Hans L. Bodlaender   Giuseppe F. Italiano (Eds.)

Algorithms – ESA 2013

21st Annual European Symposium
Sophia Antipolis, France, September 2-4, 2013
Proceedings

 Springer

Volume Editors

Hans L. Bodlaender
Utrecht University
Department of Computer Science
Princetonplein 5
3584 CC Utrecht, The Netherlands
E-mail: h.l.bodlaender@uu.nl

Giuseppe F. Italiano
Università di Roma "Tor Vergata"
Dipartimento di Ingegneria Civile e Ingegneria Informatica
via del Politecnico 1
00133 Rome, Italy
E-mail: italiano@disp.uniroma2.it

ISSN 0302-9743   e-ISSN 1611-3349
ISBN 978-3-642-40449-8   e-ISBN 978-3-642-40450-4
DOI 10.1007/978-3-642-40450-4
Springer Heidelberg New York Dordrecht London

Library of Congress Control Number: 2013945433

CR Subject Classification (1998): F.2, I.3.5, C.2, E.1, G.2, D.2, F.1

LNCS Sublibrary: SL 1 – Theoretical Computer Science and General Issues

Typesetting: Camera-ready by author, data conversion by Scientific Publishing Services, Chennai, India

Printed on acid-free paper

Springer is part of Springer Science+Business Media (www.springer.com)

shared by two papers: one by Rajesh Chitnis, László Egri and Daniél Marx for their contribution on "List H-Coloring a Graph by Removing Few Vertices" and the other by Sander P.A. Alewijnse, Quirijn W. Bouts, Alex P. ten Brink and Kevin Buchin for their contribution entitled "Computing the Greedy Spanner in Linear Space." The best student paper prize was awarded to Radu Curticapean and Marvin Künnemann for their contribution entitled "A Quantization Framework for Smoothed Analysis on Euclidean Optimization Problems." Our warmest congratulations to all of them for these achievements!

We wish to thank all the authors who submitted papers for consideration, the invited speakers, the members of the PCs for their hard work, as well as the external reviewers who assisted the PCs in the evaluation process. We are indebted to the Organizing Committee members, who helped with the local organization of the conference. We hope that the readers will enjoy the papers published in this volume, sparking their intellectual curiosity and providing inspiration for their work.

June 2013

Hans L. Bodlaender
Giuseppe F. Italiano

Preface

These proceedings contain all contributed papers presented at the 21st Annual European Symposium on Algorithms (ESA 2013), held in Sophia Antipolis, France, during September 2–4, 2013. ESA 2013 was organized as part of ALGO 2013, which also included the Workshop on Algorithms in Bioinformatics (WABI), the International Symposium on Parameterized and Exact Computation (IPEC), the Workshop on Approximation and Online Algorithms (WAOA), the International Symposium on Algorithms and Experiments for Sensor Systems, Wireless Networks and Distributed Robotics (ALGOSENSORS), the Workshop on Algorithmic Approaches for Transportation Modelling, Optimization, and Systems (ATMOS), and the Workshop on Massive Data Algorithms (MASSIVE). The previous symposia were held in Ljubljana (2012), Saarbrücken (2011), Liverpool (2010), Copenhagen (2009), Karlsruhe (2008), Eilat (2007), Zürich (2006), Palma de Mallorca (2005), Bergen (2004), Budapest (2003), Rome (2002), Aarhus (2001), Saarbrücken (2000), Prague (1999), Venice (1998), Graz (1997), Barcelona (1996), Corfu (1995), Utrecht (1994), and Bad Honnef (1993).

The ESA symposia are devoted to fostering and disseminating the results of high-quality research on the design and evaluation of algorithms and data structures. The forum seeks original algorithmic contributions for problems with relevant theoretical and/or practical applications and aims at bringing together researchers in the computer science and operations research communities. Papers were solicited in all areas of algorithmic research, both theoretical and experimental, and were evaluated by two Program Committees (PC). The PC of Track A (Design and Analysis) selected contributions with a strong emphasis on the theoretical analysis of algorithms. The PC of Track B (Engineering and Applications) evaluated papers reporting on the results of experimental evaluations and on algorithm engineering contributions for interesting applications.

In response to a call for papers, the PCs received 303 submissions from 46 countries, 229 for Track A and 74 for Track B. All submissions were reviewed by at least three PC members and were carefully evaluated on quality, originality, and relevance to the conference. Overall, the PCs wrote more than 900 reviews with the help of more than 450 external reviewers, who also participated in an extensive electronic discussion that led the committees of the two tracks to select 69 papers (53 out of 229 in Track A and 16 out of 74 in Track B), yielding an acceptance rate of about 23%. In addition to the accepted contributions, the symposium featured two invited lectures by Hannah Bast (University of Freiburg, Germany) and by Claire Mathieu (CNRS, École Normale Supérieure, France and Brown University, USA).

The European Association for Theoretical Computer Science (EATCS) sponsored a best paper award and a best student paper award. The former award was

Organization

Program Committee

David A. Bader	Georgia Institute of Technology, USA
Hans L. Bodlaender (Chair)	Universiteit Utrecht, The Netherlands
Sebastian Böcker	Friedrich-Schiller-Universität Jena, Germany
Emanuele G. Fusco	Sapienza University of Rome, Italy
Loukas Georgiadis	University of Ioannina, Greece
Roberto Grossi	University of Pisa, Italy
MohammadTaghi Hajiaghayi	University of Maryland, USA
Tobias Harks	Maastricht University, The Netherlands
Giuseppe F. Italiano (Chair)	University of Rome "Tor Vergata", Italy
Iyad Kanj	DePaul University, USA
Petteri Kaski	Aalto University, Finland
Jyrki Katajainen	University of Copenhagen, Denmark
Ken-Ichi Kawarabayashi	JST, ERATO, Kawarabayashi Large Graph Project, Japan
Guy Kortsarz	Rutgers, USA
Jan Kratochvíl	Charles University in Prague, Czech Republic
Marc van Kreveld	Universiteit Utrecht, The Netherlands
Leo Liberti	IBM T.J. Watson Research Center, USA, and LIX, École Polytechnique, France
Catherine McGeoch	Amherst College, USA
Nicole Megow	Technische Universität Berlin, Germany
Bernard M.E. Moret	EPFL, Switzerland
Gabriel Moruz	Goethe University Frankfurt am Main, Germany
Matthias Müller-Hannemann	Martin-Luther-Universität Halle-Wittenberg, Germany
Sang-Il Oum	KAIST, South Korea
Konstantinos Panagiotou	University of Munich, Germany
Andrzej Pelc	Université du Québec en Outaouais, Canada
Jeff M. Phillips	University of Utah, USA
Kunihiko Sadakane	National Institute of Informatics, Japan
Pascal Schweitzer	Forschungsinstitut für Mathematik, ETH Zürich, Switzerland
Martin Skutella	Technische Universität Berlin, Germany
Sagi Snir	University of Haifa, Israel
Kavitha Telikepalli	Tata Institute of Fundamental Research, India
Denis Trystram	Grenoble Institute of Technology, France
Dorothea Wagner	Karlsruhe Institute of Technology, Germany

Renato F. Werneck   Microsoft Research Silicon Valley, USA
Ronald de Wolf   CWI and University of Amsterdam,
                               The Netherlands
Alexander Wolff   Universität Würzburg, Germany
Norbert Zeh   Dalhousie University, Canada

Additional Reviewers

Abam, Mohammad Ali
Abdullah, Amirali
Afshani, Peyman
Aichholzer, Oswin
Alamdari, Soroush
Albers, Susanne
Aloupis, Greg
Alt, Helmut
Althaus, Ernst
Angelini, Patrizio
Angelopoulos, Spyros
Anselmi, Johnata
Anshelevich, Elliot
Antoniadis, Antonios
Araujo, Julio
Arroyuelo, Diego
Askalidis, Georgios
Attali, Dominique
Aumüller, Martin
Austrin, Per
Avron, Haim
Bampis, Evripidis
Bansal, Nikhil
Barequet, Gill
Baswana, Surender
Bateni, Mohammad
  Hossein
Batz, G. Veit
Baum, Moritz
Becchetti, Luca
Ben Avraham, Rinat
Bender, Michael
Berberich, Eric
Bereg, Sergey
Bernasconi, Anna
Bettinelli, Jeremie

Bhawalkar, Kshipra
Bienkowski, Marcin
Bille, Philip
Blaar, Holger
Blanca, Antonio
Blin, Lelia
Bonifaci, Vincenzo
Bordewich, Magnus
Borradaile, Glencora
Bose, Prosenjit
Bougeret, Marin
Brandenburg, Franz
Braverman, Vladimir
Bringmann, Karl
Brodal, Gerth Stølting
Buchin, Kevin
Buchin, Maike
Burton, Benjamin
Böckenhauer,
  Hans-Joachim
Cabello, Sergio
Cai, Leizhen
Caminiti, Saverio
Caragiannis, Ioannis
Carvalho, Margarida
Cassioli, Andrea
Casteigts, Arnaud
Chalermsook, Parinya
Chalopin, Jérémie
Chambers, Erin
Chan, Timothy
Cheong, Otfried
Chimani, Markus
Chitnis, Rajesh
Chowdhury, Rezaul
Chrobak, Marek

Cicalese, Ferdinando
Clementi, Andrea
Cohen, Johanne
Conn, Andrew
Coucheney, Pierre
Csapo, Gergely
Cygan, Marek
Czumaj, Artur
Czyzowicz, Jurek
Damaschke, Peter
Daruki, Samira
Das, Ananda Swarup
Davoodi, Pooya
Dell, Holger
Delling, Daniel
Dereniowski, Dariusz
Deshpande, Amit
Devillers, Olivier
Dibbelt, Julian
Dietzfelbinger, Martin
van Dijk, Thomas
Disser, Yann
Donatella, Firmani
Dorrigiv, Reza
Driemel, Anne
Duan, Ran
Dujmovic, Vida
Durocher, Stephane
Durr, Christoph
Dutot, Pierre
Edelkamp, Stefan
Efrat, Alon
Eirnikais, Pavlos
Elkin, Michael
Elmasry, Amr
Elsässer, Robert

Englert, Matthias
Eppstein, David
Epstein, Leah
Escoffier, Bruno
Esfandiari, Hossein
Ezra, Esther
Fagerberg, Rolf
Fairbanks, James
Feige, Uri
Fekete, Sándor
Feldman, Michal
Feldman, Moran
Felsner, Stefan
Fernau, Henning
Fiala, Jiří
Fink, Martin
Fischer, Johannes
Fleischer, Lisa
Fleszar, Krzysztof
Fogel, Efi
Fomin, Fedor
Forte, Simone
Fotakis, Dimitris
Fox, Kyle
Fraigniaud, Pierre
Friedrich, Tobias
Frongillo, Rafael
Fu, Norie
Fuchs, Fabian
Fukunaga, Alex
Fukunaga, Takuro
Fulek, Radoslav
Funke, Stefan
Fusy, Éric
Gagie, Travis
Gandhi, Rajiv
Ganian, Robert
Ganz, Maor
Garg, Naveen
Gawrychowski, Pawel
Gemsa, Andreas
Giakkoupis, George
Giannopoulos, Panos
Giegerich, Robert
Giesen, Joachim

Gilbers, Alexander
Gionis, Aristides
Gog, Simon
Goldman, Alfredo
Green, Oded
Grigoriev, Alexander
Guo, Jiong
Gutin, Gregory
Hagerup, Torben
Halldórsson, Magnús
Halperin, Dan
Hartline, Jason
Hartmann, Tanja
Haunert, Jan-Henrik
He, Meng
Heggernes, Pinar
Henze, Matthias
Henzinger, Monika
Hermelin, Danny
Hliněný, Petr
Hoefer, Martin
Holmsen, Andreas
Hon, Wing-Kai
Hong, Seok-Hee
Hornus, Samuel
Huang, Chien-Chung
Höhn, Wiebke
Høyer, Peter
Iacono, John
Imreh, Csanad
Isopi, Marco
Itai, Alon
Iwama, Kazuo
Iwata, Yoichi
Jaillet, Patrick
Jansen, Bart
Jansen, Klaus
Jansson, Jesper
Jeffery, Stacey
Jelínek, Vít
Jeż, Artur
Jeż, Lukasz
Jørgensen,
    Allan Grønlund
Kaibel, Volker

Kaiser, Tomáš
Kakimura, Naonori
Kaplan, Haim
Kappes, Andrea
Karlin, Anna
Kaufmann, Michael
Kedad-Sidhoum, Safia
de Keijzer, Bart
Kerber, Michael
Khandekar, Rohit
Khani, Reza
Khot, Subhash
Kim, Eun Jung
Kim, Ilhee
Kindermann, Philipp
Kirkpatrick, David
Klauck, Hartmut
Klavík, Pavel
Kleinberg, Robert
Klimm, Max
Knauer, Christian
Kobayashi, Yusuke
Komm, Dennis
Komusiewicz, Christian
Kontogiannis, Spyros
Koolen, Wouter M.
Kovacs, Annamaria
Kowalski, Dariusz
Kraft, Stefan
Kratsch, Dieter
Krauthgamer, Robert
Kreutzer, Stephan
Krohn, Erik
Krysta, Piotr
Krá, Daniel
Kuhnert, Sebastian
Kumar, Nirman
Kutzkov, Konstantin
Kärkkäinen, Juha
Labourel, Arnaud
Lackner, Martin
Landau, Gad M.
Lange, Carsten
Langetepe, Elmar
Lau, Lap Chi

Laue, Soeren
Laura, Luigi
Lauria, Massimo
Lengler, Johannes
Levin, Asaf
Lewenstein, Moshe
Li, Shi
Liaghat, Vahid
Liao, Chung-Shou
Ligett, Katrina
Lipták, Zsuzsanna
Lo Re, Davide
Lodi, Andrea
Loebl, Martin
Lopez-Ortiz, Alejandro
Lotker, Tzvi
Lucarelli, Giorgio
Löffler, Maarten
Lübbecke, Marco
Maehara, Takanori
Maheshwari, Anil
Mahini, Hamid
Makarychev, Konstantin
Malec, David
Maniotis, Andreas
    Milton
Mansour, Yishay
Manthey, Bodo
Manzini, Giovanni
Mareš, Martin
Mari, Federico
Marx, Dániel
Matijevic, Domagoj
Matsliah, Arie
McColl, Rob
McConnell, Ross
McGregor, Andrew
McKay, Brendan
Mendel, Manor
Mertikopoulos,
    Panayotis
Mestre, Julian
Meyer, Ulrich
Milosavljević, Nikola
Mirrokni, Vahab

Mnich, Matthias
Molfetas, Angelos
Monemizadeh, Morteza
Monti, Angelo
Morin, Pat
Moseley, Benjamin
Mount, David
Mozes, Shay
Mucherino, Antonio
Mulzer, Wolfgang
Munagala, Kamesh
Munro, Ian
Musolesi, Mirco
Muthu, Muthu
Mäkinen, Veli
Müller, Rudolf
Naamad, Yonatan
Nadarajah, Selva
Nakamoto, Atsuhiro
Narayanan, Lata
Nasre, Meghana
Navarra, Alfredo
Navarro, Gonzalo
Nederlof, Jesper
Negoescu, Andrei
Neiman, Ofer
Nicholson, Patrick K.
Niedermann, Benjamin
Niedermeier, Rolf
Nikolopoulos, Stavros
Nikou, Christophoros
Nimbhorkar, Prajakta
Nutov, Zeev
Nöllenburg, Martin
Obdržálek, Jan
Okamoto, Yoshio
Oriolo, Gianpaolo
Otachi, Yota
Ottaviano, Giuseppe
Oveis Gharan, Shayan
Pagli, Linda
Pajak, Dominik
Pajor, Thomas
Palios, Leonidas
Paluch, Katarzyna

Panigrahi, Debmalya
Papadopoulos, Charis
Paul, Christophe
Peleg, David
Peng, Richard
Perdrix, Simon
Perkovic, Ljubomir
Peter, Ueli
Pettie, Seth
Peñaranda, Luis
Piccialli, Veronica
Picone, Marco
Pierrakos, George
Pilipczuk, Michał
Piperno, Adolfo
Pokutta, Sebastian
Polishchuk, Valentin
Pralat, Pawel
Protti, Fabio
Pruhs, Kirk
Prutkin, Roman
Puglisi, Simon
Puzis, Rami
Radhakrishnan,
    Jaikumar
Rahmann, Sven
Raichel, Benjamin
Raman, Rajiv
Ramon, Jan
Rao, B.V. Raghavendra
Rawitz, Dror
Ray, Saurabh
Reidl, Felix
Riedy, Jason
Rizzi, Romeo
Roberti, Roberto
Ron, Dana
Rossmanith, Peter
Rothvoss, Thomas
Roughgarden, Tim
Roy, Sasanka
Rutter, Ignaz
Rytter, Wojciech
Sadykov, Ruslan
Sagraloff, Michael

Saha, Barna
Salavatipour,
    Mohammad R.
Sanlaville, Eric
Santaroni, Federico
Satti, Srinivasa Rao
Sau, Ignasi
Saurabh, Saket
Schapira, Michael
Schenker, Sebastian
Schmidt, Melanie
Schoenebeck, Grant
Schwartges, Nadine
Schwartz, Roy
Schäfer, Guido
Segev, Danny
Seto, Kazuhisa
Sgall, Jiří
Shachnai, Hadas
Shalom, Mordechai
Sharathkumar, Sharath
Sharir, Micha
Shen, Siqian
Shin, Chan-Su
Silvestri, Francesco
Silvestri, Simone
Sinaimeri, Blerina
Sitters, René
Skiena, Steven
Skraba, Primoz
Soma, Tasuku
Sommer, Christian
Soto, José A.
Speckmann, Bettina

Spillner, Andreas
Spoerhase, Joachim
Spöhel, Reto
Stavropoulos, Elias
van Stee, Rob
Stein, Clifford
Storandt, Sabine
Strasser, Ben
Su, Hsin-Hao
Subramani, K.
Sun, He
Suomela, Jukka
Syrgkanis, Vasilis
Talwar, Kunal
Tamir, Tami
Tatti, Nikolaj
Telle, Jan Arne
Teruyama, Junichi
Thorup, Mikkel
Todinca, Ioan
van Toll, Wouter
Ueckerdt, Torsten
Uehara, Ryuhei
Ukkonen, Antti
Vahrenhold, Jan
Vaidya, Nitin
Vakilian, Ali
Vassilvitskii, Sergei
Vatshelle, Martin
Venkatasubramanian,
    Suresh
Verbeek, Kevin
Verschae, José
Vigna, Sebastiano

Vigneron, Antoine
Vondrák, Jan
Végh, László
Wagner, Frédéric
Wagner, Uli
Wahlström, Magnus
Walen, Tomasz
Wang, Bei
Wang, Yusu
Weichert, Volker
Weimann, Oren
Welz, Wolfgang A.
Wenk, Carola
Westermann, Matthias
Wexler, Tom
White, W. Timothy J.
Wiese, Andreas
Wild, Sebastian
Winter, Sascha
Wirth, Anthony
Woeginger, Gerhard
Wollan, Paul
Wong, Prudence W.H.
Wulff-Nilsen, Christian
Xia, Ge
Xia, Lirong
Yoshida, Yuichi
Yuan, Hao
Zadimoghaddam,
    Morteza
Zakrzewska, Anita
Zastrau, David
Zhang, Qin
Zhu, Binhai

Table of Contents

The Online Replacement Path Problem

David Adjiashvili[1], Gianpaolo Oriolo[2], and Marco Senatore[2]

[1] Institute for Operations Research (IFOR)
Eidgenössische Technische Hochschule (ETH) Zürich
Rämistrasse 101, 8092 Zürich, Switzerland
david.adjiashvili@ifor.math.ethz.ch
[2] Dipartimento di Ingegneria Civile ed Ingegneria Informatica
Università di Roma Tor Vergata
Via del Politecnico 1, 00133 Rome, Italy
{oriolo,senatore}@disp.uniroma2.it

Abstract. We study a new robust path problem, the *Online Replacement Path problem (ORP)*. Consider the problem of routing a physical package through a faulty network $G = (V, E)$ from a source $s \in V$ to a destination $t \in V$ as quickly as possible. An adversary, whose objective is to maximize the latter routing time, can choose to remove a single edge in the network. In one setup, the identity of the edge is revealed to the routing mechanism (RM) while the package is in s. In this setup the best strategy is to route the package along the shortest path in the remaining network. The payoff maximization problem for the adversary becomes the *Most Vital Arc problem (MVA)*, which amounts to choosing the edge in the network whose removal results in a maximal increase of the s-t distance. However, the assumption that the RM is informed about the failed edge when standing at s is unrealistic in many applications, in which failures occur *online*, and, in particular, after the routing has started. We therefore consider the setup in which the adversary can reveal the identity of the failed edge just before the RM attempts to use this edge, thus forcing it to use a different route to t, starting from the current node. The problem of choosing the nominal path minimizing the worst case arrival time at t in this setup is ORP. We show that ORP can be solved in polynomial time and study other models naturally providing middle grounds between MVA and ORP. Our results show that ORP comprises a highly flexible and tractable framework for dealing with robustness issues in the design of RM-s.

1 Introduction

Modeling the effects of limited reliability of networks in modern routing schemes is important in many applications. It is often unrealistic to assume that the nominal network known at the stage of decision making will be available in its entirety at the stage of solution implementation. Several research directions have emerged as a result. The main paradigm in most works is to obtain a certain 'fault-tolerant' or 'redundant' solution, which takes into account a certain set of likely network realizations at the implementation phase.

H.L. Bodlaender and G.F. Italiano (Eds.): ESA 2013, LNCS 8125, pp. 1–12, 2013.

Shortest paths are often used in order to minimize routing time. In faulty network, however, simply taking the shortest path might lead to very large delays due to link failures. Two related problems that were extensively studied in the literature are the *Most Vital Arc problem (MVA)* and the *Replacement Path problem (RP)*. MVA asks given a graph $G = (V, E)$ and two nodes $s, t \in V$ to find the edge $e \in E$ whose removal results in the maximal increase in the s-t distance in G. The input to RP additionally includes a shortest path P, and the goal is to find for every $e \in P$ a shortest s-t path P_e avoiding e. In the context of robust network design both MVA and RP should be interpreted as problems in which the RM is informed about the failed edge *in advance*, namely when standing at s. This assumption is unrealistic in many situations, in which failures occur *online*, and in particular, after the routing has started. Examples of such situations range from accidents and traffic jams in road network to truly adversarial setups, in which the adversary is motivated to conceal the failure for as long as possible.

In this paper we study the *Online Replacement Path problem (ORP)*, which is motivated by such situations. We delay the formal definition of the problem to Section 3, and instead give an intuitive description. The basic assumption in ORP is that the materialized scenario is revealed to the RM 'in the last minute', namely only when the package reaches one of the endpoints of the failed edge and attempts to cross it. From this point on the package is routed through a *detour*, namely a path from the current node to the destination that avoids the failed edge. The *robust length* of a path is the maximum total travel time over all possible failure scenarios, and the goal is to find a path with minimum robust length.

ORP models online failure scenarios that occur in many situations, some of which we described before. In other applications it is only necessary to route a certain object within a certain time, called a *deadline*. As long as the object reaches its destination before the deadline, no penalty is incurred. On the other hand, if the deadline is not met, a large penalty is due. An example of such an application is organ transportation for transplants (see e.g. Moreno, Valls and Ribes [14]), in which it is critical to deliver a certain organ before the scheduled time for the surgery. In this application it does not matter how early the organ arrives at the destination, as long as it arrives in time. In such applications it is often too risky to take an unreliable shortest path, which admits only long detours in some scenarios, whereas a slightly longer path with reasonably short detours meets the deadline in *every* scenario. Thus, this domain of applications can also benefit from ORP.

Our first result is a polynomial algorithm for ORP. Concretely, we show that optimal u-t path can be found in time $O(m + n \log n)$ in undirected graphs and $O(nm + n^2 \log n)$ in directed graphs for all sources $u \in V$ and a single destination t. We prove various properties of ORP on the way to the aforementioned algorithms. In particular, we show the existence of a tree of optimal paths, and that the robust length is monotonic with respect to taking subpaths of optimal paths. These properties lead to a natural label-setting algorithm.

In Section 4 we study *Bi-objective ORP*, the optimization problem of finding a shortest path in the graph with robust length at most a given bound B. We show that this problem admits an algorithm with running time $O(m+n\log n)$ in undirected graphs (and $O(mn+n^2\log n)$ in directed graphs). We also show that the Pareto front of the latter bi-objective problem has linear size in the size of the graph, and provide a simple algorithm to compute it in time $O(m^2 + mn\log n)$, for both directed and undirected graphs. This is of course extremely nice in practical applications, as the decision maker can efficiently plot the tradeoff between the *nominal* and the *robust* length of Pareto-efficient solutions.

In Section 5 and Section 6 we study two models that provide a middle ground between MVA and ORP. In Section 5 we study the *k-Hop ORP* problem. The RM is now informed about the failed edge e as soon as it reaches a node that is k hops away from e on the nominal path. While 0-Hop ORP is simply ORP, one easily sees that $(n-1)$-Hop ORP is equivalent to MVA. For $k \in \{1, \cdots, n-2\}$ we obtain an interesting continuum of problems between ORP and MVA. We show that some of the nice properties that hold for ORP no longer hold for k-Hop ORP. In particular, while a tree of optimal paths always exists, the robust length of a subpath in this tree can be larger than the robust length of the original path. Nevertheless, we obtain a label-setting algorithm for this problem, whose running time is identical to that of our algorithms for ORP for both the directed and the undirected case (and so is independent from k). That is very interesting because, to the contrary, this is not the case with the variant where the RM is informed about the failed edge e as soon as it reaches a node that is k hops away from e in *the graph*, and not just on *the nominal path*. While this variant is equivalent to ORP for $k = 0$, we show that, already with $k = 1$, it is NP-hard to approximate within a factor of $3 - \epsilon$ for undirected graphs (and provide a simple algorithm meeting this factor), and it is strongly NP-hard to decide if there exists a nominal path with finite robust length for directed graphs.

Finally, in Section 6 we study the *ORP Game*, a two players' game related to MVA and ORP. In this game a first player, the *path builder*, is interested in arriving from s to t as quickly as possible. The second player, the *interdictor*, tries to make the latter distance as long as possible by removing a single edge from the graph. The *strategies* for the two players are the s-t paths, and the edges $e \in E$, respectively. One can see ORP and MVA as variants of the ORP Game, in which strategies are not communicated simultaneously. We show that the instances of the game which admit a pure Nash Equilibrium (NE) are exactly those where the values of the optimal solutions to ORP and MVA are equal, and build upon this fact to give an $O(m + n\log n)$-time algorithm that finds it in undirected graphs (and in time $O(mn + n^2\log n)$-time in directed graphs), or reports that no pure NE exists.

Finally, we developed a poly-time algorithm for the generalization of ORP when some fixed number of edges can fail. However, even if more involved, the algorithm goes along the same lines of that for the single-failure case, so we just defer the details to the journal version of the paper, due to space considerations.

In the next section we review related work.

2 Related Work

The Replacement Path (RP) problem was proposed by Nisan and Ronen [20] in order to study a problem in auction theory, namely that of computing *Vickrey prices*. RP is also used as a subroutine for computing the k shortest paths in a graph. The complexity of the RP problem for undirected graphs is well understood. The first paper to study this problem is due to Malik, Mittal and Gupta [13], who give a simple $O(m+n\log n)$ algorithm. A mistake in this paper was later corrected by Bar-Noy, Khuller and Schieber [4]. As a bi-product, the latter result implies an $O(m + n\log n)$-time algorithm for the Most Vital Arc (MVA) problem. This running time is asymptotically the same as a single source shortest path computation. Nardelli, Proietti and Widmayer [18] later extended the result to account for node failures. In [15] the same authors gave an algorithm that finds a detour-critical edge on a shortest path. The complexity for MVA was later improved by Nardelli, Proietti and Widmayer [17] to $O(m\alpha(m,n))$, where $\alpha(\cdot,\cdot)$ is the Inverse Ackermann function. The only general nontrivial algorithm for RP in directed graphs is due to Gotthilf and Lewenstein [10] that gave an $O(mn + n^2\log\log n)$-time algorithm. Faster algorithms for unweighted graphs (Roditty and Zwick [23]) and planar graphs (Emek, Peleg and Roditty [8], Klein, Mozes and Weimann [12] and Wulff-Nilsen [26]) were developed. The problem of approximating replacement paths was considered by Roditty [22] and Bernstein [6]. Results for bounded edge lengths were given by various authors. We refer to the paper of Vassilevska Williams [25] and references therein for details.

Some related work was carried out in the context of *routing policies* (Papadimitriou and Yannakakis [21]), the most prominent example being the Canadian Traveler Problem (Bar-Noy and Schieber [5], Nikolova and Karger [19]). In particular, in [5] the authors consider a problem that can be seen as a policy-based variant of the problem we study in Section 3. They first claim, without proof, that their problem reduces indeed to the latter one, and then claim some results that are close to those we present in Section 3. However, as we discuss later, we believe that these results are not adequately supported in [5] by rigorous arguments.

Another problem which bears resemblance to ORP is the *Stochastic Shortest Path with Recourse problem* (SSPR), studied by Andreatta and Romeo [3]. This problem can be seen as the stochastic analogue of ORP. Finally, we briefly review some related work on robust counterparts of the shortest path problem. The shortest path problem with cost uncertainty was studied by Yu and Yang [27], who consider several models for the scenario set. These results were later extended by Aissi, Bazgan and Vanderpooten [2]. These works also considered a two-stage min-max regret criterion. Dhamdhere, Goyal, Ravi and Singh [7] developed the demand-robust model and gave an approximation algorithm for the shortest path problem. A two-stage feasibility counterpart of the shortest path problem was addressed by Adjiashvili and Zenklusen [1].

3 An Algorithm for ORP

In this section we develop an algorithm for ORP. Let us establish some notation first. We are given an edge-weighted graph $G = (V, E, \ell)$, a source $s \in V$ and destination $t \in V$, and we always assume that the edge weights ℓ are nonnegative. *Unless otherwise specified, we assume that G is indifferently directed or undirected.* Let n and m be the number of nodes and edges of the input graph, respectively. For two nodes $u, v \in V$ let $\mathcal{P}_{u,v}$ be the set of simple u-v paths in G. Let $N(u)$ be the set of neighbors of u in G. For a set of edges $A \subset E$ let $\ell(A) = \sum_{e \in A} \ell(e)$. For an edge $e \in E$ and a set of edges $F \subseteq E$, let $G - e$ and $G - F$ be the graph obtained by removing the edge e and the edges in F, respectively. For a set of edges $A \subset E$ let $V(A)$ be the set of nodes incident to edges in A. Paths are always represented as sets of edges, while walks are represented as sequences of nodes. For two walks Q_1, Q_2 with the property that last node of Q_1 is the first node of Q_2 we let $Q_1 \oplus Q_2$ be their concatenation. For a path P containing nodes u and v let $P[u, v]$ be the subpath of P from u to v. For an edge $e \in E$ and $u \in V$ let Q_u^{-e} be some fixed shortest u-t path in $G - e$ and let $\pi_u^{-e} = \ell(Q_u^{-e})$. We use the convention that $Q_u^{-e} = \emptyset$ and $\pi_u^{-e} = \infty$ if u and t are in different connected components in $G - e$.

It is convenient to define the *detour* P^{-e} of a path $P \in \mathcal{P}_{v,t}$ and an edge $e = uu' \in E$ to be the walk $P[v, u] \oplus Q_u^{-uu'}$ if $uu' \in P$ (where u is the node closer to v on P), and P, otherwise. Note that we have $\ell(P^{-e}) = \ell(P[v, u]) + \pi_u^{-uu'}$ and $\ell(P^{-e}) = \ell(P)$ in the former and the latter case, respectively.

Definition 1. *Given a node $v \in V$, the robust length of the v-t path P is*

$$\text{Val}(P) = \max_{e \in E} \ell(P^{-e}).$$

ORP is to find for every $v \in V$ an optimal nominal path, namely a path P minimizing $\text{Val}(P)$ over all paths $P \in \mathcal{P}_{v,t}$.

Our algorithm uses a label-setting approach, analogous to reverse Dijkstra's algorithm for shortest paths. In every iteration, the algorithm updates certain tentative labels for the nodes of the graph, and fixes a final label to a single node u. This final label represents the connection cost of u by an optimal path to t. For $v \in V$ we define the *potential* $y(v)$ as the minimum of $\text{Val}(P)$ over all $P \in \mathcal{P}_{v,t}$. The robust length of a v-t path P is simply the maximal possible cost incurred by following P until a certain node, and then taking the best possible detour from that node to t which avoids the next edge on the path. To avoid confusion, we stress that in ORP we assume the existence of at most one failed edge in the graph. Consider next a scenario in which an edge $uu' \in P$ fails and let $u \in V$ be the node which is closer to v. Clearly, the best detour is a shortest u-t path in the graph $G - uu'$. Note that both $\text{Val}(P)$ and $y(v)$ can attain the value ∞ in case that the path P admits no detours in some scenario, and in case all v-t paths are of this sort, respectively. Furthermore, nonnegativity of ℓ implies $\text{Val}(P) \geq \text{Val}(P[v, t])$, whenever $v \in V(P)$. We can prove the following useful:

Lemma 1. *Let $P_u \in \mathcal{P}_{u,t}$ and let $v \in N(u)$ be a node, not incident to P_u. Then $\mathrm{Val}(vu \oplus P_u) = \max\{\ell(vu) + \mathrm{Val}(P_u), \pi_v^{-vu}\}$.*

Our algorithm for ORP updates the potential on the nodes of the graph, using the property established by the following lemma.

Lemma 2. *Let $U \subset V$, with $t \in U$, be the set of nodes for which the potential is known, and let vu be an edge such that:*

$$vu \in \underset{zw \in E : w \in U, z \notin U}{\arg\min} \max\{\ell(zw) + y(w), \pi_z^{-zw}\}. \tag{1}$$

Then $\mathrm{Val}(vu \oplus P_u) = y(v)$ for any optimal nominal u-t path P_u.

Lemma 2 provides the required equation for our label-setting algorithm, whose statement is given as Algorithm 1. The algorithm iteratively builds up a set U, consisting of all nodes, for which the correct potential value was already computed. The correctness of the algorithm is a direct consequence of Lemma 2.

Algorithm 1.

1: Compute π_u^{-uv} for each $uv \in E$.
2: $U = \emptyset$; $W = V$; $y'(t) = 0$; $y'(u) = \infty \; \forall u \in V - t$.
3: $successor(u) = \mathrm{NIL} \; \forall u \in V$.
4: **while** $U \neq V$ **do**
5:    Find $u = \arg\min_{z \in W} y'(z)$.
6:    $U = U + u$; $W = W - u$; $y(u) = y'(u)$.
7:    **for all** $vu \in E$ with $v \in W$ **do**
8:       **if** $y'(v) > \max\{\ell(vu) + y(u), \pi_v^{-vu}\}$ **then**
9:          $y'(v) = \max\{\ell(vu) + y(u), \pi_v^{-vu}\}$.
10:         $successor(v) = u$.

We are ready to state the main result of this section. The running time is obtained by a careful implementation that is presented in the proof. Our implementation relies on an algorithm of Nardelli, Proietti and Widmayer [16] for computing swap edges in graphs with respect to a shortest path tree. We note that Bar-Noy and Schieber [5] sketch a similar algorithm with the same time bound for a problem that can be seen as a policy-based variant of ORP. Their result are stated without proof, and we are not aware of a proof that does not build on the result of Nardelli, Proietti and Widmayer [16], which appeared afterwards.

Theorem 1. *Given an instance of ORP the potential y and the corresponding paths can be computed in time $O(m + n \log n)$ in undirected graphs, and $O(mn + n^2 \log n)$ in directed graphs.*

We end this section with a simple graph-theoretical characterization of paths with finite robust value. Let $U^2 \subset V$ be the set of nodes in G that are 2-edge connected to t, i.e., all nodes u such that there are two edge-disjoint u-t paths in G. Note that, following Theorem 2, if s and t are in different components of $G[U^2]$, there will be no path with finite robust value (and of course that happens if and only if Algorithm 1 returns $y(s) = \infty$).

Theorem 2. *A path $P \in \mathcal{P}_{s,t}$ has finite robust value if and only if $V(P) \subset U^2$.*

4   Bi-objective ORP

We turn to a natural question linking ORP and the Shortest Path problem. Consider an instance of s-t ORP for which the optimal nominal path is not unique. While all optimal paths P have the same robust length, they might differ in terms of their ordinary length $\ell(P)$. We can thus be interested in obtaining a path attaining the potential with minimum length. In general, one can consider the following bi-objective problem for any bound $B \geq y(s)$.

$$z(s, B) = \min_{P \in \mathcal{P}_{s,t}, \ \mathrm{Val}(P) \leq B} \ell(P).$$

The latter problem asks to find a *Pareto-optimal s-t* path in G with objective functions robust length and ordinary length. We call this problem *Bi-objective ORP*. Bi-objective ORP bears resemblance to the Bi-objective Shortest Path problem [11]. In the latter problem one seeks to obtain a Pareto-optimal s-t path in the graph with objective functions corresponding to ordinary length with respect to two different length functions. In this section we show that the two problems differ significantly in terms of their complexity. Concretely, we will show that a solution to bi-objective ORP and the entire Pareto front can be found in polynomial time. This contrasts to the Bi-objective Shortest Path problem, which is NP-hard, and its Pareto front can be of exponential size in the size of the graph. Our first result is:

Theorem 3. *Bi-objective ORP can be solved in time $O(m + n \log n)$ in undirected graphs and in time $O(mn + n^2 \log n)$ in directed graphs.*

Theorem 3 builds upon an algorithm that is similar to Algorithm 1. Let us turn to the problem of computing the Pareto front. Recall that a *Pareto front* F of a bi-objective optimization problem with objective functions f and g is a set of Pareto-optimal solutions to the problem with the property that, for every other solution X, there exists a solution $Y \in$ F such that $f(X) \geq f(Y)$ and $g(X) \geq g(Y)$. A Pareto front F for an instance of Bi-objective ORP is a set of paths, such that, for every Pareto-optimal path P, there exists a path in F with not longer robust length and not longer ordinary length. In general, having an entire Pareto front at hand is of course advantageous in practical applications, as it gives the decision maker a complete list of efficient strategies. The following theorem asserts that every Bi-objective ORP instance has a Pareto front with a linear number of paths. The front can be found in polynomial time using Algorithm 2 (note that, for undirected graphs, the algorithm is slightly different).

Algorithm 2.

1: $H = G$;   F $= \emptyset$.
2: **while** s and t are connected in H **do**
3:    Find a shortest s-t path P in H and add it to F.
4:    Find a critical edge $e \in E(H)$ (with $\mathrm{Val}(P) = \ell(P^{-e})$) and remove it from H.
5: Remove from F all dominated paths.
6: Return F.

Theorem 4. *Every instance of Bi-objective ORP admits a Pareto front F with at most $2m$ paths (m paths in directed graphs). The Pareto front can be found in time $O(m^2 + mn \log n)$.*

Finally, observe that Algorithm 2 is also an algorithm for ORP, as for any Pareto front F we have $y(s) = \min_{P \in F} \mathrm{Val}(P)$. This algorithm can be particularly interesting for solving ORP in sparse directed graphs, where the size of the Pareto front might compare favorably with the number of nodes in the graph.

5 k-Hop ORP

In this section we study the k-Hop ORP problem. We assume that we are given an integer k between 0 and $n - 1$ and that now the RM is informed about the failure of edge e as soon as it reaches a node that is k (or fewer) hops away from e on the nominal path P. In particular, if $e \notin P$, the RM won't be aware of the failure of e. It is easy to see that 0-Hop ORP is simply ORP and that $(n-1)$-Hop ORP is equivalent to MVA. For $k \in \{1, \cdots, n-2\}$ we obtain an interesting continuum of problems between ORP and MVA.

Apparently this new setting changes dramatically the problem. In fact, consider a (nominal) s-t path P and an edge e which belongs to P. Denote by $v(P, e)$ the first node of P that 'sees' the failure of e (note that $v(P, e) = s$ if e is at most k-hop away from s on P). Being aware of the failure of e already at $v(P, e)$ allows the RM to take a detour before getting to e, as for ORP. This justifies the following redefinition of detours. The detour P^{-e} associated with an edge $e \in E$ is defined as $P[s, v(P, e)] \oplus Q^{-e}_{v(P,e)}$ if $e \in P$, and P if $e \notin P$.

The k-Hop ORP problem is defined as finding for every $u \in V - t$ the *k-Hop potential*

$$y^k(u) = \min_{P \in \mathcal{P}_{u,t}} \mathrm{Val}^k(P),$$

where $\mathrm{Val}^k(P) = \max_{e \in E} \ell(P^{-e})$, as well as a corresponding path.

Our main result in this section is a label-setting algorithm for this problem. Obtaining this algorithm is however a more challenging task than that of obtaining such an algorithm for ORP. In particular, we will see that while there always exists a tree of optimal nominal paths, the k-hop potential needs not be a monotonic function along the paths of this tree. This property contrasts with the structure of optimal solutions to ORP. The key property of k-Hop ORP is stated in the following lemma.

Lemma 3. *Let $P_u \in \mathcal{P}_{u,t}$ be an optimal path from u, let $v \in V(P_u)$ and let $P_v \in \mathcal{P}_{v,t}$ be an optimal path from v. Then the path $P'_u = P_u[u,v] \oplus P_v$ satisfies $\mathrm{Val}^k(P'_u) = \mathrm{Val}^k(P_u)$, namely it is also optimal from u.*

Lemma 3 and the property that we state hereafter allow us to prove the correctness of a label-setting algorithm. The property follows from the fact that for any $u \in V$ and $e \in E$ one has $y^k(u) \geq \pi_u^{-e}$.

Property 1. Let $u \in V$ and $P \in \mathcal{P}_{u,t}$ be such that $\mathrm{Val}^k(P) = \pi_u^{-e}$ for some edge e seen on P by u. Then P is an optimal path from u.

Analogously to our algorithm for ORP, we proceed by incrementally computing the optimal path for every node in the graph starting from t. We maintain a set U of nodes for which a robust path was already computed. For $u \in U$ we denote this path by P_u^*. The update rule for U works as follows. First, we check if for some edge $vu \in E$ such that $v \in V \setminus U$ and $u \in U$ it holds that the path $Q = vu \oplus P_u^*$ satisfies the condition in Property 1. In other words, we check if $\mathrm{Val}^k(Q) = \pi_v^{-e}$ for some edge $e \in Q$ seen by v. If such an edge exists we set $P_v^* := Q$ and $U := U \cup \{v\}$, an update that is valid due to Property 1. Assume next that no such edge exists. We call the set U in this situation *clean*. The following lemma states an update rule for clean sets U.

Lemma 4. *Let U be clean and let $vu \in \arg\min_{qr \in E : r \in U, q \notin U} \mathrm{Val}^k(qr \oplus P_r^*)$. Then $y^k(v) = \mathrm{Val}^k(vu \oplus P_u^*)$.*

Lemmas 3 and 4 immediately imply a polynomial algorithm for k-Hop ORP. Observe that one can adopt the implementation of our label-setting algorithm for ORP to obtain the same time bounds as in Theorem 1. The details are identical, and thus omitted.

Theorem 5. *Given an instance of k-Hop ORP the potential y^k and the corresponding paths can be computed in time $O(m+n \log n)$ in undirected graphs, and $O(mn + n^2 \log n)$ in directed graphs.*

Let us make two further remarks about extensions of ORP similar to k-Hop ORP. In k-Hop ORP we assume that the RM is informed about the failed edge when it is k hops away on the nominal path. An alternative definition takes the lengths of edges on this path into account. In this problem, which we call *Radius ORP*, the integer $k \leq n-1$ is replaced by a value $R \leq \ell(E)$ called the *radius*. In this problem the RM is informed about the failed edge e on the nominal path P at the first node that is at distance at most R from its closer endpoint. The definition of $v(P, e)$ and the robust value are adapted accordingly.

We claim without proof that our algorithm for k-Hop ORP solves Radius ORP as well. Informally, this follows from fact that Lemma 3 remains correct, since it only relies on the following *monotonicity property*. The set of edges on P that a node sees is an interval on this path, and furthermore, for every two consecutive nodes $u_1, u_2 \in V(P)$, with u_2 being the closer one to t, the set of edges seen by u_1 in $P[u_2, t]$ is a subset of the set of edges seen by u_2. We defer the proof of this fact, as well as the careful treatment of Radius ORP to the journal version of the paper, due to space considerations.

We end this section with another variant of k-Hop ORP. In this variant, whose input is identical to that of k-Hop ORP, the information about the failed edge travels through the edges of the entire graph, as opposed to only the edges of the nominal path. Formally, the first node along the chosen nominal path that is informed about the failure of some edge $e \in E$ is the one closest to s that is at most k hops away from e in G. This problem, which we denote by *Strong k-Hop ORP* turns out to be NP-hard to approximate even when $k = 1$. Note that, for $k = 0$, Strong k-Hop ORP reduces to ORP, as for every path the robust value is the same in the two different problems.

Theorem 6. *for any $\epsilon > 0$ it is NP-hard to approximate Strong 1-Hop ORP within a factor of $3 - \epsilon$ in undirected graphs. In directed graphs it is strongly NP-hard to decide if there exists a nominal path with finite robust length.*

We note that in 2 s-t connected undirected graphs, every shortest path is a 3-approximation of the optimal solution to Strong 1-Hop ORP, thus the approximability of this problem is settled. The proof of this simple fact if similar to the proof of Lemma 6, and thus omitted.

6   A Two Players' Game between MVA and ORP

Let us explore next a two players' game that is the natural middle ground between the problems MVA and ORP. A first player, the *path builder*, is interested in arriving from s to t as quickly as possible. The second player, the *interdictor*, tries to make the latter distance as long as possible by removing a single edge from the graph. The *strategies* for the two players are the s-t paths, and the edges $e \in E$, respectively.

In one setup, the interdictor communicates her strategy *first*, i.e. which edge is removed from G. The path builder chooses his strategy *after*: clearly he chooses a shortest path s-t in the graph $G - e$. Therefore, the problem that the interdictor faces in this setting is clearly the MVA problem, as she will remove the edge $e \in E$ maximizing π_s^{-e}, the length of the shortest s-t path in the graph $G - e$. In the following, we let $z^*(MVA)$ be the value of an optimal solution to MVA.

In the other extreme, the path builder communicates his strategy first, i.e. an s-t path P. Then the interdictor moves, and clearly removes the edge e maximizing $\ell(P^{-e})$. Note that we assume that, if $e \in P$, the interdictor will delay the failure of the edge to the point at which the path builder attempts to cross it. Hence, the problem that the path builder faces is exactly ORP, i.e. that of choosing an s-t path with the least robust value. In the following, we let $z^*(ORP)$ be the value of an optimal solution to ORP.

The next lemma, whose simple proof we skip, shows that $z^*(ORP) \geq z^*(MVA)$.

Lemma 5. *Let P and e be an s-t path and an edge of E, respectively. Then $\mathrm{Val}(P) \geq z^*(ORP) \geq z^*(MVA) \geq \pi_s^{-e}$.*

In our two players' game, that we call the *ORP Game*, both players communicate their strategies *at the same time*. In particular, for a given s-t path P and edge $e \in E$, the payoff for the interdictor is $\ell(P^{-e})$. Lemma 5 shows that in general $z^*(ORP) \geq z^*(MVA)$. The next theorem characterizes the instances of the ORP Game admitting a pure NE as those for which $z^*(ORP) = z^*(MVA)$.

Theorem 7. *Let P and e be optimal solutions to the ORP and MVA instances on $G = (V, E)$. Then (P, e) is a pure NE of the ORP Game if and only if $\mathrm{Val}(P) = \pi_s^{-e}$. Moreover, in this case, $\mathrm{Val}(P) = z^*(ORP) = z^*(MVA) = \pi_s^{-e}$.*

Theorem 7 has also the following algorithmic implication. Recall that we can compute $z^*(MVA)$ in time $O(m + n \log n)$ [13], the same running time we obtained for unidrected ORP (Theorem 1). This clearly implies that in time $O(m + n \log n)$ we can compute a pure NE of the ORP Game in undirected graphs, if one exists, or certify that no pure NE exists. Indeed the aforementioned algorithms allow us to check the condition $z^*(ORP) = z^*(MVA)$ and compute corresponding optimal solutions, P^* and e^*, with the latter time complexity. Theorem 7 asserts that if the latter condition is satisfied, then (P^*, e^*) is a pure NE, otherwise no pure NE exists. Theorems 1 and 7 also imply a $O(mn + n^2 \log n)$ algorithm for the same problem is directed graphs.

We close this section by shortly addressing the case where $z^*(ORP) \neq z^*(MVA)$. First, in this case, Theorem 7 shows that there are no pure NE. However, the ORP Game will still admit a NE in *mixed strategies*, as for both players the sets of pure strategies is finite (s-t paths and edges). Whether it is possible to find this mixed NE in polynomial time is an interesting open question.

We conclude by analyzing the ratio $\frac{z^*(ORP)}{z^*(MVA)}$. The next lemma shows that, for undirected graphs, it is at most 3.

Lemma 6. *Let G be undirected with s-t edge-connectivity of at least two. Then $z^*(ORP) \leq 3z^*(MVA)$.*

References

1. Adjiashvili, D., Zenklusen, R.: An s - t connection problem with adaptability. Discrete Applied Mathematics 159(8), 695–705 (2011)
2. Aissi, H., Bazgan, C., Vanderpooten, D.: Approximation complexity of min-max (Regret) versions of shortest path, spanning tree, and knapsack. In: Brodal, G.S., Leonardi, S. (eds.) ESA 2005. LNCS, vol. 3669, pp. 862–873. Springer, Heidelberg (2005)
3. Andreatta, G., Romeo, L.: Stochastic shortest paths with recourse. Networks 18(3), 193–204 (1988)
4. Bar-Noy, A., Khuller, S., Schieber, B.: The complexity of finding most vital arcs and nodes. Technical report, Univ. of Maryland Institute for Advanced Computer Studies Report No. UMIACS-TR-95-96, College Park, MD, USA (1995)
5. Bar-Noy, A., Schieber, B.: The canadian traveller problem. In: SODA 1991, pp. 261–270. SIAM, Philadelphia (1991)

6. Bernstein, A.: A nearly optimal algorithm for approximating replacement paths and k shortest simple paths in general graphs. In: SODA 2010, pp. 742–755. SIAM, Philadelphia (2010)
7. Dhamdhere, K., Goyal, V., Ravi, R., Singh, M.: How to pay, come what may: Approximation algorithms for demand-robust covering problems. In: FOCS 2005, pp. 367–378. IEEE Computer Society, Washington, DC (2005)
8. Emek, Y., Peleg, D., Roditty, L.: A near-linear-time algorithm for computing replacement paths in planar directed graphs. ACM Trans. Algorithms 6, 64:1–64:13 (2010)
9. Fredman, M.L., Tarjan, R.E.: Fibonacci heaps and their uses in improved network optimization algorithms. J. ACM 34, 596–615 (1987)
10. Gotthilf, Z., Lewenstein, M.: Improved algorithms for the k simple shortest paths and the replacement paths problems. Inf. Process. Lett. 109, 352–355 (2009)
11. Hassin, R.: Approximation schemes for the restricted shortest path problem. Mathematics of Operations Research 17(1), 36–42 (1992)
12. Klein, P.N., Mozes, S., Weimann, O.: Shortest paths in directed planar graphs with negative lengths: A linear-space $O(n \log^2 n)$-time algorithm. ACM Trans. Algorithms 6, 30:1–30:18 (2010)
13. Malik, K., Mittal, A.K., Gupta, S.K.: The k most vital arcs in the shortest path problem. Operations Research Letters 8(4), 223–227 (1989)
14. Moreno, A., Valls, A., Ribes, A.: Finding efficient organ transport routes using multi-agent systems. In: Proceedings of the IEEE 3rd International Workshop on Enterprise Networking and Computing in Health Care Industry (Healthcom), pp. 233–258 (2001)
15. Nardelli, E., Proietti, G., Widmayer, P.: Finding the detour-critical edge of a shortest path between two nodes. Information Processing Letters 67(1), 51–54 (1998)
16. Nardelli, E., Proietti, G., Widmayer, P.: Swapping a failing edge of a single source shortest paths tree is good and fast. Algorithmica 35, 2003 (1999)
17. Nardelli, E., Proietti, G., Widmayer, P.: A faster computation of the most vital edge of a shortest path. Information Processing Letters 79(2), 81–85 (2001)
18. Nardelli, E., Proietti, G., Widmayer, P.: Finding the most vital node of a shortest path. In: Wang, J. (ed.) COCOON 2001. LNCS, vol. 2108, pp. 278–287. Springer, Heidelberg (2001)
19. Nikolova, E., Karger, D.R.: Route planning under uncertainty: the canadian traveller problem. In: AAAI 2008, pp. 969–974. AAAI Press (2008)
20. Nisan, N., Ronen, A.: Algorithmic mechanism design (extended abstract). In: STOC 1999, pp. 129–140. ACM, New York (1999)
21. Papadimitriou, C.H., Yannakakis, M.: Shortest paths without a map. In: Ausiello, G., Dezani-Ciancaglini, M., Della Rocca, S.R. (eds.) ICALP 1989. LNCS, vol. 372, pp. 610–620. Springer, Heidelberg (1989)
22. Roditty, L.: On the k-simple shortest paths problem in weighted directed graphs. In: SODA 2007, pp. 920–928. SIAM, Philadelphia (2007)
23. Roditty, L., Zwick, U.: Replacement paths and k simple shortest paths in unweighted directed graphs. In: Caires, L., Italiano, G.F., Monteiro, L., Palamidessi, C., Yung, M. (eds.) ICALP 2005. LNCS, vol. 3580, pp. 249–260. Springer, Heidelberg (2005)
24. Tarjan, R.E.: Efficiency of a good but not linear set union algorithm. J. ACM 22(2), 215–225 (1975)
25. Vassilevska Williams, V.: Faster replacement paths. In: SODA 2011, pp. 1337–1346. SIAM (2011)
26. Wulff-Nilsen, C.: Solving the replacement paths problem for planar directed graphs in $O(n \log n)$ time. In: SODA 2010, pp. 756–765. SIAM, Philadelphia (2010)
27. Yu, G., Yang, J.: On the robust shortest path problem. Computers & Operations Research 25(6), 457–468 (1998)

Flip Distance between Triangulations of a Simple Polygon is NP-Complete[*]

Oswin Aichholzer[1], Wolfgang Mulzer[2], and Alexander Pilz[1]

[1] Institute for Software Technology, Graz University of Technology, Austria
[2] Institute of Computer Science, Freie Universität Berlin, Germany

Abstract. Let T be a triangulation of a simple polygon. A *flip* in T is the operation of removing one diagonal of T and adding a different one such that the resulting graph is again a triangulation. The *flip distance* between two triangulations is the smallest number of flips required to transform one triangulation into the other. For the special case of convex polygons, the problem of determining the shortest flip distance between two triangulations is equivalent to determining the rotation distance between two binary trees, a central problem which is still open after over 25 years of intensive study.

We show that computing the flip distance between two triangulations of a simple polygon is NP-complete. This complements a recent result that shows APX-hardness of determining the flip distance between two triangulations of a planar point set.

1 Introduction

Let P be a simple polygon in the plane, that is, a closed region bounded by a piece-wise linear, simple cycle. A *triangulation* T of P is a geometric (straight-line) maximal outerplanar graph whose outer face is the complement of P and whose vertex set consists of the vertices of P. The edges of T that are not on the outer face are called *diagonals*. Let d be a diagonal whose removal creates a convex quadrilateral f. Replacing d with the other diagonal of f yields another triangulation of P. This operation is called a *flip*. The *flip graph* of P is the abstract graph whose vertices are the triangulations of P and in which two triangulations are adjacent if and only if they differ by a single flip. We study the *flip distance*, i.e., the minimum number of flips required to transform a given source triangulation into a target triangulation.

Edge flips became popular in the context of Delaunay triangulations. Lawson [9] proved that any triangulation of a planar n-point set can be transformed into any other by $O(n^2)$ flips. Hence, for every planar n-point set the flip graph is connected with diameter $O(n^2)$. Later, he showed that in fact every triangulation can be transformed to the Delaunay triangulation by $O(n^2)$ flips that locally

[*] O.A. partially supported by the ESF EUROCORES programme EuroGIGA - ComPoSe, Austrian Science Fund (FWF): I 648-N18. W.M. supported in part by DFG project MU/3501/1. A.P. is recipient of a DOC-fellowship of the Austrian Academy of Sciences at the Institute for Software Technology, Graz University of Technology.

H.L. Bodlaender and G.F. Italiano (Eds.): ESA 2013, LNCS 8125, pp. 13–24, 2013.

fix the Delaunay property [10]. Hurtado, Noy, and Urrutia [7] gave an example where the flip distance is $\Omega(n^2)$, and they showed that the same bounds hold for triangulations of simple polygons. They also proved that if the polygon has k reflex vertices, then the flip graph has diameter $O(n + k^2)$. In particular, the flip graph of any planar polygon has diameter $O(n^2)$. Their result also generalizes the well-known fact that the flip distance between any two triangulations of a convex polygon is at most $2n - 10$, for $n > 12$, as shown by Sleator, Tarjan, and Thurston [15] in their work on the flip distance in convex polygons. The latter case is particularly interesting due to the correspondence between flips in triangulations of convex polygons and rotations in binary trees: The dual graph of such a triangulation is a binary tree, and a flip corresponds to a rotation in that tree; also, for every binary tree, a triangulation can be constructed.

We mention two further remarkable results on flip graphs for point sets. Hanke, Ottmann, and Schuierer [6] showed that the flip distance between two triangulations is bounded by the number of crossings in their overlay. Eppstein [5] gave a polynomial-time algorithm for calculating a lower bound on the flip distance. His bound is tight for point sets with no empty 5-gons; however, except for small instances, such point sets are not in general position (i.e., they must contain collinear triples) [1]. For a recent survey on flips see Bose and Hurtado [3].

Very recently, the problem of finding the flip distance between two triangulations of a point set was shown to be NP-hard by Lubiw and Pathak [11] and, independently, by Pilz [12], and the latter proof was later improved to show APX-hardness of the problem. Here, we show that the corresponding problem remains NP-hard even for simple polygons. This can be seen as a further step towards settling the complexity of deciding the flip distance between triangulations of convex polygons or, equivalently, the rotation distance between binary trees. This variant of the problem was probably first addressed by Culik and Wood [4] in 1982 (showing a flip distance of $2n - 6$).

The formal problem definition is as follows: given a simple polygon P, two triangulations T_1 and T_2 of P, and an integer l, decide whether T_1 can be transformed into T_2 by at most l flips. We call this decision problem POLYFLIP. To show NP-hardness, we give a polynomial-time reduction from RECTILINEAR STEINER ARBORESCENCE to POLYFLIP. RECTILINEAR STEINER ARBORESCENCE was shown to be NP-hard by Shi and Su [14]. In Section 2, we describe the problem in detail. We present the well-known *double chain* (used by Hurtado, Noy, and Urrutia [7] for giving their lower bound), a major building block in our reduction, in Section 3. Finally, in Section 4, we describe our reduction and prove that it is correct. An extended abstract of this work was presented at the 29th EuroCG, 2013; for omitted proofs, see [2].

2   The Rectilinear Steiner Arborescence Problem

Let S be a set of N points in the plane whose coordinates are nonnegative integers. The points in S are called *sinks*. A *rectilinear tree* T is a connected acyclic collection of horizontal and vertical line segments that intersect only at their endpoints. The *length* of T is the total length of all segments in T (cf. [8, p. 205]). The tree T is a *rectilinear Steiner tree* for S if each sink in S appears as an endpoint of a segment in T. We call T a *rectilinear Steiner arborescence* (RSA) for S if (i) T is rooted at the origin; (ii) each leaf of T lies at a sink in S; and (iii) for each $s = (x, y) \in S$, the length of the path in T from the origin to s equals $x + y$, i.e., all edges in T point north or east, as seen from the origin [13]. In the *RSA problem*, we are given a set of sinks S and an integer k. The question is whether there is an RSA for S of length at most k. Shi and Su showed that the RSA problem is strongly NP-complete; in particular, it remains NP-complete if S is contained in an $n \times n$ grid, with n polynomially bounded in N, the number of points [14].[1]

We recall an important structural property of the RSA. Let A be an RSA for a set S of sinks. Let e be a vertical segment in A that does not contain a sink. Suppose there is a horizontal segment f incident to the upper endpoint a of e. Since A is an arborescence, a is the left endpoint of f. Suppose further that a is not the lower endpoint of another vertical edge. Take a copy e' of e and translate it to the right until e' hits a sink or another segment endpoint (this will certainly happen at the right endpoint of f); see Fig. 1. The segments e and e' define a rectangle R. The upper and left side of R are completely covered by e and (a part of) f. Since a has only two incident segments, every sink-root path in A that goes through e or f contains these two sides of R, entering the boundary of R at the upper right corner d and leaving it at the lower left corner b. We reroute every such path at d to continue clockwise along the boundary of R until it meets A again (this certainly happens at b), and we delete e and the part of f on R. In the resulting tree we subsequently remove all unnecessary segments (this happens if there are no more root-sink paths through b) to obtain another RSA A' for S. Observe that A' is not longer than A. This operation is called *sliding e to the right*. If similar conditions apply to a horizontal edge, we can *slide it upwards*. The *Hanan grid* for a point set P is the set of all vertical and horizontal lines through the points in P. In essence, the following theorem can be proved constructively by repeated segment slides in a shortest RSA.

Theorem 2.1 ([13]). *Let S be a set of sinks. There is a minimum-length RSA A for S such that all segments of A are on the Hanan grid for $S \cup \{(0,0)\}$.* □

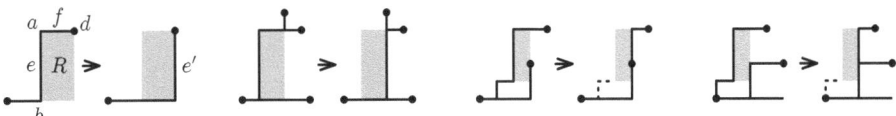

Fig. 1. The slide operation. The dots depict sinks; the rectangle R is drawn gray. The dotted segments are deleted, since they do no longer lead to a sink.

---

[1] Note that a polynomial-time algorithm was claimed [16] that later has been shown to be incorrect [13].

We use a restricted version of the RSA problem, called YRSA. An instance (S, k) of the YRSA problem differs from an instance for the RSA problem in that we require that no two sinks in S have the same y-coordinate. The NP-hardness of YRSA follows by a simple perturbation argument; see the full version for all omitted proofs.

Theorem 2.2. YRSA *is strongly* NP-*complete.*

3 Double Chains

Our definitions (and illustrations) follow [12]. A *double chain* D consists of two chains, an *upper chain* and a *lower chain*. There are n vertices on each chain, $\langle u_1, \ldots, u_n \rangle$ on the upper chain and $\langle l_1, \ldots, l_n \rangle$ on the lower chain, both numbered from left to right. Any point on one chain sees every point on the other chain, and any quadrilateral formed by three vertices of one chain and one vertex of the other chain is non-convex. Let P_D be the polygon defined by $\langle l_1, \ldots, l_n, u_n, \ldots, u_1 \rangle$; see Fig. 2 (left). We call the triangulation T_u of P_D where u_1 has maximum degree the *upper extreme triangulation*; observe that this triangulation is unique. The triangulation T_l of P_D where l_1 has maximum degree is called the *lower extreme triangulation*. The two extreme triangulations are used to show that the diameter of the flip graph is quadratic; see Fig. 2 (right).

Theorem 3.1 ([7]). *The flip distance between* T_u *and* T_l *is* $(n-1)^2$.   □

Through a slight modification of D, we can reduce the flip distance between the upper and the lower extreme triangulation to linear. This will enable us in our reduction to impose a certain structure on short flip sequences. To describe this modification, we first define the *flip-kernel* of a double chain.

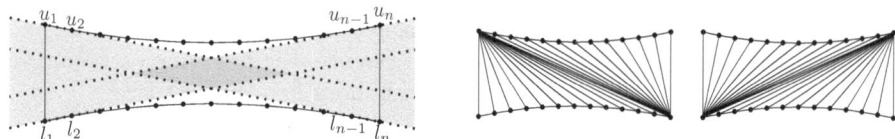

Fig. 2. Left: The polygon and the hourglass (gray) of a double chain. The diamond-shaped flip-kernel can be extended arbitrarily by flattening the chains. Right: The upper extreme triangulation T_u and the lower extreme triangulation T_l.

Let W_1 be the wedge defined by the lines through $u_1 u_2$ and $l_1 l_2$ whose interior contains no point from D but intersects the segment $u_1 l_1$. Define W_n analogously by the lines through $u_n u_{n-1}$ and $l_n l_{n-1}$. We call $W := W_1 \cup W_n$ the *hourglass* of D. The unbounded set $W \cup P_D$ is defined by four rays and the two chains. The *flip-kernel* of D is the intersection of the closed half-planes below the lines through $u_1 u_2$ and $u_{n-1} u_n$, as well as above the lines through $l_1 l_2$ and $l_{n-1} l_n$.[2]

---

[2] The flip-kernel of D might not be completely inside the polygon P_D. This is in contrast to the "visibility kernel" of a polygon.

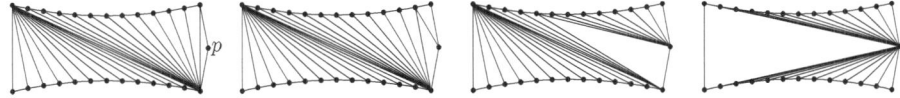

Fig. 3. The extra point p in the flip-kernel of D allows flipping one extreme triangulation of P_D^p to the other in $4n - 4$ flips

Definition 3.2. *Let D be a double chain whose flip-kernel contains a point p to the right of the directed line $l_n u_n$. The polygon P_D^p is given by the sequence $\langle l_1, \ldots, l_n, p, u_n, \ldots, u_1 \rangle$. The upper and the lower extreme triangulation of P_D^p contain the edge $u_n l_n$ and otherwise are defined in the same way as for P_D.*

The flip distance between the two extreme triangulations for P_D^p is much smaller than for P_D [17]. Fig. 3 shows how to transform them into each other with $4n - 4$ flips. The next lemma shows that this is optimal, even for more general polygons. The lemma is a slight generalization of a lemma by Lubiw and Pathak [11] on double chains of constant size.

Lemma 3.3. *Let P be a polygon that contains P_D and has $\langle l_1, \ldots, l_n \rangle$ and $\langle u_n, \ldots, u_1 \rangle$ as part of its boundary. Further, let T_1 and T_2 be two triangulations that contain the upper extreme triangulation and the lower extreme triangulation of P_D as a sub-triangulation, respectively. Then T_1 and T_2 have flip distance at least $4n - 4$.*

The following result can be seen as a special case of [12, Proposition 1].

Lemma 3.4. *Let P be a polygon that contains P_D and has $\langle u_n, \ldots, u_1, l_1, \ldots, l_n \rangle$ as part of its boundary. Let T_1 and T_2 be two triangulations that contain the upper and the lower extreme triangulation of P_D as a sub-triangulation, respectively. Consider any flip sequence σ from T_1 to T_2 and suppose there is no triangulation in σ containing a triangle with one vertex at the upper chain, the other vertex at the lower chain, and the third vertex at a point in the interior of the hourglass of P_D. Then $|\sigma| \geq (n-1)^2$.*

4   The Reduction

We reduce YRSA to POLYFLIP. Let S be a set of N sinks on an $n \times n$ grid with root at $(1, 1)$ (recall that n is polynomial in N). We construct a polygon P_D^* and two triangulations T_1, T_2 in P_D^* such that a shortest flip sequence from T_1 to T_2 corresponds to a shortest RSA for S. To this end, we will describe how to interpret any triangulation of P_D^* as a *chain path*, a path in the integer grid that starts at the origin and uses only edges that go north or east. It will turn out that flips in P_D^* essentially correspond to moving the endpoint of the chain path along the grid. We choose P_D^*, T_1, and T_2 in such a way that a shortest flip sequence between T_1 and T_2 moves the endpoint of the chain path according to an Eulerian traversal of a shortest RSA for S. To force the chain path to visit all sites, we use the observations from Section 3: the polygon P_D^* contains a double chain for each sink, so that only for certain triangulations of P_D^* it is possible to flip the double chain quickly. These triangulations will be exactly the triangulations that correspond to the chain path visiting the appropriate site.

4.1 The Construction

Our construction has two integral parameters, β and d. With foresight, we set $\beta = 2N$ and $d = nN$. We imagine that the sinks of S lie on a $\beta n \times \beta n$ grid, with their coordinates multiplied by β.

We take a double chain D with βn vertices on each chain such that the flip-kernel of D extends to the right of $l_{\beta n} u_{\beta n}$. We add a point z to that part of the flip-kernel, and we let P_D^+ be the polygon defined by $\langle l_1, \ldots, l_{\beta n}, z, u_{\beta n}, \ldots, u_1 \rangle$. Next, we add double chains to P_D^+ in order to encode the sinks. For each sink $s = (x, y)$, we remove the edge $l_{\beta y} l_{\beta y+1}$, and we replace it by a (rotated) double chain D_s with d vertices on each chain, such that $l_{\beta y}$ and $l_{\beta y+1}$ correspond to the last point on the lower and the upper chain of D_s, respectively. We orient D_s in such a way that $u_{\beta x}$ is the only point inside the hourglass of D_s and so that $u_{\beta x}$ lies in the flip-kernel of D_s; see Fig. 4. We refer to the added double chains as *sink gadgets*, and we call the resulting polygon P_D^*. For β large enough, the sink gadgets do not overlap, and P_D^* is a simple polygon. Since the y-coordinates in S are pairwise distinct, there is at most one sink gadget per edge of the lower chain of P_D^+. The precise placement of the sink gadgets is flexible, so we can make all coordinates polynomial in n; see the full version for details.

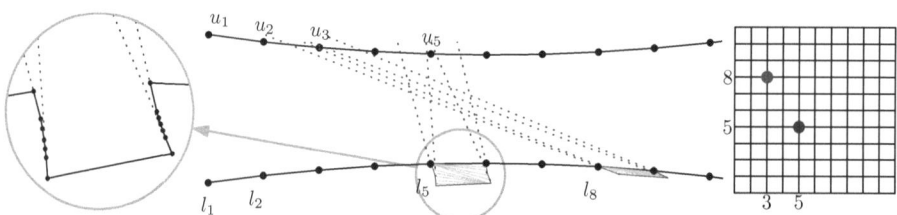

Fig. 4. The sink gadget for a site (x, y) is obtained by replacing the edge $l_{\beta y} l_{\beta y+1}$ by a double chain with d vertices on each chain. The double chain is oriented such that $u_{\beta x}$ is the only point inside its hourglass and its flip-kernel. In our example, $\beta = 1$.

Next, we describe the source and target triangulation for P_D^*. In the source triangulation T_1, the interior of P_D^+ is triangulated such that all edges are incident to z. The sink gadgets are all triangulated with the upper extreme triangulation. The target triangulation T_2 is similar, but now the sink gadgets are triangulated with the lower extreme triangulation.

To get from T_1 to T_2, we must go from one extreme triangulation to the other for each sink gadget D_s. By Lemma 3.4, this requires $(d-1)^2$ flips, unless the flip sequence creates a triangle that allows us to use the vertex in the flip-kernel of D_s. In this case, we say that the flip sequence *visits* the sink s. For d large enough, a shortest flip sequence must visit each sink, and we will show that this induces an RSA for S of similar length. Conversely, we will show how to derive a flip sequence from an RSA. The precise statement is given in the following theorem.

Theorem 4.1. *Let $k \geq 1$. The flip distance between T_1 and T_2 w.r.t. P_D^* is at most $2\beta k + (4d - 2)N$ if and only if S has an RSA of length at most k.*

We will prove Theorem 4.1 in the following sections. But first, let us show how to use it for our NP-completeness result.

Theorem 4.2. POLYFLIP *is* NP-*complete.*

Proof. As mentioned in the introduction, the flip distance in polygons is polynomially bounded, so POLYFLIP is in NP. We reduce from YRSA. Let (S, k) be an instance of YRSA such that S lies on a grid of polynomial size. We construct P_D^* and T_1, T_2 as described above. This takes polynomial time (see the full version for details). Set $l = 2\beta k + (4d - 2)N$. By Theorem 4.1, there exists an RSA for S of length at most k if and only if there exists a flip sequence between T_1 and T_2 of length at most l.   □

4.2   Chain Paths

Now we introduce the *chain path*, our main tool to establish a correspondence between flip sequences and RSAs. Let T be a triangulation of P_D^+ (i.e., the polygon P_D^* without the sink gadgets, cf. Section 4.1). A *chain edge* is an edge of T between the upper and the lower chain of P_D^+. A *chain triangle* is a triangle of T that contains two chain edges. Let $e_1, \ldots, e_m$ be the chain edges, sorted from left to right according to their intersection with a line that separates the upper from the lower chain. For $i = 1, \ldots, m$, write $e_i = (u_v, l_w)$ and set $c_i = (v, w)$. In particular, $c_1 = (1, 1)$. Since T is a triangulation, any two consecutive edges e_i, e_{i+1} share one endpoint, while the other endpoints are adjacent on the corresponding chain. Thus, c_{i+1} dominates c_i and $\|c_{i+1} - c_i\|_1 = 1$. It follows that $c_1 c_2 \ldots c_m$ is an x- and y-monotone path in the $\beta n \times \beta n$-grid, beginning at the root. It is called the *chain path* for T. Each vertex of the chain path corresponds to a chain edge, and each edge of the chain path corresponds to a chain triangle. Conversely, every chain path induces a triangulation T of P_D^+; see Fig. 5. In the following, we let b denote the upper right endpoint of the chain path.

The next lemma describes how the chain path is affected by flips; see Fig. 5.

Lemma 4.3. *Any triangulation T of P_D^+ uniquely determines a chain path, and vice versa. A flip in T corresponds to one of the following operations on the chain path: (i) move the endpoint b north or east; (ii) shorten the path at b; (iii) change an east-north bend to a north-east bend, or vice versa.*   □

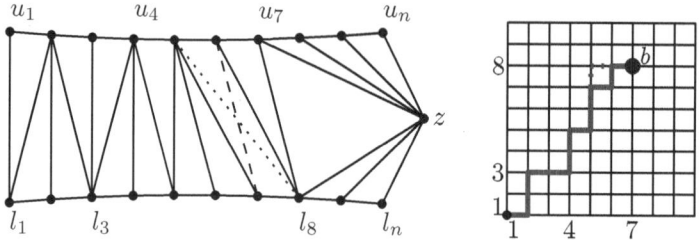

Fig. 5. A triangulation of P_D^+ and its chain path. Flipping edges to and from z moves the endpoint b along the grid. A flip between chain triangles changes a bend.

4.3 From an RSA to a Short Flip Sequence

Using the notion of a chain path, we now prove the "if" direction of Theorem 4.1.

Lemma 4.4. *Let $k \geq 1$ and A an RSA for S of length k. Then the flip distance between T_1 and T_2 w.r.t. P_D^* is at most $2\beta k + (4d - 2)N$.*

Proof. The triangulations T_1 and T_2 both contain a triangulation of P_D^+ whose chain path has its endpoint b at the root. We use Lemma 4.3 to generate flips inside P_D^+ so that b traverses A in a depth-first manner. This needs $2\beta k$ flips.

Each time b reaches a sink s, we move b north. This creates a chain triangle that allows the edges in the sink gadget D_s to be flipped to the auxiliary vertex in the flip-kernel of D_s. The triangulation of D_s can then be changed with $4d - 4$ flips; see Lemma 3.3. Next, we move b back south and continue the traversal. Moving b at s needs two additional flips, so we take $4d - 2$ flips per sink, for a total of $2\beta k + (4d - 2)N$ flips.   □

4.4 From a Short Flip Sequence to an RSA

Finally, we consider the "only if" direction in Theorem 4.1. Let σ_1 be a flip sequence on P_D^+. We say that σ_1 *visits* a sink $s = (x, y)$ if σ_1 has at least one triangulation T that contains the chain triangle $u_{\beta x} l_{\beta y} l_{\beta y + 1}$. We call σ_1 a *flip traversal* for S if (i) σ_1 begins and ends in the triangulation whose corresponding chain path has its endpoint b at the root and (ii) σ_1 visits every sink in S. The following lemma shows that every short flip sequence in P_D^* can be mapped to a flip traversal.

Lemma 4.5. *Let σ be a flip sequence from T_1 to T_2 w.r.t. P_D^* with $|\sigma| < (d-1)^2$. Then there is a flip traversal σ_1 for S with $|\sigma_1| \leq |\sigma| - (4d - 4)N$.*

Proof. We show how to obtain a flip traversal σ_1 for S from σ. Let T^* be a triangulation of P_D^*. A triangle of T^* is an *inner triangle* if all its sides are diagonals. It is an *ear* if two of its sides are polygon edges. By construction, every inner triangle of T^* must have (i) one vertex incident to z (the rightmost vertex of P_D^+), or (ii) two vertices incident to a sink gadget (or both). In the latter case, there can be only one such triangle per sink gadget. The weak (graph theoretic) dual of T^* is a tree in which ears correspond to leaves and inner triangles have degree 3.

Let D_s be a sink gadget placed between the vertices l_s and l'_s. Let u_s be the vertex in the flip-kernel of D_s. We define a triangle Δ_s for D_s. Consider the bottommost edge e of D_s, and let Δ be the triangle of T^* that is incident to e. By construction, Δ is either an ear of T^* or is the triangle defined by e and u_s. In the latter case, we set $\Delta_s = \Delta$. In the former case, we claim that T^* has an inner triangle Δ' with two vertices on D_s: follow the path from Δ in the weak dual of T^*; while the path does not encounter an inner triangle, the next triangle must have an edge of D_s as a side. There is only a limited number of such edges, so eventually we must meet an inner triangle Δ'. We then set $\Delta_s = \Delta'$; see Fig. 6. Note that Δ_s might be $l_s l'_s u_s$.

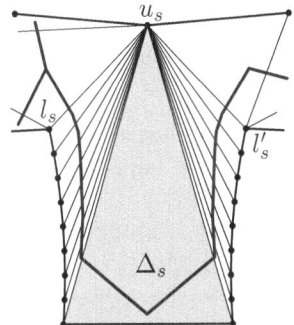 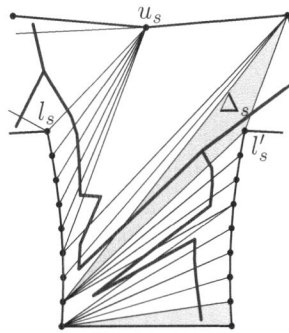

Fig. 6. Triangulations of D_s in P_D^* with $\Delta_s = \Delta$ (left), and with Δ being an ear (red) and Δ_s an inner triangle (right). The fat tree indicates the dual.

For each sink s, let the polygon $P_{D_s}^{u_s}$ consist of the D_s extended by the vertex u_s (cf. Definition 3.2). Let T^* be a triangulation of P_D^*. We show how to map T^* to a triangulation T^+ of P_D^+ and to triangulations T_s of $P_{D_s}^{u_s}$, for each s.

We first describe T^+. It contains every triangle of T^* with all three vertices in P_D^+. For each triangle Δ in T^* with two vertices on P_D^+ and one vertex on the left chain of a sink gadget D_s, we replace the vertex on D_s by l_s. Similarly, if the third vertex of Δ is on the right chain of D_s, we replace it by l'_s. For every sink s, the triangle Δ_s has one vertex at a point u_i of the upper chain. In T^+, we replace Δ_s by the triangle $l_s l'_s u_i$. No two triangles overlap, and they cover all of P_D^+. Thus, T^+ is indeed a triangulation of P_D^+.

Now we describe how to obtain T_s, for a sink $s \in S$. Each triangle of T^* with all vertices on $P_{D_s}^{u_s}$ is also in T_s. Each triangle with two vertices on D_s and one vertex not in $P_{D_s}^{u_s}$ is replaced in T_s by a triangle whose third vertex is moved to u_s in T_s (note that this includes Δ_s); see Fig. 7. Again, all triangles cover $P_{D_s}^{u_s}$ and no two triangles overlap.

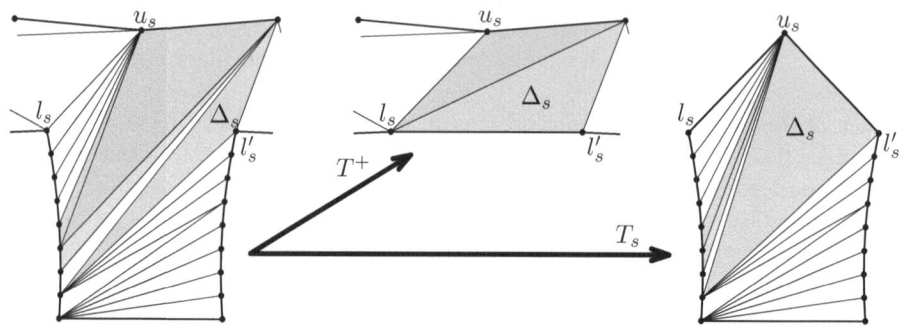

Fig. 7. Obtaining T^+ and T_s from T^*

Eventually, we show that a flip in T^* corresponds to at most one flip either in T^+ or in precisely one T_s for some sink s. We do this by considering all the possibilities for two triangles that share a common flippable edge. Note that by construction no two triangles mapped to triangulations of different polygons $P_{D_s}^{u_s}$ and $P_{D_t}^{u_t}$ can share an edge (with $t \neq s$ being another sink).

Case 1. We flip an edge between two triangles that are either both mapped to T^+ or to T_s and are different from Δ_s. This flip clearly happens in at most one triangulation.

Case 2. We flip an edge between a triangle Δ_1 that is mapped to T_s and a triangle Δ_2 that is mapped to T^+, such that both Δ_1 and Δ_2 are different from Δ_s. This results in a triangle Δ_1' that is incident to the same edge of $P_{D_s}^{u_s}$ as Δ_1(for each such triangle, the point not incident to that edge is called the *apex*), and a triangle Δ_2' having the same vertices of P_D^+ as Δ_2. Since the apex of Δ_1 is a vertex of the upper chain or z (otherwise, it would not share an edge with Δ_2), it is mapped to u_s, as is the apex of Δ_1'. Also, the apex of Δ_2' is on the same chain of D_s as the one of Δ_2. Hence, the flip affects neither T^+ nor T_s.

Case 3. We flip the edge between a triangle Δ_2 mapped to T^+ and Δ_s. By construction, this can only happen if Δ_s is an inner triangle. The flip affects only T^+, because the new inner triangle Δ_s' is mapped to the same triangle in T_s as Δ_s, since both apexes are moved to u_s.

Case 4. We flip the edge between a triangle Δ of T_s and Δ_s. Similar to Case 3, this affects only T_s, because the new triangle Δ_s' is mapped to the same triangle in T^+ as Δ_s, since the two corners are always mapped to l_s and l_s'.

Thus, σ induces a flip sequence σ_1 in P_D^+ and flip sequences σ_s in each $P_{D_s}^+$ so that $|\sigma_1| + \sum_{s \in S} |\sigma_s| \leq |\sigma|$. Furthermore, each flip sequence σ_s transforms $P_{D_s}^{u_s}$ from one extreme triangulation to the other. By the choice of d and Lemma 3.4, the triangulations T_s have to be transformed so that Δ_s has a vertex at u_s at some point, and $|\sigma_s| \geq 4d - 4$. Thus, σ_1 is a flip traversal, and $|\sigma_1| \leq |\sigma| - N(4d - 4)$, as claimed.   □

In order to obtain a static RSA from a changing flip traversal, we use the notion of a *trace*. A *trace* is a domain on the $\beta n \times \beta n$ grid. It consists of *edges* and *boxes*: an edge is a line segment of length 1 whose endpoints have positive integer coordinates; a box is a square of side length 1 whose corners have positive integer coordinates. Similar to arborescences, we require that a trace R (i) is (topologically) connected; (ii) contains the root $(1, 1)$; and (iii) from every grid point contained in R there exists an x- and y-monotone path to the root that lies completely in R. We say R is a *covering trace* for S (or, R *covers* S) if every sink in S is part of R.

Let σ_1 be a flip traversal as in Lemma 4.5. By Lemma 4.3, each triangulation in σ_1 corresponds to a chain path. This gives a covering trace R for S in the following way. For every flip in σ_1 that extends the chain path, we add the corresponding edge to R. For every flip in σ_1 that changes a bend, we add the corresponding box to R. Afterwards, we remove from R all edges that coincide with a side of a box in R. Clearly, R is (topologically) connected. Since σ_1 is a flip traversal for S, every sink is covered by R (i.e., incident to a box or edge in R). Note that every grid point p in R is connected to the root by an x- and y-monotone path on R, since at some point p belonged to a chain path in σ_1. Hence, R is indeed a trace, the unique *trace of σ_1*.

Next, we define the *cost* of a trace R, $cost(R)$, so that if R is the trace of a flip traversal σ_1, then $cost(R)$ gives a lower bound on $|\sigma_1|$. An edge has cost 2. Let B be a box in R. A *boundary side* of B is a side that is not part of another box. The cost of B is 1 plus the number of boundary sides of B. Then, $cost(R)$ is the total cost over all boxes and edges in R. For example, the cost of a tree is twice the number of its edges, and the cost of an $a \times b$ rectangle is $ab + 2(a+b)$. An edge can be interpreted as a degenerated box, having two boundary sides and no interior. The following proposition is proved in the full version.

Proposition 4.6. *Let σ_1 be a flip traversal and R the trace of σ_1. Then $cost(R) \leq |\sigma_1|$.*

Now we relate the length of an RSA for S to the cost of a covering trace for S, and thus to the length of a flip traversal. Since each sink (s_x, s_y) is connected in R to the root by a path of length $s_x + s_y$, traces can be regarded as generalized RSAs. In particular, we make the following observation.

Observation 4.7. *Let R be a covering trace for S that contains no boxes, and let A_{σ_1} be a shortest path tree in R from the root to all sinks in S. Then A_{σ_1} is an RSA for S.*   □

If σ_1 contains no flips that change bends, the corresponding trace R has no boxes. Then, R contains an RSA A_{σ_1} with $2|A_{\sigma_1}| \leq cost(R)$, by Observation 4.7. The next lemma shows that, due to the size of β, there is always a shortest covering trace for S that does not contain any boxes. See the full version for the proof.

Lemma 4.8. *Let σ_1 be a flip traversal of S. Then there exists a covering trace R for S in the $\beta n \times \beta n$ grid such that R does not contain a box and such that $cost(R) \leq |\sigma_1|$.*

Now we can finally complete the proof of Theorem 4.1 by giving the second direction of the correspondence.

Lemma 4.9. *Let $k \geq 1$ and let σ be a flip sequence on P_D^* from T_1 to T_2 with $|\sigma| \leq 2\beta k + (4d-2)N$. Then there exists an RSA for S of length at most k.*

Proof. Trivially, there always exists an RSA on S of length less than $2nN$, so we may assume that $k < 2nN$. Hence (recall that $\beta = 2N$ and $d = nN$),

$$2\beta k + 4dN - 2N < 2 \times 2N \times 2nN + 4nN^2 - 2N < 12nN^2 < (d-1)^2,$$

for $n \geq 14$ and positive N. Thus, since σ meets the requirements of Lemma 4.5, we can obtain a flip traversal σ_1 for S with $|\sigma_1| \leq 2\beta k + 2N$. By Lemma 4.8 and Observation 4.7, we can conclude that there is an RSA A for S that has length at most $\beta k + N$. By Theorem 2.1, there is an RSA A' for S that is not longer than A and that lies on the Hanan grid for S. The length of A' must be a multiple of β. Thus, since $\beta > N$, we get that A' has length at most βk, so the corresponding arborescence for S on the $n \times n$ grid has length at most k. $\square$

References

[1] Abel, Z., Ballinger, B., Bose, P., Collette, S., Dujmović, V., Hurtado, F., Kominers, S., Langerman, S., Pór, A., Wood, D.: Every large point set contains many collinear points or an empty pentagon. Graphs Combin. 27, 47–60 (2011)

[2] Aichholzer, O., Mulzer, W., Pilz, A.: Flip Distance Between Triangulations of a Simple Polygon is NP-Complete. ArXiv e-prints (2012) arXiv:1209.0579 [cs.CG]

[3] Bose, P., Hurtado, F.: Flips in planar graphs. Comput. Geom. 42(1), 60–80 (2009)

[4] Culik II, K., Wood, D.: A note on some tree similarity measures. Inf. Process. Lett. 15(1), 39–42 (1982)

[5] Eppstein, D.: Happy endings for flip graphs. JoCG 1(1), 3–28 (2010)

[6] Hanke, S., Ottmann, T., Schuierer, S.: The edge-flipping distance of triangulations. J. UCS 2(8), 570–579 (1996)

[7] Hurtado, F., Noy, M., Urrutia, J.: Flipping edges in triangulations. Discrete Comput. Geom. 22, 333–346 (1999)

[8] Hwang, F., Richards, D., Winter, P.: The Steiner Tree Problem. Annals of Discrete Mathematics (1992)

[9] Lawson, C.L.: Transforming triangulations. Discrete Math. 3(4), 365–372 (1972)

[10] Lawson, C.L.: Software for C^1 surface interpolation. In: Rice, J.R. (ed.) Mathematical Software III, pp. 161–194. Academic Press, NY (1977)

[11] Lubiw, A., Pathak, V.: Flip distance between two triangulations of a point-set is NP-complete. In: Proc. 24th CCCG, pp. 127–132 (2012)

[12] Pilz, A.: Flip distance between triangulations of a planar point set is APX-hard. ArXiv e-prints (2012) arXiv:1206.3179 [cs.CG]

[13] Rao, S.K., Sadayappan, P., Hwang, F.K., Shor, P.W.: The rectilinear Steiner arborescence problem. Algorithmica 7, 277–288 (1992)

[14] Shi, W., Su, C.: The rectilinear Steiner arborescence problem is NP-complete. In: Proc. 11th SODA, pp. 780–787 (2000)

[15] Sleator, D., Tarjan, R., Thurston, W.: Rotation distance, triangulations and hyperbolic geometry. J. Amer. Math. Soc. 1, 647–682 (1988)

[16] Trubin, V.: Subclass of the Steiner problems on a plane with rectilinear metric. Cybernetics 21, 320–324 (1985)

[17] Urrutia, J.: Algunos problemas abiertos. In: Proc. IX Encuentros de Geometría Computacional, pp. 13–24 (2001)

Empirical Evaluation of the Parallel Distribution Sweeping Framework on Multicore Architectures

Deepak Ajwani[1] and Nodari Sitchinava[2,3]

[1] Bell Laboratories Ireland, Dublin, Ireland
[2] Karlsruhe Institute of Technology, Karlsruhe, Germany
[3] University of Hawaii, Manoa

Abstract. In this paper, we perform an empirical evaluation of the Parallel External Memory (PEM) model in the context of geometric problems. In particular, we implement the parallel distribution sweeping framework of Ajwani, Sitchinava and Zeh to solve batched 1-dimensional stabbing max problem. While modern processors consist of sophisticated memory systems (multiple levels of caches, set associativity, TLB, prefetching), we empirically show that algorithms designed in simple models, that focus on minimizing the I/O transfers between shared memory and single level cache, can lead to efficient software on current multicore architectures. Our implementation exhibits significantly fewer accesses to slow DRAM and, therefore, outperforms traditional approaches based on plane sweep and two-way divide and conquer.

1 Introduction

Modern multicore architectures have complex memory systems involving multiple levels of private and/or shared caches, set associativity, TLBs, and prefetching effects. It is considered challenging to design and even engineer algorithms to directly optimize the running time on such architectures [15]. Furthermore, algorithms optimized for one architecture may not be optimal for another. To address these issues, various computational models [5,9,10,12,13] have been proposed in recent years. These computational models are simple (usually assuming only two levels of memory hierarchy, out of which one is shared) as they abstract away the messy architectural details. Also, the performance metric of these models involve a single objective function such as minimizing shared memory accesses. The simplicity of these models allows the design of practical algorithms that are expected to work well on various multicore architectures. It also allows us to compare the relative performance of algorithms theoretically.

The success of a computational model crucially depends on how well the theoretical prediction of an algorithm in that model matches the actual running time on real systems. Unfortunately so far, there has been little empirical work (such as [20]) to evaluate the predictions of algorithmic performance using these models on real multicore architectures. It is not even clear if these models can lead to the design of algorithms that are faster on current multicore systems (with 2 - 48 cores) than those designed in the traditional RAM model, external

H.L. Bodlaender and G.F. Italiano (Eds.): ESA 2013, LNCS 8125, pp. 25–36, 2013.
© Springer-Verlag Berlin Heidelberg 2013

memory model and the PRAM model. In fact, many of the algorithms designed in these models for multicores seem quite sophisticated and are likely to have high constant factors that can pay off only for architectures with hundreds of cores. This state of affairs is in sharp contrast with the sequential cache-efficient models, where a considerable empirical work (e.g., [6,11]) evaluating the algorithms on real systems exists.

At the core of the debate for the computational model is the choice of the performance metric that an algorithm designer should optimize for the current multicore systems. In the traditional RAM (and PRAM) model of computation, the algorithms are designed to minimize the number of instructions (and parallel instructions) executed by the algorithm. The *external memory (EM)* model [1] when applied to cached memories (e.g., see [16]) aims at minimizing the cache misses, ignoring the number of instructions. The *parallel external memory (PEM)* model [5] aims at minimizing the number of *parallel* cache misses.

In this work, we demonstrate that algorithms designed in simple models, that focus on minimizing the parallel I/O transfers between shared memory and a single level cache, can lead to a software performing great in practice on real multicore systems. For this purpose, we consider the algorithms to solve the problem of answering batched planar orthogonal stabbing-max queries. This problem is a fundamental geometric primitive and together with its variants is used as subroutines in solutions of many popular geometric problems such as point location in an orthogonal subdivision of the plane, orthogonal ray shooting, batched (offline) dynamic predecessor queries in 1-dimensional array and batched union-find. Also, this problem has been well-studied in various computational models and many different optimal solutions for it are known in these models. Thus, it provides a test-bed for evaluating the efficacy of theoretical analysis in various models on real multicore architectures. Another reason for selecting this non-HPC application is that the ratio of memory accesses to computation in the solutions of this problem is similar to that of many data-intensive geometric applications. For instance, our engineered PEM solution for this problem is based on the parallel distribution sweeping framework and this framework has been used for designing a wide range of other geometric algorithms in the PEM model [3,4] and a basis for PEM data structures [19].

We empirically compare the different solutions and show that a carefully engineered solution based on an algorithm in the PEM model gives the best performance on various multicore systems, outperforming traditional approaches based on plane sweep, sequential distribution sweeping and two-way divide-and-conquer. Using hardware profilers, we show that this solution exhibits significantly fewer number of accesses to slow DRAM which is correlated with the improved running time.

Since the cache line on modern systems is typically 64 bytes, I/O-efficient solutions also need to be work-efficient to compete with RAM algorithms. In other words, the total number of instructions of a cache-efficient algorithm should asymptotically match that of the best RAM solution. Therefore, we design an algorithm that is both I/O-optimal and work-efficient. To the best of our knowledge, this is the first work-efficient I/O-optimal algorithm for this problem.

2   Computational Models

External Memory Model. The widely used external memory model or the I/O model by Aggarwal and Vitter [1] assumes a two level memory hierarchy. The internal memory has a limited size and can hold at most M objects (points/line-segments) and the external memory has a conceptually unlimited size. The computation can only use the data in the internal memory, while the input and the output are stored in the external memory. The data transfer between the two memories happens in blocks of B objects. The measure of performance of an algorithm is the number of I/Os (cache misses) it performs. The number of I/Os needed to read n contiguous items from the external memory is $\mathrm{scan}(n) = \Theta(n/B)$. The number of I/Os required to sort n items is $\mathrm{sort}(n) = \Theta((n/B)\log_{M/B}(n/B))$. For all realistic values of n, B, and M, $\mathrm{scan}(n) < \mathrm{sort}(n) < n\log_2 n$.

Parallel External Memory (PEM) Model. The *parallel external memory* (PEM) model [5] is a simple parallelization of the EM model. It consists of P processors, each with a *private cache* of size M (see Figure 1). Processors communicate with each other through access to a *shared memory* of conceptually unlimited size. Each processor can use only data in its private cache for computation.

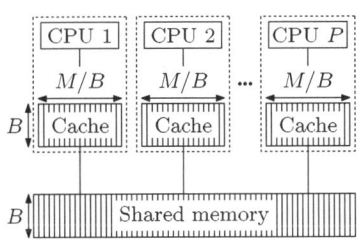

The caches and the shared memory are divided into *blocks* of size B. Data is transferred between the caches and shared memory using *parallel input-output* (I/O) operations. During each such operation, each processor can transfer one block between shared memory and its private cache. The cost of an algorithm is the number of I/Os it performs. Concurrent reading of the same block by multiple processors is allowed but concurrent block writes are disallowed (similar to a CREW PRAM). The cost of sorting

Fig. 1. The PEM model

in this model is $\mathrm{sort}_P(n) = \mathrm{O}\!\left(\frac{n}{PB}\log_{M/B}\frac{n}{B}\right)$ parallel I/Os, provided $P \le n/B^2$ and $M = B^{\mathrm{O}(1)}$ [5].

The PEM model provides the simplest possible abstraction of current multicore chips, focusing on the fundamental I/O issues that need to be addressed when designing algorithms for these architectures, similar to the I/O model [1] in the sequential setting.

3   1-D Stabbing Max Algorithms

In this section, we describe various algorithms that we implemented and used for our experimental study. We begin with formally describing the problem.

Definition 1 (Batched 1-D Stabbing-Max Problem). *Given a set of n horizontal line segments and points on the plane, report for each point the closest segment that lies directly below it.*

RAM Algorithm. In the classical RAM model, this problem is solved using the sweep line paradigm [17,7]. We sweep a hypothetical vertical line across the plane in increasing x-coordinate and perform some computation at each segment endpoint or query point. We maintain an ordered set A of *active segments* — all segments which intersect the sweep line, ordered by the y-coordinates. A segment is inserted into A when the sweep line encounters its left endpoint and removed when it encounters the right endpoint. An answer to a query point q is the segment in A with the largest y-coordinate that is smaller than the y-coordinate of q, i.e., the predecessor of q in A according to the y-ordering.

For n line segments and query points, there are O(n) insertions, deletions and predecessor searches in A. Since each of these operations can be performed in O($\log n$) time by maintaining A as a balanced binary search tree, the total complexity of this algorithm is O($n \log n$) instructions.

Sequential I/O-optimal Solution. The sequential I/O-efficient solution for this problem proceeds using the distribution sweeping framework of Goodrich et al. [14] as follows.

Let r_q be a variable associated with each query point q which we will use to store the answer. Initially r_q is initialized to a virtual horizontal line $y = -\infty$.

We partition the space into $K = \min\{M/B, n/M\}$ vertical slabs $\sigma_1, \ldots, \sigma_K$, so that each slab contains equal number of points (endpoints of horizontal segments or query points) and perform a sweep of the input by increasing y-coordinate. During the sweep we maintain for each slab σ_i a segment s_{σ_i} which is the highest segment that spans σ_i encountered by the sweep. When the sweep line encounters the query point $q \in \sigma_i$, we update r_q with s_{σ_i} iff $y(s_{\sigma_i}) > y(r_q)$. During the sweep we also generate slab lists Y_{σ_i}. A copy of a query q (resp., segment s) is added to Y_{σ_i} if q (resp., at least one of the endpoints of s) lies in slab σ_i. The sweep is followed by a recursive processing of each slab, using Y_{σ_i} as input for the recursive call. The recursion terminates when each slab contains O(M) points and the problem can be solved in internal memory, for example, by using the plane sweep algorithm.

Note, that if the initial objects are sorted by y-coordinates, we can generate the inputs Y_{σ_i} for the recursive calls sorted by y-coordinate during the sweep. Thus, the sweep at each of O($1 + \log_K(n/M)$) recursive levels takes O(n/B) I/Os and the total I/O complexity of distribution sweeping is O$\left(\frac{n}{B}(1 + \log_K n/M)\right) =$ sort(n) I/Os.

Work-optimal Solution. Note that a naive implementation of the sweep in internal memory might potentially result in updating K different variables s_{σ_i}

whenever a segment is encountered during the sweep. This could lead to $O(Kn)$ instructions at each recursive level, resulting in total $O(Kn \log_K n)$ instructions, which is larger than $O(n \log_2 n)$ instructions of the plane sweep algorithm. At the same time, the plane sweep algorithm could result in up to $O(n \log_2 n)$ I/Os, which is larger than $\mathrm{sort}(n)$ I/Os of the above algorithm.

To achieve optimal internal computation time while maintaining the optimal $\mathrm{sort}(n)$ I/O complexity we store segments s_{σ_i} in a segment tree T over K intervals defined by the slabs σ_i. Since, we are interested only in segments that fully span the slabs, each segment is stored only in one node. Also, at each node we store only the highest segment encountered up to that point in the sweep. Thus, $|T| = O(K)$, i.e. T fits in internal memory. Consider the nodes on the root to leaf path which correspond to the intervals containing q. We update r_q to the highest segment stored at these nodes. Thus, maintaining T and updating r_q takes $O(\log_2 K)$ instructions per update/query, and over $O(1 + \log_K N/M)$ recursive levels of distribution sweeping adds up to at most $O(n \log_2 n)$ instructions, which is optimal.

Parallel External Memory Solution. The PEM solution is based on the *parallel* distribution sweeping framework introduced by Ajwani et al. [3]. It differs from the sequential distribution sweeping by recursively dividing the plane into $K := \max\{2, \min\{\sqrt{n/P}, M/B, P\}\}$ vertical slabs[1] and performing the sweep in parallel using all P processors. During recursion, the slabs are processed concurrently using sets of $\Theta(P/K)$ distinct processors per slab. The parallel recursion proceeds for $O(\log_K P)$ rounds, until there are $\Theta(P)$ slabs remaining, at which point, each slab is processed concurrently using a single processor running the sequential I/O-efficient solution.

To perform the sweep of a single recursive level in parallel using multiple processors, each processor performs distribution sweeping on an equal fraction of the input. Note, that such a sweep sets the values of r_q correctly only if both the query q and the spanning segment s_{σ_i} below it are processed by the same processor. To correct the values r_q across the boundaries of the parallel sweeps we perform a round of parallel reduction on segments and queries using MAX associative operator [8]. Finally, we compact the portions of slab lists Y_{σ_i} generated by different processors into contiguous slab lists to be used as input for recursive calls. The details of the algorithm follow directly from [3] but can also be found in the full version [2].

The parallel I/O complexity of the above algorithm is $O(\mathrm{sort}_P(n))$ I/Os.

Work-optimal Solution. Similar to the sequential I/O model, we can achieve work optimality in the PEM model algorithm by maintaining a segment tree T on the K child slabs. In this case, all processors keep their own copy of T and the parallel reduction (using MAX operator) is performed over not only the K leaves, but also the $K - 1$ internal nodes of T. This does not affect the asymptotic number of parallel I/Os, but makes the scheme work-optimal, i.e. $O(\frac{n}{P} \log n)$ instructions per processor.

---

[1] The explanation for this choice of K can be found in [5].

2-way Distribution Sweeping. As a PRAM solution, we consider a recursive 2-way distribution sweeping algorithm. This framework is akin to divide-and-conquer paradigm, that is archetype for many PRAM algorithms. The 2-way distribution is continued recursively till the slab size is smaller than a fixed constant and at that stage, plane sweep algorithm is used as a base case. The distribution step is a simplified version of the corresponding step in the PEM algorithm, as the considerations of work-optimality no longer apply.

4   Implementation Details

We implemented our algorithms in C++, using OpenMP for parallelization. The engineered implementation uses some simple techniques to improve the running time of the theoretical algorithm, while trying to preserve its worst-case asymptotic guarantee on the number of shared cache accesses.

The parallel distribution sweeping calls for setting the branching parameter at $K = \max\{2, \min\{M/B, \sqrt{n/P}, P\}\}$. The parameter M also defines the size of the recursive base case. We experimentally determine the best choice of M. In particular we found that setting M to be a large fraction (e.g., $1/3$ or $1/4$) of the L3 cache results in best running times.

Having determined M, we observe that for computing K, in our compute systems the number of processors (up to 12) is far below the other two terms. Thus, the first recursive level is always a single P-way parallel distribution sweeping round, which results in P vertical slabs each of which can be processed independently of others in the consequent phases. Thus, after the parallel distribution, each of P resulting vertical slabs is assigned to a separate thread which processes it using a sequential distribution sweeping algorithm.

To perform the parallel sweep, we divide the input based on the y-coordinate among the P threads, conceptually, assigning a horizontal slab of objects to each thread. The thread with the smaller ID gets the lower y values. This can be viewed as a $P \times P$ matrix where the columns correspond to the different slabs and the rows correspond to the different threads.

We perform the prefix sum on the $P \times P$ array sequentially as the overheads associated with the synchronization barrier of OpenMP are too high to justify this operation in parallel.[2]

We combine the second scan of the data (due to reduction) with the step of compacting child slab lists into contiguous vectors. During the compaction, each processor p_j copies all partial chunks of child slab σ_j into the contiguous space. Note, the propagation of the results of the prefix sums simply needs to update the result of each query point that had been assigned the sentinel line $y = -\infty$ with the result of the prefix sums value. Thus, the propagation of the prefix sums values can be performed during this copying process.

Next, we process the P child slabs in parallel using sequential distribution sweeping. This recursively subdivides the slabs till the pre-specified threshold

---

[2] In our experiments, performing this step sequentially takes less than a millisecond, while the overall running time is in dozens or hundreds of seconds.

M is reached. When generating the input lists for the child slabs, we also store the total number of segments and query points for the child slabs. If for any slab, either the number of segments or query points is zero, we do not process it or its child invocations any further.

Space Efficiency. We carefully engineered our algorithms to reduce the space requirement of our implementations considerably. This is done while ensuring that the running time of our implementations is not affected by the space reduction. We provide more details of this in the full version [2].

Randomized vs. Deterministic Computation of Slab Boundaries. Deterministic identification of slab boundaries such that all the child slabs at each level of recursion contain the same number of objects, requires sorting the input based on the x-coordinate and storing $O(n/M)$ equally spaced entries of the sorted input in a separate array. We avoid the extra sort by instead determining the slab boundaries by partitioning the space into uniform vertical slabs. This optimization works well for random input, but in the worst case can result in the recursion depth as large as $O(\log_K \delta)$, where δ is the *spread* of the point set – the ratio between the largest and the smallest (horizontal) distance between a pair of points. In case of a large base case of the recursion and randomized input, this is not an issue. But in the case of double precision coordinates, the worse case analysis dictates that the depth of the recursion can be very large.

Constant Factors vs. EM Implementation. The I/O complexity of the sequential distribution sweeping framework is $O(n/B(1 + \log_K n/M))$, where $K = \min\{M/B, n/M\}$. Since in our experimental settings $K = n/M$, there are only 2 recursive levels: one for distribution sweeping and one for the sweep line at the base case. Thus, the implementation performs two sequential scans of the input.

In the parallel version, we have to perform two additional scans. Specifically, we perform one extra recursive step – the parallel distribution. During this step, each processor scans n/P items and writes them out into its private child slabs. After the prefix sums, which takes negligible amount of time, we must (a) propagate the result of the prefix sums to the queries that contain only sentinel values as the result and (b) construct each child slab in contiguous space. As described earlier, we combine these two tasks into a single scan.

Thus, combined with the two scans of the parallel recursive invocation of the sequential distribution sweeping, the parallel implementation performs a total of four scans of the input, i.e., twice as many as the sequential version. Since all scans are performed in parallel and in expectation each child slab contains equal number of items, the total I/Os performed by each processor is $2/P$ times the number of sequential I/Os, and (ignoring the speedup due to faster parallel internal computation) we should expect the speed up of $P/2$ on P processors.

Sorting. To perform the initial sorting of the input by the y-coordinate, we used the sorting implementation from the C++ Multicore Standard Template Library (MCSTL) [18] that is now part of the GNU libstdc++ library. For the base case of plane sweep algorithm, we use the C++ Standard Template Library (STL) sorting implementation.

5   Experiments

We performed extensive experimentation studying the performance of these algorithms (i.e., plane sweep algorithm, work-optimal I/O-optimal solution, work-optimal PEM algorithm and 2-way distribution sweeping) on various input types and on many different multicore architectures. In addition to measuring the running time of these algorithms, we used papi library and the Linux perfctr kernel module to read the hardware performance counters and measure cache misses, DRAM accesses, TLB misses, branch mispredictions, number of instructions etc.. This section summarizes the key findings of our experiments.

Our query points were generated uniformly at random inside the grid of size Grid Size × Grid Size. To elicit the asymptotic worst case cache performance of point location algorithms, we focus on long segments, whose length is chosen uniformly at random between Grid Size/4 and 3·Grid Size/4 and are at a random y-coordinate. Full discussion of the effects of segment lengths on the behaviors of various algorithms can be found in [2].

Configuration. We ran our implementation on the following multicore systems:

1. A system with a single 4-core 2.66 GHz Intel Core i7-920 processor and a total of 12.3GB RAM. Each core can run 2 threads due to hyperthreading. The processor has an L3 cache of size 8192 KB that is shared among all 4 cores. The L2 cache of 256 KB is only shared among pairs of cores.
2. A system with 4 × 12-core 1.9 GHz AMD Opteron 6168 processors and total of 264 GB of RAM. Each core contains a private L2 cache of 512 KB and groups of 6 cores share an L3 cache of 5118KB. Thus, each processor contains two L3 caches of combined size of just over 10MB.
3. A system with 2 x 16-core 2.6 GHz AMD Opteron 6282 SE processors and total of 96 GB RAM. Each core has its private L2 cache while the L3 cache is shared between 16 cores. The L2 cache size is 2 MB and L3 cache size is 16 MB.

All configurations run Linux kernels and the codebase was compiled using g++-2.4 compiler and -O3 flag.

Spatio-temporal Locality in Our Setting. The cache line size for all cache levels on all 3 systems is 64 bytes. Since our objects take 32 bytes of space, it appears that each cache line can hold only two objects. Therefore, at a first glance it is not clear if I/O efficient algorithm can utilize the spatial locality for any improvement in runtime. However, we observed that given an array that is too large to fit in cache and which contains our 32-byte objects, it takes 4-5 times faster to access the objects sequentially rather than performing access in random locations. This observation can be explained by the fact that the memory system prefetches 2-3 cache lines when performing a sequential scan. Thus, during sequential scan the prefetcher amplifies the size of the cache line by the number of lines being prefetched.[3]

---

[3] For this experiment, the array must contain the actual objects and not just pointers to the objects, which could be allocated anywhere in memory.

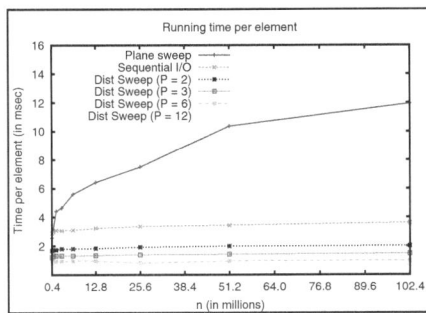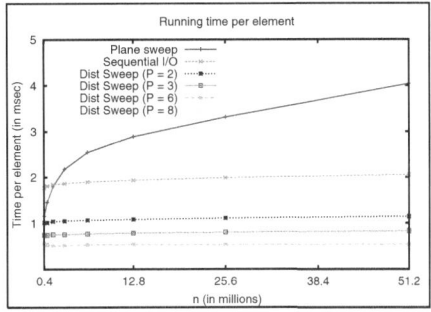

Fig. 2. Runtimes on the configuration 2 (left) and configuration 1 (right) per element. The plots exclude the times to perform initial sorting of inputs by the y-coordinate for distribution sweeping and x-coordinate for the plane sweep.

Another benefit of performing K-way distribution sweeping is that it allows us to utilize temporal locality by reducing the number of recursive calls. In particular, K is chosen as $K = \min\{n/M, M/B\}$ and the number of recursive levels is $(1 + \log_K(n/M))$. Given limit of RAM size on our systems and the large size of L3 cache, it appears from our experiments that K is set to n/M on configuration 1 and 2, resulting in a single recursive level dedicated to (sequential) distribution (with the recursive base case performing plane sweep on chunks that fit in L3 cache). On configuration 3, it requires two recursive calls. The various trade-offs involved in selecting the correct values of parameters K and M and the effect of these parameters on the actual run-time of our PEM implementation are described in the full version [2].

Random Access vs. I/O-efficient Algorithms. Figure 2 shows the absolute running times for the plane sweep and (parallel) distribution sweeping algorithms. One can see improvements in runtimes with the increase in the number of processors used. Also note the difference in the slopes in the graphs of the plane sweep algorithm compared to distribution sweeping algorithms. This is due to larger asymptotic number of cache misses of the plane sweep algorithm.

Figure 3 demonstrates this difference better. It shows the speedup of the sequential and parallel distribution sweeping algorithms relative to the plane sweep algorithm for long segments. In this figure one can see the effects of cache-efficiency on runtimes. It clearly shows that the I/O-efficient algorithms outperform the plane sweep algorithm as the input sizes increase. Recall our discussion that for the parameters of our systems $K = n/M$ and the I/O complexity of the distribution sweeping algorithm is $O((n/B)(1 + \log_K n/M)) = O(n/B)$. This explains the non-linear asymptotic speedup over plane sweep algorithm (with I/O complexity of $O((n/B)\log n/M)$) as a function of the input size.

Figure 4 shows the speedup that parallel distribution sweeping algorithm achieves relative to the sequential distribution sweeping algorithm.

PRAM vs. PEM Performance. Figure 5 (left) shows the comparative performance of the various algorithms on configuration 3. We observe that the PRAM

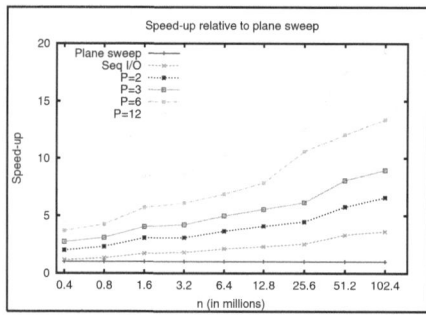 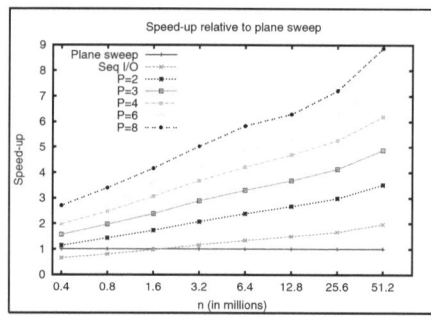

Fig. 3. Speedup of the distribution sweeping algorithms relative to the plane sweep algorithm on the configuration 2 (left) and configuration 1 (right). The plots exclude the times to perform initial sorting of inputs by the y-coordinate for distribution sweeping and x-coordinate for the plane sweep.

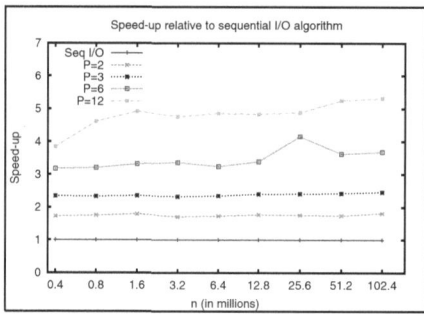 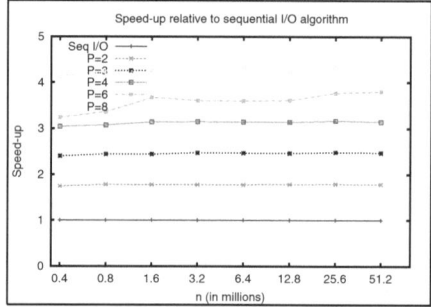

Fig. 4. Speedup of the parallel distribution sweeping algorithms relative to the sequential distribution sweeping algorithm on configuration 2 (left) and configuration 1 (right) systems. The plots exclude the times to perform initial sorting of inputs by the y-coordinate.

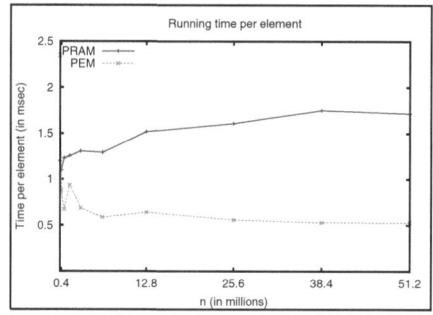 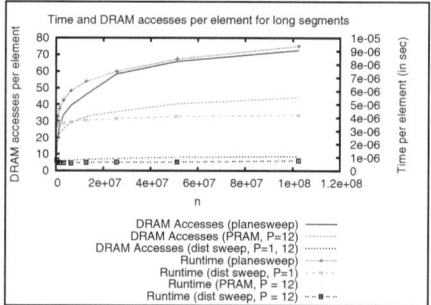

Fig. 5. Comparison of PEM and PRAM algorithms on 16 cores of configuration 3 is shown in the left figure. Running time and DRAM traffic for long segments on 12 cores of configuration 2 in the right.

implementation is significantly slower than the PEM algorithm. For instance, with 51.2 million segments and the same number of queries, PRAM implementation takes 96 seconds with 16 cores, while the PEM implementation only requires 30 seconds with the same number of cores (excluding the time for loading the input and sorting it, which is 18 seconds for both implementations). This is largely accounted for by the fact that the PRAM implementation makes poor use of temporal locality and thus, has larger number of recursive levels. In each recursive level, it scans all the segments and query points, increasing the DRAM accesses significantly.

DRAM Accesses and Cache Misses. We could not find a reliable way to measure only L3 cache misses: the `papi` library does not support measurement of shared cache events, while the hardware counters for LLC (Last Level Cache) counters returned suspiciously similar results to L2 cache misses. Instead we measured the total traffic to DRAM using `perf` tool. Figure 5 (right) shows a clear correlation between the total DRAM traffic and running times. It is interesting to note that although our algorithms are designed in simple 2-level cache model, they minimize the total traffic to DRAM, in spite of complex nature of modern memory systems.

6   Conclusions and Future Work

In this work, we explored the effects of caches on actual run-times observed on various multicore architectures in the context of the geometric stabbing-max query problem. This is used to understand how accurately the PEM model predicts the running time of combinatorial algorithms on current multicore architectures. On single-socket multicore architectures, our results show a direct correlation between traffic on DRAM memory controller and running times of implementations. Thus, the algorithms designed I/O-efficiently via the (parallel) distribution sweeping framework outperform the plane sweep algorithms which do not address the I/O-efficiency.

We chose to perfom our experiments on single-socket architectures, because the PEM model assumes uniform access latencies to shared memory. We conjecture that NUMA effects of DRAM access on multi-socket architectures might be better modeled by distributed computational models, where each processor copies data into "local" memory — DRAM address space associated with its socket — before processing it. Once the data is in the "local" DRAM banks, one can use the PEM algorithms to process it cache-efficiently. The experimental evaluation and modeling NUMA effects of multi-socket architectures is left for future investigations.

While we chose to implement an algorithm which was designed in the PEM model, it would be interesting to see how the implementations in other cache-conscious parallel models (for example, [9]) will fare in practice in similar setting.

Acknowledgments. We would like to thank Peter Sanders for encouraging to look at the work-optimality of PEM algorithms. We would also like to thank

Dennis Luxen and Dennis Schieferdecker for their extensive help with our implementations and getting `perf` and `papi` to run on our systems.

References

1. Aggarwal, A., Vitter, J.S.: The input/output complexity of sorting and related problems. Communications of the ACM 31(9), 1116–1127 (1988)
2. Ajwani, D., Sitchinava, N.: Empirical evaluation of the parallel distribution sweeping framework on multicore architectures. CoRR abs/1306.4521 (2013)
3. Ajwani, D., Sitchinava, N., Zeh, N.: Geometric algorithms for private-cache chip multiprocessors. In: de Berg, M., Meyer, U. (eds.) ESA 2010, Part II. LNCS, vol. 6347, pp. 75–86. Springer, Heidelberg (2010)
4. Ajwani, D., Sitchinava, N., Zeh, N.: I/O-optimal distribution sweeping on private-cache chip multiprocessors. In: IPDPS, pp. 1114–1123 (2011)
5. Arge, L., Goodrich, M.T., Nelson, M.J., Sitchinava, N.: Fundamental parallel algorithms for private-cache chip multiprocessors. In: SPAA, pp. 197–206 (2008)
6. Bender, M.A., Farach-Colton, M., Fineman, J.T., Fogel, Y.R., Kuszmaul, B.C., Nelson, J.: Cache-oblivious streaming B-trees. In: SPAA, pp. 81–92 (2007)
7. Bentley, J.L., Ottmann, T.A.: Algorithms for reporting and counting geometric intersections. IEEE Transactions on Computers 28(9), 643–647 (1979)
8. Blelloch, G.E.: Prefix sums and their applications. In: Reif, J.H. (ed.) Synthesis of Parallel Algorithms, pp. 35–60. Morgan Kaufmann Publishers (1993)
9. Blelloch, G.E., Chowdhury, R.A., Gibbons, P.B., Ramachandran, V., Chen, S., Kozuch, M.: Provably good multicore cache performance for divide-and-conquer algorithms. In: SODA, pp. 501–510 (2008)
10. Blelloch, G.E., Fineman, J.T., Gibbons, P.B., Simhadri, H.V.: Scheduling irregular parallel computations on hierarchical caches. In: SPAA, pp. 355–366. ACM (2011)
11. Brodal, G.S., Fagerberg, R., Vinther, K.: Engineering a cache-oblivious sorting algorithm. ACM Journal of Experimental Algorithmics 12 (2007)
12. Chowdhury, R.A., Ramachandran, V.: The cache-oblivious gaussian elimination paradigm: Theoretical framework, parallelization and experimental evaluation. In: SPAA, pp. 71–80 (2007)
13. Chowdhury, R.A., Ramachandran, V.: Cache-efficient dynamic programming for multicores. In: SPAA, pp. 207–216 (2008)
14. Goodrich, M.T., Tsay, J.J., Vengroff, D.E., Vitter, J.S.: External-memory computational geometry. In: FOCS, pp. 714–723 (1993)
15. Kang, S., Ediger, D., Bader, D.A.: Algorithm engineering challenges in multicore and manycore systems. IT - Information Technology 53(6), 266–273 (2011)
16. Mehlhorn, K., Sanders, P.: Scanning multiple sequences via cache memory. Algorithmica 35, 75–93 (2003), 10.1007/s00453-002-0993-2
17. Shamos, M.I., Hoey, D.: Geometric intersection problems. In: FOCS, pp. 208–215. IEEE Computer Society Press (1976)
18. Singler, J., Sanders, P., Putze, F.: MCSTL: The multi-core standard template library. In: Kermarrec, A.-M., Bougé, L., Priol, T. (eds.) Euro-Par 2007. LNCS, vol. 4641, pp. 682–694. Springer, Heidelberg (2007)
19. Sitchinava, N., Zeh, N.: A parallel buffer tree. In: SPAA, pp. 214–223 (2012)
20. Tang, Y., Chowdhury, R.A., Kuszmaul, B.C., Luk, C.K., Leiserson, C.E.: The Pochoir stencil compiler. In: SPAA, pp. 117–128 (2011)

Computing the Greedy Spanner in Linear Space

Sander P.A. Alewijnse, Quirijn W. Bouts,
Alex P. ten Brink, and Kevin Buchin

Eindhoven University of Technology, The Netherlands
{s.p.a.alewijnse,q.w.bouts,a.p.t.brink}@student.tue.nl, k.a.buchin@tue.nl

Abstract. The greedy spanner is a high-quality spanner: its total weight, edge count and maximal degree are asymptotically optimal and in practice significantly better than for any other spanner with reasonable construction time. Unfortunately, all known algorithms that compute the greedy spanner of n points use $\Omega(n^2)$ space, which is impractical on large instances. To the best of our knowledge, the largest instance for which the greedy spanner was computed so far has about 13,000 vertices.

We present a $O(n)$-space algorithm that computes the same spanner for points in $\mathbb{R}^d$ running in $O(n^2 \log^2 n)$ time for any fixed stretch factor and dimension. We discuss and evaluate a number of optimizations to its running time, which allowed us to compute the greedy spanner on a graph with a million vertices. To our knowledge, this is also the first algorithm for the greedy spanner with a near-quadratic running time guarantee that has actually been implemented.

1 Introduction

A t-spanner on a set of points, usually in the Euclidean plane, is a graph on these points that is a 't-approximation' of the complete graph, in the sense that shortest routes in the graph are at most t times longer than the direct geometric distance. The spanners considered in literature have only $O(n)$ edges as opposed to the $O(n^2)$ edges in the complete graph, or other desirable properties such as bounded diameter or bounded degree, which makes them a lot more pleasant to work with than the complete graph.

Spanners are used in wireless network design [7]: for example, high-degree routing points in such networks tend to have problems with interference, so using a spanner with bounded degree as network avoids these problems while maintaining connectivity. They are also used as a component in various other geometric algorithms, and are used in distributed algorithms. Spanners were introduced in network design [12] and geometry [5], and have since been subject to a considerable amount of research [9, 11].

There exists a large number of constructions of t-spanners that can be parameterized with arbitrary $t > 1$. They have different strengths and weaknesses: some are fast to construct but of low quality (Θ-graph, which has no guarantees on its total weight), others are slow to construct but of high quality (greedy spanner, which has low total weight and maximum degree), some have an extremely

H.L. Bodlaender and G.F. Italiano (Eds.): ESA 2013, LNCS 8125, pp. 37–48, 2013.

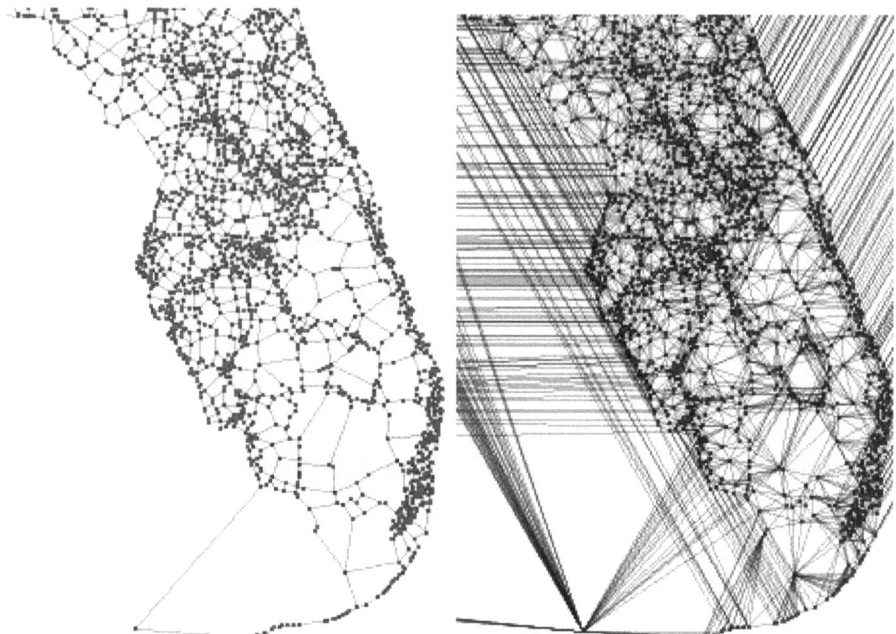

Fig. 1. The left rendering shows the greedy spanner on the USA, zoomed in on Florida, with $t = 2$. The dataset has 115,475 vertices, so it was infeasible to compute this graph before. The right rendering shows the Θ-graph on the USA, zoomed in on Florida, with $k = 6$ for which it was recently proven it achieves a dilation of 2.

low diameter (various dumbbell based constructions) and some are fast to construct in higher dimensions (well-separated pair decomposition spanners). See for example [11] for detailed expositions of these spanners and their properties.

The greedy spanner is one of the first spanner algorithms that was considered, and it has been subject to a considerable amount of research regarding its properties and more recently also regarding computing it efficiently. This line of research resulted in a $O(n^2 \log n)$ algorithm [2] for metric spaces of bounded doubling dimension (and therefore also for Euclidean spaces). There is also an algorithm with $O(n^3 \log n)$ worst-case running time that works well in practice [6]. Its running time tends to be near-quadratic in practical cases, but there are examples on which its running time is $\Theta(n^3 \log n)$. Its space usage is $\Theta(n^2)$.

Among the many spanner algorithms known, the greedy spanner is of special interest because of its exceptional quality: its size, weight and degree are asymptotically optimal, and also in practice are better than any other spanner construction algorithms with reasonable running times. For example, it produces spanners with about ten times as few edges, twenty times smaller total weight and six times smaller maximum degree as its closest well-known competitor, the Θ-graph, on uniform point sets. The contrast is clear in Fig. 1. Therefore, a method of computing it more efficiently is of considerable interest.

We present an algorithm whose space usage is $\Theta(n)$ whereas existing algorithms use $\Theta(n^2)$ space, while being only a logarithmic factor slower than the fastest known algorithm, thus answering a question left open in [2]. Our algorithm makes the greedy spanner practical to compute for much larger inputs than before: this used to be infeasible on graphs of over 15,000 vertices. In contrast, we tested our algorithm on instances of up to 1,000,000 points, for which previous algorithms would require multiple terabytes of memory. Furthermore, with the help of several optimizations we will present, the algorithm is also fast in practice, as our experiments show.

The method used to achieve this consists of two parts: a framework that uses linear space and near-linear time, and a subroutine using linear space and near-linear time, which is called a near-linear number of times by the framework. The subroutine solves the *bichromatic closest pair with dilation larger than t* problem. If there is an algorithm with a sublinear running time for this subproblem (possibly tailored to our specific scenario), then our framework immediately gives an asymptotically faster algorithm than is currently known. This situation is reminiscent to that of the minimum spanning tree, for which it is known that it is essentially equivalent to the *bichromatic closest pair* problem.

The rest of the paper is organized as follows. In Section 2 we review a number of well-known definitions, algorithms and results. In Section 3 we give the properties of the WSPD and the greedy spanner on which our algorithm is based. In Section 4 we present our algorithm and analyse its running time and space usage. In Section 5 we discuss our optimizations of the algorithm. Finally, in Section 6 we present our experimental results and compare it to other algorithms.

2   Notation and Preliminaries

Let V be a set of points in $\mathbb{R}^d$, and let $t \in \mathbb{R}$ be the intended dilation ($1 < t$). Let $G = (V, E)$ be a graph on V. For two points $u, v \in V$, we denote the Euclidean distance between u and v by $|uv|$, and the distance in G by $\delta_G(u, v)$. If the graph G is clear from the context we will simply write $\delta(u, v)$. The *dilation* of a pair of points is t if $\delta(u, v) \leq t \cdot |uv|$. A graph G has dilation t if t is an upper bound for the dilations of all pairs of points. In this case we say that G is a *t-spanner*. To simplify the analysis, we assume without loss of generality that $t < 2$.

We will often say that two points $u, v \in V$ *have a t-path* if their dilation is t. A pair of points *is without t-path* if its dilation is not t. When we say a pair of points (u, v) is the *closest* or *shortest* pair among some set of points, we mean that $|uv|$ is minimal among this set. We will talk about *a Dijkstra computation* from a point v by which we mean a single execution of the single-source shortest path algorithm known as Dijkstra's algorithm from v.

Consider the following algorithm that was introduced by Keil [10]:

Algorithm *GreedySpannerOriginal(V, t)*
1. $E \leftarrow \emptyset$
2.   **for** every pair of distinct points (u, v) in ascending order of $|uv|$
3.   **do if** $\delta_{(V,E)}(u, v) > t \cdot |uv|$
4.   **then** add (u, v) to E
5.   **return** E

Obviously, the result of this algorithm is a t-spanner for V. The resulting graph is called the *greedy spanner* for V, for which we shall present a more efficient algorithm than the above.

We will make use of the Well-Separated Pair Decomposition, or WSPD for short, as introduced by Callahan and Kosaraju in [3,4]. A WSPD is parameterized with a separation constant $s \in \mathbb{R}$ with $s > 0$. This decomposition is a set of pairs of nonempty subsets of V. Let m be the number of pairs in a decomposition. We can number the pairs, and denote every pair as $\{A_i, B_i\}$ with $1 \le i \le m$. Let u and v be distinct points, then we say that (u, v) is 'in' a well-separated pair $\{A_i, B_i\}$ if $u \in A_i$ and $v \in B_i$ or $v \in A_i$ and $u \in B_i$. A decomposition has the property that for every pair of distinct points u and v, there is exactly one i such that (u, v) is in $\{A_i, B_i\}$.

For two nonempty subsets X_k and X_l of V, we define $\min(X_k, X_l)$ to be the shortest distance between the two circles around the bounding boxes of X_k and X_l and $\max(X_k, X_l)$ to be the longest distance between these two circles. Let $diam(X_k)$ be the diameter of the circle around the bounding box of X_k. Let $\ell(X_k, X_l)$ be the distance between the centers of these two circles, also named the *length* of this pair. For a given separation constant $s \in \mathbb{R}$ with $s > 0$ as parameter for the WSPD, we require that all pairs in a WSPD are s-well-separated, that is, $\min(A_i, B_i) \ge s \cdot \max(diam(A_i), diam(B_i))$ for all i with $1 \le i \le m$.

It is easy to see that $\max(X_k, X_l) \le \min(X_k, X_l) + diam(X_k) + diam(X_l) \le (1 + 2/s)\min(X_k, X_l)$. As $t < 2$ and as we will pick $s = \frac{2t}{t-1}$ later on, we have $s > 4$, and hence $\max(X_k, X_l) \le 3/2 \min(X_k, X_l)$. Similarly, $\ell(X_k, X_l) \le \min(X_k, X_l) + diam(X_k)/2 + diam(X_l)/2 \le (1 + 1/s)\min(X_k, X_l)$ and hence $\ell(X_k, X_l) \le 5/4 \min(X_k, X_l)$.

For any V and any $s > 0$, there exists a WSPD of size $m = O(s^d n)$ that can be computed in $O(n \log n + s^d n)$ time and can be represented in $O(s^d n)$ space [3]. Note that the above four values (min, max, $diam$ and ℓ) can easily be precomputed for all pairs with no additional asymptotic overhead during the WSPD construction.

3   Properties of the Greedy Spanner and the WSPD

In this section we will give the idea behind the algorithm and present the properties of the greedy spanner and the WSPD that make it work. We assume we have a set of points V of size n, an intended dilation t with $1 < t < 2$ and a WSPD with separation factor $s = \frac{2t}{t-1}$, for which the pairs are numbered $\{A_i, B_i\}$ with $1 \le i \le m$, where $m = O(s^d n)$ is the number of pairs in the WSPD.

The idea behind the algorithm is to change the original greedy algorithm to work on well-separated pairs rather than edges. We will end up adding the edges in the same order as the greedy spanner. We maintain a set of 'candidate' edges for every well-separated pair such that the shortest of these candidates is the next edge that needs be added. We then recompute a candidate for some of these well-separated pairs. We use two requirements to decide on which pairs we perform a recomputation, that together ensure that we do not do too many recomputations, but also that we do not fail to update pairs which needed updating.

We now give the properties on which our algorithm is based. Omitted proofs are given in [1].

Observation 1 (Bose et al. [2, Observation 1]). *For every i with $1 \leq i \leq m$, the greedy spanner includes at most one edge (u, v) with (u, v) in $\{A_i, B_i\}$.*

Our definition of well-separatedness differs slightly from that in [2] but the observation still holds (see [1]). An immediate corollary is:

Observation 2 (Bose et al. [2, Corollary 1]). *The greedy spanner contains at most $O\left(\frac{1}{(t-1)^d} n\right)$ edges.*

Lemma 3. *Let E be some edge set for V. For every i with $1 \leq i \leq m$, we can compute the closest pair of points $(u, v) \in A_i \times B_i$ among the pairs of points with dilation larger than t in $G = (V, E)$ in $O(\min(|A_i|, |B_i|)(|V| \log |V| + |E|))$ time and $O(|V|)$ space.*

Proof. Assume without loss of generality that $|A_i| \leq |B_i|$. We perform a Dijkstra computation for every point $a \in A_i$, maintaining the closest point in $|B_i|$ such that its dilation with respect to a is larger than t over all these computations. To check whether a point that is considered by the Dijkstra computation is in $|B_i|$, we precompute a boolean array of size $|V|$, in which points in $|B_i|$ are marked as true and the rest as false. This costs $O(|V|)$ space, $O(|V|)$ time and achieves a constant lookup time. A Dijkstra computation takes $O(|V| \log |V| + |E|)$ time and $O(|V|)$ space, but this space can be reused between computations. □

Fact 4 (Callahan [3, Chapter 4.5]). $\sum_{i-1}^{m} \min(|A_i|, |B_i|) = O(s^d n \log n)$

Observation 5. *Let E be some edge set for V. Let $(a, b) \in E$. Let $c \in V$ and $d \in V$ be points such that $|ac|, |ad|, |bc|, |bd| > t|cd|$. Then any t-path between c and d will not use the edge (a, b).*

Proof. This directly follows from the fact that c and d are so far away from a and b that just getting to either a or b is already longer than allowed for a t-path. □

Fact 6. *Let γ and ℓ be positive real numbers, and let $\{A_i, B_i\}$ be a well-separated pair in the WSPD with length $\ell(A_i, B_i) = \ell$. The number of well-separated pairs $\{A_i', B_i'\}$ such that the length of the pair is in the interval $[\ell/2, 2\ell]$ and at least one of $R(A_i')$ and $R(B_i')$ is within distance $\gamma\ell$ of either $R(A_i)$ or $R(B_i)$ is less than or equal to $c_{s\gamma} = O\left(s^d(1 + \gamma s)^d\right)$.*

This concludes the theoretical foundations of the algorithm. We will now present the algorithm and analyze its running time.

4   Algorithm

We will now describe the algorithm in detail. The pseudocode can be found in [1]. It first computes the WSPD for V with $s = \frac{2t}{t-1}$ and sorts the resulting pairs according to their smallest distance $\min(A_i, B_i)$. It then alternates between calling the *FillQueue* procedure that attempts to add well-separated pairs to a priority queue Q, and removing an element from Q and adding a corresponding edge to E. If Q is empty after a call to *FillQueue*, the algorithm terminates and returns E.

We assume we have a procedure *ClosestPair(i)* that for the ith well-separated pair computes the closest pair of points without t-path in the graph computed so far, as presented in Lemma 3, and returns this pair, or returns **nil** if no such pair exists. For the priority queue Q, we let $\min(Q)$ denote the value of the key of the minimum of Q. Recall that $m = O(s^d n)$ denotes the number of well-separated pairs in the WSPD that we compute in the algorithm.

We maintain an index i into the sorted list of well-separated pairs. It points to the smallest untreated well-separated pair – we treat the pairs in order of $\min(A_i, B_i)$ in the *FillQueue* procedure. When we treat a pair $\{A_i, B_i\}$, we call *ClosestPair(i)* on it, and if it returns a pair (u, v), we add it to Q with key $|uv|$. We link entries in the queue, its corresponding pair $\{A_i, B_i\}$ and (u, v) together so they can quickly be requested. We stop treating pairs and return from the procedure if we have either treated all pairs, or if $\min(A_i, B_i)$ is larger than the key of the minimal entry in Q (if it exists).

After extracting a pair of points (u, v) from Q, we add it to E. Then, we update the information in Q: for every pair $\{A_j, B_j\}$ having an entry in Q for which either bounding box is at most $t|uv|$ away from $\{A_i, B_i\}$, we recompute *ClosestPair(j)* and updates its entry in Q as follows. If the recomputation returns **nil**, we remove its entry from Q. If it returns a pair (u', v'), we link the entry of j in Q with this new pair and we increase the key of its entry to $|u'v'|$.

For the full proofs of the following theorems and lemma, see [1].

Theorem 7. *Algorithm GreedySpanner computes the greedy spanner for dilation t.*

We will now analyze the running time and space usage of the algorithm. We will use the observations in Section 3 to bound the amount of work done by the algorithm.

Lemma 8. *For any well-separated pair $\{A_i, B_i\}$ $(1 \leq i \leq m)$, the number of times ClosestPair(i) is called is at most $1 + c_{st}$.*

Proof Sketch: *ClosestPair(i)* is called once for every i in the *FillQueue* procedure. *ClosestPair(i)* may also be called after an edge is added to the graph. If a well-separated pair $\{A_j, B_j\}$ causes *ClosestPair(i)* to be called, then $\ell(A_j, B_j) \in [\ell(A_i, B_i)/2, 2\ell(A_i, B_i)]$ [1]. Then, as we only perform *ClosestPair(i)* on pairs that are close by, the collection of pairs that call *ClosestPair(i)* satisfy the requirements of Fact 6 by setting $\gamma = t$, so we can conclude this happens only c_{st} times. The lemma then follows.   □

Theorem 9. *Algorithm GreedySpanner computes the greedy spanner for dilation t in $O\left(n^2 \log^2 n \frac{1}{(t-1)^{3d}} + n^2 \log n \frac{1}{(t-1)^{4d}}\right)$ time and $O\left(\frac{1}{(t-1)^d}n\right)$ space.*

Proof Sketch: We can easily precompute which well-separated pairs are near and of similar length to a given well-separated pair in $O(m^2)$ time using $O\left(\frac{1}{(t-1)^d}n\right)$ space. Other than the *ClosestPair*(i) calls, all operations performed by the algorithm stay within the time bound of the theorem. By space reuse, the space usage of all operations including the *ClosestPair*(i) calls stays within the bounds of the theorem.

By Observation 2, Lemma 3 and Lemma 8 the time taken by all *ClosestPair*(i) calls is $O\left(\sum_{i=1}^{m}(1+c_{st})\min(|A_i|,|B_i|)\left(n\log n + \frac{1}{(t-1)^d}n\right)\right)$. By Fact 4, the bound on c_{st} and after simplification, the time bound in the theorem follows. Correctness was already proven in Theorem 7. □

5 Making the Algorithm Practical

Experiments suggested that implementing the above algorithm as-is does not yield a practical algorithm. With the four optimizations described in the following sections, the algorithm attains running times that are a small constant slower than the algorithm introduced in [6] called FG-greedy, which is considered the best practical algorithm known in literature.

5.1 Finding Close-by Pairs

The algorithm at some point needs to know which pairs are 'close' to the pair for which we are currently adding an edge. In our proof above, we suggested that these pairs be precomputed in $O(m^2)$ time. Unfortunately, this precomputation step turns out to take much longer than the rest of the algorithm. If $n = 100$, then (on a uniform pointset) $m \approx 2000$ and $m^2 \approx 4000000$ while the number of edges e in the greedy spanner is about 135. Our solution is to simply find them using a linear search every time we need to know this information. This only takes $O(e \cdot m)$ time, which is significantly faster.

5.2 Reducing the Number of Dijkstra Computations

After decreasing the time taken by preprocessing, the next part that takes the most time are the Dijkstra computations, whose running time dwarfs the rest of the operations. We would therefore like to optimize this part of the algorithm. For every well-separated pair, we save the length of the shortest path found by any Dijkstra computation performed on it, that is, its source s, target t and distance $\delta(s,t)$. Then, if we are about to perform a Dijkstra computation on a vertex u, we first check if the saved path is already good enough to 'cover' all nodes in B_i. Let c be the center of the circle around the bounding box of B_i and r its radius. We check if $t \cdot |us| + \delta(s,t) + t \cdot (|tc| + r) \leq t \cdot (|uc| - r)$ and mark it as 'irrelevant for the rest of the algorithm'. This optimization roughly improves its running time by a factor three.

5.3   Sharpening the Bound of Observation 5

The bound given in Observation 5 can be improved. Let $\{A_i, B_i\}$ be the well-separated pair for which we just added an edge and let $\{A_j, B_j\}$ be the well-separated pair under consideration in our linear search. First, some notation: let X_k, X_l be sets belonging to some well-separated pair (not necessarily the same pair), then $\min(X_k, X_l)$ denotes the (shortest) distance between the two circles around the bounding boxes of X_k and X_l and $\max(X_k, X_l)$ the longest distance between these two circles. Let $\ell = \ell(A_i, B_i)$. We can then replace the condition of Lemma 5 by the sharper condition $\min(A_i, A_j) + \ell + \min(B_j, B_i) \leq t \cdot \max(A_j, B_j) \vee \min(A_i, B_j) + \ell + \min(A_j, B_i) \leq t \cdot \max(B_j, A_j)$ The converse of the condition implies that the edge just added cannot be part of a t-path between a node in $\{A_j, B_j\}$, so the correctness of the algorithm is maintained. This leads to quite a speed increase.

5.4   Miscellaneous Optimizations

There are two further small optimizations we have added to our implementation.

Firstly, rather than using the implicit linear space representation of the WSPD, we use the explicit representation where every node in the split tree stores the points associated with that node. For point sets where the ratio of the longest and the shortest distance is bounded by some polynomial in n, this uses $O(n \log n)$ space rather than $O(n)$ space. This is true for all practical cases, which is why we used it in our implementation. For arbitrary point sets, this representation uses $O(n^2)$ space. In practice, this extra space usage is hardly noticeable and it speeds up access to the points significantly.

Secondly, rather than performing Dijkstra's algorithm, we use the A^* algorithm. This algorithm uses geometric estimates to the target to guide the computation to its goal, thus reducing the search space of the algorithm [8].

We have tried a number of additional optimizations, but none of them resulted in a speed increase. We describe them here.

We have tried to replace A^* by ALT, a shortest path algorithm that uses landmarks – see [8] for details on ALT – which gives better lower bounds than the geometric estimates used in A^*. However, this did not speed up the computations at all, while costing some amount of overhead.

We have also tried to further cut down on the number of Dijkstra computations. We again used that we store the lengths of the shortest paths found so far per well-separated pair. Every time after calling $ClosestPair(i)$ we checked if the newly found path is 'good enough' for other well-separated pairs, that is, if the path combined with t-paths from the endpoints of the well-separated pairs would give t-paths for all pairs of points in the other well-separated pair. This decreased the number of Dijkstra computations performed considerably, but the overhead from doing this for all pairs was greater than its gain.

We tried to speed up finding close-by pairs using range trees. We also tried performing the optimization of the previous paragraph only to well-separated pairs 'close by' our current pair using range trees. Both optimizations sped up the core algorithm and in particular the optimization of the previous paragraph retained most of its effectiveness. The overhead of creating the range trees was greater than the gain however, in particular in terms of space usage.

6   Experimental Results

We have run our algorithm on point sets of size between 100 and 1,000,000. If the set contained at most 10,000 points, we have also run the FG-greedy algorithm to compare the two algorithms. We have recorded both space usage and running time (wall clock time). We have also performed a number of tests with decreasing values of t on datasets of size 10,000 and 50,000. Finally, as this is the first time we can compute the greedy spanner on large graphs, we have compared it to the θ-graph and WSPD-based spanners on large instances.

We have used three kinds of point distributions: a uniform distribution, a gamma distribution with shape parameter 0.75, and a distribution consisting of $\sqrt{n}$ uniformly distributed pointsets of $\sqrt{n}$ uniformly distributed points. The results from the gamma distribution were nearly identical to those of the uniform pointset, so we did not include them. All our pointsets are two-dimensional.

6.1   Experiment Environments

The algorithms have been implemented in C++. We have implemented all data structures not already in the std. The random generator used was the Mersenne Twister PRNG – we have used a C++ port by J. Bedaux of the C code by the designers of the algorithm, M. Matsumoto and T. Nishimura.

We have used two servers for the experiments. Most experiments have been run on the first server, which uses an Intel Core i5-3470 (3.20GHz) and 4GB (1600 MHz) RAM. It runs the Debian 6.0.7 OS and we compiled for 32 bits using G++ 4.7.2 with the -O3 option. For some tests we needed more memory, so we have used a second server. This server uses an Intel Core i7-3770k (3.50GHz) and 32 GB RAM. It runs Windows 8 Enterprise and we have compiled for 64 bits using the Microsoft C++ compiler (17.00.51106.1) with optimizations turned on.

6.2   Dependence on Instance Size

Our first set of tests compared FG-greedy and our algorithm for different values of n. The results are plotted in Fig. 2. As FG-greedy could only be ran on relatively small instances, its data points are difficult to see in the graph, so we added a zoomed-in plot for the bottom-left part of the plot. We have used standard fitting methods to our data points: the running time of all algorithms involved fits a quadratic curve well, the memory usage of our algorithm is linear and the memory usage of FG-greedy is quadratic. This nicely fits our theoretical analysis. In fact, the constant factors seem to be much smaller than the bound we gave in our proof. We do note a lack of 'bumps' that are often occur when instance sizes start exceeding caches: this is probably due to the cache-unfriendly

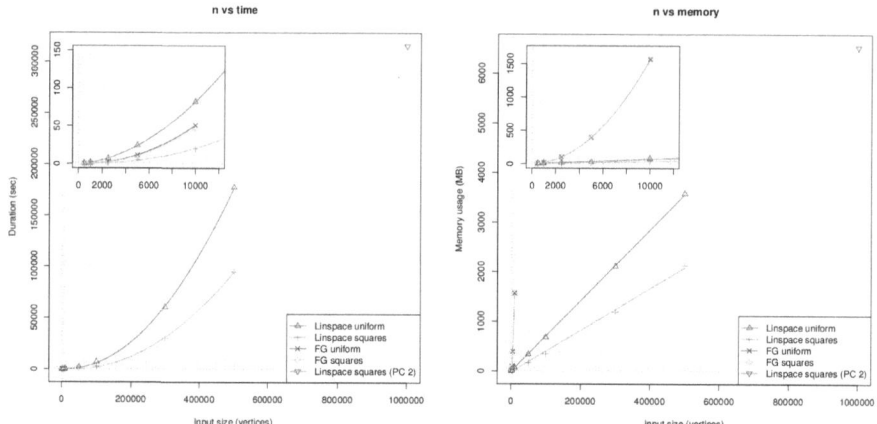

Fig. 2. The left plot shows the running time of our algorithm on uniform and clustered data for variously sized instances. The right plot shows the memory usage of our algorithm on the same data. The lines are fitted quadratic (right) and linear (left) curves. The outlier at the right side was from an experiment performed on a different server. Results for FG-greedy are also shown but were near-impossible to see, so a zoomed-in view of the leftmost corner of both plots is included in the top-left of both plots. The memory usage explosion of FG-greedy is visible in the right plot.

behavior of our algorithm and the still significant constant factor in our memory usage that will fill up caches quite quickly.

Compared to FG-greedy it is clear that the memory usage of our algorithm is vastly superior. The plot puts into perspective how much larger the instances are that our new algorithm can deal with compared to old algorithms. Furthermore, our algorithm is about twice as fast as FG-greedy on the clustered datasets, and only about twice as slow on uniform datasets. On clustered datasets the number of computed well-separated pairs is much smaller than on uniform datasets so this difference is not surprising. These plots suggest that our aim – roughly equal running times at vastly reduced space usage – is reached.

6.3   Dependence on t

We have tested our algorithms on datasets of 10,000 and 50,000 points, setting t to 1.1, 1.2, 1.4, 1.6, 1.8 and 2.0 to test the effect of this parameter. The effects of the parameter ended up being rather different between the uniform and clustered datasets. See [1] for these plots.

On uniform pointsets, our algorithm is about as fast as FG-greedy when $t = 2$, but its performance degrades quite rapidly as t decreases compared to FG-greedy. A hint to this behavior is given by the memory usage of our algorithm: it starts vastly better but as t decreases it becomes only twice as good as FG-greedy. This suggests that the number of well-separated pairs grows rapidly as t decreases, which explains the running time decrease.

On clustered pointsets, the algorithms compare very differently. FG-greedy starts out twice as slow as our algorithm when $t = 2$ and when $t = 1.1$, our algorithm is only slightly faster than FG-greedy. The memory usage of our algorithm is much less dramatic than in the uniform point case: it hardly grows with t and therefore stays much smaller than FG-greedy. The memory usage of FG-greedy only depends on the number of points and not on t or the distribution of the points, so its memory usage is the same.

6.4   Comparison with Other Spanners

We have computed the greedy spanner on the instance shown in Fig. 1, which has 115,475 points. On this instance the greedy spanner for $t = 2$ has 171,456 edges, a maximum degree of 5 and a weight of 11,086,417. On the same instance, the Θ-graph with $k = 6$ has 165,000 edges, a maximum degree of 62 and a weight of 53,341,205. The WSPD-based spanner has 16,636,489 edges, a maximum degree of 1,271 and a weight of 20,330,194,426.

As shown in Fig. 2, we have computed the greedy spanner on 500,000 uniformly distributed points. On this instance the greedy spanner for $t = 2$ has 720,850 edges, a maximum degree of 6 and a weight of 9,104,690. On the same instance, the Θ-graph with $k = 6$ has 2,063,164 edges, a maximum degree of 22 and a weight of 39,153,380. We were unable to run the WSPD-based spanner algorithm on this pointset due to its memory usage.

As shown in Fig. 2, we have computed the greedy spanner on 1,000,000 clustered points. On this instance the greedy spanner for $t = 2$ has 1,409,946 edges, a maximum degree of 6 and a weight of 4,236,016. On the same instance, the Θ-graph with $k = 6$ has 4,157,016 edges, a maximum degree of 135 and a weight of 59,643,264. We were unable to run the WSPD-based spanner algorithm on this pointset due to its memory usage.

We have computed the greedy spanner on 50,000 uniformly distributed points with $t = 1.1$. On this instance the greedy spanner has 225,705 edges, a maximum degree of 18 and a weight of 15,862,195. On the same instance, the Θ-graph with $k = 73$ (which is the smallest k for which a guarantee of $t = 1.1$ has been proven to our knowledge) has 2,396,361 edges, a maximum degree of 146 and a weight of 495,332,746. We were unable to run the WSPD-based spanner algorithm on this pointset with $t = 1.1$ due to its memory usage.

These results show that the greedy spanner really is an excellent spanner, even on large instances and for low t, as predicted by its theoretical properties.

7   Conclusion

We have presented an algorithm that computes the greedy spanner in Euclidean space in $O(n^2 \log^2 n)$ time and $O(n)$ space for any fixed stretch factor and dimension. Our algorithm avoids computing all distances by considering well-separated pairs instead. It consists of a framework that computes the greedy spanner given a subroutine for a bichromatic closest pair problem. We have presented several optimizations to the algorithm. Our experimental results show that the resulting running time is close to that of the fastest known algorithm, while massively decreasing space usage. It allowed us to compute the greedy spanner on instances of a million points, while previous algorithms were limited to at most 13,000 points. Given that our algorithm is the first algorithm with a near-quadratic running time guarantee that has been implemented, that it has linear space usage and that its running time is comparable to the best known algorithms, we think our algorithm is the method of choice to compute greedy spanners.

We leave open the problem of providing a faster subroutine for solving the *bichromatic closest pair with dilation larger than t* problem in our framework, which may allow the greedy spanner to be computed in subquadratic time. Particularly the case of the Euclidean plane seems interesting, as the closely related 'ordinary' *bichromatic closest pair* problem can be solved quickly in this setting.

References

[1] Alewijnse, S.P.A., Bouts, Q.W., ten Brink, A.P., Buchin, K.: Computing the greedy spanner in linear space. CoRR, arXiv:1306.4919 (2013)
[2] Bose, P., Carmi, P., Farshi, M., Maheshwari, A., Smid, M.: Computing the greedy spanner in near-quadratic time. Algorithmica 58(3), 711–729 (2010)
[3] Callahan, P.B.: Dealing with Higher Dimensions: The Well-Separated Pair Decomposition and Its Applications. PhD thesis, Johns Hopkins University, Baltimore, Maryland (1995)
[4] Callahan, P.B., Kosaraju, S.R.: A decomposition of multidimensional point sets with applications to k-nearest-neighbors and n-body potential fields. J. ACM 42(1), 67–90 (1995)
[5] Chew, L.P.: There are planar graphs almost as good as the complete graph. J. Comput. System Sci. 39(2), 205–219 (1989)
[6] Farshi, M., Gudmundsson, J.: Experimental study of geometric t-spanners. ACM J. Experimental Algorithmics 14 (2009)
[7] Gao, J., Guibas, L.J., Hershberger, J., Zhang, L., Zhu, A.: Geometric spanners for routing in mobile networks. IEEE J. Selected Areas in Communications 23(1), 174–185 (2005)
[8] Goldberg, A.V., Harrelson, C.: Computing the shortest path: A search meets graph theory. In: 16th ACM-SIAM Sympos. Discrete Algorithms, pp. 156–165. SIAM (2005)
[9] Gudmundsson, J., Knauer, C.: Dilation and detours in geometric networks. In: Gonzales, T. (ed.) Handbook on Approximation Algorithms and Metaheuristics, pp. 52-1–52-16. Chapman & Hall/CRC, Boca Raton (2006)
[10] Keil, J.M.: Approximating the complete euclidean graph. In: Karlsson, R., Lingas, A. (eds.) SWAT 1988. LNCS, vol. 318, pp. 208–213. Springer, Heidelberg (1988)
[11] Narasimhan, G., Smid, M.: Geometric Spanner Networks. Cambridge University Press, New York (2007)
[12] Peleg, D., Schäffer, A.A.: Graph spanners. Journal of Graph Theory 13(1), 99–116 (1989)

Friendship and Stable Matching⋆

Elliot Anshelevich[1], Onkar Bhardwaj[2], and Martin Hoefer[3]

[1] Dept. of Computer Science, Rensselaer Polytechnic Institute, Troy, NY
[2] Dept. of Electrical, Computer & Systems Engineering, RPI, Troy, NY
[3] Max-Planck-Institute for Informatics, Saarbrücken, Germany

Abstract. We study stable matching problems in networks where players are embedded in a social context, and may incorporate friendship relations or altruism into their decisions. Each player is a node in a social network and strives to form a good match with a neighboring player. We consider the existence, computation, and inefficiency of stable matchings from which no pair of players wants to deviate. When the benefits from a match are the same for both players, we show that incorporating the well-being of other players into their matching decisions significantly decreases the price of stability, while the price of anarchy remains unaffected. Furthermore, a good stable matching achieving the price of stability bound always exists and can be reached in polynomial time. We extend these results to more general matching rewards, when players matched to each other may receive different utilities from the match. For this more general case, we show that incorporating social context (i.e., "caring about your friends") can make an even larger difference, and greatly reduce the price of anarchy. We show a variety of existence results, and present upper and lower bounds on the prices of anarchy and stability for various matching utility structures.

1 Introduction

Stable matching problems capture the essence of many important assignment and allocation tasks in economics and computer science. The central approach to analyzing such scenarios is two-sided matching, which has been studied intensively since the 1970s in both the algorithms and economics literature. An important variant of stable matching is matching with cardinal utilities, when each match can be given numerical values expressing the *quality* or *reward* that the match yields for each of the incident players [5]. Cardinal utilities specify the quality of each match instead of just a preference ordering, and they allow the comparison of different matchings using measures such as social welfare. A particularly appealing special case of cardinal utilities is known as correlated stable matching, where both players who are matched together obtain the same reward. In addition to the wide-spread applications of correlated stable matching in, e.g., market sharing [16], social networks [17], and distributed computer

⋆ This work was supported by DFG grant Ho 3831/3-1 and in part by NSF grants CCF-0914782 and CCF-1101495.

H.L. Bodlaender and G.F. Italiano (Eds.): ESA 2013, LNCS 8125, pp. 49–60, 2013.

networks [27], this model also has favorable theoretical properties such as the existence of a potential function. It guarantees existence of a stable matching even in the non-bipartite case, where every pair of players is allowed to match [2,27].

When matching individuals in a social environment, it is often unreasonable to assume that each player cares only about their own match quality. Instead, players incorporate the well-being of their friends/neighbors as well, or that of friends-of-friends. Players may even be altruistic to some degree, and consider the welfare of all players in the network. Caring about friends and altruistic behavior is commonly observed in practice and has been documented in laboratory experiments [14,25]. In addition, in economics there exist recent approaches towards modeling and analyzing *other-regarding preferences* [15]. Given that other-regarding preferences are widely observed in practice, it is a fundamental question to model and characterize their influence in classic game-theoretic environments. Recently, the impact of social influence on congestion and potential games has been characterized prominently in [8, 10–12, 18–20].

We consider a natural approach to incorporate social effects into partner selection and matching scenarios by studying how social context influences stability and efficiency in matching games. Our model of social context is similar to recent approaches in algorithmic game theory and uses dyadic influence values tied to the hop distance in the graph. In this way, every player may consider the well-being of every other player to some degree, with the degree of this regardfulness possibly decaying with hop distance. The perceived utility of a player is then composed of a weighted average of player utilities. Players who only care about their neighbors or fully altruistic players are special cases of this model.

For matching in social environments, the standard model of correlated stable matching may be too constraining compared to general cardinal utilities, because matched players receive exactly the same reward. Such an *equal sharing* property is intuitive and bears a simple beauty, but other reward sharing methods might be more natural in different contexts. For instance, in theoretical computer science it is common practice to list authors alphabetically, but in other disciplines the author sequence is carefully designed to ensure a proper allocation of credit to the authors of a joint paper. The credit is often supposed to be allocated in terms of input, i.e., the first author is the one that contributed most to the project. Such input-based or proportional sharing is then sometimes overruled with sharing based on intrinsic or acquired social status, e.g., when a distinguished expert in a field is easily recognized and subconsciously credited most with authorship of an article. We are interested in how such unequal reward sharing rules affect stable matching scenarios. We consider a large class of local reward sharing rules and characterize the impact of unequal sharing on existence and inefficiency of stable matchings, both in cases when players are embedded in a social context and when they are not.

1.1 Stable Matching within a Social Context

Correlated stable matching is a prominent subclass of general ordinary stable matching. We are given a (non-bipartite) graph $G = (V, E)$ with edge weights

r_e. In a matching M, if node u is matched to node v, the reward of node u is defined to be exactly r_e. This can be interpreted as both u and v getting an identical reward from being matched together. We will also consider unequal reward sharing, where u obtains reward r_e^u and v obtains reward r_e^v with $r_e^u + r_e^v = r_e$. Therefore, the preference ordering of each node over its possible matches is implied by the rewards that this node obtains from different edges. A pair of nodes (u, v) is called a *blocking pair* in matching M if u and v are not matched to each other in M, but can both strictly increase their rewards by being matched to each other instead. A matching with no blocking pairs is called a *stable matching*.

While the matching model above has been well-studied, we are interested in stable matchings that arise in the presence of social context. Denote the reward obtained by a node v in a matching M as $R_v(M)$. When it is clear which matching we are referring to, we will simply denote this reward by R_v. We now consider the case when node v not only cares about its own reward, but also about the rewards of its friends. Specifically, the *perceived* or *friendship utility* of node v in matching M is defined as

$$U_v = R_v + \sum_{d=1}^{diam(G)} \alpha_d \sum_{u \in N_d(v)} R_u,$$

where $N_d(v)$ is the set of nodes with shortest distance exactly d from v, and $1 \geq \alpha_1 \geq \alpha_2 \geq \ldots \geq 0$ (we use $\boldsymbol{\alpha}$ to denote the vector of α_i values). In other words, for a node u that is distance d away from v, the utility of v increases by an α_d factor of the reward received by u. Thus, if $\alpha_d = 0$ for all $d \geq 2$, this means that nodes only care about their neighbors, while if all $\alpha_d > 0$, this means that nodes are altruistic and care about the rewards of everyone in the graph. The perceived utility is the quantity that the nodes are trying to maximize, and thus, in the presence of friendship, a blocking pair is a pair of nodes such that each node can increase its *perceived utility* by matching to each other. Given this definition of blocking pair, a stable matching is again defined as a matching without such a blocking pair. Note that while our definition includes α_d for all d, it is easy to see that only the values of α_1 and α_2 matter to the stability of a matching, since a deviation of a blocking pair only changes the R_v values of adjacent nodes.

Centralized Optimum and the Price of Anarchy. We study the social welfare of equilibrium solutions and compare them to an optimal centralized solution. The social welfare is the sum of rewards, i.e., a *social optimum* is a matching that maximizes $\sum_v R_v$. Notice that, while this is equivalent to maximizing the sum of player utilities when $\boldsymbol{\alpha} = 0$, this is no longer true with social context (i.e., when $\boldsymbol{\alpha} \neq 0$). Nevertheless, as in e.g. [11, 28], we believe this is a well-motivated and important measure of solution quality, as it captures the overall performance of the system, while ignoring the perceived "good-will" effects of friendship and altruism. For example, when considering projects done in pairs, the reward of an edge can represent actual productivity, while the perceived utility may not.

To compare stable solutions with a social optimum, we will often consider the price of anarchy and the price of stability. When considering stable matchings, by the price of anarchy (resp. stability) we will mean the ratio of social welfare of the social optimum and the social welfare of the worst (resp. best) stable matching.

1.2   New Results and Related Work

Our Results. In Section 2, we consider stable matching with friendship utilities and equal reward sharing. In this case, a stable matching exists and the price of anarchy (ratio of the maximum-weight matching with the worst stable matching) is at most 2, the same as in the case without friendship. The price of stability, on the other hand, improves significantly in the presence of friendship – we show a tight bound of $\frac{2+2\alpha_1}{1+2\alpha_1+\alpha_2}$. Intuitively, the bound depends only on α_1, α_2 because a deviation by a blocking pair (u, v) only affects rewards R_w for nodes w neighboring u or v. Thus, the stability of a matching depends only on the graph and α_1, α_2; changing α_i with $i \geq 3$ does not change the stability of a matching. In addition to providing a tight bound on the price of stability, we present a dynamic process that converges to a stable matching of at least this quality in polynomial time, if initiated from the maximum-weight matching. Our results imply that for socially aware players, the price of stability can greatly improve: e.g., if $\alpha_1 = \alpha_2 = \frac{1}{2}$, then the price of stability is at most $\frac{6}{5}$, and a solution of this quality can be obtained efficiently.

In Section 3 we instead study general reward sharing schemes. When two nodes matched together may receive different rewards, an integral stable matching may not exist. Thus, we focus on *fractional* stable matchings which we show to always exist, even with friendship utilities. Fractional matching is well-motivated in a social context, since the fractional amount of an edge in the matching corresponds to the strength of the link/relationship between this pair of nodes. The total relationships of any single node should add up to at most 1, modeling the fact that a single person cannot be involved in an unlimited amount of relationships. We show that for arbitrary reward sharing, prices of anarchy and stability depend on the level of inequality among reward shares. Specifically, if R is the maximum ratio over all edges $(u, v) \in E$ of the reward shares of node u and v, then the price of anarchy is at most $\frac{(1+R)(1+\alpha_1)}{1+\alpha_1 R}$. Thus, compared to the equal reward sharing case, if sharing is extremely unfair (R is unbounded), then friendship becomes even more important: changing α_1 from 0 to $\frac{1}{2}$ reduces the price of anarchy from unbounded to at most 3. In addition, for several particularly natural local reward sharing rules, we show that an integral stable matching exists, give improved price of anarchy guarantees, and show tight lower bounds.

Related Work. Stable matching problems have been studied intensively over the last few decades. On the algorithmic side, existence, efficient algorithms, and improvement dynamics for two-sided stable matchings have been of interest (for references, see standard textbooks, e.g., [29]). In this paper we address the more general stable roommates problem, in which every player can be matched to every other player. For general preference lists, there have been numerous works characterizing and algorithmically deciding existence of stable matchings [13, 30, 31]. In contrast, fractional stable matchings are always guaranteed to exist and exhibit various interesting polyhedral properties [1, 31]. For the correlated stable roommates problem, existence of (integral) stable matchings is guaranteed by a potential function argument [2, 27], and convergence time of random improvement dynamics is polynomial [3]. In [6], price of anarchy and stability bounds for *approximate* correlated stable matchings were provided. In contrast, we study friendship, altruism, and unequal reward sharing in stable roommates problems with cardinal utilities.

Another line of research closely connected to some of our results involves game-theoretic models for contribution. In [7] we consider a contribution game tied closely to matching problems. Here players have a budget of effort and contribute parts of this effort towards specific projects and relationships. For more related work on the contribution game, see [7]. All previous results for this model concern equal sharing and do not address the impact of the player's social context. As we discuss in the full version of this paper in [4], most of our results for friendship utilities can also be extended to such contribution games.

Analytical aspects of reward sharing have been a central theme in game theory since its beginning, especially in cooperative games. Recently, there have been prominent algorithmic results also for network bargaining [21, 23] and credit allocation problems [22]. A recent line of work [32, 33] treats extensions of cooperative games, where players invest into different coalitional projects. The main focus of this work is global design of reward sharing schemes to guarantee cooperative stability criteria. Our focus here is closer to, e.g., recent work on profit sharing games [9, 26]. We are interested in existence, computational complexity, and inefficiency of stable states under different reward sharing rules, with an aim to examine the impact of social context on stable matchings.

Our notion of a player's social context is based on numerical influence parameters that determine the impact of player rewards on the (perceived) utilities of other players. A recently popular model of altruism is inspired by Ledyard [24] and has generated much interest in algorithmic game theory [11, 12, 19]. In this model, each player optimizes a perceived utility that is a weighted linear combination of his own utility and the utilitarian welfare function. Similarly, for surplus collaboration [8] perceived utility of a player consists of the sum of players utilities in his neighborhood within a social network. Our model is similar to [10, 20] and smoothly interpolates between these global and local approaches.

2 Matching with Equal Reward Sharing

We begin by considering correlated stable matching in the presence of friendship utilities. In this section, the reward received by both nodes of an edge in a matching is the same, i.e., we use equal reward sharing, where every edge e has an inherent value r_e and both endpoints receive this value if edge e is in the

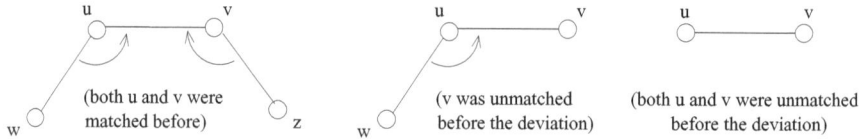

Fig. 1. (Left) Biswivel deviation (Right,Middle) Swivel deviation

matching. We consider more general reward sharing schemes in Section 3. Recall that the friendship utility of a node v increases by $\alpha_d R_u$ for every node u, where d is the shortest distance between v and u. We abuse notation slightly, and let α_{uv} denote α_d, so if u and v are neighbors, then $\alpha_{uv} = \alpha_1$.

Given a matching M, let us classify the following types of improving deviations that a blocking pair can undergo.

Definition 1. *We call an improving deviation a* **biswivel** *whenever two neighbors u and v switch to match to each other, such that both u and v were matched to some other nodes before the deviation in M.*

See Figure 1 for explanation. For such a biswivel to exist in a matching, the following necessary and sufficient conditions must hold.

$$(1 + \alpha_1)r_{uv} > (1 + \alpha_1)r_{uw} + (\alpha_1 + \alpha_{uz})\, r_{vz} \tag{1}$$

$$(1 + \alpha_1)r_{uv} > (1 + \alpha_1)r_{vz} + (\alpha_1 + \alpha_{vw})\, r_{uw} \tag{2}$$

Intuitively, the left side of Inequality (1) quantifies the utility gained by u because of getting matched to v and the right side quantifies the utility lost by u because of u and v breaking their present matchings with w and z respectively. Hence, Inequality (1) implies that u gains more utility by getting matched with v than it loses because of u and v breaking their matchings with v and z. Inequality (2) can similarly be explained in the context of node v.

Definition 2. *We call an improving deviation a* **swivel** *whenever two neighbors get matched such that at least one node among the two neighbors was not matched before the deviation.*

See Figure 1 for explanation. For a swivel to occur, it is easy to see that the reward r_{uv} of the new edge added to the matching must be strictly larger than the rewards of edges that u or v were matched to before (if any).

2.1 Existence and Social Welfare

Theorem 1. *A stable matching exists in stable matching games with friendship utilities. Moreover, the set of stable matchings without friendship (i.e., when $\alpha = 0$) is a subset of the set of stable matchings with friendship utilities on the same graph.*

Theorem 2. *The price of anarchy in stable matching games with friendship utilities is at most 2, and this bound is tight.*

2.2   Price of Stability and Convergence

The main result in this section bounds the price of stability in stable matching games with friendship utilities to $\frac{2+2\alpha_1}{1+2\alpha_1+\alpha_2}$, and this bound is tight (see Theorem 4 below). This bound has some interesting implications. It is decreasing in each α_1 and α_2, hence having friendship utilities always yields a lower price of stability than without friendship utilities. Also, note that values of $\alpha_3, \alpha_4, ..., \alpha_{diam(G)}$ have no influence. This is not surprising: after a deviation by a blocking pair (u, v), the rewards R_w remain the same for all w except those neighboring u or v. Thus, caring about players more than distance 2 away does not improve the price of stability in any way. Also, if $\alpha_1 = \alpha_2 = 1$, then PoS $= 1$, i.e., there will exist a stable matching which will also be a social optimum. Thus *loving thy neighbor and thy neighbor's neighbor but nobody beyond* is sufficient to guarantee that there exists at least one socially optimal stable matching. In fact, due to the shape of the curve, even small values of friendship quickly decrease the price of stability; e.g., setting $\alpha_1 = \alpha_2 = 0.1$ already decreases the price of stability from 2 to ~ 1.7.

We will establish the price of stability bound by defining an algorithm that creates a good stable matching in polynomial time. One possible idea to create a stable matching that is close to optimum is to use a BEST-BLOCKING-PAIR algorithm: start with the best possible matching, i.e., a social optimum, which may or may not be stable. Now choose the "best" blocking pair (u, v): the one with maximum edge reward r_{uv}. Allow this blocking pair to get matched to each other instead of their current partners. Check if the resulting matching is stable. If it is not stable then allow the best blocking pair for this matching to get matched. Repeat the procedure until there are no more blocking pairs, thereby obtaining a stable matching.

This algorithm gives the desired price of stability and running time bounds for the case of "altruism" when all α_i are the same, see Corollary 1 below. To provide the desired bound with general friendship utilities, we must alter this algorithm slightly using the concept of *relaxed* blocking pair.

Definition 3. *Given a matching M, we call a pair of nodes (u, v) a relaxed blocking pair if either (u, v) form an improving swivel, or u and v are matched to w and z respectively, with the following inequalities being true:*

$$(1 + \alpha_1)r_{uv} > (1 + \alpha_1)r_{uw} + (\alpha_1 + \alpha_2)r_{vz} \tag{3}$$

$$(1 + \alpha_1)r_{uv} > (1 + \alpha_1)r_{vz} + (\alpha_1 + \alpha_2)r_{uw} \tag{4}$$

In other words, a relaxed blocking pair ignores the possible edges between nodes u and z, and has α_2 in the place of α_{uz} (similarly, α_2 in the place of α_{vw}). It is clear from this definition that a blocking pair is also a relaxed blocking pair, since the conditions above are less constraining than Inequalities (1) and (2). Thus a matching with no relaxed blocking pairs is also a stable matching. We will call a relaxed blocking pair satisfying Inequalities (3) and (4) a *relaxed biswivel*, which may or may not correspond to an improving deviation, since a relaxed blocking pair is not necessarily a blocking pair. We define the BEST-RELAXED-BLOCKING-PAIR Algorithm to be the same as the BEST-BLOCKING-PAIR algorithm, except at each step it chooses the best *relaxed* blocking pair.

Dynamics: To establish the efficient running time of BEST-RELAXED-BLOCKING-PAIR and the price of stability bound of the resulting stable matching, we first analyze the dynamics of this algorithm and prove some helpful lemmas. We can interpret the algorithm as a sequence of swivel and relaxed biswivel deviations, each inserting one edge into M, and removing up to two edges. Note that it is not guaranteed that the inserted edge will stay forever in M, as a subsequent deviation can remove this edge from M. Let $O_1, O_2, O_3, \cdots$ denote this sequence of deviations, and $e(i)$ denote the edge which got inserted into M because of O_i. We analyze the dynamics of the algorithm by using the following key lemma.

Lemma 1. *Let O_j be a relaxed biswivel that takes place during the execution of the best relaxed blocking pair algorithm. Suppose a deviation O_k takes place before O_j. Then we have $r_{e(k)} \geq r_{e(j)}$. Furthermore, if O_k is a relaxed biswivel then $e(k) \neq e(j)$ (thus at most $|E(G)|$ relaxed biswivels can take place during the execution of the algorithm).*

It is important to note that this lemma does *not* say that $r_{e(i)} \geq r_{e(j)}$ for $i < j$. We are only guaranteed that $r_{e(i)} \geq r_{e(j)}$ for $i < j$ if O_j is a *relaxed biswivel*. Between two successive relaxed biswivels O_k and O_j, the sequence of $r_{e(i)}$ for consecutive swivels can and does increase as well as decrease, and the same edge may be added to the matching multiple times. All that is guaranteed is that $r_{e(j)}$ for a biswivel O_j will have a lower value than all the preceding $r_{e(i)}$'s. Thus, this lemma suggests a nice representation of BEST-RELAXED-BLOCKING-PAIR in terms of phases, where we define a *phase* as a subsequence of deviations that begins with a relaxed biswivel and ends with the next relaxed biswivel. Lemma 1 guarantees that at the start of each phase, the $r_{e(j)}$ value is smaller than the values in all previous phases, and that there is only a polynomial number of phases.

Theorem 3. BEST-RELAXED-BLOCKING-PAIR *outputs a stable matching after $O(m^2)$ iterations, where m is the number of edges in the graph.*

Notice that in each phase, the value of the matching only increases, since swivels only remove an edge if they add a better one. Below, we use the fact that only relaxed biswivel operations reduce the cost of the matching to bound the cost of the stable matching this algorithm produces. We do this by tracing what an edge of M^* "gets mapped to" as swivel and biswivel operations "change" this edge into another one, and showing that the image of an edge can experience at most one relaxed biswivel. The proof appears in the full version [4] of this paper.

Theorem 4. *The price of stability in stable matching games with friendship utilities is at most $\frac{2+2\alpha_1}{1+2\alpha_1+\alpha_2}$, and this bound is tight.*

From Theorems 3 and 4, we immediately get the following corollary about the behavior of best blocking pair dynamics. This corollary applies in particular to the model of altruism when $\alpha_i = \alpha$ for all $i = 1, \ldots, diam(G)$.

Corollary 1. *If $\alpha_1 = \alpha_2$ and we start from the social optimum matching, BEST-BLOCKING-PAIR converges in $O(m^2)$ time to a stable matching that is at most a factor of $\frac{2+2\alpha_1}{1+3\alpha_1}$ worse than the optimum.*

3   Matching with Friendship and General Reward Sharing

In the previous section we assumed that for $(u,v) \in M$ both u and v get the *same reward* r_{uv}. Let us now treat the more general case where u and v receive different rewards for $(u,v) \in M$. We define r_{xy}^x as the reward of x from edge $(x,y) \in M$. We interpret our model in a reward sharing context, where x and y share a total reward of $r_{xy} = r_{xy}^x + r_{xy}^y$. The correlated matching model of Section 2 can equivalently be formulated as equal sharing with nodes u and v receiving a reward of $r_{uv}/2$.

Without friendship utilities, our stable matching game reduces to the stable roommates problem (i.e., non-bipartite stable matching), since reward shares can be arbitrary and thus induce arbitrary preference lists for each node. It is well known that a stable matching may not exist in instances of the stable roommates problem. While we are able to prove existence of integral stable matching for several interesting special cases (see Section 3.1 below), the addition of friendship further complicates matters. In Section 2.1 we showed that for equal sharing, a stable matching without friendship utilities (i.e., $\boldsymbol{\alpha} = \mathbf{0}$) is also a stable matching when we have friendship utilities. This is no longer true for unequal reward sharing: adding a social context can completely change the set of stable matchings. In the full version [4] of this paper we give such examples, including an example where adding a social context (i.e., increasing $\boldsymbol{\alpha}$ above zero) destroys all stable matchings that exist when $\boldsymbol{\alpha} = \mathbf{0}$.

Although stable matchings may not exist in general non-bipartite graphs, *fractional* stable matchings are guaranteed to exist [1]. Fortunately, as we prove below, this holds even in the presence of friendship utilities with general reward sharing: A fractional stable matching always exists. By a "fractional stable matching" we simply mean a fractional matching (where the total fractional matches for a node v add up to at most 1) with no blocking pairs.

Theorem 5. *A fractional stable matching always exists, even in the case of friendship utilities and general reward sharing.*

Since an integral stable matching may not exist, we instead consider fractional matching; by price of anarchy here we mean the ratio of the total reward in a socially optimum *fractional* matching with the worst *fractional* stable matching. The corresponding ratio between the integral versions is trivially upper bounded by this amount as well. We define $R = \max_{(u,v) \in E(G)} \frac{r_{uv}^u}{r_{uv}^v}$. Note that $R \geq 1$. With this notation, we have the following theorem:

Theorem 6. *The (fractional) price of anarchy for general reward sharing with friendship utilities is at most $1 + Q$, where $Q = \frac{R+\alpha_1}{1+\alpha_1 R}$, and this bound is tight.*

Let us quickly consider the implications of the bound in Theorem 6. If $R = 1$, the bound is 2. This result implies Theorem 2, since when we have $R = 1$, then both u and v get the same reward from an edge $(u, v) \in M$. If $\alpha_1 = 0$, the bound is $1 + R$. The tightness of this bound implies that as sharing becomes more unfair, i.e., as $R \to \infty$, we can find instances where the price of anarchy is unbounded. Unequal sharing can make things much worse for the stable matching game.

Notice, however, that $\frac{R+\alpha_1}{1+\alpha_1 R}$ is a decreasing function of α_1. As α_1 goes from 0 to 1, the bound goes from $1 + R$ to 2. Without friendship utilities ($\boldsymbol{\alpha} = \mathbf{0}$), we have a tight upper bound of $1 + R$, which is extremely bad for large R. As α_1 tends to 1, however, the price of anarchy drops to 2, independent of R. For example, for $\alpha_1 = 1/2$ it is only 3. Thus, social context can drastically improve the outcome for the society, especially in the case of unfair and unequal reward sharing.

For price of stability of general reward sharing with friendship utilities, we have a lower bound within an additive factor of 1 of optimum. Specifically, define $Q' = \frac{(1+\alpha_1)(1+R)}{1+\alpha_1(R+1)}$, then we have the following theorem for the price of stability:

Theorem 7. *The price of stability of stable matching games with friendship and general reward sharing is in $[Q', Q + 1]$, with $Q < Q' \le Q + 1$.*

3.1   Specific Reward Sharing Rules

In this section we consider some particularly natural reward sharing rules and show that games with such rules have nice properties. Specifically, while for general reward sharing an (integral) stable matching may not exist, for the reward sharing rules below we show they always exist (although only if there is no social context involved) and how to compute them efficiently. We also give improved bounds on prices of anarchy for these special cases. Specifically, we consider the following sharing rules:

- *Matthew Effect sharing:* In sociology, "Matthew Effect" is a term coined by Robert Merton to describe the phenomenon which says that, when doing similar work, the more famous person tends to get more credit than other less-known collaborators. We model such phenomena for our network by associating brand values λ_u with each node u, and defining the reward that node u gets by getting matched with node v as $r^u_{uv} = \frac{\lambda_u}{\lambda_u+\lambda_v} \cdot r_{uv}$. Thus nodes u and v split the edge reward in the ratio of $\lambda_u : \lambda_v$, and a node with high λ_u value gets a disproportionate amount of reward.
- *Trust sharing:* Often people collaborate based on not only the quality of a project but also how much they trust each other. We model such a situation by associating a value β_u with each node u, which represents the *trust value* of player u, or how pleasant they are to work with. Each edge (u, v) also has an inherent quality h_{uv}. Then, the reward obtained by node u from partnering with node v is $r^u_{uv} = h_{uv} + \beta_v$.

With friendship utilities, even these intuitive special cases of reward sharing do not guarantee the existence of an integral stable matching [4]. Without friendship, however, integral stable matching exists and can be efficiently computed for Matthew Effect sharing and Trust sharing, unlike in the case of general reward sharing.

Theorem 8. *An integral stable matching always exists in stable matching games with Matthew Effect sharing and Trust sharing if $\alpha = 0$ (i.e., if there is no friendship). Furthermore, this matching can be found in $O(|V||E|)$ time.*

The price of anarchy of Matthew effect sharing can be as high as the guarantee of Theorem 6, with $R = \max_{(uv)} \frac{\lambda_u}{\lambda_v}$. For Trust sharing, however:

Theorem 9. *The price of anarchy for (fractional) stable matching games with Trust sharing and friendship utilities is at most $\max\{2 + 2\alpha_1, 3\}$.*

References

1. Abeledo, H., Rothblum, U.: Stable matchings and linear inequalities. Disc. Appl. Math. 54(1), 1–27 (1994)
2. Abraham, D., Levavi, A., Manlove, D., O'Malley, G.: The stable roommates problem with globally ranked pairs. Internet Math. 5(4), 493–515 (2008)
3. Ackermann, H., Goldberg, P., Mirrokni, V., Röglin, H., Vöcking, B.: Uncoordinated two-sided matching markets. SIAM J. Comput. 40(1), 92–106 (2011)
4. Anshelevich, E., Bhardwaj, O., Hoefer, M.: Friendship, altruism, and reward sharing in stable matching and contribution games. CoRR abs/1204.5780 (2012)
5. Anshelevich, E., Das, S.: Matching, cardinal utility, and social welfare. SIGecom Exchanges 9(1), 4 (2010)
6. Anshelevich, E., Das, S., Naamad, Y.: Anarchy, stability, and utopia: Creating better matchings. In: Mavronicolas, M., Papadopoulou, V.G. (eds.) SAGT 2009. LNCS, vol. 5814, pp. 159–170. Springer, Heidelberg (2009)
7. Anshelevich, E., Hoefer, M.: Contribution games in networks. Algorithmica 63(1-2), 51–90 (2012)
8. Ashlagi, I., Krysta, P., Tennenholtz, M.: Social context games. In: Papadimitriou, C., Zhang, S. (eds.) WINE 2008. LNCS, vol. 5385, pp. 675–683. Springer, Heidelberg (2008)
9. Augustine, J., Chen, N., Elkind, E., Fanelli, A., Gravin, N., Shiryaev, D.: Dynamics of profit-sharing games. In: Proc. 22nd Intl. Joint Conf. Artif. Intell. (IJCAI), pp. 37–42 (2011)
10. Buehler, R., Goldman, Z., Liben-Nowell, D., Pei, Y., Quadri, J., Sharp, A., Taggart, S., Wexler, T., Woods, K.: The price of civil society. In: Chen, N., Elkind, E., Koutsoupias, E. (eds.) WINE. LNCS, vol. 7090, pp. 375–382. Springer, Heidelberg (2011)
11. Chen, P.-A., de Keijzer, B., Kempe, D., Schäfer, G.: The robust price of anarchy of altruistic games. In: Chen, N., Elkind, E., Koutsoupias, E. (eds.) Internet and Network Economics. LNCS, vol. 7090, pp. 383–390. Springer, Heidelberg (2011)
12. Chen, P.-A., Kempe, D.: Altruism, selfishness, and spite in traffic routing. In: Proc. 9th Conf. Electronic Commerce (EC), pp. 140–149 (2008)

13. Chung, K.-S.: On the existence of stable roommate matchings. Games Econom. Behav. 33(2), 206–230 (2000)
14. Eshel, I., Samuelson, L., Shaked, A.: Altruists, egoists and hooligans in a local interaction model. Amer. Econ. Rev. 88(1), 157–179 (1998)
15. Fehr, E., Schmidt, K.: The economics of fairness, reciprocity and altruism: Experimental evidence and new theories. In: Handbook on the Economics of Giving, Reciprocity and Altruism. ch. 8, vol. 1, pp. 615–691. Elsevier B.V. (2006)
16. Goemans, M., Li, L., Mirrokni, V., Thottan, M.: Market sharing games applied to content distribution in ad-hoc networks. IEEE J. Sel. Area Comm. 24(5), 1020–1033 (2006)
17. Hoefer, M.: Local Matching Dynamics in Social Networks. In: Aceto, L., Henzinger, M., Sgall, J. (eds.) ICALP 2011, Part II. LNCS, vol. 6756, pp. 113–124. Springer, Heidelberg (2011)
18. Hoefer, M., Penn, M., Polukarov, M., Skopalik, A., Vöcking, B.: Considerate equilibrium. In: Proc. 22nd Intl. Joint Conf. Artif. Intell. (IJCAI), pp. 234–239 (2011)
19. Hoefer, M., Skopalik, A.: Altruism in atomic congestion games. In: Fiat, A., Sanders, P. (eds.) ESA 2009. LNCS, vol. 5757, pp. 179–189. Springer, Heidelberg (2009)
20. Hoefer, M., Skopalik, A.: Social context in potential games. In: Goldberg, P.W. (ed.) WINE 2012. LNCS, vol. 7695, pp. 364–377. Springer, Heidelberg (2012)
21. Kanoria, Y., Bayati, M., Borgs, C., Chayes, J., Montanari, A.: Fast convergence of natural bargaining dynamics in exchange networks. In: Proc. 22nd Symp. Discrete Algorithms (SODA), pp. 1518–1537 (2011)
22. Kleinberg, J., Oren, S.: Mechanisms for (mis)allocating scientific credit. In: Proc. 43rd Symp. Theory of Computing (STOC), pp. 529–538 (2011)
23. Kleinberg, J., Tardos, É.: Balanced outcomes in social exchange networks. In: Proc. 40th Symp. Theory of Computing (STOC), pp. 295–304 (2008)
24. Ledyard, J.: Public goods: A survey of experimental resesarch. In: Kagel, J., Roth, A. (eds.) Handbook of Experimental Economics, pp. 111–194. Princeton University Press (1997)
25. Levine, D.: Modeling altruism and spitefulness in experiments. Rev. Econom. Dynamics 1, 593–622 (1998)
26. Marden, J., Wierman, A.: Distributed welfare games with applications to sensor coverage. In: Proc. 47th IEEE Conf. Decision and Control, pp. 1708–1713 (2008)
27. Mathieu, F.: Self-stabilization in preference-based systems. Peer-to-Peer Netw. Appl. 1(2), 104–121 (2008)
28. Meier, D., Oswald, Y.A., Schmid, S., Wattenhofer, R.: On the windfall of friendship: Inoculation strategies on social networks. In: Proc. 9th Conf. Electronic Commerce (EC), pp. 294–301 (2008)
29. Roth, A., Sotomayor, M.O.: Two-sided Matching: A study in game-theoretic modeling and analysis. Cambridge University Press (1990)
30. Roth, A., Vate, J.V.: Random paths to stability in two-sided matching. Econometrica 58(6), 1475–1480 (1990)
31. Teo, C.-P., Sethuraman, J.: The geometry of fractional stable matchings and its applications. Math. Oper. Res. 23(5), 874–891 (1998)
32. Zick, Y., Chalkiadakis, G., Elkind, E.: Overlapping coalition formation games: Charting the tractability frontier. In: Proc. 11th Conf. Autonomous Agents and Multi-Agent Systems (AAMAS), pp. 787–794 (2012)
33. Zick, Y., Markakis, E., Elkind, E.: Stability via convexity and LP duality in OCF games. In: Proc. 26th Conf. Artificial Intelligence, AAAI (2012)

An Optimal and Practical Cache-Oblivious Algorithm for Computing Multiresolution Rasters

Lars Arge, Gerth Stølting Brodal, Jakob Truelsen,
and Constantinos Tsirogiannis

MADALGO*, Department of Computer Science, Aarhus University, Denmark
{large,gerth,jakobt,constant}@madalgo.au.dk

Abstract. In many scientific applications it is required to reconstruct a raster dataset many times, each time using a different resolution. This leads to the following problem; let $\mathcal{G}$ be a raster of $\sqrt{N} \times \sqrt{N}$ cells. We want to compute for every integer $2 \leq \mu \leq \sqrt{N}$ a raster $\mathcal{G}_\mu$ of $\lceil \sqrt{N}/\mu \rceil \times \lceil \sqrt{N}/\mu \rceil$ cells where each cell of $\mathcal{G}_\mu$ stores the average of the values of $\mu \times \mu$ cells of $\mathcal{G}$. Here we consider the case where $\mathcal{G}$ is so large that it does not fit in the main memory of the computer.

We present a novel algorithm that solves this problem in $O(\text{scan}(N))$ data block transfers from/to the external memory, and in $\Theta(N)$ CPU operations; here $\text{scan}(N)$ is the number of block transfers that are needed to read the entire dataset from the external memory. Unlike previous results on this problem, our algorithm achieves this optimal performance without making any assumptions on the size of the main memory of the computer. Moreover, this algorithm is cache-oblivious; its performance does not depend on the data block size and the main memory size.

We have implemented the new algorithm and we evaluate its performance on datasets of various sizes; we show that it clearly outperforms previous approaches on this problem. In this way, we provide solid evidence that non-trivial cache-oblivious algorithms can be implemented so that they perform efficiently in practice.

1 Introduction

Rasters are one of the most common formats for modelling spatial data. A raster is a 2-dimensional grid of square cells where each cell is assigned a real value. Among other applications, rasters are used to represent real-world terrains; in this case each cell corresponds to a region of a terrain, and the value of the cell indicates the average height of the terrain in this region. Today, it is possible to acquire massive rasters that represent terrains with very fine resolution; the size of each cell in such a raster can be less than one square meter. Yet, studying a terrain in such a small scale might lead to wrong conclusions. This happens for example when we want to identify landforms on terrains; when we study a terrain

---

* Center for Massive Data Algorithmics, a Center of the Danish National Research Foundation.

H.L. Bodlaender and G.F. Italiano (Eds.): ESA 2013, LNCS 8125, pp. 61–72, 2013.

at a scale of a few meters, we might identify many small peaks concentrated within a small area. Yet, when looking on a larger scale, these peaks may be a part of another landform; for instance a rough ridge, or a valley.

To tackle this problem, we need to have a method that can analyze the same raster in many different scales. Fisher *et al.* [4] use such a method in their landform classification algorithm; their algorithm constructs multiple rasters $\mathcal{G}_\mu$, where a cell c of $\mathcal{G}_\mu$ covers the same region as $\mu \times \mu$ cells of the original fine-resolution raster. The value assigned to c is equal to the average of the values of the original $\mu \times \mu$ cells. Given the constructed rasters $\mathcal{G}_\mu$, it is then possible to search for landforms at different scales.

Reconstructing a raster in different resolutions is an important tool for many other scientific applications; in remote sensing, Woodcock and Strahler [9] introduced an algorithm to extract the average size of tree canopies in grayscale images of forests. Here, an image is represented by a raster of square pixels, where each pixel is assigned a grayscale value. Their algorithm reconstructs many instances of a given image raster, in exactly the same way as the algorithm of Fisher *et al.* constructs different instances of a terrain raster. For their application, it is critical to construct one instance of the image for every pixel size which is an integer multiple of the pixel size in the original image, until a single pixel covers almost the entire image. This approach has been also used in other image processing algorithms [3].

Therefore, all of the different applications that we described above lead to the same algorithmic problem; let $\mathcal{G}$ be a raster that consists of $\sqrt{N} \times \sqrt{N}$ cells. For every integer $\mu \in \{2, 3, \ldots, \sqrt{N}\}$ we want to compute a raster $\mathcal{G}_\mu$ of $\lceil \sqrt{N}/\mu \rceil \times \lceil \sqrt{N}/\mu \rceil$ cells where each cell of $\mathcal{G}_\mu$ stores the average of the values of the $\mu \times \mu$ cells of $\mathcal{G}$ that cover the same region.

External Memory Algorithms. As already mentioned, today many available raster datasets are massive, and may consist of terabytes of data. A raster of this size cannot fit entirely in the main memory of a normal computer; thus, it can only be stored entirely in the hard disk. When we want to process the dataset, we have to transfer blocks of data from the disk to the main memory. We call such a block transfer an I/O-operation, or an I/O for short. Unfortunately, an I/O can take the same time as a million CPU operations. Thus, when designing an algorithm that may process such a large dataset, we want to minimise the number of block transfers that are required to process the full dataset.

For this reason, Aggarwal and Vitter [1] introduced a computational model that takes into account the number of block transfers between the disk and the main memory. This model considers two important parameters: the size of the internal memory M, and the maximum size B of a block of data that we can transfer from/to the disk. The efficiency of an algorithm in this model is equal to the number of I/Os that the algorithm requires during its execution. We call this concept of efficiency the *I/O-efficiency* of the algorithm. The I/O-efficiency of an algorithm is expressed as a function of the input size N, but also of the block size B and memory size M. To scan a set of N records stored in the disk we need $O(\text{scan}(N))$ I/Os, where $\text{scan}(N) = N/B$. To sort a set of N records we need $O(\text{sort}(N))$ I/Os, where $\text{sort}(N) = N/B \log_{M/B} N/B$.

Today computers contain several layers of memory; these include layers of cache used between the main memory of the computer and the processor. In this context, the values of parameters M and B differ for every pair of consecutive layers of cache that we consider. Then, to minimise the number of block transfers between all layers, the algorithm must be designed so that it achieves an optimal I/O-performance without knowing the parameters M and B. The algorithms that have this property are known as *cache-oblivious* algorithms [5].

Previous Results. For the problem of computing multiple resolution instances of a given raster, we study the case where the raster does not fit in the main memory of the computer. We want to design an external memory algorithm for this problem that has optimal performance both in terms of I/Os and in terms of CPU operations. In a previous paper, Arge *et al.* [2] proposed two external memory algorithms for this problem; the first algorithm requires $O(\text{sort}(N))$ I/Os and $O(N \log N)$ CPU time, and is easy to implement. Their second algorithm requires $O(\text{scan}(N))$ I/Os and $O(N)$ CPU time, which is obviously optimal. Yet, this algorithm assumes that M is at least $\Theta(B^{1+\varepsilon})$ for some selected $\varepsilon > 0$. This algorithm is *cache-aware*, which means that M and B should be known to the algorithm to achieve this performance. Moreover, this algorithm has a strong limitation when it comes to its implementation; it requires that $\Theta(B)$ files are open simultaneously during its execution. Nowadays, B can be as large as a few million units, while most operating systems can maintain only a relatively small number of files open at the same time (usually around a thousand).

Our Results. In this paper we present a new, cache-oblivious algorithm that achieves the optimal performance of $O(\text{scan}(N))$ I/Os and $O(N)$ CPU time, without making any assumptions on the size of the main memory; that is it performs $O(\text{scan}(N))$ I/Os even when $M = O(B)$. The new algorithm is very easy to implement; we have developed a purely cache-oblivious implementation of the algorithm, and we have tested its performance against an implementation of the algorithm of Arge *et al.* that requires $O(\text{sort}(N))$ I/Os. Recall that the $O(\text{scan}(N))$ algorithm of Arge *et al.* is not practically implementable due to limitations of today's operating systems. The new algorithm performs extremely well and, as expected, clearly outperforms the older approach. We consider this to be a solid proof that non-trivial cache-oblivious algorithms can be implemented to perform efficiently in practice, and be used in real-world applications in the place of standard cache-aware implementations.

2   Description of the Algorithm

Preliminaries. For a raster $\mathcal{G}$ we denote by $\mathcal{G}[i, j]$ the cell that appears in the i-th row and j-th column of $\mathcal{G}$. We use $v(i, j)$ to denote the value that is assigned to this cell. We use $|\mathcal{G}|$ to indicate the number of cells of this raster. We assume that $\mathcal{G}$ is a square; it consists of $\sqrt{N}$ rows and $\sqrt{N}$ columns of cells. Yet, it is easy to show that our analysis holds also for rasters that do not have an equal number of rows and columns. Given a cell $\mathcal{G}[i, j]$ of $\mathcal{G}$, consider the set of cells $\mathcal{G}[k, l]$ for which it holds that $1 \leq k \leq i$ and $1 \leq l \leq j$. We denote the sum of the values of these cells by $\mathrm{psum}(i, j)$, that is:

$$\mathrm{psum}(i, j) = \sum_{\substack{1 \leq k \leq i \\ 1 \leq l \leq j}} v(k, l) \ .$$

The value $\mathrm{psum}(i, j)$ is the so-called *prefix sum* of cell $\mathcal{G}[i, j]$.

Let $\mathcal{G}$ be a raster of dimensions $\sqrt{N} \times \sqrt{N}$, and let μ be an integer such that $1 < \mu \leq \sqrt{N}$. We define $\mathcal{G}_\mu$ as the raster of dimensions $\lceil \sqrt{N}/\mu \rceil \times \lceil \sqrt{N}/\mu \rceil$ such that for any cell $\mathcal{G}_\mu[i, j]$ the value $v_\mu[i, j]$ associated with this cell is equal to the average value of all cells $\mathcal{G}[k, l]$ for which we have that $(i - 1)\mu + 1 \leq k \leq i\mu$ and $(j - 1)\mu + 1 \leq l \leq j\mu$. We say that $\mathcal{G}_\mu$ is the *scale instance* of $\mathcal{G}$ at μ, and we call μ the *scale* of this instance. Considering the size of a scale instance $\mathcal{G}_\mu$, we observe that as we increase μ the number of cells of $\mathcal{G}_\mu$ decreases quadratically. In fact, Arge et al. showed that the total size of all scale instances $\mathcal{G}_\mu$ is $\Theta(N)$. We retrieve the following lemma from their paper.

Lemma 1. *Given a raster $\mathcal{G}$ of $\sqrt{N} \times \sqrt{N}$ cells, the total number of cells for all rasters $\mathcal{G}_\mu$ with $2 \leq \mu \leq \sqrt{N}$ is less than $0.65 \cdot N$.*

Proof. The total number of cells for all rasters $\mathcal{G}_\mu$ is:

$$\sum_{\mu=2}^{\sqrt{N}} \frac{N}{\mu^2} < N \sum_{\mu=1}^{\infty} \frac{1}{\mu^2} = N \cdot \zeta(2) \ ,$$

where $\zeta(x)$ is the so-called *Riemann zeta function* [6]. The value of this function is a constant for every $x > 1$. For $x = 2$ we have that $\zeta(2) \leq 1.65$. □

Let M be a 2D matrix whose entries are real numbers. We denote by $M(i, j)$ the value of the entry that appears in the i-th row and j-th column of this matrix. We denote the number of entries of this matrix by $|M|$.

2.1   A Solution Based on Prefix Sums

In the rest of this section we describe our new cache-oblivious approach for computing all scale instances of a raster $\mathcal{G}$. To describe this new approach, we first present some concepts used by Arge et al. [2]. For any scale instance $\mathcal{G}_\mu$ of a raster $\mathcal{G}$, Arge et al. observed that we can express the value of a cell $\mathcal{G}_\mu[i, j]$ using the prefix sums of the cells of $\mathcal{G}$ as $v_\mu[i, j] = \frac{Sum(i,j,\mu)}{\mu^2}$, where:

$$Sum(i, j, \mu) = \mathrm{psum}(i\mu, j\mu) - \mathrm{psum}(i\mu, (j - 1)\mu)$$
$$-\mathrm{psum}((i - 1)\mu, j\mu) + \mathrm{psum}((i - 1)\mu, (j - 1)\mu) \ .$$

Hence, to compute $\mathcal{G}_\mu$ we only need to extract the prefix sums from all cells $\mathcal{G}[i', j']$ of $\mathcal{G}$ such that *both* i' and j' are integer multiples of μ. It is easy to compute all rasters $\mathcal{G}_\mu$ if $\mathcal{G}$ fits in the main memory; first we compute a matrix that has

$\sqrt{N} \times \sqrt{N}$ entries, and which stores the prefix sums for all cells in $\mathcal{G}$. Then we can compute the value of each cell of $\mathcal{G}_\mu$ in constant time, with only four random accesses to the entries of this matrix. Since the total number of cells of all rasters $\mathcal{G}_\mu$ is $\Theta(N)$, this approach leads to an internal memory algorithm that runs in $\Theta(N)$ CPU operations. However, it is not straightforward how to compute the rasters $\mathcal{G}_\mu$ if $\mathcal{G}$ does not fit in the main memory. To solve this problem we provide the following definitions.

Let M_1 denote the 2-dimensional matrix of $\sqrt{N} \times \sqrt{N}$ entries, such that for every entry $M_1(i,j)$ of this matrix we have that $M_1(i,j) = \mathrm{psum}(i,j)$. For any $\mu \in \{2, 3, \ldots, \sqrt{N}\}$, let M_μ be the matrix that has $\lceil \sqrt{N}/\mu \rceil \times \lceil \sqrt{N}/\mu \rceil$ entries, where $M_\mu(i,j) = M_1(i\mu, j\mu)$. Thus, M_μ stores all the prefix sums that are needed for constructing $\mathcal{G}_\mu$; the value of each cell $\mathcal{G}_\mu(i,j)$ is equal to $v_\mu[i,j] = \frac{Sum_\mu(i,j)}{\mu^2}$, where: $Sum_\mu(i,j) = M_\mu[i,j] - M_\mu[i, j-1] - M_\mu[i-1, j] + M_\mu[i-1, j-1]$. Therefore, assume that we already had an efficient algorithm for computing all matrices M_μ. Then, we can extract from these matrices all scale instances $\mathcal{G}_\mu$ I/O-efficiently, in only $O(\mathrm{scan}(N))$ I/Os and $\Theta(N)$ CPU operations by simply scanning each matrix M_μ, and maintaining four pointers to access the prefix sums needed for computing each value $v_\mu(i,j)$.

Hence, we now focus on designing an efficient algorithm for computing matrices M_μ for every $\mu \in \{2, 3, \ldots, \sqrt{N}\}$. It is easy to compute M_1; we can do this by scanning $\mathcal{G}$, starting from $\mathcal{G}[1,1]$ and visiting all cells in increasing order of their row and column indices. To compute a matrix M_μ with $\mu > 1$, we could scan M_1 and extract each entry $M_1(i,j)$ such that both i and j are multiples of μ. However, in this manner we spend $O(\mathrm{scan}(N))$ I/Os to extract each matrix M_μ, leading to $O(\sqrt{N} \cdot \mathrm{scan}(N))$ I/Os for extracting all of these matrices.

To speed up the computation of the matrices M_μ, we can exploit the following property; consider two distinct integers ρ and λ such that $\rho, \lambda \in \{2, 3, \ldots, \sqrt{N}\}$, and $\rho = \nu\lambda$, for some $\nu \in \mathbb{N}$, $\nu > 1$. Then it holds that $M_\rho(i,j) = M_\lambda(i\nu, j\nu)$ for every entry $M_\rho(i,j)$ of matrix M_ρ. In other words, the entries of matrix M_ρ are a subset of the entries of M_λ if ρ is divisible by λ. Thus, we can construct M_ρ by processing a matrix that can be much smaller than M_1. To construct M_ρ faster, we want to use the smallest matrix M_λ for which ρ is a multiple of λ; we must find the largest $\lambda < \rho$ which is a divisor of ρ. We call this number the *largest distinct divisor* of ρ, and we denote it by $\mathrm{ldd}(\rho)$. Consider two matrices M_ρ and M_λ such that $\rho, \lambda \in \{1, 2, \ldots, \sqrt{N}\}$, and $\rho = \mathrm{ldd}(\lambda)$. We say that matrix M_ρ *derives* from matrix M_λ, and that M_ρ is a *derived matrix* of M_λ. In a similar manner, we say that scale instance $\mathcal{G}_\rho$ derives from instance $\mathcal{G}_\lambda$. For a matrix M_μ we denote the set of matrices that derive from M_μ by $\mathcal{D}_\mu$, that is $\mathcal{D}_\mu = \{M_\rho : \rho \in \{2, 3, \ldots, \sqrt{N}\} \text{ and } \rho = \mathrm{ldd}(\mu)\}$. To compute matrices M_μ, we first scan $\mathcal{G}$ to construct matrix M_1 that stores all prefix sums. Then, we extract all matrices $\mathcal{D}_1$ that derive from M_1; these are the matrices M_μ such that μ is a prime $\leq \sqrt{N}$. To do this, we use a function $ExtractDerived(M_\mu)$; the input of this function is a prefix sum matrix M_μ, and the output is the set of the matrices that derive from M_μ. We describe later in more detail how this function works. After constructing matrices $M_\mu \in \mathcal{D}_1$, we apply again function

ExtractDerived on these matrices to extract all sets of matrices $\mathcal{D}_\mu$. We continue this process recursively, until we have computed all matrices M_μ for the values $\mu \in \{2, 3, \ldots, \sqrt{N}\}$. We call the algorithm that we just described for computing all the scale instances of $\mathcal{G}$ as *MultirasterSpeedUp*.

It is easy to prove that *MultirasterSpeedUp* computes the scale instances of $\mathcal{G}$ correctly, assuming that function *ExtractDerived*(M_μ) computes correctly the derived matrices of any given M_μ. By Lemma 1, excluding the performance of *ExtractDerived*, the rest of the algorithm requires only $O(\text{scan}(N))$ I/Os and $\Theta(N)$ CPU operations. Next we show how we design function *ExtractDerived*.

2.2   Extracting the Derived Matrices

To compute the matrices $\mathcal{D}_\mu$ that derive from a given matrix M_μ, we first have to compute all scale values ρ such that M_ρ is a matrix that derives from M_μ. We call these values the *derived indices* of μ. We denote the set of these values by $\mathcal{S}_\mu$. Given μ, we can calculate all derived scales $\mathcal{S}_\mu$ using the following observation; let μ, ρ be two natural numbers such that $\mu = \text{ldd}(\rho)$. Then it holds that $\mu = \rho/\text{spd}(\rho)$, where $\text{spd}(\rho)$ is the *smallest prime divisor* of ρ. Since μ is the largest distinct divisor of ρ we also have that $\text{spd}(\rho) \leq \text{spd}(\mu)$. Based on the above, to compute $\mathcal{S}_\mu$ we first compute $\text{spd}(\mu)$; we go through all integers $\kappa \in \{2, \ldots, \lceil \sqrt{\mu} \rceil\}$ in increasing order, and we stop when we find the first κ that divides μ. Then we compute all prime numbers in the range $[2, \text{spd}(\mu)]$ by trivially trying all possible pairs of integers within this range, and checking if the largest of the two is divided by the smallest. For the special case $\mu = 1$ the smallest prime divisor is undefined, and we consider that $\mathcal{S}_\mu$ consists of all prime numbers smaller than $\sqrt{N}$. Thus, for $\mu > 1$ we can compute scale values $\mathcal{S}_\mu$ in $O(\mu)$ CPU operations. We need at most $O(\text{scan}(\mu))$ I/Os to store these values. For $\mu = 1$ this process requires $O(N)$ CPU operations and $O(\text{scan}(N))$ I/Os.

To extract the derived matrices $\mathcal{D}_\mu$, we will use M_μ to construct an intermediate file F_μ that contains altogether the entries of all matrices in $\mathcal{D}_\mu$. We will then process this file to extract each derived matrix I/O-efficiently. More specifically, file F_μ is organised as follows; for every prime $\rho \in \mathcal{S}_\mu$, and for every entry $M_\rho(i, j) \in M_\rho$, F_μ contains a record of the form: $< i\rho, j\rho, \rho, v_\mu(i\rho, j\rho) >$. The two first fields of the record indicate which is the entry in M_μ that has the same value as $M_\rho(i, j)$. The third field indicates the scale of M_ρ, and the last field carries the value $M_\rho(i, j)$. Most importantly, the records in F_μ appear in lexicographical order of their three first fields.

Thus, F_μ stores a record for each entry of the matrices in $\mathcal{D}_\mu$, including multiples. The number of records in F_μ is $O(|M_\mu|)$; the number of entries of M_μ is $|\mathcal{G}_\mu|$, and due to Lemma 1 the total number of cells of all the scale instances of a raster $\mathcal{G}_\mu$ cannot exceed $|\mathcal{G}_\mu|$. To construct F_μ, we create an individual file $F_{\mu,\kappa}$ for each matrix $M_\kappa \in \mathcal{D}_\mu$. File $F_{\mu,\kappa}$ contains only records of the form $\{i\kappa, j\kappa, \kappa, \otimes\}$, where $\otimes$ is a symbolic "no-data" value. Then we merge all those files into F_μ in a bottom-up manner; first we generate F_μ by merging the two files $F_{\mu,\kappa}$ and $F_{\mu,\rho}$ that correspond to the two smallest matrices M_ρ and M_κ in $\mathcal{D}_\mu$; that is

ρ, κ are the two largest values in $\mathcal{S}_\mu$. We go on merging F_μ each time with the smallest remaining file $F_{\mu,\lambda}$, until all files are merged into F_μ.

Next we fill in the prefix sum values at the last field of each record in F_μ with a single simultaneous scan of F_μ and M_μ. To extract matrices $\mathcal{D}_\mu$ from F_μ we scan F_μ once per matrix in $\mathcal{D}_\mu$. The matrices are extracted in order of decreasing size; in the first scan of F_μ we extract the largest matrix $M_\rho \in \mathcal{D}_\mu$, and so on and so forth. To extract M_ρ, we pick the records in F_μ whose third field is equal to ρ. We then throw away these records from F_μ, creating a new smaller instance of F_μ. When F_μ becomes empty we will have extracted all derived matrices in $\mathcal{D}_\mu$. The correctness of the algorithm follows from how we handle the prefix sum values in the records of file F_μ. Next we prove the efficiency of this algorithm.

Lemma 2. *Function ExtractDerived computes the set of matrices $\mathcal{D}_\mu$ that derive from M_μ in $O(\mathrm{scan}(|M_\mu| + \mu))$ I/Os and $O(|M_\mu| + \mu)$ CPU operations.*

Proof. We showed that for $\mu \geq 1$ computing the scales $\mathcal{S}_\mu$ takes $O(\mathrm{scan}(\mu))$ I/Os and $O(\mu)$ CPU time. Recall that for the case $\mu = 1$, we can compute $\mathcal{S}_\mu$ in $O(\mathrm{scan}(N))$ I/Os and $O(N)$ CPU operations. Now we prove that for any $\mu > 1$ we can construct all matrices $\mathcal{D}_\mu$ in $O(\mathrm{scan}(|M_\mu|))$ I/Os and $O(|M_\mu|)$ CPU operations. To construct file F_μ, we merge several smaller files $F_{\mu,\rho}$, one merge at a time. As soon as file $F_{\mu,\rho}$ gets merged with F_μ the records of $F_{\mu,\rho}$ become a part of F_μ; from this point and on, these records are scanned once each time we merge F_μ with another file $F_{\mu,\kappa}$. Hence, each record that initially belonged to file $F_{\mu,\rho}$ gets scanned as many times as the number of primes that are smaller or equal to ρ; this is because $\mathcal{S}_\mu$ contains all primes in the range $[2, \mathrm{spd}(\mu)]$, and because we merge files $F_{\mu,\kappa}$ in decreasing order of κ. In the mathematical literature, the number of primes that are smaller or equal to ρ is denoted by $\pi(\rho)$. As each record of $F_{\mu,\rho}$ is scanned $\pi(\rho)$ times, and as $F_{\mu,\rho}$ has $|M_{\mu\rho}|$ records, the total number of records scanned when constructing F_μ is:

$$\sum_{\rho \in \mathcal{S}_\mu} \pi(\rho)|M_{\mu\rho}| = \frac{N}{\mu^2} \sum_{\rho \in \mathcal{S}_\mu} \frac{\pi(\rho)}{\rho^2} \ . \tag{1}$$

The following upper bound is known for $\pi(\rho)$ [7]: $\pi(\rho) < \frac{1.26\rho}{\ln \rho}$. Combining this with (1) we get:

$$\frac{N}{\mu^2} \sum_{\rho \in \mathcal{S}_\mu} \frac{\pi(\rho)}{\rho^2} < 1.26 \frac{N}{\mu^2} \sum_{\rho \in \mathcal{S}_\mu} \frac{1}{\rho \ln \rho} = \frac{1.26}{\log e} \frac{N}{\mu^2} \sum_{\rho \in \mathcal{S}_\mu} \frac{1}{\rho \log \rho} \ , \tag{2}$$

where e is the base of the natural logarithm. We have that:

$$\sum_{\rho \in \mathcal{S}_\mu} \frac{1}{\rho \log \rho} \leq \sum_{i=0}^{\infty} \sum_{\substack{\rho \text{ is prime} \\ 2^{2^i} \leq \rho \leq 2^{2^{i+1}}}} \frac{1}{\rho \log \rho} \leq \sum_{i=0}^{\infty} \sum_{\substack{\rho \text{ is prime} \\ 2^{2^i} \leq \rho \leq 2^{2^{i+1}}}} \frac{1}{2^i \rho} \ . \tag{3}$$

From the mathematical literature we know that [7]:

$$\sum_{\substack{\rho \text{ is prime} \\ \rho \leq x}} \frac{1}{\rho} = O(\log \log x) \ . \tag{4}$$

Applying this on (3) we get:

$$\sum_{i=0}^{\infty} \sum_{\substack{\rho \text{ is prime} \\ 2^{2^i} \leq \rho \leq 2^{2^{i+1}}}} \frac{1}{2^i \rho} = O\left(\sum_{i=0}^{\infty} \frac{i+1}{2^i} \right) = O(1) \ .$$

Combining (2) and (4) we get that the total number of records that we need to scan in order to construct F_μ is $O(|M_\mu|)$. This requires $O(\text{scan}(|M_\mu|))$ I/Os. During the merging we do one comparison for every record that we scan, which implies that we do $O(|M_\mu|)$ operations in the CPU in total.

It remains now to show that extracting all matrices of $\mathcal{D}_\mu$ from F_μ requires $O(\text{scan}(|M_\mu|))$ I/Os and $O(|M_\mu|)$ time in the CPU. Recall that we extract the matrices M_ρ in increasing order of ρ, hence, the records of $M_{\mu\sigma}$ will get scanned as many as $\pi(\sigma)$ times each. Therefore the records scanned in this part of the algorithm are as many as the records scanned for constructing F_μ. We showed that this number is equal to $O(|M_\mu|)$, implying $O(\text{scan}(|M_\mu|))$ I/Os and $O(|M_\mu|)$ CPU operations for extracting the matrices for F_μ, and the lemma follows. □

By construction our algorithm does not require knowledge of M and B, hence it is cache-oblivious. Also, its performance does not depend on a lower bound on the size of M. We obtain the following theorem.

Theorem 1. *Given a raster $\mathcal{G}$ of $\sqrt{N} \times \sqrt{N}$ cells, we can compute all scale instances of $\mathcal{G}$ cache-obliviously in $O(\text{scan}(N))$ I/Os and $O(N)$ CPU operations.*

Proof. Function *ExtractDerived* is called only once for each matrix M_μ so, according to Lemma 2, the total number of I/Os and CPU operations required by the entire algorithm is $O(\text{scan}(\sum_\mu(|M_\mu| + \mu)))$ and $O(\sum_\mu(|M_\mu| + \mu))$ respectively. Since M_μ has the same size as $\mathcal{G}_\mu$, then according to Lemma 1 and because $\sum_\mu |M_\mu| = \Theta(N)$, the theorem follows. □

2.3 Ordering the Prefix Sum Matrices

So far, we have described an algorithm that computes efficiently all scale instances of a given raster $\mathcal{G}$. However, this algorithm does not output the scale instances of $\mathcal{G}$ in the right order. More specifically, from the description of algorithm *MultirasterSpeedUp* we can see that there can be pairs of scale instances $\mathcal{G}_\mu$ and $\mathcal{G}_\rho$ with $\mu < \rho$ such that $\mathcal{G}_\rho$ appears in the output before $\mathcal{G}_\mu$. Yet, for most practical applications, it makes sense to have those instances sorted in the output in order of increasing scale value. Fortunately, we can solve this problem while achieving the same performance as with the algorithm *MultirasterSpeedUp*. The proof of the next theorem appears in the full version of the paper.

Theorem 2. *Given a raster $\mathcal{G}$ of $\sqrt{N} \times \sqrt{N}$ cells, we can compute cache-obliviously all scale instances of $\mathcal{G}$, and output these instances in order of increasing scale using $O(\mathrm{scan}(N))$ I/Os and $O(N)$ CPU operations.*

2.4 Improving the Practical Performance of the Algorithm

Earlier in this section, we described how we can extract the prefix sum matrices $\mathcal{D}_\mu$ from a matrix M_μ by building an intermediate file F_μ. This approach requires merging several smaller files, and needs only $O(\mathrm{scan}(|M_\mu|))$ I/Os. Yet we can evade this merging process, and thus improve the I/O-performance of the algorithm by a constant factor; to build F_μ, we scan M_μ and stream the records that correspond to the entries of matrices in $\mathcal{D}_\mu$ in the form of queries to M_μ. After extracting the prefix sum value of a queried record, we append this record in F_μ.

To do this, the records are streamed to M_μ in lexico-graphical order of their three first fields. To produce the stream of the ordered records we build a min-heap structure. Each leaf node $\nu[\rho]$ of the heap corresponds to a derived matrix $M_{\mu\rho} \in \mathcal{D}_\mu$ and stores the next record of $M_{\mu\rho}$ that has to be streamed. The root of the heap stores the next record to be queried to M_μ. Figure 1 illustrates the structure of the heap; we can see that the heap is as skewed as it can get in favour of the larger derived matrices. The heap contains one leaf node for each derived matrix of M_μ, so the size of the heap is $O(\mathrm{spd}(\mu))$. Although we do not know M, we can build the heap so that at any point the nodes of the $O(M)$ topmost levels appear in memory.

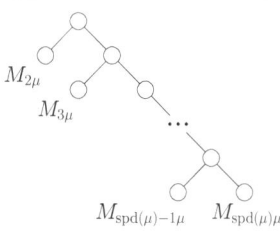

Fig. 1. The structure of the skewed heap that we use to stream the records. Each node is indicated by its corresponding derived matrix.

For the rest of the levels, a record will have to pay one I/O for every B levels that it goes up in the heap. Although this method is oblivious of M, we show that we can stream all records to M_μ so that the number of I/Os decreases as M increases. The proof of the following lemma is provided in the full version of the paper.

Lemma 3. *Let M_μ be a prefix sum matrix. We can stream all the records that correspond to the entries of the derived matrices of $\mathcal{D}_\mu$ in lexicographical order in $O(\mathrm{scan}(|M_\mu|/\log M))$ I/Os and $O(|M_\mu|)$ CPU operations.*

3   Implementation and Benchmarks

We implemented *MultirasterSpeedUp* and evaluated its efficiency on datasets of various sizes. In the experiments that we conducted, we tried several alternatives for implementing the most important routines of the algorithm, and we assessed the efficiency of the implementation for each of these alternatives. We also compared the performance of our implementation with an older implementation of the $O(\mathrm{sort}(N))$ algorithm of Arge *et al.* . Recall that it is not currently possible to implement the $O(\mathrm{scan}(N))$ algorithm of Arge *et al.* due to restrictions in standard operating systems; this algorithm requires that B files are open simultaneously, and while B today is in the order of millions of units, standard operating systems allow for about a thousand files open at the same time.

To measure the performance of our algorithm we used massive raster datasets of many sizes. The datasets that we used originate from a massive raster that consists of roughly 26 billion cells, arranged in 146974 rows and 176121 columns. This raster models the terrain surface over the entire region of Denmark. Each cell of the raster represents a square region on the terrain that has dimension of 2 meters. The elevation of each cell is stored as a 4-byte floating point number, and the entire dataset is stored in a geotif file that has 97 gigabytes size. From this dataset, we constructed all scale instances $\mathcal{G}_\mu$ for $\mu \leq 146974$, and we used the largest of these instances as input for the algorithm; we did this to evaluate the performance of the algorithm for a large range of different input sizes.

As already mentioned, we tried different options for implementing the key routines of the algorithm. These are the routines that involve merging or extracting a sequence of files from/to another larger file. For those routines we evaluated how the performance of the algorithm is affected when trying to merge/extract several files simultaneously. The routines that we tweaked are the following:

– The part of *ExtractDerived* where, given a prefix sum matrix M_μ, we merge several files to construct an intermediate file F_μ which contains the records that correspond to all the entries of the derived matrices $\mathcal{D}_\mu$.
– The part of *ExtractDerived* where we extract the derived matrices $\mathcal{D}_\mu$ from the intermediate F_μ.

For the above routines we measured how the performance of the algorithm changes if we change the number of files that are merged or extracted together. For the first routine we use f_1 to denote the number of files that we merged simultaneously at each point for constructing F_μ. For the second routine we use f_2 to denote the number of derived matrices that we extracted together each time that we performed a scan of F_μ. In the description of the algorithm, we convey that the value of each of these two parameters is equal to two. We also implemented a version of the routine that constructs the intermediate file F_μ based on the mechanism of the skewed heap described in Section 2.4. Recall that this method does not merge any files in order to construct F_μ. All versions of our implementation work in a purely cache-oblivious manner.

The algorithms were implemented in C++ using the software library TPIE (the *Templated Portable I/O Environment*) [8]. This library offers I/O-efficient algorithms for scanning and sorting large files in external memory. Our experiments where run on a machine with a 3.2GHz four-core Xeon CPU (W3565). The main memory of the computer is 12GB. This workstation has 20 disks that have a btrfs (raid 0) file system configuration. The operating system on this computer was Linux version 2.6.38. During our experiments, 8GB of memory was managed by our software, and the rest was left to the operating system for disk cache. For each of the versions of our implementation, the maximum amount of disk space used at any time during the execution was 672 GB.

In our first experiment, we ran our implementation of the algorithm on the 97GB dataset for all possible combinations of values of the two parameters $f_1, f_2 \in \{2, 3, 10, 20, 35, 50\}$. We also ran the implementation of the algorithm using the skewed heap approach for all values of parameter $f_2 \in \{2, 3, 10, 20, 40, 50\}$. From all the possible versions that we tried, the best running time was achieved by the version that uses the skewed heap approach, and parameter value $f_2 = 50$; the running time in this case was 2 hours and 15 minutes. The best running

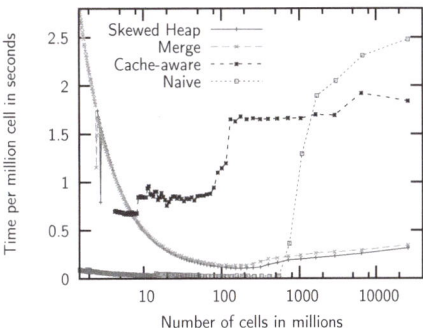

Fig. 2. The performance of the two best versions of our implementation, together with the implementation of the $O(\text{sort}(N))$ algorithm of Arge *et al.* , and the naive internal memory algorithm. The x-axis shows the input sizes using a logarithmic scale with base 10. The y-axis shows running times divided by input size.

time that we got without using the skewed heap approach was for the version with parameter values $f_1 = f_2 = 50$. In this case, the running time was 2 hours and 28 minutes. The worst running time that we got among all versions was from the version that has parameter values $f_1 = 2$, and $f_2 = 2$; the running time for this version was 3 hours and 35 minutes. In general, the running time of each version that behaved like a decreasing function on the values of paramaters f_1 and f_2. Running the implementation of the $O(\text{sort}(N))$ algorithm of Arge *et al.* on the largest dataset yielded a running time of 13 hours and 14 minutes. This running time is a bit less than four times larger than the worst running time that we got for any version of our implementation.

For our next experiment, we ran the two best versions of our implementation on the datasets that we got from extracting the 100 largest scale instances of the 97GB raster, including the initial raster itself. We also ran on these datasets the implementation of the $O(\text{sort}(N))$ algorithm of Arge *et al.*, and the naive internal-memory algorithm that uses prefix sums. Figure 2 illustrates the performance of the four implementations. There, we get a good impression on how the performance of our implementation scales with the size of the input. This is a strong indication that the theoretical bounds that we proved for the performance of the algorithm can be reflected in practice. The results of both experiments show evidently the practical efficiency of our algorithm, when also compared to the implementation of the algorithm of Arge *et al.*. Of course, it could be argued here that this result is hardly surprising; in theory, an $O(\text{sort}(N))$ algorithm has obviously worse asymptotical behaviour than an $O(\text{scan}(N))$ algorithm. However, in practice, the performance of an $O(\text{sort}(N))$ algorithm scales linearly in terms of I/Os. Figure 2 provides some evidence on this argument for the algorithm of Arge *et al.*, at least for the range of input sizes that we considered. The explanation behind this phenomenon is that the ratio M/B in most computers has a value close to one thousand, and therefore the term $\log_{M/B}(N/B)$ in sort(N) is not larger than two in all practical cases. Thus, it is not unrealistic to observe $O(\text{sort}(N))$ algorithms performing better in practice than $O(\text{scan}(N))$ algorithms. More than that, in our case, we compare a cache-aware implementation with a cache-oblivious one, and we could expect that this is an advantage for the performance of the cache-aware implementation. Yet, as we see from our experiments, this is clearly not the case; the cache-oblivious algorithm performs much better in practice. This result shows that purely cache-oblivious software can be developed to perform efficiently in real-world applications. It is interesting to see if we can get similar results also for other external memory problems.

In our last experiment, we ran the best version of our implementation on the largest of our datasets, and at every minute of the execution we measured the rate of the CPU utilisation and the I/O-throughput of this implementation. Figure 3 illustrates the results of this experiment. We see that both the I/O-throughput and CPU utilisation were fairly constant during the run. Also,

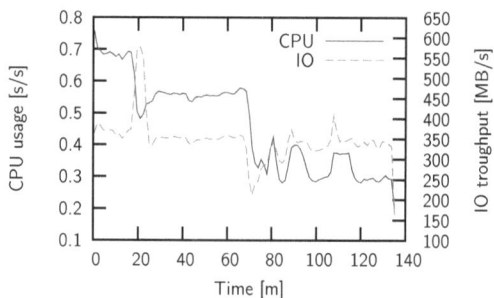

Fig. 3. The CPU utilisation and I/O-throughput of the best version of our implementation

for the largest part of the execution of the algorithm, the CPU utilisation remained above or close to 40%; hence, the running time of the algorithm was almost equally distributed between the CPU and the I/O-operations.

References

1. Aggarwal, A., Vitter, J.S.: The Input/Output complexity of sorting and related problems. Communications of the ACM 31(9), 1116–1127 (1988)
2. Arge, L., Haverkort, H., Tsirogiannis, C.: Fast Generation of Multiple Resolution Instances of Raster Data Sets. In: Proc. 20th ACM SIGSPATIAL Int. Conference on Advances in Geographic Information Systems (ACM GIS), pp. 52–60 (2012)
3. Bøcher, P.K., McCloy, K.R.: The Fundamentals of Average Local Variance - Part I: Detecting Regular Patterns. IEEE Trans. on Image Processing 15, 300–310 (2006)
4. Fisher, P., Wood, J., Cheng, T.: Where is Helvellyn? Fuzziness of Multiscale Landscape Morphometry. Trans. Inst. of British Geographers 29(1), 106–128 (2004)
5. Frigo, M., Leiserson, C.E., Prokop, H., Ramachandran, S.: Cache-Oblivious Algorithms. ACM Transactions on Algorithms 8(1), 1–22 (2012)
6. Karatsuba, A.A., Voronin, S.M.: The Riemann Zeta-Function. Walter de Gruyter, Berlin (1992)
7. Rosser, J.B., Schoenfeld, L.: Approximate Formulas for Some Functions of Prime Numbers. Illinois Journal of Mathematics 6(1), 64–94 (1962)
8. TPIE, the Templated Portable I/O Environment,
   http://www.madalgo.au.dk/tpie/
9. Woodcock, C.E., Strahler, A.H.: The Factor of Scale in Remote Sensing. Remote Sensing of Environment 21, 311–332 (1987)

Logit Dynamics with Concurrent Updates for Local Interaction Games[*]

Vincenzo Auletta[1], Diodato Ferraioli[2], Francesco Pasquale[3], Paolo Penna[4], and Giuseppe Persiano[1]

[1] Università di Salerno, Italy
{auletta,giuper}@dia.unisa.it
[2] Université Paris-Dauphine, France
diodato.ferraioli@dauphine.fr
[3] Sapienza Università di Roma, Italy
pasquale@di.unroma1.it
[4] ETH Zurich, Switzerland
paolo.penna@inf.ethz.ch

Abstract. *Logit dynamics* are a family of randomized best response dynamics based on the *logit choice function* [21] that is used to model players with limited rationality and knowledge. In this paper we study the *all-logit dynamics*, where at each time step *all* players concurrently update their strategies according to the logit choice function. In the well studied *one-logit dynamics* [7] instead at each step *only one* randomly chosen player is allowed to update.

We study properties of the all-logit dynamics in the context of *local interaction games*, a class of games that has been used to model complex social phenomena [7,23,26] and physical systems [19]. In a local interaction game, players are the vertices of a social graph whose edges are two-player potential games. Each player picks one strategy to be played for all the games she is involved in and the payoff of the player is the (weighted) sum of the payoffs from each of the games.

We prove that local interaction games characterize the class of games for which the all-logit dynamics are reversible. We then compare the stationary behavior of one-logit and all-logit dynamics. Specifically, we look at the expected value of a notable class of observables, that we call *decomposable* observables.

1 Introduction

In the last decade, we have observed an increasing interest in understanding phenomena occurring in complex systems consisting of a large number of simple networked components that operate autonomously guided by their own objectives and influenced by the behavior of the neighbors. Even though (online)

---

[*] Vincenzo Auletta and Giuseppe Persiano are supported by Italian MIUR under the PRIN 2010-2011 project *ARS TechnoMedia – Algorithmics for Social Technological Networks*. Diodato Ferraioli is supported by ANR, project COCA, ANR-09-JCJC-0066. Francesco Pasquale is supported by EU FET project MULTIPLEX 317532.

H.L. Bodlaender and G.F. Italiano (Eds.): ESA 2013, LNCS 8125, pp. 73–84, 2013.

social networks are a primary example of such systems, other remarkable typical instances can be found in Economics (e.g., markets), Physics (e.g., Ising model and spin systems) and Biology (e.g., evolution of life). A common feature of these systems is that the behavior of each component depends only on the interactions with a limited number of other components (its neighbors) and these interactions are usually very simple.

Game Theory is the main tool used to model the behavior of agents that are guided by their own objective in contexts where their gains depend also on the choices made by neighbors. Game theoretic approaches have been often proposed for modeling phenomena in a complex social network, such as the formation of the social network itself [3,5,8,16], the formation of opinions [6,11] and the spread of innovation [23,26] in the social network. Many of these models are based on *local interaction games*, where agents are represented as vertices on a *social graph* and the relationship between two agents is represented by a simple two-player game played on the edge joining the corresponding vertices.

We are interested in the *dynamics* that govern such phenomena. Several dynamics have been studied in the literature like, for example, the best response dynamics [13], the logit dynamics [7], fictitious play [12] and no-regret dynamics [15]. Any such dynamics can be seen as made of two components: (i) *Selection rule:* by which the set of players that update their state (strategy) is determined; (ii) *Update rule:* by which the selected players update their strategy. For example, the classical best response dynamics compose the *best response* update rule with a selection rule that selects one player at the time. In the best response update rule, the selected player picks the strategy that, given the current strategies of the other players, guarantees the highest utility. The Cournot dynamics [9] instead combine the best response update rule with the selection rule that selects all players. Other dynamics in which all players concurrently update their strategy are fictitious play [12] and the no-regret dynamics [15].

In this paper, we study a specific class of randomized update rules called the *logit choice function* [7,21] which is a type of noisy best response that models in a clean and tractable way the limited knowledge (or bounded rationality) of the players in terms of a parameter β called *inverse noise*. In similar models studied in Physics, β is the inverse of the temperature. Intuitively, a low value of β (that is, high temperature) models a noisy scenario in which players choose their strategies "nearly at random"; a high value of β (that is, low temperature) models a scenario with little noise in which players pick the strategies yielding higher payoffs with higher probability.

The logit choice function can be coupled with different selection rules so to give different dynamics. For example, in the *logit dynamics* [7] at every time step a single player is selected uniformly at random and the selected player updates her strategy according to the logit choice function. The remaining players are not allowed to revise their strategies in this time step.

While the logit choice function is a very natural behavioral model for approximately rational agents, the specific selection rule that selects one single player per time step avoids any form of concurrency. Therefore a natural question arises:

What happens if *concurrent* updates are allowed?

For example, it is easy to construct games for which the best response converges to a Nash equilibrium when only one player is selected at each step and does not converge to any state when more players are chosen to concurrently update their strategies.

In this paper we study how the logit choice function behaves in an extremal case of concurrency. Specifically, we couple this update rule with a selection rule by which *all* players update their strategies at every time step. We call such dynamics *all-logit*, as opposed to the classical (*one-*)logit dynamics in which only one player at a time is allowed to move. Roughly speaking, the all-logit are to the one-logit what the Cournot dynamics are to the best response dynamics.

Our Contribution. We study the all-logit dynamics for local interaction games [10,23]. Here players are vertices of a graph, called the *social graph*, and each edge is a two-player (exact) potential game. We remark that games played on different edges by a player may be different but, nonetheless, they have the same strategy set for the player. Each player picks one strategy that is used for all of her edges and the payoff is a (weighted) sum of the payoffs obtained from each game. This class of games includes coordination games on a network [10] used to model the spread of innovation in social networks [26], and the Ising model [20] for magnetism. In particular, we study the all-logit dynamics for local interaction games at every possible value of the inverse noise β and we are interested in properties of the original one-logit dynamics that are preserved by the all-logit.

We first consider *reversibility*, an important property of stochastic processes that is useful also to obtain explicit formulas for the stationary distribution. We characterize the class of games for which the all-logit dynamics (specifically, the Markov chains resulting from the all-logit dynamics) are reversible and it turns out that this class coincides with the class of local interaction games. This is to be compared with the well-known result saying that the one-logit dynamics are reversible for every potential game [7]. We find remarkable that a non-trivial property, as reversibility is, of Markov chains modeling the one-logit for potential games holds even for Markov chains modeling all-logit for a large and widely-used subclass of potential games.

Then, we focus on the *observables* of local interaction games. An observable is a function of the strategy profile (that is, the set of strategies adopted by the players) and we are interested in its expected values at stationarity for both the one-logit and the all-logit dynamics. A prominent example of observable is the difference Diff between the number of players adopting two given strategies in a game. In a local interaction game modeling the spread of innovation on a social network this observable counts the difference between the number of adopters of the new and old technology whereas in the Ising model it corresponds to the magnetic field of a magnet.

We show that there exists a class of observables whose expectation at stationarity of the all-logit is the same as the expectation at stationarity of the one-logit as long as the social network underlying the local interaction game is bipartite. Note that in many of these cases the stationary distributions of one- and all-logit dynamics are completely different. We highlight that the class of observables for which our result holds includes the Diff observable. It is interesting to note that the Ising game has been mainly studied for bipartite graphs (e.g., the two-dimensional and the three-dimensional lattice). This implies that, for the Ising model, the all-logit are dynamics that are compatible with the observations and it are arguably more natural than the one-logit dynamics (that postulate that at any given time step only one particle updates its status and then the updated strategy is instantaneously propagated). We extend this result by showing that for general graphs, the extent at which the expectations of these observables differ can be upper and lower bounded by a function of β and of the distance of the social graph from a bipartite graph.

In the full version of the paper [4] we also give preliminary bounds on the convergence time of the all-logit dynamics to their stationary distribution.

Related Works. There is a substantial body of work on the logit dynamics (see e.g. [24] and references therein). Specifically, the all-logit dynamics for strategic games have been studied in [1], where the authors consider the logit-choice function combined with general selection rules (including the selection rule of the all-logit) and investigate conditions for which a state is *stochastically stable*. A stochastically stable state is a state that has non-zero probability as β goes to infinity. We focus instead on a specific selection rule that is used by several remarkable dynamics (Cournot, fictitious play, and no-regret) and consider the whole range of values of β.

The one-logit dynamics have been actively studied starting from the work of Blume [7] that showed that for 2×2 coordination games, the risk dominant equilibria (see [14]) are stochastically stable. The one-logit for local interaction games have been analyzed in several papers with the aim of modeling and understanding the spread of innovations in a social network, see e.g. [10,26].

Remark. For readability sake, in Sections 3 and 4 most of the lemmas and theorems have "proof ideas" instead of full proofs. For full proofs and more detailed descriptions we refer the reader to the full version of the paper [4].

2   Definitions

In this section we formally define the local interaction games and the Markov chain induced by the all-logit dynamics.

Strategic Games. Let $\mathcal{G} = ([n], S_1, \ldots, S_n, u_1, \ldots, u_n)$ be a finite normal-form strategic game. The set $[n] = \{1, \ldots, n\}$ is the player set, S_i is the set of *strategies* for player $i \in [n]$, $S = S_1 \times S_2 \times \cdots \times S_n$ is the set of *strategy profiles* and $u_i \colon S \to \mathbb{R}$ is the *utility* function of player $i \in [n]$. We adopt the standard game-theoretic notation and for $\mathbf{x} = (x_1, \ldots, x_n) \in S$ and $s \in S_i$, we denote by $(\mathbf{x}_{-i}, s)$ the strategy profile $(x_1, \ldots, x_{i-1}, s, x_{i+1}, \ldots, x_n) \in S$.

Potential games [22] are an important class of games. We say that function $\Phi \colon S \to \mathbb{R}$ is an *exact potential* (or simply a *potential*) for game $\mathcal{G}$ if for every $i \in [n]$ and every $\mathbf{x} \in S$ it holds that $u_i(\mathbf{x}_{-i}, s) - u_i(\mathbf{x}_{-i}, z) = \Phi(\mathbf{x}_{-i}, z) - \Phi(\mathbf{x}_{-i}, s)$ for all $s, z \in S_i$. A game $\mathcal{G}$ that admits a potential is called a *potential game*.

Local Interaction Games. In a *local interaction game* $\mathcal{G}$, each player i, with strategy set S_i, is represented by a vertex of a graph $G = (V, E)$ (called *social graph*). For every edge $e = (i, j) \in E$ there is a two-player game $\mathcal{G}_e$ with potential function Φ_e in which the set of strategies of endpoints are exactly S_i and S_j. We denote with u_i^e the utility function of player i in the game $\mathcal{G}_e$. Given a strategy profile $\mathbf{x}$, the utility function of player i in the local interaction game $\mathcal{G}$ is $u_i(\mathbf{x}) = \sum_{e=(i,j)} u_i^e(x_i, x_j)$. It is easy to check that the function $\Phi = \sum_e \Phi_e$ is a potential function for the local interaction game $\mathcal{G}$.

Logit Choice Function. We study the interaction of n players of a strategic game $\mathcal{G}$ that update their strategy according to the *logit choice function* [21,7] described as follows: from profile $\mathbf{x} \in S$ player $i \in [n]$ updates her strategy to $s \in S_i$ with probability $\sigma_i(s \mid \mathbf{x}) = \frac{e^{\beta u_i(\mathbf{x}_{-i}, s)}}{\sum_{z \in S_i} e^{\beta u_i(\mathbf{x}_{-i}, z)}}$. In other words, the logit choice function leans towards strategies promising higher utility. The parameter $\beta \geqslant 0$ is a measure of how much the utility influences the choice of the player.

All-Logit. In this paper we consider the *all-logit* dynamics, where *all* players *concurrently* update their strategy using the logit choice function. Most of the previous works have focused on dynamics where at each step *one* player is chosen uniformly at random and she updates her strategy by following the logit choice function. We call these dynamics *one-logit*, to distinguish them from the all-logit.

The all-logit dynamics induce a Markov chain over the set of strategy profiles whose transition probability $P(\mathbf{x}, \mathbf{y})$ from profile $\mathbf{x} = (x_1, \ldots, x_n)$ to profile $\mathbf{y} = (y_1, \ldots, y_n)$ is

$$P(\mathbf{x}, \mathbf{y}) = \prod_{i=1}^n \sigma_i(y_i \mid \mathbf{x}) = \frac{e^{\beta \sum_{i=1}^n u_i(\mathbf{x}_{-i}, y_i)}}{\prod_{i=1}^n \sum_{z \in S_i} e^{\beta u_i(\mathbf{x}_{-i}, z)}}. \tag{1}$$

Sometimes it is useful to write the transition probability from $\mathbf{x}$ to $\mathbf{y}$ in terms of the *cumulative utility* of $\mathbf{x}$ with respect to $\mathbf{y}$ defined as $U(\mathbf{x}, \mathbf{y}) = \sum_i u_i(\mathbf{x}_{-i}, y_i)$ [1]. Indeed, by observing that $\prod_{i=1}^n \sum_{z \in S_i} e^{\beta u_i(\mathbf{x}_{-i}, z)} = \sum_{\mathbf{z} \in S} \prod_{i=1}^n e^{\beta u_i(\mathbf{x}_{-i}, z_i)}$, we can rewrite (1) as

$$P(\mathbf{x}, \mathbf{y}) = \frac{e^{\beta U(\mathbf{x}, \mathbf{y})}}{D(\mathbf{x})}, \tag{2}$$

where $D(\mathbf{x}) = \sum_{\mathbf{z} \in S} e^{\beta U(\mathbf{x}, \mathbf{z})}$. For a potential game $\mathcal{G}$ with potential Φ, we can define the *cumulative potential* of $\mathbf{x}$ with respect to $\mathbf{y}$ as $\Psi(\mathbf{x}, \mathbf{y}) = \sum_i \Phi(\mathbf{x}_{-i}, y_i)$. Simple algebraic manipulations show that, for a potential game, we can rewrite the transition probabilities in (2) as $P(\mathbf{x}, \mathbf{y}) = \frac{e^{-\beta \Psi(\mathbf{x}, \mathbf{y})}}{T(\mathbf{x})}$, where $T(\mathbf{x})$ is a short-hand for $\sum_{\mathbf{z} \in S} e^{-\beta \Psi(\mathbf{x}, \mathbf{z})}$.

It is easy to see that a Markov chain with transition matrix (1) is ergodic. Indeed, for example, ergodicity follows from the fact that all entries of the transition matrix are strictly positive.

Reversibility & Observables. In this work we focus on two features of the all-logit dynamics, that we formally define here: A Markov chain $\mathcal{M}$ with transition matrix P and state set S is *reversible* with respect to a distribution π if, for every pair of states $x, y \in S$, the following *detailed balance condition* holds $\pi(x)P(x, y) = \pi(y)P(y, x)$. An *observable* O is a function $O \colon S \to \mathbb{R}$, i.e. it is a function that assigns a value to each strategy profile of the game.

3 Reversibility and Stationary Distribution

It is easy to see that the one-logit dynamics for a game $\mathcal{G}$ are reversible if and only if $\mathcal{G}$ is a potential game. This does not hold for the all-logit dynamics. However, we will prove that the class of games for which the all-logit dynamics are reversible is exactly the class of local interaction games.

Reversibility Criteria. As previously stated, a Markov chain $\mathcal{M}$ is reversible if there exists a distribution π such that the detailed balance condition is satisfied. The following Kolmogorov reversibility criterion allows us to establish the reversibility of a process directly from the transition probabilities. Before stating the criterion, we introduce the following notation. A *directed path* Γ from state $x \in S$ to state $y \in S$ is a sequence of states $\langle x_0, x_1, \ldots, x_\ell \rangle$ such that $x_0 = x$ and $x_\ell = y$. The probability $\mathbf{P}(\Gamma)$ of path Γ is defined as $\mathbf{P}(\Gamma) = \prod_{j=1}^{\ell} P(x_{j-1}, x_j)$. The *inverse of path* $\Gamma = \langle x_0, x_1, \ldots, x_\ell \rangle$ is the path $\Gamma^{-1} = \langle x_\ell, x_{\ell-1}, \ldots, x_0 \rangle$. Finally, a cycle C is simply a path from a state x to itself. We are now ready to state the Kolmogorov reversibility criterion (see, for example, [17]).

Theorem 1. *An irreducible Markov chain $\mathcal{M}$ with state space S and transition matrix P is reversible if and only if for every cycle C it holds that $\mathbf{P}(C) = \mathbf{P}(C^{-1})$.*

The following lemma will be useful for proving reversibility conditions for the all-logit dynamics and for stating a closed expression for its stationary distribution.

Lemma 1. *Let $\mathcal{M}$ be an irreducible Markov chain with transition probability P and state space S. $\mathcal{M}$ is reversible if and only if for every pair of states $x, y \in S$, there exists a constant $c_{x,y}$ such that for all paths Γ from x to y, it holds that $\frac{\mathbf{P}(\Gamma)}{\mathbf{P}(\Gamma^{-1})} = c_{x,y}$.*

Proof (idea). One direction follows directly from the Kolmogorov reversibility criterion, since each cycle can be seen as a concatenation of two paths from x to y (actually, a path and the inverse of another path). As for the other direction, fix z and check that the distribution $\tilde{\pi}(x) = c_{z,x}/Z$, where Z is the normalizing constant, satisfies the detailed balance equation. □

All-Logit Reversibility Implies Potential Games. Now we prove that if the all-logit dynamics for a game $\mathcal{G}$ are reversible then $\mathcal{G}$ is a potential game.

The following lemma shows a condition on the cumulative utility of a game $\mathcal{G}$ that is necessary and sufficient for the reversibility of the all-logit for $\mathcal{G}$.

Lemma 2. *The all-logit dynamics for game $\mathcal{G}$ are reversible if and only if the following property holds for every $\mathbf{x}, \mathbf{y}, \mathbf{z} \in S$: $U(\mathbf{x}, \mathbf{y}) - U(\mathbf{y}, \mathbf{x}) = \Big(U(\mathbf{x}, \mathbf{z}) + U(\mathbf{z}, \mathbf{y})\Big) - \Big(U(\mathbf{y}, \mathbf{z}) + U(\mathbf{z}, \mathbf{x})\Big).$*

Proof (idea). One direction follows from Lemma 1. As for the other direction, the hypothesis implies that, for any fixed $\mathbf{z}$, $\tilde{\pi}(\mathbf{x}) = \frac{P(\mathbf{z},\mathbf{x})}{Z \cdot P(\mathbf{x},\mathbf{z})}$ satisfies the detailed balance equation, where Z is the normalizing constant. □

We are now ready to prove that the all-logit dynamics are reversible only for potential games.

Proposition 1. *If the all-logit dynamics for game $\mathcal{G}$ are reversible then $\mathcal{G}$ is a potential game.*

Proof (idea). We show that if the all-logit dynamics are reversible then the utility improvement over any cycle of length 4 is 0. The thesis then follows by a known characterization of potential games (Theorem 2.8 of [22]). □

A Necessary and Sufficient Condition for All-Logit Reversibility. Previously we established that the all-logit dynamics are reversible only for potential games and therefore, from now on, we only consider potential games $\mathcal{G}$ with potential function Φ. Now we present in Proposition 2 a necessary and sufficient condition for reversibility that involves the potential and the cumulative potential. The condition will then be used to prove that local interaction games are exactly the games whose all-logit dynamics are reversible.

Proposition 2. *The all-logit dynamics for a game $\mathcal{G}$ with potential Φ and cumulative potential Ψ are reversible if and only if, for all strategy profiles $\mathbf{x}, \mathbf{y} \in S$,*

$$\Psi(\mathbf{x}, \mathbf{y}) - \Psi(\mathbf{y}, \mathbf{x}) = (n - 2)\left(\Phi(\mathbf{x}) - \Phi(\mathbf{y})\right). \tag{3}$$

Proof (idea). We rewrite Lemma 2 in terms of cumulative potential as $\Psi(\mathbf{x}, \mathbf{y}) - \Psi(\mathbf{y}, \mathbf{x}) = \left(\Psi(\mathbf{x}, \mathbf{z}) + \Psi(\mathbf{z}, \mathbf{y})\right) - \left(\Psi(\mathbf{y}, \mathbf{z}) + \Psi(\mathbf{z}, \mathbf{x})\right)$. Simple algebraic manipulations shows that (3) implies the above equation. As for the other direction, we proceed by induction on the Hamming distance between $\mathbf{x}$ and $\mathbf{y}$. □

Reversibility and Local Interaction Games. Here we prove that the games for which all-logit dynamics are reversible are exactly the local interaction games.

A potential $\Phi : S_1 \times \cdots \times S_n \to \mathbb{R}$ is a *two-player potential* if there exist $u, v \in [n]$ such that, for any $\mathbf{x}, \mathbf{y} \in S$ with $x_u = y_u$ and $x_v = y_v$ we have $\Phi(\mathbf{x}) = \Phi(\mathbf{y})$. In other words, Φ is a function of only its u-th and v-th argument. It is easy to see that any two-player potential satisfies (3).

We say that a potential Φ is the sum of two-player potentials if there exist N two-player potentials $\Phi_1, \ldots, \Phi_N$ such that $\Phi = \Phi_1 + \cdots + \Phi_N$. It is easy to see that generality is not lost by further requiring that $1 \leqslant l \neq l' \leqslant N$ implies $(u_l, v_l) \neq (u_{l'}, v_{l'})$, where u_l and v_l are the two players defining potential Φ_l. At every game $\mathcal{G}$ whose potential is the sum of two-player potentials, i.e., $\Phi = \Phi_1 + \cdots + \Phi_N$, we can associate a *social graph* G that has a vertex for each player of $\mathcal{G}$ and has edge (u, v) iff there exists l such that potential Φ_l depends on players u and v. In other words, each game whose potential is the sum of two-player potentials is a local interaction game.

Observe that if two potentials satisfy (3), then such is also their sum. Hence we have the following proposition.

Proposition 3. *The all-logit dynamics for local interaction games are reversible.*

Next we prove that also the reverse implication holds.

Proposition 4. *If an n-player potential Φ satisfies (3) then it can be written as the sum of at most $N = \binom{n}{2}$ two-player potentials, $\Phi_1, \ldots, \Phi_N$ and thus it represents a local interaction game.*

Proof (idea). Let z_i^* denote the first strategy in each player's strategy set and let $\mathbf{z}^*$ be the strategy profile $(z_1^*, \ldots, z_n^*)$. Moreover, we fix an arbitrary ordering $(u_1, v_1), \ldots, (u_N, v_N)$ of the N unordered pairs of players. For a potential Φ we define the sequence $\vartheta_0, \ldots, \vartheta_N$ of potentials as follows: $\vartheta_0 = \Phi$ and, for $i = 1, \ldots, N$, set $\vartheta_i = \vartheta_{i-1} - \Phi_i$ where, for $\mathbf{x} \in S$, $\Phi_i(\mathbf{x})$ is defined as $\Phi_i(\mathbf{x}) = \vartheta_{i-1}(x_{u_i}, x_{v_i}, \mathbf{z}^*_{-u_i v_i})$. Observe that, for $i = 1, \ldots, N$, Φ_i is a two-player potential and its players are u_i and v_i. Moreover, $\sum_{i=1}^{N} \vartheta_i = \sum_{i=0}^{N-1} \vartheta_i - \sum_{i=1}^{N} \Phi_i$. Thus $\Phi - \vartheta_N = \sum_{i=1}^{N} \Phi_i$. We show that, if Φ satisfies (3), then ϑ_N is identically zero. This implies that Φ is the sum of at most N non-zero two-player potentials and thus a local interaction game.   □

We can thus conclude that if the all-logit dynamics for a potential game $\mathcal{G}$ are reversible then $\mathcal{G}$ is a local interaction game. By combining this result with Proposition 1 and Proposition 3, we obtain

Theorem 2. *The all-logit dynamics for game $\mathcal{G}$ are reversible if and only if $\mathcal{G}$ is a local interaction game.*

Stationary Distribution of the All-Logit for Local Interaction Games.

Theorem 3 (Stationary Distribution). *Let $\mathcal{G}$ be a local interaction game with potential function Φ. Then the stationary distribution of the all-logit for $\mathcal{G}$ is $\pi(\mathbf{x}) \propto e^{(n-2)\beta\Phi(\mathbf{x})} \cdot T(\mathbf{x})$, where $T(\mathbf{x}) = \sum_{\mathbf{z} \in S} e^{-\beta\Psi(\mathbf{x}, \mathbf{z})}$.*

Proof (idea). Fix any profile $\mathbf{y}$. The detailed balance equation and Proposition 2 give $\pi(\mathbf{x}) = e^{(n-2)\beta\Phi(\mathbf{x})} \cdot T(\mathbf{x}) \left(\frac{\pi(\mathbf{y})}{e^{(n-2)\beta\Phi(\mathbf{y})} \cdot T(\mathbf{y})} \right)$, for every $\mathbf{x} \in S$. Since the term in parenthesis does not depend on $\mathbf{x}$ the theorem follows.   □

For a local interaction game $\mathcal{G}$ with potential function Φ we write $\pi_1(\mathbf{x})$, the stationary distribution of the one-logit for $\mathcal{G}$, as $\pi_1(\mathbf{x}) = \gamma_1(\mathbf{x})/Z_1$ where $\gamma_1(\mathbf{x}) = e^{-\beta\Phi(\mathbf{x})}$ (also termed *Boltzmann factor*) and $Z_1 = \sum_{\mathbf{x}} \gamma_1(\mathbf{x})$ is the partition function. From Theorem 3, we derive that $\pi_A(\mathbf{x})$, the stationary distribution of the all-logit for $\mathcal{G}$, can be written in similar way, i.e., $\pi_A(\mathbf{x}) = \frac{\gamma_A(\mathbf{x})}{Z_A}$, where $\gamma_A(\mathbf{x}) = \sum_{\mathbf{y} \in S} e^{-\beta[\Psi(\mathbf{x}, \mathbf{y}) - (n-2)\Phi(\mathbf{x})]}$ and $Z_A = \sum_{\mathbf{x} \in S} \gamma_A(\mathbf{x})$ is the partition function of the all-logit. Simple algebraic manipulations show that, by setting $K(\mathbf{x}, \mathbf{y}) = 2 \cdot \Phi(\mathbf{x}) + \sum_{i \in [n]} \mathbf{d}_{\mathbf{x}, \mathbf{y}}(i) \cdot (\Phi(\mathbf{x}_{-i}, y_i) - \Phi(\mathbf{x}))$ where $\mathbf{d}_{\mathbf{x}, \mathbf{y}}$ is the characteristic vector of positions i in which $\mathbf{x}$ and $\mathbf{y}$ differ (i.e., $\mathbf{d}_{\mathbf{x}, \mathbf{y}}(i) = 1$ if $x_i \neq y_i$ and 0 otherwise), we can write $\gamma_A(\mathbf{x})$ and Z_A as

$$\gamma_A(\mathbf{x}) = \sum_{\mathbf{y} \in S} e^{-\beta K(\mathbf{x}, \mathbf{y})} \quad \text{and} \quad Z_A = \sum_{\mathbf{x}, \mathbf{y}} e^{-\beta K(\mathbf{x}, \mathbf{y})}. \tag{4}$$

4   Observables of Local Information Games

In this section we study observables of local interaction games and we focus on the relation between the expected value $\langle O, \pi_1 \rangle$ of an observable O at the stationarity of the one-logit and its expected value $\langle O, \pi_A \rangle$ at the stationarity of the all-logit dynamics. In Theorem 5, we give a sufficient condition for an observable to be invariant, that is for having the two expected values to coincide. The sufficient condition is related to the existence of a *decomposition* of the set $S \times S$ that decomposes the quantity K appearing in the expression for the stationary distribution of the all-logit for the local interaction game $\mathcal{G}$ (see Eq. 4) into a sum of two potentials. In Theorem 5 we show that if $\mathcal{G}$ admits such a decomposition μ and in addition observable O is also decomposed by μ (see Definition 2) then O has the same expected value at the stationarity of the one-logit and of the all-logit dynamics. We then show that all local interaction games on *bipartite* social graphs admit a decomposition permutation (see Theorem 4) and give an example of invariant observable.

The above finding follows from a relation between the partition functions of the one-logit and of the all-logit dynamics that might be of independent interest. More precisely, in Theorem 4 we show that if the game $\mathcal{G}$ admits a decomposition then the partition function of the all-logit is the square of the partition function of the one-logit dynamics. The partition function of the one-logit is easily seen to be equal to the partition function of the canonical ensemble used in Statistical Mechanics (see for example [18]). It is well known that a partition function of a canonical ensemble that is the union of two independent canonical ensembles is the product of the two partition functions. Thus Theorem 4 can be seen as a further evidence that the all-logit can be decomposed into two independent one-logit dynamics.

Throughout this section we assume, for the sake of ease of presentation, that each player has just two strategies available. Extending our results to any number of strategies is straightforward.

We start by introducing the concept of a *decomposition* and then we define the concept of a *decomposable* observable.

Definition 1. *A permutation* $\mu \colon (\mathbf{x}, \mathbf{y}) \mapsto (\mu_1(\mathbf{x}, \mathbf{y}), \mu_2(\mathbf{x}, \mathbf{y}))$ *of* $S \times S$ *is a decomposition for a local interaction game* $\mathcal{G}$ *with potential* Φ *if, for all* $(\mathbf{x}, \mathbf{y})$, *we have that* $K(\mathbf{x}, \mathbf{y}) = \Phi(\mu_1(\mathbf{x}, \mathbf{y})) + \Phi(\mu_2(\mathbf{x}, \mathbf{y}))$, $\mu_1(\mathbf{x}, \mathbf{y}) = \mu_2(\mathbf{y}, \mathbf{x})$ *and* $\mu_2(\mathbf{x}, \mathbf{y}) = \mu_1(\mathbf{y}, \mathbf{x})$.

Theorem 4. *If a decomposition* μ *for a local interaction game* $\mathcal{G}$ *exists, then* $Z_A = Z_1^2$.

Proof. From (4) we have $Z_A = \sum_{\mathbf{x}, \mathbf{y}} e^{-\beta K(\mathbf{x}, \mathbf{y})} = \sum_{\mathbf{x}, \mathbf{y}} e^{-\beta[\Phi(\mu_1(\mathbf{x}, \mathbf{y})) + \Phi(\mu_2(\mathbf{x}, \mathbf{y}))]}$. Since μ is a permutation of $S \times S$, we have $Z_A = \sum_{\mathbf{x}, \mathbf{y}} e^{-\beta[\Phi(\mathbf{x}) + \Phi(\mathbf{y})]} = Z_1^2$. $\square$

Definition 2. *An observable* O *is* decomposable *if there exists a decomposition* μ *such that, for all* $(\mathbf{x}, \mathbf{y})$, *it holds that* $O(\mathbf{x}) + O(\mathbf{y}) = O(\mu_1(\mathbf{x}, \mathbf{y})) + O(\mu_2(\mathbf{x}, \mathbf{y}))$.

We next prove that a decomposable observable has the same expectation at stationarity of the one-logit and the all-logit dynamics.

Theorem 5. *If observable O is decomposable then $\langle O, \pi_1 \rangle = \langle O, \pi_A \rangle$.*

Proof (idea). Suppose that O is decomposed by μ. Then we have that, for all $\mathbf{x} \in S$, $\gamma_A(\mathbf{x}) = \sum_{\mathbf{y}} \gamma_1(\mu_1(\mathbf{x}, \mathbf{y})) \cdot \gamma_1(\mu_2(\mathbf{x}, \mathbf{y}))$ and thus

$$\langle O, \pi_A \rangle = \frac{1}{2} \frac{1}{Z_A} \sum_{\mathbf{x}, \mathbf{y}} [O(\mathbf{x}) + O(\mathbf{y})] \gamma_1(\mu_1(\mathbf{x}, \mathbf{y})) \gamma_1(\mu_2(\mathbf{x}, \mathbf{y})),$$

where we used the property that $\mu_1(\mathbf{x}, \mathbf{y}) = \mu_2(\mathbf{y}, \mathbf{x})$ and $\mu_2(\mathbf{x}, \mathbf{y}) = \mu_1(\mathbf{y}, \mathbf{x})$. The theorem follows since O is decomposable. $\qquad\square$

We next prove that for all local interaction games on a bipartite social graph there exists a decomposition. We start with the following sufficient condition for a permutation to be a decomposition.

Lemma 3. *Let $\mathcal{G}$ be a social interaction game on social graph G with potential Φ and let μ be a permutation of $S \times S$ such that, for all $\mathbf{x}, \mathbf{y} \in S$, we have $\mu_1(\mathbf{x}, \mathbf{y}) = \mu_2(\mathbf{y}, \mathbf{x})$, $\mu_2(\mathbf{x}, \mathbf{y}) = \mu_1(\mathbf{y}, \mathbf{x})$ and for all edges $e = (u, v)$ of G and for all $\mathbf{x}, \mathbf{y} \in S$ either $(\tilde{x}_u, \tilde{x}_v, \tilde{y}_u, \tilde{y}_v) = (x_u, y_v, y_u, x_v)$ or $(\tilde{x}_u, \tilde{x}_v, \tilde{y}_u, \tilde{y}_v) = (y_u, x_v, x_u, y_v)$, where $\tilde{\mathbf{x}} = \mu_1(\mathbf{x}, \mathbf{y})$ and $\tilde{\mathbf{y}} = \mu_2(\mathbf{x}, \mathbf{y})$. Then μ is a decomposition of $\mathcal{G}$.*

Proof (idea). We prove by simple case analysis that the contribution of each edge $e = (u, v)$ to $K(\mathbf{x}, \mathbf{y})$ is $\Phi_e(\tilde{x}_u, \tilde{x}_v) + \Phi_e(\tilde{y}_u, \tilde{y}_v)$. The lemma is then obtained by summing over all edges e. $\qquad\square$

Theorem 6. *Let $\mathcal{G}$ be a social interaction game on a bipartite graph G. Then $\mathcal{G}$ admits a decomposition.*

Proof (idea). Let (L, R) be the set of vertices in which G is bipartite. For each $(\mathbf{x}, \mathbf{y}) \in S \times S$ we define $\tilde{\mathbf{x}} = \mu_1(\mathbf{x}, \mathbf{y})$ and $\tilde{\mathbf{y}} = \mu_2(\mathbf{x}, \mathbf{y})$ as follows: for every vertex u of G, (i) if $u \in L$ then we set $\tilde{x}_u = x_u$ and $\tilde{y}_u = y_u$; (ii) if $u \in R$ then we set $\tilde{x}_u = y_u$ and $\tilde{y}_u = x_u$.

First of all, observe that the mapping is an involution and thus it is also a permutation and that $\mu_1(\mathbf{x}, \mathbf{y}) = \mu_2(\mathbf{y}, \mathbf{x})$ and $\mu_2(\mathbf{x}, \mathbf{y}) = \mu_1(\mathbf{y}, \mathbf{x})$. From the bipartiteness of G it follows that for each edge one of the conditions of Lemma 3 is satisfied. Then we can conclude that the mapping is a decomposition. $\qquad\square$

We now give an example of decomposable observable. Consider the observable Diff that returns the (signed) difference between the number of vertices adopting strategy 0 and the number of vertices adopting strategy 1. That is, $\text{Diff}(\mathbf{x}) = n - 2 \sum_u x_u$. In local interaction games used to model the diffusion of innovations in social networks and the spread of new technology (see, for example, [26]), this observable is a measure of how wide is the adoption of the innovation. The Diff observable is also meaningful in the Ising model for ferromagnetism (see, for example, [20]) as it is the measured magnetism.

To prove that Diff is decomposable we consider the mapping used in the proof of Theorem 6 and observe that, for every vertex u and for every $(\mathbf{x}, \mathbf{y}) \in S \times S$, we have $x_u + y_u = \tilde{x}_u + \tilde{y}_u$. Whence we conclude that $O(\mathbf{x}) + O(\mathbf{y}) = O(\tilde{\mathbf{x}}) + O(\tilde{\mathbf{y}})$.

Decomposable Observables for General Graphs. We can show that for local interaction games $\mathcal{G}$ on general social graphs G the expected values of a decomposable observable O with respect to the stationary distributions of the one-logit and of the all-logit dynamics differ by a quantity that depends on β and on how far away the social graph G is from being bipartite (which in turn is related to the smallest eigenvalue of G [25]). Due to lack of space we omit the details of that result and refer the interested reader to the full version of this paper [4].

5   Future Directions

In this paper we considered the selection rule where all players play concurrently. A natural extension of this selection rule assigns a different probability to each subset of players. What is the impact of such a probabilistic selection rule on reversibility and on observables? Some interesting results along that direction have been obtained in [1,2]. Notice that if we consider the selection rule that selects player i with probability $p_i > 0$ (the one-logit dynamics set $p_i = 1/n$ for all i) then the stationary distribution is the same as the stationary distribution of the one-logit. Therefore, all observables have the same expected value and all potential games are reversible.

It is a classical result that the Gibbs distribution, that is the stationary distribution of the one-logit dynamics (the micro-canonical ensemble, in Statistical Mechanics parlance), is the distribution that maximizes the entropy among all the distributions with a fixed average potential. Can we say something similar for the stationary distribution of the all-logit? A promising direction along this line of research is suggested by the results in Section 4: at least in some cases the stationary distribution of the all-logit dynamics can be seen as a composition of simpler distributions.

References

1. Alós-Ferrer, C., Netzer, N.: The logit-response dynamics. Games and Economic Behavior 68(2), 413–427 (2010)
2. Alós-Ferrer, C., Netzer, N.: Robust stochastic stability. ECON - Working Papers 063, Department of Economics - University of Zurich (February 2012)
3. Anshelevich, E., Dasgupta, A.: Tardos, T. Wexler. Near-optimal network design with selfish agents. Theory of Computing 4(1), 77–109 (2008)
4. Auletta, V., Ferraioli, D., Pasquale, F., Penna, P., Persiano, G.: Logit dynamics with concurrent updates for local-interaction games (2012),
http://arxiv.org/abs/1207.2908
5. Bala, V., Goyal, S.: A noncooperative model of network formation. Econometrica 68(5), 1181–1229 (2000)

6. Bindel, D., Kleinberg, J.M., Oren, S.: How bad is forming your own opinion? In: In: Proc. of 52nd IEEE Ann. Symp. on Foundations of Computer Science (FOCS 2011), pp. 57–66 (2011)
7. Blume, L.E.: Blume. The statistical mechanics of strategic interaction. Games and Economic Behavior 5(3), 387–424 (1993)
8. Borgs, C., Chayes, J.T., Ding, J., Lucier, B.: The hitchhiker's guide to affiliation networks: A game-theoretic approach. In: Proc. of 2nd Symp. on Innovations in Computer Science (ICS 2011), pp. 389–400. Tsinghua University Press (2011)
9. Cournot, A.A.: Recherches sur le Principes mathematiques de la Theorie des Richesses. L. Hachette (1838)
10. Ellison, G.: Learning, local interaction, and coordination. Econometrica 61(5), 1047–1071 (1993)
11. Ferraioli, D., Goldberg, P.W., Ventre, C.: Decentralized dynamics for finite opinion games. In: Serna, M. (ed.) SAGT 2012. LNCS, vol. 7615, pp. 144–155. Springer, Heidelberg (2012)
12. Fudenberg, D., Levine, D.K.: The Theory of Learning in Games. MIT Press (1998)
13. Fudenberg, D., Tirole, J.: Game Theory. MIT Press (1992)
14. Harsanyi, J.C., Selten, R.: A General Theory of Equilibrium Selection in Games. MIT Press (1988)
15. Hart, S., Mas-Colell, A.: A general class of adaptive procedures. Journal of Economic Theory 98(1), 26–54 (2001)
16. Jackson, M.O., Wolinsky, A.: A strategic model of social and economic networks. Journal of Economic Theory 71(1), 44–74 (1996)
17. Kelly, F.: Reversibility and Stochastic Networks. Cambridge University Press (2011)
18. Landau, L.D., Lifshitz, E.M.: Statistical Physics, vol. 5. Elsevier Science (1996)
19. Levin, D.A., Luczak, M., Peres, Y.: Glauber dynamics for the mean-field Ising model: cut-off, critical power law, and metastability. Probability Theory and Related Fields 146(1-2), 223–265 (2010)
20. Martinelli, F.: Lectures on Glauber dynamics for discrete spin models. In: Lectures on Probability Theory and Statistics. Lecture Notes in Mathematics, vol. 1717, pp. 93–191. Springer, Heidelberg (1999)
21. McFadden, D.L.: Conditional logit analysis of qualitative choice behavior. In: Frontiers in Econometrics, pp. 105–142. Academic Press (1974)
22. Monderer, D., Shapley, L.S.: Potential games. Games and Economic Behavior 14, 124–143 (1996)
23. Montanari, A., Saberi, A.: Convergence to equilibrium in local interaction games. In: Proc. of 50th IEEE Ann. Symp. on Foundations of Computer Science (FOCS 2009), pp. 303–312 (2009)
24. Sandholm, W.H.: Population Games and Evolutionary Dynamics. MIT Press (2011)
25. Trevisan, L.: Max cut and the smallest eigenvalue. In: Proc. of 41st ACM Ann. Symp. on Theory of Computing (STOC 2009), pp. 263–272. ACM (2009)
26. Young, P.H.: The diffusion of innovations in social networks. Technical report (2002)

On Resilient Graph Spanners

Giorgio Ausiello[1], Paolo Giulio Franciosa[2],
Giuseppe Francesco Italiano[3], and Andrea Ribichini[1]

[1] Dipartimento di Ingegneria Informatica, Automatica e Gestionale,
Università di Roma "La Sapienza", via Ariosto 25, 00185 Roma, Italy
{ausiello,ribichini}@dis.uniroma1.it
[2] Dipartimento di Scienze Statistiche, Università di Roma "La Sapienza",
piazzale Aldo Moro 5, 00185 Roma, Italy
paolo.franciosa@uniroma1.it
[3] Dipartimento di Ingegneria Civile e Ingegneria Informatica,
Università di Roma "Tor Vergata", via del Politecnico 1, 00133 Roma, Italy
italiano@disp.uniroma2.it

Abstract. We introduce and investigate a new notion of resilience in graph spanners. Let S be a spanner of a graph G. Roughly speaking, we say that a spanner S is resilient if all its point-to-point distances are resilient to edge failures. Namely, whenever any edge in G fails, then as a consequence of this failure all distances do not degrade in S substantially more than in G (i.e., the relative distance increases in S are very close to those in the underlying graph G). In this paper we show that sparse resilient spanners exist, and that they can be computed efficiently.

1 Introduction

Spanners are fundamental graph structures that have been extensively studied in the last three decades. Given a graph G, a *spanner* is a (sparse) subgraph of G that preserves the approximate distance between each pair of vertices. More precisely, for $\alpha \geq 1$ and $\beta \geq 0$, an (α, β)-spanner of a graph $G = (V, E)$ is a subgraph $S = (V, E_S)$, $E_S \subseteq E$, that distorts distances in G up to a multiplicative factor α and an additive term β: i.e., for all vertices x, y, $d_S(x, y) \leq \alpha \cdot d_G(x, y) + \beta$, where d_G denotes the distance in graph G. We refer to (α, β) as the *distorsion* of the spanner. As a special case, $(\alpha, 0)$-spanners are known as *multiplicative spanners* (also denoted as *t-spanners*, for $t = \alpha$: $d_S(x, y) \leq t \cdot d_G(x, y)$), and $(1, \beta)$-spanners are known as *additive spanners* ($d_S(x, y) \leq d_G(x, y) + \beta$). Note that an (α, β)-spanner is trivially a multiplicative $(\alpha + \beta)$-spanner. It is known how to compute in $O(m+n)$ time a multiplicative $(2k-1)$-spanner, with $O(n^{1+\frac{1}{k}})$ edges [2,17] (which is conjectured to be optimal for any k), in $O(n^{2.5})$ time additive 2-spanners with $O(n^{\frac{3}{2}})$ edges [16] and in $O(mn^{2/3})$ time additive 6-spanners with $O(n^{\frac{4}{3}})$ edges [7], where m and n are respectively the number of edges and vertices in the original graph G. Multiplicative t-spanners are only considered for $t \geq 3$, as multiplicative 2-spanners can have as many as $\Theta(n^2)$ edges: this implies that (α, β)-spanners are considered for $\alpha + \beta \geq 3$.

Spanners have been investigated also in the fully dynamic setting, where edges may be added to or deleted from the original graph. In [4], a (2,1)-spanner and

H.L. Bodlaender and G.F. Italiano (Eds.): ESA 2013, LNCS 8125, pp. 85–96, 2013.
© Springer-Verlag Berlin Heidelberg 2013

a (3,2)-spanner of an unweighted graph are maintained under an intermixed sequence of $\Omega(n)$ edge insertions and deletions in $O(\Delta)$ amortized time per operation, where Δ is the maximum vertex degree of the original graph. The (2,1)-spanner has $O(n^{3/2})$ edges, while the (3,2)-spanner has $O(n^{4/3})$ edges. A faster randomized dynamic algorithm for general multiplicative spanners has been later proposed by Baswana [6]: given an unweighted graph, a $(2k-1,0)$-spanner of expected size $O(k \cdot n^{1+1/k})$ can be maintained in $O(\frac{m}{n^{1+1/k}} \cdot \text{polylog} \, n)$ amortized expected time for each edge insertion/deletion, where m is the current number of edges in the graph. For $k = 2, 3$ (multiplicative 3- and 5-spanners), the amortized expected time of the randomized algorithm becomes constant. The algorithm by Elkin [15] maintains a $(2k-1,0)$-spanner with expected $O(kn^{1+1/k})$ edges in expected constant time per edge insertion and expected $O(\frac{m}{n^{1/k}})$ time per edge deletion. More recently, Baswana et al. [8] proposed two faster fully dynamic randomized algorithms for maintaining $(2k-1,0)$-spanners of unweighted graphs: the expected amortized time per insertion/deletion is $O(7^{k/2})$ for the first algorithm and $O(k^2 \log^2 n)$ for the second algorithm, and in both cases the spanner expected size is optimal up to a polylogaritmic factor.

As observed in [11], this traditional fully dynamic model may be too pessimistic in several application scenarios, where the possible changes to the underlying graph are rather limited. Indeed, there are cases where there can be only temporary network failures: namely, graph edges may occasionally fail, but only for a short period of time, and it is possible to recover quickly from such failures. In those scenarios, rather than maintaining a fully dynamic spanner, which has to be updated after each change, one may be more interested in working with a static spanner capable of retaining much of its properties during edge deletions, i.e., capable of being resilient to transient failures.

Being inherently sparse, a spanner is not necessarily resilient to edge deletions and it may indeed lose some of its important properties during a transient failure. Indeed, let S be an (α, β)-spanner of G: if an edge e fails in G, then the distortion of the spanner may substantially degrade, i.e., $S \setminus e$ may no longer be an (α, β)-spanner or even a valid spanner of $G \setminus e$, where $G \setminus e$ denotes the graph obtained after removing edge e from G. In their pioneering work, Chechik et al. [11] addressed this problem by introducing the notion of *fault-tolerant spanners*, i.e., spanners that are resilient to edge (or vertex) failures. Given an integer $f \geq 1$, a spanner is said to be f-edge (resp. vertex) fault-tolerant if it preserves its original distortion under the failure of any set of at most f edges (resp. vertices). More formally, an *f-edge (resp. vertex) fault-tolerant (α, β)-spanner* of $G = (V, E)$ is a subgraph $S = (V, E_S)$, $E_S \subseteq E$, such that for any subset $F \subseteq E$ (resp. $F \subseteq V$), with $|F| \leq f$, and for any pair of vertices $x, y \in V$ (resp. $x, y \in V \setminus F$) we have $d_{S \setminus F}(x, y) \leq \alpha \cdot d_{G \setminus F}(x, y) + \beta$, where $G \setminus F$ denotes the subgraph of G obtained after deleting the edges (resp. vertices) in F. Algorithms for computing efficiently fault-tolerant spanners can be found in [5,10,11,14].

The distortion is not the only property of a spanner that may degrade because of edge failures. Indeed, even when the removal of an edge cannot change the overall distortion of a spanner (such as in the case of a fault-tolerant spanner), it may still cause a sharp increase in some of its distances. Note that while the distortion is a *global* property, distance increases are *local* properties, as they are defined for pairs of vertices. To address this problem, one would like to work with spanners that are not only *globally resilient* (such as fault-tolerant spanners) but also *locally resilient*. In other terms, we would like to make the distances between any pair of vertices in a spanner resilient to edge failures, i.e., whenever an edge fails, then the increases in distances in the spanner must be very close to the increases in distances in the underlying graph. More formally, given a graph G and an edge e in G, we define the *fragility of edge e* as the maximum relative increase in distance between any two vertices when e is removed from G:

$$\text{frag}_G(e) = \max_{x,y \in V} \left\{ \frac{d_{G \setminus e}(x, y)}{d_G(x, y)} \right\}$$

Our definition of fragility of an edge is somewhat reminiscent of the notion of *shortcut value*, as contained in [20], where the distance increase is alternatively measured by the difference, instead of the ratio, between distances in $G \setminus e$ and in G. Note that for unweighted graphs, $\text{frag}_G(e) \geq 2$ for any edge e. The fragility of edge e can be seen as a measure of how much e is crucial for the distances in G, as it provides an upper bound to the increase in distance in G between any pair of vertices when edge e fails: the higher the fragility of e, the higher is the relative increase in some distance when e is deleted.

Our Contribution. To obtain spanners whose distances are resilient to transient edge failures, the fragility of each edge in the spanner must be as close as possible to its fragility in the original graph. In this perspective, we say that a spanner S of G is σ-*resilient* if $\text{frag}_S(e) \leq \max\{\sigma, \text{frag}_G(e)\}$ for each edge $e \in S$, where σ is a positive integer. Note that in case of unweighted graphs, for $\sigma = 2$ this is equivalent to $\text{frag}_S(e) = \text{frag}_G(e)$. We remark that finding sparse 2-resilient spanners may be an overly ambitious goal, as we prove that there exists a family of dense graphs for which the only 2-resilient spanner coincides with the graph itself. It can be easily seen that in general (α, β)-spanners are not σ-resilient. Furthermore, it can be shown that even edge fault-tolerant multiplicative t-spanners are not σ-resilient, since they can only guarantee that the fragility of a spanner edge is at most t times its fragility in the graph. In fact, we exhibit 1-edge fault tolerant t-spanners, for any $t \geq 3$, with edges whose fragility in the spanner is at least $t/2$ times their fragility in G.

It seems quite natural to ask whether sparse σ-resilient spanners exist, and how efficiently they can be computed. We show that it is possible to compute σ-resilient (1,2)-spanners, (2,1)-spanners and (3,0)-spanners of optimal asymptotic size (i.e., containing $O(n^{3/2})$ edges). The total time required to compute our spanners is $O(mn)$ in the worst case. To compute our σ-resilient spanners, we start from a non-resilient spanner, and then add to it $O(n^{3/2})$ edges from a carefully chosen set of short cycles in the original graph. The algorithm is simple and thus amenable to practical implementation, while the upper bound on the number of added edges is derived from non-trivial combinatorial arguments.

The same approach can be used for turning a given (α, β)-spanner into a σ-resilient (α, β)-spanner, for any $\sigma \geq \alpha + \beta > 3$, by adding $O(n^{3/2})$ edges. Note that this result is quite general, as (α, β)-spanners contain as special cases all $(k, k-1)$-spanners, all multiplicative $(2k-1)$-spanners, for $k \geq 2$, and all additive spanners, including additive 2-spanners and 6-spanners.

All our bounds hold for undirected unweighted graphs and can be extended to the case of graphs with positive edge weights. Our results for $\sigma = \alpha + \beta = 3$ seem to be the most significant ones, both from the theoretical and the practical point of view. From a theoretical perspective, our σ-resilient (α, β)-spanners, with $\alpha + \beta = 3$, have the same asymptotic size as their non-resilient counterparts. From a practical perspective, there is empirical evidence [3] that small stretch spanners provide the best performance in terms of stretch/size trade-offs, and that spanners of larger stretch are not likely to be of practical value.

Table 1 summarizes previous considerations and compares our results with the fragility and size of previously known spanners.

Table 1. Fragility and size of spanners

Spanner S	$d_S(x,y)$	$\mathrm{frag}_S(e)$	Size	Ref.
multiplicative $(2k-1)$-spanner, $k \geq 2$	$\leq (2k-1) \cdot d_G(x,y)$	unbounded	$O\left(n^{1+\frac{1}{k}}\right)$	[2]
additive 2-spanner	$\leq d_G(x,y) + 2$	unbounded	$O\left(n^{\frac{3}{2}}\right)$	[1]
additive 6-spanner	$\leq d_G(x,y) + 6$	unbounded	$O\left(n^{\frac{4}{3}}\right)$	[7]
1-edge fault-tolerant $(2k-1)$-spanner, $k \geq 2$	$\leq (2k-1) \cdot d_G(x,y)$	$\leq (2k-1) \cdot \mathrm{frag}_G(e)$	$O\left(n^{1+\frac{1}{k}}\right)$	[11]
σ-resilient (α, β)-spanner, $\sigma \geq \alpha + \beta \geq 3$	$\leq \alpha \cdot d_G(x,y) + \beta$	$\leq \max\{\sigma, \mathrm{frag}_G(e)\}$	$O\left(n^{\frac{3}{2}}\right)$	this paper

Due to space limitations, some proofs are omitted. They will be given in the full paper.

2   Preliminaries

Let $G = (V, E)$ be an undirected unweighted graph, with m edges and n vertices. The *girth* of G, denoted by $girth(G)$, is the length of a shortest cycle in G. A *bridge* is an edge $e \in E$ whose deletion increases the number of connected components of G. Note that an edge is a bridge if and only if it is not contained in any cycle of G. Graph G is *2-edge-connected* if it does not have any bridges. The *2-edge-connected components* of G are its maximal 2-edge-connected subgraphs. Let $e \in E$, and denote by $\mathcal{C}_e$ the set of all the cycles containing edge e: if G is 2-edge-connected, then $\mathcal{C}_e$ is non-empty for each $e \in E$. A shortest cycle among all cycles in $\mathcal{C}_e$ is referred to as a *short cycle for* edge e. If G is 2-edge-connected short cycles always exist for any edge. Short cycles are not necessarily unique: for each $e \in E$, we denote by Γ_e the set of short cycles for e. Similarly, we denote by $\mathcal{P}_e(x, y)$ the set of all paths between x and y containing edge e, and by $\mathcal{P}_{\bar{e}}(x, y)$ the set of all paths between x and y avoiding edge e. We further denote by $\Pi_e(x, y)$ (respectively $\Pi_{\bar{e}}(x, y)$) the set of shortest paths in $\mathcal{P}_e(x, y)$ (respectively $\mathcal{P}_{\bar{e}}(x, y)$).

Recall that we defined the *fragility* of an edge $e = (u, v)$ in graph G as $\mathrm{frag}_G(e) = \max_{x,y \in V} \left\{ \frac{d_{G \setminus e}(x,y)}{d_G(x,y)} \right\}$. The following lemma shows that in this definition the maximum is obtained for $\{x, y\} = \{u, v\}$, i.e., exactly at the two endpoints of edge (u, v).

Lemma 1. *Let $G = (V, E)$ be a connected graph with positive edge weights, and let $e = (u, v)$ be any edge in G. Then $\mathrm{frag}_G(e) = \frac{d_{G \setminus e}(u,v)}{d_G(u,v)}$.*

Note that for unweighted graphs, Lemma 1 can be stated as $\mathrm{frag}_G(e) = d_{G \setminus e}(u, v)$.

The fragility of all edges in a graph $G = (V, E)$ with positive edge weights can be trivially computed in a total of $O(m^2 n + mn^2 \log n)$ worst-case time by simply computing all-pairs shortest paths in all graphs $G \setminus e$, for each edge $e \in E$. A faster bound of $O(mn + n^2 \log n)$ can be achieved by using either a careful modification of algorithm `fast-exclude` in [13] or by applying n times a modified version of Dijkstra's algorithm, as described in [18]. For unweighted graphs, the above bound reduces to $O(mn)$.

3   Computing σ-Resilient Subgraphs

We first show that finding sparse 2-resilient spanners may be an ambitious goal, as there are dense graphs for which the only 2-resilient spanner is the graph itself.

Theorem 1. *There is an infinite family $\mathcal{F}$ of graphs such that for each graph $G \in \mathcal{F}$ the following properties hold:*

(1) G has $\Theta(n^\delta)$ edges, with $\delta > 1.72598$, where n is the number of vertices of G.
(2) No proper subgraph of G is a 2-resilient spanner of G.
(3) There exists a 2-spanner S of G such that $\Theta(n^\delta)$ edges of $G \setminus S$, with $\delta > 1.72598$, need to be added back to S in order to make it 2-resilient.

Edge fault-tolerant spanners provide a simple way to bound distance increases under edge faults. Unfortunately, they are not σ-resilient, as the next lemma shows.

Lemma 2. *Let $G = (V, E)$ be a graph.*

(a) Let S_f be any 1-edge fault tolerant t-spanner of G. Then $\mathrm{frag}_{S_f}(e) \leq t \cdot \mathrm{frag}_G(e)$ for each $e \in S_f$.
(b) There exist 1-edge fault-tolerant t-spanners that are not σ-resilient, for any $\sigma < t/2$.

To compute a σ-resilient spanner R of graph G, without any guarantees on the number of edges in R, we may start from any (α, β)-spanner of G, with $\alpha + \beta \leq \sigma$, and add a suitable set of *backup paths* for edges with high fragility:

1. Let S be any (α, β)-spanner of G, with $\alpha + \beta \leq \sigma$: initialize R to S.
2. For each edge $e = (u, v) \in S$ such that $\mathrm{frag}_S(e) > \sigma$, select a shortest path between u and v in $G \setminus e$ and add it to R.

The correctness of our approach hinges on the following theorem.

Theorem 2. *Let S be an (α, β)-spanner of a graph G, and let R be computed by adding to S a backup path for each edge e with $\mathrm{frag}_S(e) > \sigma$. Then R is a σ-resilient (α, β)-spanner of G.*

Proof. R is trivially an (α, β)-spanner, since it contains an (α, β)-spanner S. It remains to show that R is σ-resilient. Let $e = (u, v)$ be any edge in R. We distinguish two cases, depending on whether e was in the initial (α, β)-spanner S or not:

- $e \in S$: if $\mathrm{frag}_S(e) \leq \sigma$ then also $\mathrm{frag}_R(e) \leq \sigma$. If $\mathrm{frag}_S(e) > \sigma$, a shortest path in $G \setminus e$ joining u and v has been added to S, yielding $\mathrm{frag}_R(e) = \mathrm{frag}_G(e)$;
- $e \in R \setminus S$: since S is an (α, β)-spanner of G, it must also contain a path between u and v of length at most $\alpha + \beta$. This implies that $\mathrm{frag}_R(e) \leq \mathrm{frag}_S(e) \leq \alpha + \beta \leq \sigma$. □

Note that any σ-resilient spanner R, computed by adding backup paths for high fragility edges, inherits the properties of the underlying (α, β)-spanner S, i.e., if S is fault-tolerant then R is fault-tolerant too. Let $T(m, n)$ and $S(n)$ be respectively the time required to compute an (α, β)-spanner S and the number of edges in S. A trivial implementation of the above algorithm requires a total of $O(T(m, n) + S(n) \cdot (m + n))$ time for unweighted graphs and produces σ-resilient spanners with $O(n \cdot S(n))$ edges. In the next section we will show how to improve the time complexity and how to limit the number of added edges.

3.1   Main Results: Improving Size and Running Time

Theorem 2 does not depend on how backup paths are selected. In order to bound the number of added edges, we first show that in an unweighted graph the number of edges with high fragility is small (Theorem 3), and then we show how to carefully select shortest paths to be added as backup paths, so that the total number of additional edges required is small (Theorem 4). By combining the two bounds above, we obtain σ-resilient (α, β)-spanners with $O(n^{\frac{3}{2}})$ edges in the worst case (Theorem 5). We start by bounding the number of high fragility edges in any graph. For lack of space, the following theorem is proved only for unweighted graphs. However, it holds for graphs with positive edge weights as well.

Theorem 3. *Let $G = (V, E)$ be a graph, an let σ be any positive integer. Then, the number of edges of G having fragility greater than σ is $O(n^{1 + 1/\lfloor (\sigma + 1)/2 \rfloor})$.*

Proof. Let L be the subgraph of G containing only edges whose fragility is greater than σ. If L contains no cycle, then L has at most $(n-1)$ edges and the theorem trivially follows. Otherwise, let C be a cycle in L, let ℓ be the number of edges in C, and let e be any edge in C. Note that $\text{frag}_L(e) \leq \ell - 1$. Since L is a subgraph of G, we have $\text{frag}_G(e) \leq \text{frag}_L(e)$. Thus, $\ell \geq \text{frag}_G(e) + 1$. This holds for any cycle in L, and hence $\text{girth}(L) = \min_{e \in L}\{\text{frag}_L(e)\} + 1 \geq \min_{e \in L}\{\text{frag}_G(e)\} + 1 > \sigma + 1$. As proved by Bondy and Simonovits [9], a graph with girth greater than $\sigma + 1$ contains $O(n^{1+1/\lfloor (\sigma+1)/2 \rfloor})$ edges. $\square$

We now tackle the problem of selecting the shortest paths to be added as backup paths, so that the total number of additional edges is small. Without loss of generality, we assume that G is 2-edge-connected: if it is not, all bridges in G will necessarily be included in any spanner and our algorithm can be separately applied to each 2-edge-connected component of G. Let $e = (u, v)$ be an edge of high fragility in the initial (α, β)-spanner. Note that, in order to identify a backup path for edge e, we can either refer to a shortest path between u and v in $G \setminus e$ or, equivalently, to a short cycle for e in G (i.e., the short cycle defined by the shortest path in $G \setminus e$ and the edge e itself). In the following, we will use short cycles in G rather than shortest paths in $G \setminus e$. Recall from Section 2 that we denote by Γ_e the set of short cycles for e, and by $\Pi_e(x, y)$ (respectively $\Pi_{\bar{e}}(x, y)$) the set of shortest paths in $\mathcal{P}_e(x, y)$ (respectively $\mathcal{P}_{\bar{e}}(x, y)$). The following property is immediate:

Property 1. Let e be any edge of G, let $C \in \Gamma_e$ be a short cycle for e, and let x, y be any two vertices in C. Then x and y split C into two paths C_e and $C_{\bar{e}}$, with $C_e \in \Pi_e(x, y)$ and $C_{\bar{e}} \in \Pi_{\bar{e}}(x, y)$.

Property 1 allows us to prove the following lemma:

Lemma 3. *Let e and f be any two edges in G. If two short cycles $A \in \Gamma_e$ and $B \in \Gamma_f$ share two common vertices, say x and y, then either $B_f \cup A_e$ or $B_f \cup A_{\bar{e}}$ is a short cycle for edge f, where $\{A_e, A_{\bar{e}}\}$ and $\{B_f, B_{\bar{f}}\}$ are the decompositions of A and B with respect to x and y defined in Property 1.*

Proof. Since A is a short cycle in Γ_e, edge f cannot belong to both A_e and $A_{\bar{e}}$. We distinguish two cases:

- $f \notin A_e$: in this case, $A_e \in \Pi_{\bar{f}}(x, y)$ and replacing $B_{\bar{f}}$ by A_e yields a short cycle in Γ_f.
- $f \in A_e$, which implies that $f \notin A_{\bar{e}}$. In this case, $A_{\bar{e}} \in \Pi_{\bar{f}}(x, y)$ and replacing $B_{\bar{f}}$ by $A_{\bar{e}}$ yields a short cycle in Γ_f. $\square$

Lemma 3 can be intuitively read as follows. Given two edges e and f, let C_e be a short cycle for e and let C_f be a short cycle for f. If C_e and C_f cross in two vertices, then we can compute an alternative short cycle for f, say C'_f, which has a larger intersection with C_e (i.e., such that $C_e \cup C'_f$ has fewer edges than $C_e \cup C_f$). This property allows us to select backup paths in such a way that the total number of additional edges required is relatively small. To do that efficiently, we apply a modified version of algorithm `fast-exclude` in [13].

For lack of space, we only sketch here the main modifications needed and refer to the full paper for the low-level details of the method. Algorithm `fast-exclude` is based on Dijkstra's algorithm and computes shortest paths avoiding a set of *independent paths*, from a source vertex u to any other vertex. In order to find short cycles for the set F_u of high fragility edges incident to vertex u, we would like to apply algorithm `fast-exclude` so that it avoids all edges in F_u. Unfortunately, F_u is not a proper set of independent paths (as defined in [13]), and so algorithm `fast-exclude` cannot be applied directly. We circumvent this problem by splitting each edge $(u, v) \in F_u$ into two edges (u, v'), (v', v) with the help of an extra vertex v', and by letting algorithm `fast-exclude` avoid all edges of the form (v', v), since they form a set of independent paths. A second modification of algorithm `fast-exclude` consists of giving higher priority, during the Dijkstra-like visits, to the edges that have been already used, so that whenever a short cycle for an edge e has to be output, the algorithm selects a short cycle containing fewer new edges (i.e., edges not already contained in previously output short cycles), as suggested by Lemma 3. This allows us to prove the following theorem.

Theorem 4. *Given $q > 0$ edges $e_1, e_2, \ldots, e_q$ in a 2-edge-connected graph G, there always exist short cycles $C_1, C_2, \ldots, C_q$ in G, with $C_i \in \Gamma_{e_i}$ for $1 \leq i \leq q$, such that the graph $\cup_{i=1}^{q} C_i$ has $O(\min\{q\sqrt{n} + n, n\sqrt{q} + q\})$ edges.*

Proof. Let $C_1, C_2, \ldots, C_q$ be the cycles in the order in which they are found by the modified version of algorithm `fast-exclude`, and let V_i and E_i be respectively the vertex set and the edge set of C_i, for $1 \leq i \leq q$. We partition each E_i into the following three disjoint sets:

- E_i^{old}: edges in $E_i \cap \left(\bigcup_{j=1}^{i-1} E_j \right)$, i.e., edges already in some E_j, $j < i$.
- E_i^{new}: edges with at least one endpoint not contained in $\bigcup_{j=1}^{i-1} V_j$.
- E_i^{cross}: edges not contained in E_i^{old} and with both endpoints in $\bigcup_{j=1}^{i-1} V_j$.

To prove the theorem, we have to bound the number of edges in $\bigcup_{i=1}^{q} E_i$. We only need to count the total number of edges in $\bigcup_{i=1}^{q} E_i^{\text{new}}$ and $\bigcup_{i=1}^{q} E_i^{\text{cross}}$, since each edge in E_i^{old}, for any $1 \leq i \leq q$, has been already accounted for in some E_j^{new} or E_j^{cross}, with $j < i$. Since each edge in E_i^{new} can be amortized against a new vertex, and at most two new edges are incident to each new vertex, $|\bigcup_{i=1}^{q} E_i^{\text{new}}| \leq 2 \cdot n$.

To bound the size of sets E_i^{cross}, we proceed as follows. For each cycle C_i we choose an arbitrary orientation $\overrightarrow{C_i}$, in one of the two possible directions and direct its edges accordingly. For directed edge $e = (x, y)$ we denote vertex x as tail(e). We build a bipartite graph $\mathcal{B}$ in which one vertex class represents the n vertices $v_1, v_2, \ldots, v_n$ in G, and the other vertex class represents the q directed short cycles $\overrightarrow{C_1}, \overrightarrow{C_2}, \ldots, \overrightarrow{C_q}$. There is an edge in $\mathcal{B}$ joining cycle C_i and vertex v if and only if v is the tail of an edge in E_i^{cross}. It is possible to see that the degree of C_i in $\mathcal{B}$ is the size of E_i^{cross}, since each edge in $\mathcal{B}$ corresponds to an edge in $\bigcup_{i=1}^{q} E_i^{\text{cross}}$ and vice versa.

We claim that two vertices x and y cannot be tails of two pairs of directed edges in E_i^{cross} and E_j^{cross} (see Figure 1). We prove this claim by contradiction.

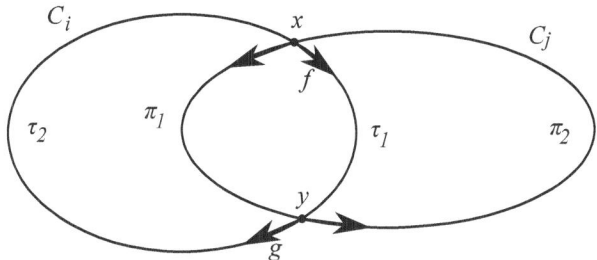

Fig. 1. On the proof of Theorem 4. If cycle C_i is detected after cycle Cj, then either edge f or g is not in E_i^{cross}. In fact, one among π_1 or π_2 should have been included in C_i in place of τ_1.

Assume without loss of generality $i > j$ (i.e., short cycle C_j has been output before short cycle C_i). Since whenever a short cycle has to be output, our algorithm selects a short cycle containing fewer new edges (i.e., edges not contained already in previously output short cycles), either the path τ_1 or τ_2 of C_i should have been replaced by portion π_1 or portion π_2 of C_j (as in Lemma 3). Let f and g be the two directed edges in E_i^{cross} with $\mathrm{tail}(f) = x$ and $\mathrm{tail}(g) = y$: by the above argument, one among f and g should be in E_i^{old} instead of E_i^{cross}, yielding a contradiction.

The previous claim implies that the bipartite graph $\mathcal{B}$ does not contain $K_{2,2}$ as a (not necessarily induced) subgraph. Determining the maximum number of edges in $\mathcal{B}$ is a special case of Zarankiewicz's problem [23]. This problem has been solved by Kővári, Sós, Turán [19] (see also [21], p. 65), who proved that any bipartite graph G with vertex classes of size m and n containing no subgraph $K_{r,s}$, with the r vertices in the class of size m and the s vertices in the class of size n, has $O\left(\min\left\{mn^{1-1/r} + n, m^{1-1/s}n + m\right\}\right)$ edges, where the constant of proportionality depends on r and s. Since in our case the bipartite graph $\mathcal{B}$ has vertex classes of size n and q, and $r = s = 2$, it follows that $\mathcal{B}$ contains $O(\min\{q\sqrt{n} + n, n\sqrt{q} + q\})$ edges.

In summary, the total number of edges in the graph $\bigcup_{i=1}^q C_i$ is bounded by

$$\left|\bigcup_{i=1}^q \left(E_i^{\mathrm{old}} \cup E_i^{\mathrm{new}} \cup E_i^{\mathrm{cross}}\right)\right| \leq 2n + O\left(\min\left\{q\sqrt{n} + n, n\sqrt{q} + q\right\}\right)$$

and thus the theorem holds. □

We observe that Theorem 4 is closely related to a result of Coppersmith and Elkin [12] on distance preservers. Given a graph G and p pairs of vertices $\{(v_1, w_1), \ldots (v_p, w_p)\}$, a pairwise distance preserver is a subgraph S of G such that $d_S(v_i, w_i) = d_S(v_i, w_i)$, for $1 \leq i \leq p$. In particular, Coppersmith and Elkin [12] showed that it is always possible to compute a pairwise distance preserver containing $O(\min\{p\sqrt{n} + n, n\sqrt{p} + p\})$ edges. Theorem 4 can be extended to weighted and directed graphs and provides an alternative (and simpler) proof of the result in [12].

We are now ready to bound the size and the time required to compute a σ-resilient (α, β)-spanner.

Theorem 5. *Let G be a graph with m edges and n vertices. Let S be any (α, β)-spanner of G, and denote by $T(m,n)$ and $S(n)$ respectively the time required to compute S and the number of edges in S. Then a σ-resilient (α, β)-spanner R of G, with $\sigma \geq \alpha + \beta$, can be computed in $O(T(m,n) + mn)$ time. Furthermore, $R \supseteq S$ and R has $O\left(S(n) + n^{3/2}\right)$ edges.*

Proof. As explained above, a σ-resilient (α, β)-spanner R can be computed by adding a set $\mathcal{C}$ of short cycles to S, one for each edge $e \in S$ with $\operatorname{frag}_G(e) > \sigma$. Let C_e be the cycle in Γ_e computed by our algorithm.

We partition egdes $e \in S$ with $\operatorname{frag}_G(e) > \sigma$ into three subsets E_ℓ, E_m and E_h, according to their fragility in G. For each subset we separately bound the number of edges in the union of cycles in $\mathcal{C}$.

low fragility edges: $E_\ell = \{e \in S \mid \sigma \leq \operatorname{frag}_G(e) \leq 5\}$. Obviously, we have $|E_\ell| \leq S(n)$, and since each cycle C_e, $e \in E_\ell$, contains at most 5 edges, we have $\left|\bigcup_{e \in E_\ell} C_e\right| = O\left(S(n)\right)$.

medium fragility edges: $E_m = \{e \in S \mid \max\{\sigma, 6\} \leq \operatorname{frag}_G(e) < \log n\}$. By Theorem 3, since the fragility of each edge in E_m is greater than 5, $|E_m| = O(n^{4/3})$. Since each cycle C_e, $e \in E_m$, contains at most $\log n$ edges, we have
$$\left|\bigcup_{e \in E_m} C_e\right| = O\left(n^{\frac{4}{3}} \cdot \log n\right).$$

high fragility edges: $E_h = \{e \in S \mid \operatorname{frag}_G(e) \geq \log n\}$. By Theorem 3, $|E_h| = O\left(n^{1 + \frac{2}{\log n}}\right) = O(n)$, and by Theorem 4 we have
$$\left|\bigcup_{e \in E_h} C_e\right| = O\left(n \cdot \sqrt{|E_h|}\right) = O\left(n^{\frac{3}{2}}\right).$$

Hence the total number of edges in R is
$$\left|\bigcup_{e \in E_\ell \cup E_m \cup E_h} C_e\right| = O\left(S(n) + n^{\frac{3}{2}}\right)$$

To bound the running time, we observe that we find the fragility of each edge in S and then we compute a set of short cycles as suggested by Lemma 3. The fragility of each edge and the set of short cycles can be computed by a proper modification of algorithm `fast-exclude` in [13] in a total of $O(mn)$ worst-case time. $\qquad\square$

Theorem 5 allows us to compute σ-resilient versions of several categories of spanners, including multiplicative $(2k - 1)$-spanners and $(k, k - 1)$-spanners for $k \geq 2$, $(1, 2)$-spanners and $(1, 6)$-spanners. Since the time required to compute all those underlying spanners is $O(mn)$, in all those cases the time required to build a σ-resilient spanner is $O(mn)$. Theorem 5 can also be applied to build σ-resilient f-spanners, where f is a general *distortion* function as defined in [22], provided that $\sigma \geq f(1)$. Furthermore, if we wish to compute a σ-resilient (α, β)-spanner with $\sigma < \alpha + \beta$, the same algorithm can still be applied starting from a σ-spanner instead of an (α, β)-spanner, yielding the same bounds given in Theorem 5.

Our results can be extended to weighted graphs, since Theorems 3 and 4 also hold for graphs with positive edge weights. Let w_{max} and w_{min} be respectively the weights of the heaviest and lightest edge in the graph, and let $W = \frac{w_{max}}{w_{min}}$. For either $\sigma > \log n$ or $\sigma \geq 5$ and $W = O\left(\left(n^{\frac{1}{2} - \frac{1}{\lceil (\sigma+1)/2 \rceil}}\right) / \log n\right)$, we can compute a σ-resilient t-spanner with $O(n^{\frac{3}{2}})$ edges in $O(mn)$ time. In the remaining cases (either $\sigma \geq 5$ and larger W or $\sigma = 3, 4$), the total number of edges becomes $O(W \cdot n^{\frac{3}{2}})$. For lack of space, the details are deferred to the full paper.

4   Conclusions and Further Work

In this paper, we have investigated a new notion of resilience in graph spanners by introducing the concept of σ-resilient spanners. In particular, we have shown that it is possible to compute small stretch σ-resilient spanners of optimal size. The techniques introduced for small stretch σ-resilient spanners can be used to turn any generic (α, β)-spanner into a σ-resilient (α, β)-spanner, for $\sigma \geq \alpha + \beta > 3$, by adding a suitably chosen set of at most $O(n^{3/2})$ edges. The same approach is also valid for graphs with positive edge weights.

We expect that in practice our σ-resilient spanners, for $\sigma \geq \alpha + \beta > 3$, will be substantially sparser than what it is implied by the bounds given in Theorem 5, and thus of higher value in applicative scenarios. Towards this aim, we plan to perform a thorough experimental study. Another intriguing question is whether our theoretical analysis on the number of edges that need to be added to an (α, β)-spanner in order to make it σ-resilient provides tight bounds, or whether it can be further improved.

References

1. Aingworth, D., Chekuri, C., Indyk, P., Motwani, R.: Fast estimation of diameter and shortest paths (without matrix multiplication). SIAM J. Comput. 28(4), 1167–1181 (1999)
2. Althofer, I., Das, G., Dobkin, D.P., Joseph, D., Soares, J.: On sparse spanners of weighted graphs. Discrete & Computational Geometry 9, 81–100 (1993)
3. Ausiello, G., Demetrescu, C., Franciosa, P.G., Italiano, G.F., Ribichini, A.: Graph spanners in the streaming model: An experimental study. Algorithmica 55(2), 346–374 (2009)
4. Ausiello, G., Franciosa, P.G., Italiano, G.F.: Small stretch spanners on dynamic graphs. Journal of Graph Algorithms and Applications 10(2), 365–385 (2006)
5. Ausiello, G., Franciosa, P.G., Italiano, G.F., Ribichini, A.: Computing graph spanner in small memory: fault-tolerance and streaming. Discrete Mathematics, Algorithms and Applications 2(4), 591–605 (2010)
6. Baswana, S.: Dynamic algorithms for graph spanners. In: Azar, Y., Erlebach, T. (eds.) ESA 2006. LNCS, vol. 4168, pp. 76–87. Springer, Heidelberg (2006)

7. Baswana, S., Kavitha, T., Mehlhorn, K., Pettie, S.: New constructions of (α, β)-spanners and purely additive spanners. In: Proc. of the 16th Annual ACM-SIAM Symposium on Discrete Algorithms (SODA 2005), pp. 672–681 (2005)

8. Baswana, S., Khurana, S., Sarkar, S.: Fully dynamic randomized algorithms for graph spanners. ACM Trans. Algorithms 8(4), 35:1–35:51 (2012)

9. Bondy, J.A., Simonovits, M.: Cycles of even length in graphs. Journal of Combinatorial Theory, Series B 16(2), 97–105 (1974)

10. Braunschvig, G., Chechik, S., Peleg, D.: Fault tolerant additive spanners. In: Golumbic, M.C., Stern, M., Levy, A., Morgenstern, G. (eds.) WG 2012. LNCS, vol. 7551, pp. 206–214. Springer, Heidelberg (2012)

11. Chechik, S., Langberg, M., Peleg, D., Roditty, L.: Fault-tolerant spanners for general graphs. In: Proc. of 41st Annual ACM Symposium on Theory of Computing (STOC 2009), pp. 435–444 (2009)

12. Coppersmith, D., Elkin, M.: Sparse sourcewise and pairwise distance preservers. SIAM J. Discrete Math. 20(2), 463–501 (2006)

13. Demetrescu, C., Thorup, M., Chowdhury, R.A., Ramachandran, V.: Oracles for distances avoiding a failed node or link. SIAM J. Comput. 37(5), 1299–1318 (2008)

14. Dinitz, M., Krauthgamer, R.: Fault-tolerant spanners: better and simpler. In: Proc. of the 30th Annual ACM Symposium on Principles of Distributed Computing (PODC 2011), pp. 169–178 (2011)

15. Elkin, M.: Streaming and fully dynamic centralized algorithms for constructing and maintaining sparse spanners. In: Arge, L., Cachin, C., Jurdziński, T., Tarlecki, A. (eds.) ICALP 2007. LNCS, vol. 4596, pp. 716–727. Springer, Heidelberg (2007)

16. Elkin, M., Peleg, D.: (1+epsilon, beta)-spanner constructions for general graphs. SIAM J. Comput. 33(3), 608–631 (2004)

17. Halperin, S., Zwick, U.: Linear time deterministic algorithm for computing spanners for unweighted graphs. Unpublished manuscript (1996)

18. Jacob, R., Koschützki, D., Lehmann, K.A., Peeters, L., Tenfelde-Podehl, D.: Algorithms for centrality indices. In: Brandes, U., Erlebach, T. (eds.) Network Analysis. LNCS, vol. 3418, pp. 62–82. Springer, Heidelberg (2005)

19. Kővári, T., Sós, V.T., Turán, P.: On a problem of K. Zarankiewicz. Colloquium Mathematicae 3(1), 50–57 (1954)

20. Koschützki, D., Lehmann, K.A., Peeters, L., Richter, S., Tenfelde-Podehl, D., Zlotowski, O.: Centrality Indices. In: Brandes, U., Erlebach, T. (eds.) Network Analysis. LNCS, vol. 3418, pp. 16–61. Springer, Heidelberg (2005)

21. Matoušek, J.: Lectures on Discrete Geometry. Springer (2002)

22. Pettie, S.: Low distortion spanners. In: Arge, L., Cachin, C., Jurdziński, T., Tarlecki, A. (eds.) ICALP 2007. LNCS, vol. 4596, pp. 78–89. Springer, Heidelberg (2007)

23. Zarankiewicz, K.: Problem p 101. Colloquium Mathematicae 2, 301 (1951)

Maximizing Barrier Coverage Lifetime
with Mobile Sensors

Amotz Bar-Noy[1], Dror Rawitz[2], and Peter Terlecky[1]

[1] The Graduate Center of the City University of New York, NY 10016, USA
amotz@sci.brooklyn.cuny.edu, pterlecky@gc.cuny.edu
[2] School of Electrical Engineering, Tel Aviv University, Tel-Aviv 69978, Israel
rawitz@eng.tau.ac.il

Abstract. Sensor networks are ubiquitously used for detection and tracking and as a result covering is one of the main tasks of such networks. We study the problem of maximizing the *coverage lifetime* of a barrier by mobile sensors with limited battery powers, where the coverage lifetime is the time until there is a breakdown in coverage due to the death of a sensor. Sensors are first deployed and then coverage commences. Energy is consumed in proportion to the distance traveled for mobility, while for coverage, energy is consumed in direct proportion to the radius of the sensor raised to a constant exponent. We study two variants which are distinguished by whether the sensing radii are given as part of the input or can be optimized, the fixed radii problem and the variable radii problem. We design parametric search algorithms for both problems for the case where the final order of the sensors is predetermined and for the case where sensors are initially located at barrier endpoints. In contrast, we show that the variable radii problem is strongly NP-hard and provide hardness of approximation results for fixed radii for the case where all the sensors are initially co-located at an internal point of the barrier.

1 Introduction

One important application of Wireless Sensor Networks is monitoring a barrier for some phenomenon. By covering the barrier, the sensors protect the interior of the region from exogenous elements more efficiently than if they were to cover the interior area. In this paper we focus on a model in which sensors are battery-powered and both moving and sensing drain energy. A sensor can maintain coverage until its battery is completely depleted. The network of sensors cover the barrier until the death of the first sensor, whereby a gap in coverage is created and the life of the network expires.

More formally, there are n sensors denoted by $\{1, \ldots, n\}$. Each sensor i has a battery of size b_i and initial position x_i. The coverage task is accomplished in two phases. In the *deployment phase*, sensors move from their initial positions to new positions, and in the *covering phase* the sensors set their sensing radii to fully cover the barrier. A sensor which moves a distance d drains $a \cdot d$ amount of battery on movement for some constant $a \geq 0$. In the coverage phase, sensing

H.L. Bodlaender and G.F. Italiano (Eds.): ESA 2013, LNCS 8125, pp. 97–108, 2013.
© Springer-Verlag Berlin Heidelberg 2013

with a radius of r drains energy per time unit in direct proportion to r^α, for some constant $\alpha \geq 1$ (see e.g., [1,11]). The lifetime of a sensor i traveling a distance d_i and sensing with a radius r_i is given by $L_i = \frac{b_i - ad_i}{r_i^\alpha}$. The coverage lifetime of the barrier is the minimum lifetime of any sensor, $\min_i L_i$. We seek to determine a destination y_i and a radius r_i, for each sensor i, that maximizes the *barrier coverage lifetime* of the network.

Many parameters govern the length of coverage lifetime, and optimizing them is hard even for simple variants. Thus, most of the past research adopted natural strategies that try to optimize the lifetime indirectly. For example, the duty cycle strategy partitions the sensors into disjoint groups that take turns in covering the barrier. The idea is that a good partition would result in a longer lifetime. Another example is the objective of minimizing the maximum distance traveled by any of the sensors. This strategy would maximize the coverage lifetime for sensors with homogeneous batteries and radii, but would fail to do so if sensors have non-uniform batteries or radii. See a discussion in the related work section.

In this paper we address the lifetime maximization problem directly. We focus on the *set-up and sense* model in which the sensors are given one chance to set their positions and sensing radii before the coverage starts. We leave the more general model in which sensors may adjust their positions and sensing radii during the coverage to future research.

Related Work. There has been previous research on barrier coverage focused on minimizing a parameter which is proportional to the energy sensors expend on movement, but not directly modeling sensor lifetimes with batteries. Czyzowicz et al. [8] assume that sensors are located at initial positions on a line barrier and that the sensors have fixed and identical sensing radii. The goal is to find a deployment that covers the barrier and that minimizes the maximum distance traveled by any sensor. Czyzowicz et al. provide a polynomial time algorithm for this problem. Chen et al. [7] extended the result to the more general case in which the sensing radii are non-uniform (but still fixed).

Czyzowicz et al.[9] considered covering a line barrier with sensors with the goal of minimizing the sum of the distances traveled by all sensors. Mehrandish et al. [12] considered the same model with the objective of minimizing the number of sensors which must move to cover the barrier. Tan and Wu [14] presented improved algorithms for minimizing the max distance traveled and minimizing the sum of distances traveled when sensors must be positioned on a circle in regular n-gon position. The problems were initially considered by Bhattacharya et al. [5]. Several works have considered the problem of covering a straight-line boundary by stationary sensors. Li et al. [11] look to choose radii for sensors for coverage which minimize the sum of the power spent. Agnetis et al. [1] seek to choose radii for coverage to minimize the sum of a quadratic cost function. Maximizing the network lifetime of battery-powered sensors that cover a barrier was previously considered for static sensors from a scheduling point of view. Buchsbaum et al. [6] and Gibson and Varadarajan [10] considered the RESTRICTED STRIP COVERING in which sensors are static and radii are fixed, but sensors may start covering at any time. Bar Noy *et al.* [2,3,4] considered the variant of this problem in which the radii are adjustable.

The only previous result we are aware of that considered a battery model with movement and transmission on a line is by Phelan *et al.* [13] who considered the problem of maximizing the transmission lifetime of a sender to a receiver on a line using mobile relays.

Our Contribution. We introduce two problems in the model in which sensors are battery-powered and both moving and sensing drain energy. In the BARRIER COVERAGE WITH VARIABLE RADII problem (abbreviated BCVR) we are given initial locations and battery powers, and the goal is to find a deployment and radii that maximizes the lifetime. In the BARRIER COVERAGE WITH FIXED RADII problem (BCFR) we are also given a radii vector ρ, and the goal is to find a deployment and a radii assignment r, such that $r_i \in \{0, \rho_i\}$, for every i, that maximizes the lifetime.

In the full version we show that the static ($a = \infty$) and fully dynamic ($a = 0$) cases are solvable in polynomial time for both BCFR and BCVR. On the negative side, we show in Section 5 that it is NP-hard to approximate BCFR (i) within any multiplicative approximation factor, or (ii) within an additive factor ε, for some $\varepsilon > 0$, in polynomial time unless P=NP, for any $a \in (0, \infty)$ and $\alpha \geq 1$, even if $x = p^n$, where $p \in (0, 1)$. We also show that BCVR is strongly NP-hard for any $a \in (0, \infty)$ and $\alpha \geq 1$.

In Section 3 we consider constrained versions of BCFR and BCVR in which the input contains a total order on the sensors that the solution is required to satisfy. We design a polynomial-time algorithm for the decision problem of BCFR in which the goal is to determine whether a given lifetime t is achievable and to compute a solution with lifetime t, if t is achievable. We design a similar algorithm for BCVR that, given t and $\epsilon > 0$, determines whether $t - \epsilon$ is achievable. Using these decision algorithms we present parametric search algorithms for constrained BCFR and BCVR. We consider the case where the sensors are initially located on the edges of the barrier (i.e., $x \in \{0, 1\}^n$) in Section 4. For both BCFR and BCVR, we show that, for every candidate lifetime t, we may assume a final ordering of the sensors. (The ordering depends only on the battery powers in the BCVR case, and it can be computed in polynomial time in the BCFR case.) Using our decision algorithms, we obtain parametric search algorithms for this special case.

Finally, we note that several proofs were omitted for lack of space.

2   Preliminaries

Model. We consider a setting in which n mobile sensors with finite batteries are located on a barrier represented by the interval $[0, 1]$. The initial position and battery power of sensor i is denoted by x_i and b_i, respectively. We denote $x = (x_1, \ldots, x_n)$ and $b = (b_1, \ldots, b_n)$. The sensors are used to cover the barrier, and they can achieve this goal by moving and sensing. In our model the sensors first move, and afterwards each sensor covers an interval that is determined by its sensing radius. In motion, energy is consumed in proportion to the distance traveled, namely a sensor consumes $a \cdot d$ units of energy by traveling a distance d, where a is a constant. A sensor i consumes r_i^α energy per time unit for sensing, where r_i is the sensor's radius and $\alpha \geq 1$ is a constant.

More formally, the system works in two phases. In the *deployment phase* sensors move from the initial positions x to new positions y. This phase is said to occur at time 0. In this phase, sensor i consumes

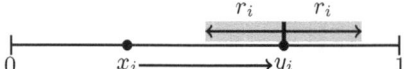

Fig. 1. Sensor i moves from x_i to y_i and covers the interval $[y_i - r_i, y_i + r_i]$

$a|y_i - x_i|$ energy. Notice that sensor i may be moved to y_i only if $a|y_i - x_i| \leq b_i$. In the *covering phase* sensor i is assigned a sensing radius r_i and covers the interval $[y_i - r_i, y_i + r_i]$. (An example is given in Figure 1.) A pair (y, r), where y is a deployment vector and r is a sensing radii vector, is called *feasible* if (i) $a|y_i - x_i| \leq b_i$, for every sensor i, and (ii) $[0, 1] \subseteq \sum_i [y_i - r_i, y_i + r_i]$. Namely, (y, r) is feasible, if the sensors have enough power to reach y and each point in $[0, 1]$ is covered by some sensor.

Given a feasible pair (y, r), the *lifetime* of a sensor i, denoted $L_i(y, r)$, is the time that transpires until its battery is depleted. If $r_i > 0$, $L_i(y, r) = \frac{b_i - a|y_i - x_i|}{r_i^\alpha}$, and if $r_i = 0$, we define $L_i(y, r) = \infty$. Given initial locations x and battery powers b, the *barrier coverage lifetime* of a feasible pair (y, r), where y is a deployment vector and r is a sensing radii vector is defined as $L(y, r) = \min_i L_i(y, r)$. We say that a t is *achievable* if there exists a feasible pair such that $L_i(y, r) = t$.

Problems. We consider two problems which are distinguished by whether the radii are given as part of the input. In the BARRIER COVERAGE WITH VARIABLE RADII problem (BCVR) we are given initial locations x and battery powers b, and the goal is to find a feasible pair (y, r) of locations and radii that maximizes $L(y, r)$. In the BARRIER COVERAGE WITH FIXED RADII problem (BCFR) we are also given a radii vector ρ, and the goal is to find a feasible pair (y, r), such that $r_i \in \{0, \rho_i\}$ for every i, that maximizes $L(y, r)$. Notice that a necessary condition for achieving non-zero lifetime is $\sum_i 2\rho_i \geq 1$.

Given a total order $\prec$ on the sensors, we consider the *constrained* variants of BCVR and BCFR, in which the deployment y must satisfy the following: $i \prec j$ if and only if $y_i \leq y_j$. That is, we are asked to maximize barrier coverage lifetime subject to the condition that the sensors are ordered by $\prec$ (this includes sensors that do not participate in the cover). Without loss of generality, we assume that the sensors are numbered according to the total order.

3   Constrained Problems and Parametric Search

In this section we present polynomial time algorithm that, given $t > 0$, decides whether t is achievable for constrained BCFR. In addition, we give a similar algorithm that, given $t > 0$ and any accuracy parameter $\epsilon > 0$, decides whether

$t - \varepsilon$ is achievable for constrained BCVR. If the answer is in the affirmative, a corresponding solution is computed by both algorithms. We use these algorithms to design parametric search algorithms for both problems.

We use the following definitions for both BCFR and BCVR. Given an order requirement $\prec$, we define:

$$l(i) \stackrel{\text{def}}{=} \max\{\max_{j \leq i} \{x_j - b_j/a\}, 0\} \quad u(i) \stackrel{\text{def}}{=} \min\{\min_{j \geq i} \{x_j + b_j/a\}, 1\}$$

$l(i)$ and $u(i)$ are the leftmost and rightmost points reachable by i.

Observation 1. *Let (y, r) be a feasible solution that satisfies an order requirement $\prec$. Then $l(i) \leq u(i)$ and $y_i \in [l(i), u(i)]$, for every i.*

Proof. If there exists i such that $u(i) < l(i)$, then there are two sensors j and k, such that where $k < j$ and $x_j + b_j/a < x_k - b_k/a$. Hence, no deployment that satisfies the total order exists. □

3.1 Fixed Radii

We start with an algorithm that solves the constrained BCFR decision problem. Given a BCFR instance and a lifetime t, we define

$$s(i) \stackrel{\text{def}}{=} \max\{x_i - (b_i - t\rho_i^\alpha)/a, l(i)\} \quad e(i) \stackrel{\text{def}}{=} \min\{x_i + (b_i - t\rho_i^\alpha)/a, u(i)\}$$

If $t\rho_i^\alpha \leq b_i$, then $s(i) \leq e(i)$. Moreover $s(i)$ and $e(i)$ are the leftmost and right-most points that are reachable by i, if i participates in the cover for t time. ($l(i)$ and $u(i)$ can be replaced by $l(i-1)$ and $u(i-1)$ in the above definitions.)

Observation 2. *Let (y, r) be a feasible pair with lifetime t that satisfies an order $\prec$. For every i, if $r_i = \rho_i$, it must be that $t\rho_i^\alpha \leq b_i$ and $y_i \in [s(i), e(i)]$.*

Algorithm **Fixed** is our decision algorithm for constrained BCFR. It first computes l, u, s, and e. If there is a sensor i such that $l(i) > u(i)$, it outputs NO. Otherwise it deploys the sensors one by one according to $\prec$. Iteration i starts with checking whether i can extend the current covered interval $[0, z]$. If it cannot, i is moved to the left as much as possible (power is used only for moving), and it is powered down (r_i is set to 0). If i can extend the current covered interval, it is assigned radius ρ_i, and it is moved to the rightmost possible position, while maximizing the right endpoint of the currently covered interval (i.e., $[0, z]$). If i is located to the left of a sensor j, where $j < i$, then j is moved to y_i.

As for the running time, l, u, s and e can be computed in $O(n)$ time. There are n iterations, each takes $O(n)$ time. Hence, the running time of Algorithm **Fixed** is $O(n^2)$. It remains to prove the correctness of the algorithm.

Theorem 1. *Given a constrained BCFR instance and t, Algorithm **Fixed** decides whether t is achievable.*

Algorithm 1. Fixed (x, b, ρ, t)

1: Compute l, u, s, and e
2: **if** there exists i such that $u(i) < l(i)$ **then return** NO
3: $z \leftarrow 0$
4: **for** $i = 1 \rightarrow n$ **do**
5:     **if** $t\rho_i^\alpha > b_i$ or $z \notin [s(i) - \rho_i, e(i) + \rho_i]$ **then**
6:         $y_i \leftarrow \max\{l(i), y_{i-1}\}$ and $r_i \leftarrow 0$ ▷ $y_0 = 0$
7:     **else**
8:         $y_i \leftarrow \min\{z + \rho_i, e(i)\}$ and $r_i \leftarrow \rho_i$
9:         $S \leftarrow \{k : k < i, y_i < y_k\}$
10:        $y_k \leftarrow y_i$ and $r_k \leftarrow 0$, for every $k \in S$
11:        $z \leftarrow y_i + r_i$
12:     **end if**
13: **end for**
14: **if** $z < 1$ **then return** NO
15: **else return** YES

Proof. If $u(i) < l(i)$ for some i, then no deployment that satisfies the order $\prec$ exists by Observation 1. Hence, the algorithm responds correctly.

We show that if the algorithm outputs YES, then the computed solution is feasible. First, notice that $y_{i-1} \leq y_i$, for every i, by construction. We prove by induction on i, that $y_j \in [l(j), u(j)]$ and that $y_j \in [s(j), e(j)]$, if $r_j = \rho_j$, for every $j \leq i$. Consider the ith iteration. If $t\rho_i^\alpha > b_i$ or $z \notin [s(i) - \rho_i, e(i) + \rho_i]$, then $y_i \in [l(i), u(i)]$, since $\max\{l(i), y_{i-1}\} \leq \max\{u(i), u(i-1)\} \leq u(i)$. Otherwise, $y_i = \min\{z + \rho_i, e(i)\} \geq s(i)$, since $z \geq s(i) - \rho_i$. Hence, if $r_i = \rho_i$, we have that $y_i \in [s(i), e(i)]$. Furthermore, if $j < i$ is moved to the left to i, then $y_j = y_i \geq s(i) \geq l(i) \geq l(j)$. Finally, let z_i denote the value of z after the ith iteration. (Initially, $z_0 = 0$.) We proof by induction on i that $[0, z_i]$ is covered. Consider iteration i. If $r_i = 0$, then we are done. Otherwise, $z_{i-1} \in [y_i - \rho_i, y_i + \rho_i]$ and $z_i = y_i + \rho_i$. Furthermore, the sensors in S can be powered down and moved, since $[y_j - r_j, y_j + r_j] \subseteq [y_i - \rho_i, y_i + \rho_i]$, for every $j \in S$.

Finally, we show that if the algorithm outputs NO, there is no feasible solution. We prove by induction that $[0, z_i]$ is the longest interval that can be covered by sensors $1, \ldots, i$. In the base case, observe that $z_0 = 0$ is optimal. For the induction step, let y' be a deployment of $1, \ldots, i$ that covers the interval $[0, z_i']$. Let $[0, z_{i-1}']$ be the interval that y' covers by $1, \ldots, i-1$. By the inductive hypothesis, $z_{i-1}' \leq z_{i-1}$. If $t\rho_i^\alpha > b_i$ or $z_{i-1} < s(i) - \rho_i$, it follows that $z_i' = z_{i-1}' \leq z_{i-1} = z_i$. Otherwise, observe that $y_i' \leq y_i$ and therefore $z_i' \leq z_i$. $\quad\square$

3.2 Variable Radii

We present an algorithm that solves the constrained BCVR decision problem.

Before presenting our algorithm, we need a few definitions. Given a BCVR instance (x, b) and $t > 0$, if sensor i moves from x_i to $p \in [l(i), u(i)]$, then we may assume without loss of generality that its radius is as large as possible, namely that $r_i(p, t) = \sqrt[\alpha]{(b_i - a|p - x_i|)/t}$.

Similarly to Algorithm **Fixed**, our algorithm tries to cover $[0,1]$ by deploying sensors one by one, such that the length of the covered prefix $[0,z]$ is maximized. This motivates the following definitions. Let $d \in [-\frac{b_i}{a}, \frac{b_i}{a}]$ denote the distance traveled by sensor i, where $d > 0$ means traveling right, and $d < 0$ means traveling left. If a sensor travels a distance d, then its lifetime t sustaining radius is given by $\sqrt[\alpha]{(b_i - a|d|)/t}$. Given t, we define: $g_i^t(d) \stackrel{\text{def}}{=} d + \sqrt[\alpha]{(b_i - a|d|)/t}$. $g_i^t(d)$ is the *right reach* of sensor i at distance d from x_i, i.e., the rightmost point that i covers when it has traveled a distance of d and the required lifetime is t. Similarly define $h_i^t(d) \stackrel{\text{def}}{=} g_i^t(-d)$ is the *left reach* of sensor i at distance d from x_i. See depiction in Figure 2. We explore these functions in the next lemma whose proof is given in the full version.

Lemma 3. *Let $t > 0$. For any i, the distance d_i^t maximizes $g_i^t(d)$, where*

$$d_i^t = \begin{cases} \frac{b_i}{a} - \frac{1}{\alpha}\sqrt[\alpha-1]{\frac{a}{\alpha t}} & \alpha > 1 \\ \frac{b_i}{a} & \alpha = 1, a < t \\ 0 & \alpha = 1, a \geq t \end{cases}$$

$$g_i^t(d_i^t) = \begin{cases} \frac{b_i}{a} + \left(1 - \frac{1}{\alpha}\right)\sqrt[\alpha-1]{\frac{a}{\alpha t}} & \alpha > 1 \\ \frac{b_i}{\min\{a,t\}} & \alpha = 1 \end{cases}$$

If $\alpha > 1$ or $a \neq t$, g_i^t is increasing for $d < d_i^t$, and decreasing for $d > d_i^t$. If $\alpha = 1$ and $a = t$, g_i^t is constant, for $d \geq 0$, and it is increasing for $d < 0$.

Given a point $z \in [0,1]$, the *attaching position* of sensor i to z, denoted by $p_i(z,t)$, is the position p for which $p - r_i(p,t) = z$ such that $p + r_i(p,t)$ is maximized, if such a position exist. If such a point does not exist we define $p_i(z,t) = \infty$. Observe that by Lemma 3 there may be at most two points

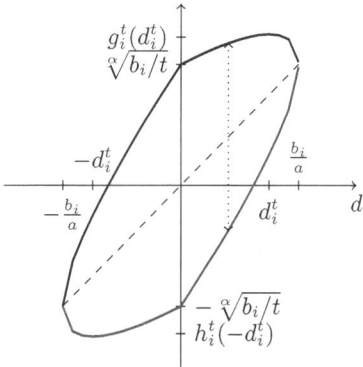

Fig. 2. Depiction of the functions $g_i^t(d)$ and $h_i^t(d)$ for $a = 2$, $\alpha = 2$, $b_i = 1$, and $t = 4$. The top curve corresponds to $g_i^t(d)$, and the bottom curve corresponds to $h_i^t(d)$. The dashed line corresponds to the location of sensor i, while the vertical interval between the curves is the interval that is covered by i at distance d from x_i.

that satisfy the equation $p - r_i(p,t) = z$. Such a position can either be found explicitly or numerically as it involves solving an equation of degree α. We ignore calculation inaccuracies for ease of presentation. These inaccuracies are subsumed by the additive factor. We omit the details.

Algorithm **Variable** is our decision algorithm for BCVR. It first computes u and l. If there is a sensor i, such that $l(i) > u(i)$, it outputs NO. Then, it deploys the sensors one by one according to $\prec$ with the goal of extending the coverage interval $[0,z]$. If i cannot increase the covering interval it is placed at $\max\{l(i), y_{i-1}\}$ so as not to block sensor $i + 1$. If i can increase coverage, it is placed in $[l(u), u(i)]$ such that z is covered and coverage to the right is maximized. It may be the case that the best place for i is to the left of previously positioned

---

Algorithm 2. Variable (x, b, t)

1: Compute l and u
2: **if** there exists i such that $u(i) < l(i)$ **then return** NO
3: $z \leftarrow 0$
4: **for** $i = 1 \rightarrow n$ **do**
5:     $q_L(i) \leftarrow \min\left\{\max\left\{x_i - d_i^t, l(i)\right\}, u(i)\right\}$
6:     $q_R(i) \leftarrow \max\left\{\min\left\{x_i + d_i^t, u(i)\right\}, l(i)\right\}$
7:     **if** $z \notin [q_L(i) - r_i(q_L(i), t), q_R(i) + r_i(q_R(i), t)]$ **then**
8:         $y_i \leftarrow \max\{l(i), y_{i-1}\}$ and $r_i \leftarrow 0$ ▷ $y_0 = 0$
9:     **else**
10:         $y_i \leftarrow \max\left\{\min\left\{p_i(z, t), u(i), x_i + d_i^t\right\}, l(i)\right\}$ and $r_i \leftarrow r_i(y_i, t)$
11:         $S \leftarrow \{k : k < i, y_i < y_k\}$
12:         $y_k \leftarrow y_i$ and $r_k \leftarrow 0$, for every $k \in S$
13:         $z \leftarrow y_i + r_i$
14:     **end if**
15: **end for**
16: **if** $z < 1$ **then return** NO
17: **else return** YES

---

sensors. In this case the algorithm moves the sensors such that coverage and order are maintained. Finally, if $z < 1$ after placing sensor n, the algorithm outputs NO, and otherwise it outputs YES.

l and u can be computed in $O(n)$ time. There are n iterations of the main loop, each taking $O(n)$ time (assuming that computing $p_i(z, t)$ takes $O(1)$ time), thus the running time of the algorithm is $O(n^2)$.

In order to analyze Algorithm **Variable** we define

$$P(i) = \{p : p \in [l(i), u(i)] \text{ and } z \in [p - r_i(p, t), p + r_i(p, t)]\} .$$

$P(i)$ is the set of points from which sensor i can cover z. Observe that $P(i)$ is an interval due to Lemma 3. Hence, we write $P(i) = [p_L(i), p_R(i)]$.

In the next two lemmas it is shown that when the algorithm checks whether $z \notin [q_L(i) - r_i(q_L(i), t), q_R(i) + r_i(q_R(i), t)]$ it actually checks whether $P(i) = \emptyset$, and that $y_i^* \overset{\text{def}}{=} \max\left\{\min\left\{p_i(z, t), u(i), x_i + d_i^t\right\}, l(i)\right\}$ is equal to $\operatorname{argmax}_{p \in P}\{p + r_i(p, t)\}$. Hence, in each iteration we check whether $[0, z]$ can be extended, and if it can, we take the best possible extension.

Lemma 4. $[p_L(i), p_R(i)] \subseteq [q_L(i), q_R(i)]$. *Moreover,* $P(i) = \emptyset$ *if and only if* $z \notin [q_L(i) - r_i(q_L(i), t), q_R(i) + r_i(q_R(i), t)]$.

Proof. By Lemma 3 $q_L(i)$ is the location that maximized coverage to the left, and $q_R(i)$ is the location that maximized coverage to the right.   □

Lemma 5. *If* $P(i) \neq \emptyset$, *then* $y_i^* = \operatorname{argmax}_{p \in P(i)}\{p + r_i(p, t)\}$.

Proof. By Lemma 3, there are three cases:

- If $x_i + d_i^t \in P(i)$, then $\operatorname{argmax}_{p \in P(i)}\{p + r_i(p, t)\} = x_i + d_i^t$. $y_i^* = x_i + d_i^t$, since $p_i(z, t) \geq x_i + d_i^t$.

- If $x_i + d_i^t > p_R(i)$, then $\operatorname{argmax}_{p \in P(i)} \{p + r_i(p,t)\} = p_R(i)$.
  $y_i^* = \min \{p_i(z,t), u(i)\}$, since $p_R(i) = \min \{p_i(z,t), u(i)\} \geq l(i)$.
- If $x_i + d_i^t < p_L(i)$, then $\operatorname{argmax}_{p \in P(i)} \{p + r_i(p,t)\} = p_L(i)$.
  $y_i^* = l(i)$, since $q_L(i) = l(i) > x_i + d_i^t \geq \min \{p_i(z,t), u(i), x_i + d_i^t\}$. $\square$

We proof of the next theorem is somewhat similar to the proof of Theorem 1.

Theorem 2. *Let $\varepsilon > 0$. Given a constrained* BCVR *instance and t, Algorithm **Variable** decides whether $t - \varepsilon$ is achievable.*

3.3   Parametric Search Algorithms

Since we have algorithm thats, given t and an order $\prec$, decides whether there exists a solution that satisfies $\prec$ with lifetime t (or $t - \varepsilon$), we can perform a binary search on t. The maximum lifetime of a given instance is bounded by the lifetime of this instance in the case where $a = 0$. In the full version we show that, for $a = 0$, the network lifetime in the fixed case is at most $\max_i \{b_i / \rho_i^\alpha\}$, and that it is $(2 \sum_j \sqrt[\alpha]{b_j})^\alpha$ in the variable radii case. These expression serve as upper bounds for the case where $a > 0$. Hence, the running time of the parametric search in polynomial in the input size and in the $\log \frac{1}{\varepsilon}$, where ε is the accuracy parameter.

4   Sensors Are Located on the Edges of the Barrier

Consider the case where the initial locations are on either edge of the barrier, namely, $x \in \{0,1\}^n$. For both BCVR and BCFR we show that, given an achievable lifetime t, there exists a solution with lifetime t in which the sensors satisfy a certain ordering. In the case of BCVR, the ordering depends only on the battery sizes, and hence we may use the parametric search algorithm for constrained BCVR from Section 3. In the case of BCFR, the ordering depends on t, and therefore may change. Even so, we may use parametric search for this special case of BCFR since, given t, the ordering can be computed in polynomial time.

Fixed Radii. We start by considering the special case of BCFR in which all sensors are located at $x = 0$. The case where $x = 1$ is symmetric. Given a BCFR instance $(0, b, \rho)$ and a lifetime t, the *maximum reach* of sensor i is defined as the farthest point from its initial position that sensors i can cover while maintaining lifetime t, and is given by: $f_t(i) = \frac{1}{a}(b_i - t\rho_i^\alpha) + \rho_i$, if $t\rho_i^\alpha \leq b_i$, and $f_t(i) = 0$, otherwise. We assume without loss of generality that the sensors are ordered according to *reach ordering*, namely that $i < j$ if and only if $f_t(i) < f_t(j)$. Also, we ignore sensors with zero reach, since they must power down. Hence, if $f_t(i) = 0$, we place i at 0 and set its radius to 0. Let t be an achievable lifetime, we show that there exists a solution (y, r) with lifetime t such that sensors are deployed according to reach ordering.

Lemma 6. *Let $(0, b, \rho)$ be a BCFR instance and let $p \in (0,1]$. Suppose that there exists a solution that covers $[0,p]$ for t time. Then, there exists a solution that covers $[0,p]$ lifetime for t time that satisfies reach ordering.*

Variable Radii. We now consider BCVR with $x = 0$. As before, the case of $x = 1$ is symmetric. Given a BCVR instance $(0, b)$ and a lifetime t, the *maximum reach* of sensor i is $g_i^t(d_i^t)$. Note that if the sensors are ordered by battery size, namely that $i < j$ if and only if $b_i < b_j$, they are also ordered by reach. Thus, we assume in the following that sensors are ordered by battery size. Let t be an achievable lifetime. We show that there exists a deployment y with lifetime t such that sensors are deployed according to the battery ordering, namely $b_i \leq b_j$ if and only if $y_i \leq y_j$.

We need the following technical lemma.

Lemma 7. *Let $c_1, c_2, d_1, d_2 \geq 0$ such that (i) $d_1 < c_1 \leq c_2 < d_2$, and (ii) $c_1 + c_2 > d_1 + d_2$. Also let $\alpha \geq 1$. Then, $\sqrt[\alpha]{c_1} + \sqrt[\alpha]{c_2} > \sqrt[\alpha]{d_1} + \sqrt[\alpha]{d_2}$*

Lemma 8. *Let $(0, b)$ be a BCVR instance and let $p \in (0, 1]$. Suppose that there exists a deployment that covers $[0, p]$ for t time. Then, there exists a deployment that covers $[0, p]$ lifetime for t time that satisfies battery ordering.*

Proof. Given a solution that covers $[0, p]$ with lifetime t, a pair of sensors is said to violate battery ordering if $b_i < b_j$ and $y_i > y_j$. Let y be a solution with lifetime t for $(0, b)$ that minimizes battery ordering violations. If there are no violations, then we are done. Otherwise, we show that the number of violations can be decreased. If y has ordering violations, then there must exist at least one violation due to a pair of adjacent sensors. Let i and j be such sensors. We assume, without loss of generality, that the batteries of both i and j are depleted at t, namely that $r_k = \sqrt[\alpha]{(b_k - a|y_k - x_k|)/t}$, for $k = i, j$.

If the barrier is covered without i, then i is moved to y_j. (Namely $y_k' = y_k$, for every $k \neq i$, and $y_i' = y_j$.) y' is feasible, since i moves to the left. Otherwise, if the barrier is covered without j, then j is moved to y_i and j's radius is decreased accordingly. Otherwise, both sensors actively participate in covering the barrier, which means that the interval $[y_j - r_j, y_i + r_i]$ is covered by i and j. In this case, we place i at y_i' with radii r_i', such that $y_i' - r_i' = y_j - r_j$. We place j at the rightmost location y_j' such that $y_j' \leq y_i$ and $y_j' - r_j' \leq y_i' + r_i'$. If $y_j' = y_i$ then we are done, as sensor j has more battery power at y_i than i does at y_i. Otherwise, we may assume that $y_j' - r_j' = y_i' + r_i'$. We show that it must be that $y_j' + r_j' \geq y_i + r_i$. We have that $y_i' < y_j$ and $y_j' < y_i$. It follows that $\beta_i' + \beta_j' > \beta_i + \beta_j$, where $\beta_i = b_i - ay_i$. Also, notice that $\beta_i < \beta_j' < \beta_j$ and $\beta_i < \beta_i' < \beta_j$. It follows that $r_i' + r_j' = \sqrt[\alpha]{\beta_i'/t} + \sqrt[\alpha]{\beta_j'/t} > \sqrt[\alpha]{\beta_i/t} + \sqrt[\alpha]{\beta_j/t} = r_i + r_j$, where the inequality is due to Lemma 7. Hence, $y_j' + r_j' = (y_j - r_j) + 2r_i' + 2r_j' > (y_j - r_j) + 2r_i + 2r_j \geq y_i + r_i$.

Since i moves to the left, it may bypass several sensors. In this case we move all sensors with smaller batteries that were bypassed by i, to y_i'. This can be done since these sensors are not needed for covering to the right of $y_i' - r_i'$. Similarly, since j moves to the right, it may bypass several sensors. As long as there is a sensor with larger reach that was bypassed by j, let k be the rightmost such sensor. Notice that k is not needed for covering to the left of y_j'. Hence, if $y_k + r_k \geq y_j' + r_j'$, we move j to y_k. Otherwise, we move k to y_j'.

In all cases, we get a deployment y' that covers $[0, p]$ with lifetime t with a smaller number of violations than y. A contradiction. $\square$

Separation. We are now ready to tackle the case where $x \in \{0,1\}^n$. We start with the fixed radii case. Given a BCFR instance (x, b, r) and a lifetime t, we assume without loss of generality that the sensors are ordered according to the following *bi-directional reach* order: first the sensors that are located at 0 according to reach order, and then the sensors that are located at 1 according to reverse reach order. We show that we may assume that the sensors are deployed using the bi-directional reach order. The first step is to show that the sensors that are located at 0 are deployed to the left of the sensors that are placed at 1.

Lemma 9. *Let (x, b, ρ) be a BCFR instance, where $x \in \{0,1\}^n$, and let t be an achievable lifetime. Then, there exists a feasible solution (y, r) with lifetime t such that $y_i \leq y_j$, for every $i < j$.*

Next we show that we may assume that the sensors are deployed using the bi-directional reach order.

Theorem 3. *Let (x, b, ρ) be a BCFR instance, and let t be an achievable lifetime. Then there exists a feasible solution (y, r) with lifetime t such that the sensors are deployed using bi-directional reach order.*

We treat the variable radii case similarly. Given a BCVR instance (x, b), we assume without loss of generality that the sensors are ordered according to a *bi-directional battery* order: first the sensors that are located at 0 according to battery order, and then the sensors that are located at 1 according to reverse battery order.

Lemma 10. *Let (x, b) be a BCVR instance, where $x \in \{0,1\}^n$, and let t be an achievable lifetime. Then, there exists a feasible solution (y, r) with lifetime t such that $y_i \leq y_j$, for every $i < j$.*

Theorem 4. *Let (x, b) be a BCVR instance, and let t be an achievable lifetime. Then there exists a feasible solution (y, r) with lifetime t such that the sensors are deployed using bi-directional battery order.*

5   Hardness Results

Theorem 5. *It is NP-hard to approximate BCFR in polynomial time (i) within any multiplicative factor, or (ii) within an additive factor ε, for some $\varepsilon > 0$, unless P=NP, for any $a \in (0, \infty)$ and $\alpha \geq 1$, even if $x = p^n$, where $p \in (0,1)$.*

Theorem 6. *BCVR is strongly NP-hard, for every $a \in (0, \infty)$ and $\alpha \geq 1$.*

6   Discussion and Open Problems

We briefly discuss some research directions and open questions. We showed that BCVR is strongly NP-Hard. Finding an approximation algorithm or showing hardness of approximation remains open. In a natural extension model, sensors could be located anywhere in the plane and asked to cover a boundary or a circular boundary. In a more general model the sensors need to cover the plane or part of the plane where their initial locations could be anywhere. Another model which can be considered is the *duty cycling* model in which sensors are partitioned into shifts that cover the barrier. Bar-Noy *et al.* [3] considered this model for stationary sensors and $\alpha = 1$. Extending it to moving sensors and $\alpha > 1$ is an interesting research direction. Finally, in the most general covering problem with the goal of maximizing the coverage lifetime, sensors could change their locations and sensing ranges at any time. Coverage terminates when all the batteries are drained.

References

1. Agnetis, A., Grande, E., Mirchandani, P.B., Pacifici, A.: Covering a line segment with variable radius discs. Computers & OR 36(5), 1423–1436 (2009)
2. Bar-Noy, A., Baumer, B.: Maximizing network lifetime on the line with adjustable sensing ranges. In: Erlebach, T., Nikoletseas, S., Orponen, P. (eds.) ALGOSEN-SORS 2011. LNCS, vol. 7111, pp. 28–41. Springer, Heidelberg (2012)
3. Bar-Noy, A., Baumer, B., Rawitz, D.: Changing of the guards: Strip cover with duty cycling. In: Even, G., Halldórsson, M.M. (eds.) SIROCCO 2012. LNCS, vol. 7355, pp. 36–47. Springer, Heidelberg (2012)
4. Bar-Noy, A., Baumer, B., Rawitz, D.: Set it and forget it: Approximating the set once strip cover problem. Technical Report 1204.1082, CoRR (2012)
5. Bhattacharya, B.K., Burmester, M., Hu, Y., Kranakis, E., Shi, Q., Wiese, A.: Optimal movement of mobile sensors for barrier coverage of a planar region. Theor. Comput. Sci. 410(52), 5515–5528 (2009)
6. Buchsbaum, A.L., Efrat, A., Jain, S., Venkatasubramanian, S., Yi, K.: Restricted strip covering and the sensor cover problem. In: SODA, pp. 1056–1063 (2007)
7. Chen, D.Z., Gu, Y., Li, J., Wang, H.: Algorithms on minimizing the maximum sensor movement for barrier coverage of a linear domain. In: Fomin, F.V., Kaski, P. (eds.) SWAT 2012. LNCS, vol. 7357, pp. 177–188. Springer, Heidelberg (2012)
8. Czyzowicz, J., Kranakis, E., Krizanc, D., Lambadaris, I., Narayanan, L., Opatrny, J., Stacho, L., Urrutia, J., Yazdani, M.: On minimizing the maximum sensor movement for barrier coverage of a line segment. In: Ruiz, P.M., Garcia-Luna-Aceves, J.J. (eds.) ADHOC-NOW 2009. LNCS, vol. 5793, pp. 194–212. Springer, Heidelberg (2009)
9. Czyzowicz, J., Kranakis, E., Krizanc, D., Lambadaris, I., Narayanan, L., Opatrny, J., Stacho, L., Urrutia, J., Yazdani, M.: On minimizing the sum of sensor movements for barrier coverage of a line segment. In: Nikolaidis, I., Wu, K. (eds.) ADHOC-NOW 2010. LNCS, vol. 6288, pp. 29–42. Springer, Heidelberg (2010)
10. Gibson, M., Varadarajan, K.: Decomposing coverings and the planar sensor cover problem. In: FOCS, pp. 159–168 (2009)
11. Li, M., Sun, X., Zhao, Y.: Minimum-cost linear coverage by sensors with adjustable ranges. In: Cheng, Y., Eun, D.Y., Qin, Z., Song, M., Xing, K. (eds.) WASA 2011. LNCS, vol. 6843, pp. 25–35. Springer, Heidelberg (2011)
12. Mehrandish, M., Narayanan, L., Opatrny, J.: Minimizing the number of sensors moved on line barriers. In: WCNC, pp. 653–658 (2011)
13. Phelan, B., Terlecky, P., Bar-Noy, A., Brown, T., Rawitz, D.: Should I stay or should I go? Maximizing lifetime with relays. In: 8th DCOSS, pp. 1–8 (2012)
14. Tan, X., Wu, G.: New algorithms for barrier coverage with mobile sensors. In: Lee, D.-T., Chen, D.Z., Ying, S. (eds.) FAW 2010. LNCS, vol. 6213, pp. 327–338. Springer, Heidelberg (2010)

Theory and Implementation of Online Multiselection Algorithms

Jérémy Barbay[1,*], Ankur Gupta[2,**], Seungbum Jo[3,***],
Satti Srinivasa Rao[3,***], and Jonathan Sorenson[2]

[1] Departamento de Ciencias de la Computación (DCC), Universidad de Chile
[2] Department of Computer Science and Software Engineering, Butler University
[3] School of Computer Science and Engineering, Seoul National University

Abstract. We introduce a new online algorithm for the *multiselection problem* which performs a sequence of selection queries on a given unsorted array. We show that our online algorithm is 1-competitive in terms of data comparisons. In particular, we match the bounds (up to lower order terms) from the optimal offline algorithm proposed by Kaligosi et al.[ICALP 2005].

We provide experimental results comparing online and offline algorithms. These experiments show that our online algorithms require fewer comparisons than the best-known offline algorithms. Interestingly, our experiments suggest that our optimal online algorithm (when used to sort the array) requires fewer comparisons than both quicksort and mergesort.

1 Introduction

Let A be an unsorted array of n elements drawn from an ordered universe. The *multiselection* problem asks for elements of rank r_i from the sequence $R = r_1, r_2, \ldots, r_q$ on A. We define $\mathcal{B}(S_q)$ as the information-theoretic lower bound on the number of comparisons required in the comparison model to answer q unique queries, where $S_q = \{s_i\}$ denotes the queries ordered by rank. We define $\Delta_i^S = s_{i+1} - s_i$, where $s_0 = 0$ and $s_{q+1} = n$. Then,

$$\mathcal{B}(S_q) = \log n! - \sum_{i=0}^{q} \log\left(\Delta_i^S!\right) = \sum_{i=0}^{q} \Delta_i^S \log \frac{n}{\Delta_i^S} - O(n).^{1}$$

As mentioned by Kaligosi et al. [10], intuitively $\mathcal{B}(S_q)$ follows from the fact that any comparison-based multiselection algorithm identifies the Δ_1^S smallest elements, Δ_2^S next smallest elements, and so on. Hence, one could sort the original array A using $\sum_i \Delta_i^S \log \Delta_i^S + O(n)$ additional comparisons.

---

* Supported in part by PROYECTO Fondecyt Regular no 1120054.

** Supported in part by the Arete Initiative at the University of Chicago.

*** Supported in part by Basic Science Research Program through the National Research Foundation of Korea (NRF) funded by the Ministry of Education, Science and Technology (Grant number 2012-0008241).

[1] We use $\log_b a$ to refer to the base b logarithm of a. By default, we let $b = 2$.

H.L. Bodlaender and G.F. Italiano (Eds.): ESA 2013, LNCS 8125, pp. 109–120, 2013.
© Springer-Verlag Berlin Heidelberg 2013

The *online multiselection* problem asks for elements of rank $r_1, r_2, \ldots, r_q$, where the sequence R is given one element at a time (in any order).

Motivation. Online multiselection is equivalent to generalized partial sorting [9]. Variants of this problem have been studied under the names *partial quicksort*, *multiple quickselect*, *interval sort*, and *chunksort*. Several applications, such as computing optimal prefix-free codes [3] and convex hulls [11], repeatedly compute medians over different ranges within an array. Online multiselection (where queries arrive one at a time) may be a key ingredient to improved results for these types of problems, whereas offline algorithms will not suffice. Most recently, Cardinal et al. [5] generalized the problem to *partial order production*, and they use multiselection as a subroutine after an initial preprocessing phase.

Previous Work. Several papers [6,12,9] have analyzed the offline multiselection problem, but these approaches must all know the queries in advance. Kaligosi et al. [10] described an algorithm performing $\mathcal{B}(S_q) + o(\mathcal{B}(S_q)) + O(n)$ comparisons.

Our Results. For the *multiselection* problem in internal memory, we describe the *first* online algorithm that supports a sequence R of q selection queries using $\mathcal{B}(S_q) + o(\mathcal{B}(S_q))$ comparisons. Our algorithm is 1-competitive in the number of comparisons performed. We match the bounds above while supporting search, insert, and delete operations, achieve similar results in the external memory model [1]. We invite readers to see [2] (or the upcoming journal version) for more details on these results.

Preliminaries. Given an unsorted array A of length n, the *median* of A is the element x such that $\lceil n/2 \rceil$ elements in A are at least x. The median can be computed in $O(n)$ comparisons [8,4,13,7], in particular, less than $3n$ comparisons [7].

Outline. In the next section, we present a simple algorithm for the online multiselection problem, and introduce some terminology to describe its analysis. In Section 2.2, we show that the simple algorithm has a constant competitive ratio. Section 3 describes modifications to the simple algorithm, and shows that the modified algorithm is optimal up to lower order terms. We describe the experimental results in Section 4.

2   A Simple Online Algorithm

Let A be an input array of n unsorted items. We describe a simple version of our algorithm for handling selection queries on array A. We call an element A[i] at position i in array A a *pivot* if $A[1 \ldots i-1] < A[i] \leq A[i+1 \ldots n]$.

Bitvector. We maintain a bitvector V of length n where $V[i] = 1$ if and only if A[i] is a pivot. During *preprocessing*, we create V and set each bit to 0. We find the minimum and maximum elements in array A, swap them into A[1] and A[n] respectively, and set $V[1] = V[n] = 1$.

Selection. The operation A.*select*(*s*) returns the *s*th smallest element of A (i.e., A[*s*] if A were sorted). To compute this result, if V[*s*] = **1** then return A[*s*] and we are done. If V[*s*] = **0**, find *a* < *s* and *b* > *s*, such that V[*a*] = V[*b*] = **1** but V[*a* + 1 ... *b* − 1] are all **0**. Perform quickselect [8] on A[*a* + 1 ... *b* − 1], marking pivots found along the way in V. This gives us A[*s*], with V[*s*] = **1**, as desired.

As queries arrive, our algorithm performs the same steps that quicksort would perform, although not necessarily in the same order. As a result, our recursive subproblems mimic those from quicksort. We can show that comparisons needed to perform *q* select queries on an array of *n* items is $O(n \log q)$. We can improve this result to $O(\mathcal{B}(S_q))$.[2] We do not prove this bound directly, since our main result is an improvement over this bound. Now, we define terminology for this improved analysis.

2.1   Terminology

Query and Pivot Sets. Let R denote a sequence of q selection queries, ordered by time of arrival. Let $S_t = \{s_1, s_2, \ldots, s_t\}$ denote the first t queries from R, sorted by position. We also include $s_0 = 1$ and $s_{t+1} = n$ in S_t for convenience of notation, since the minimum and maximum are found during preprocessing. Let $P_t = \{p_i\}$ denote the set of k pivots found by the algorithm when processing S_t, sorted by position. Note that $p_1 = 1$, $p_k = n$, V[p_i] = **1** for all i, and $S_t \subseteq P_t$.

Pivot Tree and Recursion Depth. The pivots chosen by the algorithm form a binary tree structure, defined as the *pivot tree* T of the algorithm over time.[3] Pivot p_i is the parent of pivot p_j if, after p_i was used to partition an interval, p_j was the pivot used to partition either the right or left half of that interval. The root pivot is the pivot used to partition A[2..*n* − 1] due to preprocessing. The *recursion depth*, $d(p_i)$, of a pivot p_i is the length of the path in the pivot tree from p_i to the root pivot. All leaves in the pivot tree are also selection queries, but it may be the case that a query is not a leaf.

Intervals. Each pivot was used to partition an interval in A. Let $I(p_i)$ denote the interval partitioned by pivot p_i (which may be empty), and let $|I(p_i)|$ denote its length. Intervals form a binary tree induced by their pivots. If p_i is an ancestor of pivot p_j then $I(p_j) \subset I(p_i)$. The recursion depth of an array element is the recursion depth of the smallest interval containing that element, which in turn is the recursion depth of its pivot.

Gaps. Define the query gap $\Delta_i^{S_t} = s_{i+1} - s_i$ and similarly the pivot gap $\Delta_i^{P_t} = p_{i+1} - p_i$. By telescoping we have $\sum_i \Delta_i^{S_t} = \sum_j \Delta_j^{P_t} = n - 1$.

Fact 1. *For all $\epsilon > 0$, there exists a constant c_ϵ such that for all $x \geq 4$,* $\log \log \log x < \epsilon \log x + c_\epsilon$.

---

[2] $\mathcal{B}(S_q) = n \log q$ when the q queries are evenly spaced over the input array A.

[3] Intuitively, a pivot tree corresponds to a *recursion tree*, since each node represents one recursive call made during the quickselect algorithm [8].

Proof. Since $\lim_{x \to \infty} (\log \log \log x)/(\log x) = 0$, there exists a k_ϵ such that for all $x \geq k_\epsilon$, we know that $(\log \log \log x)/(\log x) < \epsilon$. Also, in the interval $[4, k_\epsilon]$, the continuous function $\log \log \log x - \epsilon \log x$ is bounded. Let $c_\epsilon = \log \log \log k_\epsilon - 2\epsilon$, which is a constant. $\qquad\square$

2.2 Analysis of the Simple Algorithm

In this section we analyze the simple online multiselect algorithm of Section 2.

We call a pivot selection method *c-balanced* for some constant c with $1/2 \leq c < 1$ if, for all pairs (p_i, p_j) where p_i is an ancestor of p_j in the pivot tree, then $|I(p_j)| \leq |I(p_i)| \cdot c^{d(p_j) - d(p_i) + O(1)}$. If the median is always chosen as the pivot, we have $c = 1/2$ and the $O(1)$ term is zero. The pivot selection method of Kaligosi et al. [10, Lemma 8] is c-balanced with $c = 15/16$.

Lemma 1 (Entropy Lemma). *If the pivot selection method is c-balanced, then* $\mathcal{B}(P_t) = \mathcal{B}(S_t) + O(n)$.

Proof. We sketch the proof and defer the full details to the journal version of the paper. (Those results also appear in [2].) Consider any two consecutive selection queries s and s', and let $\Delta = s' - s$ be the gap between them. Let $P_\Delta = (p_l, p_{l+1}, \ldots, p_r)$ be the pivots in this gap, where $p_l = s$ and $p_r = s'$. The lemma follows from the claim that $\mathcal{B}(P_\Delta) = O(\Delta)$, since

$$
\mathcal{B}(P_t) - \mathcal{B}(S_t) = \left(n \log n - \sum_{j=0}^{k} \Delta_j^{P_t} \log \Delta_j^{P_t} \right) - \left(n \log n - \sum_{i=0}^{t} \Delta_i^{S_t} \log \Delta_i^{S_t} \right)
$$

$$
= \sum_{i=0}^{t} \Delta_i^{S_t} \log \Delta_i^{S_t} - \sum_{j=0}^{k} \Delta_j^{P_t} \log \Delta_j^{P_t} = \sum_{i=0}^{t} \mathcal{B}(P_{\Delta_i^{S_t}}) = O(n).
$$

We now sketch the proof of our claim, which proves the lemma. There must be a unique pivot p_m in P_Δ of minimal recursion depth. We split the gap Δ at p_m. Since we use a c-balanced pivot selection method, we can bound the total information content of the left-hand side by $O(\sum_{i=l}^{m-1} \Delta_i)$ and the right-hand side by $O(\sum_{i=m}^{r-1} \Delta_i)$, leading to the claim. The result follows. $\qquad\square$

3 Optimal Online Multiselection

In this section we prove the main result of our paper, Theorem 1.

Theorem 1 (Optimal Online Multiselection). *Given an unsorted array A of n elements, we provide an algorithm that supports a sequence R of q online selection queries using $\mathcal{B}(S_q)(1 + o(1)) + O(n)$ comparisons.*

Our bounds match those of the offline algorithm of Kaligosi et al. [10]. In other words, our solution is 1-competitive. We explain our proof in three main steps. In Section 3.1, we explain our algorithm and describe how it is different from Kaligosi et al. [10]. We then bound the number of comparisons resulting from merging by $\mathcal{B}(S_q)(1 + o(1)) + O(n)$ in Section 3.2. In Section 3.3, we bound the complexity of pivot finding and partitioning by $o(\mathcal{B}(S_q)) + O(n)$.

3.1   Algorithm Description

We briefly describe the deterministic algorithm from Kaligosi et al. [10]. Their result is based on tying the number of comparisons required for merging two sorted sequences to the information content of those sequences. This simple observation drives their underlying approach that both finds pivots that are "good enough" and partitions using near-optimal comparisons.

In particular, they create *runs*, which are sorted sequences from A of length roughly $\ell = \log(\mathcal{B}/n)$. Then, they compute the median μ of the medians of these sequences, and partition the runs based on μ. After partitioning, they recurse on the two sets of runs, sending *select* queries to the appropriate side of the recursion. To maintain the invariant on run length on the recursions, they merge short runs of the same size optimally until all but ℓ of the runs are again of length between ℓ and 2ℓ.

We make the following modifications to the algorithm of Kaligosi et al. [10]:

- Since the value of $\mathcal{B}(S_q)$ is not known in advance (because R is provided *online*), we cannot preset a value for ℓ, as done in Kaligosi et al. [10]. Instead, we locally set $\ell = 1 + \lfloor \log(d(p) + 1) \rfloor$ in the interval $I(p)$. Since we use only balanced pivots, $d(p) = O(\log n)$. We keep track of the recursion depth of pivots, from which it is easy to compute the recursion depth of an interval.
- We use a bitvector W to identify the endpoints of runs within each interval.
- The queries from R are processed *online*. We support online queries using the bitvector V from Section 2. Recall that a search query incurs $O(\log n)$ additional comparisons to find its corresponding interval.

To perform the operation A.*select*(s), we first use bitvector V to identify the interval I containing s. If $|I| \le 4\ell^2$, we sort the interval I (making all elements of I pivots) and answer the query s. The cost for this case is bounded by Lemma 5. Otherwise, we compute the value of ℓ for the current interval, and proceed as in Kaligosi et al. [10] to answer the query s.

We can borrow much of the analysis done in [10], but it depends heavily on the use of ℓ, which we do not know in advance. In the rest of Section 3, we modify their techniques to handle this complication.

3.2   Merging

Kaligosi et al. [10, Lemmas 5—10] count the comparisons resulting from merging. Lemmas 5, 6, and 7 do not depend on the value of ℓ and so we can use them in our analysis. Lemma 8 shows that the median-of-medians built on runs is a good pivot selection method. Although its proof uses the value of ℓ, its validity does not depend the size of ℓ. The proof merely requires that there are at least $4\ell^2$ items in each interval, which also holds for our algorithm. Lemmas 9 and 10 (from Kaligosi et al. [10]) together will bound the number of comparisons by $\mathcal{B}(S_q)(1+o(1)) + O(n)$ if we can prove Lemma 2, which bounds the information content of runs in intervals that are not yet partitioned.

Lemma 2. *Let a run r be a sorted sequence of elements from A in a gap $\Delta_i^{P_t}$, where $|r|$ is its length. Then, $\sum_{i=0}^{k} \sum_{r \in \Delta_i^{P_t}} |r| \log |r| = o(\mathcal{B}(S_t)) + O(n)$.*

Proof. In a gap of size Δ, $\ell = O(\log d)$ where d the recursion depth of the elements in the gap. This gives $\sum_{r \in \Delta} |r| \log |r| \leq \Delta \log(2l) = O(\Delta \log \log d)$, since each run has size at most 2ℓ. Because we use a good pivot selection method, we know that the recursion depth of every element in the gap is $O(\log(n/\Delta))$. Thus, $\sum_{i=0}^{k} \sum_{r \in \Delta_i^{P_t}} |r| \log |r| \leq \sum_i \Delta_i \log \log \log(n/\Delta_i)$. Recall that $\mathcal{B}(S_t) = \mathcal{B}(P_t) + O(n) = \sum_i \Delta_i \log(n/\Delta_i) + O(n)$. Fact 1 completes the proof. $\square$

3.3   Pivot Finding and Partitioning

Now we prove that the cost of computing medians and performing partitions requires at most $o(\mathcal{B}(S_q)) + O(n)$ comparisons. The algorithm computes the median m of medians of each run at a node v in the pivot tree T. Then, it partitions each run based on m. We bound the number of comparisons at each node v with more than $4\ell^2$ elements in Lemmas 3 and 4. We bound the comparison cost for all nodes with fewer elements in Lemma 5.

Let d be the current depth of the pivot tree T (defined in Section 2.1), and let the root of T have depth $d = 0$. Each node v in T is associated with some interval $I(p_v)$ corresponding to some pivot p_v. We define $\Delta_v = |I(p_v)|$ as the number of elements at node v.

Recall that $\ell = 1 + \lfloor \log(d+1) \rfloor$, and a *run* is a sorted sequence of elements in A. We define a *short run* as a run of length less than ℓ. Let βn be the number of comparisons required to compute the exact median for n elements, where β is a constant less than three [7]. Let r_v^s be the number of short runs at node v, and let r_v^l be the number of *long runs* (runs of length at least ℓ).

Lemma 3. *The number of comparisons required to find the median m of medians and partition all runs at m for any node v in the pivot tree T is at most $\beta(\ell - 1) + \ell \log \ell + \beta(\Delta_v/\ell) + (\Delta_v/\ell) \log(2\ell)$.*

Proof. We compute the cost (in comparisons) for computing the median of medians. For the $r_v^s \leq \ell - 1$ short runs, we need at most $\beta(\ell - 1)$ comparisons per node. For the $r_v^l \leq \Delta_v/\ell$ long runs, we need at most $\beta(\Delta_v/\ell)$.

Now we compute the cost for partitioning each run based on m. We perform binary search in each run. For short runs, this requires at most $\sum_{i=1}^{\ell-1} \log i \leq \ell \log \ell$ comparisons per node. For long runs, we need at most $(\Delta_v/\ell) \log(2\ell)$ comparisons per node. $\square$

Since our value of ℓ changes at each level of the recursion tree, we will sum the costs from Lemma 3 by level. The overall cost at level d is at most $2^d \beta \ell + 2^d \ell \log \ell + (n/\ell)\beta + (n/\ell) \log(2\ell)$ comparisons. Summing over all the levels, we can bound the total cost of all such nodes in the pivot tree to obtain the following lemma.

Lemma 4. *The number of comparisons required to find the median of medians and partition over all nodes v in the pivot tree T with at least $4\ell^2$ elements is within $o(\mathcal{B}(S_t)) + O(n)$.*

Proof. For all levels of the pivot tree up to level $\ell' \leq \log(\mathcal{B}(P_t)/n)$, the cost is at most

$$\sum_{d=1}^{\log(\mathcal{B}(P_t)/n)} 2^d \ell(\beta + \log \ell) + (n/\ell)(\beta + \log(2\ell)).$$

Since $\ell = \lfloor \log(d+1) \rfloor + 1$, we can easily bound the first term of the summation by $(\mathcal{B}(P_t)/n) \log \log(\mathcal{B}(P_t)/n) = o(\mathcal{B}(P_t))$. The second term can be easily upper-bounded by $n \log(\mathcal{B}(P_t)/n)(\log \log \log(\mathcal{B}(P_t)/n)/ \log \log(\mathcal{B}(P_t)/n))$, which is $o(\mathcal{B}(P_t))$. Using Lemma 1, the above two bounds are $o(\mathcal{B}(S_t)) + O(n)$.

For each level ℓ' with $\log(\mathcal{B}(P_t)/n) < \ell' \leq \log \log n + O(1)$, we bound the remaining cost. It is easy to bound each node v's cost by $o(\Delta_v)$, but this is not sufficient—though we have shown that the total number of *comparisons* for merging is $\mathcal{B}(S_t) + O(n)$, the number of *elements* in nodes with $\Delta_v \geq 4\ell^2$ could be $\omega(\mathcal{B}(S_t))$.

We bound the overall cost as follows, using the result of Lemma 3. Since node v has $\Delta_v > 4\ell^2$ elements, we can rewrite the bounds as $O(\Delta_v/\ell \log(2\ell))$. Recall that $\ell = \log d + O(1) = \log(O(\log(n/\Delta_v))) = \log \log(n/\Delta_v) + O(1)$, since we use a good pivot selection method. Summing over all nodes, we get $\sum_v (\Delta_v/\ell) \log(2\ell) \leq \sum_v \Delta_v \log(2\ell) = o(\mathcal{B}(P_t)) + O(n)$, using Fact 1 and recalling that $\mathcal{B}(P_t) = \sum_v \Delta_v \log(n/\Delta_v)$. Finally, using Lemma 1, we arrive at the claimed bound for queries. □

We now bound the comparison cost for all nodes v where $\Delta_v \leq 4\ell^2$.

Lemma 5. *For nodes v in the pivot tree T where $\Delta_v \leq 4\ell^2$, the total cost in comparisons for all operations is at most $o(\mathcal{B}(S_t)) + O(n)$.*

Proof. Nodes with no more than $4\ell^2$ elements do not incur any cost in comparisons for median finding and partitioning, unless there is (at least) one associated query within the node. Hence, we focus on nodes with at least one query.

Let z be such that $z = (\log \log n)^2 \log \log \log n + O(1)$. We sort the elements of any node v with $\Delta_v \leq 4\ell^2$ elements using $O(z)$ comparisons, since $\ell \leq \log \log n + O(1)$. We set each element as a pivot. The total comparison cost over all such nodes is no more than $O(tz)$, where t is the number of queries we have answered so far. If $t < n/z$, then the above cost is $O(n)$.

Otherwise, $t \geq n/z$. Using Jensen's inequality, we have $\mathcal{B}(P_t) \geq (n/z) \log(n/z)$, which represents the cost of sorting n/z adjacent queries. Thus, $tz = o(\mathcal{B}(P_t))$. Using Lemma 1, we know that $\mathcal{B}(P_t) = \mathcal{B}(S_t) + O(n)$, which proves the lemma. □

4   Experimental Results

In this section, we present the experimental evaluation of the online and offline multiselection algorithms. Section 4.1 describes the experimental setup. Our results are described in Section 4.2.

4.1   Experimental Setup

Our input array consists of a random permutation of the (distinct) elements from $[1, 2^{18}]$. (We also ran some experiments for larger n up to 2^{20}, and results were similar.) Our queries are generated using the indicated distribution for each experiment. We allow repetitions of queries, except in the evenly-spaced case. We only report comparisons with elements of the input array, averaged over 10 random experiments. In particular, we do *not* count comparisons between indices in the input array. Finally, we compute the *Entropy* of a query sequence S_q (defined in Section 1) by $\lfloor \log n! - \sum_{i=0}^{q} \log (\Delta_i^S!) \rfloor$ using double precision arithmetic on a 64-bit machine.

Now, we briefly describe the algorithms we considered for choosing the pivot in an unsorted interval I. The *First Element* and *Random* methods choose the corresponding element as the pivot. The *Medof3* method uses the median of the first, middle, and last elements of I as the pivot. The *Median* (using MedofMed) uses Blum et al.'s linear-time algorithm [4] as the pivot. The *MedofMed* method is the first step of Blum et al.'s algorithm [4] that computes the median of every five elements, and then uses the median of those medians as the pivot.

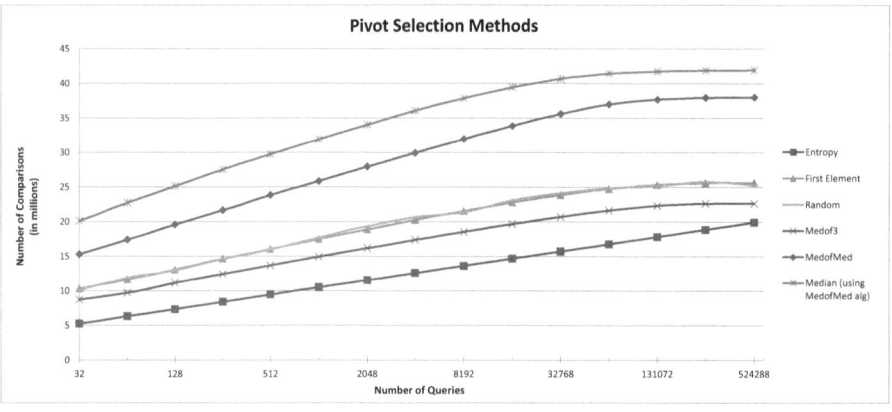

Fig. 1. Performance of various pivot selection methods on random input sequences

We compared the performance of these pivoting methods for random arrays in Figure 1 for our simple online algorithm described in Section 2. We performed similar experiments for different algorithms. The results from Figure 1 are representative of all of our findings. One can clearly see that *Medof3* uses the fewest comparisons and *Median* requires significantly more comparisons. The performance of other pivoting methods fall in between these two extremes. For the rest of the paper, we show results only for the *Medof3* pivoting method.

4.2   Results

Now, we briefly describe the algorithms we considered for multiselection. All algorithms use *Medof3* as the pivoting strategy (where applicable). The *Quicksort* algorithm is the standard quicksort, augmented by q array lookups (which require no comparisons). The *Mergesort* algorithm is the standard recursive mergesort, augmented by q array lookups (which require no comparisons). The *Simple Online* algorithm is described in Section 2. The *Optimal Online* algorithm is described in Section 3.1, where we set ℓ based on the recursion depth of the corresponding interval. The performance of the online algorithms is independent of the order of the queries. (We defer the experiments supporting this claim until the journal version of this paper.)

The *Dobkin-Munro* algorithm is described in [6]. The *Kaligosi (sorted)* algorithm is described in [10], which assumes that queries are given in sorted order. The *Kaligosi (unsorted)* algorithm first sorts the unsorted queries, and then performs the *Kaligosi (sorted)* algorithm. In some cases, sorting queries is tantamount to sorting the array. Since this algorithm is offline, one can assume that the algorithm will detect this case and revert to *Quicksort* or *Mergesort* instead.

We show our results in Figures 2 and 3. For Figure 2, the queries (in the first graph) are evenly distributed across the input array. This query distribution results in a worst-case entropy, and hence is a difficult case for multiselection algorithms. The second graph in Figure 2 has uniformly distributed queries. For Figure 3, we display results for a normal query distribution with mean $\mu = n/2$ and standard deviation $\sigma = n/8$. The second graph in Figure 3 is an exponential query distribution with $\lambda = 16/n$.

For all query distributions, our online algorithms (*Simple Online* and *Optimal Online*) outperform their offline counterparts (respectively, *Quicksort* and *Kaligosi*). The *Dobkin-Munro* algorithm requires more comparisons than *Quicksort* for any reasonably large number of queries (based on query distribution). In other words, it is usually better to sort than to use *Dobkin-Munro*. The *Kaligosi* algorithm performs quite well in terms of comparisons, but is relatively slow. The *Simple Online* algorithm converges to *Quicksort* as queries increases, highlighting that the online algorithm performs the same work as the *Quicksort*, as intuition (and the analysis) suggests.

The *Optimal Online* algorithm outperforms *Mergesort*, *Quicksort*, and *Kaligosi* (sorted and unsorted), and is even better than *Entropy* when the number of queries is large. Having an algorithm perform fewer comparisons than *Entropy* isn't a contradiction, since *Entropy* is a worst-case lower bound for an arbitrary input. Hence, the number of comparisons for an algorithm could be less than *Entropy* for a given (specific) input. Even though our algorithm is similar to *Kaligosi*, we can clearly see the value of online computation when comparing these two results. The primary reason for our improved results is due to the fact

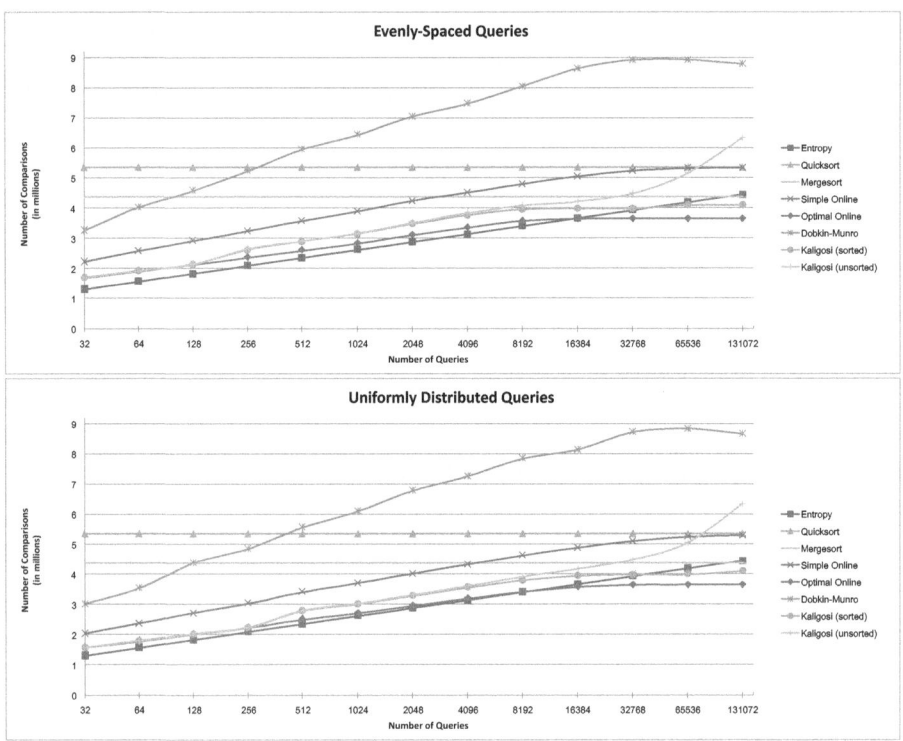

Fig. 2. Performance of multiselection algorithms on random input sequences using median of three pivot selection method, when the queries are distributed as indicated

that *Kaligosi* will pre-process runs, even for intervals that do not contain any queries. For the *Optimal Online* algorithm, since run lengths are based on the recursion depth, the algorithm will not spend comparisons generating long runs unless queries are in those intervals.

In fact, these results suggest that using the *Optimal Online* algorithm with $n/2$ queries (e.g., each odd position) can sort an array in fewer comparisons than *Mergesort*. The reason for this is that the runs computed at the beginning of the algorithm save a lot of comparisons in future recursive rounds. We are currently running experiments on tuning the length ℓ of the run to see if we can further improve this performance.

Finally, we provide similar results for a decreasing input array, since this is a best-case scenario for *Mergesort*. Notice that both *Mergesort*, *Optimal Online*, and *Kaligosi* are better than *Entropy* as queries increase. However, both multiselection algorithms outperform *Mergesort*. The sudden dip in the curve corresponding to the *Kaligosi (sorted)* algorithm after 65,536 queries corresponds to a discrete increase in the calculated value of ℓ (from 4 to 5). This sort of stair-stepping behavior is expected to continue as n increases.

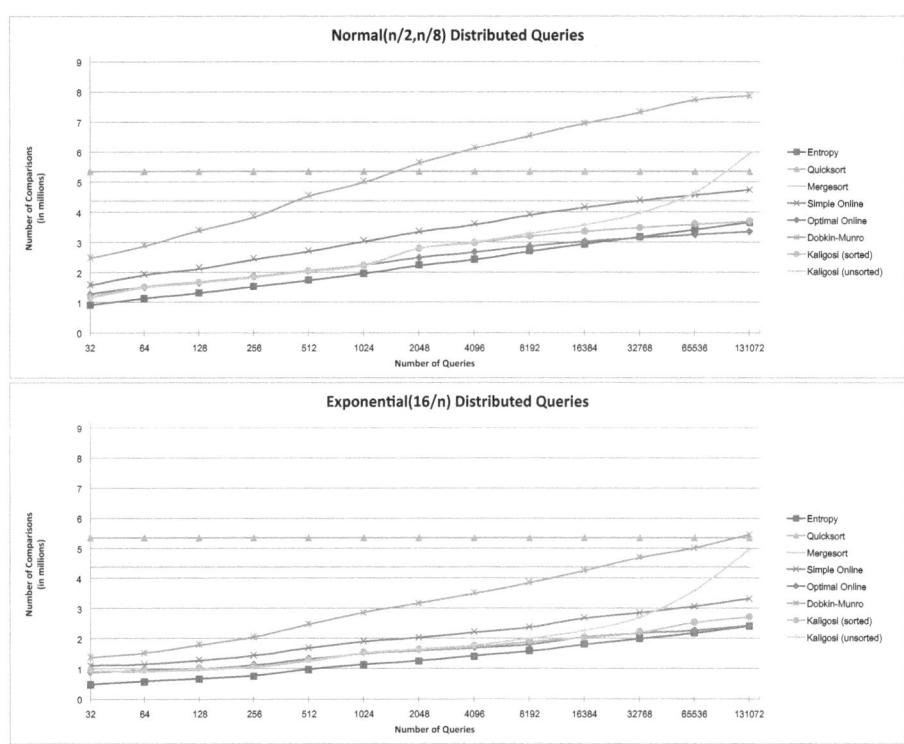

Fig. 3. Performance of multiselection algorithms on random input sequences using median of three pivot selection method, when the queries are distributed as indicated

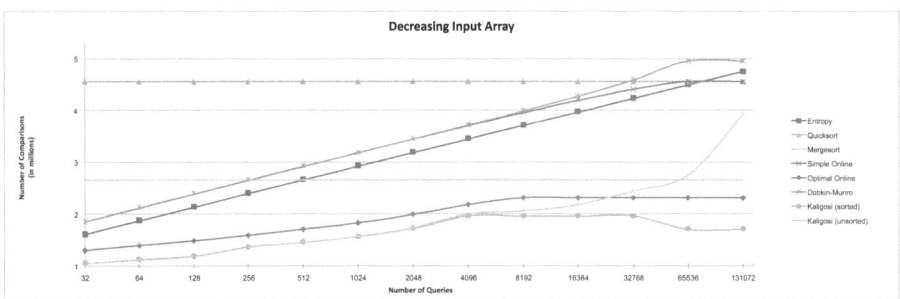

Fig. 4. Performance of multiselection algorithms on a decreasing input sequence using median of three pivot selection method, when the queries are distributed as indicated

References

1. Aggarwal, A., Vitter, J.S.: The input/output complexity of sorting and related problems. Commun. ACM 31(9), 1116–1127 (1988)
2. Barbay, J., Gupta, A., Rao, S.S., Sorenson, J.: Competitive online selection in main and external memory. CoRR, abs/1206.5336 (2012)

3. Belal, A., Elmasry, A.: Distribution-sensitive construction of minimum-redundancy prefix codes. In: Durand, B., Thomas, W. (eds.) STACS 2006. LNCS, vol. 3884, pp. 92–103. Springer, Heidelberg (2006)
4. Blum, M., Floyd, R.W., Pratt, V.R., Rivest, R.L., Tarjan, R.E.: Time bounds for selection. J. Comput. Syst. Sci. 7(4), 448–461 (1973)
5. Cardinal, J., Fiorini, S., Joret, G., Jungers, R.M., Munro, J.I.: An efficient algorithm for partial order production. In: STOC, pp. 93–100 (2009)
6. Dobkin, D.P., Munro, J.I.: Optimal time minimal space selection algorithms. J. ACM 28(3), 454–461 (1981)
7. Dor, D., Zwick, U.: Selecting the median. SICOMP 28(5), 1722–1758 (1999)
8. Hoare, C.A.R.: Algorithm 65: find. Commun. ACM 4(7), 321–322 (1961)
9. Jiménez, R.M., Martínez, C.: Interval sorting. In: Abramsky, S., Gavoille, C., Kirchner, C., Meyer auf der Heide, F., Spirakis, P.G. (eds.) ICALP 2010. LNCS, vol. 6198, pp. 238–249. Springer, Heidelberg (2010)
10. Kaligosi, K., Mehlhorn, K., Munro, J.I., Sanders, P.: Towards optimal multiple selection. In: Caires, L., Italiano, G.F., Monteiro, L., Palamidessi, C., Yung, M. (eds.) ICALP 2005. LNCS, vol. 3580, pp. 103–114. Springer, Heidelberg (2005)
11. Kirkpatrick, D.G., Seidel, R.: The ultimate planar convex hull algorithm. SIAM J. Comput. 15(1), 287–299 (1986)
12. Prodinger, H.: Multiple quickselect - Hoare's find algorithm for several elements. Inf. Process. Lett. 56(3), 123–129 (1995)
13. Schönhage, A., Paterson, M., Pippenger, N.: Finding the median. J. Comput. Syst. Sci. 13(2), 184–199 (1976)

An Implementation of I/O-Efficient Dynamic Breadth-First Search Using Level-Aligned Hierarchical Clustering*

Andreas Beckmann[1], Ulrich Meyer[2], and David Veith[2]

[1] Jülich Supercomputing Centre (JSC), Forschungszentrum Jülich,
D-52425 Jülich, Germany
[2] Institut für Informatik, Goethe-Universität Frankfurt,
Robert-Mayer-Str. 11–15, D-60325 Frankfurt am Main, Germany

Abstract. In the past a number of I/O-efficient algorithms were designed to solve a problem on a static data set. However, many data sets like social networks or web graphs change their shape frequently. We provide experimental results of the first external-memory dynamic breadth-first search (BFS) implementation based on earlier theoretical work [13] that crucially relies on a randomized clustering. We refine this approach using a new I/O-efficient deterministic clustering, which groups vertices in a level-aligned hierarchy and facilitates easy access to clusters of changing sizes during the BFS updates. In most cases the new external-memory dynamic BFS implementation is significantly faster than recomputing the BFS levels after an edge insertion from scratch.

1 Introduction

Breadth first search (BFS) is a fundamental graph traversal strategy. It can be viewed as computing single source shortest paths on unweighted graphs. BFS decomposes the input graph $G = (V, E)$ of $n = |V|$ nodes and $m = |E|$ edges into at most n levels where level i comprises all nodes that can be reached from a designated source s via a path of i edges, but cannot be reached using less than i edges.

The objective of a dynamic graph algorithm is to efficiently process an online sequence of update and query operations; see [11,15] for overviews of classic and recent results. In this paper we consider dynamic BFS for the incremental setting where additional edges are inserted one-by-one. After each edge insertion the updated BFS level decomposition has to be output.

At first sight, dynamic BFS on sparse graphs might not seem interesting since certain edge insertions could require $\Omega(n)$ updates on the resulting BFS levels, implying that the time needed to report the changes is in the same order of magnitude as recomputing the BFS levels from scratch using the standard linear time BFS algorithm. The situation, however, is completely different in the

---

* Partially supported by the DFG grant ME 3250/1-3, and by MADALGO – Center for Massive Data Algorithmics, a Center of the Danish National Research Foundation.

H.L. Bodlaender and G.F. Italiano (Eds.): ESA 2013, LNCS 8125, pp. 121–132, 2013.
© Springer-Verlag Berlin Heidelberg 2013

external-memory setting, where the currently best static BFS implementations take much more time to *compute* the BFS levels as compared to *reporting* them, i.e. writing them to disk. Thus, providing a fast dynamic alternative is an important step towards a toolbox for external-memory graph computing. In this paper we report on the engineering of an external-memory dynamic BFS implementation (based on earlier theoretical work [13]). To this end a modified external-memory clustering method tuned to our needs has been developed, too.

2   I/O-Model and Related Work

Computation model. Theoretical results on out-of-core algorithms typically rely on the commonly accepted external-memory (EM) model by Aggarwal and Vitter [1]. It assumes a two level memory hierarchy with fast internal memory having a capacity to store M data items (e.g., vertices or edges of a graph) and a slow disk of infinite size. In an I/O operation, one block of data, which can store B consecutive items, is transferred between disk and internal memory. The measure of performance of an algorithm is the number of I/Os it performs. The number of I/Os needed to read N contiguous items from disk is $\text{scan}(N) = \Theta(N/B)$. The number of I/Os required to sort N items is $\text{sort}(N) = \Theta((N/B)\log_{M/B}(N/B))$. For all realistic values of N, B, and M, $\text{scan}(N) < \text{sort}(N) \ll N$.

Review of Static and Dynamic EM BFS Algorithms. There has been a significant number of publications on external-memory graph algorithms; see [3,16] for recent overviews. In the following we shortly review the external memory MM_BFS algorithm by Mehlhorn and Meyer [12] and its dynamic extension [13]. In order to keep the description simple, we concentrate on edge insertions on already connected undirected sparse graphs with n=$|V|$ vertices and $m = |E| = \mathcal{O}(|V|)$ edges.

MM_BFS. The MM_BFS algorithm consists of two phases – a preprocessing phase and a BFS phase. In the preprocessing phase, the algorithm has to produce a clustering. This can be done with an Euler Tour technique based on an arbitrary spanning tree T of the graph G. In case T is not part of the input, it can be obtained using $\mathcal{O}((1 + \log\log(B \cdot n/m)) \cdot \text{sort}(n+m))$ I/Os [6]. Initially, each undirected edge of T is replaced by two oppositely directed edges. Then, in order to construct the Euler tour around this bi-directed tree, each node chooses a cyclic order [7] of its neighbors. The successor of an incoming edge is defined to be the outgoing edge to the next node in the cyclic order. The tour is then broken at a special node (say the root of the tree) and the elements of the resulting list are then stored in consecutive order using an external memory list-ranking algorithm; Chiang et al. [8] showed how to do this in sorting complexity. Thereafter, the Euler tour is chopped into clusters of μ nodes and duplicates are removed such that each node only remains in the *first* cluster it originally occurs; again this requires a couple of sorting steps. By construction, the distance in G between any two vertices belonging to the same cluster is bounded by $\mu - 1$ and there are $\mathcal{O}(n/\mu)$ clusters.

For the BFS phase, the key idea is to load whole preprocessed clusters into some efficient data structure (hot pool) at the expense of few I/Os, since the clusters are stored contiguously on disk and contain vertices in neighboring BFS levels. This way, the neighboring nodes $N(l)$ of some BFS level l can be computed by scanning only the hot pool. The next BFS level is obtained by removing those nodes visited in levels $l - 1$ and l from $N(l)$; see [14]. However, as the algorithm proceeds, newly discovered neighbor nodes may belong to so far unvisited clusters. Unstructured I/Os are required to import those clusters into the hot pool, from where they are gradually evicted again once they have been used to create the respective BFS levels. Maintaining the hot pool itself requires $\mathcal{O}(\mathrm{scan}(n+m) \cdot \mu)$ I/Os, whereas importing the clusters into it accounts for $\mathcal{O}(n/\mu + \mathrm{sort}(n + m))$ I/Os. Choosing $\mu = \sqrt{B}$ yields an I/O-complexity of $\mathcal{O}(n/\sqrt{B} + \mathrm{sort}(n))$ for the BFS-phase on sparse graphs.

Dynamic BFS. In the following we review the high-level ideas to computing BFS on general undirected sparse graphs in an incremental setting. Let us consider the insertion of the ith edge (u, v) and refer to the graph (and the shortest path distances from the source in the graph) before and after the insertion of this edge as G_{i-1} (d_{i-1}) and G_i (d_i), respectively.

We run the BFS phase of MM_BFS, with the difference that the adjacency list for v is added to the hot pool H when creating BFS level $\max\{0, d_{i-1}(v) - \alpha\}$ of G_i, for a certain advance $\alpha > 1$. By keeping the adjacency lists sorted according to node distances in G_{i-1} this can be done I/O-efficiently for all nodes v featuring $d_{i-1}(v) - d_i(v) \leq \alpha$. For nodes with $d_{i-1}(v) - d_i(v) > \alpha$, we import whole clusters containing their adjacency lists into H using unstructured I/Os. Each such cluster must comprise the adjacency lists of $\Omega(\alpha)$ nodes whose mutual distances in G_{i-1} are bounded by $\mu = \Theta(\alpha)$, each vertex belongs to exactly one cluster. If the BFS phase for the currently used value of α would require more than $\alpha \cdot n/B$ random cluster accesses, we increase α by a factor of two, compute a new clustering for G_{i-1} with larger chunk size μ and start a new attempt by repeating the whole approach with the increased parameters.

Meyer [13] proved an amortized high-probability bound of $\mathcal{O}(n/B^{2/3} + \mathrm{sort}(n) \cdot \log B)$ I/Os per update under a sequence of $\Theta(n)$ edge insertions. The analysis relies on the fact that there can be only be very few updates in which the BFS levels change significantly for a large number of nodes. If it can be guaranteed that each cluster loaded into the pool actually carries $\Omega(\alpha)$ vertices, most of the updates will require few cluster fetches in early attempts with small advance.

Unfortunately, the standard Euler tour based clustering method described above might produce very unbalanced clusters: in fact $\Omega(n/\mu)$ clusters may contain only a single vertex each. A randomized clustering approach [13] repairs this deficiency as follows:

Each vertex v in the spanning tree T_s is assigned an independent binary random number $r(v)$ with $\mathbf{P}[r(v) = 0] = \mathbf{P}[r(v) = 1] = 1/2$. When removing duplicates from the Euler tour, instead of storing v in the cluster related to the chunk with the *first* occurrence of a vertex v, now we only stick to its first occurrence iff $r(v) = 0$ and otherwise ($r(v) = 1$) store v in the cluster that corresponds to the *last* chunk of the Euler tour v appears in. For chunk size $\mu > 1$ and each but the last chunk, the expected number of kept vertices is at least $\mu/8$.

3   Challenges and New Results

The dynamic BFS approach in [13] applies several clusterings of *different* values for μ (say $\mu = 2^1, 2^2, 2^3, \ldots, \sqrt{B}$) for the *same* input graph. In fact, for incremental dynamic BFS on already connected graphs, these clusterings could be re-used for all subsequent edge insertions. Unfortunately, while the theoretical EM model assume external space to be of unlimited size, this is not true in reality. In fact, due to disk space limitations it may often be impossible to keep even a few different clusterings at the same time. On the other hand, even though it will not harm the theoretical worst-case bounds, re-computing the same clusterings over and over again could actually become the dominating part of the I/O numbers we see in practice: for example in improved static BFS implementations of [4,12], the preprocessing for graph clustering often takes more time than the actual BFS phase, although the latter comes with a significantly higher asymptotic I/O-bound than the preprocessing in the worst case. In addition, the old clustering method described above crucially relies on randomization. Thus, the improved deterministic clustering we propose in Section 4 features both practical and theoretical advantages.

As already mentioned in Section 2 the theoretical I/O bounds for the dynamic EM BFS algorithm in [13] are amortized over sequences of $\Theta(n)$ edge insertions. Hence, single updates could theoretically become as costly as with the static EM-BFS approach, and the hidden constants might be even worse. On the other hand, edge insertions with little effect on the resulting BFS levels should hopefully be manageable with significantly less I/O. In our practical experiments we consider such extreme cases on several graph classes. While our dynamic BFS implementation was never slower than a factor of 1.25 compared to static BFS, we have also experienced cases where dynamic BFS outperformes the static re-computation by more than a factor of 70.

4   Level-Aligned Hierarchical Clustering

The high level idea for our hierarchical clustering is rather easy: we renumber each vertex with a new bit representation $\langle b_r, \ldots, b_{q+1}, b_q, \ldots, b_1 \rangle$ that is interpreted as a combination of prefix $\langle b_r, \ldots, b_{q+1} \rangle$ and suffix $\langle b_q, \ldots, b_1 \rangle$. Different prefixes denote different clusters, and for a concrete prefix (cluster) its suffixes denote vertices within this cluster. Depending on the choice of q we get the whole spectrum between few larger clusters (q big) or many small clusters (q small). In particular we would like the following to hold:

For any $1 \leq \mu = 2^q \leq \sqrt{B}$, (1) there are $\lceil n/\mu \rceil$ clusters, (2) each cluster comprises μ vertices (one cluster may have less vertices), and (3) for any two vertices u and v belonging to the same cluster, their distance in G is $\mathcal{O}(\mu)$. In order to make this work the new vertex numbers will have to be carefully chosen. Additionally, a look-up table is built that allows to find the sequence of disk blocks for adjacency lists of the vertices associated with a concrete cluster using $\mathcal{O}(1)$ I/Os.

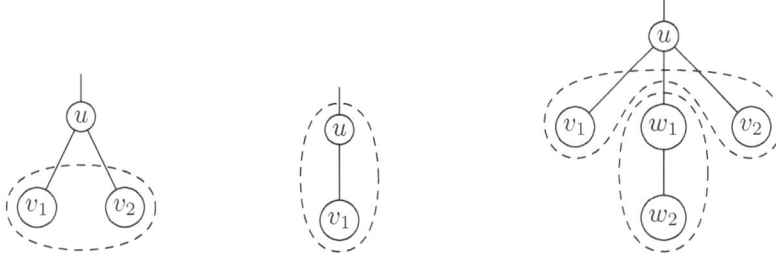

Fig. 1. Sibling-Merge   **Fig. 2.** Parent-Merge   **Fig. 3.** Complex merge example

In order to group close-by vertices into clusters (such that an appropriate renumbering can take place) we start with an arbitrary spanning tree T_s^0 rooted at source vertex s. Then we work in $p = \left\lceil \log \sqrt{B} \right\rceil$ phases, each of which transforms the current tree T_s^j into a new tree T_s^{j+1} having $\lceil |T_s^j|/2 \rceil$ vertices. (The external-memory BFS algorithms considered here only use clusters up to a size of $\sqrt{B}$ vertices so this construction is stopped after p phases. The hierarchical clustering approach imposes no limitations and can be applied for up to $\lceil \log n \rceil$ phases for other applications.) The tree shrinking is done using time-forward processing [5,8] from the leaves toward the root (for example by using negated BFS numbers for the vertices T_s^j). Consider the remaining leaves $v_1, \ldots, v_k$ with highest BFS numbers and common parent vertex u in T_s^j. If k is even then v_1 and v_2 will form a cluster (and hence a vertex in T_s^{j+1}), v_3 and v_4 will be combined, v_5 and v_6, etc. (*sibling-merge*, see Figure 1). In case k is odd, v_1 will be combined with u (*parent-merge*, Figure 2) and (if $k \geq 3$) v_2 with v_3, v_4 with v_5, etc. Merged vertices are removed from T_s^j and therefore any vertex is a leaf at the time it is reached by TFP, e.g. node w_1 shown in Figure 3 was already consumed by vertex w_2, so it is no longer available at the time v_1, v_2 get processed. Thus, each vertex of T_s^{j+1} is created out of exactly two vertices from T_s^j, except for the root which may only consist of the root from T_s^j. Note that the original graph vertices kept in a cluster are not necessarily direct neighbors but they do have short paths connecting them in the original graph. The following lemma makes this more formal:

Lemma 1. *The vertices of T_s^j form clusters in the original graph having size* $\text{size}^{(j)} = 2^j$ *(excluding the root vertex which may be smaller), maximum depth* $\overline{\text{depth}}^{(j)} = 2^j - 1$, *and maximum diameter* $\overline{\text{diam}}^{(j)} = 2^{j+1} - 2$.

Proof. By induction (obvious for $\text{size}^{(j)}$). The clusters defined by T_s^0 consist of exactly one vertex each and satisfy $\overline{\text{diam}}^{(0)} = 0$ and $\overline{\text{depth}}^{(0)} = 0$. For $j > 0$, three ways of merging a vertex v_1 have to be considered (with a sibling ($\{v_1, v_2\}$), the parent ($\{v_1, u\}$) or not merged at all ($\{v_1\}$)), resulting in

$$
\overline{\text{depth}}^{(j)} = \max \left\{
\begin{array}{l}
\overline{\text{depth}}^{(j)}(\{v_1, v_2\}), \\
\overline{\text{depth}}^{(j)}(\{v_1, u\}), \\
\overline{\text{depth}}^{(j)}(\{v_1\})
\end{array}
\right\}
$$

$$
= \max \left\{
\begin{array}{l}
\max\left\{\overline{\text{depth}}^{(j-1)}(v_1), \overline{\text{depth}}^{(j-1)}(v_2)\right\}, \\
\overline{\text{depth}}^{(j-1)}(v_1) + 1 + \overline{\text{depth}}^{(j-1)}(u), \\
\overline{\text{depth}}^{(j-1)}(v_1)
\end{array}
\right\}
$$

$$
= 2 \cdot \overline{\text{depth}}^{(j-1)} + 1 = 2 \cdot (2^{j-1} - 1) + 1 = 2^j - 1
$$

$$
\overline{\text{diam}}^{(j)} = \max \left\{
\begin{array}{l}
\overline{\text{diam}}^{(j)}(\{v_1, v_2\}), \\
\overline{\text{diam}}^{(j)}(\{v_1, u\}), \\
\overline{\text{diam}}^{(j)}(\{v_1\})
\end{array}
\right\}
$$

$$
= \max \left\{
\begin{array}{l}
\overline{\text{depth}}^{(j-1)}(v_1) + 1 + \overline{\text{diam}}^{(j-1)}(u) + 1 + \overline{\text{depth}}^{(j-1)}(v_2), \\
\overline{\text{depth}}^{(j-1)}(v_1) + 1 + \overline{\text{diam}}^{(j-1)}(u), \\
\overline{\text{diam}}^{(j-1)}(v_1)
\end{array}
\right\}
$$

$$
= 2 \cdot \overline{\text{depth}}^{(j-1)} + \overline{\text{diam}}^{(j-1)} + 2 = 2 \cdot (2^{j-1} - 1) + \overline{\text{diam}}^{(j-1)} + 2
$$

$$
= 2^j + \overline{\text{diam}}^{(j-1)} = 2^j + 2^j - 2 = 2^{j+1} - 2
$$

Note that while $\overline{\text{diam}}^{(j)}$ and $\overline{\text{depth}}^{(j)}$ denote the *maximum* diameter and depth possible for a cluster with 2^j vertices the actual values may be much smaller. □

The hierarchical approach produces a clustering with (1) $\Theta(n/\mu)$ clusters each having (2) size $\Theta(\mu)$ (excluding the root cluster) and (3) diameter $\mathcal{O}(\mu)$ for each $1 \leq \mu = 2^q \leq \sqrt{B}$.

Details on the construction of T_s^{j+1}. Two types of messages (*connect*(id) and *merged*(id)) are sent during the time-forward processing. When vertices are combined, the vertex visited first sends the ID of the new vertex in T_s^{j+1} to the other one in a *merged*() message. The *connect*() messages are used to generate edges of T_s^{j+1} using the new IDs. The *merged*() message (if any) of a vertex is sorted before the *connect*() messages of that vertex, so checking whether the current minimal entry in the priority queue has received such a message can be done in $\mathcal{O}(1)$.

Renumbering the vertices. The p phases of contracting the spanning tree each contribute one bit of the new vertex number $\langle b_r, \ldots, b_{p+1}, b_p, \ldots, b_1 \rangle$. The construction of T_s^{j+1} defines the bit b_{j+1} for the vertices from T_s^j to be 0 for the left and 1 for the right child in case of a sibling-merge, to be 0 for the parent and 1 for the child in case of a parent-merge, and to be 1 for the root vertex (unless it was already merged with another vertex). After p contraction phases T_s^p with $c = \lceil n/2^p \rceil$ vertices/clusters remain. The remaining $(r - p)$ bits are assigned by computing BFS numbers (starting with $(2^{\lceil \log_2 c \rceil} - c)$ at the root) for the vertices of T_s^p. These BFS numbers are inserted (in their binary representation) as the bits $\langle b_r, \ldots, b_{p+1} \rangle$ of the new labels for the vertices of T_s^p.

To efficiently combine the label bits from different phases and propagate them to the vertices of G again time-forward processing can be applied. The trees $T_s^p, \ldots, T_s^0$ will be revisited in that order and vertices of T_s^j can be processed e.g. in BFS order and each vertex v of T_s^j will combine the label bits received from T_s^{j+1} with the bit assigned during the construction phase and then send messages with its partial label to the vertices in T_s^{j-1} it comprises. The resulting new vertex numbering of G will cover the integers from $[n'-n, n')$ where $n' = 2^{\lceil \log_2 n \rceil}$. There will be a single gap $[0, n' - n)$ (unless n is a power of two) that can be easily excluded from storage by applying appropriate offsets when allocating and accessing arrays. Thereafter the new labeling has to be propagated to adjacency lists of G and the adjacency lists have to be reordered ($\mathcal{O}(\text{sort}(n + m))$ I/Os).

Assuming the adjacency lists are stored as an adjacency array sorted by vertex numbers (two arrays, the first with vertex information, e.g. offsets into the edge information; the second with edge information, e.g. destination vertices), there is no need for an extra index structure to retrieve any cluster in $\mathcal{O}(1 + \frac{x}{B})$ I/Os where x is the number of edges in that particular cluster. This is possible because all vertices of a cluster are numbered contiguously (for all values of $1 \leq \mu = 2^q \leq \sqrt{B}$) and the numbers of the first and last vertex in a cluster can be computed directly from the cluster number (and cluster size μ).

Lemma 2. *For an undirected connected graph G with n vertices, m edges, and given spanning tree T_s the Hierarchical Bottom-Up Clustering for all cluster sizes $1 \leq \mu = 2^q \leq n$ can be computed using $\mathcal{O}(\text{sort}(n))$ I/Os for constructing the new vertex labeling and $\mathcal{O}(\text{sort}(n + m))$ I/Os for the rearrangement of the adjacency lists of G.*

5   Implementation Details of Dynamic BFS

In the theoretical design of dynamic BFS the single parameter α was used to control the following quantities: the size of the clusters, the timer threshold (to avoid that elements are kept in the hot pool during the whole computation after a previous cluster fetch of the same element) and the number of levels that are prefetched into the hot pool. In our implementation we used two parameters α_1 and α_2 instead. The number of levels that are fetched into the hot pool is denoted by α_1 whereas the size of the clusters is controlled by α_2. The timer values are given by a simple approximation of the cluster diameter.

We split the original hot pool into two hot pools – one for the fetched levels (denoted as H) and one for the loaded clusters (denoted as HC). The elements in the two hot pools have different properties (for example the cluster id and the timer are needed in HC but not in H) and we were able to measure random I/Os from cluster fetches and sequential I/Os from feeding H with new levels and removing consumed or outdated entries separately.

For the insertion of an edge (v_1, v_2) into the graph we made the following observation. Let l_1 be the level of the vertex v_1 and l_2 the level of vertex v_2 and w.l.o.g be $l_1 \leq l_2$. The first level f_1 with possible improvements for the BFS levels is given by $f_1 = l_1 + \lfloor \frac{l_2 - l_1}{2} \rfloor + 1$. Hence, we do not need to recompute the level of any vertex in a level $l < f_1$. The distance $l_2 - l_1$ might be arbitrarily huge. The α_1 levels that could be fetched into H will never be required if $l_1 + \alpha_1 < l_2$. Therefore we start prefetching levels into H from level f_1 instead of l_1. HC will load clusters in the local neighborhood of v_2 to assign new BFS levels to adjacent vertices.

In Section 4 we argued for simplicity that the hierarchical clustering can be implemented using the time-forward processing technique and a priority queue. Since we are operating on a tree we can actually omit the priority queue in order to achieve better constant factors: we build triples $(vertex, level, parent)$ for each vertex in our tree and sort them by level and furthermore by parent. Now we scan the triples and merge two adjacent elements if they have the same parent. If there is no such adjacent element it is merged with its parent which is considered in the next level. Hence, we store a message that its parent is already clustered and then the parent is omitted later. For small levels these messages fit in internal memory, for larger levels an external sorted vector is used.

6   Experiments

Configuration. Our external-memory dynamic BFS implementation relies on the STXXL library [10]. For our static EM-BFS results we used the STXXL code from Ajwani et al. [4]. We performed our experiments on a machine with an Intel dual core E6750 processor @ 2.66 GHz, 4GB main memory (3.5 GB free), 3 hard disks with 500 GB each as external memory for STXXL, and a separate disk for the operating system, graph data, log files etc. The operating system was Debian GNU/Linux amd64 'wheezy' (testing) with kernel 3.2. We compiled with GCC 4.7.2 in C++11 mode using optimization level 3.

Graph Classes. For our experiments we used four different graph classes: one real-world graph with logarithmic diameter, two synthetic graph classes with diameters $\Theta(\sqrt{n})$ and $\Theta(n)$ and a tree graph class that was designed to elicit poor performance for the static BFS approach with standard Euler-tour clustering.

The real-world graph sk-2005 has around 50 million vertices, about 1.8 billion edges and is based on a web-crawl. It was also used by Crescenzi et al. [9] and has a known diameter of 40. The synthetic x-level graphs are similar to the B-level random graphs in [2]. They consist of x levels, each having $\frac{n}{x}$ vertices (except

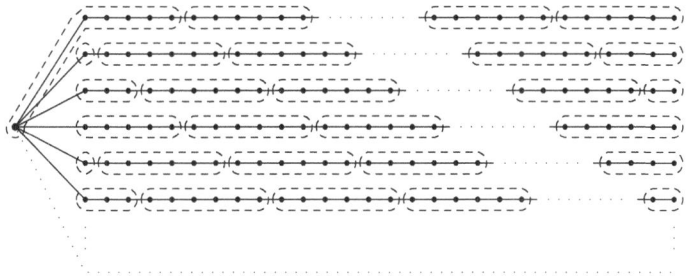

Fig. 4. Our cl_n2_29 graph has the same shape as the graph in this picture but with 1048576 lists of length 511 each. In this picture the result of an Euler tour based clustering with $\mu = 6$ is shown as it is used in MM_BFS.

the level 0 containing only one vertex). The edges are randomly distributed between consecutive levels. The $\sqrt{n}$-level graph graph features $n = 2^{28}$ nodes and $m = 1.1 \cdot 10^9$ edges, for the $\Theta(n)$-level graph we have $n = 2^{28}$ and $m = 0.9 \cdot 10^9$. The fourth graph class represents trees whose special shape are tuned to yield an Euler tour clustering that forces the static MM_BFS algorithm into $\Omega(n/\sqrt{B})$ unstructured I/Os: whenever a new BFS level is reached, many new clusters are encountered for the first time. A schematic depiction is presented in Figure 4. In our concrete case the resulting cl_n2_29 graph features about 2^{29} nodes and 2^{20} lists. The parameters were chosen in a way that prefetching heuristics will not help MM_BFS caching adjacency lists in main memory.

6.1   Results

In our experiment we inserted new edges (v_1, v_2) into the graph, where we set $v_1 = s$ (the source of the BFS tree) and select the other vertex v_2 from BFS levels $0.1 \cdot d, 0.2 \cdot d, ..., d$ where d denotes the height of the BFS tree. The experiments were executed independently. For each inserted edge the initial BFS tree / graph was the same. The source of the BFS tree was chosen to make the experiments more difficult: two vertices far away from the source might have a small distance to each other and then usually only a small fraction of the whole graph data has to be reassigned to new BFS levels in our graph data. We measured the time for dynamic BFS plus the time to write the result and the number of vertices that have been updated. Experiments during the implementation showed that for a small cluster size α_2, e. g. $\alpha_2 = 64$, its value is algorithmically never increased. This leads to a high number of random I/Os. Thus we set $\alpha_2 = 1024$ which causes a small amount of random cluster fetches. For smaller α_2 our results were slightly better for each graph class on our test machine but not for the $\sqrt{n}$-level graph. The number of elements in the hot pool H is given by the value of α_1. For large α_1 a huge hot pool H has to be scanned for each BFS level computation. We set the initial $\alpha_1 = 4$ to avoid too many sequential I/Os. Table 1 contains the time for computing static EM-BFS for each graph divided into the preprocessing and the BFS computation. The cl_n2_29 graph stands out with a slow BFS computation while the preprocessing is almost as fast as the preprocessing for the other graph classes. Table 2 contains the time for the hierarchical clustering and the time that is needed to reorganize the adjacency lists (add cluster information to edges, sort them, ...). The clustering is slower than the preprocessing of the static BFS because logarithmically many phases e. g. containing Euler Tour and list-ranking computation have to be done instead of one. Nevertheless, the computation time is still independent from the graph class and the computation time only depends on the input size. The gain of the hierarchical clustering is obtained in the dynamic BFS computation of the cl_n2_29 graph (details at the end of this section).

Table 1. Running time (in hours) of static EM-BFS with source 0

	sk-2005	$\sqrt{n}$-level graph	$\Theta(n)$-level graph	cl_n2_29 graph
Time Preprocessing	0.91	1.35	1.19	1.29
Time BFS-level computation	2.41	3.26	1.36	> 17

Table 2. Running time (in hours) of level-aligned hierarchical clustering

	sk-2005	$\sqrt{n}$-level graph	$\Theta(n)$-level graph	cl_n2_29 graph
Compute hierarchical level-aligned clustering	0.39	1.64	1.35	3.01
Reorganization of adjacency lists	1.38	0.94	0.84	0.54

Figure 5 contains the results of our dynamic BFS computing the updated BFS levels. Because one vertex of the newly inserted edge is always the source, the results mirror the local hot spots in the graph in which the update of a few BFS levels is expensive. For example, the web graph sk-2005 has many vertices in the levels close to the source 0. Vertices with a larger distance to the source seem to have a list-like path to the source. Hence, an update of the BFS-levels was very fast from vertices with large distance to the source. In our experiments the worst scenario is the $\sqrt{n}$-level graph. It is the only case in which our current implementation loses against EM-BFS for large distance between the two vertices of the new edge.

Results of experiments with cl_n2_29 graph: as expected the hot pool of static BFS had to go external and reads/writes Terabytes of data (input data set size: 8 GB). Therefore, static MM_BFS needs more than 17 hours. Each update during the dynamic BFS computation needed at most 0.23 hours.

Our results using hard disks were viable due to comparatively large α_2. In experiments on a similar machine using solid state drives we were able to improve our results. We beat the static BFS for each graph class in each test scenario by using a smaller $\alpha_2 = 256$. For our $\sqrt{n}$-level graph we were able to beat static BFS by a factor of 1.14 in our worst case. This is explained by the fact that for

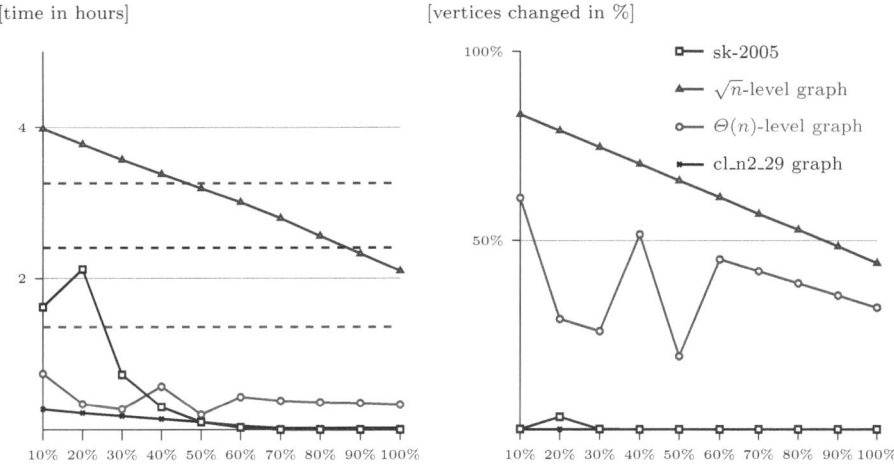

[number of BFS levels spanned by inserted edge in % of max. BFS level d]

Fig. 5. Results of dynamic BFS. The time of each static BFS is plotted in as a dashed line in the left plot for sk-2005 (2.41 h), $\sqrt{n}$-level graph (3.26 h) and $\Theta(n)$-level graph (1.36 h). The static BFS time of cl_n2_29 graph was not drawn because it is too huge with 17 hours to suit into the plot.

smaller α_2 the work on CPU is much smaller but the I/O-time increases. It seems that with SSDs the I/O time increases slower than the CPU-time decreases. We plan to present more details in a full version.

7   Conclusion

We have given initial results of the first external-memory dynamic BFS implementation using a new deterministic level-aligned hierarchical clustering. Even though we applied rather hard edge insertion scenarios our implementation was usually faster, and for some graph classes much faster, than the static BFS implementation. We investigated the interaction of the different parameters that influence the performance of our dynamic BFS in more detail.

Acknowledgements. We want to thank Asmaa Edres for her work on the implementation of the time-forward processing hierarchical clustering using a priority queue in her master's thesis.

References

1. Aggarwal, A., Vitter, J.S.: The input/output complexity of sorting and related problems. Communications of the ACM 31(9), 1116–1127 (1988)
2. Ajwani, D.: Traversing large graphs in realistic setting. PhD thesis, Saarland University (2008)

3. Ajwani, D., Meyer, U.: Design and engineering of external memory traversal algorithms for general graphs. In: Lerner, J., Wagner, D., Zweig, K.A. (eds.) Algorithmics. LNCS, vol. 5515, pp. 1–33. Springer, Heidelberg (2009)
4. Ajwani, D., Meyer, U., Osipov, V.: Improved external memory BFS implementation. In: Proc. 9th ALENEX, pp. 3–12 (2007)
5. Arge, L.: The buffer tree: A technique for designing batched external data structures. Algorithmica 37(1), 1–24 (2003)
6. Arge, L., Brodal, G., Toma, L.: On external-memory MST, SSSP and multi-way planar graph separation. J. Algorithms 53(2), 186–206 (2004)
7. Atallah, M., Vishkin, U.: Finding euler tours in parallel. Journal of Computer and System Sciences 29(30), 330–337 (1984)
8. Chiang, Y.J., Goodrich, M.T., Grove, E.F., Tamasia, R., Vengroff, D.E., Vitter, J.S.: External memory graph algorithms. In: Proceedings of the 6th Annual Symposium on Discrete Algorithms (SODA), pp. 139–149. ACM-SIAM (1995)
9. Crescenzi, P., Grossi, R., Imbrenda, C., Lanzi, L., Marino, A.: Finding the diameter in real-world graphs – experimentally turning a lower bound into an upper bound. In: de Berg, M., Meyer, U. (eds.) ESA 2010, Part I. LNCS, vol. 6346, pp. 302–313. Springer, Heidelberg (2010)
10. Dementiev, R., Sanders, P.: Asynchronous parallel disk sorting. In: Proc. 15th SPAA, pp. 138–148. ACM (2003)
11. Eppstein, D., Galil, Z., Italiano, G.: Dynamic graph algorithms. In: Atallah, M.J. (ed.) Algorithms and Theory of Computation Handbook, ch. 8. CRC Press (1999)
12. Mehlhorn, K., Meyer, U.: External-memory Breadth-First Search with sublinear I/O. In: Möhring, R.H., Raman, R. (eds.) ESA 2002. LNCS, vol. 2461, pp. 723–735. Springer, Heidelberg (2002)
13. Meyer, U.: On dynamic Breadth-First Search in external-memory. In: 25th Annual Symposium on Theoretical Aspects of Computer Science (STACS), pp. 551–560 (2008)
14. Munagala, K., Ranade, A.: I/O-complexity of graph algorithms. In: Proceedings of the 10th Annual Symposium on Discrete Algorithms (SODA), pp. 687–694. ACM-SIAM (1999)
15. Roditty, L.: Dynamic and static algorithms for path problems in graphs. PhD thesis, Tel Aviv University (2006)
16. Vitter, J.S.: Algorithms and data structures for external memory. Foundations and Trends in Theoretical Computer Science 2(4), 305–474 (2006)

Versatile Succinct Representations of the Bidirectional Burrows-Wheeler Transform*

Djamal Belazzougui, Fabio Cunial, Juha Kärkkäinen, and Veli Mäkinen

Helsinki Institute for Information Technology (HIIT),
Department of Computer Science, University of Helsinki, Finland

Abstract. We describe succinct and compact representations of the bidirectional BWT of a string $s \in \Sigma^*$ which provide increasing navigation power and a number of space-time tradeoffs. One such representation allows to extend a substring of s by one character from the left and from the right in constant time, taking $O(|s| \log |\Sigma|)$ bits of space. We then match the functions supported by each representation to a number of algorithms that traverse the nodes of the suffix tree of s, exploiting connections between the BWT and the suffix-link tree. This results in near-linear time algorithms for many sequence analysis problems (e.g. maximal unique matches), for the first time in succinct space.

1 Introduction

Suffix trees are versatile data structures on which myriads of sequence analysis tasks can be solved optimally [1, 11]. Recent progress in compressed data structures has provided alternatives that replace suffix trees verbatim with more space-efficient constructions [5, 10, 22–24]. Such black-box replacements do not always achieve optimal space-time tradeoffs, thus substantial effort has been devoted to designing the best possible setup of data structures for *specific* sequence analysis problems [6, 13, 15, 19, 27, 28]. In this paper we recognize *recurrent patterns* in the way many classical sequence analysis algorithms traverse the suffix tree of a string, and provide a corresponding set of minimal data structures that can be used to implement all such algorithms at once. The key observation is that a number of algorithms iterate over the nodes of a suffix tree either in no particular order, or explicitly in the order induced by suffix links: this allows to implement navigation using the bidirectional Burrows-Wheeler transform (BWT) [16, 18, 27, 28], often without even the need of compressed longest common prefix (LCP) arrays or range minimum query (RMQ) data structures.

Let $\mathrm{occ}_s(w)$ be a constant size representation of the occurrences of a string w in a string s over an alphabet Σ (we assume $\mathrm{occ}_s(w) = \emptyset$ if w does not occur in s). For example, $\mathrm{occ}_s(w)$ could be the locus of w in the suffix tree of s, or the lexicographical range of the suffixes of s that begin with w. In the bidirectional BWT, $\mathrm{occ}_s(w)$ consists of two lexicographical ranges of suffixes, one representing

---

* This work was partially supported by Academy of Finland under grants 250345 (CoECGR) and 118653 (ALGODAN).

H.L. Bodlaender and G.F. Italiano (Eds.): ESA 2013, LNCS 8125, pp. 133–144, 2013.

Table 1. Representations of the bidirectional BWT: summary of space, time, navigation power, and applications, for the implementations described in Section 4. Time complexities for `enumerateLeft` and `enumerateRight` are per element of output. SUS: shortest unique substrings; MR: maximal repeats; LB: longest border; QP: quasiperiod; IPS: inner product of substrings; IPK: inner product of k-mers; (N)SR: (near) supermaximal repeats; MAW: minimal absent words; BBB: bidirectional b&b (supported also by Implementation 2a).

Representation	1		2		3
Implementation	1a	1b	2a [27]	2b	3
Space (bits)	$\|s\|\log\|\Sigma\|+$ $+\|s\|+o(\|s\|)$	$\|s\|\log\|\Sigma\|+$ $+o(\|s\|\log\|\Sigma\|)$	$2\|s\|\log\|\Sigma\|+$ $+o(\|s\|)$	$2\|s\|\log\|\Sigma\|+$ $+o(\|s\|\log\|\Sigma\|)$	$O(\|s\|\log\|\Sigma\|)$
`isLeftMaximal`	$O(\log\|\Sigma\|)$	$O(1)$	$O(\log\|\Sigma\|)$	$O(1)$	$O(1)$
`isRightMaximal`	$O(1)$	$O(1)$	$O(\log\|\Sigma\|)$	$O(1)$	$O(1)$
`enumerateLeft`	$O(\log\|\Sigma\|)$	$O(1)$	$O(\log\|\Sigma\|)$	$O(1)$	$O(1)$
`enumerateRight`			$O(\log\|\Sigma\|)$	$O(1)$	$O(1)$
`extendLeft`	$O(\log\|\Sigma\|)$	$O(\|\Sigma\|)$	$O(\log\|\Sigma\|)$	$O(\|\Sigma\|)$	$O(1)$
`extendRight`			$O(\log\|\Sigma\|)$	$O(\|\Sigma\|)$	$O(1)$
Applications	MUM, SUS, MR, LB, QP, IPS, IPK		MUM, SUS, MEM, SR, NSR, MAW, IPS, IPK		BBB

the occurrences of w in s and the other representing the occurrences of the reverse of w in the reverse of s. We want to support the following operations:

- `extendLeft`$(a \in \Sigma, \mathrm{occ}_s(w)) = \mathrm{occ}_s(aw)$
- `extendRight`$(\mathrm{occ}_s(w), a \in \Sigma) = \mathrm{occ}_s(wa)$
- `enumerateLeft`$(\mathrm{occ}_s(w)) = \{\mathrm{occ}_s(aw) : a \in \Sigma, \mathrm{occ}_s(w) \neq \emptyset\}$
- `enumerateRight`$(\mathrm{occ}_s(w)) = \{\mathrm{occ}_s(wa) : a \in \Sigma, \mathrm{occ}_s(w) \neq \emptyset\}$
- `isLeftMaximal`$(\mathrm{occ}_s(w)) = $ true iff $|$`enumerateLeft`$(\mathrm{occ}_s(w))| > 1$
- `isRightMaximal`$(\mathrm{occ}_s(w)) = $ true iff $|$`enumerateRight`$(\mathrm{occ}_s(w))| > 1$

where `enumerateLeft` and `enumerateRight` produce their output in lexicographical order of a. We describe three representations of the bidirectional BWT with increasing sets of supported operations, tailored for the navigation patterns of corresponding classes of algorithms. Each representation is realized by corresponding succinct implementations, whose space–time tradeoffs are summarized in Table 1. In turn, such implementations expose a number of new tradeoffs for many problems, they achieve the first succinct-space solution for others (namely, longest border and surprising substrings), and they allow to compute maximal unique matches in succinct space and near-linear time. Our main technical contribution is to show that the $O(\log|\Sigma|)$-time bidirectional backward step operation supported by wavelet trees [27] can be performed in constant time.

2   Definitions and Notation

We use the example in Fig. 1 to introduce and illustrate the formalism of the following sections. Let Σ be a finite alphabet, let \$ $\notin \Sigma$ be a symbol lexicographically smaller than any other symbol, and let $s \in \Sigma^+$. The *suffix array*

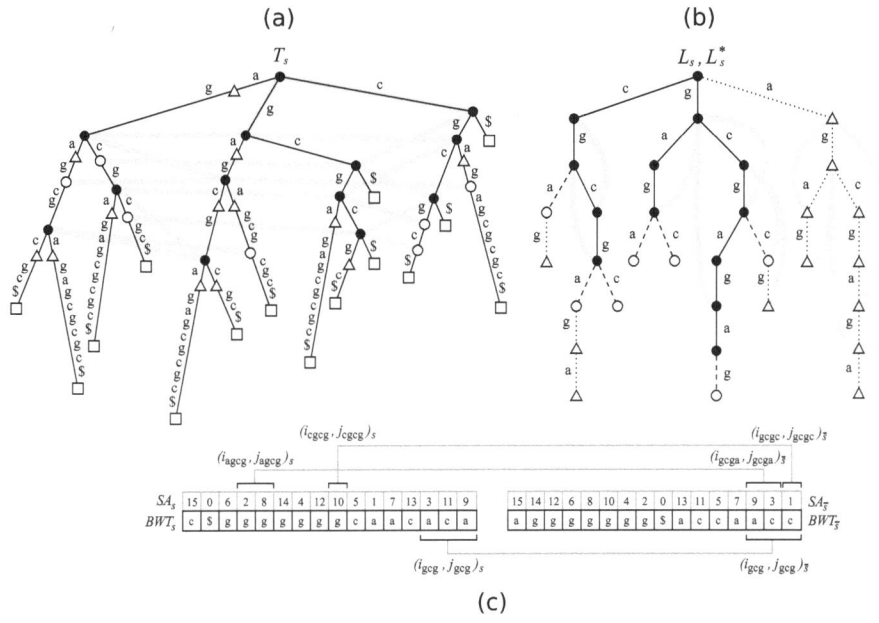

Fig. 1. Main constructs and formalisms applied to string $s = \mathtt{agagcgagagcgcgc}$ (see Sections 1 and 5.2 for more details). (a) The suffix tree T_s with extended nodes. Gray lines are implicit and explicit Weiner links. To improve clarity, the children of a node are not drawn in lexicographical order. (b) The suffix-link tree L_s and its supertree L_s^*.(c) Suffix array and BWT of s and of its reverse $\bar{s}$. Moving from the ranges of $\mathtt{gcg}$ to the ranges of $\mathtt{agcg}$ (respectively, of $\mathtt{cgcg}$) corresponds to navigating an explicit (respectively, implicit) Weiner link (see Panel b).

$SA_s[0, |s|]$ of $s\$$ is the vector of indices such that $s[SA_s[i], |s| - 1]\$$ is the i-th smallest suffix of $s\$$ in lexicographical order. The Burrows-Wheeler transform of $s\$$ is the string $BWT_s[0, |s|]$ satisfying $BWT_s[i] = s[SA_s[i] - 1]$ if $SA_s[i] > 0$, and $BWT_s[i] = \$$ otherwise. We use $\bar{s}$ to indicate the reverse of a string s, and $\mathrm{count}_s(a, i, j)$ to represent the number of occurrences of symbols lexicographically smaller than $a \in \Sigma$ in $s[i, j]$. We define the *suffix array range*, or identically the BWT *range*, of a substring w, as the maximal range $(i_w, j_w)_s$ such that the suffixes $s[SA_s[i], |s| - 1]$, $i \in [i_w, j_w]$, are prefixed by w. The range $(i_{\bar{w}}, j_{\bar{w}})_{\bar{s}}$ of $\bar{w}$ is defined analogously with suffixes $\bar{s}[SA_{\bar{s}}[i], |s| - 1]$, $i \in [i_{\bar{w}}, j_{\bar{w}}]$, prefixed by $\bar{w}$. Fig. 1c shows the ranges of $\mathtt{gcg}$, $\mathtt{agcg}$, $\mathtt{cgcg}$, and their reverses.

The *suffix tree* $T_s = (U_s \cup V_s, E_s)$ of s is a rooted tree with leaves U_s and internal nodes V_s. The edge labels $\ell(e) \in \Sigma^+$ for $e \in E_s$ induce the node labels $\ell(v) = \ell(e_0) \cdot \ell(e_1) \cdots \ell(e_{k-1})$ for $v \in U_s \cup V_s$, where $e_0, e_1, \ldots, e_{k-1}$ is the path from the root to v. Each internal node $v \in V_s$ has at least two children, the labels of the edges to the children have different first symbols, and the children are ordered by that symbol. Then all node labels are distinct and their lexicographical order corresponds to the pre-order of the nodes. The set of leaf labels is exactly the set of suffixes of s. White squares, black circles and black lines in Fig. 1a represent U_s, V_s and E_s, respectively. We denote with $W_s = \{(u,v) \mid u, v \in V_s, \ell(v) = a \cdot \ell(u), a \in \Sigma\}$ the set of *explicit Weiner links* of T_s. A Weiner link (u, v) has a label $\ell(u, v) \in \Sigma$ such that $\ell(v) = \ell(u, v) \cdot \ell(u)$. The *suffix-link tree* $L_s = (V_s, W_s)$ of s is the trie induced by W_s on V_s (black circles and solid lines in Fig. 1b). If $w = \ell(v)$ for a node $v \in U_s \cup V_s$, we say that v is the *locus* of w, $\text{locus}_s(w) = v$. For any substring w of s, let x be the shortest string such that $\text{locus}_s(wx) = v$ exists. Then v is the the *extended locus* of w, $\text{elocus}_s(w) = v$. We indicate the parent of a node v in a tree with $\text{parent}(v)$, and with $\text{lca}(u, v)$ the least common ancestor of two nodes u, v.

For a string w, let $\mathcal{L}_s(w) = \{i_0, i_1, \dots, i_{m-1}\}$ be the set of all starting positions of w in s. Consider the string sets $S = \{\ell(v) \mid v \in V_s\}$, $S' = \{a\ell(v) \mid a \in \Sigma, v \in V_s, \mathcal{L}_s(a\ell(v)) \neq \emptyset\}$ and $S'' = \{\ell(v)a \mid a \in \Sigma, v \in V_s, \mathcal{L}_s(\ell(v)a) \neq \emptyset\}$. We introduce new nodes to represent the strings in S' and S'' when necessary: $V_s' = \{v : \ell(v) = w \in S' \setminus S\}$ and $V_s'' = \{v : \ell(v) = w \in S'' \setminus (S \cup S')\}$. The suffix tree T_s, the suffix-link tree L_s, and the suffix-link tree L_s^* extended with nodes in V_s' (white circles) and in V_s'' (white triangles), are illustrated in Fig. 1. The edges $W_s' = \{(u, v) \mid u \in V_s, v \in V_s', \ell(v) = a \cdot \ell(u), a \in \Sigma\}$ are called *implicit Weiner links* and are represented as dashed lines in Fig. 1b.

3   Synchronizing the Bidirectional BWT in Constant Time

For a node $v \in V_s$, let $(i_v, j_v)_s$ and $(i_{\bar{v}}, j_{\bar{v}})_{\bar{s}}$ be shorthand notations for $(i_{\ell(v)}, j_{\ell(v)})_s$ and $(i_{\overline{\ell(v)}}, j_{\overline{\ell(v)}})_{\bar{s}}$, respectively. Assume that we know $(i_v, j_v)_s$ and $(i_{\bar{v}}, j_{\bar{v}})_{\bar{s}}$, and that we want to derive $(i_{a\ell(v)}, j_{a\ell(v)})_s$ and $(i_{\overline{\ell(v)}a}, j_{\overline{\ell(v)}a})_{\bar{s}}$ for some $a \in \Sigma$. For example, we may know the BWT ranges of node u with $\ell(u) = \text{gcg}$ in Fig. 1a, and we may want to derive the ranges for string agcg (Fig. 1c). Belazzougui and Navarro recently showed that the map $(i_v, j_v)_s \mapsto (i_{a\ell(v)}, j_{a\ell(v)})_s$ can be implemented in constant time, independent of $|\Sigma|$, using $O(|s| \log \log |\Sigma|)$ bits of space [5]. Deriving $(i_{\overline{\ell(v)}a}, j_{\overline{\ell(v)}a})_{\bar{s}}$ from $(i_{\bar{v}}, j_{\bar{v}})_{\bar{s}}$ and $(i_v, j_v)_s$ reduces to computing $\text{count}_{BWT_s}(a, i_v, j_v)$, which is typically implemented by subtracting $\text{count}_{BWT_s}(a, 0, i_v - 1)$ from $\text{count}_{BWT_s}(a, 0, j_v)$ (see e.g. [16, 28]). We prove that the running time of this approach has an intrinsic lower bound imposed by space:

Theorem 1. *Let $s \in \Sigma^+$ be a string and let k be a constant number. In the cell probe model with word size $\Theta(\log |s|)$, no data structure that uses $O(|s| \log^k |s|)$ bits of space can compute $\text{count}_s(a, 0, i)$ in $o((\log |\Sigma|)/(\log \log |s|))$ time, for any $a \in \Sigma$ and $0 \le i < |s|$.*

The proof reduces two-dimensional orthogonal range counting on an $n \times n$ grid to computing $\text{count}_s(a, 0, i)$ for a string s of length n on any alphabet size: we omit it due to lack of space. To compute $(i_{\overline{\ell(v)}a}, j_{\overline{\ell(v)}a})_{\bar{s}}$, however, we are only interested in computing $\text{count}_{BWT_s}(a, i_v, j_v)$ for intervals of BWT_s that correspond to *nodes* v of the suffix tree of s. Here we show that $\text{count}_s(a, i_v, j_v)$

for $v \in V_s$ can be computed in constant time and $O(|s| \log |\Sigma|)$ bits of space, independent of $|\Sigma|$, by applying the data structures introduced in [5]. To make the paper self-contained, we first sketch the original construction.

Theorem 2 ([5]). *There is an index of size $O(|s| \log \log |\Sigma|)$ bits that implements map $(i_v, j_v)_s \mapsto (i_{a\ell(v)}, j_{a\ell(v)})_s$ in constant time for any node $v \in V_s$ in the suffix tree of s.*

Proof sketch. We use the $O(|s|)$ representation of T_s described in [26] to commute in constant time between ranges in BWT_s and corresponding nodes in V_s, and vice versa. Assign to every $v \in V_s$ a unique identifier $\text{id}(v) \in [1, 2|s| - 1]$ by enumerating V_s in depth-first order. Note that explicit Weiner links preserve depth-first order: if $e_1 = (u_1, v_1)$ and $e_2 = (u_2, v_2)$ are explicit Weiner links with $\ell(e_1) = \ell(e_2)$, then $\text{id}(u_1) < \text{id}(u_2) \Rightarrow \text{id}(v_1) < \text{id}(v_2)$. The same holds for implicit Weiner links if we give depth-first identifiers to their destinations. For each character $a \in \Sigma$, consider set $V_a = \{\text{id}(u) : u \in V_s, e = (u, v) \in W_s \cup W'_s, \ell(e) = a\}$, and let explicit_a be a bit vector of size $|V_a|$ that marks with a one (respectively, with a zero) the explicit (respectively, implicit) Weiner links in V_a. Clearly the number of ones in explicit_a is the number of nodes in V_s whose label starts with a. Let $f_a : V_a \mapsto [1, |V_a|]$ be a *monotone minimal perfect hash function* [4]. It's easy to see that the overall space used by f_a for all $a \in \Sigma$ is $O(|s| \log \log |\Sigma|)$ bits (see [5] for additional details). The destination of an explicit Weiner link $e = (u, v)$ with label a can be obtained by first using f_a to map u to a number $x \in [1, |V_a|]$ in constant time. Then, $\text{explicit}_a[x] = 1$ implies that e is explicit, and computing $\text{id}(v)$ reduces to computing $C[a]$ and the number of ones in $\text{explicit}_a[1, x - 1]$, where $C[a] = \sum_{c \in \Sigma, c < a} \text{count}_s(c, 0, |s|)$. Note that the value of $C[a]$ for all $a \in \Sigma$ can be precomputed and stored in overall $o(n \log \sigma)$ bits. An implicit Weiner link has $\text{explicit}_a[x] = 0$, and the identifier of the extended locus of its destination is also the number of ones in $\text{explicit}_a[1, x - 1]$. Finally, since f_a returns a number in $[1, |V_a|]$ even when there is no Weiner link from u labelled by a, we need to check the existence of an a in $(i_u, j_u)_s$. In order to do so, we first convert $\text{id}(u)$ into $(i_u, j_u)_s$ and $\text{id}(v)$ into $(i_v, j_v)_s$, and then check whether $i_u \leq \text{select}(a, i_v - C[a] + 1) \leq j_u$. □

The proof of Theorem 2 extends naturally to $\text{count}_{BWT_s}(c, i_v, j_v)$.

Theorem 3. *There is an index of size $O(|s| \log |\Sigma|)$ bits that implements map $(i_v, j_v)_s \mapsto \text{count}_{BWT_s}(c, i_v, j_v)$ in constant time for any node $v \in V_s$ in the suffix tree of s.*

Proof. As above, we assume the $O(|s|)$ representation of T_s described in [26]. Let $V_c = \{u \in V_s : e = (u, v) \in W_s \cup W'_s, \ell(e) = c\}$, and let $\mathcal{T}_c = (V_c, E_c)$ be a tree induced on V_c by the set of edges E_c defined as follows: E_c contains all pairs (u, v), $u, v \in V_c$, such that u is an ancestor of v in T_s, and there is no $w \in V_c$, $w \neq v$, in the path that connects u to v in T_s. We use a monotone minimal perfect hash function f_c to map the nodes in V_c to identifiers in the depth-first order of $\mathcal{T}_c$. Let diff_c be a bit vector of size $|V_c|$ where position k corresponds to the kth node in the depth-first order of V_c. Let u_k be the kth node in the depth-first order of V_c: we set $\texttt{diff}_c[k] = \texttt{count}_{BWT_s}(c, i_{u_k}, j_{u_k}) - \sum_{(u_k,v)\in E_c} \texttt{count}_{BWT_s}(c, i_v, j_v)$. Clearly $\sum_{0\leq k<|V_c|} \texttt{diff}_c[k] \in O(|s|)$, so $\texttt{diff}_c$ can be encoded in $O(|V_c|\log(|\Sigma|/|V_c|))$ bits using a prefix-sum data structure [21]. As in the proof of Theorem 2, it follows that the total space taken by $\texttt{diff}_c$ for all $c \in \Sigma$ is $O(|s|\log|\Sigma|)$ bits. Let v be the node of V_c that corresponds to interval (i_v, j_v) in BWT_s. To compute $\texttt{count}_{BWT_s}(c, i_v, j_v)$, it suffices to retrieve the range (i'_v, j'_v) of depth-first identifiers of nodes in the subtree of $\mathcal{T}_c$ rooted at v (including v itself), and to compute $\sum_{k=i'_v}^{j'_v} \texttt{diff}_c[k]$: both such operations can be implemented in constant time using the data structure in [26] and the prefix-sum data structure in [21].

4   Succinct Representations of the Bidirectional BWT

We detail here the hierarchy of representations in Table 1[1]. *Representation 1* supports just $\texttt{enumerateLeft}$, $\texttt{isLeftMaximal}$ and $\texttt{isRightMaximal}$, and can be implemented as follows. Along the lines of [15], let $\texttt{runs}_s \in \{0,1\}^{|s|}$ be a bit vector such that $\texttt{runs}_s[i] = 1$ iff $i > 0$ and $BWT_s[i] \neq BWT_s[i-1]$. We encode $\texttt{runs}_{\bar{s}}$ as an array of $|s| + o(|s|)$ bits, and we implement $\texttt{enumerateLeft}$ and synchronize BWT_s and $BWT_{\bar{s}}$ by representing BWT_s as a wavelet tree in $|s|\log|\Sigma|+o(|s|)$ bits of space. Given $(i_w, j_w)_s$, the wavelet tree allows to enumerate all distinct characters $a \in \Sigma$ in $(i_w, j_w)_s$, to obtain $\texttt{count}_{BWT_s}(a, i_w, j)$, and to compute the number of occurrences of a in $BWT_s[1, i-1]$ and in $BWT_s[i_w, j_w]$, in $O(\log|\Sigma|)$ time per character [8, 28]. These values suffice to compute all the corresponding $(i_{aw}, j_{aw})_s$ and $(i_{\bar{w}a}, j_{\bar{w}a})_{\bar{s}}$ from $(i_w, j_w)_s$ and $(i_{\bar{w}}, j_{\bar{w}})_{\bar{s}}$ in batch. Note that this implementation supports also $\texttt{extendLeft}$ in $O(\log|\Sigma|)$ time. Operation $\texttt{isRightMaximal}$ for string w consists in checking whether the range $(i_{\bar{w}}, j_{\bar{w}})_{\bar{s}}$ contains at least two distinct characters, i.e. in counting the number of ones in $\texttt{runs}_{\bar{s}}[i_{\bar{w}}+1, j_{\bar{w}}]$. Operation $\texttt{isLeftMaximal}$ consists in checking whether the range $(i_w, j_w)_s$ contains at least two distinct characters: this can be clearly implemented in $O(\log|\Sigma|)$ time using the wavelet tree representation of BWT_s. We call *Implementation 1a* this setup of data structures that supports the functions of Representation 1.

Alternatively, we can implement $\texttt{enumerateLeft}$ and synchronize BWT_s and $BWT_{\bar{s}}$ by representing BWT_s in $|s|\log|\Sigma|$ bits as a plain sequence of characters, and by building a range minimum query data structure (RMQ, see [9]). An RMQ allows to enumerate all the distinct characters in a BWT interval in constant time per character, using $2|s| + o(|s|)$ bits of space [25]. Such characters, however, are not necessarily listed in sorted order: to put them in sorted order, we use a monotone minimal perfect hash function that stores the distinct characters for

---

[1] For brevity, we omit the space-time tradeoffs that would result from using *compressed*, rather than succinct, data structures. Some implementations described in this section can be augmented with data structures that take space within their lower order terms, and that support (albeit sometimes inefficiently) additional operations that are not required by their corresponding representations.

every interval in BWT_s. The rank of each character is thus given by the hash function in constant time per character. Finally, we build the $O(|s| \log \log |\Sigma|)$-space data structures of Theorem 2: since we know that aw exists for each $a \in \Sigma$ detected above, we can obtain $(i_{aw}, j_{aw})_s$ for each such a in constant time. As a byproduct, we also get the number of occurrences of aw: using this information we can synchronize the corresponding intervals in $BWT_{\bar{s}}$ in batch, in constant time per interval. We call *Implementation 1b* this way of supporting the functions of Representation 1.

Representation 1 suffices for applications that need to traverse the whole L_s top-down, or that just need to iterate over every node of T_s exactly once. Our second representation of the bidirectional BWT supports enumerateRight in addition to all operations of Representation 1, providing access to all the children of a node in the suffix tree of s. We implement *Representation 2* with the same data structures as in Implementations 1a and 1b, but replicating them for $BWT_{\bar{s}}$ and removing runs$_{\bar{s}}$. Thus, *Implementation 2a* supports also extendRight in $O(\log |\Sigma|)$ time, and *Implementation 2b* duplicates Implementation 1b, but without removing runs$_{\bar{s}}$. Implementation 2a was originally described in [27], and it can easily support isLeftMaximal and isRightMaximal in constant time using $2|s| + o(|s|)$ additional bits to encode runs$_s$ and runs$_{\bar{s}}$.

The third level in our hierarchy, called *Representation 3*, supports extendLeft and extendRight in addition to all operations of Representation 2, allowing selective extension by a single character in both directions. We implement Representation 3 by augmenting Implementation 2b with the data structures described in Theorem 3.

5   Applications

A number of algorithms can be expressed as iterations over the nodes of the suffix tree T_s of a string s, either in no particular order or explicitly in the order induced by a top-down navigation on the suffix-link tree L_s. In this section we implement a subset of such algorithms using the representations of the bidirectional BWT described in Section 4, and we show that the corresponding implementations allow to reach favorable regions in the space-time plane. For brevity, we waive details related to index construction time.

To warm up, consider the bidirectional branch-and-bound search used by popular read alignment tools to perform approximate string matching [16–18]: the state of the art is based on Implementation 2a or slower alternatives, thus using Representation 3 yields a speedup by a $\Theta(\log |\Sigma|)$ factor.

5.1   Maximal Repeats and Maximal Matches

We say that w is *left-maximal* (respectively, *right-maximal*) in s if $\mathcal{L}_s(w) \neq \mathcal{L}_s(aw)$ (respectively, $\mathcal{L}_s(w) \neq \mathcal{L}_s(wa)$) for all $a \in \Sigma$. We say that w is a *maximal repeat* of s if w is left- and right-maximal [11] (for example, string gag in Fig. 1 is a maximal repeat). A variety of algorithms have been proposed for discovering all the maximal repeats of s, and recently very space-efficient solutions have been achieved (see e.g. [6, 15] and references therein). Most such algorithms build an LCP array, whose construction takes $O(|s| \log^\epsilon |s|)$ time when using $O(\frac{1}{\epsilon}|s| \log |\Sigma|)$ bits of space [24]. Recently it has been shown that LCP construction is not required to solve the problem in succinct space [6]. We give here a different algorithm that also avoids LCP construction, which we later extend to other problems not considered in [6].

Lemma 1. *Assume that Implementation 1a (respectively, 1b) has already been built for a string $s \in \Sigma^+$. We can discover all the τ maximal repeats of $s \in \Sigma^+$ in $O(|s| \log |\Sigma|)$ time and $|s| \log |\Sigma| + 2|s| + o(|s|) + O(\tau \log |s|)$ bits of space (respectively, in $O(|s|)$ time and $|s| \log |\Sigma| + o(|s| \log |\Sigma|) + O(\tau \log |s|)$ bits of space).*

Proof. We navigate L_s top-down by iteratively taking all Weiner links and testing whether the string corresponding to the destination node is right-maximal: in the negative case, we stop iteration. In the positive case, we test left-maximality. To navigate L_s we keep a stack S_1 of pointers and labels of the children of each node, and to print maximal repeats to the output we keep a stack of characters S_2. Traversing L_s top-down could potentially make $|S_1| \in O(\lambda|\Sigma| \log |s|)$ bits, where λ is the depth of L_s. In the worst case $\lambda \in \Theta(|s|)$, thus it could be that $|S_1| \in \Theta(|\Sigma| \cdot |s| \log |s|)$. We keep $|S_1| \in O(|\Sigma| \log^2 |s|)$ by using the strategy of always visiting the child of a node with the largest subtree last (see e.g. [12]): this limits the depth of the navigation stack to $O(\log |s|)$. Clearly $|S_2| \in O(\lambda \log |\Sigma|) = O(|s| \log |\Sigma|)$. Reporting the positions in s of all maximal repeats would be trivial if we had the suffix array of s. In our case, we build the set of all occurrences of all maximal repeats in BWT_s, we sort them in $O(|s|)$ time using radix sort, and then we invert BWT_s. During inversion, we scan the list of occurrences and we replace each value by the corresponding position in s. This process takes overall $O(|s| \log |\Sigma|)$ time.

Note that Implementation 1a can be built in $O(|s| \log |\Sigma|)$ time and space [14], thus we can discover all maximal repeats of a string $s \in \Sigma^+$ in $O(|s| \log |\Sigma|)$ time and bits of space – the same bound as in [6]. Both algorithms extend easily to supermaximal and near-supermaximal repeats.

Let s and t be two strings on alphabet Σ. Substring w is a *maximal unique match* (MUM) between s and t iff $\mathcal{L}_s(w) = \{i\}$, $\mathcal{L}_t(w) = \{j\}$, $0 \le i < |s|$, $0 \le j < |t|$, and if $s[i-1] \ne t[j-1]$ and $s[i+|w|] \ne t[j+|w|]$. Current algorithms to detect maximal unique matches rely on LCP arrays (see e.g. [13] for the most space-efficient solution to date): using a bidirectional BWT implementation allows to reduce space.

Lemma 2. *Let s and t be two strings in Σ^+, and assume that Implementation 1a (respectively, 1b) has already been built for $s\$t$. We can discover all the τ maximal unique matches between s and t in $O((|s| + |t|) \log |\Sigma|)$ time and $(|s| + |t|) \log |\Sigma| + 3(|s| + |t|) + o(|s| + |t|) + O(\tau \log(|s| + |t|))$ bits of space (respectively, in $O(|s| + |t|)$ time and $(|s| + |t|) \log |\Sigma| + o((|s| + |t|) \log |\Sigma|) + O(\tau \log(|s| + |t|))$ bits of space).*

Proof. Let $u = s\$t$. The MUMs between s and t are precisely the nodes v of T_u with the following properties: (1) they have exactly two children; (2) one child correspond to a suffix that starts before $ in u; the other child corresponds to a suffix that starts after $ in u; (3) $a\ell(v)$ occurs in s, $b\ell(v)$ occurs in t, $a, b \in \Sigma$, and $a \neq b$. To determine MUMs it thus suffices to build a bit vector which of size $|u|$ such that $\text{which}[i] = 1$ iff the ith suffix of u in lexicographic order starts at position $|s| + 2$ or larger in u. We follow Lemma 1 to keep the depth of the navigation stack bounded by $O(\log|u|)$ and to report the positions of all MUMs in s.

This algorithm can be easily adapted to discover the shortest unique substrings of a single string s. As mentioned above, Implementation 1a is easy to construct: the following corollary ensues.

Corollary 1. *We can discover all the τ maximal unique matches between $s \in \Sigma^+$ and $t \in \Sigma^+$ in $O((|s| + |t|) \log|\Sigma|)$ time and $O((|s| + |t|) \log|\Sigma| + \tau \log(|s| + |t|))$ bits of space.*

If we have already indexed two strings s and t separately, we can compute maximal unique matches by traversing L_s and L_t synchronously:

Lemma 3. *Let s and t be two strings in Σ^+, and assume that Implementation 2a (respectively, 2b) has already been computed for s and t. We can discover all the τ maximal unique matches between s and t in $O((|s| + |t|) \log|\Sigma|)$ time and $2(|s| + |t|) \log|\Sigma| + (|s| + |t|) + o(|s| + |t|) + O(\tau \log(|s| + |t|))$ bits of space (respectively, in $O(|s| + |t|)$ time and $2(|s| + |t|) \log|\Sigma| + o((|s| + |t|) \log|\Sigma|) + O(\tau \log(|s| + |t|))$ bits of space).*

Proof. Again, let $u = s\$t$. Traversing L_u can be simulated by synchronizing the top-down traversal of L_s and L_t, as follows. Assume to be at node u_s in L_s and at node u_t in L_t, and let $(u_s, v_s) \in W_s \cup W'_s$ and $(u_t, v_t) \in W_t \cup W'_t$ be (explicit or implicit) Weiner links with label a in L_s and L_t, respectively. If $v_s \in V_s$ or $v_t \in V_t$, then $\ell(v_s)$ corresponds to a node in L_u. Otherwise, $\ell(v_s)$ is a node in L_u iff substrings $\ell(v_s)a$ of s and $\ell(v_s)b$ of t, $a, b \in \Sigma$, are such that $a \neq b$. To detect maximal unique matches, it suffices to check that $\ell(v_s)$ occurs once in s and once in t, and to retrieve the symbols that precede it in s and t. Both operations take constant time in Representation 2. To bound the depth of the navigation stack by $O(\log(|s| + |t|))$ it suffices to ensure that the largest subtree in L_s or L_t is always explored last.

Note that in this application we need `enumerateRight` just for intervals of $BWT_{\bar{s}}$ that do not correspond to nodes of T_s: we could thus use Implementation 1a or 1b with $\text{runs}_{\bar{s}}$ replaced by a plain encoding of $BWT_{\bar{s}}$ as a string of $|s| \log|\Sigma|$ bits. The proof of Lemma 3 immediately generalizes to a set of m strings:

Corollary 2. *Let $S = \{s_0, s_1, \ldots, s_{m-1}\} \subset \Sigma^+$ be a set of strings, and let $||S|| = \sum_{i=0}^{m-1} |s_i|$. Assume that Implementation 2a (respectively, 2b) has already been computed for all strings in S. We can discover all the τ maximal unique matches in S in $O(m||S|| \log|\Sigma|)$ time and $2||S|| \log|\Sigma| + ||S|| +$*

$o(||S||) + O(m\tau \log ||S||)$ *bits of space (respectively, in* $O(m||S||)$ *time and* $2||S|| \log |\Sigma| + o(||S|| \log |\Sigma|) + O(m\tau \log ||S||)$ *bits of space).*

A *maximal exact match* between strings s and t (MEM, called also *maximal pair* if $s = t$) is a quadruple (i_s, j_s, i_t, j_t) such that $s[i_s, j_s] = t[i_t, j_t]$, $s[i_s-1] \neq t[i_t-1]$ and $s[j_s + 1] \neq t[j_t + 1]$ [11, 19]. The computation of MEMs maps naturally to a navigation of the suffix-link tree of $s\$t$, and can thus be implemented with the bidirectional BWT:

Lemma 4. *Let s and t be strings on alphabet Σ, and assume that Implementation 2a or 2b has already been built for s and t. We can compute all the τ maximal exact matches between s and t in $O((|s| + |t|) \log |\Sigma| + \tau)$ time and $O((|s| + |t|) \log |\Sigma| + \tau \log(|s| + |t|))$ space.*

Proof. Again, we traverse L_u top-down, where $u = s\$t$. Assume we are at a node v of L_u with label $\ell(v)$. Build sets $P_s(v) = \{(i_{a\ell(v)b}, j_{a\ell(v)b})_s : a, b \in \Sigma, \mathcal{L}_s(a\ell(v)b) \neq \emptyset\}$ and $P_t(v) = \{(i_{c\ell(v)d}, j_{c\ell(v)d})_t : c, d \in \Sigma, \mathcal{L}_t(c\ell(v)d) \neq \emptyset\}$. Note that $\sum_{v \in V_u} |P_s(v)| + |P_t(v)| \in O(|s| + |t|)$, where V_u is the set of nodes in L_u: indeed, finding $P_s(v)$ and $P_t(v)$ coincides with exploring all explicit and implicit Weiner links $e = (v, w)$ in L_u that start from v, and then exploring the children of w in T_u if e is explicit. Linearity ensues from the folk theorem that the total number of implicit and explicit Weiner links in T_u is $O(|u|)$, and that each destination of an explicit Weiner link is explored exactly once during the traversal of L_u. We then compute $P_s(v) \otimes P_t(v) = \{((i_{a\ell(v)b}, j_{a\ell(v)b})_s, (i_{c\ell(v)d}, j_{c\ell(v)d})_t) : a \neq c, b \neq d\}$ in time linear in $|P_s(v)| + |P_t(v)|$ and in the size of the output, using a simple algorithm based on pairs of symbols that differ in both components (we omit details due to lack of space). Finally, we need to map every quadruple $(i, j, i', j') \in P_s(v) \otimes P_t(v)$ into $(j - i + 1)(j' - i' + 1)$ pairs of positions in s and t, for every $v \in L_s$. To do so, we build all pairs $(x, y) : x \in [i, j], y \in [i', j']$ of corresponding positions in BWT_s and BWT_t for all nodes of L_s, and we proceed as in Lemma 1: this takes $O((|s| + |t|) \log |\Sigma| + \tau)$ time overall.

String $w = axb$ is a *minimal absent word* of a string s, where $a, b \in \Sigma$ and $x \in \Sigma^*$, if both ax and xb occur in s, but axb does not occur in s [20]. For example, agaga is a minimal absent word in Fig. 1. Clearly only a maximal repeat of s can be the infix x of a minimal absent word axb, thus the navigation of L_s described in Lemma 4 allows to compute all the τ minimal absent words of s in $O(|s| \log |\Sigma| + \tau)$ time and $2|s| \log |\Sigma| + O(\tau \log |s|) + o(|s|)$ space, or in $O(|s| + \tau)$ time and $2|s| \log |\Sigma| + O(\tau \log |s|) + o(|s| \log |\Sigma|)$ space.

5.2 Borders and Surprising Substrings

Let $\mathcal{B}(s) \subset \Sigma^*$ be the set of nonempty borders of a string $s \in \Sigma^+$, and let $\text{bord}(s)$ be its longest border. We consider the problem of computing $|\text{bord}(w)|$ for all w in $S \cup S''$, where $S = \{\ell(v) \mid v \in V_s\}$ and $S'' = \{\ell(v)a \mid a \in \Sigma, v \in V_s, \mathcal{L}_s(\ell(v)a) \neq \emptyset\}$. In Fig. 1b, gray lines are pointers from strings in $S \cup S''$ to their longest border. Due to space constraints we omit the details of computing borders, and we just summarize the main result:

Lemma 5. *Assume that Implementation 1b has already been built for a string $s \in \Sigma^+$. We can compute $|\mathtt{bord}(w)|$ for all $w \in S \cup S''$ in overall $O(|\Sigma|\lambda \log |s|) + |s| \log |\Sigma| + o(|s| \log |\Sigma|)$ bits of space and in $O(|s|)$ time, independent of $|\Sigma|$, where $\lambda = \max\{|w| : w \in S''\}$.*

Similar lemmas hold for computing the *quasiperiod* of all strings in S [7], as well as the inner product and norm of the composition vectors of all substrings (or k-mers) of two strings s and t. Determining $|\mathtt{bord}(w)|$ for all $w \in S \cup S''$ is not just a combinatorial exercise. Consider a measure of statistical surprise $f(w)$ that scores a substring w of s with a function of $|\mathcal{L}_s(w)|$, $\mathbb{E}(w)$ and $\mathbb{V}(w)$ – respectively the expectation and variance of $|\mathcal{L}_r(w)|$ for a random string r of length $|s|$ generated by a known IID source. We call *f-cover* a set of substrings of s with the following property: for every substring v not in the cover, there is a substring w in the cover with $\mathcal{L}_s(w) = \mathcal{L}_s(v)$, $|w| > |v|$, and $f(w) \geq f(v)$ (respectively, $|w| < |v|$ and $f(w) \leq f(v)$). A large class of measures of statistical surprise is monotonic inside edges of T_s, making S (respectively, S'') an f-cover [2]. Moreover, $\mathbb{E}(w)$ and $\mathbb{V}(w)$ for all $w \in S \cup S''$ can be computed in constant time per node v of T_s using a depth-first traversal of L_s, and keeping just pointer $\mathtt{bord}_v$ and a constant amount of numerical variables in each v [3]. The following corollary is thus immediate:

Corollary 3. *Assume that Implementation 1b has already been computed for a string $s \in \Sigma^+$, and assume that $\mathbb{E}(w)$ and $\mathbb{V}(w)$ can be represented in $O(\log |s|)$ bits for any substring w of s. We can compute an f-cover of s in overall $O(|\Sigma|\lambda \log |s|) + |s| \log |\Sigma| + o(|s| \log |\Sigma|)$ bits of space and in $O(|s|)$ time, independent of $|\Sigma|$, where $\lambda = \max\{|w| : w \in B_s\}$.*

References

1. Apostolico, A.: The myriad virtues of subword trees. Technical Report 85–540, Department of Computer Science, Purdue University (1985)
2. Apostolico, A., Bock, M.E., Lonardi, S.: Monotony of surprise and large-scale quest for unusual words. In: RECOMB 2002, pp. 22–31 (2002)
3. Apostolico, A., Bock, M.E., Lonardi, S., Xu, X.: Efficient detection of unusual words. J. Comput. Biol. 7(1-2), 71–94 (2000)
4. Belazzougui, D., Boldi, P., Pagh, R., Vigna, S.: Monotone minimal perfect hashing: searching a sorted table with $o(1)$ accesses. In: SODA 2009, pp. 785–794 (2009)
5. Belazzougui, D., Navarro, G.: Alphabet-independent compressed text indexing. ACM Trans. Alg. (to appear, 2013)
6. Beller, T., Berger, K., Ohlebusch, E.: Space-efficient computation of maximal and supermaximal repeats in genome sequences. In: Calderón-Benavides, L., González-Caro, C., Chávez, E., Ziviani, N. (eds.) SPIRE 2012. LNCS, vol. 7608, pp. 99–110. Springer, Heidelberg (2012)
7. Breslauer, D.: An on-line string superprimitivity test. Inform. Process. Lett. 44(6), 345–347 (1992)
8. Ferragina, P., Manzini, G., Mäkinen, V., Navarro, G.: Compressed representations of sequences and full-text indexes. ACM T. Alg. 3(2), 20 (2007)

9. Fischer, J.: Optimal succinctness for range minimum queries. In: López-Ortiz, A. (ed.) LATIN 2010. LNCS, vol. 6034, pp. 158–169. Springer, Heidelberg (2010)

10. Fischer, J., Mäkinen, V., Navarro, G.: Faster entropy-bounded compressed suffix trees. Theor. Comput. Sci. 410(51), 5354–5364 (2009)

11. Gusfield, D.: Algorithms on strings, trees and sequences: computer science and computational biology. Cambridge University Press (1997)

12. Hoare, C.A.R.: Quicksort. The Computer Journal 5(1), 10–16 (1962)

13. Hon, W.-K., Sadakane, K.: Space-economical algorithms for finding maximal unique matches. In: Apostolico, A., Takeda, M. (eds.) CPM 2002. LNCS, vol. 2373, pp. 144–152. Springer, Heidelberg (2002)

14. Hon, W.-K., Sadakane, K., Sung, W.-K.: Breaking a time-and-space barrier in constructing full-text indices. SIAM J. Comput. 38(6), 2162–2178 (2009)

15. Kulekci, O., Vitter, J.S., Xu, B.: Efficient maximal repeat finding using the Burrows-Wheeler transform and wavelet tree. TCBB 9(2), 421–429 (2012)

16. Lam, T.W., Li, R., Tam, A., Wong, S., Wu, E., Yiu, S.: High throughput short read alignment via bi-directional BWT. In: BIBM 2009, pp. 31–36 (2009)

17. Li, H., Homer, N.: A survey of sequence alignment algorithms for next-generation sequencing. Brief. Bioinform. 11(5), 473–483 (2010)

18. Li, R., Yu, C., Li, Y., Lam, T.W., Yiu, S.-M., Kristiansen, K., Wang, J.: Soap2: An improved ultrafast tool for short read alignment. Bioinformatics 25(15), 1966–1967 (2009)

19. Ohlebusch, E., Gog, S., Kügel, A.: Computing matching statistics and maximal exact matches on compressed full-text indexes. In: Chavez, E., Lonardi, S. (eds.) SPIRE 2010. LNCS, vol. 6393, pp. 347–358. Springer, Heidelberg (2010)

20. Pinho, A.J., Ferreira, P.J.S.G., Garcia, S.P., Rodrigues, J.M.O.S.: On finding minimal absent words. BMC Bioinformatics 10(1), 137 (2009)

21. Raman, R., Raman, V., Satti, S.R.: Succinct indexable dictionaries with applications to encoding k-ary trees, prefix sums and multisets. ACM T. Alg. 3(4) (2007)

22. Russo, L.M.S., Navarro, G., Oliveira, A.L.: Dynamic fully-compressed suffix trees. In: Ferragina, P., Landau, G.M. (eds.) CPM 2008. LNCS, vol. 5029, pp. 191–203. Springer, Heidelberg (2008)

23. Russo, L.M.S., Navarro, G., Oliveira, A.L.: Fully compressed suffix trees. ACM Trans. Alg. 7(4), 53 (2011)

24. Sadakane, K.: Compressed suffix trees with full functionality. Theor. Comput. Syst. 41(4), 589–607 (2007)

25. Sadakane, K.: Succinct data structures for flexible text retrieval systems. J. Discrete Alg. 5(1), 12–22 (2007)

26. Sadakane, K., Navarro, G.: Fully-functional succinct trees. In: SODA 2010, pp. 134–149 (2010)

27. Schnattinger, T., Ohlebusch, E., Gog, S.: Bidirectional search in a string with wavelet trees. In: Amir, A., Parida, L. (eds.) CPM 2010. LNCS, vol. 6129, pp. 40–50. Springer, Heidelberg (2010)

28. Schnattinger, T., Ohlebusch, E., Gog, S.: Bidirectional search in a string with wavelet trees and bidirectional matching statistics. Inform. Comput. 213, 13–22 (2012)

Tight Lower and Upper Bounds for the Complexity of Canonical Colour Refinement

Christoph Berkholz[1], Paul Bonsma[2], and Martin Grohe[1]

[1] RWTH Aachen University, Germany
{berkholz,grohe}@informatik.rwth-aachen.de
[2] University of Twente, The Netherlands
p.s.bonsma@ewi.utwente.nl

Abstract. An assignment of colours to the vertices of a graph is *stable* if any two vertices of the same colour have identically coloured neighbourhoods. The goal of *colour refinement* is to find a stable colouring that uses a minimum number of colours. This is a widely used subroutine for graph isomorphism testing algorithms, since any automorphism needs to be colour preserving. We give an $O((m+n)\log n)$ algorithm for finding a *canonical* version of such a stable colouring, on graphs with n vertices and m edges. We show that no faster algorithm is possible, under some modest assumptions about the type of algorithm, which captures all known colour refinement algorithms.

1 Introduction

Colour refinement (also known as *naive vertex classification*) is a very simple, yet extremely useful algorithmic routine for graph isomorphism testing. It classifies the vertices by iteratively refining a colouring of the vertices as follows. Initially, all vertices have the same colour. Then in each step of the iteration, two vertices that currently have the same colour get different colours if for some colour c they have a different number of neighbours of colour c. The process stops if no further refinement is achieved, resulting in a *stable colouring* of the graph. To use colour refinement as an isomorphism test, we can run it on the disjoint union of two graphs. Any isomorphism needs to map vertices to vertices of the same colour. So, if the stable colouring differs on the two graphs, that is, if for some colour c, the graphs have a different number of vertices of colour c, then we know they are nonisomorphic, and we say that colour refinement *distinguishes* the two graphs. Babai, Erdös, and Selkow [2] showed that colour refinement distinguishes almost all graphs (in the $G(n, 1/2)$ model). In fact, they proved the stronger statement that the stable colouring is discrete on almost all graphs, that is, every vertex gets its own colour. On the other hand, colour refinement fails to distinguish any two regular graphs with the same number of vertices, such as a 6-cycle and the disjoint union of two triangles.

Colour refinement is not only useful as a simple isomorphism test in itself, but also as a subroutine for more sophisticated algorithms, both in theory and practice. For example, Babai and Luks's [1,3] $O(2^{\sqrt{n \log n}})$-algorithm — this is still

H.L. Bodlaender and G.F. Italiano (Eds.): ESA 2013, LNCS 8125, pp. 145–156, 2013.
© Springer-Verlag Berlin Heidelberg 2013

the best known worst-case running time for isomorphism testing — uses colour refinement as a subroutine, and most practical graph isomorphism tools (for example, [9,6,8,12]), starting with McKay's "Nauty" [9,10], are based on the *individualisation refinement* paradigm. The basic idea of these algorithms is to recursively compute a *canonical labelling* of a given graph, which may already have an initial colouring of its vertices, as follows. We run colour refinement starting from the initial colouring until a stable colouring is reached. If the stable colouring is discrete, then this already gives us a canonical labelling (provided the colours assigned by colour refinement are canonical, see below). If not, we pick some colour c with more than one vertex. Then for each vertex v of colour c, we modify the stable colouring by assigning a fresh colour to v (that is, we "individualise" v) and recursively call the algorithm on the resulting vertex-coloured graph. Then for each v we get a canonically labelled version of our graph, and we return the lexicographically smallest among these. (More precisely, each canonical labelling of a graph yields a canonical string encoding, and we compare these strings lexicographically.) To turn this simple procedure into a practically useful algorithm, various heuristics are applied to prune the search tree. They exploit automorphisms of the graph found during the search. However, crucial for any implementation of such an algorithm is a very efficient colour refinement procedure, because colour refinement is called at every node of the search tree.

Colour refinement can be implemented to run in time $O((n+m)\log n)$, where n is the number of vertices and m the number of edges of the input graph. To our knowledge, this was first been proved by Cardon and Crochemore [5]. Later Paige and Tarjan [11, p.982] sketched a simpler algorithm. Both algorithms are based on the partitioning techniques introduced by Hopcroft [7] for minimising finite automata. However, an issue that is completely neglected in the literature is that, at least for individualisation refinement, we need a version of colour refinement that produces a *canonical colouring*. That is, if f is an isomorphism from a graph G to a graph H, then for all vertices v of G, v and $f(v)$ should get the same colour in the respective stable colourings of G and H. However, neither of the algorithms analysed in the literature seem to produce canonical colourings. Very briefly, the reason is that these algorithms use bucketing techniques for indexing vectors with an initial segment of the natural numbers that make sure that different vectors get different indices, but do not assign indices in the lexicographical or any other canonical order. This issue can be resolved by sorting the vectors lexicographically, but this causes a logarithmic overhead in the running time. We resolve the issue differently and avoid the logarithmic overhead, thus obtaining an implementation of colour refinement that computes a canonical stable colouring in time $O((n+m)\log n)$. Ignoring the canonical part, our algorithmic techniques are similar to known results: like [11] and various other papers, we use Hopcroft's strategy of 'ignoring the largest new cell', after splitting a cell [7]. Our data structures are similar to those described by Junttila and Kaski [8]. Nevertheless, since [8] contains no complexity analysis, and [11] omits various (nontrivial) implementation details, it seems that the current paper gives the first detailed description of an $O((m+n)\log n)$ algorithm that uses this strategy. On a high level, our algorithm is also quite similar to McKay's *canonical* colour refinement algorithm [9, Alg. 2.5], but with a few key differences which enable an $O((n + m) \log n)$ implementation. McKay [9] gave an $O(n^2 \log n)$ implementation using adjacency matrices.

Now the question arises whether colour refinement can be implemented in linear time. After various attempts, we started to believe that it cannot. Of course with currently known techniques one cannot expect to disprove the existence of a linear time algorithm for the standard (RAM) computation model, or for similar general computation models. Instead, we prove the complexity lower bound for a broad class of partition-refinement based algorithms, which captures all known colour refinement algorithms, and actually every reasonable algorithmic strategy we could think of. This model can alternatively be viewed as a "proof system". We use the following assumptions. (See Sections 2 and 4 for precise definitions.) Colour refinement algorithms start with a unit partition (which has one *cell* $V(G)$), and iteratively refine this until a stable colouring is obtained. This is done using *refining operations*: choose a union of current partition cells as *refining set* R, and choose another (possibly overlapping) union of partition cells S. Cells in S are split up if their neighbourhoods in R provide a reason for this. (That is, two vertices in a cell in S remain in the same cell only if they have the same number of neighbours in every cell in R.) This operation requires considering all edges between R and S, so the number of such edges is a very reasonable and modest lower bound for the complexity of such a refining step; we call this the *cost* of the operation. We note that a naive algorithm might choose $R = S = V(G)$ in every iteration. This then requires time $\Omega(mn)$ on graphs that require a linear number of refining operations, such as paths. Therefore, all fast algorithms are based on choosing R and S smartly (and on implementing refining steps efficiently).

For our main lower bound result, we construct a class of instances such that any possible sequence of refining operations that yields the stable partition has total cost at least $\Omega((m + n) \log n)$. Note that it is surprising that a tight lower bound can be obtained in this model. Indeed, cost *upper* bounds in this model would not necessarily yield corresponding algorithms, since firstly we allow the sets R and S to be chosen nondeterministically, and secondly, it is not even clear how to refine S using R in time proportional to the number of edges between these classes. However, as we prove a lower bound, this makes our result only stronger. We formulate the lower bound result for undirected graphs and non-canonical colour refinement, so that it also holds for digraphs, and canonical colour refinement. Our proof also implies a corresponding lower bound for the coarsest relational partitioning problem considered by Paige and Tarjan [11]. Because of space constraints, some details have been omitted.

2   Preliminaries

For an undirected (simple) graph G, $N(v)$ denotes the set of neighbours of $v \in V(G)$, and $d(v) = |N(v)|$ its degree. For a digraph, $N^+(v)$ and $N^-(v)$ denote the out- and in-neighbourhoods, and $d^+(v) = |N^+(v)|$ resp. $d^-(v) = |N^-(v)|$

the out- and in-degree, respectively. A partition π of a set V is a set $\{S_1, \ldots, S_k\}$ of pairwise disjoint nonempty subsets of V, such that $\cup_{i=1}^{k} S_i = V$. The sets S_i are called *cells* of π. The *order* of π is the number of cells $|\pi|$. A partition π is *discrete* if every cell has size 1, and *unit* if it has exactly one cell. Given a partition π of V, and two elements $u, v \in V$, we write $u \approx_\pi v$ if and only if there exists a cell $S \in \pi$ with $u, v \in S$. We say that a set $V' \subseteq V$ is π-*closed* if it is the union of a number of cells of π. In other words, if $u \approx_\pi v$ and $u \in V'$ then $v \in V'$. For any subset $V' \subseteq V$, π *induces* a partition $\pi[V']$ on V', which is defined by $u \approx_{\pi[V']} v$ if and only if $u \approx_\pi v$, for all $u, v \in V'$.

Let $G = (V, E)$ be a graph. A partition π of V is *stable* for G if for every pair of vertices $u, v \in V$ with $u \approx_\pi v$ and $R \in \pi$, it holds that $|N(u) \cap R| = |N(v) \cap R|$. If G is a digraph, then $|N^+(u) \cap R| = |N^+(v) \cap R|$ should hold. For readability, all further definitions and propositions in this section are formulated for (undirected) graphs, but the corresponding statements also hold for digraphs (replace degrees/neighbourhoods by out-degrees/out-neighbourhoods). One can see that if π is stable and $d(u) \neq d(v)$, then $u \not\approx_\pi v$, which we will use throughout.

A partition ρ of V *refines* a partition π of a subset S of V if for every $u, v \in S$, $u \approx_\rho v$ implies $u \approx_\pi v$. (Usually we take $S = V$.) If ρ refines π, we write $\pi \preceq \rho$. If in addition $\rho \neq \pi$, then we also write $\pi \prec \rho$. Note that $\preceq$ is a partial order on all partitions of V.

Definition 1. *Let G be a graph, and let π and π' be partitions of $V(G)$. For vertex sets $R, S \subseteq V(G)$ that are π-closed, we say that π' is obtained from π by a refining operation (R, S) if*

- *for every $S' \in \pi$ with $S' \cap S = \emptyset$, it holds that $S' \in \pi'$, and*
- *for every $u, v \in S$: $u \approx_{\pi'} v$ if and only if $u \approx_\pi v$ and for all $R' \in \pi$ with $R' \subseteq R$, $|N(u) \cap R'| = |N(v) \cap R'|$ holds.*

Note that if π' is obtained from π by a refining operation (R, S), then $\pi \preceq \pi'$. We say that the operation (R, S) is *effective* if $\pi \prec \pi'$. In this case, at least one cell $C \in \pi$ is *split*, which means that $C \notin \pi'$. Note that an effective refining operation exists for π if and only if π is unstable. In addition, the next proposition says that if the goal is to obtain a (coarsest) stable partition, then applying any refining operation is safe.

Proposition 2 (*). [1] *Let π' be obtained from π by a refining operation (R, S). If ρ is a stable partition with $\pi \preceq \rho$, then $\pi \preceq \pi' \preceq \rho$.*

A partition π is a *coarsest* partition for a property P if π satisfies P, and there is no partition ρ with $\rho \prec \pi$ that also satisfies property P.

Proposition 3 (*). *Let $G = (V, E)$ be a graph. For every partition π of V, there is a unique coarsest stable partition ρ that refines π.*

---

[1] In the full version, (detailed) proofs are given for statements marked with a star.

3   A Fast Canonical Color Refinement Algorithm

Consider a method for obtaining a sequence $S_1^G, \ldots, S_{k(G)}^G$ of subsets of $V(G)$, for any (di)graph G. This method is called *canonical* if for any two isomorphic (di)graphs G and G', and any isomorphism $h : V(G) \rightarrow V(G')$, it holds that $k(G) = k(G')$, and $u \in S_i^G$ implies $h(u) \in S_i^{G'}$, for any $u \in V(G)$ and $i \in \{1, \ldots, k(G)\}$. In a slight abuse of terminology, we also call the sequence canonical, if the method for obtaining it is clear from the context. For instance, for simple undirected graphs G, if we define D_d to be the set of vertices of degree d, for $d \in \{0, \ldots, n-1\}$, $n = |V(G)|$, then $D_0, \ldots, D_{n-1}$ is a canonical sequence, because every isomorphism maps vertices to vertices of the same degree. (In other words: degrees are *isomorphism invariant*.) In this section we give a fast algorithm for obtaining a canonical coarsest stable partition of $V(G)$, for any digraph G. This is an *ordered partition* of V, which is a sequence of sets $C_1, \ldots, C_k$ such that $\{C_1, \ldots, C_k\}$ is a partition of V. To obtain the most general result, we formulate the algorithm for digraphs.

High-level Description and Correctness Proofs The input to our algorithm is a digraph $G = (V, E)$, with $V = \{1, \ldots, n\}$. For every vertex $v \in V$, the sets of out-neighbours $N^+(v)$ and in-neighbours $N^-(v)$ are given. Throughout, the algorithm maintains an ordered partition $\pi = C_1, \ldots, C_k$ of V, starting with the unit partition. This partition is iteratively refined using operations of the form (R, V), where $R = C_r$ for some $r \in \{1, \ldots, k\}$. We will show that when the algorithm terminates, no effective refining operations are possible on the resulting partition. So the resulting partition is the unique coarsest stable partition of G.

We now explain how to maintain a canonical order for the partition $\pi = C_1, \ldots, C_k$. To this end, indices $i \in \{1, \ldots, k\}$ are called *colours*, and the cells C_i are also called *colour classes* of the current partition. The partition π is then also viewed as a *colouring* of the vertices with colours $1, \ldots, k$. To canonically choose the next *refining colour* r, we maintain a canonical sequence (stack) S_{refine} of colours that should still be used as refining colour. When a new refining colour r should be chosen, we select r to be the last colour added to S_{refine} (i.e. r is popped from the stack). For a given refining colour class $R = C_r$ and any $x \in V$, call $d_r^+(x) := |N^+(x) \cap R|$ the *colour degree* of x. Then every colour $s \in \{1, \ldots, k\}$ will be split up according to colour degrees. More precisely, for a given refining colour r, we partition every cell C_s into new cells $C_{\sigma_1}, \ldots, C_{\sigma_p}$, such that for $x \in C_{\sigma_i}$ and $y \in C_{\sigma_j}$: (i) $i = j$ if and only if $d_r^+(x) = d_r^+(y)$, and (ii) if $i < j$ then $d_r^+(x) < d_r^+(y)$. In other words, the new colours are ordered canonically according to their colour degrees. Since we wish to have nonempty sets in our partition, we choose $\sigma_1 = s$, and $\sigma_i = k + i - 1$ for all $2 \leq i \leq p$, and then update the number of colours k. To obtain a canonical colouring, it is also important to split up the colours $s \in \{1, \ldots, k\}$ in increasing order.

It remains to explain how newly introduced colours are added to the stack S_{refine} in a canonical way. Initially, S_{refine} contains colour 1, and whenever new colours are introduced during the splitting of a colour class C_s, these are pushed onto the stack S_{refine}, in increasing order. There is however one exception: if we have already used the vertex set $S = C_s$ as refining colour class before, and this set is split up into new colours $C_{\sigma_1}, \ldots, C_{\sigma_p}$, then it is not necessary to use all of these new colours as refining colours later. Indeed, for every $i \in \{1, \ldots, p\}$ and every $x, y \in V(G)$, if $|N^+(x) \cap S| = |N^+(y) \cap S|$ and $|N^+(x) \cap C_j| = |N^+(y) \cap C_j|$ holds for every $j \neq i$, then it also holds that $|N^+(x) \cap C_i| = |N^+(y) \cap C_i|$, since $\{C_{\sigma_1}, \ldots, C_{\sigma_p}\}$ is a partition of S. Hence we may select an $i \in \{1, \ldots, p\}$, and only add the colours $\{1, \ldots, p\} \setminus \{i\}$ to the stack S_{refine}. To obtain a good complexity, we choose i such that $|C_{\sigma_i}|$ is maximised, in order to minimise the sizes of the refining colour sets used later during the computation. (This is Hopcroft's trick [7].) To be precise, for a given s, we canonically choose b to be the minimum colour degree that maximises $|\{x \in C_s \mid d_r^+(x) = b\}|$, and add all newly introduced colours to the stack, in increasing order, except the new colour that corresponds to b. On the other hand, if s was already on the stack S_{refine}, then this argument does not apply, so we have to add every new colour to the stack. The algorithm terminates when the stack S_{refine} is empty, and returns the final ordered partition $C_1, \ldots, C_k$.

Lemma 4. *For any digraph G, the above algorithm computes a canonical sequence $C_1, \ldots, C_k$, such that $\{C_1, \ldots, C_k\}$ is the coarsest stable partition of G.*

Proof sketch: The resulting partition $\pi = \{C_1, \ldots, C_k\}$ is refined by the coarsest stable partition ω of G because it is obtained from the unit partition by using refining operations (Proposition 2). It is then equal to ω since it is stable. This follows since using the argument given above, one can verify that the following invariant is maintained: if there exist colour classes C_r and C_s such that the refining operation (C_r, C_s) is effective, then the stack S_{refine} contains a colour r' such that the refining operation $(C_{r'}, C_s)$ is effective. Since the stack S_{refine} is empty when the algorithm terminates, stability follows. The final sequence is canonical since at every point during the computation, both the stack S_{refine} and the current ordered partition $C_1, \ldots, C_k$ are canonical sequences. This holds because informally, the new colours that we assign to vertices, and the order in which new colours are added to the stack, are completely determined by isomorphism-invariant values such as colour degrees with respect to sets from a canonical sequence.   □

Implementation and Complexity Bound. We now describe a fast implementation of the aforementioned algorithm. The main idea of the complexity proof is the following: one *iteration* consists of popping a refining colour r from the stack S_{refine}, and applying the refining operation (R, V), with $R = C_r$. Below we show that one such iteration takes time $O(|R| + D^-(R) + k_i \log k_i)$, where $D^-(R) = \sum_{v \in R} d^-(v)$ and k_i is the number of new colours that are introduced during iteration i. Next, we observe that for every vertex $v \in V(G)$, if $R_1^v, \ldots, R_q^v$ are the refining colour classes C_r with $v \in C_r$ that are considered throughout the computation, in chronological order, then for all $i \in \{1, \ldots, q-1\}$, $|R_i^v| \geq 2|R_{i+1}^v|$ holds. This holds because whenever a set $S = C_s$ is split up into $C_{\sigma_1}, \ldots, C_{\sigma_p}$, where S has been considered earlier as a refining colour (so it is not in S_{refine} anymore), then for all new colours σ_i that are added to the stack

S_{refine}, $|C_{\sigma_i}| \leq \frac{1}{2}|S|$ holds (since the largest colour class is not added to S_{refine}). Note that if a colour class C_{σ_i} is subsequently split up before it is considered as refining colour, the bound of course also holds. It follows that every $v \in V(G)$ appears at most $\log_2 n$ times in a refining colour class. Then we can write

$$\sum_R |R| + D^-(R) \leq \sum_{v \in V(G)} (1 + d^-(v)) \log_2 n = (n + m) \log_2 n,$$

with $m = |E(G)|$, where the first summation is over all refining colour classes $R = C_r$ considered during the computation. In addition, the total number of new colours that is introduced is at most n, since every colour class, after it is introduced, remains nonempty throughout the computation. So we may write $\sum_i k_i \log k_i \leq \sum_i k_i \log n \leq n \log n$. Combining these facts shows that the total complexity of the algorithm can be bounded by $O((n+m)\log n) + O(n \log n) = O((n+m)\log n)$.

It remains to describe an implementation such that the complexity of one iteration i of the while-loop, where refining colour class $R = C_r$ is considered, can be bounded by $O(|R| + D^-(R) + k_i \log k_i)$. The colour classes C_i are represented by doubly linked lists. For all lists, we maintain the length.

The first challenge is how to compute the colour degrees $d_r^+(v)$ efficiently for every $v \in V(G)$, with respect to the refining colour r. For this we use an array cdeg[v], indexed by $v \in \{1, \ldots, n\}$. We use the following invariant: at the beginning of every iteration, cdeg[v] = 0 for all v. Then we can compute these colour degrees by looping over all in-neighbours w of all vertices $v \in R$, and increasing cdeg[w]. At the same time, we compute a list C_{adj} of colours i that contain at least one vertex $w \in C_i$ with cdeg[w] ≥ 1, and for every such colour i, we compute a list A_i of all vertices w with cdeg[w] ≥ 1. None of these lists contain duplicates. This can all be done in time $O(|R| + D^-(R))$, assuming that at the beginning of every iteration, every A_i is an empty list, C_{adj} is an empty list, and flags are maintained for colours to keep track of membership in C_{adj}. With the same complexity, we can reset all of these data structures at the end of every iteration.

Next, we address how we can consider all colours that split up in one iteration, in canonical (increasing) order. To this end, we compute a new list C_{split}, which represents the subset of C_{adj} containing all colours that actually split up. By ensuring that all colours in C_{split} split up, we have that $|C_{\text{split}}| \leq k_i$, and therefore we can afford to sort this list. This can be done using any list sorting algorithm of complexity $O(k_i \log k_i)$, such as merge sort. To compute which colours split up, we compute for every colour in $i \in C_{\text{adj}}$ the maximum colour degree maxcdeg[i] and minimum colour degree mincdeg[i]. The value maxcdeg[i] can easily be computed while computing the colour degrees. We have mincdeg[i] = 0 if $|A_i| < |C_i|$. Otherwise, we can afford to compute mincdeg[i] by iterating over $A_i = C_i$.

Finally, we need to show how a single colour class $S = C_s$ can be split up, and how the appropriate new colours can be added to the stack S_{refine} in the proper order, all in time $O(D_R^+(S))$. Here $R = C_r$ denotes the refining colour class, and

$D_R^+(S) = \sum_{v \in S} |N^+(v) \cap R|$. Note that when summing over all $s \in C_{\text{split}}$, this indeed gives a total complexity of at most $O(D^-(R))$. Firstly, for every relevant d, we compute how many vertices in C_s have colour degree d. These values are stored in an array numcdeg$[d]$, indexed by $d \in \{0, \ldots, \text{maxcdeg}[s]\}$. (Note that maxcdeg$[s] \leq D_R^+(S)$, so we can afford to initialise an array of this size.) Using this array numcdeg, we can easily compute the (minimum) colour degree b that occurs most often in S, which corresponds to the new colour that is possibly not added to S_{refine}. Using numcdeg, we can also easily construct an array f_{newcol}, indexed by $d \in \{0, \ldots, \text{maxcdeg}[s]\}$, which represents the mapping from colour degrees that occur in S to newly introduced colours, or to the current colour s. Finally, we can loop over A_s in time $O(D_R^+(S))$, and move all vertices $v \in A_s$ from C_s to C_i, where $i = f_{\text{newcol}}[\text{cdeg}[v]]$ is the new colour that corresponds to the colour degree of v. With a proper implementation using doubly linked lists, this can be done in constant time for a single vertex. Note that looping over A_s suffices, because if there are vertices in C_s with colour degree 0, then these keep the same colour, and thus do not need to be addressed. This fact is essential since the number of such vertices may not be bounded by $O(D_R^+(S))$.

This shows how the algorithm can be implemented such that one iteration takes time $O(|R| + D^-(R) + k_i \log k_i)$. Combined with the above analysis, this shows that the algorithm terminates in time $O((n+m) \log n)$. So with Lemma 4, we obtain:

Theorem 5. *For any digraph G on n vertices with m edges, in time $O((n + m) \log n)$ a canonical coarsest stable partition can be computed.*

In individualisation refinement algorithms, one branch is as follows [9,6,8,12,10]: whenever a stable but non-discrete colouring is obtained, some new vertex v is 'individualised' by assigning it a new unique colour, v is added to S_{refine}, and the process continues. Observe that the $O((n + m) \log n)$ bound holds for this entire process.

4   Complexity Lower Bound

The *cost* of a refining operation (R, S) is $\text{cost}(R, S) := |\{(u, v) \mid u \in R, v \in S\}|$. This is basically the number of edges between R and S, except that edges with both ends in $R \cap S$ are counted twice.

Definition 6. *Let $G = (V, E)$ be a graph, and π be a partition of V.*

- *If π is stable, then $\text{cost}(\pi) := 0$.*
- *Otherwise, $\text{cost}(\pi) := \min_{R,S} \text{cost}(\pi(R, S)) + \text{cost}(R, S)$, where the minimum is taken over all effective refining operations (R, S) that can be applied to π, and where $\pi(R, S)$ denotes the partition resulting from the operation (R, S).*

A refining operation (R, S) on π is *elementary* if both $R \in \pi$ and $S \in \pi$. The following observation is useful: since non-elementary refining steps can be split up into elementary refining steps with the same total cost, we may also take the minimum over all effective *elementary* refining operations.

We can now formulate the main result of this section.

Theorem 7. *For every integer $k \geq 2$, there is a graph G_k with $n \in O(2^k k)$ vertices and $m \in O(2^k k^2)$ edges, such that $cost(\alpha) \in \Omega((m + n) \log n)$, where α is the unit partition of $V(G_k)$.*

Note that this theorem implies a complexity lower bound for all partition-refinement based algorithms for colour refinement, as discussed in the introduction. We use the following key observation to prove the theorem. For a partition π of V, denote by π_∞ the coarsest stable partition of V that refines π.

Proposition 8 (*). *Let π and ρ be partitions of V such that $\pi \preceq \rho \preceq \pi_\infty$. Then $cost(\pi) \geq cost(\rho)$.*

We say that a partition π of V *distinguishes* two sets $V_1 \subseteq V$ and $V_2 \subseteq V$ if there is a set $R \in \pi$ with $|R \cap V_1| \neq |R \cap V_2|$. This is used often for the case where $V_1 = N(u)$ and $V_2 = N(v)$ for two vertices u and v, to conclude that if π is stable, then $u \not\approx_\pi v$. If $V_1 = \{x\}$ and $V_2 = \{y\}$, then we also say that π *distinguishes* x *from* y.

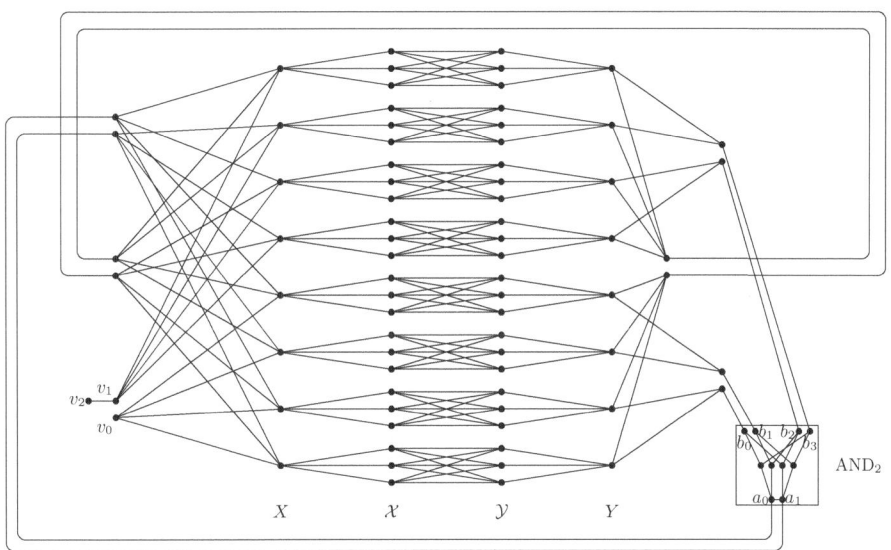

Fig. 1. The Graph G_3

Construction of the Graph For $k \in \mathbb{N}$, denote $\mathcal{B}_k = \{0, \ldots, 2^k - 1\}$. For $\ell \in \{0, \ldots, k\}$ and $q \in \{0, \ldots, 2^\ell - 1\}$, the subset $\mathcal{B}_q^\ell = \{q2^{k-\ell}, \ldots, (q+1)2^{k-\ell} - 1\}$ is called the *q-th binary block of level ℓ*. Analogously, for any set of vertices with indices in $\mathcal{B}_k$, we also consider binary blocks. For instance, if $X = \{x_i \mid i \in \mathcal{B}_k\}$, then $X_q^\ell = \{x_i \mid i \in \mathcal{B}_q^\ell\}$ is called a binary block of X. For such a set X, a partition π of X *into binary blocks* is a partition where every $S \in \pi$ is a binary block. A key fact for binary blocks that we will often use is that for any ℓ and q, $\mathcal{B}_q^\ell = \mathcal{B}_{2q}^{\ell+1} \cup \mathcal{B}_{2q+1}^{\ell+1}$.

For every integer $k \geq 2$, we will construct a graph G_k. In its core this graph consists of the vertex sets $X = \{x_i \mid i \in \mathcal{B}_k\}$, $\mathcal{X} = \{x_i^j \mid i \in \mathcal{B}_k, j \in [k]\}$, $\mathcal{Y} = \{y_i^j \mid i \in \mathcal{B}_k, j \in [k]\}$ and $Y = \{y_i \mid i \in \mathcal{B}_k\}$. Every vertex x_i is adjacent to x_i^j for all $j \in [k]$ and every y_i is adjacent to all y_i^j. Furthermore, for all i, j_1, j_2 there is an edge between $x_i^{j_1}$ and $y_i^{j_2}$. (For $\mathcal{X}$, *binary blocks* are subsets of the form $\mathcal{X}_q^\ell := \{x_i^j \mid i \in \mathcal{B}_q^\ell, j \in [k]\}$, and for $\mathcal{Y}$ the definition is analogous.)

We add gadgets to the graph to ensure that any sequence of refining operations behaves as follows. After the first step, which distinguishes vertices according to their degrees, X and Y are cells of the resulting partition. Next, X splits up into two binary blocks X_0^1 and X_1^1 of equal size. This causes $\mathcal{X}$ to split up accordingly into $\mathcal{X}_0^1$ and $\mathcal{X}_1^1$. One of these cells will be used to halve $\mathcal{Y}$ in the same way. This refining operation (R, S) is expensive because $[R, S]$ contains half of the edges between $\mathcal{X}$ and $\mathcal{Y}$. Next, Y can be split up into Y_0^1 and Y_1^1. Once this happens, there is a gadget AND_1 that causes the two cells X_0^1, X_1^1 to split up into the four cells X_q^2, for $q = 0, \ldots, 3$. Again, this causes cells in $\mathcal{X}, \mathcal{Y}$ and Y to split up in the same way and to achieve this, half of the edges between $\mathcal{X}$ and $\mathcal{Y}$ have to be considered. The next gadget AND_2 ensures that if both cells of Y are split, then the four cells of X can be halved again, etc. In general, we design a gadget AND_ℓ of level ℓ that ensures that if Y is partitioned into $2^{\ell+1}$ binary blocks of equal size, then X can be partitioned into $2^{\ell+2}$ binary blocks of equal size. By halving all the cells classes of X and Y $k = \Theta(\log n)$ times (with $n = |V(G_k)|$), this refinement process ends up with a discrete colouring of these vertices. Since every iteration uses half of the edges between $\mathcal{X}$ and $\mathcal{Y}$ (which are $\Theta(m)$), we get the cost lower bound of $\Omega(m \log n)$ (with $m = |E(G_k)|$).

We now define these gadgets in more detail. For every integer $\ell \geq 1$, we define a gadget AND_ℓ, which consists of a graph G together with two *out-terminals* a_0, a_1, and an ordered sequence of $p = 2^\ell$ in-terminals $b_0, \ldots, b_{p-1}$. For $\ell = 1$, the graph G has $V(G) = \{a_0, a_1, b_0, b_1\}$, and $E(G) = \{a_0 b_0, a_1 b_1\}$. For $\ell = 2$, the graph G is identical to the construction of Cai, Fürer and Immerman [4], but with an edge $a_0 a_1$ added (see Figure 1). The out-terminals a_0, a_1 and in-terminals $b_0, \ldots, b_3$ are indicated. For $\ell \geq 3$, AND_ℓ is obtained by taking one copy G^* of an AND_2-gadget, and two copies G' and G'' of an $\mathrm{AND}_{\ell-1}$-gadget, and adding four edges to connect the two pairs of in-terminals of G^* with the pairs of out-terminals of G' and G'', respectively. As out-terminals of the resulting gadget we choose the out-terminals of G^*. The in-terminal sequence is obtained by concatenating the sequences of in-terminals of G' and G''. For any AND_ℓ-gadget G with in-terminals $b_0, \ldots, b_{2^\ell-1}$, the *in-terminal pairs* are pairs b_{2p} and b_{2p+1}, for all $p \in \{0, \ldots, 2^{\ell-1} - 1\}$. We now state the key property for AND_ℓ-gadgets, which can be verified for $\ell = 2$, and then follows inductively for $\ell \geq 3$. We say that ρ *agrees with* ψ if $\rho[S] = \psi$.

Lemma 9 (*). *Let G be an AND_ℓ-gadget with in-terminals $B = \{b_0, \ldots, b_{2^\ell-1}\}$ and out-terminals a_0, a_1. For any partition ψ of B into binary blocks, the coarsest stable partition ρ of $V(G)$ that refines ψ agrees with ψ. Furthermore, ρ distinguishes a_0 from a_1 if and only if ψ distinguishes all in-terminal pairs.*

The graph G_k is now constructed as follows. Start with vertex sets $X, \mathcal{X}, \mathcal{Y}$ and Y, and edges between them, as defined above. For every $\ell \in \{1, \ldots, k-1\}$, we add a copy G of an AND_ℓ-gadget to the graph. Denote the in- and out-terminals of G by a_0, a_1 and $b_0, \ldots, b_{2^\ell-1}$, respectively.

- For $i = 0, 1$ and all relevant q: we add edges from a_i to every vertex in $\mathcal{X}^{\ell+1}_{2q+i}$.
- For every i, we add edges from b_i to every vertex in Y^ℓ_i.

Finally, we add a *starting gadget* to the graph, consisting of three vertices v_0, v_1, v_2, the edge $v_1 v_2$, and edges $\{v_0 x_i \mid i \in \mathcal{B}^1_0\} \cup \{v_1 x_i \mid i \in \mathcal{B}^1_1\}$. See Figure 1 for an example of this construction. (In the figure, we have expanded the terminals of AND_2 into edges, for readability. This does not affect the behavior of the graph.)

Cost Lower Bound Proof Intuitively, at level ℓ of the refinement process, the current partition contains all blocks $\mathcal{X}^{\ell+1}_q$ of level $\ell+1$ and for all $0 \le q < 2^\ell$, either $\mathcal{Y}^\ell_q$ or the two blocks $\mathcal{Y}^{\ell+1}_{2q}$ and $\mathcal{Y}^{\ell+1}_{2q+1}$. In this situation one can split up the blocks $\mathcal{Y}^\ell_q$ into blocks $\mathcal{Y}^{\ell+1}_{2q}$ and $\mathcal{Y}^{\ell+1}_{2q+1}$ using either refinement operation $(\mathcal{X}^{\ell+1}_{2q}, \mathcal{Y}^\ell_q)$ or $(\mathcal{X}^{\ell+1}_{2q+1}, \mathcal{Y}^\ell_q)$. These operations both have cost $2^{k-(\ell+1)}k^2$, and refining all the $\mathcal{Y}^\ell_q$ cells in this way costs $2^{k-1}k^2$. Once $\mathcal{Y}$ is partitioned into binary blocks of level $\ell+1$, we can partition $\mathcal{X}$ into blocks of level $\ell+2$ (using the AND_ℓ-gadget), and proceed the same way. Since there are k such refinement levels, we can lower bound the total cost of refining the graph by $2^{k-1}k^3 = \Omega(m \log n)$ and are done. What remains to show is that applying the refinement operations in this specific way is the only way to obtain a stable partition. To formalise this, we introduce a number of partitions of $V(G_k)$ that are stable with respect to the (spanning) subgraph $G'_k = G_k - [\mathcal{X}, \mathcal{Y}]$, and that partition $\mathcal{X}$ and $\mathcal{Y}$ into binary blocks. (For disjoint vertex sets S, T, we denote $[S, T] = \{uv \in E(G) \mid u \in S, v \in T\}$.) So on G_k, these partitions can only be refined using operations (R, S), where R is a binary block of $\mathcal{X}$ and S is a binary block of $\mathcal{Y}$.

Definition 10. *For any $\ell \in \{0, \ldots, k - 1\}$, and nonempty set $Q \subseteq \mathcal{B}_\ell$, by $\tau_{Q,\ell}$ we denote the partition of $\mathcal{X} \cup \mathcal{Y}$ that contains cells*

- *$\mathcal{X}^{\ell+1}_q$ for all $q \in \mathcal{B}_{\ell+1}$,*
- *$\mathcal{Y}^\ell_q$ for all $q \in Q$, and both $\mathcal{Y}^{\ell+1}_{2q}$ and $\mathcal{Y}^{\ell+1}_{2q+1}$ for all $q \in \mathcal{B}_\ell \setminus Q$.*

$\pi_{Q,\ell}$ denotes the coarsest stable partition for $G'_k = G_k - [\mathcal{X}, \mathcal{Y}]$ that refines $\tau_{Q,\ell}$.

Since $\pi_{Q,\ell}$ is stable on G'_k, any effective refining operation (with respect to G_k) should involve the edges between $\mathcal{X}$ and $\mathcal{Y}$. Using Lemma 9, it can be shown that $\pi_{Q,\ell}$ agrees with $\tau_{Q,\ell}$, and therefore any effective *elementary* refining operation has the following form:

Lemma 11 (*). *Let (R, S) be an effective elementary refining operation on $\pi_{Q,\ell}$. Then for some $q \in Q$, $R = \mathcal{X}_{2q}^{\ell+1}$ or $R = \mathcal{X}_{2q+1}^{\ell+1}$, and $S = \mathcal{Y}_q^\ell$. The cost of this operation is $k^2 2^{k-(\ell+1)}$.*

This motivates the following definition: for $q \in Q$, by $r_q(\pi_{Q,\ell})$ we denote the partition of $V(G_k)$ that results from (either of) the above refining operation(s).

Proof sketch of Theorem 7: Let G_k be the graph described above and $\pi_{Q,\ell}$ be the partitions of $V(G_k)$ from Definition 10. First, we note that using Lemma 9, it can be shown that a partition is not stable until it is discrete on $X \cup Y$. So the coarsest stable partition ω of G refines all partitions $\pi_{Q,\ell}$. For ease of notation, we define $\pi_{\emptyset,\ell} := \pi_{\mathcal{B}_{\ell+1},\ell+1}$. By Lemma 11, any effective elementary refinement operation on a partition $\pi_{Q,\ell}$ has cost $2^{k-(\ell+1)}k^2$, and results in $r_q(\pi_{Q,\ell})$ for some $q \in Q$. Denote $Q' = Q \setminus \{q\}$. Note that $r_q(\pi_{Q,\ell})$ agrees with $\tau_{Q',\ell}$ on $X \cup Y$. It can actually be shown that $r_q(\pi Q, \ell) \preceq \pi_{Q',\ell}$. So we may now apply Proposition 8 to conclude that $\mathrm{cost}(\pi_{Q,\ell}) \geq 2^{k-(\ell+1)}k^2 + \min_{q \in Q} \mathrm{cost}(\pi_{Q \setminus \{q\},\ell})$. By induction on $|Q|$ it then follows that $\mathrm{cost}(\pi_{\mathcal{B}_\ell,\ell}) \geq 2^{k-1}k^2 + \mathrm{cost}(\pi_{\mathcal{B}_{\ell+1},\ell+1})$ for all $0 \leq \ell \leq k - 1$. Hence, by induction on ℓ, $\mathrm{cost}(\pi_{\mathcal{B}_0,0}) \geq 2^{k-1}k^3$, which lower bounds $\mathrm{cost}(\alpha)$. It can be verified that $n \in O(2^k k)$ and $m \in O(2^k k^2)$, so $\log n \in O(k)$. This shows that $\mathrm{cost}(\alpha) \in \Omega((m + n) \log n)$. $\square$

References

1. Babai, L.: Moderately exponential bound for graph isomorphism. In: Gécseg, F. (ed.) FCT 1981. LNCS, vol. 117, pp. 34–50. Springer, Heidelberg (1981)
2. Babai, L., Erdős, P., Selkow, S.: Random graph isomorphism. SIAM Journal on Computing 9, 628–635 (1980)
3. Babai, L., Luks, E.: Canonical labeling of graphs. In: Proc. STOC 1983, pp. 171–183 (1983)
4. Cai, J., Fürer, M., Immerman, N.: An optimal lower bound on the number of variables for graph identifications. Combinatorica 12(4), 389–410 (1992)
5. Cardon, A., Crochemore, M.: Partitioning a graph in $O(|A| \log_2 |V|)$. Theoretical Computer Science 19(1), 85–98 (1982)
6. Darga, P., Liffiton, M., Sakallah, K., Markov, I.: Exploiting structure in symmetry detection for CNF. In: Proc. DAG 2004, pp. 530–534 (2004)
7. Hopcroft, J.: An n log n algorithm for minimizing states in a finite automaton. In: Kohavi, Z., Paz, A. (eds.) Theory of Machines and Computations, pp. 189–196. Academic Press (1971)
8. Junttila, T., Kaski, P.: Engineering an efficient canonical labeling tool for large and sparse graphs. In: Proc. ALENEX 2007, pp. 135–149 (2007)
9. McKay, B.: Practical graph isomorphism. Congressus Numerantium 30, 45–87 (1981)
10. McKay, B.: Nauty users guide (version 2.4). Computer Science Dept., Australian National University (2007)
11. Paige, R., Tarjan, R.: Three partition refinement algorithms. SIAM Journal on Computing 16(6), 973–989 (1987)
12. Piperno, A.: Search space contraction in canonical labeling of graphs. arXiv preprint arXiv:0804.4881v2 (2011)

A Faster Computation of All the Best Swap Edges of a Shortest Paths Tree[*]

Davide Bilò[1], Luciano Gualà[2], and Guido Proietti[3,4]

[1] Dipartimento di Scienze Umanistiche e Sociali, University of Sassari, Italy
[2] Dipartimento di Ingegneria dell'Impresa, University of Rome "Tor Vergata", Italy
[3] Dip. di Ingegneria e Scienze dell'Informazione e Matematica, Univ. of L'Aquila, Italy
[4] Istituto di Analisi dei Sistemi ed Informatica, CNR, Rome, Italy

Abstract. We consider a 2-edge connected, non-negatively weighted graph G, with n nodes and m edges, and a *single-source shortest paths tree* (SPT) of G rooted at an arbitrary node. If an edge of the SPT is temporarily removed, a widely recognized approach to reconnect the nodes disconnected from the root consists of joining the two resulting subtrees by means of a single non-tree edge, called a *swap edge*. This allows to reduce consistently the set-up and computational costs which are incurred if we instead rebuild a new optimal SPT from scratch. In the past, several optimality criteria have been considered to select a *best* possible swap edge, and here we restrict our attention to arguably the two most significant measures: the minimization of either the *maximum* or the *average* distance between the root and the disconnected nodes. For the former criteria, we present an $O(m \log \alpha(m, n))$ time algorithm to find a best swap edge for *every* edge of the SPT, thus improving onto the previous $O(m \log n)$ time algorithm (B. Gfeller, *ESA'08*). Concerning the latter criteria, we provide an $O(m + n \log n)$ time algorithm for the special but important case where G is *unweighted*, which compares favorably with the $O(m + n \alpha(n, n) \log^2 n)$ time bound that one would get by using the fastest algorithm known for the weighted case – once this is suitably adapted to the unweighted case.

1 Introduction

In communication networking, *broadcasting* a message from a source node to every other node of the network is one of the most common operations. Since this should be done by making use of a logical communication topology as sparser and faster as possible, then it is quite natural to resort to a *single-source shortest-paths tree* (SPT) rooted at the source node. However, despite its popularity, the SPT is highly susceptible to link malfunctioning, as any tree-based network topology: the smaller is the number of links, the higher is the average traffic for each link, and the bigger is the risk of a link overloading. Even worse, the failure of a single link may cause the disconnection of a large part of the network.

[*] This work was partially supported by the Research Grant PRIN 2010 "ARS TechnoMedia", funded by the Italian Ministry of Education, University, and Research.

H.L. Bodlaender and G.F. Italiano (Eds.): ESA 2013, LNCS 8125, pp. 157–168, 2013.
© Springer-Verlag Berlin Heidelberg 2013

In principle, two different approaches can be adopted to solve the problem of a link failure: either we rebuild a new SPT from scratch (which can be very expensive in terms of computational and set-up costs), or we quickly reconnect the two subtrees induced by the link failure by *swapping* it with a single non-tree edge (see [10,15] for some practical motivations supporting this latter approach). Quite obviously, swapping requires that the *swap edge* is wisely selected so that the resulting *swap tree* is as much efficient as possible in terms of some given post-swap distance measure from the root to the just reconnected nodes. Moreover, to be prepared to any possible failure event, it makes sense to study the problem of dealing with the failure of *each and every* single link in the network. This defines a so-called *all-best-swap edges* (ABSE) problem on the SPT [11].

Related work. The problem of swapping in spanning trees has received a significant attention from the algorithmic community. There is indeed a line of papers which address ABSE problems starting from different types of spanning trees. Just to mention a few, we recall here the *minimum spanning tree* (MST), the *minimum diameter spanning tree* (MDST), and the *minimum routing-cost spanning tree* (MRCST). For the MST, a best swap is simply a swap edge minimizing the *weight* of the swap tree, i.e., is a swap edge of minimum weight. [1] This problem is also known as the MST *sensitivity analysis* problem. Concerning the MDST, a best swap is instead an edge minimizing the *diameter* of the swap tree [9,12]. Finally, for the MRCST, a best swap is naturally an edge minimizing the *all-to-all routing cost* of the swap tree [16]. Denoting by n (resp., m) the number of nodes (resp., edges) of the given graph, the fastest solutions for solving the corresponding ABSE problems have a running time of $O(m \log \alpha(m,n))$ [14], $O(m \log n)$ [8], and $O\left(m2^{O(\alpha(2m,2m))} \log^2 n\right)$ [2], respectively, where α is the inverse of the Ackermann function originally defined in [1].

Getting back to the SPT, the appropriate definition of a best swap seems however more ambiguous, since an SPT is actually the *union* of all the shortest paths emanating from a root node, and so there is not a univocal global optimization measure we have to aim at when swapping. Thus, in [13], where the corresponding ABSE problem was initially studied, several different criteria expressing desirable features of the swap tree of an SPT were introduced in order to characterize a best possible swap edge. In particular, among all of them, two can be viewed as the most prominent ones: the *maximum* and the *average* distance from the root to the disconnected nodes. These two measures reflect a classic egalitarian *versus* utilitarian viewpoint as far as the efficiency of the swap is concerned. From the algorithmic side, the fastest known solutions for the two problems amount to $O(m \log n)$ (as a by-product of the result in [8]) and to $O(m \alpha(m,m) \log^2 n)$ time (see [4][2]), respectively.

It is worth noticing that swapping in an SPT can be reviewed as fast and good at the same time: in fact, recomputing every new optimal SPT from scratch

---

[1] Notice that a swap MST is actually a MST of the graph deprived of the failed edge.
[2] Actually, in [4] the authors claim an $O(m \log^2 n)$ time bound, but this must be augmented by an $O(\alpha(m,m))$ factor, as pointed out in [2].

would require as much as $O(mn \log \alpha(m, n))$ time [5] (no faster dynamic algorithm is indeed known). Moreover, it has been shown that in the swap tree the maximum (resp., average) distance of the disconnected nodes from the root is at most twice (resp., triple) that of the new optimal SPT, and this is tight [13].

Our Results. In this paper we focus exactly on the ABSE problem on an SPT w.r.t. these two measures. To this respect, for the former criteria we present an $O(m \log \alpha(m, n))$ time algorithm. As we will see, our result generalizes to the ABSE problem on an MDST, and thus, for both problems we improve the time complexity of $O(m \log n)$ given in [8]. It is worth noticing that in this way we beat the $O(m \log n)$ time barrier needed to sort the non-tree edges w.r.t. their weight, and we meet the time complexity of the best-known algorithm for the MST sensitivity analysis problem. In our opinion, this is particularly remarkable since this latter problem can be reduced in linear time to our problem.[3] Thus, improving the time complexity of our algorithm would provide a faster algorithm for performing a sensitivity analysis of an MST, which is one of the main open problems in the area of MST related computations.

As far as the second criteria is concerned, we focus on the special but important case where G is *unweighted*. In this case we are indeed able to first exhibit a *sparsification* technique on non-tree edges which would immediately allow to use the $O(m \, \alpha(m, m) \log^2 n)$ time algorithm known for the weighted case (see [4]) so as to obtain a faster $O(m + n \, \alpha(n, n) \log^2 n)$ time solution. However, we go beyond this improvement, by building on this sparser graph a new approach which makes use of sophisticated data structures, so as to eventually get an efficient $O(m + n \log n)$ time algorithm. Unfortunately, the extension of our machinery to weighted graphs sounds hard, and so we regard this as a challenging problem left open.

The paper is organized as follows: in Section 2 we describe a preprocessing step which will be used to guarantee the efficiency of our algorithms, while in Section 3 and 4, we present our algorithms for the maximum and the average distance criteria, respectively. Due to space limitations, some of the proofs are omitted and will be given in the full version of the paper.

2   A Preprocessing Step

In this section we present our notation, and we describe a useful preprocessing step which allows us to simplify the solution of the ABSE problems we address in this paper.

We start by defining our notation. Let $G = (V, E, w)$ be a non-negatively edge-weighted, undirected and 2-edge-connected graph, and let T be an SPT of

---

[3] Indeed, we can perform a sensitivity analysis of an MST as follows: (i) we first root the MST at any arbitrary vertex, (ii) we set the weight of every tree-edge to 0, and then (iii) we solve the ABSE problem w.r.t. the maximum distance from the root. Clearly, for every tree edge e, the best swap edge computed by the algorithm is a non-tree edge of minimum weight cycling with e.

G rooted at an arbitrary node r. Thus, T contains a shortest path from r to every other node of G. For a vertex $v \neq r$, we denote by $\bar{v}$ the *parent* of v in the tree (in particular, $\bar{\bar{v}}$ denotes the parent of $\bar{v}$). The *least common ancestor* of a given pair of nodes v, v' in T is the node of T farthest from r that is an ancestor of both v and v'; we will indicate it by $\mathtt{lca}(v, v')$. Let $e = (\bar{x}, x)$ be any tree edge. We denote by T_x the subtree of T rooted at x and containing all the descendants in T of x and by $V(T_x)$ the set of vertices in T_x. We indicate with $C_e = \{(u, y) \in E \setminus E(T) : (u \in V \setminus V(T_x)) \wedge (y \in V(T_x))\}$ the set of *swap edges* for e, i.e., the edges that may be used to replace e for reconnecting T. Since G is 2-edge connected, we have that $C_e \neq \emptyset$, $\forall e \in E(T)$. Let in the following $T_{e/f}$ denote the *swap tree* obtained from T by swapping e with $f \in C_e$. For any two vertices $v, v' \in V$, we finally denote by $d(v, v')$ and $d_{e/f}(v, v')$ the distance between v and v' in T and $T_{e/f}$, respectively.

Depending on the goal that we pursue in swapping, some swap edge may be preferable to some other one. We here focus on the following problems:

1. max-ABSE: for every $e = (\bar{x}, x) \in E(T)$, find an edge $f_e \in C_e$ s.t.:

$$f_e \in \arg \min_{f \in C_e} \left\{ \max_{v \in V(T_x)} d_{e/f}(r, v) \right\};$$

2. sum-ABSE: for every $e = (\bar{x}, x) \in E(T)$, find an edge $f_e \in C_e$ s.t.:[4]

$$f_e \in \arg \min_{f \in C_e} \left\{ \sum_{v \in V(T_x)} d_{e/f}(r, v) \right\}.$$

To solve efficiently our ABSE problems, we transform in a standard way (see for instance [17]) each non-tree edge $f = (u, y)$ into two *vertical* edges, i.e., $f' = (\mathtt{lca}(u, y), y)$ and $f'' = (\mathtt{lca}(u, y), u)$, each of them with an appropriate weight, namely $w(f') = d(r, u) + w(f) - d(r, \mathtt{lca}(u, y))$ and $w(f'') = d(r, y) + w(f) - d(r, \mathtt{lca}(u, y))$, respectively. Basically, $w(f')$ (resp., $w(f'')$) once added by $d(r, \mathtt{lca}(u, y))$, is the length of the path in $T_{e/f}$ starting from r, passing through f, and ending in y (resp., u), after that f has swapped with an edge e along the path from y (resp., u) to $\mathtt{lca}(u, y)$. In this way, we obtain an auxiliary (multi)graph G' with at most twice the number of non-tree edges of G, and which is perfectly equivalent to G as far as the study of our ABSE problems is concerned. Notice that this transformation can be performed for all non-tree edges in $O(m)$ time [6], once that the SPT is given, since essentially it only requires the computation of the least common ancestors of non-tree edges. In the rest of the paper, we will therefore assume to be working on G', unless differently stated, and that for a swap edge $f = (u, y)$, node $y \in V(T_x)$, and so $u = \mathtt{lca}(u, y)$.

3   A Faster Algorithm for max-ABSE

In this section we provide the description of an algorithm solving the max-ABSE problem in $O(m \log \alpha(m, n))$ time and $O(m)$ space. As we will see, our result

---

[4] By definition, f_e minimizes the *average* distance from r to the disconnected nodes.

generalizes to the ABSE problem on an MDST. Thus, for both problems we improve the time complexity of $O(m \log n)$ given in [8].

We need to introduce some further notation before providing the algorithm description. We denote by $P(v, v')$ the (unique) path in T between v and v' and by $w(P(v, v')) := d(v, v')$ the *length* of $P(v, v')$. A *diametral path* P of T is one of the longest simple paths in T and a *center* of T is a vertex c such that $\max_{v \in V} d(c, v) \leq \max_{v \in V} d(v', v)$, for every $v' \in V$. It is well known that a tree has either one center or two centers and, if tree has two centers, then they are adjacent. Furthermore, any diametral path of a tree passes through the tree center(s). In the rest of the paper, w.l.o.g., for every $v \in V$ and every $v' \in V(T_v)$, we will assume that there is only one longest path in T_v having v' as one of its endvertices.[5] Let v be an inner vertex of T and let P be the longest path starting at the center c of T_v and entirely contained in T_v. We denote by c_v the endvertex of the unique edge of P incident to c which is farthest from r (possibly $c_v = c$). For a leaf vertex v of T, we define $c_v := v$. Finally, we denote by U_v the vertices of the path $P(v, c_v)$ and by $\widehat{U}_v = U_v \setminus \{c_v\}$.

The following lemma shows a useful property satisfied by the vertices c_v's.

Lemma 1. *Let v and v' be two distinct vertices of T such that v' is an ancestor of v in T. Then either $c_{v'} \in U_v$ or $c_{v'} \in V \setminus V(T_v)$.*

Our algorithm traverses the tree edges in a suitable preorder and, for each tree edge $e = (\bar{x}, x)$, it computes a best swap edge f_e of e in four steps with a clever implementation of the approach used in [8]. More precisely, for each tree edge $e = (\bar{x}, x)$, our algorithm does the following:

Step 1: for every vertex $v \in V(T_x)$, it computes a best swap edge $\hat{f}_v$ of e among the set of edges in C_e which are also incident to v;[6]

Step 2: it partitions the vertices of T_x into $|U_x|$ *groups*, where each vertex z of U_x defines the group $\mathcal{G}(x, z) := \{v \in V(T_x) \mid \text{lca}(v, c_x) = z\}$;[7]

Step 3: for every vertex $z \in U_x$ it computes a *group candidate*, i.e., best swap edge f_z of e chosen among the set F_z of the best swap edges computed during Step 1 and associated to vertices of the same group z belongs to, i.e., $F_z := \{\hat{f}_v \mid v \in \mathcal{G}(x, z)\}$;

Step 4: it selects a best swap edge f_e of e among the group candidates computed in Step 3.

Let $f = (u, y) \in C_e$ be a non-tree edge and let $\text{lca}(y, c_x) = z$. Observe that $z \in U_x$. Let $\kappa_1(f) := d(r, u) + w(f) + d(r, y)$ be the *primary key* associated with f, and let $\kappa_2(f, e) := \kappa_1(f) - 2d(r, z)$ be the *secondary key* associated with the pair f and e. The primary key of f is independent of the failing tree edge and is used to compare two competing swap edges of the same set F_z. The secondary

---

[5] This property can be achieved by modifying the tree via the addition of dummy leaves, each connected with a leaf of T by a suitable cheap edge.

[6] If no such edge exists, then we assume that $\hat{f}_v$ is an imaginary edge of weight $+\infty$.

[7] Observe that, if $z = c_x$, then $\mathcal{G}(x, z) = V(T_{c_x})$; if $z \neq c_x$, then $\mathcal{G}(x, z) = V(T_z) \setminus V(T_v)$, where v is the child of z such that $c_x \in V(T_v)$.

key is used to compare two competing swap edges among the group candidates f_z's with $z \neq c_x$. The following lemmas provide useful properties suitable for an efficient implementation of Step 1, Step 3, and Step 4 of our algorithm, respectively.

Lemma 2. *Let* $e = (\bar{x}, x)$ *be a tree edge with* $\bar{x} \neq r$, *let* v *be a vertex of* T_x, *and let* f' *be a best swap edge of* $\bar{e} = (\bar{\bar{x}}, \bar{x})$ *among the set of non-tree edges in* $C_{\bar{e}}$ *which are incident to* v. *If* $(\bar{x}, v) \notin C_e$, *then* $\hat{f}_v = f'$, *otherwise* $\hat{f}_v$ *is the edge of minimum primary key between* f' *and* $(\bar{x}, v)$.

Lemma 3. *Let* $e = (\bar{x}, x)$ *be a tree edge and let* $z \in U_x$. *We have that* $f_z \in \arg\min\{\kappa_1(f) \mid f \in F_z\}$.

Lemma 4. *Let* $e = (\bar{x}, x)$ *be a tree edge and let* $\hat{f} \in \arg\min\{\kappa_2(f_z, e) \mid z \in \widehat{U}_x\}$. *Then,* f_{c_x} *or* $\hat{f}$ *is a best swap edge of* e.

The following Corollary of Lemma 1 shows that, thanks to the preorder processing of the tree edges, we can compute the new groups $\mathcal{G}(x, z)$, and thus solve Step 2, by suitably splitting some groups $\mathcal{G}(\bar{x}, z')$ which have already been computed.

Corollary 1. *Let* $\{z_1, \dots, z_k\} = U_x \setminus U_{\bar{x}}$ *such that* z_i *is the parent of* z_{i+1}, *for every* $i = 1, \dots, k-1$ *(if* $x = r$, *then* $\{z_1, \dots, z_k\} = U_r \setminus \{r\}$*) and let* z_0 *be the parent of* z_1. *We have that (a)* $\mathcal{G}(\bar{x}, z) = \mathcal{G}(x, z)$ *for every* $z \in \widehat{U}_{\bar{x}} \cap U_x$, *(b)* $\mathcal{G}(\bar{x}, c_{\bar{x}}) = \mathcal{G}(x, c_x)$ *iff* $c_{\bar{x}} = c_x$, *and (c)* $\mathcal{G}(x, z) \subseteq \mathcal{G}(\bar{x}, z_0)$ *for every* $z \in U_x \setminus U_{\bar{x}}$.

Therefore, rather than implementing our algorithm using brute force search, we make use of *split-findmin* data structures which are suitable for finding minimum-key elements of sets that can only be split. A split-findmin is a data structure that has been successfully used in [14] to solve the sensitivity analysis of an MST in $O(m \log \alpha(m, n))$ time. A split-findmin $\mathcal{S}$ maintains a set of sequences of elements each associated with a key and supports the following operations:

$\mathrm{init}(o_1, \dots, o_N)$: initializes the sequence $\mathcal{S} := \{(o_1, \dots, o_N)\}$ of N elements with key $\kappa(o_i) := +\infty$ for all i;

$\mathrm{split}(\mathcal{S}, o_i)$: let $s = (o_j, \dots, o_{i-1}, o_i, \dots, o_k)$ be the sequence in $\mathcal{S}$ containing o_i; the call of $\mathrm{split}(\mathcal{S}, o_i)$ returns $\mathcal{S} := (\mathcal{S} \setminus \{s\}) \cup \{(o_j, \dots, o_{i-1}), (o_i, \dots, o_k)\}$;

$\mathrm{findmin}(\mathcal{S}, o_i)$: let s be the sequence in $\mathcal{S}$ containing o_i; the call of $\mathrm{findmin}(\mathcal{S}, o_i)$ returns an element of minimum key in s;

$\mathrm{decreasekey}(\mathcal{S}, o_i, k)$: if $\kappa(o_i) > k$, then $\mathrm{decreasekey}(\mathcal{S}, o_i, k)$ sets $\kappa(o_i) = k$.

As proven in [14], a split-findmin data structure of N elements requires $\Theta(N)$ space, can be initialized in $\Theta(N)$ time, and supports M split, findmin, and decreasekey operations in time $O(M \log \alpha(M, N))$. The following easy-to-prove lemma is another key ingredient for the correctness of our algorithm.

Lemma 5. *Let T' be a tree rooted at r', let s be the sequence of nodes as obtained from a right-to-left preorder visit of T', and let $\widehat{S}$ be a split-findmin initialized with the sequence s. Let v be the leftmost child of r'. The execution of $\mathtt{split}(\widehat{S}, v)$ splits s into two sequences s' and s'' such that, w.l.o.g., $V(T') \setminus V(T'_v)$ contains the vertices of s' and $V(T'_v)$ contains the vertices of s''.[8]*

Our algorithm uses two split-findmin data structures $\mathcal{S}$ and $\mathcal{S}'$ both containing all the vertices of T as elements. During the visit of the tree edge $e = (\bar{x}, x)$, the first split-findmin data structure $\mathcal{S}$ is updated so as to:

($\mathcal{I}_1$): the key associated with v in $\mathcal{S}$ is $\kappa_1(\hat{f}_v)$;
($\mathcal{I}_2$): for every vertex $z \in U_x$, all the vertices in $\mathcal{G}(x, z)$ form a sequence in $\mathcal{S}$.

Thus, thanks to Lemma 3, we can compute f_z, for every $z \in U_x$, using a $\mathtt{findmin}(\mathcal{S}, z)$ query. The second split-findmin data structure $\mathcal{S}'$ is used to represent the group candidates f_z of all the groups $\mathcal{G}(x, z)$ with $z \in \widehat{U}_x$. More precisely, during the visit of the tree edge $e = (\bar{x}, x)$, $\mathcal{S}'$ is updated so as the following invariants are maintained:

($\mathcal{I}'_1$): $V(T_x)$ forms a sequence in $\mathcal{S}'$;
($\mathcal{I}'_2$): for every $v \in V(T_x)$, the key associated with v in $\mathcal{S}'$ is equal to (a) $\kappa_2(f_v, e)$, if $v \in \widehat{U}_x$ and (b) $+\infty$, if $v \notin \widehat{U}_x$.

Because of both invariants $\mathcal{I}'_1$ and $\mathcal{I}'_2$ and thanks to Lemma 4, we can compute $\hat{f} \in \arg\min\{\kappa_2(f_z, e) \mid z \in \widehat{U}_x\}$ by performing a $\mathtt{findmin}(\mathcal{S}', x)$ query.

The algorithm first arranges T in such a way that, for each non-leaf node v, the subtree rooted at the leftmost child of v contains c_v. Then, it initializes $\mathcal{S}$ and $\mathcal{S}'$ with the sequence of vertices as obtained from a right-to-left preorder visit of T and finally visits all the tree edges according to the left-to-right preorder visit of T. Let $e = (\bar{x}, x)$ be the tree edge that is visited by the algorithm. The algorithm follows the four-step approach described above to compute a best swap edge f_e of e by updating the two split-findmin data structures $\mathcal{S}$ and $\mathcal{S}'$. For the base case $x = r$, the algorithm performs only Step 2 (i.e., calls $\mathtt{split}(\mathcal{S}, z)$ for every $z \in U_r \setminus \{r\}$) to initialize $\mathcal{S}$ properly.

Step 1 of the algorithm is implemented by calling $\mathtt{decreasekey}(\mathcal{S}, v, \kappa_1(f))$ for every vertex $v \in V(T_x)$ such that $f = (u, y)$ is a swap edge of e with $u = \bar{x}$ and $v = y$. Indeed, if $\bar{x} \neq r$ and f' is a best swap edge of $\bar{e} = (\bar{\bar{x}}, \bar{x})$ among the set of non-tree edges in $C_{\bar{e}}$ which are also incident to v, then, from Lemma 2, maintaining invariant $\mathcal{I}_1$ is equivalent to comparing $\kappa_1(f)$ with $\kappa_1(f')$ which, by induction, is the key associated to vertex v in $\mathcal{S}$ at the beginning of Step 1.

Let $\{z_1, \ldots, z_k\} = U_x \setminus U_{\bar{x}}$ such that z_i is the parent of z_{i+1}, for every $i = 1, \ldots, k - 1$ (if $x = r$, then $\{z_1, \ldots, z_k\} = U_r \setminus \{r\}$) and let z_0 be the parent of z_1. Step 2 of the algorithm is implemented by simply calling $\mathtt{split}(\mathcal{S}, z_i)$ for every $i = 1, \ldots, k$. In what follows, we sketch that invariant $\mathcal{I}_2$ is maintained

---

[8] Notice that, from a topological point of view, the $\mathtt{split}(\widehat{S}, v)$ operation can be viewed as the removal of edge (r', v) in T'.

at the end of Step 2. By induction, $\mathcal{S}$ contains the sequence of the vertices in $\mathcal{G}(\bar{x}, z)$ for every $z \in U_{\bar{x}}$. From Corollary 1, we have that $\mathcal{G}(\bar{x}, z) = \mathcal{G}(x, z)$ for every $z \in \widehat{U}_{\bar{x}} \cap U_x$. Furthermore, $\mathcal{G}(\bar{x}, c_{\bar{x}}) = \mathcal{G}(x, c_x)$ iff $c_{\bar{x}} = c_x$. Finally, $\mathcal{G}(x, z) \subseteq \mathcal{G}(\bar{x}, z_0)$ for every $z \in U_x \setminus U_{\bar{x}}$. Using induction, we can prove that $\mathcal{S}$ contains the sequence s of all vertices in $\mathcal{G}(\bar{x}, z_0)$ and z_1 is the lefmost child of the subtree induced by s. Furthermore, Lemma 1 implies that z_{i+1} is the leftmost child of z_i for every $i = 1, \ldots, k-1$. Therefore, thanks to repeated applications of Lemma 5, after the execution of all split$(\mathcal{S}, z_i)$, $\mathcal{S}$ contains the sequence of all the vertices in $\mathcal{G}(x, z_j)$, for every $j = 0, \ldots, k$.

Step 3 of the algorithm can be implemented by performing the following set of operations whenever a group creation or a group modification (i.e., a split or decreasekey operation on $\mathcal{S}$) occurs during Step 1 or Step 2. Let $\mathcal{G}(x, v)$ be a group that has been created or modified. From Lemmas 2 and 3, updating $\mathcal{S}'$ is equivalent to first calling findmin$(\mathcal{S}, v)$, which returns a non-tree edge f, and, if lca$(v, c_x) \neq c_x$, to calling decreasekey$(\mathcal{S}', \text{lca}(v, c_x), \kappa_2(f, e))$. Proceeding this way, we are guaranteed that at the end of Step 3 the invariant $\mathcal{I}'_2$ holds. Observe that a decreasekey operation on $\mathcal{S}$ modifies one group while a split operation on $\mathcal{S}$ splits one group into two groups, i.e., it creates two new groups.

Step 4 of the algorithm consists of (a) computing $\hat{f}$ (the explanation of how to compute $\hat{f}$ is below), (b) calling findmin$(\mathcal{S}, c_x)$ to compute f_{c_x}, and (c) selecting a best swap edge of e between $\hat{f}$ and f_{c_x} via the explicit computation of the objective function values when e is replaced with $\hat{f}$ and with f_{c_x}, respectively.[9] To compute $\hat{f}$, the algorithm first performs split$(\mathcal{S}', x)$ which – thanks to the left-to-right preorder processing of the tree vertices and to the right-to-left preorder arrangement of the tree vertices in $\mathcal{S}'$ – maintains the invariant $\mathcal{I}_1$, and then calls findmin$(\mathcal{S}', x)$.

We can prove the following.

Theorem 1. *The* max-ABSE *problem can be solved in* $O(m \log \alpha(m, n))$ *time and* $O(m)$ *space.*

Corollary 2. *The* ABSE *problem for a MDST can be solved in* $O(m \log \alpha(m, n))$ *time and* $O(m)$ *space.*

4   A Faster Algorithm for sum-ABSE on Unweighted Graphs

In this section we solve efficiently the ABSE problem w.r.t. the criteria of minimizing the average distance from the root on *unweighted* graphs. Thus, instead of talking about an SPT, we better will refer to a rooted *breadth-first tree* (BFT) T of G.

We start by refining the "verticalization" of non-tree edges described in Section 2. Let $f' = (\text{lca}(u, y), y)$ be a vertical edge associated with a non-tree

---

[9] In [8] it is explained how the value of the objective function can be computed in $O(1)$ time for every $e \in E(T)$ and every $f \in C_e$ after a linear time preprocessing.

edge $f = (u, y)$. Since G is unweighted and T is a BFT of G, we have that either of the three following cases apply for edge f: (i) $d(r, u) = d(r, y)$, or (ii) $d(r, u) = d(r, y) - 1$, or finally (iii) $d(r, u) = d(r, y) + 1$. Depending on which of the three cases apply, we have that f' is either of *type* (i), (ii) or (iii), respectively. Clearly, a symmetric argument applies to the other vertical edge $f'' = (\text{lca}(u, y), u)$ associated with f (if any), and notice that if f' is of type (i), then the same holds for f'', while if f' is of type (ii), then f'' is of type (iii), and vice versa. Then, for any given a node $v \in V$, we can select a *representative* for any of these three types of vertical non-tree edges incident to v as follows: For each type, select a non-tree edge which swaps with a largest number of edges along the tree path from v to r. Notice that a representative edge is clearly preferable w.r.t. any other swap edge it was selected out of, since it allows for the same quality when swapping, while being usable longer. Thus, at most three vertical non-tree edges will remain associated with v, and all the other non-tree edges will be discarded. Observe that once again this refined preprocessing phase can be performed for all non-tree edges in $O(m)$ time, since it only requires to further classify vertical non-tree edges depending on their type, which is clearly doable in linear time. Notice that after the preprocessing, we have reduced to $O(n)$ the number of non-tree edges, and this immediately allows to solve the sum-ABSE problem in linear space and $O(m + n\alpha(n, n) \log^2 n)$ time, by just using the fastest algorithm known for the weighted case [4]. We will show in the following how to improve this running time to $O(m + n \log n)$.

To solve sum-ABSE, our algorithm will run in three phases, where at each phase we will only consider representative edges of either of the three types above. Indeed, for efficiency reasons, representative edges of different type need to be treated separately. This means, at each phase every node will have at most a *single* vertical edge associated, and at the end of a phase every tree edge will remain associated with a best swap edge of the current type. Finally, a best swap edge for a given tree edge will be selected as the best among those found in the three phases. Hence, in the following we assume that all non-tree edges are of the same type. We define the *level* of a node v as $\ell(v) := d(r, v)$, and the *level* of a (vertical) non-tree edge (u, y) as the level of its lowest endvertex y. Furthermore, we define the *height* of v as $h(v) := \max\{\ell(v') \mid v' \in V(T_v)\}$.

For efficiency reasons that will be clearer later, our algorithm associates two keys with each non-tree edge f, and each of these keys will be separately managed by a suitable priority queue. The first key, say $\kappa_1(f)$, is a (constant) value which depends only on f, and that can be used to compare two competing swap edges of the *same* level. Concerning the second key, this is instead a (non-constant) value which can be used to compare two competing swap edges of *different* levels. More precisely, such a key is a linear function of the form $\kappa_2(f, t) := at + b$, where a and b are constant values depending only on f, while t is a variable (called *virtual time*) that properly encodes the position in T of the failing edge e for which f must be evaluated.

The two keys of $f = (u, y)$ are defined as follows: $\kappa_1(f) := \sum_{v \in V} d(y, v)$, while $\kappa_2(f, t) := 2\ell(y)t + \kappa_1(f) - \ell(y)n$. We can prove the following.

Lemma 6. *Let f and f' be two swap edges for a tree edge $e = (\bar{x}, x)$. If $\kappa_2(f, |V(T_x)|) \leq \kappa_2(f', |V(T_x)|)$, then $\sum_{v \in V(T_x)} d_{e/f}(r, v) \leq \sum_{v \in V(T_x)} d_{e/f'}(r, v)$.*

Furthermore, Lemma 6 immediately implies the following.

Corollary 3. *Let f and f' be two non-tree edges of the same level which are also swap edges for a tree edge $e = (\bar{x}, x)$. If $\kappa_1(f) \leq \kappa_1(f')$, then $\sum_{v \in V(T_x)} d_{e/f}(r, v) \leq \sum_{v \in V(T_x)} d_{e/f'}(r, v)$.*

Thus, the main idea of our algorithm is to maintain efficiently a best swap edge for each of the levels below the failing edge. This will be done through the use of two types of priority queues, one for each type of key, according to their nature: *Fibonacci heap (F-heap)* [7] to manage constant keys of the first type, and a *kinetic heap (K-heap)* [3] to manage variable keys of the second type. Indeed kinetic heaps allow to perform findmin operations parameterized with respect to a given parameter t, i.e., the virtual time. Moreover, these operations must satisfy a *monotoniticy condition*, i.e., successive findmin operations must be performed with respect to non-decreasing values of t.

Let us now see how these heaps are built and maintained. At the beginning, for each vertex v different from r, we create an F-heap, say $F_{\ell(v)}(v)$, and if v has a non-tree edge f incident to it, we insert v in $F_{\ell(v)}(v)$ with key $\kappa_1(f)$. Then, for each leaf v, we create a K-heap, say $K(v)$, and if v has a non-tree edge f incident to it, we insert $\ell(v)$ in $K(v)$ with key $\kappa_2(f, t)$ (by also maintaining a pointer to f).

We consider the tree edges in a bottom-up fashion, by visiting the vertices of T in any post-order. When we visit node x (i.e., we consider the removal of edge $e = (\bar{x}, x)$ from T), we maintain the following invariants:

($\mathcal{I}_1$): for each level $\ell(x) \leq j \leq h(x)$, we have an F-heap $F_j(x)$ containing every node in T_x of level j having an incident non-tree edge f which can swap with e (with the corresponding key $\kappa_1(f)$);

($\mathcal{I}_2$): there is a K-heap $K(x)$ containing a subset of levels in the interval $[\ell(x), h(x)]$, with the property that a level $\ell(x) \leq j \leq h(x)$ is in $K(x)$ with key $\kappa_2(f, t)$ iff f is a best swap edge for e of level j.

Notice that from Corollary 3, ($\mathcal{I}_1$) allows to compute a best swap edge $f = (u, y)$ for e with $\ell(y) = j$. Furthermore, from Lemma 6, ($\mathcal{I}_2$) allows to compute a best swap edge for e by extracting the minimum from $K(x)$ with current virtual time $|V(T_x)|$. As will see, K-heaps are built on top of F-heaps whose use, in turn, will be only instrumental to keep the number of more expensive operations performed in the K-heaps low.

Our algorithm proceeds as follows. If x is a leaf, to get a best swap edge for e we simply perform a findmin operation on $K(x)$ with current virtual time $|V(T_x)| = 1$. Otherwise, let $q_1, \ldots, q_s$ be the children of x in T; then we do the following (notice that Steps 1-4 are performed to maintain the invariants):

1. We set $K(x) := K(q_i)$, where q_i is a child of x such that T_{q_i} is a highest subtree rooted at a child of x.

2. If x has an incident non-tree edge f such that $f \in C_e$, we insert $\ell(x)$ in $K(x)$ with key $\kappa_2(f,t)$.

3. For each non-tree edge $f = (u,y)$ with $u = x$ and $\ell(y) = j$, we first delete y from the F-heap to which y currently belongs to, say $F_j(q_z)$. Next we check whether the key of level j in $K(x)$ is associated with f. If this is the case, then (i) we delete j from $K(x)$, (ii) we perform an additional findmin on $F_j(q_z)$ to get a new best swap edge f' of level j, if any, and if this is the case (iii) we re-insert j into $K(x)$ with key $\kappa_2(f',t)$.

4. If $s = 1$, then for each level $\ell(x) \le j \le h(q_1) = h(x)$, the F-heap $F_j(x)$ is simply inherited from $F_j(q_1)$;[10] otherwise, let q_p be the child of x such that T_{q_p} is a highest subtree besides T_{q_i}. For each level $\ell(x) \le j \le h(q_p)$, we merge all F-heaps $F_j(q_k), k = 1, \ldots, s$, and we call the resulting F-heap $F_j(x)$; then, we perform a findmin operation on $F_j(x)$ to compute a best swap edge f of level j, if any. If this is the case, we first check if $K(x)$ contains j, and if so we remove it; afterwards, we re-insert j into $K(x)$, with key $\kappa_2(f,t)$. Notice that for the remaining levels $h(q_p) + 1 \le j \le h(q_i)$, the corresponding F-heaps are simply inherited from $F_j(q_i)$.

5. Finally, we perform a findmin operation on $K(x)$ with current virtual time $|V(T_x)|$ to compute a best swap edge for $e = (\bar{x}, x)$.

Theorem 2. *The* sum-ABSE *problem can be solved in $O(m + n \log n)$ time and $O(m)$ space.*

Proof. First of all, recall that the preprocessing step of the algorithm takes $O(m)$ time and space. Moreover, in order to compute the values of the two keys efficiently, we can pre-compute the value $\sum_{v' \in V} d(v, v')$ for every $v \in V$. This can be done in $O(n)$ time [4].

To derive the time complexity of the algorithm, we have to bound the number of operations performed on our data structures in each phase. Let k be the total number of merge operations on F-heaps. Notice that $k \le n$ since the number of merges is bounded by the initial number of F-heaps, namely $n - 1$ (no new F-heaps are created, since inheriting a heap is just a renaming of the heap itself). The number of insertions and deletions on K-heaps is also $O(n)$, since this is upper bounded by the number of leaves of T plus the number of merge and delete operations on F-heaps, which is $O(n)$. Concerning the findmin operations on F-heaps, it is easy to see that we have at most one such operation for each merging of F-heaps and each deletion on K-heaps, which implies that they are $O(n)$. Clearly, we perform a single findmin on a K-heap for each edge of T. From this, and from the fact that in an F-heap insert, merge and findmin operations takes each $O(1)$ time, and a delete operation takes $O(\log n)$ amortized time, while in a K-heap the amortized time for insert and delete operations is $O(\log n)$, while for findmin operations is $O(1)$ [3], and finally observing that F-heaps and K-heaps use $O(n)$ space, the claim follows.   □

---

[10] Notice that all these heaps can be inherited in $O(1)$ time by simply changing their reference from q_1 to x.

From the above discussion, it should be clear that the use of F-heaps is instrumental to reduce to $O(n)$ the number of (more expensive) operations on K-heaps, and this is exactly the key ingredient for the efficiency of our algorithm. Finally, the sparsification of non-tree edges can be used to solve the max-ABSE problem on unweighted graphs in $O(m + n \log \alpha(n, n))$ time.

References

1. Ackermann, W.: Zum hilbertschen aufbau der reellen zahlen. Mathematical Annals 99, 118–133 (1928)
2. Bilò, D., Gualà, L., Proietti, G.: Finding best swap edges minimizing the routing cost of a spanning tree. In: Hliněný, P., Kučera, A. (eds.) MFCS 2010. LNCS, vol. 6281, pp. 138–149. Springer, Heidelberg (2010)
3. Brodal, G.S., Jacob, R.: Dynamic planar convex hull. In: Proc. of the 43rd Symp. on Foundations of Computer Science (FOCS 2002), pp. 617–626. IEEE Comp. Soc. (2002)
4. Di Salvo, A., Proietti, G.: Swapping a failing edge of a shortest paths tree by minimizing the average stretch factor. Theor. Comp. Science 383(1), 23–33 (2007)
5. Gualà, L., Proietti, G.: Exact and approximate truthful mechanisms for the shortest-paths tree problem. Algorithmica 49(3), 171–191 (2007)
6. Harel, D., Tarjan, R.E.: Fast algorithms for finding nearest common ancestors. SIAM Journal on Computing 13(2), 338–355 (1984)
7. Fredman, M.L., Tarjan, R.E.: Fibonacci heaps and their uses in improved network optimization algorithms. Journal of the ACM 34(3), 596–615 (1987)
8. Gfeller, B.: Faster swap edge computation in minimum diameter spanning trees. Algorithmica 62(1-2), 169–191 (2012)
9. Italiano, G.F., Ramaswami, R.: Maintaining spanning trees of small diameter. Algorithmica 22(3), 275–304 (1998)
10. Ito, H., Iwama, K., Okabe, Y., Yoshihiro, T.: Polynomial-time computable backup tables for shortest-path routing. In: Proc. of the 10th Int. Coll. on Structural Information and Communication Complexity (SIROCCO 2003). Proceedings in Informatics, Carleton Scientific, vol. 17, pp. 163–177 (2003)
11. Nardelli, E., Proietti, G., Widmayer, P.: How to swap a failing edge of a single source shortest paths tree. In: Asano, T., Imai, H., Lee, D.T., Nakano, S.-i., Tokuyama, T. (eds.) COCOON 1999. LNCS, vol. 1627, pp. 144–153. Springer, Heidelberg (1999)
12. Nardelli, E., Proietti, G., Widmayer, P.: Finding all the best swaps of a minimum diameter spanning tree under transient edge failures. Journal of Graph Algorithms and Applications 5(5), 39–57 (2001)
13. Nardelli, E., Proietti, G., Widmayer, P.: Swapping a failing edge of a single source shortest paths tree is good and fast. Algorithmica 36(4), 361–374 (2003)
14. Pettie, S.: Sensitivity analysis of minimum spanning trees in sub-inverse-Ackermann time. In: Deng, X., Du, D.-Z. (eds.) ISAAC 2005. LNCS, vol. 3827, pp. 964–973. Springer, Heidelberg (2005)
15. Proietti, G.: Dynamic maintenance versus swapping: an experimental study on shortest paths trees. In: Näher, S., Wagner, D. (eds.) WAE 2000. LNCS, vol. 1982, pp. 207–217. Springer, Heidelberg (2001)
16. Wu, B.Y., Hsiao, C.-Y., Chao, K.-M.: The swap edges of a multiple-sources routing tree. Algorithmica 50(3), 299–311 (2008)
17. Tarjan, R.E.: Sensitivity analysis of minimum spanning trees and shortest path trees. Inf. Process. Lett. 14(1), 30–33 (1982)

Parallel String Sample Sort[*]

Timo Bingmann and Peter Sanders

Karlsruhe Institute of Technology, Karlsruhe, Germany
{bingmann,sanders}@kit.edu

Abstract. We discuss how string sorting algorithms can be parallelized on modern multi-core shared memory machines. As a synthesis of the best sequential string sorting algorithms and successful parallel sorting algorithms for atomic objects, we propose string sample sort. The algorithm makes effective use of the memory hierarchy, uses additional word level parallelism, and largely avoids branch mispredictions. Additionally, we parallelize variants of multikey quicksort and radix sort that are also useful in certain situations.

1 Introduction

Sorting is perhaps the most studied algorithmic problem in computer science. While the most simple model for sorting assumes *atomic* keys, an important class of keys are strings to be sorted lexicographically. Here, it is important to exploit the structure of the keys to avoid costly repeated comparisons of entire strings. String sorting is for example needed in database index construction, some suffix sorting algorithms, or MapReduce tools. Although there is a correspondingly large volume of work on sequential string sorting, there is very little work on parallel string sorting. This is surprising since parallelism is now the only way to get performance out of Moore's law so that any performance critical algorithm needs to be parallelized. We therefore started to look for practical parallel string sorting algorithms for modern multi-core shared memory machines. Our focus is on large inputs. This means that besides parallelization we have to take the high cost of branch mispredictions and the memory hierarchy into account. For most multi-core systems, this hierarchy exhibits many processor-local caches but disproportionately few shared memory channels to RAM.

After introducing notation and previous approaches in Section 2, Section 3 explains our parallel string sorting algorithms, in particular super scalar string sample sort (S^5) but also multikey quicksort and radix sort. These algorithms are evaluated experimentally in Section 4.

We would like to thank our students Florian Drews, Michael Hamann, Christian Käser, and Sascha Denis Knöpfle who implemented prototypes of our ideas.

---

[*] This paper is a short version of the technical report [3].

H.L. Bodlaender and G.F. Italiano (Eds.): ESA 2013, LNCS 8125, pp. 169–180, 2013.
© Springer-Verlag Berlin Heidelberg 2013

2   Preliminaries

Our input is a set $\mathcal{S} = \{s_1, \ldots, s_n\}$ of n strings with total length N. A string is a zero-based array of $|s|$ characters from the alphabet $\Sigma = \{1, \ldots, \sigma\}$. For the implementation, we require that strings are zero-terminated, i.e., $s[|s| - 1] = 0 \notin \Sigma$. Let D denote the *distinguishing prefix size* of $\mathcal{S}$, i.e., the total number of characters that need to be inspected in order to establish the lexicographic ordering of $\mathcal{S}$. D is a natural lower bound for the execution time of sequential string sorting. If, moreover, sorting is based on character comparisons, we get a lower bound of $\Omega(D + n \log n)$.

Sets of strings are usually represented as arrays of pointers to the beginning of each string. Note that this indirection means that, in general, every access to a string incurs a cache fault even if we are scanning an array of strings. This is a major difference to atomic sorting algorithms where scanning is very cache efficient. Let $\mathrm{lcp}(s, t)$ denote the length of the *longest common prefix* (LCP) of s and t. In a sequence or array of strings x let $\mathrm{lcp}_x(i)$ denote $\mathrm{lcp}(x_{i-1}, x_i)$. Our target machine is a shared memory system supporting p hardware threads (processing elements – PEs) on $\Theta(p)$ cores.

2.1   Basic Sequential String Sorting Algorithms

Multikey quicksort [2] is a simple but effective adaptation of quicksort to strings. When all strings in $\mathcal{S}$ have a common prefix of length ℓ, the algorithm uses character $c = s[\ell]$ of a pivot string $s \in \mathcal{S}$ (e.g. a pseudo-median) as a *splitter* character. $\mathcal{S}$ is then partitioned into $\mathcal{S}_<$, $\mathcal{S}_=$, and $\mathcal{S}_>$ depending on comparisons of the ℓ-th character with c. Recursion is done on all three subproblems. The key observation is that the strings in $\mathcal{S}_=$ have common prefix length $\ell + 1$ which means that compared characters found to be equal with c never need to be considered again. Insertion sort is used as a base case for constant size inputs. This leads to a total execution time of $\mathcal{O}(D + n \log n)$. Multikey quicksort works well in practice in particular for inputs which fit into the cache.

MSD radix sort [8,10,7] with common prefix length ℓ looks at the ℓ-th character producing σ subproblems which are then sorted recursively with common prefix $\ell + 1$. This is a good algorithm for large inputs and small alphabets since it uses the maximum amount of information within a single character. For input sizes $\mathrm{o}(\sigma)$ MSD radix sort is no longer efficient and one has to switch to a different algorithm for the base case. The running time is $\mathcal{O}(D)$ plus the time for solving the base cases. Using multikey quicksort for the base case yields an algorithm with running time $\mathcal{O}(D + n \log \sigma)$. A problem with large alphabets is that one will get many cache faults if the cache cannot support σ concurrent output streams (see [9] for details).

Burstsort dynamically builds a trie data structure for the input strings. In order to reduce the involved work and to become cache efficient, the trie is built lazily – only when the number of strings referenced in a particular subtree of the trie exceeds a threshold, this part is expanded. Once all strings are inserted, the relatively small sets of strings stored at the leaves of the trie are sorted recursively (for more details refer to [16,17,15] and the references therein).

LCP-Mergesort is an adaptation of mergesort to strings that saves and reuses the LCPs of consecutive strings in the sorted subproblems [11].

2.2 Architecture Specific Enhancements

Caching of characters is very important for modern memory hierarchies as it reduces the number of cache misses due to random access on strings. When performing character lookups, a caching algorithm copies successive characters of the string into a more convenient memory area. Subsequent sorting steps can then avoid random access, until the cache needs to be refilled. This technique has successfully been applied to radix sort [10], multikey quicksort [12], and in its extreme to burstsort [17].

Super-Alphabets can be used to accelerate string sorting algorithms which originally look only at single characters. Instead, multiple characters are grouped as one and sorted together. However, most algorithms are very sensitive to large alphabets, thus the group size must be chosen carefully. This approach results in 16-bit MSD radix sort and fast sorters for DNA strings. If the grouping is done to fit many characters into a machine word, this is also called *word parallelism*.

Unrolling, fission and vectorization of loops are methods to exploit out-of-order execution and super scalar parallelism now standard in modern CPUs. However, only specific, simple data in-dependencies can be detected and thus inner loops must be designed with care (e.g. for radix sort [7]).

2.3 (Parallel) Atomic Sample Sort

There is a huge amount of work on parallel sorting so that we can only discuss the most relevant results. Besides (multiway)-mergesort, perhaps the most practical parallel sorting algorithms are parallelizations of radix sort (e.g. [19]) and quicksort [18] as well as *sample sort* [4]. Sample sort is a generalization of quicksort working with $k - 1$ pivots at the same time. For small inputs sample sort uses some sequential base case sorter. Larger inputs are split into k *buckets* $b_1, \ldots, b_k$ by determining $k - 1$ splitter keys $x_1 \leq \cdots \leq x_{k-1}$ and then classifying the input elements – element s goes to bucket b_i if $x_{i-1} < s \leq x_i$ (where x_0 and x_k are defined as sentinel elements – x_0 being smaller than all possible input elements and x_k being larger). Splitters can be determined by drawing a random sample of size $\alpha k - 1$ from the input, sorting it, and then taking every α-th element as a splitter. Parameter α is the *oversampling* factor. The buckets are then sorted recursively and concatenated. "Traditional" parallel sample sort chooses $k = p$ and uses a sample big enough to assure that all buckets have approximately equal size. Sample sort is also attractive as a sequential algorithm since it is more cache efficient than quicksort and since it is particularly easy to avoid branch mispredictions (super scalar sample sort – S^4) [13]. In this case, k is chosen in such a way that classification and data distribution can be done in a cache efficient way.

2.4   More Related Work

There is some work on PRAM algorithms for string sorting (e.g. [5]). By combining pairs of adjacent characters into single characters, one obtains algorithms with work $\mathcal{O}(N \log N)$ and time $\mathcal{O}(\log N / \log \log N)$. Compared to the sequential algorithms this is suboptimal unless $D = \mathcal{O}(N) = \mathcal{O}(n)$ and with this approach it is unclear how to avoid work on characters outside distinguishing prefixes.

We found no publications on practical parallel string sorting. However, Takuya Akiba has implemented a parallel radix sort [1], Tommi Rantala's library [12] contains multiple parallel mergesorts and a parallel SIMD variant of multikey quicksort, and Nagaraja Shamsundar [14] also parallelized Waihong Ng's LCP-mergesort [11]. Of all these implementations, only the radix sort by Akiba scales fairly well to many-core architectures. For this paper, we exclude the other implementations and discuss their scalability issues in our technical report [3].

3   Shared Memory Parallel String Sorting

Already in a sequential setting, theoretical considerations and experiments [3] indicate that *the* best string string sorting algorithm does not exist. Rather, it depends at least on n, D, σ, and the hardware. Therefore we decided to parallelize several algorithms taking care that components like data distribution, load balancing or base case sorter can be reused. Remarkably, most algorithms in Section 2.1 can be parallelized rather easily and we will discuss parallel versions in Sections 3.2–3.4. However, none of these parallelizations make use of the striking new feature of modern many-core systems: many multi-core processors with individual cache levels but relatively few and slow memory channels to shared RAM. Therefore we decided to design a new string sorting algorithm based on sample sort, which exploits these properties. Preliminary result on string sample sort have been reported in the bachelor thesis of Knöpfle [6].

3.1   String Sample Sort

In order to adapt the atomic sample sort from Section 2.3 to strings, we have to devise an efficient classification algorithm. Also, in order to approach total work $\mathcal{O}(D + n \log n)$ we have to use the information gained during classification into buckets b_i in the recursive calls. This can be done by observing that

$$\forall 1 \leq i \leq k : \forall s, t \in b_i : \mathrm{lcp}(s, t) \geq \mathrm{lcp}_x(i) \ . \tag{1}$$

Another issue is that we have to reconcile the parallelization and load balancing perspective from traditional parallel sample sort with the cache efficiency perspective of super scalar sample sort. We do this by using dynamic load balancing which includes parallel execution of recursive calls as in parallel quicksort.

In our technical report [3] we outline a variant of string sample sort that uses a trie data structure and a number of further tricks to enable good asymptotic performance. However, we view this approach as somewhat risky for a first reasonable implementation. Hence, in the following, we present a more pragmatic implementation.

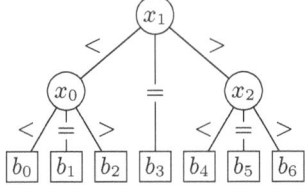

Fig. 1. Ternary search tree for $v = 3$ splitters

Super Scalar String Sample Sort (S^5) – A Pragmatic Solution. We adapt the implicit binary search tree approach used in S^4 [13] to strings. Rather than using arbitrarily long splitters as in trie sample sort [3], or all characters of the alphabet as in radix sort, we design the splitter keys to consist of *as many characters as fit into a machine word*. In the following let w denote the number of characters fitting into one machine word (for 8-bit characters and 64-bit machine words we would have $w = 8$). We choose $v = 2^d - 1$ splitters $x_0, \ldots, x_{v-1}$ from a sorted sample to construct a perfect binary search tree, which is used to classify a set of strings based on the next w characters at common prefix ℓ. The main disadvantage of this approach is that there may be many input strings whose next w characters are identical. For these strings, the classification does not reveal much information. We make the best out of such inputs by explicitly defining *equality buckets* for strings whose next w characters exactly match x_i. For equality buckets, we can increase the common prefix length by w in the recursive calls, i.e., these characters will never be inspected again. In total, we have $k = 2v + 1$ different buckets $b_0, \ldots, b_{2v}$ for a ternary search tree (see Figure 1). Testing for equality can either be implemented by explicit equality tests at each node of the search tree (which saves time when most elements end up in a few large equality buckets) or by going down the search tree all the way to a bucket b_i (i even) doing only $\leq$-comparisons, followed by a single equality test with $x_{\frac{i}{2}}$, unless $i = 2v$. This allows us to completely unroll the loop descending the search tree. We can then also unroll the loop over the elements, interleaving independent tree descents. Like in [13], this is an important optimization since it allows the instruction scheduler in a super scalar processor to parallelize the operations by drawing data dependencies apart. The strings in buckets b_0 and b_{2v} keep common prefix length ℓ. For other even buckets b_i the common prefix length is increased by $\text{lcp}_x(\frac{i}{2})$. An analysis similar to the one of multikey quicksort yields the following asymptotic running time bound.

Lemma 1. *String sample sort with implicit binary trees and word parallelism can be implemented to run in time* $\mathcal{O}\left(\frac{D}{w} \log v + n \log n\right)$.

Implementation Details. Goal of S^5 is to have a common classification data structure that fits into the cache of all cores. Using this data structure, all PEs can independently classify a subset of the strings into buckets in parallel. As most commonly done in radix sort, we first classify strings, counting how many fall into each bucket, then calculate a prefix sum and redistribute the string pointers accordingly. To avoid traversing the tree twice, the bucket index of each string is stored in an oracle. Additionally, to make higher use of super scalar parallelism, we even separate the classification loop from the counting loop [7].

Like in S^4, the binary tree of splitters is stored in level-order as an array, allowing efficient traversal using $i := 2i + \{0, 1\}$, without branch mispredictions. To perform the equality check after traversal without extra indirections, the splitters are additionally stored in order. Another idea is to keep track of the last $\leq$-branch during traversal; this however was slower and requires an extra register. A third variant is to check for equality after each comparison, which requires only an additional JE instruction and no extra CMP. The branch misprediction cost is counter-balanced by skipping the rest of the tree. An interesting observation is that, when breaking the tree traversal at array index i, then the corresponding equality bucket b_j can be calculated from i using only bit operations (note that i is an index in level-order, while j is in-order). Thus in this third variant, no additional in-order splitter array is needed.

The sample is drawn pseudo-randomly with an oversampling factor $\alpha = 2$ to keep it in cache when sorting with STL's introsort and building the search tree. Instead of using the straight-forward equidistant method to draw splitters from the sample, we use a simple recursive scheme that tries to avoid using the same splitter multiple times: Select the middle sample m of a range $a..b$ (initially the whole sample) as the middle splitter $\bar{x}$. Find new boundaries b' and a' by scanning left and right from m skipping samples equal to $\bar{x}$. Recurse on $a..b'$ and $a'..b$.

For current 64-bit machines with 256 KiB L2 cache, we use $v = 8191$. Note that the limiting data structure which must fit into L2 cache is not the splitter tree, which is only 64 KiB for this v, but is the bucket counter array containing $2v + 1$ counters, each 8 bytes long. We did not look into methods to reduce this array's size, because the search tree must also be stored both in level-order and in in-order.

Parallelization of S^5. Parallel S^5 (pS^5) is composed of four sub-algorithms for differently sized subsets of strings. For string sets $\mathcal{S}$ with $|\mathcal{S}| \geq \frac{n}{p}$, a *fully parallel version* of S^5 is run, for large sizes $\frac{n}{p} > |\mathcal{S}| \geq t_m$ a sequential version of S^5 is used, for sizes $t_m > |\mathcal{S}| \geq t_i$ the fastest sequential algorithm for medium-size inputs (caching multikey quicksort from Section 3.3) is called, which internally uses insertion sort when $|\mathcal{S}| < t_i$. The thresholds t_i and t_m depend on hardware specifics, see Section 4 for empirically determined values.

The fully parallel version of S^5 uses $p' = \lceil \frac{|\mathcal{S}|}{p} \rceil$ threads for a subset $\mathcal{S}$. It consists of four stages: selecting samples and generating a splitter tree, parallel classification and counting, global prefix sum, and redistribution into buckets. Selecting the sample and constructing the search tree are done sequentially, as these steps have negligible run time. Classification is done independently, dividing the string set evenly among the p' threads. The prefix sum is done sequentially once all threads finish counting.

In the sequential version of S^5 we permute the string pointer array in-place by walking cycles of the permutation [8]. Compared to out-of-place redistribution into buckets, the in-place algorithm uses fewer input/output streams and requires no extra space. The more complex instruction set seems to have only little negative impact, as today, memory access is the main bottleneck. However, for fully parallel S^5, an in-place permutation cannot be done in this manner. We therefore resort to out-of-place redistribution, using an extra string pointer array of size n. The string pointers are not copied back immediately. Instead, the role of the extra array and original array are swapped for the recursion.

All work in parallel S^5 is dynamically load balanced via a central job queue. Dynamic load balancing is very important and probably unavoidable for parallel string sorting, because any algorithm must adapt to the input string set's characteristics. We use the lock-free queue implementation from Intel's Thread Building Blocks (TBB) and threads initiated by OpenMP to create a light-weight thread pool.

To make work balancing most efficient, we modified all sequential sub-algorithms of parallel S^5 to use an explicit recursion stack. The traditional way to implement dynamic load balancing would be to use work stealing among the sequentially working threads. This would require the operations on the local recursion stacks to be synchronized or atomic. However, for our application fast stack operations are crucial for performance as they are very frequent. We therefore choose a different method: voluntary work sharing. If the global job queue is empty and a thread is idle, then a global atomic boolean flag is set to indicate that other threads should share their work. These then free the *bottom level* of their local recursion stack (containing the largest subproblems) and enqueue this level as separate, independent jobs. This method avoids costly atomic operations on the local stack, replacing it by a faster (not necessarily synchronized) boolean flag check. The short wait of an idle thread for new work does not occur often, because the largest recursive subproblems are shared. Furthermore, the global job queue never gets large because most subproblems are kept on local stacks.

3.2 Parallel Radix Sort

Radix sort is very similar to sample sort, except that classification is much faster and easier. Hence, we can use the same parallelization toolkit as with S^5. Again, we use three sub-algorithms for differently sized subproblems: fully parallel radix sort for the original string set and large subsets, a sequential radix sort for medium-sized subsets and insertion sort for base cases. Fully parallel radix sort consists of a counting phase, global prefix sum and a redistribution step. Like in S^5, the redistribution is done out-of-place by copying pointers into a shadow array. We experimented with 8-bit and 16-bit radixes for the full parallel step. Smaller recursive subproblems are processed independently by sequential radix sort (with in-place permuting), and here we found 8-bit radixes to be faster than 16-bit sorting. Our parallel radix sort implementation uses the same work balancing method as parallel S^5.

3.3   Parallel Caching Multikey Quicksort

Our preliminary experiments with sequential string sorting algorithms [3] showed a surprise winner: an enhanced variant of multikey quicksort by Tommi Rantala [12] often outperformed more complex algorithms. This variant employs both caching of characters and uses a super-alphabet of $w = 8$ characters, exactly as many as fit into a machine word. The string pointer array is augmented with w cache bytes for each string, and a string subset is partitioned by a whole machine word as splitter. Key to the algorithm's good performance, is that the cached characters are reused for the recursive subproblems $\mathcal{S}_<$ and $\mathcal{S}_>$, which greatly reduces the number of string accesses to at most $\lceil \frac{D}{w} \rceil + n$ in total.

In light of this variant's good performance, we designed a parallelized version. We use three sub-algorithms: *fully parallel caching multikey quicksort*, the original sequential caching variant (with explicit recursion stack) for medium and small subproblems, and insertion sort as base case. For the fully parallel sub-algorithm, we generalized a block-wise processing technique from (two-way) parallel atomic quicksort [18] to three-way partitioning. The input array is viewed as a sequence of blocks containing B string pointers together with their w cache characters. Each thread holds exactly three blocks and performs ternary partitioning by a globally selected pivot. When all items in a block are classified as $<$, $=$ or $>$, then the block is added to the corresponding output set $\mathcal{S}_<$, $\mathcal{S}_=$, or $\mathcal{S}_>$. This continues as long as unpartitioned blocks are available. If no more input blocks are available, an extra empty memory block is allocated and a second phase starts. The second partitioning phase ends with fully classified blocks, which might be only partially filled. Per fully parallel partitioning step there can be at most $3p'$ partially filled blocks. The output sets $\mathcal{S}_<$, $\mathcal{S}_=$, and $\mathcal{S}_>$ are processed recursively with threads divided as evenly among them as possible. The cached characters are updated only for the $\mathcal{S}_=$ set.

In our implementation we use atomic compare-and-swap operations for block-wise processing of the initial string pointer array and Intel TBB's lock-free queue for sets of blocks, both as output sets and input sets for recursive steps. When a partition reaches the threshold for sequential processing, then a continuous array of string pointers plus cache characters is allocated and the block set is copied into it. On this continuous array, the usual ternary partitioning scheme of multikey quicksort is applied sequentially. Like in the other parallelized algorithms, we use dynamic load balancing and free the bottom level when re-balancing is required. We empirically determined $B = 128\,\mathrm{Ki}$ as a good block size.

3.4   Burstsort and LCP-Mergesort

Burstsort is one of the fastest string sorting algorithms and cache-efficient for many inputs, but it looks difficult to parallelize. Keeping a common burst trie would require prohibitively many synchronized operations, while building independent burst tries on each PE would lead to the question how to merge multiple tries of different structure.

One would like to generalize LCP-mergesort to a parallel p-way LCP-aware merging algorithm. This looks promising in general but we leave this for future work since LCP-mergesort is not really the best sequential algorithm in our experiments.

4   Experimental Results

We implemented parallel S^5, multikey quicksort and radixsort in C++ and compare them with Akiba's radix sort [1]. We also integrated many sequential implementations into our test framework, and compiled all programs using gcc 4.6.3 with optimizations `-O3 -march=native`. In our report [3] we discuss the performance of sequential string sorters. Our implementations and test framework are available from `http://tbingmann.de/2013/parallel-string-sorting`.

Experimental results we report in this paper stem from two platforms. The larger machine, IntelE5, has four 8-core Intel Xeon E5-4640 processors containing a total of 32 cores and supporting $p = 64$ hardware threads. The second platform is a consumer-grade Intel i7 920 with four cores and $p = 8$ hardware threads. Turbo-mode was disabled on IntelE5. Our technical report [3] contains further details of these machines and experimental results from three additional platforms. We selected the following datasets, all with 8-bit alphabets. More characteristics of these instances are shown in Table 1.

URLs contains all URLs on a set of web pages which were crawled breadth-first from the authors' institute website. They include the protocol name.

Random from [16] are strings of length $[0, 20)$ over the ASCII alphabet $[33, 127)$, with both lengths and characters chosen uniform at random.

GOV2 is a TREC test collection consisting of 25 million HTML pages, PDF and Word documents retrieved from websites under the .gov top-level domain. We consider the whole concatenated corpus for line-based string sorting.

Wikipedia is an XML dump of the most recent version of all pages in the English Wikipedia, which was obtained from `http://dumps.wikimedia.org/`; our dump is dated `enwiki-20120601`. Since the XML data is not line-based, we perform *suffix sorting* on this input.

We also include the three largest inputs Ranjan **Sinha** [16] tested burstsort on: a set of **URLs** excluding the protocol name, a sequence of genomic strings of length 9 over a **DNA** alphabet, and a list of non-duplicate English words called **NoDup**. The "largest" among these is NoDup with only 382 MiB, which is why we consider these inputs more as reference datasets than as our target.

The test framework sets up a separate run environment for each test run. The program's memory is locked into RAM, and to isolate heap fragmentation, it was very important to fork() a child process for each run. We use the largest prefix $[0, 2^d)$ of our inputs which can be processed with the available RAM. We determined $t_m = 64$ Ki and $t_i = 64$ as good thresholds to switch sub-algorithms.

Figure 2 shows a selection of the detailed parallel measurements from our report [3]. For large instances we show results on IntelE5 (median of 1–3 repetitions) and for small instances on Inteli7 (of ten repetitions). The plots show the speedup of our implementations and Akiba's radix sort over the best sequential algorithm [3]. We included pS⁵-Unroll, which interleaves three unrolled descents

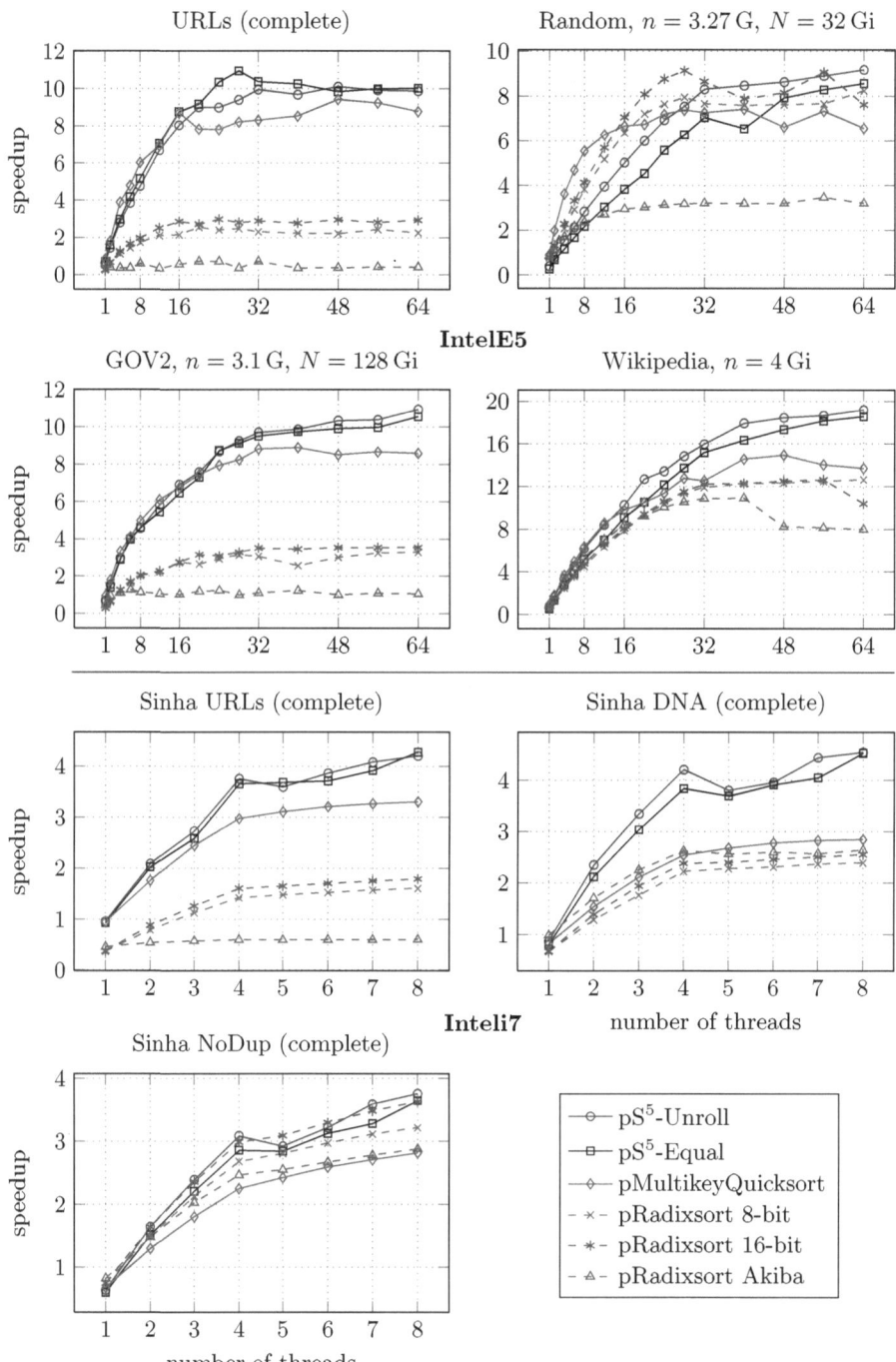

Fig. 2. Speedup of parallel algorithm implementations on IntelE5 (top four plots) and Inteli7 (bottom three plots)

Table 1. Characteristics of the selected input instances

Name	n	N	$\frac{D}{N}$ (D)	σ	avg. $|s|$
URLs	1.11 G	70.7 Gi	93.5 %	84	68.4
Random	∞	∞	–	94	10.5
GOV2	11.3 G	425 Gi	84.7 %	255	40.3
Wikipedia	83.3 G	$\frac{1}{2}n(n+1)$	(79.56 T)	213	$\frac{1}{2}(n+1)$
Sinha URLs	10 M	304 Mi	97.5 %	114	31.9
Sinha DNA	31.6 M	302 Mi	100 %	4	10.0
Sinha NoDup	31.6 M	382 Mi	73.4 %	62	12.7

of the search tree, pS^5-Equal, which unrolls a single descent testing equality at each node, our parallel multikey quicksort (pMKQS), and radix sort with 8-bit and 16-bit fully parallel steps. On all platforms, our parallel implementations yield good speedups, limited by memory bandwidth, not processing power. On IntelE5 for all four test instances, pMKQS is fastest for small numbers of threads. But for higher numbers, pS^5 becomes more efficient than pMKQS, because it utilizes memory bandwidth better. On all instances, except Random, pS^5 yields the highest speedup for both the number of physical cores and hardware threads. On Random, our 16-bit parallel radix sort achieves a slightly higher speedup. Akiba's radix sort does not parallelize recursive sorting steps (only the top-level is parallelized) and only performs simple load balancing. This can be seen most pronounced on URLs and GOV2. On Inteli7, pS^5 is consistently faster than pMKQS for Sinha's smaller datasets, achieving speedups of 3.8–4.5, which is higher than the three memory channels on this platform. On IntelE5, the highest speedup of 19.2 is gained with pS^5 for suffix sorting Wikipedia, again higher than the 4×4 memory channels. For all test instances, except URLs, the fully parallel sub-algorithm of pS^5 was run only 1–4 times, thus most of the speedup is gained in the sequential S^5 steps. The pS^5-Equal variant handles URL instances better, as many equal matches occur here. However, for all other inputs, interleaving tree descents fares better. Overall, pS^5-Unroll is currently the best parallel string sorting implementation on these platforms.

5 Conclusions and Future Work

We have demonstrated that string sorting can be parallelized successfully on modern multi-core shared memory machines. In particular, our new string sample sort algorithm combines favorable features of some of the best sequential algorithms – robust multiway divide-and-conquer from burstsort, efficient data distribution from radix sort, asymptotic guarantees similar to multikey quicksort, and word parallelism from cached multikey quicksort.

Implementing some of the refinements discussed in our report [3] are likely to yield further improvements for pS^5. To improve scalability on large machines, we may also have to look at NUMA (non uniform memory access) effects more explicitly. Developing a parallel multiway LCP-aware mergesort might then become interesting.

References

1. Akiba, T.: Parallel string radix sort in C++ (2011), http://github.com/iwiwi/parallel-string-radix-sort (git repository accessed November 2012)
2. Bentley, J.L., Sedgewick, R.: Fast algorithms for sorting and searching strings. In: ACM 8th Symposium on Discrete Algorithms, pp. 360–369 (1997)
3. Bingmann, T., Sanders, P.: Parallel string sample sort. Tech. rep. (May 2013), see ArXiv e-print arXiv:1305.1157
4. Blelloch, G.E., Leiserson, C.E., Maggs, B.M., Plaxton, C.G., Smith, S.J., Zagha, M.: A comparison of sorting algorithms for the connection machine CM-2. In: 3rd Symposium on Parallel Algorithms and Architectures. pp. 3–16 (1991)
5. Hagerup, T.: Optimal parallel string algorithms: sorting, merging and computing the minimum. In: Proceedings of the Twenty-Sixth Annual ACM Symposium on Theory of Computing, STOC 1994, pp. 382–391. ACM, New York (1994)
6. Knöpfle, S.D.: String samplesort, bachelor Thesis, Karlsruhe Institute of Technology (November 2012) (in German)
7. Kärkkäinen, J., Rantala, T.: Engineering radix sort for strings. In: Amir, A., Turpin, A., Moffat, A. (eds.) SPIRE 2008. LNCS, vol. 5280, pp. 3–14. Springer, Heidelberg (2008)
8. McIlroy, P.M., Bostic, K., McIlroy, M.D.: Engineering radix sort. Computing Systems 6(1), 5–27 (1993)
9. Mehlhorn, K., Sanders, P.: Scanning multiple sequences via cache memory. Algorithmica 35(1), 75–93 (2003)
10. Ng, W., Kakehi, K.: Cache efficient radix sort for string sorting. IEICE Transactions on Fundamentals of Electronics, Communications and Computer Sciences E90-A(2), 457–466 (2007)
11. Ng, W., Kakehi, K.: Merging string sequences by longest common prefixes. IPSJ Digital Courier 4, 69–78 (2008)
12. Rantala, T.: Library of string sorting algorithms in C++ (2007), http://github.com/rantala/string-sorting (git repository accessed November 2012)
13. Sanders, P., Winkel, S.: Super scalar sample sort. In: Albers, S., Radzik, T. (eds.) ESA 2004. LNCS, vol. 3221, pp. 784–796. Springer, Heidelberg (2004)
14. Shamsundar, N.: A fast, stable implementation of mergesort for sorting text files (May 2009), http://code.google.com/p/lcp-merge-string-sort (source downloaded November 2012)
15. Sinha, R., Wirth, A.: Engineering Burstsort: Toward fast in-place string sorting. J. Exp. Algorithmics 15(2.5), 1–24 (2010)
16. Sinha, R., Zobel, J.: Cache-conscious sorting of large sets of strings with dynamic tries. J. Exp. Algorithmics 9(1.5), 1–31 (2004)
17. Sinha, R., Zobel, J., Ring, D.: Cache-efficient string sorting using copying. J. Exp. Algorithmics 11(1.2), 1–32 (2007)
18. Tsigas, P., Zhang, Y.: A simple, fast parallel implementation of quicksort and its performance evaluation on SUN enterprise 10000. In: PDP, pp. 372–381. IEEE Computer Society (2003)
19. Wassenberg, J., Sanders, P.: Engineering a multi-core radix sort. In: Jeannot, E., Namyst, R., Roman, J. (eds.) Euro-Par 2011, Part II. LNCS, vol. 6853, pp. 160–169. Springer, Heidelberg (2011)

Exclusive Graph Searching

Lélia Blin[1], Janna Burman[2], and Nicolas Nisse[3]

[1] Université d'Evry Val d'Essonne and LIP6-CNRS, France
lelia.blin@lip6.fr
[2] LRI (CNRS/UPSud), Orsay, France
janna.burman@lri.fr
[3] COATI, Inria, I3S (CNRS/UNS), Sophia Antipolis, France
nicolas.nisse@inria.fr

Abstract. This paper tackles the well known *graph searching* problem, where a team of searchers aims at capturing an intruder in a network, modeled as a graph. All variants of this problem assume that any node can be simultaneously occupied by several searchers. This assumption may be unrealistic, e.g., in the case of searchers modeling physical searchers, or may require each individual node to provide additional resources, e.g., in the case of searchers modeling software agents. We thus investigate *exclusive* graph searching, in which no two or more searchers can occupy the same node at the same time, and, as for the classical variants of graph searching, we study the minimum number of searchers required to capture the intruder. This number is called the *exclusive search number* of the considered graph. Exclusive graph searching appears to be considerably more complex than classical graph searching, for at least two reasons: (1) it does not satisfy the *monotonicity property*, and (2) it is not *closed under minor*. Nevertheless, we design a polynomial-time algorithm which, given any tree T, computes the exclusive search number of T. Moreover, for any integer k, we provide a characterization of the trees T with exclusive search number at most k. This characterization allows us to describe a special type of exclusive search strategies, that can be executed in a distributed environment, i.e., in a framework in which the searchers are limited to cooperate in a distributed manner.

1   Introduction

Graph Searching was first introduced by Breisch [9, 10] in the context of speleology, for solving the problem of rescuing a lost speleologist in a network of caves. Alternatively, graph searching can be defined as a particular type of cops-and-robber game, as follows. Given a graph G, modeling any kind of network, design a strategy for a team of searchers moving in G resulting in capturing an intruder. There are no limitations on the capabilities of the intruder, who may be arbitrary fast, be aware of the whole structure of the network, and be perpetually aware of the current positions of the searchers. The objective is to compute the minimum number of searchers required to capture the intruder in G.

To be more formal regarding the behavior of the intruder, it is more convenient to rephrase the problem in terms of *clearing* a network of pipes contaminated by

H.L. Bodlaender and G.F. Italiano (Eds.): ESA 2013, LNCS 8125, pp. 181–192, 2013.

some gas [22]. In this framework, a team of searchers aims at clearing the edges of a graph, which are initially *contaminated*. Searchers stand on the nodes of the graph, and can *slide* along its edges. Moreover, a searcher can be removed from one node and then placed to any other node, i.e., a searcher can "jump" from node to another. Sliding of a searcher along an edge, as well as positioning one searcher at each extremity of an edge, results in clearing that edge. Nevertheless, if there is a path free of searchers between a *clear* edge and a *contaminated* edge, then the former is instantaneously *recontaminated*. Thus, to actually keep an edge clear, searchers must occupy appropriate nodes for avoiding recontamination to occur.

Informally, a *search strategy* is a sequence of movements executed by the searchers, resulting in all edges being eventually clear. The main question tackled in the context of graph searching is, given a graph G, compute a search strategy minimizing the number of searchers required for clearing G. This number, denoted by $s(G)$, is called the *search number* of the graph G. For instance, one searcher is sufficient to clear a line, while two searchers are necessary in a ring: the search number of any line is 1, while the search number of any ring is 2.

The above variant of graph searching is actually called *mixed-search* [4]. Other classical variants of graph searching are *node-search* [3], *edge-search* [21, 22], *connected-search* [2], etc. All these variants suffer from two serious limitations as far as practical applications are concerned.

– First, they all assume that any node can be simultaneously occupied by several searchers. This assumption may be unrealistic in several contexts. Typically, placing several searchers at the same node may simply be impossible in a physical environment in which, e.g., the searchers are modeling physical robots moving in a network of pipes. In the case of software agents deployed in a computer network, maintaining several searchers at the same node may consume local resources (e.g., memory, computation cycles, etc.). We investigate *exclusive* graph searching, i.e., graph searching bounded to satisfy the *exclusivity constraint* stating that no two or more searchers can occupy the same node at the same time.
– Second, most variants of graph searching also suffer from another unrealistic assumption: searcher are enabled to "jump" from one node of the graph, to another, potentially far away, node (e.g., see the classical mixed-search, defined above). We restrict ourselves to the more realistic *internal* search strategies [2], in which searchers are limited to move along the edges of the graph, that is, restricted to satisfy the *internality constraint*.

To sum up, we define *exclusive-search* as mixed-search with the additional exclusivity and internality constraints. As for all classical variants of graph searching, we study the minimum number of searchers required to clear all edges of a graph G. This number is called the *exclusive search number*, denoted by $xs(G)$.

We show that exclusive graph searching behaves very differently from classical graph searching, for at least two reasons. First, it does not satisfy the *monotonicity property* That is, there are graphs (even trees) in which every exclusive search strategy using the minimum number of searchers requires to let recontamination occurring at some step of the strategy. Second, exclusive graph searching is not *closed under minor* taking (not even under subgraph). That is, there are graphs G and H such that H is a subgraph of G, and $\mathtt{xs}(H) > \mathtt{xs}(G)$. The absence of these two properties (which will be formally established in the paper) makes exclusive-search considerably different from classical search, and its analysis requires introducing new techniques.

Our Results. First, in Sec. 2, we formally define exclusive graph searching and present basic properties for general graphs. Motivated by certain positive results for trees and inspired by the pioneering work of Parson [22] and Megiddo et al. [21], we are then essentially focussing on trees. We observe that the exclusive search number of a graph can differ exponentially from the values of classical search numbers: in a tree, the former can be linear in the number of nodes n, while all classical search numbers of trees are at most $O(\log n)$. Our main result (Sec. 3) is a polynomial-time algorithm which, given any tree T, computes the exclusive search number $\mathtt{xs}(T)$ of T, as well as an exclusive search strategy using $\mathtt{xs}(T)$ searchers for clearing T. Our algorithm is based on a characterization of the trees with exclusive search number at most k, for any given $k \leq n$.

The above characterization allows us to describe an important type of exclusive search strategies, that can be executed in a distributed environment, i.e., in a framework in which the searchers are restricted to cooperate in a distributed manner (Sec. 4). More specifically, we consider the classical (discrete) CORDA[1] (a.k.a. *Look-Compute-Move*) model [16, 24] for *autonomous* searchers moving in a network. We prove that, for any anonymous asymmetric tree T, as well as for any tree whose nodes are labeled with unique IDs, and for any $n \geq k \geq \mathtt{xs}(T)$, there exists a distributed protocol enabling k searchers to clear T.

Hence, an interesting outcome of this paper is that the minimum number of searchers needed to clear an (anonymous asymmetric or uniquely labeled) tree in a distributed manner is not larger than the one required when the searchers are coordinated and scheduled by a central entity. This is particularly surprising, especially when having in mind that, in the distributed setting, symmetry breaking becomes much more harder (even in an asymmetric network), and the scheduling of the searchers (i.e., which searchers are activated at any point in time) is under the full control of an adversary. Due to the lack of space, most of the proofs are omitted or sketched. All complete proofs can be found in [6].

Related Work. Graph searching has mainly been studied in the centralized setting for its relationship with the *treewidth* and *pathwidth* of graphs [4, 18]. The problem of computing the search number of a graph is NP-hard [21]. However, this problem is polynomial in various graph classes [17, 19, 26]. In particular, it has been widely studied in the class of trees [14, 21–23, 25].

An important property of mixed-graph searching is the monotonicity property. A strategy is *monotone* if no edges are recontaminated once they have been cleared. For any graph G, there is an optimal winning monotone (mixed-search) strategy [4]. This enables to prove that the number of steps of an optimal strategy is polynomially bounded by the number of edges. Hence, the problem to decide the mixed-search number of a graph belongs to NP. Instead, connected

---

[1] COordination of Robots in a Distributed and Asynchronous environment.

graph searching, in which the set of clear edges must always induce a connected subgraph, is not monotone in general [27] and it is not known if connected search is in NP. Connected search is monotone in trees [2]. The connectivity constraint may increase the search number of any graph by a factor up to 2 [13].

Graph searching has been intensively studied in various distributed settings (see, e.g., [7, 11, 15, 20]). Graph searching in the CORDA model has recently been studied for rings [12]. The exclusivity constraint has been already considered in the context of various coordination tasks for mobile entities in enhanced versions of the CORDA model (see, e.g., [1, 8, 12]). In the context of graph searching, the exclusivity constraint has been considered for the first time in our brief announcement [5]. Here, we present and improve some results announced in [5].

2   Exclusive Search

In this section, we provide the formal definition of exclusive graph searching, and present some basic general properties.

Given a connected graph G, an *exclusive search strategy* in G, using $k \leq n$ searchers consists in (1) placing the k searchers at k different nodes of G, and (2) performing a sequence of *moves*. A move consists in sliding one searcher from one extremity u of an edge $e = \{u, v\}$ to its other extremity v. Such a move can be performed only if v is free of searchers. That is, exclusive-search limits the strategy to place at most 1 searcher at each node, at any point in time. The edges of graph G are supposed to be initially contaminated. An edge becomes clear whenever either a searcher slides along it, or one searcher is placed at each of its extremities. An edge becomes recontaminated whenever there is a path free of searchers from that edge to a contaminated edge. A search strategy is *winning* if its execution results in all edges of the graph G being simultaneously clear. The exclusive-search number of G, denoted by $\mathrm{xs}(G)$ is the smallest k for which there exists a winning search strategy in G.

Now, we state and explain the main differences between exclusive search and all classical variants of graph searching. These differences are mainly due to the combination of the two restrictions introduced in exclusive search: two searchers cannot occupy the same node (exclusivity) and a searcher cannot "jump" (internality). Intuitively, the difficulty occurs when a searcher has to go from one node u to a far away node v, and all paths from u to v contain an occupied node.

Consider a simple example of a star with central node c and n leaves. In the classical graph searching, one searcher can occupy c, while a second searcher will sequentially clear all leaves, either by jumping from one leaf to another, or by sliding from one leaf to another, and therefore occupying several times the already occupied node c. In exclusive graph searching, such strategies are not allowed. Intuitively, if a searcher r_1 has to cross a node v that is already occupied by another searcher r_2, the latter should step aside for letting r_1 pass. However, r_2 may occupy v to preserve the graph from recontamination, and moving away from v could lead to recontaminate the whole graph. To avoid this, it may be necessary to use extra searchers (compared to the classical graph searching) that will guard several neighbors of v to prevent from recontamination when r_2 gives way to r_1. It follows that, as opposed to all classical search numbers, which differ by at most some constant multiplicative factor, the exclusive search number may be arbitrary large compared to the mixed-search number, even in trees. For instance, it is easy to check that $\mathtt{xs}(S_n) = n - 2$ for any n-node star S_n, $n \geq 3$. More generally (see [6]):

Claim 1. *For any tree T with maximum degree $\Delta \geq 2$, $\mathtt{xs}(T) > \Delta - 2$.*

This result shows an exponential increase in the number of searchers used to clear a graph since the mixed-search number of n-node trees is at most $O(\log n)$ [22]. On the positive side, we show that, for any graph G with maximum degree Δ, $\mathtt{s}(G) \leq \mathtt{xs}(G) \leq (\Delta - 1)\mathtt{s}(G)$ [6]. To prove it, we consider a classical strategy $\mathcal{S}$ for G using $\mathtt{s}(G)$ searchers. To build an exclusive strategy $\mathcal{S}_{ex}$ for G, we mimic $\mathcal{S}$ using a team of $\Delta - 1$ searchers to "simulate" each searcher in $\mathcal{S}$.

We now turn our attention to the monotonicity property. Indeed, another important difference of exclusive search compared to classical graph searching is that it is not monotone. As explained in the example of a star, when a searcher needs to cross another one, letting the former searcher pass may lead to recontaminate some edges. In spite of that, the goal of the winning strategy is to prevent an "uncontrolled" recontamination. In [6], we prove that:

Claim 2. *Exclusive graph searching is not monotone, even in trees.*

Last, but not least, contrary to classical graph searching, exclusive graph searching is not closed under minor. Indeed, even taking a subgraph can decrease the connectivity which, surprisingly, may not help the searchers (due to the exclusivity constraint). That is, there exist a graph G and a subgraph H of G such that $\mathtt{xs}(H) > \mathtt{xs}(G)$ [6]. Nevertheless, exclusive-search is closed under subgraph in trees (see [6]):

Lemma 1. *For any tree T and any subtree T' of T, $\mathtt{xs}(T') \leq \mathtt{xs}(T)$.*

Contrary to classical graph searching, the proof of this result is not trivial because of the exclusivity property. To prove it, we have to transform an exclusive strategy $\mathcal{S}$ for T into a strategy $\mathcal{S}'$ for T' using the same number of searchers, and without violating the exclusivity property. The fact that $\mathcal{S}$ may be not monotone (i.e., some recontamination may occur during $\mathcal{S}$) makes the proof technical, because one has to "control" the recontamination of T' in $\mathcal{S}'$.

3   Exclusive Search in Trees

This section is devoted to our main result. We present a polynomial-time algorithm which, given any tree T, computes the exclusive search number $\mathtt{xs}(T)$ of T and an exclusive search strategy enabling $\mathtt{xs}(T)$ searchers to clear T. Our algorithm is based on a characterization of the trees with exclusive search number at most k, for any given k. Given a node v in a tree T, a connected component of $T \setminus \{v\}$ is called a *branch* at v. Our characterization establishes a relationship between the exclusive-search number of T and the exclusive-search number of some of the branches adjacent to any node in T. More precisely, we prove that:

Theorem 1. *Let $k \geq 1$. For any tree T, $\mathrm{xs}(T) \leq k$ if and only if, for any node v, the following three properties hold:*

1. *v has degree at most $k + 1$;*
2. *for any branch B at v, $\mathrm{xs}(B) \leq k$;*
3. *for any even $i > 1$, at most i branches B at v have $\mathrm{xs}(B) \geq k - i/2 + 1$.*

To prove the theorem, we first prove (Sec. 3.1) that, for any tree T and $k \geq 1$, $\mathrm{xs}(T) \leq k$, only if the conditions of Th. 1 are satisfied. Then, we show that any tree satisfying these conditions can be decomposed in a particular way, depending on k (Fig. 1). Next, in Sec. 3.2, we describe an exclusive search strategy using at most k searchers, that clears any tree decomposed in such a way.

From the characterization of Th. 1, it follows that $\mathrm{xs}(T)$ can be computed by dynamic programming on T. Moreover, such an algorithm computes the corresponding decomposition (see Section 3.2). Hence, the following result holds:

Theorem 2. *There exists a polynomial-time algorithm that computes $\mathrm{xs}(T)$ and a corresponding exclusive search strategy for any tree T.*

We now prove Theorem 1 using the following notations. For a node $v \in T$, we denote by $N(v)$ the set of the neighbors of v. A *configuration* is a set of distinct nodes $C \subseteq V(T)$ that describes the positions of $|C|$ searchers in T.

3.1   Necessary Conditions for Theorem 1

We first show that the conditions of Theorem 1 are necessary. The fact that the first property is necessary directly follows from Claim 1. The second property is necessary by Lemma 1.

For proving that the third property is necessary, we first have to prove that, for any tree T, any branch B of T, and any exclusive strategy for T, there is a step of the strategy where at least $\mathrm{xs}(B)$ searchers are occupying the nodes of B (see [6]). While such a result is trivial in classical graph searching, it is not the case anymore subject to exclusivity and internality properties. In particular, in classical graph searching, the result is true for any subtree (not necessarily a branch) while it is not the case for the exclusive variant (see [6]). Indeed, let us consider a sub-tree T' of tree T. If T' is given independently of T, the movements of searchers are more constrained because the searchers have less "space" in T'. On the contrary, when T' is inside the tree T, the searchers can use the "extra space" provides by T to clear T'.

Lemma 2. *Let $k \geq 1$. For any tree T, if there exist $v \in V(T)$ and an even integer $i > 1$ such that there is a set $B = \{T_j : \mathrm{xs}(T_j) \geq k - i/2 + 1\}$ of branches at v and $|B| > i$, then $\mathrm{xs}(T) > k$.*

Proof. Let $\mathcal{S}$ be any exclusive strategy that clears T. By the remark above, for any $j \leq |B|$, there is a step of the strategy $\mathcal{S}$ such that at least $k - i/2 + 1$ searchers occupy simultaneously vertices in T_j. Let s_j be the last such step of $\mathcal{S}$ that occurs in T_j. W.l.o.g. assume that $s_{j-1} < s_j$, for any $1 < j \leq |B|$, and we may assume that, before step s_j, T_j is not completely clear (this means that $\mathcal{S}$ uses $k - i/2 + 1$ searchers in T_j only if it is really needed). Then, at step $s_{i/2+1}$, at least $k - i/2 + 1$ searchers are in $T_{i/2+1}$, some vertices have been cleared in T_j for any $j \leq i/2$, and T_j cannot become fully contaminated anymore until the end of the strategy (otherwise there would be another step after s_j where $k - i/2 + 1$ searchers are in T_j).

For the sake of contradiction, let us assume that $\mathcal{S}$ uses at most k searchers. Then, at step $s_{i/2+1}$, at least $k - i/2 + 1$ searchers are in $T_{i/2+1}$ and there are at most $i/2 - 1$ searchers outside $T_{i/2+1}$. That is, at that moment, there is at least one branch $X \in \{T_{i/2+2}, \ldots, T_{|B|}\}$ $(|B| > i)$ at v with still contaminated edges, and at least one branch $Y \in \{T_1, \ldots, T_{i/2}\}$ at v with (at least) some clear edges that must not be recontaminated and no searchers occupy nodes in both these branches. If there is no searcher at v, Y is fully recontaminated - a contradiction.

Otherwise, there is a searcher in v. However, since there is at least one non cleared yet branch without any searcher in it, it has to be cleared by moving there at least one searcher. For that, the searcher from v have to move. However, if this searcher moves (no matter where), there will be still at most $i/2 - 1$ searchers outside $T_{i/2+1}$ and hence, at least one cleared and one uncleared branch without any searcher, and no searcher in v. The cleared branch will be fully recontaminated - a contradiction. $\qquad\square$

Decomposition. Figure 1 presents a particular structure that we prove to exist for any tree T satisfying the properties of Theorem 1, for $k \geq 1$. Specifically, following [21], we prove that there is a unique path $A = (u_1, \cdots, u_p)$ in T called *avenue* such that $p \geq 1$ and, for any component T' of $T \setminus A$, there is an exclusive strategy that clears T' using $< k$ searchers, i.e., $\mathrm{xs}(T') < k$ (bold line in Fig. 1).

In the next section, we describe a strategy, called *ExclusiveClear*, based on this decomposition and allowing k searchers to clear T in an exclusive way. The strategy consists in clearing the subtrees of $T \setminus A$, starting with the subtrees that are adjacent to u_1, then the ones adjacent to u_2 and so on, finishing in u_p. To clear a subtree T' of $T \setminus A$, we proceed in a recursive way. That is, we recursively use *ExclusiveClear* on T' using $k' < k$ searchers. The first difficulty is to ensure that no subtrees that have been cleared are recontaminated. For this purpose, when clearing T', the remaining $k - k'$ searchers that are not needed to clear it are used to prevent recontamination. The second difficulty is to ensure exclusivity: while these $k - k'$ searchers are protecting from recontamination, k' searchers should be able to enter T' to clear it.

3.2   Exclusive Search Strategy to Clear Trees

Let $k \geq 1$ and let T be any tree satisfying Theorem 1 and thus, given with the decomposition of Figure 1. In this section, we informally describe a search

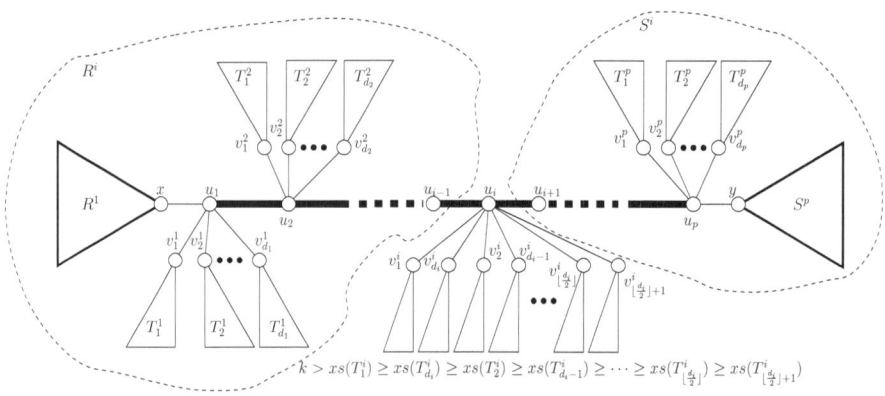

Fig. 1. A tree T with avenue $A = (u_1, \cdots, u_p)$. For any subtree X of $T \setminus A$, $\mathtt{xs}(X) < k$.

strategy that clears T using k searchers. By definition, the following strategy ensures that all moves are performed along paths free of searchers, satisfying the exclusivity and internally properties. To prove its correctness, it is sufficient to show that it uses at most k searchers (in particular, when applying the sub-procedures *bring_searchers* or *transfer* defined below). The formal proof mainly relies on the properties of the decomposition. The formal description of the strategy and the complete correctness proof are provided in [6].

Strategy *ExclusiveClear*. For ease of description, let us assume that $|V(R^1)| \geq k - 1$ (see Fig. 1). Let I be a subset of u_1 and $k - 1$ distinct nodes in R^1. The strategy starts by placing the searchers at the nodes of I. By definition of A, $\mathtt{xs}(R^1) \leq k - 1$. Then, the $k - 1$ searchers in R^1 apply *ExclusiveClear*(R^1) (such a strategy exists by induction and by the definition of A). It is important to mention that the searcher at u_1 preserves R^1 from being recontaminated by the rest of T. After this sequence of moves, all edges in $E(R^1 \cup \{(x, u_1)\})$ are cleared.

Then, we aim at clearing the remaining subtrees of $T \setminus A$ that are adjacent to u_1 ($T_1^1, \cdots, T_{d_i}^1$, in Fig. 1). Moreover, after clearing such a subtree, we need to preserve it from recontamination. Notice that, during the clearing of a subtree, u_1 will always be occupied. However, to ensure that exclusivity property is satisfied when searchers go from one subtree to another (during the *bring_searchers* procedure explained later), we need other nodes being occupied.

In order to use as few searchers as possible, the cleaning of the subtrees adjacent to u_1 must be done in a specific order. The order used to clear the subtrees is built as follows. Each subtree is considered one after the other, in the non-increasing order of $\mathtt{xs}$. In this order, we assign the first subtree to a set S_1, the second one to a set S_2, the third one to a set S_1, the fourth one to a set S_2, and we continue to divide the subtrees until each of them is assigned to one of the two sets. Note that the formula given in Figure 1 respects this order. The resulting $S_1 = \{T_1^1, \ldots, T_{\lceil d_i/2 \rceil}^1\}$ and $S_2 = \{T_{\lceil d_i/2+1 \rceil}^1, \ldots, T_{d_i}^1\}$.

The clearing of the subtrees is then divided into two phases. The subtrees in S_1 are cleared first, in the non-increasing order of their xs. Then, the subtrees in S_2 are cleared in the non-decreasing order of their xs. Each time that a subtree $T_j^1 \in S_1$ has been cleared, one searcher is left on its *root* v_j^1 (its node adjacent to u_1). That is, once a new subtree is cleared, we somehow lose a searcher to clear the next one. This is balanced by the fact that the number of searchers needed to clear the next subtree *decrease*, according to the order of clearing established above, and provided by the properties of T (Th. 1).

After clearing the subtrees in S_1, there are searchers currently "blocked" in the roots of the cleared subtrees. In order to "re-use" these searchers to clear the remaining subtrees, the strategy changes. Now, the roots of the contaminated subtrees will be occupied to prevent recontamination of the cleared subtrees. Procedure *transfer* (explained later) is used to occupy these nodes, ensuring no recontamination of the subtrees and satisfying the exclusivity property. After *transfer*, the searchers at the roots of the cleared subtrees become *free*, i.e., it is possible now to use them to clear the next subtrees.

Then, the subtrees in S_2 have to be cleared in the non-decreasing order of their xs. Each time that a subtree $T_j^1 \in S_2$ has been cleared, the searcher on its root v_j^1 becomes free. That is, we somehow gain a searcher to clear the next subtree, whose search number may *increase*, according to the properties of T.

Once all subtrees of $T \setminus A$ adjacent to u_1 are cleared, the searcher at u_1 goes to u_2 (unless it is already occupied). Now, all the searchers in R^2 (see Fig. 1) became free. Then, a similar strategy is applied for the subtrees of $T \setminus A$ adjacent to u_2, and so on, until all the subtrees adjacent to u_p are cleared.

We now describe more precisely two sub-procedures that are used to implement the strategy we have sketched above.

Procedure *bring_searchers.* It remains to detail how the searchers, once a subtree has been cleared, go to the next subtree, satisfying exclusivity and preventing recontamination. To do so, let $1 \leq i \leq p$ and let us consider the step of the strategy when the branch R^i (see Fig. 1) and all subtrees $T_1^i, \cdots, T_{j_0-1}^i$ ($1 < j_0 \leq d_i$) are cleared (the grey subtrees in Fig. 2(a)). There are two cases to be considered: $j_0 \leq \lceil \frac{d_i}{2} \rceil$ or otherwise.

Assume first that $j_0 \leq \lceil \frac{d_i}{2} \rceil$. As explained before, at this step, the nodes in $\{u_i, v_1^i, \cdots, v_{j_0-1}^i\}$ are occupied, and all other searchers are free and occupy nodes of R^i and T_j^i, for $j < j_0$. It is ensured that also u_{i-1} (if $i = 1$, set $u_{i-1} = x$) will be occupied. The process $bring\_searchers(i, j_0)$ is applied to bring $\mathrm{xs}(T_{j_0}^i)$ searchers into $T_{j_0}^i$. The searchers are brought one by one, from the clear part to $T_{j_0}^i$, without recontamination and satisfying the exclusivity property.

Fig. 2(a) depicts one phase of this process. We prove that, before each phase (but the last one, which is slightly different), there is a free searcher at some node b, either in $R^i \setminus u_{i-1}$ or in $T_j^i \setminus v_j^i$ (for some $j < j_0$). First, the searcher at u_i goes to the furthest unoccupied node in $T_{j_0}^i$ (dotted line 1 in Fig. 2(a)). Second, the searcher at v_j^i (or at u_{i-1}) goes to u_i (dotted line 2 in Fig. 2(a)). Finally, the searcher at b goes to v_j^i (or to u_{i-1}) (dotted line 3 in Fig. 2(a)). Clearly, doing

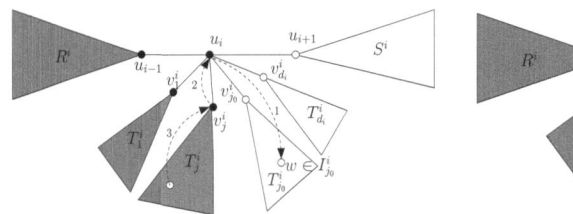
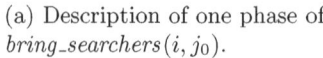
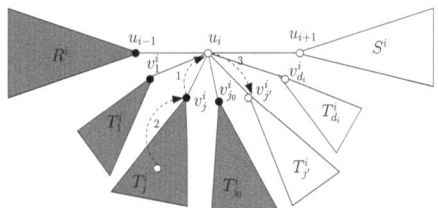

(a) Description of one phase of
$bring\_searchers(i, j_0)$.

(b) Description of one phase of
$transfer(i)$.

Fig. 2. Black nodes are occupied. Grey subtrees are cleared. Steps are depicted by dotted arrows.

so, no recontamination occurs in the cleared subtrees (but in u_i) and exclusivity is satisfied. We apply similar techniques for $j_0 > \lceil \frac{d_i}{2} \rceil$.

Procedure *transfer.* Let $1 \leq i \leq p$ and $j_0 = \lceil d_i/2 \rceil$. Just after clearing $T_{j_0}^i$, we reach a configuration where the nodes in $\{u_i, v_1^i, \cdots, v_{j_0}^i\}$ are occupied, $T_{j_0}^i$ is clear, and all other searchers are at nodes of R^i or T_j^i $(j \leq j_0)$. First, the searcher at u_i goes to u_{i-1} unless it is already occupied.

As explained before, the nodes in $\{v_{j_0+1}^i, \cdots, v_{d_i}^i\}$ must now be occupied before clearing any subtree T_j^i, for $j > j_0$. This is the role of sub-process $transfer(i)$. The searchers are brought one by one, from the clear part to $\{v_{j_0+1}^i, \cdots, v_{d_i}^i\}$, without recontamination and satisfying exclusivity.

Fig. 2(b) depicts one phase of this process. We prove that, before each phase, there is a free searcher at some node b either in $R^i \setminus \{u_{i-1}\}$ or in $T_j^i \setminus \{v_j^i\}$ (for some $j \leq j_0$). First, the searcher at v_j^i (if $b \in V(T_j^i)$) or at u_{i-1} (otherwise) goes to u_i (unless u_i is occupied) (dotted line 1 in Fig. 2(b)). Second, the searcher at b goes to v_j^i (or u_{i-1}) (dotted line 2 in Fig. 2(b)). Finally, the searcher at u_i goes to an unoccupied node in $\{v_{j_0+1}^i, \cdots, v_{d_i}^i\}$ (dotted line 3 in Fig. 2(b)). Once all these nodes are occupied, the searcher at u_{i-1} goes back to u_i. Clearly, doing so, exclusivity is satisfied and no recontamination occurs in the cleared subtrees. This, in particular, since either all the nodes $\{u_{i-1}, v_1^i, \cdots, v_{j_0-1}^i\}$, or u_i, are always occupied during $transfer(i)$.

4   Application to Distributed Graph Searching

In the previous section, we have described the *ExclusiveClear* strategy using $xs(T)$ searchers controlled by central scheduler. In this section, we briefly explain how this strategy can be adapted, and then used by the *autonomous* searchers operating in an asynchronous distributed manner (see [6] for more details).

We consider a version of the classical discrete CORDA model for autonomous mobile searchers. The searchers operate in *asynchronous* cycles of *Look-Compute-Move*. During its Look action, a searcher (instantaneously) takes a snapshot of the network map together with the relative positions of all searchers. Based on this information, during the Compute action, it computes (deterministically) the next neighboring node where to move. During the Move action, the searcher (instantaneously) changes its position according to its computation. There is a finite but unbounded delay between any two actions. Moreover, the searchers are *anonymous, uniform* (i.e., each searcher executes the same algorithm), *oblivious* (i.e., memoryless to observations and to computations performed in previous cycles) and have *no sense of direction.*

Let us consider an asymmetric tree, i.e., a tree with no non-trivial automorphisms. We show that in such a tree, each searcher can assign distinct labels to the nodes such that each node of the tree is always given the same label by all searchers during any Compute action [6]. Hence, in CORDA, an anonymous asymmetric tree can be seen as uniquely labeled.

To clear a tree T in the distributed setting, at least $\mathtt{xs}(T)$ searchers are placed in the specified different nodes of T, forming an initial configuration I. At each Compute action, a searcher computes a winning strategy S, starting from I, and using $\mathtt{xs}(T)$ searchers. Given any achievable configuration, S describes the required move (of one of the searchers). S follows the same structure as *ExclusiveClear*, inductively clearing the subtrees of $T \setminus A$, where A is the avenue. However, in contrast with *ExclusiveClear*, the main difficulty is to ensure that all configurations in S are pairwise distinct. Otherwise, since searchers are oblivious, they could enter in a loop of configurations, and the clearing would fail. In particular, an attention should be paid to the case where searchers just have cleared a subtree of $T \setminus A$, and must go back towards A to clear the next subtree. Moreover, in *ExclusiveClear*, it may happen that a searcher slides back and forth along the same edge, while no other searchers have moved. This must be avoided in S. The case of a labeled line subtree of $T \setminus A$ is particularly tricky: an extra searcher is required to clear it compared to *ExclusiveClear*. Nevertheless, we succeed to use only $\mathtt{xs}(T)$ searchers, even in the distributed case.

Theorem 3. *For any anonymous asymmetric or any uniquely labeled n-node tree T, and for any integer k with $\mathtt{xs}(T) \leq k \leq n$, there exists a distributed protocol in the discrete CORDA model enabling k searchers to clear T.*

References

1. Baldoni, R., Bonnet, F., Milani, A., Raynal, M.: Anonymous graph exploration without collision by mobile robots. Inf. Process. Lett. 109(2), 98–103 (2008)
2. Barrière, L., Flocchini, P., Fomin, F.V., Fraigniaud, P., Nisse, N., Santoro, N., Thilikos, D.M.: Connected graph searching. Inf. Comput. 219, 1–16 (2012)
3. Bienstock, D.: Graph searching, path-width, tree-width and related problems (a survey). DIMACS Ser. in Discr. Maths and Theoretical Comp. Sc. 5, 33–49 (1991)
4. Bienstock, D., Seymour, P.D.: Monotonicity in graph searching. J. Algorithms 12(2), 239–245 (1991)
5. Blin, L., Burman, J., Nisse, N.: Brief announcement: Distributed exclusive and perpetual tree searching. In: Aguilera, M.K. (ed.) DISC 2012. LNCS, vol. 7611, pp. 403–404. Springer, Heidelberg (2012)
6. Blin, L., Burman, J., Nisse, N.: Exclusive graph searching. Technical report, INRIA (2013), http://hal.archives-ouvertes.fr/hal-00837543

7. Blin, L., Fraigniaud, P., Nisse, N., Vial, S.: Distributed chasing of network intruders. Theor. Comput. Sci. 399(1-2), 12–37 (2008)
8. Blin, L., Milani, A., Potop-Butucaru, M., Tixeuil, S.: Exclusive perpetual ring exploration without chirality. In: Lynch, N.A., Shvartsman, A.A. (eds.) DISC 2010. LNCS, vol. 6343, pp. 312–327. Springer, Heidelberg (2010)
9. Breisch, R.L.: An intuitive approach to speleotopology. Southwestern Cavers 6, 72–78 (1967)
10. Breisch, R.L.: Lost in a Cave-applying graph theory to cave exploration (2012)
11. Coudert, D., Huc, F., Mazauric, D.: A distributed algorithm for computing the node search number in trees. Algorithmica 63(1-2), 158–190 (2012)
12. D'Angelo, G., Di Stefano, G., Navarra, A., Nisse, N., Suchan, K.: A unified approach for different tasks on rings in robot-based computing systems. In: 15th Workshop on Advances in Par. and Dist. Comp. Models (APDCM). IEEE (2013)
13. Dereniowski, D.: From pathwidth to connected pathwidth. SIAM J. Discrete Math. 26(4), 1709–1732 (2012)
14. Ellis, J.A., Sudborough, I.H., Turner, J.S.: The vertex separation and search number of a graph. Inf. Comput. 113(1), 50–79 (1994)
15. Flocchini, P., Huang, M.J., Luccio, F.L.: Contiguous search in the hypercube for capturing an intruder. In: 19th Int. Par. and Dist. Proc. Symp, IPDPS (2005)
16. Flocchini, P., Prencipe, G., Santoro, N., Widmayer, P.: Hard tasks for weak robots: The role of common knowledge in pattern formation by autonomous mobile robots. In: Aggarwal, A.K., Pandu Rangan, C. (eds.) ISAAC 1999. LNCS, vol. 1741, pp. 93–102. Springer, Heidelberg (1999)
17. Fomin, F.V., Heggernes, P., Mihai, R.: Mixed search number and linear-width of interval and split graphs. Networks 56(3), 207–214 (2010)
18. Fomin, F.V., Thilikos, D.M.: An annotated bibliography on guaranteed graph searching. Theor. Comput. Sci. 399(3), 236–245 (2008)
19. Heggernes, P., Mihai, R.: Edge search number of cographs in linear time. In: Deng, X., Hopcroft, J.E., Xue, J. (eds.) FAW 2009. LNCS, vol. 5598, pp. 16–26. Springer, Heidelberg (2009)
20. Ilcinkas, D., Nisse, N., Soguet, D.: The cost of monotonicity in distributed graph searching. Distributed Comp. 22(2), 117–127 (2009)
21. Megiddo, N., Hakimi, S.L., Garey, M.R., Johnson, D.S., Papadimitriou, C.H.: The complexity of searching a graph. J. Assoc. Comput. Mach. 35(1), 18–44 (1988)
22. Parsons, T.D.: The search number of a connected graph. In: 9th Southeastern Conf. on Combinatorics, Graph Theory, and Computing, Congress. Numer., XXI, Winnipeg, Man., Utilitas Math., pp. 549–554 (1978)
23. Peng, S.-L., Ho, C.-W., Hsu, T.-S., Ko, M.-T., Tang, C.Y.: Edge and node searching problems on trees. TCS 240(2), 429–446 (2000)
24. Prencipe, G.: Instantaneous actions vs. full asynchronicity: Controlling and coordinating a set of autonomous mobile robots. In: Restivo, A., Ronchi Della Rocca, S., Roversi, L. (eds.) ICTCS 2001. LNCS, vol. 2202, pp. 154–171. Springer, Heidelberg (2001)
25. Skodinis, K.: Computing optimal linear layouts of trees in linear time. J. Algorithms 47(1), 40–59 (2003)
26. Suchan, K., Todinca, I.: Pathwidth of circular-arc graphs. In: Brandstädt, A., Kratsch, D., Müller, H. (eds.) WG 2007. LNCS, vol. 4769, pp. 258–269. Springer, Heidelberg (2007)
27. Yang, B., Dyer, D., Alspach, B.: Sweeping graphs with large clique number (Extended abstract). In: Fleischer, R., Trippen, G. (eds.) ISAAC 2004. LNCS, vol. 3341, pp. 908–920. Springer, Heidelberg (2004)

Largest Chordal and Interval Subgraphs Faster Than 2^n

Ivan Bliznets[1,*], Fedor V. Fomin[2,**],
Michał Pilipczuk[2,**], and Yngve Villanger[2]

[1] St. Petersburg Academic University of the Russian Academy of Sciences, Russia
ivanbliznets@tut.by
[2] Department of Informatics, University of Bergen, Norway
{fomin,michal.pilipczuk,yngvev}@ii.uib.no

Abstract. We prove that in an n-vertex graph, induced chordal and interval subgraphs with the maximum number of vertices can be found in time $\mathcal{O}(2^{\lambda n})$ for some $\lambda < 1$. These are the first algorithms breaking the trivial $2^n n^{O(1)}$ bound of the brute-force search for these problems.

1 Introduction

The area of exact exponential algorithms is about solving intractable problems faster than the trivial exhaustive search, though still in exponential time [4]. In this paper, we give algorithms computing maximum induced chordal and interval subgraphs in a graph faster than the trivial brute-force search. These problems are interesting cases of a more general the MAXIMUM INDUCED SUBGRAPH WITH PROPERTY Π problem, where for a given graph G and hereditary property Π one asks for a maximum induced subgraph with property Π.

By the result of Lewis and Yannakakis [11], the problem is NP-complete for every non-trivial property Π. Different variants of property Π like being edgeless, planar, outerplanar, bipartite, complete bipartite, acyclic, degree-constrained, chordal etc., were studied in the literature. From the point of view of exact algorithms, as far as property Π can be tested in polynomial time, a trivial brute-force search trying all possible vertex subsets of G solves MAXIMUM INDUCED SUBGRAPH WITH PROPERTY Π in time $\mathcal{O}^*(2^n)$ on an n-vertex graph G.[1] However, many algorithms for MAXIMUM INDUCED SUBGRAPH WITH PROPERTY Π which are faster than $\mathcal{O}^*(2^n)$ can be found in the literature for explicit properties Π. Notable examples are Π being the property of being edgeless [14] (equivalent to MAXIMUM INDEPENDENT SET), acyclic [3] (equivalent to MAXIMUM INDUCED FOREST), regular [9], 2-colorable [13], planar [5], degenerate [12], cluster graph [2], or biclique [7]. A longstanding open question in the area is if

---

* Supported by The Ministry of education and science of Russian Federation, project 8216 and Russian Foundation for Basic Research (12-01-31057-mol_a).
** Supported by the European Research Council (ERC) via grant Rigorous Theory of Preprocessing, reference 267959.

[1] The $O^*(\cdot)$ notation suppresses terms polynomial in the input size.

H.L. Bodlaender and G.F. Italiano (Eds.): ESA 2013, LNCS 8125, pp. 193–204, 2013.
© Springer-Verlag Berlin Heidelberg 2013

MAXIMUM INDUCED SUBGRAPH WITH PROPERTY Π can be solved faster than the trivial $\mathcal{O}^*(2^n)$ for every hereditary property Π testable in polynomial time.

Since every hereditary class of graphs with property Π can be characterized by a (not necessarily finite) set of forbidden induced subgraphs, there is an equivalent formulation of the MAXIMUM INDUCED SUBGRAPH WITH PROPERTY Π problem. For a set of graphs $\mathcal{F}$, a graph G is called $\mathcal{F}$-free if it contains no graph from $\mathcal{F}$ as an induced subgraph. The MAXIMUM $\mathcal{F}$-FREE SUBGRAPH problem is to find a maximum induced $\mathcal{F}$-free subgraph of G. Clearly, if $\mathcal{F}$ is the set of forbidden induced subgraphs for Π, then the MAXIMUM INDUCED SUBGRAPH WITH PROPERTY Π problem and the MAXIMUM $\mathcal{F}$-FREE SUBGRAPH problem are equivalent. It is well known that when the set $\mathcal{F}$ is *finite*, then MAXIMUM $\mathcal{F}$-FREE SUBGRAPH can be solved in time $\mathcal{O}^*(2^{\lambda n})$, where $\lambda < 1$. This can be seen by applying a simple branching arguments, see Proposition 2, or by reducing to the d-HITTING SET problem, which is solvable faster than $\mathcal{O}^*(2^n)$ for every fixed d [2]. Examples of $\mathcal{F}$-free classes of graphs for some finite set $\mathcal{F}$ are split graphs, cographs, line graphs or trivially perfect graphs; see the book [1] for more information on these graph classes.

It is however completely unclear if anything faster than the trivial brute-force is possible in the case when $\mathcal{F}$ is an infinite set, even when $\mathcal{F}$ consists of very simple graphs. One of the most known and well studied classes of $\mathcal{F}$-free graphs is the class of *chordal graphs*, where $\mathcal{F}$ is the set of all cycles of length more than three. Chordal graphs form a fundamental class of graphs which properties are well understood. Another fundamental class of graphs is the class of *interval graphs*. We refer to the book of Golumbic for an overview of properties and applications of chordal and interval graphs [8]. In spite of nice structural properties of these graphs, no exact algorithms for MAXIMUM INDUCED CHORDAL SUBGRAPH and MAXIMUM INDUCED INTERVAL SUBGRAPH problems better than the trivial $\mathcal{O}^*(2^n)$ were known prior to our work.

Our Results. We define four properties of graph classes and give an algorithm that for every graph class Π satisfying these properties and for a given n-vertex graph G, returns a maximum induced subgraph of G belonging to Π in time $\mathcal{O}^*(2^{\lambda n})$ for some $\lambda < 1$, where λ depends only on the class Π. Because classes of chordal and interval graphs satisfy the required properties, as an immediate corollary of our algorithm we obtain that MAXIMUM INDUCED CHORDAL SUBGRAPH and MAXIMUM INDUCED INTERVAL SUBGRAPH can be solved in time $\mathcal{O}^*(2^{\lambda n})$ for some $\lambda < 1$. The main intention of our work was to break the trivial 2^n barrier and we did not try to optimize the constant in the exponent. There are several places where the running time of our algorithm can be improved by the cost of more involved arguments and intensive case analyses. We tried to keep the description of our algorithm as simple as possible, leaving only the ideas crucial for breaking the barrier, and postponing more complicated improvements till the full version of the paper. Moreover, pipelined with simple branching arguments, our algorithms can be used to obtain time $\mathcal{O}^*(2^{\lambda n})$ algorithms for some $\lambda < 1$ for a variety of MAXIMUM INDUCED SUBGRAPH WITH PROPERTY Π problems, where property Π is to be chordal/interval graph containing no induced subgraph from a *finite* forbidden set of graphs. Examples of such graph classes are proper interval graphs, Ptolemaic graphs, block graphs, and proper circular-arc graphs; see [1] for definitions and discussions of these graph classes.

2   Preliminaries

In the paper we use standard graph notation. A graph class Π is simply a family of graphs. We sometimes use terms Π-graph or Π-subgraph to express membership in Π. An *induced subgraph* of a graph is a subset of vertices, with all the edges between those vertices that are present in the larger graph. We say that a graph class is *hereditary* if Π is closed under taking induced subgraphs. Every hereditary graph class can be described by a (possibly infinite) list of minimum forbidden induced subgraphs $\mathcal{F}_{\Pi}$: graph G is in Π if and only if it does not contain any induced subgraph from $\mathcal{F}_{\Pi}$, and for each $H \in \mathcal{F}_{\Pi}$ every induced subgraph of H, apart from H itself, belongs to Π. The class of graphs not containing any induced subgraph from a list $\mathcal{F}$ will be denoted by $\mathcal{F}$-*free graphs*.

Chordal graphs is the class of graphs not containing any induced cycles of length more than three, that is, chordal graphs are $\mathcal{F}$-free graphs, were the set $\mathcal{F}$ consists of all cycles of length more than three. Chordal graphs are hereditary and polynomial-time recognizable. Chordal graphs admit many more characterizations, for example they are exactly graphs admitting a decomposition into a clique tree. A useful corollary of this fact is the following folklore lemma.

Proposition 1 (Folklore). *If H is a chordal graph, then there exists a clique S in H and a partition of $V(H) \setminus S$ into two subsets X_1, X_2, such that (i) $|X_1|, |X_2| \leq \frac{2}{3}|V(H)|$, and (ii) there is no edge between X_1 and X_2.*

Such a set S is called a $\frac{2}{3}$-*balanced clique separator* in H. Interval graphs form a subclass of chordal graphs admitting a decomposition into a clique path instead of less restrictive clique tree. Interval graphs are also hereditary and polynomial-time recognizable. Their characterization in terms of minimal forbidden induced subgraphs was given by Lekkerkerker and Boland [10]. The book of Golumbic [8] provides a thorough introduction to chordal and interval graphs.

We now describe the classical tools needed for the algorithm. The following folklore result basically follows from the observation that branching on forbidden structures of constant size always leads to complexity better than 2^n. Due to space constrains we omit its proof here.

Proposition 2 (Folklore). *Let $\mathcal{F}$ be a finite set of graphs and let ℓ be the maximum number of vertices in a graph from $\mathcal{F}$. Let Π be a hereditary graph class that is polynomial-time recognizable. Assume that there exists an algorithm $\mathcal{A}$ that for a given $\mathcal{F}$-free graph G on n vertices, in $\mathcal{O}^*(2^{\epsilon n})$ time finds a maximum induced Π-subgraph of G, for some $\epsilon < 1$. Then there exists an algorithm $\mathcal{A}'$ that for a given graph G on n vertices, finds a maximum induced $\mathcal{F}$-free Π-graph in G in time $\mathcal{O}^*(2^{\epsilon' n})$, where $\epsilon' < 1$ is a constant depending on ϵ and ℓ.*

We need also the following proposition from [6].

Proposition 3 ([6]). *Let $G = (V, E)$ be a graph. For every $v \in V$, and $b, f \geq 0$, the number of connected vertex subsets $B \subseteq V$ such that (i) $v \in B$, (ii) $|B| = b + 1$, and (iii) $|N(B)| = f$, is at most $\binom{b+f}{b}$. Moreover, all such subsets can be enumerated in time $\mathcal{O}^*(\binom{b+f}{b})$.*

The last necessary ingredient is the classical idea used by Schroeppel and Shamir [15] for solving SUBSET SUM by reducing it to an instance of 2-TABLE. In the 2-TABLE problem, we are given two $k \times m_i$ matrices T_i, $i = 1, 2$, and a vector $\boldsymbol{s} \in \mathbb{Q}^k$. Columns of each matrix are m_i vectors of $\mathbb{Q}^k$. The question is, if there is a column of the first matrix and a column of the second matrix such that the sum of these two vectors is $\boldsymbol{s}$. A trivial solution to the 2-TABLE problem would be to try all possible pairs of vectors; however, this problem can be solved more efficiently. We can sort columns of T_1 lexicographically in $\mathcal{O}(km_1 \log m_1)$ time, and for every column $\boldsymbol{v}$ of T_2 check whether T_1 contains a column equal to $\boldsymbol{s} - \boldsymbol{v}$ in $\mathcal{O}(k \log m_1)$ time using binary search.

Proposition 4. *The 2-TABLE problem can be solved in time $\mathcal{O}((m_1 + m_2)k \log m_1)$.*

3   Properties of the Graph Class

In this section we gather the required properties of the graph class Π for our algorithm to be applicable. We consider only hereditary subclasses of chordal graphs, hence our first property is the following.

Property (1). Π is a hereditary subclass of chordal graphs.

As Π is hereditary, it may be described by a list of vertex-minimal forbidden induced subgraphs $\mathcal{F}_\Pi$. We need the following properties of $\mathcal{F}_\Pi$:

Property (2). All graphs in $\mathcal{F}_\Pi$ are connected, and all of them do not contain a clique of size $\aleph + 1$ for some universal constant $\aleph$.

For chordal graphs $\mathcal{F}_\Pi$ consists of cycles of length at least 4, hence $\aleph = 2$. For interval graphs, an inspection of the list of forbidden induced subgraphs [10], shows that we may take $\aleph = 4$. In the following, we always treat $\aleph$ as a universal constant on which all the later constants may depend; moreover, $\aleph$ may influence the exponents of polynomial factors hidden in the O^* notation. Let us remark that connectedness of all the forbidden induced subgraphs is equivalent to requiring Π to be closed under taking disjoint union. An example of a subclass of chordal graphs not satisfying this property, is the class of strongly chordal graphs. The reason for that is that minimal forbidden subgraphs of strongly chordal graphs can contain a clique of any size, see [1] for more information on this class of graphs.

Thirdly, we need our graph class to be efficiently recognizable.

Property (3). Π is polynomial-time recognizable.

Chordal graphs and interval graphs have polynomial time recognition algorithms [8]. For our arguments to work we need one more algorithmic property.

The property that we need can be described intuitively as robustness with respect to clique separators. More precisely, we need the following statement.

Property (4). There exists a polynomial-time algorithm $\mathcal{A}$ that takes as an input a graph G together with a clique S in G. The algorithm answers YES or NO, such that the following conditions are satisfied:

- If $\mathcal{A}$ answers YES on inputs (G_1, S_1) and (G_2, S_2) where $|S_1| = |S_2|$, then graph G', obtained by taking disjoint union of G_1 and G_2 and identifying every vertex of S_1 with a different vertex of S_2 in any manner, belongs to Π.
- If $G \in \Pi$, then there exists a clique separator S in G such that $V(G) \setminus S$ may be partitioned into two sets X_1, X_2 such that (i) $|X_1|, |X_2| \leq \frac{2}{3}|V(G)|$, (ii) there is no edge between X_1 and X_2, (iii) $\mathcal{A}$ answers YES on $(G[X_1 \cup S], S)$ and on $(G[X_2 \cup S], S)$.

Observe that Property (1) and Proposition 1 already provides us with some $\frac{2}{3}$-balanced clique separator S of G. Shortly speaking, Property (4) requires that in addition belonging to Π may be tested by looking at $G[X_1 \cup S]$ and $G[X_2 \cup S]$ independently. For chordal graphs, Property (4) follows from Proposition 1 and a folklore observation that if S is a clique separator in a graph G, with (X_1, X_2) being a partition of $V(G) \setminus S$ such that there is no edge between X_1 and X_2, then G is chordal if and only if $G[X_1 \cup S]$ and $G[X_2 \cup S]$ are chordal. Hence, we may take chordality testing for the algorithm $\mathcal{A}$.

For interval graphs, we take clique path of the graph G and examine the clique separator S such that there is at most half of vertices before it and at most half after it. Let X_1 be the vertices before S on the clique path, and X_2 be the vertices after S. Clearly, S is then even a $\frac{1}{2}$-balanced clique separator, with partition (X_1, X_2) of $V(G) \setminus S$. Then it follows that $G[X_1 \cup S]$ and $G[X_2 \cup S]$ admit clique paths in which S is one of the end bags of the path. On the other hand, assume that we are given any two graphs G_1, G_2 with equally sized cliques S_1, S_2, such that G_1, G_2 admit clique paths with S_1, S_2 as the end bags. Then we may create a clique path of the graph G' obtained from the disjoint union of G_1 and G_2 and identification of S_1 and S_2 in any manner, by simply taking the clique paths for G_1 and G_2 and identifying the end bags containing S_1 and S_2, respectively. Hence, as $\mathcal{A}$ we may take an algorithm which for input (G, S) checks whether G is interval and admits a clique path with S as the end bag. Such a test may be easily done as follows: we add P_4 to G and make one end of P_4 to be adjacent to every vertex of S, thus forcing S to be the end bag, and run intervality test. Hence, interval graphs also satisfy Property (4).

4   The Algorithm

In this section we prove the main result of the paper, which is the following.

Theorem 5. *If Π satisfies Properties (1)-(4), then there exists an algorithm which, given an n-vertex graph G, returns a maximum induced subgraph of G belonging to Π in time $\mathcal{O}^*(2^{\lambda n})$ for some $\lambda < 1$, where λ depends only on $\aleph$.*

As we already observed, chordal and interval graphs satisfy Properties (1)-(4). Thus Theorem 5 implies immediately results claimed in the introduction. Our approach is based on a thorough investigation of the structure of a maximum induced subgraph. In each of the cases, we deploy a different strategy to identify possible suspects for an optimal solution. The properties we strongly rely on are the balanced separation property of chordal graphs (Property (4)), and conditions on minimal forbidden induced subgraphs for Π (Property (2)).

Let $G = (V, E)$. In the description of the algorithm we use several small positive constants: $\alpha, \beta, \gamma, \delta, \varepsilon$, and one large constant L. The final constant λ depends on the choice of $\alpha, \beta, L, \gamma, \delta, \varepsilon$; during the description we make sure that constants $(\alpha, \beta, L, \gamma, \delta, \varepsilon)$ can be chosen so that $\lambda < 1$. The choice of each constant depends on the later ones, e.g., having chosen $L, \gamma, \delta, \varepsilon$, we may find a positive upper bound on the value of β so that we may choose any positive β smaller than this upper bound.

Firstly, we observe that by Proposition 2, we may assume that the input graph does not contain any forbidden induced subgraph from $\mathcal{F}_\Pi$ of size at most ℓ for some constant ℓ, to be determined later. Indeed, if we are able to find an algorithm for maximum induced Π-subgraph running in $\mathcal{O}^*(2^{\lambda n})$ time for some $\lambda < 1$ and working in $\mathcal{F}'_\Pi$-free graphs, where $\mathcal{F}'_\Pi$ consists of graphs of $\mathcal{F}_\Pi$ of size at most ℓ, then by Proposition 2 we obtain an algorithm for maximum induced Π-subgraph working in general graphs and with running time $\mathcal{O}^*(2^{\lambda' n})$ for some $\lambda' < 1$. Hence, from now on we assume that the input graph G does not contain any forbidden induced subgraph from $\mathcal{F}_\Pi$ of size at most ℓ.

The algorithm performs a number of *steps*. After each step, depending on the result, the algorithm chooses one of the subcases.

Step 1. *Using the algorithm of Robson [14], in $\mathcal{O}^*(2^{0.276n})$ time find the largest clique K in G.*

We consider two cases: either K is large enough to finish the search directly, or K is small and we have a guarantee that the maximum induced Π-graph we are looking for contains only small cliques. The threshold for small/large is αn for a constant $\alpha > 0$, $\alpha < 1/48$, to be determined later.

Case A: $|K| \geq \alpha n$.

We show that in this case, the problem can be solved in $\mathcal{O}^*(2^{(1-(1-\kappa_0)\alpha)n})$ time for some $\kappa_0 < 1$ depending only on $\aleph$. We use the following auxiliary claim.

Lemma 6. *Let P be a subset of vertices of an n-vertex graph G that induces a graph belonging to Π, and let K be a clique in G such that $P \cap K = \emptyset$. Then in time $\mathcal{O}^*(2^{\kappa_0 \cdot |K|})$ for some $\kappa_0 < 1$ depending only on $\aleph$ it is possible to find an induced subgraph of G with the maximum number of vertices, where maximum is taken over all induced subgraphs H of G such that (i) $H \in \Pi$, (ii) $V(H) \setminus K = P$. In other words, the maximum is taken over all induced subgraphs belonging to Π which can be obtained by adding some vertices of K to P.*

Proof. For every nonempty subset W of K of size at most $\aleph$, we colour W red if $G[W \cup P] \in \Pi$. Note that this construction may be performed using at most $\aleph \cdot |K|^\aleph$ tests of belonging to Π, hence in polynomial time for constant $\aleph$.

We observe that for every subset $X \subseteq K$, $G[P \cup X]$ belongs to Π if and only if all nonempty subsets of X of size at most $\aleph$ are red. Indeed, if the latter is not the case, there is a subset $W \subseteq X$ such that $G[P \cup W] \notin \Pi$, so by Property (1) $G[P \cup X] \notin \Pi$ as well. For the opposite direction, let us assume that $G[P \cup X]$ contains some forbidden induced subgraph $F \in \mathcal{F}_\Pi$. Then $|F \cap X| > \aleph$ because otherwise, by the definition of the colouring, $F \cap X$ would not be coloured red. But since X is a clique, we conclude that F contains a clique on $\aleph + 1$ vertices, which is a contradiction with Property (2).

Hence, to obtain a maximum induced subgraph one has to find a maximum subset of X such that all its subsets of size at most $\aleph$ are coloured red. This is equivalent to finding a maximum clique in a hypergraph with hyperedges of cardinality at most $\aleph$, which can be done using a branching algorithm in $\mathcal{O}^*(2^{\kappa_0 \cdot |K|})$ time for some $\kappa_0 < 1$, depending only on $\aleph$.

We now do the following. Let H be a maximum induced subgraph of G belonging to Π. We branch into at most $2^{|V \setminus K|}$ subcases, in each fixing a different subset P of $V \setminus K$ as $V(H) \setminus K$; we discard all the branches where the subgraph induced by P does not belong to Π. For each branch, we use Lemma 6 to find a maximum induced chordal subgraph, which can be obtained from the guessed subset by adding vertices of K. This takes time $\mathcal{O}^*(2^{\kappa_0 \cdot |K|})$ for each branch. Thus the running time in this case is $\mathcal{O}^*(2^{|V \setminus K|} \cdot 2^{\kappa_0 \cdot |K|}) \leq \mathcal{O}^*(2^{(1-\alpha)n} \cdot 2^{\kappa_0 \cdot \alpha n}) = \mathcal{O}^*(2^{(1-(1-\kappa_0)\alpha)n})$. Note that $(1 - (1 - \kappa_0)\alpha) < 1$ for $\alpha > 0$ and $\kappa_0 < 1$.

Case B: *G has no clique of size αn.*

Firstly, we search for solutions that have at most $n/2 - \beta n$ or at least $n/2 + \beta n$ vertices for some $\beta > 0$, $\beta < 1/16$ to be determined later. For this, we may apply a simple brute-force check that tries all vertex subsets of size at most $\lceil n/2 - \beta n \rceil$ or at least $\lfloor n/2 + \beta n \rfloor$ in time $\mathcal{O}^*(\binom{n}{\lceil n/2 - \beta n \rceil})$, which is faster than $\mathcal{O}^*(2^n)$.

Step 2. *Iterate through all subsets of vertices of size at most $\lceil n/2 - \beta n \rceil$ or at least $\lfloor n/2 + \beta n \rfloor$, and for each of them check if it induces a graph belonging to Π. If some subset of size at least $\lfloor n/2 + \beta n \rfloor$ induces a Π-graph, output the subgraph induced by any of such subsets of maximum cardinality, and terminate the algorithm. If no subset of size exactly $\lceil n/2 - \beta n \rceil$ induces a Π-graph, output the subgraph induced by the maximum size subset inducing a Π-graph among those of size at most $\lceil n/2 - \beta n \rceil$, and terminate the algorithm.*

If execution of Step 2 did not terminate the algorithm, we know that the cardinality of the vertex set of a maximum induced subgraph belonging to Π is between $n/2 - \beta n$ and $n/2 + \beta n$. We proceed to further steps with this assumption.

Let H be a maximum induced Π-subgraph of G. We do not know how H looks like and the only information about H we have so far is that H has no clique of size αn and that $n/2 - \beta n \leq |V(H)| \leq n/2 + \beta n$. Let us note that the number of vertices of G not in H is also between $n/2 - \beta n$ and $n/2 + \beta n$.

We now use Property (4) to find a $\frac{2}{3}$-balanced clique separator in G. More precisely, there is a clique S in H such that $V(H) \setminus S$ may be partitioned into sets X_1 and X_2 so that (i) $\frac{1}{3}|V(H)| - |S| \leq |X_1|, |X_2| \leq \frac{2}{3}|V(H)|$, and (ii) there is no edge between X_1 and X_2 in G. Observe that in particular

$|X_1|, |X_2| \geq (\frac{1}{6} - \frac{\beta}{3} - \alpha)n > \frac{1}{8}n$, as $\beta < 1/16$ and $\alpha < 1/48$. As S is also a clique in G, we have that $|S| \leq \alpha n$. Property (4) gives us more algorithmic claims about the partition (X_1, S, X_2) of $V(H)$; these claims will be useful later. As α is small, we may afford the following branching step.

Step 3. *Branch into at most* $(1 + \alpha n)\binom{n}{\alpha n} \cdot (n + 1)^2$ *subproblems, in each fixing a different subset of* V *of size at most* αn *as* S, *as well as cardinalities of* X_1, X_2. *Discard all the branches where* S *is not a clique.*

From now on we focus on one subproblem; hence, we assume that the clique S is fixed and cardinalities of X_1, X_2 are known. Let $G' = G \setminus S$; to ease the notation, for $X \subseteq V(G')$ let $N'[X] = N_{G'}[X]$ and $N'(X) = N_{G'}(X)$. We now consider two cases of how the structure of the optimal solution H may look like, depending on how many connected components $H \setminus S$ has. The threshold is γn for a small constant $\gamma > 0$ to be determined later.

Step 4. *Branch into two subproblems: in the first branch assume that* $H \setminus S$ *has at most* γn *connected components, and in the second branch assume that* $H \setminus S$ *has more than* γn *connected components.*

In the branches of Step 4 the algorithm checks several cases, and for every case proceeds with further branchings. To ease the description, we do not distinguish these branchings as separate Steps, but rather explain them in the text.

Branch B.1: *Graph* $H \setminus S$ *has at most* γn *connected components.*

We first branch into at most $(n + 1)^3$ subproblems, in each guessing the sizes of sets $N'(X_1)$, $N'(X_2)$ and $N'(X_1) \cap N'(X_2)$ such that $|N'(X_1) \cap N'(X_2)| \leq |N'(X_1)|, |N'(X_2)| \leq n - (|S| + |X_1| + |X_2|)$. From now on we assume that these cardinalities are fixed. We consider a few cases depending on the sizes of $N'(X_1)$, $N'(X_2)$ and $N'(X_1) \cap N'(X_2)$; in these cases we use small constants δ, ε, to be determined later.

Case B.1.1: $||N'(X_1)| - |X_1|| \geq \delta n$, *or* $||N'(X_2)| - |X_2|| \geq \delta n$.

We concentrate only on the subcase of $||N'(X_1)| - |X_1|| \geq \delta n$, as the second is symmetric. Due to the space constraints, here we give only a brief outline. As $G[X_1]$ has only at most γn components, we can guess with $\mathcal{O}^*(\binom{n}{\gamma n})$ overhead a set that contains one element from each connected component of $G[X_1]$. Then, using Proposition 3 we can guess the whole set X_1 with $\binom{|X_1| + |N'(X_1)|}{|X_1|}$ overhead. For X_2 we perform a brute-force guess on the remaining part of $V(G')$, i.e., $V(G') \setminus N'[X_1]$, and at the end for each candidate set $X_1 \cup X_2 \cup S$ we test in polynomial time whether it induces a subgraph belonging to Π. As $||N'(X_1)| - |X_1|| \geq \delta n$, we have that $\binom{|X_1| + |N'(X_1)|}{|X_1|} = \mathcal{O}^*(2^{\kappa_2|N[X_1]|})$ for some $\kappa_2 < 1$ depending only on δ. Since $|N'[X_1]| \geq \frac{1}{8}n$, given δ we can choose α, γ small enough so that the overhead $\mathcal{O}^*(\binom{n}{\alpha n} \cdot \binom{n}{\gamma n})$ is insignificant compared to the gain obtained when guessing X_1. Hence, we produce $\mathcal{O}^*(2^{\kappa_3 n})$ candidates in total, for some $\kappa_3 < 1$.

Case B.1.2: *Case B.1.1 does not apply, but* $|N'(X_1) \cap N'(X_2)| \geq \varepsilon n$.

Again, due to the space constraints, we provide only a short description of this case. We perform a similar strategy as in Case B.1.1, but we guess both X_1 and

X_2 using Proposition 3. Observe that having guessed X_1, we can exclude $N'[X_1]$ from consideration when guessing X_2, thus removing at least εn neighbours of X_2. After the removal, the number of neighbours of X_2 differs much from $|X_2|$ (recall that δ is significantly smaller than ε), and we obtain a gain when guessing X_2. This gain depends on ε only, so we can choose α and γ small enough so that overhead $\mathcal{O}^*(\binom{n}{\alpha n} \cdot \binom{n}{\gamma n}^2)$ is insignificant compared to it.

Case B.1.3: *None of the cases B.1.1 or B.1.2 applies.*

Summarizing, sets X_1 and X_2 have the following properties:

- $\frac{1}{6}n - \frac{\beta}{3}n - \alpha n \leq |X_1|, |X_2| \leq \frac{1}{3}n + \frac{2\beta}{3}n$,
- $\frac{1}{3}n - (\alpha + \beta)n \leq |X_1| + |X_2| \leq \frac{1}{2}n + \beta n$,
- $||N'(X_i)| - |X_i|| \leq \delta n$ for $i = 1, 2$, and $|N'[X_1] \cap N'[X_2]| \leq \varepsilon n$.

Let $U_{\text{both}} = N'[X_1] \cap N'[X_2] = N'(X_1) \cap N'(X_2)$, $U_{\text{none}} = V(G') \setminus (N'[X_1] \cup N'[X_2])$, and $U = U_{\text{both}} \cup U_{\text{none}}$. We already know that $|U_{\text{both}}| \leq \varepsilon n$. We now claim that $|U_{\text{none}}| \leq \zeta n$, where $\zeta = 2\alpha + 2\beta + 2\delta + \varepsilon$. Indeed, we have that

$$|U_{\text{none}}| = |V(G')| - |X_1| - |X_2| - |N'(X_1)| - |N'(X_2)| + |N'(X_1) \cap N'(X_2)|$$
$$\leq n - 2(|X_1| + |X_2|) + 2\delta n + \varepsilon n \leq (2\alpha + 2\beta + 2\delta + \varepsilon)n$$

Given that sets U_{both} and U_{none} are small, we may fix them with $\mathcal{O}^*(\binom{n}{\varepsilon n} \cdot \binom{n}{\zeta n})$ overhead in the running time: we branch into $\mathcal{O}^*(\binom{n}{\varepsilon n} \cdot \binom{n}{\zeta n})$ subproblems, in each fixing disjoint subsets of $V \setminus S$ of sizes at most εn and ζn as $U_{\text{both}}, U_{\text{none}}$, respectively. Note that then $V(G') \setminus U$ is the symmetric difference of $N'[X_1]$ and $N'[X_2]$; let $I = V(G') \setminus U$. We are left with determining which part of I is in $X_1 \cup X_2$, and which is outside.

Observe that every vertex of I is in exactly one of the two sets: $N[X_1]$ or $N[X_2]$. Hence, by Property (4) of Π, we may look for subsets X_1, X_2 of I, such that (i) algorithm $\mathcal{A}$ run on $G[X_1 \cup S]$ and $G[X_2 \cup S]$ with clique S distinguished provides a positive answer in both of the cases, (ii) I is a disjoint union of $N[X_1]$ and $N[X_2]$. We model this situation as an instance of the 2-TABLE problem as follows. For $i = 1, 2$, enumerate all the subsets of I of size $|X_i|$ as candidates for X_i, and discard all the candidates for which the algorithm $\mathcal{A}$ does not provide a positive answer when run on the subgraph induced by the candidate plus the clique S. For each remaining candidate subset create a binary vector of length $|I|$ indicating which vertices of I belong to its closed neighbourhood. Create matrices T_1, T_2 by putting the vectors of candidates for X_1, X_2 as columns of T_1, T_2, respectively. Now, we need to check whether one can find a column of T_1 and a column of T_2 that sum up to a vector consisting only of ones.

As $|X_i| \leq \frac{1}{3}n + \frac{2\beta}{3}n$ for $i = 1, 2$, we have that tables T_1, T_2 have at most $\binom{n}{\frac{1}{3}n + \frac{2\beta}{3}n}$ columns, which is $\mathcal{O}^*(2^{\kappa_6 n})$ for some universal constant $\kappa_6 < 1$ (recall that $\beta < 1/16$, so $\frac{1}{3}n + \frac{2\beta}{3}n < \frac{3}{8}n$). Hence, by Proposition 4 we may solve the obtained instance of 2-TABLE in $\mathcal{O}^*(2^{\kappa_6 n})$ time. The total running time used by Case B.1.3, including the overheads for guessing clique S, set U and cardinalities, is $\mathcal{O}^*(\binom{n}{\alpha n} \cdot \binom{n}{\varepsilon n} \cdot \binom{n}{\zeta n} \cdot 2^{\kappa_6 n})$; note that we may choose $\alpha, \beta, \delta, \varepsilon$ small enough so that this running time is $\mathcal{O}^*(2^{\kappa_7 n})$ for some $\kappa_7 < 1$.

Branch B.2: *Graph $H \setminus S$ has more than γn connected components.*

Consider connected components of $H \setminus S$ and fix a large constant $L > 2$ depending on γ, to be determined later. We say that a component containing at most $C = L/\gamma$ vertices is *small*, and otherwise it is *large*. Let r_ℓ and r_s be the numbers of large and small components of $H \setminus S$, respectively. The number of vertices contained in large components is hence at least $\frac{L \cdot r_\ell}{\gamma}$. Thus, $\frac{L \cdot r_\ell}{\gamma} \leq n$, $r_\ell \leq \frac{\gamma n}{L}$ and, consequently, $r_s \geq \gamma n - r_\ell \geq \gamma n(1 - \frac{1}{L}) \geq \frac{\gamma n}{2}$. Since small components are nonempty, they contain at least $\frac{\gamma n}{2}$ vertices in total.

Let us wrap up the situation. The vertices of V can be partitioned into disjoint sets S, X, N_X, Y, and Z, where *(i)* S is the clique guessed in Step 3; *(ii)* X are the vertices contained in large components of $H \setminus S$; *(iii)* $N_X = N'(X)$; *(iv)* Y are the vertices contained in small components of $H \setminus S$; *(v)* Z consists of vertices not contained in H and not adjacent to X. Note that $V(H) = S \cup X \cup Y$. Unfortunately, even given X and S, the algorithm still cannot deduce the solution: we still need to split the remaining part $V \setminus (N'[X] \cup S)$ into Y that will go into the solution, and Z that will be left out. However, as we know that $G[X]$ has a small number of components, we can proceed with a branching step that guesses X using Proposition 3. Let P be a set of vertices that contains one vertex from each component of $G[X]$; we have that $|P| = r_\ell \leq \frac{\gamma n}{L}$.

Step 5. *Branch into at most $(n + 1)^4$ subbranches fixing $r_\ell, |X|, |Y|, |N'[X]|$. Then branch into $\binom{n}{r_\ell} \leq \binom{n}{\frac{\gamma n}{L}}$ cases, in each fixing a different set of size r_ℓ as a candidate for P. Add an artificial vertex v_1 adjacent to P, and using Proposition 3 in $\mathcal{O}^*(\binom{|N'[X]|}{|X|}) \leq \mathcal{O}^*(2^{|N'[X]|})$ time enumerate at most $\binom{|N'[X]|}{|X|} \leq 2^{|N'[X]|}$ vertex sets that (i) are connected, (ii) contain $P \cup \{v_1\}$, (iii) are of size $|X| + 1$ and have neighbourhood of size $|N'(X)|$. Note that we can do it by filtering out sets that do not contain P from the list given by Proposition 3. As $X \cup \{v\}$ is among enumerated candidates, branch into at most $2^{|N'[X]|}$ subcases, in each fixing a different candidate for X.*

Let $R = G[V \setminus (N'[X] \cup S)]$. Note that we need to have $|V(R)| \geq |Y| \geq r_s \geq \frac{\gamma n}{2}$, so if $|V(R)| < \frac{\gamma n}{2}$ then we may safely terminate the branch. We will now use the fact that the input graph does not contain any forbidden induced subgraphs of size bounded by some bound ℓ; recall that this assumption was justified by an application of Proposition 2. We set $\ell = 3C^2 + 1$; hence, whenever we examine an induced subgraph of G of size at most ℓ, we know that it belongs to Π. The later steps of the algorithm are encapsulated in the following lemma.

Lemma 7. *Assuming $\alpha < \frac{\gamma}{104C^3}$ and $\ell = 3C^2 + 1$, there exists a universal constant $\rho < 1$ and an algorithm working in $\mathcal{O}^*(2^{\rho|V(R)|})$ time that enumerates at most $O(2^{\rho|V(R)|})$ candidate subsets of $V(R)$, such that Y is among the enumerated candidates.*

The full proof of Lemma 7 is omitted; here, we only sketch the intuition behind the proof. However, before we proceed to this sketch, let us observe that application of Lemma 7 finishes the whole algorithm. Indeed, so far in the branching procedure we have an overhead of $\mathcal{O}^*(\binom{n}{\alpha n} \cdot \binom{n}{\frac{\gamma n}{L}} \cdot 2^{|N'[X]|})$ for guessing S and X. If we now enumerate and examine — by testing whether $G[X \cup S \cup Y] \in \Pi$ — all the candidates for Y given by Lemma 7, we arrive at running time $\mathcal{O}^*\left(\binom{n}{\alpha n} \cdot \binom{n}{\frac{\gamma n}{L}} \cdot 2^{|N'[X]|} \cdot 2^{\rho|V(R)|}\right)$.

As $|N'[X]| + |V(R)| \leq n$, $\rho < 1$ is a universal constant and $|V(R)| \geq \frac{\gamma n}{2}$, given $\gamma > 0$ we may choose L to be large enough and $\alpha > 0$ to be small enough (and smaller than $\frac{\gamma}{104C^3}$) so that this running time is $\mathcal{O}^*(2^{\kappa_8 n})$ for some $\kappa_8 < 1$. Here we exploit the fact that ρ does not depend on α, γ or L. What is really happening is that the threshold C for large components depends on γ and L, and thus the threshold ℓ for forbidden induced subgraphs on which we branch a priori using Proposition 2 depends on γ and L. Yet this branching is done outside the current reasoning and we avoid a loop in the definition of thresholds.

We now sketch the proof of Lemma 7. Firstly, as $G[Y]$ have connected components of size at most C, the degrees in $G[Y]$ are bounded by $C - 1$. Hence, whenever we see a vertex v that has high degree in R, say at least $3C$, then we can infer that if it is in Y, then at most a third of its neighbours can be also in Y. This allows us to design a branching procedure with running time $2^{O(\sigma|V(R)|)}$ for some universal $\sigma < 1$ that gets rid of high-degree vertices in R. For simplicity, assume from now on that the degrees in R are bounded by $3C$.

The crucial observation now is that Y must in fact constitute almost the whole $V(R)$, hence we can guess Y in a much more efficient manner than via a $2^{|V(R)|}$ brute-force. For the sake of contradiction, assume that $V(R) \setminus Y$ constitutes a constant fraction of $V(R)$. The strategy is to show that the assumed maximum solution H is in fact not maximum, using the fact that α is very small compared to $|V(R)|$. Let us construct an alternative solution H' as follows: we remove S from H, thus losing at most αn vertices, and add vertices of $V(R) \setminus Y$ in a greedy manner so that no component larger than $3C^2 + 1$ is created in $V(R)$. As G does not contain any forbidden induced subgraph for Π of size at most $3C^2 + 1$, all these components belong to Π. As Π is closed under taking disjoint union, H' constructed in this manner also belongs to Π. The bound on the degrees in R ensures that the greedy procedure adds at least $\frac{|V(R) \setminus Y|}{O(C^3)}$ vertices; hence if we choose $\alpha < \frac{\gamma}{O(C^3)}$, then H' is larger than H, contradicting the maximality of H.

5   Conclusion

Theorem 5 shows that for any class of graphs Π satisfying Properties (1)–(4), a maximum induced subgraph from Π of an n-vertex graph can be found in time $\mathcal{O}^*(2^{\lambda n})$ for some $\lambda < 1$. Pipelining Proposition 2 with Theorem 5 shows that we moreover may add any finite family of forbidden subgraphs on top of belonging to Π. More precisely, we have the following theorem.

Theorem 8. *Let $\mathcal{F}$ be a finite set of graphs and Π be a class of graphs satisfying Properties (1)–(4). There exists an algorithm which for a given n-vertex graph G, finds a maximum induced $\mathcal{F}$-free Π-graph in G in time $\mathcal{O}^*(2^{\lambda n})$ for some $\lambda < 1$, where λ depends only on $\aleph$ and $\mathcal{F}$.*

As mentioned in introduction, Theorem 8 covers such graph classes as proper interval graphs, i.e. claw-free interval graphs, Ptolemaic graphs, which are chordal and gem-free, block graphs, which are chordal and diamond-free; proper circular-arc graphs which are chordal, claw-free, and $\bar{S}_3$-free. We refer to [1] for the definitions and discussions on these graphs.

We conclude with the following open questions. An interesting subclass of chordal graphs that cannot be handled by our approach is the class of strongly chordal graphs. The reason is that Property (2) does not hold here and we are not aware of any algorithm for finding a maximum induced strongly chordal subgraph faster than the trivial brute-force. Secondly, our approach fails when we require the induced subgraph to be additionally *connected*, since connectivity requirements are not hereditary, and thus is Property (1) is not satisfied. Say, can a maximum induced *connected* chordal subgraph be found faster than 2^n?

References

1. Brandstädt, A., Le, V., Spinrad, J.P.: Graph Classes. A Survey, SIAM Mon. on Discrete Mathematics and Applications. SIAM, Philadelphia (1999)
2. Fomin, F.V., Gaspers, S., Kratsch, D., Liedloff, M., Saurabh, S.: Iterative compression and exact algorithms. Theor. Comput. Sci. 411, 1045–1053 (2010)
3. Fomin, F.V., Gaspers, S., Pyatkin, A.V., Razgon, I.: On the minimum feedback vertex set problem: Exact and enumeration algorithms. Algorithmica 52, 293–307 (2008)
4. Fomin, F.V., Kratsch, D.: Exact Exponential Algorithms. Springer (2010)
5. Fomin, F.V., Todinca, I., Villanger, Y.: Exact algorithm for the maximum induced planar subgraph problem. In: Demetrescu, C., Halldórsson, M.M. (eds.) ESA 2011. LNCS, vol. 6942, pp. 287–298. Springer, Heidelberg (2011)
6. Fomin, F.V., Villanger, Y.: Treewidth computation and extremal combinatorics. Combinatorica 32, 289–308 (2012)
7. Gaspers, S., Kratsch, D., Liedloff, M.: On independent sets and bicliques in graphs. Algorithmica 62, 637–658 (2012)
8. Golumbic, M.C.: Algorithmic Graph Theory and Perfect Graphs. Academic Press, New York (1980)
9. Gupta, S., Raman, V., Saurabh, S.: Maximum r-regular induced subgraph problem: Fast exponential algorithms and combinatorial bounds. SIAM J. Discrete Math. 26, 1758–1780 (2012)
10. Lekkerkerker, C.G., Boland, J.C.: Representation of a finite graph by a set of intervals on the real line. Fund. Math. 51, 45–64 (1962)
11. Lewis, J.M., Yannakakis, M.: The node-deletion problem for hereditary properties is NP-complete. J. Comput. Syst. Sci. 20, 219–230 (1980)
12. Pilipczuk, M., Pilipczuk, M.: Finding a maximum induced degenerate subgraph faster than 2^n. In: Thilikos, D.M., Woeginger, G.J. (eds.) IPEC 2012. LNCS, vol. 7535, pp. 3–12. Springer, Heidelberg (2012)
13. Raman, V., Saurabh, S., Sikdar, S.: Efficient exact algorithms through enumerating maximal independent sets and other techniques. Theory Comput. Syst. 41, 563–587 (2007)
14. Robson, J.M.: Algorithms for maximum independent sets. J. Algorithms 7, 425–440 (1986)
15. Schroeppel, R., Shamir, A.: A $T = O(2^{n/2})$, $S = O(2^{n/4})$ algorithm for certain NP-complete problems. SIAM J. Comput. 10, 456–464 (1981)

Revisiting the Problem of Searching on a Line[*]

Prosenjit Bose[1], Jean-Lou De Carufel[1], and Stephane Durocher[2]

[1] School of Computer Science, Carleton University, Ottawa, Canada
[2] Department of Computer Science, University of Manitoba, Winnipeg, Canada

Abstract. We revisit the problem of searching for a target at an unknown location on a line when given upper and lower bounds on the distance D that separates the initial position of the searcher from the target. Prior to this work, only asymptotic bounds were known for the optimal competitive ratio achievable by any search strategy in the worst case. We present the first tight bounds on the exact optimal competitive ratio achievable, parametrized in terms of the given range for D, along with an optimal search strategy that achieves this competitive ratio. We prove that this optimal strategy is unique and that it cannot be computed exactly in general. We characterize the conditions under which an optimal strategy can be computed exactly and, when it cannot, we explain how numerical methods can be used efficiently. In addition, we answer several related open questions and we discuss how to generalize these results to m rays, for any $m \geq 2$.

1 Introduction

Search problems are broadly studied within computer science. A fundamental search problem, which is the focus of this paper, is to specify how a searcher should move to find an immobile target at an unknown location on a line such that the total relative distance travelled by the searcher is minimized in the worst case [3,10,13]. The searcher is required to move continuously on the line, i.e., discontinuous jumps, such as random access in an array, are not possible. Thus, a search corresponds to a sequence of alternating left and right displacements by the searcher. This class of geometric search problems was introduced by Bellman [4] who first formulated the problem of searching for the boundary of a region from an unknown random point within its interior. Since then, many variants of the line search problem have been studied, including multiple rays sharing a common endpoint (as opposed to a line, which corresponds to two rays), multiple targets, multiple searchers, moving targets, and randomized search strategies (e.g., [1,2,3,5,6,7,8,9,12,13,14]).

For any given search strategy f and any given target location, we consider the ratio A/D, where A denotes the total length of the search path travelled by a searcher before reaching the target by applying strategy f, and D corresponds to the minimum travel distance necessary to reach the target. That is, the searcher and target initially lie a distance D from each other on a line, but the searcher

---

[*] This research has been partially funded by NSERC and FQRNT.

H.L. Bodlaender and G.F. Italiano (Eds.): ESA 2013, LNCS 8125, pp. 205–216, 2013.

knows neither the value D nor whether the target lies to its left or right. The *competitive ratio* of a search strategy f, denoted $CR(f)$, is measured by the supremum of the ratios achieved over all possible target locations. Observe that $CR(f)$ is unbounded if D can be assigned any arbitrary real value; specifically, the searcher must know a lower bound $\min \leq D$. Thus, it is natural to consider scenarios where the searcher has additional information about the distance to the target. In particular, in many instances the searcher can estimate good lower and upper bounds on D. Given a lower bound $D \geq \min$, Baeza-Yates et al. [3] show that any optimal strategy achieves a competitive ratio of 9. They describe such a strategy, which we call the *Power of Two* strategy. Furthermore, they observe that when D is known to the searcher, it suffices to travel a distance of $3D$ in the worst case, achieving a competitive ratio of 3.

We represent a search strategy by a function $f : \mathbb{N} \to \mathbb{R}^+$. Given such a function, a searcher travels a distance of $f(0)$ in one direction from the origin (say, to the right), returns to the origin, travels a distance of $f(1)$ in the opposite direction (to the left), returns to the origin, and so on, until reaching the target. We refer to $f(i)$ as the distance the searcher travels from the origin during the ith iteration. The corresponding function for the Power of Two strategy of Baeza-Yates et al. is $f(i) = 2^i \min$. Showing that every optimal strategy achieves a competitive ratio of exactly 9 relies on the fact that no upper bound on D is specified [3]. Therefore, it is natural to ask whether a search strategy can achieve a better competitive ratio when provided lower and upper bounds $\min \leq D \leq \max$.

Given R, the *maximal reach problem* examined by Hipke et al. [10] is to identify the largest bound max such that there exists a search strategy that finds any target within distance $D \leq \max$ with competitive ratio at most R. López-Ortiz and Schuierer [13] study the maximal reach problem on m rays, from which they deduce that the competitive ratio $CR(g_{opt})$ of any optimal strategy g_{opt} is at least $1 + 2m^m(m-1)^{1-m} - O(\log^{-2} \rho)$, where $\rho = \max / \min$. When $m = 2$, the corresponding lower bound becomes $9 - O(\log^{-2} \rho)$. They also provide a general strategy that achieves this asymptotic behaviour for a general m, given by $f(i) = (\sqrt{1 + i/m})(m/(m-1))^i \min$. Again, for $m = 2$ this is $f(i) = (\sqrt{1 + i/2})2^i \min$. Surprisingly, this general strategy is independent of ρ. In essence, it ignores any upper bound on D, regardless of how tight it is. Thus, we examine whether there exists a better search strategy that depends on ρ, thereby using both the upper and lower bounds on D. Furthermore, previous lower bounds on $CR(g_{opt})$ have an asymptotic dependence on ρ applying only to large values of ρ, corresponding to having only coarse bounds on D. Can we express tight bounds on $CR(g_{opt})$ in terms of ρ?

Let $g_{opt}(i) = a_i \min$ denote an optimal strategy. Since $D \leq \max$, then there must be an n such that $a_n \geq \rho$, i.e., n is the number of iterations necessary to reach the target, so that $g_{opt}(n) \geq \max$. López-Ortiz and Schuierer [13] provide an algorithm to compute the maximal reach for a given competitive ratio together with a strategy corresponding to this maximal reach. They state that the value n and the sequence $\{a_i\}_{i=0}^{n-1}$ for g_{opt} can be computed using binary search, which increases the running time proportionally to $\log \rho$. Can we find a faster algorithm for computing g_{opt}? Since in general, the values $a_0, \ldots, a_{n-1}$ are roots of a polynomial equation of unbounded degree (see Theorem 1), a binary search is equivalent to the bisection method for solving polynomial equations. However, the bisection method is a slowly converging numerical method. Can the computational efficiency be improved? Moreover, given ε, can we bound the number of steps necessary for a root-finding algorithm to identify a solution within tolerance ε of the exact value?

1.1 Overview of Results

We address all of the questions raised above. We characterize g_{opt} by computing the sequence $\{a_i\}_{i=0}^{n-1}$ for the optimal strategy. We do this by computing the number of iterations n needed to find the target and by defining a family of polynomials $p_0, \ldots, p_n$, where p_i has degree $i+1$. We can compute n in $O(1)$ time since we prove that $n \in \{\lfloor \log_2 \rho \rfloor - 1, \lfloor \log_2 \rho \rfloor\}$, where $\rho = \max / \min$. We then show that a_0 is the largest real solution to the polynomial equation $p_n(x) = \rho$. Each of the remaining elements in the sequence $\{a_i\}$ can be computed in $O(1)$ time since we prove that $a_1 = a_0(a_0 - 1)$ and $a_i = a_0(a_{i-1} - a_{i-2})$ for $2 \le i < n$. This also shows that the optimal strategy is unique. Moreover, as we show in Proposition 1, when no upper bound is known there exist infinitely many optimal strategies for any $m \ge 2$.

We give an exact characterization of g_{opt} and show that $CR(g_{opt}) = 2a_0 + 1$. This allows us to establish the following bounds on the competitive ratio of an optimal strategy in terms of ρ:

$$8 \cos^2 \left(\frac{\pi}{\lceil \log_2 \rho \rceil + 1} \right) + 1 \le CR(g_{opt}) \le 8 \cos^2 \left(\frac{\pi}{\lfloor \log_2 \rho \rfloor + 4} \right) + 1 \ .$$

López-Ortiz and Schuierer [13] show that $CR(g_{opt}) \to 9$ as $\rho \to \infty$. We show that $g_{opt} \to g_\infty$ as $\rho \to \infty$, where $g_\infty(i) = (2i + 4)2^i \min$ has a competitive ratio of 9. We thereby obtain an alternate proof of the result of Baeza-Yates et al. [3]. The strategy g_∞ is a member of the infinite family of optimal strategies in the unbounded case which we describe in Proposition 1.

We assume the Real RAM model of computation, including kth roots, logarithms, exponentiation, and trigonometric functions [15]. The computation of each term a_i in the sequence defining g_{opt} involves computing the largest real root of a polynomial equation of degree $n + 1$. We prove that $n + 1 \le 4$ if and only if $\rho \le 32 \cos^5(\pi/7) \approx 18.99761$. In this case the root can be expressed exactly using only the operations $+, -, \times, \div, \sqrt{\cdot}$ and $\sqrt[3]{\cdot}$. This implies that if $\max \le 32 \cos^5(\pi/7) \min$, then g_{opt} can be computed exactly in $O(1)$ time ($O(1)$ time per a_i for $0 \le i \le n < 4$). In general, when $n + 1 \ge 5$, Galois theory implies that the equation $p_n(x) = \rho$ cannot be solved by radicals. Since the corresponding polynomials have unbounded degree, we are required to consider approximate solutions when $\rho > 32 \cos^5(\pi/7)$. Therefore, we explain how to find a solution g_{opt}^*, such that $CR(g_{opt}^*) \le CR(g_{opt}) + \varepsilon$ for a given tolerance ε.

If $n \geq 7\varepsilon^{-1/3} - 4$, we give an explicit formula for a_0. Hence, an ε-approxima-tion can be computed in $O(n) = O(\log \rho)$ time ($O(1)$ time per a_i for $0 \leq i \leq n$). Otherwise, if $32\cos^5(\pi/7) < n < 7\varepsilon^{-1/3} - 4$, we show that a_0 lies in an interval of length at most $7^3\,(n+4)^{-3}$. Moreover, we prove that the polynomial is strictly increasing on this interval. Hence, usual root-finding algorithms work well. Given a_0, the remaining elements of the sequence $\{a_1, \ldots, a_{n-1}\}$ can be computed in $O(n)$ time ($O(1)$ time per a_i for $1 \leq i \leq n$). Finally, we explain how our technique can be generalized to m rays.

2   Searching on a Bounded Line

López-Ortiz and Schuierer [13] showed that there always exists an optimal strat-egy that is *periodic* and *monotone*. That is, the strategy alternates searching left and right between the two rays and the values in the sequence $\{a_i\}$ are increas-ing: $i > j$ implies $a_i > a_j$. Thus, it suffices to consider search strategies that are periodic and monotone. Our goal is to identify a strategy f that minimizes

$$CR(f) = \sup_{D \in [\min, \max]} \phi(f, D)/D, \quad \text{where} \quad \phi(f, D) = 2\sum_{i=0}^{f^{-1}(D)} f(i) + D$$

denotes the cost incurred by f to find a target at distance D in the worst case and $f^{-1}(D)$ is the smallest integer j such that $f(j) \geq D$.

A simple preliminary strategy is to set $g_0(i) = \max$. The corresponding com-petitive ratio is

$$CR(g_0) = \sup_{D \in [\min, \max]} (2\max + D)/D = 2\rho + 1 \ .$$

Observe that g_0 is optimal when $\rho = 1$, i.e., when D is known. A second strategy g_1 corresponds to cutting $[\min, \max]$ once at a point $a_0 \min < \rho \min = \max$. Namely, we search a sequence of two intervals, $[\min, a_0 \min]$ and $(a_0 \min, \rho \min] = (a_0 \min, \max]$, from which we define

$$g_1(i) = \begin{cases} a_0 \min & \text{if } 0 \leq i < 1, \\ \rho \min & \text{if } i \geq 1. \end{cases}$$

Therefore, a_0 needs to be chosen such that $CR(g_1)$ is minimized. We have

$$\sup_{D \in [\min, a_0 \min]} \phi(g_1, D)/D = 2a_0 + 1 \quad \text{and} \quad \sup_{D \in (a_0 \min, \rho \min]} \phi(g_1, D)/D = 3 + 2\frac{\rho}{a_0} \ .$$

Hence, to minimize $CR(g_1)$, we must select a_0, where $1 \leq a_0 \leq \rho$, such that $2a_0 + 1 = 3 + 2\rho/a_0$. Therefore, $a_0 = (1 + \sqrt{1 + 4\rho})/2$ and $CR(g_1) = 2 + \sqrt{1 + 4\rho}$. We have that $CR(g_0) \leq CR(g_1)$ if and only if $1 \leq \rho \leq 2$.

In general, we can partition the interval $[\min, \rho \min]$ into $n + 1$ subintervals whose endpoints correspond to the sequence $\min, a_0 \min, \ldots, a_{n-1} \min, \rho \min$, from which we define

$$g_n(i) = \begin{cases} a_i \min & \text{if } 0 \leq i < n, \\ \rho \min & \text{if } i \geq n. \end{cases}$$

Therefore, we must select $a_0, \ldots, a_{n-1}$, where $1 \leq a_0 \leq a_1 \leq \ldots \leq a_{n-1} \leq \rho$, such that $CR(g_n)$ is minimized. We have

$$\sup_{D \in [\min, a_0 \min]} \phi(g_n, D)/D = 2a_0 + 1 \ ,$$

$$\sup_{D \in (a_i \min, a_{i+1} \min]} \phi(g_n, D)/D = 1 + 2 \sum_{k=0}^{i+1} a_k/a_i \quad (1 \leq i \leq n-2),$$

$$\sup_{D \in (a_{n-1} \min, \rho \min]} \phi(g_n, D)/D = 1 + 2 \sum_{k=0}^{n-1} a_k/a_{n-1} + 2\rho/a_{n-1} \ .$$

Hence, the values a_i are solutions to the following system of equations:

$$\sum_{k=0}^{i+1} a_k = a_0 a_i \quad (0 \leq i \leq n-2), \quad \text{and} \quad \sum_{k=0}^{n-1} a_k + \rho = a_0 a_{n-1}. \quad (1)$$

We prove in Theorem 1 that the solution to this system of equations can be obtained using the following family of polynomials:

$$\begin{aligned} p_0(x) &= x \ , \\ p_1(x) &= x(x-1) \ , \\ p_i(x) &= x \left(p_{i-1}(x) - p_{i-2}(x) \right) \quad (i \geq 2). \end{aligned} \quad (2)$$

We apply (2) without explicitly referring to it when we manipulate the polynomials p_i. Let α_i denote the largest real root of p_i for each i.

Theorem 1. *For all $n \in \mathbb{N}$, the values a_i ($0 \leq i < n$) that define g_n satisfy the following properties:*

1. $a_i = p_i(a_0)$,
2. *a_0 is the unique solution to the equation $p_n(x) = \rho$ such that $a_0 > \alpha_n$, and*
3. $CR(g_n) = 2a_0 + 1$.

To prove Theorem 1, we use the following two formulas:

$$p_{n+1}(x) = xp_n(x) - \sum_{i=0}^{n} p_i(x), \text{ and} \quad (3)$$

$$p_n(x) = x^{\lfloor (n+1)/2 \rfloor} \prod_{k=1}^{\lfloor (n+2)/2 \rfloor} \left(x - 4\cos^2(k\pi/(n+2)) \right). \quad (4)$$

Equation (3) can be proved by induction on n. Equation (4) is a direct consequence of Corollary 10 in [11] since the p_n's are *generalized Fibonacci polynomials* (refer to [11]). We can deduce many properties of the p_n's from (4) since it provides an exact expression for all the roots of the p_n's. For instance, we have the formula $\alpha_n = 4\cos^2(\pi/(n+2))$.

Proof. 1. We can prove this theorem by induction on i, using (1) and (3).
2. From the discussion preceding Theorem 1 we know that a_0 satisfies $\sum_{k=0}^{n-1} a_k + \rho = a_0 a_{n-1}$. Therefore,

$$\begin{aligned} \rho &= a_0 a_{n-1} - \sum_{k=0}^{n-1} a_k \\ &= a_0 p_{n-1}(a_0) - \sum_{k=0}^{n-1} p_k(a_0) \qquad \text{by Theorem 1-1,} \\ &= p_n(a_0) \qquad\qquad\qquad\qquad\qquad \text{by (3).} \end{aligned}$$

Suppose $a_0 < \alpha_n$. Then, by (4), there exists an $i \in \mathbb{N}$ such that $0 \le i < n$ and $p_i(a_0) < 0$. Hence, $a_i = p_i(a_0) < 0$ by Theorem 1-1. This is impossible since all the a_i's are such that $1 \le a_i \le \rho$. Therefore, $a_0 \ge \alpha_n$. Moreover, $a_0 \ne \alpha_n$ since $p_n(a_0) = \rho \ge 1$, whereas $p_n(\alpha_n) = 0$ by the definition of α_n. Finally, this solution is unique since α_n is the biggest real root.

3. This follows directly from the discussion preceding Theorem 1.   □

From Theorem 1, the strategy g_n is uniquely defined for each n. However, this still leaves an infinite number of possibilities for the optimal strategy (one for each n). We aim to find, for a given ρ, what value of n leads to the optimal strategy. Theorem 2 gives a criterion for the optimal n in terms of ρ together with a formula that enables to compute this optimal n in $O(1)$ time.

Theorem 2.

1. *For a given ρ, if $n \in \mathbb{N}$ is such that*

$$p_n(\alpha_{n+1}) \le \rho < p_n(\alpha_{n+2}) \ , \tag{5}$$

  then g_n is the optimal strategy and $\alpha_{n+1} \le a_0 < \alpha_{n+2}$.
2. *For all $n \in \mathbb{N}$,*

$$2^n \le p_n(\alpha_{n+1}) \le \rho < p_n(\alpha_{n+2}) \le 2^{n+2} \ . \tag{6}$$

Notice that the criterion in Theorem 2-1 covers all possible values of ρ since $p_0(\alpha_1) = 1$ and $p_n(\alpha_{n+2}) = p_{n+1}(\alpha_{n+2})$ by the definition of the α_n's.

Proof. 1. Consider the strategy g_n. By Theorem 1-2 and since $p_n(\alpha_{n+1}) \le \rho < p_n(\alpha_{n+2})$, we have $\alpha_{n+1} \le a_0 < \alpha_{n+2}$.

We first prove that g_n is better than g_m for all $m < n$. Suppose that there exists an $m < n$ such that g_m is better than g_n for a contradiction. By Theorem 1-2, there exists an a_0' such that $a_0' > \alpha_m$ and $g_m(a_0') = \rho$. Moreover, since g_m is better than g_n by the hypothesis, then $2a_0' + 1 < 2a_0 + 1$ by Theorem 1-3. Therefore,

$$\alpha_m < a_0' < a_0 \ . \tag{7}$$

Also, since $m < n$, then $m+2 \le n+1$. Thus, since the α_n's are strictly increasing with respect to n, $a_0 \ge \alpha_{n+1} \ge \alpha_i$ for all $m + 2 \le i \le n + 1$. Hence, we find

$$p_m(a_0) \le p_{m+1}(a_0) \le p_{m+2}(a_0) \le \ldots \le p_{n-1}(a_0) \le p_n(a_0) \ . \tag{8}$$

But then,

$$\begin{aligned}
\rho &= p_m(a_0') \\
&< p_m(a_0) &&\text{by (7) and since } p_m \text{ is increasing on } [\alpha_m, \infty), \\
&\le p_n(a_0) &&\text{by (8),} \\
&= \rho \ ,
\end{aligned}$$

which is a contradiction. Consequently, g_n is better than g_m for all $m < n$.

We now prove that g_n is better than $g_{n'}$ for all $n' > n$. Suppose that there exists an $n' > n$ such that $g_{n'}$ is better than g_n for a contradiction. By Theorem 1-2, there exists an a_0' such that $a_0' > \alpha_{n'}$ and $g_{n'}(a_0') = \rho$. Moreover, since $g_{n'}$ is better than g_n by the hypothesis, then $2a_0' + 1 < 2a_0 + 1$ by Theorem 1-3. Therefore,

$$\alpha_n < \alpha_{n'} < a_0' < a_0 < \alpha_{n+2} \tag{9}$$

since the α_n's are strictly increasing with respect to n, from which $n' = n + 1$. But then,

$$
\begin{aligned}
\rho &= p_{n'}(a_0') \\
&= p_{n+1}(a_0') \\
&< p_n(a_0') && \text{by (9) and since } \alpha_n < \alpha_{n+1} < \alpha_{n+2}, \\
&< p_n(a_0) && \text{by (9) and since } p_n \text{ is increasing on } [\alpha_n, \infty), \\
&= \rho \ ,
\end{aligned}
$$

which is a contradiction. Consequently, g_n is better than $g_{n'}$ for all $n' > n$.
2. By standard calculus, we can prove that

$$2\cos^{n+1}(\pi/(n+3)) \geq 1 \tag{10}$$

for all $n \geq 0$. Therefore,

$$
\begin{aligned}
2^n &\leq 2^n \, 2\cos^{n+1}(\pi/(n+3)) && \text{by (10),} \\
&= \alpha_{n+1}^{(n+1)/2} && \text{by (4),} \\
&= p_n(\alpha_{n+1}) && \text{this can be proved by induction on } n, \\
&\leq \rho && \text{by (5),} \\
&< p_n(\alpha_{n+2}) && \text{by (5),} \\
&= \alpha_{n+2}^{(n+2)/2} && \text{this can be proved by induction on } n, \\
&= 2^{n+2}\cos^{n+2}(\pi/(n+4)) && \text{by (4),} \\
&\leq 2^{n+2} && \text{since } 0 < \cos(\pi/(n+4)) < 1. \qquad \square
\end{aligned}
$$

From (5), there is only one possible optimal value for n. By (6), it suffices to examine two values to find the optimal n, namely $\lfloor \log_2 \rho \rfloor - 1$ and $\lfloor \log_2 \rho \rfloor$. To compute the optimal n, let $n = \lfloor \log_2 \rho \rfloor$ and let $\gamma = 2\cos(\pi/(n+3))$. If $n+1 \leq \log_\gamma \rho$, then n is optimal. Otherwise, take $n = \lfloor \log_2 \rho \rfloor - 1$. By Theorem 2, this gives us the optimal n in $O(1)$ time.

Now that we know the optimal n, we need to compute a_i for each $0 \leq i < n$. Suppose that we know a_0. By (2) and Theorem 1-1, $a_1 = p_1(a_0) = a_0(a_0 - 1)$ and $a_i = a_0(p_{i-1}(a_0) - p_{i-2}(a_0)) = a_0(a_{i-1} - a_{i-2})$ for $2 \leq i < n$. Therefore, given a_0, each a_i can be computed in $O(1)$ time for $1 \leq i < n$. It remains to show how to compute a_0 efficiently. Since g_n is defined by n values, $\Omega(n) = \Omega(\log \rho)$ time is necessary to compute g_n. Hence, if we can compute a_0 in $O(1)$ time, then our algorithm is optimal.

By Theorem 2, for a given n, we need to solve a polynomial equation of degree $n+1$ to find the value of a_0. By Galois theory, this cannot be done by radicals if $n + 1 > 4$. Moreover, the degree of the p_n's is unbounded, so a_0 cannot be computed exactly in general. Theorem 3 explains how and why numerical methods can be used efficiently to address this issue.

Theorem 3. *Take ρ and n such that g_n is optimal for ρ.*

1. Let $a_0^* \in \mathbb{R}$ be such that $\alpha_{n+1} \leq a_0 < a_0^* \leq \alpha_{n+2}$ and define g_n^* by

$$
g_n^*(i) = \begin{cases} a_0^* \min & \text{if } i = 0, \\ p_i(a_0^*) \min & \text{if } 1 \leq i < n, \\ \rho \min & \text{if } i \geq n. \end{cases}
$$

Then $|CR(g_n) - CR(g_n^*)| \leq 7^3 (n + 4)^{-3}$.

2. The polynomial p_n is strictly increasing on $[\alpha_{n+1}, \alpha_{n+2})$ and $|\alpha_{n+2} - \alpha_{n+1}| \leq 7^3 (n + 4)^{-3}/2$.

Proof. 1. Let $a_i^* = p_i(a_0^*)$. We first prove that $CR(g_n^*) = 2a_0^* + 1$. By Theorems 1 and 2-1, there is a $\rho^* \in \mathbb{R}$ such that $p_n(\alpha_{n+1}) \leq \rho < \rho^* \leq p_n(\alpha_{n+2})$, $p_n(a_0^*) = \rho^*$ and g_n^* is optimal for ρ^*. By Theorem 1 and the discussion preceding it, we have

$$
\sup_{D \in [\min, a_0^* \min]} \frac{1}{D} \phi(g_n^*, D) = 2a_0^* + 1 \ ,
$$

$$
\sup_{D \in (a_i^* \min, a_{i+1}^* \min]} \frac{1}{D} \phi(g_n^*, D) = 1 + 2 \sum_{k=0}^{i+1} \frac{a_k^*}{a_i^*} \qquad (0 \leq i \leq n - 2)
$$

$$
= 2a_0^* + 1 \qquad (0 \leq i \leq n - 2) \ ,
$$

$$
\sup_{D \in (a_{n-1}^* \min, \rho \min]} \frac{1}{D} \phi(g_n^*, D) = 1 + 2 \sum_{k=0}^{n-1} \frac{a_k^*}{a_{n-1}^*} + 2 \frac{\rho}{a_{n-1}^*}
$$

$$
< 1 + 2 \sum_{k=0}^{n-1} \frac{a_k^*}{a_{n-1}^*} + 2 \frac{\rho^*}{a_{n-1}^*}
$$

$$
= 2a_0^* + 1 \ .
$$

This establishes that $CR(g_n^*) = 2a_0^* + 1$. Therefore,

$$
\begin{aligned}
&|CR(g_n) - CR(g_n^*)| \\
&= |(2a_0 + 1) - (2a_0^* + 1)| && \text{by Theorem 1-3 and since } CR(g_n^*) = 2a_0^* + 1, \\
&= 2(a_0^* - a_0) \\
&\leq 2(\alpha_{n+2} - \alpha_{n+1}) && \text{by the hypothesis and Theorem 2-1,} \\
&= 8 \left(\cos^2(\pi/(n + 4)) - \cos^2(\pi/(n + 3)) \right) && \text{by (4).} \\
&\leq 7^3 (n + 4)^{-3} && \text{by elementary calculus.}
\end{aligned}
$$

2. This is a direct consequence of (4) and Theorem 3-1.   □

We now explain how to compute a_0. We know the value of the optimal n. From (4) and Theorem 2-1, n satisfies $n + 1 \le 4$ if and only if $\rho \le 32 \cos^5(\pi/7) \approx 18.99761$. In this case, $p_n(x) = \rho$ is a polynomial equation of degree at most 4. Hence, by Theorem 1-2 and elementary algebra, a_0 can be computed exactly and in $O(1)$ time. Otherwise, let $\varepsilon > 0$ be a given tolerance. We explain how to find a solution g_{opt}^*, such that $CR(g_{opt}^*) \le CR(g_{opt}) + \varepsilon$.

If $n \ge 7\varepsilon^{-1/3} - 4$, then by Theorem 3, it suffices to take $a_0 = \alpha_{n+2}$ to compute an ε-approximation of the optimal strategy. But $\alpha_{n+2} = 4\cos^2(\pi/(n + 4))$ by (4). Hence, a_0 can be computed in $O(1)$ time and thus, an ε-approximation of the optimal strategy can be computed in $\Theta(n) = \Theta(\log \rho)$ time. Otherwise, if $4 \le n < 7\varepsilon^{-1/3} - 4$, then we have to use numerical methods to find the value of a_0. By Theorem 2-1, we need to solve $p_n(x) = \rho$ for $x \in [\alpha_{n+1}, \alpha_{n+2}]$. However, by Theorem 3, $|\alpha_{n+2} - \alpha_{n+1}| < 7^3(n + 4)^{-3}/2$ and p_n is strictly increasing on this interval. Hence, usual root-finding algorithms behave well on this problem.

Hence, if $n < 4$ or $n \ge 7\varepsilon^{-1/3} - 4$, then our algorithm is optimal. When $4 \le n < 7\varepsilon^{-1/3} - 4$, then our algorithm's computation time is as fast as the fastest root-finding algorithm.

It remains to provide bounds on $CR(g_n)$ for an optimal n; we present exact bounds in Theorem 4.

Theorem 4.

1. *The strategy g_0 is optimal if and only if $1 \le \rho < 2$. In this case, $CR(g_0) = 2\rho + 1$. Otherwise, if g_n is optimal $(n \ge 1)$, then*

$$8\cos^2\left(\frac{\pi}{\lceil \log_2 \rho \rceil + 1}\right) + 1 \le CR(g_n) \le 8\cos^2\left(\frac{\pi}{\lfloor \log_2 \rho \rfloor + 4}\right) + 1 . \quad (11)$$

2. *When $\max \to \infty$, the best strategy tends toward $g_\infty(i) = (2i + 4)2^i \min$ $(i \ge 0)$ and $CR(g_\infty) = 9$.*

Proof. 1. This is a direct consequence of (6), (4), and Theorems 1 and 2-1.
2. Let g_n be the optimal strategy for ρ. When $\max \to \infty$, then $\rho \to \infty$ and then, $n \to \infty$ by (6). Hence, by Theorem 2-1 and (4), $4 = \lim_{n \to \infty} \alpha_{n+1} \le \lim_{n \to \infty} a_0 \le \lim_{n \to \infty} \alpha_{n+2} = 4$. Thus, when $\max \to \infty$, $a_i = p_i(a_0) = p_i(4) = (2i + 4)2^i$ by Theorem 1-1 and (4). Hence, $g_n \to g_\infty$. □

The competitive cost of the optimal strategy is $2a_0 + 1$ by Theorem 1-3. Theorem 4-1 gives nearly tight bounds on $2a_0 + 1$. Notice that when $\rho = 1$, i.e., when D is known, then $2a_0 + 1 = 3$ which corresponds to the optimal strategy in this case. From the Taylor series expansion of $\cos^2(\cdot)$ and Theorem 4-1, we have $CR(g_n) = 9 - O(1/\log^2 \rho)$ for an optimal n. This is consistent with López-Ortiz and Schuierer' result (see [13]), although our result (11) is exact.

Letting $\rho \to \infty$ corresponds to not knowing any upper bound on D. Thus, Theorem 4-2 provides an alternate proof to the competitive ratio of 9 shown by Baeza-Yates et al. [3]. From Theorems 2 and 4, the optimal solution for a given ρ is unique. This optimal solution tends towards g_∞, suggesting that g_∞ is the canonical optimal strategy when no upper bound is given (rather than the power of two strategy).

3   Searching on m Bounded Concurrent Rays

For $m \geq 2$, when no upper bound is known, Baeza-Yates et al. [3] proved that the optimal strategy has a competitive cost of $1 + 2\,m^m/(m-1)^{m-1}$. There exist infinitely many strategies that achieve this optimal cost.

Proposition 1. *All the strategies in the following family are optimal:* $f_{a,b}(i) = (ai + b)\,(m/(m-1))^i \min$, *where* $0 \leq a \leq b/m$ *and* $(m/(m-1))^{2-m} \leq b \leq (m/(m-1))^2$.

Notice that for $m = 2$, when a and b are respectively equal to their smallest allowed value, then $f_{a,b}$ corresponds to the power of two strategy of Baeza-Yates et al. (refer to [3]). Moreover, when a and b are respectively equal to their biggest allowed value, then $f_{a,b} = g_\infty$ (refer to Theorem 4-2). This proposition can be proved by a careful computation of $CR(f_{a,b})$. For a general m, we let g_∞ be the strategy such that a and b are respectively equal to their biggest allowed value.

When we are given an upper bound $\max \geq D$, the solution presented in Section 2 partially applies to the problem of searching on m concurrent bounded rays. In this setting, we start at the crossroads and we know that the target is on one of the m rays at a distance D such that $\min \leq D \leq \max$. Given a strategy $f(i)$, we walk a distance of $f(i)$ on the $(i \bmod m)$-th ray and go back to the crossroads. We repeat for all $i \geq 0$ until we find the target. As in the case where $m = 2$, we can suppose that is the solution is periodic and monotone (refer to Section 2 or see Lemmas 2.1 and 2.2 in [13]).

Unfortunately, we have not managed to push the analysis as far as in the case where $m = 2$ because the expressions in the general case do not simplify as easily. We get the following system of equations by applying similar techniques as in Section 2

$$\sum_{k=0}^{i+m-1} a_k = a_i \sum_{k=0}^{m-2} a_k \qquad (0 \leq i \leq n - m),$$

$$\sum_{k=0}^{n-1} a_k + (i - (n-m))\rho = a_i \sum_{k=0}^{m-2} a_k \qquad (n-m+1 \leq i \leq n-1),$$

for g_n, where

$$g_n(i) = \begin{cases} a_i \min & \text{if } 0 \leq i < n, \\ \rho \min & \text{if } i \geq n. \end{cases}$$

We prove in Theorem 5 that the solution to this system of equations can be obtained using the following family of polynomials in $m - 1$ variables, where $\bar{x} = (x_0, x_1, ..., x_{m-2})$ and $|\bar{x}| = x_0 + x_1 + ... + x_{m-2}$.

$$p_n(\bar{x}) = x_n \qquad (0 \leq n \leq m - 2)$$

$$p_{m-1}(\bar{x}) = |\bar{x}|(x_0 - 1)$$

$$p_n(\bar{x}) = |\bar{x}|(p_{n-(m-1)}(\bar{x}) - p_{n-m}(\bar{x})) \qquad (n \geq m)$$

In the rest of this section, for all $n \in \mathbb{N}$, we let $\overline{a}_n = (\alpha_{n,0}, \alpha_{n,1}, ..., \alpha_{n,m-2})$ be the (real) solution to the system

$$p_n(\overline{x}) = 0, \quad p_{n+1}(\overline{x}) = 0, \quad ..., \quad p_{n+m-2}(\overline{x}) = 0$$

such that

$$0 \leq \alpha_{n,0} \leq \alpha_{n,1} \leq \cdots \leq \alpha_{n,m-2} \tag{12}$$

and $|\overline{a}_n|$ is maximized. Notice that $\overline{a}_n$ exists for any $n \in \mathbb{N}$ since $(0, 0, ..., 0)$ is a solution for any $n \in \mathbb{N}$ by the definition of the p_n's. The proof of the following theorem is similar to those of (3) and Theorem 1.

Theorem 5.

1. For all $n \in \mathbb{N}$, the values a_i $(0 \leq i < n)$ that define g_n satisfy the following properties.
   (a) $a_i = p_i(\overline{a})$.
   (b) $\overline{a}$ is a solution to the system of equations

   $$p_n(\overline{x}) = \rho, \quad p_{n+1}(\overline{x}) = \rho, \quad ..., \quad p_{n+(m-2)}(\overline{x}) = \rho.$$

   (c) $CR(g_n) = 1 + 2|\overline{a}|$.
2. The strategy g_0 is optimal if and only if $1 \leq \rho \leq m/(m-1)$. In this case, $CR(g_0) = 2(m-1)\rho + 1$.
3. For all $n \in \mathbb{N}$, $p_{n+m-1}(x) = p_n(\overline{x}) \sum_{i=0}^{m-2} x_i - \sum_{i=0}^{n+m-2} p_i(\overline{x})$.
4. For all $n \in \mathbb{N}$, $p_n(g_\infty(0), g_\infty(1), ..., g_\infty(m-2)) = g_\infty(n)$.
5. For all $0 \leq n \leq m-2$, $\overline{a}_n = (0, 0, ..., 0)$. Moreover, $\overline{a}_{m-1} = (1, 1, ..., 1)$ and $\overline{a}_m = (m/(m-1), m/(m-1), ..., m/(m-1))$.

4   Conclusion

We have generalized many of our results for searching on a line to the problem of searching on m rays for any $m \geq 2$. Even though we could not extend the analysis of the polynomials p_n as far as was possible for the case where $m = 2$, we believe this to be a promising direction for future research. By approaching the problem directly instead of studying the inverse problem (maximal reach), we were able to provide exact characterizations of g_{opt} and $CR(g_{opt})$. Moreover, the sequence of implications in the proofs of Section 2 all depend on (4), where (4) is an exact general expression for all roots of all equations p_n. As some readers may have observed, exact values of the roots of the equation p_n are not required to prove the results in Section 2; we need disjoint and sufficiently tight lower and upper bounds on each of the roots of p_n. In the case where $m > 2$, finding a factorization similar to (4) appears highly unlikely. We believe, however, that establishing good bounds for each of the roots of the p_n should be possible. Equipped with such bounds, the general problem could be solved exactly on $m > 2$ concurrent rays. We conclude with the following conjecture. It states that the strategy g_n is uniquely defined for each n, it gives a criterion for the optimal n in terms of ρ (and m) and gives the limit of g_n when max $\rightarrow \infty$.

Conjecture 1.

1. For all $n \in \mathbb{N}$, the system of equations of Theorem 5-1b has a unique solution $\overline{a}^* = (a_0^*, a_1^*, ..., a_{m-2}^*)$ satisfying (12) and such that $|\overline{a}^*| > |\overline{\alpha}_n|$. Moreover, there is a unique choice of $\overline{a}$ for g_n and this choice is $\overline{a} = \overline{a}^*$.
2. For a given ρ, if $p_n(\overline{\alpha}_{n+m-1}) \leq \rho < p_n(\overline{\alpha}_{n+m})$, then g_n is the best strategy and $|\overline{\alpha}_{n+m-1}| \leq |\overline{a}| < |\overline{\alpha}_{n+m}|$.
3. When max $\to \infty$, then the optimal strategy tends toward g_∞.
4. For all $n \in \mathbb{N}$, $0 \leq |\overline{\alpha}_n| \leq |\overline{\alpha}_{n+1}| < m^m/(m-1)^{m-1}$ with equality if and only if $0 \leq n \leq m - 3$.

References

1. Alpern, S., Baston, V., Essegaier, S.: Rendezvous search on a graph. J. App. Prob. 36(1), 223–231 (1999)
2. Alpern, S., Gal, S.: The Theory of Search Games and Rendezvous. International Series in Operations Research & Management Science. Kluwer Academic Publishers (2003)
3. Baeza-Yates, R.A., Culberson, J.C., Rawlins, G.J.E.: Searching in the plane. Inf. & Comp. 106(2), 234–252 (1993)
4. Bellman, R.: Minimization problem. Bull. AMS 62(3), 270 (1956)
5. Bender, M.A., Fernández, A., Ron, D., Sahai, A., Vadhan, S.P.: The power of a pebble: Exploring and mapping directed graphs. In: STOC, pp. 269–278 (1998)
6. Collins, A., Czyzowicz, J., Gąsieniec, L., Labourel, A.: Tell me where I am so I can meet you sooner. In: Abramsky, S., Gavoille, C., Kirchner, C., Meyer auf der Heide, F., Spirakis, P.G. (eds.) ICALP 2010. LNCS, vol. 6199, pp. 502–514. Springer, Heidelberg (2010)
7. Czyzowicz, J., Ilcinkas, D., Labourel, A., Pelc, A.: Asynchronous deterministic rendezvous in bounded terrains. Theor. Comp. Sci. 412(50), 6926–6937 (2011)
8. Dieudonné, Y., Pelc, A.: Anonymous meeting in networks. In: SODA, pp. 737–747 (2013)
9. Hammar, M., Nilsson, B.J., Schuierer, S.: Parallel searching on m rays. Comput. Geom. 18(3), 125–139 (2001)
10. Hipke, C.A., Icking, C., Klein, R., Langetepe, E.: How to find a point on a line within a fixed distance. Disc. App. Math. 93(1), 67–73 (1999)
11. Hoggatt, V., Long, C.: Divisibility properties of generalized fibonacci polynomials. Fibonacci Quart. 12(2), 113–120 (1974)
12. Koutsoupias, E., Papadimitriou, C.H., Yannakakis, M.: Searching a fixed graph. In: Meyer auf der Heide, F., Monien, B. (eds.) ICALP 1996. LNCS, vol. 1099, pp. 280–289. Springer, Heidelberg (1996)
13. López-Ortiz, A., Schuierer, S.: The ultimate strategy to search on m rays? Theor. Comp. Sci. 261(2), 267–295 (2001)
14. De Marco, G., Gargano, L., Kranakis, E., Krizanc, D., Pelc, A., Vaccaro, U.: Asynchronous deterministic rendezvous in graphs. Theor. Comp. Sci. 355(3), 315–326 (2006)
15. Preparata, F.P., Shamos, M.I.: Computational Geometry - An Introduction. Springer (1985)

On the Existence of 0/1 Polytopes with High Semidefinite Extension Complexity

Jop Briët[1,*], Daniel Dadush[1], and Sebastian Pokutta[2,**]

[1] New York University, Courant Institute of Mathematical Sciences,
New York, NY, USA
{jop.briet,dadush}@cims.nyu.edu

[2] Georgia Institute of Technology, H. Milton Stewart School of Industrial and
Systems Engineering, Atlanta, GA, USA
sebastian.pokutta@isye.gatech.edu

Abstract. Rothvoss [1] showed that there exists a 0/1 polytope (a polytope whose vertices are in $\{0,1\}^n$) such that any higher-dimensional polytope projecting to it must have $2^{\Omega(n)}$ facets, i.e., its linear extension complexity is exponential. The question whether there exists a 0/1 polytope with high PSD extension complexity was left open. We answer this question in the affirmative by showing that there is a 0/1 polytope such that any spectrahedron projecting to it must be the intersection of a semidefinite cone of dimension $2^{\Omega(n)}$ and an affine space. Our proof relies on a new technique to rescale semidefinite factorizations.

Keywords: semidefinite extended formulations, extended formulations, extension complexity.

1 Introduction

The subject of lower bounds on the size of extended formulations has recently regained a lot of attention. This is due to several reasons. First of all, essentially all NP-Hard problems in combinatorial optimization can be expressed as linear optimization over an appropriate convex hull of integer points. Indeed, many past (erroneous) approaches for proving that P=NP have proceeded by attempting to give polynomial sized linear extended formulations for hard convex hulls (convex hull of TSP tours, indicators of cuts in a graph, etc.). Recent breakthroughs of Fiorini et al. [2] and Braun et al. [3] have unconditionally ruled out such approaches for the TSP and Correlation polytope, complementing the classic result of Yannakakis [4] which gave lower bounds for symmetric extended formulations. Furthermore, even for polytopes over which optimization is in P, it is very natural to ask what the "optimal" representation of the polytope is. From this perspective, the smallest extended formulation represents the "description complexity" of the polytope in terms of a linear or semidefinite program.

---

* J.B. was supported by a Rubicon grant from the Netherlands Organisation for Scientific Research (NWO).
** Research reported in this paper was partially supported by NSF grant CMMI-1300144.

H.L. Bodlaender and G.F. Italiano (Eds.): ESA 2013, LNCS 8125, pp. 217–228, 2013.
© Springer-Verlag Berlin Heidelberg 2013

A *(linear) extension* of a polytope $P \subseteq \mathbb{R}^n$ is another polytope $Q \subseteq \mathbb{R}^d$, so that there exists a linear projection π with $\pi(Q) = P$. The *extension complexity* of a polytope is the minimum number of facets in any of its extensions. The linear extension complexity of P can be thought of as the inherent complexity of expressing P with linear inequalities. Note that in many cases it is possible to save an exponential number of inequalities by writing the polytope in higher-dimensional space. Well-known examples include the regular polygon [5, 6] or the permutahedron [7]. A *(linear) extended formulation* is simply a normalized way of expressing an extension as an intersection of the nonnegative cone with an affine space; in fact we will use these notions in an interchangeable fashion. In the seminal work of Yannakakis [8] a fundamental link between the extension complexity of a polytope and the nonnegative rank of an associated matrix, the so called *slack matrix*, was established and it is precisely this link that provided all known strong lower bounds. It states that the nonnegative rank of any slack matrix is equal to the extension complexity of the polytope.

Fiorini et al. [2] and Gouveia et al. [9] showed that the above readily generalizes to semidefinite extended formulations. Let $P \subseteq \mathbb{R}^n$ be a polytope. Then a *semidefinite extension* of P is a spectrahedron $Q \subseteq \mathbb{R}^d$ so that there exists a linear map π with $\pi(Q) = P$. While the projection of a polyhedron is polyhedral, it is open which convex sets can be obtained as projections of spectrahedra. We can again normalize the representation by considering Q as the intersection of an affine space with the cone of positive semidefinite (PSD) matrices. The *semidefinite extension complexity* is then defined as the smallest r for which there exists an affine space such that its intersection with the cone of $r \times r$ PSD matrices projects to P. We thus ask for the smallest representation of P as a projection of a spectrahedron. In both the linear and the semidefinite case, one can think of the extension complexity as the minimum size of the cone needed to represent P. Yannakakis's theorem can be generalized to this case [2, 9], and it asserts that the semidefinite extension complexity of a polytope equals the semidefinite rank (see Definition 3) of any of its slack matrices.

An important fact in the study of extended formulations is that the encoding length of coefficients is ignored: only the dimension of the required cone is measured. Also, a lower bound on the extension complexity of a polytope does not imply that building a separation oracle for the polytope is computationally hard. Indeed the perfect matching polytope is conjectured to have super polynomial extension complexity, while the associated separation problem (which allows us to compute min-cost perfect matchings) is in P. Thus standard complexity theoretic assumptions and limitations do not apply. In fact one of the main features of extended formulations is that they *unconditionally* provide lower bounds for the size of linear and semidefinite programs *independent of P vs. NP*.

The first class of polytopes with high linear extension complexity was found by Rothvoss [1]. He showed that "random" $0/1$ polytopes have exponential linear extension complexity via an elegant counting argument. Given that SDP relaxations are often more powerful than LP relaxations, an important open question is if random $0/1$ polytopes also have high PSD extension complexity.

1.1   Related Work

The basis for the study of linear and semidefinite extended formulations is the work of Yannakakis [8, 4]). The existence of a 0/1 polytope with exponential extension complexity was shown in [1] which in turn was inspired by work of Shannon [10]. The first explicit example, answering a long standing open problem of Yannakakis, was provided in [2] which, together with [9], also lay the foundation for the study of extended formulations over general closed convex cones. In [2] it was also shown that there exist matrices with large nonnegative rank but small semidefinite rank, indicating that semidefinite extended formulations can be exponentially stronger than linear ones, however falling short of giving an explicit proof. They thereby separated the expressive power of linear programs from those of semidefinite programs and raised the question:

Does every 0/1 polytope have an efficient semidefinite lift?

Other related work includes [3], where the authors study approximate extended formulations and provide examples of spectrahedra that cannot be approximated well by linear programs with a polynomial number of inequalities as well as improvements thereof in [11]. In [12] the authors prove equivalence of extended formulations to communication complexity. Recently there has been also significant progress in terms of lower bounding the linear extension complexity of polytopes by means of information theory [11, 13]. Similar techniques are not known for the semidefinite case.

1.2   Contribution

We answer the above question in the negative, i.e., we show the existence of a 0/1 polytope with exponential semidefinite extension complexity. In particular, we show that the counting argument of [1] extends to the PSD setting.

Our main technical contribution is a new rescaling technique for semidefinite factorizations of slack matrices. In particular, we show that any rank r semidefinite factorization of a slack matrix with maximum entry size Δ can be "rescaled" to a semidefinite factorization where each factor has operator norm at most $\sqrt{r\Delta}$ (Theorem 6). Our proof proceeds by a variational argument and relies on John's theorem on ellipsoidal approximation of convex bodies [14]. We note that in the linear case proving such a result is far simpler, here the only required observation is that after independent nonnegative scalings of the coordinates a nonnegative vector remains nonnegative. However, one cannot in general independently scale the entries of a PSD matrix while maintaining the PSD property.

Using our rescaling lemma, the existence proof of the 0/1 polytopes with high semidefinite extension complexity follows in a similar fashion to the linear case as presented in [1]. In addition to our main result, we show the existence of a polygon with d integral vertices and semidefinite extension complexity $\Omega((\frac{d}{\log d})^{\frac{1}{4}})$. The argument follows similarly to [6] adapting [1].

For complete proofs as well as the exact result on the polygon we refer the reader to http://arxiv.org/abs/1305.3268.

2 Preliminaries

Let $[n] := \{1, \ldots, n\}$. In the following we will consider semidefinite extended formulations. We refer to [2, 3] for a broader overview and proofs.

Let $B_2^n \subseteq \mathbb{R}^n$ denote the n-dimensional euclidean ball, and let $S^{n-1} = \partial B_2^n$ denote the euclidean sphere in $\mathbb{R}^n$. We denote by $\mathbb{S}_+^n$ the set of $n \times n$ PSD matrices which form a (non-polyhedral) convex cone. Note that $M \in \mathbb{S}_+^n$ if and only if M is symmetric ($M^\mathsf{T} = M$) and $x^\mathsf{T} M x \geq 0$ for any $x \in \mathbb{R}^n$. Equivalently, $M \in \mathbb{S}_+^n$ iff M is symmetric and has nonnegative eigenvalues.

For a matrix $A \in \mathbb{R}^{n \times n}$, we denote its trace by $\mathrm{Tr}[A] = \sum_{i=1}^n A_{ii}$. For a pair of equally-sized matrices A, B we let $\langle A, B \rangle = \mathrm{Tr}[A^\mathsf{T} B]$ denote their trace inner product and let $\|A\|_F = \sqrt{\langle A, A \rangle}$ denote the Frobenius norm of A. We denote the operator norm of a matrix $M \in \mathbb{R}^{m \times n}$ by $\|M\| = \sup_{\|x\|_2=1} \|Mx\|_2$. If M is square and symmetric ($M^\mathsf{T} = M$), then $\|M\| = \sup_{\|x\|_2=1} |x^\mathsf{T} M x|$, in which case $\|M\|$ denotes the largest eigenvalue of M in absolute value. Lastly, if $M \in \mathbb{S}_+^n$ then $\|M\| = \sup_{\|x\|_2=1} x^\mathsf{T} M x$ by nonnegativity of the inner expression.

For every positive integer ℓ and any ℓ-tuple of matrices $\mathbf{M} = (M_1, \ldots, M_\ell)$ we define $\|\mathbf{M}\|_\infty = \max\{\|M_i\| \mid i \in [\ell]\}$.

Definition 1 (Semidefinite extended formulation). *Let $K \subseteq \mathbb{R}^n$ be a convex set. A semidefinite extended formulation (semidefinite EF) of K is a system consisting of a positive integer r, an index set I and a set of triples $(a_i, U_i, b_i)_{i \in I} \subseteq \mathbb{R}^n \times \mathbb{S}_+^r \times \mathbb{R}$ such that*

$$K = \{x \in \mathbb{R}^n \mid \exists Y \in \mathbb{S}_+^r : a_i^\mathsf{T} x + \langle U_i, Y \rangle = b_i \; \forall i \in I\}.$$

The size of a semidefinite EF is the size r of the positive semidefinite matrices U_i. The semidefinite extension complexity of K, denoted $\mathrm{xc}_{SDP}(K)$, is the minimum size of a semidefinite EF of K.

In order to characterize the semidefinite extension complexity of a polytope $P \subseteq [0, 1]^n$ we will need the concept of a slack matrix.

Definition 2 (Slack matrix). *Let $P \subseteq [0, 1]^n$ be a polytope, I, J be finite sets, $\mathcal{A} = (a_i, b_i)_{i \in I} \subseteq \mathbb{R}^n \times \mathbb{R}$ be a set of pairs and let $\mathcal{X} = (x_j)_{j \in J} \subseteq \mathbb{R}^n$ be a set of points, such that*

$$P = \{x \in \mathbb{R}^n \mid a_i^\mathsf{T} x \leq b_i \; \forall i \in I\} = \mathrm{conv}(\mathcal{X}).$$

Then, the slack matrix of P associated with $(\mathcal{A}, \mathcal{X})$ is given by $S_{ij} = b_i - a_i^\mathsf{T} x_j$.

Finally, the definition of a semidefinite factorization is as follows.

Definition 3 (Semidefinite factorization). *Let I, J be finite sets, $S \in \mathbb{R}_+^{I \times J}$ be a nonnegative matrix and r be a positive integer. Then, a rank-r semidefinite factorization of S is a set of pairs $(U_i, V^j)_{(i,j) \in I \times J} \subseteq \mathbb{S}_+^r \times \mathbb{S}_+^r$ such that*

$$S_{ij} = \langle U_i, V^j \rangle$$

for every $(i, j) \in I \times J$. The semidefinite rank of S, denoted $\mathrm{rank}_{\mathrm{PSD}}(S)$, is the minimum r such that there exists a rank r semidefinite factorization of S.

The semidefinite extension complexity of a polytope can be characterized by the semidefinite rank of any of its slack matrices, which is a generalization of Yannakakis's factorization theorem [8, 4] established in [2, 9].

Theorem 4 (Yannakakis's Factorization Theorem for SDPs). *Let $P \subseteq [0, 1]^n$ be a polytope and $\mathcal{A} = (a_i, b_i)_{i \in I}$ and $\mathcal{X} = (x_j)_{j \in J}$ be as in Definition 2. Let S be the slack matrix of P associated with $(\mathcal{A}, \mathcal{X})$. Then, S has a rank-r semidefinite factorization if and only if P has a semidefinite EF of size r. That is, $\mathrm{rank}_{\mathrm{PSD}}(S) = \mathrm{xc}_{SDP}(P)$.*

Moreover, if $(U_i, V^j)_{(i,j) \in I \times J} \subseteq \mathbb{S}^r_+ \times \mathbb{S}^r_+$ is a factorization of S, then

$$P = \{x \in \mathbb{R}^n \,|\, \exists Y \in \mathbb{S}^r_+ \,:\, a_i^\mathsf{T} x + \langle U_i, Y \rangle = b_i \,\forall i \in I\}$$

and the pairs $(x_j, V^j)_{j \in J}$ satisfy $a_i^\mathsf{T} x_j + \langle U_i, V^j \rangle = b_i$ for every $i \in I$.

In particular, the extension complexity is independent of the choice of the slack matrix and the semidefinite rank of all slack matrices of P is identical.

The following well-known theorem due to [14] lies at the core of our rescaling argument. We state a version that is suitable for the later application. Recall that B_2^n denotes the n-dimensional euclidean unit ball. A *probability vector* is a vector $p \in \mathbb{R}^n_+$ such that $p(1) + p(2) + \cdots + p(n) = 1$. For a convex set $K \subseteq \mathbb{R}^n$, we let $\mathrm{aff}(K)$ denote the affine hull of K, the smallest affine space containing K. We let $\dim(K)$ denote the linear dimension of the affine hull of K. Last, we let $\mathrm{relbd}(K)$ denote the relative boundary of K, i.e., the topological boundary of K with respect to its affine hull $\mathrm{aff}(K)$.

Theorem 5 ([14]). *Let $K \subseteq \mathbb{R}^n$ be a centrally symmetric convex set with $\dim(K) = k$. Let $T \in \mathbb{R}^{n \times k}$ be such that $E = T \cdot B_2^k = \{Tx \,|\, \|x\| \leq 1\}$ is the smallest volume ellipsoid containing K. Then, there exist a finite set of points $\mathcal{Z} \subseteq \mathrm{relbd}(K) \cap \mathrm{relbd}(E)$ and a probability vector $p \in \mathbb{R}^{\mathcal{Z}}_+$ such that*

$$\sum_{z \in \mathcal{Z}} p(z) \, zz^\mathsf{T} = \frac{1}{k} TT^\mathsf{T}.$$

We will need the following lemma and corollary, whose proofs can be found in the full version of this paper.

Lemma 1. *Let r be a positive integer, $X \in \mathbb{S}^r_+$ be a non-zero positive semidefinite matrix. Let $\lambda_1 = \|X\|$, W denote the λ_1-eigenspace of X. Then for $Z \in \mathbb{R}^{r \times r}$ symmetric,*

$$\frac{d}{d\varepsilon} \|X + \varepsilon Z\| \bigg|_{\varepsilon=0} = \max_{\substack{w \in W \\ \|w\|=1}} w^\mathsf{T} Z w$$

Recall that for a square matrix X, its *exponential* is given by

$$e^X = \sum_{k=0}^{\infty} \frac{1}{k!} X^k = I + X + \frac{1}{2} X^2 + \cdots.$$

Corollary 1. *Let r be a positive integer, $X \in \mathbb{S}_+^r$ be a non-zero positive semidefinite matrices. Let $\lambda_1 = \|X\|$, W denote the λ_1-eigenspace of X. Then for $Z \in \mathbb{R}^{r \times r}$ symmetric,*

$$\frac{d}{d\varepsilon} \left\| e^{\varepsilon Z} X e^{\varepsilon Z} \right\| \Big|_{\varepsilon=0} = 2\lambda_1 \max_{\substack{w \in W \\ \|w\|_2 = 1}} w^{\mathsf{T}} Z w$$

3   Rescaling Semidefinite Factorizations

A crucial point will be the rescaling of a semidefinite factorization of a nonnegative matrix M. In the case of linear extended formulations an upper bound of Δ on the largest entry of a slack matrix S implies the existence of a minimal nonnegative factorization $S = UV$ where the entries of U, V are bounded by $\sqrt{\Delta}$. This ensures that the approximation of the extended formulation can be captured by means of a polynomial-size (in Δ) grid. In the linear case, we note that any factorization $S = UV$ can be rescaled by a nonnegative diagonal matrix D where $S = (UD)(D^{-1}U)$ and the factorization $(UD, D^{-1}V)$ has entries bounded by $\sqrt{\Delta}$. However, such a rescaling relies crucially on the fact that after independent nonnegative scalings of the coordinates a nonnegative vector remains nonnegative. However, in the PSD setting, it is not true that the PSD property is preserved after independent nonnegative scalings of the matrix entries. We circumvent this issue by showing that a restricted class of transformations, i.e. the symmetries of the semidefinite cone, suffice to rescale any PSD factorization such that the largest eigenvalue occurring in the factorization is bounded in terms of the maximum entry in M and the rank of the factorization.

Theorem 6 (Rescaling semidefinite factorizations). *Let Δ be a positive real number, I, J be finite sets, $M \in [0, \Delta]^{I \times J}$ be a nonnegative matrix and $r := \operatorname{rank}_{\mathrm{PSD}} M$. Then, there exists a semidefinite factorization $(U_i, V^j)_{(i,j) \in I \times J}$ of M (i.e., $M_{ij} = \langle U_i, V^j \rangle$ and $U_i, V_j \in \mathbb{S}_+^r$) such that $\max_{i \in I} \|U_i\| \leq \sqrt{r\Delta}$ and $\max_{j \in J} \|V^j\| \leq \sqrt{r\Delta}$.*

Proof: Let us denote by $\mathcal{E}_M^r$ the set of rank-r semidefinite factorizations $(\mathbf{U}, \mathbf{V})$ of M, where $\mathbf{U} = (U_i)_{i \in I}$ and $\mathbf{V} = (V^j)_{j \in J}$. We study the potential function $\Phi_M : \mathcal{E}_M^r \to \mathbb{R}$ defined by $\Phi_M(\mathbf{U}, \mathbf{V}) = \|\mathbf{U}\|_\infty \cdot \|\mathbf{V}\|_\infty$. In particular, we analyze how this function behaves under small perturbations of its minimizers.

To begin, we first argue that there exists a minimizer of Φ_M that satisfies $\|\mathbf{U}\|_\infty = \|\mathbf{V}\|_\infty$. For an invertible matrix $A \in \mathbb{R}^{r \times r}$ and semidefinite factorization $(\mathbf{U}, \mathbf{V}) \in \mathcal{E}_M^r$, notice that the tuple

$$(\mathbf{U}', \mathbf{V}') = \left((A^{\mathsf{T}} U_i A)_{i=1}^m, (A^{-1} V^j A^{-\mathsf{T}})_{j=1}^n \right)$$

is also a semidefinite factorization of M. To see this observe that by invariance of the trace function under similarity transformations ($\operatorname{Tr}[BWB^{-1}] = \operatorname{Tr}[W]$),

$$\langle A^{\mathsf{T}} U_i A, A^{-1} V^j A^{-\mathsf{T}} \rangle = \langle U_i, V^j \rangle = M_{ij}.$$

For $A := (\|\mathbf{V}\|_\infty / \|\mathbf{U}\|_\infty)^{1/4} \cdot I$ it is then easy to see that we obtain a factorization $(\mathbf{U}', \mathbf{V}')$ of M such that

$$\|\mathbf{U}'\|_\infty = \|\mathbf{V}'\|_\infty = \|\mathbf{U}\|_\infty^{1/2} \|\mathbf{V}\|_\infty^{1/2} .$$

It follows that $\Phi_M(\mathbf{U}', \mathbf{V}') = \Phi_M(\mathbf{U}, \mathbf{V})$. By standard compactness argument it follows that the function Φ_M has a minimizer $(\mathbf{U}, \mathbf{V})$ such that $\|\mathbf{U}\|_\infty = \|\mathbf{V}\|_\infty$ as claimed. Let us fix such a factorization $(\widetilde{\mathbf{U}}, \widetilde{\mathbf{V}})$ and let

$$\mu = \|\widetilde{\mathbf{U}}\|_\infty = \|\widetilde{\mathbf{V}}\|_\infty = \Phi_M(\widetilde{\mathbf{U}}, \widetilde{\mathbf{V}})^{1/2}.$$

Our goal is to obtain a contradiction by assuming that $\mu^2 > \Delta r + \tau$ for some $\tau > 0$. To this end we bound the value of Φ_M at infinitesimal perturbations of the point $(\widetilde{\mathbf{U}}, \widetilde{\mathbf{V}})$. For a symmetric matrix Z and parameter $\varepsilon > 0$ the type of perturbations we consider are those defined by the invertible matrix $e^{-\varepsilon Z}$, which will take the role of the matrix A above. Notice that if Z is symmetric, then so is $e^{-\varepsilon Z}$. We show that there exists a matrix Z such that for every $U \in \{\widetilde{U}_i \mid i \in I\}$ such that $\|U\| = \mu$, we have

$$\left\| e^{-\varepsilon Z} U e^{-\varepsilon Z} \right\| \leq \mu - \frac{2\mu}{r}\varepsilon + O(\varepsilon^2), \tag{1}$$

while at the same time for every $V \in \{\widetilde{V}^j \mid j \in J\}$ such that $\|V\| = \mu$, we have

$$\left\| e^{\varepsilon Z} V e^{\varepsilon Z} \right\| \leq \mu + \frac{2\Delta}{\mu}\varepsilon + O(\varepsilon^2). \tag{2}$$

This implies that there is a point $(\mathbf{U}', \mathbf{V}')$ in the neighborhood of the minimizer $(\widetilde{\mathbf{U}}, \widetilde{\mathbf{V}})$ where

$$\Phi_M(\mathbf{U}', \mathbf{V}') < \mu^2 - \frac{2\tau}{r}\varepsilon + O(\varepsilon^2).$$

Thus, for small enough $\varepsilon > 0$, we have $\Phi_M(\mathbf{U}', \mathbf{V}') < \mu^2$, a contradiction to the minimality of μ. It suffices to consider the factorization matrices with the largest eigenvalues as small perturbations cannot change the eigenvalue structure. To prove the theorem we show the existence of such a matrix Z.

Let $\mathcal{Z} \subseteq S^{r-1}$ be a finite set of unit vectors such that every $z \in \mathcal{Z}$ is a μ-eigenvector of at least one of the matrices $\widetilde{U}_i$ for $i \in I$. Let $p \in \mathbb{R}_+^{\mathcal{Z}}$ be a probability vector (i.e., $\sum_{z \in \mathcal{Z}} p(z) = 1$) and define the symmetric matrix

$$Z = \sum_{z \in \mathcal{Z}} p(z) zz^\mathsf{T}. \tag{3}$$

Claim. Let $V \in \{\widetilde{V}^j \mid j \in J\}$ be one of the factorization matrices such that $\|V\| = \mu$. Then,

$$\frac{d}{d\varepsilon} \left\| e^{\varepsilon Z} V e^{\varepsilon Z} \right\| \bigg|_{\varepsilon=0} \leq \frac{2\Delta}{\mu}. \tag{4}$$

PROOF OF CLAIM: Let $\mathcal{V} \subseteq S^{r-1}$ be the set of eigenvectors of V that have eigenvalue μ. Then, Corollary 1 gives

$$\frac{d}{d\varepsilon} \left\| e^{\varepsilon Z} V e^{\varepsilon Z} \right\|\bigg|_{\varepsilon=0} = 2\mu \max_{v \in \mathcal{V}} v^\mathsf{T} Z v = 2\mu \max_{v \in \mathcal{V}} \sum_{z \in \mathcal{Z}} p(z)(z^\mathsf{T} v)^2 \qquad (5)$$

Fix $z \in \mathcal{Z}$ and $v \in \mathcal{V}$. Let $U \in \{\widetilde{U}_1, \ldots, \widetilde{U}_m\}$ be such that z is a μ-eigenvector of U. Let $U = \sum_{k \in [r]} \lambda_k u_k u_k^\mathsf{T}$ and $V = \sum_{\ell \in [r]} \gamma_\ell v_\ell v_\ell^\mathsf{T}$ be spectral decompositions of U and V, respectively, and recall that the u_k are pairwise orthogonal as are the v_ℓ. Note that since z is an eigenvector of U and v is an eigenvector of V, we may choose spectral decompositions of U and V such that $u_1 = z$ and $v_1 = v$ respectively. Then, by our assumed bounds on the maximum entry-size of the matrix M and the fact that λ_k and γ_ℓ are nonnegative (since U and V are PSD),

$$\Delta \geq \mathrm{Tr}\left[U^\mathsf{T} V\right] = \sum_{k,\ell \in [r]} \lambda_k \gamma_\ell (u_k^\mathsf{T} v_\ell)^2 \geq \lambda_1 \gamma_1 (u_1^\mathsf{T} v_1)^2 = \mu^2 (z^\mathsf{T} v)^2$$

Putting it all together, we get that

$$2\mu \max_{v \in \mathcal{V}} \sum_{z \in \mathcal{Z}} p(z)(z^\mathsf{T} v)^2 \leq 2\mu \max_{v \in \mathcal{V}} \sum_{z \in \mathcal{Z}} p(z) \frac{\Delta}{\mu^2} = 2\mu \max_{v \in \mathcal{V}} \frac{\Delta}{\mu^2} = \frac{2\Delta}{\mu},$$

as needed.   ♦

Claim. There exists a choice of unit vectors $\mathcal{Z}$ and probabilities p such that the following holds. Let $I' = \{i \in I \mid \|\widetilde{U}_i\| = \mu\}$. Then, for Z as in (3) we have

$$\frac{d}{d\varepsilon} \left\| e^{-\varepsilon Z} \widetilde{U}_i e^{-\varepsilon Z} \right\|\bigg|_{\varepsilon=0} \leq -\frac{2\mu}{r} \quad \forall i \in I'. \qquad (6)$$

PROOF OF CLAIM: For every $i \in I'$, let $\mathcal{U}_i \subseteq \mathbb{R}^r$ be the vector space spanned by the μ-eigenvectors of $\widetilde{U}_i$. Define the convex set $K = \mathrm{conv}\left(\bigcup_{i \in I'}(\mathcal{U}_i \cap B_2^r)\right)$. Notice that K is centrally symmetric. Let $k = \dim(K)$, and let $T \in \mathbb{R}^{r \times k}$ denote a linear transformation such that that $E = TB_2^k$ is the smallest volume ellipsoid containing K. By John's Theorem, there exists a finite set $\mathcal{Z} \subseteq \mathrm{relbd}(K) \cap \mathrm{relbd}(E)$ and a probability vector $p \in \mathbb{R}_+^{\mathcal{Z}}$ such that

$$Z = \sum_{z \in \mathcal{Z}} p(z)\, z z^\mathsf{T} = \frac{1}{k} T T^\mathsf{T}. \qquad (7)$$

Notice that each $z \in \mathcal{Z}$ must be an extreme point of K (as it is one for E) and the set of extreme points of K is exactly $\bigcup_{i \in I'}(\mathcal{U}_i \cap S^{r-1})$. Hence, each $z \in \mathcal{Z}$ is a unit vector and at the same time a μ-eigenvector of some $\widetilde{U}_i$, $i \in I'$.

For $i \in I'$, by Corollary 1 and (7) we have that

$$\frac{d}{d\varepsilon} \left\| e^{-\varepsilon Z} \widetilde{U}_i e^{-\varepsilon Z} \right\| \bigg|_{\varepsilon=0} = 2\mu \max\{u^\mathsf{T}(-Z)u \mid u \in \mathcal{U}_i \cap S^{r-1}\}$$

$$= -2\mu \min\{u^\mathsf{T} Z u \mid u \in \mathcal{U}_i \cap S^{r-1}\}$$

$$= -\frac{2\mu}{k} \min\{u^\mathsf{T} TT^\mathsf{T} u \mid u \in \mathcal{U}_i \cap S^{r-1}\}$$

$$\leq -\frac{2\mu}{r} \min\{\|T^\mathsf{T} u\|_2^2 \mid u \in \mathcal{U}_i \cap S^{r-1}\}.$$

Since $E \supseteq K \supseteq (\mathcal{U}_i \cap S^{r-1})$, for any $u \in \mathcal{U}_i \cap S^{r-1}$, we have

$$\|T^\mathsf{T} u\|_2 = \sup_{x \in E} x^\mathsf{T} u \geq \sup_{y \in K} y^\mathsf{T} u \geq u^\mathsf{T} u = 1 \text{ as needed.}$$

◆

Notice that the first claim implies (2) and the second claim implies (1). Hence, our assumption $\mu^2 > \Delta r + \tau$ contradicts that μ is the minimum value of Φ_M. $\square$

4   0/1 Polytopes with High Semidefinite xc

The lower bound estimation will crucially rely on the fact that any 0/1 polytope in the n-dimensional unit cube can be written as a linear system of inequalities $Ax \leq b$ with integral coefficients where the largest coefficient is bounded by $(\sqrt{n+1})^{n+1} \leq 2^{n \log(n)}$, see e.g., [15, Corollary 26]. Using Theorem 6 the proof follows along the lines of [1]; for simplicity and exposition we chose a compatible notation. We use different estimation however and we need to invoke Theorem 6. In the following let $\mathbb{S}_+^r(\alpha) = \{X \in \mathbb{S}_+^r \mid \|X\| \leq \alpha\}$.

Lemma 2 (Rounding lemma). *For a positive integer n set $\Delta := (n+1)^{(n+1)/2}$. Let $\mathcal{X} \subseteq \{0,1\}^n$ be a nonempty set, let $r := \mathrm{xc}_{SDP}(\mathrm{conv}\,(\mathcal{X}))$ and let $\delta \leq \left(16r^3(n+r^2)\right)^{-1}$. Then, for every $i \in [n + r^2]$ there exist:*

1. *an integer vector $a_i \in \mathbb{Z}^n$ such that $\|a_i\|_\infty \leq \Delta$,*
2. *an integer b_i such that $|b_i| \leq \Delta$,*
3. *a matrix $U_i \in \mathbb{S}_+^r(\sqrt{r}\Delta)$ whose entries are integer multiples of δ/Δ and have absolute value at most $8r^{3/2}\Delta$, such that*

$$\mathcal{X} = \left\{ x \in \{0,1\}^n \,\Big|\, \exists Y \in \mathbb{S}_+^r(\sqrt{r}\Delta) : |b_i - a_i^\mathsf{T} x - \langle Y, U_i \rangle| \leq \frac{1}{4(n+r^2)} \forall i \in [n+r^2] \right\}.$$

Proof: For some index set I let $\mathcal{A} = (a_i, b_i)_{i \in I} \subseteq \mathbb{Z}^n \times \mathbb{Z}$ be a non-redundant description of $\mathrm{conv}\,(\mathcal{X})$ (i.e., $|I|$ is minimal) such that for every $i \in I$, we have $\|a_i\|_\infty \leq \Delta$ and $|b_i| \leq \Delta$. Let J be an index set for $\mathcal{X} = (x_j)_{j \in J}$ and let $S \in \mathbb{Z}_{\geq 0}^{I \times J}$ be the slack matrix of $\mathrm{conv}\,(\mathcal{X})$ associated with the pair $(\mathcal{A}, \mathcal{X})$. The largest entry of the slack matrix is at most Δ. By Yannakakis's Theorem (Theorem 4) there exists a semidefinite factorization $(U_i, V^j)_{(i,j)\in I\times J} \subseteq \mathbb{S}^r_+ \times \mathbb{S}^r_+$ of S such that

$$\operatorname{conv}(\mathcal{X}) = \{x \in \mathbb{R}^n \mid \exists Y \in \mathbb{S}^r_+ : a_i^\mathsf{T} x + \langle U_i, Y \rangle = b_i \ \forall i \in I\}.$$

By Theorem 6 we may assume that $\|U_i\| \leq \sqrt{r\Delta}$ for every $i \in I$ and $\|V^j\| \leq \sqrt{r\Delta}$ for every $j \in J$. We will now pick a subsystem of maximum volume. For a linearly independent set of vectors $x_1, \ldots, x_k \in \mathbb{R}^n$, we let $\operatorname{vol}(\{x_1, \ldots, x_k\})$ denote the k-dimensional parallelepiped volume

$$\operatorname{vol}\left(\sum_{i=1}^k a_i x_i \,\Big|\, a_1, \ldots, a_k \in [0,1]\right) = \det((x_i^\mathsf{T} x_j)_{ij})^{\frac{1}{2}}.$$

If the vectors are dependent, then by convention the volume is zero. Let $\mathcal{W} = \operatorname{span}\{(a_i, U_i) \mid i \in I\}$ and let $I' \subseteq I$ be a subset of size $|I'| = \dim(\mathcal{W})$ such that $\operatorname{vol}(\{(a_i, U_i) \mid i \in I'\})$ is maximized. Note that $|I'| \leq n + r^2$.

For any positive semidefinite matrix $U \in \mathbb{S}^r_+$ with spectral decomposition

$$U = \sum_{k \in [r]} \lambda_k u_k u_k^\mathsf{T}, \quad \text{we let} \quad \bar{U} = \sum_{k \in [r]} \bar{\lambda}_k \bar{u}_k \bar{u}_k^\mathsf{T}$$

be the matrix where for every $k \in [r]$, the value of $\bar{\lambda}_k$ is the nearest integer multiple of δ/Δ to λ_k and $\bar{u}_k$ is the vector we get by rounding each of the entries of u_k to the nearest integer multiple of δ/Δ. Since each u_k is a unit vector, the matrices $u_k u_k^\mathsf{T}$ have entries in $[-1, 1]$ and it follows that U has entries in $r\|U\|[-1, 1]$. Similarly, since each $\bar{u}_k$ has entries in $(1 + \delta/\Delta)[-1, 1]$ each of the matrices $\bar{u}_k \bar{u}_k^\mathsf{T}$ has entries in $(1 + \delta/\Delta)^2[-1, 1]$, and it follows that $\bar{U}$ has entries in $r(\|U\| + \delta/\Delta)(1 + \delta/\Delta)^2[-1, 1]$. In particular, for every $i \in I'$, the entries of $\bar{U}_i$ are bounded in absolute value by

$$r(\|U_i\| + \delta/\Delta)(1 + \delta/\Delta)^2 \leq r(\sqrt{r\Delta} + \delta/\Delta)(1 + \delta/\Delta)^2 \leq 8r^{3/2}\sqrt{\Delta}.$$

We use the following simple claim.

Claim. Let U and $\bar{U}$ be as above. Then, $\|\bar{U} - U\|_2 \leq 4\delta r^2/\sqrt{\Delta}$

Define the set

$$\bar{\mathcal{X}} = \left\{x \in \{0,1\}^n \,\Big|\, \exists Y \in \mathbb{S}^r_+(\sqrt{r\Delta}) : |b_i - a_i^\mathsf{T} x - \langle \bar{U}_i, Y \rangle| \leq \frac{1}{4(n+r^2)} \ \forall i \in I'\right\}.$$

We claim that $\bar{\mathcal{X}} = \mathcal{X}$, which will complete the proof.

We will first show that $\mathcal{X} \subseteq \bar{\mathcal{X}}$. To this end, fix an index $j \in J$. By Theorem 4 we can pick $Y = V^j \in \mathbb{S}^r_+$ such that $a_i^\mathsf{T} x_j + \langle U_i, Y \rangle = b_i$ for every $i \in I'$. Moreover, $\|Y\| = \|V^j\| \leq \sqrt{r\Delta}$. This implies that for every $i \in I'$, we have

$$|b_i - a_i^\mathsf{T} x_j - \langle \bar{U}_i, Y \rangle| = |b_i - a_i^\mathsf{T} x_j - \langle U_i, Y \rangle + \langle \bar{U}_i - U_i, Y \rangle|$$
$$\leq \|\bar{U}_i - U_i\|_F \|Y\|_F \leq 4\delta r^3,$$

where the second line follows from the Cauchy-Schwarz inequality, the above claim, and $\|Y\|_F \leq \sqrt{r}\,\|Y\| \leq r\sqrt{\Delta}$. Now, since $4\delta r^3 \leq 4r^3/(16r^3(n+r^2)) = 1/(4(n+r^2))$ we conclude that $x_j \in \bar{\mathcal{X}}$ and hence $\mathcal{X} \subseteq \bar{\mathcal{X}}$.

It remains to show that $\bar{\mathcal{X}} \subseteq \mathcal{X}$. For this we show that whenever $x \in \{0,1\}^n$ is such that $x \notin \mathcal{X}$ it follows that $x \notin \bar{\mathcal{X}}$. To this end, fix an $x \in \{0,1\}^n$ such that $x \notin \mathcal{X}$. Clearly $x \notin \operatorname{conv}(\mathcal{X})$ and hence, there must be an $i^* \in I$ such that $a_{i^*}^\mathsf{T} x > b_{i^*}$. Since x, a_{i^*} and b_{i^*} are integral we must in fact have $a_{i^*}^\mathsf{T} x \geq b_{i^*} + 1$. We express this violation in terms of the above selected subsystem corresponding to the set I'.

There exist unique multipliers $\nu \in \mathbb{R}^{I'}$ such that $(a_{i^*}, U_{i^*}) = \sum_{i \in I'} \nu_i(a_i, U_i)$. Observe that this implies that $\sum_{i \in I'} \nu_i b_i = b_{i^*}$; otherwise it would be impossible for $a_i^\mathsf{T} x + \langle U_i, Y \rangle = b_i$ to hold for every $i \in I$ and hence we would have $X = \emptyset$ (which we assumed is not the case).

Using the fact that the chosen subsystem I' is volume maximizing and using Cramer's rule,

$$|\nu_i| = \frac{\operatorname{vol}(\{(a_t, U_t) \mid t \in I' \setminus \{i\} \cup \{i^*\}\})}{\operatorname{vol}(\{(a_t, U_t) \mid t \in I'\})} \leq 1.$$

For any $Y \in \mathbb{S}_+^r(\sqrt{r\Delta})$ using $\langle U_{i^*}, Y \rangle \geq 0$ it follows thus

$$1 \leq \left|a_{i^*}^\mathsf{T} x - b_{i^*} + \langle U_{i^*}, Y \rangle\right| = \left|\sum_{i \in I'} \nu_i(a_i^\mathsf{T} x - b_i + \langle U_i, Y \rangle)\right|$$

$$\leq \sum_{i \in I'} |\nu_i|\,\left|a_i^\mathsf{T} x - b_i + \langle U_i, Y \rangle\right| \leq (n + r^2) \max_{i \in I'} \left|a_i^\mathsf{T} x - b_i + \langle U_i, Y \rangle\right|.$$

Using a similar estimation as above, for every $i \in I'$, we have

$$\left|a_i^\mathsf{T} x - b_i + \langle U_i, Y \rangle\right| = |a_i^\mathsf{T} x - b_i + \langle \bar{U}_i, Y \rangle + \langle U_i - \bar{U}_i, Y \rangle|$$

$$\leq |a_i^\mathsf{T} x - b_i + \langle \bar{U}_i, Y \rangle| + \frac{1}{4(n + r^2)}.$$

Combining this with $1 \leq (n + r^2) \max_{i \in I'} \left|a_i^\mathsf{T} x - b_i + \langle U_i, Y \rangle\right|$ we obtain

$$\frac{1}{2(n + r^2)} \leq \frac{1}{n + r^2} - \frac{1}{4(n + r^2)} \leq \max_{i \in I'} |a_i^\mathsf{T} x - b_i + \langle \bar{U}_i, Y \rangle|,$$

and so $x \notin Y$.

Via padding with empty rows we can ensure that $|I'| = n + r^2$ as claimed. $\qquad\square$

Using Lemma 2 we can establish the existence of 0/1 polytopes that do not admit any small semidefinite extended formulation following the proof of [1, Theorem 4].

Theorem 7. *For any $n \in \mathbb{N}$ there exists $\mathcal{X} \subseteq \{0,1\}^n$ such that*

$$\operatorname{xc}_{SDP}(\operatorname{conv}(\mathcal{X})) = \Omega\left(\frac{2^{n/4}}{(n \log n)^{1/4}}\right).$$

References

[1] Rothvoß, T.: Some 0/1 polytopes need exponential size extended formulations. Math. Programming, arXiv:1105.0036 (2012)
[2] Fiorini, S., Massar, S., Pokutta, S., Tiwary, H.R., de Wolf, R.: Linear vs. Semidefinite Extended Formulations: Exponential Separation and Strong Lower Bounds. In: Proceedings of STOC 2012 (2012)
[3] Braun, G., Fiorini, S., Pokutta, S., Steurer, D.: Approximation Limits of Linear Programs (Beyond Hierarchies). In: 53rd IEEE Symp. on Foundations of Computer Science (FOCS 2012), pp. 480–489 (2012)
[4] Yannakakis, M.: Expressing combinatorial optimization problems by linear programs. J. Comput. System Sci. 43(3), 441–466 (1991)
[5] Ben-Tal, A., Nemirovski, A.: On polyhedral approximations of the second-order cone. Math. Oper. Res. 26, 193–205 (2001)
[6] Fiorini, S., Rothvoß, T., Tiwary, H.R.: Extended formulations for polygons. Discrete & Computational Geometry 48(3), 658–668 (2012)
[7] Goemans, M.X.: Smallest compact formulation for the permutahedron. Manuscript (2009)
[8] Yannakakis, M.: Expressing combinatorial optimization problems by linear programs (extended abstract). In: Proc. STOC 1988, pp. 223–228 (1988)
[9] Gouveia, J., Parrilo, P.A., Thomas, R.: Lifts of convex sets and cone factorizations. Math. Oper. Res. 38(2), 248–264 (2011)
[10] Shannon, C.E.: The synthesis of two-terminal switching circuits. Bell System Tech. J. 25, 59–98 (1949)
[11] Braverman, M., Moitra, A.: An information complexity approach to extended formulations. Electronic Colloquium on Computational Complexity (ECCC) 19(131) (2012)
[12] Faenza, Y., Fiorini, S., Grappe, R., Tiwary, H.R.: Extended formulations, nonnegative factorizations, and randomized communication protocols. In: Mahjoub, A.R., Markakis, V., Milis, I., Paschos, V.T. (eds.) ISCO 2012. LNCS, vol. 7422, pp. 129–140. Springer, Heidelberg (2012)
[13] Braun, G., Pokutta, S.: Common information and unique disjointness (submitted, 2013)
[14] John, F.: Extremum problems with inequalities as subsidiary conditions. In: Studies and Essays presented to R. Courant on his 60th Birthday, pp. 187–204 (1948)
[15] Günter, M.: Lectures on 0/1-Polytopes. In: Kalai, G., Ziegler, G.M. (eds.) Polytopes — Combinatorics and Computation. DMV Seminar, vol. 29, pp. 1–41. Birkhäuser Basel (2000)

The Encoding Complexity of Two Dimensional Range Minimum Data Structures

Gerth Stølting Brodal[1], Andrej Brodnik[2,3,*], and Pooya Davoodi[4,**]

[1] MADALGO[***], Department of Computer Science, Aarhus University, Denmark
gerth@cs.au.dk
[2] University of Primorska, Department of Information Science and Technology,
Slovenia
andrej.brodnik@upr.si
[3] University of Ljubljana, Faculty of Computer and Information Science, Slovenia
[4] Polytechnic Institute of New York University, USA
pooyadavoodi@gmail.com

Abstract. In the two-dimensional range minimum query problem an input matrix A of dimension $m \times n$, $m \leq n$, has to be preprocessed into a data structure such that given a query rectangle within the matrix, the position of a minimum element within the query range can be reported. We consider the space complexity of the encoding variant of the problem where queries have access to the constructed data structure but can not access the input matrix A, i.e. all information must be encoded in the data structure. Previously it was known how to solve the problem with space $O(mn \min\{m, \log n\})$ bits (and with constant query time), but the best lower bound was $\Omega(mn \log m)$ bits, i.e. leaving a gap between the upper and lower bounds for non-quadratic matrices. We show that this space lower bound is optimal by presenting an encoding scheme using $O(mn \log m)$ bits. We do not consider query time.

1 Introduction

We study the problem of preprocessing a two dimensional array (matrix) of elements from a totally ordered set into a data structure that supports range minimum queries (RMQs) asking for the *position* of the minimum element within a range in the array. More formally, we design a data structure for a matrix $A[1..m] \times [1..n]$ with $N = mn$ elements, and each RMQ asks for the position of the minimum element within a range $[i_1..i_2] \times [j_1..j_2]$. We refer to this problem as the 2D-RMQ problem, in contrast with the 1D-RMQ problem, where the input array is one-dimensional.

Study of the 1D-RMQ and 2D-RMQ problems dates back to three decades ago, when data structures of linear size were proposed for both of the problems [8,3].

---

* Research supported by grant BI-DK/11-12-011: Algorithms on Massive Geographical LiDAR Data-sets - AMAGELDA.
** Research supported by NSF grant CCF-1018370 and BSF grant 2010437.
*** Center for Massive Data Algorithmics, a Center of the Danish National Research Foundation.

H.L. Bodlaender and G.F. Italiano (Eds.): ESA 2013, LNCS 8125, pp. 229–240, 2013.

Both problems have known applications including information retrieval, computational biology, and databases [4]. In this paper, we study the space-efficiency of the 2D-RMQ data structures.

1.1 Previous Work

There exists a long list of articles on the 1D-RMQ problem which resolve various aspects of the problem including the space-efficiency of the data structures. However, the literature of the 2D-RMQ problem is not so rich in space-efficient data structures. In the following, we review the previous results on both of the problems that deal with the space-efficiency of data structures.

1D-RMQs. Let A be an input array with n elements which we preprocess into a data structure that supports 1D-RMQs on A. A standard approach to solve this problem is to make the *Cartesian tree* of A, which is a binary tree with n nodes defined as follows: the root of the Cartesian tree is labeled by i, where $A[i]$ is the minimum element in A; the left and right subtrees of the root are recursively the Cartesian trees of $A[1..i-1]$ and $A[i+1..n]$ respectively.

The property of the Cartesian tree is that the answer to a query with range $[i..j]$ is the label of the lowest common ancestor (LCA) of the nodes with labels i and j. Indeed, the Cartesian tree stores a partial ordering between the elements that is appropriate to answer 1D-RMQs. This property has been utilized to design data structures that support 1D-RMQs in constant time [8,9].

The space usage of the 1D-RMQ data structures that rely on Cartesian trees is the amount of space required to represent a Cartesian tree plus the size of an LCA data structure. There exists a one-to-one correspondence between the set of Cartesian trees and the set of *different* arrays, where two arrays are different iff there exists a 1D-RMQ with different answers on these two arrays. The total number of binary trees with n nodes is $\binom{2n}{n}/(n+1)$, and thus the logarithm of this number yields an information-theoretic lower bound on the number of bits required to store a 1D-RMQ data structure, which is $2n - \Theta(\log n)$ bits.

Storing the Cartesian tree using the standard pointer-based representation and a linear-space LCA data structure takes $O(n \log n)$ bits. There have been a number of attempts to design 1D-RMQ data structures of size close to the lower bound. Sadakane [10] and then Fischer and Heun [7] improved the space to $4n + o(n)$ and $2n + o(n)$ bits respectively by representing the Cartesian tree using succinct encoding of the topology of the tree and utilizing the encoding to support LCA queries (in fact, Fischer and Heun [7] proposed a representation of another tree which is a transformed Cartesian tree [5]). Both of these data structures support 1D-RMQs in $O(1)$ time.

2D-RMQs. A standard two-level range tree along with a 1D-RMQ data structure can be used to design a 2D-RMQ data structure. This method was used to give a data structure of size $O(N \log N)$ that supports 2D-RMQs in $O(\log N)$ time [8]. The state of the art 1D-RMQ data structures can improve the space to $O(N)$.

The literature contains a number of results that have advanced the performance of 2D-RMQ data structures in terms of preprocessing and query times [3,1], which ended in a brilliant 2D-RMQ data structure of size $O(N)$ (that is, $O(N \log N)$ bits) which can be constructed in $O(N)$ time and supports queries in $O(1)$ time [11].

Although the complexity of the preprocessing and query times of 2D-RMQ data structures have been settled, the space complexity of 2D-RMQ data structures has been elusive.

In contrast with 1D-RMQs where the partial ordering of elements can be encoded in linear number of bits using Cartesian trees, it has been shown that encoding partial ordering of elements for 2D-RMQs would require super-linear number of bits (there is no "Cartesian tree-like" data structure for 2D-RMQs) [6]. The number of different $n \times n$ matrices is $\Omega((\frac{n}{4}!)^n/4)$, where two matrices are different if there exists a 2D-RMQ with different answers on the two matrices [6]. This counting argument implies the lower bound $\Omega(N \log N)$ on the number of bits required to store a 2D-RMQ data structure on an $n \times n$ matrix with $N = n^2$ elements.

The above counting argument was later extended to rectangular $m \times n$ matrices where $m \leq n$, yielding the information-theoretic lower bound $\Omega(N \log m)$ bits on the space of 2D-RMQ data structures [2]. This lower bound raised the question of encoding partial ordering of elements for 2D-RMQs using space close to the lower bound. In other words, the new question shifted the focus from designing 2D-RMQ data structures with efficient queries to designing encoding data structures, which we simply call encodings, that support 2D-RMQs disregarding the efficiency of queries. In fact, the fundamental problem here is to discover whether the partial ordering of elements for 2D-RMQs can be encoded in $O(N \log m)$ bits.

There exist simple answers to the above question which do not really satisfy an enthusiastic mind. On the one hand, we can ensure that each element in A takes $O(\log N)$ bits in the encoding data structure by sorting the elements and replacing each element with its rank (recall that a query looks for the position of minimum rather than its value). This provides a simple encoding of size $O(N \log N)$ bits, while this upper bound was already achieved by all the linear space 2D-RMQ data structures [3,1,11]. On the other hand, we can make a data structure of size $O(Nm)$ bits which improves the size of the first encoding for $m = o(\log n)$. This latter encoding is achieved by taking the multi-row between every two rows i and j; making an array $R[1..n]$ out of each multi-row by assigning the minimum of the k-th column of the multi-row to $R[k]$; encoding R using a 1D-RMQ data structure of size $O(n)$ bits; and making a 1D-RMQ data structure of size $O(m)$ bits for each column of A. A 2D-RMQ can be answered by finding the appropriate multi-row, computing the column of the multi-row with minimum element, and computing the minimum element in the column within the query range [2].

1.2   Our Results

The 2D-RMQ encodings mentioned above leave a gap between the lower bound $\Omega(N \log m)$ and the upper bound $O(N \cdot \min\{\log N, m\})$ on the number of bits stored in a 2D-RMQ encoding. A quick comparison between the encoding complexity of 1D-RMQ and 2D-RMQ data structures can convince the reader that the optimal number of bits per element stored in an encoding should be a function of m, at least for small enough m. For example, we should not look for functions such as $O(N \log \log N)$ as the encoding complexity of the problem. While the upper bound $O(N \cdot \min\{\log N, m\})$ would provide such a characteristic, it is often hard to believe the minimum of two functions as the complexity of a problem. In this work, we prove that the encoding complexity of 2D-RMQ data structures is $\Theta(N \log m)$ bits. As previously mentioned, we only consider the encoding complexity of the problem, and a problem that remains open from this paper is how to actually support queries efficiently, ideally in $O(1)$ time.

We describe our solution in three steps, incrementally improving the space. First we present a solution using space $O(N(\log m + \log \log n))$ bits, introducing the notion of components. We then improve this to an $O(N \log m \log^* n)$ bit solution by bootstrapping our first solution and building an $O(\log^* n)$ depth hierarchical partitioning of our components. Finally, we arrive at an $O(N \log m)$ bit solution using a more elaborate representation of a hierarchical decomposition of components of depth $O(\log n)$.

1.3 Preliminaries

We formally define the problem and then we introduce the notation used in the paper. The input is assumed to be an $m \times n$ matrix A, where $m \le n$. We let $N = m \cdot n$. The j-th entry in row i of A is denoted by $A[i, j]$, where $1 \le i \le m$ and $1 \le j \le n$. For a query range $q = [i_1, i_2] \times [j_1, j_2]$, the query RMQ($q$) reports the index (i, j) of A containing the minimum value in rows $i_1..i_2$ and columns $j_1..j_2$, i.e.

$$\mathrm{RMQ}(q) = \mathrm{argmin}_{(i,j) \in q} A[i, j] .$$

For two cells c_1 and c_2 we define Rect$[c_1, c_2]$ to be the minimal rectangle containing c_1 and c_2, i.e. c_1 and c_2 are at the opposite corners of Rect$[c_1, c_2]$.

We assume all values of a matrix to be distinct, such that RMQ(q) is always a unique index into A. This can e.g. be achieved by ordering identical matrix values by the lexicographical ordering of the indexes of the corresponding cells.

2 Tree Representation

The basic approach in all of our solutions is that we convert our problem on the matrix A into a tree representation problem, where the goal is to find a tree representation that can be represented within small space.

For a given input matrix A, we build a left-to-right ordered tree T with N leaves, where each leaf of T corresponds to a unique cell of A and each internal node of T has at least two children. Furthermore, the tree should satisfy the following crucial property:

Requirement 1. *For any rectangular query q, consider the leaves of T corresponding to the cells contained in q. The answer to the query q must be the rightmost of these leaves.*

Trivial solution. The most trivial solution is to build a tree of depth one, where all cells/leaves are the children of the root and sorted left-to-right in decreasing value order. A query q will consist of a subset of the leaves, and the rightmost of these leaves obviously stores the minimal value within q, since the leaves are sorted in decreasing order with respect to values. To represent such a tree we store for each leaf the index (i, j) of the corresponding cell, i.e. such a tree can be represented by a list of N pairs, each requiring $\lceil \log m \rceil + \lceil \log n \rceil$ bits. Total space usage is $O(N \log n)$ bits.

Note that at each leaf we explicitly store the index of the corresponding cell in A, which becomes the space bottleneck of the representation: The left to right ordering of the leaves always stores the permutation of all matrix values, i.e. such a representation will require $\Omega(N \cdot \log N)$ bits in the worst-case. In all the following representations we circumvent this bottleneck by not storing the complete total order of all matrix values and by only storing relative positions to other (already laid out) cells of A.

The structure (topology) of any tree T with N leaves can be described using $O(N)$ bits (since all internal nodes are assumed to be at least binary), e.g. using a parenthesis notation. From the known lower bound of $\Omega(N \log m)$ bits for the problem, it follows that the main task will be to find a tree T satisfying Requirement 1, and where the mapping between leaves of T and the cells of A can be stored succinctly.

3   Space $O(N(\log m + \log \log n))$ Solution

We present a solution using space $O(N(\log m + \log \log n))$ bits, i.e. it achieves optimal space $O(N \log m)$ for $\log n = m^{O(1)}$. The idea is to build a tree T of depth two in which the nodes $C_1, C_2, \ldots, C_k$ with depth one, from left-to-right, form a partitioning of A. Let C_i denote both a node in T and the subset of cells in A corresponding to the leaves/cells in the subtree of T rooted at C_i. We call such a C_i a *component*. We construct $C_1, \ldots, C_k$ incrementally from left-to-right such that Requirement 1 is satisfied. The children of C_i (i.e. leaves/cells) are sorted in decreasing value ordered left-to-right. We first describe how to construct C_1, and then describe how to construct C_i given $C_1, \ldots, C_{i-1}$, generalizing the notation and algorithm used for constructing C_1. In the following we let α denote a parameter of our construction. We will later set $\alpha = \lceil \log N / \log \log N \rceil$.

Constructing C_1. To construct C_1 we need the following notation: Two cells in A are *adjacent* if they share a side, i.e. (i, j) is adjacent to the (up to) four cells $(i-1, j)$, $(i+1, j)$, $(i, j-1)$ and $(i, j+1)$. Given a set of cells $S \subseteq A$, we define the undirected graph $G_S = (S, E)$, where $E \subseteq S \times S$ and $(c_1, c_2) \in E$ if and only if c_1 and c_2 are adjacent cells in A.

Let S be an initially empty set. We now incrementally include the cells of A in S in decreasing order of the value of the cells. While all the connected components of G_S contain less than α cells, we add one more cell to S. Otherwise the largest component C contains at least α cells, and we terminate and let $C_1 = C$. In the

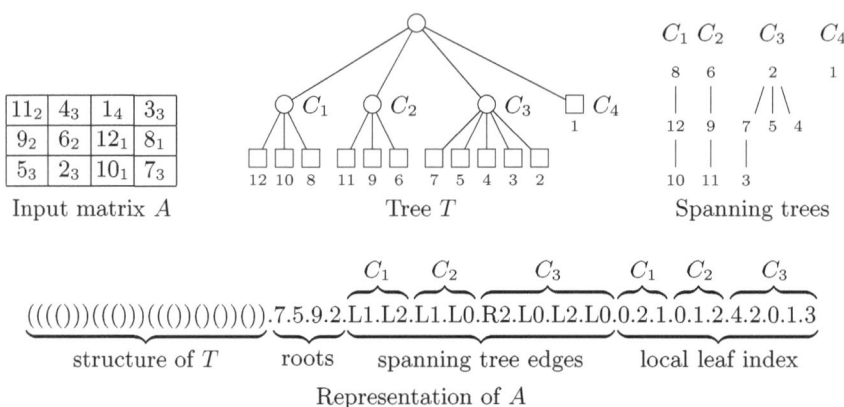

Fig. 1. Construction of $C_1, \ldots, C_4$ for a 3×4 matrix and $\alpha = 3$: In the input matrix (left) subscripts indicate the component numbers. In the tree T (middle) the leaves are labeled with the corresponding values of the cells in A. The spanning trees for $C_1, \ldots, C_4$ are shown (right). In the final representation of A (bottom) "." is not part of the stored representation.

example in Fig. 1, where $\alpha = 3$, we terminate when $S = \{12, 11, 10, 9, 8\}$ and G_S contains the two components $\{11, 9\}$ and $\{12, 10, 8\}$, and let C_1 be the largest of these two components.

We first prove the size of the constructed component C_1.

Lemma 1. $|C_1| \leq 4\alpha - 3$.

Proof. Let S be the final set in the construction process of C_1, such that S contains a connected component of size at least α. Before inserting the last cell c into S, all components in $G_{S \setminus \{c\}}$ have size at most $\alpha - 1$. Since c is adjacent to at most four cells, the degree of c in G_S is at most four, i.e. including c in S can at most join four existing components in $G_{S \setminus \{c\}}$. It follows that the component C_1 in G_S containing c contains at most $1 + 4(\alpha - 1)$ cells. □

Next we will argue that the partition of A into C_1 and $R = A \setminus C_1$ (R will be the leaves to the right of C_1 in T) supports Requirement 1, i.e. for any query q that overlaps both with C_1 and R there is an element $c \in q \cap R$ that is smaller than all elements in $q \cap C_1$, implying that the answer is not in C_1 and will be to the right of C_1 in T. Since by construction all values in $R \setminus S$ are smaller than all values in $S \supseteq C_1$, it is sufficient to prove that there exists a $c \in (q \cap R) \setminus S$.

Lemma 2. For a query q with $q \cap C_1 \neq \emptyset$ and $q \cap R \neq \emptyset$, there exists a cell $c \in (q \cap R) \setminus S$.

Proof. Assume by contradiction that $q \subseteq S$. Then each cell in q is either in C_1 or in $S \setminus C_1$, and by the requirements of the lemma and the assumption $q \subseteq S$, there exists at least one cell in both $q \cap C_1$ and $(q \cap S) \setminus C_1$. Since each cell in

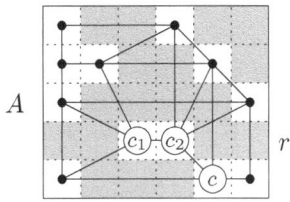

Fig. 2. The edges of $G_{A\backslash L}^{(L)}$. The cells of L are shown as grey.

q belongs to one of these sets, there must exist two adjacent cells c_1 and c_2 in q such that $c_1 \in q \cap C_1$ and $c_2 \in (q \cap S) \setminus C_1$. But since c_1 and c_2 are adjacent, both in S, and $c_1 \in C_1$, then c_2 must also be in C_1 which is a contradiction. $\square$

Constructing C_i. Given how to construct C_1, we now give a description of how to construct C_i when $C_1, \ldots, C_{i-1}$ have already been constructed. To construct C_i we let $L = C_1 \cup \cdots \cup C_{i-1}$. We generalize the notion of adjacency of cells to L-*adjacency*: Given a set L, two cells c_1 and c_2 in $A \setminus L$ are L-*adjacent* if and only if $\text{Rect}[c_1, c_2] \setminus \{c_1, c_2\} \subseteq L$. Note that the adjacency definition we used to construct C_1 is equivalent to $\emptyset$-adjacency. Finally given a set $S \subseteq A \setminus L$, we define the undirected graph $G_S^{(L)} = (S, E)$, where $(c_1, c_2) \in E$ if and only if c_1 and c_2 are L-adjacent. See Fig. 2 for an example of edges defined by L-adjacency.

The algorithm for constructing C_1 now generalizes as follows to C_i: Start with an empty set S. Consider the cells in $A \setminus L$ in decreasing value order and add the cells to S until the largest connected component C of $G_S^{(L)}$ has size at least α (or until all elements in $A \setminus L$ have been added to S). We let $C_i = C$. Note that with $L = \emptyset$ this algorithm is exactly the algorithm for constructing C_1. Fig. 1 illustrates an example of such a partitioning of a 3×4 matrix for $\alpha = 3$.

Lemma 3. *All nodes in a graph $G_S^{(L)}$ have degree at most $2m$.*

Proof. Let $G_S^{(L)} = (S, E)$. Consider a cell $c \in S$. For any row r, there exists at most one cell $c_1 \in S$ in row r to the left of c (or immediately above c), such that $(c, c_1) \in E$. Otherwise there would exist two cells c_1 and c_2 in S to the left of c in row r, where c_1 is to the left of c_2 (see Fig. 2). But then $c_2 \in \text{Rect}[c, c_1]$, i.e. c and c_1 would not be L-adjacent and $(c, c_1) \notin E$. It follows that for any row, at most one cell to the left and (symmetrically) one cell to the right of c can have an edge to c in $G_S^{(L)}$. $\square$

The following two lemmas generalize Lemmas 1 and 2 for C_1 to C_i.

Lemma 4. $|C_i| \leq m(\alpha - 1) + \alpha \leq 2m\alpha.$

Proof. Since the last cell c added to S is in C_i and by Lemma 3 has degree at most $2m$ in $G_S^{(L)}$, and before adding c to S all components had size at most $\alpha - 1$, it follows that $|C_i| \leq 1 + 2m(\alpha - 1) \leq 2m\alpha$. This bound can be improved to $1 + (m+1)(\alpha - 1)$ by observing that adding c can at most join $m+1$ existing components: if c is adjacent to two cells c_1 and c_2 in the same row, one in a column to the left and one in a column to the right of c, then c_1 and c_2 must be L-adjacent, provided if c is not in the same row as c_1 and c_2, i.e. c_1 and c_2 were already in the same component before including c. □

We next prove the generalization of Lemma 2 to C_i. Let $L = C_1 \cup \cdots \cup C_{i-1}$ be the union of the already constructed components (to the left of C_i) and let $R = A \setminus \{C_1 \cup \cdots \cup C_i\}$ be the set of cells that eventually will be to the right of C_i in T. The following lemma states that if a query q contains one element from each of C_i and R, then there exists a cell $c \in q \cap R$ with smaller value than all values in C_i. It is sufficient to show that there exists a cell $c \in q \cap R \setminus S$ since by construction all values in $R \setminus S$ are smaller than $\min(S) \leq \min(C_i)$.

Lemma 5. *For a query q with $q \cap C_i \neq \emptyset$ and $q \cap R \neq \emptyset$, there exists a $c \in q \cap R \setminus S$.*

Proof. Assume by contradiction that $q \cap R \setminus S = \emptyset$. Then $q \subseteq S \cup L = C_i \cup (S \setminus C_i) \cup L$. By the requirement of the lemma and the assumption $q \cap R \setminus S = \emptyset$, there exist $c_1 \in q \cap C_i$ and $c_2 \in q \cap S \setminus C_i$. We will now show that c_1 and c_2 are connected in $G_S^{(L)}$ (not necessarily by a direct edge), i.e. c_1 and c_2 are both in C_i, which is a contradiction.

We construct a path from c_1 to c_2 in $G_S^{(L)} = (S, E)$ as follows. If $\mathrm{Rect}[c_1, c_2] \setminus \{c_1, c_2\} \subseteq L$, then c_1 and c_2 are L-adjacent, and we are done. Otherwise let c_3 be the cell, where $\mathrm{Rect}[c_1, c_3] \setminus \{c_1, c_3\}$ contains no cell from S (c_3 is the closest cell to c_1 in $S \cap \mathrm{Rect}[c_1, c_2] \setminus \{c_1, c_2\}$). Therefore, c_1 and c_3 are L-adjacent, i.e. $\mathrm{Rect}[c_1, c_3] \setminus \{c_1, c_3\} \subseteq L$ and $(c_1, c_3) \in E$. By applying this construction recursively between c_3 and c_2 we construct a path from c_1 to c_2 in $G_S^{(L)}$. □

Representation. We now describe how to store sufficient information about the tree T such that we can decode T and answer a query. For each connected component C_i we store a *spanning tree* S_i, such that S_i consists exactly of the edges that were joining different smaller connected components while including new cells into S. We root S_i at the cell with smallest value in C_i (but any node of C_i could have been chosen as the root/anchor of the component).

In the following representation we assume n and m are already known (can be coded using $O(\log N)$ bits). Furthermore it is crucial that when we layout C_i we have already described (and can decode) what are the cells of $C_1, \ldots, C_{i-1}$, i.e. L in the construction of C_i.

Crucial to our representation of C_i is that we can efficiently represent edges in $G_S^{(L)}$: If there is an edge from c_1 to c_2 in $G_S^{(L)}$, then c_1 and c_2 are L-adjacent, i.e. to navigate from c_1 to c_2 it is sufficient to store what row c_2 is stored in, and if c_2 is to the left (L) or right (R) of c_1 (if they are in the same column, then c_2 is to the left of c_1), since we just move vertically from c_1 to the row of c_2 and then move in the left or right direction until we find the first non-L slot (here it is crucial for achieving our space bounds that we do not need to store the number of columns moved over). It follows that a directed edge from c_1 to

c_2 identifying the cell c_2 can be stored using $1 + \lceil \log m \rceil$ bits, provided we know where the cell c_1 (and L) is.

The details of the representation are the following (See Fig. 1 for a detailed example of the representation):

- We first store the structure of the spanning trees for $C_1, \ldots, C_k$ as one big tree in parenthesis notation (putting a super-root above the roots of the spanning trees); requires $2(N + 1)$ bits. This encodes the number of components, the size of each component, and for each component the tree structure of the spanning tree.
- The *global index* of the root cell for each C_i, where we enumerate the cells of A by $0..N - 1$ row-by-row, top-down and left-to-right; requires $k \lceil \log N \rceil$ bits.
- For each of the components $C_1, \ldots, C_k$, we traverse the spanning tree in a DFS traversal, and for each edge (c_1, c_2) of the spanning tree for a component C_i, where c_1 is the parent of c_2, we store a bit L/R and the row of c_2 (note that our representation ensures that the cells of $L = C_1 \cup \cdots \cup C_{i-1}$ are previously identified, which is required). Since there are $N - k$ edges in total in the k spanning trees, we have that the edges in total require $(N - k)(1 + \lceil \log m \rceil)$ bits.
- For each component C_i, we store the leaves/cells of C_i in decreasing value order (only if $|C_i| \geq 2$): For each leaf, we store its local index in C_i with respect to the following enumeration of the cells. Since the spanning tree identifies exactly the cells in C_i and $|C_i| \leq 2m\alpha$, we can enumerate these cells by $0..2m\alpha - 1$, say row-by-row, top-down and left-to-right; in total at most $N \lceil \log(2m\alpha) \rceil$ bits.

Lemma 6. *The representation of T requires $O(N(\log m + \frac{\log N}{\alpha} + \log \alpha))$ bits.*

Proof. From the above discussion we have that the total space usage in bits is

$$2(N + 1) + k \lceil \log N \rceil + (N - k)(1 + \lceil \log m \rceil) + N \lceil \log(2m\alpha) \rceil .$$

For all C_i we have $\alpha \leq |C_i| \leq 2m\alpha$, except for the last component C_k, where we have $1 \leq |C_k| \leq 2m\alpha$. It follows that $k \leq \lceil N/\alpha \rceil$, and the bound stated in the lemma follows.   □

Setting $\alpha = \lceil \log N / \log \log N \rceil$ and using $\log N \leq 2 \log n$, we get the following space bound.

Theorem 1. *There exists a 2D-RMQ encoding of size $O(N(\log m + \log \log n))$ bits.*

We can reconstruct T by reading the encoding from left-to-right, and thereby answer queries. We reconstruct the tree structure of the spanning trees of the components by reading the parenthesis representation from left-to-right. This gives us the structure of T including the size of all the components from left to right. Given the structure of T and the spanning trees, the positions of the remaining fields in the encoding are uniquely given. Since we reconstruct the components from left-to-right and we traverse the spanning trees in depth first order, we can decode each spanning tree edge since we know the relevant set L and the parent cell of the edge. Finally we decode the indexes of the leaves in each C_i, by reading their local indexes in C_i in decreasing order.

4   Space $O(N \log m \log^* n)$ Solution

In the following we describe how to recursively apply our $O(N(\log m + \log \log n))$ bit space solution to arrive at an $O(N \log m \log^* n)$ bit solution, where $\log^* n = \min\{i \mid \log^{(i)} n \leq 1\}$, $\log^{(0)} n = n$, and $\log^{(i+1)} n = \log \log^{(i)} n$.

Let K be an integer determining the depth of the recursion. At the root we have a problem of size $N_0 = N$ and apply the partitioning algorithm of Section 3 with $\alpha = \lceil \log N_0 \rceil$, resulting in sub-components $C_1, C_2, C_3, \ldots$ each of size at most $N_1 = 2m\lceil \log N_0 \rceil$. Each C_i is now recursively partitioned. A component C at level i in the recursion has size at most N_i. When constructing the partitioning of C, we initially let $L = A \setminus C$, i.e. all cells outside C act as neutral "$+\infty$" elements for the partitioning inside C_i. The component C is now partitioned as in Section 3 with $\alpha = \lceil \log N_i \rceil$ into components of size at most $N_{i+1} = 2m\lceil \log N_i \rceil$. A special case is when $|C| \leq N_{i+1}$, where we skip the partitioning of C at this level of the recursion (to avoid creating degree one nodes in T).

Let N_i denote an upper bound on the problem size at depth i of the recursion, for $0 \leq i \leq K$. We know $N_0 = N$ and $N_{i+1} \leq 2m \log N_i$. By induction it can be seen that $N_i \leq 2m \log^{(i)} N + 12m \log m$: For $i = 0$ we have $N_0 = N \leq 2m \log^{(0)} N$ and the bound is true. Otherwise we have

$$
\begin{aligned}
N_{i+1} &\leq 2m\lceil \log N_i \rceil \\
&\leq 2m\lceil \log(2m \log^{(i)} N + 12m \log m) \rceil \\
&\leq 2m\lceil \log((2m + 12m \log m) \log^{(i)} N) \rceil \quad \text{(for } \log^{(i)} N \geq 1\text{)} \\
&\leq 2m \log^{(i+1)} N + 2m \log(2m + 12m \log m) + 2m \\
&\leq 2m \log^{(i+1)} N + 12m \log m \quad \text{(for } m \geq 2\text{)} .
\end{aligned}
$$

Representation. Recall that T is an ordered tree with depth K, in which each leaf corresponds to a cell in A, and each internal node corresponds to a component and the children of the node is a partitioning of the component.

- First we have a parenthesis representation of T; requiring $2N$ bits.
- For each internal node C of T in depth first order, we store a parenthesis representation of the structure of the spanning tree for the component represented by C. Each cell c of A corresponds to a leaf ℓ of T and is a node in the spanning tree of each internal node on the path from ℓ to the root of T, except for the root, i.e. c is in at most K spanning trees. It follows that the parenthesis representation of the spanning trees takes $2KN$ bits.

- For each internal node C of T in depth first order, we store the local index of the root r of the spanning tree for C as follows. Let C' be the parent of C in T, where C' is partitioned at depth i of the recursion. We enumerate all cells in C' row-by-row, top-down and left-to-right by $0..|C'|-1$ and store the index of r among these cells. Since $|C'| \leq N_{i-1}$, this index requires $\lceil \log N_{i-1} \rceil$ bits. The total number of internal nodes in T resulting from partitionings at depth i in the recursion is at most $2N/\lceil \log N_{i-1} \rceil$ (at least half of the components in the partitionings are large enough containing at least $\lceil \log N_{i-1} \rceil$ cells), from which it follows that to store the roots we need at most $2N$ bits at each depth of the recursive partitioning, i.e. in total at most $2KN$ bits.
- For each internal node C of T in depth first order, we store the edges of the spanning tree of C in a depth first traversal of the spanning tree edges. Each edge requires $1 + \lceil \log m \rceil$ bits. Since there are less than N spanning tree edges for each level of T, the total number of spanning tree edges is bounded by NK, i.e. total space $NK(1 + \lceil \log m \rceil)$ bits.
- For each leaf ℓ of T with parent C we store the index of ℓ among the up to N_K cells in C, requiring $\lceil \log N_K \rceil$ bits. Total space $N \lceil \log N_K \rceil$ bits.

The total number of bits required becomes $2N + 2KN + 2KN + NK(1 + \lceil \log m \rceil) + N \lceil \log N_K \rceil = O(NK \log m + N \log^{(K+1)} N)$. Setting $K = \log^* N = O(\log^* n)$ we get the following theorem.

Theorem 2. *There exists a 2D-RMQ encoding using $O(N \log m \log^* n)$ bits.*

5   Space $O(N \log m)$ Solution

The $O(N \log m \log^* n)$ bits solution of the previous section can be viewed as a top-down representation, where we have a representation identifying all the cells of a component C before representing the recursive partitioning of C, and where each C_i is identified using a spanning tree local to C. To improve the space to $O(N \log m)$ bits we adopt a bottom-up representation, such that the representation of C is the concatenation of the representations of $C_1, \ldots, C_k$, prefixed by information about the sizes and the relative placements of the C_i components.

We construct T top-down as follows. Let C be the current node of T, where we want to identify its children $C_1, \ldots, C_k$. Initially C is the root of T covering all cells of A. If $|C| < m^8$, we just make all cells of (the component) C leaves below (the node) C in T in decreasing value order from left-to-right. Otherwise we create a partition $C_1, \ldots, C_k$ of C, make $C_1, \ldots, C_k$ the children of C from left-to-right, and recurse for each of the generated C_i components.

This solution which partitions a component C into a set of smaller components $C_1, \ldots, C_k$, takes a parameter $\alpha = \lfloor |C|^{1/4}/2 \rfloor$ and the set of cells L to the left of C in the final tree T, i.e. given C and L, we for C_i have the corresponding $L_i = L \cup C_1 \cup \cdots \cup C_{i-1}$. To be able to construct the partitioning, the graph $G_C^{(L)}$ must be connected and $|C| \geq \alpha$. Given such a C we can find a partition satisfying that each of the constructed C_i components has size at least α and at most $4\alpha^2 m^2$, and $G_{C_i}^{(L_i)}$ is a connected graph. It follows that $k \leq 4 \cdot |C|^{3/4}$ (since $\alpha = \lfloor |C|^{1/4}/2 \rfloor \geq |C|^{1/4}/4$) and $|C_i| \leq 4\alpha^2 m^2 \leq 4 \cdot \lfloor |C|^{1/4}/2 \rfloor^2 \cdot |C|^{1/4} \leq |C|^{3/4}$ (since $m^2 = (m^8)^{1/4} \leq |C|^{1/4}$ for $|C| \geq m^8$).

The details of how to construct the partitions and how to use it to derive a space-efficient representation is described in the full version of the paper.

Theorem 3. *There exists a 2D-RMQ encoding using $O(N \log m)$ bits.*

6   Conclusion

We have settled the encoding complexity of the two dimensional range minimum problem as being $\Theta(N \log m)$. In the worst case, a query algorithm requires to decode the representation to be able the answer queries. It remains an open problem to build a data structure with this space bound and with efficient queries.

References

1. Amir, A., Fischer, J., Lewenstein, M.: Two-dimensional range minimum queries. In: Ma, B., Zhang, K. (eds.) CPM 2007. LNCS, vol. 4580, pp. 286–294. Springer, Heidelberg (2007)
2. Brodal, G.S., Davoodi, P., Rao, S.S.: On space efficient two dimensional range minimum data structures. Algorithmica 63(4), 815–830 (2012)
3. Chazelle, B., Rosenberg, B.: Computing partial sums in multidimensional arrays. In: Mehlhorn, K. (ed.) SoCG, pp. 131–139. ACM (1989)
4. Davoodi, P.: Data Structures: Range Queries and Space Efficiency. PhD thesis, Aarhus University, Aarhus, Denmark (2011)
5. Davoodi, P., Raman, R., Rao, S.S.: Succinct representations of binary trees for range minimum queries. In: Gudmundsson, J., Mestre, J., Viglas, T. (eds.) CO-COON 2012. LNCS, vol. 7434, pp. 396–407. Springer, Heidelberg (2012)
6. Demaine, E.D., Landau, G.M., Weimann, O.: On cartesian trees and range minimum queries. In: Albers, S., Marchetti-Spaccamela, A., Matias, Y., Nikoletseas, S., Thomas, W. (eds.) ICALP 2009, Part I. LNCS, vol. 5555, pp. 341–353. Springer, Heidelberg (2009)
7. Fischer, J., Heun, V.: Space-efficient preprocessing schemes for range minimum queries on static arrays. SIAM J. Comput. 40(2), 465–492 (2011)
8. Gabow, H.N., Bentley, J.L., Tarjan, R.E.: Scaling and related techniques for geometry problems. In: DeMillo, R.A. (ed.) STOC, pp. 135–143. ACM (1984)
9. Harel, D., Tarjan, R.E.: Fast algorithms for finding nearest common ancestors. SIAM J. Comput. 13(2), 338–355 (1984)
10. Sadakane, K.: Succinct data structures for flexible text retrieval systems. J. Discrete Algorithms 5(1), 12–22 (2007)
11. Yuan, H., Atallah, M.J.: Data structures for range minimum queries in multidimensional arrays. In: Charikar, M. (ed.) SODA, pp. 150–160. SIAM (2010)

Computing the Fréchet Distance
with a Retractable Leash

Kevin Buchin[1], Maike Buchin[2], Rolf van Leusden[1],
Wouter Meulemans[1,*], and Wolfgang Mulzer[3,**]

[1] Technical University Eindhoven, The Netherlands
k.a.buchin@tue.nl, r.v.leusden@student.tue.nl, w.meulemans@tue.nl
[2] Ruhr Universität Bochum, Germany
Maike.Buchin@ruhr-uni-bochum.de
[3] Freie Universität Berlin, Germany
mulzer@inf.fu-berlin.de

Abstract. All known algorithms for the Fréchet distance between curves proceed in two steps: first, they construct an efficient oracle for the decision version; then they use this oracle to find the optimum among a finite set of critical values. We present a novel approach that avoids the detour through the decision version. We demonstrate its strength by presenting a quadratic time algorithm for the Fréchet distance between polygonal curves in $\mathbb{R}^d$ under polyhedral distance functions, including L_1 and L_∞. We also get a $(1 + \epsilon)$-approximation of the Fréchet distance under the Euclidean metric. For the exact Euclidean case, our framework currently gives an algorithm with running time $O(n^2 \log^2 n)$. However, we conjecture that it may eventually lead to a faster exact algorithm.

1 Introduction

Measuring the similarity of curves is a classic problem in computational geometry with many applications. For example, it is used for map-matching tracking data [3, 15] and moving objects analysis [5, 6]. In all these applications it is important to take the continuity of the curves into account. Therefore, the *Fréchet distance* and its variants are popular metrics to quantify (dis)similarity.

The Fréchet distance between two curves is defined by taking a homeomorphism between the curves that minimizes the maximum pairwise distance. It is commonly described using the *leash*-metaphor: a man walks on one curve and has a dog on a leash on the other curve. Both man and dog can vary their speeds, but they may not walk backwards. The Fréchet distance is the length of the shortest leash with which man and dog can walk from the beginning to the end of the respective curves.

Related work. The algorithmic study of the Fréchet distance was initiated by Alt and Godau [1]. For polygonal curves, they give an algorithm to solve the

---

[*] Supported by the Netherlands Organisation for Scientific Research (NWO) under project no. 639.022.707.
[**] Supported in part by DFG project MU/3501/1.

H.L. Bodlaender and G.F. Italiano (Eds.): ESA 2013, LNCS 8125, pp. 241–252, 2013.

decision version in $O(n^2)$ time, and then use parametric search to find the optimum in $O(n^2 \log n)$ time. Several randomized algorithms have been proposed which are based on the decision version in combination with sampling possible values for the distance, one running in $O(n^2 \log^2 n)$ time [10] and the other in $O(n^2 \log n)$ time [13]. Recently, Buchin et al. [8] showed how to solve the decision version in subquadratic time, resulting in a randomized algorithm for computing the Fréchet distance in $O(n^2 \log^{1/2} n \log \log^{3/2} n)$ time. In terms of the leash-metaphor these algorithms simply give a leash to the man and his dog to try if a walk is possible. By cleverly picking the different leash-lengths, one then finds the Fréchet distance in an efficient way. Several algorithms exist to approximate the Fréchet distance (e.g. [2, 12]). However, these rely on various assumptions of the input curve; no approximation algorithm is known for the general case.

Contribution. We present a novel approach that does not use the decision problem as an intermediate stage. We give the man a "retractable leash" which can be lengthened or shortened as required. To this end, we consider monotone paths on the *distance terrain*, a generalization of the free space diagram typically used for the decision problem. Similar concepts have been studied before, but without the monotonicity requirement (e.g., [11] or the *weak* Fréchet distance [1]).

We show that it is sufficient to focus on the boundaries of cells of the distance terrain (defined by the vertices of the curves). It seems natural to propagate through the terrain for any point on a boundary the minimal "height" (leash length) ε required to reach that point. However, this may lead to an amortized linear number of changes when moving from one boundary to the next, giving a lower bound of $\Omega(n^3)$. We therefore do not maintain these functions explicitly. Instead, we maintain sufficient information to compute the lowest ε for a boundary. A single pass over the terrain then finds the lowest ε for reaching the other end, giving the Fréchet distance.

We present the core ideas for our approach in Section 2. This framework gives a choice of distance metric, but it requires an implementation of a certain data structure. We apply this framework to the Euclidean distance (Section 3) and polyhedral distances (Section 4). We also show how to use the latter to obtain a $(1 + \epsilon)$-approximation for the former. This is the first approximation algorithm for the general case. We conclude with two open problems in Section 5.

2   Framework

2.1   Preliminaries

Curves and distances. Throughout we wish to compute the dissimilarity of two polygonal curves, P and Q. For simplicity, we assume that both curves consist of n segments. This represents the computational worst case; of course our algorithm can also cope with asymmetric cases. Both curves are given as piecewise-linear functions $P, Q: [0, n] \to \mathbb{R}^d$. That is, $P(i + \lambda) = (1 - \lambda)P(i) + \lambda P(i + 1)$ holds for any integer $i \in [0, n)$ and $\lambda \in [0, 1]$, and similarly for Q.

Let Ψ be the set of all continuous and nondecreasing functions $\psi \colon [0, n] \to [0, n]$ with $\psi(0) = 0$ and $\psi(n) = n$. Then the *Fréchet distance* is defined as

$$d_{\mathrm{F}}(P, Q) = \inf_{\psi \in \Psi} \max_{t \in [0, n]} \{\delta(P(t), Q(\psi(t)))\}.$$

Here, δ may represent any distance function between two points in $\mathbb{R}^d$. Typically, the Euclidean distance function is used; we consider this scenario in Section 3. Another option we shall consider are polyhedral distance functions (Section 4). For our framework, we require that the distance function is convex.

Distance terrain. Let us consider the joint parameter space $R = [0, n] \times [0, n]$ of P and Q. A pair $(s, t) \in R$ corresponds to the points $P(s)$ and $Q(t)$, and the distance function δ assigns a value $\delta(P(s), Q(t))$ to (s, t). We interpret this value as the "height" at point $(s, t) \in R$. This gives a *distance terrain* T, i.e., $T : R \to \mathbb{R}$ with $T(s, t) = \delta(P(s), Q(t))$. We segment T into n^2 *cells* based on the vertices of P and Q. For integers $i, j \in [0, n)$, the cell $C[i, j]$ is defined as the subset $[i, i + 1] \times [j, j + 1]$ of the parameter space. The cells form a regular grid, and we assume that i represents the column and j represents the row of each cell. An example of two curves and their distance terrain is given in Fig. 1.

A path $\pi : [0, 1] \to R$ is called *bimonotone* if it is both x- and y-monotone. For $(s, t) \in R$, we let $\Pi(s, t)$ denote the set of all bimonotone continuous paths from the origin to (s, t). The *acrophobia function* $\widetilde{T} : R \to \mathbb{R}$ is defined as

$$\widetilde{T}(s, t) = \inf_{\pi \in \Pi(s, t)} \max_{\lambda \in [0, 1]} T(\pi(\lambda)).$$

Intuitively, $\widetilde{T}(s, t)$ represents the lowest height that an acrophobic climber needs to master in order to reach (s, t) from the origin on a bimonotone path. Clearly, we have $d_{\mathrm{F}}(P, Q) = \widetilde{T}(n, n)$.

Let $x \in R$ and $\pi \in \Pi(x)$ be a bimonotone path from $(0, 0)$ to x. Let ε be a value greater than zero. We call π an ε-*witness* for x if $\max_{\lambda \in [0, 1]} T(\pi(\lambda)) \leq \varepsilon$. We call π a *witness* for x if $\max_{\lambda \in [0, 1]} T(\pi(\lambda)) = \widetilde{T}(x)$, i.e., π is an optimal path for the acrophobic climber.

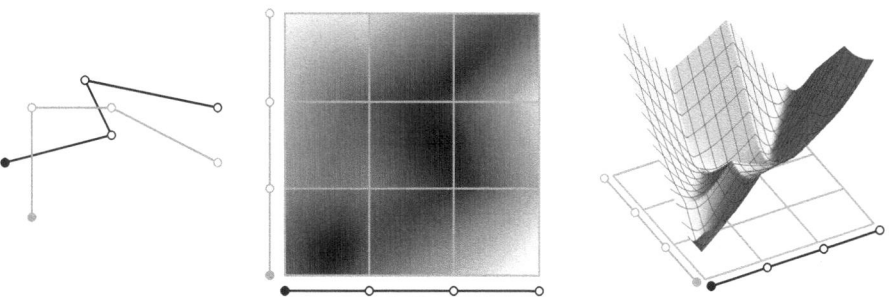

Fig. 1. Distance terrain with the Euclidean distance in $\mathbb{R}^2$. (left) Two curves. (middle) Cells as seen from above. Dark colors indicate low "height". (right) Perspective view.

2.2 Analysis of the Distance Terrain

To compute $\widetilde{T}(n,n)$, we show that it is sufficient to consider only the cell boundaries. For this, we generalize the fact that cells of the free space diagram are convex [1] to convex distance functions. For the proof, we refer to [7].

Lemma 2.1. *For a convex distance function δ and $\varepsilon \in \mathbb{R}$, the set of points (s,t) in a given cell $C[i,j]$ with $T(s,t) \leq \varepsilon$ is convex.*

Lemma 2.1 has two important consequences. First, it implies that it is indeed sufficient to consider only the cell boundaries. Second, it tells us that the distance terrain along a boundary is well-behaved. In this corollary and in the remainder of the paper, we refer to a function with a single local minimum as *unimodal*.

Corollary 2.2. *Let $C[i,j]$ be a cell of the distance terrain, and let x_1 and x_2 be two points on different boundaries of $C[i,j]$. For any y on the line segment $x_1 x_2$, we have $T(y) \leq \max\{T(x_1), T(x_2)\}$.*

Corollary 2.3. *The distance along every boundary of a cell in distance terrain T is a unimodal function.*

For any cell $C[i,j]$, we denote with $L[i,j]$ and $B[i,j]$ its left and bottom boundary respectively (and their height functions in T). The right and top boundary are given by $L[i+1,j]$ and $B[i,j+1]$.[1] With $\widetilde{L}[i,j]$ and $\widetilde{B}[i,j]$ we denote the acrophobia function along the boundary. All these restricted functions have a single parameter in the interval $[0,1]$ that represents the boundary.

Assuming that the distance function δ is symmetric, computing values for rows and columns of T is symmetric as well. Hence, we present only how to compute with rows. If δ is asymmetric, our methods still work, but some extra care needs to be taken when computing distances.

Consider a vertical boundary $L[i,j]$. We use $\widetilde{L}^*[i,j]$ to denote the minimum of the acrophobia function $\widetilde{L}[i,j]$ along $L[i,j]$. An analogous definition is used for horizontal boundaries. Our goal is to compute $\widetilde{L}^*[i,j]$ and $\widetilde{B}^*[i,j]$ for all cell boundaries of the grid. We say that an ε-witness π *passes through* an edge $B[i,j]$, if there is a $\lambda \in [0,1]$ with $\pi(\lambda) \in B[i,j]$.

Lemma 2.4. *Let $\varepsilon > 0$, and let x be a point on $L[i,j]$. Let π be an ε-witness for x that passes through $B[a,j]$, for $1 \leq a < i$. Suppose further that there exists a column b with $a < b < i$ and $\widetilde{B}^*[b,j] \leq \varepsilon$. Then there exists an ε-witness for x that passes through $B[b,j]$.*

Proof. Let y be the point on $B[b,j]$ that achieves $\widetilde{B}^*[b,j]$, and let π_y be a witness for y. Since π is bimonotone and since π passes through $B[a,j]$, it follows that π must also pass through $L[b+1,j]$. Let z be the (lowest) intersection point, and π_z the subpath of π from z to x. Let π' be the path obtained by concatenating π_y, line segment yz, and π_z. By our assumption on ε and by Corollary 2.2, path π' is an ε-witness for x that passes through $B[b,j]$. $\square$

---

[1] Note that there need not be an actual cell $C[i+1,j]$ or $C[i,j+1]$.

Lemma 2.4 implies that there are always *rightmost* witnesses for any point x on $L[i, j]$. For such witness, if it passes through $B[a, j]$ for some $a < i$, then $\widetilde{B}^*[b, j] > \widetilde{T}(x)$ for any $a < b < i$.

Corollary 2.5. *Let x be a point on $L[i, j]$. Then there is a witness for x that passes through some $B[a, j]$ such that $\widetilde{B}^*[b, j] > \widetilde{T}(x)$ for any $a < b < i$.*

Next, we argue that there is a witness for $\widetilde{L}^*[i + 1, j]$ that enters row j at or after the horizontal boundary point used by the witness for $\widetilde{L}^*[i, j]$. In other words, the rightmost witnesses behave "monotonically" in the terrain.

Lemma 2.6. *Let π be a witness for $\widetilde{L}^*[i, j]$ that passes through $B[a, j]$, for a $1 \leq a < i$. Then there exists a $a \leq b \leq i$ such that $\widetilde{L}^*[i + 1, j]$ has a witness that passes through $B[b, j]$.*

Proof. Choose b maximal such that $\widetilde{L}^*[i+1, j]$ has a witness that passes through $B[b, j]$. Suppose $b < a$. Let π' be such a witness. We know that $\widetilde{L}^*[i + 1, j] \geq \widetilde{L}^*[i, j]$, since π' passes through $L[i, j]$. However, we can now construct a witness for $\widetilde{L}^*[i + 1, j]$ that passes through $B[a, j]$: follow π to $B[a, j]$ and then switch to the intersection of π' and $L[a + 1, j]$. This contradicts the choice of b. □

We now characterize $\widetilde{L}[i, j]$ via a *witness envelope*, defined as follows. Fix a row j and two columns $a < i$. Suppose that $\widetilde{L}^*[i - 1, j]$ has a witness that passes through $B[a', j]$ with $a' \leq a$. The *witness envelope* for the column interval $[a, i]$ in row j is the upper envelope of the following functions on the interval $[0, 1]$:

(i)   the terrain function $L[i, j](\lambda)$;
(ii) the constant function $\widetilde{B}^*[a, j]$;
(iii) the constant function $\widetilde{L}^*[i - 1, j]$ if $a \leq i - 2$;
(iv) the *truncated terrain functions* $\overline{L}[b, j](\lambda) = \min_{\mu \in [0, \lambda]} L[b, j](\mu)$ for $a < b < i$.

Lemma 2.7. *Fix a row j and two columns $a < i$ as above. Let $\alpha \in [0, 1]$ and $\varepsilon > 0$. The point $x = (i, j + \alpha)$ has an ε-witness that passes through $B[a, j]$ if and only if (α, ε) lies above the witness envelope for $[a, i]$ in row j.*

Proof. Let π be an ε-witness for x that passes through $B[a, j]$. Then clearly $\varepsilon \geq \widetilde{B}^*[a, j]$ and $\varepsilon \geq L[i, j](\alpha)$. If $a \leq i - 2$, then π must pass through $L[i - 1, j]$, so $\varepsilon \geq \widetilde{L}^*[i - 1, j]$. Since π is bimonotone, it has to pass through $L[b, j]$ for $a < b < i$. Let $y_1 = (a + 1, j + \alpha_1), y_2 = (a + 2, j + \alpha_2), \ldots, y_k = (a + k, j + \alpha_k)$ be the points of intersection, from left to right. Then $\alpha_1 \leq \alpha_2 \leq \cdots \leq \alpha_k \leq \alpha$ and $\varepsilon \geq T(y_l) = L[a + l, j](\alpha_i) \geq \overline{L}[a + l, j](\alpha)$, for $l = 1, \ldots, k$. Hence (α, ε) is above the witness envelope.

Suppose (α, ε) is above the witness envelope. The conclusion follows directly if $a = i - 1$. Otherwise, $\varepsilon \geq \widetilde{L}^*[i - 1, j]$ holds. Let α' be such that the witness for $\widetilde{L}^*[i-1, j]$ that passes through $B[a', j]$ reaches $L[i-1, j]$ at point $(i-1, j+\alpha')$. If $\alpha \geq \alpha'$, we construct an appropriate ε-witness π' for x by following the witness for $\widetilde{B}^*[a, j]$, then passing to the witness for $\widetilde{L}^*[i - 1, j]$ and then taking the line segment to x. If $\alpha < \alpha'$, we construct a curve π' as before. However, π' is not bimonotone (the last line segment goes down). To fix this, let p and x be the two intersection points of π' with the horizontal line $y = j + \alpha$. We shortcut π' at the line segment px. The resulting curve π is clearly bimonotone and passes through $B[a, j]$. To see that π is an ε-witness, it suffices to check that along the segment px, the distance terrain never goes above ε. For this, we need to consider only the intersections of px with the vertical cell boundaries. Let $L[b, j]$ be such a boundary. We know that $L[b, j]$ is unimodal (Corollary 2.3) and let α^* denote the value where the minimum is obtained. By definition of the truncated terrain function, $L[b, j](\alpha) = \overline{L}[b, j](\alpha)$ if $\alpha \leq \alpha^*$. By assumption, the witness for $\widetilde{L}^*[i - 1, j]$ passes $L[b, j]$ at α or higher. Hence, if $\alpha \geq \alpha^*$, then $\widetilde{L}^*[i - 1, j] \geq L[b, j](\alpha)$. It follows that $\max\{\overline{L}[b, j](\alpha), \widetilde{L}^*[i - 1, j]\} \geq L[b, j](\alpha)$. By definition $\varepsilon \geq \max\{\overline{L}[b, j](\alpha), \widetilde{L}^*[i - 1, j]\}$ holds and thus $\varepsilon \geq L[b, j](\alpha)$. $\quad\square$

2.3 Algorithm

We are now ready to present the algorithm. We walk through the distance terrain, row by row, in each row from left to right. When processing a cell $C[i, j]$, we compute $\widetilde{L}^*[i + 1, j]$ and $\widetilde{B}^*[i, j + 1]$. For each row j, we maintain a double-ended queue (deque) Q_j that stores a sequence of column indices. We also store a data structure U_j that contains a set of (truncated) terrain functions on the vertical boundaries in j. It supports insertion, deletion, and a minimum-point query that, given up to two additional constants, returns the lowest point on the upper envelope of the terrain functions and the given constants.

The data structures fulfill the following invariant. Suppose that $\widetilde{L}^*[i, j]$ is the rightmost optimum we have computed so far in row j, and suppose that a rightmost witness for $\widetilde{L}^*[i, j]$ passes through $B[a, j]$. A point (α, β) *dominates* a point (γ, δ) if $\alpha > \gamma$ and $\beta \leq \delta$. Then Q_j stores the first coordinates of the points in the sequence $(a, \widetilde{B}^*[a, j]), (a + 1, \widetilde{B}^*[a + 1, j]), \ldots, (i - 1, \widetilde{B}^*[i - 1, j])$ that are not dominated by any other point in the sequence. Furthermore, the structure U_j stores the terrain functions for the boundaries from column $a + 1$ to i. We maintain analogous data structures for each column i.

The algorithm proceeds as follows (see Algorithm 1): since $(0, 0)$ belongs to any path through the distance terrain, we initialize $C[0, 0]$ to use $(0, 0)$ as its lowest point and compute the distance accordingly. The left- and bottommost boundaries of the distance terrain are considered unreachable. Any path to such a point also goes through the adjacent horizontal boundaries or vertical boundaries respectively. These adjacent boundaries therefore ensure a correct result.

In the body of the for-loop, we compute $\widetilde{L}^*[i + 1, j]$ and $\widetilde{B}^*[i, j + 1]$. Let us describe how to find $\widetilde{L}^*[i + 1, j]$. First, we add index i to Q_j and remove all previous indices that are dominated by it from the back of the deque. We add $L[i + 1, j]$ to upper envelope U_j. Let h and h' be the first and second element of Q_j. We perform a minimum query on U_j in order to find the smallest ε_α for which a point on $L[i + 1, j]$ has an ε_α-witness that passes through $B[h, j]$. By Lemma 2.7, this query requires the height at which the old witness enters the

---

Algorithm 1. FRECHETDISTANCE(P, Q, δ)

Input: P and Q are polygonal curves with n edges in $\mathbb{R}^d$;
$\quad$ δ is a convex distance function in $\mathbb{R}^d$
Output: Fréchet distance $d_F(P, Q)$

$\quad${We show computations only within a row, column computations are analogous}
1: $\widetilde{L}^*[0, 0] \leftarrow \delta(P(0), Q(0))$
2: $\widetilde{L}^*[0, j] \leftarrow \infty$ for all $0 < j < n$
3: For each row j, create empty deque Q_j and upper envelope structure U_j
4: **for** $j \leftarrow 0$ **to** $n - 1$; $i \leftarrow 0$ **to** $n - 1$ **do**
5: $\quad$ Remove any values x from Q_j with $\widetilde{B}^*[x, j] \geq \widetilde{B}^*[i, j]$ and append i to Q_j
6: $\quad$ **if** $|Q_j| = 1$ **then** Clear U_j
7: $\quad$ Add $L[i + 1, j]$ to U_j
8: $\quad$ Let h and h' be the first and second element in Q_j
9: $\quad$ $(\alpha, \varepsilon_\alpha) \leftarrow U_j$.MINIMUMQUERY($\widetilde{L}^*[i, j], \widetilde{B}^*[h, j]$)
10: $\quad$ **while** $|Q_j| \geq 2$ **and** $\widetilde{B}^*[h', j] \leq \varepsilon_\alpha$ **do**
11: $\quad\quad$ Remove all $L[x, j]$ from U_j with $x \leq h'$ and remove h from Q_j
12: $\quad\quad$ Let h and h' be the first and second element in Q_j
13: $\quad\quad$ $(\alpha, \varepsilon_\alpha) \leftarrow U_j$.MINIMUMQUERY($\widetilde{L}^*[i, j], \widetilde{B}^*[h, j]$)
14: $\quad$ $\widetilde{L}^*[i + 1, j] \leftarrow \varepsilon_\alpha$
15: **return** $\max\{\delta(P(n), Q(n)), \min\{\widetilde{L}^*[n - 1, n - 1], \widetilde{B}^*[n - 1, n - 1]\}\}$

---

row ($\widetilde{B}^*[h, j]$) and the value of the previous boundary $\widetilde{L}^*[i, j]$. (The latter is needed only for $h < i$, i.e. if $|Q_j| \geq 2$. For simplicity, we omit this detail in the overview.) If $\varepsilon_\alpha \geq \widetilde{B}^*[h', j]$, there is an ε_α witness for $L[i+1, j]$ through $B[h', j]$, so we can repeat the process with h' (after updating U_j). If h' does not exist (i.e., $|Q_j| = 1$) or $\varepsilon_\alpha < \widetilde{B}^*[h', j]$, we stop and declare ε_α to be optimal. We prove that this process is correct and maintains the invariant. Since the invariant is clearly satisfied at the beginning, correctness then follows by induction.

Lemma 2.8. *Algorithm 1 computes $\widetilde{L}^*[i + 1, j]$ and maintains the invariant.*

Proof. By the invariant, a rightmost witness for $\widetilde{L}^*[i, j]$ passes through $B[h_0, j]$, where h_0 is initial head of Q_j. Let h^* be the column index such that a rightmost witness for $\widetilde{L}^*[i + 1, j]$ passes through $B[h^*, j]$. Then h^* must be contained in Q_j initially, because by Lemma 2.6, we have $h_0 \leq h^* \leq i$, and by Corollary 2.5, there can be no column index a with $h^* < a \leq i$ that dominates $(h^*, B[h^*, j])$. (Note that if $h^* = i$, it is added at the beginning of the iteration.)

$\quad$ Now let h be the current head of Q_j. By Lemma 2.7, the minimum query on U_j gives the smallest ε_α for which there exists an ε_α-witness for $L[i + 1, j]$ that passes through $B[h, j]$. If the current h is less than h^*, then $\varepsilon_\alpha \geq \widetilde{L}^*[i + 1, j]$ (definition of $\widetilde{L}^*$); $\widetilde{L}^*[i + 1, j] \geq \widetilde{B}^*[h^*, j]$ (there is a witness through $B[h^*, j]$); and $\widetilde{B}^*[h^*, j] \geq \widetilde{B}^*[h', j]$ (the dominance relation ensures that the $\widetilde{B}^*$-values for the indices in Q_j are increasing). Thus, the while-loop in line 10 proceeds to the next iteration. If the current h equals h^*, then by Corollary 2.5, we have $\widetilde{B}^*[a, j] > \widetilde{B}^*[h^*, j]$ for all $h^* < a \leq i$, and the while-loop terminates with the correct value for $\widetilde{L}^*[i,j]$. It is straightforward to check that Algorithm 1 maintains the data structures Q_j and U_j according to the invariant.   □

Theorem 2.9. *Algorithm 1 computes $d_F(P,Q)$ for convex distance function δ in $\mathbb{R}^d$ in $O(n^2 \cdot f(n,d,\delta))$ time, where $f(n,d,\delta)$ represents the time to insert into, delete from, and query the upper envelope data structure.*

Proof. The correctness of the algorithm follows from Lemma 2.8. For the running time, we observe that we insert values only once into Q_j and U_j. Hence, we can remove elements at most once, leading to an amortized running time of $O(1 + f(n,d,\delta))$ for a single iteration of the loop. Since there are $O(n^2)$ cells, the total running time is $O(n^2 \cdot f(n,d,\delta))$ assuming that $f(n,d,\delta)$ is $\Omega(1)$.   □

In the generic algorithm, we must take care that U_j uses the (full) unimodal function only for $L[i+1,j]$ and the truncated versions for the other boundaries. As it turns out, we can use the full unimodal distance functions if these behave as pseudolines (i.e., they intersect at most once). Since we compare only functions in the same row (or column), functions of different rows or columns may still intersect more than once. For this approach to work, we must remove from U_j any function that is no longer relevant for our computation. This implies that U_j no longer contains all functions $L[k,j]$ with $h < k \le i+1$ but a subset of these. We prove the following (see [7] for a full proof).

Lemma 2.10. *Assume that distance functions $L[x,j]$ in row j intersect pairwise at most once. Let h denote a candidate bottom boundary. Let $(\alpha, \varepsilon_\alpha)$ denote the minimum on the upper envelope of the full unimodal distance functions in U_j. Then one of the following holds:*

(i) $(\alpha, \varepsilon_\alpha)$ is the minimum of the upper envelope of $L[i+1,j]$ and the truncated $\overline{L}[k,j]$ for $h < k \le i$.

(ii) $(\alpha, \varepsilon_\alpha)$ lies on two functions $L[a,j]$ and $L[b,j]$, one of which can be removed from U_j.

(iii) $\varepsilon_\alpha \le \widetilde{L}^*[i,j]$.

Proof (sketch). $(\alpha, \varepsilon_\alpha)$ either lies on the minimum of a function or on the intersection of two, one increasing and one decreasing. In the first case, it is easy to see that case (i) holds. In the second case, it depends on whether the increasing function is from an earlier or later column than the decreasing one. If the increasing one comes first, then we can argue that the truncated function is never part of the witness envelope for $L[i+1,j]$ or later boundaries. Hence, case (ii) is applicable. If the decreasing one comes first, then we argue that case three must hold: the given intersection is a lower bound for the minimum of the acrophobia function on the second boundary and therefore a lower bound on $\widetilde{L}^*[i,j]$.   □

From this lemma, we learn how to modify a minimum-point query. We run the query on the full unimodal functions, ignoring the given constants. If case (ii) holds, that is, the minimum lies on an increasing $L[a,j]$ and a decreasing $L[b,j]$ with $a < b$, we remove $L[a,j]$ from U_j and repeat the query. In both of the other cases, the minimum is either the computed minimum or one of the constants $\widetilde{L}^*[i,j]$ and $\widetilde{B}^*[h,j]$. We take the maximum of these three values.

3   Euclidean Distance

In this section we apply our framework to the Euclidean distance measure δ_E. Obviously, δ_E is convex (and symmetric), so our framework applies. However, instead of computing with the Euclidean distance, we use the squared Euclidean distance $\delta_E^2 = \delta_E(x,y)^2$. Squaring does not change any relative order of height on the distance terrain T, so computing the Fréchet distance with the squared Euclidean distance is equivalent to the Euclidean distance: if $\varepsilon = d_F(P,Q)$ for δ_E^2, then $\sqrt{\varepsilon} = d_F(P,Q)$ for δ_E. We now show that for δ_E^2, the distance functions in a row or column behave like pseudolines. We argue only for the vertical boundaries; horizontal boundaries are analogous.

Lemma 3.1. *For $\delta = \delta_E^2$, each distance terrain function $L[i,j]$ is part of a parabola, and any two functions $L[i,j]$ and $L[i',j]$ intersect at most once.*

Proof. Function $L[i,j]$ represents part of the distance between point $p = P(i)$ and line segment $\ell = Q(j)Q(j+1)$. Assume ℓ' is the line though ℓ, uniformly parameterized by $\lambda \in \mathbb{R}$, i.e. $\ell'(\lambda) = (1-\lambda)Q(j) + \lambda Q(j+1)$. Let λ_p denote the λ such that $\ell'(\lambda)$ is closest to p. We see that $L[i,j](\lambda) = |p - \ell'(\lambda_p)|^2 + |\ell'(\lambda) - \ell'(\lambda_p)|^2$. Since ℓ' is uniformly parameterized according to ℓ, we get that the last term is $|\ell|^2(\lambda - \lambda_p)^2$. Hence, the function is equal to $|\ell|^2\lambda^2 - 2|\ell|^2\lambda_p\lambda + |\ell|^2\lambda_p^2 + |p - \ell'(\lambda_p)|^2$, which is a parabolic function in λ. The quadratic factor depends only on ℓ. For two functions in the same row, this line segment is the same, and thus the parabolas intersect at most once.   □

By Lemma 2.10, we know that data structure U_j can use the full parabolas. The parabolas of a single row share the same quadratic term, so we can treat them as lines by subtracting $|\ell|\lambda^2$. Now we can use for U_j a standard data structure for dynamic half-plane intersections, or its dual problem: dynamic convex hulls. The fastest dynamic convex hull structure is given by Brodal and Jacob [4]. However, it does not support a query to find a minimal point for the upper envelope; it is unclear whether the structure can support such a query. Instead, we use the slightly slower structure by Overmars and Van Leeuwen [14] with $O(\log^2 n)$ time insertions and deletions. For each insertion, we also have to compute the corresponding parabola, in $O(d)$ additional time. It remains to show how to perform the minimum-point query. The data structure by Overmars and Van Leeuwen maintains a concatenable queue for the upper envelope. We assume this to be implemented via a red-black tree that maintains predecessor and successor pointers. We perform a binary for the minimum point using the intersection pattern of a node, its predecessor, and its successor (see [7] for details). To include the constants, we take the maximum of the minimum point and the constants. Hence, a single query takes $O(\log n)$ time. We obtain the following result.

Theorem 3.2. *Algorithm 1 computes the Fréchet distance under the Euclidean distance in $\mathbb{R}^2$ in $O(n^2(d + \log^2 n))$ time.*

This is slightly slower than known results for the Euclidean metric. However, we think that our framework has potential for a faster algorithm (see Section 5).

4   Polyhedral Distance

Here we consider the Fréchet distance with a (convex) polyhedral distance function δ_P, i.e., the "unit sphere" of δ_P is a convex polytope in $\mathbb{R}^d$. For instance, the L_1 and the L_∞ distance are polyhedral with the cross-polytope and the hypercube as respective unit spheres. Throughout we assume that δ_P has *complexity* k, i.e., its unit sphere has k facets. The distance terrain functions $L[i,j]$ and $B[i,j]$ are now piecewise linear with at most k parts; in each row and column the corresponding parts are parallel. Depending on the polytope, the actual maximum number k' of parts may be less. The distance δ_P has to be neither regular nor symmetric, but as before, we simplify the presentation by assuming symmetry.

We present three approaches. First, we use an upper envelope structure on piecewise linear functions. Second, we use a brute-force approach which is more efficient for small d and k. Third, we combine these methods.

Upper envelope data structure.   For piecewise linear $L[i,j]$ and $B[i,j]$, we can relax the requirements for the upper envelope data structure U_j. There are no parabolas involved, so we only need a data structure that dynamically maintains the upper envelope of lines under insertions, deletions, and minimum queries. Every function contains at most k' parts, so we insert at most nk' lines into the upper envelope. Maintaining and querying the upper envelope per row or column takes $O(nk'\log(nk'))$ time [4]. Thus, the total running time is $O(n^2k'\log(nk') + n^2g_\delta(d))$, where $g_\delta(d)$ is the time to find the parts of the function.

Brute-force approach.   We implement U_j naively. For each segment of P,Q, we sort the facets of δ_P by the corresponding slope on the witness envelope, in $O(nk(d + \log k))$ total time. For each facet $l = 1,\dots k$, we store a doubly linked list F_l of lines representing the linear parts of the unimodal functions in U_j corresponding to facet l, sorted from top to bottom (the lines are parallel). When processing a cell boundary $L[i,j]$, we update each list F_l: remove all lines below the line for $P(i)$ from the back, and append the line for $P(i)$. This takes amortized $O(d)$ time per facet, $O(kd)$ time per cell boundary. We then go through the top lines in the F_l in sorted order to determine (the minimum of) the upper envelope. This takes $O(k)$ time. The total time is $O(n^2kd + nk(d + \log k))$.

A hybrid approach.   As in the brute force approach, we maintain a list F_l for each of the k slopes. For each segment in P,Q we initialize these lists, which takes $O(nkd)$ time. But instead of sorting the slopes initially, we maintain the upper hull of the top lines in each F_l. Thus, we only need a dynamic upper hull for k lines. At each cell boundary, we only update k' lines, so we need $O(k'\log k)$ time per cell boundary, $O(n^2k'\log k)$ in total. Therefore, the total time is $O(n^2k'\log k + nkd)$, an improvement for $k' \ll k$.

Combining the previous three paragraphs, yields the following result. The method that works best depends on the relation between n, k, k', and d.

Theorem 4.1. *Algorithm 1 computes the Fréchet distance with a convex polyhedral distance function δ of complexity k in $\mathbb{R}^d$ in $O(\min\{n^2k'\log(nk') + n^2g_\delta(d), n^2kd + nk(d + \log k), n^2k'\log k + nkd\})$ time, where $g_\delta(d)$ is the time to find the parts of a distance function.*

Let us consider the implications for L_1 and L_∞. Let ℓ be the line segment and p the point defining $L[i,j]$. For L_1 at the breakpoints between the linear parts of $L[i,j]$ one of the coordinates of $\ell - p$ is zero: there are at most $k' = d + 1$ parts. The facet of the cross-polytope is determined by the signs of the coordinates. For each linear part we compute the slope in $O(d)$ time, thus $g(d) = O(d^2)$. Hence, the hybrid approach outperforms the brute-force approach.

Corollary 4.2. *Algorithm 1 computes the Fréchet distance with the L_1 distance in $\mathbb{R}^d$ in $O(\min\{n^2 d \log(nd) + n^2 d^2, n^2 d^2 + nd2^d\})$ time.*

For the L_∞ distance the facet is determined by the maximum coordinate. We have $k' \leq k = 2d$. However a facet depends on only one dimension. Hence, for the brute-force method computing the slopes does not take $O(kd)$ time, but $O(k)$. Thus the brute-force method outperforms the other methods for L_∞.

Corollary 4.3. *Algorithm 1 computes the Fréchet distance with the L_∞ distance in $\mathbb{R}^d$ in $O(n^2 d + nd \log d)$ time.*

Approximating the Euclidean distance. We can use a polyhedral distance function to approximate the Euclidean distance. A line segment and a point span exactly one single plane in $\mathbb{R}^d$ (unless they are collinear, in which case we pick an arbitrary one). On this plane, the Euclidean unit sphere is a circle. We approximate this circle with a k-regular polygon in $\mathbb{R}^2$ that has one side parallel to the line segment. Simple geometry shows that for $k = O(\epsilon^{-1/2})$, we get a $(1 + \epsilon)$-approximation. The computation is two-dimensional, but we must find the appropriate transformations, which takes $O(d)$ time per boundary. We no longer need to sort the facets of the polytope for each edge; the order is given by the k-regular polygon. This saves a logarithmic factor for the initialization. Again, the brute-force method is best, and Theorem 4.1 gives the following.

Corollary 4.4. *Algorithm 1 computes a $(1 + \epsilon)$-approximation of the Fréchet distance with the Euclidean distance in $\mathbb{R}^d$ in $O(n^2(d + \epsilon^{-1/2}))$ time.*

Alternatively we can use Corollary 4.3 to obtain a $\sqrt{d}$-approximation for the Euclidean distance. If we are willing to invoke an algorithm for the decision version, we can go to a $(1 + \epsilon)$-approximation by binary search.

Corollary 4.5. *We can calculate a $(1+\epsilon)$-approximation of the Fréchet distance with the Euclidean distance in $O(n^2 d + nd \log d + T(n) \log \frac{\sqrt{d}-1}{\epsilon})$ time, where $T(n)$ is the time needed to solve the decision problem for the Fréchet distance.*

5   Open Problems

Faster Euclidean distance. Our framework computes the Fréchet distance for polyhedral distance functions in quadratic time. For the Euclidean distance we do not achieve this running time, but we conjecture that our result can be improved to an $O(n^2)$ algorithm. Currently we use the full power of dynamic upper envelopes, which seems unnecessary as all information is available upfront.

We can for instance determine the order in which the parabolas occur on the upper envelopes, in $O(n^2)$ time for all boundaries. From the proof of Lemma 3.1, we know that the order is given by the projection of the vertices onto the line. We compute the arrangement of the lines dual to the vertices of a curve in $O(n^2)$ time. We then determine the order of the projected points by traversing the zone of a vertical line. This takes $O(n)$ for one row or column. Unfortunately, this alone is insufficient to obtain the quadratic time bound.

Locally correct Fréchet matchings. A matching between two curves that is a Fréchet matching for any two matched subcurves is called a locally correct Fréchet matching [9]. It enforces a relatively "tight" matching. The algorithm in [9] uses a linear overhead on the algorithm of Alt and Godau [1] resulting in an $O(n^3 \log n)$ execution time. We conjecture that our framework is able to avoid this overhead. However, the information we currently propagate is insufficient: a large distance early on may "obscure" the rest of the computations.

References

[1] Alt, H., Godau, M.: Computing the Fréchet distance between two polygonal curves. IJCGA 5(1-2), 78–99 (1995)
[2] Alt, H., Knauer, C.: Matching Polygonal Curves with Respect to the Fréchet Distance. In: Ferreira, A., Reichel, H. (eds.) STACS 2001. LNCS, vol. 2010, pp. 63–74. Springer, Heidelberg (2001)
[3] Brakatsoulas, S., Pfoser, D., Salas, R., Wenk, C.: On map-matching vehicle tracking data. In: Proc. 31st Int. Conf. VLDBs, pp. 853–864 (2005)
[4] Brodal, G., Jacob, R.: Dynamic planar convex hull. In: Proc. 43rd FOCS, pp. 617–626 (2002)
[5] Buchin, K., Buchin, M., Gudmundsson, J.: Constrained free space diagrams: a tool for trajectory analysis. IJGIS 24(7), 1101–1125 (2010)
[6] Buchin, K., Buchin, M., Gudmundsson, J., Löffler, M., Luo, J.: Detecting commuting patterns by clustering subtrajectories. IJCGA 21(3), 253–282 (2011)
[7] Buchin, K., Buchin, M., van Leusden, R., Meulemans, W., Mulzer, W.: Computing the Fréchet Distance with a Retractable Leash. CoRR, abs/1306.5527 (2013)
[8] Buchin, K., Buchin, M., Meulemans, W., Mulzer, W.: Four soviets walk the dog - with an application to Alt's conjecture. CoRR, abs/1209.4403 (2012)
[9] Buchin, K., Buchin, M., Meulemans, W., Speckmann, B.: Locally correct Fréchet matchings. In: Epstein, L., Ferragina, P. (eds.) ESA 2012. LNCS, vol. 7501, pp. 229–240. Springer, Heidelberg (2012)
[10] Cook, A.F., Wenk, C.: Geodesic Fréchet distance inside a simple polygon. ACM Trans. on Algo. 7(1), Art. 9, 9 (2010)
[11] de Berg, M., van Kreveld, M.J.: Trekking in the Alps Without Freezing or Getting Tired. Algorithmica 18(3), 306–323 (1997)
[12] Driemel, A., Har-Peled, S., Wenk, C.: Approximating the Fréchet distance for realistic curves in near linear time. In: Proc. 26th SoCG, pp. 365–374 (2010)
[13] Har-Peled, S., Raichel, B.: The Fréchet distance revisited and extended. In: Proc. 27th SoCG, pp. 448–457 (2011)
[14] Overmars, M., van Leeuwen, J.: Maintenance of configurations in the plane. J. Comput. System Sci. 23(2), 166–204 (1981)
[15] Wenk, C., Salas, R., Pfoser, D.: Addressing the need for map-matching speed: Localizing global curve-matching algorithms. In: Proc. 18th Int. Conf. on Sci. and Stat. Database Management, pp. 379–388 (2006)

Vertex Deletion for 3D Delaunay Triangulations

Kevin Buchin[1], Olivier Devillers[2], Wolfgang Mulzer[3],
Okke Schrijvers[4], and Jonathan Shewchuk[5]

[1] Technical University Eindhoven, The Netherlands
www.win.tue.nl/~kbuchin
[2] INRIA Sophia Antipolis – Méditerranée, France
inria.fr/sophia/members/Olivier.Devillers
[3] Freie Universität Berlin, Germany
page.mi.fu-berlin.de/mulzer
[4] Stanford University, USA
www.okke.info
[5] University of California at Berkeley, USA
www.cs.berkeley.edu/~jrs

Abstract. We show how to delete a vertex q from a three-dimensional Delaunay triangulation $\mathrm{DT}(S)$ in expected $O(C^{\otimes}(P))$ time, where P is the set of vertices neighboring q in $\mathrm{DT}(S)$ and $C^{\otimes}(P)$ is an upper bound on the expected number of tetrahedra whose circumspheres enclose q that are created during the randomized incremental construction of $\mathrm{DT}(P)$. Experiments show that our approach is significantly faster than existing implementations if q has high degree, and competitive if q has low degree.

1 Introduction

Some geometric applications require the ability to delete a vertex from a Delaunay triangulation. An early algorithm by Chew [9] for generating guaranteed-quality triangular meshes uses Delaunay vertex deletion to obtain a better bound on the minimum angle than is achieved by similar algorithms that do not use vertex deletion, and the same principle has been exploited by mesh generators that generate better-quality tetrahedra by occasionally deleting vertices from a three-dimensional Delaunay triangulation [7, Section 14.5]. Another application of Delaunay vertex deletion is interactive data cleaning, in which a user desires to remove outlier vertices from a triangulation used to interpolate data.

Let S be a finite set of points in $\mathbb{R}^2$ or $\mathbb{R}^3$, and let $\mathrm{DT}(S)$ denote its Delaunay triangulation. We study how to *delete* a vertex from $\mathrm{DT}(S)$ while maintaining the Delaunay property of the triangulation. That is, given a point $q \in S$, we wish to transform $\mathrm{DT}(S)$ into $\mathrm{DT}(S \setminus \{q\})$ quickly.

In two dimensions, Delaunay deletion is well understood. By 1990, several algorithms were known that delete a vertex q with degree d in optimal $O(d)$ time [1, 8], and fast, practical implementations are available now [13]. In three dimensions, there is still room for improvement. In theory, the best methods known are *triangulate and sew* and *ear queue*. The ear-queue algorithm has worst-case running time $O(k \log d)$ where k is the number of new tetrahedra

H.L. Bodlaender and G.F. Italiano (Eds.): ESA 2013, LNCS 8125, pp. 253–264, 2013.

created. The triangulate-and-sew algorithm triangulates the set P of vertices neighboring q in $\mathrm{DT}(S)$ with the standard randomized incremental construction (RIC) algorithm, then takes the tetrahedra adjoining q in $\mathrm{DT}(P)$ and sews them into the cavity. This method runs in expected $O(d \log d + C(P))$ time, where $d = |P|$ and $C(P)$ is the expected number of tetrahedra created during the randomized incremental construction of $\mathrm{DT}(P)$. These two complexities, $O(k \log d)$ and $O(d \log d + C(P))$, are not comparable as $\Omega(d) \leq k \leq C(P) \leq O(d^2)$.

The main innovations behind our algorithms are fast methods for point location, so all the point location steps of the incremental construction take expected total time linear in d, and a vertex insertion procedure whose cost is $C^{\otimes}(P)$, the expected number of tetrahedra *in conflict with* q created during the randomized incremental construction of $\mathrm{DT}(P)$. Our complexity $O(d + C^{\otimes}(P)) = O(C^{\otimes}(P))$ is always better than $O(d \log d + C(P))$ and better than the ear queue algorithm in the usual circumstance that $C^{\otimes}(P) = o(k \log d)$. Moreover, it uses less complicated numerical predicates than the ear queue, and therefore it is easier to implement robustly. A prototype implementation of our algorithm compares favorably with existing codes, particularly if q has high degree.

2   Related Work

Existing Algorithms. The literature contains several algorithms for Delaunay vertex deletion. All of them begin by deleting q and the simplices adjoining q, thereby evacuating a star-shaped *cavity* in the DT, which must be retriangulated. Some vertex deletion algorithms are primarily concerned with the asymptotic running time as a function of the degree d of q, but the average vertex degree in a two-dimensional DT is less than six, so some authors emphasize the speed when d is small. The cavity's Delaunay triangulation always has size $\Theta(d)$ in 2D, but in 3D its size may be as small as $\Theta(d)$ or as large as $\Theta(d^2)$.

The *gift-wrapping* or *boundary completion* algorithm constructs one triangle or tetrahedron at a time by choosing a known facet (edge or triangle) f and finding a vertex $p \in P$ so that f and p together form a Delaunay triangle or tetrahedron. Its worst-case running time is $\Theta(d^2)$ in 2D [13] and $\Theta(d^3)$ in 3D.

The *ear queue algorithm* [12] is a gift-wrapping algorithm that uses a priority queue to quickly identify an ear that can be cut off the star-shaped cavity. Each ear is assigned a priority proportional to a numerical quantity called the *power* of the circumsphere with respect to the deleted vertex q; the highest-priority ear is guaranteed to be Delaunay. The algorithm runs in $O(d \log d)$ time in 2D and $O(k \log d)$ time in 3D, where k is the number of new tetrahedra created. Unfortunately, it requires a new geometric predicate that compares the powers of two ears. This makes the code less generic and more difficult to make robust and efficient.

The *flip algorithm* connects all the facets on the cavity boundary to a single vertex in P, then performs a sequence of *flips* that replace simplices with other simplices. In 2D, the flip algorithm finds the Delaunay triangulation in $O(d^2)$ time [16], but in 3D, the flip algorithm does not always work; it can get stuck in a non-Delaunay triangulation from which no flip can make further progress [15].

In 2D, two algorithms are known that run in linear time, which is optimal. Aggarwal, Guibas, Saxe, and Shor [1] describe an algorithm that runs in deterministic $O(d)$ time, but it is complicated and not practical. In a classic paper that introduced the randomized algorithm analysis technique now known as *backward analysis*, Chew [8] proposes a simple, practical, randomized algorithm that runs in expected $O(d)$ time. The algorithm, which we call *backward reinsertion*, combines RIC with a backward point location method. The algorithm of Aggarwal et al. does not appear to generalize to 3D, but in this paper we generalize backward reinsertion to 3D.

Devillers [13] has explicitly constructed optimal algebraic decision trees for deleting vertices of degree at most 7 from 2D Delaunay triangulations. This approach, called *low-degree optimization*, obtains notable speedups for vertices of small degree. We do not foresee it being extended to 3D, where the complexity of the decision trees grows very quickly.

The *triangulate-and-sew algorithm* retriangulates the cavity by computing $DT(P)$ from scratch, taking the subset of simplices in $DT(P)$ that lie inside the cavity, and sewing them into the cavity to obtain $DT(S \setminus \{q\})$. If we use a RIC algorithm to compute $DT(P)$, triangulate-and-sew runs in $O(d \log d + C(P))$ time, whether in 2D or 3D. This approach may create many simplices that are unnecessary because they lie outside the cavity or are not helpful in computing the final simplices inside the cavity. We address this drawback below.

Existing Implementations. For Delaunay vertex deletions in 2D, CGAL [5] implements low-degree optimization for vertices of degree 7 or less. For higher degrees, it uses flipping. In 3D, the current implementation offers triangulate-and-sew. A previous version used a simplified ear queue algorithm with running time $O(dk)$ [12].

3   Preliminaries and Notation

We are given a finite point set $S \subseteq \mathbb{R}^3$ and its Delaunay triangulation $DT(S)$, and we wish to delete the vertex $q \in S$ from $DT(S)$, yielding $DT(S \setminus \{q\})$. A point $p \in S$ is a *neighbor* of q if $DT(S)$ contains the edge pq. Let P be the set of neighbors of q in $DT(S)$. Let $d = \mathrm{d}^\circ(q, DT(S)) = |P|$ be the *degree* of q in $DT(S)$.

We use a *randomized incremental construction* (RIC) algorithm to compute $DT(P)$. The standard RIC algorithm successively inserts the points in P into a Delaunay triangulation, one by one, in an order determined by a random permutation $p_1, p_2, \ldots, p_d$ of P. For $i = 1, \ldots, d$, let $P_i = \{p_1, \ldots, p_i\}$ contain the first i points of the permutation. The standard RIC constructs $DT(P)$ by successively inserting each p_i into $DT(P_{i-1})$. A tetrahedron in $DT(P_{i-1})$ is said to *conflict* with p_i if its circumsphere encloses p_i. The algorithm identifies all the *conflict tetrahedra* in three steps. First, a method called *point location* identifies one tetrahedron Δ that conflicts with p_i. Second, the algorithm finds all the other tetrahedra in $DT(P_{i-1})$ that conflict with p_i by a depth-first search from

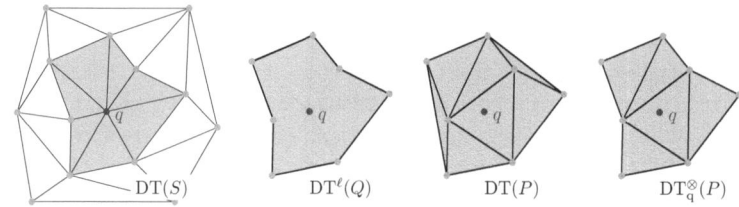

Fig. 1. A DT of S, a link DT for q, a DT of q's neighbors, and q's conflict DT

Δ. This search treats $DT(P_{i-1})$ as a graph in which each tetrahedron acts as a graph node and two nodes are connected by a graph edge if the corresponding tetrahedra share a triangular face. Third, the conflict tetrahedra are all deleted. The union of the conflict tetrahedra is a *cavity* which we retriangulate with tetrahedra adjoining p_i. This step is called *structural change*. The expected cost of the structural change, denoted $C(P)$, is obtained by summing the cost of inserting a random point into $DT(P_i)$ for $i = 1, \ldots, d$.

Point location is usually the most difficult part of incremental construction algorithms; we will discuss it at length later. We will use the special nature of Delaunay vertex deletion both to speed up point location and to reduce the number of structural changes we must make. We will also use a variant of RICs that inserts points in batches.

Let $Q = P \cup \{q\}$ and $Q_i = P_i \cup \{q\}$ for $i = 1, \ldots, d$. The *link DT*, denoted $DT^\ell(Q_i)$, is the subset of $DT(Q_i)$ containing only the tetrahedra adjoining q and their faces, as illustrated in Figure 1. The name stems from the fact that the boundary faces of $DT^\ell(Q_i)$ form a triangulation of a topological sphere. The *conflict DT*, denoted $DT_q^\otimes(P_i)$, is the subset of $DT(P_i)$ containing only the tetrahedra whose circumspheres enclose q. Observe that the boundaries of $DT_q^\otimes(P_i)$ and $DT^\ell(Q_i)$ are identical. The expected cost of the structural change restricted to the tetrahedra of $DT_q^\otimes$ is denoted $C^\otimes(P)$.

4   Algorithm

Our algorithm uses randomized incremental construction to compute $DT_q^\otimes(P)$, the tetrahedra that conflict with q in the DT of q's neighbors, and uses it to fill the cavity evacuated by q's deletion. To insert each new point p_i into $DT_q^\otimes(P_{i-1})$, we need to quickly identify a tetrahedron that conflicts with p_i. For this, we maintain the *link DT* $DT^\ell(Q_i)$: the DT of the points $P_i \cup \{q\}$, restricted to the tetrahedra adjoining q. The boundaries of $DT^\ell(Q_i)$ and $DT_q^\otimes(P_i)$ are identical, so we can use any edge of $DT^\ell(Q_i)$ adjoining p_i to find a conflict tetrahedron in $DT_q^\otimes(P_{i-1})$.

To obtain the sequence $DT^\ell(Q_4), \ldots, DT^\ell(Q_d)$, we use the *reverse deletion trick* [8]. By construction, $DT^\ell(Q) \subseteq DT(S)$. We remove the points $p_d, \ldots, p_5$ in that order from $DT^\ell(Q)$. If the deletion order is sufficiently random, this process can be implemented efficiently, as the boundary of $DT^\ell(Q)$ behaves like a 2D Delaunay triangulation. Each time we remove a point p_i from $DT^\ell(Q_i)$, we store

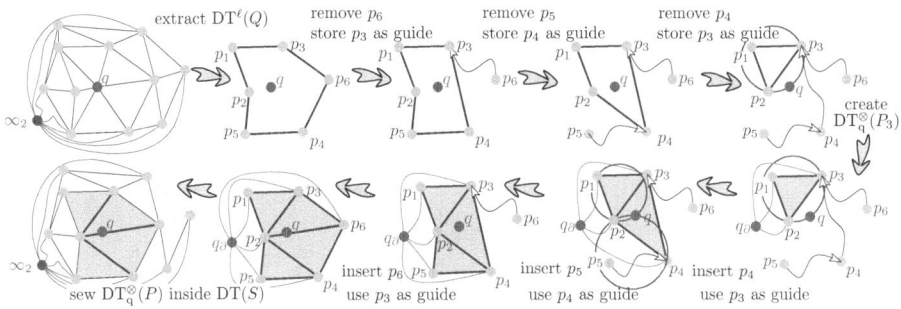

Fig. 2. An illustration of the deletion and reinsertion process in 2D

a *guide* for p_i, denoted guide(p_i), to help the point location of p_i in $\mathrm{DT}_{\mathrm{q}}^{\otimes}(P_{i-1})$; see Figure 2. The guide is usually a neighbor of p_i in $\mathrm{DT}^{\ell}(Q_i)$, but different variants of the algorithm use different guides.

We now describe how to use guide(p_i) when inserting p_i into $\mathrm{DT}_{\mathrm{q}}^{\otimes}(P_{i-1})$. The point location uses two steps: (i) finding the tetrahedron adjoining guide(p_i) that intersects the line segment guide(p_i)p_i; and (ii) walking to the tetrahedron that contains p_i. The second step visits only tetrahedra that are destroyed during insertion, so its cost can be charged to the structural change $C^{\otimes}(P)$. The time for the first step is proportional to $\mathrm{d}^{\circ}(\mathrm{guide}(p_i), \mathrm{DT}^{\ell}(Q_{i-1}))$. This depends on the exact nature of the insertion order, and we will discuss it below.

The Link Delaunay Triangulation. All the tetrahedra of $\mathrm{DT}^{\ell}(Q_i)$ share the vertex q, so the *link* of q (i.e., the triangles in $\mathrm{DT}^{\ell}(Q_i)$ that do not contain q) has the topology of a 2D sphere. Hence, we can represent $\mathrm{DT}^{\ell}(Q_i)$ as a 2D triangulation. The triangulation $\mathrm{DT}^{\ell}(Q_d)$ can be extracted from $\mathrm{DT}(S)$ in $O(d)$ time by a simple traversal of the tetrahedra adjoining q. To maintain $\mathrm{DT}^{\ell}(Q_i)$ under deletion, we can use any ordinary 2D Delaunay algorithm while replacing the in-circle test w.r.t. a triangle by the in-sphere test w.r.t. the tetrahedron formed by the triangle and q; see Section 4.1. Correctness follows because we are looking for the triangles t where the sphere passing through the vertices of t and q is empty.

The Conflict Delaunay Triangulation. Recall that we defined $\mathrm{DT}_{\mathrm{q}}^{\otimes}(P_i)$ as the set of all tetrahedra in $\mathrm{DT}(P_i)$ that have q in their circumsphere. Our goal is to prevent $\mathrm{DT}(P_i) \setminus \mathrm{DT}_{\mathrm{q}}^{\otimes}(P_i)$ from being constructed. If we were to construct all of $\mathrm{DT}(P)$, we might create tetrahedra that would be discarded when we sew the cavity back into the original triangulation. In 3D, the number of unnecessary tetrahedra can be quadratic.

Lemma 1. *For any $d \in \mathbb{N}$, there is a d-point set $P \subseteq \mathbb{R}^3$ with $C(P) = \Omega(d^2)$ and $C^{\otimes}(P) = O(d)$.*

Proof. We take P to be d points on the three-dimensional *moment curve* $t \mapsto (t, t^2, t^3)$. It is well known that $\mathrm{DT}(P)$ has complexity $\Omega(d^2)$ and that all points

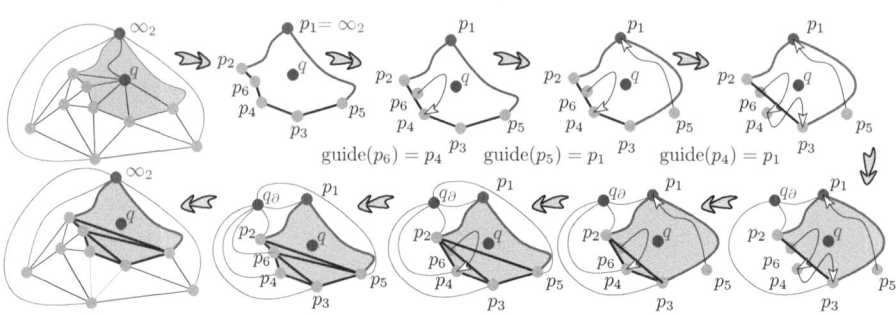

Fig. 3. 2D example of the deletion and reinsertion process with q on the convex hull

of P are on the lower part of the convex hull. Thus, if we take q sufficiently far below P, then q connects to all points of P in $DT(P \cup \{q\})$. When deleting q, $DT_q^{\otimes}(P)$ contains only $O(d)$ infinite tetrahedra, but the full triangulation $DT(P)$ consists of $\Omega(d^2)$ tetrahedra.   □

4.1   Managing the Boundaries

During the randomized incremental construction of a triangulation, we need to take care of handling the boundary and the adjacencies between boundary facets. In a full Delaunay triangulation, this boundary is the convex hull, while in an intermediate triangulation such as $DT_q^{\otimes}(P)$ it may be not convex.

To avoid complicated code for all the special cases, a classic approach adds a *dummy vertex* ∞ and creates for each facet f of the convex hull a tetrahedron between f and ∞ [3]. Thus, adjacencies between convex hull facets are managed as adjacencies between infinite tetrahedra. The circumsphere of an infinite tetrahedron is defined as the half space that is delimited by the plane through its finite facet, the side of the plane is determined by the tetrahedron orientation. The construction algorithm needs no special code for infinite tetrahedra, except inside the in-sphere predicates.

In our setting, we have three different triangulations: $DT(S)$, $DT^\ell(Q_i)$, and $DT_q^{\otimes}(P_i)$. The boundary of $DT(S)$ is managed as above, using a dummy vertex ∞_3 (the index 3 emphasizes the dimension). The triangulation $DT^\ell(Q_i)$ does not have a boundary, since it is just the link of q in some triangulation. However, note that if q lies on the convex hull of S, then ∞_3 is a vertex of $DT^\ell(Q)$. Finally, $DT_q^{\otimes}(P_i)$ is a 3D triangulation with boundary $DT^\ell(Q_i)$; to handle this boundary, we introduce another dummy vertex q_∂ (pronounced "q boundary") that forms a tetrahedron with each face of the boundary of $DT_q^{\otimes}(P_i)$.

With this approach, we can use a standard deletion algorithm for each $DT^\ell(Q_i)$ and a standard construction algorithm for each $DT_q^{\otimes}(P_i)$. All special treatment goes into the in-sphere and in-circle predicates, see below. Figure 2 shows the deletion and reinsertion process when q is not on the convex hull of S. In Figure 3, the point q is on the convex hull of S, and ∞_2 and q_∂ must interact.

In-circle Predicate for the Link DT. Let incircle(a, b, c, w) be the predicate that is true if and only if either w is inside the circumcircle for the triangle abc and abc is positively oriented or w is outside the circumcircle and abc is negatively oriented. The predicate insphere(a, b, c, d, w) is defined analogously for the point w and the tetrahedron $abcd$.

A triangle abc belongs to the link DT and is positively oriented if and only if the sphere through $qabc$ is empty and the tetrahedron $qabc$ is positively oriented. More precisely, the incircle test incircle$^\ell(a, b, c, w)$ for the 2D deletion algorithm is implemented as insphere(q, a, b, c, w). If one of a, b, c or w is ∞_3, the usual way of solving it is used, that is, insphere(q, a, b, ∞_3, w) is true if tetrahedron $qabw$ is positively oriented.

Although q initially lies inside of DT(Q_i), it might end up on the boundary during the deletion process (e.g., in Figure 2 at the deletion of p_4). We defined insphere in such a way that this does not cause a problem: if a tetrahedron has negative orientation, we want the point to be outside of its circumsphere (in Figure 2, when deleting p_4, the outside of the circle through p_2p_3q does not contain p_1, and p_2p_3 is a boundary face of DT$^\ell(Q_3)$).

In-sphere Predicate for the Conflict DT. Let $abcd$ be a positively oriented tetrahedron of DT$_q^\otimes(P_i)$. By definition, insphere(a, b, c, d, q) holds. If $q_\partial \notin \{a, b, c, d\}$, the predicate insphere$^\otimes(a, b, c, d, w)$, used to compute DT$_q^\otimes$, is defined as insphere$(a, b, c, d, w)$ (notice that a, b, c, or d might be ∞_3). If, without loss of generality, $q_\partial = d$, we consider $bacd'$, the neighboring tetrahdron of $abcq_\partial$ in DT$_q^\otimes(P_i)$. Let C be the circumsphere of $bacd'$. Then q lies inside of C. Consider moving C in the pencil of spheres through abc in the direction that places d' outside. Since abc is a face of the boundary of DT$_q^\otimes(P_i)$, the moving sphere will encounter q before any other point, so the sphere through $bacq$ is empty, and we consider it as the circumsphere of $abcq_\partial$. Again, if q is not on the same side of abc as d', the conflict zone is actually the outside of the ball. More formally, insphere$^\otimes(a, b, c, q_\partial, w)$ is defined as insphere$^\otimes(b, a, c, q, w)$. For an example in 2D refer to Figure 2: when p_5 is inserted, it is in conflict with $p_4p_2q_\partial$ and not with $p_2p_1q_\partial$ creating a non-convex angle on the boundary of DT$_q^\otimes(P_5)$. When p_4 is inserted, since p_3p_2q is counterclockwise the disk circumscribing $p_3p_2q_\partial$ is the outside of the circumscribing circle of p_2p_3q, and p_4 is in conflict.

4.2 Main Algorithm

We now present several variations of our randomized incremental algorithm. All variants use the same framework, given in Algorithm DELETEVERTEX, but they differ in the implementation of CONSTRUCTDT in line 3. Some of our schemes achieve good worst-case performance, while others yield more practical implementations.

DELETEVERTEX($DT(S), q$)

▷ Preprocessing

1    $DT^\ell(Q) \leftarrow$ CONSTRUCTLINKDT($DT(S), q$);

2    $P \leftarrow$ the neighbors of q;

▷ The actual implementation for filling the cavity

3    $DT_q^\otimes(P) \leftarrow$ CONSTRUCTDT($DT^\ell(Q), P, q$);

▷ Postprocessing

4   Sew the tetrahedra from $DT_q^\otimes(P)$ into $DT(S)$
     using the correspondence between $DT^\ell(Q)$ and the boundary of $DT_q^\otimes(P)$;

5   Delete the tetrahedra of $DT^\ell(Q)$;

It is plain to observe that the pre- and postprocessing takes time $O(|P|)$. Thus, the complexity lies in the recursive function CONSTRUCTDT.

We give two approaches for CONSTRUCTDT: *incremental backward reinsertion* (IBR) and a *biased randomized insertion order* (BRIO). IBR samples random points from P one-by-one and updates $DT^\ell(Q_i)$ before sampling a new point. BRIO uses a *gradation* of P, i.e., it batches the sampling process into *rounds*, and it updates $DT^\ell(Q_i)$ only once all points of a round have been determined.

Incremental Backward Reinsertion. The IBR-approach is given as Algorithm IBR-CONSTRUCT. It samples a point p_i from $DT^\ell(Q_i)$, recursively constructs $DT(P_{i-1})$, and then inserts p_i into $DT(P_{i-1})$ using guide(p_i). The correctness of Algorithm IBR-CONSTRUCT follows immediately by induction, but there are several choices for the implementation: how do we sample p_i in line 3, and how do we determine its guide in line 4? We discuss several variations together with the implications on the expected running time.

IBR-CONSTRUCT($DT^\ell(Q_i), P_i, q$)

1   **if** $|P_i| = 4$

2   **then return** CREATEQCONFLICTDT(P_i);

3   Sample $p_i \in P_i$;

4   guide(p_i) $\leftarrow$ one neighbor of p_i in $DT^\ell(Q_i)$;

5    $P_{i-1} \leftarrow P_i \setminus \{p_i\}$;

6    $DT^\ell(Q_{i-1}) \leftarrow$ LINKDTDELETE($DT^\ell(Q_i), p_i$);

7    $DT_q^\otimes(P_{i-1}) \leftarrow$ IBR-CONSTRUCT($DT^\ell(Q_{i-1}, P_{i-1}, q)$);

8    $DT_q^\otimes(P_i) \leftarrow$ QCONFLICTDTINSERT($DT_q^\otimes(P_{i-1}), p_i, \text{guide}(p_i)$);

9   **return** $DT_q^\otimes(P_i)$;

Uniform Sampling. A natural approach is to take p_i uniformly at random. However, if the guide is a neighbor of p_i, it is not clear how to bound the running time. If the triangulations store only incidence relations, the point location time for p_i will be proportional to $d^\circ(\text{guide}(p_i), DT^\ell(Q_{i-1}))$. If $DT^\ell(Q_{i-1})$ were a planar triangulation, choosing the nearest neighbor of p_i as guide would yield constant expected point location time [11, Lemma 1]. Unfortunately, as $DT^\ell(Q_{i-1})$ is embedded in 3D, the nearest neighbor is no longer guaranteed to be a neighbor in $DT^\ell(Q_{i-1})$.

An alternative is to use a triangle as a guide instead of a vertex. Let $\text{guide}(p_i) = t_i^\ell = p_j p_k p_l$ be a triangle created by the removal of p_i in $\text{DT}^\ell(P_i)$ (without loss of generality $j > k, l$). Then, t_i^ℓ can be matched to a tetrahedron $t_i^\otimes = p_j p_k p_l q_\partial$ created in $\text{DT}_q^\otimes$ when p_j is inserted. This matching can be efficiently computed by storing in p_j all its incident triangles in $\text{DT}^\ell(P_j)$ in counterclockwise order when p_j is deleted in DT^ℓ. After the insertion of p_j in $\text{DT}_q^\otimes$, we walk simultaneously around the edge $p_j q_\partial$ of $\text{DT}_q^\otimes(P_j)$ and through the list of triangles in order to put pointers from each t_i^ℓ to the matching $t_i^\otimes$. In a more powerful model of computation, we could also represent triangles as triples of indices in $[1, d]$ and maintain the correspondence between t_i^ℓ and $t_i^\otimes$ using universal hashing [4] in $O(1)$ expected time.

Low-degree Vertex Sampling. Instead of sampling p_i at random, another choice can be to sample it uniformly among the points with degree at most 7. Since $\text{DT}^\ell(Q_i)$ is planar, there are $\frac{2}{5}i$ candidates to choose from. The advantage is that we can delete p_i quickly using "low degree optimization". As previously, we need to take triangles as guide. Unfortunately, P_i is no longer a random subset of P of size i, and we cannot bound the expected structural change with such a sequence.

Low-degree Edge Sampling. As already pointed out, for vertex guides to be efficient, we need to control their degrees. To this aim, instead of choosing p_i first and then $\text{guide}(p_i)$ amongst its neighbors, we will choose directly the edge $p_i \text{guide}(p_i)$. The following lemma ensures that we can find an edge with $\text{d}°(p_i, \text{DT}^\ell(Q_i)) \le 8$, $\text{d}°(\text{guide}(p_i), \text{DT}^\ell(Q_i)) \le 230$, and p_i sampled at random in a subset of P_i of size greater than $\frac{i}{96}$. As for low-degree vertex sampling, we cannot guarantee a bound on the structural change for such a permutation.

Lemma 2. *Let T be a planar triangulation with n vertices such that the external face of T is a triangle. Then T contains $\Omega(n)$ edges whose both endpoints have degree $O(1)$.*

Proof. It is well known that T contains an independent set $I \subseteq P$ of vertices such that (i) $|I| \ge \frac{n}{18}$, and (ii) $\text{d}°(p, T) \le 8$ for all $p \in I$. Let N be the neighborhood of I in T. The set N induces a planar graph with at least $\frac{n}{18}$ facets (each facet contains a single point of I). Hence, $|N| \ge 2 + \frac{n}{36}$ and the average degree of a vertex in N (wrt T) is at most $3 + 3 \cdot 36 = 111$ (using the fact that $\forall v, \text{d}°(v, T) \ge 3$ if $n \ge 4$), so at least half of the vertices in N have degree at most 222. Thus, at least $\frac{n}{96}$ points in I have a neighbor of degree at most 230. □

BRIO Sampling. The *BRIO*-approach [2] uses a *gradation* to construct the permutation $\{p_i\}$. We construct a sequence $P = S_r \supseteq S_{r-1} \supseteq \cdots \supseteq S_0 = \emptyset$ of subsets such that S_{i-1} is obtained from S_i by sampling each point independently with probability $1/2$. Note that for $p \in P$, we have $\Pr[p \in S_i] = 2^{-(r-i)}$, so $r = O(\log d)$ with high probability. Now the algorithm proceeds in r rounds: in round $r - i + 1$, we have $\text{DT}^\ell(S_i \cup \{q\})$, and we compute a spanning tree T_i for

$\text{DT}^\ell(S_i \cup \{q\})$ that has maximum degree 3. This takes time $O(|S_i|)$, using an algorithm by Choi [10]. We store T_i as a guide. Then, we delete all points from $S_i \setminus S_{i-1}$ to obtain $\text{DT}^\ell(S_{i-1} \cup \{q\})$ and proceed to round $r-i+2$. This deletion takes time $O(|S_i|)$ [6]. The construction of $\text{DT}_q^\otimes(P)$ also proceeds in r rounds. In round i, we have $\text{DT}_q^\otimes(S_i)$, and we would like to obtain $\text{DT}_q^\otimes(S_{i+1})$. For this, we perform a BFS along T_{i+1}, starting from a vertex in S_i, and we insert the points as they are encountered. Since each edge of T_{i+1} must appear in in $\text{DT}_q^\otimes(S_{i+1})$, the time for walking along each edge can be charged to the structural change. However, we need to bound the time it takes to locate the tetrahedron that is intersected by the next edge. This is done in the following lemma.

Lemma 3. *The total time for locating the tetrahedra that are intersected by the edges of the bounded-degree spanning tree is $O(|S_i| + C^\otimes(S_i))$.*

Proof. The point location takes place on the boundary of $\text{DT}_q^\otimes(S_i)$. The standard BRIO analysis shows that biasing the permutation in each round increases the expected structural change by at most a constant factor [2]. Thus, throughout the round the total number of triangles that can appear on the boundary are $O(|S_i| + C^\otimes(S_i))$ (the ones present initially, plus the ones created). Since T_i has bounded degree, we scan each triangle at most $O(1)$ times for each incident vertex. The claim follows. □

We can summarize these results in the following theorem:

Theorem 4. *Backward Reinsertion takes $O(C^\otimes(P))$ expected time using*
(i) uniform sampling using triangle guides, or
(ii) BRIO sampling using vertex guides

5   Experimental Setup and Results

Variant	Sampling	Guide
IBR-Hashing	deg $p_i \leq 7$, low deg optimized delete	hash triangle
IBR-Neighbor	deg $p_i \leq 7$, low deg optimized delete	lowest degree neighbor of p_i
IBR-Edge	Random edge $\overline{uv}$ with $\text{d}^\circ(u) \leq 7, \text{d}^\circ(u) + \text{d}^\circ(v) \leq 15$	guide$(u) = v$
IBR-BRIO	Independent rounds with probability $1/2$	BDST of Choi [10]
Guide-only	Constructing $\text{DT}(P)$ and sew, edge guide and edge sampling.	
$\text{DT}_q^\otimes$-only	Constructing $\text{DT}_q^\otimes(P)$ and sew, no guide, random order.	
CGAL	CGAL: constructing $\text{DT}(P)$ and sew, no guide, "Dijkstra order" from $\text{DT}^\ell(Q)$.	
CGAL-rnd	Randomized CGAL: constructing $\text{DT}(P)$ and sew, no guide, random order.	

We implemented the above variants of our algorithm using CGAL 4.2.[1] In practice, our implementations differ a bit from theory. For IBR-hashing, we did not use a real hash table, but a balanced binary search tree. IBR-Neighbor has not been proven to be optimal in theory, taking the neighbor of p_i of lowest degree has the disadvantage of needing the computations of these degrees.

---

[1]   The experiments were performed on a 32-bit 2.53 GHz quad-core Intel i5 running Microsoft Windows 7 operating system with 3 gigabytes usable RAM. Code has been compiled using Microsoft Visual C++ in CGAL release mode.

The conditions proved in Lemma 2 are too restrictive to implement IBR-Edge, thus we are looking for edges $\overline{uv}$ such that the sum of the degrees of the two endpoints is less than 15. The vertex of smallest degree of such an edge is chosen as p_i and the other as guide. The fact that there are $\Omega(n)$ such edges is not guaranteed by our lemma but works well in practice.

We experimented on various "reasonable" datasets (several random distributions, real 3D models) where the degree of points is bounded, and we observed similar running times for all methods. Our experimental results are interesting when we use distributions with high degree points such as points on the moment curve

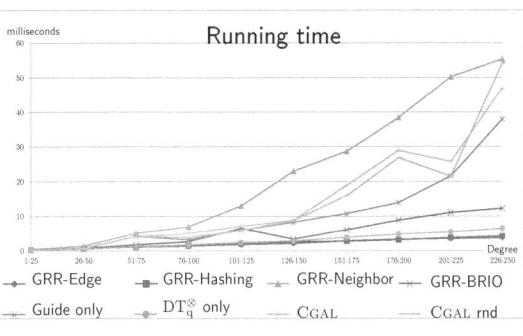

$\gamma(t) = (t, t^2, t^3)$ $(\mathrm{d}^\circ(v) = \Theta(n))$ and the *helix* distribution [14] $(\mathrm{d}^\circ(v) = \Theta(\sqrt{n}))$. On the side picture, we show the average running time per degree for degrees up

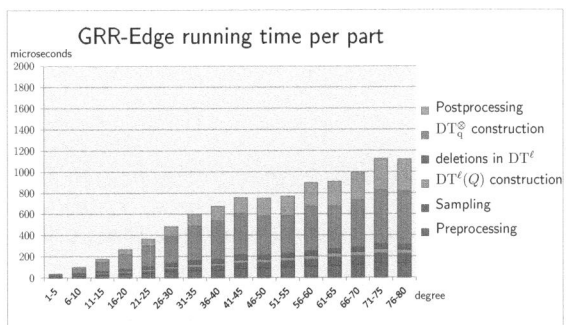

to 250 on a scale in milliseconds. The data is aggregated over all distributions, however the long term behavior is dominated by the moment curve distribution. IBR-Edge is the best variant, the good performance of $\mathrm{DT}_\mathrm{q}^\otimes$-only indicates that computing $\mathrm{DT}_\mathrm{q}^\otimes(P)$ instead of $\mathrm{DT}(P)$ is the source

of a big part of the improvement, while the improvement due to the use of guides to speed up point location is less crucial. Timings above address a complete deletion, the side figure details how this time is split in the different parts of the algorithm. More details about experiments can be found in Schrijvers's thesis [17].

In our analysis of the algorithm, we have made the general position assumption. In combination with numerical stability issues, this generally does not hold in real-world data sets. Our implementation is meant as a proof-of-concept, not production-quality code. Whenever we were unable to complete a deletion because of stability issues, we have discarded the results. This has only happened a few dozen times for the 182,051 data points we have gathered. Since our algorithm works in arbitrary dimensions, it would be possible to go to a lower-dimensional DT whenever the general position assumption is violated, as CGAL currently does for their deletion code.

6 Conclusion

Our vertex deletion algorithm has running time $O(C^\otimes(P))$, improving in theory on previous algorithms in the common circumstance that $C^\otimes(P) = o(k \log d)$ where k is the number of tetrahedra needed to retriangulate the cavity and d its number of vertices. In practice our implementation outperforms the current CGAL implementation when the deleted point has high degree (≥ 100) and remains competitive for low degree. Going from theory to practice required some compromises; our best implementations differ from the theoretical model: IBR-edge uses different degrees in the sampling method while IBR-hash does not actually use hashing but a binary search tree.

The low degree sampling schemes has not been proven to have theoretically optimal complexity. It is an open question to prove or disprove that such a permutation, where the next point is randomly chosen in a linear size subset, is random enough to obtain an expected complexity $O(C^\otimes(P))$.

References

[1] Aggarwal, A., Guibas, L., Saxe, J., Shor, P.: A linear-time algorithm for computing the Voronoi diagram of a convex polygon. Discr. Comp. Geom. 4, 591–604 (1989)

[2] Amenta, N., Choi, S., Rote, G.: Incremental constructions con BRIO. In: 19th Sympos. Comput. Geom., pp. 211–219 (2003)

[3] Boissonnat, J.-D., Devillers, O., Pion, S., Teillaud, M., Yvinec, M.: Triangulations in CGAL. Comput. Geom. 22, 5–19 (2002)

[4] Carter, L., Wegman, M.N.: Universal classes of hash functions. J. Comput. System Sci. 18(2), 143–154 (1979)

[5] CGAL. Computational Geometry Algorithms Library (2013), http://www.cgal.org

[6] Chazelle, B., Mulzer, W.: Computing hereditary convex structures. Discr. Comp. Geom. 45(4), 796–823 (2011)

[7] Cheng, S.-W., Dey, T.K., Shewchuk, J.R.: Delaunay Mesh Generation (2012)

[8] Chew, L.P.: Building Voronoi diagrams for convex polygons in linear expected time. Technical Report PCS-TR90-147, Dartmouth College (1990)

[9] Chew, L.P.: Guaranteed-Quality Mesh Generation for Curved Surfaces. In: 9th Sympos. Computat. Geom., pp. 274–280 (1993)

[10] Choi, S.: The Delaunay tetrahedralization from Delaunay triangulated surfaces. In: 18th Sympos. Comput. Geom., pp. 145–150 (2002)

[11] Devillers, O.: The Delaunay hierarchy. Int. J. Found. Comp. Sc. 13, 163–180 (2002)

[12] Devillers, O.: On deletion in Delaunay triangulations. Internat. J. Comput. Geom. Appl. 12(3), 193–205 (2002)

[13] Devillers, O.: Vertex removal in two-dimensional Delaunay triangulation: Speed-up by low degrees optimization. Comput. Geom. 44(3), 169–177 (2011)

[14] Erickson, J.: Nice point sets can have nasty Delaunay triangulations. Discr. Comp. Geo. 30(1), 109–132 (2003)

[15] Joe, B.: Three-Dimensional Triangulations from Local Transformations. SIAM Journal on Scientific and Statistical Computing 10, 718–741 (1989)

[16] Lee, D.-T., Lin, A.K.: Generalized Delaunay Triangulations for Planar Graphs. Discr. Comp. Geo. 1, 201–217 (1986)

[17] Schrijvers, O.: Insertions and deletions in Delaunay triangulations using guided point location. Master's thesis, Technische Universiteit Eindhoven (2012)

Economic 3-Colored Subdivision
of Triangulations

Lucas Moutinho Bueno and Jorge Stolfi

Institute of Computing, University of Campinas, Campinas, Brazil
lucas.bueno@students.ic.unicamp.br, stolfi@ic.unicamp.br

Abstract. We describe an algorithm to subdivide an arbitrary triangulation of a surface to produce a triangulation that is vertex-colorable with three colors. (Three-colorable triangulations can be efficiently represented and manipulated by the GEM data structure of Montagner and Stolfi.) The standard solution to this problem is the barycentric subdivision, which produces $6n$ triangles when applied to a triangulation with n faces. Our algorithm yields a subdivision with at most $2n - m + 4(2 - \chi)$ triangles, where χ is the Euler characteristic of the surface and m is the number of border edges (adjacent to only one triangle). This bound is rarely reached in practice; in particular, if the triangulation is already three-colorable the algorithm does not split any triangles.

Keywords: triangulation, 3-coloration, subdivision.

1 Introduction

We describe an algorithm to subdivide an arbitrary triangulation T of a surface, with or without border, to produce a triangulation R that is 3-vertex-colorable, namely whose vertices can be labeled with the three "colors" $\{0, 1, 2\}$ in such a way that the endpoints of every edge have distinct colors. Vertex-colorable triangulations have theoretical interest, since they can be elegantly represented by the GEM (Graph-Encoded Manifolds) graph class created by Sóstenes Lins and Arnaldo Mandel [3]. They are also relevant to geometric modeling, since they can be represented by the GEM data structure of Montagner and Stolfi [1,2] that allows efficient traversal and modification of d-dimensional triangulations, for any d, with only three topological operators.

The standard solution to this problem is the *barycentric subdivision* [4], where each edge of T is divided into two new edges and a new vertex, and each face of T into six new triangles, six new edges, and a new vertex. The procedure also yields a 3-vertex-coloration for R, where each vertex has color 0, 1 or 2 depending on whether it is contained in a vertex, edge, or face of T, respectively. The barycentric subdivision is easily implemented, but always produces a triangulation R with $6n$ faces when applied to a triangulation T with n faces, even when T is already 3-vertex-colorable.

For triangulations of the sphere, a more economical solution is to find a pairing of the dual graph, and then split the quadrilateral formed by each pair of triangles

H.L. Bodlaender and G.F. Italiano (Eds.): ESA 2013, LNCS 8125, pp. 265–276, 2013.
© Springer-Verlag Berlin Heidelberg 2013

into four triangles by adding its missing diagonal. In the resulting triangulation R every vertex has even degree, and therefore R is three-colorable [8]. This method can be adapted to work for planar triangulations with non-triangular outer face and holes. However, it always doubles the number of faces (even in cases where it would suffice to split only two of them); and it does not work for non-planar, non-spherical triangulations.

In contrast, our algorithm works for any triangulation, and yields a subdivision with at most $2n - m + 4(2 - \chi)$ triangles, where χ is the Euler characteristic of the surface and m is the number of edges on its border. This bound is $2n$ for a triangulation of the sphere with no border, and $2n - m + 4k$ for a planar triangulation with m border edges in k border cycles. Moreover, these upper bounds are only rarely attained: in experiments with Delaunay triangulations of normally distributed points, for example, the average number of triangles in R was about $1.25n$. In particular, if the triangulation T is already 3-vertex-colorable, the algorithm will not split any faces and will return $R = T$.

A lower bound to the size of R (for any algorithm) is $n + p$, where p is the number of odd-degree vertices that are not on the border. For every such vertex, at least one of the incident triangles must be bisected at that corner before those triangles can be 3-colored. Thus, if all $n/2 + 2$ vertices are odd, R will have at least $3n/2+2$ triangles. However, additional triangles may have to be subdivided to obtain a 3-colorable triangulation.

2   Definitions

For this paper, we define a *surface* as a connected compact topological space X where every point p has a neighborhood that is either homeomorphic to the plane $\mathbb{R}^2$, or homeomorphic to the closed half-plane $\mathbb{H}^2 = \{ (x,y) : x,y \in \mathbb{R} \wedge x \geq 0 \}$, with p corresponding to the origin. Points of the second type comprise the *border* of X, while the others comprise its *interior*. Note that the border is the union of zero or more connected components, each homeomorphic to the circle $\mathbb{S}^1$. See Figure 1a.

A *triangulation* T is a partition of a surface, denoted by $\mathcal{S}_T$, into a finite collection of *parts*, comprising its *vertices* $\mathcal{V}_T$, *edges* $\mathcal{E}_T$, and *faces* (or *triangles*) $\mathcal{F}_T$, such that: (i) each vertex is a singleton set; (ii) each edge is homeomorphic to $\mathbb{R}$; (iii) each face is homeomorphic to $\mathbb{R}^2$; (iv) the boundary in $\mathcal{S}_T$ of every edge is a pair of distinct vertices; (v) the boundary in $\mathcal{S}_T$ of every face is the union of three distinct vertices and three distinct edges.

With this definition, the border of $\mathcal{S}_T$ is necessarily the union of a subset of the edges and vertices of T, the *border edges* and *border vertices*. Any edge or vertex of T that is not on the border is an *interior edge* or *interior vertex*. A part p of T is said to *incide* on another part q if one of them is contained in the boundary of the other. An interior edge is incident to exactly two faces of T; whereas any border edge is incident to exactly one face. These faces are called the *wings* of the edge. See Figure 1b.

Note that a triangulation T cannot have any of the features highlighted in Figure 2. In particular, it cannot have a vertex incident to more than two border

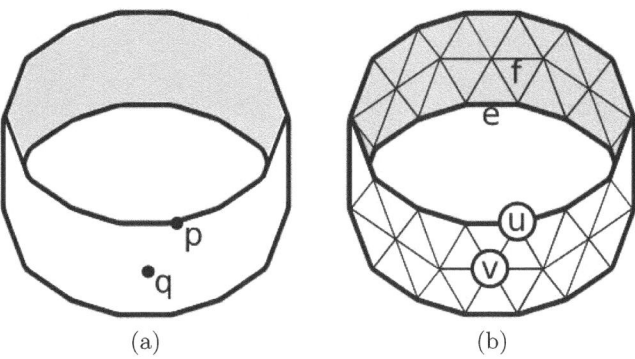

Fig. 1. A surface homeomorphic to the side surface of a cylinder, with the points p on the border and q in the interior (a); a triangulation for the surface, with a border vertex u, an interior vertex v, a border edge e and an interior edge f (b). The border has two connected components.

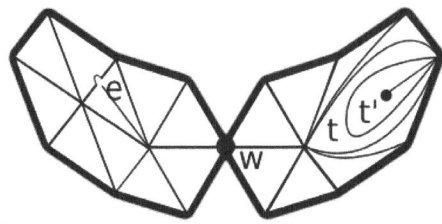

Fig. 2. A partition of a topological space that is not a triangulation. It has a local cut point w, an edge e bounded by only one vertex, a face t with only two boundary vertices and a face t' with only two boundary edges.

edges. Such a vertex would be a *local cut point* of the space $\mathcal{S}_T$, which we define as a point p in a topological space with a neighborhood X such that X is connected but $X \setminus \{p\}$ is not. Note that a surface, as defined above, cannot have any local cut points.

A *sub-triangulation* T' of a triangulation T is a subset of the triangles of T, together with all (and only) the edges and vertices that are incident to those triangles. A sub-triangulation has all the defining properties of a triangulation, except that its underlying space $\mathcal{S}_{T'}$ may not be connected, and may have local cut points. The border edges of T' are the edges of T that are incident to only one triangle of T'. The border vertices of T' are the ones incident to those edges. Note that the local cut points of $\mathcal{S}_{T'}$ are the vertices of T' that are incident to four or more border edges of T'.

A triangulation R is said to be a *subdivision* of another triangulation T if $\mathcal{S}_T = \mathcal{S}_R$ and every part of T is the union of some subset of the parts of R.

The *Euler characteristic* of a triangulation T is the number $\chi_T = |\mathcal{V}_T| - |\mathcal{E}_T| + |\mathcal{F}_T|$. It is a topological property of the surface $\mathcal{S}_T$ alone [6], being 2 for the sphere $\mathbb{S}^2$, 0 for the torus $\mathbb{T}^2 = \mathbb{S}^1 \times \mathbb{S}^1$, 1 for the closed unit disk $\mathbb{D}^2 = \{ (x,y) : x^2 + y^2 \le 1 \}$.

3   The Algorithm

3.1   Overview

Our algorithm receives a triangulation T as input, and outputs a subdivision R of T and a 3-coloration for R. The algorithm operates on one triangle of T at a time, either preserving it or replacing it (in R) by two or more new triangles, in such a way that the part of the triangulation that has already been processed is 3-vertex-colorable.

In the description of the algorithm, we will denote by R' the part of R that has already been built, and by T' the part of T that still remains to be processed. Each vertex v of R' has an assigned color $\lambda(v) \in \{0,1,2\}$. See Figure 3. Both R' and T' are valid sub-triangulations (as defined in Section 2) of R and T, respectively. Also, R' is a valid subdivision of a sub-triangulation of T.

The border edges of R' may include some edges that are contained on the border of $\mathcal{S}_T$, but also some edges that are contained in the interior of $\mathcal{S}_T$ but lie between processed and unprocessed triangles. Likewise, the border edges of T' may comprise some border edges of T, but also some interior edges of T that separate T' from R' (that is, which are incident to exactly one unprocessed triangle).

The algorithm never splits an edge of T into more than two edges. Therefore, each border edge of T' is either an unprocessed border edge of T, or a border edge of R', or the union of two border edges and a border vertex of R'.

The surface $\mathcal{S}_{R'}$ of R' is always connected once R' has been initialized. This follows from the fact that the algorithm only adds a new triangle to R' if it is incident to some edge of R' (that, by definition, is already incident to some other triangle of R').

3.2   The Old Dual Graph

Although T' may be split into two or more connected components at some stage, that may only happen, in a sense, because of the "holes" bounded by the border of $\mathcal{S}_T$. The triangulation T' would remain connected if those holes were filled in the proper way.

In order to formalize this statement (which is important for the proof of correctness) we define the *old dual graph* T'^* that represents the connectivity of T'. The vertices of T'^* are all the faces of T', plus a *null vertex* $t_\emptyset$ if T' has any border edges of T. The edges of T'^* are all the interior edges of T', plus the border edges of T' that are border edges of T. Each interior edge e of T' connects in T'^* its two wings in T'. Each border edge of T' that is a border edge of T connects in T'^* the vertex $t_\emptyset$ to its wing in T'. See Figure 4.

Throughout the algorithm, the graph T'^* remains connected. Note that the vertex $t_\emptyset$ may provide a path between two connected components of T', as long as each still has at least one border edge of T.

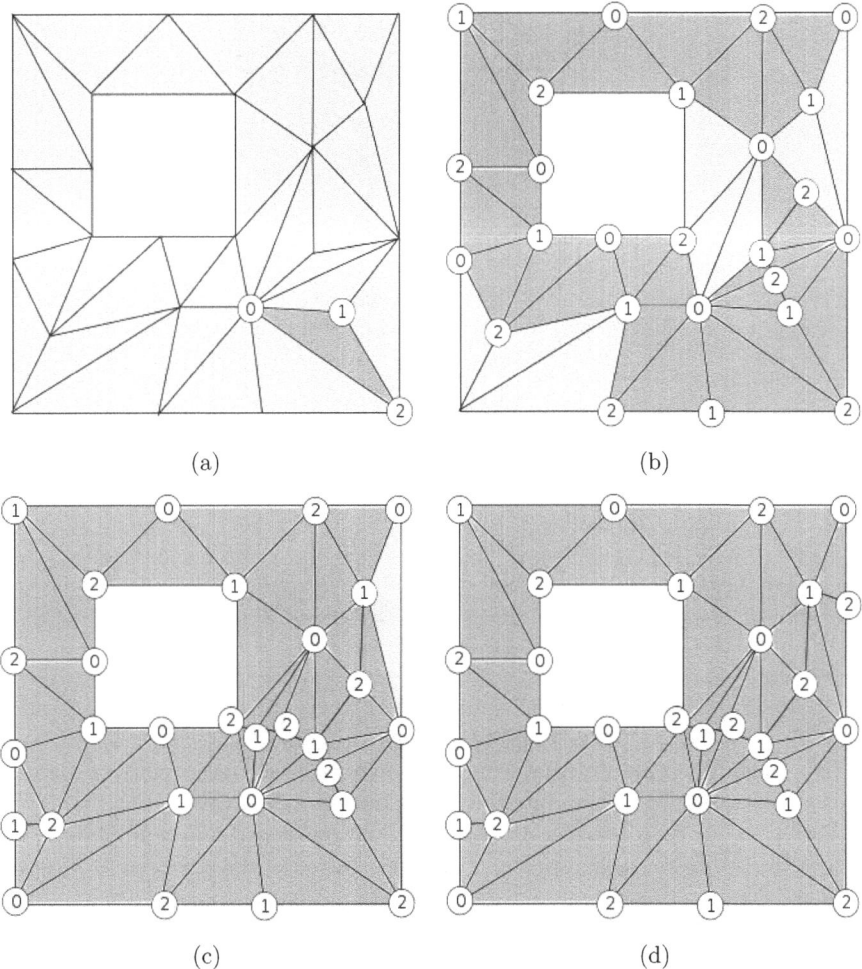

(a)   (b)

(c)   (d)

Fig. 3. Processing of a sample triangulation T, showing the situation (a) just before, (b) during, and (c) just after the main loop of the 3-coloring algorithm, and (d) the final result R. The triangles of R' are in dark gray and those of T' in light gray. The white square is a hole in $\mathcal{S}_T$.

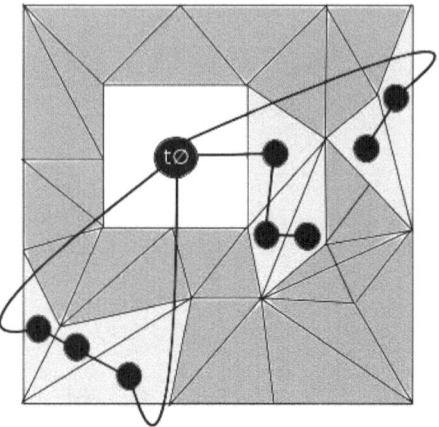

Fig. 4. The old dual graph corresponding to the situation of Figure 3(b)

3.3 Detailed Algorithm

We now show a precise description of the algorithm. It receives a triangulation T as input, and outputs a subdivision R of T and a 3-coloration λ for the vertices of R:

1. Choose any triangle t of T. Let R' and T' be the sub-triangulations of T such that $\mathcal{F}_{R'} = \{t\}$, and $\mathcal{F}_{T'} = \mathcal{F}_T \setminus \{t\}$. Let u_0, u_1, u_2 be the vertices of t. Set $\lambda(u_i) = i$ for each i (see Figure 3a).
2. While $|\mathcal{F}_{T'}| \geq 2$ do:
   (a) Find a triangle t of T' that fits any of the patterns in the left column of Figure 5, and whose removal will not disconnect the graph T'^{*}.
   (b) Apply the replacement operation shown in the second column of Figure 5, in the row corresponding to the pattern matched by t.
3. If $|\mathcal{F}_{T'}| > 0$, then $|\mathcal{F}_{T'}| = 1$. Let t be the only remaining triangle of T'. If any side of t is a border edge of T, it must fit one of the patterns in Figure 5, cases a to h; otherwise it must fit one of the patterns in Figure 6, cases n to r. Apply the corresponding replacement operation, that makes T' empty and turns R' into the final triangulation R.

On the diagrams of Figures 5 and 6, thick lines denote border edges of R'. On the left (pattern) diagrams, thin lines are edges of T' that are not in the border of R', and the only triangle shown is a triangle of T'. On the right (replacement) diagrams, all thin lines are interior edges of R', and all vertices and triangles are parts of R'. In either diagram, numbered vertices are vertices of R' (which may or may not be also vertices of T'), and the numbers are their assigned colors. The other columns of Figures 5 and 6 will only be needed in Section 5.

In steps 2a and 3, the algorithm must match a triangle t of T' to one of those pattern diagrams. Specifically, it must pair each vertex v of t with one of the three corners of the diagram, and also choose a permutation π of the three colors. If a vertex v of t is also a vertex of R', then it must be paired with a labeled vertex of the diagram, and the label of the latter should be $\pi(\lambda(v))$. Then each edge e of t will be paired with one side of the pattern diagram. Each side may be one edge of T', one edge of R', or two edges and a vertex of R'.

Note that every pattern diagram in Figures 5 and 6 has one or more thick sides. Therefore, the triangle t of T' chosen in step 2a will have at least one edge contained in the border of R'.

The replacement operation performed in steps 2b and 3 consists of: (i) remove the triangle t from T', (ii) partition it into triangles, edges and vertices as indicated by the replacement diagram, (iii) add those triangles to R', and (iv) assign colors to the new vertices of R'. In items (i) and (ii) it is understood that the sets $\mathcal{V}_{R'}$, $\mathcal{E}_{R'}$, $\mathcal{V}_{T'}$ and $\mathcal{E}_{T'}$ are adjusted as required by the definition of subtriangulation. In item (iv), each new vertex v of R' gets a color $\lambda(v) = \pi^{-1}(c)$ where c is the corresponding color on the replacement diagram.

4 Correctness

Through the execution of the algorithm the following properties can be verified: (i) the union of the parts of R' and T' is the surface $\mathcal{S}_T$; (ii) the intersection of $\mathcal{S}_{R'}$ and $\mathcal{S}_{T'}$ consists only of points in edges and vertices of R' and T'; (iii) T' is a valid sub-triangulation of T; (iv) R' is a valid sub-triangulation of R; (v) R' is a valid subdivision of a sub-triangulation of T; (vi) λ is a valid 3-coloring of R'.

These properties are true after step 1, and each of the replacements in step 2b preserves these properties. Moreover, after the replacement in step 3, T' will have no triangles and R' will be a valid subdivision of T. Therefore, R will be a valid subdivision of T and λ will be a valid 3-coloring of R. Note that the algorithm terminates, because at each iteration the triangulation T' loses one triangle.

To prove the correctness of the algorithm we must show that there is always a triangle t that satisfies either the conditions of step 2a or the conditions of step 3. For that purpose, we define the *old front* as being the set of edges and vertices of T' that are in the border of T' but not in the border of T; and, similarly, the *new front* as the edges and vertices in the border of R' that are not in the border of T. Note that the union of the parts of the old front and the union of those of the new front are the set of points of $\mathcal{S}_{R'} \cap \mathcal{S}_{T'}$, only partitioned into edges and vertices in two different ways.

The proof of correctness also rests on the following loop invariants, that will be shown to hold just before every execution of step 2a:

L0 The new front has at least one edge of R'.

L1 Each triangle of T' is incident to at most two edges of the old front.

L2 Every edge e of the old front is either an edge of the new front, or the union of two edges a, b and one vertex w of the new front. In the second case, if u, v are the ends of e, then $\lambda(u) = \lambda(v)$.

	Pattern	Replacement	$\delta\mathcal{F}_{R'}$	$\delta\mathcal{E}_{R'}^b$	$\delta\mathcal{F}_{T'}$	$\delta\chi_{R'}$	$\delta\Delta$
a			$+1$	$+1$	-1	0	$0\ (-2)$
b			$+2$	0	-1	0	$0\ (-2)$
c			$+1$	-1	-1	0	$-2\ (-4)$
d			$+2$	0	-1	0	$0\ (-2)$
e			$+2$	-2	-1	0	$-2\ (-4)$
f			$+4$	-2	-1	0	$0\ (-2)$
g			$+4$	-2	-1	0	$0\ (-2)$
h			$+4$	-2	-1	0	$0\ (-2)$
i			$+1$	$+1$	-1	-1	$-4\ (-6)$
j			$+2$	0	-1	-1	$-4\ (-6)$
k			$+4$	0	-1	-1	$-2\ (-4)$
l			$+2$	$+2$	-1	-1	$-2\ (-4)$
m			$+4$	$+2$	-1	-1	$0\ (-2)$

Fig. 5. Possible patterns (modulo color permutations) and the respective replacements for each iteration after step 1 of the algorithm

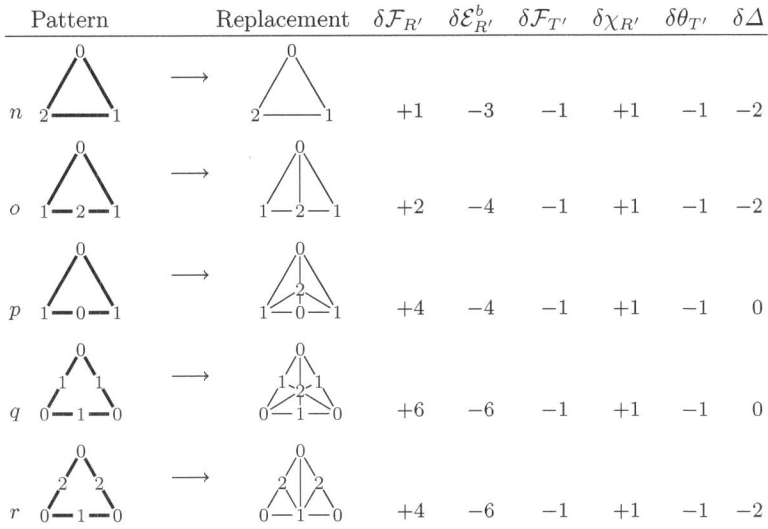

Pattern	Replacement	$\delta \mathcal{F}_{R'}$	$\delta \mathcal{E}^b_{R'}$	$\delta \mathcal{F}_{T'}$	$\delta \chi_{R'}$	$\delta \theta_{T'}$	$\delta \Delta$
n		$+1$	-3	-1	$+1$	-1	-2
o		$+2$	-4	-1	$+1$	-1	-2
p		$+4$	-4	-1	$+1$	-1	0
q		$+6$	-6	-1	$+1$	-1	0
r		$+4$	-6	-1	$+1$	-1	-2

Fig. 6. Additional possible patterns for the last operation (step 3) of the algorithm, and their replacements

The invariants **L0** and **L2** are also true just before step 3, as will be shown. The validity of **L0**, **L1**, and **L2** is established by lemmas 1–3:

Lemma 1. *The replacement in step 2b preserves invariant **L0**.*

Proof. Each replacement in Figure 5 creates at least one new edge in the border of R', so after each iteration of step 2b the border of R' is not empty.

Note that $\mathcal{F}_{T'}$ is not empty, because the main loop of the algorithm processes one triangle of T' at a time and stops if $|\mathcal{F}_{T'}| < 2$. Since $\mathcal{S}_T$ is connected, the two closed subsets $\mathcal{S}_{R'}$ and $\mathcal{S}_{T'}$ of $\mathcal{S}_T$ must have a non-empty intersection. Since $\mathcal{F}_{R'}$ and $\mathcal{F}_{T'}$ are disjoint and $\mathcal{S}_T$ has no local cut-points, the intersection must include at least one whole edge of R', which must be contained in the border of R' and the border of T'. As observed before, this edge must be part of the new front.   □

Lemma 2. *Invariant **L1** is valid just before every execution of step 2a.*

Proof. Just before each step 2a there are at least two triangles in T'. By the choice of t in step 2a, T'^* is always connected. Therefore, there is at least one edge of T'^* connecting each triangle of T'. Thus, there is no triangle of T' with all its three boundary edges in the old front and the lemma is true.   □

Lemma 3. *The replacement in step 2b preserves invariant **L2**.*

Proof. Every replacement diagram in Figure 5 has at most two edges for each corresponding edge e in the border of T'. Moreover, every time the replacement splits an edge e, the two ends of e are assigned the same same color in R'. Therefore invariant **L2** is preserved.   □

Lemmas 4, 5 and 6 guarantee the triangles chosen at steps 2a and 3 exist.

Lemma 4. *Just before every execution of step 2a, there is a triangle t of T' that has an edge e in the old front, whose removal does not disconnect the T'^* graph.*

Proof. Let $\mathcal{F}_{T'}^B$ be the set of triangles of T' that are incident to edges of the old front, and $\mathcal{F}_{T'}^I = \mathcal{F}_{T'} \setminus \mathcal{F}_{T'}^B$. Note that the vertices of T'^* are $\mathcal{F}_{T'}^B \cup \mathcal{F}_{T'}^I \cup T_\emptyset$, where $T_\emptyset$ is $\{t_\emptyset\}$, if $t_\emptyset$ exists or the empty set otherwise. By invariant **L1**, the degree of each vertex of $\mathcal{F}_{T'}^B$ in T'^* is one or two, and the degree of each vertex of $\mathcal{F}_{T'}^I$ in T'^* is three. $\mathcal{F}_{T'}^B$ cannot be empty due to invariant **L0**.

Let G be the graph obtained from T'^* by contracting all edges of $T'^* \setminus \mathcal{F}_{T'}^B$ [5], removing any loops, and coalescing any parallel edges. Note that every vertex of $\mathcal{F}_{T'}^B$ still exists in G, and its degree in G is at most two.

Since edge contraction preserves connectedness, the graph G is connected, and therefore has a spanning tree M [7]. We will now show that M has at least one leaf t that is a vertex of $\mathcal{F}_{T'}^B$. The removal of t will not disconnect T'^*, so the lemma will be proved.

Since $\mathcal{F}_{T'}^B$ is not empty and T' has at least two triangles, M has at least two vertices, and therefore at least two leaves. Even if there is one vertex $t_\emptyset$ in T'^*, there is one leaf m in M that is a vertex of $\mathcal{F}_{T'}^B$, or is the contraction of some subset X of $\mathcal{F}_{T'}^I$ (that does not include $t_\emptyset$). In the first case, we are done. In the second case, the border B of X in T is not empty and doesn't contain any border edges of T or of R'. Therefore B must consist of $k \geq 2$ edges of triangles from $\mathcal{F}_{T'}^B$, and $j \leq k$ vertices of those triangles. Since each triangle in $\mathcal{F}_{T'}^B$ has degree at most 2 in T'^* but it also has 3 distinct vertices, B must contain edges from at least two distinct triangles of $\mathcal{F}_{T'}^B$.

It follows that m has at least two neighbors in G that are in $\mathcal{F}_{T'}^B$ but only one neighbor in M. Let t be one of these neighbors that is not the neighbor of m in M. Since t has degree at most 2 in G and M is connected, t must have exactly one neighbor in M, and therefore is a leaf of M. $\qquad\square$

Lemma 5. *Just before every execution of step 2a, there is a triangle t of T' that fits one of the patterns in Figure 5 and whose removal does not disconnect the T'^* graph.*

Proof. Let t be any triangle of T' that satisfies the conditions of lemma 4. By invariants **L0** and **L1**, t has one or two sides in the old front.

By inspection one can verify that all possible cases (apart from a permutation of the colors already assigned) are covered in the left column of Figure 5. Specifically, if only one side e of t is in the old front, then t must fit one of the patterns (*a*) or (*b*) if the vertex of t that is not incident to e is not in the old front, and one of the patterns (*i*) through (*m*) otherwise. If two sides of t are in the old front, then t must fit one of the patterns (*c*) through (*h*).

Note that the color patterns that do not appear in those pattern diagrams are forbidden by invariant **L2**. $\qquad\square$

Lemma 6. *The last triangle t of T' fits one of the patterns in Figure 5 or Figure 6.*

Proof. Just before the execution of step 3, by lemma 4, t is either incident to $t_\emptyset$ in T'^* or an isolated vertex of T'^*. For the first case, t has one or two sides in the old front and, by inspection, one can verify that all possible cases are covered in the left column of Figure 5. For the second case, t has three sides in the old front and, by inspection, one can verify that all possible cases are covered in in the left column of Figure 6.   □

From these lemmas we conclude the following:

Theorem 1. *For any triangulation T, the algorithm returns a 3-colored subdivision R of T.*

5 Efficiency

For the proof of efficiency we define:

$$\Delta = |\mathcal{F}_{R'}| + 2|\mathcal{F}_{T'}| + |\mathcal{E}^b_{R'}| - |\mathcal{E}^b_T| + 4(\chi_{R'} - \chi_T) + 2\theta_{T'} \tag{1}$$

where $\chi_{R'}$ and χ_T are the Euler characteristics of R' and T, respectively; $\mathcal{E}^b_{R'}$ and $\mathcal{E}^b_T$ are the set of border edges of R' and T, respectively; and $\theta_{T'}$ is the number of connected components of T'^*, that is, 1 if T' is not empty (before step 3) and 0 otherwise (after step 3). Note that χ_T and $|\mathcal{E}^b_T|$ are constants while the other parameters vary during the algorithm.

Observe that, at each execution of steps 2b and 3, the quantity Δ decreases or stays equal. The change of its value is denoted by $\delta\Delta$. Analogously, we can define $\delta|\mathcal{F}_{R'}|$, $\delta|\mathcal{F}_{T'}|$, $\delta|\mathcal{E}^b_{R'}|$, $\delta\chi_{R'}$ and $\delta\theta_{T'}$. See Figures 5 and 6. Specifically for Figure 5, $\delta\Delta$ has two values: the first (out of the parentheses) is for step 2b; the second (in parentheses) is for step 3.

We will prove below that, after step 1, Δ is an upper bound for the number of triangles of the final triangulation R.

Lemma 7. *At any point during the execution of the algorithm, except before step 1, $\Delta \geq |\mathcal{F}_R|$*

Proof. At the end of the execution of the algorithm, $|\mathcal{F}_{R'}| = |\mathcal{F}_R|$, $|\mathcal{F}_{T'}| = 0$, $|\mathcal{E}^b_{R'}| \geq |\mathcal{E}^b_T|$ and $\chi_{R'} = \chi_T$. Substituting in formula (1), we have $\Delta \geq |\mathcal{F}_R|$.

According to lemmas 5 and 6, Figures 5 and 6 represent all possible operations during the execution of the algorithm, except for step 1. Since the change $\delta\Delta$ in Δ in all those cases is zero or negative, and $\Delta \geq |\mathcal{F}_R|$ at the end, the lemma is true.   □

Theorem 2. *For any triangulation T, the algorithm returns a subdivision R with at most $2|\mathcal{F}_T| - |\mathcal{E}^b_T| + 4(2 - \chi_T)$ faces.*

Proof. Just after step 1 we have $|\mathcal{F}_{R'}| = 1$, $|\mathcal{F}_{T'}| = |\mathcal{F}_T| - 1$, $|\mathcal{E}^b_{R'}| = 3$, $\chi_{R'} = 1$ and $\theta_{T'} = 1$. Therefore, at this point, $\Delta = 1 + 2|\mathcal{F}_T| - 2 + 3 - |\mathcal{E}^b_T| + 4(1 - \chi_T) + 2$, and then $\Delta = 2|\mathcal{F}_T| - |\mathcal{E}^b_T| + 4(2 - \chi_T)$. By lemma 7, the theorem is true.   □

In particular, if T is a triangulation of the sphere, $\mathcal{E}_T^b$ is empty and $\chi_T = 2$, so the bound reduces to $|\mathcal{F}_R| \leq 2|\mathcal{F}_T|$. Also, if T is a planar triangulation whose border consists of k separate cycles (including the outermost boundary), then $\chi_T = 2 - k$, and the bound becomes $|\mathcal{F}_R| \leq 2|\mathcal{F}_T| - |\mathcal{E}_T^b| + 4k$.

Note that these are only upper bounds, and $|\mathcal{F}_R|$ may be substantially smaller. In particular, if the triangulation T already has a 3-coloration σ, only cases (a), (c), and (i) of Figure 5 and case (n) of Figure 6 may occur. Then the algorithm will not split any triangles and will color the vertices of R' with $\lambda = \sigma$ apart from a permutation of the three colors.

References

1. Montagner, A., Stolfi, J.: Gems: A general data structure for d-dimensional triangulations. Technical Report IC-06-16, Computer Science Institute, UNICAMP (2006),
   http://www.ic.unicamp.br/~reltech/2006/06-16.pdf
2. Montagner, A.J.: A Estrutura de Dados Gema para Representação de Mapas n-Dimensionais. Masters Dissertation, Computer Science Institute, UNICAMP (2007) (in Portuguese),
   http://www.bibliotecadigital.unicamp.br/document/?code=vtls000431432
3. Lins, S., Mandel, A.: Graph-encoded 3-manifolds. Discrete Mathematics 57(3), 261–284 (1985)
4. Bing, R.H.: Geometric Topology of 3-Manifolds. Colloquium Publications vol. 40 (1983)
5. Bollobás, B.: Modern Graph Theory. Springer, New York (1998)
6. Alexandroff, P.S.: Combinatorial Topology. Dover, New York (1998)
7. Kruskal, J.B.: On the Shortest Spanning Subtree of a Graph and the Traveling Salesman Problem. Proc. American Mathematical Society 7, 48–50 (1956)
8. Tsai, M., West, D.B.: A new proof of 3-colorability of Eulerian triangulations. Mathematica Contemporanea 4(1) (2011)

Limitations of Deterministic Auction Design
for Correlated Bidders*

Ioannis Caragiannis, Christos Kaklamanis, and Maria Kyropoulou

Computer Technology Institute and Press "Diophantus" &
Department of Computer Engineering and Informatics,
University of Patras, 26504 Rio, Greece

Abstract. The seminal work of Myerson (Mathematics of OR 81) characterizes incentive-compatible single-item auctions among bidders with independent valuations. In this setting, relatively simple deterministic auction mechanisms achieve revenue optimality. When bidders have correlated valuations, designing the revenue-optimal deterministic auction is a computationally demanding problem; indeed, Papadimitriou and Pierrakos (STOC 11) proved that it is APX-hard, obtaining an explicit inapproximability factor of 99.95%. In the current paper, we strengthen this inapproximability factor to $57/58 \approx 98.3\%$. Our proof is based on a gap-preserving reduction from the problem of maximizing the number of satisfied linear equations in an over-determined system of linear equations modulo 2 and uses the classical inapproximability result of Håstad (J. ACM 01). We furthermore show that the gap between the revenue of deterministic and randomized auctions can be as low as $13/14 \approx 92.9\%$, improving an explicit gap of $947/948 \approx 99.9\%$ by Dobzinski, Fu, and Kleinberg (STOC 11).

1 Introduction

In the classical model of Auction Theory [10], a seller auctions off an item to n bidders with valuations for the item drawn independently from known but not necessarily identical probability distributions. Myerson's seminal work [14] gives an elegant characterization of revenue-maximizing auctions in this setting. Optimal revenue is achieved by simple deterministic auctions that are defined using succinct information about the probability distributions. In contrast, the case of bidders with correlated valuations has been a mystery; in spite of the vast related literature in Economics and Computer Science, no such general characterization result has been presented so far. Due to their simplicity and amenability to implement in practice, deterministic auctions are of particular importance. Recently, Papadimitriou and Pierrakos [16] provided an explanation — from the computational complexity point of view — for this lack of results, by proving that the problem of designing the optimal deterministic auction given

* This work was supported by the European Social Fund and Greek national funds through the research funding program Thales on "Algorithmic Game Theory".

H.L. Bodlaender and G.F. Italiano (Eds.): ESA 2013, LNCS 8125, pp. 277–288, 2013.

the explicit description of the joint probability distribution (with finite support) is APX-hard. Furthermore, Dobzinski et al. [7] provided a separation between randomized (truthful in expectation) and deterministic truthful auctions; there are (single-item) settings in which randomized auctions may extract strictly more revenue than any deterministic auction. Both results hold even when only three bidders participate in the auction. In this paper, we strengthen both results.

Existing approaches to single-item auctions with correlated bidders fall into three different categories. A first approach that has been mostly followed by economists (e.g, see [13,11,12,3]) assumes that each bidder has her own valuation function that depends on a shared random variable; this model is usually referred to as the *interdependent valuations model*. In a second approach, the support of the joint probability distribution is extremely large (exponentially larger than the number of players or even infinite) and an auction mechanism can obtain information about the distribution through queries (e.g, see [17,18]). The related literature focuses on the design of auctions that use only a polynomial (in terms of the number of bidders) number of queries. In the third one, the joint probability has finite support and the related work seeks for auctions that are defined in polynomial time in terms of the support size and the number of players (e.g, see [7,16]). The last two models are known as the *query model* and the *explicit model*, respectively. Among these models, the explicit one allows us to view the design of the revenue-optimal (deterministic) auction as a standard optimization problem. The auction has to define the bidder that gets the item and her payment to the auctioneer for every valuation vector of the support of the joint probability distribution. Both the allocation and the payments should be defined in such a way that no bidder has an incentive to misreport her true valuation; this constraint is known as *incentive compatibility*. The objective is to maximize the expected revenue of the auctioneer over all valuation vectors.

Since our purpose is to explore the limitations of deterministic auctions, we focus on the explicit model and the case of three bidders. Following [16], we refer to the optimization problem mentioned above (when restricted to three bidders) as 3OPTIMALAUCTIONDESIGN. The inapproximability bound presented by Papadimitriou and Pierrakos [16] is marginally smaller than 1, namely 1999/2000. It is achieved by a gap-preserving reduction from a structured maximum satisfiability problem called CATSAT. This problem has an inapproximability of 79/80; hence, the gap obtained for 3OPTIMALAUCTIONDESIGN is even closer to 1. We present a different reduction from the classical MAX-3-LIN(2) problem of approximating the number of satisfied linear equations in an over-determined system of linear equations modulo 2 (with three binary variables per equation). Our proof uses the seminal 1/2-inapproximability result of Håstad [9] and yields a significantly improved inapproximability bound of $57/58 \approx 98.3\%$ for 3OPTIMALAUCTIONDESIGN. Furthermore, we demonstrate a rather significant revenue gap between deterministic truthful mechanisms and randomized auctions that are truthful in expectation; the revenue of any deterministic auction can be at most $13/14 \approx 92.9\%$ of the optimal randomized one. This result improves the previously known explicit bound of $947/948 \approx 99.9\%$ of Dobzinski et al. [7].

Extending Myerson's work, Crémer and McLean [4,5] characterize the information structure that guarantees the auctioneer full surplus, under several settings with correlated valuations. They consider *interim* individual rationality which allows players to have negative utility for some valuation vectors. In contrast, our work focuses on *ex post* individual rationality, a design requirement that does not allow such situations. Ronen [17] and Ronen and Saberi [18] consider single-item optimal auctions in the query model. They design auctions that use queries of the form: given the valuations of a set of players, which is the conditional distribution of the remaining ones? The 1-lookahead auction in [17] yields at least half the optimal revenue to the seller. Essentially, the auction ignores the $n - 1$ lowest bids and offers the item to the remaining bidder at the price that maximizes the revenue considering the distribution of her valuations conditioned on the valuations of everybody else. Ronen and Saberi [18] present several impossibility results for auctions of particular type, e.g., no ascending auction can approximate the optimal revenue to a factor greater than $7/8$.

Dobzinski et al. [7] consider k-lookahead auctions and show that a $2/3$-approximation of the optimal revenue can be achieved by randomized auctions in the query model. For the explicit model, they show that the optimal randomized auction can be computed by linear programming while the deterministic 2-lookahead auction achieves a $3/5$-approximation of revenue. Their positive results have been strengthened by Chen et al. [2] to factors of 0.731 and 0.622, respectively. Both [7] and [16] prove that revenue-optimal auction design can be solved in polynomial time in the 2-bidder case. The 2-bidder case has also been considered in [8] and [6]. In particular, Diakonikolas et al. [6] study the tradeoff between efficiency and revenue in deterministic truthful auctions and prove that any point of the Pareto curve can be approximated with arbitrary precision.

The rest of the paper is structured as follows. We begin with preliminary definitions in Section 2. The reduction and the proof of the inapproximability of 3OPTIMALAUCTIONDESIGN are presented in Section 3. The revenue-gap construction is presented in Section 4. Due to lack of space, several details in the constructions and some proofs have been omitted.

2   Preliminaries

We study the setting in which an item is auctioned off among players (or bidders) with correlated valuations and quasi-linear utilities. Players draw their valuations from a joint probability distribution $\mathcal{D}$ over valuation vectors. A single-item auction mechanism is defined by an allocation and a payment function. For every vector of bids that are submitted by the players to the auctioneer, the allocation defines who among the players (if any) gets the item and the payment function decides the payment of the winning player to the auctioneer. Incentive compatibility, individual rationality, and the no-positive-transfers property are classically considered as important desiderata for auction mechanisms [10,15]. Incentive compatibility requires that truth-telling maximizes players' utility (i.e., valuation minus payment). Hence, each player always submits as bid her actual valuation for the item. Individual rationality encourages players to participate. In particular, we consider *ex post* individual rationality that requires that players always have non-negative utility. We also require that the players never receive payments from the auctioneer (no positive transfer). An obvious objective for auction mechanisms is the maximization of the expected revenue, i.e., the expectation over all valuation vectors of the payment received by the auctioneer.

The recent work on revenue-optimal deterministic auction mechanisms (e.g., see [16]) restricts the search space to *monotone* allocations and *threshold* payment functions. An allocation is monotone if when the item is allocated to some player i for some bidding vector b, player i is allocated the item when the bid vector is (b', b_{-i}) with $b' > b_i$. The notation (b', b_{-i}) denotes the bidding vector where player i bids b' and the remaining players keep their bids as in b. A threshold payment for a winning player i is then defined as the infimum bid b'' so that player i gets the item for the bidding vector (b'', b_{-i}).

We consider auctions with three players and assume that $\mathcal{D}$ is defined by a set S of points in $\mathbb{R}^3$ and weights associated with these points. The weight of a point indicates the probability that the corresponding valuation vector is realized. We refer to the three players as player x, y, and z; a point of S corresponds to a valuation vector where players x, y, and z have as valuation the x-, y-, and z-coordinate of the point. Naturally, we will refer to allocations of points to players. In the following, we say that two points are x-*aligned* (resp., y-aligned, z-aligned) if they have the same y- and z-coordinates (resp., x- and z-coordinates, x- and y-coordinates). Monotonicity implies that if a point p is allocated to a player (say, x), then all points that are x-aligned with p and have higher x-coordinate are *enforced* to be allocated to player x as well (similarly for the other players). The payment associated with a point p that is allocated to player x is then the lowest x-coordinate of the x-aligned points with p that are allocated to x (similarly for the other players). So, we can state the problem of designing the optimal deterministic auction mechanism as follows:

3OPTIMALAUCTIONDESIGN: Given a finite set of points $S \subset \mathbb{R}^3$ and associated weights, compute a monotone allocation of the points of S to players x, y, and z so that the weighted sum of the implied threshold payments (expected revenue) is maximized.

Randomized allocations can allocate fractions of points to players under the restriction that the total allocation fraction for a point is at most 1. Fractions correspond to allocation probabilities. In randomized auction mechanisms that are truthful-in-expectation, the allocation is monotone in the sense that the allocation probabilities to player x (similarly for the other players) are non-decreasing in the terms of the x-coordinate of x-aligned points. The payment when player x gets the item at point p depends on the x-coordinate of the points that are x-aligned to p and have lower x-coordinates and their allocation probabilities (e.g., see Chapter 13 of [15]). In any case, this payment is at least the lowest x-coordinate among the x-aligned points to p that have non-zero probability to be allocated to x.

3   The Inapproximability Result

In this section we present our gap-preserving reduction from MAX-3-LIN(2) to 3OPTIMALAUCTIONDESIGN.

MAX-3-LIN(2): Given a set of n binary variables $v_1, \ldots, v_n$ and a set of m linear equations modulo 2 each containing exactly three variables, find an assignment of the variables that maximizes the number of satisfied equations.

Consider an instance I of MAX-3-LIN(2), with n variables v_i, for $i = 1, \ldots, n$ and m linear equations modulo 2, i.e., $e(h) : v_{h_1} + v_{h_2} + v_{h_3} = \alpha_h \pmod 2$, for $h = 1, \ldots, m$, with $1 \leq h_1 < h_2 < h_3 \leq n$ and $\alpha_h \in \{0, 1\}$. Let d_i be the degree of variable v_i, i.e., the number of equations in which v_i participates. Our reduction constructs an instance $R(I)$ of 3OPTIMALAUCTIONDESIGN with a polynomial number of points. We describe the reduction giving only the relative location of most of these points; this suffices to give the intuition behind our proof. The majority of the points in $R(I)$ have coordinates in $(1-\theta, 1+\theta)$ where $\theta \in (0, 1/3600)$ is a very small constant; there are additional points called *blockers* which lie outside (and significantly far from) this region. Both the coordinates of points and their weights are rational numbers that require only polynomial precision. Without loss of generality, we consider weights that do not sum up to 1 and we consistently compute the revenue contributed by (the allocation of) a point as the product of the threshold payment and its weight.

We exploit the following property which has also been used in [1]. Given a linear equation $e(h) : v_i + v_j + v_k = \alpha$, define the four boolean clauses $e_1(h) = (v_i \vee v_j \vee v_k)$, $e_2(h) = (\neg v_i \vee \neg v_j \vee v_k)$, $e_3(h) = (\neg v_i \vee v_j \vee \neg v_k)$, and $e_4(h) = (v_i \vee \neg v_j \vee \neg v_k)$ if $\alpha = 1$ and $e_1(h) = (\neg v_i \vee \neg v_j \vee \neg v_k)$, $e_2(h) = (v_i \vee v_j \vee \neg v_k)$, $e_3(h) = (v_i \vee \neg v_j \vee v_k)$, and $e_4(h) = (\neg v_i \vee v_j \vee v_k)$ if $\alpha = 0$.

Fact 1. *If a linear equation $e(h)$ is true then the four boolean clauses $e_1(h)$, $e_2(h)$, $e_3(h)$, $e_4(h)$ are true. Otherwise, exactly three of these clauses are true.*

Instance $R(I)$ contains one variable gadget and four clause gadgets per equation (each clause gadget corresponds to one of the clauses mentioned above). The clause gadgets are carefully connected to the variable gadget. We begin by presenting the variable gadget.

The variable gadget. For every variable v_i, $i = 1 \ldots n$, the variable gadget has a *variable point* v_i with weight $d_i(1/2 + 23\theta)$. All these points are x-aligned with $(1, 1, 1)$ and their x-coordinate is in $(1, 1+\theta)$ so that the x-coordinate of v_{i+1} is higher than the x-coordinate of v_i. For each appearance of variable v_i in an equation e, there are four *connection points* $H_1^i(e)$, $H_2^i(e)$, $H_3^i(e)$, and $H_4^i(e)$ with weight θ. Two of these points are y-aligned with v_i; the other two are z-aligned with v_i. The exact location of these points will become clear after the description of the clause gadgets; for the moment, we remark that these points have coordinates in $[1, 1+\theta)$. We also add a point b with weight θ, at point $(L(b), 1, 1)$ such that $\theta L(b) = 3m(1+\theta)(1/2 + 23\theta)$.

Point b belongs to a special class of points in our construction which we call *blockers*. The idea is that a blocker can prevent the allocation of a point to a certain player. Blockers essentially play the same role that scaffolding segments play in the reduction of [16]. In particular, b x-blocks the points $v_1, ..., v_n$ in the following sense. It has weight θ (i.e., it corresponds to a highly unlikely valuation vector) but a very large x-coordinate $3m(1+\theta)(1/2+23\theta)/\theta$ to compensate for that (i.e., player x values the item greatly, unlike the other two players). In a revenue-optimal allocation, the only point among v_1, v_2, ..., v_n, and b that is allocated to player x is b. To see why this is true, observe that the contribution of b to the revenue is $3m(1+\theta)(1/2+23\theta)$ whereas by allocating some variable points and b to x, their contribution to the revenue is less than $3m(1+\theta)(1/2+23\theta)$. Furthermore, the weight $d_i(1/2+23\theta)$ of point v_i is significantly high so that in any revenue-optimal allocation, point v_i should be allocated to either player y or player z; these allocations correspond to setting variable v_i to values 1 and 0, respectively. Due to monotonicity, this will enforce the allocation of the y-aligned or z-aligned connection points; intuitively, this will propagate the fact that the variable v_i is set to a certain value to the clause gadgets.

We continue by presenting the clause gadgets and clarify the connection to the variable gadget. For each equation of I, we define four clause gadgets, one for each clause corresponding to the equation. For two x-aligned points p and q, we use the notation $p_{(+x)}q$ and $p_{(-x)}q$ to denote that q has larger and smaller x-coordinate than p, respectively (similarly for the other coordinates). We define the four clause gadgets corresponding to the equation $e(h) : v_i + v_j + v_k = 1$ (mod 2).

The clause gadget corresponding to clause $e_1(h) = (v_i \vee v_j \vee v_k)$. The clause gadget corresponding to $e_1(h)$ consists of the 5-sequence of points $[A_1(h)_{(+x)}B_1(h)_{(-z)}C_1(h)_{(+x)}D_1(h)_{(-z)}E_1(h)]$. All these points have y-coordinate in $(1-\theta, 1)$, i.e., they lie below the plane $y = 1$. Points $A_1(h)$, $C_1(h)$, and $E_1(h)$ have weight 1; points $B_1(h)$ and $D_1(h)$ have weight θ. The points $A_1(h)$, $C_1(h)$, and $E_1(h)$ are y-aligned to the connection points $H_1^i(h)$, $H_1^j(h)$, and $H_1^k(h)$, respectively. Point $A_1(h)$ is z-blocked by blocker $b_{A_1(h)}$ while point $E_1(h)$ is x-blocked by blocker $b_{E_1(h)}$. See Figure 1.

The z-blocker $b_{A_1(h)}$ has weight θ and a very high z-coordinate equal to $(1+\theta)^2/\theta$. This implies that the highest revenue allocation is the one in which $b_{A_1(h)}$ is allocated to player z and point $A_1(h)$ is not allocated to player z. To see why this is true, observe that the contribution of $b_{A_1(h)}$ to the revenue is $(1+\theta)^2$ whereas by allocating both $A_1(h)$ and $b_{A_1(h)}$ to z or neither of them to z, their contribution to the revenue is less than $(1+\theta)^2$. All blockers in the following have the same coordinate $(1+\theta)^2/\theta$ in the dimension that they block. Points with negligibly small weight θ that are part of the gadgets are crucial since they indirectly influence the allocation of unit-weight points, e.g., $B_1(h)$ prevents $A_1(h)$ and $C_1(h)$ to be allocated to x and z, respectively, at the same time. Consider a monotone allocation in which the connection points $H_1^i(h)$, $H_1^j(h)$, and $H_1^k(h)$ are allocated to players x or z. In this case, we say that the clause

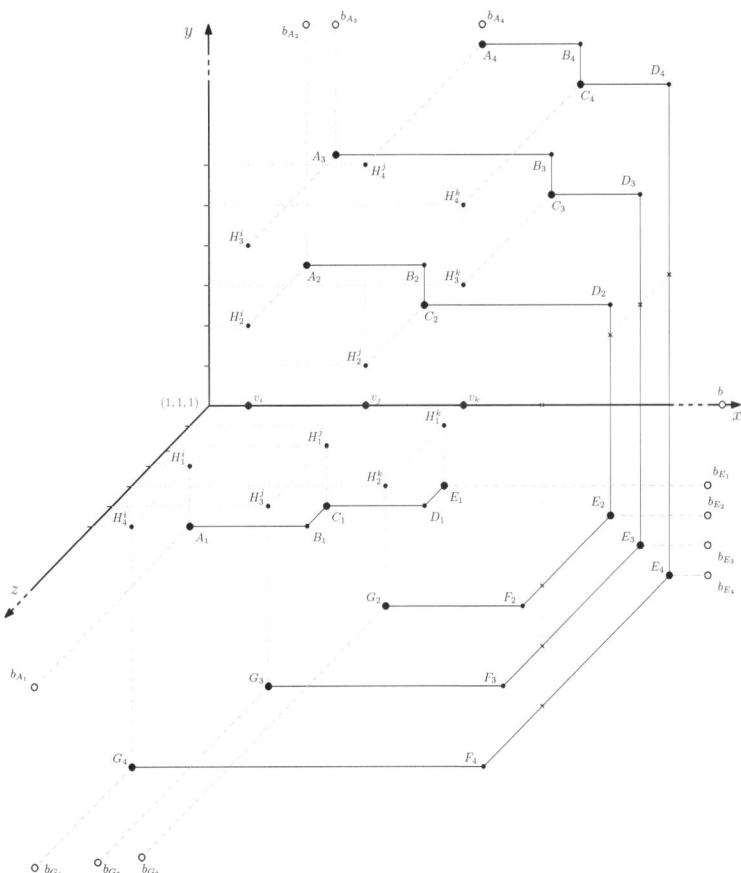

Fig. 1. The clause gadgets corresponding to equation $e(h) : v_i + v_j + v_k = 1 \pmod 2$. Large black disks represent unit-weight or variable points, smaller black disks represent connection points or θ-weight points, whereas white disk represent blockers. Note that the notation of points has been simplified by dropping the index h.

gadget is *non-breathing* in the sense that none of the points $A_1(h)$, $C_1(h)$, and $E_1(h)$ can be allocated to player y. Hence, among the monotone allocations in which the clause gadget is non-breathing, the one that maximizes revenue leaves one of $A_1(h)$, $C_1(h)$, and $E_1(h)$ unallocated. In this case, the contribution of these three unit-weight points to revenue is at least 2 and at most $2 + 2\theta$. In contrast, if some of the connection points (say $H_1^j(h)$) is not allocated to players x or z (i.e., the clause gadget is *breathing*), the contribution of $A_1(h)$, $C_1(h)$, and $E_1(h)$ can increase to at least $3 - \theta$ and at most $3 + 2\theta$ by allocating $A_1(h)$ and $B_1(h)$ to player x, $C_1(h)$ and $H_1^j(h)$ to player y, and $D_1(h)$ and $E_1(h)$ to player z. The other cases (i.e., when $H_1^i(h)$ or $H_1^k(h)$ are not allocated to any player) have similar allocations of the same improved revenue.

The clause gadget corresponding to clause $e_2(h) = (\neg v_i \vee \neg v_j \vee v_k)$.
The clause gadget corresponding to $e_2(h)$ consists of the 7-sequence of points $[A_2(h)_{(+x)}B_2(h)_{(-y)}C_2(h)_{(+x)}D_2(h)_{(-y)}E_2(h)_{(+z)}F_2(h)_{(-x)}G_2(h)]$. Points $A_2(h)$, $B_2(h)$, $C_2(h)$, $D_2(h)$, and $E_2(h)$ have z-coordinate in $(1-\theta, 1)$, i.e., they lie behind the plane $z = 1$. Points $E_2(h)$, $F_2(h)$, and $G_2(h)$ have y-coordinates in $(1 - \theta, 1)$, i.e., they lie below the plane $y = 1$. Points $A_2(h)$, $C_2(h)$, $E_2(h)$ and $G_2(h)$ have weight 1; points $B_2(h)$, $D_2(h)$, and $F_2(h)$ have weight θ. The points $A_2(h)$ and $C_2(h)$ are z-aligned to the connection points $H_2^i(h)$ and $H_2^j(h)$, respectively. Point $G_2(h)$ is y-aligned to connection point $H_2^k(h)$. Point $A_2(h)$ is y-blocked by blocker $b_{A_2(h)}$, point $E_2(h)$ is x-blocked by blocker $b_{E_2(h)}$, and point $G_2(h)$ is z-blocked by blocker $b_{G_2(h)}$. See Figure 1.

Among the monotone allocations in which the connection points $H_2^i(h)$ and $H_2^j(h)$ are allocated to players y or x, and $H_2^k(h)$ is allocated to players x or z (i.e., the clause gadget is non-breathing), the one that maximizes revenue should leave at least one of the points $A_2(h)$, $C_2(h)$, $E_2(h)$, or $G_2(h)$ unallocated. In this case, the total contribution of the unit-weight points to revenue is at least 3 and at most $3 + 3\theta$. In contrast, if some of the connection points (say $H_2^i(h)$) is not allocated to the players mentioned above, the contribution of $A_2(h)$, $C_2(h)$, $E_2(h)$, and $G_2(h)$ can increase to at least $4 - 2\theta$ and at most $4 + 2\theta$ by allocating $A_2(h)$ and $H_2^i(h)$ to player z, $C_2(h)$ and $B_2(h)$ to player y, $E_2(h)$ and $D_2(h)$ to player y, and $G_2(h)$ and $F_2(h)$ to x. The other cases (i.e., when $H_2^j(h)$ or $H_2^k(h)$ are not allocated to any player) have similar allocations of the same improved revenue.

The clause gadgets corresponding to clauses $e_3(h)$ and $e_4(h)$ are analogous. Observations about potential revenue similar to those for the clause gadget corresponding to clause $e_2(h)$ apply to these cases as well. An important property is that points in different clause gadgets are never aligned. This is achieved by dedicating a distinct xz-plane for the points in the the gadget associated with clause $e_1(h)$ and a distinct xz-plane and a distinct xy-plane for the points in each gadget associated with clause $e_2(h)$, $e_3(h)$, and $e_4(h)$, respectively. The clause gadgets corresponding to clauses of an equation $e(h') : v_i + v_j + v_k = 0 \pmod 2$ are symmetric and have identical properties.

We will show that the optimal revenue in $R(I)$ strongly depends on the maximum number of satisfied equations in I. Since approximating the second objective is hard, we will show that approximating the first objective is also hard. We exploit a particular type of monotone allocations.

Definition 1. *An allocation of instance $R(I)$ is called* simple *if for every connection point that is allocated to a player, unallocating it violates monotonicity.*

Note that the definition implies that a connection point is allocated to a player only if the allocation of its aligned variable point or clause gadget point also enforces it to be allocated to the same player. We will now explain a relation between simple allocations in $R(I)$ and assignments for I. First observe that, if the variable point v_i is not allocated, then the fact that the allocation is simple implies that all connection points aligned to v_i are not enforced by v_i and, hence, the clause gadgets corresponding to equations in which v_i participates are all breathing. Consider an equation $e(h) : v_i + v_j + v_k = \alpha \pmod 2$, one of its clauses $e_\ell(h)$, the corresponding clause gadget, and an allocation of variable points v_i, v_j, and v_k to players y and z. As mentioned above, we will associate the allocation of a variable point to player y (resp., to player z) with the assignment of value 1 (resp., 0) to its corresponding variable. Then, we can easily verify that the clause gadget associated with $e_\ell(h)$ is breathing if and only if the allocation of the variable points v_i, v_j, and v_k implies an assignment that satisfies clause $e_\ell(h)$. By Fact 1, either the four clause gadgets of $e(h)$ are breathing (if the implied assignment satisfies $e(h)$) or exactly three of them are breathing (if the implied assignment does not satisfy $e(h)$).

Furthermore, when accounting for the revenue of a simple allocation A, we will disregard the revenue obtained by connection points as well as non-blocker points in clause gadgets with weight θ; we will refer to such points as θ-weight points. We refer to the revenue obtained by the remaining points (i.e., variable points, blockers, and unit-weight points in clause gadgets) as *discounted revenue* $\mathtt{drev}(A)$.

Lemma 1. *Consider a simple allocation of maximum discounted revenue. If the four clause gadgets corresponding to an equation $e(h)$ are breathing, then their contribution to the discounted revenue is at least $26 + 15\theta + 11\theta^2$ and at most $26 + 30\theta + 11\theta^2$. If only three of them are breathing, the contribution of the four clause gadgets to the discounted revenue is at least $25 + 16\theta + 11\theta^2$ and at most $25 + 31\theta + 11\theta^2$.*

We will call a simple allocation in which all variable points are allocated to players y and z a *complete simple* allocation.

Lemma 2. *The simple allocation of maximum discounted revenue is complete.*

Lemma 3. *For every monotone allocation A with revenue $\mathtt{rev}(A)$, there is a complete simple allocation A' such that $\mathtt{drev}(A') \geq \mathtt{rev}(A) - 46m\theta$.*

Since θ has an extremely small value in our construction, it is clear that the optimal discounted revenue over complete simple allocations is a very good approximation of the optimal revenue over all monotone allocations. The proof of the next lemma exploits this observation.

Lemma 4. *If the maximum number of satisfied equations in I is K, then the revenue in the revenue-optimal monotone allocation of $R(I)$ is between $(28 - 300\theta)m + K$ and $(28 + 300\theta)m + K$.*

Proof. Let us consider an optimal assignment of values to the variables of I so that K linear equations are true. We construct a complete simple allocation A for $R(I)$ as follows. For every variable v_i that is set to 0 (resp., to 1), we allocate the variable point v_i to player z (resp., to player y). In this way, the variable points contribute $\sum_{i=1}^{n} d_i(\frac{1}{2} + 23\theta) \geq 3m/2$ to the discounted revenue of A. The blocker b is allocated to player x and contributes $3m(1 + \theta)(\frac{1}{2} + 23\theta) \geq 3m/2$ to the discounted revenue. Then, every clause gadget corresponding to a true (resp., false) clause is breathing (resp., non-breathing). The points in the clause gadgets are allocated so that their contribution to the discounted revenue is as high as possible. By Lemma 1, we have that the contribution of the four breathing clause gadgets associated with an equation to the discounted revenue is at least $26 + 15\theta + 11\theta^2 \geq 26$. For each unsatisfied equation, three of the corresponding clause gadgets are breathing and one is non-breathing. Hence, their contribution to the discounted revenue is at least $25 + 16\theta + 11\theta^2 \geq 25$. So, the total discounted revenue is at least

$$3m/2 + 3m/2 + 26K + 25(m - K) \geq (28 - 300\theta)m + K.$$

Clearly, the right-hand side of this inequality is a lower bound on the revenue of the revenue-optimal monotone allocation as well.

Now, consider a complete simple allocation of maximum discounted revenue and the assignment of values to the variables v_i this allocation implies. Consider the equations in which the four corresponding clause gadgets are all breathing. By our construction, this implies that the corresponding equations are satisfied by the assignment; so, there are at most K such quadruples and the remaining $m - K$ quadruples of clause gadgets will have three breathing and one non-breathing clause gadgets. In total, using Lemma 1, the discounted revenue of the complete simple allocation is at most

$$3m(\frac{1}{2} + 23\theta) + 3m(1 + \theta)(\frac{1}{2} + 23\theta) + K(26 + 30\theta + 11\theta^2) + (m - K)(25 + 31\theta + 11\theta^2)$$

$$< (28 + 251\theta)m + K.$$

Hence, by Lemma 3, the revenue-optimal monotone allocation has revenue at most $(28 + 300\theta)m + K$.   □

We are ready to prove our main result.

Theorem 1. *For every constant $\delta \in (0, 1/2)$, it is NP-hard to approximate* 3OPTIMALAUCTIONDESIGN *within a factor $\frac{57+\delta}{58-\delta}$.*

Proof. Let $\delta \in (0, 1/2)$, $\eta = \delta/3$ and $\theta = \delta/1800$. Since, given I, it is NP-hard to distinguish cases where the maximum number of satisfied equations is at least $(1 - \eta)m$ or at most $(1/2 + \eta)m$ [9], Lemma 4 implies that it is NP-hard to distinguish between cases where the maximum revenue among all monotone allocations of $R(I)$ is at least $(28 - 300\theta)m + (1 - \eta)m = (58 - \delta)m/2$ and at most $(28 + 300\theta)m + (1/2 + \eta)m = (57 + \delta)m/2$ is NP-hard as well.   □

4   Deterministic vs Randomized Auctions

We now present the upper bound on the revenue-gap between deterministic and randomized mechanisms.

Theorem 2. *The revenue obtained by the optimal deterministic truthful mechanism can be at most $\frac{13}{14}$ of the revenue obtained by the optimal truthful-in-expectation mechanism.*

Proof. Our construction consists of 22 points described in Table 1. The parameter θ is positive and arbitrarily small. Point $b(u)$ is a z-blocker for point u. The

Table 1. The construction in the proof of Theorem 2

Point	c_x	c_y	c_z	wgt	Point	c_x	c_y	c_z	wgt
q_1	1	$1+\theta$	1	θ	p_1	$1+\theta$	1	1	θ
q_2	1	$1+\theta$	$1-\theta$	1	p_2	$1+\theta$	$1-\theta$	1	1
$b(q_2)$	1	$(1+\theta)^2/\theta$	$1-\theta$	θ	$b(p_2)$	$(1+\theta)^2/\theta$	$1-\theta$	1	θ
q_3	$1+\theta/3$	$1+\theta$	$1-\theta$	θ	p_3	$1+\theta$	$1-\theta$	$1+\theta/2$	θ
q_4	$1+\theta/3$	$1+\theta/2$	$1-\theta$	1	p_4	$1+2\theta/3$	$1-\theta$	$1+\theta/2$	1
$b(q_4)$	$(1+\theta)^2/\theta$	$1+\theta/2$	$1-\theta$	θ	$b(p_4)$	$1+2\theta/3$	$1-\theta$	$(1+\theta)^2/\theta$	θ
q_5	$1+\theta/3$	$1+\theta/2$	$1-\theta/2$	θ	p_5	$1+2\theta/3$	$1-\theta/2$	$1+\theta/2$	θ
q_6	1	$1+\theta/2$	$1-\theta/2$	1	p_6	$1+2\theta/3$	$1-\theta/2$	1	1
$b(q_6)$	1	$(1+\theta)^2/\theta$	$1-\theta/2$	θ	$b(p_6)$	$(1+\theta)^2/\theta$	$1-\theta/2$	1	θ
q_7	1	$1+\theta/2$	1	θ	p_7	$1+2\theta/3$	1	1	θ
u	1	1	1	1	$b(u)$	1	1	$(1+\theta)^2/\theta$	θ

existence of $b(u)$ guarantees that in a revenue maximizing allocation $b(u)$ will be allocated to player z and point u should not be allocated to z. A similar observation applies for blockers $b(q_2)$, $b(q_4)$, $b(q_6)$, $b(p_2)$, $b(p_4)$, and $b(p_6)$ and their corresponding blocked points. Regarding the remaining points, we refer to the ones of weight 1 as heavy points, and to the ones of weight θ as light points.

Consider the randomized allocation in which the blockers are allocated to their preferred direction, the heavy points are allocated equiprobably between their two non-blocked directions, and the light points are allocated as follows: points q_1, q_7, and p_5 are allocated equiprobably between players y and z, points p_1, p_7, and q_3 are allocated equiprobably between players x and y, and points q_5 and p_3 are allocated equiprobably between players x and z. It can be easily verified that this is a monotone allocation with revenue at least $7(1+\theta)^2+(7+8\theta)(1-\theta) \geq 14$.

Now, let us examine the possible deterministic allocations. If u is not allocated the maximum revenue does not exceed $7(1+\theta)^2+(6+8\theta)(1+\theta) \leq 13+28\theta+15\theta^2$. Otherwise, if u is allocated to player y or z, some of the other heavy points can not be allocated to a non-blocked direction/player. To see why this is true, assume that u is allocated to player y (the case that u is allocated to player z is symmetric). Then, points q_2 and q_6 can only be allocated to player x in a monotone allocation, thus point q_4 can not be allocated to any of its non-blocked directions. Again, the maximum revenue does not exceed $13 + 28\theta + 15\theta^2$. The theorem follows since θ can take any arbitrarily small positive value. □

References

1. Chakrabarty, D., Goel, G.: On the approximability of budgeted allocations and improved lower bounds for submodular welfare maximization and GAP. In: Proceedings of the 49th Annual IEEE Symposium on Foundations of Computer Science (FOCS), pp. 687–696 (2008)
2. Chen, X., Hu, G., Lu, P., Wang, L.: On the approximation ratio of k-lookahead auction. In: Chen, N., Elkind, E., Koutsoupias, E. (eds.) Internet and Network Economics. LNCS, vol. 7090, pp. 61–71. Springer, Heidelberg (2011)
3. Constantin, F., Ito, T., Parkes, D.C.: Online Auctions for Bidders with Interdependent Values. In: Proceedings of the Sixth International Joint Conference on Autonomous Agents and Multiagent Systems (AAMAS), pp. 738–740 (2007)
4. Crémer, J., McLean, R.: Optimal selling strategies under uncertainty for a discriminating monopolist when demands are interdependent. Econometrica 53(2), 345–361 (1985)
5. Crémer, J., McLean, R.: Full extraction of the surplus in bayesian and dominant strategy auctions. Econometrica 56(6), 1247–1257 (1988)
6. Diakonikolas, I., Papadimitriou, C., Pierrakos, G., Singer, Y.: Efficiency-revenue trade-offs in auctions. In: Czumaj, A., Mehlhorn, K., Pitts, A., Wattenhofer, R. (eds.) ICALP 2012, Part II. LNCS, vol. 7392, pp. 488–499. Springer, Heidelberg (2012)
7. Dobzinski, S., Fu, H., Kleinberg, R.: Optimal auctions with correlated bidders are easy. In: Proceedings of the 43rd ACM Symposium on Theory of Computing (STOC), pp. 129–138 (2011)
8. Esö, P.: An optimal auction with correlated values and risk aversion. Journal of Economic Theory 125(1), 78–89 (2005)
9. Håstad, J.: Some optimal inapproximability results. Journal of the ACM 48(4), 798–859 (2001)
10. Krishna, V.: Auction Theory. Academic Press (2009)
11. Levin, D., Smith, J.: Optimal Reservation Prices in Auctions. The Economic Journal 106(438), 1271–1283 (1996)
12. Maskin, E.: Auctions and efficiency. In: Dewatripont, M., Hansen, L.P., Turnovsky, S.J. (eds.) Advances in Economics and Econometrics: Theory and Applications, Eighth World Congress (Econometric Society Monographs), vol. 3, pp. 1–24 (2003)
13. Milgrom, P., Weber, R.: A theory of auctions and competitive bidding. Econometrica 50(5), 1098–1122 (1982)
14. Myerson, R.: Optimal auction design. Mathematics of Operations Research 6(1), 58–73 (1981)
15. Nisan, N., Roughgarden, T., Tardos, É., Vazirani, V.V.: Algorithmic Game Theory. Cambridge University Press, New York (2007)
16. Papadimitriou, C., Pierrakos, G.: Optimal deterministic auctions with correlated priors. In: Proceedings of the 43rd ACM Symposium on Theory of Computing (STOC), pp. 119–128 (2011)
17. Ronen, A.: On approximating optimal auctions. In: Proceedings of the 3rd ACM Conference on Electronic Commerce (EC), pp. 11–17 (2001)
18. Ronen, A., Saberi, A.: On the hardness of optimal auctions. In: Proceedings of the 43rd Symposium on Foundations of Computer Science (FOCS), pp. 396–405 (2002)

Connectivity Inference in Mass Spectrometry Based Structure Determination[*]

Deepesh Agarwal[1], Julio-Cesar Silva Araujo[1,2], Christelle Caillouet[1,2], Frederic Cazals[1,**], David Coudert[1,2], and Stephane Pérennes[1,2]

[1] INRIA Sophia-Antipolis - Méditerranée
[2] Univ. Nice Sophia Antipolis, CNRS, I3S, UMR 7271,
06900 Sophia Antipolis, France

Abstract. We consider the following MINIMUM CONNECTIVITY INFERENCE problem (MCI), which arises in structural biology: given vertex sets $V_i \subseteq V, i \in I$, find a graph $G = (V, E)$ minimizing the size of the edge set E, such that the sub-graph of G induced by each V_i is connected. This problem arises in structural biology, when one aims at finding the pairwise contacts between the proteins of a protein assembly, given the lists of proteins involved in sub-complexes. We present four contributions.

First, using a reduction of the set cover problem, we establish that the MCI problem is APX-hard. Second, we show how to solve the problem to optimality using a mixed integer linear programming formulation (MILP). Third, we develop a greedy algorithm based on union-find data structures (Greedy), yielding a $2(\log_2 |V| + \log_2 \kappa)$-approximation, with κ the maximum number of subsets V_i a vertex belongs to. Fourth, application-wise, we use the MILP and the greedy heuristic to solve the aforementioned connectivity inference problem in structural biology. We show that the solutions of MILP and Greedy are more parsimonious with respect to edges than those reported by the algorithm initially developed in biophysics, which are not qualified in terms of optimality. Since MILP outputs a set of optimal solutions, we introduce the notion of *consensus solution*. Using assemblies whose pairwise contacts are known exhaustively, we show an almost perfect agreement between the contacts predicted by our algorithms and the experimentally determined ones, especially for consensus solutions.

1 Introduction

1.1 Connectivity Inference for Macro-molecular Assemblies

Macro-molecular Assemblies. Building models of macro-molecular machines is a key endeavor of biophysics, as such models not only unravel fundamental mechanisms of life, but also offer the possibility to monitor and to fix defaulting

---

[*] This work has been partially supported by ANR AGAPE and CNPq-Brazil 202049/2012-4.

[**] Correspondence to Frederic.Cazals@inria.fr or to David.Coudert@inria.fr

H.L. Bodlaender and G.F. Italiano (Eds.): ESA 2013, LNCS 8125, pp. 289–300, 2013.

systems. Example of such machines are the eukaryotic initiation factors which initiate protein synthesis by the ribosome, the ribosome which performs the synthesis of a polypeptide chain encoded in a messenger RNA derived from a gene, chaperonins which help proteins to adopt their 3D structure, the proteasome which carries out the elimination of damaged or misfolded proteins, etc. These macro-molecular assemblies involve from tens to hundreds of molecules, and range in size from a few tens of Angstroms (the size of one atom) up to 100 nanometers.

But if atomic resolution models of small assemblies are typically obtained with X-ray crystallography and/or nuclear magnetic resonance, large assemblies are not, in general, amenable to such studies. Instead, their reconstruction by *data integration* requires mixing a panel of complementary experimental data [4]. In particular, information on the hierarchical structure of an assembly, namely its decomposition into sub-complexes (complexes for short in the sequel) which themselves decompose into isolated molecules (proteins or nucleic acids) can be obtained from mass spectrometry.

Mass Spectrometry. Mass spectrometry (MS) is an analytical technique allowing the measurement of the mass-to-charge (m/z) ratio of molecules [22], based on three devices, namely a source to produce ions from samples in solution, an analyzer separating them according to their m/z ratio, and a detector to count them. The process results in a m/z spectrum, whose deconvolution yields a mass spectrum, i.e. an histogram recording the abundance of the various complexes as a function of their mass. Considering this spectrum as raw data, two mathematical questions need to be solved. The first one, known as stoichiometry determination (SD), consists of inferring how many copies of the individual molecules are needed to account for the mass of a mode of the spectrum [2,6]. The second one, known as connectivity inference, aims at finding the most plausible connectivity of the molecules involved in a solution of the SD problem.

Connectivity Inference. Given a macro-molecular assembly whose individual molecules (proteins or nucleic acids) are known, we aim at inferring the connectivity between these molecules. In other words, we are given the vertices of a graph, and we wish to figure out the edges it should have. To constrain the problem, we assume that the composition, in terms of individual molecules, of selected complexes of the assembly is known. Mathematically, this means that the vertex sets of selected *connected subgraphs* of the graph sought are known. To see where this information comes from, recall that a given assembly can be chemically denatured i.e. split into complexes by manipulating the chemical conditions prior to ionization. In extreme conditions, complete denaturation occurs, so that the individual molecules can be identified using MS. In milder conditions, multiple overlapping complexes are generated: once the masses of the proteins are known, the list of proteins in each such complex is determined by solving the aforementioned SD problem [20]. As a final comment, it should be noticed that in inferring the connectivity, *smallest-size networks* (i.e. graphs with as few edges as possible) are sought [3,25]. Indeed, due to volume exclusion constraints, a given protein cannot contact all the remaining ones, so that the minimal connectivity assumption avoids speculating on the exact (unknown) number of contacts.

Mathematical Model. Let $G = (V, E)$ be a graph, where V is the set of vertices and E the set of edges. We denote $G[V']$, respectively $G[E']$, the subgraph of G induced by $V' \subseteq V$, resp. by $E' \subseteq E$.

Consider an assembly together with the list of constituting proteins, as well as a list of associated complexes. Prosaically, we associate to each protein a vertex $v \in V$ and to each complex $i \in I \subseteq \mathbb{N}$ a subset $V_i \subseteq V$, such that if the protein v belongs to the complex i, then $v \in V_i$. Our goal is to infer the connectivity inside each complex of proteins. Therefore, we need to select a set of edges E_i between the vertices of V_i such that the graph $G_i = (V_i, E_i)$ is connected. The MINIMUM CONNECTIVITY INFERENCE problem is to find a graph $G = (V, E)$ with minimum cardinality set of edges E such that the subgraph $G[V_i]$ induced by each V_i, $i \in I$, is connected. Formally, we state the problem as follows.

Definition 1 (Minimum Connectivity Inference problem, MCI).

Inputs: *A set V of n vertices (proteins) and a set of subsets (complexes)*
$C = \{V_i \mid V_i \subseteq V \text{ and } i \in I\}$.
Constraint: *A set E of edges is feasible if $G[V_i] \subseteq G = (V, E)$ is connected, for every $i \in I$.*
Output: *A feasible set of edges E with minimum cardinality.*

Related Work. The connectivity inference problem was first addressed in [25] using a two-stage algorithm, called *network inference* (NI in the sequel). First, random graphs meeting the connectivity constraint are generated, by incrementally adding edges at random. Second, a genetic algorithm is used to reduce the number of edges, and also boost the diversity of the connectivity. Once the average size of the graphs stabilizes, the pool of graphs is analyzed to spot highly conserved edges.

From the Computer Science point of view, MCI is a network design problem in which one wants to choose a set of edges with minimum cost to connect entities (e.g., routers, antennas, etc.) subject to particular connectivity constraints. Typical examples of such constraints are that the subgraph must be k-connected, possibly with minimum degree or maximum diameter requirements (see [19] for a survey). Such network design problems are generally hard to solve. To the best of our knowledge, the problem of ensuring the connectivity for different subsets of nodes has not been addressed before.

2   Preliminaries and Hardness

2.1   Simplifying an Instance of MCI: Reduction Rules

Let (V, C) be an instance of MCI. We denote by $(V, C) \setminus u$ the instance (V', E') of MCI obtained from (V, C) by removing u from V and all the subsets of C it belongs to. So we have $V' = V \setminus \{u\}$ and $C' = \{V_i \setminus \{u\} \mid V_i \in C \text{ and } i \in I\}$. Moreover, we denote by $OPT((V, C))$ the cardinality of an optimal solution of MCI for the instance (V, C). Let us now denote $C(v) = \{i \mid V_i \ni v\} \subseteq I$, the set of complexes of the protein $v \in V$. We observe that we can apply the following reduction rules to any instance of MCI:

Lemma 1 (Reduction Rules). *Let (V, C) be an instance of MCI.*

1. *If $V_i \in C$ is such that $|V_i| = 1$, then any feasible solution for $(V, C \setminus V_i)$ is also feasible for (V, C), and we have $OPT((V, C \setminus V_i)) = OPT((V, C))$;*
2. *If $C(u) \subseteq C(v)$, for some $u, v \in V$, then a feasible solution for (V, C) is obtained from a feasible solution for $(V, C) \setminus u$ by adding the edge uv, and we have $OPT((V, C)) = OPT((V, C) \setminus u) + 1$;*

The proof is provided in the technical report [1].

By applying Lemma 1, we conclude that we can reduce the input instances of MCI to instances where every subset V_i has at least two vertices, every vertex appears in at least two subsets V_i and V_j with $i \neq j$, and the sets $C(u)$ and $C(v)$ are different, for any two vertices u and v.

2.2   Hardness

We establish that MCI is APX-hard, by showing a reduction of the SET COVER problem. The SET COVER problem is defined as follows:

Definition 2 (Set Cover problem).

Inputs: *a ground set $\mathcal{X} = \{x_1, \ldots, x_m\}$, a collection $\mathcal{F} = \{X_i \subseteq \mathcal{X}, i \in I\}$ and a positive integer k.*
Question: *does there exist $J \subseteq I$ such that $\bigcup_{i \in J} X_i = X$ and $|J| \leq k$?*

It is well-known that the SET COVER problem is NP-complete [13] and that this problem cannot be approximated in polynomial-time by a factor of $\ln n$, unless $P = NP$ [5, 17]. In order to prove our NP-completeness result, let us formally define the decision version of MCI as:

Definition 3 (Decision version of the Connectivity Inference problem, CI).

Inputs: *A set of vertices V, a set of subsets $C = \{V_i \mid V_i \subset V \text{ and } i \in I\}$ and a positive integer k.*
Constraint: *A set E of edges is feasible if $G[V_i] \subseteq G = (V, E)$ is connected, for every $i \in I$.*
Question: *Does there exists a feasible set E of edges such that $|E| \leq k$?*

Theorem 1. *The decision version of the CONNECTIVITY INFERENCE problem is NP-complete.*

The proof is provided in the technical report [1].

From the reduction used in the proof of Theorem 1 and the previous results on SET COVER problem [5, 17], we conclude that MCI is APX-hard:

Corollary 1. *There exists a constant $\mu > 0$ such that approximating MCI within $1 + \mu$ is NP-hard.*

3 Solving the Problem to Optimality Using Mixed Integer Linear Programming

3.1 Flow Based Formulation

To solve an instance (V, C) of the MCI problem, we introduce one binary variable y_e for each edge $e = uv$ of the undirected complete graph on $|V|$ vertices $K_{|V|}$, to determine whether edge e is selected in the solution. Thus, the objective function consists of minimizing the sum of the y variables, as specified by Eq. (1). To solve this problem, we form the directed graph $D = (V, A)$ in which each edge $e = uv$ of the complete graph $K_{|V|}$ is replaced by two directed arcs (u, v) and (v, u). The solution using MILP satisfies the following constraints:

▷ *Connectivity constraints.* To enforce the connectivity of each complex, we select one vertex s_i per subset $V_i \in C$ as the source of a flow that must reach all other vertices in V_i using only arcs in $D[V_i]$. We introduce continuous variables $f_a^i \in \mathbb{R}^+$ to express the quantity of flow originating from s_i and circulating along the arc $a = (u, v)$ (i.e. from node u to v), with $u, v \in V_i$. Constraint (2), the flow conservation constraint of Eq. (2), expresses that $|V_i| - 1$ units of flow are sent from s_i, and each vertex u_i collects 1 unit of flow from s_i and forwards the excess it has received from s_i to its neighbors in $D[V_i]$.

▷ *Capacity constraints.* We also introduce a continuous variable $x_a \in [0, 1]$, with $a = (u, v) \in A$ and $u, v \in V$, that is strictly positive if arc a carries some flow and 0 otherwise. In other words, no flow can use arc a when $x_a = 0$ as ensured by Constraint (3).

▷ *Symmetry constraints.* If there is some flow on arc (u, v) or (v, u) in D, then variable x is strictly positive and so the corresponding edge uv must be selected in the solution, meaning that $y_e = 1$, as ensured by Constraints (4) and (5).

Denoting $\mathcal{E}$ the edges of the complete graph $K_{|V|}$, and $A_i^+(u)$ (resp. $A_i^-(u)$) the subset of arcs of $D[V_i]$ entering (resp. leaving) node u, the formulation reads as:

$$\min \quad \sum_{e \in \mathcal{E}} y_e \tag{1}$$

$$\text{s.t.} \quad \sum_{a \in A_i^+(u)} f_a^i - \sum_{a \in A_i^-(u)} f_a^i = \begin{cases} |V_i| - 1 & \text{if } u = s_i \\ -1 & \text{if } u \neq s_i \end{cases} \quad \forall u \in V_i,\ V_i \in C \tag{2}$$

$$f_a^i \leq |V_i| \cdot x_a, \qquad\qquad \forall V_i \in C, a \in A \tag{3}$$

$$x_{(u,v)} \leq y_{uv}, \qquad\qquad \forall uv \in \mathcal{E} \tag{4}$$

$$x_{(v,u)} \leq y_{uv}, \qquad\qquad \forall uv \in \mathcal{E} \tag{5}$$

Observe that this formulation can be turned into a decision formulation, by removing the objective and adding the constraint of Eq. (6). If the formulation becomes infeasible, the optimal solution as more than k edges.

$$\sum_{e \in \mathcal{E}} y_e \leq k \quad (6) \qquad\qquad \sum_{e \in E_\ell} y_e < k \quad \forall E_\ell \in \mathcal{S} \tag{7}$$

Moreover, we can use the decision formulation to enumerate all feasible solutions for an instance (V, C, k). To do so, we use Constraints (7), where $\mathcal{S}$ is the set of feasible solutions that have already been found. This constraint prevents finding twice a solution. We first set $\mathcal{S} = \emptyset$, then we add it to all newly found solutions and repeat until the problem becomes infeasible for a solution of size k.

3.2 Implementation

The formulation has been implemented using **IBM CPLEX** solver 12.1, the corresponding software being named MILP in the sequel. Starting from the complete graph of size $|V|$, MILP allows one to compute one optimal solution, or the set of all solutions involving at most $OPT + k$ edges. For $k = 0$, one gets the set of all optimal solutions, denoted $\mathcal{S}_{MILP}$ in the sequel.

4 Approximate Solution Based on a Greedy Algorithm

4.1 Design and Properties

We now propose a greedy algorithm for MCI. Starting from the empty graph $G^0 = (V, E^0 = \emptyset)$, Algorithm 1 iteratively builds a graph $G^t = (V, E^t)$, with $E^t = E^{t-1} \cup \{e^t\}$. The edge $e^t = uv$ chosen at step t aims at reducing the number of connected components in the induced subgraphs $G^{t-1}[V_i]$ of G^{t-1}, for $i \in C(u) \cap C(v)$. More formally, at step t, we choose an edge e^t maximizing $m_t(e = uv)$ among all pairs $u, v \in V$, with $m_t(e = uv)$ the number of complexes containing u and v, and such that u and v belong to two different connected components of $G^{t-1}[V_i]$. The quantity $m_t(e = uv)$ is called the *priority* of the edge e.

Algorithm 1. Greedy algorithm for MCI

Require: $V = \{v_1, \ldots, v_n\}$ and $C = \{V_i \mid V_i \subseteq V \text{ and } i \in I\}$.
Ensure: A set E of edges such that $G[V_i] \subseteq G = (V, E)$ is connected, for every $i \in I$.
1: $t := 1$, $E^0 := \emptyset$
2: **while** there exists a disconnected graph $G^{t-1}[V_i]$, for some $i \in I$ **do**
3:     Find edge e^t maximizing the priority $m_t(e)$
4:     $E^t := E^{t-1} \cup \{e^t\}$ and $t := t + 1$
5: **return** E_{t-1}

Proposition 1. *Algorithm 1 is a $2(\log_2 |V| + \log_2 \kappa)$-approximation algorithm for MCI, with κ being the maximum number of subsets V_i a vertex belongs to.*

The proof is provided in the technical report [1].

Proposition 2. *When $\max_{v \in V} |C(v)| = 2$, Algorithm 1 always returns an optimal solution.*

The proof is provided in the technical report [1].

4.2   Implementation

In the following, we sketch an implementation of Algorithm 1, denoted `Greedy` in the sequel, which does not scan every candidate edge in E_t to find the (or a) best one, but instead maintains the priorities of all candidate edges.

Consider the following data structures:

- a priority queue Q associating to each candidate edge e its priority defined by $m_t(e)$. Note that the initial priority is given by $m_0(e = uv) = |C(u) \cap C(v)|$.
- a union-find data structure UF_i used to maintain the connected components of the induced graph $G^t[V_i]$. We assume in particular the existence of a function Find_vertices() such that UF_i.Find_vertices(u) returns the vertices of the connected component of the graph $G^t[V_i]$ containing the vertex u.

Upon popping the edge $e^t = (u, v)$ from Q, the following updates take place:

Update of the Priority Queue Q. For each complex V_i such that e^t triggers a merge between two connected components of $G^t[V_i]$, consider the two sets of vertices associated to these components, namely $K_{i,u} = UF_i$.Find_vertices(u) and $K_{i,v} = UF_i$.Find_vertices(v). The priority of all edges in the set $K_{i,u} \times K_{i,v} \setminus \{e^t\}$ is decreased by one unit.

Update of the Union-Find Data Structures. For each complex V_i such that e^t triggers a merge between two connected components of $G^{t-1}[V_i]$, the union operation UF_i.Union(UF_i.Find(u), UF_i.Find(v),) is performed.

It should be noticed that up to the logarithmic factor involved in the maintenance of Q, and up to the factor involving the inverse of Ackermann's function to run the union and find operations [24], the update complexity is output sensitive in the number of candidate edges affected in $K_{i,u} \times K_{i,v}$.

5   Experimental Results

5.1   Test Set: Assemblies of Interest

We selected three assemblies investigated by mass spectrometry, as explained in Section 1, for which we also found reference contacts between pairs of constituting proteins, against which to compare the output of our algorithms.

As explained in the appendix of [1], we classified all collected contacts into three sets, namely crystal contacts (set C_{Xtal}) observed in high resolution crystal structures, cross-linking contacts (set C_{XL}), obtained by so-called cross-linking experiments, and miscellaneous dimers (set C_{Dim}) obtained by various biophysical experiments. In case the crystal contacts are not available, we define the reference set of contacts as $C_{Exp} = C_{Dim} \cup C_{XL}$. The three systems we selected are:

▷**Yeast Exosome.** The exosome is a 3'- 5' exonuclease assembly involved in RNA processing and degradation, composed of 10 different protein types with unit stoichiometry [10]. A total of 21 complexes were determined by mass spectrometry. (See also the appendix of [1].)

▷**Yeast 19S Proteasome lid.** Proteasomes are assemblies involved in the elimination of damaged or misfolded proteins, and the degradation of short-lived regulatory proteins. The most common form of proteasome is the 26S, which involves two filtering caps (the 19S), each cap involving a peripheral lid, composed of 9 distinct protein types each with unit stoichiometry [21]. A total of 14 complexes were determined by mass spectrometry. (See also the appendix of [1].)

▷**Eukaryotic Translation Factor eIF3.** Eukaryotic initiation factors (eIF) are proteins involved in the initiation phase of the eukaryotic translation. They form a complex with the 40S ribosomal subunit, initiating the ribosomal scanning of mRNA. Among them, eIF3 consists of 13 different protein types each with unit stoichiometry [26]. A total of 27 complexes were determined by mass spectrometry. (See also the appendix of [1].)

5.2   Assessment Method

Let $\mathcal{S}_{MILP}$ be the set of optimal solutions returned by MILP, and let S_{NI} and S_G be the solutions computed by the algorithms NI [25] and Greedy respectively.

Consider an ensemble of solutions $\mathcal{S}$. The *size of a solution* $S \in \mathcal{S}$, denoted $|S|$, is its number of contacts. The *precision* of a solution S w.r.t. a reference set of contacts C is defined as the size of the intersection, i.e. $P_{\mathrm{MILP};C}(S) = |S \cap C|$. The precision is maximum if $S \subset C$, in which case no predicted contact is a false positive. The notion of precision makes sense if the reference contacts are exhaustive, which is the case for the exosome (since a crystal structure is known) and for the proteasome lid (exhaustive list of cross-links). We summarize the precision of the ensemble of solutions $\mathcal{S}$, denoted $P_{\mathrm{MILP};C}(\mathcal{S})$, by the triple (min, median, max) of the precisions of the solutions $S \in \mathcal{S}$.

The *score of a contact* appearing in a solution is the number of solutions from $\mathcal{S}$ containing it, and its *signed score* is its score multiplied by ± 1 depending on whether it is a true or false positive w.r.t C. The *score of a solution* $S \in \mathcal{S}$ is the sum of the scores of its contacts. Finally, a *consensus solution* is a solution achieving the maximum score over S, the set of all such solutions being denoted $\mathcal{S}^{cons.}$. Note that the score of a solution is meant to single out the consensus solutions from a solution set $\mathcal{S}$, while the signed score is meant to assess the solutions in $\mathcal{S}$ w.r.t a reference set.

5.3   Results

Except for the analysis of Table 1, due to the lack of space, we focus on the exosome (Fig. 1).

▷**Parsimony and Precision.** It is first observed that on the three systems, the algorithms MILP and Greedy are more parsimonious than NI (Table 1). For example, on the exosome, 10 edges are used instead of 12. The precision is excellent ($\geq 80\%$) for the three algorithms on the two systems where the reference set of contacts is exhaustive (exosome and lid).

Table 1. Size and precision of solutions. First section of the table: assembly, number of protein types, and size of the reference set C; second and third sections: size and precision for the solution returned by the algorithms NI [25] and Greedy; fourth and fifth sections: size and precision of algorithm MILP, for the whole set of optimal solutions $\mathcal{S}_{MILP}$, and for consensus solutions $\mathcal{S}_{MILP}^{cons.}$. NB: **The assignment of contacts was done manually [26]; NC^*: assembly not connected.

Complex	#types	Ref. set C	$|C|$	$|S_{NI}|$	$P_{\text{MILP};C}(S_{NI})$	$|S_G|$	$P_{\text{MILP};C}(S_G)$	$|S_{MILP}|$	$|\mathcal{S}_{MILP}|$	$P_{\text{MILP};C}(\mathcal{S}_{MILP})$	$|\mathcal{S}_{MILP}^{cons}|$	$P_{\text{MILP};C}(\mathcal{S}_{MILP}^{cons})$
Exosome	10	C_{Xtal}	26	12	12(100%)	10	10 (100%)	10	1644	(7, 9, 10)	12	(8, 9, 10)
19S Lid	9	C_{Exp}	16	9 $(NC)^*$	7(77.8%)	10	8 (80%)	10	324	(6, 7, 10)	18	(8, 8, 10)
eIF3	13	C_{Exp}	19	17**	14 (82.3%)	14	9 (64.2%)	14	2160	(8, 9, 11)	432	(8, 9, 10)

▷**Contact Scores for $\mathcal{S}_{MILP}$ on the Exosome.** Two facts emerge (Fig. 1(A)). First, four ubiquitous contacts are observed, while the remaining ones vary in the frequency. Second, there are few false positive overall. An interesting case is (Rrp42, Rrp45), which has the 7th highest count. The two polypeptide chains Rrp42 and Rrp45 are found in 16 out of 21 complexes used as input, accompanied in all cases by Rrp41. Interestingly, the point of closest approach between Rrp42 and Rrp45 in the crystal structure is circa 24Å, and this gap is precisely filled by Rrp41. That is, these three chains behave like a *rigid body*. Further inspection of the structure and of the behavior of MILP on such patterns is needed to explain why the edge (Rrp42, Rrp45) is reported.

▷**Scores for Consensus Solutions on the Exosome.** It is first observed that 12 consensus solutions amidst 1644 optimal ones are observed (Table 1 and Fig. 1(B)). In moving from $\mathcal{S}_{MILP}$ to $\mathcal{S}_{MILP}^{cons.}$, the precision increases from $(7, 9, 10)$ to $(8, 9, 10)$ — as also seen by a Pearson correlation coefficient of -0.51 between the mean false positive count per score, and the score (of a solution).

▷**Overall Assessment.** The consensus solutions from MILP are more parsimonious than those form NI, and compare favorably in terms of precision.

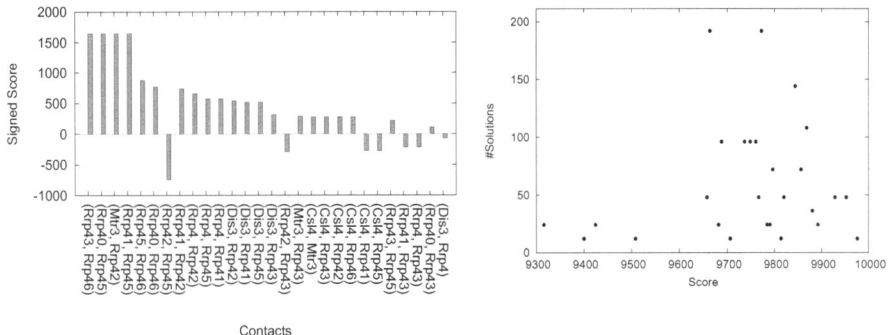

Fig. 1. Exosome (A) Signed scores for contacts in $\mathcal{S}_{MILP}$, w.r.t C_{Xtal} (B) Distribution of scores for solutions in $\mathcal{S}_{MILP}$

Fig. 2. Yeast Exosome: contacts computed by the algorithms. (A) Top and side view of the crystal structure [18, PDB 4IFD]. **(B,C,D)** Structure decorated with one edge per contact. the dash style reads as follows: *bold*: contacts in $S \cap C_{Xtal}$; *dotted*: contacts in S but not in C_{Xtal}; *dashed*: contact in C_{Xtal} but not in S (Note that only most prominent contacts in C_{Xtal} are shown to avoid cluttering). Note that a long edge i.e.

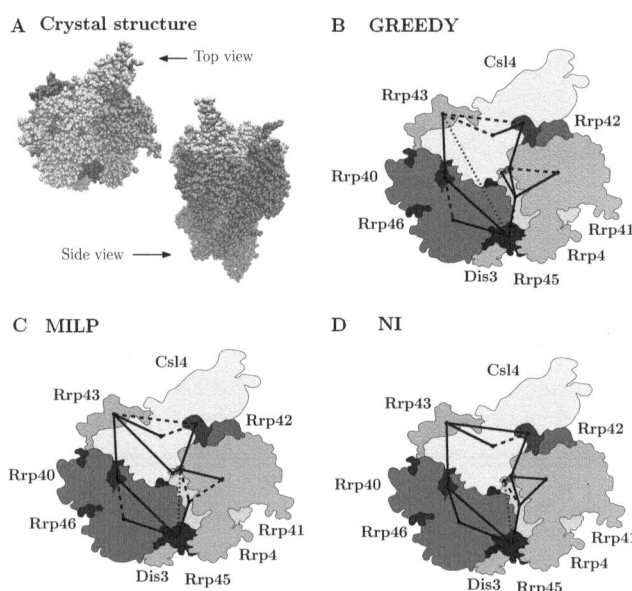

an edge between two subunits that appear distant on the top view of the assembly corresponds to a contact of these subunits located further down along the vertical direction. Also, note that part of the subunits Dis3 and Rrp42 are visible in the middle of the assembly and are trapped in between Csl4, Rrp40, Rrp41. The contact node therefore is placed there for convenience.

6   Conclusion and Outlook

A key endeavor of biophysics, for macro-molecular systems involving up to hundreds of molecules, is the determination of the pairwise contacts between these constituting molecules. The corresponding problem, known as connectivity inference, is central in mass-spectrometry based studies, which over the past five years, has proved crucial to investigate large assemblies. In this context, this paper presents a thorough study of the problem, encompassing its hardness, a greedy strategy, and a mixed integer programming algorithm. Application-wise, the key advantage of our methods w.r.t. the algorithm *network inference* developed in biophysics, is that we fully master all optimal solutions instead of a random collection of solutions which are not qualified w.r.t. the optimum. As shown by careful experiments on three assemblies recently scrutinized by other bio-physical experiments (exosome, proteasome lid, eIF3), our predictions are in excellent agreement with the experimental contacts. We therefore believe that our algorithms should leverage the interpretation of protein complexes obtained by mass spectrometry, a research vein currently undergoing major developments.

From a theoretical standpoint, a number of challenging problems deserve further work. The first one is to understand the solution space as a function of the number of input vertex sets and the structure of the unknown underlying graph. This problem is also related to the (output-sensitive) enumeration of connected subgraphs of a given graph. The second challenge is concerned with the generalization where the stoichiometry (the number of instances) of the proteins involved is more than one. In that case, complications arise since the connectivity information associated to the vertex sets of the connected subgraphs is related to protein types, while the connectivity sought is between protein instances. This extension would allow processing cases such as the nuclear pore complex, the biggest assembly known to date in eukaryotic cells, as it involves circa 450 protein instances of 30 different protein types, some of them present in 16 copies. The third one is of geometric flavor, and is concerned with the 3D embedding of the graph(s) generated. Since the nodes represent proteins and since two proteins must form a bio-physically valid interface if they touch at all, information on the shape of the proteins could be used to find plausible embeddings that would constrain the combinatorially valid solutions. This would be especially helpful to recover the edges which are known from experiments, but do not appear in exact or approximate solution to the minimal connectivity problem. Finally, the MINIMUM CONNECTIVITY INFERENCE problem also deserves investigation when the pool of candidate edges is a subset of the complete graph, which is especially relevant since pre-defined sets of edges may have been reported by a variety of experiments, some of them producing false positives.

References

1. Agarwal, D., Araujo, J., Caillouet, C., Cazals, F., Coudert, D., Pérennes, S.: Connectivity inference in mass spectrometry based structure determination. Technical Report 8320, Inria (2013), http://hal.inria.fr/hal-00837496
2. Agarwal, D., Cazals, F., Malod-Dognin, N.: Stoichiometry determination for mass-spectrometry data: the interval case (2013), http://hal.inria.fr/hal-00741491
3. Alber, F., Dokudovskaya, S., Veenhoff, L.M., Zhang, W., Kipper, J., Devos, D., Suprapto, A., Karni-Schmidt, O., Williams, R., Chait, B.T., Rout, M.P., Sali, A.: Determining the Architectures of Macromolecular Assemblies. Nature 450(7170), 683–694 (2007)
4. Alber, F., Förster, F., Korkin, D., Topf, M., Sali, A.: Integrating Diverse Data for Structure Determination of Macromolecular Assemblies. Ann. Rev. Biochem. 77, 11.1–11.35 (2008)
5. Alon, N., Moshkovitz, D., Safra, S.: Algorithmic construction of sets for k-restrictions. ACM Trans. Algorithms 2, 153–177 (2006)
6. Bocker, S., Liptak, Z.: A fast and simple algorithm for the money changing problem. Algorithmica 48(4), 413–432 (2007)
7. Cai, Q., Todorovic, A., Andaya, A., Gao, J., Leary, J.A., Cate, J.H.D.: Distinct regions of human eIF3 are sufficient for binding to the hcv ires and the 40s ribosomal subunit. Journal of Molecular Biology 403(2), 185–196 (2010)
8. Cazals, F., Proust, F., Bahadur, R., Janin, J.: Revisiting the Voronoi description of protein-protein interfaces. Protein Science 15(9), 2082–2092 (2006)
9. ElAntak, L., Tzakos, A.G., Locker, N., Lukavsky, P.J.: Structure of eIF3b rna recognition motif and its interaction with eIF3j. Journal of Biological Chemistry 282(11), 8165–8174 (2007)

10. Hernández, H., Dziembowski, A., Taverner, T., Séraphin, B., Robinson, C.V.: Subunit architecture of multimeric complexes isolated directly from cells. EMBO Reports 7(6), 605–610 (2006)
11. Janin, J., Bahadur, R.P., Chakrabarti, P.: Protein-protein interaction and quaternary structure. Quarterly Reviews of Biophysics 41(2), 133–180 (2008)
12. Kao, A., Randall, A., Yang, Y., Patel, V.R., Kandur, W., Guan, S., Rychnovsky, S.D., Baldi, P., Huang, L.: Mapping the structural topology of the yeast 19s proteasomal regulatory particle using chemical cross-linking and probabilistic modeling. Molecular & Cellular Proteomics (2012)
13. Karp, R.M.: Reducibility among combinatorial problems. In: Complexity of Computer Computations. The IBM Research Symposia Series, pp. 85–103. Plenum Press, New York (1972)
14. Lasker, K., Förster, F., Bohn, S., Walzthoeni, T., Villa, E., Unverdorben, P., Beck, F., Aebersold, R., Sali, A., Baumeister, W.: Molecular architecture of the 26s proteasome holocomplex determined by an integrative approach. Proceedings of the National Academy of Sciences 109(5), 1380–1387 (2012)
15. Levy, E., Erba, E.-B., Robinson, C., Teichmann, S.: Assembly reflects evolution of protein complexes. Nature 453(7199), 1262–1265 (2008)
16. Loriot, S., Cazals, F.: Modeling macro–molecular interfaces with intervor. Bioinformatics 26(7), 964–965 (2010)
17. Lund, C., Yannakakis, M.: On the hardness of approximating minimization problems. J. ACM 41, 960–981 (1994)
18. Makino, D.L., Baumgärtner, M., Conti, E.: Crystal structure of an rna-bound 11-subunit eukaryotic exosome complex. Nature 495(7439), 70–75 (2013)
19. Raghavan, S.: Formulations and algorithms for network design problems with connectivity requirements. PhD thesis, MIT, Cambridge, MA, USA (1995)
20. Sharon, M., Robinson, C.V.: The role of mass spectrometry in structure elucidation of dynamic protein complexes. Annu. Rev. Biochem. 76, 167–193 (2007)
21. Sharon, M., Taverner, T., Ambroggio, X.I., Deshaies, R.J., Robinson, C.V.: Structural organization of the 19s proteasome lid: insights from ms of intact complexes. PLoS Biology 4(8), e267 (2006)
22. Stengel, F., Aebersold, R., Robinson, C.V.: Joining forces: integrating proteomics and cross-linking with the mass spectrometry of intact complexes. Molecular & Cellular Proteomics 11(3) (2012)
23. Sun, C., Todorovic, A., Querol-Audí, J., Bai, Y., Villa, N., Snyder, M., Ashchyan, J., Lewis, C.S., Hartland, A., Gradia, S.: et al. Functional reconstitution of human eukaryotic translation initiation factor 3 (eIF3). Proceedings of the National Academy of Sciences 108(51), 20473–20478 (2011)
24. Tarjan, R.E.: Data Structures and Network Algorithms. CBMS-NSF Regional Conference Series in Applied Mathematics, vol. 44. Society for Industrial and Applied Mathematics, Philadelphia (1983)
25. Taverner, T., Hernández, H., Sharon, M., Ruotolo, B.T., Matak-Vinkovic, D., Devos, D., Russell, R.B., Robinson, C.V.: Subunit architecture of intact protein complexes from mass spectrometry and homology modeling. Accounts of Chemical Research 41(5), 617–627 (2008)
26. Zhou, M., Sandercock, A.M., Fraser, C.S., Ridlova, G., Stephens, E., Schenauer, M.R., Yokoi-Fong, T., Barsky, D., Leary, J.A., Hershey, J.W., Doudna, J.A., Robinson, C.V.: Mass spectrometry reveals modularity and a complete subunit interaction map of the eukaryotic translation factor eIF3. Proceedings of the National Academy of Sciences 105(47), 18139–18144 (2008)

Secluded Connectivity Problems

Shiri Chechik[1], Matthew P. Johnson[2,*], Merav Parter[3,**], and David Peleg[3]

[1] Microsoft Research Silicon Valley Center, USA
schechik@microsoft.com
[2] Department of Electrical Engineering, UCLA, Los Angeles, USA
mpjohnson@gmail.com
[3] The Weizmann Institute of Science, Rehovot, Israel
{merav.parter,david.peleg}@weizmann.ac.il[* * *]

Abstract. Consider a setting where possibly sensitive information sent over a path in a network is visible to every neighbor of (some node on) the path, thus including the nodes *on* the path itself. The *exposure* of a path P can be measured as the number of nodes adjacent to it, denoted by $N[P]$. A path is said to be *secluded* if its exposure is small. A similar measure can be applied to other connected subgraphs, such as Steiner trees connecting a given set of terminals. Such subgraphs may be relevant due to considerations of privacy, security or revenue maximization. This paper considers problems related to minimum exposure connectivity structures such as paths and Steiner trees. It is shown that on unweighted undirected n-node graphs, the problem of finding the minimum exposure path connecting a given pair of vertices is strongly inapproximable, i.e., hard to approximate within a factor of $O(2^{\log^{1-\epsilon} n})$ for any $\epsilon > 0$ (under an appropriate complexity assumption), but is approximable with ratio $\sqrt{\Delta} + 3$, where Δ is the maximum degree in the graph. One of our main results concerns the class of bounded-degree graphs, which is shown to exhibit the following interesting dichotomy. On the one hand, the minimum exposure path problem is NP-hard on *node-weighted* or *directed* bounded-degree graphs (even when the maximum degree is 4). On the other hand, we present a polynomial time algorithm (based on a nontrivial dynamic program) for the problem on unweighted undirected bounded-degree graphs. Likewise, the problem is

[*] The research of the second author was sponsored by the Army Research Laboratory and was accomplished under Cooperative Agreement Number W911NF-09-2-0053. The views and conclusions contained in this document are those of the authors and should not be interpreted as representing the official policies, either expressed or implied, of the Army Research Laboratory or the U.S. Government. The U.S. Government is authorized to reproduce and distribute reprints for Government purposes notwithstanding any copyright notation here on.
[**] Recipient of the Google Europe Fellowship in distributed computing; research supported in part by this Fellowship.
[* * *] Supported in part by the Israel Science Foundation (grant 894/09), United States-Israel Binational Science Foundation (grant 2008348), the I-CORE program of the Israel PBC and ISF (grant 4/11), Israel Ministry of Science and Technology (infrastructures grant), and Citi Foundation.

H.L. Bodlaender and G.F. Italiano (Eds.): ESA 2013, LNCS 8125, pp. 301–312, 2013.
© Springer-Verlag Berlin Heidelberg 2013

shown to be polynomial also for the class of (weighted or unweighted) bounded-treewidth graphs. Turning to the more general problem of finding a minimum exposure Steiner tree connecting a given set of k terminals, the picture becomes more involved. In undirected unweighted graphs with unbounded degree, we present an approximation algorithm with ratio $\min\{\Delta, n/k, \sqrt{2n}, O(\log k \cdot (k + \sqrt{\Delta}))\}$. On unweighted undirected bounded-degree graphs, the problem is still polynomial when the number of terminals is fixed, but if the number of terminals is arbitrary, then the problem becomes NP-hard again.

1   Introduction

The Problem. Consider a setting where possibly sensitive information sent over a path in a network is visible to every neighbor of (some node on) the path, thus including the nodes *on* the path itself. The *exposure* of a path P can be measured as the size (possibly node-weighted) of its *neighborhood* in this sense, denoted by $N[P]$. A path is said to be *secluded* if its exposure is small. A similar measure can be applied to other connected subgraphs, such as Steiner trees connecting a given set of terminals. Our interest is in finding connectivity structures with exposure as low as possible. This may be motivated by the fact that in real-life applications, a connectivity structure operates normally as part of the entire network G (and is not "extracted" from it), and so controlling the effect of its operation on the other nodes in the network may be of interest, in situations in which any "activation" of a node (by taking it as part of the structure) leads to an activation of its neighbors as well. In such settings, to minimize the set of total active nodes, we aim toward finding secluded or sufficiently private connectivity structures. Such subgraphs may be important in contexts where privacy is an important concern, or in settings where security measures must be installed on any node from which the information is visible, making it desirable to minimize their number. Another context where minimizing exposure may be desirable is when the information transferred among the participants has commercial value and overexposure to "free viewers" implies revenue loss.

This paper considers the problem of minimizing the exposure of subgraphs that satisfy some desired connectivity requirements. Two fundamental connectivity problems are considered, namely, single-path connectivity and Steiner trees, formulated as the Secluded Path and Secluded Steiner Tree problems, respectively, as follows. Given a graph $G = (V, E)$ and an s, t pair (respectively, a terminal set $\mathcal{S}$), it is required to find an $s - t$ path (respectively, a Steiner tree) of minimum exposure.

Related Work. The problems considered in this paper are variations of the classical shortest path and Steiner tree problems. In the standard versions of these problems, a cost measure is associated with edges or vertices, e.g., representing length or weight and the task is to identify a minimum cost subgraph satisfying the relevant connectivity requirement. Essentially, the cost of the solution subgraph is a *linear* sum of the solution's *constituent parts*, i.e., the sum of the weights of the edges or vertices chosen.

In contrast, in the setting of *labeled connectivity* problems, edges (and occasionally vertices) are associated with *labels* (or *colors*) and the objective is to identify a subgraph $G' \subseteq G$ that satisfies the connectivity requirements while minimizing the number of used labels. In other words, costs are now assigned to labels rather than to single edges. Such labeling schemes incorporate grouping constraints, based on partitioning the set of available edges into classes, each of which can be purchased in its entirety or not at all. These grouping constraints are motivated by applications from telecommunication networks, electrical networks, and multi-modal transportation networks. Labeled connectivity problems have been studies extensively from complexity-theoretic and algorithmic points of view [8,25,13,10]. The optimization problems in this category include, among others, the Minimum Labeled Path problem [13,25], the Minimum Labeled Spanning Tree problem [17,13], the Minimum Labeled Cut problem [26], and the Labelled Prefect Matching problem [22].

In both the traditional setting and the labeled connectivity setting, only edges or nodes that are explicitly part of the selected output structure are "paid for" in solution cost. That is, the cost of a candidate structure is a pure function of its components, ignoring the possible effects of "passive" participants, such as nodes that are "very close" to the structure in the input graph G. In contrast, in the setting considered in this paper, the cost of a connectivity structure G' is a function not only of its components but also of their immediate surroundings, namely, the manner in which G' is embedded in G plays a role as well. (Alternatively, we can say that the cost is a not necessarily a linear function of its components.)

A variant of the Secluded Path problem was recently introduced as the *Thinnest Path Problem* [11], where the focus was on directed hypergraph instances modeling transmission in wireless networks. In that application setting, each possible *transmission power* of a node yields a particular transmission range and hence a hyperedge directed from that node to the neighbors reached. The special case of that problem in which each node has a single possible transmission range is equivalent to our Secluded Path problem. They give a $\sqrt{\frac{n}{2}}$ approximation result for the DegCost algorithm (see Section 2) applied to this (more general) hypergraph setting.

The Secluded Path and Secluded Steiner Tree problems are related to several existing combinatorial optimization problems. These include the Red-Blue Set Cover problem [4,23], the Minimum Labeled Path problem [13,25] and the Steiner Tree [15] and Node Weighted Steiner Tree problems [16]. A prototypical example is the Red-Blue Set Cover problem, in which we are given a set R of red elements, a set B of blue elements and a family $S \subseteq 2^{|R| \cup |B|}$ of subsets of blue and red elements, and the objective is to find a subfamily $C \subseteq S$ covering all blue elements that minimizes the number of red elements covered. This problem is known to be strongly inapproximable.

Finally, turning to geometric settings, similarly motivated problems have been studied in the networking and sensor networks communities, where sensors are often modeled as unit disks. For example, the Maximal Breach Path problem [21] is defined in the context of traversing a region of the plane that contains sensor nodes at predetermined points, and its objective is to maximize the minimum distance between the points on the path and the sensor nodes. A dual problem studied extensively is *barrier coverage*, i.e., the (deterministic or stochastic) placement of sensors (see [18], [6]). Similarly motivated problems have been studied in the context of path planning in AI [14,19,20]. Although the motivation is similar, such problems are technically quite different from the graph-based problems studied here; those problems are typically posed in the geometric plane, amid obstacles that cause occlusion, and visibility is defined in terms of line-of-sight.

Contributions. In this paper, we introduce the concept of *secluded connectivity* and study some of its complexity and algorithmic aspects. We first state that the Secluded Path (and hence also Secluded Steiner Tree) problem is strongly inapproximable on unweighted undirected graphs with unbounded degree (more specifically, is hard to approximate with ratio $O(2^{\log^{1-\epsilon} n})$, where n is the number of nodes in the graph G, assuming $\mathcal{NP} \nsubseteq \mathcal{DTIME}\left(n^{\text{poly}\log n}\right)$). Conversely, we devise a $\sqrt{\Delta}+3$ approximation algorithm for the Secluded Path problem and a $\min\{\Delta, n/k, \sqrt{2n}, O(\log k \cdot (k+\sqrt{\Delta}))\}$ approximation algorithm for the Secluded Steiner Tree problem, where Δ is the maximum degree in the graph and k is the number of terminals.

One of our key results concerns bounded-degree graphs and reveals an interesting dichotomy. On the one hand, we show that Secluded Path is NP-hard on the class of *node-weighted* or *directed* bounded-degree graphs, even if the maximum degree is 4. In contrast, we show that on the class of unweighted undirected bounded-degree graphs, the Secluded Path problem admits an *exact* polynomial-time algorithm, which is based on a complex dynamic programming and requires some nontrivial analysis. Likewise, the Secluded Steiner Tree problem with fixed size terminal set is in P as well.

Finally, we consider some specific graph classes. We show that the Secluded Path and Secluded Steiner Tree problems are time polynomial for bounded-treewidth graphs. We also show that the Secluded Path (resp., Secluded Steiner Tree) problem can be approximated with ratio $O(1)$ (resp., $\Theta(\log k)$) in polynomial time for hereditary graph classes of bounded density. As an example, the Secluded Path problem has a 6 approximation on planar graphs. (A more careful direct analysis of the planar case yields ratio 3.)

2   Preliminaries and Notation

Consider a node-weighted graph $G(V, E, W)$, for some weight function $W : V \to \mathbb{R}_{\geq 0}$, with n nodes and maximum degree Δ. For a node $u \in V$, let $N(u) = \{v \in V \mid (u, v) \in E\}$ be the set of u's neighbors and let $N[u] = N(u) \cup \{u\}$ be u's *closed neighborhood*, i.e., including u itself. A *path* is a sequence $P = [u_1, \ldots, u_\ell]$, oriented from left to right, also termed a $u_1 - u_\ell$ path. Let $P[i] = u_i$ for $i \in \{1, \ldots, \ell\}$. Let $First(P) = u_1$ and $Last(P) = u_\ell$. For a path P and nodes x, y on it, let $P[x, y]$ be the subpath of P from x to y. For a connected subgraph

$G' \subseteq G$ and for $u_i, u_j \in V(G')$, let $\text{dist}_{G'}(u_i, u_j)$ be the distance between u_i and u_j in G'. Let $N(G') = \bigcup_{u \in G'} N(u) \setminus G'$ be the nodes that are strictly neighbors of G' nodes and $N[G'] = \bigcup_{u \in G'} N[u]$ be the set of nodes in the 1-neighborhood of G'. Define the cost of G' as

$$\text{Cost}(G') = \sum_{u \in N[G']} W(u) . \qquad (1)$$

Note that if G is unweighted, then the cost of a subgraph G' is simply the cardinality of the set of G' nodes and their neighbors, $\text{Cost}(G') = |N[G']|$.

We sometimes consider the neighbors of node $u \in V(G)$ in different subgraphs. To avoid confusion, we denote $N_{G'}(u)$ the neighbors of u restricted to graph G'.

For a subgraph $G' \subseteq G$, let $\text{DegCost}(G')$ denote the sum of the degrees of the nodes of G'. If G' is a path, then this key parameter is closely related to our problem. It is not hard to see that for any given path P, $\text{Cost}(P) \leq \text{DegCost}(P)$. The problem of finding an $s-t$ path P with minimum $\text{DegCost}(P)$ is polynomial, making it a convenient starting-point for various heuristics for the problem.

In this paper we consider two main connectivity problems. In the Secluded Path problem we are given an unweighted graph $G(V, E)$, a source node s and target node t, and the objective is to find an $s - t$ path P with minimum neighborhood size. A generalization of this problem is the Secluded Steiner Tree problem, in which instead of two terminals s and t we are given a set of k terminal nodes $\mathcal{S}$ and it is required to find a tree T in G covering $\mathcal{S}$, of minimum neighborhood size. If the given graph G is node-weighted, then the *weighted* Secluded Path and Secluded Steiner Tree problems require minimizing the neighborhood cost as given in Eq. (1). We now define these tasks formally. For an $s - t$ pair, let $\mathcal{P}_{s,t} = \{P \mid P \text{ is a s-t path}\}$ be the set of all $s - t$ paths and let $q^*_{s,t} = \min\{\text{Cost}(P) \mid P \in \mathcal{P}_{s,t}\}$ be the minimum cost among these paths. Then the objective of the Secluded Path problem is to find a path $P^* \in \mathcal{P}_{s,t}$ that attains this minimum, i.e., such that $\text{Cost}(P^*) = q^*_{s,t}$. For the Secluded Steiner Tree problem, let $\mathcal{T}(\mathcal{S}) = \{T \subseteq G \mid \mathcal{S} \subseteq V(T), T \text{ is a tree}\}$ be the set of all trees in G covering $\mathcal{S}$, and let $q^*(\mathcal{S})$ be the minimum cost among these trees, i.e.,

$$q^*(\mathcal{S}) = \min\{\text{Cost}(T) \mid T \in \mathcal{T}(\mathcal{S})\} . \qquad (2)$$

Then the solution for the problem is a tree $T^* \in \mathcal{T}(\mathcal{S})$ such that $\text{Cost}(T^*) = q^*(\mathcal{S})$. For paths P_1 and P_2, $P_1 \circ P_2$ denote the path obtained by concatenating P_2 to P_1.

3   Unweighted Undirected Graphs with Unbounded Degree

Hardness of Approximation

Theorem 1. *On unweighted undirected graphs with unbounded degree, the* Secluded Path *problem (and hence also the* Secluded Steiner Tree *problem)* *is strongly inapproximable. Specifically, unless $\mathcal{NP} \subseteq \mathcal{DTIME}(n^{poly\,\log(n)})$, the* Secluded Path *problem cannot be approximated to within a factor $O(2^{\log^{1-\epsilon} n})$ for any $\epsilon > 0$.*

Due to space limitation, missing proofs are provided in the full version of this paper [5].

Corollary 1. *The* Secluded Path *problem (and hence also the* Secluded Steiner Tree *problem) is strongly inapproximable in directed acyclic graphs.*

Approximation for the Secluded Path Problem

Theorem 2. *The* Secluded Path *problem in unweighted undirected graphs can be approximated within a ratio of $\sqrt{\Delta} + 3$.*

Proof: Given an instance of the Secluded Path problem, let P^* be an $s-t$ path that minimizes $\text{Cost}(P^*)$. Note that we may assume without loss of generality that for every node u in P^*, the only neighbors of u in G from among the nodes of $V(P^*)$ are the nodes adjacent to u in P^*. To see this, note that otherwise, if u had an edge to some neighbor $u' \in V(P^*)$ such that e is not on P^*, we could have shortened the path P^* (by replacing the subpath from u to u' with the edge e) and obtained a shorter path with at most the same cost as P^*. Recall that $\text{DegCost}(P)$ denotes the sum of the node degrees of the path P, and that $\text{Cost}(P) \leq \text{DegCost}(P)$ for any P. Recall also that the problem of finding an $s-t$ path P with minimum $\text{DegCost}(P)$ is polynomial. We claim that the algorithm that returns the path Q^* minimizing $\text{DegCost}(Q^*)$ yields a $(\sqrt{\Delta}+3)$ approximation ratio for the Secluded Path problem. In order to prove this, we show that there exists a path Q such that $\text{DegCost}(Q) \leq (\sqrt{\Delta} + 3) \cdot \text{Cost}(P^*)$. This implies that $\text{DegCost}(Q^*) \leq \text{DegCost}(Q) \leq (\sqrt{\Delta}+3) \cdot \text{Cost}(P^*)$, as required. The path Q is constructed by the following iterative process. Initially, all nodes are unmarked and we set $Q = P^*$. While there exists a node with more than $\sqrt{\Delta} + 3$ unmarked neighbors on the path Q, pick such a node x. Let y be the first (closest to s) neighbor of x in Q and let z be the last (closest to t) neighbor of x in Q. Replace the subpath $Q[y, z]$ in Q with the path $[y, x, z]$ and mark the node x. We now show that $\text{DegCost}(Q) \leq (\sqrt{\Delta} + 3) \cdot \text{Cost}(P^*)$. Let $X = \{x_1, \ldots x_\ell\}$ be the set of marked nodes in Q, where x_i is the node marked in iteration i. Note that $\ell = |X|$ is the number of iterations in the entire process. For the sake of analysis, partition $N[P^*]$ into two sets: S_1, those that have no unmarked neighbor in Q, and S_2, those that do. We claim that $|S_1| \geq \ell \cdot \sqrt{\Delta}$. For each $x_i \in X$, let $P(x_i)$ be the path that was replaced by the process of constructing Q in iteration i, and let $Q(x_i)$ be the path Q in the beginning of that iteration. Since x_i has more than $\sqrt{\Delta}+3$ unmarked neighbors in $Q(x_i)$, we get that $|P(x_i) \cap P^*| \geq \sqrt{\Delta}+4$. Let $P^-(x_i)$ be the path obtained by removing the first two nodes and last two nodes from $P(x_i)$. Note that $|P^-(x_i) \cap P^*| \geq \sqrt{\Delta}$. In addition, the sets $P^-(x_i) \cap P^*$ for $i = \{1, \ldots, \ell\}$ are pairwise disjoint, and moreover, none of the nodes in Q has a neighbor in $P^-(x_i) \cap P^*$. We thus get that $(P^-(x_i) \cap P^*) \subseteq S_1$. Note that $\text{Cost}(P^*) = |S_1| + |S_2|$ and that $\ell \leq |S_1|/\sqrt{\Delta}$.

We get that $\text{DegCost}(Q) \leq |S_2| \cdot (\sqrt{\Delta} + 3) + \ell \cdot \Delta \leq |S_2| \cdot (\sqrt{\Delta} + 3) + |S_1| \cdot \sqrt{\Delta} \leq (|S_2| + |S_1|) \cdot (\sqrt{\Delta} + 3) = \text{Cost}(P^*) \cdot (\sqrt{\Delta} + 3)$. ∎

In addition, for the Secluded Steiner Tree problem with k terminals, in the full version [5], we establish the following.

Theorem 3. *The* Secluded Steiner Tree *problem in unweighted undirected graphs can be approximated within a ratio of* $\min\{\Delta, n/k, \sqrt{2n}, O(\log k \cdot (k + \sqrt{\Delta}))\}$.

4   Bounded-Degree Graphs

Weighted/Directed Graphs. We first show that the Secluded Path problem (and thus also the Secluded Steiner Tree problem) is NP-hard on both directed graphs and weighted graphs, even if the maximum node degree is 4. We have the following.

Theorem 4. *The* Secluded Path *problem is NP-complete even for graphs of maximum degree 4 if they are either (a) node-weighted, or (b) directed.*

The Unweighted Undirected Case. In contrast, for unweighted undirected case of bounded-degree graphs, we show that the Secluded Path problem is solvable in polynomial time.

Theorem 5. *The* Secluded Path *problem is solvable in* $O(n^2 \cdot \Delta^{\Delta+1})$ *time on unweighted undirected graphs, hence in* polynomial *time for degree-bounded graphs.*

Note that in the previous section we showed that if the graph is either weighted or directed then the Secluded Path problem (i.e., the special case of Secluded Steiner Tree with two terminals) is NP-hard. In addition, it is noteworthy that the related problem of Minimum Labeled Path problem [25,13] is NP-hard even for unweighted planar graph with max degree 4 (this follows from a straightforward reduction from Vertex Cover).

We begin with notation and couple of key observations in this context. For two subpaths P_1, P_2, define their asymmetric difference as

$$\text{Diff}(P_1, P_2) = |N[P_1] \setminus N[P_2]| .$$

Observation 6. *(a) Diff$(P_1, P_2) \leq$ Diff(P_1, P') for every $P' \subseteq P_2$.*
(b) Cost$(P_1 \circ P_2) =$ Cost$(P_1) +$ Diff(P_2, P_1).

For a given path P, let $\text{dist}_P(s, u)$ be the distance in edges between s and u in the path P. Recall, that Δ is the maximum degree in graph G.

Observation 7. *Let u, v be two nodes in some optimal $s - t$ path P^* that share a common neighbor, i.e., $N(u) \cap N(v) \neq \emptyset$. Then $\text{dist}_{P^*}(u, v) \leq \Delta + 1$.*

Proof: Assume for contradiction that there exists an optimal $s - t$ path P^* such that $N(u) \cap N(v) \neq \emptyset$ where $u = P^*[i]$ and $v = P^*[j]$, $1 < i < j$ for $(i - j) \geq \Delta + 2$. Recall that due to the optimality of P^*, for every node u in P^*, the only neighbors of u in G from among the nodes of $V(P^*)$ are the nodes adjacent to u in P^* (otherwise the path can be shortcut). Let $w \in N(u) \cap N(v)$ be the mutual neighbor of u and v and consider the alternative $s - t$ path $\widehat{P}$ obtained from P^* by replacing the subpath $Q = P[i, \ldots, j]$ by the subpath $P' = [u, w, v]$. Let $Q^- = P[i + 2, \ldots, j - 2]$ be an length-ℓ' internal subpath of Q where $\ell' \geq \Delta - 1$. Then since the degree of w is at most Δ, it follows that

$$\text{Cost}(\widehat{P}) \leq N[V(P) \setminus V(Q^-)] + \Delta - 2 , \tag{3}$$

where $\Delta - 2$ is an upper bound on the number $w's$ neighbors other than u and v. In addition, note that by the optimality of P^* it contains no shortcut and hence $V(Q^-) \cap (N[V(P) \setminus V(Q^-)]) = \emptyset$ (i.e., for node $P[i]$ the only neighbors on the path P are $P[i - 1]$ and $P[i + 1]$). Thus,

$$\begin{aligned}
\text{Cost}(P^*) &\geq N[V(P) \setminus V(Q^-)] + |V(Q^-)| \\
&\geq N[V(P) \setminus V(Q^-)] + \Delta - 1 > \text{Cost}(\widehat{P}) ,
\end{aligned}$$

where the last inequality follows from Eq. (3), contradicting the optimality of P^*. The observation follows. ∎

In other words, the observation says that two vertices on the optimal path at distance $\Delta + 2$ (which is constant for bounded-degree graphs) or more have no common neighbors. This key observation is at the heart of our dynamic program, as it enables the necessary subproblem independence property. The difficulty is that this observation applies only to optimal paths, and so a delicate analysis is needed to justify why the dynamic program works. Note that the main difficulty of computing the optimal secluded path P^* is that the cost function $\text{Cost}(P^*)$ is not a linear function of path's components as in the related DegCost measure. Instead, the residual cost of the ith vertex in the path *depends* on the neighborhood of the length-$(i - 1)$ prefix of P^*. This dependency implies that the secluded path computation cannot be simply decomposed into independent subtasks. However, in contrast to suboptimal $s - t$ paths, the dependency (due to mutual neighbors) between the components of an optimal path is *limited* by the maximum degree Δ of the graph. The *limited dependency* exhibited by any $s - t'$ optimal path facilitates the correctness of the dynamic programming approach. Essentially, in the dynamic program, entries that correspond to subsolution σ of *any* optimal $s - t'$ path enjoy the limited dependency, and hence the values computed for these entries correspond to the *exact* cost of an optimal path that starts with s and ends with σ. In contrast, entries of subpaths σ that do not participate in any optimal $s - t'$ path, correspond to an *upper bound* on the cost of some path P that starts with s and ends with σ. This is due to the fact that the value of the entry is computed under the limited dependency assumption, and thus does not take into account the possible double counting of mutual neighbors between *distant* vertices in the path. Therefore, the possibly "falsified" entries cannot compete with the exact values, which are guaranteed to be computed for the entries that hold the subpaths of the optimal path. This informal intuition is formalized below.

For a path P of length $\ell \geq \Delta + 1$, let $\text{Suff}(P) = \langle v_{\ell-\Delta}, \ldots, v_\ell \rangle$ be the $(\Delta+1)$-suffix of P.

Lemma 1. *Let P^* be an optimal $s - t'$ path of length $\geq \Delta + 1$. Let $P^* = P_1 \circ P_2$ be some partition of P^* into two subpaths such that $|P_1| \geq \Delta + 1$. Then $\text{Diff}(P_2, P_1) = \text{Diff}(P_2, \text{Suff}(P_1))$.*

Proof: For ease of notation, let $\ell = |P_1|$, $P' = P_1[1, \ldots, \ell - (\Delta + 1)]$, and $\sigma = \text{Suff}(P_1)$. By definition, $\Delta(P_2, P_1) = |(N[P_2] \setminus N[\sigma]) \setminus N[P']|$. In the same manner, $\Delta(P_2, \sigma) = |(N[P_2] \setminus N[\sigma])|$. Assume for contradiction that the lemma does not hold, namely, $\Delta(P_2, \sigma) \neq \Delta(P_2, P' \circ \sigma)$. Then by Obs. 6(a) we have that $\Delta(P_2, \sigma) > \Delta(P_2, P' \circ \sigma)$. This implies that $N[P_2] \cap N[P'] \neq \emptyset$. Let $u \in N[P_2] \cap N[P']$. There are two cases to consider: (a) $u \in P'$, and (b) u has a neighbor v_1 in P'. We handle case (a) by further dividing it into two subcases: (a1) $u \in P_2$, i.e., u occurs at least twice in P^*, once in P' and once in P_2, and (a2) u has a neighbor v_2 in P_2. Note that in both subcases, there exists a shortcut of P^*, obtained in subcase (a1) by cutting the subpath between the two duplicates of u and in subcase (a2) by shortcutting from u to v_2. This shortcut results in a strictly lower cost path, in contradiction to the optimality of P^*. We proceed with case (b). Let $v_2 \in P_2$ be such that $u \in N[v_2]$. If $v_2 = u$, then clearly the path can be shortcut by going from $v_1 \in P'$ directly to $u \in P_2$, resulting in a lower cost path, in contradiction to the optimality of P^*. If $v_2 \neq u$, then $dist_{P^*}(v_1, v_2) \geq \Delta + 2$ (since $|\sigma| \geq \Delta + 1$) and $N[v_1] \cap N(v_2) \neq \emptyset$, which in contradiction to Obs. 7. The Lemma follows. ∎

Corollary 2. *Let P^* be an optimal $s - t'$ path of length $\geq \Delta + 1$. Let $P^* = P_1 \circ P_2$ be some partition of P^* into two subpaths such that $|P_1| \geq \Delta + 1$. Then $\text{Cost}(P^*) = \text{Cost}(P_1) + \text{Diff}(P_2, \text{Suff}(P_1))$.*

For clarity of representation, we describe a polynomial algorithm for the Secluded Path problem, i.e., where $\mathcal{S} = \{s, t\}$ and in addition $\Delta \leq 3$, in which case $\text{Suff}(P) = \langle v_{\ell-3}, \ldots, v_\ell \rangle$. The general case of $\Delta = O(1)$ is immediate by the description for the special case of $\Delta = 3$. The case of Secluded Steiner Tree with a fixed number of terminals $|\mathcal{S}| = O(1)$ is described in [5].

The algorithm we present is based on dynamic programming. For each $4 \leq i \leq n$, and every length-4 subpath given by a quartet of nodes $\sigma \subseteq V(G)$, it computes a length-i path $\pi(\sigma, i)$ that starts with s and ends with σ (if such exists) and an upper bound $f(\sigma, i)$ on the cost of this path, $f(\sigma, i) \geq \text{Cost}(\pi(\sigma, i))$. These values are computed inductively, using the values previously computed for other σ's and $i - 1$. In contrast to the general framework of dynamic programming, the interpretation of the computed values $f(\sigma, i)$ and $\pi(\sigma, i)$, namely, the relation between the dynamic programming values $f(\sigma, i)$ and $\pi(\sigma, i)$ and some "optimal" counterparts is more involved. In general, for arbitrary σ and i, the path $\pi(\sigma, i)$ is not guaranteed to be optimal in any sense and neither is its corresponding value $f(\sigma, i)$ (as $f(\sigma, i) \geq \text{Cost}(\pi(\sigma, i))$). However, quite interestingly, there is a subset of quartets for which a useful characterization of $f(\sigma, i)$ and $\pi(\sigma, i)$ can be established. Specifically, for every $4 \leq i \leq n$, there is a subclass of quartets Ψ_i^*, for which the computed values are in fact "optimal", in the sense that for every $\sigma \in \Psi_i^*$, the length-i path $\pi(\sigma, i)$ is of minimal cost among all other length-i paths that start with s and end with σ. We call such a path a *semi-optimal* path, since it is optimal only restricted to the length and specific suffix requirements. It turns out that for the special class of quartets Ψ_i^*, every semi-optimal path of $\sigma \in \Psi_i^*$ is also a prefix of some optimal $s - t'$ path. This property allows one to apply Cor. 2, which constitutes the key ingredient in our technique. In particular, it allows us to establish that $f(\sigma, i) = \text{Cost}(\pi(\sigma, i))$. This is sufficient for our purposes since the set $\bigcup_{i=1}^{n} \Psi_i^*$ contains *any* quartet that occurs in *some* optimal secluded $s - t$ path P^*. Specifically, for every $s - t$ optimal path P^*, it holds that the quartet $\text{Suff}(P^*) = P^*[i - 3, i]$ satisfies $\text{Suff}(P^*) \in \Psi_i^*$. The correctness of the dynamic programming is established by the fact that the values computed for quartets that occur in optimal paths in fact correspond to optimal values as required (despite the fact the these values are "useless" for other quartets).

The Algorithm. If the shortest path between s and t is less than 3, then the optimal Secluded Path can be found by an exhaustive search, so we assume throughout that $dist_G(s, t) \geq 3$. Let Ψ be the set of all length-4 subpaths σ in G. That is, for every $\sigma \in \Psi$, we have $V(\sigma) \subseteq V(G)$ and $(\sigma[i], \sigma[i + 1]) \in E(G)$ for every $i \in [1, 3]$. For $\sigma \in \Psi$, define the collection of *shifted successor* of G as $Next(\sigma) = \{\langle \sigma[2], \ldots, \sigma[4], u \rangle \mid u \in (N[\sigma[4]] \setminus \sigma)\}$. For every pair (σ, i), where $\sigma \in \Psi$ and $i \in \{1, \ldots, n\}$, the algorithm computes a value $f(\sigma, i)$ and length-i path $\pi(\sigma, i)$ ending with σ (i.e., $\pi(\sigma, i)[i - 3, \ldots, i] = \sigma$). These values are computed inductively. For $i = 4$, let $\pi(\sigma, 4) = \sigma$ and

$$f(\sigma, 4) = \begin{cases} \text{Cost}(\sigma), & \text{if } \sigma[1] = s \\ \infty, & \text{otherwise} \end{cases}$$

Once the algorithm has computed $f(\sigma, j)$ for every $4 \leq j \leq i - 1$ and every $\sigma \in \Psi$, in step $i \in \{5, n\}$ it computes

$$f(\sigma, i) = \min\{f(\sigma', i - 1) + \text{Diff}(\sigma, \sigma') \mid \sigma \in Next(\sigma')\}. \quad (4)$$

Note that

$$\text{Diff}(\sigma, \sigma') = \text{Diff}(\sigma[4], \sigma') = |N[\sigma[4]] \setminus N[\sigma']|. \quad (5)$$

Let $\sigma' \in \Psi$ such that $\sigma \in Next(\sigma')$ and σ' achieves the minimum value in Eq. (4). Then define $Pred(\sigma) = \sigma'$ and let

$$\pi(\sigma, i) = \pi(Pred(\sigma), i - 1) \circ \sigma[4]. \quad (6)$$

Let $q_{s,t}^* = \min\{f(\sigma, i) \mid i \in \{4, \ldots, n\}, Last(\pi(\sigma, i)) = t\}$, and set $P^* = \pi(\sigma^*, i^*)$ for σ^*, i^* such that $f(\sigma^*, i^*) = q_{s,t}^*$. Note that there are at most $O(n^2)$ entries $f(\sigma, i)$, each computed in constant time, and so the overall running time is $O(n^2)$. In [5], we provide a detailed analysis and establish Thm. 5. For the Secluded Steiner Tree problem, we show the following.

Theorem 8. *On unweighted undirected degree-bounded graphs, we have the following: (a) for arbitrary k, the* Secluded Steiner Tree *problem is NP-hard; (b) for $k = O(1)$, the* Secluded Steiner Tree *problem is* polynomial.

5   Secluded Connectivity for Specific Graph Families

Bounded-treewidth Graphs. For a graph $G(V, E)$, let $TW(G)$ denote the *treewidth* of G. For bounded treewidth graph, i.e. $TW(G) = O(1)$, we have the following.

Theorem 9. *The* Secluded Steiner Tree *problem (and hence also the* Secluded Path *problem) can be solved in* linear *time for graphs with fixed treewidth. In addition, given the tree decomposition of G, the* Secluded Steiner Tree *problem is solvable in $\widetilde{O}(n^3)$ if $TW(G) = O(\log n / \log \log n)$. This holds even for weighted and directed graphs.*

Bounded Density Graphs. Let $\text{DegCost}^*(\mathcal{S}) = \min\{\text{DegCost}(T) \mid T \in \mathcal{T}(\mathcal{S})\}$. In the full version [5], we show the following.

Proposition 1. *Let $\mathcal{G}$ be a hereditary class of graphs with a linear number of edges, i.e., a set of graphs such that for each $G \in \mathcal{G}$, $|E(G)| \leq \ell \cdot |V(G)|$ for some constant ℓ, and where $G \in \mathcal{G}$ implies that $G' \in \mathcal{G}$ for each subgraph G' of G. Then $\text{DegCost}^*(\mathcal{S}) \leq 2\ell \cdot q^*(\mathcal{S})$.*

An example of such a class of graphs is the family of planar graphs. For this family, the above proposition yields a 6-approximation for the Secluded Path problem. (A more careful direct analysis for planar graphs yields a 3-approximation.) Finally, note that whereas computing $\text{DegCost}^*(\mathcal{S})$ for a constant number of terminals k is polynomial, for arbitrary k, computing $\text{DegCost}^*(\mathcal{S})$ is NP-hard but can be approximated to within a ratio of $\Theta(\log k)$, see [5]. Thus, for the class of bounded density graphs, by Prop. 1, the Secluded Path problem has a constant ratio approximation and the Secluded Steiner Tree problem has an $\Theta(\log k)$ ratio approximation.

Theorem 10. *For the class of bounded-density graphs, the* Secluded Path *problem (respectively,* Secluded Steiner Tree *problem) is approximated within a ratio of $O(1)$ (respectively, $\Theta(\log k)$).*

Acknowledgment. We are grateful to Amotz Bar-Noy, Prithwish Basu and Michael Dinitz for helpful discussions.

References

1. Bodlaender, H.L.: A linear time algorithm for finding tree-decompositions of small treewidth. In: STOC, pp. 226–234 (1993)
2. Bodlaender, H.L.: A tourist guide through treewidth. Acta Cybern. 11, 1–22 (1993)

3. Bodlaender, H.L.: Treewidth: Structure and algorithms. In: Prencipe, G., Zaks, S. (eds.) SIROCCO 2007. LNCS, vol. 4474, pp. 11–25. Springer, Heidelberg (2007)

4. Carr, R.D., Doddi, S., Konjevod, G., Marathe, M.V.: On the red-blue set cover problem. In: SODA, pp. 345–353 (2000)

5. Chechik, S., Johnson, M.P., Parter, M., Peleg, D.: Secluded Connectivity Problems, http://arxiv.org/abs/1212.6176

6. Chen, A., Kumar, S., Lai, T.-H.: Local barrier coverage in wireless sensor networks. IEEE Tr. Mob. Comput. 9, 491–504 (2010)

7. Chimani, M., Mutzel, P., Zey, B.: Improved steiner tree algorithms for bounded treewidth. In: Iliopoulos, C.S., Smyth, W.F. (eds.) IWOCA 2011. LNCS, vol. 7056, pp. 374–386. Springer, Heidelberg (2011)

8. Dinur, I., Safra, S.: On the hardness of approximating label-cover. IPL 89, 247–254 (2004)

9. Feige, U.: A threshold of ln n for approximating set cover. J. ACM 45, 634–652 (1998)

10. Fellows, M.R., Guo, J., Kanj, I.A.: The parameterized complexity of some minimum label problems. JCSS 76, 727–740 (2010)

11. Gao, J., Zhao, Q., Swami, A.: The Thinnest Path Problem for Secure Communications: A Directed Hypergraph Approach. In: Proc. of the 50th Allerton Conference on Communications, Control, and Computing (2012)

12. Garey, M.R., Johnson, D.S.: The Rectilinear Steiner Tree Problem is NP-Complete. SIAM J. Appl. Math. 32, 826–834 (1977)

13. Hassin, R., Monnot, J., Segev, D.: Approximation algorithms and hardness results for labeled connectivity problems. J. Comb. Optim. 14, 437–453 (2007)

14. Johansson, A., Dell'Acqua, P.: Knowledge-based probability maps for covert pathfinding. In: Boulic, R., Chrysanthou, Y., Komura, T. (eds.) MIG 2010. LNCS, vol. 6459, pp. 339–350. Springer, Heidelberg (2010)

15. Karp, R.M.: Reducibility among combinatorial problems. In: Miller, R.E., Thatcher, J.W. (eds.) Complexity of Computer Computations, pp. 85–103. Plenum Press, NY (1972)

16. Klein, P.N., Ravi, R.: A nearly best-possible approximation algorithm for node-weighted steiner trees. J. Algo. 19, 104–115 (1995)

17. Krumke, S.O., Wirth, H.-C.: On the minimum label spanning tree problem. IPL 66, 81–85 (1998)

18. Liu, B., Dousse, O., Wang, J., Saipulla, A.: Strong barrier coverage of wireless sensor networks. In: MobiHoc, pp. 411–420 (2008)

19. Marzouqi, M., Jarvis, R.: New visibility-based path-planning approach for covert robotic navigation. Robotica 24, 759–773 (2006)

20. Marzouqi, M., Jarvis, R.: Robotic covert path planning: A survey. In: RAM, pp. 77–82 (2011)

21. Meguerdichian, S., Koushanfar, F., Potkonjak, M., Srivastava, M.B.: Coverage problems in wireless ad-hoc sensor networks. In: INFOCOM, pp. 1380–1387 (2001)

22. Monnot, J.: The labeled perfect matching in bipartite graphs. IPL 96, 81–88 (2005)

23. Peleg, D.: Approximation algorithms for the label-cover$_{max}$ and red-blue set cover problems. J. Discrete Algo. 5, 55–64 (2007)

24. Robertson, N., Seymour, P.D.: Graph minors. ii. algorithmic aspects of tree-width. J. Algo. 7, 309–322 (1986)

25. Yuan, S., Varma, S., Jue, J.P.: Minimum-color path problems for reliability in mesh networks. In: INFOCOM, pp. 2658–2669 (2005)

26. Zhang, P., Cai, J.Y., Tang, L., Zhao, W.: Approximation and hardness results for label cut and related problems. J. Comb. Optim. 21, 192–208 (2011)

List H-Coloring a Graph by Removing Few Vertices*

Rajesh Chitnis[1,**], László Egri[2], and Dániel Marx[2]

[1] Department of Computer Science, University of Maryland at College Park, USA
rchitnis@cs.umd.edu
[2] Computer and Automation Research Institute, Hungarian Academy of Sciences
(MTA SZTAKI), Budapest, Hungary
dmarx@cs.bme.hu, laszlo.egri@mail.mcgill.ca

Abstract. In the deletion version of the list homomorphism problem, we are given graphs G and H, a list $L(v) \subseteq V(H)$ for each vertex $v \in V(G)$, and an integer k. The task is to decide whether there exists a set $W \subseteq V(G)$ of size at most k such that there is a homomorphism from $G \setminus W$ to H respecting the lists. We show that DL-HOM(H), parameterized by k and $|H|$, is fixed-parameter tractable for any (P_6, C_6)-free bipartite graph H; already for this restricted class of graphs, the problem generalizes Vertex Cover, Odd Cycle Transversal, and Vertex Multiway Cut parameterized by the size of the cutset and the number of terminals. We conjecture that DL-HOM(H) is fixed-parameter tractable for the class of graphs H for which the list homomorphism problem (without deletions) is polynomial-time solvable; by a result of Feder et al. [9], a graph H belongs to this class precisely if it is a bipartite graph whose complement is a circular arc graph. We show that this conjecture is equivalent to the fixed-parameter tractability of a single fairly natural satisfiability problem, *Clause Deletion Chain-SAT*.

1 Introduction

Given two graphs G and H (without loops and parallel edges; unless otherwise stated, we consider only such graphs throughout this paper), a *homomorphism* $\phi : G \to H$ is a mapping $\phi : V(G) \to V(H)$ such that $\{u, v\} \in E(G)$ implies $\{\phi(u), \phi(v)\} \in E(H)$; the corresponding algorithmic problem *Graph Homomorphism* asks if G has a homomorphism to H. It is easy to see that G has a homomorphism into the clique K_c if and only if G is c-colorable; therefore, the algorithmic study of (variants of) Graph Homomorphism generalizes the study of graph coloring problems (cf. Hell and Nešetřil [15]). Instead of graphs, one can

---

* Supported by ERC Starting Grant PARAMTIGHT (No. 280152).
** Supported in part by NSF CAREER award 1053605, NSF grant CCF-1161626, ONR YIP award N000141110662, DARPA/AFOSR grant FA9550-12-1-0423, a University of Maryland Research and Scholarship Award (RASA) and a Summer International Research Fellowship from University of Maryland.

H.L. Bodlaender and G.F. Italiano (Eds.): ESA 2013, LNCS 8125, pp. 313–324, 2013.

consider homomorphism problems in the more general context of relational structures. Feder and Vardi [12] observed that the standard framework for Constraint Satisfaction Problems (CSP) can be formulated as homomorphism problems for relational structures. Thus variants of Graph Homomorphism form a rich family of problems that are more general than classical graph coloring, but does not have the full generality of CSPs.

List Coloring is a generalization of ordinary graph coloring: for each vertex v, the input contains a list $L(v)$ of allowed colors associated to v, and the task is to find a coloring where each vertex gets a color from its list. In a similar way, *List Homomorphism* is a generalization of Graph Homomorphism: given two undirected graphs G, H and a list function $L : V(G) \rightarrow 2^{V(H)}$, the task is to decide if there exists a list homomorphism $\phi : G \rightarrow H$, i.e., a homomorphism $\phi : G \rightarrow H$ such that for every $v \in V(G)$ we have $\phi(v) \in L(v)$. The List Homomorphism problem was introduced by Feder and Hell [8] and has been studied extensively [7, 9–11, 14, 17]. It is also referred to as List H-Coloring the graph G since in the special case of $H = K_c$ the problem is equivalent to list coloring where every list is a subset of $\{1, \dots, c\}$.

An active line of research on homomorphism problems is to study the complexity of the problem when the target graph is fixed. Let H be an undirected graph. The Graph Homomorphism and List Homomorphism problems with fixed target H are denoted by HOM(H) and L-HOM(H), respectively. A classical result of Hell and Nešetřil [16] states that HOM(H) is polynomial-time solvable if H is bipartite and NP-complete otherwise. For the more general List Homomorphism problem, Feder et al. [9] showed that L-HOM(H) is in P if H is a bipartite graph whose complement is a circular arc graph, and it is NP-complete otherwise. Egri et al. [7] further refined this characterization and gave a complete classification of the complexity of L-HOM(H): they showed that the problem is complete for NP, NL, or L, or otherwise the problem is first-order definable.

In this paper, we increase the expressive power of (list) homomorphisms by allowing a bounded number of vertex deletions from the left-hand side graph G. Formally, in the DL-HOM(H) problem we are given as input an undirected graph G, an integer k, a list function $L : V(G) \rightarrow 2^{V(H)}$ and the task is to decide if there is a *deletion set* $W \subseteq V(G)$ such that $|W| \leq k$ and the graph $G \setminus W$ has a list homomorphism to H. Let us note that DL-HOM(H) is NP-hard already when H consists of a single isolated vertex: in this case the problem is equivalent to VERTEX COVER, since removing the set W has to destroy every edge of G.

We study the parameterized complexity of DL-HOM(H) parameterized by the number of allowed vertex deletions and the size of the target graph H. We show that DL-HOM(H) is fixed parameter tractable (FPT) for a rich class of target graphs H. That is, we show that DL-HOM(H) can be solved in time $f(k, |H|) \cdot n^{O(1)}$ if H is a (P_6, C_6)-free bipartite graph, where f is a computable function that depends only of k and $|H|$ (see [5, 13, 24] for more background on fixed parameter tractability). This unifies and generalizes the fixed parameter tractability of certain well-known problems in the FPT world:

- VERTEX COVER asks for a set of k vertices whose deletion removes every edge. This problem is equivalent to DL-HOM(H) where H is a single vertex.
- ODD CYCLE TRANSVERSAL (also known as VERTEX BIPARTIZATION) asks for a set of at most k vertices whose deletion makes the graph bipartite. This problem can be expressed by DL-HOM(H) when H consists of a single edge.
- In VERTEX MULTIWAY CUT parameterized by the size of the cutset and the number of terminals, a graph G is given with terminals $t_1, \ldots, t_d$, and the task is to find a set of at most k vertices whose deletion disconnects t_i and t_j for any $i \neq j$. This problem can be expressed as DL-HOM(H) when H is a matching of d edges, in the following way. Let us obtain G' by subdividing each edge of G (making it bipartite) and let the list of t_i contain the vertices of the i-th edge e_i; all the other lists contain every vertex of H. It is easy to see that the deleted vertices must separate the terminals otherwise there is no homomorphism to H and, conversely, if the terminals are separated from each other, then the component of t_i has a list homomorphism to e_i.

Note that all three problems described above are NP-hard but known to be fixed-parameter tractable [4, 5, 21, 25].

Our Results: Clearly, if L-HOM(H) is NP-complete, then DL-HOM(H) is NP-complete already for $k = 0$, hence we cannot expect it to be FPT. Therefore, by the results of Feder et al. [9], *we need to consider only the case when H is a bipartite graph* whose complement is a circular arc graph. We focus first on those graphs H for which the characterization of Egri et al. [7] showed that L-HOM(H) is not only polynomial-time solvable, but actually in logspace: these are precisely those (bipartite) graphs that exclude the path P_6 on six vertices and the cycle C_6 on six vertices as induced subgraphs. This class of graphs admits a decomposition using certain operations (see [7]), and to emphasize this decomposition, we also call this class of graphs *skew decomposable graphs*. Let us emphasize that these graphs are bipartite by definition. Note that the class of skew decomposable graphs is a strict subclass of chordal bipartite graphs (P_6 is chordal bipartite but not skew decomposable), and bipartite cographs and bipartite trivially perfect graphs are trivially skew decomposable.

Our first result is that the DL-HOM(H) problem is fixed-parameter tractable for this class of graphs.

Theorem 1. *If H is a skew decomposable bipartite graph, then DL-HOM(H) is FPT parameterized by solution size and $|H|$.*

Observe that the graphs considered in the examples above are all skew decomposable bipartite graphs, hence Theorem 1 is an algorithmic meta-theorem unifying the fixed-parameter tractability of VERTEX COVER, ODD CYCLE TRANSVERSAL, and VERTEX MULTIWAY CUT parameterized by the size of the cutset and the number of terminals, and various combinations of these problems.

Theorem 1 shows that, for a particular class of graphs where L-HOM(H) is known to be polynomial-time solvable, the deletion version DL-HOM(H) is fixed-parameter tractable. We conjecture that this holds in general: whenever L-HOM(H) is polynomial-time solvable (i.e., the cases described by Feder et al. [9]), the corresponding DL-HOM(H) problem is FPT.

Conjecture 2. If H is a fixed graph whose complement is a circular arc graph, then DL-HOM(H) is FPT parameterized by solution size.

It might seem unsubstantiated to conjecture fixed-parameter tractability for every bipartite graph H whose complement is a circular arc graph, but we show that, in a technical sense, proving Conjecture 2 boils down to the fixed-parameter tractability of a single fairly natural problem. We introduce a variant of maximum ℓ-satisfiability, where the clauses of the formula are implication chains[1] $x_1 \to x_2 \to \cdots \to x_\ell$ of length at most ℓ, and the task is to make the formula satisfiable by removing at most k clauses; we call this problem *Clause Deletion ℓ-Chain-SAT (ℓ-CDCS)* (see Definition 14). We conjecture that for every fixed ℓ, this problem is FPT parameterized by k.

Conjecture 3. For every fixed $\ell \geq 1$, Clause Deletion ℓ-Chain-SAT is FPT parameterized by solution size.

We show that for every bipartite graph H whose complement is a circular arc graph, the problem DL-HOM(H) can be reduced to CDCS for some ℓ depending only on $|H|$. Somewhat more surprisingly, we are also able to show a converse statement: for every ℓ, there is a bipartite graph H_ℓ whose complement is a circular arc graph such that ℓ-CDCS can be reduced to DL-HOM(H_ℓ). That is, the two conjectures are equivalent. Therefore, in order to settle Conjecture 2, one necessarily needs to understand Conjecture 3 as well. Since the latter conjecture considers only a single problem (as opposed to an infinite family of problems parameterized by $|H|$), it is likely that connections with other satisfiability problems can be exploited, and therefore it seems that Conjecture 3 is a more promising target for future work.

Theorem 4. *Conjectures 2 and 3 are equivalent.*

Our Techniques: For our fixed-parameter tractability results, we use a combination of several techniques (some of them classical, some of them very recent) from the toolbox of parameterized complexity. Our first goal is to reduce DL-HOM(H) to the special case where each list contains vertices only from one side of one component of the (bipartite) graph H; we call this special case the "fixed side, fixed component" version. We note that the reduction to this special case in non-trivial: as the examples above illustrate, expressing VERTEX MULTIWAY CUT seems to require that the lists contain vertices from more than one component of H, and expressing ODD CYCLE TRANSVERSAL seems to require that the lists contain vertices from both sides of H.

We start our reduction by using the standard technique of iterative compression to obtain an instance where, besides a bounded number of precolored vertices, the graph is bipartite.

We look for obvious conflicts in this instance. Roughly speaking, if there are two precolored vertices u and v in the same component of G with colors a and b,

---

[1] The notation $x_1 \to x_2 \to \cdots \to x_\ell$ is a shorthand for $(x_1 \to x_2) \wedge (x_2 \to x_3) \wedge \cdots \wedge (x_{\ell-1} \to x_\ell)$.

respectively, such that either a and b are in different components of H, or a and b are in the same component of H but the parity of the distance between u and v is different from the parity of the distance between a and b, then the deletion set must contain a $u - v$ separator. We use the treewidth reduction technique of Marx et al. [22] to find a bounded-treewidth region of the graph that contains all such separators. As we know that this region contains at least one deleted vertex, every component outside this region can contain at most $k - 1$ deleted vertices. Thus we can recursively solve the problem for each such component, and collect all the information that is necessary to solve the problem for the remaining bounded-treewidth region. We are able to encode our problem as a Monadic Second Order (MSO) formula, hence we can apply Courcelle's Theorem [3] to solve the problem on the bounded-treewidth region.

Even if the instance has no obvious conflicts as described above, we might still need to delete certain vertices due to more implicit conflicts. But now we know that for each vertex v, there is at most one component C of H and one side of C that is consistent with the precolored vertices appearing in the component of v, otherwise a direct conflict between two precolored vertices would arise. This seems to be close to our goal of being able to fix a component C of H and a side of C for each vertex. However, there is a subtle detail here: if the deleted set separates a vertex v from every precolored vertex, then the precolored vertices do not force any restriction on v. Therefore, it seems that at each vertex v, we have to be prepared for two possibilities: either v is reachable from the precolored vertices, or not. Unfortunately, this prevents us from assigning each vertex to one of the sides of a single component. We get around this problem by invoking the "randomized shadow removal" technique introduced by Marx and Razgon [23] (and subsequently used in [1, 2, 18–20]) to modify the instance in such a way that we can assume that the deletion set does not separate any vertex from the precolored vertices, hence we can fix the components and the sides.

Note that the above reductions work for any bipartite graph H, and the requirement that H be skew decomposable is used only at the last reduction step: if H is a skew decomposable graph, then the *fixed side fixed component* version of the problem can be solved by appealing to the inductive construction of such graphs given by Egri et al. [7] and using bounded depth search.

If H is a bipartite graph whose complement is a circular arc graph (recall that this class strictly contains all skew decomposable graphs), then we show how to formulate the problem as an instance of ℓ-CDCS (showing that Conjecture 3 implies Conjecture 2). Let us emphasize that the reduction to ℓ-CDCS works only if the lists of the DL-HOM(H) instance have the "fixed side" property, and therefore our proof for the equivalence of the two conjectures (Theorem 4) needs the reduction machinery described above.

2   Preliminaries

Given a graph G, let $V(G)$ denote its vertices and $E(G)$ denote its edges. If $G = (U, V, E)$ is bipartite, we call U and V the *sides* of H. Let G be a graph and $W \subseteq V(G)$. Then $G[W]$ denotes the subgraph of G induced by the vertices in W. To simplify notation, we often write $G \setminus W$ instead of $G[V(G) \setminus W]$. The set $N(W)$ denotes the neighborhood of W in G, that is, the vertices of G which are not in W, but have a neighbor in W. Similarly to [22], we define two types of separators:

Definition 5. *A set S of vertices* separates *the sets of vertices A and B if no component of $G \setminus S$ contains vertices from both $A \setminus S$ and $B \setminus S$. If s and t are two distinct vertices of G, then an $s - t$ separator is a set S of vertices disjoint from $\{s, t\}$ such that s and t are in different components of $G \setminus S$.*

Definition 6. *Let G, H be graphs and L be a list function $V(G) \to 2^{V(H)}$. A* list homomorphism ϕ *from (G, L) to H (or if L is clear from the context, from G to H) is a homomorphism $\phi : G \to H$ such that $\phi(v) \in L(v)$ for every $v \in V(G)$. In other words, each vertex $v \in V(G)$ has a list $L(v)$ specifying the possible images of v. The right-hand side graph H is called the* target *graph.*

When the target graph H is fixed, we have the following problem:

L-Hom(H)
Input : A graph G and a list function $L : V(G) \to 2^{V(H)}$.
Question : Does there exist a list homomorphism from (G, L) to H?

The main problem we consider in this paper is the vertex deletion version of the L-Hom(H) problem, i.e., we ask if a set of vertices W can be deleted from G such that the remaining graph has a list homomorphism to H. Obviously, the list function is restricted to $V(G) \setminus W$, and for ease of notation, we denote this restricted list function $L|_{V(G) \setminus W}$ by $L \setminus W$. We can now ask the following formal question:

DL-Hom(H)
Input : A graph G, a list function $L : V(G) \to 2^{V(H)}$, and an integer k.
Parameters : k , $|H|$
Question : Does there exist a set $W \subseteq V(G)$ of size at most k such that there is a list homomorphism from $(G \setminus W, L \setminus W)$ to H?

Notice that if $k = 0$, then DL-Hom(H) becomes L-Hom(H). In the first part of the paper, we reduce DL-Hom(H) to a more restricted version of the problem where every list $L(v)$ contains vertices only from one component of H, and moreover, only from one side of that component (recall that we are assuming that H is bipartite). We call lists satisfying this property *fixed side fixed component*.

DL-Hom(H)-Fixed-Side-Fixed-Component , where H is bipartite **(FS-FC(H))**
Input : A graph G, a fixed side fixed component list function $L : V(G) \to 2^{V(H)}$, and an integer k.
Parameters : $k, |H|$
Question : Does there exist a set $W \subseteq V(G)$ such that $|W| \le k$ and $G \setminus W$ has a list homomorphism to H?

We argue that it is sufficient to solve the FS-FC(H) problem:

Theorem 7. *If the* FS-FC(H) *problem is FPT (where H is bipartite), then the* DL-HOM(H) *problem is also FPT.*

The main ideas in the reduction from DL-HOM(H) to FS-FC(H) are presented below. The proof is by induction on k, i.e., we are assuming that such a reduction is possible for $k - 1$. In the full version of the paper, we solve FS-FC(H) for skew decomposable graphs, completing the proof of Theorem 1.

Theorem 8. *If H is a skew decomposable graph, then the* FS-FC(H) *problem is FPT.*

3   The Algorithm

The algorithm proving Theorem 1 is constructed through a series of reductions. We begin with applying the standard technique of *iterative compression* [25], and this is followed by some preprocessing of the *disjoint* version of the *compression* problem.

DL-Hom(H)-Disjoint-Compression
Input : A graph G_0, a list function $L : V(G_0) \to 2^{V(H)}$, an integer k, and a set $W_0 \subseteq V(G_0)$ of size at most $k+1$ such that $G_0 \setminus W_0$ has a list homomorphism to H.
Parameters : $k, |H|$
Question : Does there exist a set $W \subseteq V(G_0)$ disjoint from W_0 such that $|W| \leq k$ and $(G_0 \setminus W, L \setminus W)$ has a list homomorphism to H?

Since the techniques related to iterative compression are folklore, we just note here that any FPT algorithm for the DL-HOM(H)-DISJOINT-COMPRESSION problem defined below translates into an FPT algorithm for DL-HOM(H) with an additional blowup factor of $O(2^{|W_0|}n)$ in the running time. The details of this reduction are given in the full version of the paper. In the rest of the paper, we concentrate on giving an FPT algorithm for the DL-HOM(H)-DISJOINT-COMPRESSION problem.

Since the new solution W can be assumed to be disjoint from W_0, for any solution set W, we must have a partial homomorphism from $G_0[W_0]$ to H. We guess all such partial list homomorphisms γ from $G_0[W_0]$ to H, and we hope that we can find a set W such that γ can be extended to a total list homomorphism from $G_0[W]$ to H. To guess these partial homomorphisms, we simply enumerate all possible mappings from W_0 to H and check whether the given mapping is a list homomorphism from $(G_0[W_0], L|_{W_0})$ to H. If not we discard the given mapping. Observe that we need to consider only $|V(H)|^{|W_0|} \leq |V(H)|^{k+1}$ mappings. Hence, in what follows we can assume that we are given a partial list homomorphism γ from $G_0[W_0]$ to H.

Recall that we are assuming that H is bipartite. Since we have a fixed partial homomorphism γ from W_0 to H, we can propagate the consequences of this homomorphism to the lists of the vertices in the neighborhood of W_0, as follows. For every $v \in W_0$, let H_v be the component of H in which $\gamma(v)$ appears. Furthermore, let S_v be the side of H_v in which $\gamma(v)$ appears, and let $\bar{S}_v$ be the other side of H_v. For each neighbor u of v in $N(W_0)$, trim $L(u)$ as $L(u) \leftarrow L(u) \cap \bar{S}_v$. The list of each vertex in $N(W_0)$ is now contained in one of the sides of a single component of H. We say that such a list is *fixed side* and *fixed component*. Note that while doing this, some of the lists might become empty. We delete those vertices from the graph, and reduce the parameter accordingly.

Recall that $G_0 \setminus W_0$ has a list homomorphism to the bipartite graph H, and therefore $G_0 \setminus W_0$ must be bipartite. We will later make use of the homomorphism from $G_0 \setminus \{W_0 \cup N(W_0)\}$ to H, so we name this homomorphism ϕ_0. To summarize the properties of the problem we have at hand, we define it formally below. Note that we will not need the graph G_0 and the set W_0 any more, only the graph $G_0 \setminus W_0$, and the neighborhood $N(W_0)$. To simplify notation, we refer to $G_0 \setminus W_0$ and $N(W_0)$ as G and N_0, respectively.

DL-Hom(H)-Bipartite-Compression (BC(H))
Input : A bipartite graph G, a list function $L : V(G) \to 2^{V(H)}$, a set $N_0 \subseteq V(G)$, where for each $v \in N_0$, the list $L(v)$ is fixed side and fixed component, a list homomorphism ϕ_0 from $(G \setminus N_0, L \setminus N_0)$ to H, and an integer k.
Parameters : $k, |H|$
Question : Does there exist a set $W \subseteq V(G)$, such that $|W| \le k$ and $(G \setminus W, L \setminus W)$ has a list homomorphism to H?

We define two types of *conflicts* between the vertices of N_0 (Definition 9). Our algorithm has two subroutines, one to handle the case when such a conflict is present, and one to handle the other case.

3.1   There Is a Conflict

If a conflict exists, its presence allows us to invoke the treewidth reduction technique of Marx et al. [22] to split the instance into a bounded-treewidth part, and into instances having parameter value strictly less than k. After solving these instances with smaller parameter value recursively, we encode the problem in Monadic Second Order logic, and apply Courcelle's theorem [3]. We outline these ideas, as follows.

Recall that the lists of the vertices in N_0 in a BC(H) instance are fixed side fixed component.

Definition 9. *Let (G, L, N_0, ϕ_0, k) be an instance of BC(H). Let u and v be vertices in the same component of G. We say that u and v are in* component conflict *if $L(u)$ and $L(v)$ are subsets of vertices of different components of H. Furthermore, u and v are in* parity conflict *if u and v are not in component conflict, and either u and v belong to the same side of G but $L(u)$ is a subset of one of the sides of a component of C of H and $L(v)$ is a subset of the other side of C, or u and v belong to different sides of G but $L(u)$ and $L(v)$ are subsets of the same side of a component of H.*

The following lemma easily follows from the definitions.

Lemma 10. *Let (G, L, N_0, ϕ_0, k) be an instance of $\mathrm{BC}(H)$. If u and v are any two vertices in N_0 that are in component or parity conflict, then any solution W must contain a set S that separates the sets $\{u\}$ and $\{v\}$.*

The result we need from [22] states that all the minimal $s - t$ separators of size at most k in G can be covered by a set C inducing a bounded-treewidth subgraph of G. In fact, a stronger statement is true: this subgraph has bounded treewidth even if we introduce additional edges in order to take into account connectivity outside C. This is expressed by the operation of taking the torso:

Definition 11. *Let G be a graph and $C \subseteq V(G)$. The graph $\mathrm{torso}(G, C)$ has vertex set C and two vertices $a, b \in C$ are adjacent if $\{a, b\} \in E(G)$ or there is a path in G connecting a and b whose internal vertices are not in C.*

Observe that by definition, $G[C]$ is a subgraph of $\mathrm{torso}(G, C)$.

Lemma 12 ([22]). *Let s and t be two vertices of G. For some $k \geq 0$, let C_k be the union of all minimal sets of size at most k that are $s - t$ separators. There is a $O(g_1(k) \cdot (|E(G) + V(G)|))$ time algorithm that returns a set $C \supset C_k \cup \{s, t\}$ such that the treewidth of $\mathrm{torso}(G, C)$ is at most $g_2(k)$, for some functions g_1 and g_2 of k.*

Lemma 10 gives us a pair of vertices that must be separated, and Lemma 12 gives us a bounded-treewidth region C of the input graph in which we know that at least one vertex must be deleted.

Courcelle's Theorem gives an easy way of showing that certain problems are linear-time solvable on bounded-treewidth graphs: it states that if a problem can be formulated in MSO, then there is a linear-time algorithm for it. This theorem also holds for relational structures of bounded-treewidth instead of just graphs, a generalization we need because we introduce new relations to encode the properties of the components of $G \setminus C$.

The following lemma formalizes the above ideas to prove that the subroutine used to handle the case when a conflict exists is correct:

Lemma 13. *Let $\mathcal{A}$ be an algorithm that correctly solves $\mathrm{DL\text{-}HOM}(H)$ for input instances in which the first parameter is at most $k - 1$. Suppose that the running time of $\mathcal{A}$ is $f(k-1, |H|) \cdot x^c$, where x is the size of the input, and c is a sufficiently large constant. Let I be an instance of $\mathrm{BC}(H)$ with parameter k that contains a component or parity conflict. Then I can be solved in time $f(k, |H|) \cdot x^c$ (where f is defined in the proof).*

Proof. Let $I = (G, L, N_0, \phi_0, k)$ be an instance of $\mathrm{BC}(H)$. Let $v, w \in N_0$ such that v and w are in component or parity conflict. Then by Lemma 10, the deletion set must contain a $v - w$ separator. Using Lemma 12, we can find a set C with the properties stated in the lemma (and note that we will also make use of the functions g_1 and g_2 in the statement of the lemma). Most importantly, C contains at least one vertex that must be removed in any solution, so the maximum number of vertices that can be removed from any connected component of $G[V(G) \setminus C]$ without exceeding the budget k is at most $k - 1$. Therefore, the outline of our strategy is the following. We use $\mathcal{A}$ to solve the problem for some slightly modified versions of the components of $G[V(G) \setminus C]$, and using these solutions, we construct an MSO formula that encodes our original problem I. Furthermore, the relational structure over which this MSO formula must be evaluated has bounded treewidth, and therefore the formula can be evaluated in linear time using Courcelle's theorem. The details of the proof are deferred to the full version of the paper. □

3.2 There Is no Conflict

In the case when there is no component or parity conflict, the problem FS-FC-IG(H) is the same as FS-FC(H) except that if the solution separates a vertex v from N_0, then we do not require that v is assigned to any vertex of H. We first trim the lists which allows us to reduce the BC(H) problem to the FS-FC-IG(H) problem. Then we use the "shadow removal" technique of Marx and Razgon [23] which allows us to reduce the FS-FC-IG(H) problem to the FS-FC(H) problem. Finally, we use the inductive construction of skew decomposable bipartite graphs [7] which allows us to solve the FS-FC(H) problem recursively. The details about this case are deferred to the full version of the paper.

4   Relation between DL-Hom(H) and Satisfiability Problems

Theorem 4 establishes the equivalence of DL-HOM(H) with the Clause Deletion ℓ-Chain SAT (ℓ-CDCS) problem, where H is restricted to be a graph for which L-HOM(H) is characterized as polynomial-time solvable by Feder et al. [9], that is, where H is restricted to be a bipartite graph whose complement is a circular arc graph. Here we only define the ℓ-CDCS problem, and the technical proof of Theorem 4 can be found in the full version of the paper.

Definition 14. *A chain clause is a conjunction of the form*

$$(x_0 \to x_1) \wedge (x_1 \to x_2) \wedge \cdots \wedge (x_{m-1} \to x_m),$$

where x_i and x_j are different variables if $i \neq j$. The length of a chain clause is the number of variables it contains. (A chain clause of length 1 is a variable, and it is satisfied by both possible assignments.) To simplify notation, we denote chain clauses of the above form as

$$x_0 \to x_1 \to \cdots \to x_m.$$

An ℓ-Chain-SAT formula consists of:

- *a set of variables V;*
- *a set of chain clauses over V such that any chain clause has length at most ℓ;*
- *a set of unary clauses (a unary clause is a variable or its negation).*

Clause Deletion ℓ-Chain-SAT   (ℓ-CDCS)

Input : An ℓ-Chain-SAT formula F.

Parameter : k

Question : Does there exist a set of clauses of size at most k such that removing these clauses from F makes F satisfiable?

---

5   Concluding Remarks

The list homomorphism problem is a widely investigated problem in classical complexity theory. In this work, we initiated the study of this problem from the perspective of parameterized complexity: we have shown that the DL-HOM(H) is FPT for any skew decomposable graph H parameterized by the solution size and $|H|$, an algorithmic meta-result unifying the fixed parameter tractability of some well-known problems. To achieve this, we welded together a number of classical and recent techniques from the FPT toolbox in a novel way. Our research suggests many open problems, four of which are:

1. If H is a fixed bipartite graph whose complement is a circular arc graph, is DL-HOM(H) FPT parameterized by solution size? (Conjecture 2.)
2. If H is a fixed digraph such that L-HOM(H) is in logspace (such digraphs have been recently characterised in [6]), is DL-HOM(H) FPT parameterized by solution size?
3. If H is a matching consisting of n edges, is DL-HOM(H) FPT, where the parameter is only the size of the deletion set?
4. Consider DL-HOM(H) for target graphs H in which both vertices with and without loops are allowed. It is known that for such target graphs L-HOM(H) is in P if and only if H is a *bi-arc* graph [10], or equivalently, if and only if H has a *majority polymorphism*. If H is a fixed bi-arc graph, is there an FPT reduction from DL-HOM(H) to ℓ-CDCS, where ℓ depends only on $|H|$?

Note that for the first problem, we already do not know if DL-HOM(H) is FPT when H is a path on 7 vertices. (If H is a path on 6 vertices, there is a simple reduction to ALMOST 2-SAT once we ensure that the instance has fixed side lists.) Observe that the third problem is a generalization of the VERTEX MULTIWAY CUT problem parameterized only by the cutset. For the fourth problem, we note that the FPT reduction from DL-HOM(H) to CDCS for graphs without loops relies on the fixed side nature of the lists involved. Since the presence of loops in H makes the concept of a fixed side list meaningless, it is not clear how to achieve such a reduction.

References

1. Chitnis, R., Cygan, M., Hajiaghayi, M., Marx, D.: Directed subset feedback vertex set is fixed-parameter tractable. In: Czumaj, A., Mehlhorn, K., Pitts, A., Wattenhofer, R. (eds.) ICALP 2012, Part I. LNCS, vol. 7391, pp. 230–241. Springer, Heidelberg (2012)

2. Chitnis, R.H., Hajiaghayi, M., Marx, D.: Fixed-parameter tractability of directed multiway cut parameterized by the size of the cutset. In: SODA (2012)
3. Courcelle, B.: Graph rewriting: An algebraic and logic approach. In: van Leeuwen, J. (ed.) Handbook of Theoretical Computer Science. Formal Models and Semantics, vol. B, pp. 193–242. Elsevier, Amsterdam (1990)
4. Cygan, M., Pilipczuk, M., Pilipczuk, M., Wojtaszczyk, J.O.: On multiway cut parameterized above lower bounds. In: Marx, D., Rossmanith, P. (eds.) IPEC 2011. LNCS, vol. 7112, pp. 1–12. Springer, Heidelberg (2012)
5. Downey, R.G., Fellows, M.R.: Parameterized Complexity. Springer
6. Egri, L., Hell, P., Larose, B., Rafiey, A.: An L vs. NL dichotomy for the digraph list homomorphism problem (2013) (manuscript in preparation)
7. Egri, L., Krokhin, A.A., Larose, B., Tesson, P.: The complexity of the list homomorphism problem for graphs. Theory of Computing Systems 51(2) (2012)
8. Feder, T., Hell, P.: List homomorphisms to reflexive graphs. J. Comb. Theory, Ser. B 72(2), 236–250 (1998)
9. Feder, T., Hell, P., Huang, J.: List homomorphisms and circular arc graphs. Combinatorica 19(4), 487–505 (1999)
10. Feder, T., Hell, P., Huang, J.: Bi-arc graphs and the complexity of list homomorphisms. Journal of Graph Theory 42(1), 61–80 (2003)
11. Feder, T., Hell, P., Huang, J.: List homomorphisms of graphs with bounded degrees. Discrete Mathematics 307, 386–392 (2007)
12. Feder, T., Vardi, M.Y.: The computational structure of monotone monadic SNP and constraint satisfaction: A study through datalog and group theory. SIAM J. Comput. 28(1), 57–104 (1998)
13. Flum, J., Grohe, M.: Parameterized Complexity Theory. Springer (2006)
14. Gutin, G., Rafiey, A., Yeo, A.: Minimum cost and list homomorphisms to semicomplete digraphs. Discrete Applied Mathematics 154, 890–897 (2006)
15. Hell, P., Nešetřil, J.: Graphs and homomorphisms. Oxford University Press
16. Hell, P., Nešetřil, J.: On the complexity of H-coloring. Journal of Combinatorial Theory, Series B 48, 92–110 (1990)
17. Hell, P., Rafiey, A.: The dichotomy of list homomorphisms for digraphs. In: SODA, pp. 1703–1713 (2011)
18. Kratsch, S., Pilipczuk, M., Pilipczuk, M., Wahlström, M.: Fixed-parameter tractability of multicut in directed acyclic graphs. In: Czumaj, A., Mehlhorn, K., Pitts, A., Wattenhofer, R. (eds.) ICALP 2012, Part I. LNCS, vol. 7391, pp. 581–593. Springer, Heidelberg (2012)
19. Lokshtanov, D., Marx, D.: Clustering with local restrictions. Inf. Comput. 222, 278–292 (2013)
20. Lokshtanov, D., Ramanujan, M.S.: Parameterized tractability of multiway cut with parity constraints. In: Czumaj, A., Mehlhorn, K., Pitts, A., Wattenhofer, R. (eds.) ICALP 2012, Part I. LNCS, vol. 7391, pp. 750–761. Springer, Heidelberg (2012)
21. Marx, D.: Parameterized graph separation problems. Theor. Comput. Sci. 351(3), 394–406 (2006)
22. Marx, D., O'Sullivan, B., Razgon, I.: Finding small separators in linear time via treewidth reduction. CoRR, abs/1110.4765 (2011)
23. Marx, D., Razgon, I.: Fixed-parameter tractability of multicut parameterized by the size of the cutset. In: STOC, pp. 469–478 (2011)
24. Niedermeier, R.: Invitation to Fixed-Parameter Algorithms. Oxford University Press (2006)
25. Reed, B.A., Smith, K., Vetta, A.: Finding odd cycle transversals. Oper. Res. Lett. 32(4), 299–301 (2004)

Rumor Spreading in Random Evolving Graphs[*]

Andrea Clementi[1], Pierluigi Crescenzi[2], Carola Doerr[3], Pierre Fraigniaud[4],
Marco Isopi[5], Alessandro Panconesi[5], Francesco Pasquale[5],
and Riccardo Silvestri[5]

[1] Università Tor Vergata di Roma
[2] Università di Firenze
[3] Université Paris Diderot and Max Planck Institute Saarbrücken
[4] CNRS and Université Paris Diderot
[5] Sapienza Università di Roma

Abstract. In this paper, we aim at analyzing the classical information spreading Push protocol in *dynamic* networks. We consider the *edge-Markovian* evolving graph model which captures natural temporal dependencies between the structure of the network at time t, and the one at time $t + 1$. Precisely, a non-edge appears with probability p, while an existing edge dies with probability q. In order to fit with real-world traces, we mostly concentrate our study on the case where $p = \Omega(\frac{1}{n})$ and q is constant. We prove that, in this realistic scenario, the Push protocol does perform well, completing information spreading in $O(\log n)$ time steps, w.h.p., even when the network is, w.h.p., disconnected at every time step (e.g., when $p \ll \frac{\log n}{n}$). The bound is tight. We also address other ranges of parameters p and q (e.g., $p+q = 1$ with arbitrary p and q, and $p = \Theta\left(\frac{1}{n}\right)$ with arbitrary q). Although they do not precisely fit with the measures performed on real-world traces, they can be of independent interest for other settings. The results in these cases confirm the positive impact of dynamism.

1 Introduction

Context and Objective. *Rumor spreading* is a well-known gossip-based distributed algorithm for disseminating information in large networks. According to the synchronous Push version of this algorithm, an arbitrary source node is initially informed, and, at each time step (a.k.a. round), each informed node u chooses one of its neighbors v uniformly at random, and this node becomes informed at the next time step.

Rumor spreading (originally called *rumor mongering*) was first introduced by [13], in the context of replicated databases, as a solution to the problem of

[*] A. Clementi is supported by Italian MIUR under the COFIN 2010-11 project *ARS TechnoMedia*. C. Doerr is supported by a F. Lynen postdoctoral research fellowship of the A. von Humboldt Foundation and by ANR project CRYQ. P. Fraigniaud is supported by ANR project DISPLEXITY, and INRIA project GANG. A. Panconesi is supported by a Google Faculty Research Award and EU FET project MULTI-PLEX 317532. F. Pasquale is supported by EU FET project MULTIPLEX 317532.

H.L. Bodlaender and G.F. Italiano (Eds.): ESA 2013, LNCS 8125, pp. 325–336, 2013.
© Springer-Verlag Berlin Heidelberg 2013

distributing updates and driving replicas towards consistency. Successively, it has been proposed in several other application areas (for a nice survey of gossip-based algorithm applications, see also [31]). Rumor spreading has also been deeply analyzed from a theoretical and mathematical point of view. Indeed, as already observed in [13], rumor spreading is just an example of an epidemic process: hence, its analysis "benefits greatly from the existing mathematical theory of epidemiology".

In particular, the *completion time* of rumor spreading, that is, the number of steps required in order to have all nodes informed with high probability[1] (w.h.p.), has been investigated in the case of several network topologies [6, 14, 17, 20–22, 30, 34], to mention just a few. Further works also derive deep connections between the completion time itself and some classic measures of graph spectral theory [7, 8, 23, 24, 35]. Recently, rumor spreading has been also analysed in the presence of transmission failures of the protocol [15, 19].

It is important to observe that the techniques and the arguments adopted in these studies strongly rely on the fact that the underlying graph is *static* and does not change over time. For instance, most of these analyses exploit the crucial fact that the degree of every node (no matter whether this is a random variable or a deterministic value) never changes during the entire execution of the rumor spreading algorithm. This paper addresses the speed of rumor spreading in the case of *dynamic* networks, where nodes and edges can appear and disappear over time (several emerging networking technologies such as ad hoc wireless, sensor, mobile networks, and peer-to-peer networks are indeed inherently dynamic).

In order to investigate the behavior of distributed protocols in the case of dynamic networks, the concept of evolving graph has been introduced in the literature. An *evolving graph* is a sequence of graphs $(G_t)_{t \geq 0}$ where $t \in \mathbb{N}$ (to indicate that we consider the graph *snapshots* at discrete time steps t, although it may evolve in a continuous manner) with the same set of n nodes.[2] This concept is rater general, ranging from *adversarial* evolving graphs [11, 32] to *random* evolving graphs [4]. In the case of *random* evolving graphs, at each time step, the graph G_t is chosen randomly according to some probability distribution over a specified family of graphs. One very well-known and deeply studied example of such a family is the set $\mathcal{G}_{n,p}$ of *Erdős-Rényi* random graphs [1, 16, 25]. In the evolving graph setting, at every time step t, each possible edge exists with probability p (independently of the previous graphs $G_{t'}$, $t' < t$, and independently of the other edges in G_t).

Random evolving graphs can exhibit communication properties which are much stronger than static networks having the same expected edge density (for a recent survey on computing over dynamic networks, see [33]). This has been proved in the case of the simplest communication protocol that implements the broadcast operation, that is, the Flooding protocol. It has been shown [3, 10, 12] that the Flooding completion time may be very fast (typically poly-logarithmic in the number of nodes) even when the network topology is, w.h.p., sparse, or

---

[1] An event holds w.h.p. if it holds with probability $1 - O(1/n^c)$ for some $c > 0$.

[2] As far as we know, this has been formally introduced for the first time in [18].

even highly disconnected at every time step. Therefore, such previous results provide analytical evidences of the fact that random network dynamics not only do not hurt, but can actually help data communication, which is of the utmost importance in several contexts, such as, e.g., delay-tolerant networking [37, 38].

The same observation has been made when the model includes some sort of *temporal* dependency, as it is in the case of the random *edge-Markovian* model. According to this model, at every time step t,

- if an edge does not exist in G_t, then it appears in G_{t+1} with probability p;
- if an edge exists in G_t, then it disappears in G_{t+1} with probability q.

For every initial graph G_0, and $0 < p, q < 1$, an edge-Markovian evolving graph will eventually converge to a (random) graph in $\mathcal{G}_{n,\tilde{p}}$ with stationary edge-probability $\tilde{p} = \frac{p}{p+q}$. However, there is a Markovian dependence between graphs at two consecutive time steps, hence, given G_t, the next graph G_{t+1} is not necessarily a random graph in $\mathcal{G}_{n,\tilde{p}}$. Interestingly enough, the edge-Markovian model has been recently subject to experimental validations, in the context of sparse opportunistic mobile networks [38], and of dynamic peer-to-peer systems [37]. These validations demonstrate a good fitting of the model with some real-world data traces.

The completion time of the Flooding protocol has been recently analyzed in the edge-Markovian model, for all possible values of $\tilde{p}$ (see [3, 12]). The Flooding protocol however generates high message complexity. Moreover, although its completion time is an interesting analog for dynamic graphs of the diameter for static graphs, it is not reflecting the kinds of gossip protocols mentioned at the beginning of this introduction, used for practical applications. Hence the main objective of this paper is to analyze the more practical Push protocol, in edge-Markovian evolving graphs.

Framework. We focus our attention on dynamic network topologies yielded by the edge-Markovian evolving graphs for parameters p (birth) and q (death) that correspond to a good fitting with real-world data traces, as observed in [37, 38]. These traces describe networks with relatively high dynamics, for which the death probability q is at least one order of magnitude greater than the birth probability p. In order to set parameters p and q fitting with these observations, let us consider the expected number of edges $\bar{m}$, and the expected node-degree $\bar{d}$ at the stationary regime, governed by $\tilde{p} = \frac{p}{p+q}$. We have $\bar{m} = \frac{p}{p+q}\binom{n}{2}$, and $\bar{d} = \frac{2\bar{m}}{n} = (n-1)\frac{p}{p+q}$. Thus, at the stationary regime, the expected number of edges ν that switch their state (from non existing to existing, or vice versa) in one time step satisfies

$$\nu = \bar{m}q + (\binom{n}{2} - \bar{m})p = \frac{n(n-1)}{2}\left(\frac{pq}{p+q} + \left(1 - \frac{p}{p+q}\right)p\right) = n(n-1)\frac{pq}{p+q} = nq\bar{d}.$$

Hence, in order to fit with the high dynamics observed in real-world data traces, we set q constant, so that a constant fraction of the edges disappear at every step, while a fraction p of the non-existing edges appear. We consider an arbitrary range for p, with the unique assumption that $p \geq \frac{1}{n}$. (For smaller p's, the completion time of any communication protocol is subject to the expected time $\frac{1}{np} \gg 1$ required for a node to acquire just one link connected to another node). To sum up, we essentially focus on the following range of parameters:

$$\frac{1}{n} \leqslant p < 1 \quad \text{and} \quad q = \Omega(1). \tag{1}$$

This range includes network topologies for a wide interval of expected edge density (from very sparse and disconnected graphs, to almost-complete ones), and with an expected number of switching edges per time step equal to some constant fraction of the expected total number of edges. Other ranges are also analyzed in the paper (e.g., $p + q = 1$ with arbitrary p and q, and $p = \Theta\left(\frac{1}{n}\right)$ with arbitrary q), but the range in Eq. (1) appears to be the most realistic one, according to the current measurements on dynamic networks.

Our Results. For the parameter range in Eq. (1), we show that, w.h.p., starting from any n-node graph G_0, the Push protocol informs all n nodes in $\Theta(\log n)$ time steps. Hence, in particular, even if the graph G_t is w.h.p. disconnected at every time step (this is the case for $p \ll \frac{\log n}{n}$), the completion time of the Push protocol is as small as it could be (the Push protocol cannot perform faster than $\Omega(\log n)$ steps in any static or dynamic graph since the number of informed nodes can at most double at every step).

We also address other ranges of parameters p and q. One such case is the sequence of independent $\mathcal{G}_{n,p}$ graphs, that is, the case where $p + q = 1$. Actually, the analysis of this special case will allow us to focus on the first important probabilistic issue that needs to be solved: spatial dependencies. Indeed, even in this case, the Push protocol induces a positive correlation among some crucial events that determine the number of new informed nodes at the next time step. This holds despite the fact that every edge is set independently from the others. For a sequence of independent $\mathcal{G}_{n,p}$ graphs, we prove that for every p (i.e., also for $p = o(\frac{1}{n})$) and $q = 1 - p$ the completion time of the Push protocol is, w.h.p., $\mathcal{O}(\log n/(\hat{p}n))$, where $\hat{p} = \min\{p, 1/n\}$. By comparing the lower bound for Flooding in [12], it turns out that this bound is tight, even for very sparse graphs.

Finally, we show that the logarithmic bound for the Push protocol holds for more "static" network topologies as well, e.g., for the range $p = \frac{c}{n}$ where $c > 0$ is a constant, and q is arbitrary. This parameter range includes edge-Markovian graphs with a small expected number of switching edges (this happens when $q = o(1)$). In this case, too, Push completes, w.h.p., in $O(\log n)$ rounds. This gives yet another evidence that dynamism helps.

Due to lack of space several proofs are omitted. We refer the interested reader to the full version of the paper [9].

2   Preliminaries

The number of vertices in the graph will always be denoted by n. We abbreviate $[n] := \{1, \ldots, n\}$ and $\binom{[n]}{2} := \{\{i, j\} \mid i, j \in [n]\}$. For any subset $E \subseteq \binom{[n]}{2}$

and any two subsets $A, B \subseteq [n]$, define $E(A) = \{$edges of E incident to $A\}$ and $E(A, B) = \{\{u, v\} \in E \mid u \in A, v \in B\}$. We consider the edge-Markovian evolving graph model $\mathcal{G}(n, p, q; E_0)$ where E_0 is the starting set of edges.

The Push Protocol over $\mathcal{G}(n, p, q; E_0)$ can be represented as a random process over the set $\mathcal{S}$ of all possible pairs (E, I) where E is a subset of edges and I is a subset of nodes. In particular, the combined Markov process works as follows

$$\ldots \to (E_t, I_t) \xrightarrow{\text{edge-Markovian}} (E_{t+1}, I_t) \xrightarrow{\text{Push protocol}} (E_{t+1}, I_{t+1}) \xrightarrow{\text{edge-Markovian}} \ldots$$

where E_t and I_t represent the set of existing edges and the set of informed nodes at time t, respectively. All events, probabilities and random variables are defined over the above random process. Given a graph $G = ([n], E)$, a node $v \in [n]$, and a subset of nodes $A \subseteq [n]$ we define $\deg_G(v, A) = |\{(v, a) \in E \mid a \in A\}|$. When we have a sequence of graphs $\{G_t = ([n], E_t) : t \in \mathbb{N}\}$ we write $\deg_t(v, A)$ instead of $\deg_{G_t}(v, A)$.

Given a graph G and an informed node $u \in I$, we define $\delta_G(u)$ as the random variable indicating the node selected by u in graph G according to the Push protocol. When G and/or t are clear from the context, they will be omitted.

Remark. It is worth noticing that analyzing the Push protocol in edge-Markovian graphs is not only subject to temporal dependencies, but also to *spatial* dependencies. To see why, consider a time step of the Push protocol. For an informed node u and a non-informed one v it is not hard to calculate the probability that $\delta(u) = v$ by conditioning on the degree of u. However, if u_1, u_2 are two informed nodes and v_1, v_2 are two non-informed ones, events "$\delta(u_1) = v_1$" and "$\delta(u_2) = v_2$" are not independent. Indeed, since the underlying graph is random, event "$\delta(u_1) = v_1$" *decreases* the probability of existence of an edge between u_1 and u_2, and so it affects the value of the random variable $\delta(u_2)$.

3   Warm Up: The Time-Independent Case

In this section we analyze the special case of a sequence of independent $G_{n,p}$ (observe that a sequence of independent $G_{n,p}$ is edge-Markovian with $q = 1 - p$). We show that the completion time of the Push protocol is $\mathcal{O}(\log n/(\hat{p}n))$ w.h.p., where $\hat{p} = \min\{p, 1/n\}$. In Theorem 1 we prove the result for $p \geqslant 1/n$ and in Theorem 2 for $p \leqslant 1/n$. From the lower bound on the flooding time for edge-Markovian graphs [12], it turns out that our bound is optimal.

As mentioned in Section 2, even though in this case there is no time-dependency in the sequence of graphs, the Push protocol introduces a kind of dependence that has to be carefully handled. The key challenge is to evaluate the probability that v receives the information from at least one of the informed nodes; i.e., $1 - \mathbf{P}\left(\cap_{u \in I}\{\delta(u) \neq v\}\right)$. We consider the Push operation on a *modified* random graph where we prove that the above events become independent and the number of new informed nodes in the original random graph is at least as large as in the modified version.

Definition 1. *Let* $G = ([n], E)$ *be a graph, let* $I \subseteq [n]$ *be a set of nodes, and let* $b \in [n]$ *be a positive integer. The* (I, b)*-modified graph* G *is the graph* $H = ([n] \cup \{v_1, \ldots, v_b\})$*, where* $\{v_1, \ldots, v_b\}$ *is a set of extra virtual nodes, obtained from* G *by the following operations: 1. For every node* $u \in I$ *with* $\deg_G(u) > b$*, remove all edges incident to* u*; 2. For every node* $u \in I$ *with* $\deg_G(u) \leqslant b$*, add all edges* $\{u, v_1\}, \ldots, \{u, v_b\}$ *between* u *and the virtual nodes; 3. Remove all edges between any pair of nodes that are both in* I*.*

Let I be the set of informed nodes performing a Push operation on a $G_{n,p}$ random graph. As previously observed, if $v \in [n] \setminus I$ is a non-informed node, then the events $\{\{\delta_G(u) = v\} : u \in I\}$ are not independent, but the events $\{\{\delta_H(u) = v\} : u \in I\}$ on the (I, b)-modified graph H are independent because of Operation 3 in Definition 1. In the next lemma we prove that, if the informed nodes perform a Push operation both in a graph and in its modified version, then the number of new informed nodes in the original graph is (stochastically) larger than the number of informed nodes in the modified one. We will then apply this result to $G_{n,p}$ random graphs.

Lemma 1. *Let* $G([n], E)$ *be a graph and let* b *an integer such that* $1 \leqslant b \leqslant n$*. Let* $I \subseteq [n]$ *be a set of nodes performing a* Push *operation in graphs* G *and* H*, where* H *is the* (I, b)*-modified* G *according to Definition 1. Let* X *and* Y *be the random variables counting the numbers of new informed nodes in* G *and* H *respectively. Then for every* $h \in [0, n]$ *it holds that* $\mathbf{P}(X \leqslant h) \leqslant \mathbf{P}(Y \leqslant h)$*.*

Proof. Consider the following coupling: Let $u \in I$ be an informed node such that $\deg_G(u) \leqslant b$ and let h and k be the number of informed and non-informed neighbors of u respectively. Choose $\delta_H(u)$ u.a.r. among the neighbors of u in H. As for $\delta_G(u)$, we do the following: If $\delta_H(u) \in [n] \setminus I$ then choose $\delta_G(u) = \delta_H(u)$; otherwise (i.e., when $\delta_H(u)$ is a virtual node) with probability $1 - x$ choose $\delta_G(u)$ u.a.r. among the informed neighbors of u in G, and with probability x choose $\delta_G(u)$ u.a.r. among the non-informed ones, where $x = \frac{k(b-h)}{(h+k)b}$. Every informed node u with $\deg_G(u) > b$ instead performs a Push operation in G independently. By construction we have that the set of new (non-virtual) informed nodes in H is a subset of the set of new informed nodes in G. Moreover, it is easy to check that, for every informed node u in I, $\delta_G(u)$ is u.a.r. among neighbors of u. $\square$

In the next lemma we give a lower bound on the probability that a non-informed node gets informed in the modified $G_{n,p}$.

Lemma 2. *Let* $I \subseteq [n]$ *be the set of informed nodes performing the* Push *operation in a* $G_{n,p}$ *random graph and let* X *be the random variable counting the number of non-informed nodes that get informed after the* Push *operation. It holds that* $\mathbf{P}(X \geqslant \lambda \cdot \min\{|I|, n - |I|\}) \geqslant \lambda$*, where* λ *is a positive constant.*

Proof. Let I be the set of currently informed nodes, let $G = ([n], E)$ be the random graph at the next time step and let H be its $(I, 3np)$-modified version. Now we show that the number of nodes that gets informed in H is at least $\lambda \cdot \min\{|I|, n - |I|\}$ with probability at least λ, for a suitable constant λ.

Let $u \in I$ be an informed node and let $v \in [n] \setminus I$ be a non-informed one. Observe that by the definition of H, u cannot choose v in H if the edge $\{u, v\} \notin E$ or if the degree of u in G is larger than $3np$ (see Operation 3 in Definition 1). Thus the probability $\mathbf{P}\left(\delta_H(u) = v\right)$ that node u chooses node v in random graph H according to the Push protocol is equal to

$$\mathbf{P}\left(\delta_H(u) = v \mid \{u, v\} \in E \wedge \deg_G(u) \leqslant 3np\right) \cdot \mathbf{P}\left(\{u, v\} \in G \wedge \deg_G(u) \leqslant 3np\right). \tag{2}$$

If $\deg_G(u) \leqslant 3np$ then node u in H has exactly $3np$ virtual neighbors plus at most other $3np$ non-informed neighbors. It follows that

$$\mathbf{P}\left(\delta_H(u) = v \mid \{u, v\} \in E \wedge \deg_G(u) \leqslant 3np\right) \geqslant 1/(6np). \tag{3}$$

We also have that

$$\mathbf{P}\left(\{u, v\} \in E, \deg_G(u) \leqslant 3np\right) = \mathbf{P}\left(\{u, v\} \in E\right) \mathbf{P}\left(\deg_G(u) \leqslant 3np \mid \{u, v\} \in E\right)$$
$$= p \cdot \mathbf{P}\left(\deg_G(u) \leqslant 3np \mid \{u, v\} \in E\right).$$

Since $\mathbf{E}\left[\deg_G(u) \mid \{u, v\} \in E\right] \leqslant np + 1$ with $np \geqslant 1$, from the Chernoff bound we can choose a positive constant c and then a positive constant $\beta < 1$ such that

$$\mathbf{P}\left(\deg_G(u) > 3np \mid \{u, v\} \in E\right) \leqslant \mathbf{P}\left(\deg_G(u) > 2np + 1 \mid \{u, v\} \in E\right)$$
$$\leqslant e^{-cnp} = \beta < 1. \tag{4}$$

By replacing Eq.s 3 and 4 into Eq. 2 we get $\mathbf{P}\left(\delta_H(u) = v\right) \geqslant \frac{\alpha}{n}$, for some constant $\alpha > 0$. Since the events $\{\{\delta_H(u) = v\}, u \in I\}$ are independent, the probability that node v is not informed in H is thus $\mathbf{P}\left(\cap_{u \in I} \delta_H(u) \neq v\right) \leqslant (1 - \alpha/n)^{|I|} \leqslant e^{-\alpha|I|/n}$. Let Y be the random variable counting the number of new informed nodes in H. The expectation of Y is $\mathbf{E}[Y] \geqslant (n - |I|) \left(1 - e^{-\alpha|I|/n}\right) \geqslant (\alpha/2)(n - |I|)|I|/n$. Hence we get

$$\mathbf{E}[Y] \geqslant \begin{cases} (\alpha/4)|I| & \text{if } |I| \leqslant n/2, \\ (\alpha/4)(n - |I|) & \text{if } |I| \geqslant n/2. \end{cases}$$

Since $Y \leqslant \min\{|I|, n - |I|\}$, it follows that $\mathbf{P}\left(Y \geqslant (\alpha/8) \cdot \min\{|I|, n - |I|\}\right) \geqslant \alpha/8$. Finally we get the thesis by applying Lemma 1. □

We can now derive the upper bound on the completion time of the Push protocol on $G_{n,p}$ random graphs.

Theorem 1. *Let $\mathcal{G} = \{G_t : t \in \mathbb{N}\}$ be a sequence of independent $G_{n,p}$ with $p \geqslant 1/n$. The completion time of the Push protocol over $\mathcal{G}$ is $\mathcal{O}(\log n)$ w.h.p.*

Proof. Consider a generic time step t of the execution of the Push protocol where $I_t \subseteq [n]$ is the set of informed nodes and $m_t = |I_t|$ is its size. For any t such that $m_t \leqslant n/2$, Lemma 2 implies that $\mathbf{P}\left(m_{t+1} \geqslant (1 + \lambda)m_t\right) \geqslant \lambda$, where λ is a positive constant. Let us define event $\mathcal{E}_t = \{m_t \geqslant (1 + \lambda)m_{t-1}\} \vee \{m_{t-1} \geqslant n/2\}$

and let $Y_t = Y_t((E_1, I_1), \ldots, (E_t, I_t))$ be the indicator random variable of that event. Observe that if $t = \frac{\log n}{\log(1+\lambda)}$ then $(1 + \lambda)^t \geqslant n/2$. Hence, if we set $T_1 = \frac{2}{\lambda} \frac{\log n}{\log(1+\lambda)}$, we get $\mathbf{P}\left(m_{T_1} \leqslant n/2\right) \leqslant \mathbf{P}\left(\sum_{t=1}^{T_1} Y_t \leqslant (\lambda/2)T_1\right)$. This probability is at most as large as the probability that in a sequence of T_1 independent coin tosses, each one giving head with probability λ, we see less than $(\lambda/2)T_1$ heads (see e.g. Lemma 3.1 in [2]). A direct application of the Chernoff bound shows that this probability is smaller than $e^{-(1/4)\lambda T_1} \leqslant n^{-c}$, for a suitable constant $c > 0$. We can thus state that, after $\mathcal{O}(\log n)$ time steps, there at least $n/2$ informed nodes w.h.p. If $m_{T_1} \geqslant n/2$, then, for every $t \geqslant T_1$, Lemma 2 implies that $\mathbf{P}\left(n - m_{t+1} \leqslant (1-\lambda)(n - m_t)\right) \geqslant \lambda$. Observe that if $t = \frac{\log n}{\lambda}$ then $(1 - \lambda)^t \leqslant 1/n$, so that for $T_2 = \frac{2}{\lambda} \cdot \frac{\log n}{\lambda} + T_1$ the probability that the Push protocol has not completed at time T_2 is $\mathbf{P}\left(m_{T_2} < n\right) \leqslant \mathbf{P}\left(m_{T_2} < n \,|\, m_{T_1} \geqslant \frac{n}{2}\right) + \mathbf{P}\left(m_{T_1} < \frac{n}{2}\right)$. As we argued in the analysis of the spreading till $n/2$, the probability $\mathbf{P}\left(m_{T_2} < n \,|\, m_{T_1} \geqslant \frac{n}{2}\right)$ is not larger than the probability that in a sequence of $\frac{2}{\lambda} \cdot \frac{\log n}{\lambda}$ independent coin tosses, each one giving head with probability λ, there are less than $\frac{\log n}{\lambda}$ heads. Again, by applying the Chernoff bound, the latter is not larger than n^{-c} for a suitable positive constant c.   □

In order to prove the bound for $p \leqslant 1/n$, we first show that one single Push operation over the union of a sequence of graphs informs (stochastically) less nodes than the sequence of Push operations performed in every single graph (this fact will also be used later in Section 4 to analyse the edge-MEG).

Lemma 3. *Let $\{G_t = ([n], E_t) : t = 1, \ldots, T\}$ be a finite sequence of graphs with the same set of nodes $[n]$. Let $I \subseteq [n]$ be the set of informed nodes in the initial graph G_1. Suppose that at every time step every informed node performs a Push operation, and let X be the random variable counting the number of informed nodes at time step T. Let $H = ([n], F)$ be such that $F = \cup_{t=1}^T E_t$ and let Y be the random variable counting the number of informed nodes when the nodes in I perform one single Push operation in graph H. Then for every $\ell = 0, 1, \ldots, n$ it holds that $\mathbf{P}(X \leqslant \ell) \leqslant \mathbf{P}(Y \leqslant \ell)$.*

Observe that if we look at a sequence of independent $G_{n,p}$ with $p \leqslant 1/n$ for a time-window of approximately $1/(np)$ time steps, then every edge appears at least once in the sequence with probability at least $1/n$. The above lemma thus allows us to reduce the case $p \leqslant 1/n$ to the case $p \geqslant 1/n$.

Theorem 2. *Let $\mathcal{G} = \{G_t : t \in \mathbb{N}\}$ be a sequence of independent $G_{n,p}$ with $p \leqslant 1/n$ and let $s \in [n]$. The Push protocol with source s over $\mathcal{G}$ completes the broadcast in $\mathcal{O}(\log n/(np))$ time steps w.h.p.*

4   Edge-Markovian Graphs with High Dynamics

In this section we prove that the Push protocol over an edge-Markovian graph $\mathcal{G}(n, p, q; E_0)$ with $p \geqslant 1/n$ and $q = \Omega(1)$ has completion time $\mathcal{O}(\log n)$ w.h.p.

As observed in the Introduction, the stationary random graph is an Erdős-Rényi $G_{n,\tilde{p}}$ where $\tilde{p} = \frac{p}{p+q}$ and the mixing time of the edge Markov chain is $\Theta(\frac{1}{p+q})$. Thus, if p and q fall into the range defined in Eq. 1, we get that the stationary random graph can be sparse and disconnected (when $p = o(\frac{\log n}{n})$) and that the mixing time of the edge Markov chain is $O(1)$. Thus, we can omit the term E_0 and assume it is random according to the stationary distribution.

The time-dependency between consecutive snapshots of the dynamic graph does not allow us to obtain directly the *increasing rate* of the number of informed nodes that we got for the independent-$G_{n,p}$ model. In order to get a result like Lemma 2 for the edge-Markovian case, we need in fact a *bounded-degree* condition on the current set of informed nodes (see Definition 2) that does not apply when the number of informed nodes is *small* (i.e., smaller than $\log n$). However, in order to reach a state where at least $\log n$ nodes are informed, we can use a different ad-hoc technique that analyzes the spreading rate yielded by the source only.

Lemma 4. *Let $\mathcal{G} = \mathcal{G}(n, p, q)$ be an edge-Markovian graph with $p \geqslant 1/n$ and $q = \Omega(1)$, and consider the Push protocol in $\mathcal{G}$ starting with one informed node. For any positive constant γ, after $\mathcal{O}(\log n)$ time steps there are at least $\gamma \log n$ informed nodes w.h.p.*

We can now start the second part of our analysis where the Push operation of all informed nodes (forming the subset I) will be considered and, thanks to the bootstrap, we can assume that $|I| = \Omega(\log n)$. As mentioned at the beginning of the section, we need to introduce the concept of *bounded-degree state* (E, I) of the Markovian process describing the information-spreading process over the dynamic graph, where E is the set of edges and I is the set of informed nodes.

Definition 2. *A state (E, I) such that $|E(I)| \leqslant (8/q)n\tilde{p}|I|$ (with $\tilde{p} = \frac{p}{p+q}$ the stationary edge probability) will be called a bounded-degree state.*

In the next lemma we show that, if I is the set of informed nodes with $|I| \geqslant \log n$, if in the starting random graph G_0 every edge exists with probability approximately $(1 \pm \varepsilon)p$, and if it evolves according to the edge-Markovian model and the informed nodes perform the Push protocol, then for a long sequence of time steps the random process is in a bounded-degree state. We will use this property in Theorem 3 by observing that, for every initial state, after $\mathcal{O}(\log n)$ time steps an edge-Markovian graph with $p \geqslant 1/n$ and $q \in \Omega(1)$ is in a state where every edge $\{u, v\}$ exists with probability $p_{\{u,v\}} \in [(1 - \varepsilon)\tilde{p}, (1 + \varepsilon)\tilde{p}]$.

Lemma 5. *Let $\mathcal{G} = \mathcal{G}(n, p, q, E_0)$ be an edge-Markovian graph starting with G_0 and consider the Push protocol in $\mathcal{G}$ where I_0 is the set of informed nodes at time $t = 0$. Then, for any constant $c > 0$, for a sequence of $c \log n$ time steps every state is a bounded-degree one w.h.p.*

Now we can bound the *increasing rate* of the number of informed nodes in an edge-Markovian graph. The proof of the following lemma combines the analysis adopted in the proof of Lemma 2 with some further ingredients required to manage the time-dependency of the edge-Markovian model.

Lemma 6. *Let (E, I) be a bounded-degree state and let X be the random variable counting the number of non-informed nodes that get informed after two steps of the Push operation in the edge-Markovian graph model. It holds that $\mathbf{P}\left(X \geqslant \varepsilon \cdot \min\{|I|, n - |I|\}\right) \geqslant \lambda$, where ε and λ are positive constants.*

Now we can prove that in $\mathcal{O}(\log n)$ time steps the Push protocol informs all nodes in an edge-Markovian graph, w.h.p.

Theorem 3. *Let $\mathcal{G} = \mathcal{G}(n, p, q, E_0)$ be an edge-Markovian graph with $p \geqslant 1/n$ and $q = \Omega(1)$ and let $s \in [n]$ be a node. The Push protocol with source s completes the broadcast over $\mathcal{G}$ in $\mathcal{O}(\log n)$ time steps w.h.p.*

Proof. Lemma 4 implies that after $\mathcal{O}(\log n)$ time steps there are $\Omega(\log n)$ informed nodes w.h.p. From Lemma 5, it follows that, after further $\mathcal{O}(\log n)$ time steps, the edge-Markovian graph reaches a bounded-degree state and remains so for further $\Omega(\log n)$ time steps. Let us rename $t = 0$ the time step where there are $\Omega(\log n)$ informed nodes and every edge $e \in \binom{[n]}{2}$ exists with probability $p_e \in [(1 - \varepsilon)\tilde{p}, (1 + \varepsilon)\tilde{p}]$. We again abbreviate $m_t := |I_t|$. Observe that if recurrence $m_{2(t+1)} \geqslant (1 + \varepsilon)m_{2t}$ holds $\log n / \log(1 + \varepsilon)$ times, then there are $n/2$ informed nodes. Let us thus name $T = \frac{2}{\lambda} \frac{\log n}{\log(1+\varepsilon)}$. If at time $2T$ there are less than $n/2$ informed nodes, then recurrence $m_{2(t+1)} \geqslant (1 + \varepsilon)m_{2t}$ held less than $\lambda T/2$ times. Since, at each time step, the recurrence holds with probability at least λ (there are less than $n/2$ informed nodes and the state is a bounded-degree one w.h.p.), the above probability is at most as large as the probability that in a sequence of T independent coin tosses, each one giving head with probability λ, we see less than $(\lambda/2)T$ heads (see, e.g., Lemma 3.1 in [2]). By the Chernoff bound such a probability is smaller than $e^{-\gamma\lambda T}$, for a suitable positive constant γ. Since γ and λ are constants and $T = \Theta(\log n)$ we have that

$$\mathbf{P}\left(m_{2T} \leqslant n/2\right) \leqslant n^{-\delta} \tag{5}$$

for a suitable positive constant δ. When m_t is larger than $n/2$ and the edge-Markovian graph is in a bounded-degree state, from Lemma 6 it follows that recurrence $n - m_{t+1} \leqslant (1 - \varepsilon)(n - m_t)$ holds with probability at least λ. If this recurrence holds $\log n / \log(1/(1 - \varepsilon))$ times then the number of informed nodes cannot be smaller than n. Hence, if we name $\tilde{T} := (2/\lambda) \log n / \log(1/(1 - \varepsilon))$, with the same argument we used to get Eq. 5, we obtain that after $2T + 2\tilde{T}$ time steps all nodes are informed w.h.p.   □

5   Edge-Markovian Graphs with Slow Dynamics

We have also considered "more static" sparse dynamic graphs. In particular, we can provide a logarithmic bound on the completion time of the Push protocol over the $\mathcal{G}(n, p, q)$ model even for $p = \Theta(1/n)$ and for $q = o(1)$. The proof of the following result combines some new coupling arguments with a previous analysis of the Push protocol for static random graphs given in [17].

Theorem 4. *Let $p = \frac{d}{n}$ for some absolute constant $d \in \mathbb{N}$ and let $q = q(n)$ be such that $q(n) = o(1)$. The* Push *protocol over edge-Markovian graphs in $\mathcal{G}(n, p, q)$ completes in $O(\log n)$ time, w.h.p.*

6   Conclusion

Completing the whole figure, i.e., for every $(p, q) \in [0, 1]^2$, is of intellectual interest. Our results obtained for the most realistic cases are however already sufficient to measure the positive impact of a certain form of network dynamics on information spreading. To go one step further, we think that the most challenging question is to analyze rumor spreading over more general classes of evolving graphs where edges may not be independent. For instance, it would be interesting to analyze the Push protocol over geometric models of mobile networks [12, 28].

References

1. Avin, C., Kouckỳ, M., Lotker, Z.: How to explore a fast-changing world (Cover time of a simple random walk on evolving graphs). In: Aceto, L., Damgård, I., Goldberg, L.A., Halldórsson, M.M., Ingólfsdóttir, A., Walukiewicz, I. (eds.) ICALP 2008, Part I. LNCS, vol. 5125, pp. 121–132. Springer, Heidelberg (2008)
2. Azar, Y., Broder, A.Z., Karlin, A.R., Upfal, E.: Balanced allocations. SIAM J. on Computing 29(1), 180–200 (1999)
3. Baumann, H., Crescenzi, P., Fraigniaud, P.: Parsimonious flooding in dynamic graphs. Distributed Computing 24(1), 31–44 (2011)
4. Bollobás, B.: Random Graphs. Cambridge University Press (2001)
5. Boyd, S., Arpita, G., Balaji, P., Devavrat, S.: Gossip algorithms: Design, analysis and applications. In: Proc. of 24th INFOCOM, pp. 1653–1664. IEEE (2005)
6. Chierichetti, F., Lattanzi, S., Panconesi, A.: Rumor Spreading in Social Networks. In: Albers, S., Marchetti-Spaccamela, A., Matias, Y., Nikoletseas, S., Thomas, W. (eds.) ICALP 2009, Part II. LNCS, vol. 5556, pp. 375–386. Springer, Heidelberg (2009)
7. Chierichetti, F., Lattanzi, S., Panconesi, A.: Almost tight bounds on rumour spreading by conductance. In: Proc. of 42nd ACM STOC, pp. 399–408 (2010)
8. Chierichetti, F., Lattanzi, S., Panconesi, A.: Rumour spreading and graph conductance. In: Proc. of 21th ACM-SIAM SODA, pp. 1657–1663 (2010)
9. Clementi, A., Crescenzi, P., Doerr, C., Fraigniaud, P., Isopi, M., Pasquale, F., Panconesi, A., Silvestri, R.: Rumor Spreading in Random Evolving Graphs, http://arxiv.org/abs/1302.3828
10. Clementi, A., Macci, C., Monti, A., Pasquale, F., Silvestri, R.: Flooding time of edge-Markovian evolving graphs. SIAM J. Discrete Math. 24(4), 1694–1712 (2010)
11. Clementi, A., Monti, A., Pasquale, F., Silvestri, R.: Broadcasting in dynamic radio networks. J. Comput. Syst. Sci. 75(4), 213–230 (2009)
12. Clementi, A., Monti, A., Pasquale, F., Silvestri, R.: Information spreading in stationary Markovian evolving graph. IEEE Trans. Parallel Distrib. Syst. 22(9), 1425–1432 (2011)
13. Demers, A., Greene, D., Hauser, C., Irish, W., Larson, J., Shenker, S., Sturgis, H., Swinehart, D., Terry, D.: Epidemic algorithms for replicated database maintenance. In: Proc. of 6th ACM PODC, pp. 1–12 (1987)

14. Doerr, B., Fouz, M., Friedrich, T.: Social networks spread rumors in sublogarithmic time. In: Proc. of 43rd ACM STOC, pp. 21–30. ACM, New York (2011)
15. Doerr, B., Huber, A., Levavi Strong, A.: robustness of randomized rumor spreading protocols. Discrete Applied Mathematics 161(6), 778–793 (2013)
16. Erdős, P., Rényi, A.: On Random Graphs. Publ. Math. 6, 290–297 (1959)
17. Feige, U., Peleg, D., Raghavan, P., Upfal, E.: Randomized broadcast in networks. Random Structures and Algorithms 1(4), 447–460 (1990)
18. Ferreira, A.: On models and algorithms for dynamic communication networks: The case for evolving graphs. In: Proc. of 4th ALGOTEL, pp. 155–161 (2002)
19. Fountoulakis, N., Huber, A., Panagiotou, K.: Reliable broadcasting in random networks and the effect of density. In: Proc. of 29th IEEE INFOCOM, pp. 2552–2560 (2010)
20. Fountoulakis, N., Panagiotou, K.: Rumor spreading on random regular graphs and expanders. In: Serna, M., Shaltiel, R., Jansen, K., Rolim, J. (eds.) APPROX and RANDOM 2010. LNCS, vol. 6302, pp. 560–573. Springer, Heidelberg (2010)
21. Fountoulakis, N., Panagiotou, K., Sauerwald, T.: Ultra-fast rumor spreading in social networks. In: Proc. of 23rd ACM-SIAM SODA, pp. 1642–1660 (2012)
22. Frieze, A., Grimmett, G.: The shortest-path problem for graphs with random arc-lengths. Discrete Applied Mathematics 10(1), 57–77 (1985)
23. Giakkoupis, G.: Tight bounds for rumor spreading in graphs of a given conductance. In: Proc. of 28th STACS. LIPIcs, vol. 9, pp. 57–68. Schloss Dagstuhl (2011)
24. Giakkoupis, G., Sauerwald, T.: Rumor spreading and vertex expansion. In: Proc. of 23rd ACM-SIAM SODA, pp. 1623–1641. SIAM (2012)
25. Gilbert, E.N.: Random graphs. Annals of Math. Statistics 30(4), 1141–1144 (1959)
26. Grindrod, P., Higham, D.J.: Evolving graphs: dynamical models, inverse problems and propagation. Proc. R. Soc. A 466(2115), 753–770 (2010)
27. Harchol-Balter, M., Leighton, T., Lewin, D.: Resource discovery in distributed networks. In: Proc. of 18th PODC, pp. 229–237. ACM, New York (1999)
28. Jacquet, P., Mans, B., Rodolakis, G.: Information Propagation Speed in Mobile and Delay Tolerant Networks. IEEE Trans. on Inf. Theory 56, 5001–5015 (2010)
29. Jelasity, M., Voulgaris, S., Guerraoui, R., Kermarrec, A.-M., van Steen, M.: Gossip-based Peer Sampling. ACM Trans. Comp. Syst. 25(3), Article 8 (2007)
30. Karp, R., Schindelhauer, C., Shenker, S., Vocking, B.: Randomized rumor spreading. In: Proc. of 41st IEEE FOCS, pp. 565–574. IEEE (2000)
31. Kermarrec, A.-M., van Steen, M.: Gossiping in distributed systems. SIGOPS Oper. Syst. Rev. 41(5), 2–7 (2007)
32. Kuhn, F., Linch, N., Oshman, R.: Distributed Computation in Dynamic Networks. In: Proc. of 42nd ACM STOC, pp. 513–522. ACM, New York (2010)
33. Kuhn, F., Oshman, R.: Dynamic networks: models and algorithms. ACM SIGACT News 42(1), 82–96 (2011)
34. Pittel, B.: On spreading a rumor. SIAM Journal on Applied Mathematics 47(1), 213–223 (1987)
35. Sauerwald, T., Stauffer, A.: Rumor spreading and vertex expansion on regular graphs. In: Proc. of 22nd ACM-SIAM SODA, pp. 462–475. SIAM (2011)
36. Van Renesse, R., Minsky, Y., Hayden, M.: A Gossip-Style Failure Detection Service. In: Proc. of Middleware, pp. 55–70 (1998)
37. Vojnovic, M., Proutier, A.: Hop limited flooding over dynamic networks. In: Proc. of 30th IEEE INFOCOM, pp. 685–693. IEEE (2011)
38. Whitbeck, J., Conan, V., de Amorim, M.D.: Performance of Opportunistic Epidemic Routing on Edge-Markovian Dynamic Graphs. IEEE Transactions on Communications 59(5), 1259–1263 (2011)

Dynamic Graphs in the Sliding-Window Model*

Michael S. Crouch, Andrew McGregor, and Daniel Stubbs

University of Massachusetts Amherst
140 Governors Drive, Amherst, MA 01003
{mcc,mcgregor,dstubbs}@cs.umass.edu

Abstract. We present the first algorithms for processing graphs in the sliding-window model. The sliding window model, introduced by Datar et al. (SICOMP 2002), has become a popular model for processing infinite data streams in small space when older data items (i.e., those that predate a sliding window containing the most recent data items) are considered "stale" and should implicitly be ignored. While processing massive graph streams is an active area of research, it was hitherto unknown whether it was possible to analyze graphs in the sliding-window model. We present an extensive set of positive results including algorithms for constructing basic graph synopses like combinatorial sparsifiers and spanners as well as approximating classic graph properties such as the size of a graph matching or minimum spanning tree.

1 Introduction

Massive graphs arise in any application where there is data about both basic entities and the relationships between these entities, e.g., web-pages and hyperlinks; papers and citations; IP addresses and network flows; phone numbers and phone calls; Tweeters and their followers. Graphs have become the de facto standard for representing many types of highly-structured data. Furthermore, many interesting graphs are dynamic, e.g., hyperlinks are added and removed, citations hopefully accrue over time, and the volume of network traffic between two IP addresses may vary depending on the time of day.

Consequently there is a growing body of work on designing algorithms for analyzing dynamic graphs. This includes both traditional data structures where the goal is to enable fast updates and queries [16, 22–24, 29] and data stream algorithms where the primary goal is to design randomized data structures of *sublinear size* that can answer queries with high probability [2, 3, 17, 18, 25, 27, 30]. The paper focuses on the latter: specifically, processing graphs using sublinear space in the sliding-window model. Although our focus isn't on update time, many of our algorithms can be made fast by using standard data structures.

Dynamic Graph Streams. Almost all of the existing work on processing graph streams considers what is sometimes referred to as the *partially-dynamic* case where the stream consists of a sequence of edges $\langle e_1, e_2, e_3, \ldots \rangle$ and the graph being monitored consists of the set of edges that have arrived so far. In other words, the graph is formed by a sequence of edge insertions. Over the last decade, it has been shown that many interesting

---

* Supported by NSF CAREER Award CCF-0953754 and associated REU supplement.

H.L. Bodlaender and G.F. Italiano (Eds.): ESA 2013, LNCS 8125, pp. 337–348, 2013.

problems can be solved using $O(n \operatorname{polylog} n)$ space, where n is the number of nodes in the graph. This is referred to as the *semi-streaming* space restriction [18]. It is only in the last year that semi-streaming algorithms for the *fully-dynamic* case, where edges can be inserted and deleted, have been discovered [2,3]. A useful example illustrating why the fully-dynamic case is significantly more challenging is testing whether a graph is connected. If there are only insertions, it suffices to track which nodes are in which connected component since these components only merge over time. In the dynamic case, connected components may also subdivide if bridge edges are deleted.

Sliding-Window Model. The sliding-window model, introduced by Datar et al. [14], has become a popular model for processing infinite data streams in small space when the goal is to compute properties of data that has arrived in the last window of time. Specifically, given an infinite stream of data $\langle a_1, a_2, \dots \rangle$ and a function f, at time t we need to return an estimate of $f(a_{t-L+1}, a_{t-L+2}, \dots, a_t)$. We refer to $\langle a_{t-L+1}, a_{t-L+2}, \dots, a_t \rangle$ as the *active* window where L is length of this window. The length of the window could correspond to hours, days, or years depending on the application. The motivation is that by ignoring data prior to the active window, we focus on the "freshest" data and can therefore detect anomalies and trends more quickly. Existing work has considered estimating various numerical statistics and geometric problems in this model [5–8,11,12,19], as well developing useful techniques such as the *exponential histogram* [14] and *smooth histogram* data structures [11,12].

1.1   Our Contributions

The paper initiates the study of processing graphs in the sliding-window model where the goal is to monitor the graph described by the last L entries of a stream of inserted edges. Note the following differences between this model and fully-dynamic model. In the sliding-window model the edge deletions are *implicit*, in the sense that when an edge leaves the active window it is effectively deleted but we may not know the identity of the deleted edge unless we store the entire window. In the case of fully-dynamic graph streams, the identity of the deleted edge is *explicit* but the edge could correspond to any of the edges already inserted but not deleted.

We present semi-streaming algorithms in the sliding-window model for various classic graph problems including testing connectivity, constructing minimum spanning trees, and approximating the size of graph matchings. We also present algorithms for constructing graph synopses including *sparsifiers* and *spanners*. We say a subgraph H of G is a $(2t - 1)$-spanner if:

$$\forall u, v \in V : \quad d_G(u, v) \le d_H(u, v) \le (2t - 1)d_G(u, v)$$

where $d_G(u, v)$ and $d_H(u, v)$ denote the distance between nodes u and v in G and H respectively. We say a weighted subgraph H of G is a $(1 + \epsilon)$ sparsifier if

$$\forall U \subset V : \quad (1 - \epsilon)\lambda_G(U) \le \lambda_H(u, v) \le (1 + \epsilon)\lambda_G(U)$$

where $\lambda_G(U)$ and $\lambda_H(U)$ denote the weight of the cut $(U, V \setminus U)$ in G and H respectively. A summary of our results can be seen in Table 1 along with the state-of-the-art results for these problems in the insert-only and insert/delete models.

Table 1. Single-Pass, Semi-Streaming Results: All the above algorithms use $O(n \operatorname{polylog} n)$ space with the exception of the spanner constructions

	Insert-Only	Insert-Delete	Sliding Window (this paper)
Connectivity	Deterministic [18]	Randomized [2]	Deterministic
Bipartiteness	Deterministic [18]	Randomized [2]	Deterministic
$(1 + \epsilon)$-Sparsifier	Deterministic [1]	Randomized [3, 21]	Randomized
$(2t - 1)$-Spanners	$O(n^{1+1/t})$ space [9, 15]	None	$O(L^{1/2} n^{(1+1/t)/2})$ space
Min. Spanning Tree	Exact [18]	$(1 + \epsilon)$-approx. [2]	$(1 + \epsilon)$-approx.
Unweighted Matching	2-approx. [18]	None	$(3 + \epsilon)$-approx.
Weighted Matching	4.911-approx. [17]	None	9.027-approx.

2   Connectivity and Graph Sparsification

We first consider the problem of testing whether the graph is k-edge connected for a given $k \in \{1, 2, 3 \ldots\}$. Note that $k = 1$ corresponds to testing connectivity. To do this, it is sufficient to maintain a set of edges $F \subseteq \{e_1, e_2, \ldots, e_t\}$ along with the time-of-arrival $\operatorname{toa}(e)$ for each $e \in F$ where F satisfies the following property:

– *Recent Edges Property.* For every cut $(U, V \setminus U)$, the stored edges F contain the most recent $\min(k, \lambda(U))$ edges across the cut where $\lambda(U)$ denotes the total number of edges from $\{e_1, e_2, \ldots, e_t\}$ that cross the cut.

Then, we can easily tell whether the graph on the active edges, $\langle e_{t-L+1}, e_{t-L+2}, \ldots, e_t \rangle$, is k-connected by checking whether F would be k-connected once we remove all edges $e \in F$ where $\operatorname{toa}(e) \leq t - L$. This follows because if there are k or more edges among the last L edges across a cut, F will include the k most recent of them.

Algorithm. The following simple algorithm maintains a set F with the above property. The algorithm maintains k disjoint sets of edges $F_1, F_2, \ldots, F_k$ where each F_i is acyclic. Initially, $F_1 = F_2 = \ldots = F_k = \emptyset$ and on seeing edge e in the stream, we update the sets as follows:

1. Define the sequence $f_0, f_1, f_2, f_3, \ldots$ where $f_0 = \{e\}$ and for each $i \geq 1$, f_i consists of the oldest edge in a cycle in $F_i \cup f_{i-1}$ if such a cycle exists and $f_i = \emptyset$ otherwise. Since each F_i is acyclic, there will be at most one cycle in each $F_i \cup f_{i-1}$.
2. For $i \in \{1, 2, \ldots, k\}$,
$$F_i \leftarrow (F_i \cup f_{i-1}) \setminus f_i$$

In other words, we add the new edge e to F_1. If it completes a cycle, we remove the oldest edge on this cycle and add that edge to F_2. If we now have a cycle in F_2, we remove the oldest edge on this cycle and add that edge to F_3. And so on. By using an existing data structure for constructing online minimum spanning trees [28], the above algorithm can be implemented with $O(k \log n)$ update time.

Lemma 1. $F = F_1 \cup F_2 \cup \ldots \cup F_k$ *satisfies the Recent Edges Property.*

Proof. Fix some $i \in [k]$ and a cut $(U, V \setminus U)$. Observe that the youngest edge $y \in F_i$ crossing a cut $(U, V \setminus U)$ is never removed from F_i since its removal would require it to be the oldest edge in some cycle C. This cannot be the case since there must be an even number of edges in C that cross the cut and so there is another edge $x \in C$ crossing the cut. This edge must have been older than y since y was the youngest.

It follows that F_1 always contains the youngest edge crossing any cut, and by induction on i, the ith youngest edge crossing any cut is contained in $\bigcup_{j=1}^{i} F_j$. This is true because this edge was initially added to $F_1 \subseteq \bigcup_{j=1}^{i} F_j$, and cannot leave $\bigcup_{j=1}^{i} F_j$. That is, for the ith youngest edge to leave F_i, there would have to be a younger crossing edge in F_i, but, inductively, any such edge is contained in $\bigcup_{j=1}^{i-1} F_j$.   □

Theorem 1. *There exists a sliding-window algorithm for monitoring k-connectivity using $O(kn \log n)$ space.*

2.1   Applications: Bipartiteness and Graph Sparsification

Bipartiteness. To monitor whether a graph is bipartite, we run the connectivity tester on the input graph and also simulate the connectivity tester on the *cycle double cover* of the input graph. The cycle double cover $D(G)$ of a graph $G = (V, E)$ is formed by replacing each node $v \in V$ by two copies v_1 and v_2 and each edge $(u, v) \in E$ by the edges (u_1, v_2) and (u_2, v_1). Note that this transformation can be performed in a streaming fashion. Furthermore, $D(G)$ has exactly twice the number of connected components as G iff G is bipartite [2].

Theorem 2. *There exists a sliding-window algorithm for monitoring bipartiteness using $O(n \log n)$ space.*

Graph Sparsification. Using the k-connectivity tester as a black-box we can also construct a $(1 + \epsilon)$-sparsifier following the approach of Ahn et al. [3]. The approach is based upon a result by Fung et al. [20] that states that sampling each edge e with probability $p_e \geq \min\{253\lambda_e^{-1}\epsilon^{-2}\log^2 n, 1\}$, where λ_e is the size of the minimum cut that includes e, and weighting the sampled edges by $1/p_e$ results in a $(1 + \epsilon)$ sparsifier with high probability. To emulate this sampling without knowing λ_e values, we subsample the graph stream to generate sub-streams that define $O(\log n)$ graphs $G_0, G_1, G_2, \ldots$ where each edge is in G_i with probability 2^{-i}. For each i, we store the set of edges $F(G_i)$ generated by the k-connectivity algorithm. If $k = \Theta(\epsilon^{-2} \log^2 n)$, then note that e is in some $F(G_i)$ with probability at least $\min\{\Omega(\lambda_e^{-1}\epsilon^{-2}\log^2 n), 1\}$ as required. See Ahn et al. [3] for further details.

Theorem 3. *There exists a sliding-window algorithm for maintaining a $(1 + \epsilon)$ sparsifier using $O(\epsilon^{-2} n \, \mathrm{polylog}\, n)$ space.*

3   Matchings

We next consider the problem of finding large matchings in the sliding-window model. We first consider the unweighted case, maximum cardinality matching, and then generalize to the weighted case.

3.1 Maximum Cardinality Matching

Our approach for estimating the size of the maximum cardinality matching combines ideas from the powerful "smooth histograms" technique of Braverman and Ostrovsky [11, 12] with the fact that graph matchings are submodular and satisfy a "smooth-like" condition.

Smooth Histograms. The smooth histogram technique gives a general framework for maintaining an estimate of a function f on a sliding window provided that f satisfies a certain set of conditions. Among these conditions are:

1. *Smoothness:* For any $\alpha \in (0, 1)$ there exists $\beta \in (0, \alpha]$ such that

$$f(B) \geq (1 - \beta)f(AB) \quad \text{implies} \quad f(BC) \geq (1 - \alpha)f(ABC) \qquad (1)$$

   where A, B, and C are disjoint segments of the stream and AB, BC, ABC denote concatenations of these segments.
2. *Approximability:* There exists a sublinear space stream algorithm that returns an estimate $\tilde{f}(A)$ for f evaluated on a (non-sliding-window) stream A, such that

$$(1 - \beta/4)f(A) \leq \tilde{f}(A) \leq (1 + \beta/4)f(A)$$

The basic idea behind smooth histograms is to approximate f on various suffixes $B_1, B_2, \ldots, B_k$ of the stream where $B_1 \supseteq W \supsetneq B_2 \supsetneq \cdots \supsetneq B_k$ and W is the active window. We refer to the B_i as "buckets." Roughly speaking, if we can ensure that $f(B_{i+1}) \approx (1 - \epsilon)f(B_i)$ for each i then $f(B_2)$ is a good approximation for $f(W)$ and we will only need to consider a logarithmic number of suffixes. We will later present the relevant parts of the technique in more detail in the context of approximate matching.

Matchings are Almost Smooth. Let $m(A)$ denote the size of the maximum matching on a set of edges A. Unfortunately, the function m does not satisfy the above smoothness condition and cannot be approximated to sufficient accuracy. It does however satisfy a "smooth-like" condition:

Lemma 2. *For disjoint segments of the stream A, B, and C and for any $\beta > 0$:*

$$m(B) \geq (1 - \beta)m(AB) \quad \text{implies} \quad m(BC) \geq \frac{1}{2}(1 - \beta)m(ABC) \qquad (2)$$

Proof. $2m(BC) \geq m(B) + m(BC) \geq (1 - \beta)m(AB) + m(BC) \geq (1 - \beta)m(ABC)$. The last step follows since $m(AB) + m(BC) \geq m(A) + m(BC) \geq m(ABC)$. $\square$

The best known semi-streaming algorithm for approximating m on a stream A is a 2-approximation and a lower bound 1.582 has recently been proved [25]. Specifically, let $\hat{m}(A)$ be the size of the greedy matching on A. Then it is easy to show that

$$m(A) \geq \hat{m}(A) \geq m(A)/2 \qquad (3)$$

Unfortunately, it is not possible to maintain a greedy matching over a sliding window.[1] However, by adjusting the analysis of [12], properties (2) and (3) suffice to show that smooth histograms can obtain an $(8 + \epsilon)$-approximation of the maximum matching in the sliding-window model. However, by proving a modified smoothness condition that takes advantage of relationships between m and $\hat{m}$, and specifically the fact that $\hat{m}$ is maximal rather than just a 2-approximation, we will show that a smooth histograms-based approach can obtain a $(3 + \epsilon)$-approximation.

Lemma 3. *Consider any disjoint segments A, B, C of a stream of edges and $\beta \in (0, 1)$.*

$$\hat{m}(B) \geq (1 - \beta)\hat{m}(AB) \quad \text{implies} \quad m(ABC) \leq \left(3 + \frac{2\beta}{1 - \beta}\right)\hat{m}(BC) .$$

Note that it the size of the *maximum* matching on ABC that is being compared with the size of the greedy matching on BC. Also to see that the above lemma is tight for any $\beta \in (0, 1)$ consider the following graph on $O(n)$ nodes:

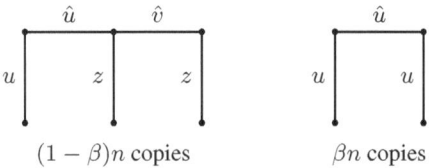

$(1 - \beta)n$ copies βn copies

and let A be the stream of the $\hat{u}$ edges (which are placed in greedy matching) followed the u edges; B are the $\hat{v}$ edges, and C are the z edges. Then $\hat{m}(AB) = n$, $\hat{m}(B) = (1 - \beta)n = \hat{m}(BC)$, and $m(ABC) = (3 - \beta)n$.

Proof (Lemma 3). Let $M(X)$ and $\hat{M}(X)$ be the set of edges in an optimal matching on X and a maximal matching on X. We say that an edge in a matching *covers* the two nodes which are its endpoints.

We first note that every edge in $M(ABC)$ covers at least one node which is covered by $\hat{M}(AB) \cup \hat{M}(BC)$; otherwise, the edge could have been added to $\hat{M}(AB)$ or $\hat{M}(BC)$ or both. Since $M(ABC)$ is a matching, no two of its edges can cover the same node. Thus $m(ABC)$ is at most the number of nodes covered by $\hat{M}(AB) \cup \hat{M}(BC)$.

The number of nodes covered by $\hat{M}(AB) \cup \hat{M}(BC)$ is clearly at most $2\hat{m}(AB) + 2\hat{m}(BC)$. But this over-counts edges in $\hat{M}(B)$. Every edge in $\hat{M}(B)$ is clearly in

---

[1] Maintaining the matching that would be generated by a greedy algorithm on the active window requires $\Omega(\min(n^2, L))$ space since it would always contain the oldest edge in the window and advancing the window allows us to recover all the edges. Similarly, it is not possible to construct the matching that would be returned by a greedy algorithm on reading the active window in reverse. This can be seen to require $\Omega(n^2)$ space even in the unbounded-stream model via reduction from INDEX. Alice considers the possible edges on an n-clique, and includes an edge iff the corresponding bit of her input is a 1. Bob then adds edges forming a perfect matching on all nodes except the endpoints of an edge of interest. The backwards greedy matching on the resulting graph consists of all of Bob's edges, plus one additional edge iff Alice's corresponding bit was a 1.

$\hat{M}(BC)$; also, every edge in $\hat{M}(B)$ shares at least one node with an edge in $\hat{M}(AB)$ since the construction was greedy. Thus we find

$$m(ABC) \leq 2\hat{m}(BC) + 2\hat{m}(AB) - \hat{m}(B)$$
$$\leq 2\hat{m}(BC) + \frac{2}{1-\beta}\hat{m}(B) - \hat{m}(B)$$
$$= 2\hat{m}(BC) + \frac{1+\beta}{1-\beta}\hat{m}(B)$$
$$\leq \left(3 + \frac{2\beta}{1-\beta}\right)\hat{m}(BC).$$

where the second inequality follows from the assumption $\hat{m}(B) \geq (1-\beta)\hat{m}(AB)$.   □

Theorem 4. *There exists a sliding-window algorithm for maintaining a $(3+\epsilon)$ approximation of the maximum cardinality matching using $O(\epsilon^{-1}n\log^2 n)$ space.*

Proof. We now use the smooth histograms technique to estimate the maximum matching size. The algorithm maintains maximal matchings over various buckets $B_1, \ldots, B_k$ where each bucket comprises of the edges in some suffix of the stream. Let W be the set of updates within the window. The buckets will always satisfy $B_1 \supseteq W \supsetneq B_2 \supsetneq \cdots \supsetneq B_k$, and thus $m(B_1) \geq m(W) \geq m(B_2)$.

Within each bucket B, we will keep a greedy matching whose size we denote by $\hat{m}(B)$. To achieve small space usage, whenever two nonadjacent buckets have greedy matchings of similar size, we will delete any buckets between them. Lemma 3 tells us that if the greedy matchings of two buckets have ever been close, then the smaller bucket's greedy matching is a good approximation of the size of the maximum matching on the larger bucket.

When a new edge e arrives, we update the buckets $B_1, \ldots, B_k$ and greedy matchings $\hat{m}(B_1), \ldots, \hat{m}(B_k)$ as follows where $\beta = \epsilon/4$:

1. Create a new empty bucket B_{k+1}.
2. Add e to the greedy matching within each bucket if possible.
3. For $i = 1 \ldots k - 2$:
   (a) Find the largest $j > i$ such that $\hat{m}(B_j) \geq (1 - \beta)\hat{m}(B_i)$
   (b) Delete B_t for any $i < t < j$ and renumber the buckets.
4. If B_2 contains the entire active window, delete B_1 and renumber the buckets.

Space Usage: Step 3 deletes "unnecessary" buckets and therefore ensures that for all $i \leq k - 2$ then $\hat{m}(B_{i+2}) < (1 - \beta)\hat{m}(B_i)$. Since the maximum matching has size at most n, this ensures that the number of buckets is $O(\epsilon^{-1}\log n)$. Hence, the total number of bits used to maintain all k greedy matchings is $O(\epsilon^{-1}n\log^2 n)$.

Approximation Factor: We prove the invariant that for any $i < k$, either $\hat{m}(B_{i+1}) \geq m(B_i)/(3 + \epsilon)$ or $|B_i| = |B_{i+1}| + 1$ (i.e., B_{i+1} includes all but the first edge of B_i) or both. If $|B_i| \neq |B_{i+1}| + 1$, then we must have deleted some bucket B which

$B_i \subsetneq B \subsetneq B_{i+1}$. For this to have happened it must have been the case that $\hat{m}(B_{i+1}) \geq (1 - \beta)\hat{m}(B_i)$ at the time. But then Lemma 3 implies that we currently satisfy:

$$m(B_i) \leq \left(3 + \frac{2\beta}{1 - \beta}\right) \hat{m}(B_{i+1}) \leq (3 + \epsilon)\hat{m}(B_{i+1}) .$$

Therefore, either $W = B_1$ and $\hat{m}(B_1)$ is a 2-approximation for $m(W)$, or we have

$$m(B_1) \geq m(W) \geq m(B_2) \geq \hat{m}(B_2) \geq \frac{m(B_1)}{3 + \epsilon}$$

and thus $\hat{m}(B_2)$ is a $(3 + \epsilon)$-approximation of $m(W)$. $\square$

3.2   Weighted Matching

We next consider the weighted case where every edge e in the stream is accompanied by a numerical value corresponding to its weight. We combine our algorithm for maximum cardinality matching with the approach of Epstein et al. [17] to give a 9.027 approximation. In this approach, we partition the set of edges into classes of geometrically increasing weights and construct a large cardinality matching in each weight class. We assume that the edge weights are polynomially bounded in n and hence there are $O(\log n)$ weight classes.

Geometrically Increasing Edge Weights. Initially, we assume that for some constants $\gamma > 1, \phi > 0$, every edge has weight $\gamma^i \phi$ for some $i \in \{0, 1, 2, \ldots\}$. Let E_i denote the set of edges with weight $\gamma^i \phi$. Our algorithm will proceed as follows:

1. For each i, use an instantiation of the maximum cardinality algorithm from the previous section to maintain a matching $A_i \subseteq E_i$ among the active edges.
2. Let R be the matching formed by greedily adding all possible edges from $A = \cup_i A_i$ in decreasing order of weight.

The next lemma bounds the total weight of edges in A in terms of the total weight of edges in R.

Lemma 4. $w(R)/w(A) \geq (\gamma - 1)/(\gamma + 1)$.

Note that the lemma is tight: consider the graph with a single edge of weight γ^k, itself adjacent to two edges of each smaller weight $\gamma^{k-1}, \gamma^{k-2}, \ldots$. If $A = E$, we have $w(R) = \gamma^k = (1 - 2/(\gamma + 1))w(A)$.

Proof (Lemma 4). Consider the process of greedily constructing R. Call an edge $e \in A$ "chosen" when it is added to R, and "discarded" if some covering edge is added to R. Edges which have not yet been chosen or discarded are said to be "in play". Note that once edges are discarded they cannot be added to R, and that the greedy construction continues until no edges remain in play.

We bound the weight of edges discarded when an edge is chosen. For an edge to be chosen, it must be the heaviest edge in play. None of its in-play neighbors can be in the same weight class, because within each weight class we have a matching. Thus, when an edge is chosen, the edges discarded are all in smaller weight classes; there are at most two edges discarded in each of these classes. If the edge $e \in A_i$ is chosen, it has weight $\gamma^i \phi$. For each $j < i$ there are at most two edges discarded with weight $\gamma^j \phi$. Let $T(e)$ be the set of edges discarded when $e \in A_i$ is chosen including e itself. Then,

$$\frac{w(e)}{w(T(e))} = \frac{\gamma^i \phi}{\gamma^i \phi + 2\sum_{j=0}^{i-1} \gamma^j \phi} = \frac{\gamma^i}{\gamma^i + 2\gamma^{i-1} \sum_{j=0}^{i-1} \gamma^{-j}} \geq \frac{1}{1 + \frac{2}{\gamma - 1}} = \frac{\gamma - 1}{\gamma + 1}$$

Since this holds for each chosen edge and all the edges appear in some $T(e)$, we conclude

$$w(R) = \sum_{e \in R} w(e) \geq \frac{\gamma - 1}{\gamma + 1} \sum_{e \in A} w(e) = \frac{\gamma - 1}{\gamma + 1} w(A)$$

as required.   □

Let OPT be the maximum-weight matching on E. $w(\text{OPT})$ is clearly at most the sum of the optimum-weight matchings on each E_i. Thus we have

Corollary 1. *If each A_i is an $(3+\epsilon)$ approximation for the maximum cardinality matching on E_i then*

$$w(\text{OPT}) \leq (3 + \epsilon)\frac{\gamma + 1}{\gamma - 1} w(R) \tag{4}$$

Arbitrary Edge Weights. We now reduce the case of arbitrary edge weights to the geometric case. Let OPT be the maximum-weight matching on $G = (V, E, w)$ and let OPT$'$ be the maximum weight matching on $G' = (V, E, w'_\phi)$ where $w'_\phi(e) = \gamma^i \phi$ for some $\gamma > 1, \phi > 0$ and i satisfies $\gamma^{i+1} \phi > w(e) \geq \gamma^i \phi$. This ensures that

$$w(\text{OPT}) < \gamma w'_\phi(\text{OPT}) \leq \gamma w'_\phi(\text{OPT}') .$$

However, Epstein et al. show that there exists $\phi \in \{\gamma^{0/q}, \gamma^{1/q}, \gamma^{2/q}, \ldots, \gamma^{1-1/q}\}$ where $q = O(\log_\gamma(1 + \epsilon))$ such that

$$w(\text{OPT}) \leq \frac{(1 + \epsilon)\gamma \ln \gamma}{\gamma - 1} w'_\phi(\text{OPT}) \leq \frac{(1 + \epsilon)\gamma \ln \gamma}{\gamma - 1} w'_\phi(\text{OPT}') .$$

And so, if we run the above algorithm with respect to w'_ϕ in parallel for each choice of ϕ, we ensure that for some ϕ,

$$w(\text{OPT}) \leq \frac{(1 + \epsilon)\gamma \ln \gamma}{\gamma - 1} w'_\phi(\text{OPT}) \leq (3 + \epsilon) \cdot \frac{(1 + \epsilon)\gamma \ln \gamma}{\gamma - 1} \cdot \frac{\gamma + 1}{\gamma - 1} w(R) ,$$

by appealing to the analysis for geometrically increasing weights (Corollary 1). This is minimized at $\gamma \approx 5.704$ to give an approximation ratio of less than 9.027 when we set ϵ to be some sufficiently small constant.

Theorem 5. *There exists a sliding-window algorithm for maintaining a 9.027 approximation for the maximum weighted matching using $O(n \log^3 n)$ space.*

4   Minimum Spanning Tree

We next consider the problem of maintaining a minimum spanning forest in the sliding-window model. We show that it is possible to maintain a spanning forest that is at most a factor $(1 + \epsilon)$ from optimal but that maintaining the exact minimum spanning tree requires $\Omega(\max(n^2, L))$ space where L is the length of the sliding window.

The approximation algorithm is based on an idea of Chazelle et al. [13] where the problem is reduced to finding maximal acyclic subgraphs, i.e., spanning forests, among edges with similar weights. If each edge weight is rounded to the nearest power of $(1+\epsilon)$, it can be shown that the minimum spanning tree in the union of these subgraphs is a $(1 + \epsilon)$ approximation of the minimum spanning tree of the original graph. The acyclic subgraphs can be found in the sliding-window model using the connectivity algorithm we presented earlier. The proof of the next theorem is almost identical to those in [2, Lemma 3.4].

Theorem 6. *There exists a sliding-window algorithm for maintaining a $(1+\epsilon)$ approximation for the minimum spanning tree using $O(\epsilon^{-1}n \log^2 n)$ bits of space.*

In the unbounded stream model, it was possible to compute the exact minimum spanning tree via a simple algorithm: 1) add the latest edge to an acyclic subgraph that is being maintained, 2) if this results in a cycle, remove the heaviest weight edge in the cycle. However, the next theorem shows that maintaining an exact minimum spanning tree in the sliding-window model is not possible in sublinear space.

Theorem 7. *Maintaining an exact minimum spanning forest in the sliding-window model requires $\Omega(\min(L, n^2))$ space.*

Proof. Let $p = \min(L, n^2/4)$. The proof is by a reduction from the communication complexity of the two-party communication problem INDEX(p) where Alice holds a binary string $a = a_1 a_2 \dots a_p$ and Bob has an index $k \in [p]$. If Alice sends a single message to Bob that enables Bob to output a_k with probability at least $2/3$, then Alice's message must contain at $\Omega(p)$ bits [26].

Alice encodes her bits on the edges of a complete bipartite graph, writing in order the edges $(u_1, v_1), (u_1, v_2), (u_1, v_3), \dots, (u_1, v_{\sqrt{p}}), (u_2, v_1), \dots, (u_2, v_{\sqrt{p}}), \dots, (u_{\sqrt{p}}, v_{\sqrt{p}})$ where the ith edge weight $2i + a_i$. Note that all these edges are in the current active window. Suppose she runs a sliding-window algorithm for exact MST on this graph and sends the memory state to Bob. Bob continues running the algorithm on an arbitrary set of $L - p + k - 1$ edges each of weight $2p + 2$. At this point any minimum spanning forest in the active window must contain the edge of weight $2k + a_k$ since it is the lowest-cost edge in the graph. Bob can thus learn a_k and hence the algorithm must have used $\Omega(p)$ bits of memory. Note that if Bob can only determine *what* the MST edges are, but not their weights, he can add an alternative path of weight $2k + 1/2$ to the node in question.   □

5   Graph Spanners

In the unbounded stream model, the following greedy algorithm constructs a $2t - 1$ stretch spanner with $O(n^{1+1/t})$ edges [4, 18]. We process the stream of edges in order;

when seeing each edge (u, v), we add it to the spanner if there is not already a path from u to v of length $2t - 1$ or less. Any path in the original graph then increases by a factor of at most $2t - 1$, so the resulting graph is a $(2t - 1)$-spanner. The resulting graph has girth at least $2t + 1$, so it has at most $O(n^{1+1/t})$ edges [10].

For graphs G_1, G_2 on the same set of nodes, let $G_1 \cup G_2$ denote the graph with the union of edges from G_1 and G_2. We will need the following simple lemma.

Lemma 5. *Let G_1 and G_2 be graphs on the same set of nodes, and let H_1 and H_2 be α-spanners of G_1 and G_2 respectively. Then $H_1 \cup H_2$ is an α-spanner of $G_1 \cup G_2$.*

Proof. Let $G = G_1 \cup G_2$ and $H = H_1 \cup H_2$. For arbitrary nodes u, v, consider a path of length $d_G(u, v)$. Each edge in this path is in G_1 or G_2 (or both). There is thus a path between the edge's endpoints in the corresponding H_1 or H_2 which is of length at most α. Thus, there is a path of length at most $\alpha d_G(u, v)$ in $H = H_1 \cup H_2$.

Theorem 8. *There exists a sliding-window algorithm for maintaining a $(2t - 1)$ spanner using $O(\sqrt{L n^{1+1/t}})$ space.*

Proof. We batch the stream into blocks $E_1, E_2, E_3, \ldots$, where each consists of s edges. We buffer the edges in each block until it has been read completely, marking each edge with its arrival time. We then run the greedy spanner construction on each block in reverse order, obtaining a spanner S_i. By Lemma 5, the union of the spanners S_i and the edges in the current block, restricted to the active edges, is a spanner of the edges in the active window. This algorithm requires space s to track the edges in the current block. Each spanner S_i has $O(n^{1+1/t})$ edges, and at most L/s past blocks are within the window. The total number of edges stored by the algorithm is thus $s + (L/s)O(n^{1+1/t})$. Setting $s = \sqrt{L n^{1+1/t}}$ gives $O(\sqrt{L n^{1+1/t}})$ edges. □

6   Conclusions

We initiate the study of graph problems in the well-studied sliding-window model. We present algorithms for a wide range of problems including testing connectivity and constructing combinatorial sparsifiers; constructing minimum spanning trees; approximating weighted and unweighted matchings; and estimating graph distances via the construction of spanners. Open problems include reducing the space required to construct graph spanners and improving the approximation ratio when estimating matching size.

References

1. Ahn, K.J., Guha, S.: Graph sparsification in the semi-streaming model. In: Albers, S., Marchetti-Spaccamela, A., Matias, Y., Nikoletseas, S., Thomas, W. (eds.) ICALP 2009, Part II. LNCS, vol. 5556, pp. 328–338. Springer, Heidelberg (2009)
2. Ahn, K.J., Guha, S., McGregor, A.: Analyzing graph structure via linear measurements. In: SODA, pp. 459–467 (2012)
3. Ahn, K.J., Guha, S., McGregor, A.: Graph sketches: sparsification, spanners, and subgraphs. In: PODS, pp. 5–14 (2012)
4. Althöfer, I., Das, G., Dobkin, D.P., Joseph, D., Soares, J.: On sparse spanners of weighted graphs. Discrete & Computational Geometry 9, 81–100 (1993)
5. Arasu, A., Manku, G.S.: Approximate counts and quantiles over sliding windows. In: PODS, pp. 286–296 (2004)

6. Ayad, A., Naughton, J.F.: Static optimization of conjunctive queries with sliding windows over infinite streams. In: SIGMOD Conference, pp. 419–430 (2004)
7. Babcock, B., Datar, M., Motwani, R.: Sampling from a moving window over streaming data. In: SODA, pp. 633–634 (2002)
8. Babcock, B., Datar, M., Motwani, R., O'Callaghan, L.: Maintaining variance and k-medians over data stream windows. In: PODS, pp. 234–243 (2003)
9. Baswana, S.: Streaming algorithm for graph spanners—single pass and constant processing time per edge. Information Processing Letters (2008)
10. Bollobás, B.: Extremal graph theory. Academic Press, New York (1978)
11. Braverman, V., Gelles, R., Ostrovsky, R.: How to catch ℓ_2-heavy-hitters on sliding windows. In: Du, D.-Z., Zhang, G. (eds.) COCOON 2013. LNCS, vol. 7936, pp. 638–650. Springer, Heidelberg (2013)
12. Braverman, V., Ostrovsky, R.: Smooth Histograms for Sliding Windows. In: 48th Annual IEEE Symposium on Foundations of Computer Science, pp. 283–293 (October 2007)
13. Chazelle, B., Rubinfeld, R., Trevisan, L.: Approximating the minimum spanning tree weight in sublinear time. SIAM J. Comput. 34(6), 1370–1379 (2005)
14. Datar, M., Gionis, A., Indyk, P., Motwani, R.: Maintaining Stream Statistics over Sliding Windows. SIAM Journal on Computing 31(6), 1794 (2002)
15. Elkin, M.: Streaming and fully dynamic centralized algorithms for constructing and maintaining sparse spanners. ACM Transactions on Algorithms 7(2), 1–17 (2011)
16. Eppstein, D., Galil, Z., Italiano, G., Nissenzweig, A.: Sparsificationa technique for speeding up dynamic graph algorithms. Journal of the ACM 44(5), 669–696 (1997)
17. Epstein, L., Levin, A., Mestre, J., Segev, D.: Improved Approximation Guarantees for Weighted Matching in the Semi-streaming Model. SIAM Journal on Discrete Mathematics 25(3), 1251–1265 (2011)
18. Feigenbaum, J., Kannan, S., McGregor, A., Suri, S., Zhang, J.: On graph problems in a semi-streaming model. Theoretical Computer Science 348(2-3), 207–216 (2005)
19. Feigenbaum, J., Kannan, S., Zhang, J.: Computing diameter in the streaming and sliding-window models. Algorithmica 41(1), 25–41 (2004)
20. Fung, W.S., Hariharan, R., Harvey, N.J.A., Panigrahi, D.: A general framework for graph sparsification. In: STOC, pp. 71–80 (2011)
21. Goel, A., Kapralov, M., Post, I.: Single pass sparsification in the streaming model with edge deletions. CoRR, abs/1203.4900 (2012)
22. Henzinger, M.R., King, V.: Randomized fully dynamic graph algorithms with polylogarithmic time per operation. J. ACM 46(4), 502–516 (1999)
23. Holm, J., de Lichtenberg, K., Thorup, M.: Poly-logarithmic deterministic fully-dynamic algorithms for connectivity, minimum spanning tree, 2-edge, and biconnectivity. J. ACM 48(4), 723–760 (2001)
24. Italiano, G., Eppstein, D., Galil, Z.: Dynamic graph algorithms. In: Algorithms and Theory of Computation Handbook. CRC Press (1999)
25. Kapralov, M.: Better bounds for matchings in the streaming model. In: SODA, pp. 1679–1697 (2013)
26. Kushilevitz, E., Nisan, N.: Communication Complexity, vol. 2006. Cambridge University Press (1997)
27. McGregor, A.: Finding graph matchings in data streams. In: Chekuri, C., Jansen, K., Rolim, J.D.P., Trevisan, L. (eds.) APPROX 2005 and RANDOM 2005. LNCS, vol. 3624, pp. 170–181. Springer, Heidelberg (2005)
28. Tarjan, R.: Data Structures and Network Algorithms. SIAM, Philadelphia (1983)
29. Thorup, M.: Near-optimal fully-dynamic graph connectivity. In: STOC, pp. 343–350 (2000)
30. Zelke, M.: Weighted matching in the semi-streaming model. Algorithmica 62(1-2), 1–20 (2012)

A Quantization Framework for Smoothed Analysis of Euclidean Optimization Problems

Radu Curticapean[1] and Marvin Künnemann[1,2]

[1] Universität des Saarlandes, Saarbrücken, Germany
curticapean@cs.uni-saarland.de
[2] Max-Planck-Institut für Informatik, Saarbrücken, Germany
marvin@mpi-inf.mpg.de

Abstract. We consider the smoothed analysis of Euclidean optimization problems. Here, input points are sampled according to density functions that are bounded by a sufficiently small smoothness parameter ϕ. For such inputs, we provide a general and systematic approach that allows to design linear-time approximation algorithms whose output is asymptotically optimal, both in expectation and with high probability.

Applications of our framework include maximum matching, maximum TSP, and the classical problems of k-means clustering and bin packing. Apart from generalizing corresponding average-case analyses, our results extend and simplify a polynomial-time probable approximation scheme on multidimensional bin packing on ϕ-smooth instances, where ϕ is constant (Karger and Onak, SODA 2007).

Both techniques and applications of our rounding-based approach are orthogonal to the only other framework for smoothed analysis on Euclidean problems we are aware of (Bläser et al., Algorithmica 2012).

1 Introduction

Smoothed analysis has been introduced by Spielman and Teng [26] to give a theoretical foundation for analyzing the practical performance of algorithms. In particular, this analysis paradigm was able to provide an explanation why the simplex method is observed to run fast in practice despite its exponential worst-case running time.

The key concept of smoothed analysis, i.e., letting an adversary choose worst-case distributions of bounded "power" to determine input instances, is especially well-motivated in a Euclidean setting. Here, input points are typically determined by physical measurements, which are subject to an inherent inaccuracy, e.g., from locating a position on a map. For clustering problems, it is often even implicitly assumed that the points are sampled from unknown probability distributions which are sought to be recovered.

Making the mentioned assumptions explicit, we call a problem *smoothed tractable* if it admits a linear-time algorithm with an approximation ratio that is bounded by $1 - o(1)$ with high probability over the input distribution specified by the adversary. Such an approximation performance is called *asymptotically*

H.L. Bodlaender and G.F. Italiano (Eds.): ESA 2013, LNCS 8125, pp. 349–360, 2013.
© Springer-Verlag Berlin Heidelberg 2013

optimal. We provide a unified approach to show that several Euclidean optimization problems are smoothed tractable, which sheds light onto the properties that render a Euclidean optimization problem likely to profit from perturbed input.

We employ the *one-step model*, a widely-used and very general perturbation model, which has been successfully applied to analyze a number of algorithms [9,10,11,15]. In this model, an adversary chooses probability densities on the input space, according to which the input instance is drawn. To prevent the adversary from modeling a worst-case instance too closely, we bound the density functions from above by a parameter ϕ. Roughly speaking, for large ϕ, we expect the algorithm to perform almost as bad as on worst-case instances. Likewise, choosing ϕ as small as possible requires the adversary to choose the uniform distribution on the input space, corresponding to an average-case analysis. Thus, the adversarial power ϕ serves as an interpolation parameter between worst and average case.

Formally, given a set of feasible distributions $\mathcal{F}$ that depends on ϕ, and a performance measure t, we define the smoothed performance of an algorithm under the perturbation model $\mathcal{F}$ as $\max_{f_1,\ldots,f_n \in \mathcal{F}} E_{(X_1,\ldots,X_n) \sim (f_1,\ldots,f_n)}[t(X_1,\ldots,X_n)]$. In this work, we will be concerned with analyzing the smoothed approximation ratio, as well as bounds on the approximation ratio that hold with high probability over the perturbations.

For given ϕ, we require the density functions chosen by the adversary to be bounded by ϕ. For real-valued input, this includes the possibility to add uniform noise in an interval of length $1/\phi$ or Gaussian noise with variance $\sigma \in \Theta(1/\phi)$. In the Euclidean case, the adversary could, e.g., specify for each point a box of volume at least $1/\phi$, in which the point is distributed uniformly.

Related Work.   Recently, Bläser, Manthey and Rao [10] established a framework for analyzing the expectation of both running times and approximation ratios for some partitioning algorithms on so-called *smooth* and *near-additive* functionals. We establish a substantially different framework for smoothed analysis on a general class of Euclidean functionals that is disjoint to the class of smooth and near-additive functionals (see Section 7 for further discussion). We contrast both frameworks by considering the maximization counterparts of two problems studied in [10], namely Euclidean matching and TSP. Our algorithms have the advantage of featuring deterministic running times and providing asymptotically optimal approximation guarantees both in expectation and with high probability.

All other related work is problem-specific and will be described in the corresponding sections. As an exception, we highlight the result of Karger and Onak [22], who studied bin packing. To the best of our knowledge, this is the only problem that fits into our framework and has already been analyzed under perturbation. In this paper, a linear-time algorithm for bin packing was given that is asymptotically optimal on instances smoothed with any constant ϕ. We provide a new, conceptually simpler rounding method and analysis that replaces a key step of their algorithm and puts the reasons for its smoothed tractability into a more general context.

Table 1. All (near) linear-time algorithms derived in our framework

problem	running time	restriction on adversary power
MaxM	$O(n)$	$\phi \in o(\sqrt[4]{n})$ or $\phi \in o(n^{\frac{1}{2}\frac{d}{d+2}-\varepsilon})$
MaxTSP	$O(n)$	$\phi \in o(\sqrt[4]{n})$ or $\phi \in o(n^{\frac{1}{2}\frac{d}{d+2}-\varepsilon})$
KMeans	$O(n)$	$k\phi \in o(n^{\frac{1}{2}\frac{1}{kd+1}\frac{d}{d+1}})$
BP$_1$	$O(n\log n)$	$\phi \in o(n^{1-\varepsilon})$
BP$_d$	$O(n)$	$\phi \in o\left(\sqrt[d(d+1)]{\log\log n/\log^{(3)} n}\right)$

Our Results. We provide very fast and simple approximation algorithms on sufficiently smoothed inputs for the following problems: The maximum Euclidean matching problem MaxM, the maximum Euclidean Traveling Salesman problem MaxTSP, the k-means clustering problem KMeans where k denotes the number of desired clusters and is part of the input, and the d-dimensional bin packing problem BP$_d$. The approximation ratio converges to one with high probability over the random inputs. Additionally, all of these algorithms can be adapted to yield asymptotically optimal expected approximation ratios as well. This generalizes corresponding average-case analysis results [14,23].

Almost all our algorithms allow trade-offs between running time and approximation performance: By choosing a parameter p within its feasible range, we obtain algorithms of running time $O(n^p)$, whose approximation ratio converges to 1 as $n \to \infty$, provided that ϕ small enough, where the restriction on ϕ depends on p. The special case of linear-time algorithms is summarized in Table 1.

2 Preliminaries

Given an n-tuple of density functions $f = (f_1, \ldots, f_n)$ and random variables $X = (X_1, \ldots, X_n)$, we write $X \sim f$ for drawing X_i according to f_i for $1 \le i \le n$. We call $Y = (Y_1, \ldots, Y_n)$ a δ-*rounding* of X if $\|X_i - Y_i\| \le \delta$ for all $1 \le i \le n$. For a given X, let $\mathcal{Y}_X^\delta := \{Y \mid \|X_i - Y_i\| \le \delta\}$ be the set of δ-roundings of X.

We will analyze Euclidean functionals $F : ([0, 1]^d)^* \to \mathbb{R}$, denoting the dimension of the input space by $d \in \mathbb{N}$. For formalizing the perturbation model, let $\phi : \mathbb{N} \to [1, \infty)$ be an arbitrary function measuring the adversary's power. For better readability, we usually write ϕ instead of $\phi(n)$. We define $\mathcal{F}_\phi$ to be the set of *feasible* probability density functions $f : [0, 1]^d \to [0, \phi]$.

Note that if $\phi = 1$, the set $\mathcal{F}_\phi$ only consists of the uniform distribution on $[0, 1]^d$. If however $\phi = n$, the adversary may specify disjoint boxes for each point. Intuitively, to obtain a particular worst-case instance, the adversary would need to specify Dirac delta functions, which corresponds figuratively to setting ϕ to infinity. Observe also that already $\phi \in \omega(1)$ suffices to let all possible locations of a fixed point X_i converge to a single point for n tending to infinity, hence we believe that a superconstant ϕ is especially interesting to analyze.

For a given Euclidean functional F, we analyze the approximation ratio ρ of approximation algorithms ALG. If the functional is induced by an optimization problem, we do not focus on constructing a feasible approximate solution, but rather on computing an approximation of the objective value. However, we adopt this simplification only for clarity of presentation. Each of the discussed algorithms can be tuned such that it also outputs a feasible approximate *solution* for the underlying optimization problem. The approximation ratio on instance X is defined as $\rho(X) = \min \left\{ \frac{\text{ALG}(X)}{F(X)}, \frac{F(X)}{\text{ALG}(X)} \right\}$, which allows to handle both maximization and minimization problems at once.

For analyzing running times, we assume the word RAM model of computation and reveal real-valued input by reading in words of at least $\log n$ bits in unit time per word. We call an approximation algorithm a *probable $g_\phi(n)$-approximation* on smoothed instances if $\rho(X) \geq g_\phi(n)$ with high probability, i.e., with probability $1 - o(1)$, when X is drawn from $\mathcal{F}_\phi^n$. The algorithms derived in our framework feature deterministic running times $t(n) \in \text{poly}(n)$ and asymptotically optimal approximation ratios $g_\phi(n)$, i.e., $g_\phi(n) \to 1$ for $n \to \infty$ if ϕ is small enough. For such choices of ϕ, each of our algorithms induces a (non-uniform) probable polynomial-time approximation scheme (PTAS) on smoothed instances.

3   Framework

Our framework builds on the notion of *quantizable* functionals. These are functionals that admit fast approximation schemes on perturbed instances using general rounding strategies. The idea is to round an instance of n points to a *quantized instance* of $\ell \ll n$ points, each equipped with a multiplicity. This quantized input has a smaller problem size, which allows to compute an approximation faster than on the original input. However, the objective function needs to be large enough to make up for the loss due to rounding.

We aim at a trade-off between running time and approximation performance. As it will turn out, varying the number $\ell(n)$ of quantized points on an instance of n points makes this possible. Thus, we keep the function ℓ variable in our definition. On instances of size n, we will write $\ell := \ell(n)$ for short.

Definition 1. *Let $d \geq 1$ and $\mathcal{F}$ be a family of probability distributions $[0, 1]^d \to \mathbb{R}_{\geq 0}$. Let $t, R : \mathbb{N} \to \mathbb{R}$ and $Q \in \mathbb{R}$. We say that a Euclidean functional $F : ([0, 1]^d)^* \to \mathbb{R}_{\geq 0}$ is t-time (R, Q)-quantizable with respect to $\mathcal{F}$, if there is a quantization algorithm A and an approximation functional $g : ([0, 1]^d \times \mathbb{N})^* \to \mathbb{R}$ with the following properties. For any function ℓ satisfying $\ell \in \omega(1)$ and $\ell \in o(n)$,*

1. *The quantization algorithm A maps, in time $O(n)$, a collection of points $X = (X_1, \ldots, X_n) \in [0, 1]^{dn}$ to a multiset $A(X) = X' = ((X_1', n_1), \ldots, (X_\ell', n_\ell))$, the quantized input, with $X_i' \in [0, 1]^d$.*
2. *The approximation functional g is computable in time $t(\ell)$ and, for any $f \in \mathcal{F}^n$, fulfills $\Pr_{X \sim f}[|F(X) - g(A(X))| \leq nR(\ell)] \in 1 - o(1)$.*
3. *For any $f \in \mathcal{F}^n$, we have $\Pr_{x \sim f}[F(X) \geq nQ] \in 1 - o(1)$.*

The following theorem states that quantizable functionals induce natural approximation algorithms on smoothed instances. We can thus restrict our attention to finding criteria that make a functional quantizable.

Theorem 2. *Let $\mathcal{F}$ be a family of probability distributions and F be $t(\ell)$-time $(R(\ell), Q)$-quantizable with respect to $\mathcal{F}$. Then for every ℓ with $\ell \in \omega(1)$ and $\ell \in o(n)$, there is an approximation algorithm ALG with the following property. For every $f \in \mathcal{F}^n$, the approximation $\text{ALG}(X)$ on the instance X drawn from f is $(1 - \frac{R(\ell)}{Q})$-close to $F(X)$ with high probability. The approximation can be computed in time $O(n + t(\ell))$.*

For all problems considered here, we also design algorithms whose *expected* approximation ratio converges to optimality in the sense that both $\text{E}[\rho] \to 1$, as already achieved by the framework algorithm, and $\text{E}[\rho^{-1}] \to 1$, the desired guarantee for *minimization* problems, which we ensure using auxiliary algorithms. A sufficient auxiliary algorithm for F is a linear-time algorithm approximating F within a constant factor $0 < c < 1$. Outputting the better solution of our framework algorithm and the c-approximation does not increase the order of the running time, but achieves an approximation ratio of $1 - o(1)$ with probability $1 - o(1)$ due to the previous theorem, yielding $\text{E}[\rho] \to 1$, and still provides a constant approximation ratio on the remaining instances sampled with probability $o(1)$. Thus, additionally $\text{E}[\rho^{-1}] \leq (1 - o(1))\frac{1}{1-o(1)} + o(1)c^{-1} \to 1$ holds.

We respresent multisets of points either as $X' = ((X_1', n_1), \ldots, (X_\ell', n_\ell)) \in ([0, 1]^d \times \mathbb{N})^\ell$ or expand this canonically to a tuple $X' \in ([0, 1]^d)^*$. By $T : ([0, 1]^d \times \mathbb{N})^* \to ([0, 1]^d)^*$ we denote the transformation that maps the former representation to the latter.

4   Grid Quantization

Our first method for verifying quantizability is grid quantization. Here, the idea is to round the input to the centers of grid cells, where the coarseness of the grid is chosen according to the desired number of distinct points. This method works well for functionals that allow for fast optimal computations on their high-multiplicity version and provide a large objective value on the chosen perturbation model.

Theorem 3. *(Grid quantization) Let $d \geq 1$, $Q \in \mathbb{R}$ and $\mathcal{F}$ be a family of probability distributions $[0, 1]^d \to \mathbb{R}_{\geq 0}$. Let $F : ([0, 1]^d)^* \to \mathbb{R}_{\geq 0}$ be a Euclidean functional with the following properties.*

1. *On the quantized input $X' = ((X_1', n_1), \ldots, (X_\ell', n_\ell))$, the value $F(T(X'))$ can be computed in time $t(\ell) + O(\sum_{i=1}^\ell n_i)$.*
2. *There is a constant C such that w.h.p., the functional differs by at most $C\delta n$ on all δ-roundings of an instance X drawn from any $f \in \mathcal{F}^n$. Formally, for each $\delta > 0$ we require $\text{Pr}_{X \sim f} \left[\forall Y \in \mathcal{Y}_X^\delta : |F(X) - F(Y)| \leq C\delta n \right] \in 1 - o(1)$.*
3. *For each $f \in \mathcal{F}^n$, it holds that $\text{Pr}_{X \sim f} \left[F(X) \geq nQ \right] \in 1 - o(1)$.*

Then F is $t(\ell)$-time $(O(\ell^{-\frac{1}{d}}), Q)$-quantizable with respect to $\mathcal{F}$.

In this section, we apply the framework to two Euclidean maximization problems, namely maximum matching and maximum TSP. Both problems have already been analyzed in the average-case world, see, e.g., an analysis of the Metropolis algorithm on maximum matching in [28]. We generalize the result of Dyer et al. [14], who proved the asymptotic optimality of two simple partitioning heuristics for maximum matching and maximum TSP on the uniform distribution in the unit square. However, in contrast to our approach, their partitioning methods typically fail if the points are not identically distributed.

4.1   Maximum Matching and Maximum TSP

Let MaxM(X) denote the maximum weight of a matching of the points $X \subseteq [0,1]^d$, where the weight of a matching M is defined as $\sum_{\{u,v\} \in M} \|u - v\|$. For simplicity, we assume that $|X|$ is even. For the more general problem of finding maximum weighted matchings on *general* graphs with non-integer weights, the fastest known algorithm due to Gabow [19] runs in time $O(mn + n^2 \log n)$.

We aim to apply Theorem 3, for which we only need to check three conditions. The rounding condition (2) is easily seen to be satisfied by a straight-forward application of the triangle inequality. The lower bound condition (3) is satisfied by the following lemma.

Lemma 4. *Let $f \in \mathcal{F}_\phi^n$. Some $\gamma > 0$ satisfies $\Pr_{X \sim f}\left[\mathrm{MaxM}(X) < \frac{\gamma n}{\sqrt[d]{\phi}}\right] \leq e^{-\Omega(n)}$.*

We call the task of computing a functional on quantized inputs *quantized version* of the functional. In the case of MaxM, an algorithm for b-matchings from [1] can be exploited, satisfying condition (1).

Lemma 5. *The quantized version of MaxM can be computed in time $O(\ell^4 + \ell^3 \log n)$, where $n = \sum_{i=1}^{\ell} n_i$.*

These observations immediately yield the following result.

Theorem 6. *MaxM is $O(\ell^4)$-time $(O(1/\sqrt[d]{\ell}), \Omega(1/\sqrt[d]{\phi}))$-quantizable w. r. t. $\mathcal{F}_\phi$. Hence, for $1 \leq p < 4$, there is a $O(n^p)$-time probable $(1 - O(\sqrt[d]{\phi/n^{p/4}}))$-approximation to MaxM for instances drawn according to some $f \in \mathcal{F}_\phi^n$. This is asymptotically optimal on smoothed instances with $\phi \in o(n^{p/4})$.*

Interestingly, the restriction on ϕ is independent of the dimension. Note that only $p < 3$ is reasonable, since deterministic cubic-time algorithms for exactly computing MaxM exist. Furthermore, as described in Section 3, an algorithm with an asymptotically optimal expected approximation ratio can be designed. E.g., we might utilize a simple greedy linear-time $\frac{1}{2}$-approximation for MaxM [4].

Similar ideas can be applied to the maximum TSP problem. For $d \geq 2$, define MaxTSP(X) as the maximum weight of a Hamiltonian cycle on $X \subseteq [0,1]^d$, where the weight of a Hamiltonian cycle C is defined as $\sum_{\{u,v\} \in C} \|u - v\|$. The problem is NP-hard (proven for $d \geq 3$ in [7], conjectured for $d = 2$,) but admits a PTAS, cf. [8,7]. According to [16], these algorithms are not practical. They stress the need for (nearly) linear-time algorithms.

Theorem 7. *Let $1 \leq p \leq 4d/(d+1)$ and $f \in \mathcal{F}_\phi^n$. On instances drawn from f, there is a $O(n^p)$-time computable probable $(1 - O(\sqrt[d]{\phi/n^{p/4}}))$-approximation for* MaxTSP. *This is asymptotically optimal for $\phi \in o(n^{p/4})$.*

Since MaxM is a $\frac{1}{2}$-approximation to MaxTSP, the greedy linear-time computable $\frac{1}{2}$-approximation to MaxM is a $\frac{1}{4}$-approximation to MaxTSP and thus provides an adapted algorithm with asymptotically optimal *expected* approximation ratio for $\phi \in o(n^{p/4})$.

5   Balanced Quantization

Grid quantization proves useful for problems in which algorithms solving the high-multiplicity version are available. To solve even more problems, this section establishes a more careful quantization step yielding *balanced* instances, i.e., instances in which each of the distinct points occurs the same number of times. This has direct applications to k-means clustering and similar problems. In general, this method can be applied to problems in which the objective scales controllably when duplicating all points.

Theorem 8. *Let $\ell : \mathbb{N} \to \mathbb{N}$ with $\ell \in \omega(1)$ and $\ell \in o(n)$. There is a function $\ell' : \mathbb{N} \to \mathbb{N}$ such that for each $n \in \mathbb{N}$ and $X = (X_1, \ldots, X_n) \in [0,1]^{dn}$, we can find, in linear time, a family of $\ell'(n)$ cells, i.e., collections of points $C_1, \ldots, C_{\ell'(n)}$ with the following properties.*

1. $\frac{\ell'(n)}{\ell(n)} \to 1$ *(we obtain ℓ cells asymptotically),*
2. $|C_i| = |C_j|$ *for all $1 \leq i, j \leq \ell'(n)$ (all cells are of equal size),*
3. $n - \sum_{i=1}^{\ell'(n)} |C_i| \in O(\frac{n}{\ell^{1/(d+1)}})$ *(almost all points are covered),*
4. $\mathrm{diam}(C_i) \in O(\frac{1}{\ell^{1/(d+1)}})$ *(each element in a cell represents this cell well).*

For some problems, an instance in which every distinct point occurs equally often can be reduced to its distinct points only. In the following, we exploit this property by applying the previous theorem to k-means clustering. The method also allows for improving the results on maximum matching and maximum TSP. We defer the details of this to a full version of this article.

5.1   K-Means Clustering

Let $d \geq 2$ and $k \in \mathbb{N}$. We define KMeans(X, k) to be the k-means objective on the points X where k is the desired number of clusters, i.e.,

$$\mathrm{KMeans}(X, k) = \min_{C_1 \,\dot\cup\, \cdots \,\dot\cup\, C_k = X} \sum_{i=1}^{k} \sum_{x \in C_i} \|x - \mu_i\|^2, \text{ where } \mu_i = \frac{1}{|C_i|} \sum_{x \in C_i} x.$$

K-Means clustering is an important problem in various areas including machine learning and data mining. If either k or d is part of the input, it is NP-hard [13,24]. However, a popular heuristic, the k-means algorithm, usually runs fast on real-world instances despite its worst-case exponential running time. This is substantiated by results proving a polynomial smoothed running time of the k-means method under Gaussian perturbations [3,2]. In terms of solution quality, however, such a heuristic can perform poorly.

Consequently, k-means clustering has also received considerable attention concerning the design of fast deterministic approximation schemes. There exist linear-time asymptotically optimal algorithms, e.g., PTASs with running time $O(nkd+d\text{poly}(k/\varepsilon)+2^{\tilde{O}(k/\varepsilon)})$ [17] and $O(ndk+2^{(k/\varepsilon)^{O(1)}}d^2 n^\sigma)$ for any $\sigma > 0$ [12]. Treating the dimension as a constant as we do, [20] showed how to compute a $(1+\varepsilon)$-approximation in time $O(n + k^{k+2}\varepsilon^{-(2d+1)k}\log^{k+1} n\log^k \frac{1}{\varepsilon})$. On a side note, a different perturbation concept has been applied to k-means clusterings in [5]. They restrict their attention to input instances which, when perturbed, maintain the same partitioning of the input points in the optimal clustering.

Define the center of mass of a set C as $\text{cm}(C) := \frac{1}{|C|}\sum_{c\in C} c$ and consider $X' = ((X_1, n_1), \ldots, (X_{\ell'}, n_{\ell'})) = ((\text{cm}(C_1), |C_1|), \ldots, (\text{cm}(C_{\ell'}), |C_{\ell'}|))$, a quantized instance using the cells $C_1, \ldots, C_{\ell'}$ obtained by applying Theorem 8. It holds that $n_1 = n_2 = \cdots = n_{\ell'} = w$. Let $Y = T(X') = (Y_1, \ldots, Y_{n'})$, where Y_i is the rounded version of X_i. Note that the number n' of points in the rounded instance is potentially smaller than n, since points may be lost in the application of Theorem 8.

Lemma 9. *There is a real Δ with $|\text{KMeans}(X, k) - \text{KMeans}(Y, k)| \leq \frac{\Delta n}{\ell^{1/(d+1)}}$.*

After establishing that rounding the input does not affect the objective value too much, the following lemma enables us to reduce the instance size significantly.

Lemma 10. *Consider $X = ((X_1, w), \ldots, (X_\ell, w))$ and $Z = ((X_1, 1), \ldots, (X_\ell, 1))$. It holds that $\text{KMeans}(T(X), k) = w\text{KMeans}(T(Z), k)$.*

It is left to give a lower bound on the objective value. Note that for other minimization functionals like minimum Euclidean matching or TSP, already the uniform distribution on the unit cube achieves an objective value of only $O(n^{(d-1)/d})$ [27], making the framework inapplicable in this case (for a more detailed discussion, refer to Section 7).

Lemma 11. *Let $f \in \mathcal{F}_\phi^n$ and $k \in o(\frac{n}{\log n})$. There exists some constant $\gamma > 0$ such that $\Pr_{X\sim f}\left[\text{KMeans}(X, k) < \frac{\gamma n}{(k\phi)^{2/d}}\right] = e^{-\Omega(n)}$.*

For solving the smaller instance obtained by quantization, two approaches are reasonable. The first is to compute an optimal solution in time $O(n^{kd+1})$ [21] and results in the following theorem.

Theorem 12. *For any $k \in o(n/\log n)$, the functional $\text{KMeans}(X, k)$ is $O(\ell^{kd+1})$-time $(O(\ell^{-1/(d+1)}), \Omega((k\phi)^{-2/d}))$-quantizable with respect to $\mathcal{F}_\phi$. Consequently, for $k \in O(\log n/\log\log n)$ and $1 \leq p \leq kd + 1$, there is a $O(n^p)$-time computable probable $\left(1 - O\left(\frac{(k\phi)^{2/d}}{n^{\frac{p}{(d+1)(kd+1)}}}\right)\right)$-approximation for $\text{KMeans}(X, k)$ on smoothed instances.*

Note that this is asymptotically optimal if $\phi \in o(\sqrt[d]{n})$ with $c = 2(1 + 1/d)(kd + 1)/p$ if $k \in O(1)$, or more generally, if $k\phi \in o(n^{\frac{pd}{2(d+1)(kd+1)}})$. Using existing linear-time approximation schemes, also an asymptotically optimal expected approximation ratio can be obtained for the same values of ϕ. Our framework algorithm applies even for large values of k, e.g., $k = \log n / \log \log n$, in which case known deterministic approximation schemes require superlinear time. However, for small k, incorporating such an approximation scheme into our algorithm yields a further improvement of the previous theorem. The details for this second approach are deferred to a full version of this article.

6   Bin Packing

For $X = (X_1, \ldots, X_n) \in [0, 1]^{dn}$, define $\mathrm{BP}_d(X)$ to be the minimum number of bins of volume one needed to pack all elements. An item $X = (x_1, \ldots, x_d)$ is treated as a d-dimensional box, where x_i is its side length in dimension i. Items must not be rotated and must be packed such that their interior is disjoint.

In the following, we extend the result of Karger and Onak [22], who gave linear-time asymptotically optimal approximation algorithms for smoothed instances with $\phi \in O(1)$ and for instances with i.i.d. points drawn from a *fixed* distribution. These tractability results are highly interesting due to the fact that there is not even an asymptotic polynomial-time approximation scheme (APTAS) solving the two-dimensional bin packing problem unless $\mathsf{P} = \mathsf{NP}$, cf. [6].

Karger and Onak's approach appears rather problem-specific, whereas our solution embeds nicely into our framework. The main difference of our approach lies in a much simpler rounding routine and analysis, after which we solve the problem exactly as in their distribution-oblivious algorithm. Note that their algorithm is supplied with a desired approximation performance $\varepsilon > 0$ and suceeds with probability of $1 - 2^{-\Omega(n)}$. Although not stated for this case, we believe that their algorithm may also apply to superconstant choices of ϕ, at a cost of decreasing the success probability. We feel that our analysis offers more insights on the reasons why bin packing is smoothed tractable by putting it into the context of our general framework.

Consider first the case that $d = 1$. Unless $\mathsf{P} = \mathsf{NP}$, BP_1 does not admit a $\frac{3}{2}$-approximation. However, asymptotic polynomial approximation schemes exist [18], i.e., $(1 - \varepsilon)$-approximations on instances with a sufficiently large objective value. These approximation schemes have an interesting connection to smoothed analysis due to the following property.

Lemma 13. *For $f \in \mathcal{F}_\phi^n$, there is a $\gamma > 0$ with $\Pr_{X \sim f}\left[\mathrm{BP}_d(X) < \gamma \frac{n}{\phi^d}\right] \leq e^{-\Omega(n)}$.*

Using this bound on the objective value, we show an example of how to transform an APTAS into a PTAS on smoothed instances. Plotkin et al. [25] have shown how to compute, in time $O(n \log \varepsilon^{-1} + \varepsilon^{-6} \log^6 \varepsilon^{-1})$, a solution with an objective value of at most $(1 + \varepsilon)\mathrm{BP}_1(X) + O(\varepsilon^{-1} \log \varepsilon^{-1})$. We derive

$$\rho = \frac{\mathrm{ALG}}{\mathrm{BP}_1(X)} \leq (1 + \varepsilon) + O\left(\frac{\varepsilon^{-1} \log \varepsilon^{-1}}{\mathrm{BP}_1(X)}\right) \leq (1 + \varepsilon) + O\left(\frac{\phi \varepsilon^{-1} \log \varepsilon^{-1}}{n}\right),$$

where the last inequality holds w.h.p. over the perturbation of the input. Setting $\varepsilon := \log n/n^\delta$ with some $\delta < 1/6$ yields a running time of $O(n \log n)$ with an approximation ratio $\leq 1 + \log n/n^\delta + O(\phi/n^{1-\delta})$. Consequently, there is a linear-time asymptotically optimal approximation algorithm on instances smoothed with $\phi \in o(n^{1-\delta})$ for any $\delta > 0$. Unfortunately, this approach is not easily generalizable to $d > 1$, since already for $d = 2$, no APTAS exists unless $P = NP$, as shown in [6]. Nevertheless, the problem is quantizable in our framework.

We say that a single item $X = (x_1, \ldots, x_d)$ fits in a box $B = (b_1, \ldots, b_d)$ if $x_i \leq b_i$ for all $1 \leq i \leq d$. In this case, we write $X \sqsubseteq B$, adopting the notation of [22]. Regarding an item as a box as well, this relation is transitive and any feasible packing containing Y induces a feasible packing by replacing Y with X. Thus, bin packing admits the monotonicity property that for each $X = (X_1, \ldots, X_n)$ and $Y = (Y_1, \ldots, Y_n)$ with $X_i \sqsubseteq Y_i$, it holds that $\mathrm{BP}_d(X) \leq \mathrm{BP}_d(Y)$.

To apply the quantization framework, we require a suitable bound on the rounding errors. Unlike for MaxM and MaxTSP, no deterministic bound of $n\delta$ is possible for a δ-rounding: Let the instance $X^{(n)}$ consist of n copies of $(\frac{1}{2}, \ldots, \frac{1}{2})$. Packing 2^d of the items per bin results in zero waste, hence $\mathrm{BP}_d(X^{(n)}) = n/2^d$. However, for *any* $\delta > 0$, the δ-rounding Y_n consisting of n copies of $(\frac{1}{2} + \frac{\delta}{\sqrt{d}}, \ldots, \frac{1}{2} + \frac{\delta}{\sqrt{d}})$ has an objective value of $\mathrm{BP}_d(Y^{(n)}) = n = 2^d \mathrm{BP}_d(X^{(n)})$. Thus, a smoothed analysis of the rounding error is necessary.

Lemma 14. *For $f \in \mathcal{F}_\phi^n$ and $t > 0$,*

$$\Pr_{X \sim f} \left[\forall Y \in \mathcal{Y}_X^t : |\mathrm{BP}_d(X) - \mathrm{BP}_d(Y)| > 2ntd\phi \right] \leq 2 \exp(-2n(dt\phi)^2).$$

Note that this probability tends to zero if $t \in \omega(\frac{1}{\phi\sqrt{n}})$. Since grid quantization rounds the points to ℓ distincts points by moving each item by at most $t = \sqrt{d}\ell^{-d}$, the requirement $\ell \in o(n)$ even implies that $t \in \omega(\frac{1}{\sqrt{n}})$ for $d \geq 2$.

Solving the high-multiplicity version of the one-dimensional case has been a key ingredient in approximation schemes for this problem since the first APTAS by [18]. The following lemma from [22] solves the multi-dimensional case.

Lemma 15. *Let $X' = ((X_1, n_1), \ldots, (X_\ell, n_\ell))$ be a quantized input with $X_i \in [\delta, 1]^d$. Then $\mathrm{BP}_d(T(X'))$ can be computed in time $O(f(\ell, \delta)\mathrm{polylog}(n))$ where $n := \sum_{i=1}^\ell n_i$, $f(\ell, \delta)$ is independent of n and $f(\ell, 1/\sqrt[d]{\ell}) \in 2^{\ell^{O(\ell)}}$.*

Observe that each coordinate of the quantized points obtained by grid quantization is at least $1/(2\sqrt[d]{\ell})$, since these points are the centroids of cubic cells of side length $1/\sqrt[d]{\ell}$. Hence, applying a slightly stronger form of the grid quantization theorem yields the following result using Lemmas 13, 14 and 15.

Theorem 16. *For $d \geq 2$, BP_d is $O(2^{\ell^{O(\ell)}})$-time $(O(\frac{\phi}{\sqrt[d]{\ell}}), \Omega(\frac{1}{\phi^d}))$-quantizable w.r.t. $\mathcal{F}_\phi$.*

Consequently, there is a linear-time probable $(1 - O(\phi^{d+1}/\sqrt[d]{\log\log n/\log^{(3)} n}))$-approximation. Hence, BP_d can be computed asymptotically exactly in time

$O(n)$ if $\phi \in o\left(\sqrt[d(d+1)]{\log \log n / \log^{(3)} n}\right)$. Here, allowing superlinear time has no effect on the admissible adversarial power. Furthermore, since BP_d can be trivially approximated by a factor of n and the success probability of our algorithm is of order $1 - \exp(-\Omega(n^{1-\varepsilon}))$, asymptotically optimal expected approximation ratios can be obtained for the same values of ϕ.

7 Concluding Remarks

Generalizing previous rounding-based approaches, we demonstrate that the general solution technique of quantization performs well on Euclidean optimization problems in the setting of smoothed analysis. We are optimistic that our framework can also be applied to disk covering and scheduling problems.

Note that our approach is orthogonal to the framework for smooth and near-additive Euclidean functionals due to Bläser et al. [10]: A smooth Euclidean functional F on n points can be bounded by $O(n^{1-1/d})$ by definition of smoothness. Hence, it can never compensate for the rounding error of at least $\Omega(\ell^{-1/d})$ per point that our quantization methods induce, as quantization is only reasonable for $\ell \leq n$ and consequently, the total rounding error amounts to $\Omega(n^{1-1/d})$. Conversely, if a functional is large enough to compensate for rounding errors induced by quantization, it cannot be smooth. Thus, for any Euclidean functional, at most one of both frameworks is applicable.

References

1. Anstee, R.P.: A polynomial algorithm for b-matchings: An alternative approach. Information Processing Letters 24(3), 153–157 (1987)
2. Arthur, D., Manthey, B., Röglin, H.: Smoothed analysis of the k-means method. Journal of the ACM 58(5), 19:1–19:31 (2011)
3. Arthur, D., Vassilvitskii, S.: Worst-case and smoothed analysis of the ICP algorithm, with an application to the k-means method. SIAM Journal on Computing 39(2), 766–782 (2009)
4. Avis, D.: A survey of heuristics for the weighted matching problem. Networks 13(4), 475–493 (1983)
5. Awasthi, P., Blum, A., Sheffet, O.: Center-based clustering under perturbation stability. Information Processing Letters 112(1-2), 49–54 (2012)
6. Bansal, N., Correa, J.É.R., Kenyon, C., Sviridenko, M.: Bin packing in multiple dimensions: Inapproximability results and approximation schemes. Mathematics of Operations Research 31, 31–49 (2006)
7. Barvinok, A., Fekete, S.P., Johnson, D.S., Tamir, A., Woeginger, G.J., Woodroofe, R.: The geometric maximum traveling salesman problem. J. ACM 50(5), 641–664 (2003)
8. Barvinok, A.I.: Two algorithmic results for the traveling salesman problem. Mathematics of Operations Research 21(1), 65–84 (1996)
9. Beier, R., Vöcking, B.: Typical properties of winners and losers in discrete optimization. SIAM Journal on Computing 35(4), 855–881 (2006)
10. Bläser, M., Manthey, B., Raghavendra Rao, B.V.: Smoothed Analysis of Partitioning Algorithms for Euclidean Functionals. Algorithmica 66(2), 397–418 (2013)

11. Boros, E., Elbassioni, K., Fouz, M., Gurvich, V., Makino, K., Manthey, B.: Stochastic mean payoff games: Smoothed analysis and approximation schemes. In: Aceto, L., Henzinger, M., Sgall, J. (eds.) ICALP 2011, Part I. LNCS, vol. 6755, pp. 147–158. Springer, Heidelberg (2011)
12. Chen, K.: On coresets for k-median and k-means clustering in metric and euclidean spaces and their applications. SIAM Journal on Computing 39(3), 923–947 (2009)
13. Dasgupta, S.: The hardness of k-means clustering. Technical report cs2007-0890, University of California, San Diego (2007)
14. Dyer, M.E., Frieze, A.M., McDiarmid, C.J.H.: Partitioning heuristics for two geometric maximization problems. Operations Research Letters 3(5), 267–270 (1984)
15. Englert, M., Röglin, H., Vöcking, B.: Worst case and probabilistic analysis of the 2-opt algorithm for the TSP: Extended abstract. In: 18th Ann. ACM-SIAM Symp. on Discrete Algorithms, SODA 2007, pp. 1295–1304. SIAM (2007)
16. Fekete, S.P., Meijer, H., Rohe, A., Tietze, W.: Solving a "hard" problem to approximate an "easy" one: Heuristics for maximum matchings and maximum traveling salesman problems. ACM J. on Experimental Algorithmics. 7, 11 (2002)
17. Feldman, D., Monemizadeh, M., Sohler, C.: A PTAS for k-means clustering based on weak coresets. In: 23rd Ann. Symp. on Computational Geometry, SCG 2007, pp. 11–18. ACM (2007)
18. Fernandez de la Vega, W., Lueker, G.: Bin packing can be solved within $1 + \epsilon$ in linear time. Combinatorica 1(4), 349–355 (1981)
19. Gabow, H.N.: An efficient implementation of Edmonds' algorithm for maximum matching on graphs. Journal of the ACM 23(2), 221–234 (1976)
20. Har-Peled, S., Mazumdar, S.: On coresets for k-means and k-median clustering. In: 36th Ann. ACM Symp. on Theory of Computing, STOC 2004, pp. 291–300 (2004)
21. Inaba, M., Katoh, N., Imai, H.: Applications of weighted Voronoi diagrams and randomization to variance-based k-clustering (extended abstract). In: 10th Annual Symp. on Computational Geometry, SCG 1994, pp. 332–339 (1994)
22. Karger, D., Onak, K.: Polynomial approximation schemes for smoothed and random instances of multidimensional packing problems. In: 18th Ann. ACM-SIAM Symp. on Discrete Algorithms, SODA 2007, pp. 1207–1216 (2007)
23. Karp, R.M., Luby, M., Marchetti-Spaccamela, A.: A probabilistic analysis of multidimensional bin packing problems. In: 16th Annual ACM Symp. on Theory of Computing, STOC 1984, pp. 289–298. ACM, New York (1984)
24. Mahajan, M., Nimbhorkar, P., Varadarajan, K.: The planar k-means problem is NP-hard. Theoretical Computer Science 442, 13–21 (2012)
25. Plotkin, S.A., Shmoys, D.B., Tardos, É.: Fast approximation algorithms for fractional packing and covering problems. Mathematics of Operations Research 20(2), 257 (1995)
26. Spielman, D.A., Teng, S.-H.: Smoothed analysis of algorithms: Why the simplex algorithm usually takes polynomial time. Journal of the ACM 51(3), 385–463 (2004)
27. Steele, J.M.: Subadditive Euclidean functionals and nonlinear growth in geometric probability. The Annals of Probability 9(3), 365–376 (1981)
28. Weber, M., Liebling, T.M.: Euclidean matching problems and the metropolis algorithm. Mathematical Methods of Operations Research 30(3), A85–A110 (1986)

Tight Kernel Bounds for Problems on Graphs with Small Degeneracy*
(Extended Abstract)**

Marek Cygan[1], Fabrizio Grandoni[2], and Danny Hermelin[3]

[1] Institute of Informatics, University of Warsaw
cygan@mimuw.edu.pl
[2] Dalle Molle Institute (IDSIA) Galleria 1, 6928, Manno - Switzerland
fabrizio@idsia.ch
[3] Department of Industrial Engineering and Management
Ben-Gurion University, Beer-Sheva, Israel
hermelin@bgu.ac.il

Abstract. Kernelization is a strong and widely-applied technique in parameterized complexity. In a nutshell, a kernelization algorithm for a parameterized problem transforms a given instance of the problem into an equivalent instance whose size depends solely on the parameter. Recent years have seen major advances in the study of both upper and lower bound techniques for kernelization, and by now this area has become one of the major research threads in parameterized complexity.

We consider kernelization for problems on d-degenerate graphs, i.e. graphs such that any subgraph contains a vertex of degree at most d. This graph class generalizes many classes of graphs for which effective kernelization is known to exist, e.g. planar graphs, H-minor free graphs, H-topological minor free graphs. We show that for several natural problems on d-degenerate graphs the best known kernelization upper bounds are essentially tight. In particular, using intricate constructions of weak compositions, we prove that unless NP$\subseteq$ coNP/poly:

- DOMINATING SET has no kernels of size $O(k^{(d-1)(d-3)-\varepsilon})$ for any $\varepsilon > 0$. The current best upper bound is $O(k^{(d+1)^2})$.
- INDEPENDENT DOMINATING SET has no kernels of size $O(k^{d-4-\varepsilon})$ for any $\varepsilon > 0$. The current best upper bound is $O(k^{d+1})$.
- INDUCED MATCHING has no kernels of size $O(k^{d-3-\varepsilon})$ for any $\varepsilon > 0$. The current best upper bound is $O(k^d)$.

We also give simple kernels for CONNECTED VERTEX COVER and CAPACITATED VERTEX COVER of size $O(k^d)$ and $O(k^{d+1})$ respectively. Both these problems do not have kernels of size $O(k^{d-1-\varepsilon})$ unless coNP/poly.

In this extended abstract we will focus on the lower bound for DOMINATING SET, which we feel is the central result of our study. The proofs of the other results can be found in the full version of the paper.

---

* Partially supported by ERC Starting Grant NEWNET 279352.
** A full version of the paper can be found in [8].

1   Introduction

Parameterized complexity is a two-dimensional refinement of classical complexity theory introduced by Downey and Fellows [14] where one takes into account not only the total input length n, but also other aspects of the problem quantified in a numerical parameter $k \in \mathbb{N}$. The main goal of the field is to determine which problems have algorithms whose exponential running time is confined strictly to the parameter. In this way, algorithms running in $f(k) \cdot n^{O(1)}$ time for some computable function $f()$ are considered feasible, and parameterized problems that admit feasible algorithms are said to be *fixed-parameter tractable*. This notion has proven extremely useful in identifying tractable instances for generally hard problems, and in explaining why some theoretically hard problems are solved routinely in practice.

A closely related notion to fixed-parameter tractability is that of kernelization. A *kernelization algorithm* (or *kernel*) for a parameterized problem $L \subseteq \{0,1\}^* \times \mathbb{N}$ is a polynomial time algorithm that transforms a given instance (x, k) to an instance (x', k') such that: (i) $(x, k) \in L \iff (x', k') \in L$, and (ii) $|x'| + k' \leq f(k)$ for some computable function f. In other words, a kernelization algorithm is a polynomial-time reduction from a problem to itself that shrinks the problem instance to an instance with size depending only on the parameter. Appropriately, the function f above is called the *size* of the kernel.

Kernelization is a notion that was developed in parameterized complexity, but it is also useful in other areas of computer science such as cryptography [21] and approximation algorithms [27]. In parameterized complexity, not only is it one of the most successful techniques for showing positive results, it also provides an equivalent way of defining fixed-parameter tractability: A decidable parameterized problem is solvable in $f(k) \cdot n^{O(1)}$ time iff it has a kernel [6]. From a practical point of view, compression algorithms often lead to efficient preprocessing rules which can significantly simplify real life instances [16,19]. For these reasons, the study of kernelization is one of the leading research frontiers in parameterized complexity. This research endeavor has been fueled by recent tools for showing lower bounds on kernel sizes [2,4,5,7,10,11,13,22,26] which rely on the standard complexity-theoretic assumption of coNP$\not\subseteq$ NP/poly.

Since a parameterized problem is fixed-parameter tractable iff it is kernelizable, it is natural to ask which fixed-parameter problems admit kernels of reasonably small size. In recent years there have been significant advances in this area. One particularly prominent line of research in this context is the development of *meta-kernelization* algorithms for problems on sparse graphs. Such algorithms typically provide compressions of either linear or quadratic size to a wide range of problems at once, by identifying certain generic problem properties that allow for good compressions. The first work in this line of research is due to Guo and Niedermeier [20], which extended the ideas used in the classical linear kernel for DOMINATING SET in planar graphs [1] to linear kernels for several other planar graph problems. This result was subsumed by the seminal paper of Bodlaender *et al.* [3], which provided meta-kernelization algorithms for problems on graphs of bounded genus, a generalization of planar graphs. Later Fomin *et al.* [17]

Table 1. Lower and upper bounds for kernel sizes for problems in d-degenerate graphs. Only the exponent of the polynomial in k is given. Results without a citation are obtained in this paper.

	Lower Bound	Upper Bound
DOMINATING SET	$(d-3)(d-1) - \varepsilon$	$(d+1)^2$ [28]
INDEPENDENT DOMINATING SET	$d - 4 - \varepsilon$	$d + 1$ [28]
INDUCED MATCHING	$d - 3 - \varepsilon$	d [15,23]
CONNECTED VERTEX COVER	$d - 1 - \varepsilon$ [9]	d
CAPACITATED VERTEX COVER	$d - \varepsilon$	$d + 1$

provided a meta-kernel for problems on H-minor free graphs which include all bounded genus graphs. Finally, a recent manuscript by Langer *et al.* [25] provides a meta-kernelization algorithm for problems on H-topological-minor free graphs. All meta-kernelizations above have either linear or quadratic size.

How far can these meta-kernelization results be extended? A natural class of sparse graphs which generalizes all graph classes handled by the meta-kernelizations discussed above is the class of d-degenerate graphs. A graph is called *d-degenerate* if each of its subgraphs has a vertex of degree at most d. This is equivalent to requiring that the vertices of the graph can be linearly ordered such that each vertex has at most d neighbors to its right in this ordering. For example, any planar graph is 5-degenerate, and for any H-minor (resp. H-topological-minor) free graph class exists a constant $d(H)$ such that all graphs in this class are $d(H)$-degenerate (see *e.g.* [12]). Note that the INDEPENDENT SET problem has a trivial linear kernel in d-degenerate graphs, which gives some hope that a meta-kernelization result yielding small degree polynomial kernels might be attainable for this graph class.

Arguably the most important kernelization result in d-degenerate graphs is due to Philip *et al.* [28] who showed a $O(k^{(d+1)^2})$ size kernel for DOMINATING SET, and an $O(k^{d+1})$ size kernel for INDEPENDENT DOMINATING SET. Erman *et al.* [15] and Kanj *et al.* [23] independently gave a $O(k^d)$ kernel for the INDUCED MATCHING problem, while Cygan *et al.* [9] showed a $O(k^{d+1})$ kernel for CONNECTED VERTEX COVER. While all these results give polynomial kernels, the exponent of the polynomial depends on d, leaving open the question of kernels of polynomial size with a fixed constant degree. This question was answered negatively for CONVC in [9] using the standard reduction from d-SET COVER. It is also shown in [9] that other problems such as CONNECTED DOMINATING SET and CONNECTED FEEDBACK VERTEX SET do not admit a kernel of any polynomial size unless coNP$\subseteq$ NP/poly. Furthermore, the results in [10,22] can be easily used to exclude a $O(k^{d-\varepsilon})$-size kernel for DOMINATING SET, for some small positive constant ε.

In the full version of the paper we show that all remaining kernelization upper bounds for d-degenerate graphs mentioned above have matching lower bounds up to some small additive constant; see Table 1 for details. Perhaps the most surprising result we obtain is the exclusion of $O(k^{(d-3)(d-1)-\varepsilon})$ size kernels for DOMINATING SET for any $\varepsilon > 0$, under the assumption of coNP $\nsubseteq$ NP/poly. This result is obtained by an intricate application of *weak compositions* which were introduced by [11], and further applied in [10,22]. What makes this result surprising is that it implies that INDEPENDENT DOMINATING SET is fundamentally easier than DOMINATING SET in d-degenerate graphs.

In the current extend abstract we focus on the lower bound obtained for DOMINATING SET, since we feel it is the most interesting and most technically challenging result we obtain. Following a brief overview of the lower bound machinery that we will use for our result, including the definition of

2   Kernelization Lower Bounds

In the following section we quickly review the main tool that we will be using for showing our kernelization lower bounds, namely compositions. A composition algorithm is typically a transformation from a classical NP-hard problem L_1 to a parameterized problem L_2. It takes as input a sequence of T instances of L_1, each of size n, and outputs in polynomial time an instance of L_2 such that (i) the output is a YES-instance iff one of the inputs is a YES-instance, and (ii) the parameter of the output is polynomially bounded by n and has only "small" dependency on T. Thus, a composition may intuitively be thought of as an "OR-gate" with a guarantee bound on the parameter of the output. More formally, for an integer $d \geq 1$, a weak d-composition is defined as follows:

Definition 1 (weak d-composition). *Let $d \geq 1$ be an integer constant, let $L_1 \subseteq \{0,1\}^*$ be a classical (non-parameterized) problem, and let $L_2 \subseteq \{0,1\}^* \times \mathbb{N}$ be a parameterized problem. A weak d-composition from L_1 to L_2 is a polynomial time algorithm that on input $x_1, \ldots, x_{t^d} \in \{0,1\}^n$ outputs an instance $(y, k') \in \{0,1\}^* \times \mathbb{N}$ such that:*

- *$(y, k') \in L_2 \iff x_i \in L_1$ for some i, and*
- *$k' \leq t \cdot n^{O(1)}$.*

The connection between compositions and kernelization lower bounds was discovered by [2] using ideas from [21] and a complexity theoretic lemma of [18]. The following particular connection was first observed in [11].

Lemma 1 ([11]). *Let $d \geq 1$ be an integer, let $L_1 \subseteq \{0,1\}^*$ be a classical NP-hard problem, and let $L_2 \subseteq \{0,1\}^* \times \mathbb{N}$ be a parameterized problem. A weak-d-composition from L_1 to L_2 implies that L_2 has no kernel of size $O(k^{d-\varepsilon})$ for any $\varepsilon > 0$, unless* coNP $\subseteq$ NP/poly.

Remark 1. Lemma 1 also holds for *compressions*, a stronger notion of kernelization, in which the reduction is not necessarily from the problem to itself, but rather from the problem to some arbitrary set.

3   Construction Overview

We next briefly sketch the main ideas behind our lower bound construction for DOMINATING SET. The general idea and framework that we use was introduced in [13], and further developed in [22]. For convenience purposes, we will show a lower bound for essentially equivalent RED BLUE DOMINATING SET problem (d-RBDS). In this problem we are given a parameter k and a n-vertex d-degenerate graph which is properly colored by two colors, red and blue, and the question is whether there are k red vertices which dominate all the blue vertices in the graph. Our goal is show how to compose any $T \approx t^d$ instances $I_1, \ldots, I_T$ of some NP-hard problem – each with the same size n – into a single instance I of RBDS which is (i) a YES-instance iff one of the T input instances is a YES-instance, and (ii) has parameter at most $t \cdot n^{O(1)}$.

As our starting problem we use the NP-hard MULTICOLORED PERFECT MATCHING problem: Given an undirected graph G with an even number of vertices n, together with a color function that assigns one of $n/2$ colors to the edges of the graph, determine whether G has a perfect matching where all edges have distinct colors. This problem easily reduces to RBDS by constructing a red vertex for each edge in G, and a blue vertex for each vertex and each edge-color, and then connecting each red vertex representing an edge to the blue vertices representing its endpoints and color. Clearly, the graph obtained by this construction is 3-degenerate.

So we want to compose a sequence $I_1, \ldots, I_T$ of MPM instances into a single instance of RBDS. As the first step, we transform each instance I_i to an instance I_i' of RBDS almost as described above. The difference is that we do not create blue vertices for edge colors, but we store the color of an edge that a given red vertex represents to use that information later on. Now the "bipartiteness" of RBDS allows for an easy start in our construction. To obtain a single RBDS instance from $I_1', \ldots, I_T'$, we identify the T sets of blue vertices in each instance into a single set. This can easily be done as all instances I_i' have the same number of blue vertices (since all instances I_i had the same number of edges and edge-colors). We then take the disjoint union of the T sets of red vertices, and connect each set to the identified set of blue vertices in the natural way. In this way, we obtain a single 2-degenerate bipartite graph H_{inst} for our instance I of RBDS. It is easy to see that if one of the instances I_i' has a solution of at most k red vertices these same red vertices form a solution in H_{inst}. The problem is that I can have a solution of size k even if all instances $I_1, \ldots I_T$ are NO-instances, since a solution in I can be composed of red vertices from more than one instance in $I_1, \ldots I_T$ and moreover we do not control the colorfulness aspect of the initial MPM instances.

Observe that we still have a lot of leeway in the parameter of our output, as we can afford a solution size of $t \cdot n^{O(1)}$, and also some leeway in the degeneracy of the output graph. Thus, we circumvent the problem above by adding an enforcement graph H_{enf} to our construction which is essentially another instance of RBDS which ensures that vertices corresponding to edges in different instances I_i will not be selected in any solution of size k', for some carefully chosen $k' = tn^{O(1)}$. This is done by connecting red vertices in H_{inst} via the edge set E_{conn} to the blue vertices of H_{enf} in an intricate manner that ensures that the resulting graph is only $(d + 2)$-degenerate.

4   Construction Details

Let us begin recalling the definition of the DOMINATING SET (DS) problem. In this problem, we are given an undirected graph $G = (V, E)$ together with an integer k, and we are asked whether there exists a set S of at most k vertices such that each vertex of G either belongs to S or has a neighbor in S (i.e. $N[S] = V$). The main result of this section is stated in Theorem 1 below.

Theorem 1. *Let $d \geq 4$. The* DOMINATING SET *problem in d-degenerate graphs has no kernel of size $O(k^{(d-1)(d-3)-\varepsilon})$ for any constant $\varepsilon > 0$ unless* NP $\subseteq$ coNP/poly.

In order to prove Theorem 1, we show a lower bound for a similar problem called the RED BLUE DOMINATING SET problem (RBDS): Given a bipartite graph $G = (R \cup B, E)$ and an integer k, where R is the set of *red* vertices and B is the set of *blue* vertices, determine whether there exists a set $D \subseteq R$ of at most k red vertices which dominate all the blue vertices (i.e. $N(D) = B$). According to Remark 1, the lemma below shows that focusing on RBDS is sufficient for proving Theorem 1 above.

Lemma 2. *There is a polynomial time algorithm, which given a d-degenerate instance $I = (G = (R \cup B, E), k)$ of* RBDS *creates a $(d+1)$-degenerate instance $I' = (G', k')$ of* DS, *such that $k' = k + 1$ and I is YES-instance iff I' is a YES-instance.*

Proof. As the graph G' we initially take $G = (R \cup B, E)$ and then we add two vertices r, r' and make r adjacent to all the vertices in $R \cup \{r'\}$. Clearly G' is $(d + 1)$-degenerate. Note that if $S \subseteq R$ is a solution in I, then $S \cup \{r\}$ is a dominating set in I'. In the reverse direction, observe that w.l.o.g. we may assume that a solution S' for I' contains r and moreover contains no vertex of B. Therefore I is a YES-instance iff I' is a YES-instance.   □

We next describe a weak $d(d + 2)$-composition from MULTICOLORED PERFECT MATCHING to RBDS in $(d + 2)$-degenerate graphs. The MULTICOLORED PERFECT MATCHING problem (MPM) is as follows: Given an undirected graph $G = (V, E)$ with even number n of vertices, together with a color function col : $E \to \{0, \ldots, n/2 - 1\}$, determine whether there exists a perfect matching in G with all the edges having distinct colors. A simple reduction from 3-DIMENSIONAL PERFECT MATCHING, which is NP-complete due to Karp [24], where we encode one coordinate using colors, shows that MPM is NP-complete when we consider multigraphs. In the full version of the paper we show that MPM is NP-complete even for simple graphs.

The construction of the weak composition is rather involved (see Fig. 1). We construct an instance graph H_{inst} which maps feasible solutions of each MPM instance into feasible solutions of the RBDS instance. Then we add an enforcement gadget (H_{enf}, E_{conn}) which prevents partial solutions of two or more MPM instances to form altogether a solution for the RBDS instance.

The overall RBDS instance will be denoted by (H, k), where H is the union of H_{inst} and H_{enf} along with the edges E_{conn} that connect between these graphs. The construction of the instance graph is relatively simple, while the enforcement gadget is rather complex. In the next subsection we describe H_{enf} and its crucial properties. In the following subsection we describe the rest of the construction, and prove the claimed lower bound on RBDS (and hence DS). Both H_{enf} and H_{inst} contain red and blue nodes. We will use the convention that R and B denote sets of red and blue nodes, respectively. We will use r and b to indicate red and blue nodes, respectively. A color is indicated by ℓ.

4.1   The Enforcement Graph

The enforcement graph $H_{enf} = (R_{enf} \cup B_{enf}, E_{enf})$ is a combination of 3 different gadgets: the *encoding gadget*, the *choice gadget*, and the *fillin gadget* (see also Fig. 1), i.e. $R_{enf} = R_{code} \cup R_{fill}$ and $B_{enf} = B_{code} \cup B_{choice} \cup B_{fill}$ (R_{choice} is empty).

Encoding Gadget: The role of this gadget is to encode the indices of all the instances by different partial solutions. It consists of nodes $R_{code} \cup B_{code}$, plus the edges among them. The set R_{code} contains one node $r_{\delta,\lambda,\gamma}$ for all integers $0 \le \delta < d + 2$, $0 \le \lambda < d$, and $0 \le \gamma < t$. In particular, $|R_{code}| = (d + 2)dt$. The set B_{code} is the union of sets B_{code}^{ℓ} for each color $0 \le \ell < n/2$. In turn, B_{code}^{ℓ} contains a node b_a^{ℓ} for each integer $0 \le a < (dt)^{d+2}$. We connect nodes $r_{\delta,\lambda,\gamma}$ and b_a^{ℓ} iff $a_{\delta} = \lambda \cdot t + \gamma$, where $(a_0, \dots, a_{d+1})$ is the expansion of a in base dt, i.e. $a = \sum_{0 \le \delta < d+2} a_{\delta}(dt)^{\delta}$. There is a subtle reason behind this connection scheme, which hopefully will be clearer soon. Note that since $0 \le \gamma < t$, pairs (λ, γ) are in one to one correspondence with possible values of digits a_{δ}.

Choice Gadget: The role of the choice gadget is to guarantee the following *choice property*: Any feasible solution to the overall RBDS instance (H, k) contains all nodes R_{code} except possibly one node $r_{\delta,\lambda,\gamma_{\delta,\lambda}}$ for each pair (δ, λ) (hence at least $(d + 2)d(t - 1)$ nodes of R_{code} altogether are taken). Intuitively, the $\gamma_{\delta,\lambda}$'s will be used to identify the index of one MPM input instance. In order to do that, we introduce a set of nodes B_{choice}, containing a node $b_{\delta,\lambda,\gamma_1,\gamma_2}$ for every pair (δ, λ) and for every $0 \le \gamma_1 < \gamma_2 < t$. We connect $b_{\delta,\lambda,\gamma_1,\gamma_2}$ to both $r_{\delta,\lambda,\gamma_1}$ and $r_{\delta,\lambda,\gamma_2}$. It is not hard to see that, in order to dominate B_{choice}, it is necessary and sufficient to select from R_{code} a subset of nodes with the choice property.

Fillin Gadget: We will guarantee that, in any feasible solution, precisely $(d + 2)d(t-1)$ nodes from R_{code} are selected. Given that, for each pair (δ, λ), there will be precisely one node $r_{\delta,\lambda,\gamma_{\delta,\lambda}}$ which is not included in the solution. Consequently, as we will prove, for each $0 \le \ell < n/2$ in B_{code}^{ℓ} there will be exactly d^{d+2} uncovered nodes, namely the nodes $b_a^{\ell} = b_{(a_0,\dots,a_{d+1})}^{\ell}$ such that for each $0 \le \delta < d+2$ and $\lambda t \le a_{\delta} < (\lambda+1)t$ one has $a_{\delta} = \lambda t + \gamma_{\delta,\lambda}$. Ideally, we would like to cover such nodes by means of red nodes in the instance graph H_{inst} (to be defined later), which encode a feasible solution to some MPM instance. However, the degeneracy of the overall graph would be too large. The role of the fillin gadget is to circumvent this problem, by leaving at most d uncovered nodes in each B_{code}^{ℓ}. The fillin gadget consists of nodes $R_{fill} \cup B_{fill}$, with some edges incident to them. The set R_{fill} is the union of sets R_{fill}^{ℓ} for each color ℓ. In turn R_{fill}^{ℓ} contains one node $r_{a,j}^{\ell}$ for each $1 \leq j \leq d^{d+2} - d$ and $0 \leq a < (dt)^{d+2}$. We connect each $r_{a,j}^{\ell}$ to b_a^{ℓ}. The set B_{fill} contains one node b_j^{ℓ}, for each color ℓ and for all $1 \leq j \leq d^{d+2} - d$. We connect b_j^{ℓ} to all nodes $\{r_{a,j}^{\ell} : 0 \leq a < (dt)^{d+2}\}$. Observe that, in order to cover B_{fill}, it is necessary and sufficient to select one node $r_{a,j}^{\ell}$ for each ℓ and j. Furthermore, there is a way to do that such that each selected $r_{a,j}^{\ell}$ covers one extra node in B_{code}^{ℓ} w.r.t. selected nodes in R_{code}. Note that we somewhat abuse notation as we denote by b_j^{ℓ} vertices of B_{fill}, while we use b_a^{ℓ} for vertices of B_{code}, hence the only distinction is by the variable name.

Lemma 3. *For any matrix $(\gamma_{\delta,\lambda})_{0 \leq \delta < d+2, 0 \leq \lambda < d}$ of size $(d+2) \times d$ with entries from $\{0, \ldots, t-1\}$, there exists a set $\tilde{R}_{enf} \subseteq R_{enf}$ of size $k' := \frac{n}{2}(d^{d+2} - d) + (d+2)d(t-1)$, such that:*

- *each vertex in $B_{choice} \cup B_{fill}$ has a neighbor in $\tilde{R}_{enf}$, and*
- *for every $0 \leq \ell < n/2$ we have $B_{code}^{\ell} \setminus N(\tilde{R}_{enf}) = \{b_a^{\ell} : 0 \leq \lambda < d, a = \sum_{0 \leq \delta < d+2}(\lambda t + \gamma_{\delta,\lambda})(dt)^{\delta}\}$.*

Proof. For each $0 \leq \delta < d+2$ and $0 \leq \lambda < d$, add to $\tilde{R}_{enf}$ the set $\{r_{\delta,\lambda,\gamma} : 0 \leq \gamma < t, \gamma \neq \gamma_{\delta,\lambda}\}$ containing $t-1$ vertices. Note that by construction $\tilde{R}_{enf}$ dominates the whole set B_{choice}. Consider a vertex $b_a^{\ell} \in B_{code}^{\ell} \setminus N(\tilde{R}_{enf})$ and observe that for each coordinate $0 \leq \delta < d+2$, there are exactly d values that a_{δ} can have, where $(a_0, \ldots, a_{d+1})$ is the (dt)-ary representation of a. Indeed, for any $0 \leq \delta < d+2$, we have $a_{\delta} \in X_{\delta} = \{\lambda t + \gamma_{\delta,\lambda} : 0 \leq \lambda < d\}$, since otherwise b_a^{ℓ} would be covered by $\tilde{R}_{enf}$ due to the δ-th coordinate. Moreover if we consider any $b_{(a_0,\ldots,a_{d+1})}^{\ell} \in B_{code}^{\ell}$ such that $a_{\delta} \in X_{\delta}$ for $0 \leq \delta < d+2$, then $b_{(a_0,\ldots,a_{d+1})}^{\ell}$ is not dominated by the vertices added to $\tilde{R}_{enf}$ so far.

Next, for each ℓ define $M^{\ell} := \{b_a^{\ell} : 0 \leq \lambda < d, a = \sum_{0 \leq \delta < d+2}(\lambda t + \gamma_{\delta,\lambda})(dt)^{\delta}\}$ and observe that M^{ℓ} are not dominated $\tilde{R}_{enf}$. For each $0 \leq \ell < n/2$, let Z^{ℓ} be the vertices of B_{code}^{ℓ} not yet covered by $\tilde{R}_{enf}$ and for each $1 \leq j \leq d^{d+2} - d$ select exactly one distinct vertex $v_j \in Z^{\ell} \setminus M^{\ell}$, where $v_j = b_a^{\ell}$, and add to $\tilde{R}_{enf}$ the vertex $r_{a,j}^{\ell}$. Observe that after this operation $\tilde{R}_{enf}$ covers B_{fill} and moreover the only vertices of B_{code} not covered by $\tilde{R}_{enf}$ are the vertices of $\bigcup_{0 \leq \ell < n/2} M^{\ell}$. Since the total size of $\tilde{R}_{enf}$ equals $d(d+2)(t-1) + \frac{n}{2}(d^{d+2} - d)$, the lemma follows. ☐

Lemma 4. *Consider an RBDS instance $(H = (R \cup B, E), k)$ containing $G_{enf} = (R_{enf} \cup B_{enf}, E_{enf})$ as an induced subgraph, with $R_{enf} \subseteq R$ and $B_{enf} \subseteq B$, such that no vertex of $B_{choice} \cup B_{fill}$ has a neighbor outside of R_{enf}. Then any feasible solution $\tilde{R}$ to (H, k) contains at least $k' := \frac{n}{2}(d^{d+2} - d) + (d+2)d(t-1)$ nodes $\tilde{R}_{enf}$*

of R_{enf}. Furthermore, for any feasible solution $\tilde{R}$ to (H, k) containing exactly k' vertices of R_{enf}, there exist a matrix $(\gamma_{\delta,\lambda})_{0 \leq \delta < d+2, 0 \leq \lambda < d}$ of size $(d+2) \times d$ with entries from $\{0, \ldots, t-1\}$, such that for each $0 \leq \ell < n/2$:

(a) there are at least d vertices in $U^\ell = B^\ell_{code} \setminus N(\tilde{R} \cap R_{enf})$, and
(b) U^ℓ is a subset of the d^{d+2} nodes $b^\ell_a = b^\ell_{(a_0,\ldots,a_{d+1})}$ such that for each $\delta \in \{0, \ldots, d+1\}$ there exists $\lambda \in \{0, \ldots d-1\}$ with $a_\delta = \lambda t + \gamma_{\delta,\lambda}$.

Proof. Let $\tilde{R}$ be any feasible solution to (H, k). Observe that since $\tilde{R}$ dominates B_{choice}, for each $0 \leq \delta < d+2$ and $0 \leq \lambda < d$ we have $|\tilde{R} \cap \{r_{\delta,\lambda,\gamma} : 0 \leq \gamma < t\}| \geq t - 1$. Moreover in order to dominate vertices of B_{fill}, the set $\tilde{R}$ has to contain at least $n/2(d^{d+2} - d)$ vertices of R_{fill}. Consequently, if $\tilde{R}$ contains exactly k' vertices of R_{enf}, then for each $0 \leq \delta < d+2$ and $0 \leq \lambda < d$, there is exactly one $\gamma_{\delta,\lambda}$ such that $r_{\delta,\lambda,\gamma_{\delta,\lambda}} \notin \tilde{R}$. By the same argument as in the proof of Lemma 3, we infer that for each ℓ, the set $B^\ell_{code} \setminus N(\tilde{R} \cap R_{code})$ contains exactly d^{d+2} vertices, and we denote them as U^ℓ_0. Observe that the set $\tilde{R} \setminus R_{code}$ dominates at most $d^{d+2} - d$ vertices of U^ℓ_0, for each $0 \leq \ell < n/2$, which proves properties (a) and (b) of the lemma. $\qquad\square$

4.2   The Overall Graph

The construction of $H_{inst} = (R_{inst} \cup B_{inst}, E_{inst})$ is rather simple. Let $(G_i = (V, E_i), \text{col}_i)$ be the input MPM instances, with $0 \leq i < T = t^{d(d+2)}$. By standard padding arguments we may assume that all the graphs G_i are defined over the same set V of even size n, i.e. $G_i = (V, E_i)$. For each $v \in V$, we create a blue node $b_v \in B_{inst}$. For each $e_i = \{u, v\} \in E_i$, we create a red node $r_{e,i} \in R^i_{inst}$ and connect it to both b_u and b_v. We let $R_{inst} := \bigcup_{0 \leq i < T} R^i_{inst}$. Intuitively, we desire that a RBDS solution, if any, selects exactly $n/2$ nodes from one set R^i_{inst}, corresponding to edges of different colors, which together dominate all nodes B_{inst}: This induces a feasible solution to MPM for the i-th instance.

It remains to describe the edges E_{conn} which connect H_{enf} with H_{inst}. This is the most delicate part of the entire construction. We map each index $i, 0 \leq i < T$, into a distinct $(d+2) \times d$ matrix M_i with entries $M_i[\delta, \lambda] \in \{0, \ldots, t-1\}$, for all possible values of δ and λ. Consider an instance G_i. We connect $r_{e,i}$ to b^ℓ_a iff $\ell = \text{col}_i(e_i)$ and there exists $0 \leq \lambda < d$ such that the expansion $(a_0, \ldots, a_{d+1})$ of a in base dt satisfies $a_\delta = M_i[\delta, \lambda] + \lambda \cdot t$ on each coordinate $0 \leq \delta < d+2$. The final graph $H := (R \cup B, E)$ we construct for our instance RBDS is then given by $R := R_{inst} \cup R_{enf}$ and $E := E_{inst} \cup E_{enf} \cup E_{conn}$. See Fig. 1.

Lemma 5. *H is $(d+2)$-degenerate.*

Proof. Observe that each vertex of $\bigcup_{0 \leq i < T} R^i_{inst}$ is of degree exactly $d+2$ in H, so we put all those vertices first to our ordering. Next, we take vertices of B_{inst}, as those have all neighbors already put into the ordering. Therefore it is enough to argue about the $(d+2)$-degeneracy of the enforcement gadget. We order vertices of $R_{fill} \cup B_{choice}$, since those are of degree exactly two in H.

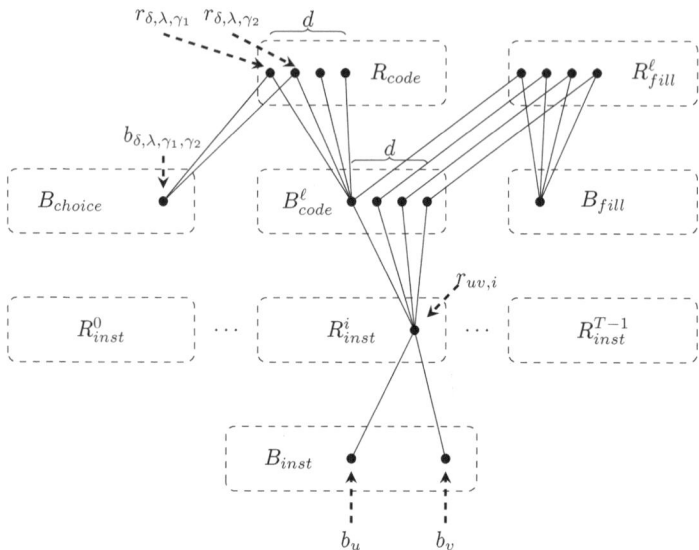

Fig. 1. Construction of the graph H. For simplicity the figure does not include sets $R^{\ell'}_{fill}$ and $B^{\ell'}_{code}$ for $\ell' \neq \ell$.

In $H \setminus R_{fill}$ the vertices of B_{fill} become isolated, so we put them next to our ordering. We are left with the vertices of the encoding gadget. Observe, that each blue vertex of the encoding gadget has exactly $d+2$ neighbors in R_{code}, one due to each coordinate, hence we put the vertices of B_{code} next and finish the ordering with vertices of R_{code}.   □

Lemma 6. *Let* $k := (d+2)d(t-1) + n/2(d^{d+2} - d) + n/2 = k' + n/2$. *Then* (H, k) *is a YES-instance of* RBDS *iff* (G_i, col_i) *is a YES-instance of* MPM *for some* $i \in \{0, \ldots, T-1\}$.

Proof. Let us assume that for some i_0 the instance (G_{i_0}, col_{i_0}) is a YES-instance and $E' \subseteq E_{i_0}$ is the corresponding solution. We use Lemma 3 with the matrix M_{i_0} assigned to the instance i_0 to obtain the set $\tilde{R}_{enf}$ of size $(d+2)d(t-1) + \frac{n}{2}(d^{d+2} - d)$. As the set $\tilde{R}$ we take $\tilde{R}_{enf} \cup \{r_{e,i_0} : e \in E'\}$. Clearly $|\tilde{R}| = k$. Since E' is a perfect matching, $\tilde{R}$ dominates B_{inst}. By Lemma 3, $\tilde{R}$ dominates $B_{fill} \cup B_{choice}$ and all but d vertices of each B^{ℓ}_{code}, so denote those d vertices by M^{ℓ}. Consider each $0 \leq \ell < n/2$, and observe that since E' is multicolored and by the construction of H, the set of neighbors of r_{e,i_0} in B_{code} is exactly $M^{col_{i_0}(e)}$; and hence $\tilde{R}$ is a solution for (H, k).

In the other direction, assume that (H, k) is a YES-instance and let $\tilde{R}$ be a solution of size at most k. By Lemma 4, the set $\tilde{R}$ contains at least $k' = \frac{n}{2}(d^{d+2} - d) + (d+2)d(t-1)$ vertices of R_{enf} and since $\tilde{R}$ needs to dominate also B_{inst} it contains at least $\frac{n}{2}$ vertices of $\bigcup_{0 \leq i < T} R^i_{inst}$, since no vertex of H dominates more than two vertices of B_{inst}. Consequently $|\bigcup_{0 \leq i < T} R^i_{inst} \cap \tilde{R}| = n/2$ and

$|R_{enf} \cap \tilde{R}| = k'$. We use Lemma 4 to obtain a matrix $M = (\gamma_{\delta,\lambda})$ of size $(d+2) \times d$. Moreover, by property (a) of Lemma 4, there are at least d vertices in U^ℓ, and consequently for each color ℓ the set $\tilde{R}$ contains exactly one vertex of the set $\{r_{e,i} : 0 \leq i < T, \text{col}_i(e) = \ell\}$. Our goal is to show that for each $0 \leq i < T$, such that a matrix different than M is assigned to the i-th instance, we have $\tilde{R} \cap R^i_{inf} = \emptyset$, which is enough to finish the proof of the lemma. Consider any such i and assume that the matrices M_i and M differ in the entry $M_i[\delta', \lambda'] \neq \gamma_{\delta',\lambda'}$. Let ℓ be a color such that $r_{e,i} \in \tilde{R}$ and $\text{col}_i(e) = \ell$. By property (b) of Lemma 4, the set of at least d vertices of B^ℓ_{code} not dominated by $\tilde{R} \cap R_{enf}$ is contained in $U^\ell_0 = \{b^\ell_{(a_0,\ldots,a_{d+1})} : \forall_{0 \leq \delta < d+2} \text{ if } \lambda t \leq a_\delta < (\lambda+1)t \text{ then } a_\delta = \lambda t + \gamma_{\delta,\lambda}\}$. However, by our construction of edges of H between R^i_{inst} and B^ℓ_{code}, we have $(N_H(r_{e,i}) \cap B^\ell_{code}) \not\subseteq U^\ell_0$ since the vertex $b^\ell_{(a_0,\ldots,a_{\delta+1})} \in N_H(r_{e,i}) \cap B^\ell_{code}$ with $a_\delta = \lambda' t + M_i[\delta', \lambda']$ does not belong to U^ℓ_0 and consequently does not belong to U^ℓ, which leaves at least one vertex of B^ℓ_{code} not dominated by $\tilde{R}$; a contradiction.

$\square$

Acknowledgements. We thank anonymous referees for their helpful remarks and comments.

References

1. Alber, J., Fellows, M.R., Niedermeier, R.: Polynomial-time data reduction for dominating set. Journal of the ACM 51(3), 363–384 (2004)
2. Bodlaender, H.L., Downey, R.G., Fellows, M.R., Hermelin, D.: On problems without polynomial kernels. Journal of Computer and System Sciences 75(8), 423–434 (2009)
3. Bodlaender, H.L., Fomin, F.V., Lokshtanov, D., Penninkx, E., Saurabh, S., Thilikos, D.M. (Meta) kernelization. In: FOCS, pp. 629–638 (2009)
4. Bodlaender, H.L., Jansen, B.M.P., Kratsch, S.: Cross-composition: A new technique for kernelization lower bounds. In: STACS, pp. 165–176 (2011)
5. Bodlaender, H.L., Thomassé, S., Yeo, A.: Kernel bounds for disjoint cycles and disjoint paths. In: Fiat, A., Sanders, P. (eds.) ESA 2009. LNCS, vol. 5757, pp. 635–646. Springer, Heidelberg (2009)
6. Cai, L., Chen, J., Downey, R.G., Fellows, M.R.: Advice classes of paramterized tractability. Annals of Pure and Applied Logic 84(1), 119–138 (1997)
7. Chen, Y., Flum, J., Müller, M.: Lower bounds for kernelizations and other preprocessing procedures. Theory of Computing Systems 48(4), 803–839 (2011)
8. Cygan, M., Grandoni, F., Hermelin, D.: Tight kernel bounds for problems on graphs with small degeneracy. CoRR, abs/1305.4914 (2013)
9. Cygan, M., Pilipczuk, M., Pilipczuk, M., Wojtaszczyk, J.O.: Kernelization hardness of connectivity problems in d-degenerate graphs. Discrete Applied Mathematics 160(15), 2131–2141 (2012)
10. Dell, H., Marx, D.: Kernelization of packing problems. In: SODA, pp. 68–81 (2012)
11. Dell, H., van Melkebeek, D.: Satisfiability allows no nontrivial sparsification unless the polynomial-time hierarchy collapses. In: STOC, pp. 251–260 (2010)
12. Diestel, R.: Graph Theory, 3rd edn. Springer-Verlag (2005)

13. Dom, M., Lokshtanov, D., Saurabh, S.: Incompressibility through colors and IDs. In: Albers, S., Marchetti-Spaccamela, A., Matias, Y., Nikoletseas, S., Thomas, W. (eds.) ICALP 2009, Part I. LNCS, vol. 5555, pp. 378–389. Springer, Heidelberg (2009)

14. Downey, R.G., Fellows, M.R.: Parameterized Complexity. Springer-Verlag (1999)

15. Erman, R., Kowalik, L., Krnc, M., Walen, T.: Improved induced matchings in sparse graphs. Discrete Applied Mathematics 158(18), 1994–2003 (2010)

16. Fellows, M.R.: The lost continent of polynomial time: Preprocessing and kernelization. In: Bodlaender, H.L., Langston, M.A. (eds.) IWPEC 2006. LNCS, vol. 4169, pp. 276–277. Springer, Heidelberg (2006)

17. Fomin, F.V., Lokshtanov, D., Saurabh, S., Thilikos, D.M.: Bidimensionality and kernels. In: SODA, pp. 503–510 (2010)

18. Fortnow, L., Santhanam, R.: Infeasibility of instance compression and succinct PCPs for NP. In: STOC, pp. 133–142 (2008)

19. Guo, J., Niedermeier, R.: Invitation to data reduction and problem kernelization. SIGACT News 38(1), 31–45 (2007)

20. Guo, J., Niedermeier, R.: Linear problem kernels for NP-hard problems on planar graphs. In: Arge, L., Cachin, C., Jurdziński, T., Tarlecki, A. (eds.) ICALP 2007. LNCS, vol. 4596, pp. 375–386. Springer, Heidelberg (2007)

21. Harnik, D., Naor, M.: On the compressibility of NP instances and cryptographic applications. SIAM Journal on Computing 39(5), 1667–1713 (2010)

22. Hermelin, D., Wu, X.: Weak compositions and their applications to polynomial lower-bounds for kernelization. In: SODA, pp. 104–113 (2012)

23. Kanj, I.A., Pelsmajer, M.J., Schaefer, M., Xia, G.: On the induced matching problem. Journal of Computer and System Sciences 77(6), 1058–1070 (2011)

24. Karp, R.M.: Reducibility among combinatorial problems. In: Miller, R.E., Thatcher, J.W. (eds.) Complexity of Computer Computations, pp. 85–103. Plenum Press (1972)

25. Kim, E.J., Langer, A., Paul, C., Reidl, F., Rossmanith, P., Sau, I., Sikdar, S.: Linear kernels and single-exponential algorithms via protrusion decompositions. In: Fomin, F.V., Freivalds, R., Kwiatkowska, M., Peleg, D. (eds.) ICALP 2013, Part I. LNCS, vol. 7965, pp. 613–624. Springer, Heidelberg (2013)

26. Kratsch, S.: Co-nondeterminism in compositions: a kernelization lower bound for a Ramsey-type problem. In: SODA, pp. 114–122 (2012)

27. Nemhauser, G.L., Trotter Jr., L.E.: Vertex packings: Structural properties and algorithms. Mathematical Programming 8(2), 232–248 (1975)

28. Philip, G., Raman, V., Sikdar, S.: Polynomial kernels for dominating set in graphs of bounded degeneracy and beyond. ACM Transactions on Algorithms 9(1), 11 (2012)

Labeling Moving Points with a Trade-Off between Label Speed and Label Overlap

Mark de Berg and Dirk H.P. Gerrits

Department of Mathematics and Computer Science
Technische Universiteit Eindhoven, The Netherlands

Abstract. Traditional map-labeling algorithms ensure that the labels being placed do not overlap each other, either by omitting labels or scaling them. This is undesirable in applications where the points to be labeled are moving. We develop and experimentally evaluate a heuristic for labeling moving points. Our algorithm labels all the points with labels of a fixed size, while trying to minimize the number of overlapping labels and ensuring smoothly moving labels. It allows a trade-off between label speed and label overlap.

1 Introduction

To do their jobs air-traffic controllers need a real-time visualization of the airplanes in their designated air space. Similarly, companies may want to track their fleet of taxis, trucks, or ships, and biologists may want to track wildlife tagged with GPS devices. A natural visualization for these kinds of applications is to represent each object as a moving point, and to place a *label* with each point that gives information about the object—an identifier, velocity and/or altitude, and so on. This leads to the *dynamic point-labeling problem*, which is the topic of our paper: how can we maintain a suitable labeling of a set of moving points in the plane?

Dynamic point labeling generalizes *static point labeling*, a problem in automated cartography that has attracted much attention; see the online Map Labeling Bibliography [9]. Here a static set of points (representing cities, say) is to be labeled (with their names). Such a *static labeling* should have readable labels, and an unambiguous association between labels and points. Readability is typically formalized by regarding the labels as axis-aligned rectangles slightly larger than the text they contain, and requiring that they be placed so that their interiors are disjoint. To associate labels with their points one usually requires each label to be placed so that it contains the point on its boundary. This can be done in several different ways, for example by requiring that the point is one of the four corners of the label (the 4-position model), or by allowing the point to be anywhere on the boundary (the 4-slider model). Other models, such as the 2-position model or the 1-slider model, have been studied as well.

Given a label model, one would like to label all of the points with non-overlapping labels. Unfortunately, this is not always possible and deciding

H.L. Bodlaender and G.F. Italiano (Eds.): ESA 2013, LNCS 8125, pp. 373–384, 2013.

whether this is the case is strongly NP-complete for most label models [5]. This led to the investigation of two optimization problems: the *size-maximization problem*, which asks for the maximum scaling factor for the labels that allows them to be placed without overlap, and the *number-maximization problem*, which asks for a maximum-cardinality subset of the points that allows a non-overlapping labeling. Results have been obtained for many variants of these problems.

One way to extend these results to dynamic point labeling would be to recompute a static labeling in real time—say 50 times per second. While several algorithms for static labeling are fast enough for this, and while the quality of the static solution is very good in practice, it still does not lead to satisfactory results. Indeed, algorithms for number-maximization would lead to labels appearing and disappearing between consecutive time steps, which is distracting for the user—note that heuristics for the dynamic number-maximization problem (see, for instance, [1,7,8] and their references) suffer from the same problem. Algorithms for size-maximization would lead to labels changing size (possibly in a non-continuous manner), which is disturbing as well. In an earlier paper [2], we therefore studied the *free-label maximization problem* which asks to label all points with labels of a given size, while maximizing the number of labels that are *free* (that is, interior-disjoint with all other labels). This avoids the downsides of the size- and number-maximization problems mentioned above. In our earlier paper we studied the static variant of the free-label maximization problem. In this paper we turn our attention to the dynamic version of the problem.

Our results. The formal problem setting is as follows. The input is a *dynamic point set P*, which specifies for each point $p \in P$ the time $birth(p)$ at which it is added to the point set, the time $death(p)$ at which it is removed, and its continuous trajectory between those times. For simplicity we assume that the trajectories are polygonal, although the results can be extended to curved trajectories. We refer to the time interval during which p is present in the point set as its *lifespan* and denote it by $life(p) = [birth(p), death(p)]$. Whenever $t \in life(p)$ we say that p is *alive* at time t, and then denote its position by $p(t)$. For such a dynamic point set P we compute a dynamic labeling $\mathcal{L}$, which for all t assigns a static labeling $\mathcal{L}(t)$ to the point set $P(t)$. Here $P(t) = \{p(t) \mid p \in P, t \in life(p)\}$ denotes the set of points that are alive at time t, at their respective locations. The label model we use is a slider model, where we require the axis-aligned label to be "behind" the point (with respect to its direction of movement) at all times. More precisely, the ray from the point to the center of its label should make an angle of at least 90° with the ray from the point in its direction of movement. This way, the label placement does not obscure the movement of the points.

Our global approach is simple: we compute a static labeling at regular intervals, and then interpolate between successive static labelings to obtain the dynamic labeling. For the static labeling we use an algorithm from our previous paper [2]; the crucial and novel part lies in the interpolation. Our solution has several attractive properties. Firstly, it is fast. The static labelings can be computed in $O(n \log n)$ time for n points, and interpolating takes time linear in the combined complexity of the point trajectories. Secondly, the static labelings contain many free labels and the interpolation is such that it minimizes both the maximum speed of the labels as well as their average speed. Thirdly, the user can vary Δt, the time between successive time steps, to obtain a trade-off between label speed and label overlap. The trade-off turns out be to be very favorable in practice: a small sacrifice in label freeness can yield greatly reduced label speeds. Finally, the algorithm is "semi-online", in the sense that it only needs to know the trajectories and the times at which points are added or removed Δt time in advance.

2   A Heuristic Algorithm

As mentioned above, we propose a dynamic labeling algorithm that computes a series of static labelings and then moves the labels smoothly from their positions in one static labeling to their positions in the next. In computing the static labelings we try to maximize the number of free labels, in computing the interpolation between labelings we try to minimize the movement speed of the labels. Thus we hope to achieve a good dynamic labeling according to all criteria mentioned above. The algorithm can be expressed by the following pseudocode, where $\Delta t > 0$ is the time between successive static labelings, and $\mathcal{L}(t, t')$ refers to the part of a dynamic labeling $\mathcal{L}$ between times t and t'.

Algorithm INTERPOLATIVELABELING(P, $t_{\max}$)
1. $\mathcal{L} \leftarrow \emptyset;\ t \leftarrow 0$
2. $\mathcal{L}(t) \leftarrow$ STATICLABELING(P, t)
3.   **while** $t < t_{\max}$
4.   **do** $t_{\text{next}} \leftarrow \min(t + \Delta t, t_{\max})$
5. $\mathcal{L}(t_{\text{next}}) \leftarrow$ STATICLABELING(P, t_{next})
6. $\mathcal{L}(t, t_{\text{next}}) \leftarrow$ INTERPOLATE($\mathcal{L}(t), t,\ \mathcal{L}(t_{\text{next}}), t_{\text{next}}$)
7. $t \leftarrow t_{\text{next}}$
8.   **return** $\mathcal{L}$

This method has three parameters. Firstly, the size of the time step Δt. Smaller time steps may lead to a greater number of free labels over time. Larger time steps may lead to less or slower label movement, and fewer calls to the STATIC-LABELING subroutine. Secondly, the algorithm used for STATICLABELING. We shall use a simple, greedy algorithm called FOURGREEDYSWEEPS, which was described in an earlier paper [2]. For labels of equal dimensions that can be placed anywhere around their point it yields a constant-factor approximation to free-label maximization. The algorithm still works in our setting where labels may only be placed behind their points, but the proof of its approximation ratio unfortunately does not. Lastly, the algorithm used for INTERPOLATE. We shall use the simple, linear-time algorithm described below. It minimizes both the average and the maximum movement speed of each label.

Interpolation algorithm. We are given two moments in time t_1 and t_2, and static labelings $\mathcal{L}(t_1)$ and $\mathcal{L}(t_2)$ of the dynamic point set at those times. We then wish to compute a dynamic labeling $\mathcal{L}(t_1, t_2)$ which *respects* them, that is, whose static labelings at times t_1 and t_2 equal $\mathcal{L}(t_1)$ and $\mathcal{L}(t_2)$. We will do this independently for each moving point $p \in P$ with $life(p) \cap [t_1, t_2] \neq \emptyset$. For the rest of this section, let p be such a point and let $[t'_1, t'_2] = life(p) \cap [t_1, t_2]$.

Recall that the position of a label is restricted in that it must contain the point it labels on its boundary. Furthermore, the labels must trail "behind" the points. Specifically, a ray from the point in its direction of movement must make an angle of at least $90°$ with a ray from the point through its label's center. Now suppose we translate the coordinate system so that point p is always at the origin. The allowed positions for the center of p's label then trace out part of the surface of a box in 3-dimensional space-time. We refer to this *configuration space* for point p over the interval $[t'_1, t'_2]$ as $\mathcal{C}$. Figure 1(b) shows an example of such a configuration space for the point trajectory shown in Figure 1(a).

A label trajectory for p corresponds to a path through $\mathcal{C}$, monotone with respect to the time axis. To compute one, it will be convenient to "unfold" $\mathcal{C}$ into a rectilinear polygon R as in Figure 1(c). If point trajectory p has k vertices, then R is the union of $k - 1$ closed, axis-aligned rectangles. The intersection of any two consecutive rectangles is a vertical line segment which we refer to as a *portal*—see Figure 1(c). If a portal's coordinate along the time axis is t we refer to it as the *portal at t*, denoted by $\Psi(t)$. In addition to these $k - 2$ portals we define two extra portals at t'_1 and t'_2. These are simply the sets of label positions that respect $\mathcal{L}(t_1)$ and $\mathcal{L}(t_2)$. If $t'_1 = t_1$ then $\Psi(t'_1)$ is a single point representing the label given to p by $\mathcal{L}(t_1)$. If $t'_1 \neq t_1$ then $\mathcal{L}(t_1)$ specifies nothing about p's label position at time t'_1, and $\Psi(t'_1)$ is simply the leftmost edge of R.

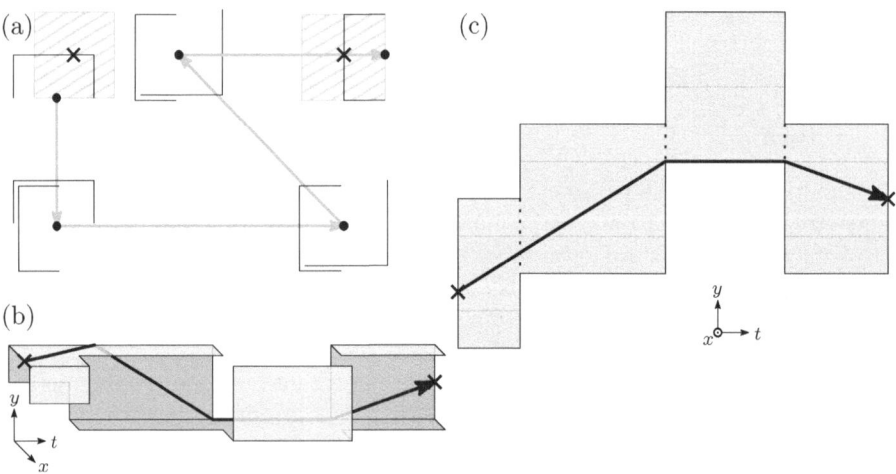

Fig. 1. (a) A piecewise linear point trajectory (in gray, with dots as vertices) with given labels (in gray, hatched, centers marked by crosses) at the endpoints. (b) The corresponding configuration space of allowed label positions. (c) An unfolding of (b) and a shortest path through it. Portals are shown as dotted line segments.

Analogously, $\Psi(t_2')$ is either the rightmost edge of R, or a point representing p's label in $\mathcal{L}(t_2)$. With these definitions, a label trajectory for p corresponds to a time-monotone path through R from portal $\Psi(t_1')$ to portal $\Psi(t_2')$, passing through all intermediate portals in sequence. We will now argue that the shortest such path produces the most desirable label trajectory.

Lemma 1. *A shortest path from $\Psi(t_1')$ to $\Psi(t_2')$ through R yields a label trajectory minimizing both (i) the average speed and (ii) the maximum speed of p's label relative to p.*

Proof. (i) Let π be such a shortest path, which must be a t-monotone polygonal chain. Let $T(\pi)$ be the summed length of the projections onto the t-axis of the links of π, and let $Y(\pi)$ the same quantity for the y-axis. Every t-monotone path π' from $\Psi(t_1')$ to $\Psi(t_2')$ must traverse the same distance in the t-direction, namely $T(\pi') = T(\pi) = t_2' - t_1'$. Since π is the shortest such path, it minimizes the distance traveled in y-direction, that is, $Y(\pi) \leqslant Y(\pi')$ for all π'. Thus π minimizes the average speed $Y(\pi)/T(\pi)$ of p's label.

 (ii) Let ab be a steepest link of π, that is, ab has the maximum absolute slope among the links of π. Suppose that ab's projection onto the t-axis is $[t', t'']$ and that ab has negative slope (the case where it has positive slope is similar). Then π cannot make a left turn at a, as that would make ab less steep than the link preceding it. So π must either make a right turn at a, or start at a. If π makes a right turn at a, then a must be the bottom endpoint of portal $\Psi(t')$, as we could otherwise shorten π by a downward deformation at a. Since a lies above b, this makes ab the shortest path not only from a to b, but also from $\Psi(t')$ to b. If π instead starts at a, then $t_1' = t'$, so the same condition holds. Reasoning symmetrically for b yields that ab must also be the shortest path from a to $\Psi(t'')$. We conclude that ab is the shortest path from $\Psi(t')$ to $\Psi(t'')$. This implies that any t-monotone path from $\Psi(t')$ to $\Psi(t'')$ must be at least as steep as ab somewhere between t' and t''. Thus π minimizes the maximum speed $Y(ab)/T(ab)$ of p's label. $\qquad\square$

Theorem 1. *Suppose we are given a dynamic point set P with n points, along with static labelings $\mathcal{L}(t_1)$ and $\mathcal{L}(t_2)$ of it at times t_1 and t_2. For each $p \in P$ let k_p be the number of vertices in its polygonal trajectory between times t_1 and t_2. In $O(\sum_{p \in P} k_p)$ time and $O(\max\{k_p \mid p \in P\})$ space we can then compute a dynamic labeling $\mathcal{L}(t_1, t_2)$ respecting $\mathcal{L}(t_1)$ and $\mathcal{L}(t_2)$, that for each point minimizes both its average and its maximum label speed.*

Proof. Lemma 1 shows that $\mathcal{L}(t_1, t_2)$ can be obtained by computing a shortest path through the unfolded configuration space for each point. It remains to show that this can be done in $O(k_p)$ time and space for a point $p \in P$ with $life(p) \cap [t_1, t_2] = [t_1', t_2'] \neq \emptyset$. As before, let R denote the simple, rectilinear polygon with $O(k_p)$ vertices that is p's unfolded configuration space. Through R we seek either a shortest point-to-point path (if $life(p) \supseteq [t_1, t_2]$), a shortest edge-to-edge path (if $life(p) \subset [t_1, t_2]$), or a shortest point-to-edge path (otherwise). This can be done in linear time using an algorithm by Lee and Preparata [6], with some additional techniques by Guibas et al. [4]. $\qquad\square$

Hourglass trimming. In the previous section we described an algorithm to minimize both the average and the maximum speed of all labels in a dynamic labeling
that interpolates between two given static labelings. As the experimental evaluation in the next section will show, however, high label speeds can still occur
when a poor choice of static labelings is made. To see why this occurs, consider
the example in Figure 2. Recall that we compute static labelings at regular intervals of Δt time. In the example, point p's trajectory makes a sharp left turn
at time $t + \varepsilon$ for some small $\varepsilon > 0$. The static labeling algorithm completely
disregards this when computing the labeling for time t, and comes up with the
depicted leftmost label position for p. Now whatever labeling is picked for time
$t + \Delta t$, the label for p will have to move substantially within ε time units. The
problem here is that the chosen label position is just in the range of positions
considered to be behind the point at time t, but the same position lies quite a
bit in front of the point at time $t + \varepsilon$.

To fix this, we shall restrict the range of label positions that the static labeling
algorithm is allowed to use, based on the trajectories of the points. We will
need some definitions first. Consider a fixed point p, and let R be its unfolded
configuration space. Let $\pi(a, b)$ denote the shortest path in R from a to b, for any
$a, b \in R$. For a time interval $[t', t'']$ in which p is alive, we define the hourglass
$H(t', t'')$ as the region enclosed by $\pi(a, c)$ and $\pi(b, d)$, where the segments ab
and cd are the intersections of R with the vertical lines at t' and t''. Of the two
paths $\pi(a, c)$ and $\pi(b, d)$ we call the upper one the hourglass's *upper chain*, and
the other its *lower chain*. If the two chains intersect, then their intersection is
a common subchain called the *string*. The hourglass is then called *closed*, and
removal of the string leaves two connected components called *funnels*. If the
hourglass is not closed, it is called *open*. Figure 3(a) illustrates these concepts.

Now consider a time instance $t \in life(p)$. Let $t_- = \max(birth(p), t - \Delta t)$, let
$t_+ = \min(t + \Delta t, death(p))$, and consider the hourglasses $H(t_-, t)$ and $H(t, t_+)$.
Whatever label positions are chosen at times t_-, t, and t_+, the slowest interpolation between them (that is, the shortest path connecting them in R) must stay
within the union of these two hourglasses. If $H(t_-, t)$ or $H(t, t_+)$ has steep edges,
as in the example of Figure 3, then a fast moving label may result. We shall therefore narrow the ranges of valid label positions in such a way that these steep edges
are "trimmed off" the hourglasses. If $H(t_-, t)$ is closed, then let $F_-(t)$ denote

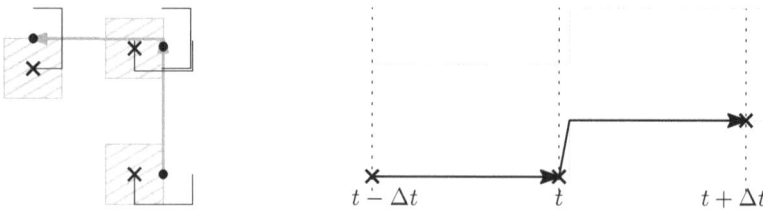

Fig. 2. An example of the static labeling algorithm making a poor choice with regards
to the dynamic labeling

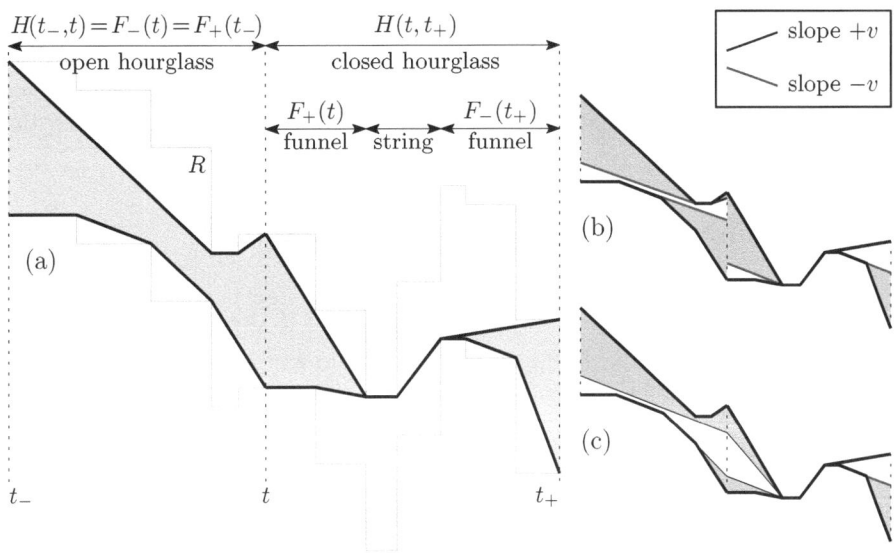

Fig. 3. (a) An example of an open hourglass (left) and a closed hourglass (right), the latter consisting of two funnels connected by a string. (b) The same hourglasses after trimming, shown in white. (c) When necessary (as in this case), the trimmed hourglasses are modified so that they connect to each other.

the rightmost funnel of $H(t_-,t)$, and let $F_-(t) = H(t_-,t)$ otherwise. Similarly, let $F_+(t)$ denote either $H(t,t_+)$ (if open) or its leftmost funnel (if closed). We assume we are given a parameter v that denotes a speed deemed reasonable for labels. We now translate a line with slope $+v$ down from positive infinity along the y-axis until it has become tangent to one of the two chains defining $F_-(t)$. Similarly, we translate a line with slope $-v$ up from negative infinity until it has become tangent to one of the two chains defining $F_-(t)$. These two lines define a narrower interval on the vertical line at t—see Figure 3(b). If we apply the same procedure at time t_- (using $F_+(t_-)$), then the new hourglass $H'(t_-,t)$ between the two narrowed intervals at t_- and t is the trimmed hourglass we are after. Specifically, the slowest label interpolation through $H'(t_-,t)$ cannot exceed the speed v, except in two cases. Firstly, this occurs when $H'(t_-,t)$ is closed, and its string contains an edge steeper than v, as on the right in Figure 3(b). In this case there is nothing that can be done to avoid exceeding the speed v with p's label. Secondly, this occurs when $H'(t_-,t)$ is open, and a bi-tangent of its upper and lower chain is steeper than v, as on the left in Figure 3(b). In this case it might be possible to trim the hourglass further, but sometimes this will simply result in a closed hourglass, making the previous case apply. Hence, we decided not to trim the hourglass further. Our experiments described in the next section show that our current method works quite well in practice.

In the same way, we compute the trimmed hourglass $H'(t,t_+)$, using $F_+(t)$ and $F_-(t_+)$. Typically, the narrowed interval at t that defines $H'(t_-,t)$ will differ from the narrowed interval at t that defines $H'(t,t_+)$. If they overlap this poses no problem, as we may then simply take their intersection. In Figure 3(b), however, they are disjoint. In that case, there is nothing we can do to avoid high label speeds on both sides of t. We then take the interval in between the two, as in Figure 3(c). Note that this can undo some of the trimming, making the hourglass wider again.

3   Experimental Evaluation

We will now evaluate the effect of varying the time-step parameter Δt on the quality of the produced labeling, that is, on the number of free labels and on the speeds at which labels move. Intuitively, one would expect both the number of free labels and the label speeds to increase as the time step approaches 0. Thus there should be a trade-off between how many labels are free and how slow the labels move. Our main goal is to quantify this trade-off experimentally.

Computation time. We have measured the computation time of our C++ implementation only informally, just to ensure it was fast enough for use in interactive applications. On the modest hardware used for our experiments (an Intel Q6600 2.40GHz with 3GB RAM running Ubuntu 12.10), INTERPOLATION takes about 0.4 milliseconds to interpolate between two 100-point labelings, and 2.2 milliseconds between two 1,000-point labelings. Producing such labelings with FOUR-GREEDYSWEEPS takes about 5 milliseconds for 100 points, and 189 milliseconds for 1,000 points. Note that this is with a simple $O(n^2)$-time implementation, even though a more sophisticated $O(n \log n)$-time implementation is possible [2]. The latter could no doubt label 1,000 points much more quickly, but such large numbers of labels cannot be legibly displayed on a reasonably sized screen anyway. The extra engineering effort was therefore deemed unnecessary.

Experimental setup. To evaluate the quality of the produced dynamic labelings, we have used a network of streets in the Dutch city of Eindhoven (see Figure 5). Moving points were created to move along five polygonal paths through the network, at constant and equal speeds of 35 px/s. The arrival times of the points on each route were created by a Poisson process with parameter 5 s. This makes the time between arrivals of successive points on a route an exponentially distributed random variable with a mean of 5 s. Different seeds for the random number generator thus create different problem instances. We used 100 different seeds to create 100 problem instances. For each problem instance we used our algorithm to produce dynamic labelings from $t = 0$ to $t = t_{\max} = 60$ s for several different values of the time step Δt. To determine the quality of such a dynamic labeling $\mathcal{L}$ we did not compute the exact intervals of time during which each label was free. Instead, $\mathcal{L}$ was sampled at regular times at a rate of 25.6 samples per second (1537 samples total over 60 seconds). This is roughly the same framerate as used in movies (24 frames per second), but with the time between samples changed to have a finite binary floating point representation (from $1/24$ s $= 5/120$ s to $5/128$ s). For each sample we recorded the number of free and non-free labels, as well as the amount each label moved (relative to its point) since the last sample. In addition, we recorded the *free label area* for each sample: the area covered by the union of the labels, minus the area covered by more than one label. In practice, this might be a more useful measure to maximize than the number of completely free labels. All of the above was done for several different time steps, the lowest being $\Delta t = 1/25.6$ s ≈ 0, so that each sample is labeled independently without regard for label speed, and the highest being $\Delta t = 61$ s > 60 s, so that label speeds are minimized without regard for label freeness. The graphs below offer a summary of the resulting data. The software used to generate the data and the graphs can be downloaded from http://dirkgerrits.com/programming/flying-labels/.

Results at a glance. The graphs in Figure 4 provides a high-level overview of the quality of labelings computed by our algorithm. The top graph shows how the label speed (measured in pixels per second), averaged over all moving points and over the whole time interval $[0, t_{\max}]$, decreases as we pick higher and higher values for Δt. The middle graph shows the same for the fraction of the labels that are completely free. The bottom graph, lastly, shows the free label area divided by the total area that would be spanned by the labels if they did not overlap. Each graph shows the minimum and maximum (dotted lines), 25% and 75% quantiles (dashed lines), and mean (solid line) over the 100 problem instances. The red lines show the results of the algorithm without hourglass trimming, the black lines show the results when hourglass trimming is used with a parameter of $v = 10$ px/s. In both cases we see a very sharp decline in label speeds until around $\Delta t = 2$ s, with a more modest corresponding decrease in label freeness. For higher Δt the decrease in label speeds slows down substantially while the decrease in label freeness continues. Thus, these preliminary results suggest that a time step of around 2 seconds should yield good labelings.

Detailed analysis of one instance. The effect of turning hourglass trimming on is similar to that of increasing the time step: the label speeds and freenesses both decrease. In this sense it forms an alternative to raising the time step. Hourglass trimming does something more, however, as can be seen when we examine a single problem instance in more details. We have selected an instance with particularly few free labels, but the effect can also be seen in other instances.

In Figure 6, seven graphs show how the label speeds of the individual moving points develop over time in this problem instance, for seven different values of Δt. Each of the moving points is drawn as a polyline, showing its speeds at all time samples. The red lines show the situation when hourglass trimming is not used. In the case of $\Delta t \approx 0$ almost all labels move rather violently (as was to be expected). For the other values of Δt, and especially the higher ones, a curious pattern appears. Label movement tends to be slow overall and decrease when Δt increases, but there are "spikes" of very high label movement near times that are a multiple of Δt. This effect is caused by a point changing direction near a multiple of Δt, as was explained in Section 2. This was our motivation for introducing hourglass trimming. The black lines show what happens when hourglass trimming is employed with the parameter $v = 10$ px/s. The regularly occurring spikes vanish, leaving label speeds with less variance. Note that the resulting label speeds do still exceed v, and by quite a bit for smaller time steps.

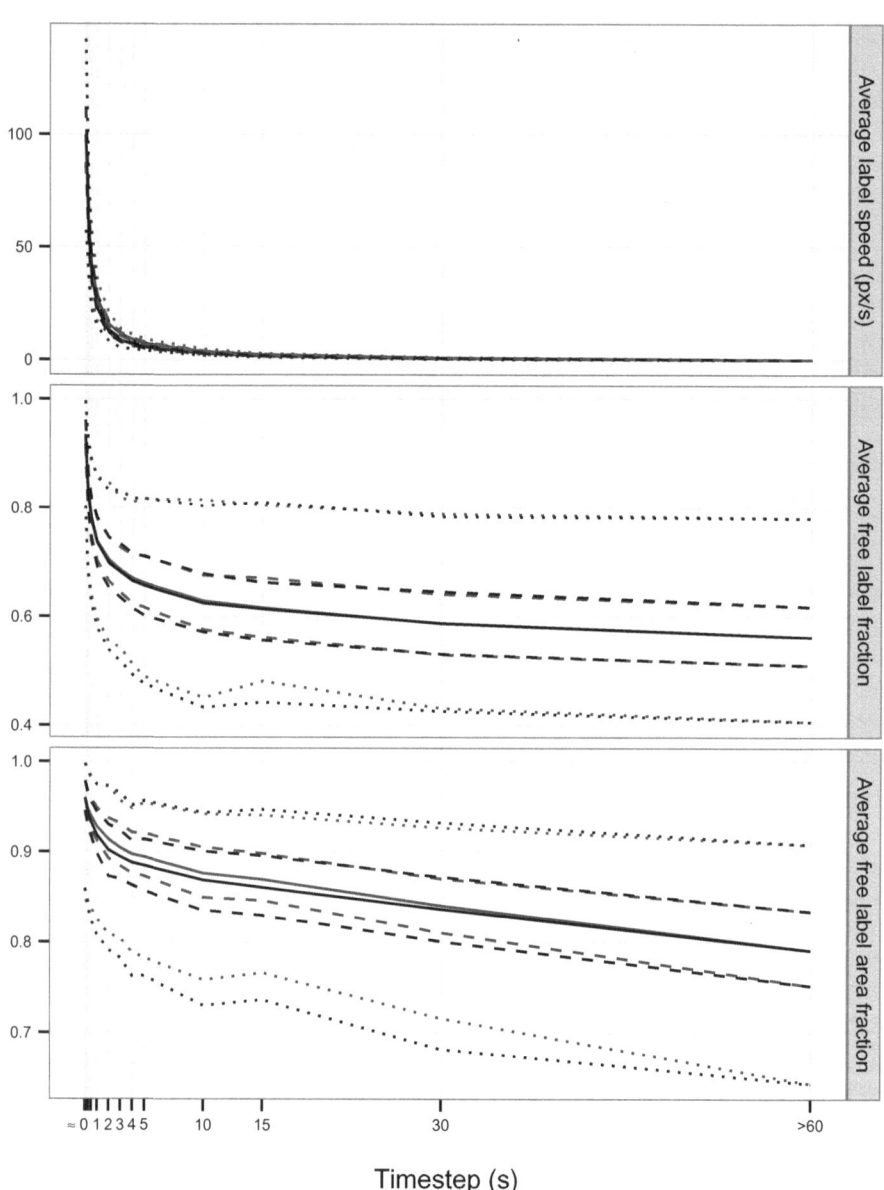

Fig. 4. The effects of varying the time step from $\Delta t \approx 0$ to $\Delta t > 60$ s on label speeds, number of free labels, and free label area, both with (black) and without (red) hourglass trimming. Shown are the minimum and maximum (dotted lines), 25% and 75% quantiles (dashed lines), and mean (solid line) over the 100 problem instances.

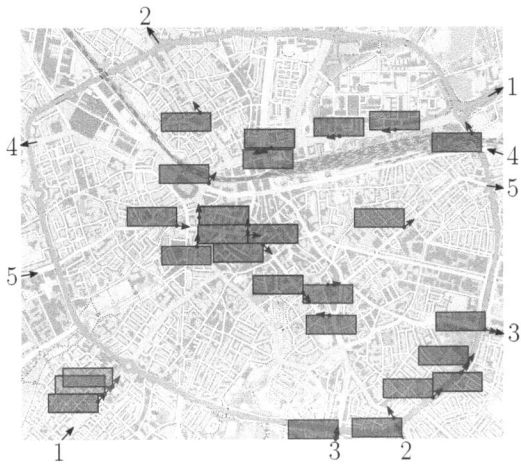

Fig. 5. The road network used for our experiments, and some labeled points moving across it along five routes. Blue labels are free, red labels are not.

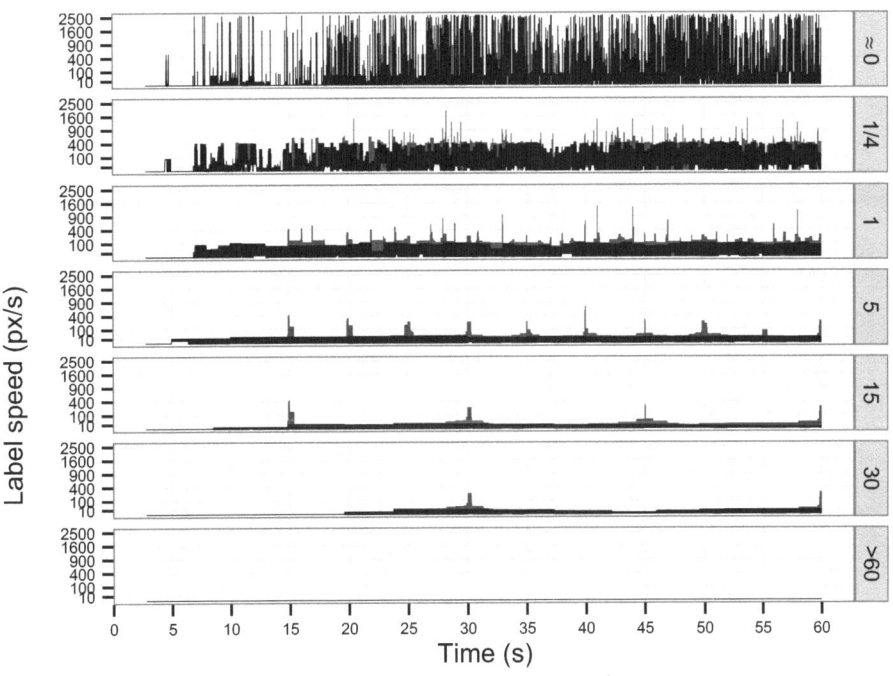

Fig. 6. The progression of label speeds over one particular problem instance, for several different time steps, both with (black) and without (red) hourglass trimming. Note that the y-axis uses a square root scale. With a linear scale the higher time steps would have indistinguishably low speeds, while with a logarithmic scale the lower time steps would have indistinguishably high speeds.

4   Conclusion

We developed a heuristic algorithm for free-label maximization on dynamic point sets, and evaluated it experimentally. The algorithm has been presented with the assumption that all points move on polygonal trajectories, but could be implemented just as well for curved trajectories. Instead of operating on polygons our algorithms will then work with curved *splinegons*. This change can be effected using techniques due to García-López and Ramos [3]. Our algorithm works by computing static labelings with many free labels at regular intervals, and then interpolating between these static labelings in a way that minimizes both the average and the maximum label speed. By varying the time between static labelings one obtains a trade-off between the number of free labels over time, and the speeds of the labels. The trade-off seemed favorable in experiments: a small increase in label speeds can yield a great increase in the number of free labels.

With these preliminary results we have only scratched the surface. From a theoretical point of view, algorithms with proven approximation ratios are still sorely lacking for dynamic labeling. From a practical point of view, there is room for improvement in other directions. While hourglass trimming was a step in the right direction, our method for choosing static labelings can undoubtedly be improved further. Testing on real-world data is also needed to get a more realistic picture of our method's performance.

References

1. Bell, B., Feiner, S., Höllerer, T.: View management for virtual and augmented reality. In: Proc. 14th ACM Sympos. User Interface Software and Technology (UIST 2001), pp. 101–110. ACM (2001)
2. de Berg, M., Gerrits, D.H.P.: Approximation algorithms for free-label maximization. Comput. Geom. Theory Appl. 45(4), 153–168 (2012)
3. García-López, J., Ramos, P.A.: A unified approach to conic visibility. Algorithmica 28(3), 307–322 (2000)
4. Guibas, L., Hershberger, J., Leven, D., Sharir, M., Tarjan, R.E.: Linear-time algorithms for visibility and shortest path problems inside triangulated simple polygons. Algorithmica 2(1), 209–233 (1987)
5. van Kreveld, M., Strijk, T., Wolff, A.: Point labeling with sliding labels. Comput. Geom. Theory Appl. 13, 21–47 (1999)
6. Lee, D.T., Preparata, F.P.: Euclidean shortest paths in the presence of rectilinear barriers. Networks 14(3), 393–410 (1984)
7. Rosten, E., Reitmayr, G., Drummond, T.: Real-time video annotations for augmented reality. In: Bebis, G., Boyle, R., Koracin, D., Parvin, B. (eds.) ISVC 2005. LNCS, vol. 3804, pp. 294–302. Springer, Heidelberg (2005)
8. Vaaraniemi, M., Treib, M., Westermann, R.: Temporally coherent real-time labeling of dynamic scenes. In: Proc. 3rd Internat. Conf. Computing for Geospatial Research and Applications, COM.Geo 2012, article no. 17 (2012)
9. Wolff, A., Strijk, T.: The Map Labeling Bibliography (2009), http://liinwww.ira.uka.de/bibliography/Theory/map.labeling.html

Inefficiency of Standard Multi-unit Auctions

Bart de Keijzer[1], Evangelos Markakis[3,*], Guido Schäfer[1,2],
and Orestis Telelis[3,**]

[1] CWI Amsterdam, The Netherlands
[2] VU Amsterdam, The Netherlands
{b.de.keijzer,g.schaefer}@cwi.nl
[3] Athens University of Economics and Business, Greece
{markakis,telelis}@gmail.com

Abstract. We study two standard multi-unit auction formats for allocating multiple units of a single good to multi-demand bidders. The first one is the Discriminatory Auction, which charges every winner his winning bids. The second is the Uniform Price Auction, which determines a uniform price to be paid per unit. Variants of both formats find applications ranging from the allocation of state bonds to investors, to online sales over the internet. For these formats, we consider two bidding interfaces: (i) standard bidding, which is most prevalent in the scientific literature, and (ii) uniform bidding, which is more popular in practice. In this work, we evaluate the economic inefficiency of both multi-unit auction formats for both bidding interfaces, by means of upper and lower bounds on the Price of Anarchy for pure Nash equilibria and mixed Bayes-Nash equilibria. Our developments improve significantly upon bounds that have been obtained recently for submodular valuation functions. Also, for the first time, we consider bidders with subadditive valuation functions under these auction formats. Our results signify near-efficiency of these auctions, which provides further justification for their use in practice.

1 Introduction

We study standard multi-unit auction formats for allocating multiple units of a single good to multi-demand bidders. Multi-unit auctions are one of the most widespread and popular tools for selling identical units of a good with a single auction process. In practice, they have been in use for a long time, one of their most prominent applications being the auctions offered by the U.S. and U.K. Treasuries for selling bonds to investors, see e.g., the U.S. treasury report [21]. In more recent years, they are also implemented by various online brokers [17].

* Supported by the project AGT of the research funding program THALES (co-financed by the EU and Greek national funds).
** Supported by project "DDCOD" (PE6-213). The project is implemented within the framework of the Action "Supporting Postdoctoral Researchers" of the Operational Program "Education and Lifelong Learning" (Action's Beneficiary: General Secretariat for Research and Technology), and is co-financed by the European Social Fund (ESF) and the Greek State.

H.L. Bodlaender and G.F. Italiano (Eds.): ESA 2013, LNCS 8125, pp. 385–396, 2013.

Table 1. Upper bounds on the (Bayesian) economic inefficiency of multi-unit auctions

Valuation Functions	Auction Format (bidding: standard | uniform)	
	Discriminatory Auction	*Uniform Price Auction*
Submodular	$e/(e-1)$	3.1462
Subadditive	2 | $2e/(e-1)$	4 | 6.2924

In the literature, multi-unit auctions have been a subject of study ever since the seminal work of Vickrey [22] (although the need for such a market enabler was conceived even earlier, by Friedman, in [8]) and the success of these formats has led to a resurgence of interest in auction design.

There are three simple *Standard Multi-Unit Auction* formats that have prevailed and are being implemented; these are the *Discriminatory Auction*, the *Uniform Price Auction* and the *Vickrey Multi-Unit Auction*. All three formats share a common allocation rule and bidding interface and have seen extensive study in auction theory [12,15]. Each bidder under these formats is asked to issue a sequence of non-increasing marginal bids, one for each additional unit. For an auction of k units, the k highest marginal bids win, and each grants its issuing bidder a single unit. The formats differ in the way that payments are determined for the winning bidders. The Discriminatory Auction prescribes that each bidder pays the sum of his winning bids. The Uniform Price Auction charges the lowest winning or highest losing marginal bid per allocated unit. The Vickrey auction charges according to an instance of the Clarke payment rule (thus being a generalization of the well known single-item Second-Price auction).

Except for the Vickrey auction, which is truthful and efficient, the others suffer from a *demand reduction* effect [1], whereby bidders may have incentives to understate their value, so as to receive less units at a better price. This effect is amplified when bidders have non-submodular valuation functions, since the bidding interface forces them to encode their value within a submodular bid vector. Even worse, in many practical occasions bidders are asked for a *uniform bid* per unit together with an upper bound on the number of desired units. In such a setting, each bidder is required to "compress" his valuation function into a bid that scales linearly with the number of units. The mentioned allocation and pricing rules apply also in this *uniform bidding* setting, thus yielding different versions of Discriminatory and Uniform Price Auctions. Despite the volume of research from the economics community [1,16,6,3,4] and the widespread popularity of these auction formats, the first attempts of quantifying their economic efficiency are only very recent [14,19]. There has also been no study of these auction formats for non-submodular valuations, as noted by Milgrom [15].

Our Contributions. We study the inefficiency of the Discriminatory Auction and Uniform Price Auction under the standard and uniform bidding interfaces. Our main results are improved inefficiency bounds for bidders with submodular valuation functions and new bounds for bidders with subadditive valuation functions.[1] The results are summarized in Table 1. Our bounds indicate that these auctions are nearly efficient, which, paired with their simplicity, provides further justification for their use in practice.

Our focus is on the inefficiency of Bayes-Nash equilibria; we refer the reader to the full version of the paper [11] for a discussion of pure Nash equilibria. For submodular valuation functions, we derive upper bounds of $\frac{e}{e-1}$ and $3.1462 < \frac{2e}{e-1}$ for the Discriminatory and the Uniform Price Auctions, respectively. These improve upon the previously best known bounds of $\frac{2e}{e-1}$ and $\frac{4e}{e-1}$ [19]. For the Uniform Price Auction, our bound is less than a factor 2 away from the known lower bound of $\frac{e}{e-1}$ [14]. We also prove lower bounds of $\frac{e}{e-1}$ and 2 for the Discriminatory Auction and Uniform Price Auction, with respect to the currently known proof techniques [19,7,5,2,10]. As a consequence, unless the upper bound of $\frac{e}{e-1}$ for the Discriminatory Auction is tight, its improvement requires the development of novel tools; the same holds for reducing the Uniform Price Auction upper bound below 2 (if $\frac{e}{e-1}$ from [14] is indeed worst-case). For subadditive valuations, we obtain bounds of $\frac{2e}{e-1}$ and $6.2924 < \frac{4e}{e-1}$ for Discriminatory and Uniform Price Auctions respectively, independent of the bidding interface. Further, for the standard bidding interface we are able to derive improved bounds of 2 and 4, respectively, by adapting a recent technique from [7]. We also give a lower bound of almost 2 for uniform pricing and subadditive valuations. In Section 4 we discuss further applications of our results in connection with the smoothness framework of [19]. In particular, some of our bounds carry over to simultaneous and sequential compositions of such auctions (Table 2).

Related Work. The multi-unit auction formats that we examine here present technical and conceptual resemblance to the *Simultaneous Auctions* format that has received significant attention recently [7,5,2,10,19]. However, upper bounds in this setting do not carry over to our format. Simultaneous auctions were first studied by Christodoulou, Kovacs and Schapira [5]. The authors proposed that each of a collection of distinct goods, with one unit available for each of them, is sold in a distinct *Second Price Auction*, simultaneously and independently of the other goods. Bidders in this setting may have combinatorial valuation functions over the subsets of goods, but they are forced to bid separately for each good. For bidders with fractionally subadditive valuation functions, they proved a tight upper bound of 2 on the mixed Bayesian Price of Anarchy of the Simultaneous Second Price Auction. Bhawalkar and Roughgarden [2] extended the study of inefficiency for subadditive bidders and showed an upper bound of $O(\log m)$ which was recently reduced to 4 by Feldman *et al.* [7]. For arbitrary valuation functions, Fu, Kleinberg and Lavi [9] proved an upper bound of 2 on the inefficiency of pure Nash equilibria, when they exist.

Hassidim *et al.* [10] studied *Simultaneous First Price Auctions*. They showed that pure Nash equilibria in this format are always efficient, when they exist. They proved constant upper bounds on the inefficiency of mixed Nash equilibria for (fractionally) subadditive valuation functions and $O(\log m)$ and $O(m)$ for

---

[1] To the best of our knowledge, for subadditive valuation functions our bounds provide the first quantification of the inefficiency of these auction formats.

the inefficiency of mixed Bayes-Nash equilibria for subadditive and arbitrary valuation functions. Syrgkanis showed in [20] that this format has Bayesian Price of Anarchy $\frac{e}{e-1}$ for fractionally subadditive valuation functions. Feldman *et al.* [7] proved an upper bound of 2 for subadditive ones.

Recently, Syrgkanis and Tardos [19] and Roughgarden [18] independently developed extensions of the *smoothness technique* for games of incomplete information. In [19], these ideas are further developed for analyzing the inefficiency of simultaneous and sequential *compositions* of simple auction mechanisms. They demonstrate applications of their techniques on welfare analysis of standard multi-unit auction formats *and* their compositions. For submodular valuation functions, they prove inefficiency upper bounds of $\frac{2e}{e-1}$ and $\frac{4e}{e-1}$ for the Discriminatory Auction and Uniform Price Auction, respectively. Here, we improve upon these results, also regarding simultaneous and sequential compositions.

2   Definitions and Preliminaries

We consider auctioning k units of a single good to a set $[n] = \{1, ..., n\}$ of n bidders, indexed by $i = 1, \ldots, n$. Every bidder $i \in [n]$ has a non-negative non-decreasing private valuation function $v_i : (\{0\} \cup [k]) \mapsto \Re^+$ over quantities of units, where $v_i(0) = 0$. We denote by $\mathbf{v} = (v_1, \ldots, v_n)$ the *valuation function profile* of bidders. We consider in particular (symmetric) *submodular* and *subadditive* functions:

Definition 1. *A valuation function $f : (\{0\} \cup [k]) \mapsto \Re^+$ is called:*
- *submodular iff for every $x < y$, $f(x) - f(x-1) \geq f(y) - f(y-1)$.*
- *subadditive iff for every x, y, $f(x+y) \leq f(x) + f(y)$.*

The class of submodular functions is strictly contained in the class of subadditive ones [13]. For any non-negative non-decreasing function $f : (\{0\} \cup [k]) \mapsto \Re^+$ and any integers $x, y \in [k], x < y$, the following are known to hold: If f is submodular, then $f(x)/x \geq f(y)/y$. If f is subadditive, then $f(x)/x \geq f(y)/(x+y)$.

Standard Multi-unit Auctions. The *standard format*, as described in auction theory [12,15], prescribes that each bidder $i \in [n]$ submits a vector of k non-negative non-increasing *marginal bids* $\mathbf{b}_i = (b_i(1), \ldots, b_i(k))$ with $b_i(1) \geq \cdots \geq b_i(k)$. We will often refer to these simply as *bids*. In the *uniform bidding format*, each bidder i submits only a single bid $\bar{b}_i$ along with a quantity $q_i \leq k$, the interpretation being that i is willing to pay at most $\bar{b}_i$ per unit for up to q_i units.

The allocation rule of standard multi-unit auctions grants the issuer of each of the k highest (marginal) bids a distinct unit per winning bid. The pricing rule differentiates the formats. Let $x_i(\mathbf{b})$ be the number of units won by bidder i under profile $\mathbf{b} = (\mathbf{b}_1, \ldots, \mathbf{b}_n)$. We study the following two pricing rules:

(i) Discriminatory Pricing. Every bidder i pays for every unit a price equal to his corresponding winning bid, i.e., the utility of i is
$$u_i^{v_i}(\mathbf{b}) = v_i(x_i(\mathbf{b})) - \sum_{j=1}^{x_i(\mathbf{b})} b_i(j).$$

(ii) Uniform Pricing. Every bidder i pays for every unit a price equal to the *highest losing bid* $p(\mathbf{b})$, i.e., the utility of i is
$$u_i^{v_i}(\mathbf{b}) = v_i(x_i(\mathbf{b})) - x_i(\mathbf{b})p(\mathbf{b}).$$

For a bidding profile $\mathbf{b}$, the produced allocation $\mathbf{x}(\mathbf{b}) = (x_1(\mathbf{b}), x_2(\mathbf{b}), \dots, x_n(\mathbf{b}))$ has a *social welfare* equal to the bidders' total value: $SW(\mathbf{v}, \mathbf{b}) = \sum_{i=1}^{n} v_i(x_i(\mathbf{b}))$. The (pure) Price of Anarchy is the worst case ratio, over all pure Nash equilibrium profiles $\mathbf{b}$, of the optimal social welfare over $SW(\mathbf{v}, \mathbf{b})$.

Incomplete Information. Under the incomplete information model of Harsanyi, the valuation function $\mathbf{v}_i$ of bidder i is drawn from a finite set V_i according to a discrete probability distribution $\pi_i : V_i \to [0, 1]$ (independently of the other bidders); we will write $\mathbf{v}_i \sim \pi_i$. The actual drawn valuation function of every bidder is *private*. A valuation profile $\mathbf{v} = (\mathbf{v}_1, \dots, \mathbf{v}_n) \in \mathcal{V} = \times_{i \in [n]} V_i$ is drawn from a *publicly known distribution* $\pi : \mathcal{V} \to [0, 1]$, where π is the product distribution of $\pi_1, \dots, \pi_n$, i.e., $\pi(\mathbf{v}) \mapsto \prod_{i \in [n]} \pi_i(\mathbf{v}_i)$. Every bidder i knows his own valuation function but does not know the valuation function $\mathbf{v}_{i'}$ drawn by any other bidder $i' \neq i$. Bidder i may only use his knowledge of π to estimate $\mathbf{v}_{-i}$. Given the publicly known distribution π, the (possibly mixed) strategy of every bidder is a function of his own valuation $\mathbf{v}_i$, denoted by $B_i(\mathbf{v}_i)$. B_i maps a valuation function $\mathbf{v}_i \in V_i$ to a *distribution* $B_i(\mathbf{v}_i) = B_i^{v_i}$, over all possible bid vectors for i. In this case we will write $\mathbf{b}_i \sim B_i^{v_i}$, for any particular bid vector $\mathbf{b}_i$ drawn from this distribution. We also use the notation $\mathbf{B}_{-i}^{\mathbf{v}_{-i}}$, to refer to the vector of randomized strategies of bidders other than i, under profile $\mathbf{v}_{-i}$.

A *Bayes-Nash equilibrium* (BNE) is a strategy profile $\mathbf{B} = (B_1, \dots, B_n)$ such that for every bidder i and for every valuation $\mathbf{v}_i$, $B_i(\mathbf{v}_i)$ maximizes the utility of i in expectation, over the distribution of the other bidders' valuations $\mathbf{w}_{-i}$ *given* $\mathbf{v}_i$ and over the distribution of i's own and the other bidders' strategies, $\mathbf{B}^{(\mathbf{v}_i, \mathbf{w}_{-i})}$, i.e., for every pure strategy $\mathbf{c}_i$ of i:

$$\mathbb{E}_{\mathbf{w}_{-i}|\mathbf{v}_i, \ \mathbf{b} \sim \mathbf{B}^{(\mathbf{v}_i, \mathbf{w}_{-i})}} \left[u_i^{v_i}(\mathbf{b}) \right] \geq \mathbb{E}_{\mathbf{w}_{-i}|\mathbf{v}_i, \ \mathbf{b}_{-i} \sim \mathbf{B}^{\mathbf{w}_{-i}}} \left[u_i^{v_i}(\mathbf{c}_i, \mathbf{b}_{-i}) \right]$$

where $\mathbb{E}_{\mathbf{v}}$ and $\mathbb{E}_{\mathbf{w}_{-i}|\mathbf{v}_i}$ denote expectation over the distributions π and $\pi(\cdot|\mathbf{v}_i)$.

Fix a valuation profile $\mathbf{v} \in \mathcal{V}$ and consider a (mixed) bidding configuration $\mathbf{B}^{\mathbf{v}}$ under $\mathbf{v}$. The Social Welfare $SW(\mathbf{v}, \mathbf{B}^{\mathbf{v}})$ under $\mathbf{B}^{\mathbf{v}}$ when the valuations are $\mathbf{v}$ is defined as the expectation over the bidding profiles chosen by the bidders from their randomized strategies, i.e., $SW(\mathbf{v}, \mathbf{B}^{\mathbf{v}}) = \mathbb{E}_{\mathbf{b} \sim \mathbf{B}^{\mathbf{v}}} \left[\sum_i v_i(x_i(\mathbf{b})) \right]$. The *expected* Social Welfare in Bayes-Nash equilibrium $\mathbf{B}^{\mathbf{v}}$ is then $\mathbb{E}_{\mathbf{v} \sim \pi}[SW(\mathbf{v}, \mathbf{B}^{\mathbf{v}})]$. The socially optimum assignment under valuation profile $\mathbf{v} \in \mathcal{V}$ will be denoted by $\mathbf{x}^{\mathbf{v}}$. The *expected* optimum social welfare is then $\mathbb{E}_{\mathbf{v}}[SW(\mathbf{v}, \mathbf{x}^{\mathbf{v}})]$. Under these definitions, we will study the *Bayesian Price of Anarchy*, i.e., the worst case ratio $\mathbb{E}_{\mathbf{v}}[SW(\mathbf{v}, \mathbf{x}^{\mathbf{v}})]/\mathbb{E}_{\mathbf{v}}[SW(\mathbf{v}, \mathbf{B}^{\mathbf{v}})]$ over all possible product distributions π and Bayes-Nash equilibria $\mathbf{B}$ for π.[2] Similarly to previous works, when analyzing the Uniform Price Auction we assume *no-overbidding*, i.e., each bidder never bids more than his value for every number of units; formally, for every $s \in [k]$, $\sum_{j \in [s]} b_i(j) \leq v_i(s)$. In our analysis, we will use $\beta_j(\mathbf{b})$ to refer to the j-th lowest winning bid under profile $\mathbf{b}$; thus $\beta_1(\mathbf{b}) \leq \dots \leq \beta_k(\mathbf{b})$.

---

[2] As in previous works [5,7], we ensure existence of Bayes-Nash equilibria in our auction formats by assuming that bidders have bounded and finite strategy spaces, e.g., derived through discretization. Our bounds on the auctions' Bayesian inefficiency hold for sufficiently fine discretizations (see also Appendix D of [7]).

Due to space limitations, we omit several proofs from this extended abstract; all missing proofs can be found in the full version of the paper [11].

3   Bayes-Nash Inefficiency

Our main results concern the inefficiency of Bayes-Nash equilibria (we defer a discussion of pure Nash equilibria to the full version [11]). We derive bounds on the (mixed) Bayesian Price of Anarchy for the Discriminatory and the Uniform Price Auctions with submodular and subadditive valuation functions. For the latter class our bounds are the first results to appear in the literature of standard multi-unit auctions (see also the commentary in [15, Chapter 7]).

Theorem 1. *The Bayesian Price of Anarchy (under the standard or uniform bidding format) is at most*

(i) $\frac{e}{e-1}$ *and* $\frac{2e}{e-1}$ *for the Discriminatory Auction with submodular and subadditive valuation functions, respectively,*

(ii) $|W_{-1}(-1/e^2)| \approx 3.1462 < \frac{2e}{e-1}$ *and* $2|W_{-1}(-1/e^2)| \approx 6.2924 < \frac{4e}{e-1}$ *for the Uniform Price Auction with submodular and subadditive valuation functions, respectively,* W_{-1} *being the lower branch of the Lambert W function.*

This theorem improves on the currently best known upper bounds of $\frac{2e}{e-1}$ and $\frac{4e}{e-1}$ for the Discriminatory Auction and the Uniform Price Auction, respectively, with submodular valuation functions due to Syrgkanis and Tardos [19]. For the Uniform Price Auction, this further reduces the gap from the known lower bound of $\frac{e}{e-1}$ [14].

Syrgkanis and Tardos [19] obtained their bounds through an adaptation of the *smoothness framework* for games with incomplete information ([18,20]). The bounds of Theorem 1 and some additional results can also be obtained through this framework. We comment on this in more detail in Section 4.

For subadditive valuation functions and the standard bidding format, however, better bounds can be obtained by adapting a technique recently introduced by Feldman *et al.* [7], which does not fall within the smoothness framework. We were unable to derive these bounds via a smoothness argument and believe that this is due to the additional flexibility provided by this technique.

Theorem 2. *The Bayesian Price of Anarchy is at most* 2 *and* 4 *for the Discriminatory Auction and the Uniform Price Auction, respectively, with subadditive valuation functions under the standard bidding format.*

3.1   Proof Template for Bayesian Price of Anarchy

In order to present all our bounds from Theorem 1 and Theorem 2 in a self-contained and unified manner, we make use of a proof template, formalized in Theorem 3, below. Adaptations of it have been used in previous works [14,5,2].

Theorem 3. *Let V be a class of valuation functions. Suppose that for every valuation profile $\mathbf{v} \in V^n$, for every bidder $i \in [n]$, and for every distribution $\mathcal{P}_{-i}$ over non-overbidding profiles $\mathbf{b}_{-i}$, there is a bidding profile $\mathbf{b}'_i$ such that the following inequality holds for some $\lambda > 0$ and $\mu \geq 0$:*

$$\mathbb{E}_{\mathbf{b}_{-i} \sim \mathcal{P}_{-i}} \left[u_i^{\mathbf{v}_i}(\mathbf{b}'_i, \mathbf{b}_{-i}) \right] \geq \lambda \cdot v_i(x_i^{\mathbf{v}}) - \mu \cdot \mathbb{E}_{\mathbf{b}_{-i} \sim \mathcal{P}_{-i}} \left[\sum_{j=1}^{x_i^{\mathbf{v}}} \beta_j(\mathbf{b}_{-i}) \right]. \quad (1)$$

Then the Bayesian Price of Anarchy is at most
(i) $\max\{1, \mu\}/\lambda$ for the Discriminatory Auction,
(ii) $(\mu + 1)/\lambda$ for the Uniform Price Auction.

In this theorem we make no assumptions regarding the bidding interface; proving a bound for the uniform bidding interface only requires that we exhibit a uniform bidding strategy $\mathbf{b}'_i$ for each bidder i and for any distribution $\mathcal{P}_{-i}$ over uniform non-overbidding profiles $\mathbf{b}_{-i}$. In Section 3.3 we show that our bound of $\frac{e}{e-1}$ for the Discriminatory Auction is best possible, for the proof template of Theorem 3; this rules out achievement of better bounds via the techniques in [19,7].

3.2   Key Lemma and Proofs of Theorem 1 and Theorem 2

The following is our key lemma to prove Theorem 1. We point out that it applies to arbitrary valuation functions and to any multi-unit auction which is *discriminatory price dominated*, i.e., the total payment $P_i(\mathbf{b})$ of bidder i under profile $\mathbf{b}$ satisfies $P_i(\mathbf{b}) \leq \sum_{j \in [x_i(\mathbf{b})]} b_i(j)$. Note that every multi-unit auction guaranteeing *individual rationality* must satisfy this condition.

Lemma 1 (Key Lemma). *Let $\mathbf{v}$ be a valuation profile and suppose that the pricing rule is discriminatory price dominated. Define $\tau_i = \arg\min_{j \in [x_i^{\mathbf{v}}]} v_i(j)/j$ for every $i \in [n]$. Then for every bidder $i \in [n]$ and every bidding profile $\mathbf{b}_{-i}$ there exists a randomized uniform bidding profile $\mathbf{b}'_i$ such that for every $\alpha > 0$*

$$\mathbb{E}[u_i^{\mathbf{v}_i}(\mathbf{b}'_i, \mathbf{b}_{-i})] \geq \alpha \left(1 - \frac{1}{e^{1/\alpha}} \right) x_i^{\mathbf{v}} \frac{v_i(\tau_i)}{\tau_i} - \alpha \sum_{j=1}^{x_i^{\mathbf{v}}} \beta_j(\mathbf{b}_{-i}). \quad (2)$$

Proof. Define $B = (1 - e^{-1/\alpha})$ and let $\mathbf{c}_i$ be the vector that is $v_i(\tau_i)/\tau_i$ on the first $x_i^{\mathbf{v}}$ entries, and is 0 everywhere else. Let t be a random variable drawn from $[0, B]$ with probability density function $f(t) = \alpha/(1 - t)$. Define the random deviation of bidder i as $\mathbf{b}'_i = t\mathbf{c}_i$. Note that $\mathbf{b}'_i$ is always a uniform bid vector.

Let k^* be the number of items that bidder i would win in profile $(B\mathbf{c}_i, \mathbf{b}_{-i})$, i.e., the number of items won by i, when i would deviate to bid vector $B\mathbf{c}_i$. For $j = 0, \ldots, k^*$, let γ_j refer to the infimum value in $[0, B]$ such that bidder i would win j items if he would deviate to bid vector $\gamma_j \mathbf{c}_i$. Note that this definition is equivalent to defining γ_j as the least value in $[0, B]$ that satisfies $\gamma_j v_i(\tau_i)/\tau_i = \beta_j(\mathbf{b}_{-i})$. For notational convenience, we define $\gamma_{k^*+1} = B$.

Let $x_i(\mathbf{b}'_i, \mathbf{b}_{-i})$ be the random variable that denotes the number of units allocated to bidder i under $(\mathbf{b}'_i, \mathbf{b}_{-i})$. It always holds that $x_i(\mathbf{b}'_i, \mathbf{b}_{-i}) \leq k^* \leq x_i^{\mathsf{v}}$, because bidder i bids $b'_i(j) = 0$ for all $j = x_i^{\mathsf{v}} + 1, \ldots, k$. More precisely, we have $x_i(\mathbf{b}'_i, \mathbf{b}_{-i}) = j$ if $t \in (\gamma_j, \gamma_{j+1}]$ for $j = 0, \ldots, k^*$. By assumption, the payment of bidder i under profile $(\mathbf{b}'_i, \mathbf{b}_{-i})$ is at most $t x_i(\mathbf{b}'_i, \mathbf{b}_{-i}) v_i(\tau_i)/\tau_i$. Also note that, by definition of τ_i, it holds that $v_i(j) \geq j v_i(\tau_i)/\tau_i$ for $j \leq x_i^{\mathsf{v}}$. Using these two facts, we can bound the expected utility of bidder i as follows:

$$\mathbb{E}[u_i^{\mathsf{v}}(\mathbf{b}'_i, \mathbf{b}_{-i})] \geq \sum_{j=1}^{k^*} \int_{\gamma_j}^{\gamma_{j+1}} \left(v_i(j) - tj\frac{v_i(\tau_i)}{\tau_i} \right) f(t)\, dt$$

$$\geq \sum_{j=1}^{k^*} \int_{\gamma_j}^{\gamma_{j+1}} j\frac{v_i(\tau_i)}{\tau_i}(1-t)f(t)\, dt = \alpha \sum_{j=1}^{k^*} j\frac{v_i(\tau_i)}{\tau_i} \int_{\gamma_j}^{\gamma_{j+1}} 1\, dt$$

$$= \alpha \sum_{j=1}^{k^*} j\frac{v_i(\tau_i)}{\tau_i}(\gamma_{j+1} - \gamma_j) = \alpha B k^* \frac{v_i(\tau_i)}{\tau_i} - \alpha \sum_{j=1}^{k^*} \gamma_j \frac{v_i(\tau_i)}{\tau_i}$$

$$= \alpha B k^* \frac{v_i(\tau_i)}{\tau_i} - \alpha \sum_{j=1}^{k^*} \beta_j(\mathbf{b}_{-i}) \geq \alpha B x_i^{\mathsf{v}} \frac{v_i(\tau_i)}{\tau_i} - \alpha \sum_{j=1}^{x_i^{\mathsf{v}}} \beta_j(\mathbf{b}_{-i}).$$

The last inequality holds because $B v_i(\tau_i)/\tau_i \leq \beta_j(\mathbf{b}_{-i})$, for $k^* + 1 \leq j \leq x_i^{\mathsf{v}}$, by the definition of k^*. The above derivation implies (2).   □

The deviation $\mathbf{b}'_i$ defined in Lemma 1 is a distribution on uniform bidding strategies. That is, the lemma applies to both the standard and the uniform bidding format. Observe also that $\mathbf{b}'_i$ satisfies the no-overbidding assumption.

Proof (of Theorem 1). First consider the case of submodular valuation functions. In this case, $\tau_i = x_i^{\mathsf{v}}$ for every $i \in [n]$, as explained in Section 2. Using our Key Lemma, we conclude that Theorem 3 holds for $(\lambda, \mu) = (\alpha(1 - e^{-1/\alpha}), \alpha)$. The stated bounds are obtained by choosing $\alpha = 1$ for the Discriminatory Auction and $\alpha = -1/(W_{-1}(-1/e^2) + 2) \approx 0.87$ for the Uniform Price Auction.

Next consider the case of subadditive valuation functions. The following lemma shows that subadditive valuation functions can be approximated by uniform ones, thereby losing at most a factor 2.

Lemma 2. *If v_i is subadditive, $\tau_i = \arg\min_{j \in [x_i^{\mathsf{v}}]} \frac{v_i(j)}{j}$ yields $\frac{v_i(\tau_i)}{\tau_i} \geq \frac{1}{2}\frac{v_i(x_i^{\mathsf{v}})}{x_i^{\mathsf{v}}}$.*

By combining Lemma 2 with our Key Lemma, it follows that Theorem 3 holds for $(\lambda, \mu) = (\frac{\alpha}{2}(1 - e^{-1/\alpha}), \alpha)$. The bounds stated in Theorem 1 are obtained by the same choices of α as for the submodular valuation functions.   □

Next, consider subadditive valuations under the *standard* bidding format. We derive improved bounds of 2 and 4 for the Discriminatory and Uniform Price Auction, respectively. To this end, we adapt an approach recently developed by Feldman *et al.* [7] to establish an analog of our Key Lemma. The main idea is to construct the bid $\mathbf{b}'_i$ by using the distribution $\mathcal{P}_{-i}$ on the profiles $\mathbf{b}_{-i}$. Theorem 2 then follows from Theorem 3 in combination with Lemma 3 below.

Lemma 3. *Let V be the class of subadditive valuation functions. Then Theorem 3 holds true with $(\lambda, \mu) = (\frac{1}{2}, 1)$ for the Discriminatory and $(\lambda, \mu) = (\frac{1}{2}, 1)$ for the Uniform Price Auction (under the standard bidding format).*

3.3   Lower Bounds

A lower bound of approximately $\frac{e}{e-1}$ for Uniform Price Auctions with submodular bidders was given in [14]; our upper bound is less than a factor 2 away. For subadditive valuation functions, we prove a lower bound of almost 2:

Theorem 4. *The Price of Anarchy is at least $\frac{2k}{k+1}$ for the Uniform Price Auction with subadditive valuations (under the uniform bidding interface).*

No lower bound is known for the Discriminatory Auction, although *Demand Reduction* (which is responsible for welfare loss in this format) has been observed previously [12,1]. In light of this, we prove here an *impossibility result* showing that for the Discriminatory Auction no bound better than $\frac{e}{e-1}$ on the Price of Anarchy can be achieved via the proof template given in Theorem 3. Similarly, for the Uniform Price Auction we rule out that a bound better than 2 on the Price of Anarchy can be derived through this template.

Theorem 5. *There is a lower bound of $\frac{e}{e-1}$ and 2 on the Bayesian Price of Anarchy for the Discriminatory Auction and the Uniform Price Auction, respectively, with submodular valuation functions that can be derived through the proof template given in Theorem 3.*

Theorem 5 rules out the possibility of obtaining better bounds by means of the smoothness framework of [19], or by means of *any* approach aiming at identifying the b'_i required by Theorem 3, including [7]. These are essentially the only known techniques for obtaining upper bounds on the Bayesian Price of Anarchy. Thus, any improvement on our upper bound for the Discriminatory Auction must use either specific properties of the (Bayes-Nash equilibrium) distribution $\mathcal{D}$, or a completely new approach altogether. The same holds for improvements of the upper bound for the Uniform Price Auction below 2 – and towards the only known lower bound of $\frac{e}{e-1}$ from [14] (should it be worst-case).

4   Smoothness and Its Implications

We elaborate on the connections of our results to the smoothness framework for auction mechanisms, which has very recently been developed by Syrgkanis and Tardos [19]. We first review the smoothness definitions introduced in [19] (adapted to our multi-unit auction setting). As introduced earlier, let $P_i(\mathbf{b})$ refer to the payment of bidder i under bidding profile $\mathbf{b}$.

Definition 2 ([19]). *A mechanism $\mathcal{M}$ is (λ, μ)-smooth for $\lambda > 0$ and $\mu \geq 0$ if for any valuation profile $\mathbf{v}$ and for any bidding profile $\mathbf{b}$ there exists a randomized bidding profile $\mathbf{b}'_i = \mathbf{b}'_i(\mathbf{v}, \mathbf{b}_i)$ for each i such that*

$$\sum_{i \in [n]} \mathbb{E}[u_i^{\mathbf{v}_i}(\mathbf{b}'_i, \mathbf{b}_{-i})] \geq \lambda SW(\mathbf{v}, \mathbf{x}^{\mathbf{v}}) - \mu \sum_{i \in [n]} P_i(\mathbf{b}).$$

Table 2. Upper bounds on the Bayesian Price of Anarchy for compositions

| Valuation Functions | Discriminatory Auction | | Uniform Price Auction |
	Simultaneous	Sequential	Simultaneous/Sequential
Submodular	$e/(e-1)$	$2e/(e-1)$	3.1462
Subadditive	$2e/(e-1)$	$4e/(e-1)$	6.2924

In [19] it is shown that if a mechanism is (λ, μ)-smooth, then several results follow automatically. One such result concerns upper bounds on the Price of Anarchy. Another result is that the smoothness property is retained under *simultaneous and sequential compositions*. In these compositions there are m mechanisms with separate allocation and payment rules. Every bidder specifies for each mechanism a bidding profile. In the simultaneous composition, these profiles are submitted simultaneously, while in the sequential composition, they are submitted sequentially. A bidder expresses his valuation for the m-tuples of outcomes of the mechanisms in a restricted way.[3] We summarize the main composition results of Syrgkanis and Tardos [19] in the theorem below.

Theorem 6 (Theorems 4.2, 4.3, 5.1, and 5.2 in [19]).
(i) If $\mathcal{M}$ is (λ, μ)-smooth, then the correlated (or mixed Bayesian) Price of Anarchy of $\mathcal{M}$ is at most $\max\{1, \mu\}/\lambda$.
(ii) If $\mathcal{M}$ is a simultaneous (respectively, sequential) composition of m (λ, μ)-smooth mechanisms, then $\mathcal{M}$ is (λ, μ)-smooth (resp., $(\lambda, \mu + 1)$-smooth).

By exploiting our Key Lemma, we can show that the Discriminatory Auction is smooth. Theorem 7 in combination with Theorem 6 leads to the composition results stated in Table 2 (these bounds are achieved for $\alpha = 1$).

Theorem 7. *The Discriminatory Auction is (λ, μ)-smooth (both in the standard and uniform bidding format) with*
(i) $(\lambda, \mu) = (\alpha \left(1 - e^{-1/\alpha}\right), \alpha)$ for submodular valuation functions, and
(ii) $(\lambda, \mu) = (\frac{\alpha}{2} \left(1 - e^{-1/\alpha}\right), \alpha)$ for subadditive valuation functions.

For auction mechanisms where one needs to impose a no-overbidding assumption, a different smoothness notion is introduced in [19]. Given a mechanism $\mathcal{M}$, define bidder i's *willingness-to-pay* as the maximum payment he could ever pay conditional to being allocated x units, i.e., $B_i(\mathbf{b}_i, x) = \max_{\mathbf{b}_{-i}:x_i(\mathbf{b})=x} P_i(\mathbf{b})$.

Definition 3 ([19]). *A mechanism $\mathcal{M}$ is weakly (λ, μ_1, μ_2)-smooth for $\lambda > 0$ and $\mu_1, \mu_2 \geq 0$ if for any valuation profile $\mathbf{v}$ and for any bidding profile $\mathbf{b}$ there exists a randomized bidding profile $\mathbf{b}'_i = \mathbf{b}'_i(\mathbf{v}, \mathbf{b}_i)$ for each bidder i such that*

$$\sum_{i \in [n]} \mathbb{E}[u_i^{\mathbf{v}_i}(\mathbf{b}'_i, \mathbf{b}_{-i})] \geq \lambda SW(\mathbf{v}, \mathbf{x}^{\mathbf{v}}) - \mu_1 \sum_{i \in [n]} P_i(\mathbf{b}) - \mu_2 \sum_{i \in [n]} B_i(\mathbf{b}_i, x_i(\mathbf{b})).$$

---

[3] More precisely, in the simultaneous composition it is assumed that the valuation function of each bidder is *fractionally subadditive* across the m mechanisms (see [19] for formal definitions). In the sequential composition, the valuation function of each bidder is defined as the maximum of his valuations over these mechanisms.

Syrgkanis and Tardos [19] establish the following results.

Theorem 8 (Theorems 7.4, C.4 and C.5 in [19]).
(i) If $\mathcal{M}$ is (λ, μ_1, μ_2)-weakly smooth, then the correlated (or mixed Bayesian) Price of Anarchy of $\mathcal{M}$ is at most $(\mu_2 + \max\{1, \mu_1\})/\lambda$.
(ii) If $\mathcal{M}$ is a simultaneous (resp., sequential) composition of m (λ, μ_1, μ_2)-weakly smooth mechanisms, then $\mathcal{M}$ is (λ, μ_1, μ_2)-weakly smooth (resp., $(\lambda, \mu_1 + 1, \mu_2)$-weakly smooth).

Using our Key Lemma, we can show that the Uniform Price Auction is weakly smooth. As a consequence, we obtain the composition results stated in Table 2 (these bounds are achieved for $\alpha = -1/(W_{-1}(-1/e^2) + 2) \approx 0.87$).

Theorem 9. *The Uniform Price Auction is weakly (λ, μ_1, μ_2)-smooth (both in the standard and uniform bidding format) with*
(i) $(\lambda, \mu_1, \mu_2) = (\alpha \left(1 - e^{-1/\alpha}\right), 0, \alpha)$ for submodular valuation functions, and
(ii) $(\lambda, \mu_1, \mu_2) = (\frac{\alpha}{2} \left(1 - e^{-1/\alpha}\right), 0, \alpha)$ for subadditive valuation functions.

Some additional results on mechanisms with *budgets* (see [19]) can be inferred from Theorems 7 and 9. We defer further details to the full version of the paper.

5 Conclusions

We derived inefficiency upper bounds in the incomplete information model for the widely popular Discriminatory and Uniform Price Auctions, when bidders have submodular or subadditive valuation functions. Notably, our bounds for subadditive valuation functions already improve upon the ones that were known for submodular bidders [14,19]. Moreover, for each of the two formats and valuation function classes we considered both the *standard* bidding interface [12,15] and a practically motivated *uniform* bidding interface.

To derive our results, we elaborated on several techniques from the recent literature on *Simultaneous Auctions* [19,7,5,2]. By the recent developments of [19], our bounds for submodular bidders yield improved inefficiency bounds for *simultaneous* and *sequential* compositions of the considered formats. In absence of an indicative lower bound in the incomplete information model, we showed that our upper bound of $\frac{e}{e-1}$ for the Discriminatory Auction with submodular valuation functions is best possible, w.r.t. the currently known proof techniques. Additionally, for the Uniform Price Auction (with submodular bidders), we showed that, proving an upper bound of less than 2, also requires novel techniques; this poses a particularly challenging problem, given the lower bound of $\frac{e}{e-1}$ from [14].

References

1. Ausubel, L., Cramton, P.: Demand Reduction and Inefficiency in Multi-Unit Auctions. Tech. rep., University of Maryland (2002)
2. Bhawalkar, K., Roughgarden, T.: Welfare Guarantees for Combinatorial Auctions with Item Bidding. In: Proc. of the 22nd ACM-SIAM Symposium on Discrete Algorithms (SODA), pp. 700–709 (2011)

3. Binmore, K., Swierzbinski, J.: Treasury auctions: Uniform or discriminatory? Review of Economic Design 5(4), 387–410 (2000)
4. Bresky, M.: Pure Equilibrium Strategies in Multi-unit Auctions with Private Value Bidders. Tech. Rep. 376, CERGE Economics Institute (2008)
5. Christodoulou, G., Kovács, A., Schapira, M.: Bayesian Combinatorial Auctions. In: Aceto, L., Damgård, I., Goldberg, L.A., Halldórsson, M.M., Ingólfsdóttir, A., Walukiewicz, I. (eds.) ICALP 2008, Part I. LNCS, vol. 5125, pp. 820–832. Springer, Heidelberg (2008)
6. Engelbrecht-Wiggans, R., Kahn, C.M.: Multi-unit auctions with uniform prices. Economic Theory 12(2), 227–258 (1998)
7. Feldman, M., Fu, H., Gravin, N., Lucier, B.: Simultaneous Auctions are (almost) Efficient. In: Proc. of the 45th ACM Symposium on Theory of Computing (STOC), pp. 201–210 (2013)
8. Friedman, M.: A Program for Monetary Stability. Fordham University Press, New York (1960)
9. Fu, H., Kleinberg, R., Lavi, R.: Conditional equilibrium outcomes via ascending price processes with applications to combinatorial auctions with item bidding. In: Proc. of the 13th ACM Conference on Electronic Commerce, p. 586 (2012)
10. Hassidim, A., Kaplan, H., Mansour, Y., Nisan, N.: Non-price equilibria in markets of discrete goods. In: Proc. of the 12th ACM Conference on Electronic Commerce, pp. 295–296 (2011)
11. de Keijzer, B., Markakis, E., Schäfer, G., Telelis, O.: On the Inefficiency of Standard Multi-Unit Auctions. arXiv:1303.1646 [cs.GT] (2013)
12. Krishna, V.: Auction Theory. Academic Press (2002)
13. Lehmann, B., Lehmann, D.J., Nisan, N.: Combinatorial auctions with decreasing marginal utilities. Games and Economic Behavior 55(2), 270–296 (2006)
14. Markakis, E., Telelis, O.: Uniform Price Auctions: Equilibria and Efficiency. In: Serna, M. (ed.) SAGT 2012. LNCS, vol. 7615, pp. 227–238. Springer, Heidelberg (2012)
15. Milgrom, P.: Putting Auction Theory to Work. Cambridge University Press (2004)
16. Noussair, C.: Equilibria in a multi-object uniform price sealed bid auction with multi-unit demands. Economic Theory 5, 337–351 (1995)
17. Ockenfels, A., Reiley, D.H., Sadrieh, A.: Economics and Information Systems. In: Handbooks in Information Systems, ch. 12. Online Auctions, vol. 1, pp. 571–628. Elsevier (2006)
18. Roughgarden, T.: The price of anarchy in games of incomplete information. In: Proc. of the 13th ACM Conference on Electronic Commerce, pp. 862–879 (2012)
19. Syrgkanis, V., Tardos, E.: Composable and efficient mechanisms. In: Proc. of the 45th ACM Symposium on Theory of Computing (STOC), pp. 211–220 (2013)
20. Syrgkanis, V.: Bayesian games and the smoothness framework. arXiv:1203.5155 [cs.GT] (2012)
21. U.S. Department of Treasury: Uniform-Price Auctions: Update of the Treasury Experience (1998),
http://www.treasury.gov/press-center/press-releases/Documents/upas.pdf
22. Vickrey, W.: Counterspeculation, auctions, and competitive sealed tenders. Journal of Finance 16(1), 8–37 (1961)

FPTAS for Minimizing Earth Mover's Distance under Rigid Transformations[*]

Hu Ding and Jinhui Xu

Department of Computer Science and Engineering
State University of New York at Buffalo
{huding,jinhui}@buffalo.edu

Abstract. In this paper, we consider the problem (denoted as EMDRT) of minimizing the earth mover's distance between two sets of weighted points A and B in a fixed dimensional $\mathbb{R}^d$ space under rigid transformation. EMDRT is an important problem in both theory and applications and has received considerable attentions in recent years. In this paper, we present the first FPTAS algorithm for EMDRT. Our algorithm runs roughly in $O((nm)^{d+2}(\log nm)^{2d})$ time which matches the order of magnitude of the degree of a lower bound for any PTAS of this problem, where n and m are the sizes of A and B, respectively. Our result is based on several new techniques, such as the *Sequential Orthogonal Decomposition (SOD)* and *Optimum Guided Base (OGB)*. Our technique can also be extended to several related problems, such as the alignment problem, and achieves FPTAS for each of them.

1   Introduction

In this paper, we study the problem (denoted as EMDRT) of minimizing the earth mover's distance between two sets of weighted points A and B (with size n and m, respectively) in a fixed dimensional $\mathbb{R}^d$ space under rigid transformation. In EMDRT, each point in A and B is associated with a nonnegative weight, and the objective is to determine the best rigid transformation $\mathcal{T}$ for B so that the earth mover's distance (EMD) between A and $\mathcal{T}(B)$ is minimized, where EMD is the minimum transportation cost between the two point sets. EMDRT is an important problem in both theory and applications. In theory, it generalizes the bipartite matching problem (*i.e.*, from one-to-one matching to many-to-many matching) and is a powerful model for a number of other matching or partial matching problems. For instance, if all points in A and B have unit weight, EMDRT becomes a one-to-one matching problem (*e.g.*, the congruent and alignment problem). If all points in A have unit weight and all points in B have infinity weight, EMDRT becomes a many-to-one matching problem (*i.e.*, the Hausdorff distance matching problem). In applications, EMDRT has connections to many EMD based problems in pattern recognition and computer vision [5, 12], and can be used to solve the challenging alignment problem for rigid objects and detect the similarity between multi-dimensional point sets.

---

[*] This research was supported in part by NSF under grant IIS-1115220.

H.L. Bodlaender and G.F. Italiano (Eds.): ESA 2013, LNCS 8125, pp. 397–408, 2013.
© Springer-Verlag Berlin Heidelberg 2013

A number of results exist for EMDRT and its related problems. Cabello *et. al* [7] presented several approximation results in $\mathbb{R}^2$ space; particularly, they gave a $(2 + \epsilon)$-approximation solution for the 2D EMDRT problem, and a $(1 + \epsilon)$-approximation solution for some special cases. Later, Klein and Veltkamp [11] introduced a few improved results by using *reference points*, and achieved an $O(2^{d-1})$-approximation for EMDRT in $\mathbb{R}^d$ space. Andoni *et al.* presented several algorithms for computing EMD (without transformation) [2,3]. For the one-to-one and many-to-one matching problems, a number of existing results (mainly on lower dimensional space) can be found in the survey paper by Alt and Guibas [1]. For the related alignment problem, there is a long and rich history [4, 8–10].

In this paper, we present the first FPTAS algorithm for EMDRT in any fixed dimensional space. Our result is based on a few new techniques, such as *Sequential Orthogonal Decomposition (SOD)* and *Optimum-Guided-Base (OGB)*. SOD decomposes a rigid transformation into a sequence of primitive operations which enables us to accurately analyze how the transportation flow changes in a step-by-step fashion during the whole process of rigid transformation and therefore have a better estimation on the quality of solution. OGB enables us to use some information of the unknown optimal solution to select some critical points which partially define the rigid transformation. We show that although OGB cannot be explicitly implemented, its result can actually be implicitly obtained. A major advantage of OGB is that it can help us to significantly reduce the search space. Consequently, our FPTAS runs in roughly $O((nm)^{d+2}(\log nm)^{2d})$ time, which is close (*i.e.*, matches the order of magnitude of the degree) to the lower bound $\Omega(mn^{O(d)})$ (where $m \leq n$) for any FPTAS algorithm of EMDRT [6].

Our technique for EMDRT can be extended to several related problems, such as the problem of minimizing EMD for weighted point sets under similarity transformation (i.e., rigid transformation plus scaling), and the alignment problem, and achieve an FPTAS for each of them. For the alignment problem, we consider one-to-one matching and many-to-one matching (Hausdorff distance). Our result for the Hausdorff distance metric is an FPTAS, while existing results [8, 9] are only pseudo-polynomial (depending on the spread ratio).

2   Preliminaries

This section introduces several definitions to be used throughout the paper.

Definition 1 (Rigid Transformation). *Let P be a set of points in $\mathbb{R}^d$. A rigid transformation $\mathcal{T}(P)$ of P is a transformation (e.g., rotation, translation, reflection, or their combinations) preserving the pairwise distances of its points. After a rigid transformation, the new point set $\mathcal{T}(P)$ is called an image of P.*

Definition 2 (Earth Mover's Distance (EMD)). *Let $A = \{p_1, \cdots, p_n\}$ and $B = \{q_1, \cdots, q_m\}$ be two sets of weighted points in $\mathbb{R}^d$ with nonnegative weights α_i and β_j for each $p_i \in A$ and $q_j \in B$ respectively, and W^A and W^B be their respective total weights. The earth mover's distance between A and B is*

$$EMD(A, B) = \frac{\min_F \sum_{i=1}^n \sum_{j=1}^m f_{ij} \|p_i - q_j\|}{\min\{W^A, W^B\}}, \tag{1}$$

where $F = \{f_{ij}\}$ is a feasible flow satisfying the following conditions. (1) $f_{ij} \geq 0$, for any $1 \leq i \leq n$ and $1 \leq j \leq m$; (2) $\sum_{j=1}^{m} f_{ij} \leq \alpha_i$ for any $1 \leq i \leq n$; (3) $\sum_{i=1}^{n} f_{ij} \leq \beta_j$ for any $1 \leq j \leq m$; (4) $\sum_{i=1}^{n} \sum_{j=1}^{m} f_{ij} = \min\{W^A, W^B\}$.

Definition 3 (Earth Mover's Distance Under Rigid Transformation (EMDRT)). *Given two weighted point sets A and B in $\mathbb{R}^d$, the problem of the earth mover's distance between A and B under rigid transformation is to determine a rigid transformation $\mathcal{T}$ for B so as to minimize the earth mover's distance $EMD(A, \mathcal{T}(B))$.*

Orientation and Reflection: For simplicity, we do not consider reflection in the rigid transformation, as it can be captured by performing our algorithm twice, one for the original point set and the other for its mirror image.

3   Overview of Our Approach

From Definition 3, we know that our main task for achieving an FPTAS is to find a good approximation of the optimal rigid transformation $\mathcal{T}_{opt}$ for B. In Lemma 1, we show that this is equivalent to identify d points R (called *reference system*) from A and another d sorted points (called *base*) from B and determine a rigid transformation $\mathcal{T}$ to map points in the base to the neighborhoods of R.

Our approach consists of two main steps: (1) Design a polynomial-approximation algorithm to compute an upper bound of the optimal objective value, and (2) use the upper bound to derive an FPTAS. In both steps, directly searching for $\mathcal{T}$ in the rigid transformation space could be very costly. Our idea is to introduce a new technique called *Sequential Orthogonal Decomposition (SOD)*, which decomposes the rigid transformation into a sequence of d primitive operations (*i.e.*, a translation and $d - 1$ one-dimensional rotations). One important property of SOD is that its outcome is independent of the initial position of B and depends only on the choice of R and the base. This enables us to assume that B is initially located at $\mathcal{T}_{opt}(B)$. Another important property of SOD is that it allows us to analyze, in a step by step fashion, how the transportation flow (*e.g.*, bottleneck flow) changes when B changes from $\mathcal{T}_{opt}(B)$ to $SOD(B)$. This gives us an accurate estimation of the quality of solution.

The quality of the rigid transformation $\mathcal{T}$ determined by SOD depends on the reference system R and the base. To find a good R, we first build a grid in the neighborhood of each point in A and then consider as R all possible subsets (with cardinality d) of A and its grid points. To find a good base, a key problem we need to avoid is that a small error in the rotation could cause some point in B to move a long distance and therefore introduce a large error. To avoid this problem, we select the base as a set of d ordered points in B which are as "dispersed" as possible (since other choices cause larger error). We consider two types of dispersions. Type 1 dispersion is based on the weighted distance (with weight β_j) between each point q_j and the flat spanned by all points appearing before q_j in the sorted order of the base (see *Algorithm Base-Selection* in Section 5.1). Type 2 dispersion considers not only the weighted distance, but also the distance between each q_j and all points in A in the optimal solution (see *Optimum-Guided-Base (OGB)* in Section 6). We show that although OGB cannot be explicitly implemented (as it depends on the optimal solution), it can actually be implicitly obtained. A major advantage of the second type of dispersion is that it enables us to significantly improve the running time. **Note** that since type 2 is modified from type 1, we present both of them for better understanding of the readers.

4   Sequential Orthogonal Decomposition

To solve the EMDRT problem, we first introduce the sequential orthogonal decomposition which will be used as a key technique in our algorithms. We start with the following lemma on rigid structure.

Lemma 1. *For any rigid structure in $\mathbb{R}^d$ with $m \geq d$ vertices (or points), its position is completely fixed if the locations of any d vertices, which are not contained in any $(d-2)$-dimensional flat, are fixed.*

Sequential Orthogonal Decomposition (SOD). Let $P = \{p_1, p_2, \cdots, p_n\}$ be a set of points in $\mathbb{R}^d$ with $n \geq d$, and $R = \{r_1, \cdots, r_d\}$ be another set of fixed points (i.e. whose locations will not change) which spans a $(d-1)$-dimensional flat in $\mathbb{R}^d$, called a *reference system*. Then for any injective mapping f from $\{1, \cdots, d\}$ to $\{1, \cdots, n\}$, we define a sequential orthogonal decomposition, $SOD(P, R, f)$, as follows.

1. In step 1, perform a translation on P such that $p_{f(1)}$ coincides with r_1. Let p_i^1 be $p_i \in P$ in its new position, and P^1 be the new image of P.
2. In the j-th step for $2 \leq j \leq d$
   (a) Let $\mathcal{H}_{j-1}$ be the flat spanning $\{r_1, \cdots, r_{j-1}\}$, and $\tilde{p}_{f(j)}^{j-1}$ and $\tilde{r}_j$ be the projections of $p_{f(j)}^{j-1}$ and r_j on $\mathcal{H}_{j-1}$, respectively.
   (b) Let Δ_j be the two dimensional subspace determined by the two vectors, $p_{f(j)}^{j-1} - \tilde{p}_{f(j)}^{j-1}$ and $r_j - \tilde{r}_j$, and $\Delta_j^\perp$ be the $(d-2)$-dimensional flat orthogonal to Δ_j and containing $\tilde{p}_{f(j)}^{j-1}$.
   (c) Perform a one-dimensional rotation $\mathcal{T}$ on P^{j-1} about $\Delta_j^\perp$ such that the vector $\mathcal{T}(p_{f(j)}^{j-1}) - \tilde{p}_{f(j)}^{j-1}$ is parallel to the vector $r_j - \tilde{r}_j$ (see Fig. 1).
   (d) Let p_i^j denote p_i in its new position, and P^j denote the new image of P.

We first prove the feasibility of SOD.

Theorem 1 (SOD feasibility). *In Step j ($2 \leq j \leq d$) of SOD, there exists a one-dimensional rotation $\mathcal{T}$ on P^{j-1} about $\Delta_j^\perp$ such that the vector $\mathcal{T}(p_{f(j)}^{j-1}) - \tilde{p}_{f(j)}^{j-1}$ is parallel to the vector $r_j - \tilde{r}_j$.*

Next, we study some important properties of SOD.

Lemma 2. *Let $I(P)$ be any image of P (i.e., $I(P)$ is the new P after a rigid transformation). Then the output of $SOD(P, R, f)$ and $SOD(I(P), R, f)$ are the same.*

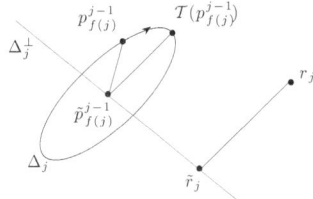 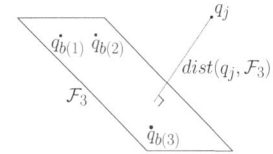

Fig. 1. Rotation in SOD

Fig. 2. An illustration for Algorithm Base-Selection (for $l = 3$)

Lemma 3. *For* $2 \leq j \leq d$, $\|p^j_{f(j)} - p^{j-1}_{f(j)}\| \leq 2\|p^{j-1}_{f(j)} - r_j\|$.

As for the running time of SOD, we know that there are d steps in the process, and each step involves computing the projection of one point to the corresponding flat, which costs $O(d^3)$ time. Thus, we have the following lemma.

Lemma 4. *SOD can be performed in* $O(|P|d^4)$ *time.*

5 FPTAS for EMDRT

In this section, we present an FPTAS for EMDRT. Our algorithm first applies SOD to obtain an upper bound on the optimal objective value of EMDRT, and then use it to determine a proximity region (called grid-ball) for each point in A containing its possible match in B. By searching a grid in each such grid-ball, we show that an FPTAS can be attained for EMDRT.

To solve EMDRT, a basic problem is to determine the earth mover's distance (EMD) between two sets of fixed points without considering any rigid transformation. In [7], Cabello *et al.* introduced a $(1 + \epsilon)$-approximation algorithm for computing EMD in a plane, and generalized it to any d-dimensional space. Below is a lemma proved in [7].

Lemma 5 ([7]). *Given two weighted point sets A and B in $\mathbb{R}^d$ and a small $\epsilon > 0$, there exists an algorithm which outputs a $(1 + \epsilon)$-approximation of EMD between A and B in $O((n^2/\epsilon^{2(d-1)}) \log^2(n/\epsilon))$ time, where $n = \max\{|A|, |B|\}$.*

5.1 Upper Bound

For the upper bound, we notice that although [11] provides an $O(2^{d-1})$-approximate solution, it cannot be used as an upper bound as it assumes that the two input point sets have equal total weight, which may not be the case in our problem. To obtain an upper bound for EMDRT, we need the following definition.

Definition 4 (Bottleneck). *Let $F = \{f_{ij}\}$ be any feasible flow between A and B in Definition 2. Then the Bottleneck of F is defined as $\mathcal{BN}(A, B, F) = \max_{i,j}\{\frac{f_{ij}\|p_i - q_j\|}{\min\{W^A, W^B\}}\}$.*

From the above definition, we immediately have the following lemma.

Lemma 6. *For any feasible flow F between A and B,*

$$\frac{1}{nm} \frac{\sum_{i=1}^{n} \sum_{j=1}^{m} f_{ij} \|p_i - q_j\|}{\min\{W^A, W^B\}} \leq \mathcal{BN}(A, B, F) \leq \frac{\sum_{i=1}^{n} \sum_{j=1}^{m} f_{ij} \|p_i - q_j\|}{\min\{W^A, W^B\}}.$$

Our main idea for obtaining an upper bound is to first identify a good base from B, then enumerate all possible subsets of A with cardinality d as the reference system R, and finally use SOD to obtain a rigid transformation for B. The criterium for selecting the base is to make them as "dispersed" as possible, where the dispersiveness is measured by the weighted distance between each point $q_j \in B$ and the flat spanned by all determined base points.

Algorithm Upper-Bound-for-EMDRT
Input: Two weighted point sets $A = \{p_1, \cdots, p_n\}$ and $B = \{q_1, \cdots, q_m\}$ in $\mathbb{R}^d$ with weight $\alpha_i \geq 0$ and $\beta_j \geq 0$ for p_i and q_j respectively, and $W^A \geq W^B$.
Output: An upper bound on $\min_{\mathcal{T}} EMD(A, \mathcal{T}(B))$.

1. Call Base-Selection on B, and let $\{q_{b(1)}, \cdots, q_{b(d)}\}$ be the output base.
2. Enumerate all d-point tuples from $[A]^d = A \times \cdots \times A$. For each tuple $R = \{p_{i(1)}, \cdots, p_{i(d)}\}$, Do
   (a) Define the mapping such that $f(i(j)) = b(j)$ for each $1 \leq j \leq d$.
   (b) Emulate the execution of the procedure $SOD(B, R, f)$, and stop at the final step or at the l-th step if $p_{i(l)}$ locates on the flat $span\{p_{i(1)}, \cdots, p_{i(l-1)}\}$. Then compute the $(1 + \epsilon)$-approximation of EMD between A and the output image of B by using the algorithm in Lemma 5.
3. Output the image of B which has the minimum EMD to A among all images of B corresponding to the tuples of $[A]^d$.

Algorithm Base-Selection
Input: A weighted point set $B = \{q_1, \cdots, q_m\}$ in $\mathbb{R}^d$ with nonnegative weight β_j for each q_j.
Output: A base which is an ordered subset $\{q_{b(1)}, \cdots, q_{b(d)}\}$ of points in B.

1. Select the point with largest weight from B, and denote it as $q_{b(1)}$. Let $l = 1$, and repeat the following steps until $l = d$.
   (a) Let $\mathcal{F}_l$ be the flat spanned by $\{q_{b(1)}, \cdots, q_{b(l)}\}$. See Fig. 2
   (b) Find the point realizing $\max\{\beta_j \cdot dist(q_j, \mathcal{F}_l) \mid 1 \leq j \leq m\}$, and denote it as $q_{b(l+1)}$, where $dist(q_j, \mathcal{F}_l)$ is the distance between q_j and $\mathcal{F}_l$.
   (c) Let $l = l + 1$.
2. Output $\{q_{b(1)}, \cdots, q_{b(d)}\}$.

Theorem 2. *The algorithm of Upper-Bound-for-EMDRT yields in $O(n^{d+2}(\log n)^2 md^4)$ time an upper bound T which is a $((1 + \epsilon)nm(n + 1)(2n + 1)^{d-1})$-approximation of the optimal objective value.*

Let $\mathcal{T}_{opt}$ be an optimal rigid transformation (i.e., the one realizing the value of $\min_{\mathcal{T}} EMD(A, \mathcal{T}(B))$), and $\overline{F} = \{\overline{f}_{ij}\}$ be the corresponding optimal flow. Since the algorithm enumerates all d-point tuples in $[A]^d$, we just need to focus on the tuple $\{p_{i(1)}, \cdots, p_{i(d)}\}$ which has $\overline{f}_{i(j)b(j)} = \max\{\overline{f}_{ib(j)} \mid 1 \leq i \leq n\}$ for each $1 \leq j \leq d$. With a slight abuse of notation, we use $R = \{p_{i(1)}, \cdots, p_{i(d)}\}$ to denote the tuple which satisfies this requirement. Before proving Theorem 2, we first introduce the following lemma.

Lemma 7. *Let $I(B)$ be the final output image of B by $SOD(B, R, f)$ in the above algorithm. Then $\mathcal{BN}(A, I(B), \overline{F}) \leq (n+1)(2n+1)^{d-1}\mathcal{BN}(A, \mathcal{T}_{opt}(B), \overline{F})$.*

Proof (of Theorem 2). Let $T_{opt} = EMD(A, \mathcal{T}_{opt}(B))$, and T be the objective value returned by the algorithm. Obviously, $T_{opt} \leq T$. By Lemma 6, we know

$$T \leq nm\mathcal{BN}(A, I(B), \overline{F}), \tag{2}$$

$$\mathcal{BN}(A, \mathcal{T}_{opt}(B), \overline{F}) \leq T_{opt}. \tag{3}$$

By Lemma 7, we immediately have $T \leq nm(n+1)(2n+1)^{d-1}T_{opt}$. Since the algorithm in Lemma 5 only yields a $(1+\epsilon)$-approximation of EMD, we have an additional $(1+\epsilon)$ factor in the approximation ratio.

Running Time: It is easy to see that the Base-Selection algorithm takes $O(md^4)$ time. Since we enumerate all n^d tuples in $[A]^d$ and each tuple corresponds to a call to the procedure of SOD which costs $O(md^4)$ time (by Lemma 4), the total running time is thus $O(n^{d+2}(\log n)^2 md^4)$ (including the time by the algorithm in Lemma 5), where the hidden constant depends on ϵ and d. $\square$

5.2   The FPTAS Algorithm

The upper bound determined in last section enables us to achieve a $(1+\epsilon)$-approximation for EMDRT in the following way. We first draw d balls (called *Grid-Ball*) centered at the d points of A respectively, and then build a grid inside each ball. For each d-grid-point tuple (one from each grid-ball), emulate the SOD procedure on B. The purpose of using grid points is for determining more "accurate" rigid transformation for B. To implement this approach, we need to resolve two major issues: (1) What is the radius of each grid-ball and (2) what is the density of the grids. Below we discuss our ideas on each of them.

Lemma 8. *If $W^A \geq W^B$ in EMDRT, then for any $1 \leq j \leq m$,*

$$\min_{1 \leq i \leq n} \|p_i - \mathcal{T}_{opt}(q_j)\| \leq \frac{nW^B}{\beta_j} EMD(A, \mathcal{T}_{opt}(B)). \tag{4}$$

Lemma 8 indicates that for each q_j, there exists some p_i whose grid-ball contains $\mathcal{T}_{opt}(q_j)$ if its radius is set to be $\frac{nW^B}{\beta_j} EMD(A, \mathcal{T}_{opt}(B))$. However, since $EMD(A, \mathcal{T}_{opt}(B))$ is unknown, we can use the upper bound to approximate it. The following lemma determines the density of the grid.

Lemma 9. *If $W^A \geq W^B$ and $\mathcal{T}$ is a rigid transformation for B such that for each $1 \leq j \leq m$, at least one of the following two conditions hold,*

1. $\|\mathcal{T}(q_j) - \mathcal{T}_{opt}(q_j)\| \leq c\epsilon \min_{1 \leq i \leq n} \|p_i - \mathcal{T}_{opt}(q_j)\|$;
2. $\|\mathcal{T}(q_j) - \mathcal{T}_{opt}(q_j)\| \leq c\epsilon \frac{W^B}{m\beta_j} EMD(A, \mathcal{T}_{opt}(B))$,

where $c > 0$ is a constant, then $EMD(A, \mathcal{T}(B)) \leq (1 + 2c\epsilon)EMD(A, \mathcal{T}_{opt}(B))$.

Algorithm FPTAS-for-EMDRT

Input: Two weighted point sets $A = \{p_1, \cdots, p_n\}$ and $B = \{q_1, \cdots, q_m\}$ in $\mathbb{R}^d$ with weight $\alpha_i \geq 0$ and $\beta_j \geq 0$ for p_i and q_j respectively, and $W^A \geq W^B$; a small number $\epsilon > 0$.
Output: A rigid transformation $\mathcal{T}$ for B with $EMD(A, \mathcal{T}(B)) \leq (1+\epsilon)EMD(A, \mathcal{T}_{opt}(B))$.

1. Call Upper-Bound-for-EMDRT in Section 5.1, and let T be the yielded upper bound. Construct the set $\Gamma = \{\frac{T}{2^t} \mid t = 0, 1, \cdots, 2d\log(\max\{n, m\})\}$.
2. Call Base-Selection on B in Section 5.1, and let $\{q_{b(1)}, \cdots, q_{b(d)}\}$ be the base.
3. Enumerate all d-point tuples from $[A]^d = A \times \cdots \times A$. For each tuple $R = \{p_{i(1)}, \cdots, p_{i(d)}\}$, do
   (a) Call Grid-Construction algorithm, and let $\{G_1, \cdots, G_d\}$ be the output.
   (b) For each tuple $\{g_1, \cdots, g_d\} \in G_1 \times G_2 \times \cdots \times G_d$, emulate SOD on B, $\{g_1, \cdots, g_d\}$, and the mapping f, where $f(j) = b(j)$. Compute the EMD between A and the output image of B.
4. Output the image of B among all the obtained images which has the minimum EMD to A.

Algorithm Grid-Construction

Input: Γ, $\{q_{b(1)}, \cdots, q_{b(d)}\}$ and $R = \{p_{i(1)}, \cdots, p_{i(d)}\}$.
Output: Grids $\{G_1, \cdots, G_d\}$

1. For each $1 \leq j \leq d$, do:
   (a) Build the set of radius candidates $\{\frac{nW^B}{\beta_{b(j)}}\gamma \mid \gamma \in \Gamma\}$.
   (b) For each radius candidate r, construct a grid-ball centered at $p_{i(j)}$ and with radius r, and build a grid inside the ball with grid length $\frac{r\epsilon}{8 \cdot 3^d nm\sqrt{d}}$.
   (c) Let G_j denote the union of the grids inside all the grid-balls.
2. Output $\{G_1, \cdots, G_d\}$.

Theorem 3. *The above FPTAS-for-EMDRT algorithm yields a $(1 + \epsilon)$-approximation for the EMDRT problem in $O((nm)^{O(d^2)}(\sqrt{d}/\epsilon)^{d^2} 3^{d^3} d^4)$ time.*

Sketch for the Proof: From Lemma 8 and 9, we know that for each $\mathcal{T}_{opt}(q_{b(j)})$, there is one grid point close to it. We denote these d grid points as $\{g_1, \cdots, g_d\}$ (see Figure 3). We construct an **implicit** set of points $B' = \{q'_1, \cdots, q'_m\}$ called 'relayer', where $q'_{b(j)} = g_j$ for $1 \leq j \leq d$, and $q'_l = \mathcal{T}_{opt}(q_l)$ for any $l \notin \{b(j) \mid 1 \leq j \leq d\}$, and assign each q'_l a weight β_l. B' is used as a bridge to show the quality of solution from the above algorithm. Particularly, we first prove that

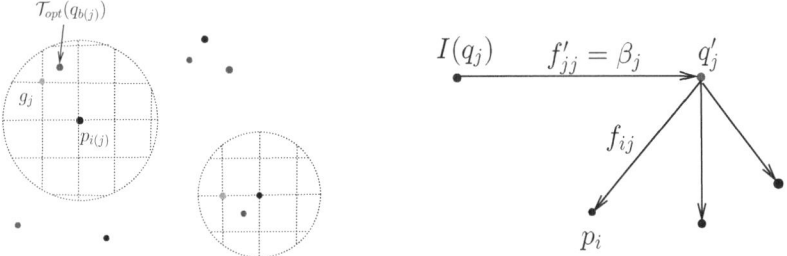

Fig. 3. Black points are A, red points are $\mathcal{T}_{opt}(B)$, and green points are grid points

Fig. 4. An example showing the flow from $I(q_j)$ to p_i passing through the q'_j

$EMD(A, B')$ is close to the optimal objective value, since B' is close to $\mathcal{T}_{opt}(B)$. Then, we show that $I(B)$ is close to B', where $I(B)$ is the output from the execution of SOD on B and $\{g_1, \cdots, g_d\}$. Finally, we view B' as the one who relays all flow from $I(B)$ to A, which implies the $EMD(A, I(B))$ is also close to the optimal objective value.

Proof. First, by Theorem 2 and Step 1 of the FPTAS-for-EMDRT algorithm, we know that there must exist $\gamma_0 \in \Gamma$ such that

$$EMD(A, \mathcal{T}_{opt}(B)) \leq \gamma_0 \leq 2EMD(A, \mathcal{T}_{opt}(B)). \tag{5}$$

Since the algorithm enumerates all the d-tuples from $[A]^d$, we can focus on the tuple $\{p_{i(1)}, \cdots, p_{i(d)}\}$, such that $||p_{i(j)} - \mathcal{T}_{opt}(q_{b(j)})|| = \min_{1 \leq i \leq n} ||p_i - \mathcal{T}_{opt}(q_{b(j)})||$ for each $1 \leq j \leq d$. We know that if let $r_j = \frac{nW^B}{\beta_{b(j)}}\gamma_0$ for each $1 \leq j \leq d$, then the grid length satisfies the following inequality,

$$\frac{r_j\epsilon}{8 \cdot 3^d nm\sqrt{d}} \leq \frac{\epsilon W^B}{4 \cdot 3^d m\sqrt{d}\beta_{b(j)}}EMD(A, \mathcal{T}_{opt}(B)). \tag{6}$$

From Lemma 8, we know that $\mathcal{T}_{opt}(q_{b(j)})$ locates inside the grid-ball centered at $p_{i(j)}$ and with radius r_j. Thus, there exists one grid point $g_j \in G_j$ such that

$$||g_j - \mathcal{T}_{opt}(q_{b(j)})|| \leq c\epsilon\frac{W^B}{m\beta_{b(j)}}EMD(A, \mathcal{T}_{opt}(B)), \tag{7}$$

where $c = 1/(4 \cdot 3^d)$ (see Figure 3). Now, we construct an implicit point set $B' = \{q'_1, \cdots, q'_m\}$ as the *relayer* point set, where $q'_{b(j)} = g_j$ for $1 \leq j \leq d$, and $q'_l = \mathcal{T}_{opt}(q_l)$ for any $l \notin \{b(j) \mid 1 \leq j \leq d\}$, and assign each q'_l a weight β_l. From Lemma 9 and inequality (7), we have

$$EMD(A, B') \leq (1 + 2c\epsilon)EMD(A, \mathcal{T}_{opt}(B)). \tag{8}$$

This means that $EMD(A, B')$ is close to the optimal objective value.

Consider executing the SOD procedure on B and $\{g_1, \cdots, g_d\}$ (*i.e.*, $\{q'_{b(1)}, \cdots, q'_{b(d)}\}$), and let $I(B)$ be the output image of B. We define the flow $F' = \{f'_{ij}\}$

from $\mathcal{T}_{opt}(B)$ to B' as $f'_{jj} = \beta_j$ for each $1 \leq j \leq m$ and $f'_{ij} = 0$ for all $i \neq j$. Note that from (7), we have

$$\mathcal{BN}(\mathcal{T}_{opt}(B), B', F') \leq \frac{c\epsilon}{m} EMD(A, \mathcal{T}_{opt}(B)). \tag{9}$$

Using a similar idea in the proof of Lemma 7, we have

$$\mathcal{BN}(I(B), B', F') \leq 2 \cdot 3^{d-1} \mathcal{BN}(\mathcal{T}_{opt}(B), B', F') \leq \frac{\epsilon}{2m} EMD(A, \mathcal{T}_{opt}(B)). \tag{10}$$

The only difference from Lemma 7 is that we replace the factor of $(n+1)(2n+1)^{d-1}$ in Lemma 7 by $2 \cdot 3^{d-1}$ (due to the difference in the flow F').

In the above construction, B' can be viewed as a point set which **relays** all the flow from $I(B)$ to A. More specifically, any flow from $I(q_j)$ to p_i can be thought as a flow arriving q'_j first, and then flowing to p_i (see Figure 4). From (8), (10), and triangle inequality, we have

$$EMD(A, I(B)) \leq EMD(A, B') + \sum_{j=1}^{m} \mathcal{BN}(I(B), B', F') \leq (1+\epsilon)EMD(A, \mathcal{T}_{opt}(B)).$$

This means that $I(B)$ is the desired image of B which yields a $(1+\epsilon)$-approximation for EMDRT. Since the algorithm enumerates all tuples in $[A]^d$ and $G_1 \times G_2 \times \cdots \times G_d$, $I(B)$ must be one of the output images.

Running Time: The most time consuming step is enumerating all tuples in $G_1 \times G_2 \times \cdots \times G_d$. Since each $|G_j| = O(d \log(\max\{n, m\})(\frac{8 \cdot 3^d nm \sqrt{d}}{\epsilon})^d)$, the total running time is $O((nm)^{O(d^2)} (\sqrt{d}/\epsilon)^{d^2} 3^{d^3} d^4)$ (including the time for SOD). $\qquad \square$

6   FPTAS with Improved Running Time

The FPTAS algorithm given in Section 5.2 can be further improved in its running time. To achieve this, we notice that the most time-consuming part of the algorithm is for examining all combinations of the grid points inside the d grid-balls. To speed up the computation, it is desirable to reduce the size of the grids. For this purpose, we first observe that the contribution of a point $q_j \in B$ to the optimal objective value is $\sum_{i=1}^{n} f_{ij} \|p_i - \mathcal{T}_{opt}(q_j)\| / \min\{W^A, W^B\}$, where $\sum_{i=1}^{n} f_{ij}$ can be viewed as the weight β_j of q_j (by Definition 2) and $\|p_i - \mathcal{T}_{opt}(q_j)\|$ is the distance between $p_i \in A$ and the position of q_j in the optimal solution. This means that, to avoid the error caused by the rigid transformation, we need to consider both β_j and $\|p_i - \mathcal{T}_{opt}(q_j)\|$ when selecting the base. The Base-Selection Algorithm in Section 5.1 maximizes the dispersion of the base using only the weight (*i.e.*, type 1 dispersion; see Section 3). Thus, a better way for maximizing the dispersion of the base is to consider both factors (*i.e.*, type 2 dispersion). This motivates us to consider the following algorithm.

Algorithm Optimum-Guided-Base (OGB)
Input: A weighted point set $B = \{q_1, \cdots, q_m\}$ in $\mathbb{R}^d$ with weight $\beta_j \geq 0$ for q_j.
Output: A base which is an ordered subset of points in B.

1. For each $1 \leq j \leq m$, define a value $v_j = \max\{\min_{1 \leq i \leq n} ||p_i - \mathcal{T}_{opt}(q_j)||,$ $\frac{W^B}{m\beta_j} EMD(A, \mathcal{T}_{opt}(B))\}$.

2. Select the point in B with the minimum v_j value, and denote it as $q_{b(1)}$.

3. Set $l = 1$, and repeat the following steps until $l = d$.

   (a) Let $\mathcal{F}_l$ the flat spanning $\{q_{b(1)}, \cdots, q_{b(l)}\}$.

   (b) Find the point in B realizing the value of $\max\{\frac{1}{v_j} dist(q_j, \mathcal{F}_l) \mid 1 \leq j \leq m\}$, and denote it as $q_{b(l+1)}$, where $dist(q_j, \mathcal{F}_l)$ is the distance between q_j and $\mathcal{F}_l$.

   (c) Let $l = l + 1$.

4. Output $\{q_{b(1)}, \cdots, q_{b(d)}\}$.

Note: The above OGB algorithm differs from the Base-Selection algorithm mainly on the use of v_j value for selecting the base. However, since v_j depends on the optimal solution, OGB cannot be directly implemented. To resolve this issue, we can enumerate all tuples in $[B]^d$, and find the one corresponding to the output of OGB. Thus, for ease of analysis, we can assume that OGB is available.

With OGB, we can now discuss our improved FPTAS algorithm. Since most part of the improved algorithm is similar to the original FPTAS algorithm given in Section 5.2, below we just list the main differences.

1. Replace the Base-Selection algorithm by the OGB algorithm.

2. In the Grid-Construction algorithm, replace the set of radius candidates by $\{\frac{nW^B}{2^t \beta_{b(j)}} \gamma \mid \gamma \in \Gamma, t = 1, 2, \cdots, \log(nm)\}$ for each $1 \leq j \leq d$, and change the grid length to $\frac{r\epsilon}{8 \cdot 3^d \sqrt{d}}$.

Why is the Grid Size Reduced? To see why the above algorithm is an improved FPTAS, we first analyze the grid size. Similar to the original FP-TAS, we can build a grid-ball centered at $p_{i(j)}$ and with radius $v_{b(j)}$ (*i.e.,* $\max\{\min_{1 \leq i \leq n} ||p_i - \mathcal{T}_{opt}(q_{b(j)})||, \frac{W^B}{m\beta_{b(j)}} EMD(A, \mathcal{T}_{opt}(B))\}$). With this, we know that (1) $\mathcal{T}_{opt}(q_{b(j)})$ locates inside the grid-ball of $p_{i(j)}$ and (2) the grid length can be set as $\frac{\epsilon}{\sqrt{d}} v_{b(j)}$ (by Lemma 9). Thus, the grid size of each grid-ball is $O((\sqrt{d}/\epsilon)^d)$, which is independent of n and m, and thus a significant reduction on its size. Note that although the exact value of $v_{b(j)}$ is unknown, it is possible to find a good approximation (less than $2v_{b(j)}$ by Lemma 8) from the set of new radius candidates, which increases the grid size only by a constant factor.

With the above understanding, we can show the following theorem using a similar argument given in the proof of Theorem 3.

Theorem 4. *The improved FPTAS algorithm yields a* $(1+\epsilon)$*-approximation for the EMDRT problem in* $O((nm)^{d+2}(\log(nm))^{2d}(\sqrt{d}/\epsilon)^{d^2} 3^{d^3} d^4)$ *time.*

Lower Bound on Running Time. Any PTAS for EMDRT has a lower bound of $\Omega(mn^{O(d)})$ on its running time, where $m \leq n$ [6]. Since our algorithm takes roughly $O((nm)^{d+2} (\log(nm))^{2d})$-time, it is close to the limit.

References

1. Alt, H., Guibas, L.: Discrete geometric shapes: matching, interpolation, and approximation. In: Sack, J.-R., Urrutia, J. (eds.) Handbook of Computational Geometry, pp. 121–153. Elsevier, Amsterdam (1999)
2. Andoni, A., Indyk, P., Krauthgamer, R.: Earth mover distance over high-dimensional spaces. In: SODA, pp. 343–352 (2008)
3. Andoni, A., Do Ba, K., Indyk, P., Woodruff, D.P.: Efficient Sketches for Earth-Mover Distance, with Applications. In: FOCS, pp. 324–330 (2009)
4. Benkert, M., Gudmundsson, J., Merrick, D., Wolle, T.: Approximate one-to-one point pattern matching. J. Discrete Algorithms 15, 1–15 (2012)
5. Cohen, S.: Finding Color and Shape Patterns in Images. PhD thesis, Stanford University, Department of Compute Science (1999)
6. Cabello, S., Giannopoulos, P., Knauer, C.: On the parameterized complexity of d-dimensional point set pattern matching. Inf. Process. Lett. 105(2), 73–77 (2008)
7. Cabello, S., Giannopoulos, P., Knauer, C., Rote, G.: Matching Point Sets with Respect to the Earth Mover's Distance. Computational Geometry: Theory and Applications 39(2), 118–133 (2008)
8. Cardoze, D.E., Schulman, L.J.: Pattern Matching for Spatial Point Sets. In: FOCS, pp. 156–165 (1998)
9. Gavrilov, M., Indyk, P., Motwani, R., Venkatasubramanian, S.: Combinatorial and Experimental Methods for Approximatie Point Pattern Matching. Algorithmica 38(1), 59–90 (2004)
10. Goodrich, M.T., Mitchell, J.S.B., Orletsky, M.W.: Approximate Geometric Pattern Matching Under Rigid Motions. IEEE PAMI 21(4), 371–379 (1999)
11. Klein, O., Veltkamp, R.C.: Approximation Algorithms for Computing the Earth Mover's Distance Under Transformations. In: Deng, X., Du, D.-Z. (eds.) ISAAC 2005. LNCS, vol. 3827, pp. 1019–1028. Springer, Heidelberg (2005)
12. Rubner, Y., Tomasi, C., Guibas, L.J.: The Earth Movers Distance as a Metric for Image Retrieval. Int. J. of Comp. Vision 40(2), 99–121 (2000)

Maximizing a Submodular Function
with Viability Constraints

Wolfgang Dvořák[1], Monika Henzinger[1], and David P. Williamson[2]

[1] Universität Wien, Fakultät für Informatik, Währingerstraße 29, A-1090 Vienna,
Austria
[2] School of Operations Research and Information Engineering, Cornell University,
Ithaca, New York, 14853, USA

Abstract. We study the problem of maximizing a monotone submodular function with viability constraints. This problem originates from computational biology, where we are given a phylogenetic tree over a set of species and a directed graph, the so-called food web, encoding viability constraints between these species. These food webs usually have constant depth. The goal is to select a subset of k species that satisfies the viability constraints and has maximal phylogenetic diversity. As this problem is known to be NP-hard, we investigate approximation algorithm. We present the first constant factor approximation algorithm if the depth is constant. Its approximation ratio is $(1 - \frac{1}{\sqrt{e}})$. This algorithm not only applies to phylogenetic trees with viability constraints but for arbitrary monotone submodular set functions with viability constraints. Second, we show that there is no $(1 - 1/e + \epsilon)$-approximation algorithm for our problem setting (even for additive functions) and that there is no approximation algorithm for a slight extension of this setting.

1 Introduction

We consider the problem of maximizing a monotone submodular set function f over subsets of a ground set X, subject to a restriction on what subsets are allowed. As discussed below, this problem has been well-studied with constraints on the allowed sets that are *downward-closed*; that is, if S is allowed subset, then so is any $S' \subset S$. Here we study the problem of maximizing such a function with a constraint that is not downwards-closed. Specifically, we assume that there exists a directed acyclic graph D with the elements of X as nodes in the graph and only consider so-called *viable* sets of a certain size. A set S is viable if each element either has no outgoing edges in D or it has a path P to such an element with $P \subseteq S$. Such viability constraints are a natural way to model dependencies between elements, where an element can only contribute to the function if it appears together with specific other elements.

We are motivated by a problem arising in conservation biology. The problem is given as a rooted phylogenetic tree $\mathcal{T}$ with nonnegative weights on the edges, where the leaves of the tree represent species, and the weights represent genetic distance. Given a conservation limit k, we would like to select k species so as

H.L. Bodlaender and G.F. Italiano (Eds.): ESA 2013, LNCS 8125, pp. 409–420, 2013.
© Springer-Verlag Berlin Heidelberg 2013

to maximize the overall phylogenetic diversity of the set, which is equivalent to maximizing the weight of the induced subtree on the k selected leaves plus the root. This problem, known as *Noah's Ark problem* [16], can be solved in polynomial time via a greedy algorithm [3,12,14]. Moulton, Semple, and Steel [10] introduced a more realistic extension of the problem which takes into account the dependence of various species upon one another in a food web. In this food web, an arc is directed from species a towards species b if a's survival depends on species b's. Moulton et al. now consider selecting viable subsets given by the food web of size k, i.e. a species is viable if also at least one of its successors in the food web is preserved. Note that in real life the *depth* of the food web, i.e., the longest shortest path between any node in D and the "nearest" node with no out-edge, is constant (usually no larger than 30). Faller et al. [4] show that the problem of maximizing phylogenetic diversity with viability constraints is NP-hard, even in simple special cases with constant depth (e.g. the food web is a directed tree of constant depth).

Since phylogenetic diversity induces a monotone, submodular function on a set of species, this problem is a special case of the problem of maximizing a submodular function with viability constraints. There exists a long line of research on approximately maximizing monotone submodular functions with constraints. This line of work was initiated by Nemhauser et al. [11] in 1978; they give a greedy $(1 - \frac{1}{e})$-approximation algorithm for maximizing a monotone submodular function subject to a cardinality constraint. Fisher et al. [6] introduced approximation algorithms for maximizing a monotone submodular function subject to matroid constraints (in which the set S must be an independent set in a single or multiple matroids). In recent work other types of constraints have been studied, as well as nonmonotone submodular functions; see the surveys by Vondrak [15] and Goundan and Schulz [7].

In our case, the viability constraints are *not downwards-closed* while most of the prior work studies downwards-closed constraints. One notable exception, where not downwards-closed constraints are considered, are matroid base constraints [9]. The viability constraint could be extended to be downwards-closed by simply defining every subset of a viable set to be allowable. But then it would be NP-hard to test whether a given set satisfies the constraint (this is equivalent to the acyclic directed Steiner tree problem). Moreover this extension violates the exchange property of matroids and thus viability constraints also differ from matroid base constraints. Hence we consider a new type of constraint in submodular function maximization. We show how variants on the standard greedy algorithm can be used to derive approximation algorithms for maximizing a monotone, submodular function with viability constraints; thus we show that a new type of constraint can be handled in submodular function maximization.

Specifically we first present a scheme of $(1 - \frac{1}{e^{p/(p+d-1)}})/2$ - approximation algorithms for monotone submodular set functions with viability constraints, where d is the depth of the food web and p is a parameter of the algorithm, such that the running time is exponential in p but is polynomial for any fixed p. For instance if we set $p = d$ we achieve a $(1 - \frac{1}{\sqrt{e}})/2$ - approximation algorithm.

We further present a variant of these algorithms which are $(1 - \frac{1}{e^{p/(p+d-1)}})$ - approximations, but whose running time is exponential in both d and p. For fixed $d=p$ this is polynomial and provides an $(1 - \frac{1}{\sqrt{e}})$ - approximation algorithm.

Next by a reduction from the maximum coverage problem, we show that there is no $(1-1/e+\epsilon)$-approximation algorithm for the phylogenetic diversity problem with viability constraints (unless $\mathsf{P} = \mathsf{NP}$). Finally we consider a generalization of our problem where we additionally allow AND-constraints such as "species a is only viable if we preserve *both* species b and species c" and show that this generalization has no approximation algorithm (assuming $\mathsf{P} \neq \mathsf{NP}$) by a reduction from 3-SAT.

We define the problem more precisely in Section 2, introduce our algorithms in Section 3, and give the hardness results in Section 4. All omitted proofs are available at: http://eprints.cs.univie.ac.at/3736/.

2 Phylogenetic Diversity with Viability Constraints

We first give a formal definition of the problem.

Definition 1. *A (rooted)* phylogenetic tree $\mathcal{T} = (T, E_{\mathcal{T}})$ *is a rooted tree with root r and each non-leaf node having at least 2 child-nodes together with a weight function w assigning non-negative integer weights to the edges. Let X denote the set of leaf nodes of $\mathcal{T}$. For any set $A \subseteq X$ the operator $\mathcal{T}(A)$ yields the spanning tree of the set $A \cup \{r\}$, and by $\mathcal{T}_E(A)$ we denote the edges of this spanning tree. Then for any set $S \subseteq X$ the* phylogenetic diversity *is defined as*

$$\mathcal{PD}(S) = \sum_{e:e\in\mathcal{T}_E(S)} w(e)$$

A food web D *for the phylogenetic tree $\mathcal{T} = (T, E_{\mathcal{T}})$ is an acyclic directed graph (X, E). A set $S \subseteq X$ is called* viable *if each $s \in S$ is either a sink (a node with out-degree 0) in D or there is a $s' \in S$ such that $(s, s') \in E$.*

Now our problem of interest is defined as follows.

Definition 2. *The problem of* Optimizing Phylogenetic Diversity with Viability Constraints *(OptPDVC) is defined as follows. You are given a phylogenetic tree $\mathcal{T}$ and a food web $D = (X, E)$, and a positive integer k. Find a viable subset $S \subseteq X$ of size (at most) k maximizing $\mathcal{PD}(S)$.*

OptPDVC is known to be NP-hard [4], even for restricted classes of phylogenetic trees and dependency graphs.

First we study fundamental properties of the function $\mathcal{PD}$.

Definition 3. *The set function $\mathcal{PD}(.|.) : 2^X \times 2^X \mapsto \mathbb{N}_0$ for each $A, B \subseteq X$ is defined as $\mathcal{PD}(A|B) = \mathcal{PD}(A \cup B) - \mathcal{PD}(B)$.*

The intuitive meaning of $\mathcal{PD}(A|B)$ is the gain of diversity we get by adding the set A to the already selected species B.

We next recall the definition of submodular set functions. We call a set function *submodular* if

$$\forall A, B, C \subseteq \Omega : A \subseteq B \Rightarrow f(A \cup C) - f(A) \geq f(B \cup C) - f(B).$$

Proposition 1. $\mathcal{PD}$ *is a non-negative, monotone, and submodular function [2].*

Now consider the function $\mathcal{PD}(.|.)$. As $\mathcal{PD}(.)$ is monotone also $\mathcal{PD}(.|.)$ is monotone in the first argument and because of the submodularity of $\mathcal{PD}(.)$ the function $\mathcal{PD}(.|.)$ is anti-monotone in the second argument.

In the remainder of the paper we will not refer to the actual definition of the functions $\mathcal{PD}(.)$, $\mathcal{PD}(.|.)$, but only exploit monotonicity, submodularity and the fact that these functions can be efficiently computed.

Moreover we will consider a function VE (*viable extension*) which, given a set of species S, returns a viable set S' of minimum size containing S. In the simplest case where S consist of just one species it computes a shortest path to any sink node in the food web. We define the *depth* d of a food web (X, E) as

$$d(X, E) = \max_{s \in X} |\operatorname{VE}(\{s\})|.$$

If the food web is clear from the context we just write d instead of $d(X, E)$. Note that the problem remains NP-hard for instances with $d = 2$, even if $\mathcal{PD}$ is additive [4]. Finally we define the *costs* of adding a set of species A to a set B

$$c(A|B) = |\operatorname{VE}(B \cup A)| - |B|.$$

We will tacitly assume that $d \leq k$. Otherwise we can eliminate species s with $c(\{s\}|\emptyset) > k$ by polynomial time preprocessing, using a shortest path algorithm.

3   Approximation Algorithms

In this section we assume that a non-negative, monotone submodular function $\mathcal{PD}(.)$ is given as an oracle and we want to maximize $\mathcal{PD}(.)$ under viability constraints together with a cardinality constraint. We first review the greedy algorithm given by Faller et al. [4] presented in Algorithm 1. The idea is that, in each step, one considers only species which either have no successors in the food web or for which one of the successors has already been selected (adding one of these species will keep the set viable). Then one adds the species that gives the

---

Algorithm 1. Greedy, Faller et al.

1: $S \leftarrow \emptyset$
2: **while** $|S| < k$ **do**
3:     $s \leftarrow \underset{c(s|S)=1}{\operatorname{argmax}} \, v(\{s\}|S)$
4:     $S \leftarrow S \cup \{s\}$
5: **end while**

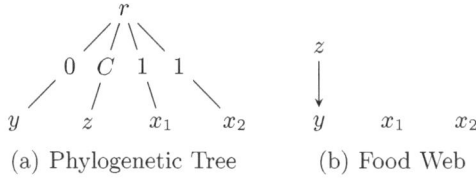

(a) Phylogenetic Tree   (b) Food Web

Fig. 1. An illustration of Example 1

largest gain of preserved diversity. By the restriction on the considered species the constructed set is always viable, but we might miss highly valuable species which is illustrated by the following example.

Example 1. Consider the set of species $X = \{y, z, x_1, x_2\}$ the phylogenetic tree $\mathcal{T} = (\{r\} \cup X, \{(r, y), (r, z), (r, x_1), (r, x_2)\}$, weights $w(r, x_i) = 1$, $w(r, y) = 0$, $w(r, z) = C$ with $C > 2$ and the food web $(X, \{(z, y)\})$ (see Fig. 1). Assume a budget $k = 2$, now as the species y has weight 0, Algorithm 1 would pick x_1, and x_2. Hence Algorithm 1 results a viable set with diversity 2. But the set $\{z, y\}$ is viable and has diversity C, which can be made arbitrarily large.

This example shows that the greedy solution might have an approximation ratio that is arbitrarily bad, because it ignores highly weighted species if they are "on the top of" less valuable species.

Hence, to get approximation ratio, we have to consider all subsets of species up to a certain size which can be made viable and pick the most valuable subset. Algorithm 2 deals with this observation. It generalizes concepts from [1], which itself builds on [8]. Lines 5 - 10 of the algorithm implement a greedy algorithm that in each step selects the most "cost efficient" subset of species of size p, i.e. the subset S of species that maximizes the ratio of the increase in PD over the cost of adding S, and adds it to the solution. But Algorithm 2 does not solely run the greedy algorithm, it first computes the set with maximal $\mathcal{PD}$ among

---

Algorithm 2.

1: $\mathcal{B} \leftarrow \{B \subseteq X \mid |B| \leq p, 1 \leq c(B|\emptyset) \leq k\}$
2: $S \leftarrow \underset{B \in \mathcal{B}}{\mathrm{argmax}}\ \mathcal{PD}(B)$
3: $\mathcal{G} = \mathrm{VE}(S)$
4: $G \leftarrow \emptyset$
5: **while** $\mathcal{B} \neq \emptyset$ **do**
6:      $S \leftarrow \underset{B \in \mathcal{B}}{\mathrm{argmax}}\ \frac{\mathcal{PD}(B|G)}{c(B|G)}$
7:      $G = \mathrm{VE}(G \cup S)$
8:      $\mathcal{B} \leftarrow \{B \in \mathcal{B} \mid |B| \leq p, 1 \leq c(B|G) \leq k - |G|\}$
9: **end while**
10: **if** $\mathcal{PD}(G) > \mathcal{PD}(\mathcal{G})$ **then**
11:      $\mathcal{G} \leftarrow G$
12: **end if**

---

all sets of size $\leq p$ that can be made viable. In certain cases this set is better than the viable set obtained by the greedy algorithm, a fact that we exploit in the proof of Theorem 1. In the algorithm G denotes the current set of selected species; $\mathcal{B}$ contains the species we might add to G; $\mathcal{G}$ denotes the best viable set we have found so far.

The next theorem will analyze the approximation ratio of Algorithm 2.

Theorem 1. *Algorithm 2 is a* $(1 - \frac{1}{e^{p/(p+d-1)}})/2$ *approximation (for any* $p \in \{1, \dots, \lfloor k/3 \rfloor\}$).

To prove Theorem 1, we introduce some notation. First let $O \subseteq X$ denote the optimal solution. We will consider a decomposition $\mathcal{D}_O$ of O into sets $O_1, \dots, O_{\lceil k/p \rceil}$ of size $\leq p$. By decomposition we mean that (i) $\bigcup_{i=1}^{\lceil k/p \rceil} O_i = O$ and (ii) $O_i \cap O_j = \emptyset$ if $i \neq j$. Moreover we require that $|\operatorname{VE}(O_i)| \leq p + d - 1$ and $\sum_i |\operatorname{VE}(O_i)| \leq \frac{k}{p}(p + d - 1)$. Next we show that such a decomposition $\mathcal{D}_O$ always exists.

Lemma 1. *There exist* $\lceil \frac{k}{p} \rceil$ *many pairs* $(O_1, B_1), \dots, (O_{\lceil \frac{k}{p} \rceil}, B_{\lceil \frac{k}{p} \rceil})$ *such that* $O = \bigcup_{1 \leq i \leq \lceil \frac{k}{p} \rceil} O_i$, $O_i \cup B_i$ *is viable,* $|O_i| \leq p$, $|B_i| \leq d - 1$ *and* $\sum_i |O_i \cup B_i| \leq \frac{k}{p}(p + d - 1)$.

Proof. The optimal solution O is a viable subset of size at most k. Consider the reverse graph G of D projected on the set O i.e. $G = (O, E^- \cap (O \times O))$, and add an artificial root r that has an edge to all roots of G. Start a depth-first-search in r and with the empty sets O_1, B_1. Whenever the DFS removes a node from the stack, we add this node to the current set O_i, $i \geq 1$. When $|O_i| = p$ then we add the nodes on the stack, except r, to the set B_i, but do not change the stack itself. Then we continue the DFS with the next pair (O_{i+1}, B_{i+1}), again initialized by empty sets. Eventually the DFS stops, then the stack is empty and thus $(O_{\lceil \frac{k}{p} \rceil}, \emptyset)$ is the last pair. Notice that by the definition of d there are at most d nodes on the stack, one being the root r and hence $|B_i| \leq d - 1$. Since the DFS removes each node exactly once from the stack, all the sets $O_i \subseteq O$ are disjoint and all except the last one are of size p. Hence the DFS produces $\lceil \frac{k}{p} \rceil$ many sets O_i satisfying $O_i \cup B_i$ is viable, $|O_i| \leq p$ and $|B_i| \leq d - 1$. Finally as, by construction, $B_{\lceil \frac{k}{p} \rceil} = \emptyset$ and $|O_{\lceil \frac{k}{p} \rceil}| = k \mod p$ we obtain that $\sum_{i=1}^{\lceil \frac{k}{p} \rceil} |O_i \cup B_i| = \sum_{i=1}^{\lfloor \frac{k}{p} \rfloor} |O_i \cup B_i| + (k \mod p) \leq \lfloor \frac{k}{p} \rfloor (p + d - 1) + (k \mod p) \leq \frac{k}{p}(p + d - 1)$. □

Now consider the greedy algorithm and the value l where the l-th iteration is the first iteration such that after executing the loop body, $\max_{O_j \in \mathcal{D}_O} \frac{PD(O_j|G)}{c(O_j|G)} > \max_{B \in \mathcal{B}} \frac{PD(B|G)}{c(B|G)}$. If $\mathcal{G} \neq O$ the inequality holds at least for the last iteration of the loop where $\mathcal{B} = \emptyset$. We define $O'_{l+1} = \operatorname*{argmax}_{O_j \in \mathcal{D}_O} \frac{PD(O_j|G)}{c(O_j|G)}$, i.e. O'_{l+1} is in the optimal viable set and would be a better choice than the selection of the algorithm, but the greedy algorithm cannot make $S \cup O'_{l+1}$ viable without violating the cardinality constraint.

Let S_i denote the set S added to G in iteration i of the while loop in Line 6. Moreover, for $i \leq l$ we denote the set G after the i-th iteration by G_i, with $G_0 = \emptyset$, the set $G \cup S$ from Line 7 as $G_i^* = G_{i-1} \cup S_i$ and the "costs" of adding set S_i by $c_i = c(S_i|G_{i-1}) = c(G_i|G_{i-1})$. With a slight abuse of notation we will use G_{l+1} to denote the viable set $\text{VE}(G_l \cup O'_{l+1})$, c_{l+1} to denote $c(O'_{l+1}|G_l)$ and G_{l+1}^* to denote the set $G_l \cup O'_{l+1}$ (G_{l+1} is not a feasible solution as $|G_{l+1}| > k$). Notice that while the sets G_i^* are not necessarily viable sets, all the $G_i, i \geq 0$ are viable sets and moreover $\mathcal{PD}(\mathcal{G}) \geq \mathcal{PD}(G_i), i \leq l$.

First we show that in each iteration of the algorithm the set S_i gives us a certain approximation of the missing part of the optimal solution.

Lemma 2. *For $1 \leq i \leq l+1$, $p \in \{1, \ldots, \lfloor k/3 \rfloor\}$:*

$$\frac{\mathcal{PD}(S_i|G_{i-1})}{c_i} \geq \frac{p}{(p+d-1)k} \cdot \mathcal{PD}(O|G_{i-1})$$

Proof. By the definition of S_i for each $O_j \in \mathcal{D}_O$ the following holds:

$$\frac{\mathcal{PD}(O_j|G_{i-1})}{c(O_j|G_{i-1})} \leq \frac{\mathcal{PD}(S_i|G_{i-1})}{c(G_i|G_{i-1})}$$

Next we use the monotonicity and submodularity of $\mathcal{PD}$ (for the first inequality) and the inequality from above (for the second inequality).

$$\mathcal{PD}(O|G_{i-1}) \leq \sum_{O_j \in \mathcal{D}_O} \mathcal{PD}(O_j|G_{i-1}) = \sum_{O_j \in \mathcal{D}_O} \frac{\mathcal{PD}(O_j|G_{i-1})}{c(O_j|G_{i-1})} c(O_j|G_{i-1})$$

$$\leq \sum_{O_j \in \mathcal{D}_O} \frac{\mathcal{PD}(S_i|G_{i-1})}{c_i} c(O_j|G_{i-1}) \leq \frac{\mathcal{PD}(S_i|G_{i-1})}{c_i} \frac{p+d-1}{p} \cdot k$$

The last step exploits that by Lemma 1 $\sum_{O_j \in \mathcal{D}_O} c(O_j|G_{i-1}) \leq \frac{k}{p} \cdot (p+d-1)$. □

Lemma 3. *For $1 \leq i \leq l+1$:*

$$\mathcal{PD}(G_i^*) \geq \left[1 - \prod_{j=1}^{i} \left(1 - \frac{p \cdot c_j}{(d+p-1) \cdot k} \right) \right] \mathcal{PD}(O)$$

Proof. The proof is by induction on i. The base case for $i = 1$ is by Lemma 2. For the induction step we show that if the claim holds for all $i' < i$ then it must also hold for i. For convenience we define $C_i = \frac{p \cdot c_i}{(d+p-1) \cdot k}$.

$$\mathcal{PD}(G_i^*) = \mathcal{PD}(G_{i-1}) + \mathcal{PD}(G_i^*|G_{i-1}) \geq \mathcal{PD}(G_{i-1}) + C_i \cdot \mathcal{PD}(O|G_{i-1})$$

$$= \mathcal{PD}(G_{i-1}) + C_i \cdot (\mathcal{PD}(O \cup G_{i-1}) - \mathcal{PD}(G_{i-1}))$$

$$\geq (1 - C_i) \cdot \mathcal{PD}(G_{i-1}) + C_i \cdot \mathcal{PD}(O) \geq (1 - C_i) \cdot \mathcal{PD}(G_{i-1}^*) + C_i \cdot \mathcal{PD}(O)$$

$$\geq (1 - C_i) \left[1 - \prod_{j=1}^{i-1} (1 - C_j) \right] \mathcal{PD}(O) + C_i \cdot \mathcal{PD}(O)$$

$$= \left[1 - \prod_{j=1}^{i} (1 - C_j) \right] \mathcal{PD}(O)$$

□

Proof (Theorem 1). Towards a bound for G_{l+1}^* consider $\sum_{m=1}^{l+1} c_m$. As G_{l+1} exceeds the cardinality constraint $\sum_{m=1}^{l+1} c_m > k$ it follows that:

$$1 - \prod_{j=1}^{l+1} \left(1 - \frac{p \cdot c_j}{(d+p-1) \cdot (k)}\right) \geq 1 - \prod_{j=1}^{l+1} \left(1 - \frac{p \cdot c_j}{(d+p-1) \cdot \sum_{m=1}^{l+1} c_m}\right)$$

$$\geq 1 - \left(1 - \frac{p}{(d+p-1) \cdot (l+1)}\right)^{l+1} \geq 1 - \frac{1}{e^{p/(d+p-1)}}$$

To obtain Line 2 we used the fact the term $1 - \prod_{j=1}^{l+1}\left(1 - \frac{c_j'}{C}\right)$ with constant C and the constraint $\sum_{j=1}^{l+1} c_j' = 1$ has its maximum at $c_j' = 1/(l+1)$.

By Lemma 3 we obtain $\mathcal{PD}(G_{l+1}^*) \geq \left(1 - \frac{1}{e^{p/(d+p-1)}}\right) \cdot \mathcal{PD}(O)$, thus it only remains to relate $\mathcal{PD}(G_{l+1}^*)$ to $\mathcal{PD}(G_l)$. To this end we consider the optimal set of size p computed in Line 3 and denote it by S_o. If the greedy solution has higher $\mathcal{PD}$ than S_o the algorithm returns a superset of G_l^* otherwise a superset of S_o. Hence, $\mathcal{PD}(G)$ is larger or equal to the maximum of $\mathcal{PD}(G_l^*)$ and $\mathcal{PD}(S_o)$. From the definitions of G_l^* and S_o it follows that

$$\mathcal{PD}(G_{l+1}^*) \leq \mathcal{PD}(G_l^*) + \mathcal{PD}(O_{l+1}') \leq \mathcal{PD}(G_l^*) + \mathcal{PD}(S_o).$$

With the above result for $\mathcal{PD}(G_{l+1}^*)$ we obtain that:

$$\mathcal{PD}(G) \geq \max(\mathcal{PD}(G_l^*), \mathcal{PD}(S_o)) \geq \left(1 - \frac{1}{e^{p/(d+p-1)}}\right) \cdot \frac{\mathcal{PD}(O)}{2}$$

Hence Algorithm 2 provides an $\left(1 - \frac{1}{e^{p/(d+p-1)}}\right)/2$ - approximation. □

Theorem 2. *Algorithm 2 is in time $\mathcal{O}\left(k \cdot (3^p n^{p+2} + n^{p+1} m)\right)$.*

Proof. First notice that computing the function VE can be reduced to a Steiner tree problem by (i) taking all the species in S that are already connected (via nodes in S) to a sink node in S, and merging these nodes into a single terminal node t, and (ii) connecting the remaining sink nodes to t. As starting nodes for the Steiner tree problem we use the remaining species in S. A viable set S is reduced to a single vertex t and thus the number of terminal nodes in the Steiner tree problems is bounded by a constant, hence we can solve them in polynomial time: The Steiner tree problem on acyclic directed graphs can be solved in time $\mathcal{O}(3^j n^2 + nm)$ [13], where j is the number of starting and terminal nodes. In Line 2 we have to consider $\mathcal{O}(n^p)$ sets S and for each of them we solve a Steiner tree problem. with at most p starting nodes. So this first loop can be done in time $\mathcal{O}(3^p n^{p+2} + n^{p+1} m)$. The number of iterations of the while loop is bounded by k and in each iteration, in Line 6, we have to solve $\mathcal{O}(n^p)$ Steiner tree problems with at most p starting nodes. Now as each iteration takes time $\mathcal{O}(3^p n^{p+2} + n^{p+1} m)$ we get a total running time of $\mathcal{O}(k \cdot (3^p n^{p+2} + n^{p+1} m))$. □

Finally, notice that one can use a modification of the enumeration technique as described in [8], to get rid of the factor $1/2$ in the approximation ratio. The idea is to consider all (viable) sets of a certain size and for each of them to run the greedy algorithm starting with this set. Finally one chooses the best of the produced solutions. These sets typically have to contain three objects of interest, in the case of the maximum coverage problem [8] (cf. Def. 4 below) just three sets from the collection SC. However, in our setting an object of interest is a pair (O_i, B_i), i.e. a set O_i of size $\leq p$ and a set B_i of size $< d$ making O_i viable. Thus three objects result in a set of size of $3p + 3d - 3$. This increases the running time by a factor of $n^{3p+3d-3}$. The proof of the following theorem is very similar to the above analysis for Algorithm 2 (details are provided in the appendix).

Theorem 3. *There exists an* $\left(1 - \frac{1}{e^{p/(d+p-1)}}\right)$ *- approximation algorithm for OptPDVC which runs in time* $\mathcal{O}\left(k \cdot (3^p n^{4p+3d-1} + n^{4p+3d-2} m)\right)$.

4   Impossibility Results

If we allow arbitrary monotone submodular functions it is easy to see that no $1 - \frac{1}{e} + \epsilon$-approximation algorithm exists (unless P = NP). This is immediate by the corresponding result for Max Coverage (with cardinality constraints). Here we show that, when considering viability constraints, this also holds for additive functions and in particular for the phylogenetic diversity $\mathcal{PD}$.

Definition 4. *The input to the* Max Coverage *problem is a set of domain elements* $D = \{1, 2, \ldots, n\}$, *together with non-negative integer weights* $w_1, \ldots w_n$, *a collection of subsets of* D SC $= \{S_1, \ldots S_m\}$ *and a positive integer* k. *The goal is to find a set* SC$' \subseteq$ SC *of cardinality* k *maximizing* $\sum_{i \in \bigcup_{S \in SC'} S} w_i$.

Proposition 2. *There is no* α-*approximation algorithm for Max Coverage with* $\alpha > 1 - \frac{1}{e}$ *(unless* P = NP*)* [5,8].

Reduction 1. *Given an instance* (D, SC, k) *of the Max Coverage problem, we build an instance of OptPDVC as follows (cf. Fig. 2)*

$$X = D \cup \{S_{i,j} \mid S_i \in SC, 1 \leq j \leq n\}$$
$$E = \{(j, S_{i,n}) \mid j \in S_i\} \cup \{(S_{i,j+1}, S_{i,j}) \mid 1 \leq i \leq m, 1 \leq j < n\}$$
$$\mathcal{T} = (\{r\} \cup X, \{(r, s) \mid s \in X\})$$
$$w_e = \begin{cases} w_i & e = (r, i), i \in D \\ 0 & otherwise \end{cases}$$
$$k' = (k + 1) \cdot n$$

Lemma 4. *Let* (D, SC, k) *be an instance of the Max Coverage problem and let* $(TC, (X, E), k')$ *be an instance of OptPDVC given by Reduction 1. Let* $W > 0$. *Then there exists a cover* $C \subseteq SC$ *of size* k *with* $w(C) \geq W$ *for* (D, SC, k) *iff there exists a viable set* A *of size* $k' = (k + 1) \cdot n$ *with* $PD(A) \geq W$.

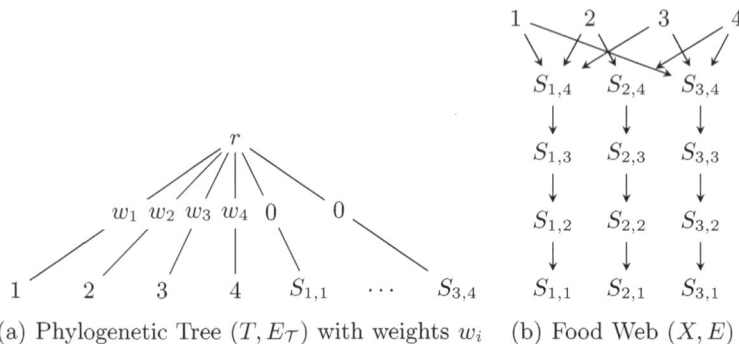

(a) Phylogenetic Tree (T, E_T) with weights w_i (b) Food Web (X, E)

Fig. 2. An illustration of Reduction 1, applied to $D = \{1, 2, 3, 4\}$, $\mathsf{SC} = \{S_1, S_2, S_3\}$, $S_1 = \{1, 2, 3\}$, $S_2 = \{2, 4\}$, $S_3 = \{1, 3, 4\}$

Proof. $\Rightarrow$: First assume that there is a cover C of size k with $w(C) = W$. Then $A' = \{S_{i,j} \mid S_i \in C, 1 \le j \le n\} \cup \bigcup_{S_i \in C} S_i$ is a viable set of size $\le k \cdot n + n$. Clearly $PD(A') = W$ and thus we have a viable set A of size $(k+1) \cdot n$ with $PD(A) \ge W$ by adding arbitrary viable species.

$\Leftarrow$: Assume there is a viable set A of size $(k+1) \cdot n$ with $PD(A) = W$. There are at most $k + 1$ elements $S_i \in \mathsf{SC}$ such that $S_{i,n} \in A$. This is by the fact that if $S_{i,n} \in A$ then also $S_{i,1}, \ldots, S_{i,n-1} \in A$. Now consider the case where there are exactly $k+1$ such elements. Then we already have $(k+1) \cdot n$ species in A and thus no $x \in D$ is contained in A. But then $PD(A) = 0$ as only the edges (r, x) with $x \in D$ have non-zero weight. Assuming $W > 0$ we thus have at most k elements $S_i \in E$ such that $S_{i,n} \in A$ and further as A is viable for each $x \in A$ there is an $S_{i,n} \in A$ such that $x \in S_i$. Hence $C' = \{S_i \mid S_{i,n} \in A\}$ is of size at most k and covers all $x \in A \cap D$, i.e. $w(C') = W$. Now by adding arbitrary $S_i \in \mathsf{SC}$ we can construct a cover C of size k with $w(C) \ge W$. □

Theorem 4. *There is no α-approximation algorithm for OptPDVC with $\alpha > 1 - \frac{1}{e}$ (unless $\mathsf{P} = \mathsf{NP}$), even if $\mathcal{PD}$ is an additive function.*

Proof. Immediate by Proposition 2, Lemma 4 and the fact that Reduction 1 can be performed in polynomial time. □

Finally let us consider a straightforward generalization of the viability constraints. So far we assumed that a species is viable iff at least one of its successors survives, but one can also imagine cases where one node needs several or even all of its successors to survive to be viable. In the following we consider food webs where we allow two types of nodes: (i) nodes that are viable if at least one successors survives and (ii) nodes that are only viable if all successors survive. We will show that in this setting no approximation algorithm is possible using a reduction from the NP-hard problem of deciding whether a propositional formula in 3-CNF is satisfiable. A 3-CNF formula is a propositional formula which is the conjunction of clauses, and each clause is the disjunction of exactly three literals, e.g. $\phi = (x_1 \vee x_2 \vee x_3) \wedge (x_2 \vee \neg x_3 \vee \neg x_4) \wedge (x_2 \vee x_3 \vee x_4)$.

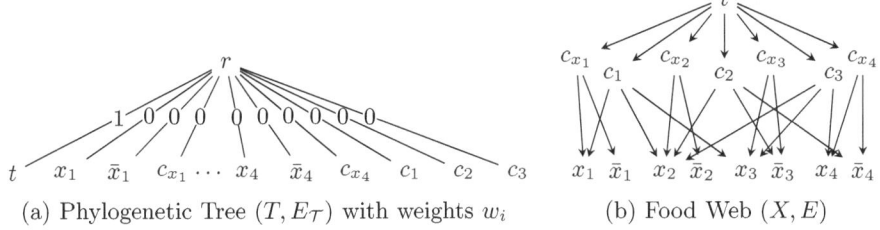

(a) Phylogenetic Tree $(T, E_{\mathcal{T}})$ with weights w_i (b) Food Web (X, E)

Fig. 3. An illustration of Reduction 2, applied to the propositional formula $\phi = (x_1 \vee x_2 \vee x_3) \wedge (x_2 \vee \neg x_3 \vee \neg x_4) \wedge (x_2 \vee x_3 \vee x_4)$

Reduction 2. *Given a propositional formula ϕ in 3-CNF over propositional variables $\mathcal{V} = \{x_1, \ldots, x_n\}$ with clauses $c_1, \ldots, c_m$ build the following instance $(T, E_{\mathcal{T}})$, (X, E) and weight w_e (cf. Fig. 3) :*

$$X = \{c_1, \ldots, c_m\} \cup \{x, \bar{x}, c_x \mid x \in \mathcal{V}\} \cup \{t\}$$
$$\mathcal{T} = (\{r\} \cup X, \{(r, s) \mid s \in X\})$$
$$w_e = \begin{cases} 1 & e = (r, t) \\ 0 & otherwise \end{cases}$$
$$E = \{(c_x, x), (c_x, \bar{x}) \mid x \in \mathcal{V}\} \cup \{(c_i, x) \mid x \in c_i\} \cup \{(c_i, \bar{x}) \mid \neg x \in c_i\}$$
$$\cup \{(t, c_i), (t, c_x) \mid 1 \leq i \leq m, x \in \mathcal{V}\}$$
$$k = 2 \cdot |\mathcal{V}| + m + 1$$

The species $\{c_1, \ldots c_m\} \cup \{x, \bar{x}, c_x \mid x \in \mathcal{V}\}$ are viable in the traditional sense and t is viable iff all its successors survive. More formally, a set $S \subseteq X$ is viable if (i) for each $s \in S$ either s is a sink or there is a $s' \in S$ with $(s, s') \in E$ and (ii) if $t \in S$ it holds for all s' with $(t, s') \in E$ that $s' \in S$.

Lemma 5. *Given a propositional formula ϕ and the instance $(\mathsf{TC}, (X, E), k)$ of OptPDVC given by Reduction 2. Then ϕ is satisfiable iff there exists a viable set A of size $\leq k$ with $PD(A) > 0$.*

Now assuming that there is an approximation algorithm for OptPDVC with generalized viability constraints we would immediately get an procedure deciding 3-CNF formulas: apply Reduction 2, compute $\mathcal{PD}$ using the α-approximation algorithm, and return satisfiable if $\mathcal{PD}$ is positive.

Theorem 5. *It is NP-hard to decide whether an instance of OptPDVC with generalized viability constraints has a viable set S with $\mathcal{PD}(S) > 0$. Thus no approximation algorithm for the problem can exist unless P = NP.*

Proof. Immediate by Lemma 5, and the fact that Reduction 2 can be performed in polynomial time.

References

1. Bordewich, M., Semple, C.: Nature reserve selection problem: A tight approximation algorithm. IEEE/ACM Transactions on Computational Biology and Bioinformatics 5(2), 275–280 (2008)
2. Bordewich, M., Semple, C.: Budgeted nature reserve selection with diversity feature loss and arbitrary split systems. Journal of Mathematical Biology 64(1-2), 69–85 (2012)
3. Faith, D.P.: Faith. Conservation evaluation and phylogenetic diversity. Biological Conservation 61(1), 1–10 (1992)
4. Faller, B., Semple, C., Welsh, D.: Optimizing Phylogenetic Diversity with Ecological Constraints. Annals of Combinatorics 15, 255–266 (2011)
5. Feige, U.: A threshold of ln n for approximating set cover. J. ACM 45(4), 634–652 (1998)
6. Fisher, M.L., Nemhauser, G.L., Wolsey, L.A.: An analysis of approximations for maximizing submodular set functions – II. Mathematical Programming Study 8, 73–87 (1978)
7. Goundan, P.R., Schulz, A.S.: Revisiting the greedy approach to submodular set function maximization. Working Paper, Massachusetts Institute of Technology (2007), http://www.optimization-online.org/DB_HTML/2007/08/1740.html
8. Khuller, S., Moss, A., Naor, J.: The budgeted maximum coverage problem. Inf. Process. Lett. 70(1), 39–45 (1999)
9. Lee, J., Mirrokni, V.S., Nagarajan, V., Sviridenko, M.: Non-monotone submodular maximization under matroid and knapsack constraints. In: Mitzenmacher, M. (ed.) Proceedings of the 41st Annual ACM Symposium on Theory of Computing, STOC 2009, Bethesda, MD, USA, May 31-June 2, pp. 323–332. ACM (2009)
10. Moulton, V., Semple, C., Steel, M.: Optimizing phylogenetic diversity under constraints. Journal of Theoretical Biology 246(1), 186–194 (2007)
11. Nemhauser, G.L., Wolsey, L.A., Fisher, M.L.: An analysis of approximations for maximizing submodular set functions — I. Mathematical Programming 14, 265–294 (1978)
12. Pardi, F., Goldman, N.: Species choice for comparative genomics: being greedy works. PLoS Genetics 71, 71 (2005)
13. Hsu, T.S., Tsai, K.-H., Wang, D.-W., Lee, D.T.: Two variations of the minimum steiner problem. J. Comb. Optim. 9(1), 101–120 (2005)
14. Steel, M.: Phylogenetic diversity and the greedy algorithm. Systematic Biology 54(4), 527–529 (2005)
15. Vondrák, J.: Submodular functions and their applications. In: SODA 2013 Plenary Talk (2013) Slides available at, http://theory.stanford.edu/~jvondrak/data/SODA-plenary-talk.pdf
16. Weitzman, M.L.: The Noah's ark problem. Econometricay 66, 1279–1298 (1998)

Table Cartograms

William Evans[1], Stefan Felsner[2], Michael Kaufmann[3], Stephen G. Kobourov[4],
Debajyoti Mondal[5], Rahnuma Islam Nishat[6], and Kevin Verbeek[7]

[1] Department of Computer Science, University of British Columbia
[2] Institut für Mathematik, Technische Universität Berlin
[3] Wilhelm-Schickard-Institut für Informatik, Universität Tübingen
[4] Department of Computer Science, University of Arizona
[5] Department of Computer Science, University of Manitoba
[6] Department of Computer Science, University of Victoria
[7] Department of Computer Science, University of California, Santa Barbara

Abstract. A table cartogram of a two dimensional $m \times n$ table A of
non-negative weights in a rectangle R, whose area equals the sum of the
weights, is a partition of R into convex quadrilateral faces corresponding
to the cells of A such that each face has the same adjacency as its corresponding cell and has area equal to the cell's weight. Such a partition
acts as a natural way to visualize table data arising in various fields of
research. In this paper, we give a $O(mn)$-time algorithm to find a table
cartogram in a rectangle. We then generalize our algorithm to obtain table cartograms inside arbitrary convex quadrangles, circles, and finally,
on the surface of cylinders and spheres.

1 Introduction

A *cartogram*, or *value-by-area diagram*, is a thematic cartographic visualization,
in which the areas of countries are modified in order to represent a given set
of values, such as population, gross-domestic product, or other geo-referenced
statistical data. Red-and-blue population cartograms of the United States were
often used to illustrate the results in the 2000 and 2004 presidential elections.
While geographically accurate maps seemed to show an overwhelming victory
for George W. Bush, population cartograms effectively communicated the near
50-50 split, by deflating the rural and suburban central states.

The challenge in creating a good cartogram is thus to shrink or grow the
regions in a map so that they faithfully reflect the set of pre-specified area values,
while still retaining their characteristic shapes, relative positions, and adjacencies
as much as possible. In this paper we introduce a new *table cartogram* model,
where the input is a two dimensional $m \times n$ table of non-negative weights, and the
output is a rectangle with area equal to the sum of the input weights partitioned
into $m \times n$ convex quadrilateral faces each with area equal to the corresponding
input weight. Fig. 1 shows two such examples. Such a visualization preserves
both area and adjacencies, furthermore, it is simple, visually attractive, and
applicable to many fields that require visualization of data table.

H.L. Bodlaender and G.F. Italiano (Eds.): ESA 2013, LNCS 8125, pp. 421–432, 2013.
© Springer-Verlag Berlin Heidelberg 2013

4.5	4.5	16	2.5
4	3	4.5	3
2.5	6	4.5	10.5
7	9	9	6

B (2.34)	C (2.26)	N(1.25)	O(1.45)
Al (2.70)	Si (2.33)	P (1.82)	S (2.07)
Ga (5.91)	Ge (5.32)	As (5.72)	Se(4.79)
In (7.31)	Sn (7.31)	Sb (6.68)	Te(4.93)

Fig. 1. A 4×4 table, its table cartogram, and a cartogram of some elements of the periodic table according to their density in grams per cubic centimeter (for solids) or per liter (for gases).

The solution to the problem is not obvious even for a 2×2 table. For example, Fig. 2(a) shows a table A. One attempt to find the cartogram of A in a unit square R may be to first split R horizontally according to the sum of each row, and then to find a good split in each subrectangle to realize the correct areas. But this approach does not work, because the first split prevents the creation of the two convex quadrilaterals with area ϵ in opposite corners that share a boundary vertex, Fig. 2(b). Fig. 2(c) shows a possible cartogram.

The following little argument shows that 2×2 table cartograms exist. The argument contains some elements that will be reused for the general case. The input is a 2×2 table with four positive reals a, b, c, d with $a + b + c + d = 1$, as shown in Fig. 2(d). Rotational symmetry of the problem allows us to assume that $a + b \leq 1/2$. Fix the unit square R with corners $(0,0), (0,1), (1,1), (1,0)$ as the frame for the table cartogram. Now consider the horizontal line ℓ with the property that every triangle $T(p)$ with top side equal to the top side of R and one corner p on ℓ has area $a + b$. Since $a + b \leq 1/2$, the line ℓ intersects R in a horizontal segment. For $p \in \ell \cap R$, the vertical line through p partitions $R \backslash T(p)$ into a left 4-gon S^- and a right 4-gon S^+. The areas of these two 4-gons depend continuously on the position of point p but their sum is always $c + d$. If p is on the left boundary, $Area\,(S^+) = c + d$, and if p is on the right boundary, $Area\,(S^+) = 0$. Hence, it follows from the intermediate value theorem that there is a position for p on $\ell \cap R$ such that $Area\,(S^-) = c$ and $Area\,(S^+) = d$. By

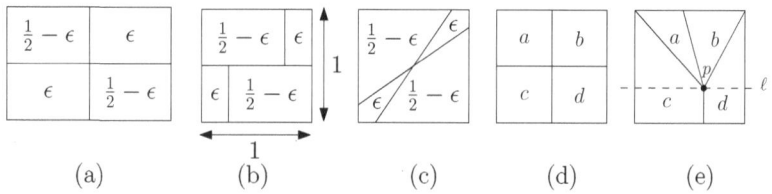

Fig. 2. (a) A 2×2 table A. (b) R. (c) An attempt to find a cartogram. (d) A cartogram of A in R. (e) A 2×2 table A. (f) The cartogram showing ℓ as a dashed line.

rotating a line around this p, we find a line that partitions $T(p)$ such that the left triangle has area a and thus the right triangle has area b, this again uses the intermediate value theorem. The resulting partition of R into four parts is a table cartogram for the input table, see Fig. 2(e). The critical reader may object that two of the 4-gons have a degenerate side. This can be avoided by perturbing the cartogram slightly to make a very short edge instead of a point. The result is an ε approximate cartogram without degeneracies. Another approach is to modify the construction rules so that degeneracies are avoided. We take this approach in Section 2 to show the existence of non-degenerate table cartograms in general.

Related Work. The problem of representing additional information on top of a geographic map dates back to the 19th century, and highly schematized rectangular cartograms can be found in the 1934 work of Raisz [1]. Recently, van Kreveld and Speckmann describe automated methods to produce rectangular cartograms [2]. With such rectangular cartograms it is not always possible to represent all adjacencies and areas accurately [3,2]. However, in many "simple" cases, such as France, Italy and the USA, rectangular cartograms and even table cartograms offer a practical and straightforward schematization, e.g., Fig. 3. *Grid maps* are a special case of single-level spatial treemaps: the input is a geographic map mapped onto a grid of equal-sized rectangles, in such a way as to preserve as well as possible the relative positions of the corresponding regions [4,5]. As we show, such maps can always be visualized as table cartograms.

Eppstein *et al.* studied area-universal rectangular layouts and characterized the class of rectangular layouts for which all area-assignments can be achieved with combinatorially equivalent layouts [6]. If the requirement that rectangles are used is relaxed to allow the use of rectilinear regions then de Berg *et al.* [7] showed that all adjacencies can be preserved and all areas can be realized with 40-sided regions. In a series of papers the polygon complexity that is sufficient to realize any rectilinear cartogram was decreased from 40 sides down to 8 sides [8], which is best possible due to an earlier lower bound [9].

More general cartograms without restrictions to rectangular or rectilinear shapes have also been studied. For example adjacencies can be preserved and

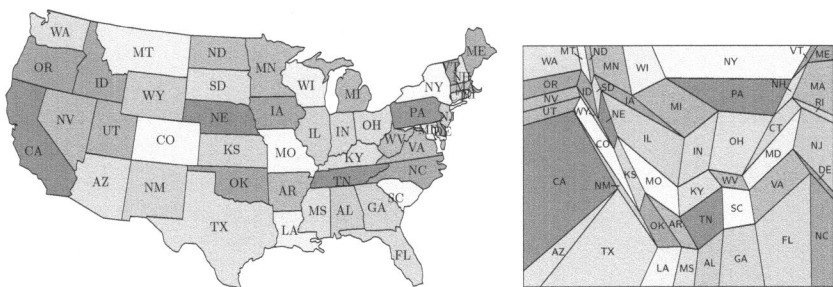

Fig. 3. A table cartogram of USA according to the population of the states in 2010, using the grid map of [5].

areas represented perfectly using convex quadrilaterals if the dual of the map is an outerplanar graph [10]. Dougenik *et al.* introduced a method based on force fields where the map is divided into cells and every cell has a force related to its data value which affects the other cells [11]. Dorling used a cellular automaton approach, where regions exchange cells until an equilibrium has been achieved, i.e., each region has attained the desired number of cells [12]. This technique can result in significant distortions, thereby reducing readability and recognizability. Keim *et al.* defined a distance between the original map and the cartogram with a metric based on Fourier transforms, and then used a scan-line algorithm to reposition the edges so as to optimize the metric [13]. Gastner and Newman [14] project the original map onto a distorted grid, calculated so that cell areas match the pre-defined values. The desired areas are then achieved via an iterative diffusion process inspired by physical intuition. The cartograms produced this way are mostly readable but the complexity of the polygons can increase significantly. Edelsbrunner and Waupotitsch [15] generated cartograms using a sequence of homeomorphic deformations. Kocmoud and House [16] described a technique that combines the cell-based approach of Dorling [12] with the homeomorphic deformations of Edelsbrunner and Waupotitsch [15].

There are many papers, spanning over a century, and covering various aspects of cartograms, from geography to geometry and from interactive visualization to graph theory and topology. The above brief review is woefully incomplete; the survey by Tobler [17] provides a more comprehensive overview.

Our Results. The main construction is presented in Section 2. We start with a simple constructive algorithm that realizes any table inside a rectangle in which each cell is represented by a convex quadrilateral with its prescribed weight. The approach relies on making many of the regions be triangles. We then modify the method to remove such degeneracies. The construction can be implemented to run in $O(mn)$ time, i.e., in time linear in the input size.

In Section 3 we find table cartograms inside arbitrary triangles or convex quadrilaterals, which is best possible, because regular n-gons, $n \geq 5$, do not always support table cartograms (e.g., consider a table with some cell value larger than the maximum-area convex quadrangle that can be drawn inside the n-gon). We also realize table cartograms inside circles, using circular-arcs, and on the surface of a sphere via a transformation from a realization on the cylinder.

2   Table Cartograms in Rectangles

We first construct a cartogram with degenerate 4-gons. The input is a table A with m rows and n columns of non-negative numbers $A_{i,j}$. Let $S = \sum_{i,j} A_{i,j}$ and let S_i be the sum of the numbers in row i, i.e., $S_i = \sum_{1 \leq j \leq n} A_{i,j}$. Assume, by scaling, that $S > 4$. Let R be the rectangle with corners $(0,0)$, $(S/2, 0)$, $(S/2, 2)$, $(0, 2)$. We construct the cartogram within R and later generalize the construction to all rectangles with area S. Let k be the largest index such that the sum of the numbers in rows $1, 2, \ldots, k-1$ is less than $S/2$. We may then

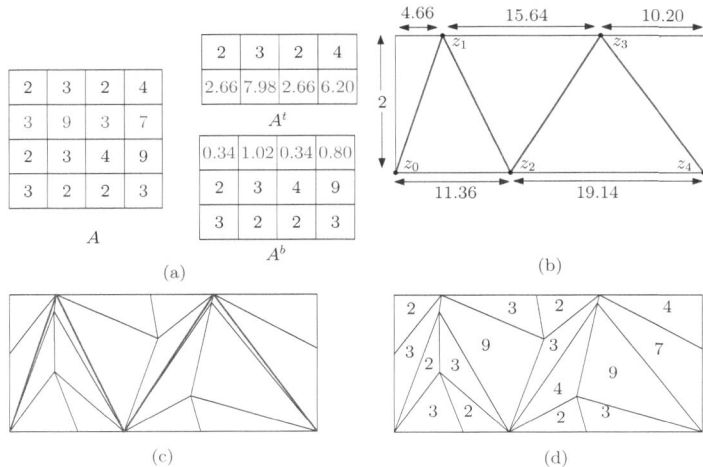

Fig. 4. (a) Illustration for A, A^t and A^b, where $k = 2$ and $\lambda \approx 0.886$. (b) The zigzag path Z. We have distorted the aspect ratio of the figure to increase readability. (c) The subdivision of triangles, where Z is shown in red, and (d) the complete cartogram.

choose $\lambda \in (0, 1]$ such that $\sum_{1 \leq i \leq k-1} S_i + \lambda S_k = S/2$. We split the table A into two tables A^t and A^b. Table A^t consists of k rows and n columns. The first $k - 1$ rows are taken from A, i.e., $A_{i,j}^t = A_{i,j}$ for $1 \leq i \leq k - 1$ and $1 \leq j \leq n$. The last row is a λ-fraction of row k from A, i.e., $A_{k,j}^t = \lambda \cdot A_{k,j}$ for all j. Table A^b consists of $m - k + 1$ rows and n columns. The first row accommodates the remaining portion of row k from A, i.e., $A_{1,j}^b = (1 - \lambda) \cdot A_{k,j}$. All the other rows are taken from A, i.e., $A_{i,j}^b = A_{i+k-1,j}$ for $i > 1$ and all j. An example is shown in Fig. 4(a). If $\lambda = 1$, then A^b contains a top row of zeros.

Let D_j^t be the sum of entries in columns $2j - 2$ and $2j - 1$ from A^t, where $1 \leq j \leq \lfloor m/2 + 1 \rfloor$. Note that D_1^t is only responsible for one column. The same may hold for the last D_j^t depending on the parity of m. Similarly, D_l^b is the sum of entries in columns $2l - 1$ and $2l$ from A^b, where $1 \leq l \leq \lceil m/2 \rceil$. Again, depending on the parity of m the last D_l^b may only be responsible for one column.

We now define a zig-zag Z in R (formally, Z is a polygonal line) such that the areas of the triangles defined by Z are the numbers $D_1^t, D_1^b, D_2^t, D_2^b, D_3^t, \ldots$ in this order. The zig-zag starts at $z_0 = (0, 0)$. Since the height of R is 2, the first segment ends at $z_1 = (D_1^t, 2)$ and the second segment goes down to $z_2 = (D_1^b, 0)$. In general, for i odd, $z_i = (\sum_{j=1}^{\lceil i/2 \rceil} D_j^t, 2)$ and for i even, $z_i = (\sum_{l=1}^{i/2} D_l^b, 0)$. An important property of Z is that it ends at one of the two corners on the right side of R. This is because $\sum_j D_j^t = S/2 = \sum_l D_l^b$.

Lemma 2 shows that we can partition each triangle created by the zig-zag Z into triangles whose areas are the corresponding entries in A^t or A^b. It relies on the following lemma which is a consequence of properties of barycentric coordinates. We omit the proof due to space constraints.

Lemma 1 (Triangle Lemma). *Let $\triangle abc$ be a triangle and let α, β, γ be non-negative numbers, where $\alpha + \beta + \gamma = Area(\triangle abc)$. Then we can find a point p in $\triangle abc$, where $Area(\triangle pbc) = \alpha$, $Area(\triangle apc) = \beta$, $Area(\triangle abp) = \gamma$, in $O(1)$ arithmetic operations.*

Lemma 2. *Let A be an $m \times 2$ table such that each cell is assigned a non-negative number. Let $\triangle abc$ be a triangle such that the area of $\triangle abc$ is equal to the sum of the numbers of A. Then A admits a cartogram inside $\triangle abc$ such that all cells of A are represented by triangles and the boundary between those triangles representing cells in the left column and those representing cells in the right column is a polygonal path connecting point a to some point on the segment bc.*

Proof. The proof is by induction on m. The case $m = 1$ is obvious. If $m > 1$ we define $\alpha = \sum_{1 \leq i \leq m-1} A_{i,1} + A_{i,2}$, $\beta = A_{m,1}$ and $\gamma = A_{m,2}$. Using Lemma 1 we find a point p in $\triangle abc$ that partitions the triangle into triangles of areas α, β and γ. We keep the triangles $\triangle apc$ and $\triangle abp$ as representatives for $A_{m,1}$ and $A_{m,2}$ and construct the cartogram for the first $m - 1$ rows of A in the triangle $\triangle pbc$ by induction. $\qquad\square$

To partition triangle $\triangle z_{2j-2}, z_{2j-3}, z_{2j-1}$, for $1 \leq j \leq \lfloor m/2 + 1 \rfloor$ (where $z_{-1} = (0,2)$ and $z_{m+1} = (S/2, 2)$ if needed), we appeal to Lemma 2 with A (in the lemma) being the two columns from A^t whose sum is D_j^t. To make Lemma 2 applicable to cases like D_1^t which represent only one column from A^t, we simply add a column of zeros to A. Similarly, we can partition triangle $\triangle z_{2l-1}, z_{2l-2}, z_{2l}$, for $1 \leq l \leq \lceil m/2 \rceil$.

This yields a table cartogram of the $(m + 1) \times n$ table A^+ that is obtained by stacking A^t on A^b. Note, however, that all triangles representing cells from the last row of A^t have a side that equals one of the edges of Z. Symmetrically, all triangles representing cells from the first row of A^b have a side on Z. Hence, by removing the edge of Z we glue two triangles of area $\lambda A_{k,j}$ and $(1 - \lambda)A_{k,j}$ into a 4-gon of area $A_{k,j}$. The 4-gons obtained by removing edges of Z are convex because they have crossing diagonals. This completes the construction.

To complete the proof of the following theorem, in which R is a $w \times h$ rectangle with area S, we scale the above cartogram by a factor of $h/2$ vertically and a factor of $2/h$ horizontally.

Theorem 1. *Let A be an $m \times n$ table of non-negative numbers $A_{i,j}$. Let R be a rectangle with width w, height h and area equal to the sum of the numbers of A. Then there exists a cartogram of A in R such that every face in the cartogram is convex. The construction requires $O(mn)$ arithmetic operations.*

Removing degeneracies. The construction of the proof of Theorem 1 creates faces of degenerate shape, i.e., some faces may not be perfect quadrangles, as shown in Fig. 4(d). We modify this construction to avoid the degeneracies. Of course we have to make a stronger assumption on the input: All entries $A_{i,j}$ of the table are strictly positive. The first part of the construction remains unaltered.

- Determine k and λ such that $\sum_{1 \leq i \leq k-1} S_i + \lambda S_k = S/2$.
- Define A^t and A^b and the two-column sums D_j^t and D_l^b for these tables.
- Compute the zig-zag in the rectangle R of height 2 and width $S/2$.

Let $z_0, z_1, \ldots, z_n$ be the corner points of the zig-zag Z. For i even we define $z_i' = z_i + (0, v)$ and for i odd $z_i' = z_i - (0, v)$, i.e., z_i' is obtained by shifting z_i vertically a distance of v into R. We will choose this positive value v to obey conditions (B1) and (B2) required by the construction (these conditions have been specified later). Let Z' be the zig-zag with corners $z_0', z_1', \ldots, z_n'$. The segment z_i', z_i is the *leg at* z_i'. The union of all the legs and Z' is the *skeleton* G' of a partition of R into 5-gons. We refer to the 5-gons with corners $z_{i-1}, z_{i+1}, z_{i+1}', z_i', z_{i-1}'$ as F_i. We abstain from introducing extra notation for the two 4-gons at the ends of Z' and just think of them as degenerate 5-gons.

Lemma 3. *A 5-gon in R with vertices $(x_1, 0), (x_3, 0), (x_3, v), (x_2, 2 - v), (x_1, v)$ has the same area $x_3 - x_1$ as the triangle with corners $(x_1, 0), (x_3, 0), (x_2, 2)$.*

Proof. First note that changing the value of x_2 (shear) preserves the area of the 5-gon and of the triangle. Hence we may assume that $x_2 = x_3$. Now let P be the parallelogram with corners $(x_1, 0), (x_1, v), (x_2, 2), (x_2, 2 - v)$. Both, the 5-gon and the triangle can be partitioned into the triangle $(x_1, 0), (x_2, 2 - v), (x_3, 0)$ and a triangle that makes a half of P.   □

Some of the 5-gons F_i may not be convex. However, concave corners can only be at z_{i+1}' or z_{i-1}'. To get rid of concave corners we deal with corners at $z_1', z_2', \ldots, z_{n-1}'$ in this order. At each z_i' we may slightly shift z_i' horizontally and bend the leg to rebalance the areas. This can be done so that the concave corner at z_i' is resolved. We then say that z_i' has been *convexified*. Fig. 5 shows an example of the process.

The vertex z_i' has a concave corner in at most one of F_{i-1} and F_{i+1} In the first case we move z_i' to the right in the second case we move z_i' to the left. By symmetry, we only detail the second case, i.e. z_i' has a concave corner in F_{i+1}.

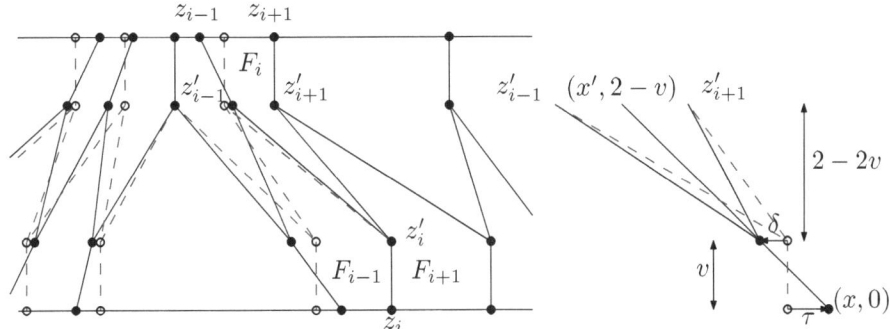

Fig. 5. Before and after convexifying z_i'. The dashed lines represent the original Z'.

Shifting z_i' horizontally keeps the area of F_i invariant, only the areas of F_{i-1} and F_{i+1} are affected by the shift. By shifting z_i' a distance of δ to the left while keeping z_i at its place the increase in area of F_{i+1} is $\delta(2-v)/2$. To balance the increase we move z_i, the other end of the leg, to the right by an amount τ, where $\tau v/2 = \delta(2-v)/2$. To make sure that the corners at z_i' after shifting are convex we choose δ and τ so that the line connecting the new positions of z_i and z_i' contains the midpoint of z_{i-1}' and z_{i+1}'. If the new position of z_i is $(x,0)$ and $(x',2-v)$ is the midpoint of z_{i-1}' and z_{i+1}' then $v/(\tau+\delta) = 2/(x-x')$.

We do not want the shift of z_i to introduce a crossing. We ensure this with a bound on v. For all j, let $T_j = Area(F_j)$ and this is the distance between z_{j-1} and z_{j+1} before shifting (since the height of the strip is 2). If $\tau \le T_{i+1}$, then the leg z_i', z_i does not intersect leg z_{i+2}', z_{i+2}. The absolute value of the slope of the leg z_i', z_i after convexifying z_i' is less than v/τ. The slope of the leg is also between the slopes of z_i', z_{i-1}' and z_i', z_{i+1}'. The absolute value of these slopes is larger than $(2-2v)/(S/2)$ which is the minimum possible slope of a segment of Z' in R. Define $T = \min_j T_j$. Hence, if $v/T < (2-2v)/(S/2) = 4(1-v)/S$, then $\tau < T$. We thus have an inequality that we want to be true for v:

$$v \le \frac{4T}{S+4T}. \tag{B1}$$

Observe that convexifying z_{i+1}' may require a shift of z_{i+1}' by δ' (and a compensating shift of z_{i+1} by τ') after z_i' has been convexified. However, if $v \le 1/4$, then balancing area and (B1) imply $\frac{1}{4}T > v\tau' = \delta'(2-v) \ge \delta'\frac{7}{4}$ whence $\delta' \le T/7$. This shows that z_{i+1}' stays on the right side of the old midpoint of z_{i-1}' and z_{i+1}' so that the corners at z_i' stay convex.

The next step of the construction is to place equidistant points on each of the legs. The segments between two consecutive points on the leg z_i', z_i will serve as sides for quadrangles of the quadrangular subdivision of F_{i-1} and F_{i+1}. Specifically, a leg z_i', z_i with i odd is subdivided into $k-1$ segments of equal length and a leg z_i', z_i with i even is subdivided into $m-k$ segments. Recall that k is the number of rows in A^t. For the partition of F_i into 4-gons with the prescribed areas we proceed inductively as in Lemma 2. We again need a partition lemma, whose proof is omitted due to space constraints.

Lemma 4. *Consider a convex 5-gon F as shown in Fig. 6(a). Let α, β, γ be positive numbers with $\alpha + \beta + \gamma = Area(F)$. If $\alpha > Area(\square p_0, q_0, q_j, p_j)$,*

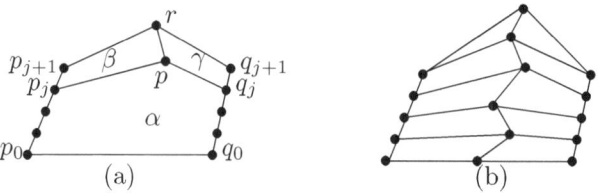

Fig. 6. (a) The α, β and γ partition of F. (b) A final partition of F_i.

$\beta > Area(\triangle p_j, r, p_{j+1})$, $\gamma > Area(\triangle q_j, q_{j+1}, r)$, then there exists $p \in F$ such that $\alpha = Area(\bigcirc p_0, q_0, q_j, p, p_j)$, $\beta = Area(\square p_j, p, r, p_{j+1})$, $\gamma = Area(\square q_j, q_{j+1}, r, p)$.

To ensure that the conditions for Lemma 4 are satisfied throughout the inductive partition of the regions F_i we need to bound v. Let $M = \min_{i,j} A_{i,j}$ be the minimum value in the table. Recall that $S/2$ is the width of R, and the y-distance of p_j and p_{j+1} is at most v. Hence, $vS/2$ is a generous upper bound on $Area(\triangle p_j, r, p_{j+1})$, $Area(\triangle q_j, q_{j+1}, r)$, and $Area(\square p_0, q_0, q_j, p_j)$. We ensure that these areas are less than β, γ, and α respectively by requiring

$$v < \frac{2M}{S} \tag{B2}$$

Theorem 2. *Let A be an $m \times n$ table of non-negative numbers $A_{i,j}$. Let R be a rectangle of width w and height h such that $w \cdot h = \sum_{i,j} A_{i,j}$. Then there exists a non-degenerate cartogram of A in R such that every face in the cartogram is convex. The construction requires $O(mn)$ arithmetic operations.*

Proof. The steps of the construction are:

- Construct the table cartogram with degeneracies.
- Compute the bounds and fix an appropriate value for v, compute the skeleton G' and its regions F_i, and convexify the legs in order of increasing index.
- Subdivide each of the regions F_i into convex 4-gons (and two triangles).
- Remove the edges of the zig-zag to get the cells of the middle row as unions of two triangles, which must generate convex quadrangles, since these triangles are contained in rectangle R and the common side of a pair of triangles connects opposite sides of R.

All can be done with $O(mn)$ arithmetic operations. Regarding the degeneracies, however, there is an issue that remains. To break A into A^t and A^b, we split row k so that the last row of A^t is a λ-fraction of row k from A while the rest of this row becomes the first row of A^b. Degeneracies occur if $\lambda = 1$. However, rather than splitting row k in this case, we can treat cells of row k as generic cells and assign a section of a leg to each of them. The construction is almost as before. Two details have to be changed. The first partition of each F_i into three pieces now produces two 4-gons and a 5-gon, before (see Fig. 6(b)) we had two triangles and a 5-gon in this step. The other change is that we don't remove zig-zag edges belonging to Z' to merge triangles to 4-gons at the end of the construction.   □

Instead of just knowing that there are no degeneracies, it would be nice to have a lower bound on the *feature size*, that is the minimum side-length of a 4-gon in the table cartogram. The segments subdividing the legs have length at least v/m. Because these leg segments have length at most v and $vS/2 < M$ (by (B2)), the opposite edges in a generic 4-gon (the blue edges in Fig. 6(b)) have length at least v. However, the triangles whose composition creates the 4-gons representing cells of row k can have area smaller than M. These triangles may have area $\hat{\lambda} M$ where $\hat{\lambda} = \min\{\lambda, 1 - \lambda\}$. This may lead to a very small feature size. To improve on this, another degree of freedom in the construction can be used. Instead of breaking each cell $A_{k,j}$ into a λ and a $1 - \lambda$ fraction, we can use individual values λ_j to define $A_{k,j}^t = \lambda_j A_{k,j}$. The choice of the values λ_j must satisfy two conditions: (1) $\sum_j \lambda_j A_{k,j} = \lambda S_k = \lambda \sum_j A_{k,j}$ and (2) if $\lambda_i = 0$ and $\lambda_j = 1$ then $|i - j| > 1$. By choosing most of the λ_j to be 0 or 1, and avoiding degeneracies, we may be able to have a substantial improvement in feature size.

3   Generalizations

We generalize the notion of "area" by specifying the weight of a region as an integral over some density function $w : R \to \mathbb{R}^+$. The density function should be *positive*, meaning that the integrals over triangular regions with nonempty interiors exist and are positive. The following generalizes Lemma 1 for any positive density function, allowing us to compute cartograms on weighted $\mathbb{R}^2$.

Lemma 5 (Weighted Triangle Lemma). *Let $\triangle abc$ be a triangle and $w :$ $\triangle abc \to \mathbb{R}^+$ be a positive density function on $\triangle abc$. Let $Area(\triangle abc)$ be the w-weighted area of the triangle $\triangle abc$. Given three non-negative real numbers α, β, γ, where $\alpha + \beta + \gamma = Area(\triangle abc)$, there exists a unique point p inside $\triangle abc$ such that $Area(\triangle pbc) = \alpha$, $Area(\triangle apc) = \beta$, and $Area(\triangle abp) = \gamma$.*

We now discuss some scenarios where the outerface of the cartogram has a more general shape. The following theorem considers the case when the outerface is a convex quadrangle $\square pqrs$. In such a case, we use a binary search to find the zigzag path that starts at q and ends at r or s.

Theorem 3. *Let A be an $m \times n$ table of non-negative numbers. Let $\square pqrs$ be an arbitrary convex quadrilateral with area equal to the sum, S, of the numbers of A. Then there exists a cartogram of A in $\square pqrs$ (with degeneracies).*

Next, we show how to compute a table cartogram inside a circle. A *circular triangle* $\bigcirc abc$ is a region in the plane bounded by three circular arcs (called *arms*) that pairwise meet at the points a, b, and c (called *vertices*), such that for every vertex $v \in \{a, b, c\}$ and for every point x on the arc that is not incident to v, one can draw a circular arc between v and x inside $\bigcirc abc$ that does not cross the boundary of $\bigcirc abc$. An arm is *convex* if the straight line joining any two

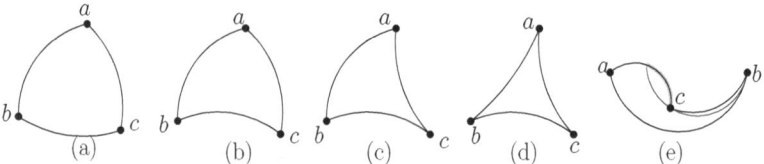

Fig. 7. (a–d) A circular triangle of Type i, $0 \le i \le 3$, i.e., a circular triangle with i concave arms. (e) A region bounded by circular arcs, but not a circular triangle.

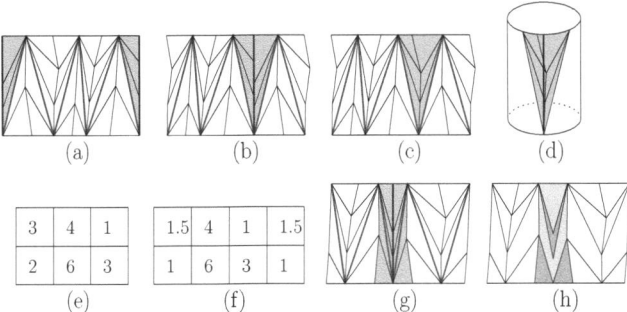

Fig. 8. Cylindrical cartogram construction. (a–d) n is even, (e–h) n is odd. Note that when n is even, the faces are convex quadrilaterals. However, when n is odd, the faces with areas from the leftmost column of A may be concave hexagons.

points on the arm is interior to the region bounded by the triangle. Otherwise, the arm is *concave*. We distinguish four types of circular triangles, see Figs. 7(a–d). We generalize Lemma 1 for circular triangles, i.e., given a circular triangle, one can split it into three other circular triangles with prescribed areas, and find the following generalization of Theorem 1.

Theorem 4. *Let A be an $m \times n$ table of non-negative numbers. Let C be an arbitrary circle with area equal to the sum of the numbers of A. Then there exists a cartogram of A in C where every face is a circular triangle.*

One can compute a cartogram of a table on the surface of a sphere using area preserving map projection techniques. For example, here we use Lambert's cylindrical equal-area projection. The construction used is shown in Fig. 8.

Theorem 5. *Every $m \times n$ table of non-negative numbers admits a cartogram on a sphere.*

4   Conclusions and Future Work

We have presented a simple constructive algorithm that realizes any table inside a rectangle in which each cell is represented by a convex quadrilateral with its prescribed weight. If all weights are strictly positive, then we can also obtain non-degenerate realizations. This method can be further extended to realize any table inside an arbitrary convex quadrilateral, inside a circle using circular arcs, or even on a sphere. From a practical point of view, the cartograms obtained by our method may not be visually pleasing, but by using additional straightforward heuristics that improve the visual quality while keeping the areas the same, we can obtain cartograms of practical relevance, as shown in Figs. 1 and 3. Our theoretical solution plays a vital role in this context, since heuristics used directly may get stuck, being unable to obtain the correct areas. Whether there exists a method that can gradually change the areas to provably obtain the correct areas remains an interesting open problem. It would also be interesting to examine table cartograms for other types of tables, such as triangular or hexagonal grids. From a theoretical point of view, finding algorithms for table cartograms on a sphere with less distortion, and generalizing our result to 3D table cartograms (inside a box) are further interesting open problems.

Acknowledgments. Initial work on this problem began at Dagstuhl Seminar 12261 "Putting Data on the Map" in June 2012 and most of the results of this paper were obtained at the Barbados Computational Geometry workshop in February 2013. We would like to thank the organizers of these events, as well as many participants for fruitful discussions and suggestions.

References

1. Raisz, E.: The rectangular statistical cartogram. Geographical Review 24(2), 292–296 (1934)
2. van Kreveld, M., Speckmann, B.: On rectangular cartograms. Comput. Geom. Theory Appl. 37(3), 175–187 (2007)
3. Heilmann, R., Keim, D.A., Panse, C., Sips, M.: Recmap: Rectangular map approximations. In: Proc. of InfoVis 2004, pp. 33–40 (2004)
4. Wood, J., Dykes, J.: Spatially ordered treemaps. IEEE Transactions on Visualization and Computer Graphics 14(6), 1348–1355 (2008)
5. Eppstein, D., van Kreveld, M., Speckmann, B., Staals, F.: Improved grid map layout by point set matching. In: Proc. of PacificVis 2013 (to appear, 2013)
6. Eppstein, D., Mumford, E., Speckmann, B., Verbeek, K.: Area-universal rectangular layouts. In: SoCG 2009, pp. 267–276. ACM (2009)
7. de Berg, M., Mumford, E., Speckmann, B.: On rectilinear duals for vertex-weighted plane graphs. Discrete Mathematics 309(7), 1794–1812 (2009)
8. Alam, M.J., Biedl, T., Felsner, S., Kaufmann, M., Kobourov, S.G., Ueckerdt, T.: Computing cartograms with optimal complexity. In: SoCG 2012, pp. 21–30 (2012)
9. Yeap, K.H., Sarrafzadeh, M.: Floor-planning by graph dualization: 2-concave rectilinear modules. SIAM J. Comput. 22(3), 500–526 (1993)
10. Alam, M.J., Biedl, T., Felsner, S., Kaufmann, M., Kobourov, S.G.: Proportional contact representations of planar graphs. Journal of Graph Algorithms and Applications 16(3), 701–728 (2012)
11. Dougenik, J.A., Chrisman, N.R., Niemeyer, D.R.: An algorithm to construct continuous area cartograms. The Professional Geographer 37(1), 75–81 (1985)
12. Dorling, D.: Area cartograms: their use and creation. Number 59 in Concepts and Techniques in Modern Geography. University of East Anglia (1996)
13. Keim, D.A., North, S.C., Panse, C.: Cartodraw: A fast algorithm for generating contiguous cartograms. IEEE Trans. Vis. Comput. Graph. 10(1), 95–110 (2004)
14. Gastner, M.T., Newman, M.E.J.: Diffusion-based method for producing density-equalizing maps. National Academy of Sciences 101(20), 7499–7504 (2004)
15. Edelsbrunner, H., Waupotitsch, R.: A combinatorial approach to cartograms. Computational Geometry: Theory and Applications 7(5-6), 343–360 (1997)
16. House, D.H., Kocmoud, C.J.: Continuous cartogram construction. In: Proc. of VIS 1998, pp. 197–204 (1998)
17. Tobler, W.: Thirty five years of computer cartograms. Annals Assoc. American Geographers 94(1), 58–73 (2004)

Network Bargaining with General Capacities

Linda Farczadi, Konstantinos Georgiou, and Jochen Könemann

University of Waterloo, Waterloo, Canada

Abstract. We study *balanced solutions* for *network bargaining games* with general capacities, where agents can participate in a fixed but arbitrary number of contracts. We provide the first polynomial time algorithm for computing balanced solutions for these games. In addition, we prove that an instance has a balanced solution if and only if it has a stable one. Our methods use a new idea of reducing an instance with general capacities to a network bargaining game with unit capacities defined on an auxiliary graph. This represents a departure from previous approaches, which rely on computing an allocation in the intersection of the *core* and *prekernel* of a corresponding *cooperative game*, and then proving that the solution corresponding to this allocation is balanced. In fact, we show that such cooperative game methods do not extend to general capacity games, since contrary to the case of unit capacities, there exist allocations in the intersection of the core and prekernel with no corresponding balanced solution. Finally, we identify two sufficient conditions under which the set of balanced solutions corresponds to the intersection of the *core* and *prekernel*, thereby extending the class of games for which this result was previously known.

1 Introduction

Exchanges in networks have been studied for a long time in both sociology and economics. In sociology, they appear under the name of *network exchange theory*, a field which studies the behaviour of agents who interact across a network to form bilateral relationships of mutual benefit. The goal is to determine how an agent's location in the network influences its ability to negotiate for resources [1992]. In economics, they are known as *cooperative games* and have been used for studying the distribution of resources across a network, for example in the case of two-sided markets [1971] [1984].

From a theoretical perspective the most commonly used framework for studying such exchanges is that of *network bargaining games*. The model consists of an undirected graph $G = (V, E)$ with edge weights $w : E(G) \to \mathrm{R}_+$ and vertex capacities $c : V(G) \to \mathrm{Z}_+$. The vertices represent the agents, and the edges represent possible pairwise contracts that the agents can form. The weight of each edge represents the value of the corresponding contract. If a contract is formed between two vertices, its value is divided between them, whereas if the contract is not formed neither vertex receives any profit from this specific contract. The capacity of each agent limits the number of contracts it can form. This constraint, together with an agent's position in the network determine its bargaining power.

H.L. Bodlaender and G.F. Italiano (Eds.): ESA 2013, LNCS 8125, pp. 433–444, 2013.

A *solution* for the network bargaining model specifies the set of contracts which are formed, and how each contract is divided. Specifically, a solution consists of a pair (M, z), where M is a c-matching of the underlying graph G, and z is a vector which assigns each edge uv two values $z_{uv}, z_{vu} \geq 0$ corresponding to the profit that agent u, respectively agent v, earn from the contract uv. To be a valid solution, the two values z_{uv} and z_{vu} must add up to the value of the contract whenever the edge uv belongs to the c-matching M, and must be zero otherwise.

Solutions to network bargaining games are classified according to two main concepts: *stability* and *balance*. A solution is stable if the profit an agent earns from any formed contract is at least as much as its outside option. An agent's *outside option*, in this context, refers to the maximum profit that the agent can rationally receive by forming a new contract with one of its neighbours, under the condition that the newly formed contract would benefit both parties. The notion of balance, first introduced in [1983], [1984], is a generalization of the Nash bargaining solution to the network setting. Specifically, in a balanced solution the value of each contract is split according to the following rule: both endpoints must earn their outside options, and any surplus is to be divided equally among them. Balanced solutions have been shown to agree with experimental evidence, even to the point of picking up on subtle differences in bargaining power among agents [1999]. This is an affirmation of the fact that these solutions are natural and represent an important area of study.

Our Contribution and Results. Our main result is providing the first polynomial time algorithm for computing balanced solutions for network bargaining games with general capacities and fully characterizating the existence of balanced solutions for these games. Specifically we show the following results in sections 4.3 and 4.4 respectively:

RESULT 1. *There exists a polynomial time algorithm which given an instance of a network bargaining game with general capacities and a maximum weight c-matching M, computes a balanced solution (M, z) whenever one exists.*

RESULT 2. *A network bargaining game with general capacities has a balanced solution if and only if it has a stable one.*

Our method relies on a new approach of reducing a general capacity instance to a network bargaining game with unit capacities defined on an auxiliary graph. This allows us to use existing algorithms for obtaining balanced solutions for unit capacities games, which we can then transform to balanced solutions of our original instance. This represents a departure from previous approaches of [2010] which relied on proving an equivalence between the set of balanced solutions and the intersection of the core and prekernel of the corresponding matching game. In section 3.1 we show that such an approach cannot work for our case, since this equivalence does not extend to all instances of general capacity games:

RESULT 3. *There exists an instance of a network bargaining game with general capacities for which we can find an allocation in the intersection of the core and prekernel such that there is no corresponding balanced solution for this allocation.*

Despite this result, we provide two necessary conditions which ensure that the correspondence between the set of balanced solutions and allocations in the intersection of the core and prekernel is maintained. Using the definition of *gadgets* from section 3.2 we have the following result given in section 3.3:

RESULT 4. *If the network bargaining game has no gadgets and the maximum c-matching M is acyclic, the set of balanced solutions corresponds to the intersection of the core and prekernel.*

Related Work. Kleinberg and Tardos [2008] studied network bargaining games with unit capacities and developed a polynomial time algorithm for computing the entire set of balanced solutions. They also show that such games have a balanced solution whenever they have a stable one and that a stable solution exists if and only if the linear program for the maximum weight matching of the underlying graph has an integral optimal solution.

Bateni et al. [2010] consider network bargaining games with unit capacities, as well as the special case of network bargaining games on bipartite graphs where one side of the partition has all unit capacities. They approach the problem of computing balanced solutions from the perspective of cooperative games. In particular they use the matching game of Shapley and Shubik [1971] and show that the set of stable solutions corresponds to the core, and the set of balanced solutions corresponds to the intersection of the core and prekernel.

Like we do here, Kanoria et al. [2009] also study network bargaining games with general capacities. They show that a stable solution exists for these games if and only the linear program for the maximum weight c-matching of the underlying graph has an integral optimal solution. They are also able to obtain a partial characterization of the existence of balanced solutions by proving that if this integral optimum is unique, then a balanced solution is guaranteed to exist. They provide an algorithm for computing balanced solutions in this case which uses local dynamics but whose running time is not polynomial.

2   Preliminaries and Definitions

An instance of the *network bargaining game* is a triple (G, w, c) where G is a undirected graph, $w \in \mathrm{R}_+^{|E(G)|}$ is a vector of edge weights, and $c \in \mathrm{Z}_+^{|V(G)|}$ is a vector of vertex capacities. A set of edges $M \subseteq E(G)$ is a c-matching of G if $|\{v : uv \in M\}| \leq c_u$ for all $u \in V(G)$. Given a c-matching M, we let d_u denote the degree of vertex u in M. We say that vertex u is *saturated* in M if $d_u = c_u$.

A *solution* to the network bargaining game (G, w, c) is a pair (M, z) where M is a c-matching of G and $z \in \mathrm{R}_+^{2|E(G)|}$ assigns each edge uv a pair of values z_{uv}, z_{vu} such that $z_{uv} + z_{vu} = w_{uv}$ if $uv \in M$ and $z_{uv} = z_{vu} = 0$ otherwise.

The *allocation* associated with the solution (M, z) is the vector $x \in \mathrm{R}^{|V(G)|}$ where x_u represents the total payoff of vertex u, that is for all $u \in V(G)$ we have $x_u = \sum_{v:uv \in M} z_{uv}$. The *outside option* of vertex u with respect to a solution (M, z) is defined as

$$\alpha_u(M, z) := \max\left(0, \max_{v:uv \in E(G)\setminus M}\left(w_{uv} - \mathbf{1}_{[d_v=c_v]}\min_{vw \in M} z_{vw}\right)\right),$$

where $\mathbf{1}_E$ is the indicator function for the event E, which takes value one whenever the event holds, and zero otherwise. If $\{v : uv \in E(G)\setminus M\} = \emptyset$ then we set $\alpha_u(M, z) = 0$. We write α_u instead of $\alpha_u(M, z)$ whenever the context is clear.

A solution (M, z) is *stable* if for all $uv \in M$ we have $z_{uv} \geq \alpha_u(M, z)$, and for all unsaturated vertices u we have $\alpha_u(M, z) = 0$.

A solution (M, z) is *balanced* if it is stable and in addition for all $uv \in M$ we have $z_{uv} - \alpha_u(M, z) = z_{vu} - \alpha_v(M, z)$.

2.1 Special Case: Unit Capacities

The definitions from the previous section simplify in the case where all vertices have unit capacities. Specifically a *solution* to the unit capacity game (G, w) is a pair (M, x) where M is now a matching of G and $x \in \mathbb{R}_+^{|V(G)|}$ assigns a value to each vertex such that for all edges $uv \in M$ we have $x_u + x_v = w_{uv}$ and for all $u \in V(G)$ not covered by M we have $x_u = 0$. Since each vertex has at most one unique contract, the vector x from the solution (M, x) is also the allocation vector in this case.

The *outside option* of vertex u can now be expressed as

$$\alpha_u(M, x) := \max_{v:uv \in E(G)\setminus M}(w_{uv} - x_v),$$

where as before we set $\alpha_u(M, x) = 0$ whenever $\{v : uv \in E(G)\setminus M\} = \emptyset$.

A solution (M, x) is *stable* if for all $u \in V(G)$ we have $x_u \geq \alpha_u(M, x)$ and *balanced* if it is stable and in addition $x_u - \alpha_u(M, x) = x_v - \alpha_v(M, x)$ for all $uv \in M$.

2.2 Cooperative Games

Given an instance (G, w, c) of the network bargaining game we let $N = V(G)$ and define the *value* $\nu(S)$ of a set of vertices $S \subseteq N$ as

$$\nu(S) := \max_{M \text{ } c\text{-matching of } G[S]} w(M).$$

Then the pair (N, v) denotes an instance of the *matching game* of Shapley and Shubik [1971]. We will refer to this as the matching game associated with the instance (G, w, c).

Given $x \in \mathbb{R}_+^{|N|}$ and two vertices $u, v \in V(G)$ we define the *power* of vertex u over vertex v with respect to the vector x as

$$s_{uv}(x) := \max_{S \subseteq N: u \in S, v \notin S} \nu(S) - x(S),$$

where $x(S) = \sum_{u \in S} x_u$. We write s_{uv} instead of $s_{uv}(x)$ whenever the context is clear. The *core* of the game is defined as the set

$$\mathcal{C} := \left\{ x \in R_+^{|N|} : x(S) \geq \nu(S), \forall S \subset N, x(N) = \nu(N) \right\}.$$

The *prekernel* of the game is the set

$$\mathcal{K} := \left\{ x \in R_+^{|N|} : s_{uv}(x) = s_{vu}(x) \quad \forall u, v \in N \right\}.$$

3 Balanced Solutions via Cooperative Games

The first attempt towards computing balanced solutions for the network bargaining game with general capacities is to use the connection to cooperative games presented in [2010]. For the special class of unit capacity and constrained bipartite games, Bateni et al. show that the set of stable solutions corresponds to the core, and the set of balanced solutions corresponds to the intersection of the core and prekernel of the associated matching game. This implies that efficient algorithms, such as the one of [1998], can be used to compute points in the intersection of the core and prekernel from which a balanced solution can be uniquely obtained.

3.1 Allocations in $\mathcal{C} \cap \mathcal{K}$ with No Corresponding Balanced Solutions

The first question of interest is whether this equivalence between balanced solutions and the intersection of the core and prekernel extends to network bargaining games with arbitrary capacities. The following lemma proves that this is not always the case.

Lemma 1. *There exists an instance (G, w, c) of the network bargaining game and a vector $x \in \mathcal{C} \cap \mathcal{K}$ such that there exists no balanced solution (M, z) satisfying*
$x_u = \sum_{v:uv \in M} z_{uv}$ *for all $u \in V(G)$.*

Proof. Consider the following graph where every vertex has capacity 2 and the edge weights are given above each edge

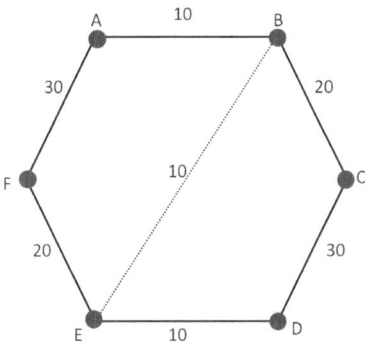

Consider the vector x defined as $x_u = 20$ for all $u \in V(G)$. We claim that the vector x is in the intersection of the core and prekernel and there exists no balanced solution (M, z) corresponding to x. The proof is deferred to the full version of this paper [2013].   □

In view of Lemma 1, we cannot hope to extend the correspondence between balanced solutions and allocations in the intersection of the core and prekernel to all network bargaining games. However we can generalize the results of [2010] by characterizing a larger class of network bargaining games, including unit capacity and constrained bipartite games, for which this correspondence holds. We achieve this by defining a certain gadget whose absence, together with the fact that the c-matching M is acyclic, will be sufficient for the correspondence to hold.

3.2   Gadgets

Let (G, w, c) be an instance of the network bargaining game and (M, z) a solution. Consider a vertex $u \in V(G)$ with $\alpha_u(M, z) > 0$ and let v be a neighbour of u in M. Let v' be vertex u's best outside option and if v' is saturated in M, let u' be its weakest contract. Using these definitions we have

$$\alpha_u(M, z) = w_{uv'} - \mathbf{1}_{[d_{v'} = c_{v'}]} z_{v'u'}.$$

We say that u is a *bad* vertex in the solution (M, z) if at least one of the following two conditions holds:

1. There is a $v - v'$ path in M,
2. There is a $u - u'$ path in M, that does not pass through vertex v'.

We refer to such $v - v'$ or $u - u'$ paths as *gadgets* of the solution (M, z). The following figure depicts these two types of gadgets, solid lines denote edges in M and dashed lines denote edges in $E \setminus M$.

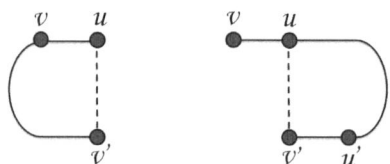

3.3   Sufficient Conditions for Correspondence between Set of Balanced Solutions and $\mathcal{C} \cap \mathcal{K}$

We can now state our main theorem of this section.

Theorem 1. *Let (G, w, c) be an instance of the network bargaining game. Let $x \in \mathcal{C}$ and (M, z) be a corresponding stable solution so that $x_u = \sum_{v:uv \in M} z_{uv}$ for all $u \in V(G)$. If the following two conditions are satisfied*

1. M is acyclic,
2. there are no bad vertices in the solution (M, z),

then, the following statement holds

$$(M, z) \text{ is a balanced solution if and only if } x \in \mathcal{K}.$$

Proof. Fix $uv \in M$. Note that it suffices to show $s_{uv} = -z_{uv} + \alpha_u$, since this would imply that $s_{uv} = s_{vu}$ if and only if $z_{uv} - \alpha_u = z_{vu} - \alpha_v$. Our strategy is to first show that s_{uv} is upper bounded $-z_{uv} + \alpha_u$, after which it will be sufficient to find a set T for which $\nu(T) - x(T)$ achieves this upper bound. We start with the following lemma whose proof is deferred to the full version [2013].

Lemma 2. $s_{uv} \leq -z_{uv} + \alpha_u$.

Hence it suffices to find a set $T \subseteq V(G)$ such that $u \in T$, $v \notin T$ and show that $\nu(T) - x(T) \geq -z_{uv} + \alpha_u$. Given a set of vertices S we let M_S denote the edges of M which have both endpoints in S. Note that for any set of vertices S we have

$$w(M_S) - x(S) = - \sum_{ab \in M: a \in S, b \notin S} z_{ab}. \tag{1}$$

We define $\mathcal{C}$ to be the set of components of G induced by the edges in M. Since u and v are neighbours in M they will be in the same component, call it C. Now suppose we remove the edge uv from C. Since M is acyclic, this disconnects C into two components C_u and C_v, containing vertices u and v respectively. Now M_{C_u} is a valid c-matching of C_u hence applying equation (1) to the vertex set the component C_u we obtain

$$\nu(C_u) - x(C_u) \geq w(M_{C_u}) - x(C_u) = -z_{uv}.$$

If $\alpha_u = 0$ then setting T to be the vertex set of component C_u completes the proof for this case. Hence it remains to consider the case where $\alpha_u > 0$. Then by stability of the solution (M, z) vertex u must be saturated in M. Let v' be vertex u's best outside option. We split the analysis into four cases. Figure 3.3 shows an example of the first two cases.

Case 1: $v' \in C_u$ and v' is not saturated in M. Since $uv \notin C_u$ and v' is not saturated in M the set of edges $M_{C_u} \cup \{uv'\}$ is a valid c-matching of C_u and therefore

$$\begin{aligned}
\nu(C_u) - x(C_u) &\geq w(M_{C_u} \cup \{uv'\}) - x(C_u) \\
&= w(M_{C_u}) - x(C_u) + w_{uv'} \\
&= -z_{uv} + w_{uv'} \qquad \text{by applying (1) to } C_u.
\end{aligned}$$

Case 2: $v' \in C_u$ and v' is saturated in M. Let u' be the weakest contract of v' and suppose we remove the edge $v'u'$ from C_u. Since M is acyclic, this disconnects C_u into two components. From condition (2) we know that u' is not on the $u - v'$ path in M. Hence u and v' are in the same component of $C_u \setminus \{v'u'\}$. Denote this component by D_u. Now $M_{D_u} \cup \{uv'\}$ is a c-matching of D_u and thus

$$
\begin{aligned}
\nu(D_u) - x(D_u) &\geq w\left(M_{D_u} \cup \{uv'\}\right) - x(D_u) \\
&= w(M_{D_u}) - x(D_u) + +w_{uv'} \\
&= -z_{uv} - z_{v'u''} + +w_{uv'} \qquad \text{by applying (1) to } D_u \\
&= -z_{uv} + \alpha_u \qquad\qquad\qquad \text{by choice of } v' \text{ and } u'.
\end{aligned}
$$

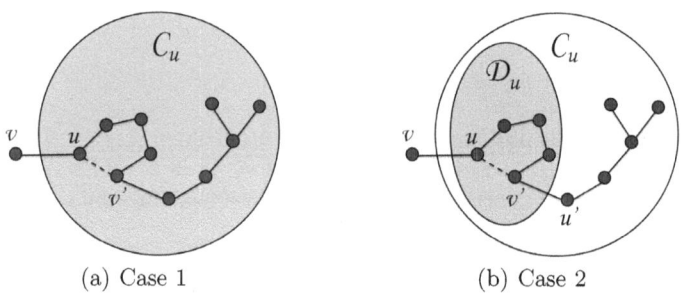

(a) Case 1   (b) Case 2

Fig. 1. Cases 1 and 2 in the proof Theorem 1

It remains to consider the following two cases whose proof is deferred to the full version [2013]:

Case 3: $v' \notin C_u$ and v' is not saturated in M.
Case 4: $v' \notin C_u$ and v' is saturated in M. □

We note that all network bargaining games studied in [2010] satisfy conditions (1) and (2) of Theorem 1. In addition to these, Theorem 1 also covers the case of network bargaining games where the underlying graph is a tree, but the vertices are allowed to have arbitrary capacities. Hence starting with a maximum weight c-matching M we can use the polynomial time algorithm of [1998] to compute a point in the intersection of the core and prekernel for these games, from which we can obtain a corresponding solution (M, z). Then using Theorem 1 we know that (M, z) will be balanced.

4   Balanced Solutions via Reduction to the Unit Capacity Games

While we were able to generalize the class of network bargaining games for which balanced solutions can be obtained by computing a point in the intersection of the core and prekernel, we were not able to apply this technique to all network bargaining games. In this section we show that balanced solutions can be obtained to any network bargaining game (G, w, c) by a reduction to a unit capacity game defined on an auxiliary graph.

4.1 Construction of the Instance (G', w') and Matching M'

Suppose we are given an instance (G, w, c) of the network bargaining game together with a c-matching M of G. We describe below how to obtain an instance (G', w') of the unit capacity game together with a matching M of G.

Construction: $[(G, w, c), M] \rightarrow [(G', w'), M']$

1. for each $u \in V(G)$: fix a labelling $\sigma_u : \{v : uv \in M\} \rightarrow \{1, \cdots, d(u)\}$ and create c_u copies $u_1, \cdots, u_{c_u}$ in $V(G')$.
2. for each uv in $E(G) \cap M$: add the edge $u_{\sigma_v(u)} v_{\sigma_u(v)}$ to $E(G') \cap M'$ and set its weight to w_{uv}.
3. for each edge $uv \in E(G) \setminus M$: add all edges $u_i v_j$ to $E(G')$ for all $i \in [c_u]$ and $j \in [c_v]$, and set all their weights to w_{uv}.

Example 1: Consider the instance depicted on the left hand side of the figure below. The solid edges are in the c-matching M and the dotted edges are in $E \setminus M$. Node u has capacity four, nodes x and y have capacity two and all other nodes have capacity one. All edges have unit weight.

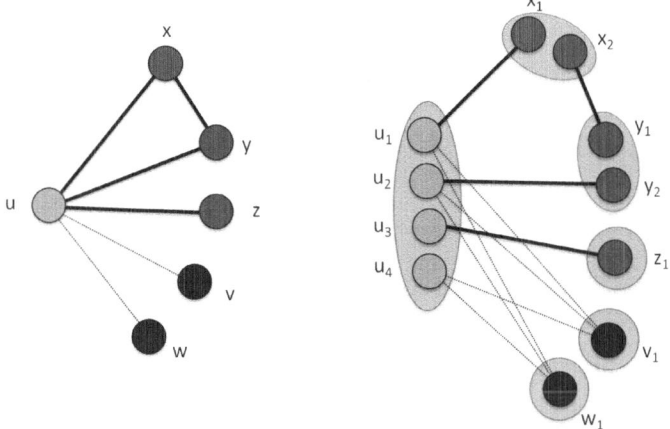

We make four copies of u in G', two copies of x and y, and one copy of every other node. Each edge in M corresponds to a unique edge in M'. For the edges uv and uw which are not in M, we connect every copy of u to every copy of v and w with edges in $E(G') \setminus M'$. The resulting graph is on the right.

4.2 Mapping between the Two Solution Sets

Suppose we are given an instance of the network bargaining game (G, w, c) with a c-matching M. Let $[(G', w'), M']$ be obtained using the construction given in section 4.1. Note that M and M' have the same number of edges and each edge $uv \in M$ is mapped to the unique edge $u_i v_j \in M'$ where $i = \sigma_v(u)$ and $j = \sigma_u(v)$.

This allows us to go back and forth between solutions on M and M' by dividing the weight of each edge in the same way as its corresponding pair.

We define the two solution sets:

$$\mathcal{X} := \left\{ x \in \mathrm{R}^{|V(G')|} : (M', x) \text{ is a solution to } (G', w') \right\}$$

$$\mathcal{Z} := \left\{ z \in \mathrm{R}^{2|E(G)|} : (M, z) \text{ is a solution to } (G, w, c) \right\}.$$

And the two mappings:

1. $\phi : \mathcal{X} \to \mathcal{Z}$
   For all $uv \in E$ define

$$(\phi(x)_{uv}, \phi(x)_{vu}) := \begin{cases} \left(x_{u_{\sigma_v(u)}}, x_{v_{\sigma_u(v)}} \right) & \text{if } uv \in M, \\ (0, 0) & \text{otherwise.} \end{cases}$$

2. $\phi^{-1} : \mathcal{Z} \to \mathcal{X}$.
   For all $u_i \in V(G')$ define

$$\phi^{-1}(z)_{u_i} := \begin{cases} z_{uv} & \text{if } i = \sigma_u(v), \\ 0 & \text{otherwise.} \end{cases}$$

Note that $z = \phi(x)$ if and only if $x = \phi^{-1}(z)$. The following lemma, whose proof is deferred to the full version [2013], shows that the mapping given by the function ϕ and its inverse ϕ^{-1} defines a bijection between the $\mathcal{X}$ and $\mathcal{Z}$

Lemma 3.   *1. If $x \in \mathcal{X}$ and $z = \phi(x)$, then $z \in \mathcal{Z}$.*
2. If $z \in \mathcal{Z}$ and $x = \phi^{-1}(z)$ then $x \in \mathcal{X}$.

From now on we write $(M, z) \sim (M', x)$ whenever $z = \phi(x)$ or equivalently $x = \phi^{-1}(z)$. The next lemma is the key step in showing that certain properties of a solution are preserved under our mapping.

Lemma 4. *Let (G, w, c) be an instance of the network bargaining game and M a c-matching on G. Suppose the auxiliary instance (G', w') and the matching M' were obtained using the construction given in section 4.1. Let (M, z) be a solution to (G, w, c) and (M', x') a solution to (G', w') such that $(M, z) \sim (M', x)$. Then for any $u \in V(G)$ and any $i \in [d_u]$ we have*

$$\alpha_u(M, z) = \alpha_{u_i}(M', x).$$

The proof is deferred to the full version [2013]. Using Lemma 4 we can now prove that stability and balance are preserved when mapping between solutions of the network bargaining game and the corresponding unit capacity game of the auxiliary instance.

Theorem 2. *Let (G, w, c) be an instance of the network bargaining game and M a c-matching on G. Suppose the auxiliary instance (G', w') and the matching M' were obtained using the construction given in section 4.1. Let (M, z) be a solution to (G, w, c) and (M', x') a solution to (G', w') such that $(M, z) \sim (M', x)$. Then:*

1. (M, z) is stable if and only if (M', x) is stable.
2. (M, z) is balanced if and only if (M', x) is balanced.

Proof. Let $uv \in M$. Suppose that $i = \sigma_v(u)$. Then $z_{uv} = x_{u_i}$ and using Lemma 4 we have

$$z_{uv} \geq \alpha_u(M, z) \quad \text{if and only if} \quad x_{u_i} \geq \alpha_{u_{\sigma_v(u)}}(M', x).$$

It remains to show that if (M', x) is stable then $\alpha_u(M, z) = 0$ for any unsaturated vertices u of G. Suppose u is such a vertex. Then the vertex u_{d_u+1} is not covered in M' and therefore $x'_{d_u+1} = 0$. If (M', x) is stable then $\alpha_{u_{d_u+1}} = 0$ and by Lemma 4 we have $\alpha_u(M, z) = 0$ as desired. This completes the proof of the first statement. To prove the second statement let $uv \in M$ and suppose that $i = \sigma_v(u)$ and $j = \sigma_u(v)$. Then $z_{uv} = x_{u_i}$, $z_{vu} = x_{v_j}$ and by Lemma 4 we have:

$$z_{uv} - \alpha_u(M, z) = z_{vu} - \alpha_v(M, z) \quad \Leftrightarrow \quad x_{u_i} - \alpha_{u_i}(M', x) = x_{v_j} - \alpha_{v_j}(M', x).$$

This completes the proof. □

4.3 Algorithm for Computing Balanced Solutions

Using Theorem 2 we have the following algorithm for finding a balanced solution to the network bargaining game (G, w, c):

1. Find a maximum c-matching M in G.
2. Obtain unit capacity game (G', w') with matching M' using the construction from section 4.1.
3. Find a balanced solution x on the matching M' in G'.
4. Set $z = \phi(x)$ and return (M, z).

We note that step 3 of the algorithm can be implementing using the existing polynomial time algorithm of Kleinberg and Tardos [2008]. Given any instance of a network bargaining game with unit capacities together with a maximum weight matching, their algorithm returns a balanced solution on the given matching, whenever one exists.

4.4 Existence of Balanced Solutions

Using Theorem 2 we know that stable solutions of the original problem map to stable solutions of the matching problem and viceversa. Since any stable solution must occur on a c-matching, respectively matching, of maximum weight we have the following corollary

Corollary 1. *Let (G, w, c) be an instance of the network bargaining game and M a c-matching on G. Suppose the auxiliary instance (G', w') and the matching M' were obtained using the given construction. Then*

1. *M is a maximum weight c-matching for (G, w, c) if and only if M' is a maximum weight matching for (G', w').*

2. *There exists a balanced solution for (G, w, c) on the c-matching M if and only if there exists a balanced solution for (G', w') on the matching M'.*

It was previously shown in [2008] that a unit capacity game possesses a balanced solution if and only if it has a stable solution, which in turn happens if and only if the linear program for the maximum weight matching of the underlying graph has an integral optimal solution. For the case of network bargaining game with general capacities, [2009] have shown that a stable solution exists if and only if the linear program for the maximum weight c-matching of the underlying graph has an integral optimal solution. In terms of existence of balanced solutions, they only obtain a partial characterization by proving that if this integral optimum is unique, then a balanced solution is guaranteed to exist. Our results imply the following full characterization for the existence of balanced solutions, thus extending the results of [2009]:

Theorem 3. *An instance (G, w, c) of the network bargaining game has a balanced solution if and only if it has a stable one.*

References

[2008]   Kleinberg, J., Tardos, É.: Balanced outcomes in social exchange networks. In: Proceedings of the 40th Annual ACM Symposium on Theory of Computing, pp. 295–304. ACM (2008)

[2010]   Azar, Y., Devanur, N., Jain, K., Rabani, Y.: Monotonicity in bargaining networks. In: Proceedings of the Twenty-First Annual ACM-SIAM Symposium on Discrete Algorithms. Society for Industrial and Applied Mathematics, pp. 817–826 (2010)

[2010]   Bateni, M., Hajiaghayi, M., Immorlica, N., Mahini, H.: The cooperative game theory foundations of network bargaining games. Automata, Languages and Programming, 67–78 (2010)

[1983]   Cook, K., Emerson, R., Gillmore, M., Yamagishi, T.: The distribution of power in exchange networks: Theory and experimental results. American Journal of Sociology, 275–305 (1983)

[1992]   Cook, K., Yamagishi, T.: Power in exchange networks: A power-dependence formulation. Social Networks 14(3), 245–265 (1992)

[1998]   Faigle, U., Kern, W., Kuipers, J.: An efficient algorithm for nucleolus and prekernel computation in some classes of tu-games (1998)

[2009]   Kanoria, Y., Bayati, M., Borgs, C., Chayes, J., Montanari, A.: A natural dynamics for bargaining on exchange networks. arXiv preprint arXiv:0911.1767 (2009)

[1950]   Nash Jr., J.: The bargaining problem. Econometrica: Journal of the Econometric Society, 155–162 (1950)

[1984]   Rochford, S.: Symmetrically pairwise-bargained allocations in an assignment market. Journal of Economic Theory 34(2), 262–281 (1984)

[1971]   Shapley, L., Shubik, M.: The assignment game i: The core. International Journal of Game Theory 1(1), 111–130 (1971)

[1999]   Willer, D.: Network exchange theory. Praeger Publishers (1999)

[2013]   Farczadi, L., Georgiou, K., Könemann, J.: Network bargaining with general capacities. arXiv preprint arXiv:1306.4302 (2013)

Nearly Optimal Private Convolution

Nadia Fawaz[1], S. Muthukrishnan[2], and Aleksandar Nikolov[2]

[1] Technicolor, Palo Alto, CA
[2] Rutgers University

Abstract. We study algorithms for computing the convolution of a private input x with a public input h, while satisfying the guarantees of (ε, δ)-differential privacy. Convolution is a fundamental operation, intimately related to Fourier Transforms. In our setting, the private input may represent a time series of sensitive events or a histogram of a database of confidential personal information. Convolution then captures important primitives including linear filtering, which is an essential tool in time series analysis, and aggregation queries on projections of the data. We give an algorithm for computing convolutions which satisfies (ε, δ)-differentially privacy and is nearly optimal *for every public h*, i.e. is instance optimal with respect to the public input. We prove optimality via spectral lower bounds on the *hereditary discrepancy* of convolution matrices. Our algorithm is very efficient – it is essentially no more computationally expensive than a Fast Fourier Transform.[1]

1 Introduction

Much useful data contains sensitive information about individuals (or the actions they take): typical examples are census data, data from medical studies, and financial data. While analyzing such sensitive datasets is valuable for scientific studies, policy and decision making, care must be taken to protect the privacy of the individuals represented in the data. Simple measures such as removing personally identifying attributes, replacing names with pseudonyms and publishing only aggregate statistics have proved inadequate protection from sophisticated linkage attacks [27,25,26]. An extreme solution would be to remove all sensitive information from the datasets, but this approach can destroy the utility of the data: a medical study without disease incidence rates would be useless, for example. In recent years *differential privacy* [10] has become a standard framework in which to reason about trade offs between privacy and utility, and this is the framework we adopt in this paper.

We study the *noise complexity* of a special class of queries. Consider a database representing users of N different types, or a time series of events that occurred over N time steps. We may encode the database as a vector $\mathbf{x}$ indexed by $\{1, \ldots, N\}$, where x_i gives the number of users of type i, in the database example, or x_i is the count of events that occurred at time step i. We say that two such vectors $\mathbf{x}$ and $\mathbf{x}'$ are *neighbors* when $\|\mathbf{x} - \mathbf{x}'\|_1 \leq 1$. Neighboring input vectors

---

[1] The full version of this paper can be found at http://arxiv.org/abs/1301.6447

H.L. Bodlaender and G.F. Italiano (Eds.): ESA 2013, LNCS 8125, pp. 445–456, 2013.
© Springer-Verlag Berlin Heidelberg 2013

correspond to databases that differ in at most a single user/event. Informally, an algorithm is differentially private if its output distribution is almost identical for neighboring inputs. More precisely, a randomized algorithm $\mathcal{A}$ satisfies (ε, δ)-differential privacy if for all neighbors $\mathbf{x}, \mathbf{x}' \in [0,1]^n$, and all measurable subsets T of the range of $\mathcal{A}$, we have

$$\Pr[\mathcal{A}(\mathbf{x}) \in T] \leq e^{\varepsilon} \Pr[\mathcal{A}(\mathbf{x}') \in T] + \delta,$$

where probabilities are taken over the randomness of $\mathcal{A}$.

In this work we are interested in *workloads* of M *linear queries*, given as a matrix $\mathbf{A}$; the intended output for the workload is $\mathbf{Ax}$. Differential privacy necessitates randomization and approximation for all non-trivial workloads; we discuss accuracy in terms of *mean squared error* (MSE) as a measure of approximation: the expected average of squared error over all M queries. The MSE achieved by an algorithm is the worst MSE the algorithm achieves on any input database.

The queries in a workload $\mathbf{A}$ can have different degrees of correlation, and this poses different challenges for the private approximation algorithm. In one extreme, when $\mathbf{A}$ is a set of $\Omega(N)$ independently sampled random $\{0, 1\}$ (i.e. counting) queries differentially private algorithm needs to incur at least $\Omega(N)$ squared error per query on average [9]. On the other hand, if $\mathbf{A}$ consists of the same counting query repeated M times, we only need to add $O(1)$ noise per query [10]. While these two extremes are well understood, relatively less is known about workloads of queries with some, but not perfect, correlation.

The *convolution*[2] $y = x * y$ of the private input sequence x with a public sequence h is defined as

$$y_i = \sum_{j=0}^{N-1} h_j x_{i-j \bmod N}.$$

If we view the input sequence as a vector $\mathbf{x}$, and define the circulant *convolution matrix* $\mathbf{H} = (h_{N+j-i \bmod N})_{i,j \in \{0,\dots,N-1\}}$, we see the convolution map is equivalent to computing the N linear queries $\mathbf{Hx}$. Each query is a circular shift of the previous one, and, therefore, the queries are far from independent but not identical either. Convolution is a fundamental operation that arises in algebraic computations such as polynomial multiplication, in signal analysis, and has well known connection to Fourier transforms. Of primary interest to us, it is a natural primitive in various applications:

- linear filters in the analysis of time series data can be cast as convolutions; as example applications, linear filtering can be used to isolate cycle components in time series data from spurious variations, and to compute time-decayed statistics of the data;
- when user type in the database is specified by d binary attributes, aggregate queries such as k-wise marginals and generalizations to other predicate queries can be represented as convolutions.

---

[2] Here we define circular convolution, but, as discussed in the paper, our results generalize to other types of convolution, which are defined similarly.

Privacy concerns arise naturally in these applications: the time series data can contain records of sensitive events, such as financial transactions, records of user activity, etc.; some of the attributes in a database can be sensitive, for example when dealing with databases of medical data.

We give the first (ε, δ)-differentially private algorithm which is nearly *query-optimal*: it achieves MSE which is not much smaller than the smallest MSE that any (ε, δ)-differentially private algorithm can achieve on the given convolution query.[3]

To prove the optimality of our algorithm, we need to prove *optimal lower bounds* on the noise complexity of private algorithms for computing convolutions. We use the recent discrepancy-based noise lower bounds of Muthukrishnan and Nikolov [24]. We use a characterization of combinatorial discrepancy in terms of determinants of submatrices discovered by Lovász, Spencer, and Vesztergombi [23], together with ideas by Hardt and Talwar [18]. A main technical ingredient in the proof of our lower bound is a connection between the discrepancy of a matrix $\mathbf{A}$ and the discrepancy of $\mathbf{PA}$ where $\mathbf{P}$ is an orthogonal projection operator.

Related Work. The problem of computing private convolutions has not been considered in the literature before. However, there is a fair amount of work on the more general problem of computing arbitrary linear queries, as well as some work on special cases of convolution maps.

Bolot et al. [4] give algorithms for various decayed sum queries: window sums, exponentially and polynomially decayed sums. Any decayed sum function is a type of linear filter, and, therefore, a special case of convolution. Thus, our current work gives a nearly optimal (ε, δ)-differentially private approximation for *any decayed sum function*. Moreover, as far as mean squared error is concerned, our algorithms give improved error bounds for the window sums problem: constant squared error per query. However, unlike [4], we only consider the offline batch-processing setting, as opposed to the online continual observation setting.

The work of Barak et al. [1] on computing k-wise marginals concerns a restricted class of convolutions (see Section 5). Moreover, Kasiviswanathan [19] show a noise lower bound for k-wise marginals which is tight in the worst case. Our work is a generalization: we are able to give nearly optimal approximations to a wider class of queries, and our lower and upper bounds nearly match for any convolution.

Li and Miklau [21,22] proposed the class of extended matrix mechanisms, building on prior work on the matrix mechanism [20], and showed how to efficiently compute the optimal mechanism from the class. Since our mechanism is a special instance of the extended matrix mechanism, the algorithms of Li and Miklau have at most as much error as our algorithm. They also derived a spectral lower bound [22] on the extended matrix mechanism; their results further imply that the spectral lower bound is tight *for the extended mechanism* for workloads corresponding to convolutions. However, unlike our lower bounds, this has no

---

[3] We note that while our algorithm is instance optimal with respect to queries, the measure of error we use is still worst-case over databases.

direct implication for private algorithms which are not an instantiation of the matrix mechanism.

Independently and concurrently with our work, Cormode et al. [8] considered adding optimal non-uniform noise to a fixed transform of the private database. Similarly to [8], we gain significantly in efficiency over the general extended matrix mechanism by fixing a specific transform (in our case the Fourier transform) of the data and computing a closed form expression for the optimal noise magnitudes. Our lower bounds show that, somewhat surprisingly, this simplification of Cormode et al. in fact comes without loss of generality for *any* set of convolution queries.

In the setting of $(\epsilon, 0)$-differential privacy, [18,2] prove nearly optimal upper and lower bounds on approximating $\mathbf{A}\mathbf{x}$ for any matrix $\mathbf{A}$. Prior to our work a similar result was not known for the weaker notion of approximate privacy, i.e. (ε, δ)-differential privacy. After a preliminary version of this paper was made available, our results were generalized by Nikolov, Talwar, and Zhang [28] to give nearly optimal algorithms for computing any linear map A under (ε, δ)-differential privacy. However, this comes at the cost of higher computational complexity: even the algorithm from [28], which is more efficient than the algorithms from [18,2], has running time $\Omega(N^3)$, as it needs to approximate the minimum enclosing ellipsoid of an N-dimensional convex body. By contrast our algorithm's running time is dominated by the running time of the Fast Fourier Transform, i.e. $O(N \log N)$, making it suitable for practical applications.

A related line of research exploits sparsity assumptions on the private database in order to reduce error [3,11,16,28]. Using techniques from learning theory, more efficient algorithms for sparse databases have been designed for the set of marginal queries [15,17,7,29,6]. As we do not limit the database size, our results are not directly comparable. Also, our lower bounds already hold when the database size (which in our notation corresponds to $\|\mathbf{x}\|_1$) is at least the number of linear queries, and in that regime our algorithm is nearly optimal, and cannot be significantly improved in terms of noise complexity. Finally, note that the optimal error for *a subset of all marginal queries* may be less than linear in database size, and our algorithms will give near optimal error for the *specific subset* of interest.

Recent work [17,7,29] on privately answering marginal queries has taken the approach of treating the database as a function from queries to the reals, and approximating this function by a small degree polynomial. This technique bears some resemblance to our approach for generalized marginals: we compute the Fourier transform of the database privately and spend most of the privacy budget on lower order Fourier coefficients, since they carry the most information.

Organization. We begin with preliminaries on differential privacy and convolution operators. In section 3 we derive our main lower bound result, and in section 4 we describe and analyze our nearly optimal algorithm. In section 5 we describe applications of our main results.

2   Preliminaries

Notation: $\mathbb{N}$, $\mathbb{R}$, and $\mathbb{C}$ are the sets of non-negative integers, real, and complex numbers respectively. By log we denote the logarithm in base 2 while by ln we denote the logarithm in base e. Matrices and vectors are represented by boldface upper and lower cases, respectively. $\mathbf{A}^T$, $\mathbf{A}^*$, $\mathbf{A}^H$ stand for the transpose, the conjugate and the transpose conjugate of $\mathbf{A}$, respectively. The trace and the determinant of $\mathbf{A}$ are respectively denoted by $\text{tr}(\mathbf{A})$ and $\det(\mathbf{A})$. $\mathbf{A}_{m:}$ denotes the m-th row of matrix $\mathbf{A}$, and $\mathbf{A}_{:n}$ its n-th column. $\mathbf{A}|_S$, where $\mathbf{A}$ is a matrix with N columns and $S \subseteq [N]$, denotes the submatrix of $\mathbf{A}$ consisting of those columns corresponding to elements of S. $\lambda_{\mathbf{A}}(1), \ldots, \lambda_{\mathbf{A}}(n)$ represent the eigenvalues of an $n \times n$ matrix $\mathbf{A}$. $\mathbf{I}_N$ is the identity matrix of size N. $\text{E}[\cdot]$ is the statistical expectation operator. $\text{Lap}(x, s)$ denotes the Laplace distribution centered at x with scale s, i.e. the distribution of the random variable $x + \eta$ where η has probability density function $p(y) \propto \exp(-|y|/s)$.

2.1   Fourier Eigen-Decomposition of Convolution

In this section, we recall the definition of the Fourier basis, and the eigen-decomposition of circular convolution in this basis.

Definition 1. *The normalized Discrete Fourier Transform (DFT) matrix of size N is defined as*

$$\mathbf{F}_N = \left(\frac{1}{\sqrt{N}} \exp\left(-\frac{j2\pi\, m\, n}{N} \right) \right)_{m,n \in \{0,\ldots,N-1\}}. \tag{1}$$

Note that $\mathbf{F}_N$ is symmetric ($\mathbf{F}_N = \mathbf{F}_N^T$) and unitary ($\mathbf{F}_N \mathbf{F}_N^H = \mathbf{F}_N^H \mathbf{F}_N = \mathbf{I}_N$).

We denote by $\mathbf{f}_m = N^{-1/2}(1, e^{\frac{j2\pi\, m}{N}}, \ldots, e^{\frac{j2\pi\, m\,(N-1)}{N}})^T \in \mathbb{C}^N$ the m-th column of the inverse DFT matrix $\mathbf{F}_N^H$. Or alternatively, $\mathbf{f}_m^H$ is the m-th row of $\mathbf{F}_N$. The normalized DFT of a vector $\mathbf{h}$ is simply given by $\hat{\mathbf{h}} = \mathbf{F}_N \mathbf{h}$.

Theorem 1 ([14]). *Any circulant matrix $\mathbf{H}$ can be diagonalized in the Fourier basis $\mathbf{F}_N$: the eigenvectors of $\mathbf{H}$ are given by the columns $(\mathbf{f}_m)_{m \in \{0,\ldots,N-1\}}$ of the inverse DFT matrix $\mathbf{F}_N^H$, and the associated eigenvalues $\{\lambda_m\}_{m \in \{0,\ldots,N-1\}}$ are given by $\sqrt{N}\hat{\mathbf{h}}$, i.e. by the DFT of the first column $\mathbf{h}$ of $\mathbf{H}$:*

$$\forall m \in \{0, \ldots, N-1\}, \quad \mathbf{H}\mathbf{f}_m = \lambda_m \mathbf{f}_m$$

$$\text{where} \quad \lambda_m = \sqrt{N}\hat{h}_m = \sum_{n=0}^{N-1} h_n e^{-\frac{j2\pi\, m\, n}{N}}.$$

Equivalently, in the Fourier domain, the circular convolution matrix $\mathbf{H}$ becomes a diagonal matrix $\hat{\mathbf{H}} = \text{diag}\{\sqrt{N}\hat{\mathbf{h}}\}$.

Corollary 1. *Consider the circular convolution $\mathbf{y} = \mathbf{H}\mathbf{x}$ of $\mathbf{x}$ and $\mathbf{y}$. Let $\hat{\mathbf{x}} = \mathbf{F}_N \mathbf{x}$ and $\hat{\mathbf{h}} = \mathbf{F}_N \mathbf{h}$ denote the normalized DFT of $\mathbf{x}$ and $\mathbf{h}$. In the Fourier domain, the circular convolution becomes a simple entry-wise multiplication of the components of $\sqrt{N}\hat{\mathbf{h}}$ with the components of $\hat{\mathbf{x}}$: $\hat{\mathbf{y}} = \mathbf{F}_N \mathbf{y} = \hat{\mathbf{H}}\hat{\mathbf{x}}$.*

2.2   Accuracy

Definition 2. *Given a vector* $\mathbf{h} \in \mathbb{R}^N$ *which defines a convolution matrix* $\mathbf{H}$, *the mean (expected) squared error (MSE) of an algorithm* $\mathcal{A}$ *is defined as*

$$\text{MSE} = \sup_{\mathbf{x} \in \mathbb{R}^N} \frac{1}{N} \mathrm{E}[\|\mathcal{A}(\mathbf{x}) - \mathbf{H}\mathbf{x}\|_2^2].$$

Note that MSE measures the mean expected squared error *per output component*. Note further that MSE is a function of both the algorithm and the public convolution matrix, but is defined to be worst-case over private inputs.

3   Lower Bounds

In this section we derive a spectral lower bound on mean squared error of differentially private approximation algorithms for circular convolution. We prove that this bound is nearly tight for every fixed $\mathbf{h}$ in the Section 4. The lower bound is stated as Theorem 2.

Theorem 2. *Let* $\mathbf{h} \in \mathbb{R}^N$ *be an arbitrary real vector and let us relabel the Fourier coefficients of* $\mathbf{h}$ *so that* $|\hat{h}_0| \geq \ldots \geq |\hat{h}_{N-1}|$. *For all sufficiently small* ε *and* δ, *the expected mean squared error MSE of any* (ε, δ)-*differentially private algorithm* $\mathcal{A}$ *that approximates* $\mathbf{h} * \mathbf{x}$ *is at least*

$$\text{MSE} = \Omega \left(\max_{K=1}^{N} \frac{K^2 \hat{h}_{K-1}^2}{N \log^2 N} \right). \tag{2}$$

For the remainder of the paper, we define the notation $\text{specLB}(\mathbf{h})$ for the right hand side of (2), i.e. $\text{specLB}(\mathbf{h}) = \max_{K=1}^{N} \frac{K^2 \hat{h}_{K-1}^2}{N \log^2 N}$.

3.1   Discrepancy Preliminaries

We define (ℓ_2) hereditary discrepancy as

$$\text{herdisc}(\mathbf{A}) = \max_{W \subseteq [N]} \min_{\mathbf{v} \in \{-1,+1\}^W} \|\mathbf{A}\mathbf{v}\|_2.$$

The following result connects discrepancy and differential privacy:

Theorem 3 ([24]). *Let* $\mathbf{A}$ *be an* $M \times N$ *complex matrix and let* $\mathcal{A}$ *be an* (ε, δ)-*differentially private algorithm for sufficiently small constant* ε *and* δ. *There exists a constant* C *and a vector* $\mathbf{x} \in \{0, 1\}^N$ *such that* $\mathrm{E}[\|\mathcal{A}(\mathbf{x}) - \mathbf{A}\mathbf{x}\|_2^2] \geq C \frac{\text{herdisc}(\mathbf{A})^2}{\log^2 N}$.

The determinant lower bound for hereditary discrepancy due to Lovász, Spencer, and Vesztergombi gives us a spectral lower bound on the noise required for privacy.

Theorem 4 ([23]). *There exists a constant C' such that for any complex $M \times N$ matrix $\mathbf{A}$, $\mathrm{herdisc}(\mathbf{A}) \geq C' \max_{K,\mathbf{B}} \sqrt{K} |\det(\mathbf{B})|^{1/K}$, where K ranges over $[\min\{M,N\}]$ and $\mathbf{B}$ ranges over $K \times K$ submatrices of $\mathbf{A}$.*

Corollary 1. *Let $\mathbf{A}$ be an $M \times N$ complex matrix and let $\mathcal{A}$ be an (ε, δ)-differentially private algorithm for sufficiently small constant ε and δ. There exists a constant C and a vector $\mathbf{x} \in \{0,1\}^N$ such that, for any $K \times K$ submatrix $\mathbf{B}$ of $\mathbf{A}$, $\mathrm{E}[\|\mathcal{A}(\mathbf{x}) - \mathbf{Ax}\|_2^2] \geq C \frac{K |\det(\mathbf{B})|^{2/K}}{\log^2 N}$.*

3.2   Proof of Theorem 2

We exploit the power of the determinant lower bound of Corollary 1 by combining the simple but very useful observation that projections do not increase mean squared error with a lower bound on the maximum determinant of a submatrices of a rectangular matrix. We present these two ingredients in sequence and finish the section with a proof of Theorem 2.

Lemma 1. *Let $\mathbf{A}$ be an $M \times N$ complex matrix and let $\mathcal{A}$ be an (ε, δ)-differentially private algorithm for sufficiently small constant ε and δ. There exists a constant C and a vector $\mathbf{x} \in \{0,1\}^N$ such that for any $L \times M$ projection matrix $\mathbf{P}$ and for any $K \times K$ submatrix $\mathbf{B}$ of $\mathbf{PA}$, $\mathrm{E}[\|\mathcal{A}(\mathbf{x}) - \mathbf{Ax}\|_2^2] \geq C \frac{K |\det(\mathbf{B})|^{2/K}}{\log^2 N}$.*

The proof of the lemma is based on the observation that $\mathcal{A}$ can be used to answer linear queries $\mathbf{Bx}$ by computing $\mathbf{y} = \mathcal{A}(x)$ and outputting (a subset of the coordinates of) $\mathbf{Px}$. The MSE of this new mechanism is no larger than the error of $\mathcal{A}$. Details can be found in the full version of the paper.

Our main technical tool is a linear algebraic fact connecting the determinant lower bound for $\mathbf{A}$ and the determinant lower bound for any projection of $\mathbf{A}$.

Lemma 2. *Let $\mathbf{A}$ be an $M \times N$ complex matrix with singular values $\lambda_1 \geq \ldots \geq \lambda_N$ and let $\mathbf{P}$ be a projection matrix onto the span of the left singular vectors corresponding to $\lambda_1, \ldots, \lambda_K$. There exists a constant C and $K \times K$ submatrix $\mathbf{B}$ of $\mathbf{PA}$ such that*

$$|\det(\mathbf{B})|^{1/K} \geq C \sqrt{\frac{K}{N}} \left(\prod_{i=1}^K \lambda_i \right)^{1/K}$$

Proof. Let $\mathbf{C} = \mathbf{PA}$ and consider the matrix $\mathbf{D} = \mathbf{CC}^H$. It has eigenvalues $\lambda_1^2, \ldots, \lambda_K^2$, and therefore $\det(\mathbf{D}) = \prod_{i=1}^K \lambda_i^2$. On the other hand, by the Binet-Cauchy formula for the determinant, we have

$$\det(\mathbf{D}) = \det(\mathbf{CC}^H) = \sum_{S \in \binom{[N]}{K}} \det(\mathbf{C}|_S)^2 \leq \binom{N}{K} \max_{S \in \binom{[N]}{K}} \det(\mathbf{C}|_S)^2.$$

Rearranging and raising to the power $1/2K$, we get that there exists a $K \times K$ submatrix of $\mathbf{C}$ such that $|\det(\mathbf{B})|^{1/K} \geq \binom{N}{K}^{-1/2K} \left(\prod_{i=1}^K \lambda_i \right)^{1/K}$. Using the bound $\binom{N}{K} \leq \left(\frac{Ne}{K} \right)^K$ completes the proof.

We can now prove our main lower bound theorem by combining Lemma 1 and Lemma 2.

Proof (of Theorem 2). As usual, we will express $\mathbf{h} * \mathbf{x}$ as the linear map $\mathbf{Hx}$, where $\mathbf{H}$ is the convolution matrix for $\mathbf{h}$. By Lemma 1, it suffices to show that for each K, there exists a projection matrix $\mathbf{P}$ and a $K \times K$ submatrix $\mathbf{B}$ of $\mathbf{PH}$ such that $|\det(B)|^{1/K} \geq \Omega(\sqrt{K}|\hat{h}_K|)$. Recall that the eigenvalues of $\mathbf{H}$ are $\sqrt{N}\hat{h}_0, \ldots, \sqrt{N}\hat{h}_{N-1}$, and, therefore, the i-th singular value of $\mathbf{H}$ is $\sqrt{N}|\hat{h}_{i-1}|$. By Lemma 2, there exists a constant C, a projection matrix P, and a submatrix $\mathbf{B}$ of $\mathbf{PH}$ such that

$$|\det(\mathbf{B})|^{1/K} \geq C\sqrt{\frac{K}{N}} \left(\prod_{i=0}^{K-1} \sqrt{N}|\hat{h}_i| \right)^{1/K} \geq C\sqrt{K}|\hat{h}_K|.$$

This completes the proof.

4 Upper Bounds

Next we describe an algorithm which is nearly optimal for (ε, δ)-differential privacy. This algorithm is derived by formulating the error of a natural class of private algorithms as a convex program and finding a closed form solution. The class of private algorithms we consider is those which add independent Gaussian noise to the Fourier coefficients of the private input $\mathbf{x}$. This is a special case of the extended matrix mechanism [21]; working with a less general algorithm is what allows us to derive a closed form for the optimal algorithm. At the same time, the error of our algorithm matches the lower bound on extended matrix mechanisms from [22].

Consider the class of algorithms, which first add independent Laplacian noise variables $z_i = \mathrm{Lap}(0, b_i)$ to the Fourier coefficients $\hat{x}_i$ to compute $\tilde{x}_i = \hat{x}_i + z_i$, and then output $\tilde{\mathbf{y}} = \mathbf{F}_N^H \hat{\mathbf{H}} \tilde{\mathbf{x}}$. This class of algorithms is parameterized by the vector $\mathbf{b} = (b_0, \ldots, b_{N-1})$; a member of the class will be denoted $\mathcal{A}(\mathbf{b})$ in the sequel. The question we address is: For given $\varepsilon, \delta > 0$, how should the noise parameters $\mathbf{b}$ be chosen such that the algorithm $\mathcal{A}(\mathbf{b})$ achieves (ε, δ)-differential privacy in $\mathbf{x}$ for ℓ_1 neighbors, while minimizing the mean squared error MSE? It turns out that by convex programming duality we can derive a closed form expression for the optimal $\mathbf{b}$, and moreover, the optimal $\mathcal{A}(\mathbf{b})$ is nearly optimal *among all (ε, δ)-differentially private algorithms*. The optimal parameters are used in Algorithm 1.

Theorem 5. *Algorithm 1 satisfies (ε, δ)-differential privacy, and achieves expected mean squared error*

$$\mathrm{MSE} = 4\frac{\ln(1/\delta)}{\varepsilon^2 N}\|\hat{\mathbf{h}}\|_1^2. \tag{3}$$

Moreover, Algorithm 1 runs in time $O(N \log N)$.

Algorithm 1. FOURIER MECHANISM

Set $\gamma = \frac{2\ln(1/\delta)\|\hat{\mathbf{h}}\|_1}{\varepsilon^2 N}$

Compute $\hat{\mathbf{x}} = \mathbf{F}_N\mathbf{x}$ and $\hat{\mathbf{h}} = \mathbf{F}_N\mathbf{x}$.

for all $i \in \{0,\ldots,N-1\}$ **do**

  if $|\hat{h}_i| > 0$ **then**

    Set $z_i = \mathrm{Lap}\left(\sqrt{\frac{\gamma}{|\hat{h}_i|}}\right)$

  else if $|\hat{h}_i| = 0$ **then**

    Set $z_i = 0$

  end if

  Set $\tilde{x}_i = \hat{x}_i + z_i$.

  Set $\bar{y}_i = \sqrt{N}\hat{h}_i\tilde{x}_i$.

end for

Output $\tilde{\mathbf{y}} = \mathbf{F}_N^H\bar{\mathbf{y}}$

The proof of Theorem 5 is omitted from the current version of the paper. Next, we show that it implies that Algorithm 1 is almost optimal *for any given* **h**.

Theorem 6. *For any* **h**, *Algorithm 1 satisfies* (ε, δ)-*differential privacy and achieves expected mean squared error* $O\left(\mathrm{specLB}(\mathbf{h})\frac{\log^2 N \log^2 |I| \ln(1/\delta)}{\varepsilon^2}\right)$.

Proof. Assume that $|\hat{h}_0| > |\hat{h}_1| > \ldots > |\hat{h}_{N-1}|$. Then, by definition of $I = \{0 \le i \le N-1 : |\hat{h}_i| > 0\}$, we have $|\hat{h}_j| = 0$, for all $j > |I| - 1$. Thus,

$$\|\hat{\mathbf{h}}\|_1 = \sum_{i=0}^{|I|-1} |\hat{h}_i| = \sum_{i=1}^{|I|} \frac{1}{i}i|\hat{h}_{i-1}| \le \left(\sum_{i=1}^{|I|} \frac{1}{i}\right)\sqrt{N}\log N\sqrt{\mathrm{specLB}(\mathbf{h})}$$

$$= H_{|I|}\sqrt{N}\log N\sqrt{\mathrm{specLB}(\mathbf{h})}, \qquad (4)$$

where $H_m = \sum_{i=1}^{m}\frac{1}{i}$ denotes the m-th harmonic number. Recalling that $H_m = O(\log m)$, and combining the bound (4) with the expression of the MSE (3) yields the desired bound.

5   Generalizations and Applications

In this section we describe some generalizations and applications of our lower bounds and algorithms for private convolution. Next we sketch applications to computing running sums, and linear filters motivated by analysis of time series data. Applications to computing compressible convolution maps, and computing generalized marginals on data cubes (which are an example of compressible convolutions) are described in the full version of the paper.

5.1   Running Sum

Running sums can be defined as the circular convolution $x' * h$ of the sequences $h = (1,\ldots,1,0,\ldots,0)$, where there are N ones and N zeros, and

$x' = (x, 0, \ldots, 0)$, where the private input x is padded with N zeros. An elementary computation reveals that $\hat{h}_1 = \sqrt{N}$ and $\hat{h}_i = O(N^{-1/2})$ for all $i > 1$. By Theorem 5, Algorithm 1 computes running sums with mean squared error $O(1)$ (ignoring dependence on ϵ and δ), improving on the bounds of [5,12,30] in the mean squared error regime.

5.2   Linear Filters in Time Series Analysis

Linear filtering is a fundamental tool in analysis of time-series data. A time series is modeled as a sequence $x = (x_t)_{t=-\infty}^{\infty}$, supported on a finite set of time steps. A filter converts the time series into another time series. A linear filter does so by computing the convolution of x with a series of *filter coefficients* w, i.e. computing $y_t = \sum_{i=-\infty}^{\infty} w_i x_{t-i}$. For a finitely supported x, y can be computed using circular convolution by restricting x to its support set and padding with zeros on both sides.

We consider the case where x is a time series of sensitive *events*. Each element x_i is a count of events or sum of values of individual transactions that have occurred at time step i. When we deal with values of transactions, we assume that individual transactions have much smaller value than the total. We emphasize that the definition of differential privacy with respect to x defined this way corresponds to *event-level privacy*.

We consider applications to financial analysis, but our methods are applicable to other instances of time series data, e.g. we may also consider network traffic logs or a time series of movie ratings on an online movie streaming service. We can perform almost optimal differentially private linear filtering by casting the filter as a circular convolution. For more references and detailed description, we refer the reader to the full version of our paper and the book of Gençan, Selçuk, and Whitcher [13].

6   Conclusion

We derive nearly tight upper and lower bounds on the error of (ε, δ)-differentially private algorithms for computing convolutions. Our lower bounds rely on recent general lower bounds based on discrepancy theory and elementary linear algebra; our upper bound is a simple computationally efficient algorithm. We also sketch several applications of private convolutions, in time series analysis and in computing generalizes marginal queries on a d-attribute database.

Since our algorithm for computing convolutions has running time $O(N \log N)$, we conjecture that there exists an $\tilde{O}(Nn)$ time algorithm for computing convolutions with optimal error when the database size is at most n. This would improve on the more general algorithm from [28], which has running time $O(M^2 Nn)$.

References

1. Barak, B., Chaudhuri, K., Dwork, C., Kale, S., McSherry, F., Talwar, K.: Privacy, accuracy, and consistency too: a holistic solution to contingency table release. In: Proceedings of the Twenty-Sixth ACM SIGMOD-SIGACT-SIGART Symposium on Principles of Database Systems, pp. 273–282. ACM (2007)

2. Bhaskara, A., Dadush, D., Krishnaswamy, R., Talwar, K.: Unconditional differentially private mechanisms for linear queries. In: Proceedings of the 44th Symposium on Theory of Computing, STOC 2012, pp. 1269–1284. ACM, New York (2012)

3. Blum, A., Ligett, K., Roth, A.: A learning theory approach to non-interactive database privacy. In: STOC 2008: Proceedings of the 40th Annual ACM Symposium on Theory of Computing, pp. 609–618. ACM, New York (2008)

4. Bolot, J., Fawaz, N., Muthukrishnan, S., Nikolov, A., Taft, N.: Private decayed sum estimation under continual observation. Arxiv preprint arXiv:1108.6123 (2011)

5. Hubert Chan, T.-H., Shi, E., Song, D.: Private and continual release of statistics. In: Abramsky, S., Gavoille, C., Kirchner, C., Meyer auf der Heide, F., Spirakis, P.G. (eds.) ICALP 2010. LNCS, vol. 6199, pp. 405–417. Springer, Heidelberg (2010)

6. Chandrasekaran, K., Thaler, J., Ullman, J., Wan, A.: Faster private release of marginals on small databases. arXiv preprint arXiv:1304.3754 (2013)

7. Cheraghchi, M., Klivans, A., Kothari, P., Lee, H.: Submodular functions are noise stable. In: Proceedings of the Twenty-Third Annual ACM-SIAM Symposium on Discrete Algorithms, pp. 1586–1592. SIAM (2012)

8. Cormode, G., Procopiuc, C.M., Srivastava, D., Yaroslavtsev, G.: Accurate and efficient private release of datacubes and contingency tables

9. Dinur, I., Nissim, K.: Revealing information while preserving privacy. In: Proceedings of the Twenty-Second ACM SIGMOD-SIGACT-SIGART Symposium on Principles of Database Systems, pp. 202–210. ACM (2003)

10. Dwork, C., McSherry, F., Nissim, K., Smith, A.: Calibrating noise to sensitivity in private data analysis. In: Halevi, S., Rabin, T. (eds.) TCC 2006. LNCS, vol. 3876, pp. 265–284. Springer, Heidelberg (2006)

11. Dwork, C., Naor, M., Reingold, O., Rothblum, G.N., Vadhan, S.: On the complexity of differentially private data release: efficient algorithms and hardness results. In: Proceedings of the 41st Annual ACM Symposium on Theory of Computing, pp. 381–390. ACM (2009)

12. Dwork, C., Pitassi, T., Naor, M., Rothblum, G.: Differential privacy under continual observation. In: STOC (2010)

13. Gençay, R., Selçuk, F., Whitcher, B.: An Introduction to Wavelets and Other Filtering Methods in Finance and Economics. Elsevier Academic Press (2002)

14. Gray, R.M.: Toeplitz and circulant matrices: a review. Foundations and Trends in Communications and Information Theory 2(3), 155–239 (2006)

15. Gupta, A., Hardt, M., Roth, A., Ullman, J.: Privately releasing conjunctions and the statistical query barrier. In: Proceedings of the 43rd Annual ACM Symposium on Theory of Computing, pp. 803–812. ACM (2011)

16. Hardt, M., Rothblum, G.: A multiplicative weights mechanism for privacy-preserving data analysis. In: Proc. 51st Foundations of Computer Science (FOCS). IEEE (2010)

17. Hardt, M., Rothblum, G., Servedio, R.: Private data release via learning thresholds. In: Proceedings of the Twenty-Third Annual ACM-SIAM Symposium on Discrete Algorithms, pp. 168–187. SIAM (2012)

18. Hardt, M., Talwar, K.: On the geometry of differential privacy. In: Proceedings of the 42nd ACM Symposium on Theory of Computing (2010)

19. Kasiviswanathan, S., Rudelson, M., Smith, A., Ullman, J.: The price of privately releasing contingency tables and the spectra of random matrices with correlated rows. In: Proceedings of the 42nd ACM Symposium on Theory of Computing, pp. 775–784. ACM (2010)

20. Li, C., Hay, M., Rastogi, V., Miklau, G., McGregor, A.: Optimizing linear counting queries under differential privacy. In: Proceedings of the Twenty-Ninth ACM SIGMOD-SIGACT-SIGART Symposium on Principles of Database Systems, PODS 2010, pp. 123–134. ACM, New York (2010)
21. Li, C., Miklau, G.: An adaptive mechanism for accurate query answering under differential privacy. PVLDB 5(6), 514–525 (2012)
22. Li, C., Miklau, G.: Measuring the achievable error of query sets under differential privacy. CoRR abs/1202.3399 (2012)
23. Lovász, L., Spencer, J., Vesztergombi, K.: Discrepancy of set-systems and matrices. European Journal of Combinatorics 7(2), 151–160 (1986)
24. Muthukrishnan, S., Nikolov, A.: Optimal private halfspace counting via discrepancy. In: Proceedings of the 44th ACM Symposium on Theory of Computing (2012)
25. Narayanan, A., Shi, E., Rubinstein, B.: Link prediction by de-anonymization: How we won the kaggle social network challenge. In: The 2011 International Joint Conference on Neural Networks (IJCNN), pp. 1825–1834. IEEE (2011)
26. Narayanan, A., Shmatikov, V.: Robust de-anonymization of large sparse datasets. In: IEEE Symposium on Security and Privacy, SP 2008, pp. 111–125. IEEE (2008)
27. Narayanan, A., Shmatikov, V.: De-anonymizing social networks. In: 2009 30th IEEE Symposium on Security and Privacy, pp. 173–187. IEEE (2009)
28. Nikolov, A., Talwar, K., Zhang, L.: The geometry of differential privacy: the sparse and approximate cases
29. Thaler, J., Ullman, J., Vadhan, S.: Faster algorithms for privately releasing marginals. In: Czumaj, A., Mehlhorn, K., Pitts, A., Wattenhofer, R. (eds.) ICALP 2012, Part I. LNCS, vol. 7391, pp. 810–821. Springer, Heidelberg (2012)
30. Xiao, X., Wang, G., Gehrke, J.: Differential privacy via wavelet transforms

Tractable Parameterizations for the Minimum Linear Arrangement Problem

Michael R. Fellows[1], Danny Hermelin[2],
Frances A. Rosamond[1], and Hadas Shachnai[3]

[1] School of Engineering and IT, Charles Darwin University, Darwin, Australia
{michael.fellows,frances.rosamond}@cdu.edu.au
[2] Department of Industrial Engineering and Management,
Ben-Gurion University, Beer-Sheva, Israel
hermelin@bgu.ac.il
[3] Computer Science Department, Technion, Haifa 32000, Israel
hadas@cs.technion.ac.il

Abstract. The MINIMUM LINEAR ARRANGEMENT (MLA) problem asks to embed a given graph on the integer line so that the sum of the edge lengths of the embedded graph is minimized. Most layout problems are either intractable, or not known to be tractable, parameterized by the *treewidth* of the input graphs. We investigate MLA with respect to three parameters that provide more structure than treewidth. In particular, we give a factor $(1+\varepsilon)$-approximation algorithm for MLA parameterized by (ε, k), where k is the *vertex cover number* of the input graph. By a similar approach, we describe two FPT algorithms that exactly solve MLA parameterized by, respectively, the *max leaf* and *edge clique cover* numbers of the input graph.

1 Introduction

Given a graph $G = (V, E)$, a *linear arrangement* is a linear ordering on the set of vertices V of G which is specified by a permutation $\pi : V \to \{1, 2, \ldots, |V|\}$. The *cost* of the arrangement is defined by $cost(\pi) = \sum_{(u,v) \in E} |\pi(u) - \pi(v)|$; that is, the cost of the ordering is the sum total of the edge lengths under the ordering. In typical applications one is interested in linear arrangements of low cost. The MINIMUM LINEAR ARRANGEMENT (MLA) problem is the problem of finding a linear arrangement of minimum cost, and the *standard parameterization* of the problem is to determine if an input graph G has a layout of cost at most k (the parameter).

MLA is one of the most important and well studied graph ordering problems. It is in some sense closely related to the BANDWIDTH problem, which seeks to minimize the maximum edge length of an ordering of the vertices. It has various applications, most of which stem from the domain of VLSI circuit design. As it was known to be NP-complete already from the mid 70's [12], most of the early work on this problem focused on designing heuristics and approximation algorithms. For a graph with n vertices, the best known approximation ratio for MLA

H.L. Bodlaender and G.F. Italiano (Eds.): ESA 2013, LNCS 8125, pp. 457–468, 2013.

is $O(\sqrt{\log n}\,\log\log n)$, due to Feige and Lee [5], and Charikar, Hajiaghayi, Karloff and Rao [4]. Earlier notable work includes the paper by Rao and Richa [18] who presented an $O(\log n)$-approximation algorithm for the problem, along with another algorithm for planar graphs that achieves a ratio of $O(\log\log n)$. Recently, Ambühl, Mastrolilli, and Svensson [1] showed that MLA does not admit a PTAS unless NP-complete problems can be solved in randomized subexponential time. We refer the reader to [18] for a further account of earlier work on MLA.

In terms of parameterized complexity, the standard parameterization of MLA is trivially *fixed-parameter tractable (FPT)*. Gutin *et al.* presented in [14] an FPT algorithm which, given a graph on n vertices and m edges, and a parameter k, outputs a linear arrangement of cost at most $m + k$, if one exists. Fernau in [11] described efficient bounded search tree FPT algorithms for MLA under its standard parameterization. Bodlaender *et al.* [3] investigated exact exponential-time algorithms for several ordering problems, including MLA.

Most parameterized problems are FPT parameterized by the treewidth of the input graph. However, graph layout and width problems are a notable exception (but see also [6] for further examples of parameterized problems that are $W[1]$-hard when parameterized by the input treewidth, or even the input vertex cover number). Parameterized by the treewidth of the input graph, BANDWIDTH is known to be hard for $W[t]$ for all t (this follows from the results in [2]). Whether something similar holds for MLA is unknown. This general situation motivates studying the complexity of these problems, parameterized by structural parameters even stronger than treewidth, a program that is sometimes called *parameter ecology*. See [8] for a recent survey of this area.

In this paper, we consider the complexity of MLA parameterized by three structural parameters that have a certain commonality: all three, when bounded, force the input graph G to have a structure that essentially consists of an elaboration of some small (parametrically bounded) "seed" graph H, that gives us sufficient information about the entire graph G to be able to derive efficient algorithms. The three graph structural parameters we consider are:

(*i*) The *vertex cover number* of G, denoted $vc(G)$, which is the size of the smallest set of vertices intersecting every edge in G.

(*ii*) The *maximum leaf number* of G, denoted $m\ell(G)$, which is the maximum number of leaves in any spanning tree of G.

(*iii*) The *edge clique cover number* of G, denoted $ecc(G)$, which is defined to be the minimum number of cliques required to cover all the edges of G.

The question of whether MLA is FPT by the vertex cover number of the input graph has been prominently raised in [10], where it is shown that a number of graph layout and width problems such as BANDWIDTH, CUTWIDTH and DISTORTION are FPT by this parameter, and this question has been a noted open problem in parameterized algorithmics. Here we offer a partial positive answer *by taking an approach that combines parameterization and approximation.* One of our main results shows that MLA can be approximated to within a factor of $(1 + \varepsilon)$ of optimal, in FPT time for the aggregate parameter $(1/\varepsilon, k)$, where k is the vertex cover number of the input graph. (Whether the MLA problem can be exactly solved in FPT time for the parameter k alone still remains open.) In [9] it is shown that BANDWIDTH is FPT, parameterized by the *max leaf number* of the input graph. Here we obtain a matching result for MLA; it can be exactly solved in FPT time for this parameter. While the techniques used in both results look similar from a bird's eye, there are several major differences hiding in the details. Finally, our last result shows that the *edge clique cover number* can potentially be a useful parameterization for other graph layout problems.

The paper is organized as follows. In Section 2 we give a $(1+\varepsilon)$-approximation for MLA parameterized by $vc(G)$. In Sections 3 and 4 we present FPT algorithms for MLA using the parameters $m\ell(G)$ and $ecc(G)$, respectively. We conclude in Section 5 with some open problems. Due to space constraints some of the proofs are omitted. The detailed results appear in [7].

2   MLA Parameterized by Vertex-Cover Number

In this section we present an algorithm which yields a $(1+\varepsilon)$-approximation for MLA in FPT-time with respect to $k = vc(G)$ and $1/\varepsilon$, where G is the input graph and $\varepsilon > 0$. Our algorithm proceeds as follows.

W.l.o.g., we assume that G has no isolated vertices, and let $m = |E|$. The first step of our algorithm is to compute a vertex cover $V' \subseteq V$ of G of size k; let $I = V \setminus V'$. Note that each vertex in I has neighbors only in V'. We define the type of node $u \in I$ to be its set of neighbors, $N(u)$. Clearly, there are $T \leq 2^k$ different types of vertices in I. Let n_t denote the number of vertices of type t, $1 \leq t \leq T$. The main idea of our algorithm is as follows. We put together vertices of the same type into groups of an appropriately chosen size, and then compute an optimal linear arrangement for the graph obtained by merging each group into a single *mega-vertex*. The analysis of our algorithm relies on an interesting homogeneity lemma, which relates to the behavior of vertices of identical type inside "gaps" formed by the vertices of V' in an optimal arrangement for G. A detailed description of the algorithm is given below (see Algorithm 1).

2.1   Analysis

We now prove that the algorithm yields a solution that is within factor $(1 + \varepsilon)$ from the optimal.

Theorem 1. *Algorithm 1 computes a $(1 + \varepsilon)$-approximate linear arrangement for G in FPT time with respect to k and $1/\varepsilon$.*

We use in the proof a few lemmas. Given a layout π for G, let $u_1, \ldots, u_k$ denote the vertices in V' as ordered in π. We say that a vertex $v \in I$ is in *gap* i, $1 \leq i \leq k-1$, if $\pi(u_i) < \pi(v) < \pi(u_{i+1})$. Similarly, v is in *gap 0* if $\pi(v) < \pi(u_1)$, and in *gap k* if $\pi(u_k) < \pi(v)$. The following lemma shows that the vertices of $V' \setminus V$ appear homogenously according to their type in some optimal linear arrangement of G.

Algorithm 1. MLA parameterized by $vc(G)$

Input: (G, ε, k).
Output: A linear ordering $\pi = (\pi(1), \ldots, \pi(n))$ for the vertices in G.

1: Set $s = \frac{\varepsilon m}{4k^2 (k+1) 2^k}$.
2: Apply an FPT algorithm to find a minimum vertex cover $V' \subseteq V$ of G. If $|V'| > k$ then STOP.
3: Partition the vertices in I into T types.
4: **for all** $1 \le t \le T$ **do**
5:    Partition the vertices of type t arbitrarily to groups (*mega-vertices*), where each group is of size s (except, maybe, for the last group).
6:    Set the neighbors of each mega-vertex to be the neighbors (in V') of a vertex of type t.
7: Let $G' = (V_M \cup V', E_M)$ be the graph formed by the mega-vertices, where V_M is the set of mega-vertices and E_M is the set of edges connecting V_M and V'.
8: **for all** linear arrangements π' of G' **do**
9:    Lift π' to a linear arrangement π of G, replacing each mega-vertex by the corresponding set of vertices in I; calculate the cost of π.
10: **return** the layout π found in Step 9 yielding the minimum cost.

Lemma 1 (Homogeneity for $vc(G)$). *There exists an optimal solution in which the vertices of each type appear in gap i consecutively, for all $1 \le i \le k-1$.*

Proof. Let $\pi : V \to \{1, \ldots, n\}$ denote an optimal linear arrangement of G, and let $\delta(v)$ denote the *force* of v (with respect to π) defined by

$$\delta(v) = |\{u \in N(v) \mid \pi(u) > \pi(v)\}| - |\{u \in N(v) \mid \pi(u) < \pi(v)\}|.$$

We note that for any gap i, $0 \le i \le k$, placing the vertices in I from left to right in non-decreasing order by $\delta(v)$ gives an optimal ordering within this gap. Indeed, if there exists a pair of vertices $u, v \in I$ which are adjacent according to π in gap i with $\delta(v) > \delta(u)$ and $\pi(v) < \pi(u)$, we can swap v and u and obtain a linear arrangement of smaller cost. It follows that all of the vertices v with equal force in gap i will be placed consecutively. We can thus place all of the vertices of type t as a contiguous block in gap i without harming the optimality of the arrangement, since all of these vertices have the same force in gap i. $\qquad\square$

Lemma 2. *The cost of any linear arrangement for G is at least $\frac{m^2}{4k^2}$.*

Proof. It is not difficult to see that a graph G with $vc(G) = k$ and m edges attaining the minimum possible cost of a linear arrangement is the disjoint union of k stars. Consider such a graph G^*, and let ℓ_r denote the number of edges in the r-th star of G^*, $1 \le r \le k$. Then the cost of a minimum linear arrangement of G^* is lower-bounded by

$$MLA(G^*) \ge \sum_{r=1}^{k} \sum_{d=1}^{\lfloor \ell_r / 2 \rfloor} 2d \ge \sum_{r=1}^{k} \frac{\ell_r^2}{4}.$$

The function $\sum_{r=1}^{k} \frac{\ell_r^2}{4}$ above is minimized when $\ell_r = \frac{m}{k}$ for all $1 \leq r \leq k$. Thus, $MLA(G^*) \geq \frac{m^2}{4k^2}$, and the lemma follows.   □

Lemma 3. *Let $MLA(G)$ denote the cost of the optimal arrangement for G, and let $MLA(G')$ denote the cost of the layout returned by Algorithm 1. Then $MLA(G') \leq MLA(G) + \frac{\varepsilon m^2}{4k^2}$.*

Proof. We note that Algorithm 1 considers only assignments of integral numbers of mega-vertices in each gap. However, it may be the case that any optimal ordering for G contains fractional assignments of mega-vertices in some of the gaps. Consider such an optimal ordering π_o for G. Let π_o^I be an *integral assignment* in which the number of mega-vertices of each type in any gap of π_o is rounded up/down to the next integral value. Thus, the total increase in the number of vertices in any gap is at most $s \cdot 2^k = \frac{\varepsilon m}{4k^2(k+1)}$. Let $length_o(e)$ be the cost incurred by $e \in E$ (or, the *length* of e) in π_o. Consider an iteration of Algorithm 1, in which it considers an integral assignment that corresponds to π_o. Then, $length_o(e)$ is stretched by the additional vertices in π_o^I, in each gap that e crosses, i.e., at most by $s \cdot 2^k \cdot (k+1) = \frac{\varepsilon m}{4k^2}$. Since Algorithm 1 outputs an integral assignment of minimum cost, we have the statement of the lemma.

Proof of Theorem 1: Observe that the running time of Algorithm 1 can be roughly bounded by the number of linear arrangements of G'. Note that $|E'| = \lceil m/s \rceil = O(2^k k^3)$ and $|V'| \leq 2|E'|$, as G' has no isolated vertices. Thus, the total running time of the algorithm can be bounded by $O^*((2^k k^3)!)$. Furthermore, by Lemmas 2 and 3, we have

$$\frac{MLA(G')}{MLA(G)} \leq 1 + \frac{\varepsilon m^2}{4k^2 \cdot MLA(G)} \leq 1 + \varepsilon,$$

where $MLA(G')$ denotes the minimum cost of the layout returned by the algorithm, and $MLA(G)$ denotes the optimal cost. Thus, Algorithm 1 returns a $(1+\varepsilon)$-approximate solution in FPT time with respect to (ε, k).   □

3   MLA Parameterized by Max-Leaf Number

In this section we give an FPT algorithm for MLA parameterized by $k = m\ell(G)$, for $k > 1$. We start with some definitions. Given a graph $H = (V', E')$, a *subdivision* of an edge $e' = (u, v) \in E'$ replaces e' by $E_{sub}(e') = \{(u, v_1)(v_1, v_2) \ldots (v_{n_{e'}}, v)\}$, for some $n_{e'} \geq 1$. Thus, the edge e' becomes an *edge-path*, and we add to H' $n_{e'}$ vertices. We say that a graph G is *a subdivision* of a graph H if G was generated by subdivision operations on the edges of H.

As shown in [16], if $m\ell(G) = k$ then G is a subdivision of a graph H on at most $4k - 2$ vertices. We call $H = (V', E')$, where $V' \subseteq V$ and $|V'| = k' \leq 4k - 2$, the *seed* graph of G. Let $S \subseteq E'$ denote the set of subdivided edges in H. Then, $V = V' \cup (\bigcup_{e' \in S} \{v_1, \ldots, v_{n_{e'}}\})$, and $E = E' \setminus S \cup (\bigcup_{e' \in S} E_{sub}(e'))$. We say that a vertex $w \in V \setminus V'$ belongs to e', where $e' = (u, v) \in E'$, if w is on the path $(u, v_1, \ldots v_{n_{e'}}, v)$ in G.

Our algorithm for MLA, parameterized by $m\ell(G)$, branches on an exhaustive set of solution patterns which has size bounded by a function of k. This set of patterns contains at least one optimal solution for the graph G, if one exists. Then, to determine when a solution pattern can be realized, we solve a linear integer program which outputs placements for all vertices of G in a permutation that is consistent with the given pattern. A detailed description is given in Algorithm 2.

Algorithm 2. MLA parametrized by $m\ell(G)$

Input: (G, k).
Output: A linear ordering $\pi = (\pi(1), \ldots, \pi(n))$ for the vertices in G.
1: Let $H = (V', E')$ be the seed graph of G, where $E' = \{e_1, \ldots, e_{m'}\}$
2: **for all** permutations $\sigma \in S_{k'}$ of the vertices in V' **do**
3:    **for all** configurations of E', $\bar{C} = (\bar{C}_{e_1}, \ldots, \bar{C}_{e_{m'}})$ **do**
4:        Solve an integer linear program to determine the position of vertices in $V \setminus V'$ among vertices in V', such that the total cost is minimized (see details below).

5: **return** a permutation σ of V' which yields minimum cost for G.

We define a permutation π of the vertices in G, by extending a permutation σ of the vertices in H. Given such a permutation σ, we now formulate an integer linear program to find the position of vertices in $V \setminus V'$ among the vertices of H. Recall that any permutation $\sigma = (u_1, \ldots, u_{k'})$ defines $(k' + 1)$ gaps. Let π be a permutation of V that is consistent with σ. Let $\pi(v) \in [n]$ denote the position of vertex $v \in V$ in π, where $n = |V|$. We say that a vertex $w \in V \setminus V'$ is placed in gap i in π, $1 \leq i \leq k' - 1$, if $\pi(u_i) < \pi(w) < \pi(u_{i+1})$, where $u_i, u_{i+1} \in V'$; w is placed in gap 0 (k') if $\pi(w) < \pi(u_1)$ $(\pi(w) > \pi(u_{k'}))$.

A *configuration for an edge-path* of $e' \in E'$ is a $(k' + 1)$-vector, $\bar{C}_{e'} = (c_{e',0}, \ldots, c_{e',k'})$, in which the i-th entry is equal to '1' if gap i contains a vertex in e', and '0' otherwise. Let $|E'| = m'$. Then, a *configuration for E'* is a set of m' configuration vectors for all edge-paths in E'. We use in the integer program the variables $x_{e',i}$ to indicate the number of vertices on the edge-path of $e' \in E'$ in the i-th gap, $0 \leq i \leq k'$. Let $\sigma \in S_{k'}$ be a permutation of the vertices in V'. We denote by $Cost(G|\sigma, \bar{x})$ the total cost of the linear arrangement of the vertices in G, with the given permutation σ and the variable values. This cost can be computed in FPT time, using Lemma 5 (see Section 3.1). Then our program can be formulated as $LP_{m\ell}$ below.

3.1   Analysis

We now prove that Algorithm 2 yields an optimal solution. Our analysis crucially relies on the next two lemmas.

$(LP_{m\ell}):$ minimize $Cost(G|\,\sigma, \bar{x})$

subject to: $\displaystyle\sum_{i=0}^{k'} x_{e',i} = n_{e'}$ $\forall\, e' \in E'$

$x_{e',i} = 0$ if $c_{e',i} = 0$

$x_{e',i} \in \{1, \ldots, n_{e'}\}$ if $c_{e',i} = 1$

Lemma 4. *Given a permutation $\sigma \in S_{k'}$ for V', and the number of vertices of edge-path e' in gap i, $x_{e',i}$, $0 \le i \le k'$, there exists an optimal order for V where each gap i for which $x_{e',i} > 0$ contains exactly one or two contiguous segments of e'.*

Lemma 5 (Homogeneity for $m\ell(G)$). *There exists an optimal layout in which vertices in $V \setminus V'$ that belong to the same edge-path appear consecutively in gap i, for all $0 \le i \le k'$.*

Proof. Given a permutation π of the vertices in G, we show below that if the ordering of vertices in gap i does not satisfy the contiguity property in the lemma then we can modify the order of vertices to satisfy the property, without increasing the total cost. Given the permutation π, we say that a vertex $v \in V \setminus V'$ is a *right-bend* of an edge-path if its two neighbors, u, w, satisfy $\pi(u) < \pi(v)$ and $\pi(w) < \pi(v)$, i.e., both u and w appear to the left of v in π. Similarly, v is a *left-bend* if its two neighbors appear to its right in π.

In proving the lemma, we consider pairs of vertices in gap i and show that we can swap the order of vertices in these pairs until we obtain a consecutive ordering of the vertices in each edge-path. Let $z, v \in V \setminus V'$ be a pair of vertices that appear consecutively in π. Suppose that z and v belong to two different edge-paths, e'_1 and e'_2, respectively. Also, assume that the two neighbors of z on e'_1 are $r, x \in V$, and the two neighbors of v on e'_2 are $u, w \in V$. We distinguish between four cases.

(i) Neither z nor v is a bend. Let π' be the permutation resulting from a swap of z and v in $\pi = (v_{i_1}, \ldots, x, \ldots, u, \ldots v, z, \ldots, w, \ldots r, \ldots, v_{i_n})$. Then, due to the swap, the distance between z and r (x) increases (decreases) by one; similarly, the distance between v and u (w) increases (decreases) by one. Thus, the overall change in the permutation cost is equal to zero.

(ii) Exactly one of z, v is a bend. W.l.o.g, assume that v is a right bend (the proof is similar for the case where v is a left bend). Suppose that $\pi(w) < \pi(u) < \pi(v)$, and assume that the vertex z precedes v in gap i, i.e., we have $\pi = (v_{i_1}, \ldots, w \ldots, u, \ldots, z, v, \ldots, v_{i_n})$. We show that swapping z and v does not increase the total cost. Let π' be the resulting permutation. Since z is not a bend, it has two neighbors x, r, such that $\pi(x) < \pi(z) < \pi(r)$.

Thus, after swapping z and v, z gets closer (by one) to r, while its distance from x increases by one. On the other hand, v gets closer to both u and w. Hence, overall, we have that $cost(\pi') - cost(\pi) < 0$.

(iii) Suppose that z, v are bends of the same direction. W.l.o.g., assume that both are right bends, and we have that $\pi(x) < \pi(r) < \pi(z)$, $\pi(w) < \pi(u) < \pi(v)$. More precisely, let $\pi = (v_{i_1}, \ldots, x \ldots, w, \ldots, r, \ldots, u, \ldots z, v, \ldots, v_{i_n})$. Then, it is easy to verify that $cost(\pi') - cost(\pi) = 0$.

(iv) Suppose that z, v are bends having opposite directions. W.l.o.g., assume that v is a right bend and z is a left bend. We further assume that $\pi(z) < \pi(r) < \pi(x)$, and $\pi(w) < \pi(u) < \pi(v)$, i.e., $\pi = (v_{i_1}, \ldots, w \ldots, u, \ldots, z, v, \ldots r, \ldots, x, \ldots, v_{i_n})$. Indeed, as a result of the swap, both v and z become closer to their neighbors. Hence, $cost(\pi') - cost(\pi) < 0$. $\qquad\square$

We can perform a sequence of the above swaps for pairs of vertices that belong to different edge-paths, until we have that all of the vertices on the same edge-path appear consecutively in gap i, without increasing the cost.

Lemma 6. *Given a permutation σ and the vector $\bar{x}$, $Cost(G| \sigma, \bar{x})$ is linear and can be computed in FPT time.*

Proof. We can write $Cost(G| \sigma, \bar{x}) = \sum_{i=0}^{k'} Cost(\text{gap } i|\sigma, \bar{x})$, where $Cost(\text{gap } i|\sigma, \bar{x})$ is the contribution of gap i to the total cost. Specifically, suppose that the two ends of gap i are $x, y \in V'$, where $\pi(x) < \pi(y)$. Let $(v_1, \ldots, v_h)$ be a segment of edge-path e' in gap i. W.l.o.g., we assume that v_1 is closer to x. Then we compute the contribution of this segment to the cost of gap i as follows. We compute the cost incurred by each internal edge on this segment and add to that a term that depends on the type of the segment. By Lemma 4, we may assume that the segment $(v_1, \ldots, v_h)$ is contiguous.

(a) If the segment is straight (i.e., has no bend in gap i), we add the distance from v_1 to x and from v_h to y. This term would then partially account for the cost incurred by the edges connecting the two ends of the segment, v_1 and v_h, to their neighbors in other gaps.

(b) If the segment has a right bend, we add the distance between v_1 and x plus the distance between v_h and x.

(c) If the segment has a left bend, we add the distance between v_1 and y and the distance between v_h and y.

Thus, it suffices to show that $Cost(\text{gap } i|\sigma, \bar{x})$ is linear and can be computed in FPT time. We now show how to order optimally the segments in gap i. Let B_ℓ, B_r and S denote the sets of segments of types *left-bend*, *right-bend* and *straight* in gap i, respectively. We denote by $z_i = \sum_{e' \in E'} x_{e',i}$ the number of vertices assigned to gap i.

Claim 2. *The cost incurred by any segment of type S in gap i is equal to z_i.*

Claim 3. *Given a set of segments in B_r in gap i, of lengths $x_{e_1,i} \leq x_{e_2,i} \leq \cdots \leq x_{e_r,i}$, the minimum total cost incurred by these segments in gap i is given* by $r \cdot x_{e_1,i} + (r-1)x_{e_2,i} + \cdots + x_{e_r,i}$. *Similarly, given a set of segments in* B_ℓ, *of lengths* $x_{e_1,i} \leq x_{e_2,i} \leq \cdots \leq x_{e_\ell,i}$, *the minimum cost incurred by these segments is* $\ell \cdot x_{e_1,i} + (\ell-1)x_{e_2,i} + \cdots + x_{e_\ell,i}$.

Thus, letting $|B_r| = r$ and $|B_\ell| = \ell$, by Claims 2 and 3 we have that the total cost of gap i is given by

$$Cost(\text{gap } i|\sigma, \bar{x}) = |S| \cdot \sum_{e' \in E'} x_{e',i} + \sum_{j=1}^{r}(r-j+1)x_{e_j,i} + \sum_{j=1}^{\ell}(\ell-j+1)x_{e_j,i}, \quad (1)$$

where $e_1, \ldots, e_r$ are the edge-paths having in gap i segments in B_r, and $e_1, \ldots, e_\ell$ are the edge-paths having in gap i segments in B_ℓ. For a vector $\bar{x}$ that is consistent with a given configuration of E', we have that the values of r and ℓ in (1) are fixed, thus $Cost(\text{gap } i|\sigma, \bar{x})$ is linear, and so is $Cost(G|\sigma, \bar{x})$. Hence, we can solve the integer program in FPT time (see, e.g., [17,15]).   □

We summarize in the next result.

Theorem 4. *There is an FPT algorithm for MLA parameterized by the maximum leaf number.*

Proof. Clearly, by Lemmas 4 and 5, our algorithm exhaustively searches over the set of solution patterns which contains an optimal one. It remains to show that the algorithm runs in FPT time. We note that the two outer loops of Algorithm 2 require $O(k'! \cdot 2^{k^3})$ iterations, each requires solving an integer linear program having $O(|E'| \cdot k') = O(k^3)$ variables. This gives the statement of the theorem. ∎

4   MLA Parameterized by Edge-Clique-Cover Number

In this section we show that MLA parameterized by the edge clique cover number of the input graph is in FPT. More precisely, we prove the following.

Theorem 5. *There is an* $O^*(2^k!)$-*time algorithm for MLA where* $k = eec(G)$ *is the edge clique cover number of the input graph* G.

We define the type of node $u \in V$ to be $N[u] = N(u) \cup \{u\}$. Note that our notion of type here is different from the one we use in Section 2, as vertices of the same type are necessarily adjacent. Nevertheless, the two different definitions are conceptually very similar, as we can prove a certain homogeneity lemma for this notion of type as well.

Lemma 7 (Homogeneity for $ecc(G)$**).** *There exists an optimal vertex ordering which is* homogenous. *That is, an ordering in which vertices of each type appear consecutively.*

Proof of Theorem 5 [assuming Lemma 7]: By Lemma 7, there exists an optimal solution in which vertices of each type appear consecutively. Observe that in any such homogenous ordering, the ordering of vertices of the same type can be arbitrary. That is, reordering vertices of a given type does not affect the

---

Algorithm 3. MLA parametrized by $ecc(G)$

Input: (G, k).
Output: A linear ordering π for the vertices of G.

1: Compute the type of each vertex of G.
2: Compute the cost of each homogenous ordering of types.
3: **return** a homogenous ordering with minimum cost.

---

total edge lengths of the ordering. Now, it is well known that a graph with edge clique cover number at most k has at most 2^k different types [13]. Thus, our algorithm searches through all $O(2^k!)$ homogenous vertex orderings and outputs the best one.   □

To prove Lemma 7, we introduce some additional notation. For an ordering $\pi = V \to \{1,\ldots,n\}$ and a pair of vertices $u, v \in V$ such that $\pi(u) < \pi(v)$, we let $\overleftarrow{\pi_{u,v}}$ and $\overrightarrow{\pi_{u,v}}$ denote the following permutations:

$$\overleftarrow{\pi_{u,v}}(x) = \begin{cases} \pi(x) & \text{if } \pi(x) \leq \pi(u) \text{ or } \pi(v) < \pi(x), \\ \pi(u) + 1 & \text{if } x = v, \\ \pi(x) + 1 & \text{if } \pi(u) < \pi(x) < \pi(v), \end{cases}$$

and

$$\overrightarrow{\pi_{u,v}}(x) = \begin{cases} \pi(x) & \text{if } \pi(x) < \pi(u) \text{ or } \pi(v) \leq \pi(x), \\ \pi(v) - 1 & \text{if } x = u, \\ \pi(x) - 1 & \text{if } \pi(u) < \pi(x) < \pi(v). \end{cases}$$

Thus, $\overleftarrow{\pi_{u,v}}$ is the vertex ordering obtained from π by placing v directly after u, and $\overrightarrow{\pi_{u,v}}$ is the ordering obtained from π by placing u directly before v.

Lemma 8. *Let $\pi = V \to \{1,\ldots,n\}$ be an optimal vertex ordering, and let u and v be two vertices of the same type with $\pi(u) < \pi(v)$. Then both $\overleftarrow{\pi_{u,v}}$ and $\overrightarrow{\pi_{u,v}}$ are optimal as well.*

Proof. Define three set of vertices $A = \{x \in V : \pi(x) < \pi(u)\}$, $B = \{x \in V : \pi(u) < x < \pi(v)\}$, and $C = \{x \in V : \pi(v) < \pi(x)\}$, and consider the two quantities $\overleftarrow{\Delta} = cost(\overleftarrow{\pi_{u,v}}) - cost(\pi)$ and $\overrightarrow{\Delta} = cost(\overrightarrow{\pi_{u,v}}) - cost(\pi)$. As π is optimal, both these quantities are non-negative, *i.e.* $\overleftarrow{\Delta} \geq 0$ and $\overrightarrow{\Delta} \geq 0$. If the lemma were false, at least one of these quantities would be strictly positive, *i.e.* we would either have $\overleftarrow{\Delta} > 0$ or $\overrightarrow{\Delta} > 0$. Aiming towards a contradiction let us assume that $\overleftarrow{\Delta} > 0$ (the proof in case $\overrightarrow{\Delta} > 0$ is symmetric).

Let us examine which edges contribute to $\overleftarrow{\Delta}$, *i.e.* which edges $\{x, y\} \in E$ have different lengths in $\overrightarrow{\pi_{u,v}}$ and π ($(|\overrightarrow{\pi_{u,v}}(x) - \overrightarrow{\pi_{u,v}}(y)| \neq |\pi(x) - \pi(y)|)$. Clearly, edges in $E(A, A)$, $E(B, B)$, $E(C, C)$ and $E(A, C)$ do note contribute to $\overleftarrow{\Delta}$. All other edge lengths are different in the two orderings. The length of each edge in $E(v, C)$ increases by $|B|$ in $\overleftarrow{\pi_{u,v}}$ when compared to its length in π, while the length of each edge in $E(v, A)$ decreases by $|B|$. Similarly, the length of each edge in $E(A, B)$ increases by 1, while the length of each edge in $E(B, C)$ decreases by 1.

What remains to account for are the edges involving u and v. Let $\ell(u, B) = \sum_{b \in B} |\pi(u) - \pi(b)|$ and $\ell(v, B) = \sum_{b \in B} |\pi(v) - \pi(b)|$ denote the total length of the edges of $E(u, B)$ and $E(v, B)$ with respect to π. Observe that as u and v are twins (and are thus adjacent to the same set of vertices in B), the total length of all edges of $E(v, B)$ becomes $\ell(u, B)$ in $\overleftarrow{\pi}_{u,v}$, while the length of each edge of $E(u, B)$ increases by 1. Thus, the total contribution of edges in $E(u, B) \cup E(v, B)$ to $\overleftarrow{\Delta}$ is $|E(u, B)| + \ell(u, B) - \ell(v, B)$. Finally, the last edge contributing to $\overleftarrow{\Delta}$ is the edge $\{u, v\}$ itself (which necessarily exists as u and v are from the same type) whose length decreases by $|B|$ in $\overleftarrow{\pi}_{u,v}$ when compared to π.

Summing up all these contributions, we get the following equality for $\overleftarrow{\Delta}$:

$$\overleftarrow{\Delta} = |E(v, C)| \cdot |B| + |E(A, B)| + \ell(u, B) + |E(u, B)|$$
$$- |E(v, A)| \cdot |B| - |E(B, C)| - \ell(v, B) - |B|. \tag{2}$$

Symmetrically, a similar calculation will give us the following equation for $\overrightarrow{\Delta}$:

$$\overrightarrow{\Delta} = |E(u, A)| \cdot |B| + |E(B, C)| + \ell(v, B) + |E(v, B)|$$
$$- |E(u, C)| \cdot |B| - |E(A, B)| - \ell(u, B) - |B|. \tag{3}$$

Now as u and v are twins, we know that $|E(u, A)| = |E(v, A)|$ and $|E(u, C)| = |E(v, C)|$. Furthermore, we also have $|B| \geq |E(u, B)| = |E(v, B)|$. Thus, using equations (2) and (3), and our assumption that $\overleftarrow{\Delta} > 0$ and $\overrightarrow{\Delta} \geq 0$, we get

$$|E(v, C)| \cdot |B| + |E(A, B)| + \ell(u, B) + |E(u, B)| >$$
$$|E(v, A)| \cdot |B| + |E(B, C)| + \ell(v, B) + |B| \geq$$
$$|E(u, A)| \cdot |B| + |E(B, C)| + \ell(v, B) + |E(v, B)| \geq$$
$$|E(u, C)| \cdot |B| + |E(A, B)| + \ell(u, B) + |B| \geq$$
$$|E(v, C)| \cdot |B| + |E(A, B)| + \ell(u, B) + |E(u, B)|$$

our desired contradiction.   □

Proof of Lemma 7: Let π be an optimal vertex ordering. If π is not homogenous we can use Lemma 8 repeatedly to transform it into one. The lemma thus follows.

5   Summary and Open Problems

We have shed light on the complexity of MLA for structural parameterizations stronger than treewidth, including a nice example of a successful combination of parameterization and approximation. We believe our algorithmic strategy in this may be applicable elsewhere, but can only so far show its applicability for the three parameters above. Can our results be extended to the aggregate parameterization based on treewidth? Furthermore, the aim of this paper was only at qualitative FPT results. We leave the best running times for such FPT algorithms as an open problem.

References

1. Ambühl, C., Mastrolilli, M., Svensson, O.: Inapproximability results for maximum edge biclique, minimum linear arrangement, and sparsest cut. SIAM J. Comput. 40(2), 567–596 (2011)
2. Bodlaender, H.L., Fellows, M.R., Hallett, M.T.: Beyond NP-completeness for problems of bounded width: hardness for the W hierarchy. In: Proceedings of the 26th Annual Symposium on the Theory of Computing (STOC), pp. 449–458 (1994)
3. Bodlaender, H.L., Fomin, F.V., Koster, A.M.C.A., Kratsch, D., Thilikos, D.M.: A note on exact algorithms for vertex ordering problems on graphs. Theory of Computing Systems 50(3), 420–432 (2012)
4. Charikar, M., Hajiaghayi, M.T., Karloff, H.J., Rao, S.: ℓ_2^2 spreading metrics for vertex ordering problems. In: SODA, pp. 1018–1027 (2006)
5. Feige, U., Lee, J.R.: An improved approximation ratio for the minimum linear arrangement problem. Inf. Process. Lett. 101(1), 26–29 (2007)
6. Fellows, M.R., Fomin, F.V., Lokshtanov, D., Rosamond, F.A., Saurabh, S., Szeider, S., Thomassen, C.: On the complexity of some colorful problems parameterized by treewidth. Information and Computation 209(2), 143–153 (2011)
7. Fellows, M.R., Hermelin, D., Rosamond, F., Shachnai, H.: Tractable parameterizations for the minimum linear arrangement problem, full version http://www.cs.technion.ac.il/~hadas/PUB/MLA_FHRS.pdf/
8. Fellows, M.R., Jansen, B.M.P., Rosamond, F.A.: Towards fully multivariate algorithmics: Parameter ecology and the deconstruction of computational complexity. European Journal of Combinatorics 34(3), 541–566 (2013)
9. Fellows, M.R., Lokshtanov, D., Misra, N., Mnich, M., Rosamond, F.A., Saurabh, S.: The complexity ecology of parameters: An illustration using bounded max leaf number. Theory Comput. Syst. 45(4), 822–848 (2009)
10. Fellows, M.R., Lokshtanov, D., Misra, N., Rosamond, F.A., Saurabh, S.: Graph layout problems parameterized by vertex cover. In: Hong, S.-H., Nagamochi, H., Fukunaga, T. (eds.) ISAAC 2008. LNCS, vol. 5369, pp. 294–305. Springer, Heidelberg (2008)
11. Fernau, H.: Parameterized algorithmics for linear arrangement problems. Discrete Applied Mathematics 156(17), 3166–3177 (2008)
12. Garey, M.R., Johnson, D.S.: Computers and intractability: a guide to the theory of NP-completeness. W. H. Freeman (1979)
13. Gramm, J., Guo, J., Hüffner, F., Niedermeier, R.: Data reduction and exact algorithms for clique cover. ACM Journal of Experimental Algorithmics 13 (2008)
14. Gutin, G., Rafiey, A., Szeider, S., Yeo, A.: The linear arrangement problem parameterized above guaranteed value. Theory Comput. Syst. 41(3), 521–538 (2007)
15. Kannan, R.: Minkowski's convex body theorem and integer programming. Mathematics of Operations Research 12, 415–440 (1987)
16. Kleitman, D.J., West, D.B.: Spanning trees with many leaves. SIAM J. Discrete Math. 4(1), 99–106 (1991)
17. Lenstra, H.: Integer programming with a fixed number of variables. Mathematics of Operations Research 8, 538–548 (1983)
18. Rao, S., Richa, A.W.: New approximation techniques for some linear ordering problems. SIAM J. Comput. 34(2), 388–404 (2004)

Compressed Cache-Oblivious String B-tree*

Paolo Ferragina and Rossano Venturini

Dipartimento di Informatica, Università di Pisa, Italy

Abstract. In this paper we address few variants of the well-known prefix-search problem in a dictionary of strings, and provide solutions for the cache-oblivious model which improve the best known results.

1 Introduction

The Prefix Search problem is probably the most well-known problem in data-structural design for strings. It asks for preprocessing a set $\mathcal{S}$ of K strings, having total length N, in such a way that, given a query-pattern P, it can return efficiently in time and space (the range of) all strings in $\mathcal{S}$ having P as a prefix. This easy-to-formalize problem is the backbone of many other algorithmic applications, and recently it received a revamped interest because of its Web-search (e.g., auto-completion search) and Internet-based (e.g., IP-lookup) applications.

In order to prove our surprising statement, and thus contextualize the contribution of this paper, we need to survey the main achievements in this topic and highlight their missing algorithmic points. The first solution to the prefix-search problem dates back to Fredkin [13], who introduced in 1960 the notion of (Compacted) trie to solve it. The trie structure became very famous in the '80s-'90s due to its suffix-based version, known as the Suffix Tree, which dominated first the String-matching scene [1], and then Bio-Informatics [15]. Starting from the Oxford English Dictionary initiative [12], and the subsequent advent of the Web, the design of tries managing large sets of strings became mandatory. It turned immediately clear that laying out a trie in a disk memory with page size B words, requiring optimal space and path traversals in $O(|P|/B)$ I/Os was not an easy task. And indeed Demaine *et al.* [6] showed that any layout of an arbitrary tree (and thus a trie) in external memory needs a poor number of I/Os to traverse a downward path spelling the pattern P.

The turning point in disk-efficient prefix-searching was the design of the String B-tree data structure [9], which was able to achieve $O(\log_B K + \mathsf{Scan}(P))$ I/Os, where $\mathsf{Scan}(P) = O(1 + \frac{|P|}{B \cdot \log N})$ indicates the number of I/Os needed to examine the input pattern, given that each disk page consists of B memory-words each of $\Theta(\log N)$ bits, and $|P|$ denotes the length of the binary representation of the pattern P. String B-trees provided I/O-efficient analogues of tries and

---

* This work was partially supported by MIUR of Italy under projects PRIN ARS Technomedia 2012 and FIRB Linguistica 2006, the Midas EU Project, and the eCloud EU Project.

H.L. Bodlaender and G.F. Italiano (Eds.): ESA 2013, LNCS 8125, pp. 469–480, 2013.

suffix trees, with the specialty of introducing some *redundancy* in the representation of the classic trie, which allowed the author to surpass the lower bounds
in [6]. The I/O-bound is optimal whenever the alphabet size is large and the
data structure is required to support the search for the lexicographic position of
P among the strings $\mathcal{S}$ too. The space usage is $O(K \log N + N)$ bits, which is
uncompressed, given that strings and pointers are stored explicitly. The string
B-tree is based upon a careful orchestration of a B-tree layout of string pointers, plus the use of a Patricia Trie [19] in each B-tree node which organizes its
strings in optimal space and supports prefix searches in $O(1)$ string accesses.
Additionally, the string B-tree is *dynamic* in that it allows the insertion/deletion
of individual strings from $\mathcal{S}$. As for B-trees, the data structure needs to know
B in advance so depending from this parameter crucially for its design. Brodal
and Fagerberg [4] made one step further by removing the dependance on B, and
thus designing a static *cache-oblivious* trie-like data structure [14]. Unlike string
B-trees, this structure is independent of B and does store, basically, a trie over
the indexed strings *plus* few paths which are *replicated multiple times*. This *redundancy* is the essential feature that gets around the lower bound in [6], and it
comes essentially at no additional asymptotic space-cost. Overall this solution
solves the prefix-search in $O(\log_B K + \mathsf{Scan}(P))$ I/Os by using $O(K \log N + N)$
bits of space, simultaneously over all values of B and thus cache-obliviously, as
currently said in the literature. In order to reduce the space-occupancy, Bender *et al.* [3] designed the (randomized) cache-oblivious string B-tree (shortly,
COSB). It achieves the improved space of $(1 + \epsilon)\mathsf{FC}(\mathcal{S}) + O(K \log N)$ bits, where
$\mathsf{FC}(\mathcal{S})$ is the space required by the Front-coded storage of the strings in $\mathcal{S}$ (see
Section 2), and ϵ is a user-defined parameter that controls the trade-off between
space occupancy and I/O-complexity of the query/update operations. COSB
supports searches in $O(\log_B K + (1 + \frac{1}{\epsilon})(\mathsf{Scan}(P) + \mathsf{Scan}(succ(P))))$ I/Os, where
$succ(P)$ is the successor of P in the ordered $\mathcal{S}$.[1] The solution is randomized
so I/O-bounds holds with high probability. Furthermore, observe that the term
$O((1 + \frac{1}{\epsilon})\mathsf{Scan}(succ(P)))$ may degenerate becoming $\Theta((1 + \frac{1}{\epsilon})\sqrt{N}/B)$ I/Os for
some sets of strings. Subsequently, Ferragina *et al.* [10] proposed an improved
cache-oblivious solution for the static-version of the problem regarding the space
occupancy. They showed that there exists a static data structure which takes
$(1 + \epsilon)\mathsf{LT}(\mathcal{S}) + O(K)$ bits, where $\mathsf{LT}(\mathcal{S})$ is a lower-bound to the storage complexity of $\mathcal{S}$. Searches can be supported in $O(\log_2 K + \mathsf{Scan}(P))$ I/Os or in
$O(\log_B K + (1 + \frac{1}{\epsilon})(\mathsf{Scan}(P) + \mathsf{Scan}(succ(P))))$ I/Os. Even if this solution is deterministic, its query complexity still has the costly dependency on $\mathsf{Scan}(succ(P))$.
For completeness, we notice that the literature proposes many other compressed
solutions but their searching algorithms are not suitable for the Cache-oblivious
model (see e.g., [11,21,17]).

Recently, Belazzougui *et al.* [2] introduced the *weak* variant of the problem
that allows for a *one-side* answer, namely the answer is requested to be correct

---

[1] This index can be also dynamized to support insertion and deletion of a string P
in $O(\log_B K + (\log^2 N)(1 + \frac{1}{\epsilon})\mathsf{Scan}(P))$ I/Os plus the cost of identifying P's rank
within $\mathcal{S}$.

only in the case that P prefixes some of the strings in $\mathcal{S}$; otherwise, it leaves to the algorithm the possibility to return a un-meaningful solution to the query. The *weak*-feature allowed the authors of [2] to reduce the space occupancy from $O(N)$, which occurs when strings are incompressible, to the surprisingly succinct bound of $O(K \log \frac{N}{K})$ bits which was indeed proved to be a space lower-bound, regardless of the query complexity. In the cache-oblivious model the query complexity of their solution is $O(\log_2 |P| + \mathsf{Scan}(P))$ I/Os. Their key contribution was to propose a solution which do not store the string set $\mathcal{S}$ but uses only $O(\log \frac{N}{K})$ bits per string. This improvement is significant for very-large string sets, and we refer the reader to [2] for a discussion on its applications. Subsequently, Ferragina [8] proposed a very simple (randomized) solution for the weak-prefix search problem which matches the best known solutions for the prefix-search and the weak prefix-search, by obtaining $O(\log_B N + \mathsf{Scan}(P))$ I/Os within $O(K \log \frac{N}{K})$ bits of space occupancy. The searching algorithm is randomized, and thus its answer is correct with high probability.

In this paper we attack three problems of increasing sophistication by posing ourselves the challenging question: *how much redundancy we have to add to the classic trie space occupancy in order to achieve $O(\log_B K + \mathsf{Scan}(P))$ I/Os in the supported search operations.*[2]

Weak-Prefix Search. Returns the (lexicographic) range of strings prefixed by P, or an arbitrary value whenever such strings do not exist.

Full-Prefix Search. Returns the (lexicographic) range of strings prefixed by P, or $\perp$ if such strings do not exist.

Longest-Prefix Search. Returns the (lexicographic) range of strings sharing the longest common prefix with P.

We get the above I/O-bound for Weak-Prefix Search Problem, for the other problems we achieve $O(\log_B K + (1 + \frac{1}{\epsilon})\mathsf{Scan}(P))$ I/Os, for any constant $\epsilon > 0$. The space complexities are asymptotically optimal, in that they match the space lower bound for tries up to constant factors. The query complexity is optimal for the latter problem. This means that for Weak-Prefix Search Problem we improve [8] via a *deterministic* solution (rather than randomized) with a better space occupancy and a better I/O-complexity; for Longest-Prefix Search Problem we improve both [3] and [10] via a space-I/O optimal *deterministic* solution (rather than randomized, space sub-optimal, or I/O-inefficient solutions in [3] and [10]). Technically speaking, our results are obtained by adopting few technicalities plus a *new storage scheme* that extends the *Locality Preserving Front-Coding* scheme, at the base of COSB, in such a way that prefixes of the compressed strings can be decompressed in optimal I/Os, rather than just the entire strings. This scheme is surprisingly simple and it can be looked as a compressed version of the Blind-trie, backbone of the String B-tree [9].

---

[2] We remark that this query bound can be looked at as nearly optimal for the first two problems because it has not been proved yet that the term $\log_B K$ is necessary within the space bounds obtained in this paper.

Table 1. A summary of our notation and terminology

$\mathcal{S}$	The set of strings
N	Total length of the strings in $\mathcal{S}$
K	Number of strings in $\mathcal{S}$
$\mathcal{T_S}$	The compact trie built on $\mathcal{S}$
t	Number of nodes in $\mathcal{T_S}$, it is $t \leq 2K - 1$
$\mathsf{p}(u)$	The parent of the node u in $\mathcal{T_S}$
$\mathsf{string}(u)$	The string spelled out by the path in $\mathcal{T_S}$ reaching u from the root
$\mathsf{label}(u)$	The label of the edge $(\mathsf{p}(u), u)$
$\hat{\mathcal{S}}$	The set $\mathcal{S}$ augmented with all strings $\mathsf{string}(u)$
$\mathsf{Trie}(\mathcal{S})$	The sum of edge-label lengths in $\mathcal{T_S}$
$\mathsf{LT}(\mathcal{S})$	Lower bound (in bits) to the storage complexity of the set of strings $\mathcal{S}$ (it is $\mathsf{LT}(\mathcal{S}) = \mathsf{Trie}(\mathcal{S}) + \log \binom{\mathsf{Trie}(\mathcal{S})}{t-1}$)

2   Notation and Background

In order to simplify the following presentation of our results, we assume to deal with *binary* strings. In the case of a larger alphabet Σ, it is enough to transform the strings over Σ into binary strings, and then apply our algorithmic solutions. The I/O complexities do not change because they depend only on the number of strings K in $\mathcal{S}$ and on the number of bits that fit in a disk block (hence $\Theta(B \log N)$ bits). As a further simplifying assumption we take $\mathcal{S}$ to be *prefix free*, so that no string in the set is prefix of another string; condition that is satisfied in practice because of the null-character terminating each string.

Table 1 summarizes all our notation and terminology. Below we will briefly recall few algorithmic tools that we will deploy to design our solutions to the prefix-search problem. We start with Front Coding, a compression scheme for strings which represents $\mathcal{S}$ as the sequence $\mathsf{FC}(\mathcal{S}) = \langle n_1, L_1, n_2, L_2, \ldots, n_K, L_K \rangle$, where n_i is the length of longest common prefix between S_{i-1} and S_i, and L_i is the suffix of S_i remaining after the removal of its first n_i (shared) characters, hence $L_i = |S_i| - n_i$. The first string S_1 is represented in its entirety, so that $L_1 = S_1$ and $n_1 = 0$. FC is a well established practical method for encoding a string set [22], and we will use interchangeably FC to denote either the algorithmic scheme or its output size in bits.

In order to estimate the space-cost of $\mathsf{FC}(\mathcal{S})$ in bits, the authors of [10] introduced the so called *trie measure* of $\mathcal{S}$, defined as: $\mathsf{Trie}(\mathcal{S}) = \sum_{i=1}^{K} |L_i|$, which accounts for the number of characters outputted by $\mathsf{FC}(\mathcal{S})$. And then, they devised a lower-bound $\mathsf{LT}(\mathcal{S})$ to the storage complexity of $\mathcal{S}$ which adds to the trie measure the cost, in bits, of storing the lengths $|L_i|$s. We have $\mathsf{LT}(\mathcal{S}) = \mathsf{Trie}(\mathcal{S}) + \log \binom{\mathsf{Trie}(\mathcal{S})}{t-1}$ bits.

In the paper we will often obtain bounds in terms of $\log \binom{\mathsf{Trie}(\mathcal{S})}{t-1}$, so the following fact is helpful:

Fact 1. *For any dictionary of strings $\mathcal{S}$, we have* $\log \binom{\text{Trie}(\mathcal{S})}{t-1} = O(K \log \frac{N}{K})$. *Nevertheless there exist dictionaries for which $K \log \frac{N}{K}$ may be up to $\log K$ times larger than* $\log \binom{\text{Trie}(\mathcal{S})}{t-1}$. *Finally, it is* $O(\log \binom{\text{Trie}(\mathcal{S})}{t-1}) = o(\text{Trie}(\mathcal{S})) + O(K)$.

Despite its simplicity $\mathsf{FC}(\mathcal{S})$ is a good compressor for $\mathcal{S}$, and indeed [10] showed that the representation obtained via Front-Coding takes the following number of bits:
$$\mathsf{LT}(\mathcal{S}) \leq \mathsf{FC}(\mathcal{S}) \leq \mathsf{LT}(\mathcal{S}) + O(K \log \frac{N}{K}). \tag{1}$$

It is possible to show that there exist pathological cases in which Front Coding requires space close to that upper bound. The main drawback of Front-coding is that decoding a string S_j might require the decompression of the entire sequence $\langle 0, L_1, \ldots, n_j, L_j \rangle$. In order to overcome this drawback, Bender *et al.* [3] proposed a variant of FC, called *locality-preserving front coding* (shortly, LPFC), that, given a parameter ϵ, adaptively partitions $\mathcal{S}$ into blocks such that decoding any string S_j takes $O((1 + \frac{1}{\epsilon})|S_j|/B)$ optimal time and I/Os, and requires $(1+\epsilon)\mathsf{FC}(\mathcal{S})$ bits of space. This adaptive scheme offers a clear space/time tradeoff in terms of the user-defined parameter ϵ and it is agnostic in the parameter B.

A different linearization, called *rear-coding* (RC), is a simple variation of FC which *implicitly* encodes the length n_i by specifying the length of the suffix of S_{i-1} to be removed from it in order to get the longest common prefix between S_{i-1} and S_i. This change is crucial to avoid *repetitive* encodings of the same longest common prefixes, the space inefficiency in FC. And indeed RC is able to come very close to LT, because we can encode the lengths of the suffixes to be dropped via a binary array of length $\text{Trie}(\mathcal{S})$ with $K - 1$ bits set to 1, as indeed those suffixes partition $\mathcal{T}_\mathcal{S}$ into K disjoint paths from the leaves to the root. So RC can be encoded in
$$\mathsf{RC}(\mathcal{S}) \leq \text{Trie}(\mathcal{S}) + 2 \log \binom{\text{Trie}(\mathcal{S})}{K - 1} + O(\log \text{Trie}(\mathcal{S})) = (1+o(1))\mathsf{LT}(\mathcal{S}) \text{ bits,} \tag{2}$$

where the latter equality follows from the third statement in Fact 1. Comparing eqn. (2) and (1), the difference between RC and FC is in the encoding of the n_i, so $\text{Trie}(\mathcal{S}) \leq N$ (in practice $\text{Trie}(\mathcal{S}) \ll N$).

In the design of our algorithms and data structures we will need two other key tools which are nowadays the backbone of every compressed index: namely, Rank/Select data structures for binary strings. Their complexities are recalled in the following theorems.

Theorem 1 ([7]). *A binary vector $B[1..m]$ with n bits set to 1 can be encoded within $\log \binom{m}{n} + O(n)$ bits so that we can solve in $O(1)$ time the query $\mathsf{Select}_1(B, i)$, with $1 \leq i \leq n$, which returns the position in B of the ith occurrence of 1.*

Theorem 2 ([20]). *A binary vector $B[1..m]$ with n bits set to 1 can be encoded within $m + o(m)$ bits so that we can solve in $O(1)$ time the queries $\mathsf{Rank}_1(B, i)$, with $1 \leq i \leq m$, which returns the number of 1s in the prefix $B[1..i]$, and $\mathsf{Select}_1(B, i)$, with $1 \leq i \leq n$, which returns the position in B of the ith occurrence of 1.*

3   A Key Tool: Cache-Oblivious Prefix Retrieval

The novelty of our paper consists of a surprisingly simple representation of $\mathcal{S}$ which is compressed and still supports the cache-oblivious retrieval of any prefix of any string in this set in optimal I/Os and space (up to constant factors). The striking news is that, despite its simplicity, this result will constitute the backbone of our improved algorithmic solutions.

In this section we instantiate our solution on tries even if it is sufficiently general to represent any (labeled) tree in compact form still guaranteeing optimal traversal in the cache-oblivious model of any root-to-a-node path. We assume that the trie nodes are numbered accordingly to the time of their DFS visit. Any node u in $\mathcal{T}_\mathcal{S}$ is associated with label(u) which is (the variable length) string on the edge $(p(u), u)$, where $p(u)$ is the parent of u in $\mathcal{T}_\mathcal{S}$. Observe that any node u identifies uniquely the string string(u) that prefixes all strings of $\mathcal{S}$ descending from u. Obviously, string(u) can be obtained by juxtaposing the labels of the nodes on the path from the root to u. Our goal is to design a storage scheme whose space occupancy is $(1+\epsilon)\mathsf{LT}(\mathcal{S})+O(K)$ bits and supports in optimal time/IO the operation Retrieval(u, ℓ) which asks for the prefix of the string string(u) having length $\ell \in (|\mathsf{string}(p(u))|, |\mathsf{string}(u)|]$. Note that the returned prefix ends up in the edge $(p(u), u)$. In other words, we want to be able to access the labels of the nodes in any root-to-a-node path and any of their prefixes. Formally, we aim to prove the following theorem.

Theorem 3. *Given a set $\mathcal{S}$ of K binary strings having total length N, there exists a storage scheme for $\mathcal{S}$ that occupies $(1+\epsilon)\mathsf{LT}(\mathcal{S})+O(K)$ bits, where $\epsilon > 0$ is any fixed constant, and solves the query Retrieval(u, ℓ) in $O(1+(1+\frac{1}{\epsilon})\frac{\ell}{B \log N})$ optimal I/Os.*

Before presenting a proof, let us discuss efficiency of two close relatives of our solution: *Giraffe tree decomposition* [4] and *Locality-preserving front Coding* (LPFC) [3]. The former solution has the same time complexity of our solution but has a space occupancy of at least $3 \cdot \mathsf{LT}(\mathcal{S})+O(K)$ bits. The latter approach has (almost) the same space occupancy of our solution but provides no guarantee on the number of I/Os required to access *prefixes* of the strings in $\mathcal{S}$.

Our goal is to accurately lay out the labels of nodes of $\mathcal{T}_\mathcal{S}$ so that any string(u) can be retrieved in optimal $O((1 + \frac{1}{\epsilon})\mathsf{Scan}(\mathsf{string}(u)))$ I/Os. This suffices for obtaining the bound claimed in Theorem 3 because, once we have reconstructed string$(p(u))$, Retrieval(u, ℓ) is completed by accessing the prefix of label(u) of length $j = \ell - |\mathsf{string}(p(u))|$ which is written consecutively in memory. One key feature of our solution is a proper replication of some labels in the lay out, whose space is bounded by $\epsilon \cdot \mathsf{LT}(\mathcal{S})$ bits.

The basis of our scheme is the amortization argument in LPFC [3] which represents $\mathcal{S}$ by means of a variant of the classic front-coding in which some strings are stored explicitly rather than front-coded. More precisely, LPFC writes the string S_1 explicitly, whereas all subsequent strings are encoded in accordance with the following argument. Suppose that the scheme already compressed $i - 1$ strings and has to compress string S_i. It scans back $c|S_i|$ characters in the current representation to check if it is possible to decode S_i, where $c = 2 + 2/\epsilon$. If this is the case, S_i is compressed by writing its suffix L_i, otherwise S_i is fully written. A sophisticated amortization argument in [3] proves that LPFC requires $(1+\epsilon)$LT$+O(K \log(N/K))$ bits. This bound can be improved to $(1+\epsilon)$LT$+O(K)$ bits by replacing front-coding with rear-coding (see eqn. 2). By construction, this scheme guarantees an optimal decompression time/IO of any string $S_i \in \mathcal{S}$, namely $O((1 + \frac{1}{\epsilon})$Scan$(S_i))$ I/Os. But unfortunately, this property does not suffice to guarantee an optimal decompression for prefixes of the strings: the decompression of *a prefix* of a string S_i may cost up to $\Theta((1+\frac{1}{\epsilon})$Scan$(S_i))$ I/Os.

In order to circumvent this limitation, we modify LPFC as follows. We define the superset $\hat{\mathcal{S}}$ of $\mathcal{S}$ which contains, for each node u in $\mathcal{T}_{\mathcal{S}}$ (possibly a leaf), the string string(u). This string is a prefix of string(v), for any descendant v of u in $\mathcal{T}_{\mathcal{S}}$, so it is lexicographically smaller than string(v). The lexicographically ordered $\hat{\mathcal{S}}$ can thus be obtained by visiting the nodes of $\mathcal{T}_{\mathcal{S}}$ according to a DFS visit. In our solution we require that all the strings emitted by LPFC$(\hat{\mathcal{S}})$ are self-delimited. Thus, we prefix each of them with its length coded with Elias' Gamma coding. Now, let R be the compressed output obtained by computing LPFC$(\hat{\mathcal{S}})$ with rear-coding. We augment R with two additional data structures:

- The binary array $E[1..|R|]$ which sets to 1 the positions in R where the encoding of some string(u) starts. E contains $t - 1$ bits set to 1. Array E is enriched with the data structure in Theorem 1 so that Select$_1$ queries can be computed in constant time.
- The binary array $V[1..t]$ that has an entry for each node in $\mathcal{T}_{\mathcal{S}}$ according to their (DFS-)order. The entry $V[u]$ is sets to 1 whenever string(u) has been copied in R, 0 otherwise. We augment V with the data structure of Theorem 2 to support Rank and Select queries.

In order to answer Retrieval(u, ℓ) we first implement the retrieval of string(u). The query Select$_1(E, u)$ gives in constant time the position in R where the encoding of string(u) starts. Now, if this string has been fully copied in R then we are done; otherwise we have to reconstruct it. This has some subtle issues that have to be addressed efficiently, for example, we do not even know the length of string(u) since the array E encodes the individual edge-labels and not their lengths from the root of $\mathcal{T}_{\mathcal{S}}$. We reconstruct string$(u)$ forward by starting from the first copied string (say, string(v)) that precedes string(u) in R. The node index v is obtained by computing Select$_1(V, $Rank$_1(V, u))$ which identifies the position of the first 1 in V that precedes the entry corresponding to u (i.e., the closer copied strings preceding u in the DFS-visit of $\mathcal{T}_{\mathcal{S}}$).

Assume that the copy of string(v) starts at position p_v, which is computed by selecting the v-th 1 in the E. By the DFS-order in processing $\mathcal{T}_{\mathcal{S}}$ and by the fact that string(u) is not copied, it follows that string(u) can be reconstructed by copying characters in R starting from position p_v up to the occurrence of string(u). We recall that rear-coding augments each emitted string with a value: let w and w' two nodes consecutive in the DFS-visit of $\mathcal{T}_{\mathcal{S}}$, rear-coding writes the value $|$string$(w)| - lcp($string$(w), $string$(w'))$ (namely, the length of the suffix of string(w) that we have to remove from w in order to obtain the length of its longest common prefix with string(w')). This information is exploited in reconstructing string(u). We start by copying string(v) in a buffer by scanning R forward from position p_v. At the end, the buffer will contain string(u). For every value m written by rear-coding, we overwrite the last m characters of the buffer with the characters in R of the suffix of the current string (delimited by E's bits set to 1). By LPFC-properties, we are guaranteed that this scan takes $O((1 + \frac{1}{\epsilon})\mathsf{Scan}(\mathsf{string}(u))$ I/Os.

Let us now come back to the solution of Retrieval(u, ℓ). First of all we reconstruct string($\mathsf{p}(u)$), then determine the edge-label $(\mathsf{p}(u), u)$ in E given the DFS-numbering of u and a select operation over E. We thus take from this string its prefix of length $\ell - |\mathsf{string}(\mathsf{p}(u))|$, the latter is known because we have indeed reconstructed that string.

To conclude the proof of Theorem 3 we are left with bounding the space occupancy of our storage scheme. We know that R requires no more than $(1 + \epsilon)\mathsf{LT}(\hat{\mathcal{S}}) + O(K)$ bits, since we are using rear-coding. The key observation is then the trie measure of $\hat{\mathcal{S}}$ coincides with the one of $\mathcal{S}$ (i.e., $\mathsf{Trie}(\hat{\mathcal{S}}) = \mathsf{Trie}(\mathcal{S})$), so that $|R| = (1 + \epsilon)\mathsf{LT}(\hat{\mathcal{S}}) + O(K) = (1 + \epsilon)\mathsf{LT}(\mathcal{S}) + O(K)$. The space occupancy of E is $\log \binom{|R|}{t-1}$ bits (Theorem 1), therefore $|E| \leq t \log(|R|/t) + O(t) = o(\mathsf{Trie}(\mathcal{S})) + O(K)$ bits. It is easy to see that the cost of self-delimiting the strings emitted by LPFC with Elias' Gamma coding is within the same space bound. The vector V requires just $O(K)$ bits, by Theorem 2.

The query Retrieval(u, ℓ) suffices for the aims of this paper. However, it is more natural an operation that, given a string index i and a length ℓ, returns the prefix of $S_i[1..\ell]$. This can be supported by using a variant of the ideas presented later for the Weak-prefix Search problem, which adds $O(\log \binom{\mathsf{Trie}(\mathcal{S})}{t-1} + K) = o(\mathsf{Trie}(\mathcal{S})) + O(K)$ bits to space complexity (hidden by the other terms) and a term $O(\log_B K)$ I/Os to query time. Alternatively, it is possible to keep an I/O-optimal query time by adding $O(K \log \frac{N}{K})$ bits of space.

4   Searching Strings: Three Problems

In this section we address the three problems introduced in the Introduction, they allow us to frame the wide spectrum of algorithmic difficulties and solutions related with the search for a pattern within a string set.

Problem 1 (Weak-Prefix Search Problem). Let $\mathcal{S} = \{S_1, S_2, \ldots, S_K\}$ be a set of K binary strings of total length N. We wish to preprocess $\mathcal{S}$ in such a way that, given a pattern P, we can efficiently answer the query weakPrefix(P) which asks for the range of strings in $\mathcal{S}$ prefixed by P. An arbitrary answer could be returned whenever P is not a prefix of any string in $\mathcal{S}$.   □

The lower bound in [2] states that $\Omega(K \log \frac{N}{K})$ bits are necessary regardless the query time. We show the following theorem.

Theorem 4. *Given a set of $\mathcal{S}$ of K binary strings of total length N, there exists a deterministic data structure requiring $2 \cdot \log \binom{\mathsf{Trie}(\mathcal{S})}{t-1} + O(K)$ bits of space* *that solves the* Weak-Prefix Search Problem *for any pattern P with $O(\log_B K +$ Scan$(P))$ I/Os.*

The space occupancy is optimal up to constant factor since $\log \binom{\text{Trie}(\mathcal{S})}{t-1}$ is always at most $K \log \frac{N}{K}$ (see Fact 1). Moreover, our refined estimate of the space occupancy, by means of Trie$(\mathcal{S})$, shows that it can go below the lower bound in [2] even by a factor $\Theta(\log K)$ depending on the characteristics of the indexed dictionary (see Fact 1). The query time instead is almost optimal, because it is not clear whether the term $\log_B K$ is necessary within this space bound. Summarizing, our data structure is smaller, deterministic and faster than previously known solutions.

We follow the solution in [8] by using a two-level indexing. We start by partitioning $\mathcal{S}$ into $s = K/\log N$ groups of (contiguous) strings defined as follows: $\mathcal{S}_i = \{S_{1+i \log N}, S_{2+i \log N}, \ldots, S_{(i+1) \log N}\}$ for $i = 0, 1, 2, \ldots, s - 1$. We then construct a subset $\mathcal{S}_{\text{top}}$ of $\mathcal{S}$ consisting of $2s = \Theta(\frac{n}{\log n})$ representative strings obtained by selecting the smallest and the largest string in each of these groups. The index in the first level is responsible for searching the pattern P within the set $\mathcal{S}_{\text{top}}$, in order to identify an approximated range. This range is guaranteed to contain the range of strings prefixed by P. A search on the second level suffices to identify the correct range of strings prefixed by P. We have two crucial differences w.r.t. the solution in [8]: 1) our index is deterministic; 2) our space-optimal solution for the second level is the key for achieving Theorem 4.

First level. As in [8] we build the Patricia Trie PT_{top} over the strings in $\mathcal{S}_{\text{top}}$ with the speciality that we store in each node u of PT_{top} a fingerprint of $O(\log N)$ bits computed for string(u) according to Karp-Rabin fingerprinting [18]. The crucial difference w.r.t. [8] is the use of a (deterministic) injective instance of Karp-Rabin that maps any prefix of any string in $\mathcal{S}$ into a distinct value in a interval of size $O(N^2)$.[3] Given a string $S[1..s]$, the Karp-Rabin fingerprinting rk(S) is equal to $\sum_{i=1}^{s} S[i] \cdot t^i \pmod{M}$, where M is a prime number and t is a randomly chosen integer in $[1, M-1]$. Given the set of strings $\mathcal{S}$, we can obtain an instance rk() of the Karp-Rabin fingerprinting that maps all the prefixes of all the strings in $\mathcal{S}$ to the first $[M]$ integers without collisions, with M chosen among the first $O(N^2)$ integers. It is known that a value of t that guarantees no collisions can be found in expected $O(1)$ attempts. In the cache-oblivious setting, this implies that finding a suitable function requires $O(\text{Sort}(\mathsf{N}) + N/B)$ I/Os in expectation, where Sort(N) is the number of I/Os required to sort N integers.

Given PT_{top} and the pattern P, our goal is that of finding the lowest edge $e = (v, w)$ such that string(v) is a prefix of P and string(w) is not. This edge can be found with a standard blind search on PT_{top} and by also comparing fingerprints of P with the ones stored in the traversed nodes (see [8] for more details). A cache-oblivious efficient solution is obtained by laying out PT_{top} via the centroid trie decomposition [3]. This layout guarantees that the above search requires $O(\log_B K + \text{Scan}(P))$ I/Os. However, in [8] the edge e is correctly identified

---

[3] Notice that we require the function to be injective for prefixes of strings in $\mathcal{S}$ not $\mathcal{S}_{\text{top}}$.

only with high probability. The reason is that a prefix of P and a prefix of a string in $\mathcal{S}$ may have the same fingerprint even if they are different. Our use of the injective Karp-Rabin fingerprints avoids this situation guaranteeing that the search is always correct[4].

Second level. For each edge $e = (u, v)$ of $\mathsf{PT}_{\mathsf{top}}$ we define the set of strings $\mathcal{S}_e$ as follows. Assume that each node v of $\mathsf{PT}_{\mathsf{top}}$ points to its leftmost/rightmost descending leaves, denoted by $L(v)$ and $R(v)$ respectively. We call $\mathcal{S}_{L(v)}$ and $\mathcal{S}_{R(v)}$ the two groups of strings, from the grouping above, that contain $\mathcal{S}_{L(v)}$ and $\mathcal{S}_{R(v)}$. Then $\mathcal{S}_e = \mathcal{S}_{L(v)} \cup \mathcal{S}_{R(v)}$. We have a total of $O(K/\log n)$ sets, each having $O(\log N)$ strings. The latter is the key feature that we exploit in order to index these small sets efficiently by resorting to the following lemma. We remark that $\mathcal{S}_e$ will not be constructed and indexed explicitly, rather we will index the sets $\mathcal{S}_{L(v)}$ and $\mathcal{S}_{R(v)}$ individually, and keep two pointers to each of them for every edge e. This avoids duplication of information and some subtle issues in the storage complexities. But poses the problem of how to weak-prefix search in $\mathcal{S}_e$ which is only virtually available. The idea is to search in $\mathcal{S}_{L(v)}$ and $\mathcal{S}_{R(v)}$ individually, three cases may occur. Either we find that the range is totally within one of the two sets, and in this case we return that range; or we find that the range includes the rightmost string in $\mathcal{S}_{L(v)}$ and the leftmost string in $\mathcal{S}_{R(v)}$, and in this case we merge them. The correctness comes from the properties of trie's structure and the first-level search, as one can prove by observing that the trie built over $\mathcal{S}_{L(v)} \cup \mathcal{S}_{R(v)}$ is equivalent to the two tries built over the two individual sets except for the rightmost path of $\mathcal{S}_{L(v)}$ and the leftmost path of $\mathcal{S}_{R(v)}$ which are merged in the trie for $\mathcal{S}_e$. This merging is not a problem because if the range is totally within $\mathcal{S}_{R(v)}$, then the dominating node is within the trie for this set and thus the search for P would find it by searching both $\mathcal{S}_{R(v)}$ or $\mathcal{S}_e$. Similarly this holds for a range totally within $\mathcal{S}_{L(v)}$. The other case comes by exclusion, so the following lemma allows to establish the claimed I/O and space bounds.

Lemma 1. *Let $\mathcal{S}_i$ be a set of $K_i = O(\log N)$ strings of total length at most N. The Patricia trie of $\mathcal{S}_i$ can be represented by requiring $\log \binom{\mathsf{Trie}(\mathcal{S}_i)}{t_i - 1} + O(K_i)$ bits of space so that the blind search of any pattern P with $O((\log K_i)/B + \mathsf{Scan}(P))$ I/Os, where t_i is the number of nodes in the trie of the set $\mathcal{S}_i$.*

To conclude the proof of Theorem 4, we distinguish two cases based on the value of K. If $K = O(\log N)$, we do not use the first level since Lemma 1 with $K_i = K$ already matches the bounds in Theorem 4. Otherwise $K = \Omega(\log N)$, and so searching P requires $O(\log_B K + \mathsf{Scan}(P))$ I/Os on the first level and $O((\log \log N)/B + \mathsf{Scan}(P)) = O(\log_B K + \mathsf{Scan}(P))$ I/Os on the second level. For the space occupancy, we observe that the first level requires $O(K)$ bits, and the second level requires $\sum_i (\log \binom{\mathsf{Trie}(\mathcal{S}_i)}{t_i - 1} + K_i)$ bits (Lemma 1). Note that $\sum_i t_i \le t = O(K)$ because each string of $\mathcal{S}$ belongs to at most one $\mathcal{S}_i$.

---

[4] Recall that in the **Weak-Prefix Search Problem** we are searching under the assumption that P is a prefix of at least a string in $\mathcal{S}$.

Problem 2 (Full-Prefix Search Problem). Let $\mathcal{S} = \{S_1, S_2, \ldots, S_K\}$ be a set of K binary strings of total length N. We wish to preprocess $\mathcal{S}$ in such a way that, given a pattern P, we can efficiently answer the query Prefix(P) which asks for the range of strings in $\mathcal{S}$ prefixed by P, the value $\perp$ is returned whenever P is not a prefix of any string in $\mathcal{S}$.   □

This the classic prefix-search which requires to recognize whether P is or is not the prefix of any string in $\mathcal{S}$. By combining Theorems 3 and 4 we get:

Theorem 5. *Given a set of $\mathcal{S}$ of binary strings of size K of total length N, there exists a data structure requiring $(1+\epsilon)\mathsf{LT}(\mathcal{S})+O(K)$ bits of space that solves the* Full-Prefix Search Problem *for any pattern P with $O(\log_B K + (1 + \frac{1}{\epsilon})\mathsf{Scan}(P))$ I/Os, where $\epsilon > 0$ is any constant.*

We use the solution of Theorem 4 to identify the highest node u from which descends the largest range of strings that are prefixed by P. Then, we use Theorem 3 to check I/O-optimally whether Retrieval($u, |P|$) equals P. The space occupancy of this solution is optimal up to a constant factor; the query complexity is almost optimal being unclear whether it is possible to remove the $\log_B K$ term and still maintain optimal space.

Problem 3 (Longest-Prefix Search Problem). Let $\mathcal{S} = \{S_1, S_2, \ldots, S_K\}$ be a set of K binary strings of total length N. We wish to preprocess $\mathcal{S}$ in such a way that, given a pattern P, we can efficiently answer the query LPrefix(P) which asks for the range of strings in $\mathcal{S}$ sharing the longest common prefix with P.   □

This problem waives the requirement that P is a prefix of some string in $\mathcal{S}$, and thus searches for the longest common prefix between P and $\mathcal{S}$'s strings. If P prefixes some strings in $\mathcal{S}$, then this problem coincides with the classic prefix-search. Possibly the identified lcp is the *null* string, and thus the returned range of strings is the whole set $\mathcal{S}$. We will prove the following result.

Theorem 6. *Given a set of $\mathcal{S}$ of K binary strings of total length N, there exists a data structure requiring $(1+\epsilon)\mathsf{LT}(\mathcal{S})+O(K)$ bits of space that solves the* Longest-Prefix Search Problem *for any pattern P with $O(\log_B K+(1+\frac{1}{\epsilon})\mathsf{Scan}(P))$ I/Os, where $\epsilon > 0$ is any constant.*

First we build the data structures of Theorem 3 with a constant ϵ' to be fixed later, in order to efficiently access prefixes of strings in $\mathcal{S}$ but also as a basis to partition the strings. It is convenient to observe this process on $\mathcal{T}_\mathcal{S}$. Recall that the data structure of Theorem 3 processes nodes of $\mathcal{T}_\mathcal{S}$ in DFS-order. For each visited node u, it encodes string(u) either by copying string(u) or by writing label(u). In the former case we will say that u is marked. Let $\mathcal{S}_{\text{copied}}$ be the set formed by the string(u) of any marked node u. The goal of a query LPrefix(P) is to identify the lowest node w in $\mathcal{T}_\mathcal{S}$ sharing the longest common prefix with P. We identify the node w in two phases. In the first phase we solve the query LPrefix(P) on the set $\mathcal{S}_{\text{copied}}$ in order to identify the range of all the (consecutive) marked nodes $[v_l, v_r]$ sharing the longest common prefix with P. Armed with this information, we start a second phase that scans appropriate portions of the compressed representation R of Theorem 3 to identify our target node w. (For space reasons we defer the description of our solution to the full paper.)

References

1. Apostolico, A.: The myriad virtues of subword trees. In: Combinatorial Algorithms on Words, pp. 85–96 (1985)
2. Belazzougui, D., Boldi, P., Pagh, R., Vigna, S.: Fast prefix search in little space, with applications. In: de Berg, M., Meyer, U. (eds.) ESA 2010, Part I. LNCS, vol. 6346, pp. 427–438. Springer, Heidelberg (2010)
3. Bender, M.A., Farach-Colton, M., Kuszmaul, B.C.: Cache-oblivious string B-trees. In: PODS, pp. 233–242. ACM (2006)
4. Brodal, G.S., Fagerberg, R.: Cache-oblivious string dictionaries. In: SODA, pp. 581–590. ACM Press (2006)
5. Brodnik, A., Munro, J.I.: Membership in constant time and almost-minimum space. SIAM J. Comput. 28(5), 1627–1640 (1999)
6. Demaine, E.D., Iacono, J., Langerman, S.: Worst-case optimal tree layout in a memory hierarchy. CoRR, cs.DS/0410048 (2004)
7. Elias, P.: Efficient storage and retrieval by content and address of static files. J. ACM 21(2), 246–260 (1974)
8. Ferragina, P.: On the weak prefix-search problem. In: Giancarlo, R., Manzini, G. (eds.) CPM 2011. LNCS, vol. 6661, pp. 261–272. Springer, Heidelberg (2011)
9. Ferragina, P., Grossi, R.: The string B-tree: A new data structure for string search in external memory and its applications. J. ACM 46(2), 236–280 (1999)
10. Ferragina, P., Grossi, R., Gupta, A., Shah, R., Vitter, J.S.: On searching compressed string collections cache-obliviously. In: PODS, pp. 181–190 (2008)
11. Ferragina, P., Venturini, R.: The compressed permuterm index. ACM Transactions on Algorithms 7(1), 10 (2010)
12. Frakes, W., Baeza-Yates, R.: Information Retrieval: Data Structures and Algorithms. Prentice-Hall (1992)
13. Fredkin, E.: Trie memory. Communication of the ACM 3(9), 490–499 (1960)
14. Frigo, M., Leiserson, C.E., Prokop, H., Ramachandran, S.: Cache-oblivious algorithms. ACM Transactions on Algorithms 8(1), 4 (2012)
15. Gusfield, D.: Algorithms on Strings, Trees, and Sequences - Computer Science and Computational Biology. Cambridge University Press (1997)
16. He, M., Munro, J.I., Rao, S.S.: Succinct ordinal trees based on tree covering. In: Arge, L., Cachin, C., Jurdziński, T., Tarlecki, A. (eds.) ICALP 2007. LNCS, vol. 4596, pp. 509–520. Springer, Heidelberg (2007)
17. Hon, W.-K., Shah, R., Vitter, J.S.: Compression, Indexing, and Retrieval for Massive String Data. In: Amir, A., Parida, L. (eds.) CPM 2010. LNCS, vol. 6129, pp. 260–274. Springer, Heidelberg (2010)
18. Karp, R.M., Rabin, M.O.: Efficient randomized pattern-matching algorithms. IBM Journal of Research and Development 31(2), 249–260 (1987)
19. Morrison, D.R.: PATRICIA - practical algorithm to retrieve coded in alphanumeric. J. ACM 15(4), 514–534 (1968)
20. Munro, J.I.: Tables. In: Foundations of Software Technology and Theoretical Computer Science (FSTTCS), pp. 37–42 (1996)
21. Navarro, G., Mäkinen, V.: Compressed full-text indexes. ACM Comput. Surv. 39(1) (2007)
22. Witten, I.H., Moffat, A., Bell, T.C.: Managing Gigabytes: Compressing and Indexing Documents and Images. Morgan Kaufmann Publishers (1999)

BICO: BIRCH Meets Coresets
for *k*-Means Clustering[*]

Hendrik Fichtenberger, Marc Gillé, Melanie Schmidt,
Chris Schwiegelshohn, and Christian Sohler

Efficient Algorithms and Complexity Theory, TU Dortmund, Germany
`{firstname.lastname}@tu-dortmund.de`

Abstract. We design a data stream algorithm for the *k*-means problem, called BICO, that combines the data structure of the SIGMOD Test of Time award winning algorithm BIRCH [27] with the theoretical concept of coresets for clustering problems. The *k*-means problem asks for a set C of k centers minimizing the sum of the squared distances from every point in a set P to its nearest center in C. In a data stream, the points arrive one by one in arbitrary order and there is limited storage space.

BICO computes high quality solutions in a time short in practice. First, BICO computes a summary S of the data with a provable quality guarantee: For every center set C, S has the same cost as P up to a $(1 + \varepsilon)$-factor, i.e., S is a *coreset*. Then, it runs *k*-means++ [5] on S.

We compare BICO experimentally with popular and very fast heuristics (BIRCH, MacQueen [24]) and with approximation algorithms (Stream-KM++ [2], StreamLS [16, 26]) with the best known quality guarantees. We achieve the same quality as the approximation algorithms mentioned with a much shorter running time, and we get much better solutions than the heuristics at the cost of only a moderate increase in running time.

1 Introduction

Clustering is the task to partition a set of objects into groups such that objects in the same group are similar and objects in different groups are dissimilar. There is a huge amount of work on clustering both in practice and in theory. Typically, theoretic work focuses on exact solutions or approximations with guaranteed approximation factors, while practical algorithms focus on speed and results that are reasonably good on the particular data at hand.

We study the *k*-means problem, which given a set of points P from $\mathbb{R}^d$ asks for a set of k centers such that the cost defined as the sum of the squared distances of all points in P to their closest center is minimized. The centers induce a clustering defined by assigning every point to its closest center.

For this problem, the algorithm most used in practice is Lloyd's algorithm, an iterative procedure that converges to a local optimum after a possibly exponential number of iterations. An improved algorithm known as *k*-means++ by Arthur and Vassilvitskii [6] has a $\mathcal{O}(\log k)$ approximation guarantee.

---

[*] This research was partly supported by DFG grants BO 2755/1-1 and SO 514/4-3 and within the Collaborative Research Center SFB876, project A2.

H.L. Bodlaender and G.F. Italiano (Eds.): ESA 2013, LNCS 8125, pp. 481–492, 2013.

Big Data is an emerging area of computer science. Nowadays, data sets arising from large scale physical experiments or social networks analytics are far too large to fit in main memory. Some data, e. g., produced by sensors, additionally arrives one by one, and it is desirable to filter it at arrival without intermediately storing large amounts of data. Considering k-means in a data stream setting, we assume that points arrive in arbitrary order and that there is limited storage capacity. Neither Lloyd's algorithm nor k-means++ work in this setting, and both would be too slow even if the data already was on a hard drive [2].

A vast amount of approximation algorithms were developed for the k-means problem in the streaming setting. They usually use the concept of *coresets*. A coreset is a small weighted set of points S, that ensures that if we compute the weighted clustering cost of S for any given set of centers C, then the result will be a $(1 + \varepsilon)$-approximation of the cost of the original input.

Approximation algorithms are rather slow in practice. Algorithms fast in practice are usually heuristics and known to compute bad solutions on occasions. The best known one is BIRCH [27]. It also computes a summary of the data, but without theoretical quality guarantee.

This paper contributes to the field of interlacing theoretical and practical work to develop an algorithm good in theory and practice. An early work in this direction is StreamLS [16, 26] which applies a local search approach to chunks of data. StreamLS is significantly outperformed by Stream-KM++[2] which computes a coreset and then solves the k-means problem on the coreset by applying k-means++. Stream-KM++ computes very good solutions and is reasonably fast for small k. However, especially for large k, its running time is still far too large for big data sets.

We develop BICO, a data stream algorithm for the k-means problem which also computes a coreset and achieves very good quality in practice, but is significantly faster than Stream-KM++.

Related Work. To solve k–means in a stream, there are three basic approaches. First, one can apply an online gradient descent such as MacQueen's algorithm [24]. Such an algorithm is usually the fastest available, yet the computed solution has poor performance in theory and (usually) also in practice.

The other two approaches compute summaries of the data for further processing. Summaries should be small to speed up the optimization phase, they should have a good quality, and their computation should be fast. Summary computing algorithms either update the summary one point at a time or read a batch of points and process them together.

Theoretical analysis has mostly focused on the latter scenario and then relies on the *merge & reduce* framework originally by Bentley and Saxe [7] and first applied to clustering in [4]. Informally speaking, a coreset construction can be defined in a non-streaming manner and then be embedded into the merge & reduce framework. The computed coreset is by a factor of $\log^{t+1} n$ larger than the non-streaming version where t is the exponent of ε^{-1} in the coreset size. The running time is increased to $2C(m) \cdot n/m$ where m is the batch size and $C(m)$ is the time of the non-streaming version of the coreset construction. Thus, the asymptotic running time is not increased (for at least linear $C(m)$), but from a practical point of view this overhead is not desirable. Another drawback is that the size of the computed coresets usually highly depends on $\log n$. Streaming algorithms using summaries include [8, 10, 11, 12, 20, 21, 22]. In particular, StreamLS uses a batch approach, and Stream-KM++ uses merge & reduce. There is usually a high dependency on $\log n$ in merge & reduce constructions. If d is not a constant, the smallest dependency on the dimension is achieved by [12] and this coreset has a size of $\mathcal{O}(k^2 \cdot \varepsilon^{-4} \log^5 n)$ (which is independent of d).

For pointwise updates, in particular notice the construction in [15] computing a streaming coreset of size $\mathcal{O}(k \cdot \log n \cdot \varepsilon^{-(d+2)})$. For low dimensions, i. e., if d is a constant, this is the lowest dependency on k and $\log n$ of any coreset construction. This is due to the fact that the coreset is maintained without merge & reduce. The time to compute the streaming coreset is $\tilde{\mathcal{O}}(n \cdot \rho + k \cdot \log n \cdot \rho \cdot \epsilon^{-(d+2)})$ with $\rho = \log(n\Delta/\epsilon)$ where Δ is the spread of the points, i. e., the maximal distance divided by the smallest distance of two distinct points.

Pointwise updates are usually preferable for practical purposes. Probably the best known practical algorithm is BIRCH [27]. It reads the input only once and computes a summary by pointwise updates. Then, it solves the k-means problem on the summary using agglomerative clustering.

The summary that BIRCH computes consists of a tree of so-called Clustering Features. A Clustering Feature (CF) summarizes a set of points by the sum of the points, the number of points and the sum of the squared lengths of all points. BIRCH has no theoretical quality guarantees and does indeed sometimes perform badly in practice [18, 19].

We are not aware of other very popular data stream algorithms for the k-means problem. There is a lot of work on related problems, for example CURE [18] which requires more than one pass over the data, DBSCAN [9] which is not center based, CLARANS [25] which is typically used when centers have to be chosen from the dataset and is not particularly optimized for points in Euclidean space and ROCK [17] and COBWEB [14] which are designed for categorical attributes.

Our Contribution. We develop BICO, a data stream algorithm based on the data structure of BIRCH. Both algorithms compute a summary of the data, but while the summary computed by BIRCH can be arbitrarily bad as we show at the start of Section 3, we show that BICO computes a coreset S, so for every set of centers C, the cost of the input point set P can be approximated by computing the cost of S. For constant dimension d, we bound the size m of our coreset by $\mathcal{O}(k \cdot \log n \cdot \varepsilon^{-(d+2)})$ and show that BICO needs $\mathcal{O}\left(N(m) \cdot (n + m \log n\Delta)\right)$ time to compute it where $N(m)$ is the time needed to solve a nearest neighbor problem within m points. Trivially, $N(m) = \mathcal{O}(m)$. By using range query data structures, $N(m) = \mathcal{O}(\log^{d-1} m)$ can be achieved at the cost of $\mathcal{O}(m \log^{d-1} m)$ additional space [3]. Notice that the size of the coreset is asymptotically equal to [15].

We implement BICO and show how to realize the algorithm in practice by introducing heuristic enhancements. Then we compare BICO experimentally to two heuristics, BIRCH and MacQueen's k-means algorithm, and to two algorithms designed for high quality solutions, Stream-KM++ and StreamLS. BICO computes solutions of the same high quality as Stream-KM++ and StreamLS which we believe to be near optimal. For small k, BICOs running time is only beaten by MacQueen's and in particular, BICO is 5-10 times faster than Stream-KM++ (and more for StreamLS). For larger k, BICO needs to maintain a larger coreset to keep the quality up. However, BICO can trade quality for speed. We do additional testruns showing that with different parameters, BICO still beats the cost of MacQueen and BIRCH in similar running time. We believe that BICO provides the best quality-speed trade-off in practice.

2   Preliminaries

The k-means Problem. Let $P \subseteq \mathbb{R}^d$ be a set of points in d-dimensional Euclidean space with $|P| = n$. For two points $p, q \in P$, we denote their Euclidean distance by $||p - q|| := \sqrt{\sum_{i=1}^{d} (p_i - q_i)^2}$. The k-means problem asks for a set C of k points in $\mathbb{R}^d$ (called *centers*) that minimizes the sum of the squared distances of all points in P to their closest point in C, i.e., the objective is $\min_{C \subset \mathbb{R}^d, |C|=k} \sum_{p \in P} \min_{c \in C} ||p - c||^2 =: \min_{C \subset \mathbb{R}^d, |C|=k} \text{cost}(P, C)$.

The weighted k-means cost is defined by $\text{cost}_w(P, C) := \sum_{p \in P} w(p) \min_{c \in C} ||p-c||^2$ for any weight function $w : P \to \mathbb{R}^+$, and the weighted k-means objective then is $\min_{C \subset \mathbb{R}^d, |C|=k} \text{cost}_w(P, C)$. For a point set P, we denote the centroid of P as $\mu(P) := \frac{1}{|P|} \sum_{p \in P} p$. The k-means objective function satisfies the following well-known equation that allows to compute $\text{cost}(P, \{c\})$ via $\mu(P)$.

Fact 1. *Let $P \subset \mathbb{R}^d$ be a finite point set. Then the following equation holds:* $\sum_{p \in P} ||p - c||^2 = \sum_{p \in P} ||p - \mu(P)||^2 + |P| ||\mu(P) - c||^2.$

BIRCH. We only describe the main features of BIRCH's preclustering phase. The algorithm processes given points on the fly, storing them in a so called *CF Tree* where CF is the abbreviation of *Clustering Feature*.

Definition 1 (Clustering Feature). *Let $P := \{p_1, \dots, p_n\} \subset \mathbb{R}^d$ be a set of n d-dimensional points. The* Clustering Feature CF_P *of P is a 3-tuple (n, s_1, S_2) where n is number of points, $s_1 = \sum_{i=1}^{n} p_i$ is the linear sum of the points, and $S_2 = \sum_{i=1}^{n} ||p_i||^2$ is the sum of the squared lengths of the points.*

The usage of Clustering Features are the main space reduction concept of BIRCH. Notice that given a Clustering Feature $CF_P = (n, s_1, S_2)$, the squared distances of all points in a point set P to *one* given center c can be calculated *exactly* by $\text{cost}(P, \{c\}) = \sum_{i=1}^{n} ||p_i||^2 - n \cdot ||\mu(P) - 0||^2 + n \cdot ||c - \mu_i||^2 = S_2 - \frac{1}{n} ||s_1||^2 + n ||c - s_1/n||^2.$

When using Clustering Features to store a small summary of points, the quality of the summary decreases when storing points together in one CF that should be assigned to different centers later on. If we summarize points in a CF and later on get centers where all these points are closest to the same center, then their clustering cost can be computed with the CF without any error. Thus, the idea of BIRCH is to heuristically identify points that are likely to be clustered together. For this purpose, they use the following insertion process.

The first point in the input opens the first CF, i. e., a CF only containing the first point is created. Then, iteratively, the next points are added. For a new point p, BIRCH first looks for the CF which is 'closest' to p. Let CF_S be an arbitrary existing CF in the CF tree of BIRCH and recall that CF_S represents the set of points S. The distance between p and CF_S is defined as

$$\sum_{q \in S \cup \{p\}} (q - (\sum_{q' \in (S \cup \{p\})} q')/(|S| + 1))^2 - \sum_{q \in S} (q - (\sum_{q' \in S} q')/|S|)^2. \qquad (1)$$

Let CF_{S^*} be the CF closest to p. Then, p is added to CF_{S^*} if the radius $\sqrt{(\sum_{p \in (S^* \cup \{p\})} (q - \mu_{S^*})^2)/(|S| + 1)}$ is smaller than a given threshold t. If the radius exceeds the threshold, then p opens a new CF.

BIRCH works with increasing thresholds when processing the input data. It starts with threshold $t = 0$ and then increases t whenever the number of CFs exceeds a given space bound, calling a rebuilding algorithm to shrink the tree. This algorithm ensures that the number of CFs is decreased sufficiently. Notice that CFs cannot be split again, so the rebuilding might return a different tree than the one computed directly with the new threshold.

Coresets. BIRCH decides heuristically how to group the points into subclusters. We aim at refining this process such that the reduced set is not only small but is also guaranteed to approximate the original point set. More precisely, we aim at constructing a coreset:

Definition 2 ([21]). *A (k, ε)-coreset is a subset $S \subset \mathbb{R}^d$ weighted with a weight function $w' : S \to \mathbb{R}^d$ such that for all $C \subset \mathbb{R}^d$, $|C| = k$, it holds that $|\mathrm{cost}_{w'}(S, C) - \mathrm{cost}(P, C)| \leq \varepsilon \cdot \mathrm{cost}(P, C)$.*

3   BICO: Combining BIRCH and Coresets

The main problem with the insertion procedure of BIRCH is that the decision whether points are added to a CF or not is based on the increase of the radius of the candidate CF. Figure 1 shows a point set that was generated with two rather close but clearly distinguishable clusters plus randomly added points serving as noise. The problem for BIRCH is that the distance between the two clusters is not much larger than the average distance between the points in the noise. We see that BIRCH merges the clusters together and thus later computes only one center for them while the second center is placed inside the noise.

Our lower bound example for the quality guarantee of BIRCH follows this intuition. It looks similar to Figure 1, but is multi-dimensional and places the points deterministically in a structured way useful for the theoretical analysis. Let $d > 1$ and define the point sets $P_1 := \{(7, 3i_2, \ldots, 3i_d) \mid i_2, \ldots, i_d \in \{-2, -1, 0, 1, 2\}\}$ and $P_2 := \{(-7, 3i_2, \ldots, 3i_d) \mid i_2, \ldots, i_d \in \{-2, -1, 0, 1, 2\}\}$. Let R be the set of $(n - |P_1| - |P_2|)/2$ points at position $(1, 0, \ldots, 0)$ and equally many points at position $(-1, 0, \ldots, 0)$ and set $P = P_1 \cup P_2 \cup R$.

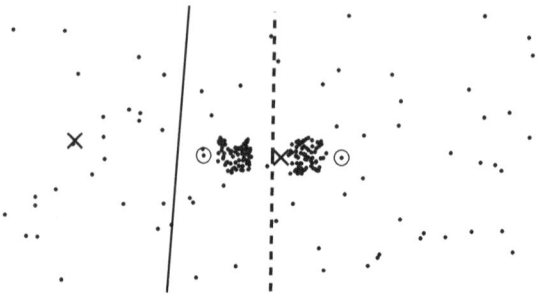

Fig. 1. An example created by drawing 150 points uniformly at random from the areas around $(-0.5, 0)$ and $(0, 0.5)$ and 75 points from $[-4, -2] \times [4, 2]$. BIRCH computed the centers marked by x leading to the partitioning by the solid line. BICO computed the same centers in 10 independent runs, marked by circles and the dashed line partitioning.

Theorem 1. *For $c > 0$, input $\tilde{P}$ and threshold $T > 0$, BIRCH either has $\Omega(n^{1/c})$ CFs or the computed solution has cost of at least $\Omega\left((n^{1-\frac{1}{c}}/\log n) \cdot OPT\right)$.*

The Basic Algorithm. Like BIRCH, BICO uses a tree whose nodes correspond to CFs. The tree has no distinguished root, but starts with possibly many CFs in level 1 (which can be seen as children of an imaginary root node). When we open a CF, we keep the first point in the CF as its *reference point*. The first point in the stream just opens a CF in level 1. Now, for each new point, we first try to add the point into an existing CF. We start on the first level $i = 1$. We try to insert the current point to the nearest CF in level i. The insertion fails if all CFs are far away, i.e. the distance between the current point and all CF reference points is larger than the *radius* R_i of CFs on level i, or the nearest CF is full, i.e. the cost of the point set represented by the CF is greater or equal than a threshold parameter T. In the first case we open a new CF with the current point as reference point in level i and in the second case we recurse and try to insert the point into the children of the nearest CF which are on level $i + 1$.

The algorithm is given in Algorithm 1. We assume that $n_{\max}$ is large enough to ensure that line 13 is never executed. We denote the CF that a point r is reference point of as $CF(r)$. By nearest(p, S) we refer to the reference point closest to p of all CFs in S. By children$(CF(r))$ we denote the set of CFs that are children of $CF(r)$ in the tree. For sake of shorter notation, we also use a virtual point ρ with virtual CF $CF(\rho)$ for the root node of the tree.

Theorem 2. *Let $\varepsilon > 0$, $f(\varepsilon) = (2 \cdot (\log n) \cdot 4^d \cdot \sqrt{40}^{d+2})/(\varepsilon^{d+2})$, $OPT/(k \cdot f(\varepsilon)) \leq T \leq 2 \cdot OPT/(k \cdot f(\varepsilon))$ and $R_i = \sqrt{T/(8 \cdot 2^i)}$. The set of centroids s_1/n', where (n', s_1, S_2) is a CF resulting from Algorithms 1, weighted with n' is a (k, ε)-coreset of size $O(k \cdot \log n \cdot \varepsilon^{-(d+2)})$ if the dimension d is a constant.*

The Rebuilding Algorithm. Above, we assumed that we know the cost of an optimal solution beforehand. To get rid of this assumption, we start with a small threshold, increase it if necessary and use the rebuilding algorithm given

---

Algorithm 1. Update mechanism where T and R_i are fixed parameters

---

input: $p \in \mathbb{R}^d$

1 Set $f = \rho$, $S = $ children $(CF(\rho))$ and $i = 1$;
2 **if** $S = \emptyset$ *or* $||p - \text{nearest}(p, S)|| > R_i$ **then**
3 | Open new CF with reference point p in level i as child of $CF(f)$;
4 **else**
5 | Set $r := \text{nearest}(p, S)$;
6 | **if** $\text{cost}_{CF}(CF(r) \cup \{p\}) \leq T$ **then**
7 | | Insert p in $CF(r)$;
8 | **else**
9 | | Set $S := $ children $(CF(r))$;
10 | | Set $f := r$ and $i := i + 1$;
11 | | Goto line 2;

12 **if** *number of current CFs* $> n_{\max}$ **then**
13 | Start rebuilding algorithm;

---

in Algorithm 2 to adjust the tree to the new threshold. If we start with a T smaller than $OPT/(k \cdot f(\varepsilon))$ and keep doubling it, then at some point T will be in $[OPT/(k \cdot f(\varepsilon)), 2 \cdot OPT/(k \cdot f(\varepsilon))]$, and at this point in time our coreset size of $O(k \cdot \log n \cdot \varepsilon^{-(d+2)})$ is sufficient to store a (k, ε)-coreset. Until this point, we will need rebuilding steps, but we will not lose quality.

The aim of the rebuilding algorithm is to create a tree which is similar to the tree which would have resulted from using the new threshold in the first place. Let R_i' be the radii before and let R_i be the radii after one iteration of the rebuilding algorithm. Notice that $R_i = \sqrt{2 \cdot T'/(8 \cdot 2^i)} = \sqrt{T'/(8 \cdot 2^{i-1})} = R_{i-1}'$. Thus, if a CF is not moved up and thus its level is increased by one, the radius will remain the same. This is nice as thus the CF in the same level automatically satisfy that the reference points are not within the radii of their neighbors. However, other properties of the tree do no longer hold in the original way: (1) If a CF $CF(r)$ on level i is inserted to a CF $CF(r')$ on level $i - 1$, i.e., $CF(r)$ becomes a new child of $CF(r')$ or they are merged, it is possible that some points which are represented by $CF(r)$ are not within the radius R_{i-1} of r'. But the distance can be bounded by $R_{i-1} + R_i$ such that the CFs on each level do not overlap too much. (2) The rebuilding algorithm can not split CFs. Thus, the set of CFs is different compared to a run where the new threshold was used in the first place. In total, these changes slightly increase the coreset size compared to Theorem 2, but it can be still bounded by $O(k \cdot \log n \cdot \varepsilon^{-(d+2)})$, and the weighted set of centroids also remains a (k, ε)-coreset.

Running Time. When trying to insert a point on level i, we need to decide whether the point is within distance R_i of its nearest neighbor, and if so, we need to locate the nearest neighbor. Let $N(m)$ denote the time needed for this, depending on the coreset size $m \in \mathcal{O}(k \cdot \log n \cdot \varepsilon^{-(d+2)})$. The running time of BICO is $\mathcal{O}(N(m) \cdot n)$ plus the time needed for the rebuilding steps.

Algorithm 2. Rebuilding algorithm when number of CFs gets too large

---

1 Set $T := 2 \cdot T$;
2 Create a new empty level 1 (which implicitly increases the number of all existing levels);
3 Let S_1 be the empty set of Clustering Features in level 1;
4 Let S_2 be the set of all Clustering Features in level 2;
5 **for** all Clustering Features $X \in S_2$ with reference point p **do**
6   **if** $S_1 = \emptyset$ or $||p - nearest(p, S_1)||^2 > R_1$ **then**
7   Move X from S_2 to S_1;
8   Notice that this implicitly moves all children of X one level up;
9   **else**
10   **if** $cost\,(X \cup CF(nearest(p, S_1))) \leq T$ **then**
11   Insert X into $CF(nearest(p, S_1))$;
12   **else**
13   Make X a child of $CF(nearest(p, S_1))$;

14 Traverse through the CF tree and, if possible, merge CFs into parent CFs

---

A rebuilding step needs to go through all m elements of the coreset and to insert them into the new-build tree. This takes $\mathcal{O}(m \cdot N(m))$ time. The number of rebuilding steps depends on how we choose the start value for T. We proved that BICO computes a (k, ε)-coreset for large enough m. In particular, this means that for any $m + 1$ points, BICO contracts at least two of them during the process. We use this observation by scanning through the first $m + 1$ points and calculating the minimal distance d_0 between two points (here, just ignore multiple points at the same position). Notice that if $T < d_0^2$, we are not able to merge any two points into one Clustering Feature. We set $T = d_0^2$. Then, T cannot be too small, because otherwise we cannot contract the m points.

The cost of any clustering is bounded from above by $n \cdot \Delta_{\max}$ where $\Delta_{\max}$ is the maximal distance between any two points. Our start value for T is bounded from below by the smallest distance $\Delta_{\min}$ between any two points. Thus, the factor between the start and end value of T is bounded by $n \cdot \frac{\Delta_{\max}}{\Delta_{\min}}$. The fraction $\Delta := \frac{\Delta_{\max}}{\Delta_{\min}}$ is called the *spread* of the points. With each rebuilding step we double T, and thus the number of rebuilding steps is bounded by $\log(n \cdot \Delta)$.

Corollary 1. *BICO computes a coreset for a set of n points in $\mathbb{R}^d$ given as an input only data stream in time bounded by $\mathcal{O}(N(m)(n + \log(n\Delta)m))$ using $\mathcal{O}(m)$ space where $m \in \mathcal{O}(k \cdot \log n \cdot \varepsilon^{-(d+2)})$ is the coreset size and d is constant.*

4 Experiments

Algorithms. We compare BICO with Stream-KM++, StreamLS, BIRCH and MacQueen's k-means algorithm. Stream-KM++ also aims at a trade-off between quality and speed which makes it most relevant for our work. BIRCH is the most relevant practical algorithm. We include MacQueen because it performed very well on one data set and is very fast. We use the author's implementations for Stream-KM++ [2], BIRCH [27] and StreamLS [16, 26] and an open source implementation of MacQueen's k-means [13]. We use the same parameters for BIRCH as in [2] except that we increase the memory to 26% on BigCross and 8% on Census in order to enable BIRCH to compute solutions for our larger k.

Setting. All computations were performed on seven identical machines with the same hardware configuration (2.8 Ghz Intel E7400 with 3 MB L2 Cache and 8 GB main memory). BICO and k-means++ are implemented in C++ and compiled with gcc 4.5.2. The source code for the algorithms, the testing environment and links to the other algorithms' source codes will appear at our website[1].

After computing the coreset, we determine the final solution via five weighted k-means++ runs (until convergence) and chose the solution with best cost on the coreset. The implementation of BICO differs from Section 3 in two points.

Coreset Size. Our theoretic analysis gives a worst case bound on the space needed by BICO even for adversarial inputs. On average instances, we expect that $\mathcal{O}(k)$ Clustering Features suffice to get a very good solution. The authors in [2] used the size $200k$ for Stream-KM++ which we also opted to use for both Stream-KM++ and BICO in our line of experiments. This leads to a asymptotic running time of $\mathcal{O}(k \cdot (n + k \log n\Delta))$ for BICO.

Filtering. A large part of the running time of BICO is spent for nearest neighbor queries on the first level of the tree. We speed it up by an easy heuristic: All CF reference points are projected to d one-dimensional subspaces chosen uniformly at random at the start of BICO. Let p be a new point. Since only reference points that are close to p in each subspace are candidates for the cluster feature we are searching for, we take the subspace where the number of points within distance R_1 of p is smallest and only iterate through these to find the nearest neighbor.

Datasets. We used the four largest data sets evaluated in [2], *Tower, Covertype* and *Census* from the UCI Machine Learning Repository [1] and *BigCross*, which is a subset of the Cartesian product of *Tower* and *Covertype*, created by the authors of [2] to have a very large data set. Additionally, we use a data set we call CalTech128 which is also large and has higher dimension. It consists of 128 SIFT descriptors [23] computed on the Caltech101 object database.

	BigCross	CalTech128	Census	CoverType	Tower
Number of Points (n)	11620300	3168383	2458285	581012	4915200
Dimension (d)	57	128	68	55	3
Total size ($n \cdot d$)	662357100	405553024	167163380	31955660	14745600

Experiments. On Census, Tower and BigCross, we ran tests with all values for k from [2], and $k = 250$ and $k = 1000$ in addition. On CalTech128, we tested $k = 50, 100, 250$ and $k = 1000$. We repeated all randomized algorithms 100 times and the diagrams show the mean values. In all diagrams, the bar of BICO is composed of two bars on top of each other corresponding to the core BICO part and the k-means++ part of BICO. We did not find parameters that enabled BIRCH to compute centers on CalTech128. Due to tests that we did

---

[1] http://ls2-www.cs.uni-dortmund.de/bico/

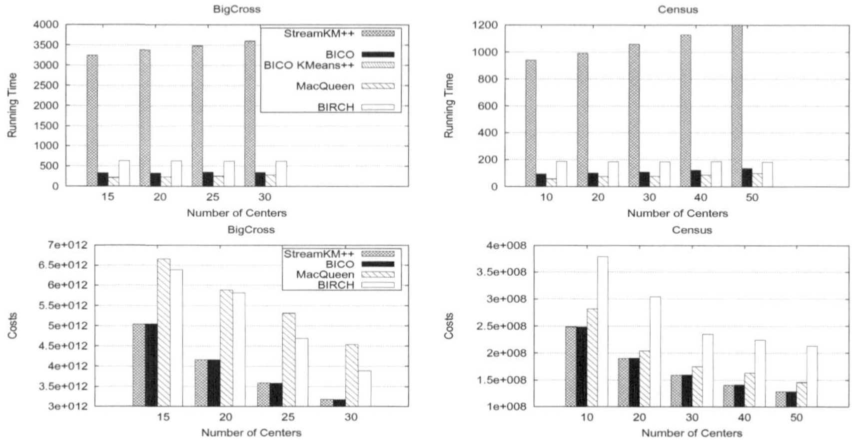

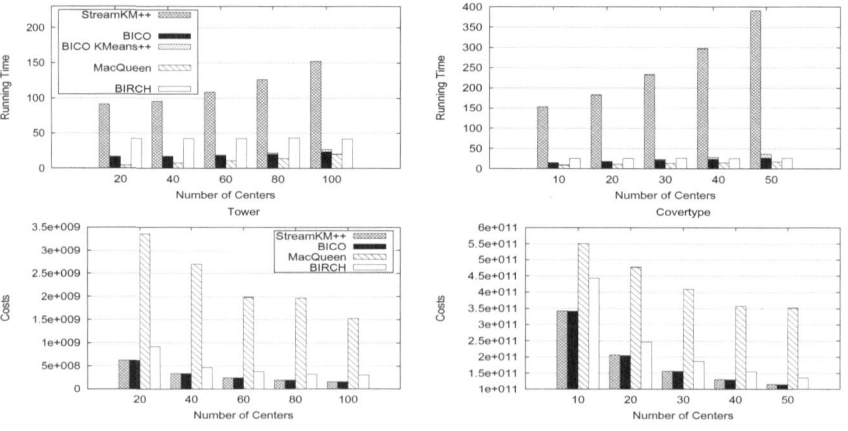

Fig. 2. Running times (in seconds) and costs for datasets BigCross and Census

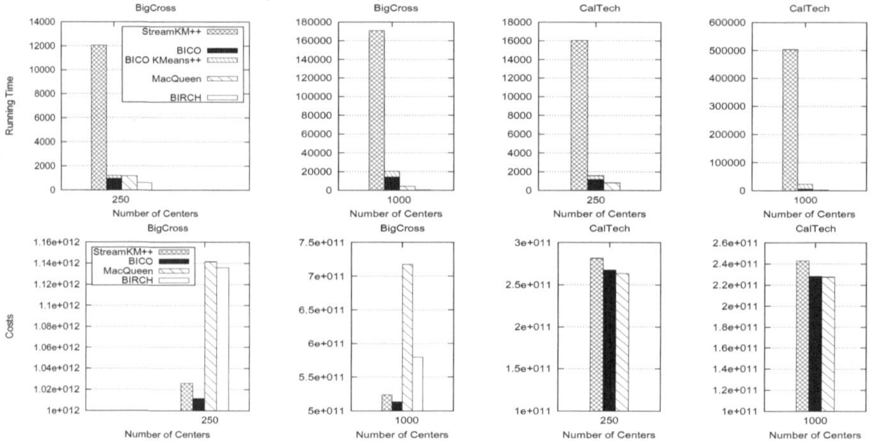

Fig. 3. Running times and costs for datasets BigCross and CalTech for large k

Fig. 4. Running times and costs for datasets Tower and Covertype

with modified versions of CalTech128 we believe that the implementation is not able to handle data with a dimension like CalTech128.

Comparison with Stream-KM++ and StreamLS. StreamLS, Stream-KM++ and BICO all have comparable solution quality (see Figures 2, 4, 3). StreamLS, however, is rather slow such that we did not include it in the diagrams and did not include Stream-LS in the tests for larger k. The running times of Stream-KM++ and BICO both depend on the number of centers which is reasonable because more centers require a larger coreset size which induces more effort to keep the coreset up to date. However, BICO is 5-10 times faster and applicable to much higher values of k (see Figure 2, 3, 4).

Comparison with BIRCH and MacQueen. BIRCH and MacQueen both tend to compute much worse solutions. MacQueen performs okay on Census, really well on CalTech128, but badly on BigCross and worse on Tower and Covertype (see Figures 2, 3, 4). MacQueen starts being faster than BIRCH, but is slower for larger k (compare BigCross in Figure 2 and 3) because BIRCH does not adjust for larger k. The running time of BICO is nearly always lower than that of BIRCH for small k (see Figures 2, 4) and within two times the running time of MacQueen in most experiments. For large k, BICO has significantly larger running time. If the dataset is high-dimensional like CalTech128, this is mainly due to k-means++, while on data sets with a lot of points of lower dimension like BigCross, the core part of BICO is dominating.

BICO for Large k. We point out that BICO is still practical for large k despite the large running times when adjusted. We chose BigCross because a running time of 4.6 hours is unfavourable and because here, the running time is due to core BICO and cannot be tackled by improving the k-means++ implementation (which is implemented without any speed-ups). By reducing the coreset size, the running time of BICO decreases. We lower it until BICO runs in 619 seconds compared to a running time of 616 seconds by BIRCH and 4241 seconds by MacQueen. The solution computed by BICO is still significantly better than the solutions by MacQueen and BIRCH.

Remark. Notice that many proof and experiment details are omitted due to space restrictions and will appear in a long version of the paper.

Acknowledgements. We thank René Grzeszick for providing the data set Cal-Tech128 and for pointing out the implementation of MacQueen's k-means algorithm. We also thank Frank Hellweg, Daniel R. Schmidt and Lukas Pradel for their help with the implementation and the example in Figure 1, and anonymous referees for their valuable comments concerning the presentation of the paper.

References

[1] Asuncion, A., Newman, D.J.: UCI machine learning repository (2007)
[2] Ackermann, M.R., Märtens, M., Raupach, C., Swierkot, K., Lammersen, C., Sohler, C.: Streamkm++: A clustering algorithm for data streams. ACM Journal of Experimental Algorithmics 17(1) (2012)
[3] Agarwal, P.K., Erickson, J.: Geometric range searching and its relatives. Contemporary Mathematics 223, 1–56 (1999)

[4] Agarwal, P.K., Har-Peled, S., Varadarajan, K.R.: Approximating extent measures of points. Journal of the ACM 51(4), 606–635 (2004)

[5] Arthur, D., Vassilvitskii, S.: How slow is the k-means method? In: Proc. of the 22nd SoCG, pp. 144–153 (2006)

[6] Arthur, D., Vassilvitskii, S.: k-means++: the advantages of careful seeding. In: Proc. of the 18th SODA, pp. 1027–1035 (2007)

[7] Bentley, J.L., Saxe, J.B.: Decomposable searching problems i: Static-to-dynamic transformation. J. Algorithms 1(4), 301–358 (1980)

[8] Chen, K.: On coresets for k-median and k-means clustering in metric and euclidean spaces and their applications. SIAM Journal on Computing 39(3), 923–947 (2009)

[9] Ester, M., Kriegel, H.P., Sander, J., Xu, X.: A density-based algorithm for discovering clusters in large spatial databases with noise. In: KDD, pp. 226–231 (1996)

[10] Feldman, D., Langberg, M.: A unified framework for approximating and clustering data. In: Proc. of the 43rd STOC, pp. 569–578 (2011)

[11] Feldman, D., Monemizadeh, M., Sohler, C.: A PTAS for k-means clustering based on weak coresets. In: Proc. 23rd SoCG, pp. 11–18 (2007)

[12] Feldman, D., Schmidt, M., Sohler, C.: Constant-size coresets for k-means, pca and projective clustering. In: Proc. of the 24th SODA, pp. 1434–1453 (2012)

[13] Fink, G.A., Plötz, T.: Open source project ESMERALDA

[14] Fisher, D.H.: Knowledge acquisition via incremental conceptual clustering. Machine Learning 2(2), 139–172 (1987)

[15] Frahling, G., Sohler, C.: Coresets in dynamic geometric data streams. In: Proc. of the 37th STOC, pp. 209–217 (2005)

[16] Guha, S., Meyerson, A., Mishra, N., Motwani, R., O'Callaghan, L.: Clustering data streams: Theory and practice. IEEE TKDE 15(3), 515–528 (2003)

[17] Guha, S., Rastogi, R., Shim, K.: Rock: A robust clustering algorithm for categorical attributes. Inform. Systems 25(5), 345–366 (2000)

[18] Guha, S., Rastogi, R., Shim, K.: Cure: An efficient clustering algorithm for large databases. Inform. Systems 26(1), 35–58 (2001)

[19] Halkidi, M., Batistakis, Y., Vazirgiannis, M.: On clustering validation techniques. Journal of Intelligent Inform. Systems 17(2-3), 107–145 (2001)

[20] Har-Peled, S., Kushal, A.: Smaller coresets for k-median and k-means clustering. Discrete & Computational Geometry 37(1), 3–19 (2007)

[21] Har-Peled, S., Mazumdar, S.: On coresets for k-means and k-median clustering. In: Proc. of the 36th STOC, pp. 291–300 (2004)

[22] Langberg, M., Schulman, L.J.: Universal epsilon-approximators for integrals. In: Proc. of the 21st SODA, pp. 598–607 (2010)

[23] Lowe, D.G.: Distinctive image features from scale-invariant keypoints. International Journal of Computer Vision 60(2), 91–110 (2004)

[24] MacQueen, J.B.: Some methods for classification and analysis of multivariate observations. In: Proc. 5th Berkeley Symp. on Math. Stat. and Prob., pp. 281–297 (1967)

[25] Ng, R.T., Han, J.: Clarans: A method for clustering objects for spatial data mining. IEEE TKDE 14(5), 1003–1016 (2002)

[26] O'Callaghan, L., Meyerson, A., Motwani, R., Mishra, N., Guha, S.: Streaming-data algorithms for high-quality clustering. In: Proc. 18th ICDE, pp. 685–694 (2002)

[27] Zhang, T., Ramakrishnan, R., Livny, M.: Birch: A new data clustering algorithm and its applications. Data Mining and Knowledge Discovery 1(2), 141–182 (1997)

Long Circuits and Large Euler Subgraphs[*]

Fedor V. Fomin and Petr A. Golovach

Department of Informatics, University of Bergen, PB 7803, 5020 Bergen, Norway
{fomin,petr.golovach}@ii.uib.no

Abstract. An undirected graph is Eulerian if it is connected and all its vertices are of even degree. Similarly, a directed graph is Eulerian, if for each vertex its in-degree is equal to its out-degree. It is well known that Eulerian graphs can be recognized in polynomial time while the problems of finding a maximum Eulerian subgraph or a maximum induced Eulerian subgraph are NP-hard. In this paper, we study the parameterized complexity of the following Euler subgraph problems:

- LARGE EULER SUBGRAPH: For a given graph G and integer parameter k, does G contain an induced Eulerian subgraph with at least k vertices?
- LONG CIRCUIT: For a given graph G and integer parameter k, does G contain an Eulerian subgraph with at least k edges?

Our main algorithmic result is that LARGE EULER SUBGRAPH is fixed parameter tractable (FPT) on undirected graphs. We find this a bit surprising because the problem of finding an induced Eulerian subgraph with exactly k vertices is known to be W[1]-hard. The complexity of the problem changes drastically on directed graphs. On directed graphs we obtained the following complexity dichotomy: LARGE EULER SUBGRAPH is NP-hard for every fixed $k > 3$ and is solvable in polynomial time for $k \leq 3$. For LONG CIRCUIT, we prove that the problem is FPT on directed and undirected graphs.

1 Introduction

One of the oldest theorems in Graph Theory is attributed to Euler, and it says that a (undirected) graph admits an *Euler circuit*, i.e., a closed walk visiting every edge exactly once, if and only if the graph is connected and all its vertices are of even degrees. Respectively, a directed graph has a *directed Euler circuit* if and only if the graph is (weakly) connected and for each vertex, its in-degree is equal to its out-degree. While checking if a given directed or undirected graph is Eulerian is easily done in polynomial time, the problem of finding k edges (arcs) in a graph to form an Eulerian subgraph is NP-hard. We refer to the book of Fleischner [12] for a thorough study of Eulerian graphs and related topics.

In [5], Cai and Yang initiated the study of parameterized complexity of subgraph problems motivated by Eulerian graphs. Particularly, they considered the following parameterized subgraph and induced subgraph problems:

---

[*] Supported by the European Research Council (ERC) via grant Rigorous Theory of Preprocessing, reference 267959.

H.L. Bodlaender and G.F. Italiano (Eds.): ESA 2013, LNCS 8125, pp. 493–504, 2013.

---

k-CIRCUIT   **Parameter:** k
Input: A (directed) graph G and non-negative integer k
Question: Does G contain a circuit with k edges (arcs)?

---

and

EULER k-SUBGRAPH   **Parameter:** k
Input: A (directed) graph G and non-negative integer k
Question: Does G contain an induced Euler subgraph with k vertices?

The decision versions of both k-CIRCUIT and EULER k-SUBGRAPH are known to be the NP-complete [5]. Cai and Yang in [5] proved that k-CIRCUIT on undirected graphs is FPT. On the other hand, the authors have shown in [14] that EULER k-SUBGRAPH is W[1]-hard. The variant of the problem $(m-k)$-CIRCUIT, where one asks to remove at most k edges to obtain an Eulerian subgraph was shown to be FPT by Cygan et al. [7] on directed and undirected graphs. The problem of removing at most k vertices to obtain an induced Eulerian subgraph, namely EULER $(n-k)$-SUBGRAPH, was shown to be W[1]-hard by Cai and Yang for undirected graphs [5] and by Cygan et al. for directed graphs [7]. Dorn et al. in [8] provided FPT algorithms for the weighted version of Eulerian extension.

In this work we extend the set of results on the parameterized complexity of Eulerian subgraph problems by considering the problems of finding an (induced) Eulerian subgraph with *at least* k (vertices) edges. We consider the following problems:

LARGE EULER SUBGRAPH   **Parameter:** k
Input: A (directed) graph G and non-negative integer k
Question: Does G contain an induced Euler subgraph with at least k vertices?

---

and

LONG CIRCUIT   **Parameter:** k
Input: A (directed) graph G and non-negative integer k
Question: Does G contain a circuit with at least k edges (arcs)?

The decision version of LONG CIRCUIT was shown to be NP-complete by Cygan et al. in [7] and it is not difficult to see that the same is true for LARGE EULER SUBGRAPH. Let us note that by plugging-in these observations into the framework of Bodlaender et al. [4], it is easy to conclude that on undirected graphs both problems have no polynomial kernels unless NP $\subseteq$ coNP/poly.

However, the parameterized complexity of these problems appears to be much more interesting.

Our Results. We show that LARGE EULER SUBGRAPH behaves differently for directed and undirected cases. For undirected graphs, we prove that the problem is FPT. We find this result surprising, because the very closely related EULER

k-SUBGRAPH problem is known to be W[1]-hard [14]. The proof is based on a structural result interesting in its own. Roughly speaking, we show that large treewidth certifies containment of a large induced Euler subgraph. For directed graphs, LARGE EULER SUBGRAPH is NP-complete for each $k \geq 4$, and this bound is tight—the problem is polynomial-time solvable for each $k \leq 3$. We also prove that EULER k-SUBGRAPH is W[1]-hard for directed graphs. LONG CIRCUIT is proved to be FPT for directed and undirected graphs. Our algorithm is based on the results by Gabow and Nie [15] about the parameterized complexity of finding long cycles. The known and new results about Euler subgraph problems are summarized in Table 1.

Table 1. Parameterized complexity of Euler subgraph problems

	Undirected	Directed
k-CIRCUIT	FPT [5]	FPT, Prop. 2
EULER k-SUBGRAPH	W[1]-hard [14]	W[1]-hard, Thm 3
$(m-k)$-CIRCUIT	FPT [7]	FPT [7]
EULER $(n-k)$-SUBGRAPH	W[1]-hard [5]	W[1]-hard [7]
LONG CIRCUIT	FPT, Thm 7	FPT, Thm 7
LARGE EULER SUBGRAPH	FPT, Thm 2	NP-complete for $\forall\, k \geq 4$, Thm 4; in P for $k \leq 3$

This paper is organised as follows. Section 2 contains basic definitions and preliminaries. In Section 3.1 we show that LARGE EULER SUBGRAPH is FPT on undirected graphs. In Section 3.2 we prove that on directed graphs, EULER k-SUBGRAPH is W[1]-hard while LARGE EULER SUBGRAPH is NP-complete for each $k \geq 4$. In Section 4 we treat LONG CURCUIT and show that it is FPT on directed and undirected graphs.

2   Basic Definitions and Preliminaries

Graphs. We consider finite directed and undirected graphs without loops or multiple edges. The vertex set of a (directed) graph G is denoted by $V(G)$, the edge set of an undirected graph and the arc set of a directed graph G is denoted by $E(G)$. To distinguish edges and arcs, the edge with two end-vertices u, v is denoted by $\{u, v\}$, and we write (u, v) for the corresponding arc. For a set of vertices $S \subseteq V(G)$, $G[S]$ denotes the subgraph of G induced by S, and by $G - S$ we denote the graph obtained form G by the removal of all the vertices of S, i.e., the subgraph of G induced by $V(G) \setminus S$. Let G be an undirected graph. For a vertex v, we denote by $N_G(v)$ its *(open) neighborhood*, that is, the set of vertices which are adjacent to v. The *degree* of a vertex v is denoted by $d_G(v) = |N_G(v)|$, and $\Delta(G)$ is the maximum degree of G. Let now G be a directed graph. For a vertex $v \in V(G)$, we say that u is an *in-neighbor* of v if $(u, v) \in E(G)$. The set of all in-neighbors of v is denoted by $N_G^-(v)$. The *in-degree* $d_G^-(v) = |N_G^-(v)|$.

Respectively, u is an *out-neighbor* of v if $(v,u) \in E(G)$, the set of all out-neighbors of v is denoted by $N_G^+(v)$, and the *out-degree* $d_G^+(v) = |N_G^+(v)|$.

For a (directed) graph G, a (directed) *trail* of *length* k is a sequence $v_0, e_1, v_1, e_2, \ldots, e_k, v_k$ of vertices and edges (arcs resp.) of G such that $v_0, \ldots, v_k \in V(G)$, $e_1, \ldots, e_k \in E(G)$, the edges (arcs resp.) $e_1, \ldots, e_k$ are pairwise distinct, and for $i \in \{1, \ldots, k\}$, $e_i = \{v_{i-1}, v_i\}$ ($e_i = (v_{i-1}, v_i)$ resp.). A trail is said to be *closed* if $v_0 = v_k$. A closed (directed) trail is called a (directed) *circuit*, and it is a (directed) *cycle* if all its vertices except $v_0 = v_k$ are distinct. Clearly, any cycle is a subgraph of G, and it is said that C is an *induced cycle* of G if $C = G[V(C)]$. A (directed) path is a trail such that all its vertices are distinct. For a (directed) walk (trail, path resp.) $v_0, e_1, v_1, e_2, \ldots, e_k, v_k$, v_0 and v_k are its *end-vertices*, and $v_1, \ldots, v_{k-1}$ are its *internal* vertices. For a (directed) walk (trail, path resp.) with end-vertices u and v, we say that it is an (u,v)-*walk* (*trail*, *path* resp.). We omit the word "directed" if it does not create a confusion. Also we write a trail as a sequence of its vertices $v_0, \ldots, v_k$.

A connected (directed) graph G is an *Euler* (or *Eulerian*) graph if it has a (directed) circuit that contains all edges (arcs resp.) of G. By the celebrated result of Euler (see, e.g., [12]), a connected graph G is an Euler graph if and only if all its vertices have even degrees. Respectively, a connected directed graph G is an Euler directed graph if and only if for each vertex $v \in V(G)$, $d_G^-(v) = d_G^+(v)$.

Ramsey Numbers. The *Ramsey number* $R(r,s)$ is the minimal integer n such that any graph on n vertices has either a clique of size r or an independent set of size s. By the famous paper of Erdös and Szekeres [10], $R(r,s) \leq \binom{r+s-2}{r-1}$.

Parameterized Complexity. Parameterized complexity is a two dimensional framework for studying the computational complexity of a problem. One dimension is the input size n and another one is a parameter k. It is said that a problem is *fixed parameter tractable* (or FPT), if it can be solved in time $f(k) \cdot n^{O(1)}$ for some function f, and it is said that a problem is in XP, if it can be solved in time $O(n^{f(k)})$ for some function f. One of the basic assumptions of the Parameterized Complexity theory is the conjecture that the complexity class W[1] $\neq$ FPT, and it is unlikely that a W[1]-hard problem could be solved in FPT time. A problem is Para-NP-*hard*(*complete*) if it is NP-hard (complete) for some fixed value of the parameter k. Clearly, a Para-NP-hard problem is not in XP unless P=NP. We refer to the books of Downey and Fellows [9], Flum and Grohe [13], and Niedermeier [19] for detailed introductions to parameterized complexity.

Treewidth. A *tree decomposition* of a graph G is a pair (X,T) where T is a tree and $X = \{X_i \mid i \in V(T)\}$ is a collection of subsets (called *bags*) of $V(G)$ such that:

1. $\bigcup_{i \in V(T)} X_i = V(G)$,
2. for each edge $\{x,y\} \in E(G)$, $x, y \in X_i$ for some $i \in V(T)$, and
3. for each $x \in V(G)$ the set $\{i \mid x \in X_i\}$ induces a connected subtree of T.

The *width* of a tree decomposition $(\{X_i \mid i \in V(T)\}, T)$ is $\max_{i \in V(T)} \{|X_i| - 1\}$. The *treewidth* of a graph G (denoted as $\mathbf{tw}(G)$) is the minimum width over all tree decompositions of G.

3   Large Euler Subgraphs

3.1   Large Euler Subgraphs for Undirected Graphs

In this section we show that LARGE EULER SUBGRAPH is FPT for undirected graphs. Using Ramsey arguments, we prove that if a graph G has sufficiently large treewidth, then G has an induced Euler subgraph on at least k vertices. Then if the input graph has large treewidth, we have a YES-answer. Otherwise, we use the fact that LARGE EULER SUBGRAPH can be solved in FPT time for graphs of bounded treewidth. All graphs considered here are undirected.

For a given positive integer k, we define the function $f(\ell)$ for integers $\ell \geq 2$ recursively as follows:

- $f(2) = R(k, k-1) + 1$,
- $f(\ell) = (k-1)(2(\ell-1)(f(\lfloor \frac{\ell}{2} \rfloor + 1) - 1) + 1) + 1$ for $\ell > 2$.

We need the following two lemmas.

Lemma 1. *Let G be a graph, and suppose that s, t are distinct vertices joined by at least $f(\ell)$ internally vertex-disjoint paths of length at most ℓ in G for some $\ell \geq 2$. Then G has an induced Euler subgraph on at least k vertices.*

Proof. Consider the minimum value of ℓ such that G has $f(\ell)$ internally vertex disjoint (s, t)-paths. We have at least $r = f(\ell) - 1$ such paths $P_1, \ldots, P_r$ that are distinct from the trivial (s, t)-path with one edge. We assume that each path P_i has no chords that either join two internal vertices or an internal vertex and one of the end-vertices, i.e., each internal vertex is adjacent in $G[V(P_i)]$ only to its two neighbors in P_i. Otherwise, we can replace P_i by a shorter path with all vertices in $V(P_i)$ distinct from the path s, t. We consider two cases.

Case 1. $\ell = 2$. The paths $P_1, \ldots, P_r$ are of length two and therefore have exactly one internal vertex. Assume that $u_1, \ldots, u_r$ are internal vertices of these paths. Because $r = f(2) - 1 = R(k, k-1)$, the graph $G[\{u_1, \ldots, u_r\}]$ either has a clique K of size k or an independent set I of size at least $k - 1$. Suppose that G has a clique K. If k is odd, then $G[K]$ is an induced Euler subgraph on k vertices. If k is even, then $G[K \cup \{s\}]$ is an induced Euler subgraph on $k + 1$ vertices. Assume now that that $I \subseteq \{u_1, \ldots, u_{r-1}\}$ is an independent set of size $k - 1$. Let $v \in I$. If $\{s, t\} \in E(G)$ and k is even or $\{s, t\} \notin E(G)$ and k is odd, then $G[I \cup \{s, t\}]$ is an induced Euler subgraph on $k + 1$ vertices. Else if $\{s, t\} \notin E(G)$ and k is even or $\{s, t\} \in E(G)$ and k is odd, then $G[I \cup \{s, t\} \setminus \{v\}]$ is an induced Euler subgraph on k vertices.

Case 2. $\ell \geq 3$. We say that paths P_i and P_j are *adjacent* if they have adjacent internal vertices. Let $p = f(\lfloor \ell/2 \rfloor + 1)$. Suppose that there is an internal vertex v of one of the paths $P_1, \ldots, P_r$ that is adjacent to at least $2p - 1$ internal vertices of some other distinct $2p - 1$ paths. Then there are $p = f(\lfloor \ell/2 \rfloor + 1)$ paths $P_{i_1}, \ldots, P_{i_p}$ that have respective internal vertices $v_1, \ldots, v_p$ such that i) v is adjacent to $v_1, \ldots, v_p$ and ii) either each v_j is at distance at most $\lfloor \ell/2 \rfloor$ from s in P_{i_j} for all $j \in \{1, \ldots, p\}$ or each v_j is at distance at most $\lfloor \ell/2 \rfloor$ from t in P_{i_j}

for all $j \in \{1, \dots, p\}$. But then either the vertices s, v or v, t are joined by at least $f(\lfloor \ell/2 \rfloor + 1)$ internally vertex-disjoint paths of length at most $\lfloor \ell/2 \rfloor + 1 < \ell$. This contradicts our choice of ℓ. Hence, for each $i \in \{1, \dots, r\}$, any internal vertex of P_i has adjacent internal vertices in at most $2p - 2$ other paths, and P_i is adjacent to at most $2(\ell - 1)(p - 1)$ other paths. As $r = (k-1)(2(\ell-1)(p-1)+1)$, there are $k - 1$ distinct paths $P_{i_1}, \dots, P_{i_{k-1}}$ that are pairwise non-adjacent, i.e., they have no adjacent internal vertices.

Let $H = G[V(P_{i_1}) \cup \dots \cup V(P_{i_{k-1}})]$ and $H' = G[V(P_{i_1}) \cup \dots \cup V(P_{i_{k-2}})]$. Notice that by our choice of the paths, $H = P_{i_1} \cup \dots \cup P_{i_{k-1}}$ and $H' = P_{i_1} \cup \dots \cup P_{i_{k-2}}$ if $\{s, t\} \notin E(G)$, and $P_{i_1} \cup \dots \cup P_{i_{k-1}}$ ($P_{i_1} \cup \dots \cup P_{i_{k-2}}$ resp.) can be obtained from H (H' resp.) by the removal of $\{s, t\}$ if s, t are adjacent. If $\{s, t\} \in E(G)$ and k is even or $\{s, t\} \notin E(G)$ and k is odd, then H is an induced Euler subgraph on at least $k + 1$ vertices. Else if $\{s, t\} \notin E(G)$ and k is even or $\{s, t\} \in E(G)$ and k is odd, then H' is an induced Euler subgraph on at least k vertices. $\qquad\square$

For $k \geq 4$, let

$$\Delta_k = 1 + \frac{(f(3k-8)-1)((f(3k-8)-2)^{3(k-3)}-1)}{f(3k-8)-3}.$$

Lemma 2. *For $k \geq 4$, any 2-connected graph G with $\Delta(G) > \Delta_k$ has an induced Euler subgraph on at least k vertices.*

Proof. Let G be a 2-connected graph and let u be a vertex of G with $d_G(u) = \Delta(G)$. As G is 2-connected, $G' = G - u$ is connected. Let v be an arbitrary vertex of $N_G(u)$. Denote by T a tree of shortest paths from v to all other vertices of $N_G(u)$ in G'.

Claim A. *If there is a (v, w)-path P of length at least $3(k-3)+1$ in T for some $w \in N_G(u)$, then G has an induced Euler subgraph on at least k vertices.*

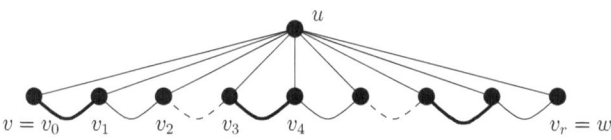

Fig. 1. The path P and the graphs Q_1 (shown by the thick lines), Q_2 (shown by the thin lines), and Q_3 (shown by the dashed lines)

Proof (of Claim A). Denote by $v_0, \dots, v_r$ the vertices of $N_G(u)$ in P in the path order, $v_0 = v$ and $v_r = w$. Let Q_1 be the union of the $(v_0, v_1), (v_3, v_4), \dots, (v_{\lfloor r/3 \rfloor}, v_{\lfloor r/3 \rfloor + 1})$-subpaths of P, let Q_2 be the union of the $(v_1, v_2), (v_4, v_5), \dots, (v_{\lfloor r/3 \rfloor + 1}, v_{\lfloor r/3 \rfloor + 2})$-subpaths of P, and let Q_3 be the union of the $(v_2, v_3), (v_5, v_6), \dots, (v_{\lfloor r/3 \rfloor - 1}, v_{\lfloor r/3 \rfloor})$-subpaths of P as it is sown in Fig. 1. Notice that some subpaths can be empty depending whether r modulo 3 is 0, 1 or 2. Observe that Q_1, Q_2, Q_3 are edge-disjoint induced subgraphs of G. Since $Q_1 \cup Q_2 \cup Q_3 = P$, there is Q_i for $i \in \{1, 2, 3\}$ with at least $k - 2$ edges. Then Q_i has at least $k - 1$ vertices. Let $H = G[V(Q_i) \cup \{u\}]$. By the definition of Q_i, H is a union of induced cycles with one common vertex u such that for different cycles C_1, C_2 in the union, $V(C_1) \cap V(C_2) = \{u\}$ and $\{x, y\} \notin E(G)$ whenever $x \in V(C_1) \setminus \{u\}$ and $y \in V(C_2) \setminus \{u\}$. Hence, H is an Euler graph with at least k vertices. $\qquad\square$

From now we assume that all (v, w)-paths in T have length at most $3(k - 3)$ for $w \in N_G(u)$.

Claim B. *If there is a vertex $w \in V(T)$ with $d_T(w) \geq f(3(k - 3) + 1)$, then G has an induced Euler subgraph on at least k vertices.*

Proof (of Claim B). Recall that T is a tree of shortest paths from v to all other vertices of $N_G(u)$ in G'.. We assume that T is rooted in v. Then the root defines the parent-child relation on T. Let x_0 be a parent of w (if exists) and let $x_1, \ldots, x_r$ be the children of w. If w has no parent, then $w = v$ and $r \geq f(3(k - 3) + 1)$. Otherwise, $r \geq f(3(k - 3) + 1) - 1$. Let $y_0 = v$. Because each leaf of T is a vertex of $N_G(u)$, for each $i \in \{1, \ldots, r\}$, there is a descendant $y_i \in V(T) \cap N_G(u)$ of x_i in T. Denote by P_i the unique (w, y_i)-path in T for $i \in \{0, \ldots, r\}$. As all (v, w)-paths in T have length at most $3(k - 3)$ for $w \in N_G(u)$, the paths $P_0, \ldots, P_r$ have length at most $3(k - 3)$. Notice that these paths have no common vertices except w. Observe now that $y_0, \ldots, y_r$ are adjacent to u in G. Therefore, we have at least $f(3(k - 3) + 1)$ internally vertex-disjoint (u, w)-paths in G. By Lemma 1, it implies that G has an induced Euler subgraph on at least k vertices. $\qquad\square$

To complete the proof of the lemma, it remains to observe that if $\Delta(T) < f(3(k - 3) + 1)$ and all (v, w)-paths in T have length at most $3(k - 3)$ for $w \in N_G(u)$, then

$$d_G(u) \leq |V(T)| \leq 1 + \frac{(f(3k - 8) - 1)((f(3k - 8) - 2)^{3(k-3)} - 1)}{f(3k - 8) - 3} = \Delta_k.$$

$\qquad\square$

Kosowski et al. [18] obtained the following bound for treewidth.

Theorem 1 ([18]). *Let G be a graph without induced cycles with at least $k \geq 3$ vertices and let $\Delta(G) \geq 1$. Then $\mathbf{tw}(G) \leq k(\Delta(G) - 1) + 2$.*

This theorem together with Lemma 2 immediately imply the next lemma.

Lemma 3. *Let G be a graph and let $k \geq 4$. If $\mathbf{tw}(G) > k(\Delta_k - 1) + 2$, then G has an induced Euler subgraph on at least k vertices.*

Proof. Suppose that $\mathbf{tw}(G) > k(\Delta_k - 1) + 2$. Then G has a 2-connected component G' with $\mathbf{tw}(G') > k(\Delta_k - 1) + 2$. If $\Delta(G') > \Delta_k$, then G' has an induced Euler subgraph on at least k vertices by Lemma 2. Otherwise, by Theorem 1, G' has an induced cycle on at least k vertices, i.e., an induced Euler subgraph. $\qquad\square$

Now we observe that LARGE EULER SUBGRAPH is FPT for graphs of bounded treewidth.

Lemma 4. *For any positive integer* t, LARGE EULER SUBGRAPH *can be solved in linear time for graphs of treewidth at most* t.

Now we can prove the main result of this section.

Theorem 2. *For any positive integer* k, LARGE EULER SUBGRAPH *can be solved in linear time for undirected graphs.*

Proof. Clearly, we can assume that $k \geq 3$, as any Euler graph has at least three vertices. If $k = 3$, then we can find any shortest cycle in the input graph G. It is straightforward to see that if G has no cycles, then we have no Euler subgraph, and any induced cycle is an induced Euler subgraph on at least three vertices. Hence, it can be assumed that $k \geq 4$. We check in linear time whether $\mathbf{tw}(G) \leq k(\Delta_k - 1) + 2$ using the Bodlaender's algorithm [3]. If it is so, we solve our problem using Lemma 4. Otherwise, by Lemma 3, we conclude that G has an induced Euler subgraph on at least k vertices and return a YES-answer. □

Notice, that the proof of Theorem 2 is not constructive. Next, we sketch the algorithm that produces an induced Euler subgraph on at least $k \geq 4$ vertices if it exists.

First, for each $\ell \geq 2$, we can test the existence of two vertices s, t such that the input graph G has at least $f(\ell)$ internally vertex-disjoint (s, t)-paths of length at most ℓ in FPT time with the parameter ℓ using the color coding technique [17]. If we find such a structure for $\ell \leq 3k - 8$, we find an induced Euler subgraph with at least k vertices that is either a clique or a union of (s, t)-paths as it is explained in the proof of Lemma 1.

Otherwise, we find all 2-connected components. If there is a 2-connected component G' with a vertex u with $d_{G'}(u) > \Delta_k$, then we find an induced Euler subgraph with at least k vertices that is a union of cycles with the common vertex u using the arguments form the proof of Lemma 2.

If all 2-connected components have bounded maximum degrees, we use the algorithm of Kosowski et al. [18] that in polynomial time either finds an induced cycle on at least k vertices or constructs a tree decomposition of width at most $k(\Delta_k - 1) + 2$. In the fist case we have an induced Euler subgraph on at least k vertices. In the second case the treewidth is bounded, and LARGE EULER SUBGRAPH is solved by a dynamic programming algorithm instead of applying Lemma 4.

3.2   Large Euler Subgraphs for Directed Graphs

In this section we show that EULER k-SUBGRAPH and LARGE EULER SUBGRAPH are hard for directed graphs. Due the space restrictions, the proofs are omitted.

First, we consider EULER k-SUBGRAPH. It is straightforward to see that this problem is in XP, since we can check for every subset of k vertices, whether it induces an Euler subgraph. We prove that this problem cannot be solved in FPT time unless FPT = W[1].

Theorem 3. *The* EULER *k*-SUBGRAPH *is* W[1]*-hard for directed graphs.*

For LARGE EULER SUBGRAPH for directed graphs, we prove that this problem is Para-NP-complete.

Theorem 4. *For any* $k \geq 4$, LARGE EULER SUBGRAPH *is* NP*-complete for directed graphs.*

We proved that LARGE EULER SUBGRAPH is NP-complete for directed graphs for $k \geq 4$. In the conclusion of this section we observe that the bound $k \geq 4$ is tight unless P=NP.

Proposition 1. LARGE EULER SUBGRAPH *can be solved in polynomial time for* $k \leq 3$.

4 Long Circuits

In this section we show that the LONG CIRCUIT problem is FPT for directed and undirected graphs.

We need the following auxiliary problem:

AT LEAST k AND AT MOST k'-CIRCUIT   **Parameter:** k'
Input: A (directed) graph G and non-negative integers k, k', $k \leq k'$
Question: Does G contain a circuit with at least k and at most k' edges (arcs)?

Clearly, we can solve this problem in FPT time for undirected graphs applying the algorithm by Cai and Yang [5] for r-CIRCUIT for each $r \in \{k, \ldots, k'\}$. For directed graphs, we can use the same approach based on the color coding technique introduced by Alon, Yuster and Zwick [1]. For completeness, we sketch the proof here.

Lemma 5. *The* AT LEAST k AND AT MOST k'-CIRCUIT *problem can be solved in* $2^{O(k')} \cdot nm$ *expected time and in* $2^{O(k')} \cdot nm \log n$ *worst-case time for (directed) graphs with* n *vertices and* m *edges (arcs).*

Proof. As the algorithms for directed and undirected graphs are basically the same, we consider here the directed case. For simplicity, we solve the decision problem, but the algorithm can be easily modified to obtain a circuit of prescribed size if it exists.

Let G be a directed graph with n vertices and m arcs.

First, we describe the randomized algorithm. We color the arcs of G by k' colors $1, \ldots, k'$ uniformly at random independently from each other. Now we are looking for a *colorful* circuit in G that has at least k arcs, i.e., for a circuit such that all arcs are colored by distinct colors.

To do it, we apply the dynamic programming across subsets. We choose an initial vertex u and try to construct a circuit that includes u. For a set of colors $X \subseteq \{1, \ldots, k'\}$, denote by $U(X)$ the set of vertices $v \in V(G)$ such that there is a (u, v)-trail with $|X|$ edges colored by distinct colors from X. It is straightforward to see that $U(\emptyset) = \{u\}$. For $X \neq \emptyset$, $v \in U(X)$ if and only if v has an in-neighbor $w \in N_G^-(v)$ such that (w, v) is colored by a color $c \in X$ and $w \in U(X \setminus \{c\})$. We consequently construct the sets $U(X)$ for X with $1, 2, \ldots, k'$ elements. We stop and return a YES-answer if $u \in U(X)$ for some X of size at least k. Notice that the sets $U(X)$ can be constructed in time $O(k'2^{k'} \cdot m)$. Since we try all possibilities to select u, the running time is $O(k'2^{k'} \cdot mn)$.

Now we observe that for any positive number $p < 1$, there is a constant c_p such that after running our randomized algorithm $c_p 2^{O(k')}$ times, we either get a YES-answer or can claim that with probability p G has no directed circuit with at least k and at most k' arcs.

This algorithm can be derandomized by the technique proposed by Alon, Yuster and Zwick [1]. To do it, we replace random colorings by a family of at most $2^{O(k')} \log n$ hash functions that can be constructed in time $2^{O(k')} \cdot m \log n$.   □

If we set $k' = k$, then Lemma 5 immediately implies the following proposition. Notice that it was proved in [5] for undirected graphs.

Proposition 2. *The k-CIRCUIT problem can be solved in $2^{O(k)} \cdot nm$ expected time and in $2^{O(k)} \cdot nm \log n$ worst-case time for (directed) graphs with n vertices and m edges (arcs).*

Gabow and Nie in [15] considered the LONG CYCLE problem:

LONG CYCLE **Parameter:** k
Input: A (directed) graph G and a positive integer k
Question: Does G contain a cycle with at least k edges (arcs)?

In particular, they proved the following theorem.

Theorem 5 ([15]). *The LONG CYCLE problem can be solved in $2^{O(k \log k)} \cdot nm$ expected time and in $2^{O(k \log k)} \cdot nm \log n$ worst-case time for directed graphs with n vertices and m arcs.*

Let us recall that a fundamental cycle in undirected graph is formed from a spanning tree and a nontree edge. For undirected graphs, it is slightly more convenient to use the structural result by Gabow and Nie.

Theorem 6 ([15]). *In a connected undirected graph having a cycle with k edges, either every depth-first search tree has a fundamental cycle with at least k edges or some cycle with at least k edges has at most $2k - 4$ edges.*

We need the following observation.

Lemma 6. *Let G be a (directed) graph without cycles of length at least k. If G has a circuit with at least k edges (arcs resp.), then G has a circuit with at least k and at most $2k - 2$ edges (arcs resp.).*

Proof. Let C be a circuit in G. It is well-known (see, e.g., [12]) that C is a union of edge-disjoint (arc-disjoint) cycles $C_1, \ldots, C_r$. Moreover, it can be assumed that for any $i \in \{1, \ldots, r\}$, the circuit $C_1 \cup \ldots \cup C_i$ is connected. Suppose now that C is a circuit with at least k edges (arcs resp.) that has minimum length. Then the circuit $C' = C_1 \cup \ldots \cup C_{r-1}$ has at most $k - 1$ edges (arcs resp.). Since G has no cycles of length at least k, C_r has at most $k - 1$ edges (arcs resp.). Thus C has at most $2k - 2$ edges (arcs resp.). □

Now we are ready to prove the main result of this section.

Theorem 7. *The LONG CIRCUIT problem can be solved in $2^{O(k \log k)} \cdot nm$ expected time and in $2^{O(k \log k)} \cdot nm \log n$ worst-case time for directed graphs with n vertices and m arcs, and in $2^{O(k)} \cdot nm$ expected time and in $2^{O(k)} \cdot nm \log n$ worst-case time for undirected graphs with n vertices and m edges.*

Proof. First, we consider directed graphs. Let G be a directed graph. By Theorem 5, we can check whether G has a cycle with at least k arcs. If we find such a cycle C, then C is a circuit with at least k arcs, and we have a YES-answer. Otherwise, we conclude that each cycle in G is of length at most $k - 1$. Then by Lemma 6, if G has a circuit with at least k arcs, then it has a circuit with at least k and at most $2k - 2$ arcs. We find such a circuit (if it exist) by solving AT LEAST k AND AT MOST k'-CIRCUIT for $k' = 2k - 2$ by making use of Lemma 5. Combining the running times, we have that LONG CIRCUIT can be solved in $2^{O(k \log k)} \cdot nm$ expected time and in $2^{O(k \log k)} \cdot nm \log n$ worst-case time.

For the undirected case, we assume that the input graph G is connected, as otherwise we can solve the problem for each component. We choose a vertex v arbitrarily and perform the depth-first search from v. In this way we find the fundamental cycles for the dfs-tree rooted in v, and check whether there is a fundamental cycle of length at least k. If we have such a cycle, then it is a circuit with at least k edges, and we have a YES-answer. Otherwise, by Theorem 6, either G has no cycles of length at least k or G has a cycle with at least k and at most $2k - 4$ edges. If G has no cycles with at least k edges, then by Lemma 6, if G has a circuit with at least k edges, it contains a circuit with at least k and at most $2k - 2$ edges. We conclude that if the constructed fundamental cycles have lengths at most $k - 1$, then G either has a circuit with at least k and at most $2k - 2$ edges or has no circuit with at least k edges. We check whether G has a circuit with at least k and at most $2k - 2$ edges by solving AT LEAST k AND AT MOST k'-CIRCUIT for $k' = 2k - 2$ using Lemma 5. Since the depth-first search runs in linear time, we have that on undirected graphs LONG CIRCUIT can be solved in $2^{O(k)} \cdot nm$ expected time and in $2^{O(k)} \cdot nm \log n$ worst-case time. □

References

1. Alon, N., Yuster, R., Zwick, U.: Color-coding. J. ACM 42(4), 844–856 (1995)
2. Berman, P., Karpinski, M., Scott, A.D.: Approximation hardness of short symmetric instances of MAX-3SAT. Electronic Colloquium on Computational Complexity, ECCC (049) (2003)
3. Bodlaender, H.L.: A linear-time algorithm for finding tree-decompositions of small treewidth. SIAM J. Comput. 25(6), 1305–1317 (1996)
4. Bodlaender, H.L., Downey, R.G., Fellows, M.R., Hermelin, D.: On problems without polynomial kernels. J. Comput. Syst. Sci. 75(8), 423–434 (2009)
5. Cai, L., Yang, B.: Parameterized complexity of even/odd subgraph problems. J. Discrete Algorithms 9(3), 231–240 (2011)
6. Courcelle, B.: The monadic second-order logic of graphs III: tree-decompositions, minor and complexity issues. ITA 26, 257–286 (1992)
7. Cygan, M., Marx, D., Pilipczuk, M., Pilipczuk, M., Schlotter, I.: Parameterized complexity of Eulerian deletion problems. In: Kolman, P., Kratochvíl, J. (eds.) WG 2011. LNCS, vol. 6986, pp. 131–142. Springer, Heidelberg (2011)
8. Dorn, F., Moser, H., Niedermeier, R., Weller, M.: Efficient algorithms for eulerian extension and rural postman. SIAM J. Discrete Math. 27(1), 75–94 (2013)
9. Downey, R.G., Fellows, M.R.: Parameterized complexity. Monographs in Computer Science. Springer, New York (1999)
10. Erdös, P., Szekeres, G.: A combinatorial problem in geometry. Compositio Math. 2, 463–470 (1935)
11. Fellows, M.R., Hermelin, D., Rosamond, F.A., Vialette, S.: On the parameterized complexity of multiple-interval graph problems. Theor. Comput. Sci. 410(1), 53–61 (2009)
12. Fleischner, H.: Eulerian Graphs and Related Topics, Part 1, Amsterdam. Annals of Discrete Mathematics 45, vol. 1 (1990)
13. Flum, J., Grohe, M.: Parameterized complexity theory. Texts in Theoretical Computer Science. An EATCS Series. Springer, Berlin (2006)
14. Fomin, F.V., Golovach, P.A.: Parameterized complexity of connected even/odd subgraph problems. In: STACS 2012. LIPIcs, vol. 14, pp. 432–440. Schloss Dagstuhl - Leibniz-Zentrum fuer Informatik (2012)
15. Gabow, H.N., Nie, S.: Finding a long directed cycle. ACM Transactions on Algorithms 4(1) (2008)
16. Garey, M.R., Johnson, D.S., Tarjan, R.E.: The planar hamiltonian circuit problem is NP-complete. SIAM J. Comput. 5(4), 704–714 (1976)
17. Golovach, P.A., Thilikos, D.M.: Paths of bounded length and their cuts: Parameterized complexity and algorithms. Discrete Optimization 8(1), 72–86 (2011)
18. Kosowski, A., Li, B., Nisse, N., Suchan, K.: k-Chordal graphs: From cops and robber to compact routing via treewidth. In: Czumaj, A., Mehlhorn, K., Pitts, A., Wattenhofer, R. (eds.) ICALP 2012, Part II. LNCS, vol. 7392, pp. 610–622. Springer, Heidelberg (2012)
19. Niedermeier, R.: Invitation to fixed-parameter algorithms, Oxford Lecture Series in Mathematics and its Applications, vol. 31. Oxford University Press, Oxford (2006)

Subexponential Parameterized Algorithm for Computing the Cutwidth of a Semi-complete Digraph*

Fedor V. Fomin and Michał Pilipczuk

Department of Informatics, University of Bergen, Norway
{fomin,michal.pilipczuk}@ii.uib.no

Abstract. Cutwidth of a digraph is a width measure introduced by Chudnovsky, Fradkin, and Seymour [4] in connection with development of a structural theory for tournaments, or more generally, for semi-complete digraphs. In this paper we provide an algorithm with running time $2^{O(\sqrt{k \log k})} \cdot n^{O(1)}$ that tests whether the cutwidth of a given n-vertex semi-complete digraph is at most k, improving upon the currently fastest algorithm of the second author [18] that works in $2^{O(k)} \cdot n^2$ time. As a byproduct, we obtain a new algorithm for FEEDBACK ARC SET in tournaments (FAST) with running time $2^{c\sqrt{k}} \cdot n^{O(1)}$, where $c = \frac{2\pi}{\sqrt{3} \cdot \ln 2} \leq 5.24$, that is simpler than the algorithms of Feige [9] and of Karpinski and Schudy [16], both also working in $2^{O(\sqrt{k})} \cdot n^{O(1)}$ time. Our techniques can be applied also to other layout problems on semi-complete digraphs. We show that the OPTIMAL LINEAR ARRANGEMENT problem, a close relative of FEEDBACK ARC SET, can be solved in $2^{O(k^{1/3} \cdot \sqrt{\log k})} \cdot n^{O(1)}$ time, where k is the target cost of the ordering.

1 Introduction

A directed graph is *simple* if it contains no multiple arcs or loops; it is moreover *semi-complete* if for every two vertices v, w, at least one of the arcs (v, w) or (w, v) is present. An important subclass of semi-complete digraphs is the class of *tournaments*, where we require that exactly one of these arcs is present. Tournaments are extensively studied both from combinatorial and computational point of view; see the book of Bang-Jensen and Gutin [2] for an overview.

One reason why semi-complete digraphs are interesting, is that for this class it is possible to construct a structural theory, resembling the theory of minors for undirected graphs. This theory has been developed recently by Chudnovsky, Fradkin, Kim, Scott, and Seymour [4,5,6,12,13,17]. In particular, two natural notions of digraph containment, namely immersion and minor orders, have been proven to well-quasi-order the set of semi-complete digraphs [6,17]. The developed structural theory has many algorithmic consequences, including fixed-parameter tractable algorithms for containment testing problems [4,11,18].

* Supported by the European Research Council (ERC) via grant Rigorous Theory of Preprocessing, reference 267959.

H.L. Bodlaender and G.F. Italiano (Eds.): ESA 2013, LNCS 8125, pp. 505–516, 2013.

In both theories, for undirected graphs and semi-complete digraphs, width parameters play crucial roles. While in the theory of undirected graph minors the main parameter is treewidth, for semi-complete digraphs *cutwidth* becomes one of the key notions. Given a semi-complete digraph T and a vertex ordering $\sigma = (v_1, v_2, \ldots, v_n)$ of $V(T)$, the *width* of σ is the maximum number of arcs that are directed from a suffix of the ordering to the complementary prefix. The *cutwidth* of T, denoted $\mathbf{ctw}(T)$, is the smallest possible width of an ordering of $V(T)$. It turns out that excluding a fixed digraph as an immersion implies an upper bound on cutwidth of a semi-complete digraph [4]. Hence, the claim that the immersion relation is a well-quasi-ordering of semi-complete digraphs can be easily reduced to the case of semi-complete digraphs of bounded cutwidth; there, a direct reasoning can be applied [6].

From the computational point of view, Chudnovsky, Fradkin, and Seymour give an approximation algorithm that, given a semi-complete digraph T on n vertices and an integer k, in time $O(n^3)$ either outputs an ordering of width $O(k^2)$ or concludes that $\mathbf{ctw}(T) > k$ [4]. It also follows from the work of Chudnovsky, Fradkin, and Seymour [4,5] that the value of cutwidth can be computed exactly by a non-uniform fixed-parameter algorithm working in $f(k) \cdot n^3$ time: as the class of semi-complete digraphs of cutwidth at most k is characterized by a finite set of forbidden immersions, we can approximate cutwidth and test existence of any of them using dynamic programming on the approximate ordering. These results were further improved by the second author [18]: he gives an $O(OPT)$-approximation in $O(n^2)$ time by proving that any ordering of $V(T)$ according to outdegrees has width at most $O(\mathbf{ctw}(T)^2)$, and a fixed-parameter algorithm that in $2^{O(k)} \cdot n^2$ time finds an ordering of width at most k or concludes that it is not possible.

Our results and techniques. In this work we present an algorithm that, given a semi-complete digraph T on n vertices and an integer k, in $2^{O(\sqrt{k \log k})} \cdot n^{O(1)}$ time computes an ordering of width at most k or concludes that $\mathbf{ctw}(T) > k$. In other words, we prove that the cutwidth of a semi-complete digraph can be computed in subexponential parameterized time.

The idea behind our approach is inspired by the recent work of a superset of the current authors on clustering problems [10]. The algorithm in [10] is based on a combinatorial result that in every YES instance of that problem the number of k-cuts, i.e., partitions of the vertex set into two subsets with at most k edges crossing the partition, is bounded by a subexponential function of k. We apply a similar strategy to compute the cutwidth of a semi-complete digraph. A k-*cut* in a semi-complete digraph T is a partition of its vertices into two sets X and Y, such that only at most k arcs are directed from Y to X. Our algorithm is based on a new combinatorial lemma that the number of k-cuts in a semi-complete digraph of cutwidth at most k is at most $2^{O(\sqrt{k \log k})} \cdot n$. The crucial ingredient of its proof is to relate k-cuts of a transitive tournament to partition numbers: a notion extensively studied in classical combinatorics and which subexponential asymptotics is very well understood. Then we roughly do the following. It is possible to show that all k-cuts can be enumerated with polynomial time delay.

We enumerate all k-cuts and if we exceed the combinatorial bound, we are able to say that the cutwidth of the input digraph is more than k. Otherwise, we have a bounded-size family of objects on which we can employ a dynamic programming routine. The running time of this step is up to a polynomial factor proportional to the number of k-cuts, and thus is subexponential.

As a byproduct of the approach taken, we also obtain a new algorithm for FEEDBACK ARC SET (FAS) in semi-complete digraphs, with running time $2^{c\sqrt{k}}$. $n^{O(1)}$ for $c = \frac{2\pi}{\sqrt{3}\cdot\ln 2} \leq 5.24$. The FAS problem was the first problem in tournaments shown to admit subexponential parameterized algorithms. The first algorithm with running time $2^{O(\sqrt{k}\log k)} \cdot n^{O(1)}$ is due to Alon, Lokshtanov, and Saurabh [1]. This has been further improved by Feige [9] and by Karpinski and Schudy [16], who have independently shown two different algorithms with running time $2^{O(\sqrt{k})} \cdot n^{O(1)}$. The algorithm of Alon et al. introduced a new technique called *chromatic coding* that proved to be useful also in other problems in dense graphs [14]. The algorithms of Feige and of Karpinski and Schudy were based on the degree ordering approach, and the techniques developed there were more specific to the problem.

In our approach, the $2^{O(\sqrt{k})} \cdot n^{O(1)}$ algorithm for FAS on semi-complete digraphs follows immediately from relating k-cuts of a transitive tournament to partition numbers, and an application of the general framework. It is also worth mentioning that the explicit constant in the exponent obtained using our approach is much smaller than the constants in the algorithms of Feige and of Karpinski and Schudy; however, optimizing these constants was not the purpose of these works. Similarly to the algorithm of Karpinski and Schudy, our algorithm works also in the weighted setting.

Lastly, we show that our approach can be also applied to other layout problems in semi-complete digraphs. For example, we consider a natural variant of the well-studied OPTIMAL LINEAR ARRANGEMENT problem [3,7], and we prove that one can compute in $2^{O(k^{1/3}\cdot\sqrt{\log k})} \cdot n^{O(1)}$ time an ordering of cost at most k, or conclude that it is impossible (see Section 2 for precise definitions). Although such a low complexity may be explained by the fact that the optimal cost may be even cubic in the number of vertices, we find it interesting that results of this kind can be also obtained by making use of our techniques.

Organization of the paper. In Section 2 we introduce basic notions and problem definitions. In Section 3 we prove combinatorial lemmata concerning k-cuts of semi-complete digraphs. In Section 4 we apply the results of the previous section to obtain the algorithmic results. Section 5 is devoted to concluding remarks.

2   Preliminaries

We use standard graph notation. For a digraph D, we denote by $V(D)$ and $E(D)$ the vertex and arc sets of D, respectively. A digraph is *simple* if it has no loops and no multiple arcs, i.e., for every pair of vertices v, w, the arc (v, w) appears in $E(D)$ at most once. Note that we do not exclude existence of arcs (v, w) and (w, v) at the same time. All the digraphs considered in this paper will be simple.

A digraph is *acyclic* if it contains no cycle. It is known that a digraph is acyclic if and only if it admits a *topological ordering* of vertices, i.e., an ordering $(v_1, v_2, \ldots, v_n)$ of $V(D)$ such that arcs are always directed from a vertex with a smaller index to a vertex with a larger index.

A simple digraph T is *semi-complete* if for every pair (v, w) of vertices **at least** one of the arcs (v, w), (w, v) is present. A semi-complete digraph T is moreover a *tournament* if for every pair (v, w) of vertices **exactly** one of the arcs (v, w), (w, v) is present. A *transitive tournament* with ordering $(v_1, v_2, \ldots, v_n)$ is a tournament T defined on vertex set $\{v_1, v_2, \ldots, v_n\}$ where $(v_i, v_j) \in E(T)$ if and only if $i < j$.

We use Iverson notation: for a condition φ, $[\varphi]$ denotes value 1 if φ is true and 0 otherwise. We also use $\exp(t) = e^t$.

2.1   Feedback Arc Set

Definition 1. *Let T be a digraph. A subset $F \subseteq E(T)$ is called a* feedback arc set *if $T \setminus F$ is acyclic.*

The FAS problem in semi-complete digraphs is defined as follows.

FEEDBACK ARC SET **Parameter:** k
Input: A semi-complete digraph T, an integer k
Question: Is there a feedback arc set of T of size at most k?

We have the following easy observation that enables us to view FAS as a graph layout problem.

Lemma 2. *Let T be a digraph. Then T admits a feedback arc set of size at most k if and only if there exists an ordering $(v_1, v_2, \ldots, v_n)$ of $V(T)$ with at most k arcs of $E(T)$ directed backward in this ordering, i.e., of form (v_i, v_j) for $i > j$.*

Proof. If F is a feedback arc set in T then the ordering can be obtained by taking any topological ordering of $T \setminus F$. On the other hand, given the ordering we may simply define F to be the set of backward edges.

2.2   Cutwidth

Definition 3. *Let T be a digraph. For an ordering $\sigma = (v_1, v_2, \ldots, v_n)$ of $V(T)$, the width of σ is $\max_{1 \leq t \leq n-1} |E(\{v_{t+1}, v_{t+2}, \ldots, v_n\}, \{v_1, v_2, \ldots, v_t\})|$. Cutwidth of T, denoted $\mathbf{ctw}(T)$, is the smallest possible width of an ordering of $V(T)$.*

The CUTWIDTH problem can be hence defined as follows:

CUTWIDTH **Parameter:** k
Input: A semi-complete digraph T, an integer k
Question: Is $\mathbf{ctw}(T) \leq k$?

Note that the introduced notion reverses the ordering with respect to the standard literature on cutwidth [4,18]. We chose to do so to be consistent within the paper, and compatible with the literature on FAS in tournaments.

2.3   Optimal Linear Arrangement

Definition 4. *Let T be a digraph and $(v_1, v_2, \ldots, v_n)$ be an ordering of its vertices. Then the cost of this ordering is defined as*

$$\sum_{(v_i, v_j) \in E(T)} (i - j) \cdot [i > j],$$

that is, every arc directed backwards in the ordering contributes to the cost with the distance between the endpoints in the ordering.

Whenever the ordering is clear from the context, we also refer to the contribution of a given arc to its cost as to the *length* of this arc. By a simple reordering of the computation we obtain the following:

Lemma 5. *For a digraph T and ordering $(v_1, v_2, \ldots, v_n)$ of $V(T)$, the cost of this ordering is equal to:*

$$\sum_{t=1}^{n-1} |E(\{v_{t+1}, v_{t+2}, \ldots, v_n\}, \{v_1, v_2, \ldots, v_t\})|.$$

Proof. Observe that

$$\sum_{(v_i, v_j) \in E(T)} (i - j) \cdot [i > j] = \sum_{(v_i, v_j) \in E(T)} \sum_{t=1}^{n-1} [j \le t < i]$$

$$= \sum_{t=1}^{n-1} \sum_{(v_i, v_j) \in E(T)} [j \le t < i] =$$

$$= \sum_{t=1}^{n-1} |E(\{v_{t+1}, v_{t+2}, \ldots, v_n\}, \{v_1, v_2, \ldots, v_t\})|.$$

The problem OLA (OPTIMAL LINEAR ARRANGEMENT) in semi-complete digraphs is defined as follows:

OPTIMAL LINEAR ARRANGEMENT	**Parameter: k**
Input: A semi-complete digraph T, an integer k	
Question: Is there an ordering of $V(T)$ of cost at most k?	

3 k-Cuts of Semi-complete Digraphs

In this section we provide all the relevant observations on k-cuts of semi-complete digraphs. We start with the definitions, and then proceed to bounding the number of k-cuts when the given semi-complete digraph is close to a structured one.

3.1 Definitions

Definition 6. *A k-cut of a digraph T is a partition (X, Y) of $V(T)$ with the following property: there are at most k arcs $(u, v) \in E(T)$ such that $u \in Y$ and $v \in X$.*

The following lemma will be needed to apply the general framework.

Lemma 7. *k-cuts of a digraph T can be enumerated with polynomial-time delay.*

Proof. Let $\sigma = (v_1, v_2, \ldots, v_n)$ be an arbitrary ordering of vertices of T. We perform a classical branching strategy: we start with empty X and Y, and consider the vertices in order σ, at each step branching into one of the two possibilities: vertex v_i is to be incorporated into X or into Y. However, after assigning each consecutive vertex we run a max-flow algorithm from Y to X to find the size of a minimum edge cut between Y and X. If this size is more than k, we terminate the branch as we know that it cannot result in any solutions found. Otherwise we proceed. We output a partition after the last vertex, v_n, is assigned a side; note that the last max-flow check ensures that the output partition is actually a k-cut. Moreover, as during the algorithm we consider only branches that can produce at least one k-cut, the next partition will be always found within polynomial waiting time, proportional to the depth of the branching tree times the time needed for computations at each node of the branching tree.

3.2 k-Cuts of a Transitive Tournament and Partition Numbers

For a nonnegative integer n, a partition of n is a multiset of positive integers whose sum is equal to n. The partition number $p(n)$ is equal to the number of different partitions of n. Partition numbers are studied extensively in analytic combinatorics, and there are sharp estimates on their value. In particular, we will use the following:

Lemma 8 ([8,15]). *There exists a constant A such that for every nonnegative k it holds that $p(k) \leq \frac{A}{k+1} \cdot \exp(C\sqrt{k})$, where $C = \pi\sqrt{\frac{2}{3}}$.*

We remark that the original proof of Hardy and Ramanujan [15] shows moreover that the optimal constant A tends to $\frac{1}{4\sqrt{3}}$ as k goes to infinity. From now on, we adopt constants A, C given by Lemma 8 in the notation. We use Lemma 8 to obtain the following result, which is the core observation of this paper.

Lemma 9. *Let T be a transitive tournament with n vertices and k be a nonnegative integer. Then T has at most $A \cdot \exp(C\sqrt{k}) \cdot (n+1)$ k-cuts, where A, C are defined as in Lemma 8.*

Proof. We prove that for any number a, $0 \leq a \leq n$, the number of k-cuts (X, Y) such that $|X| = a$ and $|Y| = n - a$, is bounded by $A \cdot \exp(C\sqrt{k})$; summing through all the possible values of a proves the claim.

We naturally identify the vertices of T with numbers $1, 2, \ldots, n$, such that arcs of T are directed from smaller numbers to larger. Let us fix some k-cut (X, Y) such that $|X| = a$ and $|Y| = n - a$. Let $x_1 < x_2 < \ldots < x_a$ be the vertices of X.

Let $m_i = x_{i+1} - x_i - 1$ for $i = 0, 1, \ldots, a$; we use convention that $x_0 = 0$ and $x_{a+1} = n + 1$. In other words, m_i is the number of elements of Y that are between two consecutive elements of X. Observe that every element of Y between x_i and x_{i+1} is the tail of exactly $a - i$ arcs directed from Y to X: the heads are $x_{i+1}, x_{i+2}, \ldots, x_a$. Hence, the total number of arcs directed from Y to X is equal to $k' = \sum_{i=0}^{a} m_i \cdot (a - i) = \sum_{i=0}^{a} m_{a-i} \cdot i \leq k$.

We define a partition of k' as follows: we take m_{a-1} times number 1, m_{a-2} times number 2, and so on, up to m_0 times number a. Clearly, a k-cut of T defines a partition of k' in this manner. We now claim that knowing a and the partition of k', we can uniquely reconstruct the k-cut (X, Y) of T, or conclude that this is impossible. Indeed, from the partition we obtain all the numbers $m_0, m_1, \ldots, m_{a-1}$, while m_a can be computed as $(n - a) - \sum_{i=0}^{a-1} m_i$. Hence, we know exactly how large must be the intervals between consecutive elements of X, and how far is the first and the last element of X from the respective end of the ordering, which uniquely defines sets X and Y. The only possibilities of failure during reconstruction are that (i) the numbers in the partition are larger than a, or (ii) computed m_a turns out to be negative; in these cases, the partition does not correspond to any k-cut. Hence, we infer that the number of k-cuts of T having $|X| = a$ and $|Y| = n - a$ is bounded by the sum of partition numbers of nonnegative integers smaller or equal to k, which by Lemma 8 is bounded by $(k + 1) \cdot \frac{A}{k+1} \cdot \exp(C\sqrt{k}) = A \cdot \exp(C\sqrt{k})$.

3.3 k-Cuts of Semi-complete Digraphs with a Small FAS

We have the following simple fact.

Lemma 10. *Assume that T is a semi-complete digraph with a feedback arc set F of size at most k. Let T' be a transitive tournament on the same set of vertices, with vertices ordered as in any topological ordering of $T \setminus F$. Then every k-cut of T is also a $2k$-cut of T'.*

Proof. The claim follows directly from the observation that if (X, Y) is a k-cut in T, then at most k additional arcs directed from Y to X can appear after introducing arcs in T' in place of deleted arcs from F.

From Lemmata 9 and 10 we obtain the following corollary.

Corollary 11. *Every semi-complete digraph with n vertices and with a feedback arc set of size at most k, has at most $A \cdot \exp(C\sqrt{2k}) \cdot (n + 1)$ k-cuts.*

3.4 k-Cuts of Semi-complete Digraphs of Small Cutwidth

To bound the number of k-cuts of semi-complete digraphs of small cutwidth, we need the following auxiliary combinatorial result.

Lemma 12. *Let (X, Y) be a partition of $\{1, 2, \ldots, n\}$ into two sets. We say that a pair (a, b) is bad if $a < b$, $a \in Y$ and $b \in X$. Assume that for every integer t there are at most k bad pairs (a, b) such that $a \leq t < b$. Then the total number of bad pairs is at most $k(1 + \ln k)$.*

Proof. Let $y_1 < y_2 < \ldots < y_p$ be the elements of Y. Let m_i be equal to the total number of elements of X that are greater than y_i. Note that m_i is exactly equal to the number of bad pairs whose first element is equal to y_i, hence the total number of bad pairs is equal to $\sum_{i=1}^{p} m_i$. Clearly, sequence (m_i) is non-increasing, so let p' be the last index for which $m_{p'} > 0$. We then have that the total number of bad pairs is equal to $\sum_{i=1}^{p'} m_i$. Moreover, observe that $p' \leq k$, as otherwise there would be more than k bad pairs (a, b) for which $a \leq y_{p'} < b$: for a we can take any y_i for $i \leq p'$ and for b we can take any element of X larger than $y_{p'}$.

We claim that $m_i \leq k/i$ for every $1 \leq i \leq p'$. Indeed, observe that there are exactly $i \cdot m_i$ bad pairs (a, b) for $a \leq y_i$ and $b > y_i$: a can be chosen among i distinct integers $y_1, y_2, \ldots, y_i$, while b can be chosen among m_i elements of X larger than y_i. By the assumption we infer that $i \cdot m_i \leq k$, so $m_i \leq k/i$. Concluding, we have that the total number of bad pairs is bounded by $\sum_{i=1}^{p'} m_i \leq \sum_{i=1}^{p'} k/i = k \cdot H(p') \leq k \cdot H(k) \leq k(1 + \ln k)$, where $H(k) = \sum_{i=1}^{k} 1/i$ is the harmonic function. $\qquad \square$

We now apply Lemma 12 to the setting of semi-complete digraphs.

Lemma 13. *Assume that T is a semi-complete digraph with n vertices that admits an ordering of vertices $(v_1, v_2, \ldots, v_n)$ of width at most k. Let T' be a transitive tournament on the same set of vertices, where $(v_i, v_j) \in E(T')$ if and only if $i < j$. Then every k-cut of T is a $2k(1 + \ln 2k)$-cut of T'.*

Proof. Without loss of generality we assume that T is in fact a tournament, as deleting any of two opposite arcs connecting two vertices can only make the set of k-cuts of T larger, and does not increase the width of the ordering.

Identify vertices $v_1, v_2, \ldots, v_n$ with numbers $1, 2, \ldots, n$. Let (X, Y) be a k-cut of T. Note that arcs of T' directed from Y to X correspond to bad pairs in the sense of Lemma 12. Therefore, by Lemma 12 it suffices to prove that for every integer t, the number of arcs $(a, b) \in E(T')$ such that $a \leq t < b$, $a \in Y$, and $b \in X$, is bounded by $2k$. We know that the number of such arcs in T is at most k, as there are at most k arcs directed from Y to X in T in total. Moreover, as the considered ordering of T has cutwidth at most k, at most k arcs between vertices from $\{1, 2, \ldots, t\}$ and $\{t + 1, \ldots, n\}$ can be directed in different directions in T and in T'. We infer that the number of arcs $(a, b) \in E(T')$ such that $a \leq t < b$, $a \in Y$, and $b \in X$, is bounded by $2k$, and so the lemma follows. $\qquad \square$

From Lemmata 9 and 13 we obtain the following corollary.

Corollary 14. *Every tournament with n vertices and of cutwidth at most k, has at most $A \cdot \exp(2C\sqrt{k(1 + \ln 2k)}) \cdot (n + 1)$ k-cuts.*

3.5 k-Cuts of Semi-complete Digraphs with an Ordering of Small Cost

We firstly show the following lemma that proves that semi-complete digraphs with an ordering of small cost have even smaller cutwidth.

Lemma 15. *Let T be a semi-complete digraph on n vertices that admits an ordering $(v_1, v_2, \ldots, v_n)$ of cost at most k. Then the width of this ordering is at most $(4k)^{2/3}$.*

Proof. We claim that for every integer $t \geq 0$, the number of arcs in T directed from the set $\{v_{t+1}, \ldots, v_n\}$ to $\{v_1, \ldots, v_t\}$ is at most $(4k)^{2/3}$. Let ℓ be the number of such arcs; without loss of generality assume that $\ell > 0$. Observe that at most one of these arcs may have length 1, at most 2 may have length 2, etc., up to at most $\lfloor\sqrt{\ell}\rfloor - 1$ may have length $\lfloor\sqrt{\ell}\rfloor - 1$. It follows that at most $\sum_{i=1}^{\lfloor\sqrt{\ell}\rfloor - 1} i \leq \ell/2$ of these arcs may have length smaller than $\lfloor\sqrt{\ell}\rfloor$. Hence, at least $\ell/2$ of the considered arcs have length at least $\lfloor\sqrt{\ell}\rfloor$, so the total sum of lengths of arcs is at least $\frac{\ell \cdot \lfloor\sqrt{\ell}\rfloor}{2} \geq \frac{\ell^{3/2}}{4}$. We infer that $k \geq \frac{\ell^{3/2}}{4}$, which means that $\ell \leq (4k)^{2/3}$. $\square$

Lemma 15 ensures that only $(4k)^{2/3}$-cuts are interesting from the point of view of dynamic programming. Moreover, from Lemma 15 and Corollary 14 we can derive the following statement that bounds the number of states of the dynamic program.

Corollary 16. *If T is a semi-complete digraph with n vertices that admits an ordering of cost at most k, then the number of $(4k)^{2/3}$-cuts of T is bounded by $A \cdot \exp(2C \cdot (4k)^{1/3} \cdot \sqrt{1 + \ln(2 \cdot (4k)^{2/3})}) \cdot (n+1)$.*

4   The Algorithms

In this section we apply the results of the previous section to obtain algorithmic results. We begin with the algorithm for FEEDBACK ARC SET, and then proceed to CUTWIDTH and OLA. All three applications are very similar and based on the principle of dynamic programming. Therefore we give in detail only the first algorithm, which contains all the necessary ingredients. The remaining two algorithms (marked with (†)) we defer to the full version of the paper.

For a semi-complete digraph T, let $\mathcal{N}(T, k)$ denote the family of k-cuts of T.

Theorem 17. *There exists an algorithm that, given a semi-complete digraph T on n vertices and an integer k, in time $\exp(C\sqrt{2k}) \cdot n^{O(1)}$ either finds a feedback arc set of T of size at most k or correctly concludes that this is impossible, where $C = \pi\sqrt{\frac{2}{3}}$.*

Proof. Using Lemma 7, we enumerate all the k-cuts of T. If we exceed the bound of $A \cdot \exp(C\sqrt{2k}) \cdot (n+1)$ during enumeration, by Corollary 11 we may safely terminate the computation providing a negative answer; note that this happens after using at most $\exp(C\sqrt{2k})\cdot n^{O(1)}$ time, as the cuts are output with polynomial time delay. Hence, from now on we assume that we have the set $\mathcal{N} := \mathcal{N}(T, k)$ and we know that $|\mathcal{N}| \leq A \cdot \exp(C\sqrt{2k}) \cdot (n+1)$.

We now describe a dynamic programming procedure that computes the size of optimal feedback arc set from the set $\mathcal{N}$; the dynamic program is based on the approach presented in [18]. We define an auxiliary weighted digraph D with vertex set $\mathcal{N}$. Intuitively, a vertex from $\mathcal{N}$ corresponds to a partition into prefix and suffix of the ordering.

Formally, we define arcs of D as follows. We say that cut (X_2, Y_2) *extends* cut (X_1, Y_1) if there is one vertex $v \in Y_1$ such that $X_2 = X_1 \cup \{v\}$ and, hence, $Y_2 = Y_1 \setminus \{v\}$. We put an arc in D from cut (X_1, Y_1) to cut (X_2, Y_2) if (X_2, Y_2) extends (X_1, Y_1); the weight of this arc is equal to $|E(\{v\}, X_1)|$, that is, the number of arcs that cease to be directed from the right side to the left side of the partition when moving v between these parts. Note that thus each vertex of D has at most n outneighbours, so $|E(D)|$ is bounded by $O(|\mathcal{N}| \cdot n)$. Moreover, the whole graph D can be constructed in $|\mathcal{N}| \cdot n^{O(1)}$ time by considering all the vertices of D and examining each of at most n candidates for outneighbours in polynomial time.

Observe that a path from vertex $(\emptyset, V(T))$ to a vertex $(V(T), \emptyset)$ of total weight ℓ defines an ordering of vertices of T that has exactly ℓ backward arcs — each of these edges was taken into account while moving its tail from the right side of the partition to the left side. On the other hand, every ordering of vertices of T that has exactly $\ell \leq k$ backward arcs defines a path from $(\emptyset, V(T))$ to $(V(T), \emptyset)$ in D of total weight ℓ; note that all partitions into prefix and suffix in this ordering are k-cuts, so they constitute legal vertices in D. Hence, we need to check whether vertex $(V(T), \emptyset)$ can be reached from $(\emptyset, V(T))$ by a path of total length at most k. This, however, can be done in time $O((|V(D)| + |E(D)|)\log|V(D)|) = O(\exp(C\sqrt{2k})\cdot n^{O(1)})$ using Dijkstra's algorithm. The feedback arc set of size at most k can be easily retrieved from the constructed path in polynomial time.

We remark that it is straightforward to adapt the algorithm of Theorem 17 to the weighted case, where all the arcs are assigned a real weight larger or equal to 1 and we parametrize by the target total weight of the solution. As the minimum weight is at least 1, we may still consider only k-cuts of the digraph where the weights are forgotten. On this set we employ a modified dynamic programming routine, where the weights of arcs in digraph D are not simply the number of arcs in $E(\{v\}, X_1)$, but their total weight. We omit the details here.

We now proceed to the main result of this paper, i.e., the subexponential algorithm for cutwidth of a semi-complete digraph. The following theorem essentially follows from Lemma 7, Corollary 14 and the dynamic programming algorithm from [18].

Theorem 18 (†). *There exists an algorithm that, given a semi-complete digraph T on n vertices and an integer k, in time $2^{O(\sqrt{k \log k})} \cdot n^{O(1)}$ either computes a vertex ordering of width at most k or correctly concludes that this is impossible.*

Similarly as in the proof of Theorem 17, the algorithm builds an auxiliary digraph D on the set $\mathcal{N}(T, k)$. Lemma 7 can be used to enumerate $\mathcal{N}(T, k)$ with polynomial delay, while Corollary 14 ensures that the enumeration can be terminated after finding $2^{O(\sqrt{k \log k})} \cdot n$ cuts. Arcs of D are defined via the same notion of extension, and this time D is unweighted. Then paths in D from $(\emptyset, V(T))$ to $(V(T), \emptyset)$ correspond to orderings of $V(T)$ of cutwidth at most k. Details are omitted.

Finally, we present how the framework can be applied to the OLA problem.

Theorem 19 (†). *There exists an algorithm that, given a semi-complete digraph T on n vertices and an integer k, in time $2^{O(k^{1/3}\sqrt{\log k})} \cdot n^{O(1)}$ either computes a vertex ordering of cost at most k, or correctly concludes that it is not possible.*

Here, Lemma 15 ensures that we may work on family $\mathcal{N}(T, (4k)^{2/3})$, while Corollary 16 gives us a bound after which enumeration of cuts may be terminated. Weights of arcs in the digraph D are set according to the formula from Lemma 5.

Similarly to Theorem 17, it is also straightforward to adapt the algorithm of Theorem 19 to the natural weighted variant of the problem, where each arc is assigned a real weight larger or equal to 1, each arc directed backward in the ordering contributes to the cost with its weight multiplied by the length of the arc, and we parametrize by the total target cost.

5   Conclusions

In this paper we have showed that a number of vertex ordering problems on tournaments, and more generally, on semi-complete digraphs, admit subexponential parameterized algorithms. We believe that our approach provides a new insight into the structure of problems on semi-complete digraphs solvable in subexponential parameterized time: in instances with a positive answer the space of naturally relevant objects, namely k-cuts, is of subexponential size. We hope that such an algorithmic strategy may be applied to other problems as well.

Clearly, it is possible to pipeline the presented algorithm for FAS on semi-complete digraphs with a simple kernelization algorithm, which can be found, e.g., in [1], to separate the polynomial dependency on n from the subexponential dependency on k in the running time. This can be done also for OLA, as this problem admits a simple linear kernel; we omit the details here.

However, we believe that a more important challenge is to investigate whether the $\sqrt{\log k}$ factor in the exponent of the running times of the algorithms of Theorems 18 and 19 is necessary. At this moment, appearance of this factor is a result of pipelining Lemma 9 with Lemma 12 in the proof of Lemma 13. A closer examination of the proofs of Lemmata 9 and 12 shows that bounds given by them are essentially optimal on their own; yet, it is not clear whether the bound given by pipelining them is optimal as well. Hence, we would like to pose the following open problem: is the number of k-cuts of a semi-complete digraph on n vertices and of cutwidth at most k bounded by $2^{O(\sqrt{k})} \cdot n^{O(1)}$? If the answer to this combinatorial question is positive, then the $\sqrt{\log k}$ factor could be removed.

References

1. Alon, N., Lokshtanov, D., Saurabh, S.: Fast FAST. In: Albers, S., Marchetti-Spaccamela, A., Matias, Y., Nikoletseas, S., Thomas, W. (eds.) ICALP 2009, Part I. LNCS, vol. 5555, pp. 49–58. Springer, Heidelberg (2009)
2. Bang-Jensen, J., Gutin, G.: Digraphs. In: Springer Monographs in Mathematics, 2nd edn. Theory, algorithms and applications, Springer-Verlag London Ltd., London (2009)
3. Chinn, P.Z., Chvátalová, J., Dewdney, A.K., Gibbs, N.E.: The bandwidth problem for graphs and matrices — a survey. J. Graph Theory 6, 223–254 (1982)
4. Chudnovsky, M., Fradkin, A., Seymour, P.: Tournament immersion and cutwidth. J. Comb. Theory Ser. B 102, 93–101 (2012)
5. Chudnovsky, M., Scott, A., Seymour, P.: Vertex disjoint paths in tournaments (2011) (manuscript)
6. Chudnovsky, M., Seymour, P.D.: A well-quasi-order for tournaments. J. Comb. Theory, Ser. B 101, 47–53 (2011)
7. Díaz, J., Petit, J., Serna, M.J.: A survey of graph layout problems. ACM Comput. Surv. 34, 313–356 (2002)
8. Erdős, P.: On an elementary proof of some asymptotic formulas in the theory of partitions. Annals of Mathematics (2) 43, 437–450 (1942)
9. Feige, U.: Faster FAST (Feedback Arc Set in Tournaments), CoRR, abs/0911.5094 (2009)
10. Fomin, F.V., Kratsch, S., Pilipczuk, M., Pilipczuk, M., Villanger, Y.: Tight bounds for parameterized complexity of cluster editing. In: Proceedings of the 30th International Symposium on Theoretical Aspects of Computer Science (STACS). LIPIcs, vol. 20, pp. 32–43. Schloss Dagstuhl - Leibniz-Zentrum fuer Informatik (2013)
11. Fomin, F.V., Pilipczuk, M.: Jungles, bundles, and fixed parameter tractability. In: Proceedings of the 24th ACM-SIAM Symposium on Discrete Algorithms (SODA), pp. 396–413. SIAM (2012)
12. Fradkin, A., Seymour, P.: Edge-disjoint paths in digraphs with bounded independence number (2010) (manuscript)
13. Fradkin, A., Seymour, P.: height 2pt depth -1.6pt width 23pt, Tournament pathwidth and topological containment. J. Comb. Theory Ser. B (in press, 2013)
14. Ghosh, E., Kolay, S., Kumar, M., Misra, P., Panolan, F., Rai, A., Ramanujan, M.S.: Faster parameterized algorithms for deletion to split graphs. In: Fomin, F.V., Kaski, P. (eds.) SWAT 2012. LNCS, vol. 7357, pp. 107–118. Springer, Heidelberg (2012)
15. Hardy, G.H., Ramanujan, S.: Asymptotic formulae in combinatory analysis. Proceedings of the London Mathematical Society s2-17, 75–115 (1918)
16. Karpinski, M., Schudy, W.: Faster algorithms for feedback arc set tournament, Kemeny rank aggregation and betweenness tournament. In: Cheong, O., Chwa, K.-Y., Park, K. (eds.) ISAAC 2010, Part I. LNCS, vol. 6506, pp. 3–14. Springer, Heidelberg (2010)
17. Kim, I., Seymour, P.: Tournament minors, CoRR, abs/1206.3135 (2012)
18. Pilipczuk, M.: Computing cutwidth and pathwidth of semi-complete digraphs via degree orderings. In: Proceedings of the 30th International Symposium on Theoretical Aspects of Computer Science (STACS). LIPIcs, vol. 20, pp. 197–208. Schloss Dagstuhl - Leibniz-Zentrum fuer Informatik (2013)

Binary Jumbled Pattern Matching
on Trees and Tree-Like Structures

Travis Gagie[1], Danny Hermelin[2], Gad M. Landau[3,4,*], and Oren Weimann[3]

[1] University of Helsinki
`travis.gagie@cs.helsinki.fi`
[2] Ben-Gurion University
`hermelin@bgu.ac.il`
[3] University of Haifa
`{landau,oren}@cs.haifa.ac.il`
[4] NYU poly

Abstract. Binary jumbled pattern matching asks to preprocess a binary string S in order to answer queries (i, j) which ask for a substring of S that is of length i and has exactly j 1-bits. This problem naturally generalizes to vertex-labeled trees and graphs by replacing "substring" with "connected subgraph". In this paper, we give an $O(n^2 / \log^2 n)$-time solution for trees, matching the currently best bound for (the simpler problem of) strings. We also give an $O(g^{2/3} n^{4/3} / (\log n)^{4/3})$-time solution for strings that are compressed by a grammar of size g. This solution improves the known bounds when the string is compressible under many popular compression schemes. Finally, we prove that the problem is fixed-parameter tractable with respect to the treewidth w of the graph, even for a constant number of different vertex-labels, thus improving the previous best $n^{O(w)}$ algorithm.

1 Introduction

Jumbled pattern matching is an important variant of classical pattern matching with several applications in computational biology, ranging from alignment [4] and SNP discovery [6], to the interpretation of mass spectrometry data [9] and metabolic network analysis [21]. In the most basic case of strings, the problem asks to determine whether a given pattern P can be rearranged so that it appears in a given text T. That is, whether T contains a substring of length $|P|$ where each letter of the alphabet occurs the same number of times as in P. Using a straightforward sliding window algorithm, such a jumbled occurrence can be found optimally in $O(n)$ time on a text of length n. While jumbled pattern matching has a simple efficient solution, its *indexing* problem is much more challenging. In the indexing problem, we preprocess a given text T so that on queries P we can determine quickly whether T has a jumbled occurrence of P. Very little is known about this problem besides the trivial naive solution.

Most of the interesting results on indexing for jumbled pattern matching relate to binary strings (where a query pattern (i, j) asks for a substring of T that

---

[*] Partially supported by the National Science Foundation Award 0904246, Israel Science Foundation grant 347/09, and Grant No. 2008217 from the United States-Israel Binational Science Foundation (BSF).

H.L. Bodlaender and G.F. Italiano (Eds.): ESA 2013, LNCS 8125, pp. 517–528, 2013.

is of length i and has j 1s). Given a binary string of length n, Cicalese, Fici and Lipták [13] showed how one can build in $O(n^2)$ time an $O(n)$-space index that answers jumbled pattern matching queries in $O(1)$ time. Their key observation was that if one substring of length i contains fewer than j 1s, and another substring of length i contains more than j 1s, then there must be a substring of length i with exactly j 1s. Using this observation, they construct an index that stores the maximum and minimum number of 1s in any i-length substring, for each possible i. Burcsi et al. [9] (see also [10,11]) and Moosa and Rahman [22] independently improved the construction time to $O(n^2/\log n)$, then Moosa and Rahman [23] further improved it to $O(n^2/\log^2 n)$ in the RAM model. Currently, faster algorithms than $O(n^2/\log^2 n)$ exist only when the string compresses well using run-length encoding [3,19] or when we are willing to settle for approximate indexes [14].

The natural extension of jumbled pattern matching from strings to trees is much harder. In this extension, we are asked to determine whether a vertex-labeled input tree has a connected subgraph where each label occurs the same number of times as specified by the input query. The difficulty here stems from the fact that a tree can have an exponential number of connected subgraphs as opposed to strings. Hence, a sliding window approach becomes intractable. Indeed, the problem is NP-hard [21], even if our query contains at most one occurrence of each letter [17]. It is not even fixed-parameter tractable when parameterized by the alphabet size [17]. The fixed-parameter tractability of the problem was further studied when extending the problem from trees to graphs [2,5,15,16]. In particular, the problem (also known as the *graph motif problem*) was recently shown by Fellows et al. [17] to be polynomial-time solvable when the number of letters in the alphabet as well as the treewidth of the graph are both fixed.

Our results. In this paper we extend the currently known state-of-the-art for binary jumbled pattern matching. Our results focus on trees, and tree-like structures such as grammars and bounded treewidth graphs. The problem on such trees turns out to be more challenging than on strings and requires substantially different ideas and techniques.

- **Trees:** For a tree T of size n, we present an index of size $O(n)$ bits that is constructed in $O(n^2/\log^2 n)$ time and answers binary jumbled pattern matching queries in $O(1)$ time. This matches the performance of the best known index for binary strings. In fact, our index for trees is obtained by multiple applications of an efficient algorithm for strings [23] under a more careful analysis. This is combined with both a micro-macro [1] and centroid decomposition of the input tree. Our index can also be used as an $O(ni/\log^2 n)$-time algorithm for the pattern matching (as opposed to the indexing) problem, where i denotes the size of the pattern. Finally, by increasing the space of our index to $O(n\log n)$ bits, we can output in $O(\log n)$ time a node of T that is part of the pattern occurrence.

- **Grammars:** For a binary string S of length n derived by a grammar of size g, we show how to construct in $O(g^{2/3}n^{4/3}/\log^{4/3} n)$ time an index of size $O(n)$ bits that answers jumbled pattern matching queries on S in $O(1)$ time. The size of the grammar g can be exponentially smaller than

n and is always at most $O(n/\log n)$. This means that our time bound is $O(n^2/\log^2 n)$ even when S is not compressible. If S is compressible but with other compression schemes such as the LZ-family, then we can transform it into a grammar-based compression with little or no expansion [12,24].

- **Bounded Treewidth Graphs:** For a graph G with treewidth bounded by w, we show how to improve on the $O(n^{O(w)})$ time algorithm of Fellows *et al.* [17] to an algorithm which runs in $2^{O(w^3)}n + w^{O(w)}n^{O(1)}$ time. Thus, we show that for a binary alphabet, jumbled pattern matching is fixed-parameter tractable when parameterized only by the treewidth. This result extends easily to alphabets of constant sizes.

We present our results for trees, grammars, and bounded treewidth graphs in sections 2, 3 and 4 respectively. Proofs of all lemmas are given in the full version of this paper.

2   Jumbled Pattern Matching on Trees

In this section we consider the natural extension of binary jumbled pattern matching to trees. Recall that in this extension we are given a tree T with n nodes, where each node is labeled by either 1 or 0. We will refer to the nodes labeled 1 as *black* nodes, and the nodes labeled 0 as *white* nodes. Our goal is to construct a data structure that on query (i,j) determines whether T contains a connected subgraph with exactly i nodes, j of which are black. Such a subgraph of T is referred to as a *pattern* and (i,j) is said to *appear* in T. The main result of this section is stated below.

Theorem 1. *Given a tree T with n nodes that are colored black or white, we can construct in $O(n^2/\log^2 n)$ time a data structure of size $O(n)$ bits that given a query (i,j) determines in $O(1)$ time if (i,j) appears in T.*

Notice that the bounds of Theorem 1 match the currently best bounds for the case where T is a string [22,23]. This is despite the fact that a string has only $O(n^2)$ substrings while a tree can have $\Omega(2^n)$ connected subgraphs. The following lemma indicates an important property of string jumbled pattern matching that carries on to trees. It gives rise to a simple index described below.

Lemma 1. *If (i,j_1) and (i,j_2) both appear in T, then for every $j_1 \leq j \leq j_2$, (i,j) appears in T.*

2.1   A Simple Index

As in the case of strings, the above lemma suggests an $O(n)$-size data structure: For every $i = 1,\ldots,n$, store the minimum and maximum values i_{min} and i_{max} such that (i,i_{min}) and (i,i_{max}) appear in T. This way, upon query (i,j), we can report in constant time whether (i,j) appears in T by checking if $i_{min} \leq j \leq i_{max}$. However, while $O(n^2)$ construction-time is trivial for strings (for every $i = 0,\ldots,n$, slide a window of length i through the text in $O(n)$ time) it is harder on trees.

To obtain $O(n^2)$ construction time, we begin by converting our tree into a rooted binary tree. We arbitrarily root the tree T. To convert it to a binary tree, we duplicate each node with more than two children as follows: Let v be a node with children $u_1, \ldots, u_k$, $k \geq 3$. We replace v with $k-1$ new nodes $v_1, \ldots, v_{k-1}$, make u_1 and u_2 be the children of v_1, and make $v_{\ell-1}$ and $u_{\ell+1}$ be the children of v_ℓ for each $\ell = 2, \ldots, k-1$. We call the nodes $v_2, \ldots, v_k$ *dummy nodes*. This procedure at most doubles the size of T. To avoid cumbersome notation, we henceforth use T and n to denote the resulting binary rooted tree and its number of nodes. For a node v, we let T_v denote the subtree of T rooted at v (*i.e.* the connected subgraph induced by v and all its descendants).

Next, in a bottom-up fashion, we compute for each node v of T an array A_v of size $|T_v| + 1$. The entry $A_v[i]$ will store the maximum number of black nodes that appear in a connected subgraph of size i that includes v and another $i-1$ nodes in T_v. Computing the minimum (rather than maximum) number of black nodes is done similarly. Throughout the execution, we also maintain a global array A such that $A[i]$ stores the maximum $A_v[i]$ over all nodes v considered so far. Notice that in the end of the execution, $A[i]$ holds the desired value i_{max} since every connected subgraph of T of size i includes some node v and $i-1$ nodes in T_v.

We now show how to compute $A_v[i]$ for a node v and a specific value $i \in \{1, \ldots, |T_v|\}$. If v has a single child u, then v is necessarily not a dummy node and we set $A_v[i] = col(v) + A_u[i-1]$, where $col(v) = 1$ if v is black and $col(v) = 0$ otherwise. If v has two children u and w, then any pattern of size i that appears in T_v and includes v is composed of v, a pattern of size ℓ in T_u that includes u, and a pattern of size $i-1-\ell$ in T_w that includes w. We therefore set $A_v[i] = col(v) + \max_{0 \leq \ell \leq i-1}\{A_u[\ell] + A_w[i-1-\ell]\}$ and $A_v[i] = \max_{1 \leq \ell \leq i-1}\{A_u[\ell] + A_w[i-1-\ell]\}$ when v is a dummy node. Observe that in the latter the ℓ index starts with 1 to indicate that the non-dummy copy of v is already included in the pattern.

Lemma 2. *The above algorithm runs in $O(n^2)$ time.*

Note that if at any time the algorithm only stores arrays A_v which are necessary for future computations, then the total space used by the algorithm is $O(n)$. The space can be made $O(n)$ *bits* by storing the A_v arrays in a succinct fashion (this will also prove useful later for improving the running time): Observe that $A[i]$ is either equal to $A_v[i]$ or to $A_v[i] + 1$. This is because any pattern of size i with b black nodes can be turned into a pattern of size $i-1$ with at least $b-1$ black nodes by removing a leaf. We can therefore represent A_v as a binary string B_v of n bits, where $B_v[0] = 0$, and $B_v[i] = A_v[i] - A_v[i-1]$ for all $i = 1, \ldots, n-1$. Notice that since $A_v[i] = \sum_{\ell=0}^{i} B_v[\ell]$, each entry of A_v can be retrieved from B_v in $O(1)$ time using *rank* queries [20].

2.2   Pattern Matching

Before improving the above algorithm, we show that it can already be analyzed more carefully to get a bound of $O(n \cdot i)$ when the pattern size is known to be at most i. This is useful for the *pattern matching* problem: Without preprocessing, decide whether a given pattern (i, j) appears in T.

In the case of strings, this problem can trivially be solved in $O(n)$ time by sliding a window of length i through the string thus effectively considering every substring of length i. This sliding-window approach however does not extend to trees since we cannot afford to examine all connected subgraphs of T. We next show that, in trees, searching for a pattern of size i can be done in $O(n \cdot i)$ time by using our above indexing algorithm. This is useful when the pattern is small (i.e., when $i = o(n)$). Obtaining $O(n)$ time remains our main open problem.

Lemma 3. *Given a tree T with n nodes that are colored black or white and a query pattern (i, j), we can check in $O(n \cdot i)$ time and $O(n)$ space if T contains the pattern (i, j).*

2.3 An Improved Index

In this subsection, we will gradually improve the construction time from $O(n^2)$ to $O(n^2 / \log^2 n)$. For simplicity of the presentation, we will assume the input tree T is a rooted binary tree. This extends to arbitrary trees using a similar dummy-nodes trick as above.

From trees to strings. Recall that we can represent every A_v by a binary string B_v where $B_v[i] = A_v[i] - A_v[i - 1]$. We begin by showing that if v has two children u, w then the computation of B_v can be done by solving a variant of jumbled pattern matching on the string $S_v = X_v \circ col(v) \circ Y_v$ (here $\circ$ denotes concatenation) of length $|S_v| = |T_u| + |T_w| + 1$, where X_v is obtained from B_u by reversing it and removing its last bit, and Y_v is obtained from B_w by removing its first bit. We call the position in S_v with $col(v)$ the *split position* of S_v. Recall that $A_v[i] = col(v) + \max_{0 \le \ell \le i-1} \{A_u[\ell] + A_w[i - 1 - \ell]\}$. This is equal to the maximum number of 1s in a window of S that is of length i and includes the split position of S_v.

We are therefore interested only in windows including the split position, and this is the important distinction from the standard jumbled pattern matching problem on strings. Clearly, using the fastest $O(n^2 / \log^2 n)$-time algorithm [23] for the standard string problem we can also solve our problem and compute A_v in $O(|S|^2 / \log^2 n)$ time. However, recall that for our total analysis (over all nodes v) to give $O(n^2 / \log^2 n)$ we need the time to be $O(|X_v| \cdot |Y_v| / \log^2 n)$ and not $O((|X_v| + |Y_v|)^2 / \log^2 n)$.

First Speedup. The $O(\log^2 n)$-factor speedup for jumbled pattern matching on strings [23] is achieved by a clever combination of lookup tables. One log factor is achieved by computing the maximum 1s in a window of length i only when i is a multiple of $s = (\log n)/6$. Using a lookup table over all possible windows of length s, a sliding window of size i can be extended in $O(1)$ time to all windows of sizes $i + 1, \ldots, i + s - 1$ that start at the same location (see [23] for details). Their algorithm can output in $O(n^2 / \log n)$ time an array of $O(n / \log n)$ words. For each i that is a multiple of s, the array keeps one word storing the maximum number of 1s over all windows of length i and another word storing the binary increment vector for the maximum number of 1s in all windows of length $i + 1, \ldots, i + s - 1$.

By only considering windows that include the split position of S_v, this idea easily translates to an $O(|X_v| \cdot |Y_v| / \log n)$-time algorithm to compute A_v and implicitly store it in $O((|X_v| + |Y_v|) / \log n)$ words. From this it is also easy to obtain an $O((|X_v| + |Y_v|) / \log n)$-words representation of B_v. Notice that if v has a single child then the same procedure works with $|X_v| = 0$ in time $O(|Y_v| / \log n) = O(n / \log n)$. Summing over all nodes v, we get an $O(n^2 / \log n)$-time solution for binary jumbled indexing on trees.

Second Speedup. In strings, an additional logarithmic improvement shown in [23] can be obtained as follows: When sliding a window of length i (i is a multiple of s) the window is shifted s locations in $O(1)$ time using a lookup table over all pairs of binary substrings of length $\leq s$ (representing the leftmost and rightmost bits in all these s shifts). This further improvement yields an $O(n^2 / \log^2 n)$-time algorithm for strings. In trees however this is not the case. While we can compute A_v in $O((|X_v| + |Y_v|)^2 / \log^2 n)$ time, we can guarantee $O(|X_v| \cdot |Y_v| / \log^2 n)$ time only if both $|X_v|$ and $|Y_v|$ are greater than s. Otherwise, say $|X_v| < s$ and $|Y_v| \geq s$, we will get $O(|X_v| \cdot |Y_v| / |X_v| \log n) = O(|Y_v| / \log n)$ time. This is because our windows must include the $col(v)$ index and so we never shift a window by more than $|X_v|$ locations. Overcoming this obstacle is the main challenge of this subsection. It is achieved by carefully ensuring that the $O(|Y_v| / \log n) = O(n / \log n)$ costly constructions will be done only $O(n / \log n)$ times.

A Micro-Macro Decomposition. A *micro-macro decomposition* [1] is a partition of T into $O(n / \log n)$ disjoint connected subgraphs called *micro trees*. Each micro tree is of size at most $\log n$, and at most two nodes in a micro tree are adjacent to nodes in other micro trees. These nodes are referred to as *top* and *bottom boundary* nodes. The top boundary node is chosen as the root of the micro tree. The *macro tree* is a rooted tree of size $O(n / \log n)$ whose nodes correspond to micro trees as follows (See Fig 1): The top boundary node $t(C)$ of a micro tree C is connected to a boundary node in the parent micro tree $parent(C)$ (apart from the root). The boundary node $t(C)$ might also be connected to a top boundary node of a child micro tree $child(C)$.[1] The bottom boundary node $b(C)$ of C is connected to top boundary nodes of at most two child micro trees $\ell(C)$ and $r(C)$ of C.

A Bottom Up Traversal of the Macro Tree. With each micro tree C we associate an array A_C. Let T_C denote the union of micro tree C and all its descendant micro trees. The array A_C stores the maximum 1s (black nodes) in every pattern that includes the boundary node $t(C)$ and other nodes of T_C. We also associate three auxiliary arrays: A_b, A_t and A_{tb} The array A_b stores

---

[1] The root of the macro tree is unique as it might have a top boundary node connected to two child micro trees. We focus on the other nodes. Handling the root is done in a very similar way.

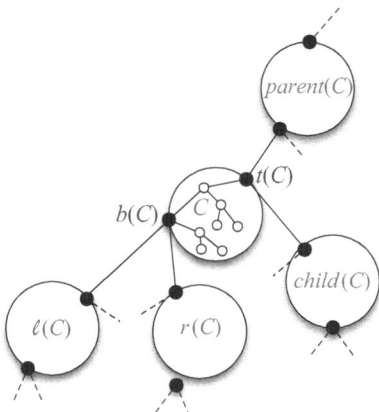

Fig. 1. A micro tree C and its neighboring micro trees in the macro tree. Inside each micro tree, the black nodes correspond to boundary nodes and the white nodes to non-boundary nodes.

the maximum 1s in every pattern that includes the boundary node $b(C)$ and other nodes of C, $T_{\ell(C)}$, and $T_{r(C)}$. The array A_t stores the maximum 1s in every pattern that includes the boundary node $t(C)$ and other nodes of C and $T_{child(C)}$. Finally, the array A_{tb} stores the maximum 1s in every pattern that includes *both* boundary nodes $t(C)$ and $b(C)$ and other nodes of C, $T_{\ell(C)}$, and $T_{r(C)}$.

We initialize for every micro tree C its $O(|C|) = O(\log n)$ sized arrays. Arrays A_C and A_t are initialized to hold the maximum 1s in every pattern that includes $t(C)$ and nodes of C. This can be done in $O(|C|^2)$ time for each C by rooting C at $t(C)$ and running the algorithm from the previous subsection. Similarly, we initialize the array A_b to hold the maximum 1s in every pattern that includes $b(C)$ and nodes of C. The array A_{tb} is initialized as follows: First we check how many nodes are 1s and how many are 0s on the unique path between $t(C)$ and $b(C)$. If there are i 1s and j 0s we set $A_{tb}[k] = 0$ for every $k < i + j$ and we set $A_{tb}[i + j] = i$. We compute $A_{tb}[k]$ for all $k > i + j$ in total $O(|C|^2)$ time by contracting the $b(C)$-to-$t(C)$ path into a single node and running the previous algorithm rooting C in this contracted node. The total running time of the initialization step is therefore $O(n \cdot |C|^2 / \log n) = O(n \log n)$ which is negligible. Notice that during this computation we have computed the maximum 1s in all patterns that are completely inside a micro tree. We are now done with the leaf nodes of the macro tree.

We next describe how to compute the arrays of an internal node C of the macro tree given the arrays of $\ell(C), r(C)$ and $child(C)$. We first compute the maximum 1s in all patterns that include $b(C)$ and vertices of $T_{\ell(C)}$ and $T_{r(C)}$. This can be done using the aforementioned string speedups in $O(|T_{\ell(C)}| \cdot |T_{r(C)}| / \log^2 n)$ time when both $|T_{\ell(C)}| > \log n$ and $|T_{r(C)}| > \log n$ and in $O(n / \log n)$ time otherwise. Using this and the initialized array A_b of C (that is of size $|C| \le \log n$) we can compute the final array A_b of C in time $O((|T_{\ell(C)}| + |T_{r(C)}|)/\log n) = O(n/\log n)$. Similarly, using the initialized A_{tb} of C, we can compute the final array A_{tb} of C in $O(n/\log n)$ time. Next, we compute the array A_t using the initialized array A_t of C and the array A_t of $child(C)$ in time $O(n/\log n)$. Finally, we compute A_C of C using A_{tb} of C and A_t of $child(C)$ in $O((|T_{\ell(C)}|+|T_{r(C)}|+ |C|)\cdot|T_{child(C)}|/\log^2 n)$ time if both $|T_{\ell(C)}|+|T_{r(C)}|+|C| > \log n$ and $|T_{child(C)}| > \log n$ and in $O(n/\log n)$ otherwise. To finalize A_C we must then take the entry-wise maximum between the computed A_C and A_t. This is because a patten in T_C may or may not include $b(C)$.

To bound the total time complexity over all clusters C, notice that some computations required $O(\alpha(v)\cdot\beta(v)/\log^2 n)$ when $\alpha(v) > \log n$ and $\beta(v) > \log n$ are the subtree sizes of two children of some node $v \in T$. We have already seen that the sum of all these terms over all nodes of T is $O(n^2/\log^2 n)$. The other type of computations each require $O(n/\log n)$ time but there are at most $O(n/\log n)$ such computations ($O(1)$ for each micro tree) for a total of $O(n^2/\log^2 n)$. This completes the proof of Theorem 1.

2.4   Finding the Query Pattern

In this subsection we extend the index so that on top of identifying in $O(1)$ time if a pattern (i, j) appears in T, it can also locate in $O(\log n)$ time a node $v \in T$ that is part of such a pattern appearance. We call this node an *anchor* of the appearance. This extension increases the space of the index from $O(n)$ bits to $O(n \log n)$ bits (i.e., $O(n)$ words).

Recall that given a tree T we build in $O(n^2/\log^2 n)$ time an array A of size $n = |T|$ where $A[i]$ stores the minimum and maximum values i_{min} and i_{max} such that (i, i_{min}) and (i, i_{max}) appear in T. Now consider a *centroid decomposition* of T: A centroid node c in T is a node whose removal leaves no connected component with more than $n/2$ nodes. We first construct the array A of T in $O(n^2/\log^2 n)$ time and store it in node c. We then recurse on each remaining connected component. This way, every node $v \in T$ will compute the array corresponding to the connected component whose centroid was v. Notice that this array is not the array A_v since we do not insist the pattern uses v. Observe that since each array A is implicitly stored in an n-sized bit array B, and since the recursion tree is balanced the total space complexity is $O(n \log n)$ bits. Furthermore, the time to construct all the arrays is bounded by $T(n) = 2T(n/2) + O(n^2/\log^2 n) = O(n^2/\log^2 n)$.

Let c denote the centroid of T whose removal leaves at most three connected components T_1, T_2, and T_3 (recall we assume degree at most 3). Upon query (i, j) we first check the array of c if pattern (i, j) appears in T (i.e., if $i_{min} \le j \le i_{max}$). If it does then we check the centroids of T_1, T_2 and T_3. If (i, j) appears in any of them then we continue the search there. This way, after at most $O(\log n)$ steps we reach the first node v whose connected component includes (i, j) but none of its child components do. We return v as the anchor node since such a pattern must include v. Finally, we note that the above can be extended so that for *every* occurrences of (i, j) one node that is part of this occurrence is reported.

3 Jumbled Pattern Matching on Grammars

In grammar compression, a binary string S of length n is compressed using a context-free grammar $G(S)$ in Chomsky normal form that generates S and only S. Such a grammar has a unique *parse tree* that generates S. Identical subtrees of this parse tree indicate substring repeats in S. The size of the grammar $g = |G(S)|$ is defined as the total number of variables and production rules in the grammar. Note that g can be exponentially smaller than $n = |S|$, and is always at most $O(n/\log n)$. We show how to solve the jumbled pattern matching problem on S by solving it on the parse tree of $G(S)$, taking advantage of subtree repeats. We obtain the following bounds. The proof is given in the appendix.

Theorem 2. *Given a binary string S of length n compressed by a grammar $G(S)$ of size g, we can construct in $O(g^{2/3}n^{4/3}/(\log n)^{4/3})$ time a data structure of size $O(n)$ bits that on query (i, j) determines in $O(1)$ time if S has a substring of length i with exactly j 1s.*

We also note that similarly to the case of trees (subsection 2.4), if we are willing to increase our index space to $O(n \log n)$ bits, then it is not difficult to turn indexes for *detecting* jumbled pattern matches in grammars into indexes for *locating* them. To obtain this, we build an index for S and recurse (build indexes) on $S_1 = B_1 \circ \cdots \circ B_k$ and $S_2 = B_{k+1} \circ \cdots \circ B_d$ where $|S_1|$ and $|S_2|$ are roughly $n/2$. This way, like in the centroid decomposition for trees, we can get in $O(\log n)$ time an anchor index of S. That is, an index of S that is part of a pattern appearance. Furthermore, as opposed to trees, we can then find the actual appearance (not just the anchor) in additional $O(i)$ time by sliding a window of size i that includes the anchor.

4 Jumbled Pattern Matching on Bounded Treewidth Graphs

In this section we consider the extension of binary jumbled pattern matching to the domain of graphs: Given a graph G whose vertices are colored either black and white, and a query (i, j), determine whether G has a connected subgraph G' with i white vertices and j black vertices[2]. This problem is also known as the (binary) graph motif problem in the literature. Fellows *et al.* [17] provided an $n^{O(w)}$ algorithm for this problem, where w is the treewidth of the input graph. Here we will substantially improve on this result by proving the following theorem, asserting that the problem is fixed-parameter tractable in the treewidth of the graph.

Theorem 3. *Binary jumbled pattern matching can be solved in $f(w) \cdot n^{O(1)}$ time on graphs of treewidth w.*

---

[2] The difference between the meaning of the query here and elsewhere in the paper is for ease of the presentation.

The function $f(w)$ in the theorem above can be replaced with $w^{O(w)}$ in case a tree decomposition of width w (see below) is provided with the input graph, and otherwise it can be replaced by $2^{O(w^3)}$. Also, the algorithm in the theorem actually computes all queries (i, j) that appear in G, and can thus be easily converted to an index for the input graph.

We begin by first introducing some necessary notation and terminology. Let $G = (V(G), E(G))$ be a graph. A *tree decomposition* of G is a tree $\mathcal{T}$ whose nodes are subsets of $V(G)$, called *bags*, with the following two properties: (i) the union of all subgraphs induced by the bags of $\mathcal{T}$ is G, and (ii) for any vertex v, the set of all bags including v induces a connected subgraph in $\mathcal{T}$. We use $\mathcal{X}$ to denote the set of bags in a given tree decomposition. The *width* of the decomposition is defined as $\max_{X \in \mathcal{X}} |X| - 1$. The *treewidth* of G is the smallest possible width of any tree decomposition of G. Given a bag X of a given tree decomposition $\mathcal{T}$, we let G_X denote the subgraph induced by the union of all bags in $\mathcal{T}_X$. Bodlaender [7] gave an algorithm for computing a width-w tree decomposition of a given graph with treewidth w in $2^{O(w^3)}n$ time.

We next describe the information we store for each bag in the tree decomposition of G. Let X be an arbitrary bag. A partition $\Pi_X = \{X_0, X_1, \ldots, X_x\}$ of X is *positive* for a given query (i, j) in G_X if there are x disjoint connected subgraphs $G_1, \ldots, G_x$ of G_X such that (1) the total number of black and white vertices in $G' = G_1 \cup \cdots \cup G_x$ is i and j respectively, and (2) $V(G') \cap X_0 = \emptyset$ and $V(G_\ell) \cap X = X_\ell$ for each $\ell = 1, \ldots, x$. Here we slightly abuse our terminology and allow X_0 to be the empty set. The information we compute for each bag X is an array A_X which has an entry for each possible query (i, j), where the entry $A_X[i, j]$ contains the set of all positive partitions of X for (i, j) in G_X. Note that the query (i, j) appears in G_X iff there exists some partition into two sets $\{X_0, X_1\}$ that is positive for (i, j). Since (i, j) appears in G iff (i, j) appears in G_X for some bag $X \in \mathcal{X}$, computing the arrays A_X for all bags allows us to determine whether (i, j) appears in G. Our algorithm computes all arrays A_X in a bottom-top fashion from the leaves to the root of $\mathcal{T}$. It is easy to verify that the size of each array A_X is bounded by $w^{O(w)}n^2$. To get a similar term in our running time, we will show that computing the array A_X from the arrays of the children of X can be done in polynomial-time with respect to the child array sizes.

We will work with a specific kind of tree decompositions, namely *nice tree decompositions* [8]. A nice tree decomposition is a binary rooted tree decomposition $\mathcal{T}$ with four types of bags: *Leaf, forget, introduce*, and *join*. Leaf bags are the leaves of $\mathcal{T}$ and include a single vertex of G, and so computing A_X for leaf bags is trivial. A forget bag X has a single child Y with $X = Y \setminus \{v\}$ for some vertex v of G. Computing A_X from A_Y in this case amounts to converting each positive partition Π_Y of Y to a corresponding positive partition Π_X of X by removing v from the class it belongs to in Π_Y. An introduce bag X also has a single child Y, but this time we have $X = Y \cup \{v\}$ for some vertex $v \notin Y$ of G. By the properties of a tree decomposition, we know that v is only adjacent to vertices of Y in G_X. Computing A_X from A_Y in this case requires the consideration of all partitions of X which are formed from positive partitions Π_Y of Y by adding

v to a class in Π_Y with one of its neighbors (or adding $\{v\}$ as a new singleton class). We leave the precise details to the full version of this paper, but it should be easy to see that computing A_X in this case, as well as in all cases above, can be done in $w^{O(w)} n^{O(1)}$.

The more challenging case is when X is a join bag. A join bag X has two children Y and Z in $\mathcal{T}$, with $X = Y = Z$. Consider two partitions $\Pi_Y = \{Y_0, \ldots, Y_y\}$ and $\Pi_Z = \{Z_0, \ldots, Z_z\}$ for Y and Z. We define the partition $\Pi_Y \oplus \Pi_Z$ as follows: First we set X_0 to be $Y_0 \cap Z_0$. The remaining classes are constructed such that any pair of vertices in X belong to the same class in $\Pi_X \setminus \{X_0\}$ iff they belong to the same class in $\Pi_Y \setminus \{Y_0\}$ or to the same class in $\Pi_Z \setminus \{Z_0\}$.

Let i_0 and j_0 respectively denote the number of white and black vertices in X. We claim that if (i_1, j_1) and (i_2, j_2) are two queries for which Π_Y and Π_Z are respectively positive in G_Y and G_Z, then $\Pi_X = \Pi_Y \oplus \Pi_Z$ is positive for $(i_1 + i_2 - i_0, j_1 + j_2 - j_0)$. This can be verified by considering the connected components in $G_1^Y \cup \cdots \cup G_y^Y \cup G_1^Z \cdots \cup G_z^Z$, where $G_1^Y, \ldots, G_y^Y$ and $G_1^Z, \ldots, G_z^Z$ are sets of graphs witnessing that Π_Y and Π_Z are positive for (i_1, j_1) in G_Y and (i_2, j_2) in G_Z. It is easy to see that the total number of white and black vertices in these components is $i_1 + i_2 - i_0$ and $j_1 + j_2 - j_0$, where i_0 white vertices and j_0 black vertices are subtracted due to double counting the vertex colors in X. Moreover, it can be verified that these components intersect X as required by Π_X.

On the other hand, it can also be seen on the same lines that if (i, j) is a query for which Π_X is positive in G_X, then $(i, j) = (i_1 + i_2 - i_0, j_1 + j_2 - j_0)$ for some pair of queries (i_1, j_1) and (i_2, j_2) for which Π_Y and Π_Z are positive in G_Y and G_Z. We can therefore compute $A_X[i, j]$ by examining all such pairs (i_1, j_1) and (i_2, j_2), and computing the partition $\Pi_Y \oplus \Pi_Z$ for each pair of positive partitions $\Pi_Y \in A_Y[i_1, j_1]$ and $\Pi_Z \in A_Z$. This requires $w^{O(w)} n^{O(1)}$ time.

To summarize we compute each array A_X in $w^{O(w)} n^{O(1)}$ time. As the total number of bags is $O(wn)$, we obtain an algorithm whose total running time is $w^{O(w)} n^{O(1)}$, excluding the time required to compute the nice tree decomposition $\mathcal{T}$. We note that the running time of our algorithm can be improved slightly by using an extension of Lemma 1 to graphs. Also, our result straightforwardly extends to an $w^{O(w)} n^{O(c)}$ time algorithm for the case where the vertices of G are colored with c colors.

References

1. Alstrup, S., Secher, J., Sporkn, M.: Optimal on-line decremental connectivity in trees. Information Processing Letters 64(4), 161–164 (1997)
2. Ambalath, A.M., Balasundaram, R., Rao H., C., Koppula, V., Misra, N., Philip, G., Ramanujan, M.S.: On the kernelization complexity of colorful motifs. In: Raman, V., Saurabh, S. (eds.) IPEC 2010. LNCS, vol. 6478, pp. 14–25. Springer, Heidelberg (2010)
3. Badkobeh, G., Fici, G., Kroon, S., Lipták, Z.: Binary jumbled string matching for highly run-length compressible texts. Inf. Process. Lett. 113(17), 604–608 (2013)
4. Benson, G.: Composition alignment. In: Benson, G., Page, R.D.M. (eds.) WABI 2003. LNCS (LNBI), vol. 2812, pp. 447–461. Springer, Heidelberg (2003)
5. Betzler, N., van Bevern, R., Fellows, M.R., Komusiewicz, C., Niedermeier, R.: Parameterized algorithmics for finding connected motifs in biological networks. IEEE/ACM Trans. Comput. Biology Bioinform. 8(5), 1296–1308 (2011)

6. Böcker, S.: Simulating multiplexed SNP discovery rates using base-specific cleavage and mass spectrometry. Bioinformatics 23(2), 5–12 (2007)
7. Bodlaender, H.L.: A linear time algorithm for finding tree-decompositions of small treewidth. SIAM Journal on Computing 25, 1305–1317 (1996)
8. Bodlaender, H.L.: Treewidth. Algorithmic techniques and results. In: Privara, I., Ružička, P. (eds.) MFCS 1997. LNCS, vol. 1295, pp. 19–36. Springer, Heidelberg (1997)
9. Burcsi, P., Cicalese, F., Fici, G., Lipták, Z.: On table arrangement, scrabble freaks, and jumbled pattern matching. In: Boldi, P. (ed.) FUN 2010. LNCS, vol. 6099, pp. 89–101. Springer, Heidelberg (2010)
10. Burcsi, P., Cicalese, F., Fici, G., Lipták, Z.: Algorithms for jumbled pattern matching in strings. International Journal of Foundations of Computer Science 23(2), 357–374 (2012)
11. Burcsi, P., Cicalese, F., Fici, G., Lipták, Z.: On approximate jumbled pattern matching in strings. Theory of Computing Systems 50(1), 35–51 (2012)
12. Charikar, M., Lehman, E., Liu, D., Panigrahy, R., Prabhakaran, M., Sahai, A., Shelat, A.: The smallest grammar problem. IEEE Transactions on Information Theory 51(7), 2554–2576 (2005)
13. Cicalese, F., Fici, G., Lipták, Z.: Searching for jumbled patterns in strings. In: Proc. of the Prague Stringology Conference, pp. 105–117 (2009)
14. Cicalese, F., Laber, E., Weimann, O., Yuster, R.: Near linear time construction of an approximate index for all maximum consecutive sub-sums of a sequence. In: Kärkkäinen, J., Stoye, J. (eds.) CPM 2012. LNCS, vol. 7354, pp. 149–158. Springer, Heidelberg (2012)
15. Dondi, R., Fertin, G., Vialette, S.: Complexity issues in vertex-colored graph pattern matching. J. Discrete Algorithms 9(1), 82–99 (2011)
16. Dondi, R., Fertin, G., Vialette, S.: Finding approximate and constrained motifs in graphs. In: Giancarlo, R., Manzini, G. (eds.) CPM 2011. LNCS, vol. 6661, pp. 388–401. Springer, Heidelberg (2011)
17. Fellows, M.R., Fertin, G., Hermelin, D., Vialette, S.: Upper and lower bounds for finding connected motifs in vertex-colored graphs. J. Comput. Syst. Sci. 77(4), 799–811 (2011)
18. Gawrychowski, P.: Faster algorithm for computing the edit distance between SLP-compressed strings. In: Calderón-Benavides, L., González-Caro, C., Chávez, E., Ziviani, N. (eds.) SPIRE 2012. LNCS, vol. 7608, pp. 229–236. Springer, Heidelberg (2012)
19. Giaquinta, E., Grabowski, S.: New algorithms for binary jumbled pattern matching. Inf. Process. Lett. 113(14-16), 538–542 (2013)
20. Jacobson, G.: Space-efficient static trees and graphs. In: Proc. of the 30th Annual Symposium on Foundations of Computer Science (FOCS), pp. 549–554 (1989)
21. Lacroix, V., Fernandes, C.G., Sagot, M.-F.: Motif search in graphs: Application to metabolic networks. IEEE/ACM Trans. Comput. Biology Bioinform. 3(4), 360–368 (2006)
22. Moosa, T.M., Rahman, M.S.: Indexing permutations for binary strings. Information Processing Letters 110(18-19), 795–798 (2010)
23. Moosa, T.M., Rahman, M.S.: Sub-quadratic time and linear space data structures for permutation matching in binary strings. Journal of Discrete Algorithms 10, 5–9 (2012)
24. Rytter, W.: Application of Lempel-Ziv factorization to the approximation of grammar-based compression. Theoretical Computer Science 302(1-3), 211–222 (2003)

Kernelization Using Structural Parameters on Sparse Graph Classes*

Jakub Gajarský[1], Petr Hliněný[1], Jan Obdržálek[1], Sebastian Ordyniak[1],
Felix Reidl[2], Peter Rossmanith[2],
Fernando Sánchez Villaamil[2], and Somnath Sikdar[2]

[1] Faculty of Informatics, Masaryk University,
Brno, Czech Republic
`{gajarsky,hlineny,obdrzalek,ordyniak}@fi.muni.cz`
[2] Theoretical Computer Science, Department of Computer Science,
RWTH Aachen University, Aachen, Germany
`{reidl,rossmani,fernando.sanchez,sikdar}@cs.rwth-aachen.de`

Abstract. Meta-theorems for polynomial (linear) kernels have been the subject of intensive research in parameterized complexity. Heretofore, there were meta-theorems for linear kernels on graphs of bounded genus, H-minor-free graphs, and H-topological-minor-free graphs. To the best of our knowledge, there are no known meta-theorems for kernels for any of the larger sparse graph classes: graphs of bounded expansion, locally bounded expansion, and nowhere dense graphs. In this paper we prove meta-theorems for these three graph classes. More specifically, we show that graph problems that have finite integer index (FII) admit linear kernels on hereditary graphs of bounded expansion when parameterized by the size of a modulator to constant-treedepth graphs. For hereditary graph classes of locally bounded expansion, our result yields a quadratic kernel and for hereditary nowhere dense graphs, a polynomial kernel. While our parameter may seem rather strong, a linear kernel result on graphs of bounded expansion with a weaker parameter would for some problems violate known lower bounds. Moreover, we use a relaxed notion of FII which allows us to prove linear kernels for problems such as LONGEST PATH/CYCLE and EXACT s, t-PATH which do not have FII in general graphs.

1 Introduction

Data preprocessing has always been a part of algorithm design. The last decade has seen steady progress in the area of *kernelization*, an area which deals with the design of polynomial-time preprocessing algorithms. These algorithms compress an input instance of a parameterized problem into an equivalent output instance whose size is bounded by some function of the parameter. Parameterized

* Research funded by DFG-Project RO 927/12-1 "Theoretical and Practical Aspects of Kernelization", by the Czech Science Foundation, project P202/11/0196, and by Employment of Newly Graduated Doctors of Science for Scientific Excellence (CZ.1.07/2.3.00/30.0009).

H.L. Bodlaender and G.F. Italiano (Eds.): ESA 2013, LNCS 8125, pp. 529–540, 2013.

complexity theory guarantees the existence of such *kernels* for problems that are *fixed-parameter tractable*. Some problems admit stronger kernelization in the sense that the size of the output instance is bounded by a polynomial (or even linear) function of the parameter, the so-called *polynomial (or linear) kernels*.

Of great interest are *algorithmic meta-theorems*, results that focus on problem classes instead of single problems. In the area of graph algorithms, such meta-theorems usually have the following form: all problems with a specific property admit an algorithm of a specific type on a specific graph class. In this paper we focus on meta-theorems for linear or polynomial kernels on sparse graph classes. After early results such as [2, 20], the first such meta-theorem due to Bodlaender et al. [4] states that problems that have *finite integer index* (FII) and are *quasi-compact* admit linear kernels on graphs of bounded genus. Fomin et al. [18] extended the result to the strictly larger class of H-minor-free graphs, for problems which have FII, are *bidimensional* and satisfy the *separation property*. This result was, in turn, generalized in [21] to H-topological-minor-free graphs, which strictly contain H-minor-free graphs. Here, the problems are required to have FII and to be *treewidth-bounding*.

The keystone of all these meta-theorems is *finite integer index*. This property is the basis of the *protrusion replacement rule* whereby protrusions (pieces of the input graph satisfying certain requirements) are replaced by members of a finite set of canonical graphs. The protrusion replacement rule is a crucial ingredient to obtaining small kernels.

Although these meta-theorems (viewed in chronological order) steadily covered larger graph classes, the set of problems captured in their framework diminished as the second precondition became stricter. For H-topological-minor-free graphs this precondition is to be treewidth bounding. A (parameterized) graph problem is treewidth-bounding if YES-instances have a vertex set of size linear in the parameter, deletion of which results in a graph of bounded treewidth. Such a vertex set is called a *modulator to bounded treewidth*. While treewidth-boundedness is a strong prerequisite, it is important to note that the combined properties of bidimensionality and separability (used to prove the result on H-minor-free graphs) imply treewidth-boundedness [18]. Quasi-compactness on bounded genus graphs imply the same [4]. This demonstrates that all three previous meta-theorems on linear kernels implicitly or explicitly used treewidth-boundedness.

Another way of viewing the meta-theorem in [21] is as follows: when parameterized by a treewidth modulator, problems that have FII have linear kernels in H-topological-minor-free graphs. A natural problem therefore is to identify the least restrictive parameter that can be used to prove a meta-theorem for linear kernels for the next well-known class in the sparse-graph hierarchy, namely, graphs of bounded expansion. This class was defined by Nešetřil and Ossona de Mendez [26] and subsumes the class of H-topological-minor-free graphs. However, a modulator to bounded treewidth does not seem to be a useful parameter for this class. Any graph class $\mathcal{G}$ can be transformed into a class $\tilde{\mathcal{G}}$ of bounded expansion by replacing every graph $G \in \mathcal{G}$ with $\tilde{G}$, obtained in turn by replacing each edge of G by a path on $|V(G)|$ vertices. For problems like TREEWIDTH t-VERTEX DELETION and, in particular, FEEDBACK VERTEX SET this operation neither changes the instance membership nor does it increase the parameter. As both problems do not admit kernels of size $O(k^{2-\epsilon})$ unless coNP $\subseteq$ NP/poly by a result of Dell and Melkebeek [10], a linear kernelization result on this class of graph and under this parameterization must necessarily exclude both problems.

This suggests that to encompass these problems, the chosen parameter must not be invariant under edge subdivision. If we assume that the parameter does not increase for subgraphs, it must necessarily attain high values on paths. *Treedepth* [26] is precisely a parameter that enforces this property, since graphs of bounded treedepth are essentially degenerate graphs with no long paths. Note that bounded treedepth implies bounded treewidth.

In terms of parameters, a modulator to bounded treedepth is a generalization of vertex cover. This is easy to see as a vertex cover leaves a graph of treedepth one. The vertex cover number has often been used as a parameter for problems that are W-hard or otherwise difficult to parameterize such as LONGEST PATH [5], CUTWIDTH [8], BANDWIDTH, IMBALANCE, DISTORTION [14], LIST COLORING, PRECOLORING EXTENSION, EQUITABLE COLORING, L(p,1)-LABELING, CHANNEL ASSIGNMENT [15]. Other generalizations of vertex cover have also been used as a parameter [12, 19]. Treedepth thus seems to be a good compromise.

Our contribution. We show that, for the class of problems with FII, a parameterization by the size of a modulator to bounded treedepth allows for linear kernels in linear time on graphs of bounded expansion. The same parameter yields quadratic kernels in graphs of locally bounded expansion and polynomial kernels in nowhere dense graphs, both strictly larger classes. In particular, nowhere dense graphs are the largest class that may still be called sparse [26]. In these results we do not require a treedepth modulator to be supplied as part of the input, as we show that it can be approximated to within a constant factor.

Furthermore, we only need FII to hold on graphs of bounded treedepth, thus including problems which do not have FII in general. This relaxation enables us to obtain kernels for LONGEST PATH/CYCLE none of which have polynomial kernels even on sparse graphs with respect to their standard parameters since they admit simple AND/OR-Compositions [3]. Problems covered by our framework include HAMILTONIAN PATH/CYCLE, several variants of DOMINATING SET, (CONNECTED) VERTEX COVER, CHORDAL VERTEX DELETION, FEEDBACK VERTEX SET, INDUCED MATCHING, and ODD CYCLE TRANSVERSAL. In particular, we cover all problems having FII in general graphs that were considered in earlier frameworks [4, 18, 21]. We wish to emphasize, however, that this paper does not subsume these results because of our usage of a structural parameter.

Finally, notice that a kernelization result for LONGEST CYCLE on graphs of bounded expansion with a parameter closed under edge subdivision would automatically imply the same result for general graphs. This forms the crux of our belief that any relaxation of the treedepth parameter to prove a meta-theorem for linear kernels on graphs of bounded expansion will exclude problems such as this one.

2   Preliminaries

We use standard graph-theoretic notation (see [11] for any undefined terminology). All our graphs are finite and simple. Given a graph G, we use $V(G)$ and $E(G)$ to denote its vertex and edge sets. For convenience we assume that $V(G)$ is a totally ordered set, and use uv instead of $\{u, v\}$ to denote an edge of G. Since we primarily consider sparse graphs, we let $|G|$ denote the number of vertices in the graph G. The distance $d_G(v, w)$ between two vertices $v, w \in V(G)$ is the length (number of edges) of a shortest v, w-path in G and ∞ if v and w lie in different connected components. By $\omega(G)$ we denote the size of the largest complete subgraph of G. For a class $\mathcal{G}$ of graphs we denote $\omega(\mathcal{G})$ the $\max\{\omega(G) | G \in \mathcal{G}\}$ and set $\omega(\mathcal{G}) = \infty$ if the maximum does not exist.

For $S \subseteq V(G)$, we let $N^G(S)$ denote the set of vertices in $V(G) \setminus S$ that have at least one neighbor in S, and for a subgraph H of G we define $N^G(H) := N^G(V(H))$. If X is a subset of vertices disjoint from S, then $N_X^G(S)$ is the set $N^G(S) \cap X$ (and similarly for $N_X^G(H)$). Given a graph G and a set $W \subseteq V(G)$, we define $\partial_G(W)$ as the set of vertices in W that have a neighbor in $V \setminus W$. Note that $N^G(W) = \partial_G(V(G) \setminus W)$. A graph G is d-degenerate if every subgraph G' of G contains a vertex $v \in V(G')$ with $deg^G(v) \leqslant d$. The degeneracy of G is the smallest d such that G is d-degenerate. In the rest of the paper we drop the index G from all the notation if it is clear which graph is being referred to.

A *graph problem* Π is a set of pairs (G, ξ), where G is a graph and $\xi \in \mathbf{N}_0$, such that for all graphs G_1, G_2 and all $\xi \in \mathbf{N}_0$, if G_1 is isomorphic to G_2, then $(G_1, \xi) \in \Pi$ iff $(G_2, \xi) \in \Pi$. For a graph class $\mathcal{G}$, we define $\Pi_\mathcal{G}$ as the set of pairs $(G, \xi) \in \Pi$ such that $G \in \mathcal{G}$.

Graph Classes. We denote the treewidth of a graph G by $\mathbf{tw}(G)$ and its pathwidth by $\mathbf{pw}(G)$. As treedepth is not a well-known measure, we provide its definition here. In this context, a *rooted forest* is a disjoint union of rooted trees. For a vertex x in a tree T of a rooted forest, the depth of x in the forest is the number of vertices in the path from the root of T to x. The *height of a rooted forest* is the maximum depth of a vertex of the forest. The *closure* $\mathrm{clos}(\mathcal{F})$ of a rooted forest $\mathcal{F}$ is the graph with vertex set $\bigcup_{T \in \mathcal{F}} V(T)$ and edge set $\{xy \mid x \neq y \text{ is an ancestor of } y \text{ in } \mathcal{F}\}$. A *treedepth decomposition* of a graph G is a rooted forest $\mathcal{F}$ such that $G \subseteq \mathrm{clos}(\mathcal{F})$.

Definition 1 (Treedepth). *The* treedepth $\mathbf{td}(G)$ *of a graph G is the minimum height of any treedepth decomposition of G.*

We list some well-known facts about graphs of bounded treedepth. Proofs are omitted and can be found in [26]. If a graph has no path with more than d vertices, then its treedepth is at most d. For any graph G with $\mathbf{td}(G) \leqslant d$, it holds that (1) G has no paths with 2^d vertices and, in particular, any DFS-tree of G has height at most $2^d - 1$; (2) G is d-degenerate and hence has at most $d \cdot |V(G)|$ edges; (3) $\mathbf{tw}(G) \leqslant \mathbf{pw}(G) \leqslant d - 1$. A useful way of thinking about graphs of bounded treedepth is that they are (sparse) graphs with no long paths. Similar to treewidth, a treedepth decomposition can be computed in linear time.

Definition 2 (Shallow minor [26]). *For $d \in \mathbf{N}_0$, a graph H is a* shallow *minor at depth d of G if there exist disjoint subsets $V_1, \ldots, V_p$ of $V(G)$ such that*

1. *each graph $G[V_i]$ has radius at most d, meaning that there exists $v_i \in V_i$ (a center) such that every vertex in V_i is within distance at most d in $G[V_i]$;*
2. *there is a bijection $\psi \colon V(H) \to \{V_1, \ldots, V_p\}$ such that for every $u, v \in V(H)$, if $uv \in E(H)$ then there is an edge in G with an endpoint each in $\psi(u)$ and $\psi(v)$.*

Note that if $u, v \in V(H)$, $\psi(u) = V_i \ni v_i$, and $\psi(v) = V_j \ni v_j$ then $d_G(v_i, v_j) \leqslant (2d + 1) \cdot d_H(u, v)$. The class of shallow minors of G at depth d is denoted by $G \triangledown d$. This notation is extended to graph classes $\mathcal{G}$ as well: $\mathcal{G} \triangledown d = \bigcup_{G \in \mathcal{G}} G \triangledown d$.

The class of graphs of bounded expansion is defined using the notion of *greatest reduced average density (grad)* (see [23, 27] for details). Let $\mathcal{G}$ be a graph class. Then the greatest reduced average density of $\mathcal{G}$ with *rank d* is defined as $\nabla_d(\mathcal{G}) = \sup_{H \in \mathcal{G} \triangledown d}(|E(H)|/|V(H)|)$. This notation is extended to graphs via the convention $\nabla_d(G) := \nabla_d(\{G\})$. In particular, note that $G \triangledown 0$ denotes the set of subgraphs of G and hence $2\nabla_0(G)$ is the maximum average degree of all subgraphs of G—i.e. its *degeneracy*.

Definition 3 (Bounded expansion [23]). *A graph class $\mathcal{G}$ has* bounded expansion *if there exists a function $f \colon \mathbf{N} \to \mathbf{R}$ (called the* expansion function*) such that for all $d \in \mathbf{N}$, $\nabla_d(\mathcal{G}) \leqslant f(d)$. We say that $\mathcal{G}$ has expansion bounded by f.*

An important relation we make use of later is: $\nabla_d(G) = \nabla_0(G \triangledown d)$, i.e. the grad of G with rank d is precisely one half the maximum average degree of subgraphs of its depth d shallow minors.

3 The Protrusion Machinery

We state the main definitions of the protrusion machinery developed in [4, 18], in some cases modifying them to suit our purpose.

Definition 4 (r-protrusion [4]). *Given a graph $G = (V, E)$, a set $W \subseteq V$ is an r-protrusion of G if $|\partial_G(W)| \leqslant r$ and $\mathbf{tw}(G[W]) \leqslant r - 1$.* [1] *We call $\partial_G(W)$ the* protrusion boundary *and $|W|$ the* size *of the protrusion W.*

A *t-boundaried graph* is a pair $(G, bd(G))$, where $G = (V, E)$ is a graph and $bd(G) \subseteq V$ is a set of t vertices with distinct labels from the set $\{1, \ldots, t\}$. The graph G is called the *underlying unlabeled graph* and $bd(G)$ is called the *boundary*[2]. Given a graph class $\mathcal{G}$, we let $\mathcal{G}_t$ denote the class of t-boundaried graphs $(G, bd(G))$ where $G \in \mathcal{G}$. For t-boundaried graphs $(H, bd(H))$ and $(G, bd(G))$, we say that $(H, bd(H)$ is a subgraph of $(G, bd(G))$ if $H \subseteq G$ and $bd(H) = bd(G)$. We say that $(H, bd(H))$ is an induced subgraph of $(G, bd(G))$ if for some $X \subseteq V(G)$, $H = G[X]$

---

[1] We want the bags in a tree-decomposition of $G[W]$ to be of size at most r.
[2] Usually denoted by $\partial(G)$, but this collides with our usage of ∂.

and $bd(H) = bd(G)$. We say that the boundaries of two t-boundaried graphs $\widetilde{G} = (G, bd(G))$ and $\widetilde{H} = (H, bd(H))$ are *identical* if the function mapping each vertex of $bd(G)$ to that vertex of $bd(H)$ with the same label is an isomorphism between $G[bd(G)]$ and $H[bd(H)]$. Note that in the case of $\widetilde{H}$ being an induced subgraph of $\widetilde{G}$, the boundaries are identical. By slightly abusing notation, we often denote a t-boundaried graph by the underlying unlabeled graph when the boundary is clear from the context.

If $W \subseteq V(G)$ is an r-protrusion of G, we let G_W be the r-boundaried graph $(G[W], B)$, where B is the labeled set of vertices of $\partial(W)$, each vertex being assigned a unique label from the set $\{1, \ldots, r\}$ according to its order in G. For t-boundaried graphs $(G_1, bd(G_1))$ and $(G_2, bd(G_2))$, we let $G_1 \oplus G_2$ denote the graph obtained by taking the disjoint union of G_1 and G_2 and identifying each vertex in $bd(G_1)$ with the vertex in $bd(G_2)$ with the same label, and then making the graph simple, if necessary. The resulting order of vertices is an arbitrary extension of the orderings on $V(G_1)$ and $V(G_2) \setminus V(G_1)$. Note that the $\oplus$ "destroys" the boundaries of two t-boundaried graphs and creates a simple graph. In the opposite direction, let $H \subseteq G$ such that $\partial(H)$ has t vertices. Let B be the labeled set of vertices from $\partial(H)$ such that each vertex is assigned a unique label from $\{1, \ldots, t\}$ according to its order in G. We define $G \ominus_B H$ to be the t-boundaried graph $(G - (V(H) \setminus B), B)$. The $\ominus$ operation, therefore, creates a t-boundaried graph from a simple graph. To make things clear we sometimes annotate the $\oplus$ operator with the boundary as well.

Definition 5 (Replacement). *Let W be a t-protrusion of a graph G and let B be the labeled set of vertices of $\partial(W)$, where the labels are assigned to vertices from $\{1, \ldots, t\}$ according to their order in G. For a t-boundaried graph H, replacing W by H is defined by the operation $(G \ominus_B G[W]) \oplus_B H$.*

The following definition concerns the centerpiece of our framework.

Definition 6 (Finite integer index; FII). *Let Π be a graph problem and let $\widetilde{G}_1 = (G_1, bd(G_1))$, $\widetilde{G}_2 = (G_2, bd(G_2))$ be two t-boundaried graphs. We say that $\widetilde{G}_1 \equiv_{\Pi,t} \widetilde{G}_2$ if there exists an integer constant $\Delta_{\Pi,t}(\widetilde{G}_1, \widetilde{G}_2)$ such that for all t-boundaried graphs $\widetilde{H} = (H, bd(H))$ and for all $\xi \in \mathbf{N}$: $(\widetilde{G}_1 \oplus \widetilde{H}, \xi) \in \Pi$ iff $(\widetilde{G}_2 \oplus \widetilde{H}, \xi + \Delta_{\Pi,t}(\widetilde{G}_1, \widetilde{G}_2)) \in \Pi$. We say that Π has finite integer index in the class $\mathcal{F}$ if, for every $t \in \mathbf{N}$, the number of classes of $\equiv_{\Pi,t}$ which have a non-empty intersection with $\mathcal{F}$ is finite.*

Note that the constant $\Delta_{\Pi,t}(\widetilde{G}_1, \widetilde{G}_2)$ depends on Π, t, and the *ordered pair* $(\widetilde{G}_1, \widetilde{G}_2)$ so that $\Delta_{\Pi,t}(\widetilde{G}_1, \widetilde{G}_2) = -\Delta_{\Pi,t}(\widetilde{G}_2, \widetilde{G}_1)$. On most occasions, the problem Π and the class $\mathcal{F}$ will be clear from the context and in such situations, we use $\equiv_t$ and Δ_t instead of $\equiv_{\Pi,t}$ and $\Delta_{\Pi,t}$, respectively.

The next lemma shows that if we assume that a graph problem Π has FII in a graph class $\mathcal{F}$, then we can choose representatives for the equivalence classes of $\equiv_{\Pi,t}$ from a (possibly different) graph class $\mathcal{G}$ under certain circumstances.

Lemma 1. *Let $\preceq$ be a relation on graphs, let $\mathcal{F}, \mathcal{G}$ be graph classes and Π a graph problem such that Π has FII in $\mathcal{F}$ and $\mathcal{G}$ is well quasi-ordered by $\preceq$. Then for each $t \in \mathbf{N}$, there exists a finite set $\mathcal{R}(t, \mathcal{F}, \mathcal{G}, \preceq) \subseteq \mathcal{F}_t \cap \mathcal{G}_t$ with the following property: for every $\widetilde{G} = (G, bd(G)) \in \mathcal{F}_t \cap \mathcal{G}_t$ there exists $\widetilde{H} = (H, bd(H)) \in \mathcal{R}(t, \mathcal{F}, \mathcal{G}, \preceq)$ such that $\widetilde{G} \equiv_{\Pi, t} \widetilde{H}$; $bd(G)$ and $bd(H)$ are identical; and $H \preceq G$.*

Let us explain how we use Lemma 1. The graph problems Π that we consider in this paper have FII on the class of general graphs or, for all $p \in \mathbf{N}$, in the class of graphs of treedepth at most p. In accordance with the notation in Lemma 1, the class $\mathcal{F}$ corresponds to the class where Π has FII. The choice of our parameter now ensures that our kernelization replaces protrusions of treedepth at most a previously fixed constant d: choosing $\mathcal{G}$ to be the graphs of treepdepth at most d, all protrusions (actually the graphs induced by them) are members of $\mathcal{F} \cap \mathcal{G}$. As $\mathcal{G}$ is well-quasi ordered under the induced subgraph relation [26, Chapter 6, Lemma 6.13], we choose $\preceq$ to be $\subseteq_{\text{ind}}$.

Now consider a restriction of the graph problem Π to a class $\mathcal{K}$ that is closed under taking induced subgraphs. In this paper, the class $\mathcal{K}$ is a hereditary graph class of bounded expansion or locally bounded expansion or a hereditary nowhere dense class. This ensures that $\emptyset \neq \mathcal{K} \cap \mathcal{G} \subseteq \mathcal{F} \cap \mathcal{G}$. Given an instance (G, ξ) of Π with $G \in \mathcal{K}$, one can replace a protrusion of G by a representative (of constant size) that is an *induced subgraph* of that protrusion, ensuring that this replacement creates a graph that still resides in $\mathcal{K}$. To summarize, Lemma 1 guarantees that the protrusion replacement rule (described next) preserves the graph class $\mathcal{K}$ and the parameter.

As preparation for the kernelization theorems of the next section, let $\mathfrak{P}$ denote the set of all graph problems that have FII on general graphs or, for each $p \in \mathbf{N}$, in the set of graphs of treedepth at most p. Our reduction rule is formalized as follows.

Reduction Rule 1 (Protrusion replacement) *Let $t, d \in \mathbf{N}$ and let $\Pi \in \mathfrak{P}$. Let (G, ξ) be an instance of Π and suppose that $W \subseteq V(G)$ is a t-protrusion of treedepth at most d such that $\beta(t, d) < |W| \leqslant \beta(t, d) + t$, where β is some function fixed in advance. Let $R \in \mathcal{R}(t, d)$ be the representative of G_W. The protrusion replacement rule is the following: Reduce (G, ξ) to $(G', \xi') := ((G \ominus G[W]) \oplus R, \xi + \Delta_t(G_W, R))$.*

The existence of a finite set of representatives $\mathcal{R}(t, d)$ is guaranteed by Lemma 1. The safety of the protrusion replacement follows from the definition of FII.

As usual, we let $\mathcal{F}$ denote the class on which the problem has FII. For a problem $\Pi \in \mathfrak{P}$ and the class $\mathcal{G}$ of graphs of treedepth at most d, we let $\mathcal{R}(t, d)$ denote the finite set $\mathcal{R}(t, \mathcal{F}, \mathcal{G}, \subseteq_{\text{ind}})$ from Lemma 1 and $\rho(t, d)$ to denote the size of the largest member of $\mathcal{R}(t, d)$. In what follows, when applying the protrusion replacement rule, we will assume that for each $t, d \in \mathbf{N}$, we are given the set $\mathcal{R}(t, d)$ of representatives of the equivalence classes of $\equiv_{\Pi, t}$. Note that previous work on meta-kernels implicitly made this assumption [4, 16–18].

4   Linear Kernels on Graphs of Bounded Expansion

In this section we show that graph problems that have FII on general graphs or in the class of graphs with bounded treedepth admit linear kernels on hereditary graph classes with bounded expansion, when parameterized by the size of a modulator to constant treedepth. On hereditary graph classes of locally bounded expansion we obtain quadratic vertex kernels and on hereditary nowhere dense classes, polynomial kernels. Our main theorem is the following.

Theorem 1. *Let $\mathcal{K}$ be a hereditary graph class of bounded expansion and let $d \in \mathbf{N}$ be a constant. Let $\Pi \in \mathfrak{P}$. Then there is an algorithm that takes as input $(G, \xi) \in \mathcal{K} \times \mathbf{N}$ and, in time $\mathcal{O}(|G| + \log \xi)$, outputs (G', ξ') such that*

1. $(G, \xi) \in \Pi$ *if and only if* $(G', \xi') \in \Pi$;
2. G' *is an induced subgraph of G; and*
3. $|G'| = \mathcal{O}(|S|)$, *where S is an optimal treedepth-d modulator of the graph G.*

In the rest of this section we let $\Pi \in \mathfrak{P}$ and let $\mathcal{K}$ be a hereditary graph class whose expansion is bounded by f.

We proceed as follows. Because an optimal treedepth-d modulator cannot be assumed as part of the input, we obtain an approximate modulator $S \subseteq V(G)$ to partition $V(G)$ into sets $Y_0 \uplus Y_1 \uplus \cdots \uplus Y_\ell$ such that $S \subseteq Y_0$ and $|Y_0| = \mathcal{O}(|S|)$ and for $1 \leqslant i \leqslant l$, Y_i induces a collection of connected components of $G - Y_0$ that have exactly the same *small* neighborhood in Y_0. We then use bounded expansion to show that $\ell = \mathcal{O}(|S|)$ and use protrusion reduction to replace each $G[Y_i]$, $1 \leqslant i \leqslant l$, by an *induced subgraph* of $G[Y_i]$ of constant size. Every time the protrusion replacement rule is applied, ξ is modified. This results in an equivalent instance (G', ξ') such that $G' \subseteq G$ and $|G'| = \mathcal{O}(|S|)$, as claimed in Theorem 1.

Lemma 2. *Fix $d \in \mathbf{N}$. Given a graph G, one can in $O(|G|^2)$ time compute a subset $S \subseteq V(G)$ such that $\mathbf{td}(G - S) \leqslant d$ and $|S|$ is at most 2^d times the size of an optimal treedepth-d modulator of G. For graphs of bounded expansion, S can be computed in linear time.*

To prove the size bounds on decompositions of $V(G)$ into vertex-disjoint sets $Y_0 \uplus Y_1 \uplus \cdots \uplus Y_\ell$, we use the following lemma about grads of bipartite graphs.

Lemma 3. *Let $G = (X, Y, E)$ be a bipartite graph, and $p \geqslant \nabla_1(G)$. Then there are at most $2p|X|$ vertices in Y with degree greater than $2p$; and at most $(4^p + 2p) \cdot |X|$ subsets $X' \subseteq X$ such that $X' = N(u)$ for some $u \in Y$.*

The proof of the next lemma tells us how to find clusters of connected components with a small neighborhood, which will be targeted by the reduction.

Lemma 4. *Let $G \in \mathcal{K}$ and $S \subseteq V(G)$ be a set of vertices such that $\mathbf{td}(G - S) \leqslant d$ $(d \in \mathbf{N}$ is a constant). There is an algorithm that runs in time $\mathcal{O}(|G|)$ and computes a partition, called protrusion-decomposition, of $V(G)$ into sets $Y_0 \uplus Y_1 \uplus \cdots \uplus Y_\ell$ such that the following hold:*

1. $S \subseteq Y_0$ and $|Y_0| = \mathcal{O}(|S|)$;
2. for $1 \leqslant i \leqslant \ell$, Y_i induces a set of connected components of $G - Y_0$ that have the same neighborhood in Y_0 of size at most $2^{d+1} + 2 \cdot f(2^d)$;
3. $\ell \leqslant \left(4^{f(2^d)} + 2f(2^d)\right) \cdot |S| = \mathcal{O}(|S|)$.

Proof Sketch. We first construct a DFS-forest $\mathcal{D}$ of $G - S$. Assume that there are q trees $T_1, \ldots, T_q$ in this forest rooted at $r_1, \ldots, r_q$, respectively. Since $\mathbf{td}(G - S) \leqslant d$, the height of every tree in $\mathcal{D}$ is at most $2^d - 1$. Next we construct for each T_i, $1 \leqslant i \leqslant q$, a path decomposition of the subgraph of $G[V(T_i)]$. Suppose that T_i has leaves $l_1, \ldots, l_s$ ordered according to their DFS-number. For $1 \leqslant j \leqslant s$, create a bag B_j containing the vertices on the unique path from l_j to r_i and string these bags together in the order $B_1, \ldots, B_s$. Clearly, this is a path decomposition $\mathcal{P}_i$ of $G[V(T_i)]$ with width at most $2^d - 2$. Note that the root r_i is in every bag of $\mathcal{P}_i$.

We now use a marking algorithm similar to the one in [21] to mark $\mathcal{O}(|S|)$ bags in the path decompositions $\mathcal{P}_1, \ldots, \mathcal{P}_q$ with the property that each marked bag can be uniquely identified with a connected subgraph of $G - S$ that has a large neighborhood in S. We use Lemma 3 to show that the set $\mathcal{M}$ of marked bags has at most $2 \cdot f(2^d - 1 + 1) \cdot |S| = \mathcal{O}(|S|)$ members, allowing us to put $Y_0 := V(\mathcal{M}) \cup S$. We also show that each connected component in $G - Y_0$ has less than $2^{d+1} + 2 \cdot f(2^d)$ neighbors in Y_0. As the marking stage of the algorithm runs through $\mathcal{P}_1, \ldots, \mathcal{P}_q$ exactly once, this phase takes only linear time.

Finally, we cluster the connected components of $G - Y_0$ according to their neighborhoods in Y_0 to obtain the sets $Y_1, \ldots, Y_\ell$. We again use Lemma 3 to show that the number ℓ of clusters is at most $\left(4^{f(2^d)} + 2f(2^d)\right) \cdot |S| = \mathcal{O}(|S|)$, as claimed. To accomplish this in linear time, we use a radix sort on the constant-sized neighborhoods in Y_0 of the components of $G - Y_0$. Thus the clustering and therefore the whole decomposition is linear-time computable. $\qquad\square$

All which is left to show is that each cluster Y_i, $1 \leqslant i \leqslant \ell$, can be reduced to constant size. Note that each cluster is separated from the rest of the graph via a small set of vertices in Y_0 and that each component of $G - Y_0$ has constant treedepth. These facts enable us to use the protrusion reduction rule. Lemma 1 assures us that FII either on general graphs or in the class of graphs with bounded treedepth, implies the existence of a finite set of representatives such that every protrusion can be replaced by an induced subgraph.

Lemma 5. *For fixed $d, h \in \mathbf{N}$, let (G, ξ) be an instance of Π with $G \in \mathcal{K}$ and let $S \subseteq V(G)$ be a treedepth-d modulator of G. Let $Y_0 \uplus Y_1 \uplus \cdots \uplus Y_\ell$ be a protrusion-decomposition of G, where $S \subseteq Y_0$ and for $1 \leqslant i \leqslant \ell$, $|N_{Y_0}(Y_i)| \leqslant h$. Then one can in $\mathcal{O}(|G| + \log \xi)$ time obtain (G', ξ') and a protrusion-decomposition $Y_0' \uplus Y_1' \uplus \cdots \uplus Y_\ell'$ of G' such that*

1. *$(G, \xi) \in \Pi$ if and only if $(G', \xi') \in \Pi$;*
2. *G' is an induced subgraph of G with $Y_0' = Y_0$; and*
3. *for $1 \leqslant i \leqslant \ell$ it is $|N_{Y_0'}(Y_i')| \leqslant h$, $|Y_i'| \leqslant \rho(d + h, d) = \mathcal{O}(1)$ and $\mathbf{td}(Y_i) \leqslant d$.*

Proof (Theorem 1). Given an instance (G, ξ) of Π with $G \in \mathcal{K}$ having fixed a constant $d \in \mathbf{N}$, we calculate a 2^d-approximate modulator S using Lemma 2. Using the algorithm outlined in the proof of Lemma 4, we compute the decomposition $Y_0 \uplus Y_1 \uplus \cdots \uplus Y_\ell$. Each cluster $Y_i, 1 \leqslant i \leqslant \ell$ forms a protrusion with boundary size $|N(Y_i)| \leqslant 2^{d+1} + 2f(2^d) =: h$ and treedepth (and thus treewidth) $\leqslant d$. Applying Lemma 5 now yields an equivalent instance (G', ξ') with $|V(G')| = |Y_0| + \sum_{i=1}^{\ell} |Y_i'|$ vertices, where Y_i' denote the clusters obtained through applications of the reduction rule. This quantity is at most $\mathcal{O}(|S|) + \ell \cdot \rho(d + 2^{d+1} + 2f(2^d), d) = \mathcal{O}(|S|)$. As G' is an induced subgraph of G, the above implies that $|V(G')| + |E(G')| = \mathcal{O}(|S|)$ by the degeneracy of G and that $G' \in \mathcal{K}$. □

Some problems do not have FII in general (see [9]) but only when restricted to, say, graphs of bounded treedepth.

Lemma 6. *Let $\mathcal{G}(d)$ be the class of all graphs that have treedepth at most d. The problems* LONGEST PATH, LONGEST CYCLE, EXACT s, t-PATH, EXACT CYCLE *have FII in $\mathcal{G}(d)$ for each $d \in \mathbf{N}$.*

Corollary 1. *The following graph problems either have FII in general or in the class of graphs with bounded treedepth, and hence have linear kernels in hereditary graph classes of bounded expansion, when the parameter is the size of a modulator to constant treedepth:* (CONNECTED) DOMINATING SET, r-DOMINATING SET, EFFICIENT DOM. SET, (CONNECTED) VERTEX COVER, LONGEST PATH/CYCLE *and hence also* HAMILTONIAN PATH/CYCLE, INDEPENDENT SET, FEEDBACK VERTEX SET, EDGE DOM. SET, INDUCED MATCHING, CHORDAL VERTEX DELETION, ODD CYCLE TRANSVERSAL, INDUCED d-DEGREE SUBGRAPH, MIN LEAF SPANNING TREE, MAX FULL DEGREE SPANNING TREE, EXACT s, t-PATH, EXACT CYCLE.

For a more comprehensive list of problems that have FII in general graphs (and hence fall under the purview of the above corollary), see [4].

Extension to Larger Graph Classes. We can lift our results to prove polynomial kernels in graphs of *locally bounded expansion* and the even larger class of *nowhere dense* graphs. Let $N_d(v)$ denote the neighborhood of a vertex v up to distance d.

A graph class $\mathcal{K}$ has *locally bounded expansion* if there exists a function $f \colon \mathbf{N} \times \mathbf{N} \to \mathbf{R}$ such that for every graph $G \in \mathcal{K}$ and all $r, d \in \mathbf{N}$ and each $v \in V(G)$, it holds $\nabla_r(G[N_d(v)]) \leqslant f(d, r)$ [13]. A graph class $\mathcal{K}$ is *nowhere dense* if for all $r \in \mathbf{N}$ it holds that $\omega(\mathcal{K} \triangledown r) < \infty$ [24,25].

The two kernelization results that we are about to state apply to all problems listed in Section 4. Note that the running time of the kernelization algorithm is quadratic only because it starts by computing an approximate treedepth-d modulator which takes quadratic time. If we assume that we are given such a modulator as part of the input, then the kernelization algorithm runs in linear time.

Theorem 2. *Let $\mathcal{K}$ be a hereditary class of locally bounded expansion and let $d \in \mathbf{N}$ be a constant. Let $\Pi \in \mathfrak{P}$. Then there is an algorithm that takes as input $(G, \xi) \in \mathcal{K} \times \mathbf{N}$ and, in time $\mathcal{O}(|G|^2 + \log \xi)$, outputs (G', ξ') such that*

1. *$(G, \xi) \in \Pi$ if and only if $(G', \xi') \in \Pi$;*
2. *G' is an induced subgraph of G; and*
3. *$|G'| = \mathcal{O}(|S|^2)$, where S is an optimal treedepth-d modulator of G.*

Theorem 3. *Let $\mathcal{K}$ be hereditary and nowhere-dense and let $d \in \mathbf{N}$ be a constant. Let $\Pi \in \mathfrak{P}$. Then there is a constant $c > 0$ and an algorithm that takes as input $(G, \xi) \in \mathcal{K} \times \mathbf{N}$ and, in time $\mathcal{O}(|G|^2 + \log \xi)$, outputs (G', ξ') such that*

1. *$(G, \xi) \in \Pi$ if and only if $(G', \xi') \in \Pi$;*
2. *G' is an induced subgraph of G; and*
3. *$|G'| = \mathcal{O}(|S|^c)$, where S is an optimal treedepth-d modulator of G.*

5   Further Research

We conclude with some open problems. Which graph problems admit linear kernels on graphs of bounded expansion with the *natural* parameter? Is there a meta-theorem on graphs of bounded expansion for problems that are not closed under edge-subvisions with a weaker parameterization?

References

1. IPEC 2011. LNCS, vol. 7112. Springer (2011)
2. Alber, J., Fellows, M.R., Niedermeier, R.: Polynomial-time data reduction for Dominating Set. J. ACM 51, 363–384 (2004)
3. Bodlaender, H.L., Downey, R.G., Fellows, M.R., Hermelin, D.: On problems without polynomial kernels. Journal of Computer and System Sciences 75(8), 423–434 (2009)
4. Bodlaender, H.L., Fomin, F.V., Lokshtanov, D., Penninkx, E., Saurabh, S., Thilikos, D.M.: (Meta) Kernelization. In: Proc. of 50th FOCS, pp. 629–638. IEEE Computer Society (2009)
5. Bodlaender, H.L., Jansen, B.M.P., Kratsch, S.: Kernel bounds for path and cycle problems. In: IPEC 2011 [1], pp. 145–158
6. Bodlaender, H.L., Kloks, T.: Better algorithms for the pathwidth and treewidth of graphs. In: Leach Albert, J., Monien, B., Rodríguez-Artalejo, M. (eds.) ICALP 1991. LNCS, vol. 510, pp. 544–555. Springer, Heidelberg (1991)
7. Courcelle, B.: The monadic second order logic of graphs I: Recognizable sets of finite graphs. Inform. and Comput. 85, 12–75 (1990)
8. Cygan, M., Lokshtanov, D., Pilipczuk, M., Pilipczuk, M., Saurabh, S.: On cutwidth parameterized by vertex cover. In: IPEC 2011 [1], pp. 246–258
9. de Fluiter, B.: Algorithms for Graphs of Small Treewidth. PhD thesis, Utrecht University (1997)
10. Dell, H., van Melkebeek, D.: Satisfiability allows no nontrivial sparsification unless the polynomial-time hierarchy collapses. In: Proceedings of the 42nd ACM Symposium on Theory of Computing, STOC 2010, pp. 251–260. ACM (2010)

11. Diestel, R.: Graph Theory, 4th edn. Springer, Heidelberg (2010)
12. Doucha, M., Kratochvíl, J.: Cluster vertex deletion: a parameterization between vertex cover and clique-width. In: Rovan, B., Sassone, V., Widmayer, P. (eds.) MFCS 2012. LNCS, vol. 7464, pp. 348–359. Springer, Heidelberg (2012)
13. Dvořák, Z., Král, D.: Algorithms for classes of graphs with bounded expansion. In: Paul, C., Habib, M. (eds.) WG 2009. LNCS, vol. 5911, pp. 17–32. Springer, Heidelberg (2010)
14. Fellows, M.R., Lokshtanov, D., Misra, N., Rosamond, F.A., Saurabh, S.: Graph layout problems parameterized by vertex cover. In: Hong, S.-H., Nagamochi, H., Fukunaga, T. (eds.) ISAAC 2008. LNCS, vol. 5369, pp. 294–305. Springer, Heidelberg (2008)
15. Fiala, J., Golovach, P.A., Kratochvíl, J.: Parameterized complexity of coloring problems: Treewidth versus vertex cover. In: Chen, J., Cooper, S.B. (eds.) TAMC 2009. LNCS, vol. 5532, pp. 221–230. Springer, Heidelberg (2009)
16. Fomin, F.V., Lokshtanov, D., Misra, N., Philip, G., Saurabh, S.: Hitting forbidden minors: Approximation and kernelization. In: Proc. of 28th STACS. LIPIcs, vol. 9, pp. 189–200. Schloss Dagstuhl–Leibniz-Zentrum für Informatik (2011)
17. Fomin, F.V., Lokshtanov, D., Misra, N., Saurabh, S.: Planar $\mathcal{F}$-Deletion: Approximation and Optimal FPT Algorithms. In: FOCS 2012, pp. 470–479. IEEE Computer Society (2012)
18. Fomin, F.V., Lokshtanov, D., Saurabh, S., Thilikos, D.M.: Bidimensionality and kernels. In: Proc. of 21st SODA, pp. 503–510. SIAM (2010)
19. Ganian, R.: Twin-cover: beyond vertex cover in parameterized algorithmics. In: IPEC 2011 [1], pp. 259–271
20. Guo, J., Niedermeier, R.: Linear problem kernels for NP-hard problems on planar graphs. In: Arge, L., Cachin, C., Jurdziński, T., Tarlecki, A. (eds.) ICALP 2007. LNCS, vol. 4596, pp. 375–386. Springer, Heidelberg (2007)
21. Kim, E.J., Langer, A., Paul, C., Reidl, F., Rossmanith, P., Sau, I., Sikdar, S.: Linear kernels and single-exponential algorithms via protrusion decomposition. In: Fomin, F.V., Freivalds, R., Kwiatkowska, M., Peleg, D. (eds.) ICALP 2013, Part I. LNCS, vol. 7965, pp. 613–624. Springer, Heidelberg (2013)
22. Nešetřil, J., de Mendez, P.O.: Linear time low tree-width partitions and algorithmic consequences. In: STOC 2006, pp. 391–400. ACM (2006)
23. Nešetřil, J., Ossona de Mendez, P.: Grad and classes with bounded expansion I. Decompositions. European J. Combin. 29(3), 760–776 (2008)
24. Nešetřil, J., Ossona de Mendez, P.: First order properties on nowhere dense structures. The Journal of Symbolic Logic 75(3), 868–887 (2010)
25. Nešetřil, J., Ossona de Mendez, P.: On nowhere dense graphs. European J. Combin. 32(4), 600–617 (2011)
26. Nešetřil, J., Ossona de Mendez, P.: Sparsity: Graphs, Structures, and Algorithms. Algorithms and Combinatorics, vol. 28. Springer (2012)
27. Nešetřil, J., Ossona de Mendez, P., Wood, D.R.: Characterisations and examples of graph classes with bounded expansion. Eur. J. Comb. 33(3), 350–373 (2012)
28. Wood, D.: On the maximum number of cliques in a graph. Graphs and Combinatorics 23, 337–352 (2007)

On the Computational Complexity
of Erdős-Szekeres and Related Problems in $\mathbb{R}^3$

Panos Giannopoulos[1,*], Christian Knauer[1], and Daniel Werner[2]

[1] Institut für Informatik, Universität Bayreuth, Bayreuth, Germany
{panos.giannopoulos,christian.knauer}@uni-bayreuth.de
[2] Institut für Informatik, Freie Universität Berlin, Berlin, Germany
werner@zedat.fu-berlin.de

Abstract. The Erdős-Szekeres theorem states that, for every k, there is a number n_k such that every set of n_k points in general position in the plane contains a subset of k points in convex position. If we ask the same question for subsets whose convex hull does not contain any other point from the set, this is not true: as shown by Horton, there are sets of arbitrary size that do not contain an empty convex 7-gon.

These problems have been studied also from a computational point of view, and, while several polynomial-time algorithms are known for finding a largest (empty) convex subset in the planar case, the complexity of the problems in higher dimensions has been, so far, open. In this paper, we give the first non-trivial results in this direction. First, we show that already in dimension 3 (the decision versions of) both problems are NP-hard. Then, we show that when an empty convex subset is sought, the problem is even W[1]-hard w.r.t. the size of the solution subset.

1 Introduction

Let P be a set of points. A set $P' \subseteq P$ is in *convex position*, if no point $p' \in P'$ is contained in the convex hull of $P' \smallsetminus \{p'\}$. It is in *emtpy convex position* if it is in convex position and does not contain any other point of P in its convex hull.

The Erdős-Szekeres theorem [7], one of the major theorems from combinatorial geometry and one of the earliest results in geometric Ramsey theory, states that, for every k, there is a number n_k such that every set of n_k points in general position in the plane contains a subset of k points in convex position. In particular, if n_k^* is the smallest such number, it holds that $2^{k-2} + 1 \leq n_k^* \leq 4^k$.

A closely related question is whether the theorem still holds when we ask for subsets that are in empty convex position. As shown by Horton [9], this is not the case: there are arbitrarily large sets in the plane that do not contain empty 7-gons; see also Matoušek [10, Chapter 3] for further details and references. Both questions generalize to dimension larger than 2 in the obvious way, and, clearly, the numbers n_k^* do not increase when the dimension gets higher (proof:

* Research supported by the German Science Foundation (DFG) under grant Kn 591/3-1.

H.L. Bodlaender and G.F. Italiano (Eds.): ESA 2013, LNCS 8125, pp. 541–552, 2013.
© Springer-Verlag Berlin Heidelberg 2013

project to some plane); see the surveys by Bárány and Károly [2] and Morris and Soltan [12] for (more or less) recent progress in the subject.

In this paper, we study the corresponding computational problems. More specifically, we study the following decision problems:

Given a set P of n points in $\mathbb{R}^d$ and $k \in \mathbb{N}$,

(i) (ERDŐS-SZEKERES) decide whether there is a set $Q \subset P$ of k points in convex position;

(ii) (LARGEST-EMPTY-CONVEX-SUBSET) decide whether there is a set $Q \subset P$ of k points in empty convex position.

1.1 Previous Results

For the planar case, the computational problems have received a lot of attention in the past. Chvátal and Klincsek [5] gave an $O(n^3)$-time algorithm for the problem of finding a largest convex subset. This algorithm was used by Avis and Rappaport [1] for finding a largest *empty* convex subset. Several years later, Dobkin et al. [6] improved this algorithm to run in time $O(\gamma_3(P))$, where $\gamma_3(P)$ is the number of empty triangles in P, which lies between n^2 and n^3. These algorithms are all based on dynamic programming and do not generalize to higher dimensions. Another approach that enumerates all empty convex sets of size k in time that is polynomial in both n and k was given by Mitchell et al. [11]. On the other hand, in higher dimensions, the only computational result appears in [6], where it was shown that, in $\mathbb{R}^3$, all subsets of size k in empty convex position can be found in time $O(k!n\log^3 n)$ per set. As there can be as many as n^k such subsets, this is (at best) a small improvement over the trivial $O(n^{k+1})$-time algorithm. Consequently, in that paper, the question was raised whether it is possible to find a largest subset in (empty) convex position in $\mathbb{R}^3$ in polynomial time.

1.2 Our Contributions

We first show, in Section 2, that, already in $\mathbb{R}^3$, both the ERDŐS-SZEKERES and LARGEST-EMPTY-CONVEX-SUBSET problems are NP-hard, therefore giving a negative answer to the question of Dobkin et al. [6]. We reduce from the independent set problem in penny graphs [4]. The reduction also uses a lifting transform to the elliptic paraboloid, which is well known in computational geometry, but whose only other application in an NP-hardness proof that we are aware of is due to Buchin et al. [3] for the problem of approximating polyhedral objects by spherical caps.

Having established NP-hardness, it it natural to ask whether any of these problems is fixed-parameter tractable with respect to the solution size k, i.e., solvable in time $O(f(k) \cdot n^c)$ for some computable function f and constant c (NP-hardness does not rule out this possibility). Observe here that due to the Erdős-Szekeres theorem itself, this is trivially true for the ERDŐS-SZEKERES problem. The theorem states that any set of at least 4^k points admits a convex subset of size k. Therefore, given a point set P and a $k \in \mathbb{N}$, if $n > 4^k$, we simply answer yes, while if $n := |P| \leq 4^k$, we use a brute force algorithm, i.e., simply try all subsets of size k; this takes time $\binom{n}{k} \approx n^k \leq (4^k)^k$.

In Section 3, we show that, in contrast to ERDŐS-SZEKERES, the LARGEST-EMPTY-CONVEX-SUBSET problem in $\mathbb{R}^3$ is W[1]-hard with respect to the solution size k, under the extra condition that the solution set is *strictly* convex, i.e., the interior of the convex hull of any of its subsets is empty. This implies that (under standard complexity-theoretic assumptions) the problem is not fixed-parameter tractable with respect to k; see Flum and Grohe [8] for basic notions of parameterized complexity theory.

2   NP-Hardness

We first prove NP-hardness for the LARGEST-EMPTY-CONVEX-SUBSET problem and then show how to adapt the reduction to the ERDŐS-SZEKERES problem.

2.1   Largest Empty Convex Subset

We use a reduction from a slight modification of the following problem: Given a set of pairwise non-overlapping unit disks in $\mathbb{R}^2$, decide whether there are k disks such that no two of them touch.

Here, non-overlapping means that the interiors of the disks are pairwise disjoint. The intersection graphs of non-intersecting unit disks are also called *penny graphs*. As shown by Cerioli et al. [4], this problem is NP-hard by a simple reduction from a variant of the VERTEX-COVER problem. However, the reduction has a little flaw, because some of the centers of the disks have irrational coordinates.

We overcome this obstacle by perturbing each disk center a little by at most some small ε, and enlarging the diameter of each disk by 2ε. This way, we get an instance of unit disks that *almost* forms a penny graph—some of the disks may now overlap a little. Moreover, in the original construction, the angle between any two intersections along a circle is always at least $\pi/2$. Thus, by choosing ε appropriately, after the perturbation and enlargement of the circles, the angle between the two closest points of two circles intersecting a common circle is still $\pi/2 \pm \alpha^*$, for some small $\alpha^* < \pi/2 - \pi/100$, and the minimum distance between two centers is still at least $\delta^* = 2 - 4\varepsilon$. Observe that the value of this ε does not depend on the input, and thus can be chosen as some arbitrarily small constant, say $\varepsilon = 10^{-6}$. We call instances of unit disks that arise this way *quasi non-intersecting*.

In particular, instances that arise this way do not induce any new incidences, and thus there is an independent set of size k among the perturbed disks if and only if there is a set of k independent disks in the original instance. Combining this with the reduction from [4] gives the following corollary:

Corollary 1. *The following problem is NP-hard: Given a set of quasi non-intersecting unit disks and $k \in \mathbb{N}$, decide whether there are k disks such that no two of them intersect.*

For a given instance $\mathcal{D}$ of unit disks in the plane, we will create a set of points in $\mathbb{R}^3$. All these points will lie close to the elliptic paraboloid, in a sense to be made precise later.

For a point $x = (x_1, x_2) \in \mathbb{R}^2$, let lift: $(x_1, x_2) \mapsto (x_1, x_2, x_1^2 + x_2^2)$ denote the standard lifting transform to the paraboloid. Also, let D_c denote the n centers of the disks in $\mathcal{D}$, and let L denote the set of all points $\hat{x} = \text{lift}(x)$, for $x \in D_c$. Finally, let $\text{ch}(P)$ be the convex hull of a set P.

We want to forbid certain pairs of points to lie in empty convex positions, namely, those for which the corresponding disks intersect. Thus, for a pair of intersecting disks d, d' and their centers $c_d, c_{d'}$, we add a blocking point

$$b_{dd'} = \frac{1}{2} \left(\text{lift}(c_{d'}) + \text{lift}(c_d) \right).$$

Let $B = \{b_{dd'} \mid d \cap d' \neq \varnothing\}$. Note that B lies slightly above the paraboloid. Finally, we set $P = L \cup B$.

We have created $O(|\mathcal{D}|)$ points, and as the underlying geometric graph is planar, the size of the reduction is linear in the input size.

Lemma 1. *A blocking point $b_{dd'}$ is contained in the convex hull of a set $Q \subseteq L$ if and only if both $\hat{c}_d$ and $\hat{c}_{d'}$ are contained in Q.*

Proof. "$\Leftarrow$" By definition.

"$\Rightarrow$" We show that there is a plane that contains $b_{dd'}$, $\hat{c}_d$, and $\hat{c}_{d'}$ and is strictly below all other points. Here we will make use of the fact that our instance consists of quasi non-intersecting unit disks—otherwise, the claim would not hold.

Let C be the circle whose center is the projection of $b_{dd'}$ to the first two coordinates that passes through c_d and $c_{d'}$. Since the centers of the disks have a distance at least $\delta^* > 3/2$, and every point in C is at most at distance $\sqrt{2} < 3/2$ from either c_d or $c_{d'}$, this circle does not contain any other points from D_c. Further, since all intersection points are at least an angle of α^* apart, the circle does not contain any (projection of) a blocking point. See Fig. 1. Define h to be the unique plane whose intersection with the paraboloid projects to the circle C. This plane contains all three points, and because C does not contain any other points, all other points from L and thus B lie strictly above h. Thus $b_{dd'}$ is contained in $\text{ch}(Q)$ if and only if $b_{dd'} \in \text{ch}(Q \cap h) = \text{ch}(\{\hat{c}_d, \hat{c}_{d'}\})$.   □

The following lemma states that whether or not a set is in empty convex position will depend only on which points we choose from L. Set B can always be added without destroying this property.

Lemma 2. *Sets L and B are in empty convex position, and $\text{ch}(L) = \text{ch}(L \cup B)$.*

Proof. By construction, all points of L lie on the paraboloid. The points from B can be separated from each other by the plane defined in the previous proof. As all of them are convex combinations of points in L, we have $\text{ch}(B) \subseteq \text{ch}(L)$.   □

Combining Lemmas 1 and 2, we get the following.

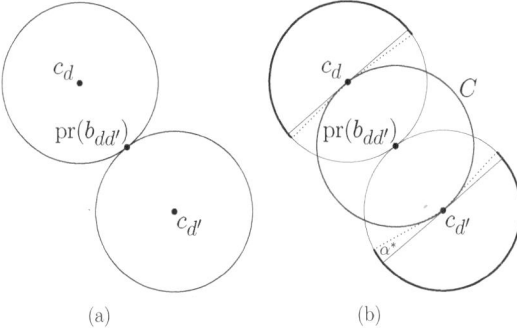

Fig. 1. Finding an empty circle: (a) Two intersecting disks and the projected blocking point; (b) Since all intersections with other disks lie at the bold arcs, no projected blocking points lies inside C

Corollary 2. *A set $L' \cup Q' \subseteq P$ is in empty convex position if and only if no point of $Q' \subseteq Q$ is contained in the convex hull of $L' \subseteq L$.*

Lemma 3. *There is an independent set of size m among the unit disks if and only if there are $m + |B|$ points in empty convex position.*

Proof. "$\Rightarrow$" Let I, $|I| = m$, be an independent set among the set of disks. Let $\hat{I} \subseteq L$ denote the corresponding lifted centers. We claim that $S = \hat{I} \cup B$ is in empty convex position. Indeed, by Corollary 2, no point of $L \setminus \hat{I}$ is in the convex hull of S. Further, by Lemma 1, if some point $b \in B$ was in $\mathrm{ch}(S)$, this would mean that there are two points in $\hat{I}$ that contained b in their convex hull. Thus, by Lemma 1, the corresponding disks would touch, and I would not be an independent set. This means that there are $m + |B|$ points in empty convex position.
"$\Leftarrow$" Now assume that there is no independent set of size m. This means that for any choice of m disks, two of them touch. Now take any set S of $m + |B|$ points. As there are only $|L| + |B|$ points in total, this must contain at least m points from L. Thus, at least two of them belong to disks that intersect. By Lemma 1, their convex hull contains a point of B. Thus, S is not in empty convex position. □

Theorem 1. *Problem* LARGEST-EMPTY-CONVEX-SUBSET *is NP-hard in $\mathbb{R}^3$.*

2.2 Erdős-Szekeres

In order to show NP-hardness of this problem, we use the same construction as in the previous section. We need only the following analogue of Lemma 3.

Lemma 4. *There is an independent set of size m among the unit disks if and only if there are $m + |B|$ points in convex position.*

Proof. "$\Rightarrow$" By Lemma 3, the existence of an independent set of size m corresponds to an empty convex set of size $m + |B|$.

"⇐" We show that every set of points in convex position can be modified to yield a set in empty convex position.

Let S be a set of $m+|B|$ points in convex position with $|S \cap B| < |B|$, let $I = S \cap L$, and let $\mathcal{D}_I$ be the corresponding set of disks. Observe that, if $|S \cap B| < |B|$, then $|I| > m$, and thus if all disks from $\mathcal{D}_I$ are independent, we are done.

Otherwise, we show how to construct a set S' in convex position of the same size such that $|S' \cap B| = |S \cap B| + 1$. Let d and d' be two disks from $\mathcal{D}_I$ that intersect. The point $b_{dd'}$ cannot be part of S, for otherwise S would not be in convex position. If we thus set $S' = I \smallsetminus \{\hat{d}\} \cup B \cup \{b_{dd'}\}$, by Corollary 2 the set is still in convex position, and we have $|S'| = |S|$ and $|S' \cap B| > |S \cap B|$.

Thus, after finitely many steps we end up with a set of $m + |B|$ points in convex position which contains all points from B. In particular, no point of B is contained in the convex hull of $S \cap L$. By Lemma 3, this means that the disks corresponding to the m points from L do not intersect. Thus, we have an independent set of size m.   □

Theorem 2. *Problem* ERDŐS-SZEKERES *is NP-hard in* $\mathbb{R}^3$.

3   Fixed-Parameter Intractability

In this section, we show that, under the strict convexity condition, LARGEST-EMPTY-CONVEX-SUBSET is W[1]-hard by an *fpt*-reduction from the W[1]-hard k-CLIQUE problem [8]: Given a graph $G([n], E)$ and $k \in \mathbb{N}$, decide whether G contains a k-size clique. More precisely, we will construct a set P of $\Theta(k^2 n^2)$ points in $\mathbb{R}^3$ such that there exists a set $Q \subset P$ in empty convex position and $|Q| = f(k)$, for some function $f(k) \in \Theta(k^2)$, if and only if G has a clique of size k; a clique will actually correspond to $k!$ emtpy convex subsets of P.

3.1   High-Level Description

We begin with a high-level description of the construction, see Fig. 2. Initially, the construction will lie on the plane; later on, it will be lifted to the elliptic paraboloid with a (more or less) standard transform. The construction is organized as the upper diagonal part of a grid with k rows and k columns. The ith row and column represent a choice for the ith vertex of a clique in G and are made of $4i - 2$ and $4(k - i + 1) - 2$ gadgets respectively. There are n choices and each choice is represented by a collection of empty convex subsets of points – one subset with a constant number of points from each gadget.

Each gadget consists of $\Theta(n)$ points within a rectangular region, which are organized in sets (of pairs) of collinear points. There is a constant number of such sets and, since we are looking for strictly convex subsets, only one pair of consecutive points per set can be chosen at any time. Certain choices are rendered invalid by additional points. Neighboring gadgets share the points on their common rectangle edge, see the zoomed-in area in Fig 2. Through these common points, the choice of subsets is made consistent among the gadgets. In

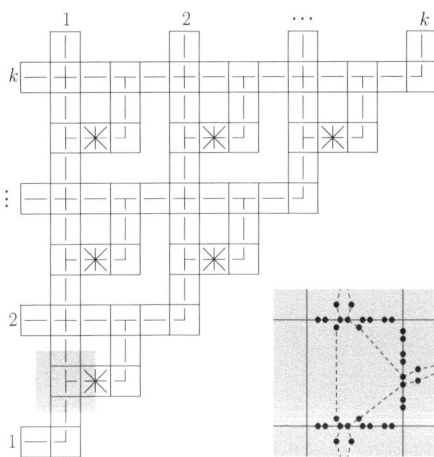

Fig. 2. High-level schematic of the construction. The zoomed-in area shows points shared among gadgets on their common boundaries and some pairs of points inside each gadget that take part in a choice of an empty convex set (in dashed).

particular, the choice in the ith row is made consistent with the choice in the ith column via the 'diagonal', ⊟ gadget in their intersection corner; consistency here means that they both correspond to the same choice of a vertex of G. On the other hand, in the intersection of the ith column with the jth row, for every $j \geq i+1$, there is a 'cross', ⊞ gadget, which ensures that the choice in the column is propagated independently of the choice in the row and vice versa. Finally, the jth column is 'connected' to the ith row, for every $j + 1 \leq i \leq k$, by three additional gadgets. One of them, the 'star', ⊛ gadget, encodes graph G, i.e., it allows only for combinations of choices (in the column and the row) that are consistent with the edges in G.

Locally, every valid subset from a gadget consists of points that are in strictly convex position and whose convex hull is empty. By lifting the whole construction to the paraboloid appropriately, we make sure that this property is true globally, i.e., for any set constructed from the local choices in a consistant manner.

3.2   Gadgets

There are five different types of gadgets, and each type has a specific function.

⊟ **gadget.** This gadget propagates a choice of pairs of points horizontally, see Fig. 3. It has n pairs of points L_i, R_i on the left and right side of its rectangle respectively, which correspond to the vertices of G; pairs corresponding to the same vertex are aligned horizontally. It also has $n(n-1) + 2$ points inside the rectangle on a vertical line ℓ as follows. For every two pairs L_i and R_j, with $i \neq j$, a point is placed such that it is inside the parallelogram $L_i R_j$ formed by the pairs but outside the parallelogram formed by any other two pairs. Effectively, this

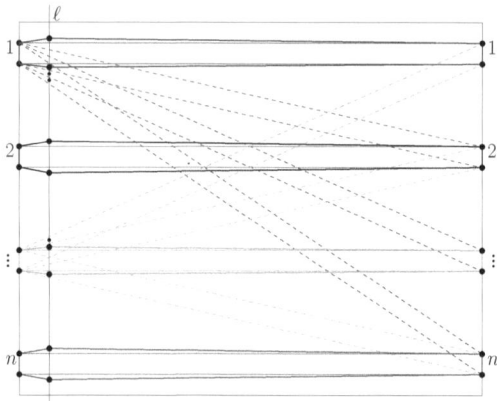

Fig. 3. The ⊟ gadget. Dashed parallelograms represent choices cancelled by points on
ℓ. Parallelograms in full are not cancelled. There are n choices of empty convex 6-gons.

point cancels a choice of pairs that correspond to different vertices of G. One
point is placed above L_1R_1 on ℓ and close to its boundary; similarly one point
is placed below L_nR_n. Due to the strict convexity condition, at most one pair
per rectangle side and at most one pair on ℓ can be chosen. Thus, there are
n maximum size empty convex subsets. Each subset contains six points and
is formed by three pairs: L_i, R_i, for some i, and the pair of points on ℓ that
are closest to the parallelogram L_iR_i; this latter pair is formed by the point
that cancels L_iR_{i-1} and the point that cancels L_iR_{i+1}. The ⊡ gadget is just a
$90°$-rotated copy of the ⊟ gadget.

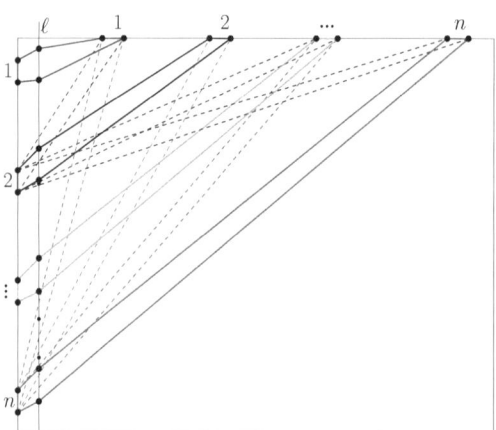

Fig. 4. The ⊡ gadget

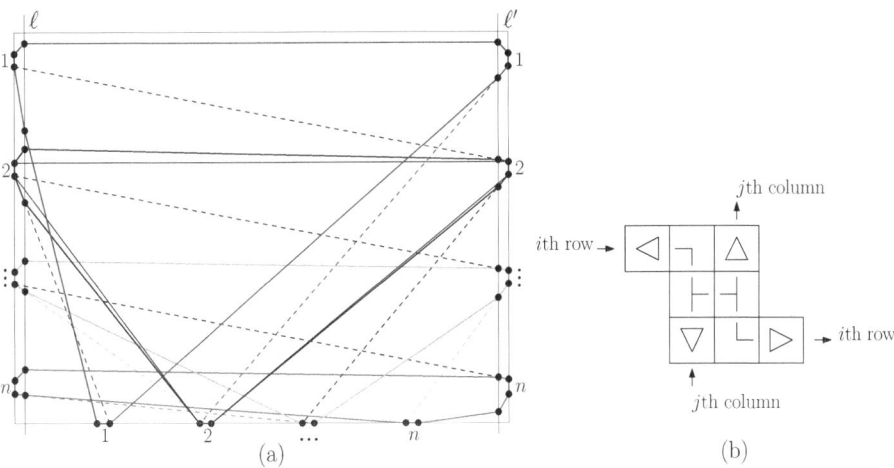

Fig. 5. (a) The ⊟ gadget: there are n choices of empty convex 10-gons; an empty convex 6-gon, as described in the text, is shown only for $i = 2$. (b) The ⊞ gadget: a high level schematic.

⊟ **gadget.** This gadget has basically the same structure as the ⊟ gadget but propagates information diagonally. See Fig. 4.

⊞ **gadget.** This gadget propagates information both horizontally and diagonally, see Fig. 5(a). It has n pairs of points L_i, R_i, and B_i, on the left, right, and bottom side of its rectangle respectively. As before, an ith pair corresponds to the ith vertex of G. There will be only n valid choices and three pairs per choice, namely, the ith pair from each side. This is enforced by the strict convexity condition and by placing, for every i, four additional points on two vertical lines ℓ and ℓ' inside the rectangle. These points are placed *outside* the convex 6-gon $L_iR_iB_i$ that is formed by the points in the corresponding pairs. (The points are also outside every other 6-gon for $j \neq i$.) See the example for $i = 2$ in Fig. 5(a).

More specifically, looking at the gadget from top to bottom and from left to right, a point is placed on the intersection of ℓ and the line through the second point of L_i and the second point of B_{i-1}; for the case of $i = 1$, the point is placed just below the 6-gon. A second point is placed on ℓ just above the 6-gon. At the right side, two points are placed on ℓ' as follows. One point is placed on the intersection of ℓ' and the line through the second point of L_i and the first point of R_{i+1}; for $i = 1$, the point is placed just above the 6-gon. A second point is placed on the intersection of ℓ' and the line through the second point of R_i and the first point of B_{i+1}; for $i = n$, the point is placed just below the 6-gon.

There are n maximum size empty convex subsets with 10 points each. A subset is formed by the points in the pairs L_i, R_i, B_i, and the four points on ℓ and ℓ' that are closest to the corresponding 6-gon.

⊞ **gadget.** This gadget consists of eight subgadgets, which are very similar to the ones we have already described above. See Fig. 5(b). It can be thought of as having two inputs (at the upper and lower left corner) and two outputs (at the upper and lower right corner). It propagates the inputs (choices) independently from each other, one horizontally and one vertically. The input subgadgets ◁ and ▽ are respectively connected (through their left and bottom rectangle sides) to the ⊟ and ⊡ gadgets of the ith row and jth column of the global construction (Fig. 2). The output subgadgets ▷ and △ are similarly connected to a ⊟ and ⊡ gadget. Note that the subgadgets ⊞ and ⊞ in the middle of the ⊞ gadget (Fig. 5(b)) as well as the subgadgets ⊟ and ⊔ are *not* connected directly to any row or column of the global construction. Roughly speaking, the ⊞ gadget has the following function: it multiplexes the two inputs, then it mirrors them (vertically and horizontally), and then demultiplexes them. Next, we describe the subgadgets in more detail.

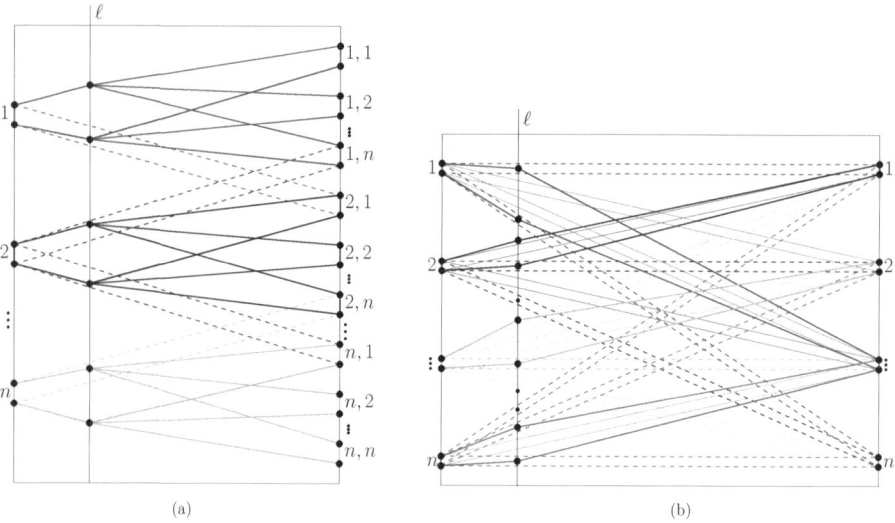

Fig. 6. (a) The ◁ subgadget. (b) The ⊛ gadget: several examples of cancelled choices (in dashed) and of empty convex 6-gons (in bold) are shown.

The ◁ subgadget is shown in Fig. 6(a). It is similar to the previously described ⊟ gadget in Fig. 3. It has again n pairs of points L_i on the left side. The difference now is that there are n^2 pairs of points $R_{i,j}$, $1 \le i, j \le n$, on the right side of the gadget. The second index j basically encodes the choice coming from the input subgadget ▽ at the lower left corner, which is communicated through the ⊞ and ⊟ subgadgets inbetween. Only the n^2 pairs L_i and $R_{i,j}$ constitute valid choices. The rest are cancelled by additional points as usually. Together with pairs of canceling points (which are also chosen as before) there are exactly n^2 empty convex 6-gons, and these are of maximum size.

The input ▽ subgadget propagates information vertically and is defined similarly to the ◁ subgadget. It has n pairs of points B_j on the bottom side and n^2 pairs T_{ij} on the top side. Note that now only the n^2 pairs B_j and $T_{i,j}$ are valid.

The output subgadgets ▷ and △ are just mirrored images of their input counterparts. The ⊟ and ⊞ subgadgets are constructed in the same way as the ⊟ gadget in Fig. 4 but have n^2 valid 6-gons, while the ⊞ and ⊞ subgadgets are constructed in the same way as the ⊟ gadget in Fig. 5 and have n^2 valid 10-gons.

⊛ **gadget.** This gadget encodes the edges of the input graph G. See Fig. 6(b). It is similar to gadget ⊟ (Fig. 3) and allows only combinations of pairs that correspond to edges of the graph: for every *non*-edge ij of G, a point is placed inside the parallelogram L_iR_j.

3.3   Lifting to $\mathbb{R}^3$ and Correctness

Every corner of a gadget rectangle is lifted to the paraboloid with the map $(x_1, x_2) \mapsto (x_1, x_2, x_1^2 + x_2^2)$. The images of the corners of each rectangle lie on one distinct plane (since the corners lie on a circle). The points in the gadget are projected orthogonally on the corresponding plane. As this is an affine map, collinearity and convexity within the gadget are preserved. Each gadget lies on a distinct facet (a parallelogram) of a convex polyhedron.

The total number of (sub)gadgets of each type together with the size of a largest valid (i.e., empty and convex) subset in a gadget of the type is

$$
\begin{array}{ll}
⊟ : k^2,\, 6; & ⊞,\, ⊞,\, ⊞ : 2k(k-1),\, 10; \\
\square : k(3k-1)/2,\, 6; & ◁,\, ▷,\, △,\, ▽ : 2k(k-1),\, 6; \\
⊟,\, ⊟,\, ⊞ : k(3k-1)/2,\, 6; & ⊛ : k(k-1)/2,\, 6.
\end{array}
$$

A global valid subset is formed by locally choosing one valid subset from every gadget in a consistent manner. When a largest locally possible subset (as given above) can be chosen, the global subset has size $k(35k - 23)$. As we will now prove, such a global subset corresponds to a k-size clique of G. Let P be the set of all the points in our construction.

Lemma 5. *There exists an empty convex subset of P with $k(35k-23)$ points if and only if G has a clique of size k.*

Proof. "⇒" Suppose there exists a global valid subset of size $k(35k-23)$. Then, a largest locally possible valid subset must be chosen from every gadget. Consider such a choice of subsets and let v_i be the vertex of G corresponding to the choice from the leftmost gadget of the ith grid row. By construction, the subset corresponding to the same vertex v_i must be chosen from every other gadget in this row as well as every gadget in the ith column. Consider the jth row, for some $j \neq i$. Through the ⊛ gadget that connects the ith column to the jth row, when $i + 1 \leq j$, or the jth column to the ith row, when $j < i$, the subset chosen from the jth row must correspond to a vertex v_j such that v_iv_j is an edge of G. Hence $\{v_1, \ldots, v_k\}$ is a clique in G.

"⇐" Obvious.   □

Therefore, we have shown the following.

Theorem 3. LARGEST-EMPTY-CONVEX-SUBSET *is $W[1]$-hard, under the condition of strict convexity.*

4 Open Problems

The approximability status of both problems studied in the paper is open: none of our reductions imply hardness while no constant-factor approximation algorithms are known.

References

1. Avis, D., Rappaport, D.: Computing the largest empty convex subset of a set of points. In: Proceedings of the first Annual Symposium on Computational Geometry, SCG 1985, pp. 161–167. ACM (1985)
2. Bárány, I., Károlyi, G.: Problems and Results around the Erdős-Szekeres Convex Polygon Theorem. In: Akiyama, J., Kano, M., Urabe, M. (eds.) JCDCG 2000. LNCS, vol. 2098, pp. 91–105. Springer, Heidelberg (2001)
3. Buchin, K., Plantinga, S., Rote, G., Sturm, A., Vegter, G.: Convex approximation by spherical patches. In: Proceedings of the 23rd EuroCG, pp. 26–29 (2007)
4. Cerioli, M., Faria, L., Ferreira, T., Protti, F.: On minimum clique partition and maximum independent set on unit disk graphs and penny graphs: complexity and approximation. Electronic Notes in Discrete Mathematics 18, 73–79 (2004)
5. Chvátal, V., Klincsek, G.: Finding largest convex subsets. In: Congresus Numeratium, pp. 453–460 (1980)
6. Dobkin, D., Edelsbrunner, H., Overmars, M.: Searching for empty convex polygons. Algorithmica 5(4), 561–571 (1990)
7. Erdős, P., Szekeres, G.: A combinatorial problem in geometry. Compositio Math. 2, 463–470 (1935)
8. Flum, J., Grohe, M.: Parameterized Complexity Theory. Texts in Theoretical Computer Science. Springer (2006)
9. Horton, J.D.: Sets with no empty convex 7-gons. C. Math. Bull. 26, 482–484 (1983)
10. Matoušek, J.: Lectures on Discrete Geometry. Graduate Texts in Mathematics, vol. 212. Springer (2002)
11. Mitchell, J.S.B., Rote, G., Sundaram, G., Woeginger, G.: Counting convex polygons in planar point sets. Information Processing Letters 56(1), 45–49 (1995)
12. Morris, W., Soltan, V.: The Erdős-Szekeres problem on points in convex position – a survey. Bull. Amer. Math. Soc. 37, 437–458 (2000)

Encodings for Range Selection and Top-k Queries

Roberto Grossi[1], John Iacono[2], Gonzalo Navarro[3,*], Rajeev Raman[4],
and Satti Srinivasa Rao[5,**]

[1] Dip. di Informatica, Univ. of Pisa, Italy
[2] Dept. of Comp. Sci. and Eng., Polytechnic Institute of New York Univ., USA
[3] Dept. of Comp. Sci., Univ. of Chile, Chile
[4] Dept. of Comp. Sci., Univ. of Leicester, UK
[5] School of Comp. Sci. and Eng., Seoul National Univ.

Abstract. We study the problem of *encoding* the positions the top-k elements of an array $A[1..n]$ for a given parameter $1 \le k \le n$. Specifically, for any i and j, we wish create a data structure that reports the positions of the largest k elements in $A[i..j]$ in decreasing order, *without* accessing A at query time. This is a natural extension of the well-known encoding range-maxima query problem, where only the position of the maximum in $A[i..j]$ is sought, and finds applications in document retrieval and ranking. We give (sometimes tight) upper and lower bounds for this problem and some variants thereof.

1 Introduction

We consider the problem of *encoding* range top-k queries over an array of distinct values $A[1..n]$. Given an integer $1 \le k \le n$, we wish to preprocess A and create a data structure that can answer top-k-pos queries: given two indices i and j, return the *positions* where the largest k values in $A[i..j]$ occur.

The encoding version of the problem requires this query to be answered *without* accessing A: this is useful when the values in A are intrinsically uninteresting and only the indices where the top-k values occur are of interest. An example is auto-completion search in databases and search engines [13, 15]. Here, as the user types in a query, the system presents the user with the k most popular completions, chosen from a lexicon of phrases, based on the text entered so far. Viewing the lexicon as a sorted sequence of strings with popularity scores stored in A, the indices i and j can specify the range of phrases prefixed by the text typed in so far. Similarly, in document search engines, A could contain the (virtual) sequence of PageRank values of the pages in an inverted list sorted by URL. Then we could efficiently retrieve the k most highly ranked documents that contain a query term, restricted to a range of page identifiers (which can model a domain

* Partially funded by Millennium Nucleus Information and Coordination in Networks ICM/FIC P10-024F, Mideplan, Chile.
** Research partly supported by Basic Science Research Program through the National Research Foundation of Korea (NRF) funded by the Ministry of Education, Science and Technology (Grant number 2012-0008241).

H.L. Bodlaender and G.F. Italiano (Eds.): ESA 2013, LNCS 8125, pp. 553–564, 2013.

of any granularity). An encoding data structure for top-k queries will allow us to reduce the amount of space needed to perform these searches in main memory.

One can always use a non-encoding data structure (i.e., one that accesses A during the execution of queries) for top-k-pos, such as that of Brodal et al. [4], on an array A' that contains the sorted permutation of the elements in A, and thus trivially avoid access to A at query time. This yields an encoding that uses $O(n)$ words, or $O(n \lg n)$ bits, of memory and answers top-k-pos queries in optimal $O(k)$ time. We aim to find non-trivial encodings of size $o(n \lg n)$ bits (from which, of course, it is not possible to recover the sorted permutation, but one can still answer any top-k-pos query). As we prove a lower bound of $\Omega(n \lg k)$ bits on the space of any top-k encoding, non-trivial encodings can exist only if $\lg k = o(\lg n)$. This is a reasonable assumption for the aforementioned applications.

Related Work. The encoding top-k problem is closely related to the problem of encoding *range maximum query (RMQ)*, which is the particular case with $k = 1$: $\mathsf{RMQ}_A(i, j) = \mathrm{argmax}_{i \le p \le j} A[p]$. The RMQ problem has a long history and many applications [2, 3, 12], and the problem of encoding RMQs has been studied in [8, 19]. In particular, Fischer and Heun [8] gave an encoding of A that uses $2n + o(n)$ bits and answers RMQ in $O(1)$ time; their space bound is asymptotically optimal to within lower-order terms. We are not aware of any work on top-k encoding for $k > 1$.

Our work is also related to *range selection*, which is to preprocess A to find the kth largest element in a range $A[i..j]$, with i, j, k given at query time. This problem has recently been studied intensively in its non-encoding version [5, 6, 9, 10, 14]. Jørgensen and Larsen [14] obtained a query time of $O(\lg k/ \lg \lg n + \lg \lg n)$, very recently improved to $O(\lg k/ \lg \lg n)$ by Chan and Wilkinson [6], both using $\Theta(n)$ words, i.e., $\Omega(n \lg n)$ bits. Jørgensen and Larsen [14] introduced the κ-*capped* range selection problem, where a parameter κ is given at preprocessing time, and the data structure only supports selection for ranks $k \le \kappa$. They showed that even the one-sided κ-capped range selection problem requires query time $\Omega(\lg k/ \lg \lg n)$ for structures using $O(n \,\mathrm{polylog}\, n)$ words, and the result of Chan and Wilkinson is therefore the best possible. Although the problems we consider are different in essential ways, we borrow some techniques, most notably that of *shallow cuttings* [6, 14], in some of our results.

Contributions. We present new lower and upper bounds shown in Table 1, where we assume that the word RAM model has word size of $w = \Omega(\lg n)$ bits, for the following operations on encoding data structures for one- and two-sided range selection and range top-k queries and for any $1 \le i \le j \le n$.

1. kth-pos(i) returns the position of the k-th largest value in range $A[1..i]$ for any array A, and
2. top-k-pos(i) returns the top-k largest positions in $A[1..i]$ (one-sided variant) or top-k-pos(i, j) for the top-k largest positions in $A[i..j]$ (two-sided variant).

We make heavy use of rank and select operations on bitmaps. Given a bitmap $B[1..n]$, operation $\mathsf{rank}_b(B, i)$ is the number of occurrences of bit b in the prefix

Table 1. Our lower and upper bounds for encodings for one- and two-sided range selection and range top-k queries, simplified for the interesting case $\lg k = o(\lg n)$, and valid for any function $\lg \lg k / \lg k \preceq \epsilon(k) \preceq 1$

Problem	Lower bound space (bits)	Upper bound space (bits)	Upper bound time
kth-pos(i)	$n \lg k - O(n)$	$n \lg k + O(\epsilon(k)n \lg k)$	$O(1/\epsilon(k))$
top-k-pos(i)	$n \lg k - O(n)$	$n \lg k + O(n)$	$O(k)$
top-k-pos(i,j)	$n \lg k - O(n)$	$O(n \lg k)$	$O(k)$

$B[1..i]$, whereas $\mathsf{select}_b(B, j)$ is the position in B of the j-th occurrence of bit b. These operations generalize in the obvious way to sequences over an alphabet $[\sigma]$. We also make use of the Cartesian tree [21] of an array $A[1..n]$, which is fundamental for RMQ solutions. The root of the Cartesian tree represents the position m of a maximum of $A[1..n]$, thus $m = \mathrm{RMQ}_A(1, n)$. Its left and right subtrees are the Cartesian trees of $A[1..m - 1]$ and $A[m + 1..n]$, respectively.

2   One-Sided Queries

We start by considering one-sided queries. We are given an array A of n integers and a fixed value k. We can assume w.l.o.g. that A is a permutation in $[n]$, otherwise we can replace each $A[i]$ by its rank in $\{A[1], \ldots, A[n]\}$, breaking ties as desired, and obtain the same results from queries. We want to preprocess and encode A to support the one-sided operations of Table 1 efficiently.

2.1   Lower Bounds

For queries kth-pos(i) and top-k-pos(i), we give a lower bound of $\Omega(n \lg k)$ bits on the size of the encoding. Assume for simplicity that $n = \ell k$, for some integer ℓ. Consider an array A of length n, with $A[i]$ initialized to i, for $1 \leq i \leq n$, and re-order its elements as follows: take $\ell - 1$ permutations π_j of size k, $0 \leq j < \ell - 1$, and permute the elements in the subarray $A[jk + 1..(j + 1)k]$ according to permutation π_j, $A[jk + i] = jk + \pi_j(i)$ for $0 \leq j < \ell - 1$ and $1 \leq i \leq k$. Observe that, for each $1 \leq j < \ell$, $A[x] < A[y]$ for any $x \leq jk$ and $y > jk$. We now show how to reconstruct the $\ell - 1$ permutations by performing several kth-pos queries on the array A. By the above property of A, the position of the k-th value in the prefix $A[1..jk + i - 1]$ is the position of value $(j - 1)k + i$, for any $1 \leq j < \ell$ and $1 \leq i \leq k$. This position is $(j - 1)k + \pi_{j-1}^{-1}(i)$. Then, any π_{j-1} can be easily computed with the k queries, kth-pos($jk + i - 1$) for $1 \leq i \leq k$. Since the $\ell - 1$ permutations require $(\ell - 1) \lg k! = (n - k) \lg k - O(n)$ bits, the claim follows.

Similar arguments apply to top-k-pos(i) as well, even if it gives the results not in order: using the array A above, kth-pos(i) is precisely the element that disappears from the answer when we move from top-k-pos(i) and top-k-pos($i+1$).

Theorem 1. *Any encoding of an array $A[1..n]$ answering kth-pos or top-k-pos queries requires at least $(n - k) \lg k - O(n)$ bits of space.*

A	12	18	17	20	14	19	22	11	25	21	28	16	23	13	15	24	29	27
X	1	2	3	1	-	3	2	-	3	1	1	-	2	-	-	2	2	3
P	1	1	1	1	0	1	1	0	1	1	1	0	1	0	0	1	1	1

Fig. 1. Encoding of an array A of length n to support kth-pos and top-k-pos queries, for $k = 3$. The encoding consists of a bitmap P of length $n = 18$ with $n' = 13$ ones, and a string X of length n' over the alphabet $[1..k]$.

2.2   Upper Bounds and Encodings

We first consider query kth-pos(i). We scan the array from left to right, and keep track of the top-k values in the prefix seen so far. At any position $i > k$, if we insert $A[i]$ into the top-k values, then we have to remove the k-th largest value of the prefix $A[1..i-1]$. The idea to solve these queries is to record the position of that leaving k-th largest value, so that to solve kth-pos(i) we find the next $i' > i$ where the top-k list changes, and then find the value leaving the list when $A[i']$ enters it. This one was the k-th largest value in $A[1..i]$. We wish to store this data using $O(n \lg k)$ bits.

We store a bit vector P of length n, where $P[i] = 1$ iff a new element is inserted into the top-k values at position i (or equivalently, the k-th largest value of $A[1..i-1]$ is deleted at position i). Let $n' \leq n$ be the number of ones in P. The first k bits of P are set to 1. We encode P using $n + o(n)$ bits, while supporting constant-time rank and select operations on it [7, 16].

Further, we store string X of length n', such that $X[j] = j$ for $1 \leq j \leq k$, and $X[j] = X[\text{rank}_1(P, \text{kth-pos}(\text{select}_1(P, j) - 1))]$, for $k < j \leq n'$. String X encodes the positions of the top-k values in $A[1..i]$ as follows. Let $j = \text{rank}_1(P, i) > k$. Then the last occurrence of $X[j] = \alpha$ in $X[1..j-1]$ marks the position of the element that was the k-th in the segment $A[1..i-1]$. This is because the last occurrence of each distinct symbol α in $X[1..j]$ is the position of a top-k value in $A[1..i]$. This is obviously true for $i = j = k$, and by induction it stays true when, at a position $P[i] = 1$, we set $X[j+k] = \alpha$, where α marked the position of the k-th maximum in $A[1..i-1]$. Figure 1 shows an example.

Note that X is a string of length n' over the alphabet $[k]$. Hence, X can be encoded using $(1 + \epsilon(k))n' \lg k$ bits, so that select is supported in $O(1)$ time and access in $O(1/\epsilon(k))$ time [11], for any $\epsilon(k) = O(1)$ (including functions in $o(1)$). On top of this we add $O(n' \lg \lg k)$ bits. to support in constant time *partial* rank queries, of the form $\text{rank}_{X[i]}(X, i)$. This is obtained by storing one monotone minimum perfect hash function (mmphf) [1] per distinct symbol c appearing $n_c > 0$ times in X. The space for each c is $O(n_c \lg \lg(n'/n_c))$ bits, which adds up to $O(n' \lg \lg k)$ by Jensen's inequality.

By the discussion above, for $i < n$ we compute $j = \text{rank}_1(P, i) + 1$, and $\alpha = X[j]$. Then it holds kth-pos(i) $= \text{select}_1(P, \text{select}_\alpha(X, \text{rank}_\alpha(X, j) - 1))$. This is correct because the top-k list changes when $A[i]$ enters the list, and we find the next time $X[j] = \alpha$ is mentioned, which is where $A[i]$ is finally displaced from the top-k list. Thus, this operation can be supported in $O(1/\epsilon(k))$ time, where the time to access X dominates. Theorem 2 follows.

Theorem 2. *Given an array $A[1..n]$ and a value k, there is an encoding of A and k on a RAM machine of $w = \Omega(\lg n)$ bits that uses $n \lg k + O(\epsilon(k) n \lg k)$ bits and support* kth-pos(i) *queries in $O(1/\epsilon(k))$ time, for any function $\epsilon(k) \in O(1)$ and $\epsilon(k) \in \Omega(\lg \lg k / \lg k)$.*

For supporting top-k-pos(i) queries we need a different query on X: Given a position j, find the rightmost occurrence preceding j of every symbol in $[k]$. This can be done in $O(k)$ time using our representation of X [11]: The string is cut into chunks of size k. We can traverse the chunk of j in time $O(k)$ to find all the last occurrences, preceding j, of distinct symbols[1]. For each symbol not found in the chunk, we use constant-time rank and select on bitmaps already present in the representation to find the previous chunk where it appears, and finally find in constant time its last occurrence in that previous chunk (as we have already chosen, for our purposes, constant-time select inside the chunks).

By the discussion above on the meaning of X, it is clear that the rightmost occurrences, up to position $j = \mathsf{rank}_1(P, i) + 1$, of the distinct symbols in $[k]$, form precisely the answer to top-k-pos(i). Thus we find all those positions p in time $O(k)$ and remap them to the original array using $\mathsf{select}_1(P, p)$. Since we need only select queries on X, we need only $n \lg k + O(n)$ bits for it [11].

Note the top-k positions do not come sorted by largest value. By the same properties of X, if the first occurrence of α after $X[j]$ precedes the first occurrence of β after $X[j]$, then the value associated to α in our answer is smaller than that associated to β, as it is replaced earlier. Thus we find the first occurrence, after j, of all the symbols in $[k]$, analogously as before, and sort the results according to the positions found to the right, in $O(k \lg k)$ time. Thus Theorem 3 follows.

Theorem 3. *Given an array $A[1..n]$ and a value k, there is an encoding of A and k that uses $n \lg k + O(n)$ bits and supports* top-k-pos(i) *queries in $O(k)$ time on a RAM machine of $w = \Omega(\lg n)$ bits. The result can be sorted from largest to lowest value in $O(k \lg k)$ time. The encoding is a subset of that of Theorem 2.*

3   Two-Sided Range Top-k Queries

We now consider the problem of encoding the array $A[1..n]$ so as to answer the query top-k-pos(i, j). We will also consider solving top-k queries for any $k \le \kappa$, where κ is set at construction time. Clearly, a lower bound on the encoding size of $\Omega(n \lg \kappa)$ bits follows from Section 2.1.

Corollary 1. *Any encoding of an array $A[1..n]$ answering* top-k-pos(i, j) *queries requires at least $(n - k) \lg k - O(n)$ bits of space.*

We give now two upper bounds for query top-k-pos(i, j). The first is weaker, but it is used to obtain the second.

---

[1] Although we do not have constant-time access to the symbols, we can select all the (overall) k positions of all the k distinct symbols in the chunk, in time $O(k)$.

3.1   Using $O(kn)$ Bits and $O(k^2)$ Time

Let $A[1..n] = a_1 \ldots a_n$. We define, for each element a_j, κ pointers $j > P_1[j] > \ldots > P_\kappa[j]$, to the last κ elements to the left of j that are larger than a_j.

Definition 1. *Given a sequence $a_1, \ldots, a_n$, we define arrays of pointers $P_0[1..n]$ to $P_\kappa[1..n]$ as $P_0[j] = j$, and $P_{k+1}[j] = \max(\{i, \ i < P_k[j] \land a_i > a_j\} \cup \{0\})$.*

These pointers allow us to answer top-k queries without accessing A. We now prove a result that is essential for their space-efficient representation.

Lemma 1. *Let $1 \le j_1, j_2 \le n$ and $0 < k \le \kappa$, and let us call $i_1 = P_{k-1}[j_1]$ and $i_2 = P_{k-1}[j_2]$. Then, if $i_1 < i_2$ and $P_k[j_2] < i_1$, it holds $P_k[j_1] \ge P_k[j_2]$.*

Proof. It must hold $a_{i_1} < a_{i_2}$, since otherwise $P_k[j_2] \ge i_1$ by Definition 1 (as it would hold $a_{j_2} < a_{i_2} \le a_{i_1}$ and $0 < i_1 < i_2$), contradicting the hypothesis.

Now let us call $r_1 = P_k[j_1]$ and $r_2 = P_k[j_2] < i_1$. If it were $r_1 < r_2$ (and thus $r_2 > 0$), then we would have the following contradiction: (1) $a_{j_1} \ge a_{r_2}$ (because otherwise it would be $r_1 = P_k[j_1] \ge r_2$, as implied by Definition 1 since $r_2 = P_k[j_2] < i_1 = P_{k-1}[j_1]$ and $a_{r_2} > a_{j_1}$); (2) $a_{r_2} > a_{j_2}$ (because $r_2 = P_k[j_2]$); (3) $a_{j_2} \ge a_{i_1}$ (because otherwise it would be $r_2 = P_k[j_2] \ge i_1$, as implied by Definition 1 since $i_1 < i_2 = P_{k-1}[j_2]$ and $a_{i_1} > a_{j_2}$, and $r_2 \ge i_1$ contradicts the hypothesis); (4) $a_{i_1} > a_{j_1}$ (because $i_1 = P_{k-1}[j_1]$). □

This lemma shows that if we draw, for a given k, all the arcs starting at $P_{k-1}[j]$ and ending at $P_k[j]$ for all j, then no arc "crosses" another. This property enables a space-efficient implementation of the pointers.

Pointer Representation. We represent each "level" $k > 0$ of pointers separately, as a set of arcs leading from $P_{k-1}[j]$ to $P_k[j]$. For a level $k > 0$ and for any $0 \le i \le n$, let $p_k[i] = |\{j, \ P_k[j] = i\}|$ be the number of pointers of level k that point to position i. We store a bitmap

$$T_k[1..2n+1] \quad = \quad 10^{p_k[0]} \, 10^{p_k[1]} \, 10^{p_k[2]} \ldots 10^{p_k[n]},$$

where we mark the number of times each position is the target of pointers from level k. Each 1 corresponds to a new position and each 0 to the target of an arc. Note that the sources of those arcs correspond to the 0s in bitmap T_{k-1}, that is, to arcs that go from $P_{k-2}[j]$ to $P_{k-1}[j]$. Arcs that enter the same position i are sorted according to their source position, so that we associate the leftmost 0s of $0^{p_k[i]}$ to the arcs with the rightmost sources. Conversely, we associate the rightmost 0s of $0^{p_{k-1}[i]}$ to the arcs with the leftmost targets. This rule ensures that those arcs entering, or leaving from, a same position do not cross in T_k. Then we define a balanced sequence of parentheses

$$B_k[1..2n] \quad = \quad \left(^{p_{k-1}[0]}\right)^{p_{k-1}[0]} \left(^{p_k[0]-p_{k-1}[0]}\right)^{p_{k-1}[1]} \left(^{p_k[1]}\right)^{p_{k-1}[2]} \left(^{p_k[2]} \ldots \right)^{p_{k-1}[n]}.$$

This sequence matches arc targets (opening parentheses) and sources (their corresponding closing parentheses). The arcs that leave from and enter at the special position 0 receive special treatment to make the sequence balanced.

Calling $findopen(B_k, i)$ the position of the opening parenthesis matching the closing parenthesis at $B_k[i]$, the following algorithm computes the position z_k of the 0 of T_k corresponding to $P_k[j]$, given the position z_{k-1} of the 0 of T_{k-1} corresponding to $P_{k-1}[j]$.

1. $p \leftarrow rank_0(T_{k-1}, z_{k-1})$ 5. $c \leftarrow select_)(B_k, p)$
2. $z \leftarrow select_1(T_{k-1}, z_{k-1} - p)$ 6. $o \leftarrow findopen(B_k, c)$
3. $z' \leftarrow select_1(T_{k-1}, z_{k-1} - p + 1)$ 7. $r \leftarrow rank_((B_k, o)$
4. $p \leftarrow z' - (z_{k-1} - z)$ 8. $z_k \leftarrow select_0(T_k, r)$

The code works as follows. Given the position $T_{k-1}[z_{k-1}] = 0$ corresponding to the target of pointer $P_{k-1}[j]$, we first compute in p the number of 0s up to z_{k-1} in T_{k-1}. This position is corrected so as to (virtually) reverse the 0s that form the run where z_{k-1} lies (between the 1s at positions z and z'), in order to convert entering into leaving arcs. Then we find c, the p-th closing parenthesis in B_k, which is the target of this arc, and find its source o, the matching opening parenthesis. Finally we compute the rank r of o among the opening parentheses to the left, and match it with the corresponding 0 in T_k, z_k.

We use the code as follows. Starting with $P_0[j] = j$ and $z_0 = 2j + 1$, we use the code up to κ times in order to find, consecutively, $z_1, z_2, \ldots, z_\kappa$. At any point we have that $P_k[j] = rank_1(T_k, z_k) - 1$. Finally, since we know $T_0 = 1(10)^n$, we can avoid storing it, and replace lines 1–4 by $p \leftarrow j + 1$, when computing z_1.

By using a representation of T_k that supports constant-time $rank$ and $select$ operations in $2n + O(n/\lg^2 n)$ bits [17], and a representation of B_k that in addition supports operation $findopen$ in constant time and the same space [20], we have that the overall space is $4\kappa n + o(n)$ bits for any $\kappa = O(\lg n)$. With such a representation, we can compute any $P_k[j]$, for any j and any $1 \le k \le \kappa$, in time $O(k)$ and, more precisely, in time $O(1)$ after having computed $P_{k-1}[j]$ using the same procedure.

Top-k Algorithm. To find the k largest elements of $A[i..j]$ we use the structure of Fischer and Heun [8] that takes $2n + o(n)$ bits and answers RMQs in constant time. Our algorithm reconstructs the top part of the Cartesian tree [21] of $A[i..j]$ that contains the top-k elements, and also their children. The invariant of the algorithm is that, at any time, the internal nodes of the reconstructed tree are top-k elements already reported, whereas the next largest element is one of the current leaves. The tree that is reconstructed is of size at most $3k$.

The nodes p of the tree will be associated with an interval $[i_p..j_p]$ of $[1..n]$, and with a position m_p where the maximum of $A[i_p..j_p]$ occurs. In internal nodes it will hold $i_p = j_p = m_p$. Those intervals will form a cover of $[i..j]$ (i.e., will be disjoint and their union will be $[i, j]$), and values i_p (and j_p) will increase as we traverse the tree in inorder form.

We start taking RMQ(i, j), which gives the position m of the maximum (top-1); this would be enough for $k = 1$. In general, we initialize a tree of just one leaf and no internal nodes. The leaf is associated to position m and interval $[i..j]$. This establishes the invariants.

To report the next largest element, we take the position m_l of the maximum of the rightmost leaf l, and traverse the interval $[i..m_l]$ backwards using $P_1[m_l]$, $P_2[m_l]$, and so on. Each position (larger than 0) we arrive at contains an element larger than $A[m_l]$. However, if those are elements we have already reported, our candidate m_l is still good to be the next one to report. To determine this in constant time, we traverse the tree in reverse inorder at the same time we do the backward interval traversal. When we are at an internal node, we know that the backward traversal will stop there, as the element is larger than $A[m_l]$. Leaves, instead, are not yet reported and their interval may be skipped by the traversal.

If, however, the backward traversal stops at a position $P_r[m_l]$ that falls within the interval of another leaf p, then m_l is not the next largest element, since $P_r[m_l]$ is not yet reported. Instead of continuing with the new candidate at position $P_r[m_l]$, we take the leaf position m_p, which is indeed the largest of the interval. We restart the backward traversal from $l \leftarrow p$, using again $P_1[m_l]$, $P_2[m_l]$, and so on. When the backward traversal surpasses the left limit i, the current candidate is the next largest element to report. We split its area into two, compute RMQs to define the two new leaves of l, and restart the process.

For example, the first thing that happens when we start this algorithm for $k > 1$ is that $P_1[m] < i$, thus we report m and create a left child with interval $[i..m - 1]$ and position RMQ$(i, m - 1)$ and a right child with interval $[m + 1..j]$ and position RMQ$(m + 1, j)$. Then we go on to report the second element.

Since each step can be carried out in constant time, and our backward traversal performs k to $3k$ steps to determine the k-th answer, it follows that the time complexity of the algorithm is $O(k^2)$. We are able to run this algorithm for any $k \leq \kappa + 1$. By renaming κ we have our final result, that with $\kappa = 1$ matches the RMQ lower bounds.

Theorem 4. *Given an array $A[1..n]$ and a value κ, there is an encoding of A and κ that uses $(4\kappa - 2)n + o(n)$ bits and supports top-k-pos(i, j) queries for any $k \leq \kappa$, in $O(k^2)$ time on a RAM machine of $w = \Omega(\lg n)$ bits. The positions are given sorted by value.*

3.2   Using $O(n \lg k)$ Bits and $O(k)$ Time

Our final solution achieves asymptotically optimal time and space, building on the results of Section 3.1. It uses Jørgensen and Larsen's "shallow cuttings" idea [14], which we now outline. Imagine the values of $A[1..n]$, already mapped to the interval $[n]$, as a grid of points $(i, A[i])$. Now sweep a horizontal line from $y = n$ to $y = 1$. Include all the points found along the sweep in a *cell*, that is, a rectangle $[1, n] \times [y, n]$. Once we reach a point y_0 such that the cell reaches the threshold of containing $2\kappa + 1$ points, create two new cells by splitting the current cell. Let (x, y) be the point whose x-coordinate is the median in the current cell.

This will be called a *split point*; note it is not necessarily the point (x_0, y_0) that caused the split. Then the two new cells are initialized as $[1, x] \times [y_0, n]$ and $[x, n] \times [y_0, n]$ (note the vertical limit is y_0, that of the point causing the split, which now belongs to one of the two cells). Both cells now contain κ points, and the sweep continues, further splitting the new cells as we include more points. We create a *binary tree of cells* T_C, where the new cells are the left and right children, respectively, of the current cell.

In general, at any point in time, we will have a sequence of split points already determined, $x_1, x_2, \ldots$, and the cells that are leaves in the current T_C cover an x-coordinate interval of the form $[x_i, x_{i+1}]$ (we implicitly add split points 1 and n at the extremes). When the next split occurs at a point (x_0, y_0) within the cell covering the interval $[x_i, x_{i+1}]$, we will split it into two new cells covering $[x_i, x] \times [y_0, n]$ and $[x, x_{i+1}] \times [y_0, n]$, for some x. We will associate to those cells the *keys* $[x_i, x]$ and $[x, x_{i+1}]$, respectively, and the *extents* $[x_{i-1}, x_{i+1}]$ and $[x_i, x_{i+2}]$, respectively. Finally, once we have finished the sweep on the plane, we are left with a final set of split points $x_1, x_2, \ldots, x_t$ (from now on x_i will refer to this final sequence of split points). We add t further *keyless* cells with extents $[x_{i-1}, x_{i+1}]$ for all $1 \leq i \leq t$.

Jørgensen and Larsen prove various interesting properties of this process: (i) it creates $O(t) = O(n/\kappa)$ cells, each containing κ to 2κ points (if $n \geq \kappa$); (ii) if $c = [x_i, x_j] \times [y_0, n]$ is the cell with maximum y_0 value whose key is contained in a query range $[l, r]$, then $[l, r]$ is contained in the extent of c and (iii) the top-κ values in $[l, r]$ belong to the union of the 3 cells that comprise the extent of c.

We give now an encoding of this data structure that contains two parts. The first part uses $O((n/\kappa) \lg \kappa) + o(n) = o(n)$ bits[2] and identifies in constant time the desired cell whose extent contains $[l, r]$. The second part uses $O(n \lg \kappa)$ bits and gives us the first κ elements in the extent, in $O(\kappa)$ time.

Finding the Cell. We mark the final split points x_i in a bitmap $S[1..n]$ with constant-time rank and select support. As there are t bits set, S can be implemented in $O((n/\kappa) \lg \kappa) + o(n)$ bits [18]. It allows us finding in constant time the range $[m, M]$ of split points $l \leq x_m < \ldots < x_M \leq r$ contained in $[l, r]$. If this range contains zero or one split point (i.e., $m \geq M$), then $[l, r]$ is contained in the extent of the keyless cell number m and we are done for the first part.

Otherwise, the following procedure finds the desired key [14]. Find the split point x_i with maximum associated y_0-coordinate (this is the y_0 coordinate given to the two cells created by split point x_i). Find the split point x_j with second maximum. If $j < i$ (i.e., x_j is to the left of x_i), then the desired key is $[x_j, x_i]$, else it is $[x_i, x_j]$.

We map, using rank_1 on S, the range $[l, r]$ to the range $[m, M]$. Consider the array $Y[1..t]$ of y_0 values associated to the t split points. We store a range top-2 encoding T on the array Y, using the result of Section 3.1. This requires $O(n/\kappa)$ bits and returns the positions of the first and second maxima in $Y[m, M]$, x_i and

---

[2] It is not $o(n)$ if $\kappa = O(1)$, yet in this case the results of Section 3.1 are asymptotically equivalent.

x_j, in $O(1)$ time. Assume w.l.o.g. that $i < j$ and thus the desired key is $[x_i, x_j]$; the case $[x_j, x_i]$ is symmetric.

Now the final problem is to find the extent associated to the key $[x_i, x_j]$. For this we need to find the split points that, at the moment when the key $[x_i, x_j]$ was created, preceded x_i and followed x_j. Since, at the time we created split point x_j, the split points that existed were precisely those with y_0 larger than that associated to x_j, it follows that the split point that preceded x_i is $x_{i'}$, where $i' = P_2[j]$, as defined in Section 3.1 (Def. 1). Similarly, the split point that followed x_j was $x_{j'}$, where $j' = N_1[j]$ (N_k is defined symmetrically to P_k but pointing to the right, and it can be represented analogously). Those structures still require $O(n/\kappa)$ bits and answer in $O(1)$ time.

Thus, using $O((n/\kappa) \lg \kappa) + o(n)$ bits and $O(1)$ time, we find the extent that contains query $[l, r]$. The actual x-coordinates of the extent, $[x_{i'}, x_{j'}]$, are found using select_1 on S.

Traversing the Maxima. For the second structure, we represent the tree of cells T_C using $O(n/\kappa)$ bits, so that a number of operations are supported in constant time [20]. Since the key $[x_i, x_j]$ was created with the split point x_j, the corresponding node of T_C is the left child of the j-th node of T_C in inorder. This node with inorder j is computed in constant time, and then we can compute the preorder of its left child also in constant time (note that leaves do not have inorder number, but all nodes have a preorder position) [20].

Associated to the preorder index of each node of T_C we store an array $M[1..O(\kappa)]$ over $[O(\kappa)]$, using $O(\kappa \lg \kappa)$ bits (and $O(n \lg \kappa)$ overall). This array stores the information necessary to find the successive maxima of the extent of the node. We use RMQ queries on $A[x_{i'}, x_{j'}]$. Clearly the maximum in the extent is $m_1 = \mathrm{RMQ}_A(x_{i'}, x_{j'})$. The second maximum is either $m_2 = \mathrm{RMQ}_A(x_{i'}, m_1 - 1)$ or $m_2 = \mathrm{RMQ}_A(m_1 + 1, x_{j'})$. Which of the two is greater is stored (in some way to be specified soon) in $M[1]$. Assume $M[1]$ indicates that $m_2 = \mathrm{RMQ}_A(x_{i'}, m_1 - 1)$ (the other case is symmetric). Then the third maximum is either $m_3 = \mathrm{RMQ}_A(x_{i'}, m_2 - 1)$, $m_3 = \mathrm{RMQ}_A(m_2 + 1, m_1 - 1)$, or $m_3 = \mathrm{RMQ}_A(m_1 + 1, x_{j'})$. Which of the three is the third maximum is indicated by $M[2]$, and so on. Note that we cannot store directly the maxima positions in M because we would need $O(\kappa \lg n)$ bits. Rather, we use M to guide the search across the Cartesian tree slice that covers the extent.

To achieve $O(\kappa)$ time we will encode the values in M in the following way. At query time, will initialize an array $I[1..k]$ and start with the interval $I[1] = [x_{i'}, x_{j'}]$. Now $M[1] = 1$ will tell us that we must now split the interval at $I[1]$ using an RMQ query, $m_1 = \mathrm{RMQ}_A(I[M[1]]) = \mathrm{RMQ}_A(I[1]) = \mathrm{RMQ}_A(x_{i'}, x_{j'})$. We write the two resulting subintervals in $I[2] = [x_{i'}, m_1 - 1]$ and $I[3] = [m_1 + 1, x_{j'}]$. Now $M[2] \in \{2, 3\}$ will tell us in which of those intervals is the second maximum. Assume again it is in $M[2] = 2$. Then we compute $m_2 = \mathrm{RMQ}_A(I[2]) = \mathrm{RMQ}_A(x_{i'}, m_1 - 1)$, and write the two resulting subintervals in $I[4] = [x_{i'}, m_2 - 1]$ and $I[5] = [m_2 + 1, m_1 - 1]$. Now $M[3] \in \{3, 4, 5\}$ tells which interval contains the third maximum, and so on. Note the process is deterministic, so we can precompute the M values.

Therefore, we obtain in $O(\kappa)$ time the $O(\kappa)$ elements that belong to the extent, that is, the union of the three cells. By the properties of the shallow cutting, those contain the κ maxima of the query interval $[l, r]$. Therefore, in $O(\kappa)$ time we obtain the successive maxima of the extent, and filter out those not belonging to $[l, r]$. We are guaranteed to have seen the κ maxima of $[l, r]$, in order, after examining $O(\kappa)$ maxima of the extent. Note that if we want only the top-k, for $k \leq \kappa$, we also need $O(\kappa)$ time.

Theorem 5. *Given an array $A[1..n]$ and a value κ, there is an encoding of A and κ that uses $O(n \lg \kappa)$ bits and supports* top-k-pos(i, j) *queries for any $k \leq \kappa$, in $O(\kappa)$ time on a RAM machine of $w = \Omega(\lg n)$ bits. The positions are given sorted by value.*

By building this structure for $\lceil \lg \kappa \rceil$ successive powers of 2, we can use one where the search cost is $O(k)$, for any $k \leq \kappa$.

Corollary 2. *Given an array $A[1..n]$ and a value κ, there is an encoding of A and κ that uses $O(n \lg^2 \kappa)$ bits and supports* top-k-pos(i, j) *queries for any $k \leq \kappa$, in $O(k)$ time on a RAM machine of $w = \Omega(\lg n)$ bits. The positions are given sorted by value.*

4   Conclusions

We have given lower and upper bounds to several extensions of the RMQ problem, considering the encoding scenario. Some variants of range selection and range top-k queries were considered in the simpler one-sided version, where the interval starts at the beginning of the array. For those, we have obtained optimal or nearly-optimal time, and matched the space lower bound up to lower-order terms. For the general two-sided version of the problem we have largely focused on the range top-k query, where we have obtained optimal time and asymptotically optimal space (up to constant factors). Several problems remain open, especially handling range selection queries in the two-sided case, which we have not addressed. Tightening the constant space factors is also possible. Finally, most of our results fix k at construction time (although for two-sided queries we can fix a maximum k at construction time, at the price of an $O(\lg k)$ extra space factor). Removing these restrictions is also of interest.

References

1. Belazzougui, D., Boldi, P., Pagh, R., Vigna, S.: Monotone minimal perfect hashing: searching a sorted table with $o(1)$ accesses. In: Proc. SODA, pp. 785–794 (2009)
2. Bender, M., Farach-Colton, M.: The level ancestor problem simplified. Theor. Comp. Sci. 321(1), 5–12 (2004)
3. Berkman, O., Vishkin, U.: Recursive star-tree parallel data structure. SIAM J. Comp. 22(2), 221–242 (1993)

4. Brodal, G.S., Fagerberg, R., Greve, M., López-Ortiz, A.: Online sorted range reporting. In: Dong, Y., Du, D.-Z., Ibarra, O. (eds.) ISAAC 2009. LNCS, vol. 5878, pp. 173–182. Springer, Heidelberg (2009)

5. Brodal, G., Gfeller, B., Jørgensen, A., Sanders, P.: Towards optimal range medians. Theor. Comp. Sci. 412(24), 2588–2601 (2011)

6. Chan, T., Wilkinson, B.: Adaptive and approximate orthogonal range counting. In: Proc. SODA, pp. 241–251 (2013)

7. Clark, D.: Compact Pat Trees. Ph.D. thesis, Univ. of Waterloo, Canada (1996)

8. Fischer, J., Heun, V.: Space-efficient preprocessing schemes for range minimum queries on static arrays. SIAM J. Comp. 40(2), 465–492 (2011)

9. Gagie, T., Navarro, G., Puglisi, S.: New algorithms on wavelet trees and applications to information retrieval. Theor. Comp. Sci., 426–427, 25–41 (2012)

10. Gagie, T., Puglisi, S., Turpin, A.: Range quantile queries: another virtue of wavelet trees. In: Karlgren, J., Tarhio, J., Hyyrö, H. (eds.) SPIRE 2009. LNCS, vol. 5721, pp. 1–6. Springer, Heidelberg (2009)

11. Golynski, A., Munro, I., Rao, S.: Rank/select operations on large alphabets: a tool for text indexing. In: Proc. SODA, pp. 368–373 (2006)

12. Harel, D., Tarjan, R.: Fast algorithms for finding nearest common ancestors. SIAM J. Comp. 13(2), 338–355 (1984)

13. Hsu, P., Ottaviano, G.: Space-efficient data structures for top-k completion. In: Proc. WWW, pp. 583–594 (2013)

14. Jørgensen, A., Larsen, K.: Range selection and median: Tight cell probe lower bounds and adaptive data structures. In: Proc. SODA, pp. 805–813 (2011)

15. Li, G., Ji, S., Li, C., Feng, J.: Efficient type-ahead search on relational data: a tastier approach. In: Proc. SIGMOD, pp. 695–706. ACM (2009)

16. Munro, I.: Tables. In: Chandru, V., Vinay, V. (eds.) FSTTCS 1996. LNCS, vol. 1180, pp. 37–42. Springer, Heidelberg (1996)

17. Pătraşcu, M.: Succincter. In: Proc. FOCS, pp. 305–313 (2008)

18. Raman, R., Raman, V., Rao, S.S.: Succinct indexable dictionaries with applications to encoding k-ary trees, prefix sums and multisets. ACM Trans. Alg. 2(4), 43:1–43:25 (2007)

19. Sadakane, K.: Succinct representations of lcp information and improvements in the compressed suffix arrays. In: Proc. SODA, pp. 225–232 (2002)

20. Sadakane, K., Navarro, G.: Fully-functional succinct trees. In: Proc. SODA, pp. 134–149 (2010)

21. Vuillemin, J.: A unifying look at data structures. Comm. ACM 23(4), 229–239 (1980)

Fréchet Queries in Geometric Trees

Joachim Gudmundsson[1] and Michiel Smid[2]

[1] School of IT, University of Sydney, Australia
[2] School of Computer Science, Carleton University, Ottawa, Canada

Abstract. Let T be a tree that is embedded in the plane and let $\Delta, \varepsilon > 0$ be real numbers. The aim is to preprocess T into a data structure, such that, for any query polygonal path Q, we can decide if T contains a path P whose Fréchet distance $\delta_F(P, Q)$ to Q is less than Δ. We present an efficient data structure that solves an approximate version of this problem, for the case when T is c-packed and each of the edges of T and Q has length $\Omega(\Delta)$ (not required if T is a path): If the data structure returns NO, then there is no such path P. If it returns YES, then $\delta_F(P, Q) \leq \sqrt{2}(1 + \varepsilon)\Delta$ if Q is a line segment, and $\delta_F(P, Q) \leq 3(1 + \varepsilon)\Delta$ otherwise.

1   Introduction

The Fréchet distance [14] is a measure of similarity between two curves P and Q that takes into account the location and ordering of the points along the curves. Let p and p' be the endpoints of P and let q and q' be the endpoints of Q. Imagine a dog walking along P and, simultaneously, a person walking along Q. The person is holding a leash that is attached to the dog. Neither the dog nor the person is allowed to walk backwards along their curve, but they can change their speeds. The Fréchet distance between P and Q is the length of the shortest leash such that the dog can walk from p to p' and the person can walk from q to q'. To define this formally, let $|xy|$ denote the Euclidean distance between two points x and y. Then the Fréchet distance $\delta_F(P, Q)$ between P and Q is

$$\delta_F(P, Q) = \inf_{f} \max_{z \in P} |z f(z)|,$$

where f is any orientation-preserving homeomorphism $f : P \to Q$ with $f(p) = q$ and $f(p') = q'$.

Measuring the similarity between curves has been extensively studied in the last 20 years in computational geometry [3,13,16], as well as in other areas.

Alt and Godau [3] showed that the Fréchet distance between two polygonal curves P and Q can be computed in $O(n^2 \log n)$ time, where n is the total number of vertices of P and Q. Buchin et al. [8] improved the running time to $O(n^2 \sqrt{\log n}(\log \log n)^{3/2})$ expected time. A lower bound of $\Omega(n \log n)$ was given by Buchin et al. [7]. Alt [1] conjectured the decision problem to be 3SUM-hard.

Until recently, subquadratic time algorithms were only known for restricted cases, such as closed convex curves and κ-bounded curves [4]. In 2010, Driemel et al. [13] introduced a new class of realistic curves, the so-called c-packed curves.

H.L. Bodlaender and G.F. Italiano (Eds.): ESA 2013, LNCS 8125, pp. 565–576, 2013.

A polygonal path is c-packed if the total length of the edges inside any ball is bounded by c times the radius of the ball. The definition generalizes naturally to geometric graphs. Driemel et al. showed that a $(1 + \varepsilon)$-approximation of the Fréchet distance between two c-packed curves with a total of n vertices can be computed in $O(cn/\varepsilon + cn \log n)$ time. The notion of c-packedness has been argued [9,13] to capture many realistic settings for geometric paths and graphs.

In many applications, it is important to find the path in a geometric graph G that is most similar to a polygonal curve Q [2]. Given a planar geometric graph G with n vertices and a polygonal curve Q with n' vertices, the problem is to find the path P on G that has the smallest Fréchet distance to Q. They presented an algorithm that finds such a path P, with both endpoints being vertices of G, in $O(nn' \log(nn') \log n)$ time using $O(nn')$ space; see also [6,17]. The bound on the running time is close to quadratic in the worst case. Chen et al. [9] considered the case when the embedding of G is so-called ϕ-low density and the curve Q is c-packed. In $\mathbb{R}^d$, they presented a $(1+\varepsilon)$-approximation algorithm for the problem with running time $O((\phi n' + cn) \log(nn') \log(n + n') + (\phi n'/\varepsilon^d + cn/\varepsilon) \log(nn'))$.

Little is known about query variants. The aim is to preprocess a given geometric graph G, such that, for a polygonal path Q and a real value $\Delta > 0$ as a query, it can be decided if there exists a path P in G whose Fréchet distance to Q is less than Δ; the path P does not have to start or end at a vertex of G. In [5], de Berg et al. studied the case when G is a polygonal path with n vertices and Q is a straight-line segment. For any fixed Δ and any s with $n \le s \le n^2$, they show how to build, in $O(n^2 + s \operatorname{polylog}(n))$ time, a data structure of size $O(s \operatorname{polylog}(n))$, such that for any query segment Q of length more than 6Δ, the query algorithm associated with the data structure returns YES or NO in $O((n/\sqrt{s}) \operatorname{polylog}(n))$ time. If the output is YES, then there exists a path P in G such that $\delta_F(Q, P) \le (2 + 3\sqrt{2})\Delta$. (In fact, their algorithm counts all such "minimal" paths.) On the other hand, if the output is NO, then $\delta_F(Q, P) \ge \Delta$ for any path P in G. By increasing the preprocessing time to $O(n^3 \log n)$, the same result holds for the case when the threshold Δ is part of the query.

We consider the same problem as de Berg et al. [5] for the case when the graph is a tree and a query consists of a polygonal path Q. Let T be a tree (not necessarily plane) with n vertices that is embedded in the plane. Thus, any vertex of T is a point in $\mathbb{R}^2$ and any edge is the line segment joining its two vertices. A point x in $\mathbb{R}^2$ is said to be *on* T, if either x is a vertex of T or x is in the relative interior of some edge of T. If x and y are two points on T, then $T[x, y]$ denotes the path on T from x to y.

Let $\Delta > 0$ be a fixed real number. We want to preprocess T such that given a (possibly crossing) polygonal path Q with n' vertices, decide if there exist two points x and y on T, for which $\delta_F(Q, T[x, y]) < \Delta$.

Assume that the tree T is c-packed, for some constant c. Also, assume that each edge of T has length $\Omega(\Delta)$. For any constant $\varepsilon > 0$, we show that a data structure of size $O(n \operatorname{polylog}(n))$ can be built in $O(n \operatorname{polylog}(n))$ time. For any polygonal path Q with n' vertices, each of whose edges has length $\Omega(\Delta)$, the query algorithm associated with the data structure returns YES or NO in $O(n' \operatorname{polylog}(n))$ time.

1. If the output is YES and $n' = 2$ (i.e., Q is a segment), then the algorithm also reports two points x and y on T such that $\delta_F(Q, T[x, y]) \le \sqrt{2}(1+\varepsilon)\Delta$.
2. If the output is YES and $n' > 2$, then the algorithm also reports two points x and y on T such that $\delta_F(Q, T[x, y]) \le 3\Delta$.
3. If the output is NO, then $\delta_F(Q, T[x, y]) \ge \Delta$ for any x and y on T.

Compared to the structure in [5], the main drawbacks are that we require the input tree T to be c-packed and all its edges to have length $\Omega(\Delta)$ (not required if T is a path). However, the advantages are that (1) our structure can report a path $T[x, y]$, (2) the approximation bound is improved from $(2 + 3\sqrt{2})$ to $\sqrt{2}(1 + \varepsilon)$ for query segments, (3) our query algorithm can handle polygonal query paths Q, (4) the preprocessing time is improved from nearly quadratic to $O(n \operatorname{polylog}(n))$, and (5) the input graph can be a tree and is not restricted to being a polygonal path.

The rest of this paper is organized as follows. In Section 2, we present the general approach for querying a tree with a line segment, by giving a generic query algorithm and proving its correctness. Since the running time of this algorithm can be $\Omega(n)$ for arbitrary trees, we recall c-packed trees and μ-simplifications in Section 3. In Section 4, we use μ-simplifications to show how the generic algorithm can be implemented efficiently for querying a polygonal c-packed path with a query segment. In Section 5, we generalize the result of Section 4 to c-packed trees, all of whose edges have length $\Omega(\Delta)$. In Section 6, we use the previous results to query polygonal paths and trees with a polygonal path.

2   A Generic Algorithm for Query Segments

We first state two technical lemmas. Let T be a tree embedded in the plane and let $\Delta > 0$ be a real number. We assume that a query consists of a line segment $Q = [a, b]$ with endpoints a and b. We also assume that $|ab| > 2\Delta$.

Let $R(a, b)$ be the rectangle with sides of length $|ab|$ and 2Δ as indicated in Figure 1. Let D_a be the disk with center a and radius Δ, and let C_{ab} be the part of the boundary of this disk that is contained in $R(a, b)$. Define D_b and C_{ba} similarly with respect to b.

Lemma 1. *Let $\varepsilon > 0$ be a sufficiently small real number, let $Q = [a, b]$ be a line segment of length more than 2Δ, and assume that the angle between Q and the positive X-axis is at most ε. Assume there exist two points x and y on T, such that $\delta_F(Q, T[x, y]) < \Delta$. Then, there exist two points x' and y' on $T[x, y]$, such that the following are true: 1. x' and y' are on the half-circles C_{ab} and C_{ba}, respectively, and x' is on the path $T[x, y']$. 2. The path $T[x', y']$ is completely contained in the region $R(a, b) \setminus (D_a \cup D_b)$. 3. $\delta_F(Q, T[x', y']) \le \Delta$. 4. Let p and q be the first and last vertices of T on the path $T[x', y']$, respectively. For each vertex r on the path $T[p, q]$, let L_r be the vertical line through r, and let L'_r be the vertical line that is obtained by translating L_r by a distance $2(1+\varepsilon)\Delta$ to the left. Then, for each such r, the path $T[r, q]$ does not cross the line L'_r.*

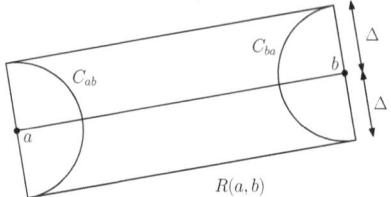

Fig. 1. The rectangle $R(a, b)$ and the half-circles C_{ab} and C_{ba} corresponding to the line segment ab

Lemma 2. *Let $\varepsilon > 0$ be a sufficiently small real number, let $Q = [a, b]$ be a line segment of length more than 2Δ, and assume that the angle between Q and the positive X-axis is at most ε. Assume there exist two points x' and y' on T, such that the following are true: (i) x' and y' are on the half-circles C_{ab} and C_{ba}, respectively, and (ii) the second and fourth item in Lemma 1 hold. Then, $\delta_F(Q, T[x', y']) \le \sqrt{2}(1 + 3\varepsilon)\Delta$.*

2.1 The Generic Algorithm

As before, let T be a tree embedded in the plane and let $\Delta > 0$. We define $T(x, y)$ to be the "open" path obtained by removing x and y from $T[x, y]$. We choose a small $\varepsilon > 0$. For each m with $0 \le m < \pi/\varepsilon$, consider the (X_m, Y_m)-coordinate system that is obtained by rotating the X-axis and Y-axis by an angle of $2m\varepsilon$.

Consider a query segment $Q = [a, b]$ with $|ab| > 2\Delta$. We present a generic algorithm that approximately decides if there exist two points x and y on T such that $\delta_F(Q, T[x, y]) < \Delta$. The idea is to choose a coordinate system in which Q is approximately horizontal. Using this, we find all pairs (x, y) of points on T for which the three conditions in Lemma 2 hold.

Step 1: Compute the set A of intersection points between T and the half-circle C_{ab}, and the set B of intersection points between T and the half-circle C_{ba}.

Step 2: Compute $I = \{(x, y) \in A \times B : T(x, y) \cap A = \emptyset$ and $T(x, y) \cap B = \emptyset\}$.

Step 3: Compute $J = \{(x, y) \in I : T[x, y]$ is completely contained in $R(a, b)\}$.

Step 4: Let m be an index such that the angle between Q and the positive X_m-axis is at most ε. For each pair (x, y) in J, do the following: Let p and q be the first and last vertices of T on the path $T[x, y]$, respectively. For each vertex r on $T[p, q]$, let L_r be the line through r that is parallel to the Y_m-axis, and let L'_r be the line obtained by translating L_r by a distance $2(1 + \varepsilon)\Delta$ in the negative X_m-direction. Decide if, for each vertex r on $T[p, q]$, the path $T[r, q]$ does not cross the line L'_r. If this is the case, then return YES together with the two points x and y, and terminate the algorithm. If, at the end of this fourth step, the algorithm did not terminate yet, then return NO and terminate.

If T has n vertices, then the worst-case running time of this algorithm will be $\Omega(n)$, because the sets A and B in Step 1 may have size $\Theta(n)$.

Lemma 3. *Let $Q = [a, b]$ be a line segment of length more than 2Δ and consider the output of the generic algorithm on input Q. If the output is YES, then there exist two points x and y on T such that $\delta_F(Q, T[x, y]) \leq \sqrt{2}(1 + 3\varepsilon)\Delta$. If the output is NO, then for all x and y on T, $\delta_F(Q, T[x, y]) \geq \Delta$.*

3 c-Packed Trees and μ-Simplifications

Following Driemel et al. [13], we say that a tree T is *c-packed* if for any $\rho > 0$ and any disk D of radius ρ, the total length of $D \cap T$ is at most $c\rho$.

Lemma 4. *The arrangement of a c-packed tree with n vertices has size $O(cn)$.*

Lemma 5. *Let $\Delta > 0$ and $c' \geq 1$, and let T be a c-packed tree, all of whose edges have length at least Δ/c'. Then, for any circle C of radius Δ, the number of intersection points between T and C is $O(cc')$.*

In Section 4, we consider polygonal paths P that are c-packed, but do not have the property that each edge has length $\Omega(\Delta)$. Since Lemma 5 does not hold for such a path, we would like to simplify P, resulting in a path P', such that P' is $O(c)$-packed, each of its edges has length $\Omega(\Delta)$, and P' is close to P with respect to the Fréchet distance. This leads to the notion of a μ-simplification:

Let $P = (p_1, p_2, \ldots, p_n)$ be a polygonal path in the plane and let $\mu > 0$ be a real number. A *μ-simplification* of P is a polygonal path $P' = (p_{i_1}, p_{i_2}, \ldots, p_{i_k})$ such that $1 = i_1 < i_2 < \ldots < i_k = n$ and each edge of P', except possibly the last one, has length at least μ.

Lemma 6 (Driemel et al. [13]). *In $O(n)$ time, a μ-simplification P' of P can be computed such that $\delta_F(P, P') \leq \mu$. If the polygonal path P is c-packed, then this μ-simplification P' is $(6c)$-packed.*

4 Polygonal Paths and Query Segments

In this section, we assume that the tree T is a polygonal path and write $P = (p_1, p_2, \ldots, p_n)$ instead of T. For two points x and y on P, we write $x \leq_P y$ if x is on the subpath $P[p_1, y]$. We choose positive real numbers c, c', and ε. We assume that P is c-packed and that each edge of P, except possibly the last one, has length at least Δ/c'. At the end of this section, we will show how to remove the latter assumption. We start by describing the preprocessing algorithm.

Arrangement $\mathcal{A}(P)$: We construct the arrangement $\mathcal{A}(P)$ induced by the path P and preprocess it for point-location queries. By Lemma 4, $\mathcal{A}(P)$ has size $O(cn)$.

Circular ray shooting structures: For each face F of $\mathcal{A}(P)$, we construct the data structure $CRS(F)$ in Cheng et al. [10], which supports circular ray shooting queries for the fixed radius Δ. If m is the number of vertices on F, then $CRS(F)$ has query time $O(\log m)$, size $O(m)$, and can be constructed in $O(m \log m)$ time.

Subpath rectangle intersection structure: We construct a balanced binary search tree storing the points $p_1, p_2, \ldots, p_n$ at its leaves (sorted by their indices). At each node v of this tree, we store the convex hull CH_v of all points stored in the subtree of v. The resulting structure $SRI(P)$ can be used to answer the following type of query in $O(\log^2 n)$ time: Given any rectangle R (not necessarily axes-parallel) and any two indices i and j with $1 \leq i \leq j \leq n$, decide if the subpath $P[p_i, p_j] = (p_i, p_{i+1}, \ldots, p_j)$ of P is completely contained in R.

Priority search trees: In Section 2.1, we defined, for each m with $0 \leq m < \pi/\varepsilon$, the coordinate system with axes X_m and Y_m. For each m with $0 \leq m < \pi/\varepsilon$, we do the following: For any k with $1 \leq k \leq n$, let L_{km} be the line through the point p_k that is parallel to the Y_m-axis, and let L'_{km} be the line obtained by translating L_{km} by a distance $2(1 + \varepsilon)\Delta$ in the negative X_m-direction. Define

$$f_m(k) = \min\{\ell : k < \ell \leq n, \; p_\ell \text{ is to the left (w.r.t. the } X_m\text{-direction) of } L'_{km}\}.$$

Consider the set $S_m = \{(k, f_m(k)) : 1 \leq k \leq n\}$ of n points in the plane. We construct a priority search tree $PST_m(P)$ for the set S_m; see McCreight [15]. We can use $PST_m(P)$ to answer the following query in $O(\log n)$ time: Given any two indices i and j with $1 \leq i \leq j \leq n$, decide if, for each k with $i \leq k \leq j$, the subpath $P[p_k, p_j]$ does not cross the line L'_{km}. Indeed, observe that this is the case if and only if no point of S_m is in the three-sided rectangle $[i, j] \times (-\infty, j]$.

Lemma 7. *The entire preprocessing algorithm takes $O((c + 1/\varepsilon)n \log n)$ time and produces a data structure of size $O(n \log n + (c + 1/\varepsilon)n)$.*

The query algorithm: Let $Q = [a, b]$ be a query line segment of length more than 2Δ. Recall that we assume that the polygonal path P is c-packed and each edge of P, except possibly the last one, has length at least Δ/c'. We show how the data structures can be used to implement the four steps of the generic algorithm.
Step 1: Consider the half-circle C_{ab} and let z be one of its endpoints. Find the face F in the arrangement $\mathcal{A}(P)$ that contains z. Then use $CRS(F)$ to find the first intersection between the boundary of F and the circular ray along C_{ab} that starts at z. This gives us the first intersection, say x, between P and C_{ab}. Set z to x and repeat this procedure. We obtain the set A of intersection points between P and C_{ab}. Similarly we obtain the set B of intersection points between P and the half-circle C_{ba}. Using Lemma 5, this step takes $O(cc' \log n)$ time.
Step 2: We sort the points of $A \cup B$ in the order in which they occur along the path P. By scanning the sorted order, we obtain the set

$$I = \{(x, y) \in A \times B : x \leq_P y, P(x, y) \cap A = \emptyset \text{ and } P(x, y) \cap B = \emptyset\}.$$

Since $|A| + |B| = O(cc')$, this step takes $O(cc' \log(cc'))$ time.
Step 3: For each (x, y) in I, we use the data structure $SRI(P)$ to decide if the subpath $P[x, y]$ is completely contained in the rectangle $R(a, b)$. If this is the case, then we add the pair (x, y) to an initially empty set. In this way, we obtain

$$J = \{(x, y) \in I : P[x, y] \text{ is completely contained in } R(a, b)\}.$$

Since $|I| = O(cc')$, this step takes $O(cc' \log^2 n)$ time.

Step 4: Let m be an index such that the angle between Q and the positive X_m-axis is at most ε. For each (x, y) in the set J, do the following: If x and y are on the same edge of P, then return YES together with the two points x and y, and terminate the algorithm. Otherwise, let p_i and p_j be the first and last vertices of P on the path $P[x, y]$, respectively. Use the priority search tree $PST_m(P)$ to decide if, for each k with $i \leq k \leq j$, the subpath $P[p_k, p_j]$ does not cross the line L'_{km}. If this is the case, then return YES together with the two points x and y, and terminate the algorithm. If the algorithm did not terminate yet, then return NO and terminate. Since $|J| = O(cc')$, this step takes $O(cc' \log n)$ time.

Lemma 8. *The query algorithm takes $O(cc' \log^2 n)$ time and correctly implements the generic algorithm of Section 2.1. Thus, the two claims in Lemma 3 hold for the output of this algorithm.*

This result assumes that each edge of P, except possibly the last one, has length at least Δ/c'. We now remove this assumption.

Let P be a polygonal path and assume that P is c-packed for some real number $c > 0$. We choose real numbers Δ, ε, and c'. Let $\mu = \Delta/c'$. In $O(n)$ time, we compute a μ-simplification P' of P; see Lemma 6. Then we run the preprocessing algorithm on P'. For any given query segment $Q = [a, b]$ of length more than 2Δ, we run the above query algorithm on the data structure for P'.

Assume the output of the query algorithm is YES. By Lemma 3, there exist two points x' and y' on P' such that $x' \leq_{P'} y'$ and $\delta_F(Q, P'[x', y']) \leq \sqrt{2}(1 + 3\varepsilon)\Delta$. Since, by Lemma 6, $\delta_F(P, P') \leq \mu = \Delta/c'$, it follows that there exist two points x and y on P such that $x \leq_P y$ and

$$\delta_F(Q, P[x, y]) \leq \sqrt{2}(1 + 3\varepsilon)\Delta + \Delta/c' = \sqrt{2}\Delta(1 + 3\varepsilon + 1/(c'\sqrt{2})). \quad (1)$$

On the other hand, if the output of the query algorithm is NO, then we have, by Lemma 3, $\delta_F(Q, P'[x', y']) \geq \Delta$ for any two points x' and y' on P' with $x' \leq_{P'} y'$. Therefore, we have $\delta_F(Q, P[x, y]) \geq \Delta - \mu = (1 - 1/c')\Delta$ for any two points x and y on P with $x \leq_P y$.

By taking $c' = 1/\varepsilon$ and defining $\Delta_0 = (1 - 1/c')\Delta$, the right-hand side in (1) becomes $\sqrt{2}\Delta_0(1 + O(\varepsilon))$. This proves the following result (rename Δ_0 as Δ).

Theorem 1. *Let P be a polygonal path in the plane with n vertices, let c and Δ be positive real numbers, and assume that P is c-packed. For any $\varepsilon > 0$, we can construct a data structure of size $O(n \log n + (c + 1/\varepsilon)n)$ in $O((c + 1/\varepsilon)n \log n)$ time. Given any segment Q of length more than 2Δ, the query algorithm corresponding to this data structure takes $O((c/\varepsilon) \log^2 n)$ time and outputs either YES or NO. If the output is YES, then there exist two points x and y on P with $x \leq_P y$ such that $\delta_F(Q, P[x, y]) \leq \sqrt{2}(1 + \varepsilon)\Delta$. If the output is NO, then $\delta_F(Q, P[x, y]) \geq \Delta$ for any two points x and y on P with $x \leq_P y$.*

Remark 1. Using a recent result of Driemel and Har-Peled [12], the approximation factor can be improved from $\sqrt{2}(1 + \varepsilon)$ to $1 + \varepsilon$. For constant c and ε, the space and query time improve to $O(n)$ and $O(\log n \log \log n)$, respectively, whereas the preprocessing time increases to $O(n \log^2 n)$. It is not clear if this can be used to improve the approximation factors in the rest of this paper.

5   Trees and Query Segments

Let T be a tree with n vertices that is embedded in the plane. We fix a real number $\Delta > 0$ and consider query segments $Q = [a, b]$ of length more than 2Δ. We choose positive real numbers c, c', and ε, and assume that T is c-packed and each edge of T has length at least Δ/c'.

Cole and Vishkin [11] show how to decompose the tree T into a collection of paths; we refer to this collection as the *path decomposition* $PD(T)$ of T. For any two points x and y on T, the path $T[x, y]$ overlaps $O(\log n)$ paths in $PD(T)$. More precisely, given x and y, in $O(\log n)$ time, a sequence $v_1, \ldots, v_k$ can be computed, such that (i) $k = O(\log n)$, (ii) $v_1 = x$ and $v_k = y$, (iii) for each i with $2 \leq i \leq k-1$, v_i is an endpoint of some path in $PD(T)$ (and, thus, a vertex of T), (iv) for each i with $1 \leq i < k$, the path $T[v_i, v_{i+1}]$ is contained in some path in $PD(T)$, (v) the path $T[x, y]$ in T between x and y is equal to the concatenation of the paths $T[v_1, v_2], T[v_2, v_3], \ldots, T[v_{k-1}, v_k]$. Using this, the following query can be answered in $O(\log n)$ time: Given points x, y, and z on T, together with the edges that contains them, decide if z is on the path $T[x, y]$.

We first describe the preprocessing algorithm. Construct the arrangement $\mathcal{A}(T)$ induced by the tree T and preprocess it for point-location queries. For each face F of $\mathcal{A}(T)$, construct the data structure $CRS(F)$ for circular ray shooting queries, as in Section 4. Compute the path decomposition $PD(T)$ of T. For each path P in $PD(T)$, construct the data structure $SRI(P)$ of Section 4. Also, for each path P in $PD(T)$ and for each m with $0 \leq m < \pi/\varepsilon$, construct two data structures $PST_M^{\rightarrow}(P)$ and $PST_M^{\leftarrow}(P)$ one structure for each direction in which the path P can be traversed. Additionally, we do the following:

Leftmost and rightmost structures: For each (X_m, Y_m)-coordinate systems, $0 \leq m < \pi/\varepsilon$, and each path P in $PD(T)$, construct a balanced binary search tree $LR_m(P)$ storing the points of P at its leaves, in the order in which they appear along P. At each node v of this tree, we store the point in the subtree of v that is extreme in the positive X_m-direction and the point in the subtree of v that is extreme in the negative X_m-direction. For any given path P in $PD(T)$, any index m, and any two points x and y on P, we can use $LR_m(P)$ to compute, in $O(\log n)$ time, the point on $P[x, y]$ that is extreme in the positive X_m-direction and the point on $P[x, y]$ that is extreme in the negative X_m-direction.

The query algorithm: Consider a query segment $Q = [a, b]$ of length more than 2Δ. We show how to implement the four steps of the generic algorithm.

Step 1: As in Section 4, we compute the sets A and B of intersection points between T and the half-circles C_{ab} and C_{ba}, respectively.

Step 2: For each x in A and each y in B: Consider all points $z \in (A \cup B) \setminus \{x, y\}$ and, for each such point, decide if it is on the path $T[x, y]$. If this is not the case for all such z, then we add the pair (x, y) to an initially empty set. At the end of this step, we have computed

$$I = \{(x, y) \in A \times B : T(x, y) \cap A = \emptyset \text{ and } T(x, y) \cap B = \emptyset\}.$$

Step 3: For each (x, y) in I: Use the path decomposition $PD(T)$ to compute the sequence of $O(\log n)$ paths in $PD(T)$ that overlap $T[x, y]$. For each such path P, use the data structure $SRI(P)$ to decide if the maximal subpath of P that is on $T[x, y]$ is completely contained in the rectangle $R(a, b)$. If this is the case for each such P, then we add the pair (x, y) to an initially empty set. At the end of this step, we have computed the set

$$J = \{(x, y) \in I : T[x, y] \text{ is completely contained in } R(a, b)\}.$$

Step 4: Let m be an index such that the angle between Q and the positive X_m-axis is at most ε. Consider any pair (x, y) in the set J. Let p and q be the first and last vertices on the path $T[x, y]$, respectively. For each vertex r on $T[p, q]$, let L_r be the line through r that is parallel to the Y_m-axis, and let L_r' be the line obtained by translating L_r by a distance of $2(1 + \varepsilon)\Delta$ in the negative X_m-direction. We have to decide if for each such vertex r, the path $T[r, q]$ does not cross the line L_r'.

We compute the sequence $v_1, \ldots, v_k$ as explained in the beginning of this section. For each i with $1 \le i < k$, let P_i be the path in the path decomposition $PD(T)$ that contains $T[v_i, v_{i+1}]$.

Assume there are two vertices r and s on $T[p, q]$, such that s is on $T[r, q]$ and s is to the left of L_r' (with respect to the X_m-direction). Let i and j be the indices such that r is on the path P_i and s is on the path P_j. Observe that $i \le j$. There are two possibilities: If $i = j$, then we use one of the two priority search trees $PST_M^{\rightarrow}(P_i)$ and $PST_M^{\leftarrow}(P_i)$ to find such vertices r and s. If $i < j$, then we may assume that r is the vertex on $P_i \cap T[p, q]$ that is extreme in the positive X_m-direction, whereas s is the vertex on the concatenation of the paths $P_{i+1} \cap T[p, q], \ldots, P_{k-1} \cap T[p, q]$ that is extreme in the negative X_m-direction. Thus, we handle this case in the following way: For each i with $1 \le i < k$, use the data structure $LR_m(P_i)$ to find the two points r_i and s_i on $P_i \cap T[p, q]$ that are extreme in the positive and negative X_m-direction, respectively. Then for $i = k - 2, k - 3, \ldots, 1$, find the point s_i' on the concatenation of the paths $P_{i+1} \cap T[p, q], \ldots, P_{k-1} \cap T[p, q]$ that is extreme in the negative X_m-direction. Finally, for each i, $1 \le i < k$, check if the point s_i' is to the left of the line L_{r_i}'.

Theorem 2. *Let T be a tree with n vertices that is embedded in the plane, and let c, c', and Δ be positive real numbers. Assume that T is c-packed and each of its edges has length at least Δ/c'. For any $\varepsilon > 0$, we can construct a data structure of size $O(n \log n + (c + 1/\varepsilon)n)$ in $O((c + 1/\varepsilon)n \log n)$ time. Given any segment Q of length more than 2Δ, the query algorithm corresponding to this data structure takes $O((cc')^2 \log^3 n + (cc')^3 \log n)$ time and outputs either YES* or NO. If the output is YES, then there exist two points x and y on T such that $\delta_F(Q, T[x, y]) \leq \sqrt{2}(1 + \varepsilon)\Delta$. If the output is NO, then $\delta_F(Q, T[x, y]) \geq \Delta$ for any x and y on T.

Unfortunately, the simplification technique of Section 3 cannot be used to remove the assumption that each edge of T has length at least Δ/c': The number of paths in the path decomposition $PD(T)$ of the tree T can be $\Omega(n)$. Therefore, if we apply the simplification technique to each such path, as we did in Section 4, we may get $\Omega(n)$ simplified paths, each of which contains one edge of length less than $\mu = \Delta/c'$. As a result, the sets A and B that are computed in Step 1 of the query algorithm may have linear size.

6   Querying with a Polygonal Path

Until now, we have considered querying polygonal paths or trees with a line segment. In this section, we generalize our results to queries Q consisting of a polygonal path. Unfortunately, the approximation factor increases from $\sqrt{2}(1+\varepsilon)$ to 3, and we need the requirement that edge lengths of Q are more than 4Δ.

Let T be a tree embedded in the plane and let $\Delta > 0$ be a real number. We consider polygonal query paths $Q = (q_1, q_2, \ldots, q_{n'})$, all of whose edges have length more than 4Δ. As before, we fix a sufficiently small real number $\varepsilon > 0$. The following two lemmas generalize Lemmas 1 and 2.

Lemma 9. *Let $Q = (q_1, q_2, \ldots, q_{n'})$ be a polygonal path, all of whose edges have length more than 4Δ. Assume there exist two points x and y on T, such that $\delta_F(Q, T[x, y]) < \Delta$. Then, there exist points $x'_1, y'_1, \ldots, x'_{n'-1}, y'_{n'-1}$ on $T[x, y]$, such that the following are true:*

1. *By traversing the path $T[x, y]$, we visit $x'_1, y'_1, \ldots, x'_{n'-1}, y'_{n'-1}$ in this order.*
2. *For each i with $1 \leq i < n'$,*
   (a) *x'_i and y'_i are on the half-circles $C_{q_i q_{i+1}}$ and $C_{q_{i+1} q_i}$, respectively,*
   (b) *the path $T[x'_i, y'_i]$ is completely contained in the region $R(q_i, q_{i+1}) \backslash (D_{q_i} \cup D_{q_{i+1}})$.*
3. *For each i with $1 \leq i < n' - 1$, the path $T[y'_i, x'_{i+1}]$ is completely contained in the disk with center q_{i+1} and radius 3Δ.*
4. *$\delta_F(Q, T[x'_1, y'_{n'-1}]) \leq \Delta$.*
5. *For each i with $1 \leq i < n'$, let m_i be an index such that the angle between the line segment $[q_i, q_{i+1}]$ and the positive X_{m_i}-axis is at most ε. Let p_i and q_i be the first and last vertices of T on the path $T[x'_i, y'_i]$, respectively. For each vertex r on the path $T[p_i, q_i]$, let L_r be the line through r that is parallel to the Y_{m_i}-axis, and let L'_r be the line that is obtained by translating L_r by a distance $2(1 + \varepsilon)\Delta$ in the negative X_{m_i}-direction. Then, for each such r, the path $T[r, q_i]$ does not cross the line L'_r.*

Lemma 10. *Let $Q = (q_1, q_2, \ldots, q_{n'})$ be a polygonal path, all of whose edges have length more than 4Δ. Assume there exist points $x'_1, y'_1, \ldots, x'_{n'-1}, y'_{n'-1}$ on T, such that the first (with $x = x'_1$ and $y = y'_{n'-1}$), second, third, and fifth items in Lemma 9 hold. Then, $\delta_F(Q, T[x'_1, y'_{n'-1}]) \leq 3\Delta$.*

Our algorithm will run the query algorithms of the previous sections separately on each edge of Q. Afterwards, we check if the partial paths in T obtained for the edges of Q can be combined into one global path that is close to the entire path Q with respect to the Fréchet distance. The following lemma implies that this combining step is possible and, in fact, unique, if such a global path exists. This is where we need that each edge of Q has length more than 4Δ.

Lemma 11. *Let $P = (p_1, p_2, \ldots, p_n)$ and $Q = (q_1, q_2, \ldots, q_{n'})$ be polygonal paths and assume that each edge of Q has length more than 4Δ. Let i be any index with $1 \le i < n' - 1$ and assume that there exists a point y on P, such that y is on the half-circle $C_{q_i q_{i+1}}$. Then there exists at most one pair (x', y') of points on P such that (i) $y \le_P x' \le_P y'$, (ii) x' and y' are on the half-circles $C_{q_{i+1} q_{i+2}}$ and $C_{q_{i+2} q_{i+1}}$, respectively, (iii) the path $P[x', y']$ is completely contained in the region $R(q_{i+1}, q_{i+2}) \setminus (D_{q_{i+1}} \cup D_{q_{i+2}})$, (iv) the path $P[y, x']$ is completely contained in the disk with center q_{i+1} and radius 3Δ.*

Assume that the tree T is a polygonal path $P = (p_1, p_2, \ldots, p_n)$. We choose Δ, c, c', and ε, and assume that P is c-packed. We start by assuming that each edge of P, except possibly the last one, has length at least Δ/c'. We run the preprocessing algorithm of Section 4. Additionally, we do the following:

Subpath furthest point structure: We construct a balanced search tree storing the vertices of P at its leaves, sorted by their indices. At each node v, we store the furthest-point Voronoi diagram of all points in the subtree of v. Let $SFP(P)$ be the resulting data structure. For any disk D in the plane and any two indices i and j with $1 \le i \le j \le n$, we can use $SFP(P)$ to decide if the subpath $P[p_i, p_j]$ is completely contained in D. The time to answer such a query is $O(\log^2 n)$.

Consider a query path $Q = (q_1, q_2, \ldots, q_{n'})$ all of whose edges have length more than 4Δ. As mentioned before, we run the query algorithm of the previous section on each edge $[q_i, q_{i+1}]$ of Q. If (x_i, y_i) and (x_{i+1}, y_{i+1}) are returned for the i-th and $(i+1)$-st edges, respectively, then we check if $y_i \le_P x_{i+1}$ and the path $P[y_i, x_{i+1}]$ is completely contained in the disk with center q_{i+1} and radius 3Δ. By doing this for $i = 1, 2, \ldots, n' - 2$, we answer the query.

To remove the assumption that each edge of P has length at least Δ/c', we compute a μ-simplification P' of P, and proceed as in Section 4.

Theorem 3. *Let P be a polygonal path in the plane with n vertices, let c and Δ be positive real numbers, and assume that P is c-packed. For any $\varepsilon > 0$, we can construct a data structure of size $O(n \log n + cn)$ in $O(n \log^2 n + cn \log n)$ time. Given any polygonal path Q with n' vertices, all of whose edges have length more than 4Δ, the query algorithm corresponding to this data structure takes $O((c/\varepsilon)^2 n' \log^2 n)$ time and outputs either YES or NO. If the output is YES, then there exist two points x and y on P with $x \le_P y$ such that $\delta_F(Q, P[x, y]) \le 3(1 + \varepsilon)\Delta$. If the output is NO, then $\delta_F(Q, P[x, y]) \ge \Delta$ for any two points x and y on P with $x \le_P y$.*

Theorem 4. *Let T be a tree with n vertices that is embedded in the plane, and let c, c', and Δ be positive real numbers. Assume that T is c-packed and each* *of its edges has length at least Δ/c'. We can construct a data structure of size $O(n \log n + cn)$ in $O(n \log^2 n + cn \log n)$ time. Given any polygonal path Q with n' vertices, all of whose edges have length more than 4Δ, the query algorithm corresponding to this data structure takes $O((cc')^4 n' \log^3 n)$ time and outputs either YES or NO. If the output is YES, then there exist two points x and y on T such that $\delta_F(Q, T[x, y]) \leq 3\Delta$. If the output is NO, then $\delta_F(Q, T[x, y]) \geq \Delta$ for any x and y on T.*

References

1. Alt, H.: The computational geometry of comparing shapes. In: Albers, S., Alt, H., Näher, S. (eds.) Festschrift Mehlhorn. LNCS, vol. 5760, pp. 235–248. Springer, Heidelberg (2009)

2. Alt, H., Efrat, A., Rote, G., Wenk, C.: Matching planar maps. Journal of Algorithms 49(2), 262–283 (2003)

3. Alt, H., Godau, M.: Computing the Fréchet distance between two polygonal curves. International Journal of Computational Geometry & Applications 5, 75–91 (1995)

4. Alt, H., Knauer, C., Wenk, C.: Comparison of distance measures for planar curves. Algorithmica 38(1), 45–58 (2003)

5. de Berg, M., Cook, A.F., Gudmundsson, J.: Fast Fréchet queries. Computational Geometry – Theory and Applications 46(6), 747–755 (2013)

6. Brakatsoulas, S., Pfoser, D., Salas, R., Wenk, C.: On map-matching vehicle tracking data. In: VLDB, pp. 853–864 (2005)

7. Buchin, K., Buchin, M., Knauer, C., Rote, G., Wenk, C.: How difficult is it to walk the dog? In: EuroCG, pp. 170–173 (2007)

8. Buchin, K., Buchin, M., Meulemans, W., Mulzer, W.: Four soviets walk the dog - with an application to Alt's conjecture. CoRR abs/1209.4403 (2012)

9. Chen, D., Driemel, A., Guibas, L.J., Nguyen, A., Wenk, C.: Approximate map matching with respect to the Fréchet distance. In: ALENEX, pp. 75–83 (2011)

10. Cheng, S.W., Cheong, O., Everett, H., van Oostrum, R.: Hierarchical decompositions and circular ray shooting in simple polygons. Discrete & Computational Geometry 32, 401–415 (2004)

11. Cole, R., Vishkin, U.: The accelerated centroid decomposition technique for optimal parallel tree evaluation in logarithmic time. Algorithmica 3, 329–346 (1988)

12. Driemel, A., Har-Peled, S.: Jaywalking your dog: computing the Fréchet distance with shortcuts. In: SODA, pp. 318–337 (2012)

13. Driemel, A., Har-Peled, S., Wenk, C.: Approximating the Fréchet distance for realistic curves in near linear time. Discrete & Computational Geometry 48, 94–127 (2012)

14. Fréchet, M.: Sur quelques points du calcul fonctionnel. Rendiconti del Circolo Mathematico di Palermo 22, 1–74 (1906)

15. McCreight, E.M.: Priority search trees. SIAM Journal on Computing 14, 257–276 (1985)

16. Wenk, C.: Shape matching in higher dimensions, Dissertation, Freie Universität Berlin, Germany (2003)

17. Wenk, C., Salas, R., Pfoser, D.: Addressing the need for map-matching speed: Localizing global curve-matching algorithms. In: SSDBM, pp. 379–388 (2006)

A Computationally Efficient FPTAS for Convex Stochastic Dynamic Programs

Nir Halman[1], Giacomo Nannicini[2,*], and James Orlin[3]

[1] Jerusalem School of Business Administration, The Hebrew University, Jerusalem, Israel, and Dept. of Civil and Environmental Engineering, MIT, Cambridge, MA
halman@mit.edu

[2] Singapore University of Technology and Design, Singapore
nannicini@sutd.edu.sg

[3] Sloan School of Management, MIT, Cambridge, MA
jorlin@mit.edu

Abstract. We propose a computationally efficient Fully Polynomial-Time Approximation Scheme (FPTAS) for convex stochastic dynamic programs using the technique of K-approximation sets and functions introduced by Halman et al. This paper deals with the convex case only, and it has the following contributions: First, we improve on the worst-case running time given by Halman et al. Second, we design an FPTAS with excellent computational performance, and show that it is faster than an exact algorithm even for small problem instances and small approximation factors, becoming orders of magnitude faster as the problem size increases. Third, we show that with careful algorithm design, the errors introduced by floating point computations can be bounded, so that we can provide a guarantee on the approximation factor over an exact infinite-precision solution. Our computational evaluation is based on randomly generated problem instances coming from applications in supply chain management and finance.

1 Introduction

We consider a finite horizon stochastic dynamic program (DP), as defined in [1]. Our model has an underlying discrete time dynamic system, and a cost function that is additive over time. We now introduce the type of problems addressed in this paper. We postpone a rigorous definition of each symbol until Sect. 2, without compromising clarity. The system dynamics are of the form:
$I_{t+1} = f(I_t, x_t, D_t), \quad t = 1, \ldots, T$, where:

$$t : \text{discrete time index},$$
$$I_t \in \mathcal{S}_t : \text{state of the system at time } t$$
$$(\mathcal{S}_t \text{ is the } state \ space \text{ at stage } t),$$
$$x_t \in \mathcal{A}_t(I_t) : \text{action or decision to be selected at time } t$$
$$(\mathcal{A}_t(I_t) \text{ is the } action \ space \text{ at stage } t \text{ and state } I_t),$$
$$D_t : \text{discrete random variable over the sample space } \mathcal{D}_t,$$
$$T : \text{number of time periods}.$$

---

* Corresponding author.

H.L. Bodlaender and G.F. Italiano (Eds.): ESA 2013, LNCS 8125, pp. 577–588, 2013.

The cost function $g_t(I_t, x_t, D_t)$ gives the cost of performing action x_t from state I_t at time t for each possible realization of the random variable D_t. The random variables are assumed independent but not necessarily identically distributed. The total incurred cost is $\sum_{t=1}^{T} g_t(I_t, x_t, D_t) + g_{T+1}(I_{T+1})$, where g_{T+1} is the terminal cost function. The problem is that of choosing a sequence of actions $x_1, \ldots, x_T$ that minimizes the expectation of the total incurred cost. This problem is called a *stochastic dynamic program*. Formally, we want to determine:

$$z^*(I_1) = \min_{x_1,\ldots,x_T} E\left[g_1(I_1, x_1, D_1) + \sum_{t=2}^{T} g_t(f(I_{t-1}, x_{t-1}, D_{t-1}), x_t, D_t) + g_{T+1}(I_{T+1}) \right],$$

where I_1 is the initial state of the system and the expectation is taken with respect to the joint probability distribution of the random variables D_t.

Theorem 1.1. *[2] For every initial state I_1, the optimal value $z^*(I_1)$ of the DP is given by $z_1(I_1)$, where z_1 is the function defined by the recursion:*

$$z_t(I_t) = \begin{cases} g_{T+1}(I_{T+1}) & \text{if } t = T+1 \\ \min_{x_t \in \mathcal{A}_t(I_t)} E_{D_t}\left[g_t(I_t, x_t, D_t) + z_{t+1}(f(I_t, x_t, D_t)) \right] & \text{if } t = 1, \ldots, T. \end{cases}$$

Assuming that $|\mathcal{A}_t(I_t)| = |\mathcal{A}|$ and $|\mathcal{S}_t| = |\mathcal{S}|$ for every t and $I_t \in \mathcal{S}_t$, this gives a pseudopolynomial algorithm that runs in time $O(T|\mathcal{A}||\mathcal{S}|)$.

Halman et al. [3] give an FPTAS for three classes of problems that fall into this framework. This FPTAS is not problem-specific, but relies solely on structural properties of the DP. The three classes of [3] are: convex DP, nondecreasing DP, nonincreasing DP. In this paper we propose a modification of the FPTAS for the convex DP case that achieves better running time. Several applications of convex DPs are discussed in [4]. Two examples are:

1. **Stochastic single-item inventory control** [5]: we want to find replenishment quantities for a warehouse in each time period to minimize the expected procurement and holding/backlogging costs. This is a classic problem in supply chain management.
2. **Cash Management** [6]: we want to manage the cash flow of a mutual fund. At the beginning of each time period we can buy or sell stocks, thereby changing the cash balance. At the end of each time period, the net value of deposits/withdrawals is revealed. If the cash balance of the mutual fund is positive, we incur some opportunity cost because the money could have been invested somewhere. If the cash balance is negative, we must borrow money from the bank at some cost.

Assuming convex cost functions, these problems fall under the Convex DP case.

Our modification of the FPTAS is designed to improve the theoretical worst-case running time while making the algorithm a computationally useful tool. We show that our algorithm, unlike the original FPTAS of [3], has excellent performance on randomly generated problem instances of the two applications described above: it is faster than an exact algorithm even on small instances where no large numbers are involved and for low approximation factors (0.1%), becoming orders of magnitude faster on larger instances. To the best of our knowledge, this is the first time that a framework for the automatic generation of FPTASes is shown to be a practically as well as theoretically useful tool. The only computational evaluation with positive results of an FPTAS we could find in the literature is [7], which tackles a specific problem (multiobjective 0-1 knapsack), whereas our algorithm addresses a whole *class* of problems, without explicit knowledge of problem-specific features. We believe that this paper is a first step in showing that FPTASes do not have to be looked at as a purely theoretical tool. Another novelty of our approach is that the algorithm design allows bounding the errors introduced by floating point computations, so that we can guarantee the specified approximation factor with respect to the optimal infinite-precision solution under reasonable assumptions.

The rest of this paper is organized as follows. In Sect. 2 we introduce our notation and the original FPTAS of [3]. In Sect. 3, we improve the worst-case running time of the algorithm. In Sect. 4, we discuss our implementation of the FPTAS. Sect. 5 contains an experimental evaluation of the proposed method.

An important difference between our approach and traditional Approximate DP (ADP) methods (see e.g. [1]) is that we provide an approximation guarantee. For this reason, a fair comparison with ADP is difficult and is not undertaken in this work, but a brief discussion is presented in Sect. 5. Note that at this stage, our comparison with ADP methods is only preliminary.

2   Preliminaries and Algorithm Description

Let $\mathbb{N}, \mathbb{Z}, \mathbb{Q}, \mathbb{R}$ be the sets of nonnegative integers, integers, rational numbers and real numbers respectively. For $\ell, u \in \mathbb{Z}$, we call any set of the form $\{\ell, \ell+1, \dots, u\}$ a *contiguous* interval. We denote a contiguous interval by $[\ell, \dots, u]$, whereas $[\ell, u]$ denotes a real interval. Given $D \subset \mathbb{R}$ and $\varphi : D \to \mathbb{R}$ such that φ is not identically zero, we denote $D^{\min} := \min\{x \in D\}$, $D^{\max} := \max\{x \in D\}$, $\varphi^{\min} := \min_{x \in D}\{|\varphi(x)| : \varphi(x) \neq 0\}$, and $\varphi^{\max} := \max_{x \in D}\{|\varphi(x)|\}$. Given a finite set $D \subset \mathbb{R}$ and $x \in [D^{\min}, D^{\max}]$, for $x < D^{\max}$ let $\text{next}(x, D) := \min\{y \in D : y > x\}$, and for $x > D^{\min}$ let $\text{prev}(x, D) := \max\{y \in D : y < x\}$. Given a function defined over a finite set $\varphi : D \to \mathbb{R}$, we define $\sigma_\varphi(x) := \frac{\varphi(\text{next}(x,D)) - \varphi(x)}{\text{next}(x,D) - x}$ as the slope of φ at x for any $x \in D \setminus \{D^{\max}\}$, $\sigma_\varphi(D^{\max}) := \sigma_\varphi(\text{prev}(D^{\max}, D))$. Let $\mathcal{S}_{T+1}$ and $\mathcal{S}_t$ be contiguous intervals for $t = 1, \dots, T$. For $t = 1, \dots, T$ and $I_t \in \mathcal{S}_t$, let $\mathcal{A}_t$ and $\mathcal{A}_t(I_t) \subseteq \mathcal{A}_t$ be contiguous intervals. For $t = 1, \dots, T$ let $\mathcal{D}_t \subset \mathbb{Q}$ be a finite set, let $g_t : \mathcal{S}_t \times \mathcal{A}_t \times \mathcal{D}_t \to \mathbb{N}$ and $f_t : \mathcal{S}_t \times \mathcal{A}_t \times \mathcal{D}_t \to \mathcal{S}_{t+1}$. Finally, let $g_{T+1} : \mathcal{S}_{T+1} \to \mathbb{N}$.

In this paper we deal with a class of problems labeled "convex DP" for which an FPTAS is given by [3]. [3] additionally defines two classes of monotone DPs, but in this paper we address the convex DP case only. The definition of a convex DP requires the notion of an integrally convex set, see [8].

Definition 2.1. *[3] A DP is a* Convex DP *if: The terminal state space $\mathcal{S}_{T+1}$ is a contiguous interval. For all $t = 1, \dots, T + 1$ and $I_t \in \mathcal{S}_t$, the state space*

S_t and the action space $\mathcal{A}_t(I_t)$ are contiguous intervals. g_{T+1} is an integer-valued convex function over S_{T+1}. For every $t = 1, \ldots, T$, the set $S_t \otimes \mathcal{A}_t$ is integrally convex, function g_t can be expressed as $g_t(I, x, d) = g_t^I(I, d) + g_t^x(x, d) + u_t(f_t(I, x, d))$, and function f_t can be expressed as $f_t(I, x, d) = a(d)I + b(d)x + c(d)$ where $g_t^I(\cdot, d), g_t^x(\cdot, d), u_t(\cdot)$ are univariate integer-valued convex functions, $a(d) \in \mathbb{Z}, b(d) \in \{-1, 0, 1\}$, and $c(d) \in \mathbb{Z}$.

Let $U_S := \max_{t=1,\ldots,T+1} |S_t|$, $U_{\mathcal{A}} := \max_{t=1,\ldots,T} |\mathcal{A}_t|$ and $U_g := \frac{\max_{t=1,\ldots,T+1} g_t^{\max}}{\min_{t=1,\ldots,T+1} g_t^{\min}}$. Given $\varphi : D \to \mathbb{R}$, let $\sigma_\varphi^{\max} := \max_{x \in D} \{|\sigma_\varphi(x)|\}$ and $\sigma_\varphi^{\min} := \min_{x \in D} \{|\sigma_\varphi(x)| : |\sigma_\varphi(x)| > 0\}$. For $t = 1, \ldots, T$, we define $\sigma_{g_t}^{\max} := \max_{x_t \in \mathcal{A}_t, d_t \in D_t} \sigma_{g_t(\cdot, x_t, d_t)}^{\max}$ and $\sigma_{g_t}^{\min} := \min_{x_t \in \mathcal{A}_t, d_t \in D_t} \sigma_{g_t(\cdot, x_t, d_t)}^{\min}$. Let $U_\sigma := \frac{\max_{t=1,\ldots,T+1} \sigma_{g_t}^{\max}}{\min_{t=1,\ldots,T+1} \sigma_{g_t}^{\min}}$. We require that $\log U_S$, $\log U_{\mathcal{A}}$ and $\log U_g$ are polynomially bounded in the input size. This implies that $\log U_\sigma$ is polynomially bounded.

Under these conditions, it is shown in [3] that a Convex DP admits an FPTAS, using a framework that we review later in this section. Even though these definitions may look burdensome, the conditions cannot be relaxed. In particular, [3] shows that a Convex DP where $b(d) \notin \{-1, 0, 1\}$ in Def. 2.1 does not admit an FPTAS unless P = NP.

The input data of a DP problem consists of the number of time periods T, the initial state I_1, an oracle that evaluates g_{T+1}, oracles that evaluate the functions g_t and f_t for each time period $t = 1, \ldots, T$, and the discrete random variable D_t. For each D_t we are given n_t, the number of different values it admits with positive probability, and its support $\mathcal{D}_t = \{d_{t,1}, \ldots, d_{t,n_t}\}$, where $d_{t,i} < d_{t,j}$ for $i < j$. Moreover, we are given positive integers $q_{t,1}, \ldots, q_{t,n_t}$ such that $P[D_t = d_{t,i}] = \frac{q_{t,i}}{\sum_{j=1}^{n_t} q_{t,j}}$ (see [9] for an extension). For every $t = 1, \ldots, T$ and $i = 1, \ldots, n_t$, we define the following values:

$$p_{t,i} := P[D_t = d_{t,i}] : \text{probability that } D_t \text{ takes value } d_{t,i},$$
$$n^* := \max_t n_t : \text{maximum number of different values that } D_t \text{ can take.}$$

For any function $\varphi : D \to \mathbb{R}$, t_φ denotes an upper bound on the time needed to evaluate φ.

The basic idea underlying the FPTAS of Halman et al. is to approximate the functions involved in the DP by keeping only a logarithmic number of points in their domain. We then use a step function or linear interpolation to determine the function value at points that have been eliminated from the domain.

Definition 2.2. *[10] Let $K \geq 1$ and let $\varphi : D \to \mathbb{R}^+$, where $D \subset \mathbb{R}$ is a finite set. We say that $\tilde{\varphi} : D \to \mathbb{R}^+$ is a K-approximation function of φ (or more briefly, a K-approximation of φ) if for all $x \in D$ we have $\varphi(x) \leq \tilde{\varphi}(x) \leq K\varphi(x)$.*

Definition 2.3. *[10] Let $K \geq 1$ and let $\varphi : D \to \mathbb{R}^+$, where $D \subset \mathbb{R}$ is a finite set, be a unimodal function. We say that $W \subseteq D$ is a K-approximation set of φ if the following three properties are satisfied: (i) $D^{\min}, D^{\max} \in W$. (ii) For every $x \in W \setminus \{D^{\max}\}$, either $next(x, W) = next(x, D)$ or $\max\{\varphi(x), \varphi(next(x, W))\} \leq K \min\{\varphi(x), \varphi(next(x, W))\}$. (iii) For every $x \in D \setminus W$, we have $\varphi(x) \leq \max\{\varphi(prev(x, W)), \varphi(next(x, W))\} \leq K\varphi(x)$.*

An algorithm to construct K-approximations of functions with special structure (namely, convex or monotone) in polylogarithmic time was first introduced in [5]. In this paper we only deal with the convex case, therefore when presenting results from [3,10] we try to avoid the complications of the two monotone cases. In the rest of this paper it is assumed that the conditions of Def. 2.1 are met.

Definition 2.4. *[10] Let $\varphi : D \to \mathbb{R}$. $\forall E \subseteq D$, the convex extension of φ induced by E is the function $\hat{\varphi}$ defined as the lower envelope of the convex hull of $\{(x, \varphi(x)) : x \in E\}$.*

The main building block of the FPTAS is the routine APXSET (see [10]), which computes a K-approximation set of a function φ over the domain D. The idea is to only keep points in D such that the function value "jumps" by less than a factor K between adjacent points. For brevity, in the rest of this paper the algorithms to compute K-approximation sets are presented for convex nondecreasing functions. They can all be extended to general convex functions by applying the algorithm to the right and to the left of the minimum, which can be found in $O(\log |D|)$ time by binary search. Hence, theorems are presented for the general case.

Theorem 2.5. *[10] Let $\varphi : D \to \mathbb{R}^+$ be a convex function over a finite domain of real numbers, and let $x^* = \arg\min_{x \in D}\{\varphi(x)\}$. Then for every $K > 1$ the following holds. (i) APXSET computes a K-approximation set W of cardinality $O(1 + \log_K \frac{\varphi^{\max}}{\varphi^{\min}})$ in $O(t_\varphi(1 + \log_K \frac{\varphi^{\max}}{\varphi^{\min}}) \log |D|)$ time. (ii) The convex extension $\hat{\varphi}$ of φ induced by W is a convex K-approximation of φ whose value at any point in D can be determined in $O(\log |W|)$ time for any $x \in [D^{\min}, D^{\max}]$ if W is stored in a sorted array $(x, \varphi(x)), x \in W$. (iii) $\hat{\varphi}$ is minimized at x^*.*

Proposition 2.6. *[10] Let $1 \le K_1 \le K_2, 1 \le t \le T, I_t \in \mathcal{S}_t$ be fixed. Let $\hat{g}_t(I_t, \cdot, d_{t,i})$ be a convex K_1-approximation of $g_t(I_t, \cdot, d_{t,i})$ for every $i = 1, \ldots, n_t$. Let $\hat{z}_{t+1}$ be a convex K_2-approximation of z_{t+1}, and let:*

$$\hat{G}_t(I_t, \cdot) := E_{D_t}\left[\hat{g}_t(I_t, \cdot, D_t)\right], \quad \hat{Z}_{t+1}(I_t, \cdot) := E_{D_t}\left[\hat{z}_{t+1}(f_t(I_t, \cdot, D_t))\right].$$

Then

$$\bar{z}_t(I_t) := \min_{x_t \in \mathcal{A}(I_t)}\left\{\hat{G}_t(I_t, x_t) + \hat{Z}_{t+1}(I_t, x_t)\right\} \tag{1}$$

is a K_2-approximation of z_t that can be computed in $O(\log(|A_t|)n_t(t_{\hat{g}} + t_{\hat{z}_t} + t_{f_t}))$ time for each value of I_t.

We now have all the necessary ingredients to describe the FPTAS for convex DPs given in [10]. The algorithm is given in Algorithm 1. It is shown in [10] that $\bar{z}_t, \hat{z}_t$ are convex for every t.

Theorem 2.7. *[10] Given $0 < \epsilon < 1$, for every initial state I_1, $\hat{z}_1(I_1)$ is a $(1+\epsilon)$-approximation of the optimal cost $z^*(I_1)$. Moreover, Algorithm 1 runs in $O((t_g + t_f + \log(\frac{T}{\epsilon}\log(TU_g)))\frac{n^*T^2}{\epsilon}\log(TU_g)\log U_{\mathcal{S}}\log U_{\mathcal{A}})$ time, which is polynomial in $1/\epsilon$ and the input size.*

Algorithm 1. FPTAS for convex DP.

1: $K \leftarrow 1 + \frac{\epsilon}{2(T+1)}$
2: $W_{T+1} \leftarrow \text{ApxSet}(g_{T+1}, \mathcal{S}_{T+1}, K)$
3: Let $\hat{z}_{T+1}$ be the convex extension of g_{T+1} induced by W_{T+1}
4: **for** $t = T, \ldots, 1$ **do**
5:     Define $\bar{z}_t$ as in (1) with $\hat{g}_t$ set equal to g_t
6:     $W_t \leftarrow \text{ApxSet}(\bar{z}_t, \mathcal{S}_t, K)$
7:     Let $\hat{z}_t$ be the convex extension of $\bar{z}_t$ induced by W_t
8: **return** $\hat{z}_1(I_1)$

Algorithm 2. $\text{ApxSetSlope}(\varphi, D, K)$

1: $x \leftarrow D^{\min}$
2: $W \leftarrow \{D^{\min}\}$
3: **while** $x < D^{\max}$ **do**
4:     $x \leftarrow \max\{\text{next}(x, D), \max\{y \in D : (y \leq D^{\max}) \wedge (\varphi(y) \leq K(\varphi(x) + \sigma_\varphi(x)(y - x))\}\}$
5:     $W \leftarrow W \cup \{x\}$
6: **return** W

---

3 Improved Running Time

In this section we show that for the convex DP case, we can improve the running time given in Thm. 2.7.

In the framework of [10], monotone functions are approximated by a step function, and Def. 2.3 guarantees the K-approximation property for this case. However, ApxSet greatly overestimates the error committed by the convex extension induced by W. For the convex DP case we propose the simpler Def. 3.1 of K-approximation set, that preserves correctness of the FPTAS and the analysis carried out in [10].

Definition 3.1. *Let $K \geq 1$ and let $\varphi : D \to \mathbb{R}^+$, where $D \subset \mathbb{R}$ is a finite set, be a convex function. Let $W \subseteq D$ and let $\hat{\varphi}$ be the convex extension of φ induced by W. We say that W is a K-approximation set of φ if: (i) $D^{\min}, D^{\max} \in W$; (ii) For every $x \in D$, $\hat{\varphi}(x) \leq K\varphi(x)$.*

Note that a K-approximation set according to the new definition is not necessarily such under the original Def. 2.3 as given in [10]. E.g.: $D = \{0, 1, 2\}, \varphi(0) = 0, \varphi(1) = 1, \varphi(2) = 2K$; the only K-approximation set according to the original definition is D itself, whereas $\{0, 2\}$ is also a K-approximation set in the sense of Def. 3.1. An algorithm to compute a K-approximation set in the sense of Def. 3.1 is given in Algorithm 2 (for nondecreasing functions).

Theorem 3.2. *Let $\varphi : D \to \mathbb{N}^+$ be a convex function over a finite domain of integers. Then for every $K > 1$, $\text{ApxSetSlope}(\varphi, D, K)$ computes*

a K-approximation   set W of   size $O(\log_K \min\{\frac{\sigma_\varphi^{\max}}{\sigma_\varphi^{\min}}, \frac{\varphi^{\max}}{\varphi^{\min}}\})$ in $O(t_\varphi \log_K \min\{\frac{\sigma_\varphi^{\max}}{\sigma_\varphi^{\min}}, \frac{\varphi^{\max}}{\varphi^{\min}}\} \log|D|)$ *time.*

We can improve on Thm. 2.7 by replacing each call to ApxSet with a call to ApxSetSlope in Algorithm 1.

Theorem 3.3. *Given* $0 < \epsilon < 1$, *for every initial state* I_1, $\hat{z}_1(I_1)$ *is a* $(1 + \epsilon)$-*approximation of the optimal cost* $z^*(I_1)$. *Moreover, Algorithm 1 runs in* $O((t_g + t_f + \log(\frac{T}{\epsilon} \log \min\{U_\sigma, TU_g\}))\frac{n^* T^2}{\epsilon} \log \min\{U_\sigma, TU_g\} \log U_S \log U_A)$ *time, which is polynomial in both* $1/\epsilon$ *and the (binary) input size.*

4   From Theory to Practice

In this section we introduce an algorithm that computes smaller K-approximation sets than ApxSetSlope in practice, although we do not improve over Thm. 3.2 from a theoretical standpoint, and the analysis is more complex. Given two points $(x, y), (x', y') \in \mathbb{R}^2$ we denote by $\textsc{Line}((x, y), (x', y'), \cdot) : \mathbb{R} \to \mathbb{R}$ the straight line through them. We first discuss how to exploit convexity of φ to compute a bound on the approximation error $\frac{\textsc{Line}((x, \varphi(x)), (x', \varphi(x')), w)}{\varphi(w)}$ $\forall w \in [x, \ldots, x']$.

Proposition 4.1. *Let* $\varphi : [\ell, u] \to \mathbb{R}^+$ *be a nondecreasing convex function. Let* $h \geq 3$, $E = \{x_i : x_i \in [\ell, u], i = 1, \ldots, h\}$ *with* $\ell = x_1 < x_2 < \cdots < x_{h-1} < x_h = u$, *let* $y_i := \varphi(x_i) \forall i$, $(x_0, y_0) := (x_1 - 1, y_1)$ *and* $(x_{h+1}, y_{h+1}) := (x_h + 1, 2y_h - f(x_h - 1))$. *For all* $i = 1, \ldots, h - 1$, *define* $LB_i(x)$ *as:*

$$LB_i(x) := \max\{\textsc{Line}((x_{i-1}, y_{i-1}), (x_i, y_i), x), \textsc{Line}((x_{i+1}, y_{i+1}), (x_{i+2}, y_{i+2}), x)\}$$

for $x \in [x_i, x_{i+1}]$, *and* $LB_i(x) := 0$ *otherwise. Define* $\underline{\varphi}(x) := \sum_{i=1}^{h-1} LB_i(x)$. *Observe that* $LB_i(x)$ *is the maximum of two linear functions, so it has at most one breakpoint over the interval* (x_i, x_{i+1}). *Let* B *be the set of these breakpoints. For* $1 \leq j < k \leq h$ *let*

$$\gamma_E(x_j, x_k) := \max_{x_e \in [x_j, x_k] \cap (E \cup B)} \left\{\frac{\textsc{Line}((x_j, y_j), (x_k, y_k), x_e)}{\underline{\varphi}(x_e)}\right\}. \quad (2)$$

Then $\left|\frac{\textsc{Line}((x_j, y_j), (x_k, y_k), w)}{\varphi(w)}\right| \leq \gamma_E(x_j, x_k) \leq \frac{y_k}{y_j}$ $\forall w \in [x_j, x_k].$

The set B of Prop. 4.1 allows the computation of a bound $\gamma_E(x_j, x_k)$ on the error committed by approximating φ with a linear function between x_j, x_k. We use this bound in ApxSetConvex, see Algorithm 3 (for nondecreasing functions). In the description of ApxSetConvex, $\Lambda > 1$ is a given constant. We used $\Lambda = 2$ in our experiments. The running time of ApxSetConvex can be a factor $\log|D|$ slower than ApxSetSlope, but its practical performance is superior for two reasons: it produces smaller approximation sets, and evaluates $\bar{z}$ fewer times (each evaluation of $\bar{z}$ is expensive, see below). We experimented with applying Prop. 4.1 in conjunction with ApxSetSlope, but this did not improve the algorithm's performance, therefore we omit the discussion.

Algorithm 3. APXSETCONVEX(φ, D, K).

1: $W \leftarrow \{D^{\min}, D^{\max}\}$
2: $x \leftarrow D^{\max}$
3: $E \leftarrow \{D^{\min}, \lfloor (D^{\min} + D^{\max})/2 \rfloor, D^{\max}\}$
4: **while** $x > D^{\min}$ **do**
5:     $\ell \leftarrow D^{\min}, r \leftarrow x, \text{counter} \leftarrow 0, z \leftarrow x$
6:     **while** $r > \text{next}(\ell, D)$ **do**
7:         $w \leftarrow \min\{y \in E : (y > D^{\min}) \wedge (\gamma_E(y, x) \leq K)\}$
8:         $\ell \leftarrow \text{prev}(w, E), r \leftarrow w, \text{counter}++$
9:         $E \leftarrow E \cup \{\lfloor (\ell + r)/2 \rfloor\} \cap [D^{\min}, \max\{\text{next}(r, E), \arg \gamma_E(r, x)\}]$
10:        **if** $\text{counter} > \Lambda \log(|D|)$ **then**
11:            **if** $\varphi(z) > K\varphi(r)$ **then**
12:                $\text{counter} \leftarrow 0, z \leftarrow r$
13:            **else**
14:                $\ell \leftarrow \text{prev}(r, D)$
15:        $x \leftarrow \min\{\text{prev}(x, D), r\}$
16:    $W \leftarrow W \cup \{x\}, E \leftarrow E \cap [D^{\min}, x]$
17: **return** W

Theorem 4.2. *Let $\varphi : D \to \mathbb{N}^+$ be a convex function defined over a set of integers. Then* APXSETCONVEX(φ, D, K) *computes a K-approximation set of φ of size $O(\log_K \min\{\frac{\sigma_\varphi^{\max}}{\sigma_\varphi^{\min}}, \frac{\varphi^{\max}}{\varphi^{\min}}\})$ in time $O(t_\varphi \log_K \min\{\frac{\sigma_\varphi^{\max}}{\sigma_\varphi^{\min}}, \frac{\varphi^{\max}}{\varphi^{\min}}\} \log^2 |D|)$.*

We now briefly discuss our implementation of the FPTAS, omitting details for space reasons. By (1), each evaluation of $\bar{z}_t$ requires the minimization of a convex function. Hence, evaluating $\bar{z}_t$ is expensive. We briefly discuss our approach to perform the computation of a K-approximation set of $\bar{z}$ efficiently, which is crucial because such computation is carried out T times in the FPTAS. First, we use a dictionary to store function values of $\bar{z}_t$ to make sure that the minimization in (1) is performed at most once for each state I_t. Second, we use golden section search [11] instead of binary search to minimize convex functions, in all cases where a function evaluation requires more than $O(1)$ time.

Our implementation uses floating point arithmetics for the sake of speed. Most modern platforms provide both floating point (fixed precision) arithmetics and rational (arbitrary precision) arithmetics. The latter is not implemented in hardware and is considerably slower, but does not incur into numerical errors and can handle arbitrarily large numbers. In particular, only by using arbitrary precision one can guarantee to find the optimal solution of a problem instance. Our implementation guarantees that the final result satisfies the desired approximation guarantee, by bounding the error introduced by the floating point computations, and rounding key calculations appropriately.

Finally, at each stage of the dynamic program, we adaptively compute an approximation factor K_t that guarantees the approximation factor $(1 + \epsilon)$ for the value function at stage 1 taking into account the errors in the floating point

---

Algorithm 4. Implementation of the FPTAS for convex DP.

1: $K \leftarrow (1 + \epsilon)^{\frac{1}{T+1}}$
2: $W_{T+1} \leftarrow \textsc{ApxSetConvex}(g_{T+1}, \mathcal{S}_{T+1}, K)$
3: Let K'_{T+1} be the maximum value of γ_E recorded by $\textsc{ApxSetConvex}$
4: Let $\hat{z}_{T+1}$ be the convex extension of g_{T+1} intluced by W_{T+1}
5: **for** $t = T, \ldots, 1$ **do**
6:    Define $\bar{z}_t$ as in (1) with $\hat{g}_t$ set equal to g_t
7:    $K_t \leftarrow \frac{K^{T+2-t}}{\prod_{j=t+1}^{T+1} K'_j}.$
8:    $W_t \leftarrow \textsc{ApxSetConvex}(\bar{z}_t, \mathcal{S}_t, K_t)$
9:    Let K'_t be the maximum value of γ_E recorded by $\textsc{ApxSetConvex}$
10:   Let $\hat{z}_t$ be the convex extension of $\bar{z}_t$ induced by W_t
11: **return** $\hat{z}_1(I_1)$

---

computations. The choice of K_t is based on the actual maximum approximation factor recorded at all stages $\tau > t$. We summarize the FPTAS for convex DPs in Algorithm 4.

5 Computational Experiments

We implemented the FPTAS in Python 3.2. Better performance could be obtained using a faster language such as C. However, in this context we are interested in comparing different approaches, and because all the tested algorithms are implemented in Python, the comparison is fair. The tests were run on Linux on an Intel Xeon E5-4620@2.20 Ghz (HyperThreading and TurboBoost disabled).

5.1 Generation of Random Instances

We now give an overview of how the problem instances used in the computational testing phase are generated. We consider two types of problems: stochastic single-item inventory control problems, and cash management problems. For each instance we require 4 parameters: the number of time periods T, the state space size parameter M, the size of the support of the random variable N, the degree of the cost functions d. The state space size parameter determines the maximum demand in each period for single-item inventory control instances, and the maximum difference between cash deposit and cash withdrawal for cash management instances. The actual values of these quantities for each instance are determined randomly, but we ensure that they are beetween $M/2$ and M, so that the difficulty of the instance scales with M. Each instance requires the generation of some costs: procurement, storage and backlogging costs for inventory control instances; opportunity, borrowing and transaction costs for cash management instances. We use polynomial functions $c_t x^d$ to determine these costs, where c_t is a coefficient that is drawn uniformly at random in a suitable set of

Fig. 1. Average CPU time on INVENTORY$(T, M, N, 1)$ instances. On the x-axis we identify the group of instances with the label M, T, N. The y-axis is on a logarithmic scale. The label of the approximation algorithms indicates the value of ϵ and the subroutine used to compute the K-approximation sets, namely: "Apx" uses APXSET, "Slope" uses APXSETSLOPE, "Convex" uses APXSETCONVEX.

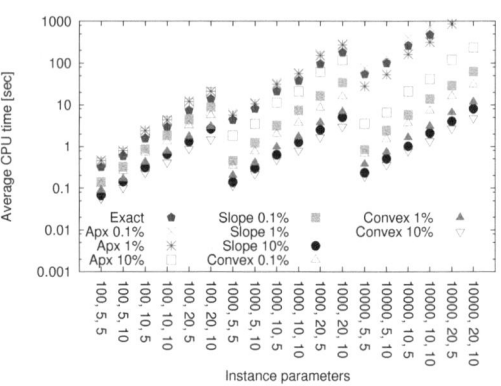

values for each stage, and $d \in \{1, 2, 3\}$. The difficulty of the instances increases with d, as the numbers involved and U_σ grow.

The random instances are labeled INVENTORY(T, M, N, d) and CASH(T, M, N, d) to indicate the values of the parameters. In the future we plan to experiment on real-world data. At this stage, we limit ourselves to random instances with "reasonable" numbers.

5.2 Analysis of the Results

We generated 50 random INVENTORY and CASH instances for each possible combination of the following values: $T \in \{5, 10, 20\}, M \in \{100, 1000, 10000\}$, $N \in \{5, 10\}$. We applied to each instance the following algorithms: an exact DP algorithm (label EXACT), and the FPTAS for $\epsilon \in \{0.001, 0.01, 0.1\}$ as described in Alg. 4 using one of the subroutines APXSET, APXSETSLOPE, APXSETCONVEX. This allows us to verify whether or not the modifications proposed in this paper are beneficial. We remark that our implementation of EXACT runs in $O(T|\mathcal{S}| \log |\mathcal{A}|)$ time exploiting convexity.

For each group of instances generated with the same parameters, we look at the sample mean of the running time for each algorithm. Because the sample standard deviation is typically low, comparing average values is meaningful. When the sample means are very close, we rely on additional tests – see below.

The maximum CPU time for each instance was set to 1000 seconds. Part of the results are reported in Fig. 1, the rest are omitted for space reasons.

We summarize the results. The FPTAS with APXSET is typically slower than EXACT, whereas APXSETSLOPE and APXSETCONVEX are always faster than EXACT. Surprisingly, they are faster than EXACT by more than a factor 2 even on the smallest instances in our test set: $T = 5, M = 100, N = 5, d = 1$ (on INVENTORY, 0.32 seconds for EXACT vs. 0.12 seconds for APXSETCONVEX with $\epsilon = 0.001$). As the size of the instances and the numbers involved grows, the difference increases in favor of the FPTAS. For INVENTORY instances with $d = 1$ the FPTAS with APXSETCONVEX and $\epsilon = 1\%$ can be faster than EXACT by more than 2 orders of magnitude (482.8 seconds for EXACT, 3.0 seconds for

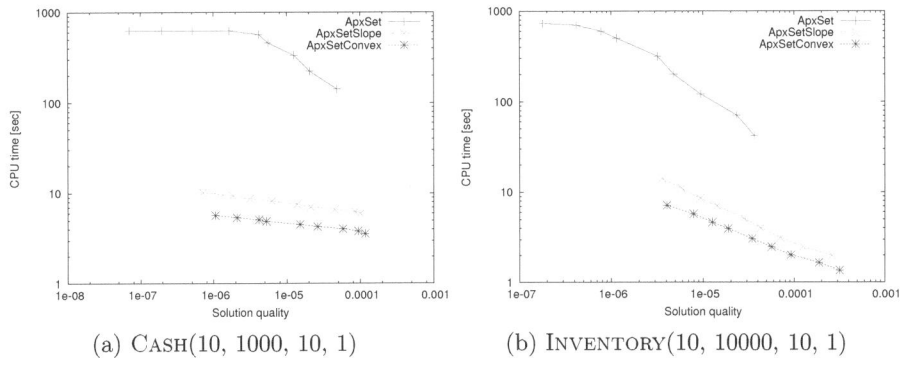

(a) CASH(10, 1000, 10, 1)   (b) INVENTORY(10, 10000, 10, 1)

Fig. 2. Average solution quality (computed as $f^{\mathrm{apx}}/f^* - 1$, where f^{apx} is the cost of the approximated policy, and f^* the cost of the optimal policy) vs. average CPU time.

APXSETCONVEX with $\epsilon = 1\%$). For the largest problems in our test set, EXACT does not return a solution before the time limit whereas APXSETSLOPE ($\epsilon = 0.1$) and APXSETCONVEX ($\epsilon \in \{0.01, 0.1\}$) terminate on *all* instances. The results suggest that the improvements of the FPTAS can be over 3 orders of magnitude and increase with the problem size. The CASH problem seems to heavily favor the FPTAS over EXACT, with consistent speed-ups of two (resp. three) orders of magnitude on medium (resp. large) instances. On the largest CASH instance that can be solved by EXACT with $d = 1$ ($T = 5, M = 10000, N = 5$), APXSETCON-VEX takes 0.9 seconds, compared to 717.6 for EXACT. The difference increases in favor of the FPTAS for larger d. This may be due to our random instance generator, and has to be investigated further.

APXSETCONVEX is faster than APXSETSLOPE despite its overhead per iteration on all groups of instances except two (out of 108). This was verified with a Wilcoxon rank-sum test at the 95% significance level. For two groups of INVEN-TORY instances with $d = 2$, APXSETCONVEX and APXSETSLOPE yield comparable CPU time. For $d = 1, 3$ APXSETCONVEX is clearly the fastest algorithm, whereas for INVENTORY and $d = 2$ APXSETSLOPE is occasionally comparable with APXSETCONVEX, see below the analysis on solution quality. At this stage we do not have a clear explanation for this behavior.

We analyze the quality of the approximated solutions using the three possible APXSET routines. We run the FPTAS with 9 different values of K (equally spaced on a log scale between 10^{-3} and 10^{-1}) on INVENTORY and CASH instances with $T = 10, N = 10$, and varying M, d. We then compare on a graph the average cost of the policy (i.e. the sequence of actions) obtained with the different APXSET routines, and the respective average running times. This can only be done for instances on which EXACT finds a solution. The graphs for CASH with $M = 1000, d = 1$ and for INVENTORY with $M = 10000, d = 1$ are reported in Fig. 2. They show very clearly that APXSETCONVEX is faster than APXSETSLOPE for equal solution quality, and both algorithms are much faster than APXSET. Graphs for other instances with $d = 1, 3$ yield similar conclusions and do not contribute further insight. On INVENTORY problems with $d = 2$, in some cases the performance of APXSETSLOPE is comparable to APXSETCONVEX, as stated above. However, APXSETCONVEX gives a better approximation factor guarantee for equal CPU time, and seems therefore the best algorithm.

We conclude with a short comparison with the ADP approach of [6] for CASH problems. The instances and approach discussed in [6] are significantly different from ours, hence we cannot provide a fair comparison. However, to give a rough idea, we observe that [6] reports speed-ups of three orders of magnitude over an exact DP approach with an average approximation factor of 0.28% and no approximation guarantee, whereas our FPTAS achieves similar speed-ups on large instances guaranteeing an approximation factor of 0.1% and achieving an actual approximation factor $< 0.001\%$ in most cases.

Acknowledgements. Partial support was provided by: EU FP7/2007-2013 grant 247757 and Recanati Fund of Jerusalem School of Business Administration (N. Halman), SUTD grant SRES11012 and IDC grants IDSF1200108, IDG21300102 (G. Nannicini), ONR grant N000141110056 (J. Orlin).

References

1. Bertsekas, D.P.: Dynamic Programming and Optimal Control. Athena Scientific, Belmont (1995)
2. Bellman, R.: Dynamic Programming. Princeton University Press, Princeton (1957)
3. Halman, N., Klabjan, D., Li, C.L., Orlin, J., Simchi-Levi, D.: Fully polynomial time approximation schemes for stochastic dynamic programs. In: Teng, S.H. (ed.) SODA, pp. 700–709. SIAM (2008)
4. Nascimento, J.M.: Approximate dynamic programming for complex storage problems. PhD thesis, Princeton University (2008)
5. Halman, N., Klabjan, D., Mostagir, M., Orlin, J., Simchi-Levi, D.: A fully polynomial time approximation scheme for single-item inventory control with discrete demand. Mathematics of Operations Research 34(3), 674–685 (2009)
6. Nascimento, J., Powell, W.: Dynamic programming models and algorithms for the mutual fund cash balance problem. Management Science 56(5), 801–815 (2010)
7. Bazgan, C., Hugot, H., Vanderpooten, D.: Implementing an efficient FPTAS for the 0-1 multiobjective knapsack problem. European Journal of Operations Research 198(1), 47–56 (2009)
8. Murota, K.: Discrete Convex Analysis. SIAM, Philadelphia (2003)
9. Halman, N., Orlin, J.B., Simchi-Levi, D.: Approximating the nonlinear newsvendor and single-item stochastic lot-sizing problems when data is given by an oracle. Operations Research 60(2), 429–446 (2012)
10. Halman, N., Klabjan, D., Li, C.L., Orlin, J., Simchi-Levi, D.: Fully polynomial time approximation schemes for stochastic dynamic programs. Technical Report 3918, Optimization Online (June 2013)
11. Kiefer, J.: Sequential minimax search for a maximum. Proceedings of the American Mathematical Society 4(3), 502–506 (1953)

An Optimal Online Algorithm for Weighted Bipartite Matching and Extensions to Combinatorial Auctions

Thomas Kesselheim[1,*], Klaus Radke[2,**],
Andreas Tönnis[2,***], and Berthold Vöcking[2]

[1] Department of Computer Science, Cornell University, Ithaca, NY, USA
kesselheim@cs.cornell.edu
[2] Department of Computer Science, RWTH Aachen University, Germany
{radke,toennis,voecking}@cs.rwth-aachen.de

Abstract. We study online variants of weighted bipartite matching on graphs and hypergraphs. In our model for online matching, the vertices on the right-hand side of a bipartite graph are given in advance and the vertices on the left-hand side arrive online in random order. Whenever a vertex arrives, its adjacent edges with the corresponding weights are revealed and the online algorithm has to decide which of these edges should be included in the matching. The studied matching problems have applications, e.g., in online ad auctions and combinatorial auctions where the right-hand side vertices correspond to items and the left-hand side vertices to bidders.

Our main contribution is an optimal algorithm for the weighted matching problem on bipartite graphs. The algorithm is a natural generalization of the classical algorithm for the secretary problem achieving a competitive ratio of e ≈ 2.72 which matches the well-known upper and lower bound for the secretary problem. This shows that the classic algorithmic approach for the secretary problem can be extended from the simple selection of a best possible singleton to a rich combinatorial optimization problem.

On hypergraphs with $(d+1)$-uniform hyperedges, corresponding to combinatorial auctions with bundles of size d, we achieve competitive ratio $O(d)$ in comparison to the previously known ratios $O(d^2)$ and $O(d \log m)$, where m is the number of items. Additionally, we study variations of the hypergraph matching problem representing combinatorial auctions for items with bounded multiplicities or for bidders with submodular valuation functions. In particular for the case of submodular valuation functions we improve the competitive ratio from $O(\log m)$ to e.

* Supported by a fellowship within the Postdoc-Programme of the German Academic Exchange Service (DAAD).
** Supported by the Studienstiftung des deutschen Volkes.
*** Supported by the DFG GRK/1298 "AlgoSyn".

H.L. Bodlaender and G.F. Italiano (Eds.): ESA 2013, LNCS 8125, pp. 589–600, 2013.

1   Introduction

Consider the following natural generalization of the classical secretary problem: Suppose an administrator wants to hire people for a set of open positions (rather than only one secretary for a single position). The applicants are interviewed one at a time and in every interview the interviewer learns weights representing the degree of qualification of the current candidate for each of the possible positions. Now, immediately after an interview, the administrator has to either assign the applicant to one of the open positions or the candidate will leave the room and take a job at another company. The administrator is interested in maximizing the sum of the assigned weights.

The described problem corresponds to the weighted online bipartite matching problem. In general, the jobs correspond to the offline vertices on the right-hand side which are known in advance. The vertices on the left-hand side arrive online one by one, each with its incident edges and their respective weights. The decision whether and how to assign the current vertex has to be made online. Unfortunately, for general weights and when the vertices arrive in adversarial order, every algorithm can perform arbitrarily bad. To achieve any reasonable competitive ratio, it is necessary to make additional assumptions on the model. In this work, we assume that the vertices arrive in random order, analogously to the original *secretary problem*.

The weighted online matching problem in the secretary model was introduced by Korula and Pál [19]. It is a generalization of the matroid secretary problem on transversal matroids which was introduced by Babaioff et al. [3] and later improved by Dimitrov and Plaxton [9]. This, respectively, is a generalization of the classical secretary problem. Here, we present the first optimal algorithm for weighted online matching which also matches the lower bound for the secretary problem.

The online matching problem is closely related to combinatorial auctions. Let, e.g., the right-hand side of the graph represent the items and the left-hand side correspond to the bidders. Then the weighted online bipartite matching corresponds to an online combinatorial auction where every bidder can buy at most one item. Now, we extend the graph towards $(d+1)$-uniform hyperedges so that every edge contains exactly one bidder and d items. Thus, every hyperedge represents a bid on a bundle of items in a combinatorial auction. This setting was first analyzed by Korula and Pál [19] on whose results we improve. Additionally, we allow for multiplicities on the items, the right-hand side vertices of the hypergraph, which has applications in ad auctions. Furthermore, we consider submodular[1] weight functions which is a reasonable assumption for these economically motivated problems. Like Korula and Pál, we analyze our algorithms in terms of competitive ratio, i.e. the ratio of the value of an optimal offline solution to the expected weight achieved by the online algorithm.

---

[1] A set function $f: 2^{\Omega} \to \mathbb{R}$ is submodular if for every $X, Y \subseteq \Omega$ we have that $f(X \cup Y) + f(X \cap Y) \leq f(X) + f(Y)$.

Our Contribution: We provide algorithms for several variants of weighted bipartite matching in the secretary model, i.e. with random arrival order. All our algorithms are generalizations of the classical approach to the secretary problem. First, they gather information on the instance via sampling. Then, in every later step, they solve the known part of the instance optimally and treat the left-hand side vertex that has just arrived according to that locally optimal solution. The most important feature of our analysis is to interpret the random arrival order as a sequence of stochastically independent experiments.

For online bipartite matching we obtain an e-competitive algorithm. This improves on the 8-competitive algorithm by Korula and Pál [19] and matches the lower bound on the classical secretary problem, see e.g. Buchbinder et al. [5]. While Korula and Pál follow a similar approach, their analysis requires them to use a greedy approximation algorithm for the online decision making instead of locally optimal solutions.

When we apply the algorithmic approach to online bipartite hypermatching, we use randomized rounding on a fractional LP solution. Therefore, the obtained competitive ratios are with respect to the fractional offline optimum. When the online bidders are interested in sets of size at most d and every item is available at least b times, we obtain an expected competitive ratio of $O(d^{1/b})$. Thus, for classical combinatorial auctions this translates to a $O(d)$-approximation in contrast to the previously known $O(d^2)$ by Korula and Pál. For multiplicities $b \geq \log(d)$, a common assumption in ad auctions, the competitive ratio becomes a constant $O(1)$. For general valuations on sets of unbounded size, our randomized algorithm is $O(m^{1/(b+1)})$-competitive, where m is the number of items. Furthermore, if valuation functions are submodular, the competitive ratio of our algorithm is again e even for multiplicity one and thus optimal.

All these results are based on a random order of arrivals. Using this assumption, we beat the lower bound of $\Omega(b \cdot d^{1/b})$ for any deterministic set packing algorithm in the online adversary model by Azar and Regev [2]. We show that for $b = 1$, every randomized online algorithm in the secretary model, even with unlimited computational power, is $\Omega(\ln(d)/\ln\ln(d))$-competitive.

Related Work: When analyzing online bipartite matching, it is necessary to make additional assumptions on the model as no algorithm can handle adversarial arrival with general edge weights; see Aggarwal et al. [1] for a proof. A common choice is to assume a random order of the vertices on the left-hand side. Another option is to admit arbitrary order but to make restrictions on the edge weights. Some recent, loosely related papers adopt slight changes to the model and assume budgeted allocations with stochastic arrivals, see e.g. [7,12,22].

The random order model has its origins in the famous secretary problem, where n candidates for a job arrive online in random order and the goal is to pick the best one with maximal probability. This is identical to edge-weighted bipartite matching with only one vertex on the right-hand side. Although the problem was folklore, it was not published until 1960 by Gardner and it was solved many times. See Ferguson [11] for historical details. The optimal algorithm for the secretary problem is e-competitive in expectation.

A generalization of the classical secretary problem is the matroid secretary problem, introduced by Babaioff et al. [3]. Here, the elements of a matroid arrive online in random order and the objective is to select an independent set of maximum weight. For general matroids they gave an $O(\log(\rho))$-competitive algorithm, where ρ is the rank of the matroid. This result was later improved to $O(\sqrt{\log(\rho)})$ by Chakraborty and Lachish [6]. Various results are known for special kinds of matroids, see [3,13,14,18,19]. Note that transversal matroids are a special case of bipartite matching, where all edges incident to the same left-hand side vertex have identical weight. Babaioff et al. [3] presented a $4d$-competitive algorithm for the case of transversal matroids with bounded left degree d. This was improved by Dimitrov and Plaxton [9] who gave a 16-competitive algorithm for transversal matroids. The first result on general bipartite matching in the secretary model is by Korula and Pál [19] who presented an 8-competitive online algorithm.

The research on online bipartite matching with adversarial arrival was initiated by Karp et al. [17] who analyzed the unweighted case. They presented a randomized algorithm obtaining an expected competitive ratio of $e/(e-1)$ and a matching lower bound. The proof was later simplified by Goel and Mehta [12] and Birnbaum and Mathieu [4]. A primal-dual analysis was given by Devanur et al. [8]. Karande et al. [16] and independently Mahdian and Yan [21] showed that the lower bound of $e/(e-1)$ does not hold when the left-hand side vertices arrive in random, instead of arbitrary, order. Aggarwal et al. [1] were the first to analyze online matching with adversarial order in a general weighted setting. They obtained an expected competitive ratio of $e/(e-1)$ as long as all edges incident to the same vertex on the right-hand side have identical weight. Kalyanasundaram and Pruhs [15] presented a deterministic 3-competitive algorithm when the edge weights represent a metric space.

Bipartite hypermatching in the secretary model was first analyzed by Korula and Pál [19]. They obtained an expected competitive ratio of $O(d^2)$ when the hyperedges have bounded size $d + 1$. Krysta and Vöcking [20] investigated online combinatorial auctions with random arrival of the bidders and developed randomized mechanisms that are incentive compatible. For valuations on sets of bounded size d and when each of the m items is available b times, they showed an expected competitive ratio of $O(d^{1/b} \log(bm))$. In the case of general valuations, they obtained an expected competitive ratio of $O(m^{1/(b+1)} \log(bm))$. When the valuation functions are XOS and $b = 1$ the achieved competitive ratio is $O(\log(m))$. Feldmann et al. [10] provide constant competitive algorithms for different variants of the secretary problem using submodular weight functions. E.g., they consider a submodular secretary problem on partition matroids.

2 Edge-Weighted Bipartite Online Matching

In the *bipartite online matching* problem, we are initially given the set R of an edge-weighted bipartite graph $G = (L \cup R, E)$ and the cardinality $n := |L|$ of the set L. At every step, a new vertex $v \in L$ arrives together with the weights

$w(e) \in \mathbb{R}_{\geq 0}$ of its incident edges. Most importantly, the vertices in L are revealed online and in random order. The algorithm always has to either assign the current vertex to one of its unmatched neighbors in R, or decide to leave it unassigned.

Our algorithm is a generalization of the classical approach to the secretary problem. There, a constant fraction of the candidates is ignored. Then, when an online candidate arrives that is better than all previous ones, it is selected. We also start by sampling a constant fraction of the vertices on the left-hand side. Afterwards, whenever a new vertex is presented to the algorithm, we compute an optimum solution on the revealed part of the graph. If, in this local solution, the current vertex on the left-hand side is assigned to an unmatched vertex, we add this edge to our matching.

Algorithm 1. Bipartite online matching

Input : vertex set R and cardinality $n = |L|$
Output: matching M

Let L' be the first $\lfloor n/e \rfloor$ vertices of L;
$M := \emptyset$;
for each subsequent vertex $\ell \in L - L'$ **do**   // steps $\lceil n/e \rceil$ to n
$\quad$ $L' := L' \cup \ell$;
$\quad$ $M^{(\ell)} :=$ optimal matching on $G[L' \cup R]$;   // e.g. by Hungarian method
$\quad$ Let $e^{(\ell)} := (\ell, r)$ be the edge assigned to ℓ in $M^{(\ell)}$;
$\quad$ **if** $M \cup e^{(\ell)}$ is a matching **then**
$\quad\quad$ add $e^{(\ell)}$ to M;

For convenience of notation, we will number the vertices in L from 1 to n in the (random) order they are presented to the algorithm. Hence, we will use the variable ℓ synonymously as an integer, the name of an iteration and the name of the current vertex.

Lemma 1. *Let the random variable A_v denote the contribution of the vertex $v \in L$ to the output, i.e. the weight of the edge (v, r) assigned to v in M. And let OPT be the value of a maximum-weight matching in the full graph G. For the vertices $\ell \in \{\lceil n/e \rceil, \ldots, n\}$ we have,*

$$\mathbf{E}[A_\ell] \geq \frac{\lfloor n/e \rfloor}{\ell - 1} \cdot \frac{OPT}{n} .$$

Proof. First, we will show that the expected weight of $e^{(\ell)}$, i.e. of the edge assigned to vertex ℓ in the matching $M^{(\ell)}$, is a significant fraction of OPT. Then, we will analyze the probability of adding this edge to the matching M.

The proof relies on the fact that in any step k of the algorithm the choice of the random permutation up to this point can be modeled as a sequence of the following *independent* random experiments: First choose a set of size k from L. Then determine the order of these k vertices by iteratively selecting a vertex at random and removing it. We need this interpretation to exploit the randomness in each of these experiments separately.

Now in step ℓ we have $|L'| = \ell$ and the algorithm calculates an optimal matching $M^{(\ell)}$ on $G[L' \cup R]$. As explained, the current vertex ℓ can be seen as being selected uniformly at random from the set L'. Hence, the expected weight of the edge $e^{(\ell)}$ in $M^{(\ell)}$ is $w(M^{(\ell)})/\ell$. Also, since L' can be seen as being uniformly selected from L with size ℓ we know $\mathbf{E}\left[w\left(M^{(\ell)}\right)\right] \geq \ell/n \cdot OPT$. Together we have,

$$\mathbf{E}\left[w(e^{(\ell)})\right] \geq \frac{OPT}{n} . \tag{1}$$

Note that the above expectation is only over the random choice of the set L' and the choice of the element to be last in their order. The rest of the proof will exploit the randomness in the order of the remaining $\ell - 1$ vertices in L'.

The edge $e^{(\ell)} = (\ell, r)$ can only be added to the matching M if r has not already been matched in an earlier step. Consider the vertex r. In any of the preceding steps $k \in \{\lceil n/e \rceil, \ldots, \ell - 1\}$ the vertex r was only matched if it was in $e^{(k)}$, i.e. if in $M^{(k)}$ the vertex r was assigned to the left-hand side vertex k. Again, the last vertex in the order can be seen as being chosen uniformly at random from the k participating vertices on the left-hand side. Hence, the probability of r being matched in step k was at most $1/k$. As before, the order of the vertices $1, \ldots, k - 1$ is irrelevant for this event. Therefore, also the respective events if some vertex $k' < k$ was matched to r can be regarded as independent. Following this argument inductively from $k = \ell - 1$ down to $\lceil n/e \rceil$, we get,

$$\mathbf{Pr}\left[r \text{ unmatched in step } \ell\right] = \mathbf{Pr}\left[\bigwedge_{k=\lceil n/e \rceil}^{\ell-1} r \notin e^{(k)}\right] \geq \prod_{k=\lceil n/e \rceil}^{\ell-1} \frac{k-1}{k} = \frac{\lceil \frac{n}{e} \rceil - 1}{\ell - 1} .$$

Thus we have $\mathbf{Pr}\left[M \cup e^{(\ell)} \text{ is a matching}\right] \geq \frac{\lceil n/e \rceil}{\ell - 1}$. Together with inequality (1) we obtain the lemma. $\qquad\square$

Theorem 2. *The online matching algorithm is* e-*competitive in expectation.*

Proof. The weight of the matching M is obtained by summing the variables A_ℓ. Using Lemma 1 we get,

$$\mathbf{E}\left[w(M)\right] = \mathbf{E}\left[\sum_{\ell=1}^{n} A_\ell\right] \geq \sum_{\ell=\lceil n/e \rceil}^{n} \frac{\lfloor n/e \rfloor}{\ell - 1} \cdot \frac{OPT}{n} = \frac{\lfloor n/e \rfloor}{n} \cdot \sum_{\ell=\lfloor n/e \rfloor}^{n-1} \frac{1}{\ell} \cdot OPT .$$

We have $\frac{\lfloor n/e \rfloor}{n} \geq \frac{1}{e} - \frac{1}{n}$ and $\sum_{\ell=\lfloor n/e \rfloor}^{n-1} \frac{1}{\ell} \geq \ln\left(\frac{n}{\lfloor n/e \rfloor}\right) \geq 1$ which gives,

$$\mathbf{E}\left[w(M)\right] \geq \frac{\lfloor n/e \rfloor}{n} \cdot \sum_{\ell=\lfloor n/e \rfloor}^{n-1} \frac{1}{\ell} \cdot OPT \geq \left(\frac{1}{e} - \frac{1}{n}\right) \cdot OPT .$$

$\qquad\square$

3 Packing Sets of Size at Most d with Capacity b

A common generalization of bipartite online matching is the *bipartite online b-hypermatching* problem. Here, the underlying structure is an edge-weighted hypergraph $H = (L \cup R, E)$. We assume that the hyperedges in E are of the form $e = (v, S)$, with $v \in L$, $S \subseteq R$ and $|S| \leq d$. Again, we are initially given the vertex set R together with the size n of the vertex set L and the capacity b. The vertices in L are presented to the algorithm online and in a random order. At each step, when a vertex $v \in L$ is revealed, the algorithm also observes all its incident hyperedges $\delta(v) := \{e \in E \mid v \in e\}$ together with their respective weights[2] $w(e) \in \mathbb{R}_{\geq 0}$. As before, the algorithm decides online whether to assign one of the edges in $\delta(v)$, or to leave v unmatched. The objective is a b-hypermatching in H of maximum weight, i.e. every vertex $r \in R$ may be contained in up to b edges of the matching but every vertex in L may be matched only once.

In every step ℓ we will solve the LP-relaxation of max-weight b-hypermatching on the revealed part of the graph computing a fractional solution $x^{(\ell)}$. Note that for a particular subset $L' \subseteq L$ of the left-hand side vertices the restricted hypergraph $H[L' \cup R]$ has exactly the edge set $E' := \{(v, S) \in E \mid v \in L'\}$.

In a matching, a vertex on the left-hand side is assigned to at most one hyperedge. Hence, for every vertex $v \in L'$ a feasible solution to the LP-relaxation satisfies the constraint $\sum_{e=(v,S) \in E'} x_e \leq 1$. Therefore we can interpret the restricted vector $x\big|_{\delta(v)}$ as a probability distribution over all hyperedges incident to v. The second LP-constraint is $\sum_{e=(v,S) \in E', r \in S} x_e \leq b$ for every vertex $r \in R$.

Algorithm 2. Bipartite online b-hypermatching

Input : vertex set R, cardinality $n = |L|$ and parameter $p < 1$
Output: b-hypermatching M

Let L' be the first $p \cdot n$ vertices of L;
$M := \emptyset$;
for each subsequent vertex $\ell \in L - L'$ **do**   // steps $pn + 1$ to n
    $L' := L' \cup \ell$;
    $x^{(\ell)} :=$ optimal fractional solution of LP-relaxation on $H[L' \cup R]$;
    Choose $e^{(\ell)}$ randomly according to the distribution $x^{(\ell)}\big|_{\delta(\ell)}$;
    if $M \cup e^{(\ell)}$ is a b-hypermatching **then**
        add $e^{(\ell)}$ to M;

The parameter $p < 1$ will be set later. In line with the analysis of bipartite matching, we will number the vertices in L from 1 to n in their online order.

Note that the linear program and the randomized rounding in the above algorithm are only required to maintain polynomial runtime. Furthermore, all following competitive ratios are with respect to the optimal fractional solution.

---

[2] The weight functions are generally represented implicitly, e.g. by demand oracles, which allows to solve the LP-relaxation in polynomial time, see [23].

Lemma 3. *Let the random variable A_v denote the contribution of the vertex $v \in L$ to the output, i.e. the weight of the edge (v, S) assigned to v in M. And let OPT_{LP} be the value of a fractional offline optimum, i.e. of the LP-relaxation on the full hypergraph H. For the vertices $\ell \in \{pn + 1, \ldots, n\}$ we have,*

$$\mathbf{E}\left[A_\ell\right] \geq \left(1 - d \cdot \left(\frac{e(1-p)}{p}\right)^b\right) \frac{OPT_{LP}}{n} \, .$$

Proof. In analogy to the proof of Lemma 1, we interpret the random permutation up to any step k as multiple independent experiments: Choose k vertices out of L, then pick one of these to be the last in the ordering and remove it. To determine the ordering of the other $k - 1$ elements, iteratively select and remove the remaining vertices. Here we have to consider one additional independent random experiment due to the randomized rounding.

In step ℓ, the algorithm calculates an optimal fractional solution $x^{(\ell)}$ to the LP-relaxation on $H[L' \cup R]$ with value $OPT_{LP^{(\ell)}}$. Since $e^{(\ell)}$ is chosen according to the restricted vector $x^{(\ell)}|_{\delta(\ell)}$, we have $\mathbf{E}\left[w\left(e^{(\ell)}\right)\right] = \sum_{e \in \delta(\ell)} w(e) \cdot x_e^{(\ell)}$. Exactly as in the proof of Lemma 1 one can show

$$\mathbf{E}\left[w\left(e^{(\ell)}\right)\right] \geq \frac{OPT_{LP}}{n} \, . \tag{2}$$

The expectation is taken over the choice of the set L', the choice of the vertex to be last in their order and the randomized rounding.

The rounded hyperedge $e^{(\ell)} = (\ell, S)$ is only added to M if every vertex in S is covered by at most $b - 1$ other edges in M.

We will first bound the probability of covering a vertex $r \in R$ in any preceding step $k \in \{pn + 1, \ldots, \ell - 1\}$. Assume for the moment that, within step k, all participating left-hand side vertices did randomized rounding according to their respective restriction of $x^{(k)}$. Let us denote these tentative hyperedges by h_1 to h_k and remember that $e^{(k)}$ corresponds to the last one. For $r \in R$ the probability of being covered in step k is at most

$$\mathbf{Pr}\left[r \in e^{(k)}\right] = \mathbf{Pr}\left[\bigvee_{v \in \{1, \ldots, k\}} ((v \text{ is last in the order}) \wedge (r \in h_v))\right]$$

$$\leq \sum_{v \in \{1, \ldots, k\}} \mathbf{Pr}\left[(v \text{ is last in the order}) \wedge (r \in h_v)\right] \, .$$

The randomized rounding is stochastically independent of the order and we know that the last vertex in the order is chosen uniformly out of k vertices. Hence,

$$\mathbf{Pr}\left[r \in e^{(k)}\right] \leq \frac{1}{k} \cdot \sum_{v \in \{1, \ldots, k\}} \mathbf{Pr}\left[r \in h_v\right] \, .$$

The hyperedge h_v was drawn according to the distribution $x^{(k)}|_{\delta(v)}$. This gives $\mathbf{Pr}\left[r \in h_v\right] = \sum_{e \in \delta(v), \, r \in e} x_e^{(k)}$. Since $x^{(k)}$ is a feasible LP solution and thus satisfies $\sum_{e \in \delta(v), \, r \in e} x_e^{(k)} \leq b$ for all $r \in R$, we have,

$$\mathbf{Pr}\left[r \in e^{(k)}\right] \leq \frac{1}{k} \cdot \sum_{v \in \{1,\dots,k\}} \sum_{\substack{e \in \delta(v), \\ r \in e}} x_e^{(k)} \leq \frac{b}{k}, \tag{3}$$

which bounds the probability of $r \in R$ being covered in step k.

Finally, we can bound the probability of adding $e^{(\ell)} = (\ell, S)$ to M. The attempt fails if any of the vertices in S was already covered b times in previous steps. For any vertex $r \in S$, we have by inequality (3),

$$\mathbf{Pr}\left[r \text{ is covered at least } b \text{ times}\right] \leq \sum_{\substack{C \subseteq \{pn+1,\dots,\ell-1\}, \\ |C|=b}} \left(\prod_{k \in C} \frac{b}{k}\right) \tag{4}$$

$$\leq \binom{(1-p)n}{b} \cdot \left(\frac{b}{pn}\right)^b \leq \left(\frac{e(1-p)}{p}\right)^b.$$

Using a union bound over all vertices $r \in S$, and as $|S| \leq d$, we get,

$$\mathbf{Pr}\left[M \cup e^{(\ell)} \text{ is a } b\text{-hypermatching}\right] \geq 1 - d \cdot \left(\frac{e(1-p)}{p}\right)^b.$$

Together with inequality (2) we obtain the result.   □

Theorem 4. *Set the parameter p to $\frac{e(2d)^{1/b}}{1+e(2d)^{1/b}}$. Then the expected competitive ratio of the b-hypermatching algorithm for edges of size at most $d+1$ is $O(d^{1/b})$.*

Proof. The weight $w(M)$ of the b-hypermatching is equal to $\sum_{\ell=1}^{n} A_\ell$. Using Lemma 3 we get,

$$\mathbf{E}\left[w(M)\right] \geq \sum_{\ell=pn+1}^{n} \left(1 - d \cdot \left(\frac{e(1-p)}{p}\right)^b\right) \frac{OPT_{LP}}{n}.$$

The sum yields a factor of $(1-p) \cdot n$, substituting $p = \frac{e(2d)^{1/b}}{1+e(2d)^{1/b}}$, we obtain,

$$\mathbf{E}\left[w(M)\right] \geq \frac{OPT_{LP}}{1 + e(2d)^{1/b}} \cdot \left(1 - d \cdot \left(\frac{e}{e(2d)^{1/b}}\right)^b\right) \geq \frac{OPT_{LP}}{2 + 4ed^{1/b}}.$$

□

A tighter analysis for the case of $b = 1$ gives a competitive ratio of ed.

The above result for hyperedges of bounded size can be generalized to hyperedges of unbounded size using a technique by Krysta and Vöcking [20]. Flip a fair coin to choose one out of two algorithms. In case one, apply the algorithm for hyperedges of bounded size where the instance is restricted to those edges covering at most $d = \lfloor |R|^{b/(b+1)} \rfloor$ vertices on the right-hand side. In the other case, restrict the hyperedges of every vertex on the left-hand side to the single incident hyperedge of maximum weight. Now, apply Algorithm 2 as for sets of size $d = 1$ and with only one vertex on the right-hand side which is available b times. For a proof see full version.

Corollary 5. *For online b-hypermatching with general weight functions the described randomized algorithm has an expected competitive ratio of $O(|R|^{1/(b+1)})$.*

4   Lower Bound

In Section 3, we presented an $O(d)$-competitive algorithm for online hypermatching with edges of size at most $d+1$. Here, we will complement this result with a lower bound of $\Omega\left(\ln(d)/\ln\ln(d)\right)$. Note that this bound is due to the online nature of the problem and holds even when we admit unbounded computational power. The result is inspired by a lower bound from Babaioff et al. [3].

We will construct a set system with the following more easily imaginable conflict graph, i.e. where we have a vertex for every set and an edge between intersecting sets. The conflict graph is d-partite and all partitions are completely connected to each other. Therefore, choosing a single set (resp. vertex in the conflict graph) precludes the selection of any other set whose corresponding vertex in the conflict graph is in a different partition.

Proposition 6. *For every prime number q, there is a hypergraph $H = (V, E)$ with $|E| = |V| = q^2$ and $|e| = q$ ($\forall e \in E$), satisfying the following properties:*

1. *the edges E can be partitioned into q disjoint sets C_i each containing q edges,*
2. *the q hyperedges in each set C_i are pairwise disjoint,*
3. *every edge in C_i intersects all $q \cdot (q-1)$ edges that belong to any C_j, $j \neq i$.*

For a proof see full version. To turn the graph in Proposition 6 into a lower bound instance for $d = q$, we set $R := V$ and let every vertex in L be incident to exactly one of the hyperedges. So, the edges of the graph arrive online in random order. Every hyperedge is independently assigned the weight 1 with probability $1/d$, and 0 otherwise.

Theorem 7. *Any online algorithm obtains a matching of expected weight less than 2. With high probability there is a matching of weight $\Omega\left(\frac{\ln(d)}{\ln\ln(d)}\right)$.*

Proof. When an algorithm assigns the first hyperedge e, say $e \in C_i$, all the edges that do not belong to C_i are blocked by Property 3. The only edges disjoint to e are those in C_i. There are at most $d-1$ such edges, each having an expected weight of $1/d$. So their accumulated expected weight is less than 1.

By Property 2 the d edges of a set C_i form a hypermatching. For any i we have $\mathbf{Pr}$ [at least λ edges in C_i have weight 1] $= \binom{d}{\lambda}(\frac{1}{d})^\lambda \geq (\frac{d}{\lambda})^\lambda(\frac{1}{d})^\lambda = \lambda^{-\lambda}$. Choosing $\lambda := \ln(d)/2\ln\ln(d)$, the last term is at least $1/\sqrt{d}$. The probability that every set C_i has less than λ heavy edges is at most $(1 - 1/\sqrt{d})^d \leq e^{-d/\sqrt{d}} = e^{-\sqrt{d}}$. Hence, w.h.p. there is a matching of weight $\Omega\left(\ln(d)/\ln\ln(d)\right)$.   □

5   Submodular Weight Functions

Let us assume that the hypergraph is complete, i.e. $H = (L \cup R, E)$ with $E = L \times 2^R$. Then we can define a weight function $w_v \colon 2^R \to \mathbb{R}_{\geq 0}$ for every vertex $v \in L$ by setting $w_v(S) := w\big((v, S)\big)$, $\forall S \subseteq R$. Now, if all these weight functions are normalized monotone submodular, then we can modify Algorithm 2 to obtain a constant competitive ratio. Note that this setting corresponds to on-line combinatorial auctions with submodular valuations, where the bidders arrive in random order.

Let us first analyze the case when the vertices in R have multiplicity one. In every online step $\ell \in \{\lceil n/e \rceil, \ldots, n\}$ we solve the LP-relaxation of the revealed part of the instance and randomly round the vector to obtain $e^{(\ell)}$. Hence we have $\mathbf{E}\left[w\left(e^{(\ell)}\right)\right] \geq \frac{OPT_{LP}}{n}$. Remember that in Algorithm 2 we had to completely reject a hyperedge $e^{(\ell)} = (\ell, S)$ if any of the vertices in S was already covered. Here, we can still add the hyperedge $e^{(\ell)\prime} := (\ell, S')$, where $S' \subseteq S$ are those vertices in S that are not yet covered by the matching. For every $r \in R$ the probability of still being unmatched at the beginning of step ℓ can be analyzed exactly as in Lemma 1. Thus, we again have $\mathbf{Pr}\left[r \text{ was still unmatched in step } \ell\right] \geq \frac{\lfloor n/e \rfloor}{\ell - 1}$.

A known property of submodular functions is the following, see e.g. [10].

Proposition 8. *Given a normalized monotone submodular function $f : 2^R \to \mathbb{R}_{\geq 0}$, a set $S \subseteq R$ and a random set $S' \subseteq S$, where every element of S is contained in S' with probability at least p (not necessarily independently). Then $\mathbf{E}[f(S')] \geq p \cdot f(S)$.*

Combining the above observations with Proposition 8 we get,

$$\mathbf{E}\left[w\left(e^{(\ell)\prime}\right)\right] \geq \frac{\lfloor n/e \rfloor}{\ell - 1} \cdot \frac{OPT_{LP}}{n} \ .$$

This inequality is identical to the one in Lemma 1. By the same calculations as in Theorem 2 we obtain our result.

Theorem 9. *For online combinatorial auctions with submodular weight functions the algorithm is e-competitive.*

Note that for b-hypermatching, i.e. when the vertices in R are available with multiplicity b, we can obtain the same competitive ratio. Simply replace every vertex in R by b copies, each with multiplicity one, and expand the valuation function in the obvious way. This equivalent instance can then be handled with the above algorithm.

References

1. Aggarwal, G., Goel, G., Karande, C., Mehta, A.: Online vertex-weighted bipartite matching and single-bid budgeted allocations. In: SODA, pp. 1253–1264 (2011)
2. Azar, Y., Regev, O.: Combinatorial algorithms for the unsplittable flow problem. Algorithmica 44(1), 49–66 (2006)
3. Babaioff, M., Immorlica, N., Kleinberg, R.D.: Matroids, secretary problems, and online mechanisms. In: SODA, pp. 434–443 (2007)
4. Birnbaum, B.E., Mathieu, C.: On-line bipartite matching made simple. SIGACT News 39(1), 80–87 (2008)

5. Buchbinder, N., Jain, K., Singh, M.: Secretary problems via linear programming. In: Eisenbrand, F., Shepherd, F.B. (eds.) IPCO 2010. LNCS, vol. 6080, pp. 163–176. Springer, Heidelberg (2010)
6. Chakraborty, S., Lachish, O.: Improved competitive ratio for the matroid secretary problem. In: SODA, pp. 1702–1712 (2012)
7. Devanur, N.R., Hayes, T.P.: The adwords problem: online keyword matching with budgeted bidders under random permutations. In: ACM Conference on Electronic Commerce, pp. 71–78 (2009)
8. Devanur, N.R., Jain, K., Kleinberg, R.D.: Randomized primal-dual analysis of ranking for online bipartite matching. In: SODA, pp. 101–107 (2013)
9. Dimitrov, N.B., Plaxton, C.G.: Competitive weighted matching in transversal matroids. Algorithmica 62(1-2), 333–348 (2012)
10. Feldman, M., Naor, J., Schwartz, R.: Improved competitive ratios for submodular secretary problems (extended abstract). In: Goldberg, L.A., Jansen, K., Ravi, R., Rolim, J.D.P. (eds.) RANDOM/APPROX 2011. LNCS, vol. 6845, pp. 218–229. Springer, Heidelberg (2011)
11. Ferguson, T.S.: Who solved the secretary problem? Statistical Science, 282–289 (1989)
12. Goel, G., Mehta, A.: Online budgeted matching in random input models with applications to adwords. In: SODA, pp. 982–991 (2008)
13. Im, S., Wang, Y.: Secretary problems: Laminar matroid and interval scheduling. In: SODA, pp. 1265–1274 (2011)
14. Jaillet, P., Soto, J.A., Zenklusen, R.: Advances on matroid secretary problems: Free order model and laminar case. CoRR abs/1207.1333 (2012)
15. Kalyanasundaram, B., Pruhs, K.: Online weighted matching. J. Algorithms 14(3), 478–488 (1993)
16. Karande, C., Mehta, A., Tripathi, P.: Online bipartite matching with unknown distributions. In: STOC, pp. 587–596 (2011)
17. Karp, R.M., Vazirani, U.V., Vazirani, V.V.: An optimal algorithm for on-line bipartite matching. In: STOC, pp. 352–358 (1990)
18. Kleinberg, R.D.: A multiple-choice secretary algorithm with applications to online auctions. In: SODA, pp. 630–631 (2005)
19. Korula, N., Pál, M.: Algorithms for secretary problems on graphs and hypergraphs. In: Albers, S., Marchetti-Spaccamela, A., Matias, Y., Nikoletseas, S., Thomas, W. (eds.) ICALP 2009, Part II. LNCS, vol. 5556, pp. 508–520. Springer, Heidelberg (2009)
20. Krysta, P., Vöcking, B.: Online mechanism design (randomized rounding on the fly). In: Czumaj, A., Mehlhorn, K., Pitts, A., Wattenhofer, R. (eds.) ICALP 2012, Part II. LNCS, vol. 7392, pp. 636–647. Springer, Heidelberg (2012)
21. Mahdian, M., Yan, Q.: Online bipartite matching with random arrivals: an approach based on strongly factor-revealing lps. In: STOC, pp. 597–606 (2011)
22. Mehta, A., Saberi, A., Vazirani, U.V., Vazirani, V.V.: Adwords and generalized online matching. J. ACM 54(5) (2007)
23. Nisan, N., Roughgarden, T., Tardos, E., Vazirani, V.V.: Algorithmic game theory. Cambridge University Press (2007)

Balls into Bins Made Faster

Megha Khosla

Max Planck Institute for Informatics, Saarbrücken, Germany
mkhosla@mpi-inf.mpg.de

Abstract. Balls-into-bins games describe in an abstract setting several multiple-choice scenarios, and allow for a systematic and unified theoretical treatment. In the process that we consider, there are n bins, initially empty, and $m = \lfloor cn \rfloor$ balls. Each ball chooses independently and uniformly at random $k \geq 3$ bins. We are looking for an allocation such that each ball is placed into one of its chosen bins and no bin has load greater than 1. How quickly can we find such an allocation? We present a simple and novel algorithm that finds such an allocation (if it exists) and runs in *linear time* with high probability.

Our algorithm finds applications in finding perfect matchings in a special class of sparse random bipartite graphs, orientation of random hypergraphs, load balancing and hashing.

1   Introduction

Balls and bins games offer a powerful model with various applications in computer science and mathematics, e.g., the analysis of hashing, the modeling of load balancing strategies and matchings in bipartite graphs. The typical aim of such games is to allocate a collection of m balls into a set of n bins such that the maximum load is minimized. A simple strategy would be to place each ball randomly into one of the bins. For $m = n$ the maximum load then grows like $\frac{\log n}{\log \log n}$. Now suppose that balls arrive sequentially and each ball chooses two bins independently and uniform at random. Azar et al. [1] showed that if each ball is placed into the one which is least loaded at the time of placement then with high probability[1] the maximum load is $\frac{\log \log n}{\log 2} + \Theta(1)$, which is an exponential improvement over the previous case.

In this work we restrict the maximum capacity of any bin to one, i.e., each bin can hold at most 1 ball. Additionally we provide that each ball, in addition to having $k \geq 3$ choices, can also be moved among its choices on demand. An important example of an allocation strategy in this direction is *cuckoo hashing* [14,4]. Cuckoo hashing is a collision resolution scheme used in building large hash tables. Here bins are the locations on the hash table and balls represent the items. In this scheme when a ball arrives, it chooses its k random bins (chosen using k random hash functions) and is allocated to one of them. In case the bin

---

[1] We use *with high probability* to mean with probability $1 - n^{-\zeta}$ for some constant $\zeta > 0$.

H.L. Bodlaender and G.F. Italiano (Eds.): ESA 2013, LNCS 8125, pp. 601–612, 2013.

is full, the previously allocated ball is moved out and placed in one of its other $k - 1$ choices. This process may be repeated indefinitely or until a free bin is found. We give a simple linear time algorithm that builds on the idea of cuckoo hashing and is asymptotically optimal. Roughly speaking we propose an efficient strategy to choose the bin in case all the choices of the incoming ball are full.

We model the k-choice balls-into-bins game by a directed graph $G = (V, E)$ such that the set of vertices V corresponds to bins. We say a vertex is *occupied* if there is a ball assigned to the corresponding bin, otherwise it is *free*. Let $\mathcal{I}$ be the set of m balls. We represent each ball $x \in \mathcal{I}$ as a tuple of its k chosen bins, so we say a vertex $v \in x$ if v corresponds to one of the chosen bins of ball x. For vertices $u, v \in V$, a directed edge $e = (u, v) \in E$ if and only if there exists a ball $y \in \mathcal{I}$ so that the following two conditions hold, (i) $u, v \in y$, and (ii) u is occupied by y. Note that a vertex with outdegree 0 is a free vertex. We denote the set of free vertices by F and the minimum of the distance of vertices in F from v by $d(v, F)$. Since G represents an allocation we call G an *allocation graph*.

Assume that at some instance a ball z arrives such that all its k choices are occupied. Let $v \in z$ be the vertex chosen to place z. The following are the main observations.

1. The necessary condition for ball z to be successfully assigned to v is the existence of a path from v to F. This condition remains satisfied as long as some allocation is possible.
2. A free location will be found in the minimum number of steps if for all $u \in z$ the distance $d(v, F) \leq d(u, F)$.

With respect to our first observation, a natural question would be the following. Is it possible to place each of the $m = \lfloor cn \rfloor$ balls into one of their chosen bins such that each bin holds at most one ball? From [11,6,8] we know that there exists a critical size $c_k^* n$ such that if $c < c_k^* n$ then such an allocation is possible with high probability, otherwise this is not the case. In particular the following is known.

Theorem 1. *For integers $k \geq 3$ let ξ^* be the unique solution of the equation*

$$k = \frac{\xi(1 - e^{-\xi})}{1 - e^{-\xi} - \xi e^{-\xi}}. \tag{1}$$

Let $c_k^* = \frac{\xi^*}{k(1 - e^{-\xi^*})^{k-1}}$. Then

$$\Pr \text{ allocation of } m = \lfloor cn \rfloor \text{ balls to } n \text{ bins is possible} \overset{(n \to \infty)}{=} \begin{cases} 0, & \text{if } c > c_k^* \\ 1, & \text{if } c < c_k^* \end{cases}. \tag{2}$$

The proof of the above theorem is non-constructive, i.e., it does not give us an algorithm to find such an allocation.

Our second observation suggests that the allocation time depends on the selection of the bin, which we make for each assignment, from among the k possible bins. One can in principle use breadth first search (BFS) to always make assignments over the shortest path (in the allocation graph). BFS is analyzed in [4] and is shown to run in linear time only in expectation. One can also select uniformly at random a bin from the available bins. This resembles a random walk on the vertices of the allocation graph and is called the *random walk insertion*. In [7,9] the authors analyzed the random walk insertion method and gave a polylogarithmic bound (with high probability) on the maximum allocation time, i.e., the maximum time it can take to allocate a single ball. The random walk method does not provide any guarantees for the total allocation time. In fact it might run for ever in some worst case. In this paper we propose a novel allocation algorithm which runs in linear time with probability $1 - n^{-\zeta}$ for some constant $\zeta > 0$. Moreover it is guaranteed to find an allocation if it exists. Through simulations we demonstrate that the our allocation method requires drastically less number of selections (to place or replace an item) when compared to the random walk method (which to our knowledge is also the state of art method for the process under consideration). For instance the number of selections in the worst case is reduced by a factor of 10 when using our method.

1.1 More Related Work

Questions in the multiple choice balls-into-bins games can also be phrased in terms of *orientation of graphs* or more generally *orientations of k-uniform hypergraphs*. The n bins are represented as vertices and each of the m balls form an edge with its k-vertices representing the k random choices of the ball. In fact, this is a random hypergraph $H_{n,m,k}$ (or random graph $G_{n,m}$ for $k = 2$) with n vertices and m edges where each edge is drawn uniformly at random (with replacement) from the set of all k-multisubsets of the vertex set. An s-orientation of a graph then amounts to a mapping of each edge to one of its vertices such that no vertex receives more than one edge. For $k = 2$ several allocation algorithms and their analysis are closely connected to the cores of the associated graph. The s-core of a graph is the maximum vertex induced subgraph with minimum degree at least s. For instance Czumaj and Stemann [8] give a linear time algorithm achieving maximum load $O(m/n)$ based on computation of all cores. Fernholz and Ramachandran [3] and Cain, Sanders, and Wormald [2] gave linear time algorithms for computing an optimal allocation (asymptotically almost surely). Their analysis also determined the threshold for s-orientability of sparse random graphs which is also the threshold for the $s + 1$ core to have density (ratio of number of edges to that of vertices) less than s.

Another closely related problem is that of finding maximum cardinality matchings in random bipartite graphs. Consider a bipartite graph $G = (E, V; \mathcal{E})$ where E corresponds to the m balls and V represents the n bins. For $e \in E$ and $v \in V$, we have $(e, v) \in E$ if and only if v represent one of the k choices of the ball represented by e. This represents a k-*left regular* random (as the k- neighbors of left vertex set are chosen randomly) bipartite graph. The problem of finding an optimal allocation now reduces to finding a left perfect matching in G. One can refer [13,4] for expected linear time algorithms for maximum matchings for k-left regular random bipartite graphs. More results and techniques in multiple-choice allocation schemes can be found in [12].

1.2   Applications

Load Balancing. We are given a set of $m = \lfloor cn \rfloor$ identical jobs, and n machines on which they can be executed. Suppose that each job may choose randomly among k different machines. Our result implies that as long as $c < c^*_{k,\ell}$, we can find an assignment of the jobs to their preferred machines, such that each machine is assigned at most one task, in time $O(n)$ with high probability.

Parallel Access to Hard Disks. We are given n hard disks (or any other means of storing large amounts of information), which can be accessed independently of each other. The goal would be to store large data sets allowing for a level of redundancy for fault tolerance, and at the same time minimize the number of I/O steps needed to retreive the data (see [15] for more details). Using our allocation algorithm we can allocate k copies of each of the $m = \lfloor cn \rfloor$ blocks randomly on n hard disks in linear time (with high probability) as long as $c < c^*_k$. The m different data blocks can now be read with at most 1 parallel query on each disk.

Efficient Dictionaries. In the data structure language one can understand the above problem as one of designing efficient dictionaries. We are given $m = \lfloor cn \rfloor$ items (balls) and n locations (bins). Each item chooses its preferred k locations by using k random hash functions. Our aim is to assign each item to one of its chosen locations such that all items are assigned and no location receives more than 1 item. Assuming that such an allocation is possible our proposed algorithm constructs the dictionary in time $O(n)$ with high probability.

1.3   Notation

Throughout the paper we use n to denote the number of bins, m for the number of balls and k denotes the number of random choices of any ball. For an allocation graph $G = (V, E)$ and any two vertices $u, v \in V$, the shortest distance from u to v is denoted by $d(u, v)$. We denote the set of free vertices by F. We denote the shortest distance from a vertex $v \in V$ to any set of vertices say S by $d(v, S)$ which is defined as

$$d(v, S) := \min_{u \in S} d(v, u).$$

We use R to denote the set of vertices furthest to F, i.e.,

$$R := \{v \in V | d(v, F) \geq \max_{u \in V} d(u, F)\}.$$

For an integer $t \in \{0, 1, \ldots, n\}$ and a subset of vertex set $V' \subseteq V$ we use $N_t(u)$ and $N_t(V')$ to denote the set of vertices at distance at most t from the vertex $u \in V$ and the set V' respectively. Mathematically,

$$N_t(u) := \{v \in V \mid d(u,v) \leq t\} \quad \text{and} \quad N_t(V') := \{v \in V \mid d(v,V') \leq t\}.$$

We say an allocation is *proper* if each of the m balls is allocated to one of its chosen bins and no bin holds more than one ball. We use a common definition of *with high probability*, namely the event occurs with probability $1 - O(n^{-\zeta})$ for some $\zeta > 0$. log refers to the natural logarithms.

1.4   Our Contribution

We propose a simple and efficient algorithm to find a proper allocation. In a nutshell we provide a deterministic strategy of how to select a vertex (bin) for placing a ball when all its choices are occupied. We assign to each vertex $v \in V$ an integer label, $L(v)$. Initially all vertices have 0 as their labels. Note that at this stage, for all $v \in V$, $L(v) = d(v, F)$, i.e., the labels on the vertices represent their shortest distances from F. When a ball x appears, it chooses the vertex with the least label from among its k choices. If the vertex is free, the ball is placed on it. Otherwise, the previous ball is kicked out. The label of the vertex is then updated and set to one more than the minimum label of the remaining $k - 1$ choices of the ball x. The kicked out ball chooses the bin with minimum label from its k choices and the above procedure is repeated till an empty bin is found. Note that to maintain the labels of the vertices as their shortest distances to F we would require to update labels of the neighbors of the selected vertex and the labels of their neighbors and so on. This corresponds to performing a breadth first search starting from the selected vertex. We avoid the BFS and perform only local updates. Therefore, we also call our method as *local search allocation*.

Previous work [4] on total allocation time for the case $k \geq 3$ has analysed breadth first search, and it was shown to be linear only in *expectation*. A simple reduction suggests that to match the probability bounds given by our algorithm, BFS would require $O(n \log(n))$ run time. Our algorithm finds an allocation (with probability 1) whenever it exists. This is in contrast to the random walk insertion method which might run indefinitely for a solvable instance. In the last section we present experimental results comparing the performance of these two algorithms. The results reveal that local search allocation is at least 10 times faster even for the worst case allocation. We now state our main result.

Theorem 2. *Let $k \geq 3$. For any fixed $\varepsilon > 0$, set $m = (1 - \varepsilon)c_k^* n$. Assume that each of the m balls chooses k random bins from a total of n bins. Then with high probability, a proper allocation of these balls can be found in time $O(n)$.*

We prove the above theorem in two steps. First we show that the algorithm is correct and finds an allocation in polynomial time. To this end we prove that, at any instance, the label of a vertex is at most its shortest distance to the set of free vertices. Therefore, no vertex can have a label greater than $n - 1$.

This would imply that the algorithm could not run indefinitely and would stop after making at most n changes at each location. We then show that the local search allocation method will find an allocation in a time proportional to the sum of distances of the n vertices to F (in the resulting allocation graph). We complete the proof by showing that (i) if for $\varepsilon > 0$, $m = (1 - \varepsilon)c_k^*$ balls are placed in n bins using k random choices for each ball then the corresponding allocation graph has two special structural properties with high probability, and (ii) if the allocation graph has these two properties, then the sum of distances of its vertices to F is linear in n. In the next section we give a formal description of our algorithm and its analysis.

2   Local Search Allocation and Its Analysis

Assume that we are given balls in an online fashion, i.e., each balls chooses its k random bins whenever it appears. Moreover, balls appear in an arbitrary order. The allocation using local search method goes as follows. For each vertex $v \in V$ we maintain a label. Initially each vertex is assigned a label 0. To assign a ball x at time t we select one of its chosen vertices v such that its label is minimum and assign x to v. We assign a new label to v which is one more than the minimum label of the remaining $k - 1$ choices of x. However, v might have already been occupied by a previously assigned ball y. In that case we kick out y and repeat the above procedure. Let $\mathbf{L} = \{L(v) \mid v \in V\}$ and $\mathbf{T} = \{T(v) \mid v \in V\}$ where $L(v)$ denotes the label of vertex v and $T(v)$ denotes the ball assigned to vertex v. We initialize $\mathbf{L}$ with all 0s and $\mathbf{T}$ with $\emptyset$, i.e., all vertices are free. We then use Algorithm 1 to assign an arbitrary ball when it appears. In the next subsection

---

Algorithm 1. AssignBall $(x, \mathbf{L}, \mathbf{T})$

1: Choose a bin v among the k choices of x with minimum label $L(v)$.
2: **if** $(L(v) == n - 1)$ **then**
3:    **EXIT**   ▷**Allocation does not exist**
4: **else**
5:    $L(v) \leftarrow 1 + \min(L(w)|w \neq v \text{ and } w \in x)$
6:    **if** $(T(v) \neq \emptyset)$ **then**
7:       $y \leftarrow T(v)$ ▷**Move that replaces a ball**
8:       $T(v) \leftarrow x$
9:       **CALL** AssignBall$(y, \mathbf{L}, \mathbf{T})$
10:   **else**
11:     $T(v) \leftarrow x$ ▷**Move that places a ball**

---

we first prove the correctness of the algorithm, i.e., it finds an allocation in a finite number of steps whenever an allocation exists. We show that the algorithm takes a maximum of $O(n^2)$ time before it obtains a bin for each ball. We then proceed to give a stronger bound on the running time.

2.1 Analysis of the Local Search Allocation

We need some additional notation. In what follows a *move* denotes either placing a ball in a free bin or replacing a previously allocated ball. Let M be the total number of moves performed by the algorithm. For $p \in [M]$ we use $L_p(v)$ to denote the label of vertex v at the end of the pth move. Similarly we use F_p to denote the set of free vertices at the end of pth move. The corresponding allocation graph is denoted as $G_p = (V, E_p)$. We need the following proposition.

Proposition 1. *For all $p \in [M]$ and all $v \in V$, the shortest distance of v to F_p is at least the label of v, i.e., $d(v, F_p) \geq L_p(v)$.*

Proof. We first note that the label of a free vertex always remain 0, i.e.,

$$\forall p \in [M], \forall w \in F_p, \qquad L_p(w) = 0. \tag{3}$$

We will now show that throughout the algorithm the label of a vertex is at most one more than the label of any of its immediate neighbors (neighbors at distance 1). More precisely,

$$\forall p \in [M], \forall (u, v) \in E_p, \qquad L_p(u) \leq L_p(v) + 1. \tag{4}$$

We prove (4) by induction on the number of moves performed by the algorithm. Initially when no ball has appeared all vertices have 0 as their labels. When the first ball is assigned, i.e., there is a single vertex say u such that $L_1(u) = 1$. Clearly, (4) holds after the first move. Assume that (4) holds after p moves.

For the $(p+1)$th move let $w \in V$ be some vertex which is assigned a ball x. Consider an edge $(u, v) \in E_p$ such that $u \neq w$ and $v \neq w$. Note that the labels of all vertices $v \in V \setminus w$ remain unchanged in the $(p+1)$th move. Therefore by induction hypothesis, (4) is true for all edges which does not contain w. By Step 2 of Algorithm 1 the new label of w is one more than the minimum of the labels of its $k-1$ neighbors, i.e,

$$L_{p+1}(w) = \min_{w' \in x \setminus w} L_{p+1}(w') + 1.$$

Therefore (4) holds for all edges originating from w. Now consider a vertex $u' \in V$ such that $(u', w) \in E_p$. Now by induction hypothesis we have $L_{p+1}(u) = L_p(u) \leq L_p(w) + 1$. Note that the vertex w was chosen because it had the minimum label among the k possible choices for the ball x, i.e.,

$$L_p(w) \leq \min_{w' \in x} L_p(w') = \min_{w' \in x \setminus w} L_{p+1}(w') < L_{p+1}(w).$$

We therefore obtain $L_{p+1}(u) \leq L_p(w) + 1 < L_{p+1}(w) + 1$, thereby completing the induction step. We can now combine (3) and (4) to obtain the desired result. To see this, consider a vertex v at distance $s < n$ to a free vertex $f \in F_p$ such that s is also the shortest distance from v to F_p. By iteratively applying (4) we obtain $L_p(v) \leq s + L_p(f) = d(v, F_p)$, which completes the proof.

We know that whenever the algorithm visits a vertex, it increases its label by at least 1. Trivially the maximum distance of a vertex from a free vertex is $n-1$ (if an allocation exists), and so is the maximum label. Therefore the algorithm will stop in at most $n(n-1)$ steps, i.e., after visiting each vertex at most $n-1$ times, which implies that the algorithm is correct and finds an allocation in $O(n^2)$ time. In the following we show that the total running time is proportional to the sum of labels of the n vertices.

Lemma 1. *Let* $\mathbf{L}^*$ *be the array of labels of the vertices after all balls have been allocated using Algorithm 1. Then the total time required to find an allocation is* $O(\sum_{v \in V} L^*(v))$.

Proof. Now each invocation of Algorithm 1 increases the label of the chosen vertex by at least 1. Therefore, a vertex with label ℓ has been selected (for any move) at most ℓ times. Now the given number of balls can be allocated in a time proportional to the number of steps required to obtain the array $\mathbf{L}^*$ (when the initial set consisted of all zeros) and hence is $O(\sum_{v \in V} L^*(v))$.

For notational convenience let $F := F_M$ and $G := G_M$ denote the set of free vertices and the allocation graph (respectively) at the end of the algorithm. By Proposition 1 we know that for each $v \in V$, $L^*(v) \leq d(v, F)$. Moreover, by Step 2 of Algorithm 1 the maximum value of a label is n. Thus the total sum of labels of all vertices is bounded as follows.

$$\sum_{v_i \in V} L^*(v_i)) \leq \min \left(\sum_{v_i \in V} d(v, F), n^2 \right).$$

To compute the desired sum, i.e., $\sum_{v_i \in V} d(v, F)$, we study the structure of the allocation graph. The following lemma states that, with probability $1 - o(1)$, a fraction of the vertices in the allocation graph are at a constant distance to the set of free vertices, F. This would imply that the contribution for the above sum made by these vertices is $O(n)$.

Lemma 2. *For any fixed* $\varepsilon > 0$, *let* $m = (1 - \varepsilon)c_k^* n$ *balls are assigned to* n *bins using* k *random choices for each ball. Then the corresponding allocation graph* $G = (V, E)$ *satisfies the following with probability* $1 - O(1/n)$: *for every* $\alpha > 0$ *there exist* $C = C(\alpha, \varepsilon) > 0$ *and a set* $S \subseteq V$ *of size at least* $(1 - \alpha)n$ *such that every vertex* $v \in S$ *satisfies* $d(v, F) \leq C$.

We remark that the above lemma has been originally proved in [7]. With respect to an allocation graph recall that we denote the set of vertices furthest from F by R. Also for an integer s, $N_s(R)$ denotes the set of vertices at distance at most s from R. The next lemma states that the neighborhood of R expands suitably with high probability. We remark that the estimate, for expansion factor, presented here is not the best possible but nevertheless suffices for our analysis. Our proof is similar to the proof of Proposition 2.4 in [7] and is deferred to the full version.

Lemma 3. *For any fixed $\varepsilon > 0$, let $m = (1 - \varepsilon)c_k^* n$ balls are assigned to n bins using k random choices for each ball and $G = (V, E)$ be the corresponding allocation graph. Then for any $0 < \alpha < \frac{1}{k-1}$ and every integer s such that $1 \leq |N_s(R)| \leq \alpha n$, there exists a constant $\zeta > 0$ such that G satisfies the following with probability $1 - n^{-\zeta}$.*

$$|N_{s+1}(R)| > \left(k - 1 - \frac{\log e^k(k-1)}{\log \frac{1}{\alpha(k-1)}}\right) |N_s(R)|.$$

The following corollary easily follows from the above two lemmas.

Corollary 1. *With high probability, the maximum label of any vertex in the allocation graph is $O(\log n)$.*

Due to space constraints, the missing proofs are deferred to the full version. We now prove our main theorem.

Proof (Proof of Theorem 2). Let for $\gamma > 0$

$$\alpha = \min\left\{\frac{0.1}{k-1}, \ \exp\left(-1 - \frac{k}{k-2-\gamma}\right) \cdot (k-1)^{\frac{1}{k-2-\gamma}}\right\} \text{ and } \delta = \frac{\log e^k(k-1)}{\log \frac{1}{\alpha(k-1)}}.$$

Then by Lemma 2, with probability $1 - O(1/n)$, there exists a $C = C(\alpha, \varepsilon)$ and a set S such that $|S| \geq (1 - \alpha)n$ and every vertex $v \in S$ satisfies $d(v, F) \leq C$. Let $T + 1$ be the maximum of the distances of vertices in R to S, i.e.,

$$T = \max_{v \in R} d(v, S) - 1.$$

Clearly the number of vertices at distance at most T from R is at most αn, i.e., $|N_T(R)| \leq \alpha n$. Moreover for all $t < T$, $|N_t(R)| < |N_T(R)|$. Then by Lemma 3, for all $t \leq T$ the following holds with high probability,

$$|N_{t+1}(R)| > (k - 1 - \delta) |N_t(R)|.$$

One can check that for $\gamma > 0$ and α as chosen above, $\delta < k - 2 - \gamma$. The total distance of all vertices from F is then given by

$$D = \sum_{v \in N_T(R)} d(v, F) + \sum_{v \in S} d(v, F).$$

As every vertex in S is at a constant distance from F, we obtain $\sum_{v \in S} d(v, F) = O(n)$. Note that for every $i > 0$, $|N_i(R)| - |N_{i-1}(R)|$ is the number of vertices at distance i from R. Therefore,

$$\sum_{v \in N_T(R)} d(v, F) = (T + C)|N_0(R)| + \sum_{i=1}^{T}(T + C - i)(|N_i(R)| - |N_{i-1}(R)|)$$

$$= (T + C)|N_0(R)| + \sum_{i=1}^{T}(T - i)(|N_i(R)| - |N_{i-1}(R)|) + C\sum_{i=1}^{T}(|N_i(R)| - |N_{i-1}(R)|)$$

$$= (T + C)|N_0(R)| + \sum_{i=1}^{T}(T - i)(|N_i(R)| - |N_{i-1}(R)|) + C(|N_T(R)| - |N_0(R)|)$$

$$= \sum_{i=1}^{T}\left((T - i)(|N_i(R)| - |N_{i-1}(R)|) + |N_0(R)|\right) + C \cdot |N_T(R)| = \sum_{i=0}^{T-1}|N_i(R)| + O(n).$$

Now with high probability, we have $|N_{T-j}(R)| < \frac{|N_T(R)|}{(k-1-\delta)^j}$. Therefore,

$$\sum_{i=0}^{T-1}|N_i(R))| < |N_T(R)|\sum_{j=1}^{T}\frac{1}{(k-1-\delta)^j} < |N_T(R)|\sum_{j=1}^{T}\frac{1}{(1+\gamma)^j} = O(n),$$

which completes the proof of Theorem 2.

2.2   Experimental Results and Discussion

We present some simulations to compare the performance of local search allocation with the random walk method which (to the best of our knowledge) is currently the state-of-art method and so far considered to be the fastest algorithm for the case $k \geq 3$. We recall that in the random walk method we choose a bin at random from among the k possible bins to place the ball. If the bin is not free, the previous ball is moved out. The moved out ball again chooses a random bin from among its choices and the procedure goes on till an empty bin is found. In our experiments we consider $n \in [10^5, 5 \times 10^6]$ balls and $\lfloor cn \rfloor$ bins. The k random bins are chosen when the ball appears. All random numbers in our simulations are generated by *ranlxs2* generator of GNU Scientific Library [10].

Recall that a move is either placing an item at a free location or replacing it with other item. In Figure 1 we give a comparison of the total number of moves (averaged over 50 random instances) performed by local search and random walk methods for $k = 3$ and $k = 4$. Figure 2 compares the maximum number of moves (averaged over 50 random instances) for a single insertion performed by local search and random walk methods. Figure 3 shows a comparison when the number of balls are fixed and density (ratio of number of balls to that of bins) approaches the threshold density. Note that the time required to obtain an allocation by random walk or local search methods is directly proportional to the number of moves performed.

We remark that local search allocation has some additional cost, i.e., the extra space required to store the labels. Though this space is $O(n)$, local search allocation is still useful for the applications where the size of objects (representing the balls) to be allocated is much larger than the labels which are integers. Moreover,

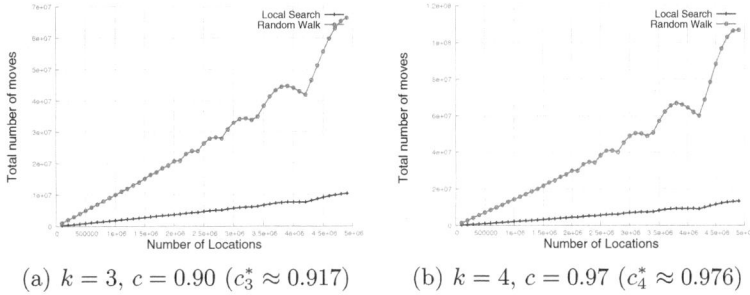

(a) $k = 3$, $c = 0.90$ ($c_3^* \approx 0.917$)   (b) $k = 4$, $c = 0.97$ ($c_4^* \approx 0.976$)

Fig. 1. Comparison of total number of moves performed by local search and random walk methods

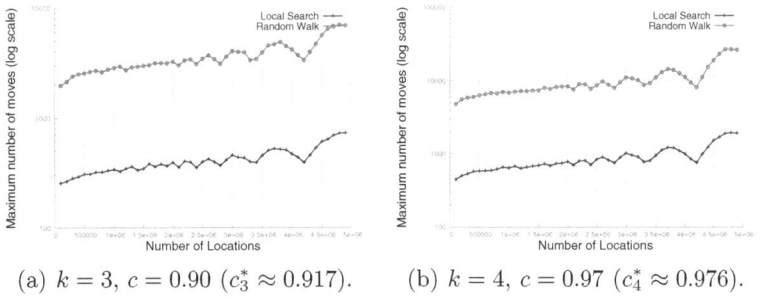

(a) $k = 3$, $c = 0.90$ ($c_3^* \approx 0.917$).   (b) $k = 4$, $c = 0.97$ ($c_4^* \approx 0.976$).

Fig. 2. Comparison of maximum number of moves performed by local search and random walk methods

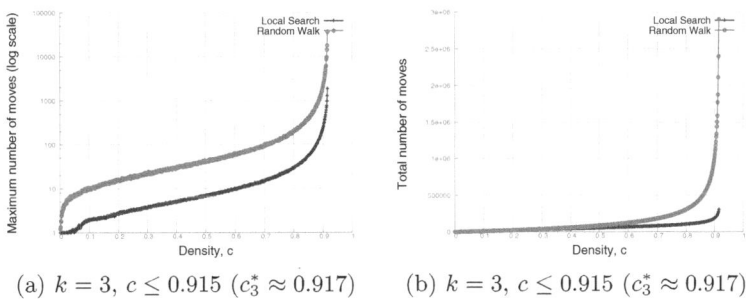

(a) $k = 3$, $c \leq 0.915$ ($c_3^* \approx 0.917$)   (b) $k = 3$, $c \leq 0.915$ ($c_3^* \approx 0.917$)

Fig. 3. Comparison of total number of moves and maximum number of moves (for fixed number of locations, $n = 10^5$) performed by local search and random walk methods when density c approaches c_k^*.

with high probability, the maximum label of any vertex is $O(\log n)$. Many integer compression methods [16] have been proposed for compressing small integers and can be potentially useful for further optimizations. Also in most of the load balancing problems, the speed of assignment is a much desired and the most important requirement. Though we have not provided any theoretical guarantees for the maximum allocation time, our simulations suggest that the local search allocation is at least 10 times faster than the random walk. As future work it would be interesting to extend our algorithm for the case where each bin can hold more than one ball [5,11].

Acknowledgements. The author would like to thank Kurt Mehlhorn and Ali Pourmiri for their valuable suggestions to improve the presentation of the paper.

References

1. Azar, Y., Broder, A., Karlin, A., Upfal, E.: Balanced allocations. SIAM Journal on Computing 29(1), 180–200 (1999)
2. Cain, J.A., Sanders, P., Wormald, N.: The random graph threshold for k-orientiability and a fast algorithm for optimal multiple-choice allocation. In: ACM-SIAM Symposium on Discrete Algorithms, SODA 2007, pp. 469–476 (2007)
3. Fernholz, D., Ramachandran, V.: The k-orientability thresholds for gn, p. In: ACM-SIAM Symposium on Discrete Algorithms, SODA 2007, pp. 459–468 (2007)
4. Fotakis, D., Pagh, R., Sanders, P., Spirakis, P.: Space efficient hash tables with worst case constant access time. In: Alt, H., Habib, M. (eds.) STACS 2003. LNCS, vol. 2607, pp. 271–282. Springer, Heidelberg (2003)
5. Fountoulakis, N., Khosla, M., Panagiotou, K.: The multiple-orientability thresholds for random hypergraphs. In: ACM-SIAM Symposium on Discrete Algorithms, SODA 2011, pp. 1222–1236 (2011)
6. Fountoulakis, N., Panagiotou, K.: Sharp load thresholds for cuckoo hashing. Random Structures & Algorithms 41(3), 306–333 (2012)
7. Fountoulakis, N., Panagiotou, K., Steger, A.: On the insertion time of cuckoo hashing. CoRR, abs/1006.1231 (2010)
8. Frieze, A., Melsted, P.: Maximum matchings in random bipartite graphs and the space utilization of cuckoo hash tables. Random Structures & Algorithms 41(3), 334–364 (2012)
9. Frieze, A., Melsted, P., Mitzenmacher, M.: An analysis of random-walk cuckoo hashing. SIAM Journal on Computing 40(2), 291–308 (2011)
10. Galassi, M., Davies, J., Theiler, J., Gough, B., Jungman, G., Booth, M., Rossi, F.: Gnu scientific library reference manual (2003), http://www.gnu.org/software/gsl
11. Lelarge, M.: A new approach to the orientation of random hypergraphs. In: ACM-SIAM Symposium on Discrete Algorithms, SODA 2012, pp. 251–264 (2012)
12. Mitzenmacher, M., Richa, A., Sitaraman, R.: The power of two random choices: A survey of techniques and results. In: Handbook of Randomized Algorithms, pp. 255–312 (2000)
13. Motwani, R.: Average-case analysis of algorithms for matchings and related problems. J. ACM 41(6), 1329–1356 (1994)
14. Pagh, R., Rodler, F.F.: Cuckoo hashing. In: Meyer auf der Heide, F. (ed.) ESA 2001. LNCS, vol. 2161, pp. 121–133. Springer, Heidelberg (2001)
15. Sanders, P., Egner, S., Korst, J.: Fast concurrent access to parallel disks. In: ACM-SIAM Symposium on Discrete Algorithms, pp. 849–858 (1999)
16. Schlegel, B., Gemulla, R., Lehner, W.: Fast integer compression using simd instructions. In: Workshop on Data Management on New Hardware, DaMoN 2010, pp. 34–40. ACM (2010)

An Alternative Approach to Alternative Routes: HiDAR

Moritz Kobitzsch

Karlsruhe Institute of Technology
kobitzsch@kit.edu

Abstract. Alternatives to a shortest path are a common feature for modern navigation providers. In contrast to modern speed-up techniques, which are based on the unique distance between two locations within the map, computing alternative routes that might include slightly sub-optimal routes seems a way more difficult problem. Especially testing a possible alternative route for its quality can so far only be done utilizing considerable computational overhead. This forces current solutions to settle for any viable alternative instead of finding the best alternative routes possible. In this paper we show a way on how to deal with this overhead in an effective manner, allowing for the computation of high quality alternative routes while maintaining competitive query times.

1 Introduction

The process of finding a single shortest path in a graph with respect to some weight function has come a long way from Dijkstra's algorithm [1], which allows for solutions in almost linear time. Speed-up techniques [2–5] offer real time shortest path computation through extensive use of preprocessing. These techniques are used in many web-based navigation services that provide driving directions to the general public.

Even though the computed paths do not rely on heuristics, the optimality of a shortest path remains a subjective measure. Navigation services offer multiple choices, hoping one of them might reflect the preferences of the user. These choices are referred to as alternative routes. The first formal mention of alternative routes, aside from the k-shortest path problem or blocked vertex/arc routes, can be found in [6]. Still, until Abraham et al. [7] introduced the idea of via-node alternatives, which describe an alternative route using source, target, and an additional vertex, only proprietary algorithms were used by navigation providers. This approach was later on improved by Luxen and Schieferdecker through use of extensive preprocessing [8], while Bader et al. [9] proposed an entirely different approach using penalties.

In this paper we focus on alternative routes using the definition of Abraham et al. [7]. So far, via-node alternatives [7,8] focus on reporting any alternative route, satisfying some minimal quality[1] requirements, as current techniques require

---

[1] We will focus on the exact definition of quality later on.

H.L. Bodlaender and G.F. Italiano (Eds.): ESA 2013, LNCS 8125, pp. 613–624, 2013.
© Springer-Verlag Berlin Heidelberg 2013

three full shortest path queries, which we will describe later on, to check a potential route for its quality.

After the introduction of our terminology in Section 2 and a more extensive discussion of the work related to this paper in Section 3, we introduce a new algorithm to solve the problem of alternative routes in Section 4. Our algorithm allows for the selection of the best possible among the found candidates by reducing the problem to a small graph representing all potentially viable alternative routes. We evaluate our algorithm in Section 5 where we show not only the possibility of calculating higher quality routes but also show the competitiveness regarding query times.

2   Definitions and Terminology

Every road network can be viewed as a directed and weighted graph:

Definition 1 (Graph,Restricted Graph). *A weighted **graph** $G = (V, A, c)$ is described as a set of vertices $V, |V| = n$, a set of arcs $A \subseteq V \times V, |A| = m$ and a cost function $c : A \mapsto \mathbb{N}_{>0}$. We might choose to restrict G to a subset $\tilde{V} \subseteq V$. This **restricted graph** $G_{\tilde{V}}$ is defined as $G_{\tilde{V}} = (\tilde{V}, \tilde{A}, c)$, with $\tilde{A} = \{a = (u, v) \in A \mid u, v \in \tilde{V}\}$.*

All methods described within this work are targeted at shortest path computations. We define paths and the associated distances as follows:

Definition 2 (Paths,Length,Distance). *Given a graph $G = (V, A, c)$: We call a sequence $p_{s,t} = \langle s = v_0, \ldots, v_k = t \rangle$ with $v_i \in V, (v_i, v_{i+1}) \in A$ a **path** from s to t. Its **length** $\mathcal{L}(p_{s,t})$ is given as the combined weights of the represented arcs: $\mathcal{L}(p_{s,t}) = \sum_{i=0}^{k-1} c(v_i, v_{i+1})$. If the length of a path $p_{s,t}$ is minimal over all possible paths between s and t with respect to c, we call the path a **shortest path** and denote $p_{s,t} = \mathcal{P}_{s,t}$. The length of such a shortest path is called the **distance** between s and t: $\mathcal{D}(s, t) = \mathcal{L}(\mathcal{P}_{s,t})$. Furthermore, we define $\mathcal{P}_{s,v,t}$ as the **concatenated path** $\mathcal{P}_{s,v,t} = \mathcal{P}_{s,v} \cdot \mathcal{P}_{v,t}$.*

In the context of multiple graphs or paths, we denote the desired restriction via subscript. For example $\mathcal{D}_G(s, t)$ denotes the distance between s and t in G, with $\mathcal{D}_{p_{s,t}}(a, b)$ we denote the distance between a and b when following $p_{s,t}$.

Our metric of choice is the average travel time. Therefore, we might choose to omit the cost function c when naming a graph and simply give $G = (V, A)$.

3   Related Work

The large amount of research available that addresses shortest paths in road networks cannot be appropriately discussed within the limits of this paper. We therefore focus on the work most closely related to ours. Existing overview papers like [4] can give a better impression of available speed-up techniques and their different properties. In the following, we first discuss speed-up techniques and focus on alternative routes afterwards.

3.1   Speed-Up Techniques

While Dijkstra's algorithm [1] still is the fastest algorithm to calculate a single shortest path on a single CPU, the same does not hold true any more in the case of static graph data and multiple queries. The assumption that a large number of queries will be directed at a static network allows to invest a lot of computing power in advance to speed up queries later on. This idea has led to a large number of speed-up techniques. Due to the close relation to our work we cover Reach [2, 10], Contraction Hierarchies [3], and PHAST [11].

Contraction Hierarchies: Contraction Hierarchies [3] are a hierarchical speed-up technique. The hierarchy itself is defined through some strict order of the vertices $\prec: V \times V \mapsto \{\texttt{true}, \texttt{false}\}$. Vertices are processed, or *contracted*, in this order. During the contraction of a vertex v we inspect all pairs of $u \in V_i = \{u \in V \mid (u, v) \in A^+, v \prec u\}$ and $w \in V_o = \{w \in V \mid (v, w) \in A^+, v \prec w\}$, where A^+ is initially given as A but continually augmented with additional arcs. So to speak, we check whether the distance between two neighbouring vertices of v depends on v; whenever $\mathcal{D}_{G^+_{v \prec}}(u, w) > \mathcal{D}_{G^+_{v \preccurlyeq}}(u, w)$, we add the arc (u, w) with cost $\mathcal{D}_{G^+_{v \prec}}(u, w)$ to the current A^+, where $G^+_{v \prec} = (V_{v \prec}, A^+_{v \prec})$, which represents G augmented with additional arcs and restricted to vertices $u \in V$ with $v \prec u$. These new arcs are called *shortcuts* and preserve shortest path distances between vertices of the restricted graph $G^+_{v \prec}$. Each $G^+_{v \prec}$ can be interpreted as an overlay graph to G in which added shortcuts represent whole shortest paths containing vertices $u \prec v$. When all vertices have been processed, the graph G^+ is split up into two new graphs: $G^\uparrow$ and $G^\downarrow$, which represent sets of arcs leading away from ($G^\uparrow$) or to ($G^\downarrow$) the vertices.

Formally, $G^\uparrow$ is defined as $G^\uparrow = (V, A^\uparrow)$ with $A^\uparrow = \{a = (v, u) \in A^+ \mid v \prec u\}$. $G^\downarrow$ is defined the same way: $G^\downarrow = (V, A^\downarrow)$ with $A^\downarrow = \{a = (u, v) \in A^+ \mid v \prec u\}$.

A search within a Contraction Hierarchy consists of a forward search from s in $G^\uparrow$ and a backwards search from t in $G^\downarrow$. Geisberger et al. [3] prove the correctness of this approach by showing that for any $\mathcal{P}_{s,t}$ in the original graph, a concatenated path $\mathcal{P}_{s,v,t}$ can be found with $\mathcal{P}_{s,v,t} = \mathcal{P}_{G^\uparrow,s,v} \cdot \mathcal{P}_{G^\downarrow,v,t}$ with identical length to $\mathcal{P}_{s,t}$ within the original, unprocessed graph. This specific form of path is called an *up-down* path, as it climbs up the hierarchy to v and then descends to t. To retrieve the full shortest path information the concatenated path has to be *unpacked*: All shortcuts are of the form (u, v, w) with (u, v) and (v, w) potentially describing shortcuts as well. By remembering the middle vertex v for every shortcut, the full shortest path in the original graph can be extracted recursively.

PHAST: This technique, though not really a speed-up technique itself, is a method to exploit Contraction Hierarchies and modern hardware architectures to compute full shortest path trees [11]. It exploits the up-down characteristic of shortest paths within a Contraction Hierarchy as well as the fact that, though originally designed as an n-*level* hierarchy (compare the strict ordering), the actual hierarchical information of a Contraction Hierarchy order represents a rather shallow hierarchy; whenever two vertices cannot reach each other within the hierarchy, they are completely independent from one another and can be viewed as if they were in the same level. The method performs two consecutive steps: In a first step Delling et al. perform a full upwards search from s in $G^\uparrow$. In the second step, vertices are processed (or *swept*) in reverse contraction order, updating distances whenever a better distance value than the current one can be found by using an arc from $G^\downarrow$ in combination with the by now correct distance values of higher level vertices. The nature of a Contraction Hierarchy guarantees the distance value of the vertex with the highest level to be set correctly after this initial search. Inductively, the correctness of lower level vertices follows. This principle can be localized [12,13] by restricting the set of swept vertices to those reachable in backwards direction in $G^\downarrow$, starting at one of the desired targets.

Reach: Reach is a goal-directed technique introduced by Gutman [10]. Gutman defines the *reach* of a vertex v as $r(v) = \max_{\mathcal{P}_{s,t}, v \in \mathcal{P}_{s,t}} \min(\mathcal{P}_{s,v}, \mathcal{P}_{v,t})$. During a search between s and t, the reach of a vertex v allows for pruning of v whenever $\mathcal{D}(s,v) > r(v)$ and $\mathcal{D}(v,t) > r(v)$. Goldberg et al. [2] improved Gutman's approach through the inclusion of shortcuts, allowing for better pruning.

3.2   Alternative Routes

Alternative routes in road networks have first been formally studied by Abraham et al. [7]. By now two general approaches can be found in the literature: via-node alternative routes or the related plateaux-method [6–9], and penalty-based approaches [9]. Due to the large differences between the two approaches and the limited scope of this paper, we only cover the via-node approach at this point.

Via-Node Alternative Routes: Within a graph $G = (V, A)$, a via-node alternative to a shortest path $\mathcal{P}_{s,t}$ can be described by a single vertex $v \in V \setminus \mathcal{P}_{s,t}$. The alternative route is described as $\mathcal{P}_{s,v,t}$. As this simple description can result in arbitrarily bad paths, for example paths containing loops, Abraham et al. [7] define a set of criteria to be fulfilled for an alternative route to be *viable*. A viable alternative route provides the user with a real alternative, not just minimal variations (*limited sharing*), is not too much longer (*bounded stretch*) and does not contain obvious flaws, i.e. sufficiently small sub-paths have to be optimal (*local optimality*). Formally, we define these criteria as follows:

Definition 3 (Viable Alternative Route). *Given a graph $G = (V, A)$, a source s, a target t, and a via-candidate v as well as three tuning parameters γ, ϵ, α; v is a viable via-candidae, and thus defines a viable via-node alternative route $\mathcal{P}_{s,v,t} = \mathcal{P}_{s,v} \cdot \mathcal{P}_{v,t}$, if following three criteria are fulfilled:*

1. $\mathcal{L}(\mathcal{P}_{s,t} \cap \mathcal{P}_{s,v,t}) \leq \gamma \cdot \mathcal{D}(s,t)$ (*limited sharing*)
2. $\forall a, b \in \mathcal{P}_{s,v,t}, \mathcal{D}_{\mathcal{P}_{s,v,t}}(s,a) < \mathcal{D}_{\mathcal{P}_{s,v,t}}(s,b)$:
   $\mathcal{D}_{\mathcal{P}_{s,v,t}}(a,b) \leq (1+\epsilon) \cdot \mathcal{D}(a,b)$ (*bounded stretch*)
3. $\forall a, b \in \mathcal{P}_{s,v,t}, \mathcal{D}_{\mathcal{P}_{s,v,t}}(s,a) < \mathcal{D}_{\mathcal{P}_{s,v,t}}(s,b)$,
   $\mathcal{D}_{\mathcal{P}_{s,v,t}}(a,b) \leq \alpha \cdot \mathcal{D}(s,t) : \mathcal{D}_{\mathcal{P}_{s,v,t}}(a,b) = \mathcal{D}(a,b)$ (*local optimality*)

The usual choice of γ, ϵ, and α is to allow at most $\gamma = 80$ % overlap between $\mathcal{P}_{s,v,t}$ and $\mathcal{P}_{s,t}$. Furthermore the user should never travel more than $\epsilon = 25$ % longer than necessary between any two points on its track, and every subpath that is at most $\alpha = 25$ % as long as the original shortest path should be an optimal path.

These criteria require a quadratic number of shortest path queries to be fully evaluated. Therefore, Abraham et al. propose a possibility to approximate these criteria [7]. The *approximated viability test* requires three shortest path queries. The first two queries are required to calculate the actual candidate itself, which is necessary to test the amount of sharing; this is unless we use Dijkstra's algorithm, which already computes the shortest path information. For the stretch, we have to check the distance between the two vertices d_1, d_2 where both $\mathcal{P}_{s,t}$ and the alternative candidate $\mathcal{P}_{s,v,t}$ deviate/meet up. As long as $\mathcal{D}_{\mathcal{P}_{s,v,t}}(d_1, d_2)$ does not violate the stretch criteria in respect to $\mathcal{D}(d_1, d_2)$, the path passes the test. Subpath optimality gives the distance necessary for the comparison. The third query is used to approximate the local optimality and is performed between the two vertices o_1, o_2 which are closest to v on $\mathcal{P}_{s,v,t}$ and still fulfil $\mathcal{D}_{\mathcal{P}_{s,v,t}}(o_1, v) \geq T$ and $\mathcal{D}_{\mathcal{P}_{s,v,t}}(v, o_2) \geq T$ respectively, where $T = \alpha \cdot \mathcal{D}(d_1, d_2)$. This check is called the *T-test*. Note that for a correct approximation of the above mentioned criteria, $T = \alpha \cdot \mathcal{D}(s, t)$ would be required. Due to the nature of the test, that would make it impossible to allow for alternatives with up to 80% sharing without setting α to less than 10%. Therefore Abraham et al. [7] propose to check against α, and for consistency reasons ϵ, in relation to $\mathcal{D}(d_1, d_2)$.

Definition 3 can be directly extended to allow for second or third alternatives (alternative routes of *degree* 2, 3 or even n). Only the limited sharing parameter has to be tested against the full set of alternatives already known. Bounded stretch and local optimality are only checked against the original shortest path and therefore translate directly.

Abraham et al. [7] give multiple algorithms to compute alternative routes. The reference algorithm (X-BDV) is based on a bidirectional implementation of Dijkstra's algorithm and is used as the gold standard. To avoid the long query time of Dijkstra's algorithm on continental-sized networks, they also give techniques based on Reach and Contraction Hierarchies. Due to the strong restrictions of the search spaces caused by the speed-up techniques, they present weakened search criteria which they call relaxation. For example, in the Contraction Hierarchy they allow to look downwards the hierarchy under certain conditions. The relaxation can be applied in multiple levels. The respective algorithms are referred to by X-CHV-3/X-CHV-5 for the Contraction Hierarchy-based methods with relaxation levels of 3/5 and X-REV-1/X-REV-2 for the Reach-based variants and their respective relaxation levels[2].

Luxen and Schieferdecker [8] improved the algorithm of Abraham et al. in terms of query times by storing a precomputed small set of via-candidates for pairs of regions within the graph. While their method is faster than our algorithm, we refrain from comparing ours to their implementation as they require

---

[2] We refer to these in the experimental section.

significantly more preprocessing and storage overhead. Additionally, the small set of candidates lowers the chance to find high quality routes even more.

Plateaux and Via-Nodes: A *plateau* is defined with respect to two shortest paths $\mathcal{P}_{s,u}$ and $\mathcal{P}_{v,t}$. If $\mathcal{P}_{s,u} \cap \mathcal{P}_{v,t} \neq \emptyset$, the common segment is called a plateau. It is easy to see that if two paths share a plateau of appropriate length, any vertex on said plateau can be chosen as via-candidate and will fulfil the local optimality criterion. It is therefore equally valid to find plateaux of length T (compare *approximated viability*) instead of performing the T-test. Even prior to [7], Cambridge Vehicle Information Technology Ltd. published a method based on full shortest path trees and plateaux [6] called *Choice Routing*. Our algorithm takes a rather similar approach to [6] while not relying on full shortest path trees.

4   Hierarchy Decomposition for Alternative Routes

One of the most unsatisfying properties of the known algorithms for via-node alternatives is the selection process. Due to the three necessary shortest path queries, testing all potential candidates proves expensive. The relaxation process, which is necessary to increase the chance of finding alternative routes in the first place, does not only increase the number of valid via candidates, it also increases the number of candidates that do not provide viable alternative routes immensely. Therefore, current algorithms order the candidates heuristically. The candidates are tested in this order until a first candidates passes the viability test; the respective candidate is reported as the result. Potentially better candidates are discarded.

For our new algorithm, we want to take a different approach. The main goal is to make the check for viability fast enough to test all potential candidates. Our algorithm (HiDAR), makes the approach of [6] viable:

1. Compute (*pruned*) shortest path DAGs from s and directed at t
2. Utilize plateaux to create a compact and meaningful alternative graph
3. Extract alternative routes from this graph

The main task is the calculation of the alternative graph. In this alternative graph $H = (V_H, A_H)$ we want to encode the following information: $\mathcal{D}(s, v)$ and $\mathcal{D}(v, t)$ for any $v \in V_H$. Furthermore, we encode for any $a \in A_H$ whether a forms a plateau with respect to the paths in the shortest path DAGs from the first step. No vertex in V_H, except for s, t or the first/last vertex of a plateau should have indegree and outdegree equal to one. It is obvious that alternative graph extraction on such graphs is simple, as all the information for the (approximated) viability check is present. Sufficiently small graphs allow for very fast alternative graph extraction. This process is illustrated graphically in Figure 2.

4.1   High Level Algorithm Description

The first step of our algorithm computes shortest path trees within a Contraction Hierarchy, instead of the full graph (compare Figure 1). Due to possible overlaps hidden within shortcuts, these shortest path trees actually describe a shortest path DAG in the original graph.

Shortest Path Search: Our approach resembles the behaviour described in [12, 13]. In a first step, we perform two exhaustive upwards searches from s in $G^\uparrow$ and backwards from t in $G^\downarrow$. These searches describe two sets of vertices V_s and V_t which contain those vertices which were reached by the two respective searches, together with their (tentative) distance values. We aim at computing the correct distance values from s to $V_s \cup V_t$ as well as from $V_s \cup V_t$ to t. In the following, we describe the process for s, t is handled analogously. Following the principles of [12, 13], we can compute $\mathcal{D}(s, v)$ for any $v \in V_t$ by sweeping (see Section 3) V_t in reverse contraction order. To compute $\mathcal{D}(s, v)$ for all $v \in V_s$, we need to extract what we call the *backwards hull* $H_{G^\downarrow}(V_s)$ of V_s: the union of all vertices reachable by backwards arcs in $G^\downarrow$ from any vertex in V_s. Note that in an undirected graph $H_{G^\downarrow}(V_s) = V_s$. By sweeping $H_{G^\downarrow}(V_s)$ in reverse contraction order, we can calculate $\mathcal{D}(s, v)$ for all $v \in V_s$.

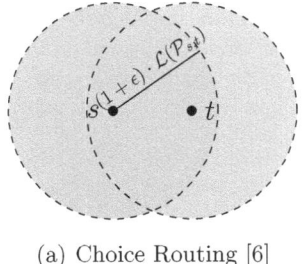

(a) Choice Routing [6]

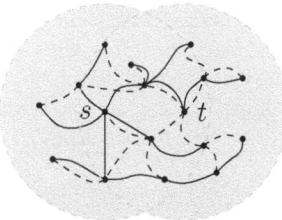

(b) HiDAR; dots mark $V_s \cup V_t$, forward tree painted solid, backward tree dashed

Fig. 1. Schematic representation of graph exploration for shortest path tree calculation

For the further steps, we restrict the processing to vertices $v \in V_s \cup V_t$ with $\mathcal{D}(s, v) + \mathcal{D}(v, t) \leq (1 + \epsilon) \cdot \mathcal{D}(s, t)$ (compare Definition 3, Figure 2(b)).

Hierarchy Decomposition: The data calculated in the first step contains all necessary distance information for our alternative graph. The alternative graph itself has to be calculated by finding all vertices that actually provide relevant information. Those vertices might currently be encoded within the shortcuts of the shortest path trees: any two shortcuts, whether they belong to the tree rooted at s or the one rooted at t, can have common segments, as each of the shortcuts represents a shortest path in the original graph. We need to find the vertices at which two of these paths meet up or deviate from one another and decide whether they form a plateau, i.e. do not belong to the same DAG.

We perform a Hierarchy Decomposition [14] to unpack all shortcuts at once. During the process, we handle shortcuts in reverse contraction order of their middle vertices, ordering them with a priority queue. The order enables us to

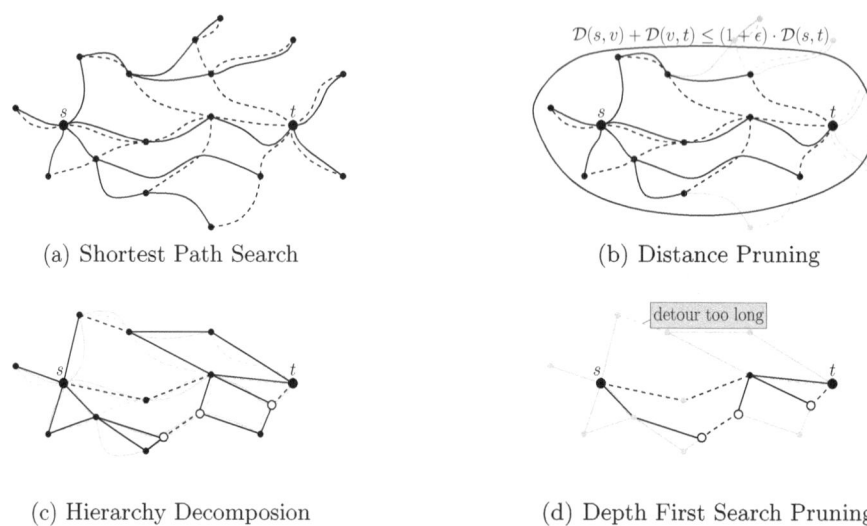

(a) Shortest Path Search

(b) Distance Pruning

(c) Hierarchy Decomposion

(d) Depth First Search Pruning

Fig. 2. Schematic representation of the alternative graph generation process of HiDAR. $H(V_s, V_t)$ is marked with solid dots. The forward shortest path tree is marked with solid curves, the backward tree with dashed ones. Hollowed circles represent newly found important vertices. Solid straight lines (c,d) mark arcs in the alternative graphs. Dashed straight lines mark plateaux within the alternative graph. Grayed out information represents pruned information.

decide locally whether a given middle vertex is important by just storing its predecessor and successor in the fully unpacked shortcut, as all shortcuts containing the same middle vertex are processed subsequently. All vertices with *indegree + outdegree* > 2 are deemed important for the alternative graph.

We can avoid special cases in this process by including some dummy arcs and vertices. By handling the shortcuts in a carefully chosen order, we also manage to calculate some additional data during the unpacking process that allows for direct generation of a compacted version of the alternative graph, consisting only of important vertices as well as source/target vertices of the input shortcuts. This step is illustrated schematically in Figure 2(c).

Alternative Graph Extraction: While the graph calculated in the previous step already contains all necessary information desired for our alternative graph, it still contains a lot of unwanted information, too. Actually important for the alternative route extraction are only paths that connect plateaux of sufficient size to the source and the target vertex. To finalize our alternative graph, we perform a depth first search (DFS) from the source. Whenever this DFS encounters a plateau large enough to justify the necessary detour to reach it, which we approximate by trivial lower bounds, we mark the paths from s to the plateau and from the plateau to t to be included in the alternative graph.

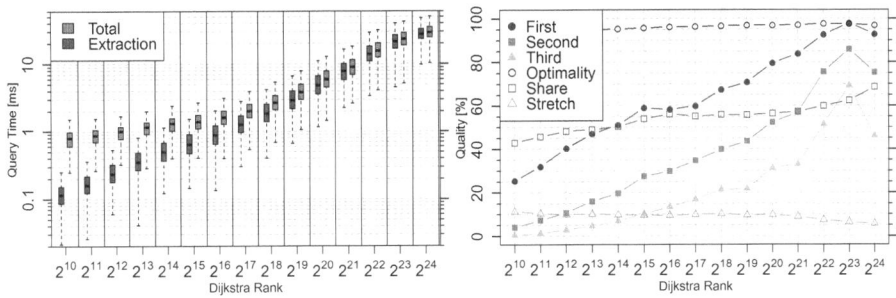

Fig. 3. Query time, in total and restricted to the graph generation process (left). Quality measurements in terms of success rates for the first three alternatives and specifically for the first alternative (right).

All other plateaux, compare Figure 2(d), are discarded. Finally, we compact the graph to get the alternative graph we report as the result of our algorithm. A more in depth discussion of our algorithm can be found in [15].

5   Experiments

Our experimental configuration is as follows: All our experiments were performed on a single core of an Intel(R) Core(TM) i7 920 CPU clocked at 2.67 GHz. The machine is equipped with 12GB DDR3-1333 RAM, runs SUSE Linux (kernel 2.6.34.10-0.6-default), and approximates the machine used in [7]. The code was developed in c++ and compiled using g++ in version 4.5.0 using -std=c++0x -O3 -mtune=native as parameters. We use two kinds of inputs to test our algorithm. The first set T_n is composed of 10 000 source and target pairs, the second set T_r of 1 000 source and target pairs per Dijkstra rank. All queries are chosen uniformly at random. We use classic parallel Contraction Hierarchies preprocessing, allowing to process the European road network within a few minutes. In this section, we focus only on the runtime and the quality of our algorithm. *Success rates* given describe the number of times an alternative route of a given degree was found. For alternative route selection, we follow the approach of [6] and select long plateaux first.

Runtime: First, we compare the query time and success rates of our algorithm to the techniques presented in [7]. For the first alternative, our algorithm is already competitive in its query time. Higher degree alternative route extraction does not really affect our algorithm. In fact, we can extract all alternatives within less than 0.1 ms. In contrast, the other algorithms, which provide similar success rates, require significantly longer query times for higher degree alternatives. The algorithm by Luxen and Schieferdecker would outperform ours in pure query time, but requires additional sets of via candidates for any possible further alternative and a significant amount of additional storage and preprocessing time.

Table 1. Comparison of runtime and success rates for multiple alternatives based on the test set T_n. Numbers for X-BDV, X-CHV-3/X-CHV-5, X-REV-1/X-REV-2 are based upon the data given in [7]. Query times are accumulated times up to the n-th alternative.

	first		second		third	
algorithm	time [ms]	success rate [%]	time [ms]	success rate [%]	time [ms]	success rate [%]
X-BDV	11 451.5	94.5	12 225.9	80.6	13 330.9	59.6
X-CHV-3	16.9	90.7	20.3	70.1	22.1	42.3
X-CHV-5	55.2	92.9	65.0	77.0	73.2	53.3
X-REV-1	20.4	91.3	33.6	70.3	42.6	43.0
X-REV-2	34.3	94.2	50.3	79.0	64.9	56.9
HiDAR	18.2	91.5	18.2	75.7	18.2	55.9

Our algorithm not only allows for finding second and third degree alternatives with a high success rate; we observe a fourth alternative in about 36.4% of all cases and a fifth/sixth alternative in 22.4% and 12.18% of the cases while still only requiring 18.2 *ms* on average.

The runtime of our algorithm is dominated by the graph generation process, as can be seen in Figure 3. While unpacking shortcuts in advance does speed up most techniques, it does not so in ours. On the contrary, we measured significantly longer query times for our algorithm, well beyond the 80 *ms* mark. But due to the large amount of overlaps present in our data and due to the faster compression of our new technique, our live extraction manages to outperform pre-unpacked shortcuts.

Quality: As our method to alternative routes differs greatly from the approaches presented in the literature so far, we also evaluate the quality of the alternative routes generated by our algorithm based on stretch, sharing and local optimality. The quality values are given in Table 2. In the worst case, we report stretch values worse than previous algorithms. These values originate from a very small set of outliers that would have been deemed viable by the other algorithms as well. On the other hand, we outperform the classic algorithms by far by providing better stretch values than previously known algorithms. What strikes the eye the most are the values for local optimality. The worst case performance of our algorithm can nearly compete with the average case performance of the other algorithms. On average, we can report local optimality of 90 %. These routes are probably not reported by X-BDV, as the selection method prefers other alternatives first that share some parts with the high quality route, thus making it inadmissible later on. However, the computational overhead of X-BDV does not allow to justify this assumption by fully exploring all potential candidate selection schemes. Only in terms of sharing our first alternative performs worse than other algorithms.

Table 2. Quality of HiDAR alternatives (based on test set T_n) compared to X-BDV, X-CHV and X-REV. For optimality we report local optimality restricted to the detour (as also done in [7]). For second and third alternatives, no quality measures are given for the relaxed variants of X-CHV and X-REV.

#	algorithm	success rate [%]	UBS avg	max	sharing [%] avg	max	loc opt [%] avg	min
first	X-BDV	94.5	9.4	35.8	47.2	79.9	73.1	30.3
	X-CHV-3	90.7	11.5	45.4	45.4	80.0	67.7	30.0
	X-REV-2	94.2	9.7	31.6	46.6	79.9	71.3	27.6
	HiDAR	91.5	6.6	64.2	63.0	80.0	97.4	70.5
second	X-BDV	80.6	11.8	38.5	62.4	80.0	71.8	29.6
	HiDAR	75.7	10.6	57.7	63.2	80.0	94.4	64.0
third	X-BDV	59.6	13.2	41.2	68.9	80.0	68.7	30.6
	HiDAR	55.9	12.9	76.2	62.5	80.0	92.4	65.3

The quality for higher degree alternatives still remains high. The uniformly bounded stretch gradually increases to 21.4 % on average for the 10th alternative, while the amount of sharing averages out at around 60 % and local optimality remains high with an average of around 82.9 %. Figure 3 shows that we can achieve this high quality over the full range of queries.

6   Conclusion and Future Work

We have presented a new method to compute alternative routes in a competitive way, allowing for high success rates and the extraction of a maximum number of alternative routes without additional effort. By this, we present the only algorithm providing reasonable means for extracting up to 10 alternative routes. The high performance during the actual alternative route extraction allows for a precise evaluation of the alternative candidates and for extraction schemes based on personal preferences instead of accepting any acceptable candidate.

Still, the runtime of our algorithm itself could possibly be improved. The full extraction of shortcuts does seem like an unnecessary amount of effort compared to the low number of important vertices. Finding methods to restrict the extractions to only shortcuts promising some results would greatly speed up our algorithm.

Although every plateau defines a viable via-node alternative route, this does not hold true the other way around. Therefore, it would be interesting to see how many more alternative routes can be found through a combination of both via-node alternatives and our method or even to incorporate other approaches like [13].

Acknowledgements. We would like to thank Dennis Schieferdecker for interesting discussions and insight into the world of alternative routes.

References

1. Dijkstra, E.W.: A Note on Two Problems in Connexion with Graphs. Numerische Mathematik 1, 269–271 (1959)
2. Goldberg, A.V., Kaplan, H., Werneck, R.F.: Reach for A*: Efficient Point-to-Point Shortest Path Algorithms. In: Proceedings of the 8th Workshop on Algorithm Engineering and Experiments (ALENEX 2006), pp. 129–143. SIAM (2006)
3. Geisberger, R., Sanders, P., Schultes, D., Delling, D.: Contraction Hierarchies: Faster and Simpler Hierarchical Routing in Road Networks. In: McGeoch, C.C. (ed.) WEA 2008. LNCS, vol. 5038, pp. 319–333. Springer, Heidelberg (2008)
4. Delling, D., Sanders, P., Schultes, D., Wagner, D.: Engineering Route Planning Algorithms. In: Lerner, J., Wagner, D., Zweig, K.A. (eds.) Algorithmics. LNCS, vol. 5515, pp. 117–139. Springer, Heidelberg (2009)
5. Delling, D., Goldberg, A.V., Pajor, T., Werneck, R.F.: Customizable Route Planning. In: Pardalos, P.M., Rebennack, S. (eds.) SEA 2011. LNCS, vol. 6630, pp. 376–387. Springer, Heidelberg (2011)
6. Cambridge Vehicle Information Technology Ltd. (Choice Routing)
7. Abraham, I., Delling, D., Goldberg, A.V., Werneck, R.F.: Alternative Routes in Road Networks. ACM Journal of Experimental Algorithmics 18(1), 1–17 (2013)
8. Luxen, D., Schieferdecker, D.: Candidate Sets for Alternative Routes in Road Networks. ACM Journal of Experimental Algorithmics (submitted, 2013)
9. Bader, R., Dees, J., Geisberger, R., Sanders, P.: Alternative Route Graphs in Road Networks. In: Marchetti-Spaccamela, A., Segal, M. (eds.) TAPAS 2011. LNCS, vol. 6595, pp. 21–32. Springer, Heidelberg (2011)
10. Gutman, R.J.: Reach-Based Routing: A New Approach to Shortest Path Algorithms Optimized for Road Networks. In: Proceedings of the 6th Workshop on Algorithm Engineering and Experiments (ALENEX 2004), pp. 100–111. SIAM (2004)
11. Delling, D., Goldberg, A.V., Nowatzyk, A., Werneck, R.F.: PHAST: Hardware-accelerated shortest path trees. Journal of Parallel and Distributed Computing (2012)
12. Delling, D., Goldberg, A.V., Werneck, R.F.: Faster Batched Shortest Paths in Road Networks. In: Proceedings of the 11th Workshop on Algorithmic Approaches for Transportation Modeling, Optimization, and Systems (ATMOS 2011). OpenAccess Series in Informatics (OASIcs), vol. 20, pp. 52–63 (2011)
13. Delling, D., Kobitzsch, M., Luxen, D., Werneck, R.F.: Robust Mobile Route Planning with Limited Connectivity. In: Proceedings of the 14th Meeting on Algorithm Engineering and Experiments (ALENEX 2012), pp. 150–159. SIAM (2012)
14. Luxen, D., Sanders, P.: Hierarchy Decomposition for Faster User Equilibria on Road Networks. In: Pardalos, P.M., Rebennack, S. (eds.) SEA 2011. LNCS, vol. 6630, pp. 242–253. Springer, Heidelberg (2011)
15. Kobitzsch, M.: An Alternative to Alternative Routes: HiDAR. Technical Report

Efficient Indexes for Jumbled Pattern Matching with Constant-Sized Alphabet

Tomasz Kociumaka[1], Jakub Radoszewski[1], and Wojciech Rytter[1,2,*]

[1] Faculty of Mathematics, Informatics and Mechanics,
University of Warsaw, Warsaw, Poland
{kociumaka,jrad,rytter}@mimuw.edu.pl
[2] Faculty of Mathematics and Computer Science,
Copernicus University, Toruń, Poland

Abstract. We introduce efficient data structures for an indexing problem in non-standard stringology — jumbled pattern matching. Moosa and Rahman [J. Discr. Alg., 2012] gave an index for jumbled pattern matching for the case of binary alphabets with $O(\frac{n^2}{\log^2 n})$-time construction. They posed as an open problem an efficient solution for larger alphabets. In this paper we provide an index for any constant-sized alphabet. We obtain the first $o(n^2)$-space construction of an index with $o(n)$ query time. It can be built in $O(n^2)$ time. Precisely, our data structure can be implemented with $O(n^{2-\delta})$ space and $O(m^{(2\sigma-1)\delta})$ query time for any $\delta > 0$, where m is the length of the pattern and σ is the alphabet size ($\sigma = O(1)$). We also break the barrier of quadratic construction time for non-binary constant alphabet simultaneously obtaining poly-logarithmic query time.

1 Introduction

The problem of *jumbled pattern matching* is a variant of the standard pattern matching problem. The *match* between a given pattern and a factor of the word is defined in a nonstandard way. In this paper by a *match* of two words we mean their commutative (Abelian) equivalence: one word can be obtained from the other by permuting its symbols. This relation can be conveniently described using *Parikh vectors*, which show frequency of each symbol of the alphabet in a word: u and v are commutatively equivalent (denoted as $u \approx v$) if and only if their Parikh vectors are equal. In the jumbled pattern matching the query pattern is given as a Parikh vector, which in our case is of a constant size (due to small alphabet).

Several results related to indexes for jumbled pattern matching in binary words have been obtained recently. Cicalese et al. [8] proposed an index with $O(n)$ size and $O(1)$ query time and gave an $O(n^2)$ time construction algorithm for the index. The key observation used in this index is that if a word contains two factors of length ℓ containing i and j ones, $i < j$, respectively, then it

---

* Supported by grant no. N206 566740 of the National Science Centre.

H.L. Bodlaender and G.F. Italiano (Eds.): ESA 2013, LNCS 8125, pp. 625–636, 2013.
© Springer-Verlag Berlin Heidelberg 2013

must contain a factor of length ℓ with any intermediate number of ones. The construction time was improved independently by Burcsi et al. [4] (see also [5,6]) and Moosa and Rahman [20] to $O(n^2/\log n)$ and then by Moosa and Rahman [21] to $O(n^2/\log^2 n)$. An index for trees vertex-labeled with $\{0, 1\}$ achieving the same complexity bounds ($O(n^2/\log^2 n)$ construction time, $O(n)$ size and $O(1)$ query time) was given in [17]. The general problem of computing an index for jumbled pattern matching in trees and graphs is known to be NP-complete [13,19].

Moosa and Rahman [20,21] posed an open problem for a construction of an $o(n^2)$ indexing scheme for general alphabet (with $o(n)$ query time). In particular, even for a ternary alphabet none was known, the basic observation used to obtain a binary index does not hold for any larger alphabet. We prove that the answer for this problem is positive for any constant-sized alphabet. We show an $O(n^2(\log\log n)^2/\log n)$ time and space construction of an index that enables queries in $O((\frac{\log n}{\log\log n})^{2\sigma-1})$ time. We also give a solution with $O(n^{2-\delta})$ size of the index, $O(n^2)$ construction time and $O(m^{(2\sigma-1)\delta})$ query time. Here σ is the size of the alphabet and m is the length of the pattern, that is, the sum of components of the Parikh vector of the pattern. Our construction algorithms are randomized (Las Vegas) and work in word-RAM model with word size $\Omega(\log n)$. The latter index is described in Section 3 and in Section 4 (improvement from $n^{(2\sigma-1)\delta}$ to $m^{(2\sigma-1)\delta}$ in query time) and the former index is given in Section 5 (auxiliary tools) and in Section 6.

A notion closely related to jumbled pattern matching are Abelian periods, first defined in [9]. The pair (i, p) is an Abelian period of w if $w = u_0 u_1 \ldots u_k u_{k+1}$ where $u_1 \approx u_2 \approx \ldots \approx u_k$, u_0 and u_{k+1} contain a subset of letters of u_1, and $|u_0| = i$, $|u_1| = p$. Recently there have been a number of results related to efficient algorithms for Abelian periods [9,14,15,18,10]. There have also been several combinatorial results on Abelian complexity in words [1,7,11,12] and partial words [2,3].

2 Preliminaries

In this paper we assume that the alphabet Σ is $\{1, 2, \ldots, \sigma\}$ for $\sigma = O(1)$. For a word $w \in \Sigma^n$ by $w[i \mathinner{.\,.} j]$ denote a *factor* of w equal to $w_i \ldots w_j$. We say that the factor $w[i \mathinner{.\,.} j]$ *occurs* at the position i.

Let $\#_s(x)$ denote the number of occurrences of the letter s in x. The Parikh vector $\Psi(x)$ of the word $x \in \Sigma^*$ is defined as:

$$\Psi(x) = (\#_1(x),\ \#_2(x),\ \#_3(x),\ \ldots, \#_\sigma(x)).$$

We say that the words x and y are *commutatively equivalent* (denoted as $x \approx y$) if y can be obtained from x by a permutation of its letters. Observe that we have:

$$x \approx y \iff \Psi(x) = \Psi(y).$$

Example 1. $1221 \approx 2211$, since $\Psi(1221) = (2, 2) = \Psi(2211)$.

We define $Occ(p)$ as the set of all positions where factors of w with the Parikh vector p occur. If $Occ(p) \neq \emptyset$, we say that p *occurs* in w.

The problem of *indexing for jumbled pattern matching* (also called indexing for permutation matching [20,21]) is defined as follows:

Preprocessing: build an index for a given word w of length n;

Query: for a given Parikh vector (a pattern) p decide whether p occurs in w.

$$1 \quad 2 \quad 3 \quad 1 \quad 2 \quad 3 \quad 4 \quad 3 \quad 2 \quad 1 \quad 3 \quad 4 \quad 1 \quad 2 \quad 2 \quad 3 \quad 1 \quad 3 \quad 4 \quad 3$$

Fig. 1. Let $w = 1\,2\,3\,1\,2\,3\,4\,3\,2\,1\,3\,4\,1\,2\,2\,3\,1\,3\,4\,3$. The factor $2\,3\,1\,2\,3\,4$ occurs (as a word) starting at position 2. We have $\Psi(2\,3\,1\,2\,3\,4) = (1,2,2,1)$, hence the jumbled pattern $p = (1,2,2,1)$ (Parikh vector) also occurs at position 2. There are several occurrences of the jumbled pattern p: $Occ(p) = \{2,4,5,11,14\}$.

Define the *norm* of a Parikh vector $p = (p_1, p_2, \ldots, p_\sigma)$ as:

$$|p| = \sum_{i=1}^{\sigma} |p_i|.$$

For two Parikh vectors p, q, by $p + q$ and $p - q$ we denote their component-wise sum and difference. We define the *extension* sets of Parikh vectors:

$$Ext_{<r}(p) = \{p + p' : |p'| < r\}, \quad Ext_r(p) = \{p + p' : |p'| = r\}.$$

Also define

$$Ext'_{<r}(p) = Ext_{<r}(p) \cap \{p' : Occ(p) \cap Occ(p') \neq \emptyset\}.$$

For a set X of Parikh vectors, we define:

$$Ext_{<r}(X) = \bigcup_{p \in X} Ext_{<r}(p), \quad Ext'_{<r}(X) = \bigcup_{p \in X} Ext'_{<r}(p).$$

Lemma 2. *For any Parikh vector p and integer $r \geq 0$, $|Ext_r(p)| = O(r^{\sigma-1})$, $|Ext_{<r}(p)| = O(r^\sigma)$ and $|\{q : p \in Ext_r(q)\}| = O(r^{\sigma-1})$.*

Proof. Both $|Ext_r(p)|$ and $|\{q : p \in Ext_r(q)\}|$ are bounded by the number of Parikh vectors of norm r, which is $\binom{r+\sigma-1}{\sigma-1}$, since each Parikh vector corresponds to a placement of r indistinguishable balls ('positions') into σ distinguishable urns ('letters').

To bound $|Ext_{<r}(p)|$ it suffices to observe that

$$Ext_{<r}(p) = \bigcup_{k=0}^{r-1} Ext_k(p).$$

$\square$

Let us also introduce an efficient tool for determining Parikh vectors of factors of a given word.

Lemma 3. *After $O(n)$ time preprocessing, the Parikh vector $\Psi(w[i \mathinner{.\,.} j])$ for any $1 \le i \le j \le n$ can be computed in $O(1)$ time.*

Proof. For each $k \in \{0, \ldots, n\}$ we precompute $\Psi(w[1 \mathinner{.\,.} k])$ in $O(n)$ time. Then

$$\Psi(w[i \mathinner{.\,.} j]) = \Psi(w[1 \mathinner{.\,.} j]) - \Psi(w[1 \mathinner{.\,.} i-1]).$$

$\square$

3 Index with Sublinear Time Queries

A Parikh vector which occurs in w is called an *Abelian factor* of w. Observe that the *zero vector* is an Abelian factor of every word, since it corresponds to the empty word.

Define $Abelian(w)$ as the set of all Abelian factors of w. For example:

$$Abelian(a^k b^k) = \{(i,j) \ : \ 0 \le i, j \le k\}.$$

For a positive integer L we say that a pair $(\mathcal{A}, \mathcal{B})$ of two disjoint subsets of $Abelian(w)$ is *L-good* if the following conditions are satisfied:

(1) $\sum_{p \in \mathcal{A} \cup \mathcal{B}} |Occ(p)| \le n^2/L$;

(2) $|Ext_{<L}(\mathcal{B})| = O(n^2/L)$;

(3) $|Occ(p)| \le L^\sigma$ for each $p \in \mathcal{A}$;

(4) for each $z \in Abelian(w)$ we have:

$$z \in Ext'_{<L}(\mathcal{B}) \quad \text{or} \quad \exists_{p \in \mathcal{A}, \, 0 \le |z| - |p| < L} \ Occ(z) \cap Occ(p) \ne \emptyset.$$

Note that condition (4) could also be stated as $z \in Ext'_{<L}(\mathcal{B}) \cup Ext_{<L}(\mathcal{A})$, however we choose the above statement due to operational reasons (see the following Query algorithm).

Let

$$\mathcal{F}_L = \{p \in Abelian(w) \ : \ |p| \bmod L = 0\}.$$

Elements of $\mathcal{F}_L$ are called *L-factors*, clearly $|\mathcal{F}_L| \le \frac{n^2}{L}$.

We partition the set of L-factors into the set $\mathcal{L}_L$ of *light* Abelian factors (small number of occurrences) and the set $\mathcal{H}_L$ of *heavy* Abelian factors (more occurrences in w). More formally:

$$\mathcal{L}_L = \{p \in \mathcal{F}_L : |Occ(p)| \le L^\sigma\},$$
$$\mathcal{H}_L = \{p \in \mathcal{F}_L : |Occ(p)| > L^\sigma\} \cup \{\bar{0}\}.$$

Lemma 4. *The pair* $(\mathcal{L}_L, \mathcal{H}_L)$ *is* L-good.

Proof. The total size of $Occ(p)$ for all p, such that $|p| = k \cdot L$ for a fixed k, is at most n. Hence, the total size of $Occ(p)$ for all $p \in \mathcal{F}_L$ is at most $\frac{n^2}{L}$, which gives part (1) of the definition of an L-good pair.

Part (3) follows from the definition of $\mathcal{L}_L$. As for part (2), by Lemma 2, for each $p \in \mathcal{H}_L$ we have

$$|Ext_{<L}(p)| = O(L^\sigma) = O(|Occ(p)|).$$

The latter inequality follows from the definition of $\mathcal{H}_L$. This shows that the size of $Ext_{<L}(\mathcal{H}_L)$ is bounded by the total size of $Occ(p)$ for $p \in \mathcal{H}_L$, which we have already shown (part (1)) to be bounded by $\frac{n^2}{L}$.

As for property (4), consider any $z \in Abelian(w)$ and its occurrence at position i. Let $p = \Psi(w[i .. i + k \cdot L - 1])$, where $k \cdot L \le |z| < (k+1) \cdot L$. Then either $p \in \mathcal{H}_L$ and therefore $z \in Ext'_{<L}(p)$ or $p \in \mathcal{L}_L$ and clearly $i \in Occ(p)$. $\square$

Data Structure. The index consists of the following parts:

1. the sets $\mathcal{L}_L$, $\mathcal{H}_L$ and $Ext'_{<L}(\mathcal{H}_L)$;

2. $Occ(p)$ for each $p \in \mathcal{L}_L$.

All the components are stored in hash tables indexed by p. With perfect hashing [16] we obtain $O(1)$ access time and $O(n^2/L)$ space. The following algorithm realizes a query. Note that the only Abelian factor of length 0 is heavy. Hence, if $|z| < L$ then z is an Abelian factor of w if and only if $z \in Ext'_{<L}(\mathcal{H}_L)$.

```
Algorithm QUERY(z)
    if z ∈ Ext'<L(HL) then
        return true;
    r := |z| mod L;
    foreach p : z ∈ Extr(p) do
        if p ∈ LL then
            foreach i ∈ Occ(p) do
                if w[i .. i + |z| − 1] ≈ z then
                    return true;
    return false;
```

In the query algorithm we check if $z \in Ext'_{<L}(\mathcal{H}_L)$ or, otherwise, if there exists $p \in \mathcal{L}_L$, $0 \le |z| - |p| < L$, such that z occurs as an extension of p. Hence, the correctness of the query algorithm follows from property (4) of an L-good pair. Let us analyze the complexity of the data structure.

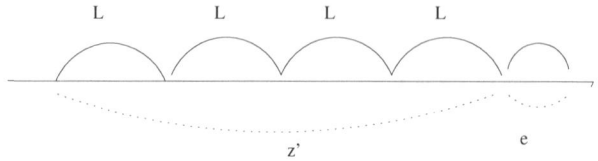

Fig. 2. When searching for the Abelian factor z we look for any L-factor z' and (short) factor e such that $z = z' + e$ and $|e| < L$

Theorem 5. *For any integer $L > 0$ there exists an index for jumbled pattern matching with $O(n^2/L)$ space and $O(L^{2\sigma-1})$ query time. The preprocessing time is $O(n^2)$.*

Proof. Assume that the elements of $\mathcal{F}_L$ are indexed using a hash table. Let us consider the complexity of the main iteration of a single query. By Lemma 2, $|\{p : p \in Ext_r(z)\}| \leq L^{\sigma-1}$, and for any $p \in \mathcal{L}_L$, by definition, $|Occ(p)| \leq L^\sigma$. Thus, using constant-time equivalence queries from Lemma 3, we obtain the desired $O(L^{2\sigma-1})$ query time.

The index size is bounded by

$$|\mathcal{F}_L| + \sum_{p \in \mathcal{F}_L} |Occ(p)| + |Ext'_{<L}(\mathcal{H}_L)|$$

which is $O(n^2/L)$ by the conditions (1-2) of an L-good pair.

Finally consider the preprocessing time. The sets $\mathcal{F}_L$, $\mathcal{L}_L$ and $\mathcal{H}_L$, and also $Occ(p)$ for all $p \in \mathcal{F}_L$, can be computed in $O(n^2/L)$ time. To compute $Ext'_{<L}(\mathcal{H}_L)$, we consider each $z \in \mathcal{H}_L$, all the elements of $Occ(z)$ and all the extensions $z + e$ of the corresponding occurrences of z by at most L letters. This yields $O(n^2/L \cdot L) = O(n^2)$ time. $\square$

Corollary 6. *For any $0 < \delta < 1$ there exists an index for jumbled pattern matching with $O(n^{2-\delta})$ space and $O(n^{(2\sigma-1)\delta})$ query time.*

Proof. We take $L = n^\delta$ and apply Theorem 5. $\square$

4   The Case of Small Patterns

While $O(n^{(2\sigma-1)\delta})$ is sublinear in n for small δ, it is still rather large, and, especially for very small patterns, might be considered unsatisfactory. We modify the data structure to handle such patterns much more efficiently, in $O(m^{(2\sigma-1)\delta})$ time for patterns of length m. We start with an auxiliary data structure.

Lemma 7. *For any $0 < \delta < 1$ there exists an index for jumbled pattern matching with $O(n \cdot k^{1-\delta})$ space and $O(k^{(2\sigma-1)\delta})$ query time for patterns of size that is at most k.*

Proof. We slightly change the definition of L-factors, we only take the L-factors of size at most k. Let $\mathcal{F}_{L,k}$ denote the set of these factors. Similarly as in the case of $\mathcal{F}_L$ we obtain $|\mathcal{F}_{L,k}| \leq \frac{n \cdot k}{L}$.

Now we take $L = k^\delta$ and the rest of the construction is essentially the same as before. The size of the data structure is $O(\frac{n \cdot k}{L})$, which is $O(n \cdot k^{1-\delta})$. The query time is $O(L^{2\sigma-1})$, hence of order $O(k^{(2\sigma-1)\delta})$. $\square$

Theorem 8. *For any $\delta > 0$ there exists an index for jumbled pattern matching with $O(n^{2-\delta})$ space and $O(m^{(2\sigma-1)\delta})$ query time, where m is the size of the pattern.*

Proof. Let
$$K = \{2^i : 0 \leq i \leq \lfloor \log n \rfloor\} \cup \{n\}.$$

We can precompute the data structures from Lemma 7 for each $k \in K$. The total size will be of order:
$$n^{2-\delta} + \sum_{i=0}^{\lfloor \log n \rfloor} n \cdot 2^{i(1-\delta)} = n^{2-\delta} + n \cdot O(2^{(1-\delta)\log n}) = O(n^{2-\delta}).$$

To answer a query about a pattern p of size m we take
$$k = \min\{j \in K : j \geq m\}.$$

Then we apply the query algorithm from Lemma 7 (using only the part of the data structure relevant to k). This completes the proof. $\square$

In particular if we take $\delta = 1/((2\sigma - 1)a)$ then we have a more concrete result.

Observation 9. *For any integer $a > 1$ there exists an index for jumbled pattern matching with $O(n^{2-1/((2\sigma-1)a)})$ space and $O(\sqrt[a]{m})$ query time.*

5   Efficient Merging of Packed Sets

In this section by merging we mean computing a set-theoretic union, i.e. at most one copy of each element is preserved. We merge large families of sets whose union is relatively small. We aim at sublinear time in the total size of those families, which requires suitable compact encoding of sets. The algorithm that we develop in this section is used to obtain an $o(n^2)$ time construction algorithm for an index for jumbled pattern matching, which is shown in the following section.

Let $U = \{1, 2, \ldots, N\}$, where $N = \left\lceil \left(\frac{\log n}{\log \log n}\right)^\sigma \right\rceil$ and $M(n) = \delta \cdot \frac{\log n}{(\sigma \log \log n)}$.

The set U is called the *universe*, and its subsets of size not larger than $M(n)$ are called here *small sets*. We have $\log|U| = \sigma \log \log n$ and $|U|^{M(n)} = 2^{\log|U| \cdot M(n)} = n^\delta$. Consequently, we obtain a bound for the number of small sets.

Observation 10. *There are $O(n^\delta)$ small sets.*

Assume all small subsets are listed in lexicographic order $(t = O(n^\delta))$:

$$\mathcal{S} = \{S_0, S_1, \dots, S_t\}.$$

Then each small subset can be identified by its rank in the list above. A set of identifiers $\mathcal{X} = \{\gamma_1, \gamma_2, \dots, \gamma_m\} \subseteq \{0, 1, \dots, t\}$ represents a family $\mathcal{R} = S_{\gamma_1}, S_{\gamma_2}, \dots, S_{\gamma_m}$. We say that each identifier in $\mathcal{X}$ is a *packed* representation of the subset of U and $\mathcal{X}$ is the *packed* version of a family $\mathcal{R}$ of small sets. We denote:

$$\mathcal{R} = UNPACK(\mathcal{X}) \text{ and } \mathcal{X} = PACK(\mathcal{R}).$$

For a family $\mathcal{R}$ of subsets of U denote by $Merge(\mathcal{R})$ the sorted set of all elements of $\bigcup_{S \in \mathcal{R}} S$, with duplicates removed (each element has unique occurrence in the merge). Denote

$$PackedMerge(\mathcal{X}) = Merge(UNPACK(\mathcal{X})).$$

Define the following **Packed Merging Problem:**

   Input: a family $\mathcal{X} = \{\gamma_1, \gamma_2, \dots, \gamma_m\}$ of integers (packed small sets);

   Output: $PackedMerge(\mathcal{X})$.

Example 11. Let $U = \{1, 2, 3, 4\}$ and $M(n) = 2$. Then:
$$PackedMerge(\{2, 4, 7\}) = Merge(S_2, S_4, S_7) = \{1, 2, 4\}.$$
In this case the lexicographically ordered ordered list of small sets is:
$S_0 = \emptyset$, $S_1 = \{1\}$, $S_2 = \{1, 2\}$, $S_3 = \{1, 3\}$, $S_4 = \{1, 4\}$, $S_5 = \{2\}$,
$S_6 = \{2, 3\}$, $S_7 = \{2, 4\}$, $S_8 = \{3\}$, $S_9 = \{3, 4\}$, $S_{10} = \{4\}$.

Lemma 12. [PackedMerge Lemma] *Let $\delta = \frac{1}{2} - \varepsilon$ for any $0 < \varepsilon < \frac{1}{2}$. Then after $O(n)$ time preprocessing, for each packed family $\mathcal{X}$ of m integers $S = PackedMerge(\mathcal{X})$ can be computed in $O(|S| + (m + \log^\sigma n) \log \log n)$ time.*

Proof. For a set S of integers we write $S \leq K$ if all elements of S are smaller than or equal to K, similarly we write $S > K$ if all the elements are greater than K. For two identifiers i, j denote $SmallSplit(i, j, K) = (p, q)$, where (p, q) is any pair of indices of subsets such that:

$$S_p \cup S_q = S_i \cup S_j \text{ and } (S_p \leq K \text{ or } S_p > K).$$

For an identifier i and an integer K also denote $Split(i, K) = (p, q)$, where (p, q) is any pair of indices of subsets such that:

$$S_p \cup S_q = S_i \text{ and } S_p \leq K \text{ and } S_q > K.$$

Note that the number of triples (i, j, K) is $o(n)$. Hence:

Claim. After $o(n)$-time preprocessing each $SmallSplit$ and $Split$ query can be answered in constant time.

Using *SmallSplit* and *Split* operations we can *scan* the sequence $\mathcal{X}$, then each time we process two current consecutive sets S_i, S_{i+1}, S_{i+1} is replaced by S_q and we have $S_p \leq K$ or $S_p > K$, where $SmallSplit(i, j, K) = (p, q)$.

Algorithm LARGESPLIT($\mathcal{X}, K$)

    Assume $\mathcal{X} = \{\gamma_1, \gamma_2, \ldots, \gamma_m\}$;

    $\mathcal{X}_1 := \mathcal{X}_2 := \emptyset$;

    for $i := 1$ **to** $m - 1$ **do**

        $(p, q) := SmallSplit(\gamma_i, \gamma_{i+1}, K)$;

        $\gamma_{i+1} := q$;

        if $S_p \leq K$ **then** add p to $\mathcal{X}_1$ **else** add p to $\mathcal{X}_2$;

    $(p, q) := Split(\gamma_m, K)$;

    add p to $\mathcal{X}_1$;

    add q to $\mathcal{X}_2$;

    return $(\mathcal{X}_1, \mathcal{X}_2)$;

In this way we have shown:

Claim. We can compute in $O(|\mathcal{X}|)$ time two families $\mathcal{X}_1, \mathcal{X}_2$ of packed sets such that:

- $PackedMerge(\mathcal{X}_1) \cup PackedMerge(\mathcal{X}_2) = PackedMerge(\mathcal{X})$;
- $PackedMerge(\mathcal{X}_1) \leq K, PackedMerge(\mathcal{X}_2) > K$.
- The total number of packed sets in $\mathcal{X}_1, \mathcal{X}_2$ is at most $|\mathcal{X}| + 1$.

After at most $\log |U|$ operations *LargeSplit*, each time applied to a smaller range of integers in U, we arrive at the situation when $\mathcal{X}$ is transformed into a series of nonempty packed families, each of them contains packed subsets included in the subrange of U of size at most $M(n)$.

Algorithm GENERATE($\mathcal{X}$)

    $Queue := \{(\mathcal{X}, [0, N])\}$; $OutputList := \emptyset$;

    while $Queue \neq \emptyset$ **do**

        $(\mathcal{X}', \Delta) := delete(Queue)$; $middle := mid(\Delta)$;

        $(\mathcal{X}_1, \mathcal{X}_2) := $ LARGESPLIT($\mathcal{X}', middle$);

        if $|\Delta|/2 \leq M(n)$ **then** add $\mathcal{X}_1, \mathcal{X}_2$ to $OutputList$;

        else add $(\mathcal{X}_1, left(\Delta))$, $(\mathcal{X}_2, right(\Delta))$ to $Queue$;

    return $OutputList$;

The algorithm above returns the set of families of packed subsets, for each family all its sets are subsets of the same interval of size $M(n)$. For an interval

Δ let $mid(\Delta)$ be the middle point in Δ and $left(\Delta)$, $(right(\Delta))$ be the left (resp. right) half of Δ.

Similarly as for $SmallSplit$ queries, we obtain the following claim that enables us to efficiently merge packed sets belonging to the same short range.

Claim. After $O(n)$ time preprocessing for each two packed sets S_i, S_j which are subsets of some range Δ', $|\Delta'| \leq M(n)$, we can compute the packed version of their union (the identifier p such that $S_p = S_i \cup S_j$) in $O(1)$ time.

Now we can compute union of each subfamily in $OutputList$ in time proportional to the number of returned elements. Since returned sets are disjoint for different subfamilies the time is proportional to the total number of returned elements.

The total time of all operations $LargeSplit$ is $O((m + |U|) \log |U|) = O((m + \log^\sigma n) \log \log n)$ since we perform operations on $O(\log |U|)$ *levels* (a level corresponds iterations with the same $|\Delta|$) and at each level we spend $O(m + |U|)$ time. □

6   Reducing Preprocessing Time

In Section 3 we have presented a subquadratic-space and sublinear query-time index. However the construction time was quadratic. Now we give an index which can be built slightly faster. The following theorem solves an open problem stated by Moosa and Rahman [20,21] for the case of constant-sized alphabet.

Theorem 13. *Each query in the index for jumbled pattern matching can be answered in $O((\frac{\log n}{\log \log n})^{2\sigma - 1})$ time after preprocessing in $O\left(\frac{(\log \log n)^2}{\log n} \cdot n^2\right)$ time and space.*

Proof. The queries work as in Theorem 5 with $L = \lfloor M(n) \rfloor$. Recall that $M(n) = \delta \cdot \log n / (\sigma \log \log n)$, where $\delta = \frac{1}{2} - \varepsilon$ for any $0 < \varepsilon < \frac{1}{2}$. For simplicity we extend the word w with L trailing sentinel letters. The index itself is also the same, its space complexity is $O(n^2/L) = O\left(\frac{\log \log n}{\log n} \cdot n^2\right)$. As described in the proof of Theorem 5, all the parts of the preprocessing excluding the computation of $Ext'_{<L}(\mathcal{H}_L)$ work in $O(n^2/L)$ time. The missing component is constructed by a reduction to Packed Merging Problem.

Let N be the number of distinct Abelian factors e of the input word such that $|e| < L$. Then $N \leq L^\sigma = O(\log^\sigma n)$. All such e's can be computed and assigned different identifiers from $\{0, 1, \ldots, N\}$ in $O(nL)$ time. These identifiers form the universe in the Packed Merging Problem.

Denote by A the number of distinct (ordinary) factors u of the input word such that $|u| = L$. We have $A \leq \sigma^L = O(n^\delta)$. All such factors can be computed and assigned different identifiers from $\{0, 1, \ldots, A\}$ in $O(n \cdot (L + \log n)) = O(n \log n)$ time.

For each factor $u \in \{0, 1, \ldots, A\}$ we can also compute in $O(n^\delta \cdot L)$ total time the set $S(u)$ of identifiers of all Parikh vectors corresponding to prefixes of u.

Note that the size of $S(u)$ is $L = \lfloor M(n) \rfloor$. Hence, the sets $S(u)$ form the *small sets* from the PackedMerge problem.

Now the extension sets $Ext'_{<L}(p)$ for each $p \in \mathcal{H}_L$ are computed separately. We consider all positions in $Occ(p)$, for each such position i we take the identifier u of the L-letter word coming after the corresponding occurrence of p. To compute $Ext'_{<L}(p)$, it suffices to find all the distinct elements among all the sets $S(u)$ and add p to each of them. By the PackedMerge Lemma, this can be performed in $O(|S| + (\log n)^\sigma \log \log n + |Occ(p)| \cdot \log \log n)$ time, where $S = Ext'_{<L}(p)$. In total $|S| > (\log n)^\sigma$ both sum up to at most $O(n^2/L)$ and $|Occ(p)| \cdot \log \log n$ sum up to $O(n^2 \log \log n/L)$, which yields the time complexity of the construction. $\square$

Acknowledgement. The authors would like to thank several researchers present at the Stringmasters 2013 workshop in Verona for introducing the problem and comments on the preliminary solution: Péter Burcsi, Ferdinando Cicalese, Gabriele Fici, Travis Gagie, Arnaud Lefebvre and Zsuzsanna Lipták. We are especially grateful to Ferdinando Cicalese and Travis Gagie for very valuable remarks.

References

1. Avgustinovich, S.V., Glen, A., Halldórsson, B.V., Kitaev, S.: On shortest crucial words avoiding Abelian powers. Discrete Applied Mathematics 158(6), 605–607 (2010)
2. Blanchet-Sadri, F., Kim, J.I., Mercaş, R., Severa, W., Simmons, S.: Abelian square-free partial words. In: Dediu, A.-H., Fernau, H., Martín-Vide, C. (eds.) LATA 2010. LNCS, vol. 6031, pp. 94–105. Springer, Heidelberg (2010)
3. Blanchet-Sadri, F., Simmons, S.: Avoiding Abelian powers in partial words. In: Mauri, G., Leporati, A. (eds.) DLT 2011. LNCS, vol. 6795, pp. 70–81. Springer, Heidelberg (2011)
4. Burcsi, P., Cicalese, F., Fici, G., Lipták, Z.: On table arrangements, scrabble freaks, and jumbled pattern matching. In: Boldi, P. (ed.) FUN 2010. LNCS, vol. 6099, pp. 89–101. Springer, Heidelberg (2010)
5. Burcsi, P., Cicalese, F., Fici, G., Lipták, Z.: Algorithms for jumbled pattern matching in strings. Int. J. Found. Comput. Sci. 23(2), 357–374 (2012)
6. Burcsi, P., Cicalese, F., Fici, G., Lipták, Z.: On approximate jumbled pattern matching in strings. Theory Comput. Syst. 50(1), 35–51 (2012)
7. Cassaigne, J., Richomme, G., Saari, K., Zamboni, L.Q.: Avoiding Abelian powers in binary words with bounded Abelian complexity. Int. J. Found. Comput. Sci. 22(4), 905–920 (2011)
8. Cicalese, F., Fici, G., Lipták, Z.: Searching for jumbled patterns in strings. In: Holub, J., Žďárek, J. (eds.) Proceedings of the Prague Stringology Conference 2009, Czech Technical University in Prague, Czech Republic, pp. 105–117 (2009)
9. Constantinescu, S., Ilie, L.: Fine and Wilf's theorem for Abelian periods. Bulletin of the EATCS 89, 167–170 (2006)
10. Crochemore, M., Iliopoulos, C., Kociumaka, T., Kubica, M., Pachocki, J., Radoszewski, J., Rytter, W., Tyczyński, W., Waleń, T.: A note on efficient computation of all Abelian periods in a string. Information Processing Letters 113(3), 74–77 (2013)

11. Currie, J.D., Aberkane, A.: A cyclic binary morphism avoiding Abelian fourth powers. Theor. Comput. Sci. 410(1), 44–52 (2009)
12. Currie, J.D., Visentin, T.I.: Long binary patterns are Abelian 2-avoidable. Theor. Comput. Sci. 409(3), 432–437 (2008)
13. Fellows, M.R., Fertin, G., Hermelin, D., Vialette, S.: Upper and lower bounds for finding connected motifs in vertex-colored graphs. J. Comput. Syst. Sci. 77(4), 799–811 (2011)
14. Fici, G., Lecroq, T., Lefebvre, A., Prieur-Gaston, É.: Computing Abelian periods in words. In: Holub, J., Žďárek, J. (eds.) Proceedings of the Prague Stringology Conference 2011, Czech Technical University in Prague, Czech Republic, pp. 184–196 (2011)
15. Fici, G., Lecroq, T., Lefebvre, A., Prieur-Gaston, E., Smyth, W.: Quasi-linear time computation of the abelian periods of a word. In: Holub, J., Žďárek, J. (eds.) Proceedings of the Prague Stringology Conference 2012, Czech Technical University in Prague, Czech Republic, pp. 103–110 (2012)
16. Fredman, M.L., Komlós, J., Szemerédi, E.: Storing a sparse table with O(1) worst case access time. J. ACM 31(3), 538–544 (1984)
17. Gagie, T., Hermelin, D., Landau, G.M., Weimann, O.: Binary jumbled pattern matching on trees and tree-like structures. In: Bodlaender, H.L., Italiano, G.F. (eds.) ESA 2013. LNCS, vol. 8125, pp. 517–528. Springer, Heidelberg (2013)
18. Kociumaka, T., Radoszewski, J., Rytter, W.: Fast algorithms for abelian periods in words and greatest common divisor queries. In: Portier, N., Wilke, T. (eds.) STACS. LIPIcs, vol. 20, pp. 245–256. Schloss Dagstuhl - Leibniz-Zentrum fuer Informatik (2013)
19. Lacroix, V., Fernandes, C.G., Sagot, M.-F.: Motif search in graphs: Application to metabolic networks. IEEE/ACM Trans. Comput. Biology Bioinform. 3(4), 360–368 (2006)
20. Moosa, T.M., Rahman, M.S.: Indexing permutations for binary strings. Inf. Process. Lett. 110(18-19), 795–798 (2010)
21. Moosa, T.M., Rahman, M.S.: Sub-quadratic time and linear space data structures for permutation matching in binary strings. J. Discrete Algorithms 10, 5–9 (2012)

Better Approximation Algorithms
for Technology Diffusion

Jochen Könemann, Sina Sadeghian, and Laura Sanità

University of Waterloo, Faculty of Mathematics,
Department of Combinatorics & Optimization, Waterloo, ON N2L 3G1
{jochen,s3sadegh,laura.sanita}@uwaterloo.ca

Abstract. Motivated by cascade effects arising in network technology upgrade processes in the Internet, Goldberg and Liu [SODA, 2013] recently introduced the following natural *technology diffusion* problem. Given a graph $G = (V, E)$, and thresholds $\theta(v)$, for all $v \in V$. A vertex *u activates* if it is adjacent to a connected component of active nodes of size at least $\theta(v)$. The goal is to find a seed set $\mathcal{A}$ whose initial activation would trigger a cascade activating the entire graph.

Goldberg and Liu presented an algorithm for this problem that returns a seed set of size $O(rl \log(n))$ times that of an optimum seed set, where r is the diameter of the given graph, and l is the number of distinct thresholds used in the instance. We improve upon this result by presenting an $O(\min\{r, l\} \log(n))$-approximation algorithm. Our algorithm is simple and combinatorial, in contrast with the previous approach that is based on randomized rounding applied to the solution of a linear program.

Keywords: Approximation Algorithms, Technology Diffusion, Combinatorial Optimization.

1 Introduction

Networks connecting autonomous entities are pervasive in today's world, and it is not surprising that their various properties are the subject of a vast and growing body of research (e.g., see the two books [2, 9]). In this paper, we focus on the study of algorithmic aspects of diffusion processes and cascade effects in such networks. How does a virus spread through a population of individuals, and how quickly is a rumor propagated through the members of a group of friends? Modeling dynamic network aspects like this has been an increasingly active sub-area in its own right and we refer the reader to Kleinberg's survey in [14] for an introduction.

Our work here is specifically motivated by a question addressed by Domingos and Richardson [1, 15] and their work on viral marketing. The authors studied "word-of-mouth" strategies in advertising a new product. The authors posed the following question: given a social network that connects potential customers, can we identify a small *seed* set of influential individuals that, if initially convinced to adopt a product, will eventually persuade all other customers to follow? The authors propose a probabilistic model, and heuristics to address this question.

H.L. Bodlaender and G.F. Italiano (Eds.): ESA 2013, LNCS 8125, pp. 637–646, 2013.
© Springer-Verlag Berlin Heidelberg 2013

Kempe et al. [10] later recast the questions of Domingos and Richardson in the language of discrete optimization. The authors single out two diffusion models: the *linear threshold* [7, 16], and *independent cascade* [6] models. In this paper, we will focus on adaptations of the former, where an individual's inclination to adopt the product is a function of the behaviour of her *immediate* network neighbours. Each edge $uv \in E$ has a weight b_{uv} such that the sum of weights incident to a node v is at most 1. A vertex v adopts the product (*activates*) if the total weight of active neighbours is at least her *threshold* θ_v. The goal in the *influence maximization* problem is to find a minimum-size set of initially active *seed* vertices $\mathcal{A}$ that cause all vertices to eventually adopt the product. In [10], the authors proved that this problem is NP-hard, and presented several approximation algorithms.

The threshold model described above is inherently *local* as a player's behaviour is only impacted by immediate network neighbours. Goldberg and Liu [5] recently pointed out that locality may not adequately model network externalities [12]; the authors argue that the standard threshold model used by Kempe et al. is a particularly unsuitable model for cascade effects arising in technology upgrade processes in the Internet [4, 8]. The authors propose a generalized threshold model in which a vertex' utility is influenced by the size of its connected component in the graph of active nodes. We describe this model formally next.

1.1   Non-local Threshold Model

Consider a network $G = (V, E)$ connecting a population of individuals, each of which has a *threshold* $\theta(v) \in \{\theta_1, \ldots, \theta_l\}$. Choose a *seed set* $\mathcal{A}_0 \subseteq V$ of vertices that are initially *active*. Goldberg and Liu describe the following diffusion process: in any step $i \geq 1$, the set of active vertices $\mathcal{A}_i$ consists of all previously active vertices $\mathcal{A}_{i-1}$ and vertices v whose connected component in the graph $G[\mathcal{A}_{i-1} \cup \{v\}]$ induced by $\mathcal{A}_{i-1}$ and v has size at least $\theta(v)$. The smallest t such that $v \in \mathcal{A}_t$ is called the *activation time* of v.

In the *technology diffusion* (TD) problem, we want to find a minimum-cardinality seed set $\mathcal{A}_0$ such that the above process yields the activation of all vertices in V.

1.2   Goldberg-Liu and Our Results

Following the notation of [5], we let r be the *diameter* of G; i.e., if $P(u, v)$ is the smallest number of edges on any u, v-path in G, than we let r be the maximum of $P(u, v)$ over all pairs u, v of vertices. We also use l for the number of thresholds of the given TD instance, and assume that

$$\theta_1 < \theta_2 < \ldots < \theta_l.$$

We use $\mathcal{A}_0^*$ for the seed set of an optimum solution. Goldberg and Liu showed that TD is as hard to approximate as *set-cover*, and hence, no $o(\log(n))$-approximation may exist unless NP has $n^{O(\log \log(n))}$-time deterministic algorithms [3]. In the same paper the authors propose an $O(rl \log(n))$-approximation. Our main result is the following improvement over Goldberg and Liu's work.

Theorem 1. *There is an $O(\min\{r, l\} \log(n))$-approximation algorithm for TD.*

The algorithm is obtained in two steps. We first describe an $O(r \log(n))$-approximation by reducing an instance of TD to one of *submodular set-cover* [17]. In order to obtain an $O(l \log(n))$-approximation, we reduce the problem to an instance of the *quota* version of the *node-weighted Steiner tree* problem [11, 13]. Both of our algorithms are substantially simpler than those presented in [5], and are deterministic and combinatorial while Goldberg and Liu's methods relied on randomized rounding applied to the solution of a linear program. We complement our algorithmic improvements with the following negative result.

Theorem 2. *TD is as hard to approximate as the quota-version of the unit-weight node-weighted Steiner tree problem.*

The above theorem is as strong as that given in [5] as the quota-version of the unit-weight node-weighted Steiner tree problem generalizes set cover, and at the same time it is based on a simpler reduction. Although the above theorem does not immediately imply any new hardness result for TD, it show that TD with only two thresholds is already as hard to approximate as the quota-version of the unit-weight node-weighted Steiner tree problem, and this may indicate the need for new ideas to shave off the $\min\{r, l\}$ factor in our result in Theorem 1.

2 An $O(r\,log(n))$ Approximation Algorithm for TD

In this section we develop an $O(r \log n)$ approximation algorithm for TD. In the following, we use $\mathcal{A}_0^*$ for an optimum seed set, and let $\mathcal{A}_t^*$ be the set of vertices that are activated at the end of round t of the dynamics described in Section 1.1.

We begin by obtaining a good lower-bound on the cardinality of $\mathcal{A}_0^*$. For a threshold θ_i and a vertex v, let $G_v^{\theta_i}$ be the subgraph induced by v and all vertices of G with threshold at most θ_i:

$$G_v^{\theta_i} = G[\{v\} \cup \{u | \theta(u) \leq \theta_i\}].$$

Note that $G_v^{\theta_i}$ may in general be disconnected. In the following, let $\Gamma(\theta_i, v)$ be the vertex set of the connected component of $G_v^{\theta_i}$ containing v. Note that, by definition, $v \in \Gamma(\theta_i, v)$. Similarly, for a general set of vertices $S \subseteq V$, we define $\Gamma(\theta_i, S) = \bigcup_{v \in S} \Gamma(\theta_i, v)$. For ease of notation, we let $\theta_0 = 0$, so that $\Gamma(\theta_0, S) = S$ for every $S \subseteq V$.

Lemma 21. *For all $\theta_i \in \{\theta_0, \ldots, \theta_{l-1}\}$, $|\Gamma(\theta_i, \mathcal{A}_0^*)| \geq \theta_{i+1} - 1$.*

Proof. Consider a threshold θ_i. If for all vertices $v \notin \mathcal{A}_0^*$ we have $\theta(v) \leq \theta_i$, then by definition $\Gamma(\theta_i, \mathcal{A}_0^*) = V$ and therefore $|\Gamma(\theta_i, \mathcal{A}_0^*)| = |V| \geq \theta_{i+1} - 1$.

Now assume that there is a vertex $v \notin \mathcal{A}_0^*$ with threshold at least θ_{i+1}. In particular, among such vertices, let v be one with smallest activation time t;

i.e., $v \in \mathcal{A}_t^* \setminus \mathcal{A}_{t-1}^*$. Directly from the definition we now see that v's connected component in $G[\mathcal{A}_{t-1}^* \cup \{v\}]$ has at least $\theta(v) = \theta_{i+1}$ vertices, and among these, only v has threshold larger than θ_i. Thus $\mathcal{A}_{t-1}^* \subseteq \Gamma(\theta_i, \mathcal{A}_0^*)$, and $|\mathcal{A}_{t-1}^*| \geq \theta_{i+1} - 1$. The lemma follows.

By the previous lemma, we can conclude that the size of the minimum cardinality subset of vertices S satisfying $|\Gamma(\theta_i, S)| \geq \theta_{i+1} - 1$ for all $1 \leq i < l$, gives us a lower bound on $|\mathcal{A}_0^*|$.

Corollary 22. *An optimal solution S^* to the following minimum threshold problem*

$$\min_{S \subseteq V}\{|S| \ : \ |\Gamma(\theta_i, S)| \geq \theta_{i+1} - 1, \ \forall \, 0 \leq i < l\}. \tag{MT}$$

has size at most $\mathcal{A}_0^*$.

The above corollary suggests the strategy to follow in designing an approximation algorithm. First, we will search for a vertex set that is a good approximate solution to the minimization problem (MT), and then we will slightly adjust it to turn it into a feasible solution for the technology diffusion problem. As a first step, we present an $O(\log(n))$-approximation algorithm for (MT) by reducing it to the *submodular set-cover* (SSC) problem. The input to an instance of SSC is a universe U, and a *submodular* function f defined over the subsets of U. Recall that f is submodular, whenever it satisfies

$$f(A \cup \{e\}) - f(A) \geq f(B \cup \{e\}) - f(B)$$

for all $A \subseteq B \subseteq U$ and for all $e \in U \setminus B$. Given a cost c_e for all $e \in U$, the goal is now to find a minimum-cost set $T \subseteq U$ such that $f(T) = f(U)$. The problem clearly generalizes the set-cover problem, and admits an $O(\log(\max_{e \in U} f(\{e\})))$-approximation [17].

Theorem 23. *There is an $O(\log n)$-approximation algorithm for (MT).*

Proof. As promised, we achieve the result by reducing (MT) to SSC. Let

$$U = V = \{v_1, \ldots, v_n\}$$

be the universe of the SSC instance. For each threshold θ_i, $0 \leq i \leq l$, we define function f_i by letting

$$f_i(D) = \min\{|\Gamma(\theta_i, D)|, \theta_{i+1} - 1\},$$

for all $D \subseteq U$. It is an easy exercise to show that f_i is indeed submodular for all i, and hence, so is their sum f, defined by

$$f(D) = \sum_{i=0}^{l-1} f_i(D),$$

ALGORITHM 1. SelectSeedSet1

Input: Graph $G = (V, E)$, thresholds θ.
Output: A seed set $\mathcal{A}_0^r$.
$\mathcal{A}_0^r = \emptyset$
Compute a solution D to the instance of (MT) according to Theorem 23
  While $D \neq \emptyset$ Choose $v \in D$ and add it to $\mathcal{A}_0^r$
    If $(G[S]$ is not connected)
      Compute a shortest v, S-path P in G
      Add the vertices of the path P to $\mathcal{A}_0^r$
    Remove v from D

for all $D \subseteq U$. Consider the SSC instance with groundset U, submodular function f, and unit cost for each $e \in U$. Suppose that S^* is an optimal solution to (MT) and note that

$$f(S^*) = \sum_{i=0}^{l-1} f_i(S^*) = \sum_{i=0}^{l-1} \theta_{i+1} - 1 = f(U).$$

Thus, an optimal solution D^* of the SSC instance defined above has cardinality at most $|S^*|$. Finally note that

$$f(U) = \sum_{i=0}^{l-1} (\theta_{i+1} - 1) \leq n^2,$$

and the algorithm of [17] therefore returns a set $D \subseteq U$ of size at most $O(\log(n))|S^*|$. By definition, $f(D) = f(U) = \sum_{i=0}^{l-1} \theta_{i+1} - 1$, and therefore $f_i(D) = \theta_{i+1} - 1$, for all $i = 0, \ldots, l-1$. But this implies that $|\Gamma(\theta_i, D)| \geq \theta_{i+1} - 1$ for all such i, and thus, D is a feasible solution to the instance of our minimum threshold problem.

We are now ready to give our $O(r \log n)$-approximation algorithm for the technology diffusion problem.

Theorem 24. *Algorithm 1 is an $O(r \log n)$-approximation algorithm for TD.*

Proof. Given an instance of the technology diffusion problem, Algorithm 1 first computes a solution D to the instance of the minimum threshold problem defined by G and θ, according to Theorem 23. Note that Theorem 23 and Corollary 22 imply $|D| = O(\log n)|\mathcal{A}_0^*|$. Then, the algorithm constructs the connected seed set $\mathcal{A}_0^r$ from D as follows.

The algorithm adds to $\mathcal{A}_0^r$, one by one, each vertex $v \in D$: after each addition, if the vertex v is not adjacent to any vertex in the current set $\mathcal{A}_0^r$, the algorithm also adds to $\mathcal{A}_0^r$ the vertices of a shortest v, S-path P. Note that P has at most r new vertices. Therefore, eventually the set $\mathcal{A}_0^r$ output by the algorithm has size $\leq r|D| \leq O(r \log n)|\mathcal{A}_0^*|$.

Finally, we argue that $\mathcal{A}_0^r$ is a feasible solution to our instance of the technology diffusion problem.

By induction, we prove that if we start the technology diffusion process with the seed set $\mathcal{A}_0^r$, all of the vertices in $\Gamma(\theta_i, \mathcal{A}_0^r)$ will be activated at some time $t < \infty$, for all $1 < i \leq l$. Since $\Gamma(\theta_l, \mathcal{A}_0^r) = V$, this will prove feasibility for our seed set $\mathcal{A}_0^r$.

By definition, $\Gamma(\theta_0, \mathcal{A}_0^r) = \mathcal{A}_0^r$ and $G[\mathcal{A}_0^r]$ is connected and contains at least $\theta_1 - 1$ vertices. Assume now that all vertices in $\Gamma(\theta_i, \mathcal{A}_0^r)$ are activated, for some $i \geq 0$. By the definition of our minimum threshold problem, the set D computed by the algorithm in its first step is such that $|\Gamma(\theta_i, D)| \geq \theta_{i+1} - 1$ and since $D \subseteq S$, $|\Gamma(\theta_i, \mathcal{A}_0^r)| \geq \theta_{i+1} - 1$. Therefore, $\Gamma(\theta_i, \mathcal{A}_0^r)$ is a set that induces a connected subgraph of activated vertices of size at least $\theta_{i+1} - 1$. This implies that all the vertices in $\Gamma(\theta_{i+1}, \mathcal{A}_0^r)$ will be activated after some finite time.

3   An $O(l \, log(n))$ Approximation Algorithm

In this section, we give an $O(l \, \log n)$-approximation algorithm for the technology diffusion problem. We recall from [5] that we may, w.l.o.g., limit our search to seed sets that *induce connected activation sequences*.

Definition 1. *A seed set $\mathcal{A}_0$ induces a connected activation sequence with anchor s if there is a permutation $v_1, \ldots, v_n$ of vertices so that (i) $s = v_1$, (ii) for all $1 \leq t \leq n$, vertices $v_1, \ldots, v_{t-1}$ induce a connected component in G and v_t is adjacent to it, and (iii) v_t is in $\mathcal{A}_0$ whenever $t < \theta(v_t)$.*

[5] showed that there is a choice of anchor such that the size of a minimum cardinality seed set inducing a connected activation sequence is at most twice the size of an optimal seed set.

Lemma 31 ([5]). *Given an instance of TD, let $\mathcal{A}_0^*$ be the optimal seed set. There is a choice of anchor $s \in \mathcal{A}_0^*$ such that the minimum-cardinality seed set $\mathcal{A}^c$ that induces a connected activation sequence anchored at s has cardinality at most $2|\mathcal{A}_0^*|$.*

Given the above lemma, we will from now on assume that we know anchor vertex s, and will search for a seed set that induces a connected activation sequence anchored at s.

The key insight in our algorithm is its connection with the *quota-constrained node-weighted Steiner Tree* (qNST) problem. In an instance of qNST we are given an undirected graph $G = (V, E)$, a root vertex $s \in V$, vertex weights $w(v)$ for all $v \in V$, and a quota $Q \in \mathbb{Z}_+$. The goal is to find a tree T containing $s \in T$, that spans at least Q vertices, and has smallest weight $w(T) = \sum_{v \in V(T)} w(v)$. This problem is known to have an $O(\log n)$-approximation algorithm.

Theorem 3 ([11, 13]). *There is a polynomial time $O(\log n)$ approximation algorithm for qNST.*

We now show how to use the above theorem to give an $O(l \, \log n)$-approximation for the technology diffusion problem. Algorithm 2 runs in l steps. In each step

---

ALGORITHM 2. SelectSeedSet2

Input: Graph $G = (V, E)$, thresholds θ and a vertex s.
Output: A seed set $\mathcal{A}_0^l$.
$\mathcal{A}_0^l = \emptyset$
for $i = 1 \ldots l$ **do**
    For each $v \in V$, set $w_i(v) = 0$ if $\theta(v) < \theta_i$ and $w_i(v) = 1$, otherwise
    Find an $O(\log n)$-approximate tree T_i for the instance $(G, s, w_i, \theta_i - 1)$ of
    qNST problem
    Add to $\mathcal{A}_0^l$ each vertex $v \in T_i$ with $w_i(v) = 1$
end

---

i, we define a weight function $w_i(v) = 1$ for all $v : \theta(v) \geq \theta_i$, and $w_i(v) = 0$ otherwise, and find an $O(\log n)$-approximate minimum w_i-weight tree containing s and covering at least $\theta_i - 1$ vertices. This can be done in polynomial time as stated in Theorem 3.

We now argue that (i) the set $\mathcal{A}_0^l$ output by the algorithm is a feasible seed set and (ii) the cardinality of $\mathcal{A}_0^l$ is $O(l \log n)|\mathcal{A}_0^*|$.

Lemma 32. *Algorithm 2 outputs a feasible seed set $\mathcal{A}_0^l$.*

Proof. By induction on i, we prove that $\mathcal{A}_0^l$ activates all the vertices in T_i at some time $t < \infty$. This implies that $\mathcal{A}_0^l$ activates $\theta_i - 1$ vertices which form a connected component. For $i = l$, this would lead to the result since if there is a connected component C of activated vertices of size at least $\theta_l - 1$ at some time t, clearly all non-activated vertices adjacent to C will become activated at time $t + 1$, and therefore eventually all vertices in V will be activated.

In the first step, the algorithm finds a set T_1 of $\theta_1 - 1$ vertices and adds all of them to the seed set $\mathcal{A}_0^l$, since $w_1(v) = 1$ for every vertex v. For $i > 1$, assume that all vertices in T_{i-1} are activated at time t. Clearly they form a connected component of size at least $\theta_{i-1} - 1$. Consider the set W_t of non-activated vertices in $T_i \setminus T_{i-1}$ at time t. If $W_t = \emptyset$, we are done. If not, clearly there is a subset $W' \subseteq W_t$ of vertices that are adjacent to the connected component formed by the vertices in T_{i-1}: this is because $T_i \cap T_{i-1} \neq \emptyset$, since both contains the vertex s. Note that a vertex $u \in W'$ cannot have $w_i(u) = 1$, since otherwise the algorithm would have added it to $\mathcal{A}_0^l$, and therefore u would be activated, a contradiction. It follows that $w_i(u) = 0$, i.e. $\theta(v) \leq \theta_{i-1}$ and therefore it becomes activated at time $t + 1$. If we now consider the set W_{t+1} we have that $W_{t+1} \subset W_t$: so we can repeat this process till eventually all vertices in W_t become activated.

Lemma 33. *In each phase i, Algorithm 2 adds to $\mathcal{A}_0^l$ a number of vertices equals to $w_i(T_i) = O(\log n)|\mathcal{A}_0^*|$.*

Proof. Consider the connected activation process starting with the seed set $\mathcal{A}^c$. Let t be the first time in which at least $\theta_i - 1$ vertices forming a connected component are activated. If a vertex v with $\theta(v) \geq \theta_i$ is active at time t, then v must be in the seed set $\mathcal{A}^c$. Note that $\mathcal{A}_t^c$ contains a tree T_i^* of at least $\theta_i - 1$

vertices and $w_i(T_i^*)$ is exactly the number of vertices in T_i^* with threshold $\geq \theta_i$. Since T_i is an $O(\log n)$-approximate minimum w_i-weight tree, it follows that $w_i(T_i) \leq O(\log n)w_i(T_i^*) \leq O(\log n)|\mathcal{A}^c|$. The lemma follows from Lemma 31.

Lemma 33, Lemma 32 and Lemma 31 imply:

Theorem 34. *Algorithm 2 is an $O(l \log n)$ approximation algorithm for technology diffusion problem.*

Theorems 24 and 34 together provide a proof of Theorem 1.

4   Complexity

We now provide a proof of Theorem 2, and show that TD with only two thresholds is as hard to approximate as $0, 1$-cost qNST. Our reduction is simpler and at the same time as strong as that given in [5] as $0, 1$-cost qNST generalizes set cover.

Consider an unrooted instance of qNST on graph $G = (V, E)$ with weights $w(v) \in \{0, 1\}$ for all $v \in V$, and quota Q. We define an instance of TD as follows. For every vertex $v \in V(G)$, let $\theta(v) = Q$ if $w(v) = 1$ and let $\theta(v) = 1$ if $w(v) = 0$.

First assume that T is a solution for the given qNST instance of cost k. Then let $\mathcal{A}_0$ be the set of k weight 1 vertices of T. Since all other vertices of T have threshold 1, using seed set $\mathcal{A}_0$ will lead to the activation of all vertices of T. Since T spans at least Q vertices, and all thresholds are at most Q, all other vertices will eventually be activated. Hence, the constructed TD instance has a solution of size at most k.

Now let $\mathcal{A}_0$ be a valid seed set of size k for the TD instance. If there is no vertex in $V \setminus \mathcal{A}_0$ with threshold Q then all vertices not in $\mathcal{A}_0$ have weight 0. Thus, any spanning tree T of G has weight at most $|\mathcal{A}_0| = k$.

Now assume that there is a vertex $v \in V \setminus \mathcal{A}_0$ with threshold Q. Pick such a vertex with smallest activation time t. By definition, the connected component T containing v in $G[\mathcal{A}_{t-1} \cup \{v\}]$ has at least Q vertices. Moreover, all but one of the weight-1 vertices in T are also in $\mathcal{A}_0$. Thus, the weight of T is at most $k + 1$.

As we discussed in previous section, we can solve the technology diffusion problem using an algorithm for qNST which together with the discussion above shows that the technology diffusion problem with two threshold is essentially equivalent to $0, 1$-weight qNST.

5   Further Work

Using the threshold model proposed by Goldberg and Liu, we studied the problem of finding a smallest seed set whose activation triggers a cascade that eventually activates the entire population. We presented improved approximation algorithms for this problem, and pointed out challenges standing in the way of further improvements.

Many open problems remain. For example, given a target value Q, can we find a small seed set whose activation triggers the activation of at least Q vertices? Or, given a parameter k, we could be looking for a seed set of size at most k such that the largest number of vertices are activated.

Another natural generalization is the weighted version of TD where each vertex $v \in V$ has an associated cost c_v, and the goal is to find a minimum-cost seed set activating all vertices. Unlike the unit-cost version in which the optimal solution inducing connected activation sequence approximates the optimal solution within factor of two, there is an arbitrarily large gap between these two for version of the problem with costs.

This can be easily seen by way of the following example. Consider a star graph with $n + 1$ vertices such that the threshold of every vertex is $n + 1$. Let the cost of each leaf vertex be $\frac{1}{n}$ while the cost of the vertex at center of the star be an arbitrary large value C. Then, the optimal solution is the set of leaf vertices with total cost of 1 while the cost of any optimal solution inducing a connected activation sequence is at least C as it must contain the vertex at the center. This suggests that one should seek other approaches for solving the version of problem with vertex costs.

References

[1] Domingos, P., Richardson, M.: Mining the network value of customers. In: Proceedings of the Knowledge Discovery and Data Mining, pp. 57–66 (2001)

[2] Easley, D.A., Kleinberg, J.M.: Networks, Crowds, and Markets - Reasoning About a Highly Connected World. Cambridge University Press (2010)

[3] Feige, U.: A threshold of ln n for approximating set cover. J. ACM 45 (1998)

[4] Gill, P., Schapira, M., Goldberg, S.: Let the market drive deployment: a strategy for transitioning to bgp security. SIGCOMM Comput. Commun. Rev. 41(4), 14–25 (2011)

[5] Goldberg, S., Liu, Z.: Technology diffusion in communication networks. In: Proceedings of the ACM-SIAM Symposium on Discrete Algorithms, pp. 233–240 (2013)

[6] Goldenberg, J., Libai, B., Muller, E.: Talk of the network: A complex systems look at the underlying process of word-of-mouth. Marketing letters 12(3), 211–223 (2001)

[7] Granovetter, M.: Threshold models of collective behavior. American Journal of Sociology, 1420–1443 (1978)

[8] Guérin, R., Hosanagar, K.: Fostering ipv6 migration through network quality differentials. ACM SIGCOMM Computer Communication Review 40(3), 17–25 (2010)

[9] Jackson, M.: Social and Economic Networks. Princeton University Press (2008)

[10] Kempe, D., Kleinberg, J., Tardos, E.: Maximizing the spread of influence through a social network. In: Proceedings, Knowledge Discovery and Data Mining, KDD 2003, pp. 137–146 (2003)

[11] Könemann, J., Sadeghian, S., Sanità, L.: An LMP O(log n)-approximation algorithm for node weighted prize collecting Steiner tree. Technical Report 1302.2127, arXiv (2013)

[12] Metcalfe, B.: Metcalfe's law: A network becomes more valuable as it reaches more users. In: InfoWorld (1995)
[13] Moss, A., Rabani, Y.: Approximation algorithms for constrained node weighted steiner tree problems. SIAM J. Comput. 37(2), 460–481 (2007)
[14] Nisan, N., Roughgarden, T., Tardos, E., Vazirani, V.: Algorithmic Game Theory. Cambridge University Press (2007)
[15] Richardson, M., Domingos, P.: Mining knowledge-sharing sites for viral marketing. In: Proceedings of the Knowledge Discovery and Data Mining (2002)
[16] Schelling, T.: Micromotives and Macrobehavior. Norton (1978)
[17] Wolsey, L.A.: An analysis of the greedy algorithm for the submodular set covering problem. Combinatorica 2(4), 385–393 (1982)

On Polynomial Kernels for Integer Linear Programs: Covering, Packing and Feasibility[*]

Stefan Kratsch[**]

Technical University Berlin, Germany
stefan.kratsch@tu-berlin.de

Abstract. We study the existence of polynomial kernels for the problem of deciding feasibility of integer linear programs (ILPs), and for finding good solutions for covering and packing ILPs. Our main results are as follows: First, we show that the ILP FEASIBILITY problem admits no polynomial kernelization when parameterized by both the number of variables and the number of constraints, unless NP ⊆ coNP/poly. This extends to the restricted cases of bounded variable degree and bounded number of variables per constraint, and to covering and packing ILPs. Second, we give a polynomial kernelization for the COVER ILP problem, asking for a solution to $Ax \geq b$ with $c^T x \leq k$, parameterized by k, when A is row-sparse; this generalizes a known polynomial kernelization for the special case with 0/1-variables and coefficients (d-HITTING SET).

1 Introduction

This work seeks to extend the theoretical understanding of preprocessing and data reduction for Integer Linear Programs (ILPs). Our motivation lies in the fact that ILPs encompass many important problems, and that ILP solvers, especially CPLEX, are known for their preprocessing to simplify (and shrink) input instances before running the main solving routines (cf. [2]). When it comes to NP-hard problems, then, formally, being able to reduce *every* instance of some problem would give an efficient algorithm for solving it entirely, and prove P = NP (cf. [3]). We avoid this issue by studying the question for efficient preprocessing via the notion of *kernelization* from parameterized complexity [4], which relates the performance of the data reduction to one or more problem-specific parameters, like the number n of variables of an ILP.

A kernelization with respect to some parameter n is an efficient algorithm that given an input instance returns an equivalent instance of size depending only on n; a *polynomial kernelization* guarantees size polynomial in the parameter (see Section 2 for formal definitions). This notion has been successfully applied to a wide range of problems (see Lokshtanov et al. [5] for a recent survey). A breakthrough result by Bodlaender et al. [6] (using [7]) gave a framework for ruling out polynomial kernels for certain problems, assuming NP ⊄ coNP/poly (else the polynomial hierarchy collapses).[1]

---

[*] Some proofs are omitted from this extended abstract and can be found in [1].

[**] Supported by the DFG, research project PREMOD, KR 4286/1.

[1] All kernelization lower bounds mentioned in this work are modulo this assumption.

H.L. Bodlaender and G.F. Italiano (Eds.): ESA 2013, LNCS 8125, pp. 647–658, 2013.

ILP Feasibility. Let us first discuss the INTEGER LINEAR PROGRAM FEASI-
BILITY (ILPF) problem: Given a set of m linear (in)equalities in n variables with
integer coefficients, decide whether some integer point $x \in \mathbb{Z}^n$ fulfills all of them.
A well-known result of Lenstra [8] gives an $\mathcal{O}(\alpha^{n^3} m^c)$ time algorithm for this
problem, later improved, e.g., by Kannan [9] to $\mathcal{O}(n^{\mathcal{O}(n)} m^c)$. We can trivially
"reduce" to size $N = \mathcal{O}(n^{\mathcal{O}(n)})$ by observing that Kannan's algorithm solves
all larger instances in polynomial time $\mathcal{O}(Nm^c) = N^{\mathcal{O}(1)}$. Can actual reduction
rules give smaller kernels, for example with size polynomial in n?

It is clear that we can store an ILPF instance, for example, (A, b), with $A \in
\mathbb{Z}^{m \times n}$, $b \in \mathbb{Z}^m$, asking for $x \geq 0$ with $Ax \leq b$, by encoding all $\mathcal{O}(nm)$ coefficients,
which takes $\mathcal{O}(nm \log C)$ bits where C is the largest absolute value among co-
efficients. Let us check what can be said about polynomial kernels with respect
to these parameters for ILPF and the r-row-sparse[2] variant r-ILPF:

If the row-sparseness is unrestricted, then $\text{ILPF}(n + C)$ and $\text{ILPF}(m + C)$
encompass HITTING SET(n) and HITTING SET(m)[3], respectively, which admit
no polynomial kernels [10,11]. What about $\text{ILPF}(n + m)$, which is the maximal
open case below the trivial parameter $n + m + \log C$?

With bounded row-sparseness r, things turn out differently: For r-$\text{ILPF}(n+C)$
and r-$\text{ILPF}(m + \log C)$ there are polynomial kernels: The former is not hard
and a sketch is provided in the full version [1]; the latter is trivial since row-
sparseness r entails $n \leq r \cdot m$ and hence the above encoding uses $\mathcal{O}(nm \log C) =
\mathcal{O}(rm^2 \log C) = (m + \log C)^{\mathcal{O}(1)}$ bits. It was showed previously that r-$\text{ILPF}(n)$
admits no polynomial kernel [12], and it can be seen that the proof works also
for r-$\text{ILPF}(n + \log C)$.[4] Again, this leaves parameter $n + m$ open.

Our contribution for ILPF is the following theorem which, unfortunately, settles both $\text{ILPF}(n + m)$ and r-$\text{ILPF}(n + m)$ negatively (see Section 3). It can
be seen that this completes the picture regarding the existence of polynomial
kernels for ILPF and r-ILPF for parameterization by any subset of n, m, $\log C$,
and C (see [1]). The same is true for the column-sparse case q-ILPF, but we
omit a detailed discussion since it is quite similar to the row-sparse case.

Theorem 1. *ILPF$(n + m)$ does not admit a polynomial kernelization or compression, unless* $\text{NP} \subseteq \text{coNP/poly}$. *This holds also if each constraint has at most
three variables and each variable is in at most three constraints.*

It appears that $\text{ILPF}(n + m)$ is the first problem for which we know that a
polynomial kernelization fails solely due to the encoding size of large numbers in
the input data (and taking into account our proof that no reduction is possible);
an additional parameter $\log C$ would trivially give a polynomial kernel. This
of course fits into the picture of hardness results for weight(ed) problems, e.g.,

---

[2] Row-sparseness r: at most r variables per constraint; column-sparseness q: each
variable occurs in at most q constraints; we use r and q throughout this work.

[3] HITTING SET: Given a base set U of size n, a set $\mathcal{F}$ of m subsets of U, and an
integer k, find a set of k elements of U that intersects each set in $\mathcal{F}$ (if possible). ILP
formulation: Is there $(x)_{u \in U}$ with $\sum_{u \in U} x_u \leq k$, and $\sum_{u \in F} x_u \geq 1$ for all $F \in \mathcal{F}$?

[4] The used cross-composition with t input instances creates coefficients of value $\mathcal{O}(t^2)$
with encoding size $\log C = \mathcal{O}(\log t)$ which is permissible for a cross-composition.

W[1]-hardness of SMALL SUBSET SUM(k) where the task is to pick a subset of at most k numbers to match some target sum [13] and the kernelization lower bound for SMALL SUBSET SUM($k+d$) where the value of numbers is upper bound by 2^d [10]; however, in both cases the *number* of weights is not bounded in the parameters. Furthermore, there are lower bounds for weighted graph problems (e.g., [14]), but there the used weights have *value* polynomial in the instance size and hence negligible *encoding size*. We also point out two contrasting positive results: A randomized polynomial compression of SUBSET SUM(n) [3], and a randomized reduction of KNAPSACK(n) to many instances of size polynomial in n [15] (the number of instances depends on the bit size of the largest weight).

Covering and Packing ILPs. Given the overwhelming amount of negative results for ILPF, we turn to the more restricted cases of covering and packing ILPs (cf. [16]) with the hope of identifying some positive cases:

$$\text{(covering ILP:)} \quad \min \quad c^T x \qquad\qquad \text{(packing ILP:)} \quad \max \quad c^T x$$
$$\text{s.t.} \quad Ax \geq b \qquad\qquad\qquad\qquad\qquad \text{s.t.} \quad Ax \leq b$$
$$x \geq 0 \qquad\qquad\qquad\qquad\qquad\qquad\qquad x \geq 0$$

Here A, b, and c have non-negative integer entries (coefficients). Feasibility for these ILPs is usually trivial (e.g., $x = 0$ is feasible for packing), and the more interesting question is whether there exist feasible solutions x with small (resp. large) value of $c^T x$. Encompassing many well-studied problems from parameterized complexity, we ask whether $c^T x \leq k$ (resp. $c^T x \geq k$), and parameterize by k; instances are given as (A, b, c, k). Unsurprisingly, there are a couple of special cases contained in this setting that have been studied before (e.g., with 0/1-variables and coefficients); some of those are W[1]-hard (and are hence unlikely to have polynomial kernels), whereas others have positive results that we could hope to generalize to the more general ILP setting. To capture the different cases, we use the column-sparseness q and row-sparseness r of the matrix A: taking q and r as constants, additional parameters, or unrestricted values defines different problems. Our main result in this part is a polynomial kernelization for r-COVER ILP(k) (see Section 4); the special case of only 0/1 variables and coefficients is known as r-HITTING SET(k) and admits kernels of size $\mathcal{O}(k^r)$ [17,18]. Our result also uses the Sunflower Lemma (like [17]), but the reduction arguments for sunflowers of linear constraints are of course more involved than for sets.

Theorem 2. *The r-COVER ILP(k) problem admits a reduction to $\mathcal{O}(k^{r^2+r})$ variables and constraints, and a polynomial kernelization.*

Furthermore, we show how to preprocess instances of PACKING ILP($k+q+r$) and COVER ILP($k+q+r$) to equivalent instances with polynomial in kqr many variables and constraints (see Section 5). For r-PACKING ILP($k + q$) this is extended to a polynomial kernelization, generalizing that for the special case of bounded degree INDEPENDENT SET(k). To put these results into context, we provide an overview containing also the inherited hard cases in Tables 1 and 2; a brief discussion of these cases can be found in the full version [1].

Table 1. "PK" stands for polynomial kernel, "no PK" stands for no polynomial kernel unless NP ⊆ coNP/poly. All normal-font entries are implied by boldface entries.

Parameterized complexity of Packing ILP(k)			
		row-sparseness r	
	constant	parameter	unrestricted
column-sparse. q — constant	(PK)	(FPT)	**W[1]-hard from Subset Sum(k)**
column-sparse. q — parameter	**PK (Theorem 4)**	**FPT;** $n = \mathcal{O}(kqr)$ and $m = \mathcal{O}(kq^2r)$ **(Theorem 4)**	(W[1]-hard)
column-sparse. q — unrestricted	**W[1]-hard from Independent Set(k)**	(W[1]-hard)	(W[1]-hard)

Table 2. "PK" stands for polynomial kernel, "no PK" stands for no polynomial kernel unless NP ⊆ coNP/poly. All normal-font entries are implied by boldface entries.

Parameterized complexity of Cover ILP(k)			
		row-sparseness r	
	constant	parameter	unrestricted
column-sparse. q — constant	(PK)	(FPT)	**W[1]-hard from Subset Sum(k)**
column-sparse. q — parameter	(PK)	(FPT); $n = \mathcal{O}(kqr)$ and $m = \mathcal{O}(kq)$ **(Theorem 5)**	(W[1]-hard)
column-sparse. q — unrestricted	**PK (Theorem 2)**	**FPT but no PK** (cf. [1, Prop. 4])	**W[2]-hard from Hitting Set(k)**

2   Preliminaries

A *parameterized problem* over some finite alphabet Σ is a language $\mathcal{P} \subseteq \Sigma^* \times \mathbb{N}$. The problem $\mathcal{P}$ is *fixed-parameter tractable* if $(x, k) \in \mathcal{P}$ can be decided in time $f(k) \cdot (|x| + k)^{\mathcal{O}(1)}$, where f is an arbitrary computable function. A *kernelization* for $\mathcal{P}$ is a polynomial-time algorithm that, given input (x, k), computes an equivalent instance (x', k') with $|x'|, k' \leq h(k)$ where h is some computable function; K is a *polynomial* kernelization if $h(k)$ is polynomially bounded in k. By relaxing the restriction that the created instance (x', k') must be of the same problem and allow the output to be an instance of any language (i.e., any decision problem) we get the notion of *(polynomial) compression*. Almost all lower bounds for kernelization apply also for this weaker notion.

We also use the concept of an (OR-)cross-composition of Bodlaender et al. [19] which builds on the breakthrough results of Bodlaender et al. [6] and Fortnow and Santhanam [7] for proving lower bounds for kernelization.

Definition 1 ([19]). *An equivalence relation $\mathcal{R}$ on Σ^* is called a* polynomial equivalence relation *if the following two conditions hold:*

1. *There is a polynomial-time algorithm that decides whether two strings belong to the same equivalence class (time polynomial in $|x| + |y|$ for $x, y \in \Sigma^*$).*
2. *For any finite set $S \subseteq \Sigma^*$ the relation $\mathcal{R}$ partitions the elements of S into a number of classes that is polynomial in the size of the largest element of S.*

Definition 2 ([19]). *Let $L \subseteq \Sigma^*$ be a language, let $\mathcal{R}$ be a polynomial equivalence relation on Σ^*, and let $\mathcal{P} \subseteq \Sigma^* \times \mathbb{N}$ be a parameterized problem. An OR-cross-composition of L into $\mathcal{P}$ (with respect to $\mathcal{R}$) is an algorithm that, given t instances $x_1, x_2, \ldots, x_t \in \Sigma^*$ of L that are $\mathcal{R}$-equivalent, takes time polynomial in $\sum_{i=1}^{t} |x_i|$ and outputs an instance $(y, k) \in \Sigma^* \times \mathbb{N}$ such that:*
1. *The parameter value k is polynomially bounded in $\max_i |x_i| + \log t$.*
2. *The instance (y, k) is **yes** for $\mathcal{P}$ iff at least one instance x_i is **yes** for L.*
We then say that L OR-cross-composes into $\mathcal{P}$.

Theorem 3 ([19]). *If an NP-hard language L OR-cross-composes into the parameterized problem $\mathcal{P}$, then $\mathcal{P}$ does not admit a polynomial kernelization or compression unless $NP \subseteq coNP/poly$ and the polynomial hierarchy collapses.*

3   A Kernel Lower Bound for ILPs with Few Coefficients

In this section, we prove Theorem 1, i.e., that INTEGER LINEAR PROGRAM FEASIBILITY$(n+m)$ admits no polynomial kernelization unless $NP \subseteq coNP/poly$. We begin with a technical lemma that expresses multiplication by powers of two in an ILP; its proof can be found in [1]. The crucial point is that we need multiplication by t different powers of two, but can afford only $\mathcal{O}(\log^c t)$ variables and coefficients (direct products of variables are not legal in linear constraints).

Lemma 1. *Let a, b, and p be variables, and let $b_{\max}, p_{\max} \geq 0$ be integers. Let $\ell = \lceil \log p_{\max} \rceil$. There is a system of $6\ell + 7$ linear constraints with $2\ell - 1$ auxiliary variables such that all integer solutions have $0 \leq b \leq b_{\max}$, $0 \leq p \leq p_{\max}$, and $a = b \cdot 2^p$. Conversely, if these three conditions hold then feasible values for the auxiliary variables exist. The system uses coefficients with bit size $\mathcal{O}(p_{\max} \log b_{\max})$ and all variables have range at most $\{0, \ldots, b_{\max} \cdot 2^{p_{\max}}\}$.*

Now we are set up to prove the first part of Theorem 1.

Lemma 2. *INTEGER LINEAR PROGRAM FEASIBILITY$(n + m)$ admits no polynomial kernelization or compression unless $NP \subseteq coNP/poly$.*

Proof. We give an OR-cross-composition from the NP-hard INDEPENDENT SET problem. The input instances are of the form $(G = (V, E), k)$ where G is a graph and $k \leq |V|$ is an integer, asking whether G contains an independent set of size at least k. For the polynomial equivalence relation $\mathcal{R}$ we let two instances be equivalent if they have the same number of vertices and the same solution size k. It is easy to check that this fulfills Definition 1. For convenience we consider all input graphs to be on vertex set $V = \{1, \ldots, n\}$, for some integer n.

Let t $\mathcal{R}$-equivalent instances $(G_0 = (V, E_0), k), \ldots, (G_{t-1} = (V, E_{t-1}), k)$ be given. Without loss of generality we assume that $t = 2^\ell$ for some integer ℓ since otherwise we could copy one instance sufficiently often (at most doubling the input size and not affecting whether at least one of the given instances is **yes**).

Construction. We now describe an instance of INTEGER LINEAR PROGRAM FEASIBILITY that is **yes** if and only if at least one of the instances $(G_0 = (V, E_0), k), \ldots, (G_{t-1} = (V, E_{t-1}), k)$ is **yes** for INDEPENDENT SET. We first define an encoding of the t edge sets E_p into $\binom{n}{2}$ integer constants $D(i, j)$, one for each possible edge $\{i, j\} \in \binom{V}{2}$ (throughout the proof we use $1 \le i < j \le n$):

$$D(i, j) := \sum_{p=0}^{t-1} 2^p \cdot D(i, j, p), \quad \text{where} \quad D(i, j, p) := \begin{cases} 1 & \text{if } \{i, j\} \in E_p, \\ 0 & \text{else.} \end{cases}$$

In other words, the p-th bit of $D(i, j)$ is one if and only if $\{i, j\}$ is an edge of $G_p = (V, E_p)$. The values $0 \le D(i, j) \le 2^t - 1$ will be used as constants in the ILP that we construct next.

1. We start with a single variable p that is intended for choosing an instance number; its range is $\{0, \ldots, t-1\}$. The rest of the ILP will be made in such a way that it ensures that a feasible solution for the ILP will imply that instance (G_p, k) is **yes** for INDEPENDENT SET, and vice versa.
2. Now we will add constraints that allow us to extract the necessary information regarding which of the possible edges $\{i, j\}$ are present in graph G_p. Recall that the p-th bit of the constant $D(i, j)$ encodes this. For convenience, we will derive the needed constraints and argue their correctness right away. For each possible edge $\{i, j\}$ with $1 \le i < j \le n$ we introduce a variable $e_{i,j}$ with the goal of enforcing $e_{i,j} = 1$ if $\{i, j\} \in E_p$, and $e_{i,j} = 0$ else.
   Let i, j with $1 \le i < j \le n$ be fixed (we apply the following for all these choices). Clearly,

$$D(i, j) = \sum_{s=0}^{p-1} 2^s D(i, j, s) + 2^p D(i, j, p) + \sum_{s=p+1}^{t-1} 2^s D(i, j, s).$$

We are of course interested in the $2^p D(i, j, p)$ term, which takes value either 0 or 2^p. Since the first term $\sum_{s=0}^{p-1} 2^s D(i, j, s)$ is at most $2^p - 1$ and the last is a multiple of 2^{p+1}, we will extract it via a constraint

$$D(i, j) = \alpha + \beta + \gamma, \tag{1}$$

assuming that we can enforce the following conditions: (i) $0 \le \alpha \le 2^p - 1$, (ii) $\beta \in \{0, 2^p\}$, and (iii) $\gamma \in \{0, 2^{p+1}, 2 \cdot 2^{p+1}, \ldots\}$. (Note that we use new variables $\alpha_{i,j}, \beta_{i,j}, \gamma_{i,j}, \delta_{i,j}, \varepsilon_{i,j}$ for each choice of i and j, but in the construction the indices are omitted for readability.) This is where Lemma 1 comes into the picture as it permits us to enforce the creation of the required values and range restrictions without using overly many variables and constraints. For (i) we add a variable δ and enforce $\delta = 2^p$ by using Lemma 1 on $a = \delta$, $b = b_{\max} = 1$, and p with $p_{\max} = t - 1$. We then add constraints $\alpha \ge 0$ and $\alpha \le \delta - 1$. For (ii) we apply the lemma on $a = \beta$, $b = e_{i,j}$

with $b_{\max} = 1$, and p with $p_{\max} = t - 1$, enforcing $\beta = 2^p e_{i,j}$. (Note that we want to get $e_{i,j} = D(i,j,p)$ from $\beta = 2^p \cdot D(i,j,p)$ anyway; this way it is already enforced.) For (iii) we add a new variable $\varepsilon \geq 0$ and apply the lemma on $a = \gamma$, $b = 2\varepsilon$ with $b_{\max} = 2^t - 1$ (formally, this requires a new variable b' and constraint $b' = 2\varepsilon$), and p with $p_{\max} = t - 1$, enforcing $\gamma = 2^p 2\varepsilon = 2^{p+1}\varepsilon$. The fact that there are no further restrictions on ε effectively allows γ to take on any multiple of 2^{p+1} (and no other values). The upper bound $b_{\max}$ on $b = \varepsilon$ for Lemma 1 comes from $D(i,j) \leq 2^t - 1$ (we do not require larger values of γ and ε since $\gamma \leq D(i,j)$ and $\gamma = 2^p\varepsilon$).

The most costly application of Lemma 1 is for (iii) where we have $b_{\max} = 2^t - 1$ and $p_{\max} = t - 1$. This incurs $\mathcal{O}(\log p_{\max}) = \mathcal{O}(\log t)$ additional variables and constraints, and uses coefficient bit size $\mathcal{O}(p_{\max} \log b_{\max}) = \mathcal{O}(t^2)$ (same bounds suffice also for (i) and (ii)). Thus, over all choices of $1 \leq i < j \leq n$, i.e., for getting all the needed edge information, we use $\mathcal{O}(\binom{n}{2} \log t)$ additional variables and constraints. For each variable $e_{i,j}$ our constraints ensure that it is equal one if $\{i,j\}$ is an edge in G_p; else it has value zero.

3. Now we can finally add the actual edge constraints needed to express the INDEPENDENT SET problem. We add n variables $x_1, \ldots, x_n$ with range $\{0,1\}$, one variable x_i for each vertex $i \in V$. For each possible edge $\{i,j\} \in \binom{V}{2}$, i.e., for all $1 \leq i < j \leq n$ we add the following constraint.

$$x_i + x_j + e_{i,j} \leq 2 \tag{2}$$

Finally, we add a constraint

$$\sum_{i=1}^{n} x_i \geq k \tag{3}$$

to ensure that we select at least k vertices. This completes the construction.

Let us now check the number of variables and constraints in our created ILP. The number of variables is dominated by the $\mathcal{O}(\binom{n}{2} \log t)$ variables added in Step 2, which come from $\mathcal{O}(\binom{n}{2})$ applications of Lemma 1. The same is true for the number of constraints used. Thus both parameters of our target instance are polynomially bounded in the largest input instance plus $\log t$. Since we postulated no further restrictions on the target ILP (e.g., A and b may have negative coefficients), we will omit a discussion of how to write all constraints as $Ax \leq b$ with $x \geq 0$ (here x is the vector of all variables used) since that is straightforward. The largest bit size of a coefficient is $\mathcal{O}(t^2)$ hence it is easy to see that the whole ILP can be generated in time polynomial in the total input size (which is roughly $\mathcal{O}(t \cdot n^2)$ from the t INDEPENDENT SET instances with n vertices each).

It remains to argue correctness of the construction. Due to space restrictions, and since most of the needed arguments were already given, this is deferred to the full version [1].

Thus we have an OR-cross-composition from the NP-hard INDEPENDENT SET problem to the INTEGER LINEAR PROGRAM FEASIBILITY$(n + m)$ problem. By Theorem 3, this implies that INTEGER LINEAR PROGRAM FEASIBILITY$(n + m)$ has no polynomial kernel or compression unless NP $\subseteq$ coNP/poly. $\qquad\square$

The second part of Theorem 1 is now an easy corollary. Similarly, we get lower bounds for covering and packing ILP with parameter $n + m$. Getting the sparseness for Corollary 1 is easier than what was needed for r-ILPF(n) in [12] since both number of constraints and number of variables are bounded in the parameter value. Proofs for both corollaries are given in the full version [1].

Corollary 1. INTEGER LINEAR PROGRAM FEASIBILITY *(n + m) restricted to instances that have at most 3 variables per constraint (row-sparseness) and with each variable occurring in at most 3 constraints (column-sparseness) admits no polynomial kernel or compression unless* NP $\subseteq$ coNP/poly.

Corollary 2. COVER ILP *(n + m)* and PACKING ILP *(n + m) do not admit polynomial kernelizations or compressions unless* NP $\subseteq$ coNP/poly.

4   Polynomial Kernelization for Row-Sparse Cover ILP(k)

In this section, we prove Theorem 2, by giving a polynomial kernelization for r-COVER ILP(k), generalizing polynomial kernelizations for r-HITTING SET(k) [17,18]. Our result uses the sunflower lemma (stated below) that can also be used for r-HITTING SET (as in [17]). However, the application is complicated by the fact that the replacement rules for a sunflower of constraints are not as clear-cut as for sets in the hitting set case: Constraints forming a sunflower pairwise overlap on the same variables but with (in general) different coefficients; hence no small replacement is implied. Additionally, we have to bound the number of constraints that have exactly the same set of variables with nonzero coefficients, called *scope*, since the sunflower lemma will only be applied to the set of different scopes. The main work lies in the proof of the following lemma; the polynomial kernelization is given as a corollary.

Lemma 3. *The r-*COVER ILP *(k) problem admits a polynomial-time preprocessing to an equivalent instance with $\mathcal{O}(k^{r^2+r})$ constraints and variables.*

Before we turn to the proof, we give a lemma that captures some initial reduction arguments including a bound on the number of constraints with the same scope. The full proof is given in [1].

Lemma 4. *Given an instance (A, b, c, k) of r-*COVER ILP *(k) we can in polynomial time reduce to an equivalent instance (A', b', c', k) such that:*
1. *No constraint is satisfied if all variables in its scope are zero, i.e., $b'_i \geq 1$.*
2. *The cost function c'^T is restricted to $1 \leq c'_i \leq k$.*
3. *All feasible solutions with $c'^T x \leq k$ have $x_i \in \{0, \ldots, k\}$ for all i.*
4. *There are at most $(k + 1)^d$ constraints for any scope of at most d variables.*

Proof (Part 4.). Consider any set of d variables with more than $(k + 1)^d$ constraints having exactly this scope. It is clear that there are at most $(k + 1)^d$ possible assignments to these variables that do not violate the maximum cost of k. Each constraint can rule out some of those. It is therefore sufficient to keep only one constraint for each infeasible assignment, as all further constraints are redundant and may be deleted, giving the claimed bound. Note that each constraint has at most r variables in its scope, hence we can perform this reduction in time polynomial in n. □

We recall sunflowers and the sunflower lemma of Erdős and Rado [20].

Definition 3 ([20]). *Let $\mathcal{F}$ denote a family of sets. A* sunflower *in $\mathcal{F}$ of cardinality t and with* core C *is a collection of t sets $\{F_1, \ldots, F_t\} \subseteq \mathcal{F}$ such that $F_i \cap F_j = C$ for all $i \neq j$. The sets $F_1 \setminus C, \ldots, F_t \setminus C$ are called the* petals *of the sunflower; they are pairwise disjoint. The core C may be empty.*

Lemma 5 (Sunflower Lemma [20]). *Let $\mathcal{F}$ denote a family of sets each of size d. If the cardinality of $\mathcal{F}$ is greater than $d! \cdot (t-1)^d$ then $\mathcal{F}$ contains a sunflower of cardinality t, and such a sunflower can be found in polynomial time.*

Proof (Lemma 3). To begin with, we apply Lemma 4 in polynomial time. Afterwards, for each constraint scope with $d \leq r$ variables, there are at most $(k+1)^d = \mathcal{O}(k^r)$ constraints with that scope. Now, we will apply the sunflower lemma to the set of all scopes of size d, for each $d \in \{1, \ldots, r\}$. If there are more than $d! \cdot (t-1)^d = \mathcal{O}(t^r)$ sets then we find a sunflower consisting of t sets (scopes). We will show how to remove at least one constraint matching one of the scopes, when t is sufficiently large (and polynomially bounded in k).

If, instead, the number of scopes is at most $d! \cdot (t-1)^d$ (for all d) then we can bound the total number of constraints as follows: We have r choices for $d \in \{1, \ldots, r\}$. For each d we have at most $d! \cdot (t-1)^d = \mathcal{O}(t^r)$ constraint scopes. For each scope there are at most $\mathcal{O}(k^r)$ constraints. In total this gives a bound of $r \cdot \mathcal{O}(t^r) \cdot \mathcal{O}(k^r) = \mathcal{O}(t^r \cdot k^r)$. Since each constraint has at most r variables we get the same bound (in $\mathcal{O}$-notation) for the number of variables.

Removing a Constraint. Let us now see how to find a redundant constraint when there are more than $d! \cdot (t-1)^d$ constraint scopes for some $1 \leq d \leq r$. We will also derive an appropriate value for t (at least $t > k$). Consider a t-sunflower in the set of constraint scopes of size d (which we can get in polynomial time from the sunflower lemma). Let its core be denoted by $C = \{x_1, \ldots, x_s\}$, with $0 \leq s < d \leq r$, and its pairwise disjoint petals by $\{y_{1,s+1}, \ldots, y_{1,d}\}, \ldots, \{y_{t,s+1}, \ldots, y_{t,d}\}$. (Note that $s < d$ is needed since all sets are different, which requires nonempty petals; else they would all equal the core.) Thus there must be constraints in the ILP matching these scopes. We arbitrarily pick one constraint for each scope:

$$a_{1,1}x_1 + a_{1,2}x_2 + \ldots + a_{1,s}x_s + a_{1,s+1}y_{1,s+1} + \ldots + a_{1,d}y_{1,d} \geq b_1$$
$$a_{2,1}x_1 + a_{2,2}x_2 + \ldots + a_{2,s}x_s + a_{2,s+1}y_{2,s+1} + \ldots + a_{2,d}y_{2,d} \geq b_2$$
$$\vdots$$
$$a_{t,1}x_1 + a_{t,2}x_2 + \ldots + a_{t,s}x_s + a_{t,s+1}y_{t,s+1} + \ldots + a_{t,d}y_{t,d} \geq b_t$$

Note that to keep notation simple the indexing of the variables and coefficients is only with respect to the sunflower and makes no assumption about the actual numbering within $Ax \geq b$; all arguments are local to the sunflower.

First, let us note that we may return **no** (or a dummy **no**-instance) if the core is empty and $t > k$: That would give us more than k constraints on disjoint sets of variables that each require at least one nonzero variable; this is impossible at maximum cost k. In the following, assume that $s \geq 1$.

Since each variable takes values from $\{0, 1, \ldots, k\}$ (by Lemma 4), there are at most $(k+1)^s$ possible assignments for the s core variables (in fact there are even less since the sum is at most k). It is clear that assigning zero to all core variables does not lead to a feasible solution since each of the t constraints requires at least one nonzero variable and the petal variables are disjoint. However, unlike for r-HITTING SET, assigning one to a single core variable is not necessarily sufficient to satisfy all constraints, and the value of each variable for a constraint might be quite different. Thus, we may not simply replace the constraints of the sunflower by the restriction of one constraint to the core variables.

To cope with this difficulty we employ a marking strategy. We check all $(k+1)^s$ assignments of choosing a value from $\{0, 1, \ldots, k\}$ for each core variable. It is possible that some of the constraints are already satisfied by this core assignment due to the monotone behavior of covering constraints. If more than k constraints need an additional nonzero variable (which would be a petal variable) then clearly this core assignment is infeasible. In this case we arbitrarily mark $k + 1$ constraints that are not satisfied by the core assignment alone (i.e., if all their other variables would be zero); these serve to prevent the core assignment from being chosen. If at most k constraints are not yet satisfied then we mark all of them. Clearly, in total we mark at most $(k + 1) \cdot (k + 1)^s$ constraints.

We will now argue that all unmarked constraints can be deleted. Clearly, deletion can only create false positives, so consider a solution to the instance obtained by deleting any unmarked constraint. If the assignment does not also satisfy the removed constraint (where we take value zero for variables that do not occur after deletion) then, in particular, the same is true for the core assignment made by this assignment. Hence, while marking constraints with respect to this core assignment we must have marked $k + 1$ other constraints (that are also not satisfied by the core assignment alone), which are hence not deleted. However, these other constraints cannot all be satisfied with a budget of at most k, a contradiction. Thus, deleting all unmarked constraints is safe.

Therefore, if we have a sunflower with core size s and more than $(k + 1) \cdot (k + 1)^s$ constraints then our marking procedure allows us to delete at least one constraint. Thus, allowing for core size s up to $r - 1$, we set $t = (k + 1) \cdot (k + 1)^{r-1} + 1 = \mathcal{O}(k^r)$. While we can find t-sunflowers (via the sunflower lemma) we can always delete at least one constraint. Our earlier discussion at the start of the proof now gives a bound of $\mathcal{O}(t^r \cdot k^r) = \mathcal{O}(k^{r^2+r})$ on the number of constraints and variables achieved by this reduction process. This completes the proof. $\square$

Corollary 3. *The r-COVER ILP(k) problem admits a polynomial compression to size $\mathcal{O}(k^{r^2+2r})$ and a polynomial kernelization.*

Proof (sketch). We know how to reduce to $\mathcal{O}(k^{r^2+r})$ constraints in polynomial time. Since each constraint can equivalently be described by the infeasible assignments that it defines for the variables in its scope, we may encode the instance by replacing each constraint on $d \leq r$ variables by a 0/1-table of dimension $(k+1)^d$: Each 0 entry means that the assignment corresponding to this coordinate is infeasible (e.g., entry at $(2,3,5)$ tells whether $x_1 = 2$, $x_2 = 3$, and $x_3 = 5$ is feasible for this constraint, when (x_1, x_2, x_3) is its scope); each 1 stands for a feasible entry. For each constraint this requires $(k+1)^d = \mathcal{O}(k^r)$ bits for the table plus $r \cdot \log(k^{r^2+r}) = \mathcal{O}(\log k)$ bits for encoding the scope (there are at most r variables, each one encoded by a number from 1 to $\mathcal{O}(k^{r^2+r})$). The cost function $c^T x$ requires only the storing of $\mathcal{O}(k^{r^2+r})$ integers with values from 0 to k, taking $\mathcal{O}(\log k)$ bits each. In total this gives the claimed size for the compression.

To get a polynomial kernelization let us first observe that the problem described above is clearly in NP since feasible solutions (certificates) can be verified in polynomial time. Thus, by an argument of Bodlaender et al. [21], we get an encoding as an instance of r-COVER ILP(k) by using the implicit Karp reduction whose existence follows from the fact that r-COVER ILP(k) is complete for NP. The size of this instance is polynomial in $\mathcal{O}(k^{r^2+2r})$, and hence polynomial in k. This completes the polynomial kernelization. $\square$

5   Further Results

In this section we state some results for PACKING ILP(k) and COVER ILP(k) with column-sparseness q and row-sparseness r as additional parameters or constants; proofs are deferred to [1].

Theorem 4. PACKING ILP $(k + q + r)$ is fixed-parameter tractable and admits a polynomial-time preprocessing to $\mathcal{O}(kqr)$ variables and $\mathcal{O}(kq^2r)$ constraints. For r-PACKING ILP $(k + q)$ this gives a polynomial kernelization.

Theorem 5. COVER ILP $(k + q + r)$ is fixed-parameter tractable and admits a polynomial-time preprocessing to $\mathcal{O}(kqr)$ variables and $\mathcal{O}(kq)$ constraints. (A polynomial kernelization for r-COVER ILP $(k + q)$ follows from Corollary 3.)

6   Conclusion

We have studied different problem settings on integer linear programs. For INTEGER LINEAR PROGRAM FEASIBILITY parameterized by the numbers n of variables and m of constraints we ruled out polynomial kernelizations, assuming NP $\not\subseteq$ coNP/poly. This still holds when both column- and row-sparseness are at most three. Adding further new and old results, this settles the existence of polynomial kernels for ILPF and r-ILPF for parameterization by any subset of $\{n, m, \log C, C\}$ where C is the maximum absolute value of coefficients.

Regarding covering and packing ILPs, we gave polynomial kernels for r-COVER ILP(k), generalizing r-HITTING SET(k), and for r-PACKING ILP$(k+q)$. Further results and observations give an almost complete picture regarding q and r except for the question about polynomial kernels for parameter $k + q + r$. We recall that for both problems one can reduce to $n, m \leq (kqr)^{\mathcal{O}(1)}$ but parameterization by n and m only admits no polynomial kernelization (Corollary 2).

References

1. Kratsch, S.: On polynomial kernels for integer linear programs: Covering, packing and feasibility. CoRR abs/1302.3496 (2013)
2. Atamtürk, A., Savelsbergh, M.W.P.: Integer-programming software systems. Annals OR 140(1), 67–124 (2005)
3. Harnik, D., Naor, M.: On the compressibility of $\mathcal{NP}$ instances and cryptographic applications. SIAM J. Comput. 39(5), 1667–1713 (2010)
4. Downey, R.G., Fellows, M.R.: Parameterized Complexity. Monographs in Computer Science. Springer (1998)
5. Lokshtanov, D., Misra, N., Saurabh, S.: Kernelization – preprocessing with a guarantee. In: Bodlaender, H.L., Downey, R., Fomin, F.V., Marx, D. (eds.) Fellows Festschrift. LNCS, vol. 7370, pp. 129–161. Springer, Heidelberg (2012)
6. Bodlaender, H.L., Downey, R.G., Fellows, M.R., Hermelin, D.: On problems without polynomial kernels. J. Comput. Syst. Sci. 75, 423–434 (2009)
7. Fortnow, L., Santhanam, R.: Infeasibility of instance compression and succinct PCPs for NP. J. Comput. Syst. Sci. 77(1), 91–106 (2011)
8. Lenstra, H.W.: Integer programming with a fixed number of variables. Mathematics of Operations Research 8, 538–548 (1983)
9. Kannan, R.: Minkowski's convex body theorem and integer programming. Mathematics of Operations Research 12(3), 415–440 (1987)
10. Dom, M., Lokshtanov, D., Saurabh, S.: Incompressibility through colors and ids. In: Albers, S., Marchetti-Spaccamela, A., Matias, Y., Nikoletseas, S., Thomas, W. (eds.) ICALP 2009, Part I. LNCS, vol. 5555, pp. 378–389. Springer, Heidelberg (2009)
11. Hermelin, D., Kratsch, S., Soltys, K., Wahlström, M., Wu, X.: Hierarchies of inefficient kernelizability. CoRR abs/1110.0976 (2011)
12. Kratsch, S.: On polynomial kernels for sparse integer linear programs. In: Portier, N., Wilke, T. (eds.) STACS. LIPIcs, vol. 20, pp. 80–91. Schloss Dagstuhl - Leibniz-Zentrum fuer Informatik (2013)
13. Downey, R.G., Fellows, M.R.: Fixed-parameter tractability and completeness II: On completeness for W[1]. Theor. Comput. Sci. 141(1&2), 109–131 (1995)
14. Jansen, B.M.P., Bodlaender, H.L.: Vertex cover kernelization revisited - upper and lower bounds for a refined parameter. Theory Comput. Syst. 53(2), 263–299 (2013)
15. Nederlof, J., van Leeuwen, E.J., van der Zwaan, R.: Reducing a target interval to a few exact queries. In: Rovan, B., Sassone, V., Widmayer, P. (eds.) MFCS 2012. LNCS, vol. 7464, pp. 718–727. Springer, Heidelberg (2012)
16. Plotkin, S.A., Shmoys, D.B., Tardos, É.: Fast approximation algorithms for fractional packing and covering problems. In: FOCS, pp. 495–504 (1991)
17. Flum, J., Grohe, M.: Parameterized Complexity Theory. Texts in Theoretical Computer Science. An EATCS Series. Springer (2006)
18. Abu-Khzam, F.N.: A kernelization algorithm for d-hitting set. J. Comput. Syst. Sci. 76(7), 524–531 (2010)
19. Bodlaender, H.L., Jansen, B.M.P., Kratsch, S.: Cross-composition: A new technique for kernelization lower bounds. In: Schwentick, T., Dürr, C. (eds.) STACS. LIPIcs, vol. 9, pp. 165–176 (2011)
20. Erdős, P., Rado, R.: Intersection theorems for systems of sets. J. London Math. Soc. 35, 85–90 (1960)
21. Bodlaender, H.L., Thomassé, S., Yeo, A.: Kernel bounds for disjoint cycles and disjoint paths. Theor. Comput. Sci. 412(35), 4570–4578 (2011)

Balanced Neighbor Selection
for BitTorrent-Like Networks*

Sándor Laki and Tamás Lukovszki

Faculty of Informatics, Eötvös Loránd University
Pázmány Péter sétány 1/C, H-1117 Budapest, Hungary
{lakis,lukovszki}@inf.elte.hu

Abstract. In this paper we propose a new way of constructing the evolving graph in BitTorrent (BT) as new peers join the system one by one. The maximum degree in the constructed graph will be $O(1)$ while the diameter will remain $O(\ln n)$, with high probability, where n is the number of nodes. Considering a randomized upload policy, we prove that the distribution of b blocks on the overlay generated by our neighbor selection strategy takes $O(b + \ln n)$ phases only, with high probability, which is optimal up to a constant factor. It improves the previous upper bound of $O(b + (\ln n)^2)$ by Arthur and Panigrahy (SODA'06). Besides theoretical analysis, thorough simulations have been done to validate our algorithm and demonstrate its applicability in the BT network.

1 Introduction

In the past decade, tracker-based peer-to-peer networks like BitTorrent (BT) [1] and Tribler [2] have emerged as popular solutions in the area of not only simple file-sharing, but video-on-demand services as well. These applications are still showing an increasing interest, generating a significant part of the overall Internet traffic.

The selection of neighbors is an important design decision of peer-to-peer systems. In tracker-based peer-to-peer networks, each peer that enters the network, first has to connect to a central component called tracker to obtain a peer set representing the initial neighborhood of the joining client. The tracker maintains a list of all nodes in the system, called the swarm, and returns a random subset of the existing nodes. This random neighbor selection may lead to suboptimal overlay topologies. In order to optimize the network, various neighbor selection strategies can be found in the literature that considers different aspects from locality [14] and load balancing [3,8] to quality of experience [12].

The performance of BT like peer-to-peer systems has been widely analyzed in the past few years from theoretical and practical aspects as well. The empirical

---

* Supported by the EU EIT project SmartUC, the grant EITKIC-12-1-2012-0001 of the National Development Agency, and the EU FP7 OpenLab project (Grant No. 287581). Performed in cooperation with the EIT ICT Labs Budapest Associate Partner Group (www.ictlabs.elte.hu).

H.L. Bodlaender and G.F. Italiano (Eds.): ESA 2013, LNCS 8125, pp. 659–670, 2013.

results [9,5] show that the simple routing policy applied by the original BT is quite effective even in case of a flash crowd setting when a great deal of peers join the network almost at the same time. Besides empirical evidence, this heavy loaded case has already been investigated from a theoretical perspective as well. In [3], several algorithms are demonstrated that share b data blocks among n clients in a network of diameter d and degree D in $O(D(b + d))$ steps with high probability[1], where in one time step, each client can upload one data block to, and download one block from one of its neighbor. For a network used by BT it results in a time bound of $O(b \ln n)$ time steps. They propose a neighbor selection strategy which improves this bound to a near-optimal $O(b + (\ln n)^2)$ steps.

In this paper, we improve the neighbor selection strategy resulting in a time bound of $O(b + \ln n)$ steps, which is optimal up to a constant factor. Our method uses the idea of multiple choice [4] and takes into account not only the actual load of the peers, but the possibility as well that a client will be selected in the future. This will ensure an overlay network of constant degree and logarithmic diameter with high probability. The constructed overlay topologies are examined from both theoretical and practical perspectives as well. We model these overlay networks as a graph, whose vertices are the peers and neighboring peers are connected by an edge. We analyze the key graph properties of the proposed network, showing that the maximum degree in our overlay topology is $O(1)$, with high probability, while its diameter still remains logarithmic in the number of peers n. In such a network, the randomized upload policy will share b data blocks among n clients in $O(b + \ln n)$ time steps with high probability, which is optimal in networks of n vertices, in which the degree of the vertices is bounded by a constant. Besides the theoretical analysis thorough simulations have been performed to validate the different properties of the constructed overlay networks.

The rest of the paper is organized as follows: in Section 2, we briefly overview the related works. In Section 3, we describe our model and the different neighbor selection methods. Section 4 details our theoretical analysis and includes the proof of our theoretical upper bounds. In Section 5, we experimentally analyze the different properties of the constructed overlay networks. Finally, we include some concluding remarks in Section 6.

2   Related Works

A great deal of theoretical and empirical studies have emerged in the past decade to analyze the performance of existing BT like peer-to-peer networks and propose new neighbor selection methods to optimize the constructed overlay topology. At the empirical end, Izal et al. [9] and Pouwelse et al. [13] present measurement based studies of BT which are based on tracker logs of different torrents. Their analysis show that the simple mechanisms that can be found in the original BT makes this file-sharing system very efficient.

---

[1] An event E is said to occur with high probability, if given $n > 1$, $Pr[E] > 1 - 1/n^c$, where $c > 1$ is a constant [10].

Bharambe et al. [6] conducted simulations to confirm that BT performs near-optimally in terms of uplink bandwidth utilization, and download time except under certain extreme conditions. They have also found that the rate-based tit-for-tat policy is not effective in preventing unfairness, which means that low bandwidth peers can download more than they upload to the network when high bandwidth peers are present. To solve this issue, they propose some slight changes to the tracker and a stricter tit-for-tat policy.

Bindal et al. [7] examine a new approach to enhance BT traffic locality, in which a peer chooses the majority, but not all, of its neighbors from peers within the same ISP. In this way the traffic costs at ISPs can significantly be reduced.

Besides empirical works there are several theoretical ones as well. Zhang et al. [15] formulate an optimization problem to solve for the optimal peer selection strategy to maximize the global system-wide performance. They also derive a purely distributed algorithm that is provably globally optimal. Besides the tracker, the proposed solution requires some changes in the ordinary peers as well.

Arthur and Panigrahy [3] propose a mathematical framework to model the distribution of individual data blocks. They examine several properties of BT like networks and discuss a number of extensions to them, including a new neighbor selection strategy that can easily be implemented in the trackers to achieve near-optimal performance in the distribution of data blocks. In our work, we use the same analytical framework introduced in [3].

3   System Model

In this section, we briefly outline our system model, which is in accordance with what is proposed by Arthur and Panigrahy in [3], and describe the neighbor selection strategies to be examined.

For the sake of simplicity, we model the constructed overlay topologies as directed graphs where each vertex represents a client in the peer-to-peer network. We also assume equal bandwidth and delay among all the peers, and ignore other BT specific mechanisms such as tit-for-tat and optimistic unchoke. This simplified model does not take care of the process of clients joining and leaving the network during the file-sharing. Furthermore, the file sharing process can be considered as routing data blocks on the directed graph over discrete time steps. To this end, Arthur and Panigrahy [3] proposed the randomized upload policy where each vertex attempts to upload a block to a random neighboring client during each time step. They proved the following theorem.

Theorem 1 ([3]). *Suppose a vertex u begins with a copy of every block. Let D denote the maximum out-degree in a directed graph consisting of n vertices, and suppose the distance from u to every other vertex is at most d. If we route on this graph using the randomized upload policy, then $T \leq 4D(4d + b)$ with probability at least $1 - 2n \exp\left(-\frac{d}{2}\right)$, where b denotes the number of distinct blocks to be distributed and T is the number of time steps before the routing completes.*

We introduce a neighbor selection strategy, which results in a network of constant degree and logarithmic diameter. Applying Theorem 1 to this network we obtain a bound of $O(b + \log n)$ time steps for the completion of the routing of the b blocks, with high probability. Note that every graph of n vertices with constant degree has a diameter $\Omega(\log n)$. Thus the routing of a block to the farthest vertex takes $\Omega(\log n)$ steps. Since it can receive one block in a step receiving all the blocks takes $\Omega(b + \log n)$ steps. Thus, our result is optimal up to a constant factor.

As we mentioned before, in BT each joining client first sends a request to the tracker that returns a peer set chosen uniformly at random among the existing peers. These nodes form the initial neighborhood of the new client. To model such a peer-to-peer network, Arthur and Panigrahy [3] propose the BitTorrent-C graph ($C \geq 2$) which is a directed graph constructed by the following method.

BitTorrent-C graph (abbr. **BT-C graph**)[3]:

1. At the beginning it consists of C vertices, $v_1, ..., v_C$ and edges from v_j to v_i if and only if $j < i$.
2. While the total number of vertices is less than n, add a vertex and add directed edges from C existing vertices chosen uniformly at random to the new vertex.

This graph has been analyzed thoroughly in [3]. It has been shown that with high probability the maximum out-degree in a BT-C graph is at most $3C(1 + \ln N)$, while the diameter is at most $3 \lg n$. Furthermore, based on Theorem 1 they also proved that the required time steps for distributing b blocks with the randomized upload policy can be bounded by $O(\ln n(b + \lg n))$, with high probability.

To improve the performance of the original BT, especially in case of flash crowd settings, they recommend a practical variant of BT-C, called Smoothed-BT-C graph that can be constructed as follows.

Smoothed-BT-C graph [3]:

1. At the beginning it consists of C vertices, $v_1, ..., v_C$ and edges from v_j to v_i if and only if $j < i$.
2. While the total number of vertices is less than n, add a vertex and add directed edges from C existing vertices to the new vertex, but instead of choosing each previous vertex uniformly at random, select two previous nodes and connect the one with higher index to the new vertex.

In [3] it has been proved that by using the randomized upload policy in a Smoothed-BT-C graph with n vertices, the routing completes in at most $O(b + (\ln n)^2)$ time steps, with high probability.

In this paper, we propose two novel neighbor selection strategies that use the idea of multiple choice to ensure an overlay network of constant out-degree, with high probability. We first introduce the MultipleChoice-BT-C graph that can be built up in the following manner.

MultipleChoice-BT-C graph:

1. At the beginning it consists of C vertices, $v_1, ..., v_C$ and edges from v_j to v_i if and only if $j < i$.
2. We add the remaining vertices in order. Let $t > 2$ be a constant. When we add the ith vertex, $C < i \leq n$, we choose $t \log n$ vertex from the vertex set $\{v_j : j \in [\min(i/2, i - C), i - 1]\}$ uniformly at random and connect the lowest degree vertex to the new vertex. We repeat this C times in order to add C directed edges to the new vertex.

Estimating $\log n$ can be done in various ways, some simple methods are discussed e.g. in [11]. We will show in Section 4 that the multiple choice policy guarantees constant out-degrees with high probability. The disadvantage of this strategy is that at the insertion of the ith vertex it tends to select the neighbors from the last $\frac{1}{2t \log n}$ fraction of the previous vertices, since the newer vertices have more likely a lower degree. The consequence of this will be a super-logarithmic diameter.

To remedy this problem, we present a modification of this multiple choice neighbor selection policy, which guaranties an overlay topology with $O(C)$ maximal degree and logarithmic diameter with high probability. This improved variant of BT-C is called Balanced-BT-C graph. It is similar to the MultipleChoice-BT-C graph with the following modification. Instead of choosing the vertex with the smallest out-degree from $t \log n$ previous vertices, we also take into account the expected number of times it will be selected in the future. Before the insertion of the i-th node, for a vertex v_j, $j < i$, let $\delta(v_j)$ denote the out-degree of v_j and $w(v_j) := \sum_{i < \ell \leq 2j} \frac{2C}{\ell}$. The value $w(v_j) \cdot t \log n$ expresses the expected number of times v_j will be chosen after the insertion of v_i. For $j < i/2$, $w(v_j) = 0$ and for $j \geq i$, $w(v_j)$ is the expected number of future edges. Let $\delta^*(v_j) := \delta(v_j) + w(v_j)$.

Balanced-BT-C graph:

1. At the beginning the graph consists of C vertices, $v_1, ..., v_C$ and edges from v_j to v_i if and only if $j < i$.
2. We add the remaining vertices in order. Let $t > 2$ be a constant. When we add the ith vertex, $C < i \leq n$, we choose $t \log n$ vertex from the vertex the set $\{v_j : j \in [\min(i/2, i - C), i - 1]\}$ uniformly at random and connect the vertex v_j with the lowest $\delta(v_j) + w(v_j)$ to the new vertex. We repeat this C times in order to add C directed edges to the new vertex.

Using this neighbor selection strategy ensures that at the moment we insert the ith vertex, the expected values of $\delta^*(v_j)$, $j \in [i/2, i - 1]$ are approximately the same. Thus the selected neighbor is distributed evenly among $\{v_j : j \in [i/2, i-1]\}$. As a consequence, the median neighbor is selected expectedly from the middle of the interval $[i/2, i - 1]$, which results a network with logarithmic diameter.

4   Theoretical Analysis

In order to give a bound on the diameter of the network the term of median depth has been introduced in [3], and defined as follows.

Median Depth: For any i, consider the set of integers j such that there is an edge from v_j to v_i. Let $m(i)$ denote a median element in this set. Recursively, let $m^k(i) = m(m^{k-1}(i))$ for $k > 0$ and $m^0(i) = i$. The median depth of v_i is defined as the the smallest k such that $m^k(i) = 1$.

Clearly, the distance from v_1 to v_i is at most the median depth of v_i.

Lemma 1. *For $t \geq 2$, the maximum out-degree in a MultipleChoice-BT-C graph is at most $2C$ with probability $1 - \frac{C}{n}$.*

Proof. Consider the insertion of the ith vertex. To create a maximum out-degree greater than $2C$, at least one edge to be inserted from a previous vertex with out-degree of at least $2C$. The neighborhood of the ith vertex consists of peers from the interval $I = [\frac{i}{2}, i-1]$. We also know that peers from I may previously have been selected by at most $\frac{i}{2} - 1$ vertices as neighbors, so the number of edges originated from I is at most $C(\frac{i}{2} - 1) < C\frac{i}{2}$. Consequently, the interval contains less than $\frac{i}{4}$ peers with out-degree $2C$. Thus the probability that we select a vertex from I having out-degree $2C$ is at most $\frac{1}{2}$. The probability that all the $t \log n$ vertices have out-degree $2C$ is at most $(\frac{1}{2})^{t \log(n)} = \frac{1}{n^t}$. Since all the vertices have C ingress edges, the probability that after inserting the ith vertex there is a peer with out-degree $2C+1$ is at most $\frac{C}{n^t}$. Since $t \geq 2$ using the union bound over the error probabilities of all the vertices proves the lemma:

$$\sum_{i=1}^{n} \frac{C}{n^t} < \frac{C}{n^{t-1}} \leq \frac{C}{n}. \qquad \square$$

Although this neighbor selection strategy guarantees a constant degree with high probability, the median depth of the resulting graph will be super-logarithmic. The reason is that newer vertices tend to have lower degree and the neighbors of the new vertex are preferentially selected among the last $\frac{1}{2t \log n}$ fraction of the previous vertices. When we insert the nth vertex, a vertex v_j, $j \in [n/2, n-1]$ has been chosen before the nth insertion expectedly $t \log n \sum_{j < i < n} \frac{2}{i}$ times. Thus newer vertices have been chosen less frequently and more likely have lower degree.

Lemma 2. *For $t \geq 2$, the maximum out-degree in a Balanced-BT-C graph is at most $6C$ with probability $1 - \frac{C}{n}$.*

Proof. Consider the insertion of the ith vertex. The neighborhood of the ith vertex consists of peers from the interval $I = [\frac{i}{2}, i-1]$. We prove that after the insertion of the ith vertex, for each v_j, $j \in I$, $\delta^*(v_j) \leq 6C$, with high probability. It will imply the claim of the lemma. To create a vertex v_j with $\delta^*(v_j) > 6C$, all of the $t \log n$ chosen vertices must have a $\delta^*(.)$ value of at least $6C$. We know that peers from I may previously have been selected by at most $\frac{i}{2} - 1$ vertices as neighbors, so the number of edges originated from I is at most $C(\frac{i}{2} - 1) < C\frac{i}{2}$. Furthermore, we know that, for each $j \in I$, $w(v_j) = \sum_{i < \ell \leq 2j} \frac{2C}{\ell} \leq (2j - i)\frac{2C}{i} \leq 2C$. Since I contains $\frac{i}{2}$ vartices, $\sum_{j \in I} w(v_j) \leq Ci$. Therefore, $\sum_{j \in I} \delta^*(v_j) < C\frac{3i}{2}$. Consequently, less less than $\frac{i}{4}$ peers with a $\delta^*(.)$ value $6C$. Thus the probability that we select a vertex from I having a $\delta^*(.)$ value $6C$ is at most $\frac{1}{2}$. The probability that all the $t \log n$ vertices have a $\delta^*(.)$

value $6C$ is at most $(\frac{1}{2})^{t \log(n)} = \frac{1}{n^t}$. Since all vertices have C ingress edges, the probability that, after inserting the ith vertex, there is a peer with a $\delta^*(.)$ value greater than $6C$ is at most $\frac{C}{n^t}$. Since $t \geq 2$ using the union bound over the error probabilities of all the vertices proves the lemma: $\sum_{i=1}^{n} \frac{C}{n^t} < \frac{C}{n^{t-1}} \leq \frac{C}{n}$.   □

One can observe that our balanced neighbor selection strategy ensures that at the moment we insert the ith vertex, the expected values of $\delta^*(v_j)$, $j \in [i/2, i-1]$ are approximately the same, resulting that the selected neighbor is distributed evenly among $\{v_j : j \in [i/2, i-1]\}$. In this case, the median neighbor is expectedly from the middle of the interval $[i/2, i-1]$, which leads to a network with logarithmic median-depth.

Applying Theorem 1, it can be claimed that in a Balanced-BT-C graph with n vertices, routing b data blocks with the randomized upload policy completes in at most $O(b + \log n)$ time steps, with high probability, which is provable optimal up to constants.

5   Experimental Results

In the previous section, high probability upper bounds have been proved for the maximum out-degree and the diameter of the different networks. In addition, simulations enable us not only to validate these theoretical results but to reveal more information on the distributions themselves and explore the practical strength of the above bounds.

The parameters of the constructed overlay networks have been varying in a reasonable wide range: C was chosen from the range $[2, 90]$, while n from $[500, 100000]$. Due to page limitation, results for $C = 2$ and 30 are only presented in this paper. $C = 2$ demonstrates well the behavior of different strategies for constant size neighbor set, while $C = 30$ corresponds to the neighbor set whose size is rather logarithmic in the number of peers in our simulations. For each setting, the simulations have been repeated 100 times to obtain statistically enough data for further analysis. Note that the constant t in our multiple choice methods has been fixed to 2 during the analysis.

First, we pay attention to the maximum out degrees observed in the different networks. In a file sharing network, the out-degree of a peer determines the maximum load on it, indicating the maximum number of other nodes that the given peer can upload data to. Figure 1 illustrates the maximum out-degrees for two different C values. Each dot in these figures indicates the maximum out-degree observed in a given experiment. Since for each setting and algorithm, the simulations have been repeated 100 times, we can see 100 individual dots for each n value and graph construction. The fitted curves for the different networks are marked by different colors, expressing the connection between the maximum out-degree values and the number of clients. Looking at Figure 1(a), one can observe that, not surprisingly, the original BT-C results the highest maximum out-degrees and shows a logarithmic correlation between the maximum values and the network size. In case of a such small C, Smoothed-BT-C behaves very

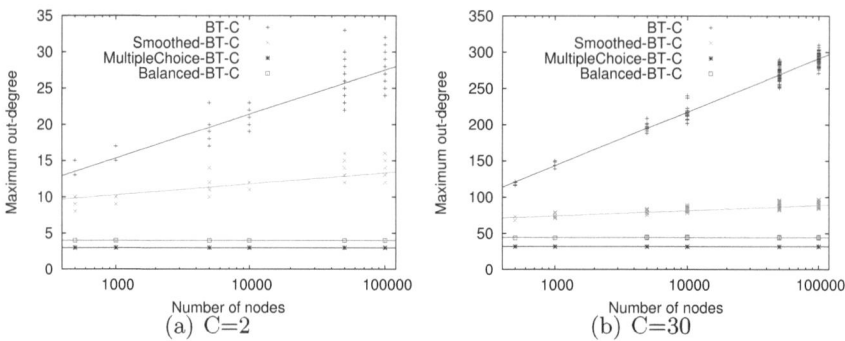

Fig. 1. Maximum out-degrees for the different overlay networks. Each dot represents a maximum out-degree value after the given simulation has been done. For each parameterization and algorithm, the simulations have been repeated 100 times. The fitted curves/lines indicate the relationship between the maximum out-degrees and the number of clients in the network.

similarly to the previous overlay construction. Although it provides less maximum out-degrees for large n values, it is almost the half of what we can see in BT-C, but the above relationship is still logarithmic. In accordance with the theory, both our MultipleChoice-BT-C and Balanced-BT-C algorithms aim at keeping the out-degrees at a constant level. For $C = 2$, the maximum values are 3 and 4, respectively. These values correspond to the theoretical high probability upper bounds. Considering larger neighborhood sizes in Figure 1(b), where C is 30, similar correlations can be identified, but there are some slight differences we have to shed light on. First of all, for large peer set sizes (C), the maximum out-degrees in the networks constructed by MultipleChoice-BT-C and Balanced-BT-C never reach the theoretical upper bound and, as it is expected, the former method results a bit lower load level on the peers than the latter one. From a practical perspective, when the network size is within a reasonable range (e.g. less than 1 million), the out-degrees in a Smoothed-BT-C graph are only slightly influenced by the number of peers.

Besides the analysis of the maximum values, our experiments enable us to examine the out-degree distributions themselves. Figure 2 shows the complementer cumulative distribution function (CCDF) of the out-degrees for the different algorithms on a semi-log plot. First, we consider $C = 2$, in case of BT-C the out-degrees seem to follow an exponential decay, but for larger C values the results show slightly better load distribution, for the tail of the distribution decays sharply after a certain point. For Smoothed-BT-C, the figures indicate that the CCDF of the out-degrees decreases faster than exponential, for both C values. Not surprisingly, our MultipleChoice-BT-C and Balanced-BT-C methods result a sudden drop in the CCDF plots at a constant value which is less than or equal to $2C$, indicating that the majority of the nodes have the same constant out-degree, and the load of the peers are much more balanced.

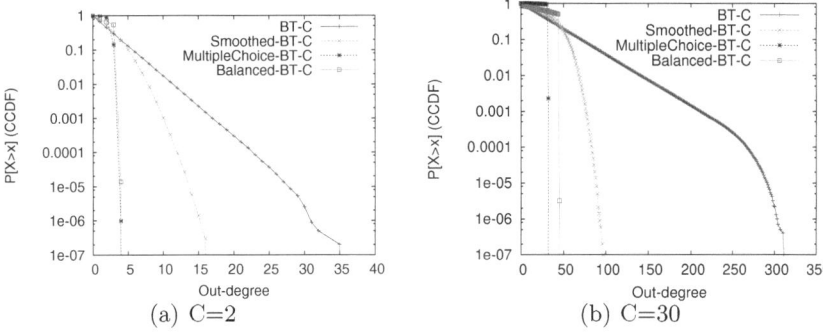

Fig. 2. The CCDF of the out-degrees in a network containing $n = 100000$ peers. Each method has been evaluated 100 times.

Besides the maximum out-degree, the diameter of the overlay topology also plays a crucial role in data distribution. For example, if we consider a peer-to-peer network containing only one seeder node having the whole file at the beginning, the data blocks need to propagate through the whole network to reach all the peers. It has already shown in Section 4 that besides the number of blocks to be distributed and the maximum out-degree, the diameter of the network has also significant effect on the time required for spreading the blocks of a given file.

Figure 3 depicts the relationship between the maximum median-depth and the number of peers in the different overlay networks. Each dot in the figures represents an individual experiment where the network size can be seen on the horizontal axis, while the observed maximum median-depth on the vertical ones. The fitted curves show the trend of this relationship for the different methods. Looking at the case $C = 2$, the smallest maximum median-depth values are produced by the simple BT-C and our Balanced-BT-C approaches. In accordance with our theoretical results, in both cases a logarithmic correlation can be identified, with a slight difference between them. Considering larger C values, this difference increases a bit, but it is not significant. For $C = 30$, in BT-C and Balanced-BT-C networks with 100000 clients, the maximum median-depths are 20 and 30, respectively, and they are even less in smaller networks. Smoothed-BT-C also produces a logarithmic relationship with a bit larger base. In the previous example, the maximum diameter resulted by this approach is less than 65. Taking into account the resulted maximum out-degrees and the ease of its implementability, this method could also perform well in practice, providing a good trade-off between reduced load and a bit longer diameter. In addition, we have seen that MultipleChoice-BT-C results constant maximum out-degree which is less than what we can get in the case of our Balanced-BT-C method. However, the correlation between the diameter and the network size is much worse than logarithmic and seems to be proportional to $(\ln n)^2$. According to Theorem 1, both the diameter and the maximum out-degree of a network have significant influence on how much time it takes to fully distribute the data blocks of a given

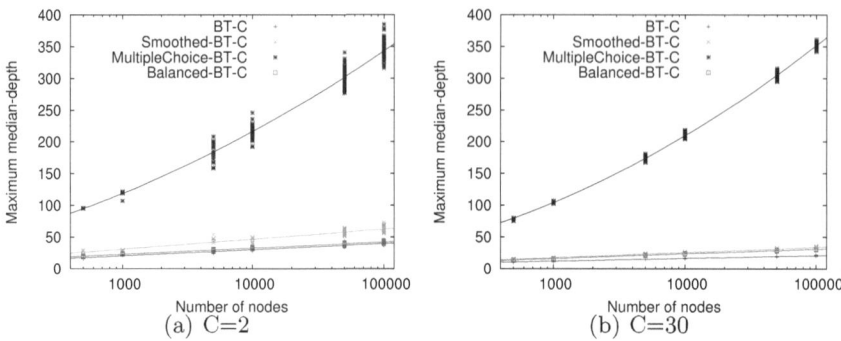

Fig. 3. Maximum median-depth for the different overlay networks. Each dot represents a maximum median-depth value after the given simulation has been done. For each parameterization and algorithm, the simulations have been repeated 100 times. The fitted curves/lines indicate the relationship between the maximum median-depth and the number of clients in the network.

file in the network. In this respect, our improved Balanced-BT-C method surpasses all the other examined approaches, keeping the load of peers at a constant level and producing short paths in the network whose lengths can be bounded by $O(\ln n)$ with high probability.

It can be said that though the multiple choice can reduce the maximum outdegree in the overlay topology, but in itself it results a too regular network with unmanageably huge diameter. It shows the practical meaning of our balancing technique that can remedy this phenomenon.

The histograms of the maximum median-depth values observed in our simulations are presented in Figure 4. The first conspicuous difference we can recognize is that our balanced algorithm shows significantly narrower distribution compared to the other cases. All the values fall between 34 and 47, while in the other networks they show much higher variance, resulting twice or more wider ranges. We can also see that the difference between BT-C and Balanced-BT-C is even less than what can be derived from Figure 3. The most likely maximum median-depth values are 39 and 42, respectively. In case of Smoothed-BT-C and MultipleChoice-BT-C the observed values cover significantly larger ranges.

Besides the maximum values, we have also examined the distribution of median-depths in the different overlays. Figure 5 depicts the CCDF of the median-depth values where each network consists of 100000 peers. As it is expected, MultipleChoice-BT-C provides significantly higher median-depths with a much slower decay than the others. Meanwhile, the other three methods show very similar distributions. In Figure 5(a), where $C = 2$, the slope in the CCDF plot of Balanced-BT-C is much sharper than in the case of the other two methods, which indicates that the most likely values are coming from a much wider range. It means that not only the maximum diameters have smaller variance, but the median distances from the source to ordinary peers as well. The CCDF of Balanced-BT-C

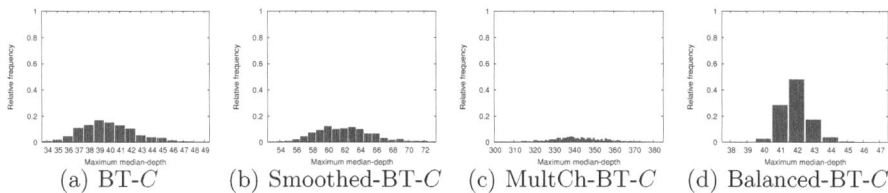

(a) BT-C (b) Smoothed-BT-C (c) MultCh-BT-C (d) Balanced-BT-C

Fig. 4. The empirical distribution of the maximum median-depths observed in our simulations. Each algorithm has been performed 500 times with the parameters $n = 100000$ and $C = 2$.

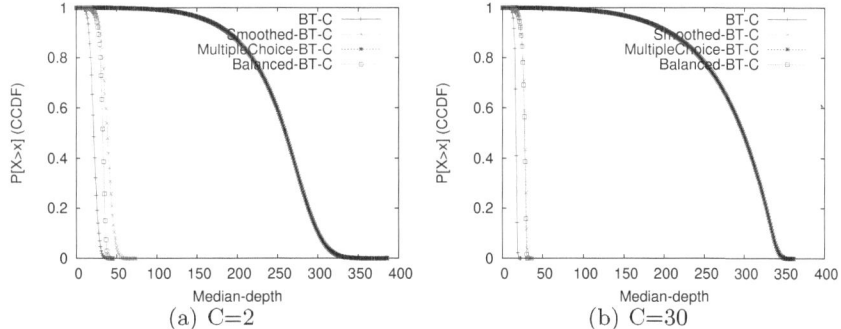

(a) C=2   (b) C=30

Fig. 5. The CCDF of the median-depths for different overlay constructions with fixed network sizes $n = 100000$. Each method has been evaluated 100 times.

takes place between BT-C and Smoothed-BT-C in all three figures ($C = 2$ and 30). One can also recognize that for larger C, the median-depths in our balanced overlay follow almost the same distribution as what can be seen in the case of Smoothed-BT-C. As C is increasing, the difference between the two distribution is basically disappearing which suggests that, in terms of median-depths, Balanced-BT-C could be at most as worse as the Smoothed-BT-C method.

6   Conclusion

In this paper, we have introduced a novel neighbor selection strategy which uses the idea of multiple choice to improve the performance of spreading blocks in a BT like peer-to-peer network. Our multiple choice algorithm takes into account not only the current load of a given peer, but the expected value that it will be selected as uploading neighbor in the future. The constructed overlay topology has been analyzed from both theoretical and experimental aspects and it has been proved that this topology has constant degree and logarithmic diameter with high probability. We have also shown that considering a randomized upload policy, routing of b data blocks in the proposed network requires at most

$O(b + \log n)$ time steps with high probability, which is optimal up to a constant factor. Besides the theoretical analysis, thorough simulations has been performed to examine the graph properties of the constructed networks and validate the theoretical results as well.

References

1. Bittorrent, http://bittorrent.com
2. Tribler, http://www.tribler.org
3. Arthur, D., Panigrahy, R.: Analyzing bittorrent and related peer-to-peer networks. In: Proc. of the Seventeenth Annual ACM-SIAM Symposium on Discrete Algorithms, pp. 961–969. ACM (2006)
4. Azar, Y., Broder, A.Z., Karlin, A.R., Upfal, E.: Balanced allocations. SIAM J. Comput. 29(1), 180–200 (1999)
5. Bharambe, A.R., Herley, C., Padmanabhan, V.N.: Understanding and Deconstructing BitTorrent Performance. Technical Report MSR-TR-2005-03, Microsoft Research (2005)
6. Bharambe, A.R., Herley, C., Padmanabhan, V.N.: Analyzing and improving a bittorrent networks performance mechanisms. In: Proceedings of the 25th IEEE Int. Conference on Computer Communications, INFOCOM 2006, pp. 1–12 (April 2006)
7. Bindal, R., Cao, P., Chan, W., Medved, J., Suwala, G., Bates, T., Zhang, A.: Improving traffic locality in bittorrent via biased neighbor selection. In: ICDCS 2006, p. 1 (2006)
8. Hecht, F.V., Bocek, T., Stiller, B.: B-tracker: Improving load balancing and efficiency in distributed p2p trackers. In: 2011 IEEE Int. Conference on Peer-to-Peer Computing, P2P, pp. 310–313 (2011)
9. Izal, M., Urvoy-Keller, G., Biersack, E.W., Felber, P., Al Hamra, A., Garcés-Erice, L.: Dissecting bittorrent: Five months in a torrent's lifetime. In: Barakat, C., Pratt, I. (eds.) PAM 2004. LNCS, vol. 3015, pp. 1–11. Springer, Heidelberg (2004)
10. Motwani, R., Raghavan, P.: Randomized algorithms. ACM Comput. Surv. 28(1), 33–37 (1996)
11. Naor, M., Wieder, U.: Novel architectures for p2p applications: The continuous-discrete approach. ACM Transactions on Algorithms 3(3) (2007)
12. Picone, M., Amoretti, M., Zanichelli, F.: An evaluation criterion for adaptive neighbor selection in heterogeneous peer-to-peer networks. In: Pfeifer, T., Bellavista, P. (eds.) MMNS 2009. LNCS, vol. 5842, pp. 144–156. Springer, Heidelberg (2009)
13. Pouwelse, J., Garbacki, P., Epema, D., Sips, H., Epema, D.H.J., Sips, H.J.: A measurement study of the bittorrent peer-to-peer file-sharing system (2004)
14. Wang, H., Liu, J., Chen, B., Xu, K., Ma, Z.: On tracker selection for peer-to-peer traffic locality. In: 2010 IEEE Tenth Int. Conference on Peer-to-Peer Computing, P2P, pp. 1–10 (2010)
15. Zhang, H., Shao, Z., Chen, M., Ramchandran, K.: Optimal neighbor selection in bittorrent-like peer-to-peer networks. In: Proc. ACM SIGMETRICS Joint Int. Conference on Measurement and Modeling of Computer Systems, pp. 141–142 (2011)

Parameterized Complexity of Directed Steiner Tree on Sparse Graphs

Mark Jones[1], Daniel Lokshtanov[2], M.S. Ramanujan[3], Saket Saurabh[2,3], and Ondřej Suchý[4*]

[1] Royal Holloway University of London, United Kingdom
markj@cs.rhul.ac.uk
[2] University of Bergen, Norway
daniello@ii.uib.no
[3] Institute of Mathematical Sciences, India
{msramanujan,saket}@imsc.res.in
[4] Czech Technical University in Prague, Czech Republic
ondrej.suchy@fit.cvut.cz

Abstract. We study the parameterized complexity of the directed variant of the classical STEINER TREE problem on various classes of directed sparse graphs. While the parameterized complexity of STEINER TREE parameterized by the number of terminals is well understood, not much is known about the parameterization by the number of non-terminals in the solution tree. All that is known for this parameterization is that both the directed and the undirected versions are W[2]-hard on general graphs, and hence unlikely to be fixed parameter tractable (**FPT**). The undirected STEINER TREE problem becomes **FPT** when restricted to sparse classes of graphs such as planar graphs, but the techniques used to show this result break down on directed planar graphs.

In this article we precisely chart the tractability border for DIRECTED STEINER TREE (DST) on sparse graphs parameterized by the number of non-terminals in the solution tree. Specifically, we show that the problem is fixed parameter tractable on graphs excluding a topological minor, but becomes W[2]-hard on graphs of degeneracy 2. On the other hand we show that if the subgraph induced by the terminals is required to be acyclic then the problem becomes **FPT** on graphs of bounded degeneracy.

We also show that our algorithm achieves the best possible running time dependence on the solution size and degeneracy of the input graph, under standard complexity theoretic assumptions. Using the ideas developed for DST, we also obtain improved algorithms for DOMINATING SET on sparse undirected graphs. These algorithms are asymptotically optimal.

---

* Part of this work was done while with the Saarland University, supported by the DFG Cluster of Excellence on Multimodal Computing and Interaction and the DFG project DARE (GU 1023/1-2), while at TU Berlin, supported by the DFG project AREG (NI 369/9), and while visiting IMSc, Chennai, supported by the Indo-German Max Planck Center for Computer Science.

H.L. Bodlaender and G.F. Italiano (Eds.): ESA 2013, LNCS 8125, pp. 671–682, 2013.
© Springer-Verlag Berlin Heidelberg 2013

1   Introduction

In the STEINER TREE problem we are given as input a n-vertex graph $G = (V, E)$ and a set $T \subseteq V$ of terminals. The objective is to find a subtree ST of G spanning T that minimizes the number of vertices in ST. STEINER TREE is one of the most intensively studied graph problems in Computer Science. Steiner trees are important in various applications such as phylogenetic tree reconstruction [11] and network routing [16]. We refer to the book of Prömel and Steger [20] for an overview of the results on, and applications of the STEINER TREE problem. The STEINER TREE problem is known to be NP-hard [8], and remains hard even on planar graphs [7]. The minimum number of non-terminals can be approximated to within $O(\log n)$, but cannot be approximated to $o(\log t)$, where t is the number of terminals, unless $P \subseteq DTIME[n^{\text{polylog } n}]$ (see [14]). Furthermore the weighted variant of STEINER TREE remains APX-complete, even when the graph is complete and all edge costs are either 1 or 2 (see [2]).

In this paper we study a natural generalization of STEINER TREE to directed graphs, from the perspective of parameterized complexity. The goal of parameterized complexity is to find ways of solving NP-hard problems more efficiently than by brute force. The aim is to restrict the combinatorial explosion in the running time to a parameter that is much smaller than the input size for many input instances occurring in practice. Formally, a *parameterization* of a problem is the assignment of an integer k to each input instance and we say that a parameterized problem is *fixed-parameter tractable* (FPT) if there is an algorithm that solves the problem in time $f(k) \cdot |I|^{O(1)}$, where $|I|$ is the size of the input instance and f is an arbitrary computable function depending only on the parameter k. Above FPT, there exists a hierarchy of complexity classes, known as the W-hierarchy. Just as NP-hardness is used as an evidence that a problem is probably not polynomial time solvable, showing that a parameterized problem is hard for one of these classes gives evidence that the problem is unlikely to be fixed-parameter tractable. The main classes in this hierarchy are:

$$FPT \subseteq W[1] \subseteq W[2] \subseteq \cdots \subseteq W[P] \subseteq XP.$$

The principal analogue of the classical intractability class NP is W[1]. In particular, this means that an FPT algorithm for any W[1]-hard problem would yield an $f(k)n^c$ time algorithm for every problem in the class W[1]. For more background on parameterized complexity the reader is referred to the monograph [6]. We consider the following directed variant of STEINER TREE.

DIRECTED STEINER TREE (DST)   Parameter: k
Input: A directed graph $D = (V, A)$, a root vertex $r \in V$, a set $T \subseteq V \setminus \{r\}$ of terminals and an integer $k \in \mathbb{N}$.
Question: Is there a set $S \subseteq V \setminus (T \cup \{r\})$ of at most k vertices such that the digraph $D[S \cup T \cup \{r\}]$ contains a directed path from r to every terminal $t \in T$?

The DST problem is well studied in approximation algorithms, as the problem generalizes several important connectivity and domination problems on undirected as well as directed graphs [4,5,10,21,22]. These include GROUP STEINER TREE, NODE WEIGHTED STEINER TREE, TSP and CONNECTED DOMINATING SET. However, this problem has so far largely been ignored in the realm of parameterized complexity. The aim of this paper is to fill this gap.

It follows from the reduction presented in [18] that DST is W[2]-hard on general digraphs. Hence we do not expect FPT algorithms to exist for these problems, and so we turn our attention to classes of *sparse* digraphs. Our results give a nearly complete picture of the parameterized complexity of DST on sparse digraphs. Specifically, we prove the following results. We use the O^* notation to suppress factors polynomial in the input size.

1. There is a $O^*(2^{O(hk)})$-time algorithm for DST on digraphs excluding K_h as a minor[1]. Here K_h is a clique on h vertices.
2. There is a $O^*(f(h)^k)$-time algorithm for DST on digraphs excluding K_h as a topological minor. This algorithm is quite involved and uses the recently developed Grohe-Marx decomposition for graphs excluding K_h as a topological minor [9] and our algorithm for the problem on K_h-minor free digraphs as a subroutine. The proof is given in the appended full version of the paper.
3. There is a $O^*(2^{O(hk)})$-time algorithm for DST on digraphs excluding K_h as a topological minor if the graph induced by the set of terminals (also referred to as the graph induced on terminals) is acyclic.
4. DST is W[2]-hard on 2-degenerate digraphs if the graph induced on terminals is allowed to contain directed cycles.
5. There is a $O^*(2^{O(dk)})$-time algorithm for DST on d-degenerate graphs if the graph induced on terminals is acyclic, implying that DST is FPT parameterized by k on $o(\log n)$-degenerate graph classes. This yields the first FPT algorithm for STEINER TREE on *undirected* d-degenerate graphs.
6. For any constant $c > 0$, there is no $f(k)n^{o(\frac{k}{\log k})}$-time algorithm on graphs of degeneracy $c \log n$ even if the graph induced on terminals is acyclic, unless the Exponential Time Hypothesis [12] (ETH) fails.

Our algorithms for DST hinge on a novel branching which exploits the domination-like nature of the DST problem. The branching is based on a new measure which seems useful for various connectivity and domination problems on both directed and undirected graphs of bounded degeneracy. We demonstrate the versatility of the new branching by applying it to the DOMINATING SET problem on graphs excluding a topological minor and more generally, graphs of bounded degeneracy. In the DOMINATING SET problem, we are given an undirected graph $G = (V, E)$ and a positive integer k and the objective is to check whether there exists a subset $S \subseteq V$ of size at most k such that every vertex in G is either in S or adjacent to a vertex in S. Our $O^*(2^{O(dk)})$-time algorithm for DOMINATING SET on d-degenerate graphs improves over the $O^*(k^{O(dk)})$-time algorithm by Alon and Gutner [1]. It turns out that our algorithm is essentially optimal – we show

---

[1] When we say that a digraph excludes a fixed (undirected) graph as a minor or a topological minor, or that the digraph has degeneracy d we mean that the statement is true for the underlying undirected graph.

that assuming the ETH, the running time dependence of our algorithm on the degeneracy of the input graph and solution size k can not be significantly improved. Using these ideas we also obtain a polynomial time $O(d^2)$ factor approximation algorithm for DOMINATING SET on d-degenerate graphs. The results related to DOMINATING SET as well as proofs missing in this paper appear in the full version [13]. We believe that our new branching and corresponding measure will turn out to be useful for several other problems on sparse (di)graphs.

2   Preliminaries

Given a digraph $D = (V, A)$, for each vertex $v \in V$, we define $N^+(v) = \{w \in V | (v, w) \in A\}$ and $N^-(v) = \{w \in H | (w, v) \in A\}$. In other words, the sets $N^+(v)$ and $N^-(v)$ are the set of out-neighbors and in-neighbors of v, respectively.

The degeneracy of an undirected graph $G = (V, E)$ is defined as the least number d such that every subgraph of G contains a vertex of degree at most d. The degeneracy of a digraph is defined to be the degeneracy of the underlying undirected graph. We say that a class of (di)graphs $\mathcal{C}$ is $o(\log n)$-degenerate if there is a function $f(n) = o(\log n)$ such that every (di)graph $G \in \mathcal{C}$ is $f(|V(G)|)$-degenerate.

In a directed graph, we say that a vertex u dominates a vertex v if there is an arc (u, v) and in an undirected graph, we say that a vertex u dominates a vertex v if there is an edge (u, v) in the graph.

Given a vertex v in a directed graph D, we define the operation of short-circuiting across v as follows. We add an arc from every vertex in $N^-(v)$ to every vertex in $N^+(v)$ and delete v.

For a set of vertices $X \subseteq V(G)$ such that $G[X]$ is connected we denote by G/X the graph obtained by contracting edges of a spanning tree of $G[X]$ in G.

Given an instance (D, r, T, k) of DST, we say that a set $S \subseteq V \setminus (T \cup \{r\})$ of at most k vertices is a solution to this instance if in the digraph $D[S \cup T \cup \{r\}]$ there is a directed path from r to every terminal $t \in T$.

Minors and Topological Minors. For a graph $G = (V, E)$, a graph H is a minor of G if H can be obtained from G by deleting vertices, deleting edges, and contracting edges. We denote that H is a minor of G by $H \preceq G$. A mapping $\varphi : V(H) \to 2^{V(G)}$ is a model of H in G if for every $u, v \in V(H)$ with $u \neq v$ we have $\varphi(u) \cap \varphi(v) = \emptyset$, $G[\varphi(u)]$ is connected, and, if $\{u, v\}$ is an edge of H, then there are $u' \in \varphi(u)$ and $v' \in \varphi(v)$ such that $\{u', v'\} \in E(G)$. It is known, that $H \preceq G$ iff H has a model in G. A subdivision of a graph H is obtained by replacing each edge of H by a non-trivial path. We say that H is a topological minor of G if some subgraph of G is isomorphic to a subdivision of H and denote it by $H \preceq_T G$. In this paper, whenever we make a statement about a directed graph having (or being) a minor of another graph, we mean the underlying undirected graph. A graph G excludes graph H as a (topological) minor if H is not a (topological) minor of G. We say that a class of graphs $\mathcal{C}$ excludes $o(\log n)$-sized (topological) minors if there is a function $f(n) = o(\log n)$ such that for every graph $G \in \mathcal{C}$ we have that $K_{f(|V(G)|)}$ is not a (topological) minor of G.

3 DST on Sparse Graphs

In this section, we introduce our main idea and use it to design algorithms for the DIRECTED STEINER TREE problem on classes of sparse graphs. We begin by giving a $O^*(2^{O(hk)})$-time algorithm for DST on K_h-minor free graphs. Then, we show that in general, even in 2-degenerate graphs, we cannot expect to have an FPT algorithm for DST parameterized by the solution size. Finally, we show that when the graph induced on the terminals is acyclic, then our ideas are applicable and we can give a $O^*(2^{O(hk)})$-time algorithm on K_h-topological minor free graphs and a $O^*(2^{O(dk)})$-time algorithm on d-degenerate graphs.

DST on minor free graphs. We begin with a polynomial time preprocessing which will allow us to identify a *special* subset of the terminals with the property that it is enough for us to find an arborescence from the root to these terminals.

Rule 1. *Given an instance (D, r, T, k) of DST, let C be a strongly connected component with at least 2 vertices in the graph $D[T]$. Then, contract C to a single vertex c, to obtain the graph D' and return the instance $(D', r, T' = (T \setminus C) \cup \{c\}, k)$.*

Lemma 1. *Rule 1 is sound, i.e., the instance (D', r, T', k) produced by the rule is a yes-instance of DST if and only if (D, r, T, k) is a yes-instance of DST.*

Proof. Suppose S is a solution to (D, r, T, k). Then there is a directed path from r to every terminal $t \in T$ in the digraph $D[S \cup T \cup \{r\}]$. Contracting the vertices of C will preserve this path. Hence, S is also a solution for (D', r, T', k).

Conversely, suppose S is a solution for (D', r, T', k). If the path P from r to some $t \in T' \setminus C$ in $D'[S \cup T' \cup \{r\}]$ contains c, then there must be a path from r to some vertex x of C and a path (possibly trivial) from some vertex $y \in C$ to t in $D[S \cup T \cup \{r\}]$. As there is a path between any x and y in $D[C]$, concatenating these three paths results in a path from r to t in $D[S \cup T \cup \{r\}]$. Hence, S is also a solution to (D, r, T, k). □

Proposition 1. *Given an undirected graph $G = (V, E)$ which excludes K_h as a minor for some h, and a vertex subset $X \subseteq V$ inducing a connected subgraph of G, the graph G/X also excludes K_h as a minor.*

We call an instance *reduced* if Rule 1 cannot be applied to it. Given an instance (D, r, T, k), we first apply Rule 1 exhaustively to obtain a reduced instance. Since the resulting graph still excludes K_h as a minor (by Proposition 1), we have not changed the problem and hence, for ease of presentation, we denote the reduced instance also by (D, r, T, k). We call a terminal vertex $t \in T$ a *source-terminal* if it has no in-neighbors in $D[T]$. We use T_0 to denote the set of all source-terminals. Since for every terminal, the graph $D[T]$ contains a path from some source terminal to this terminal, we have the following observation.

Observation 1. *Let (D, r, T, k) be a reduced instance and let $S \subseteq V$. Then the digraph $D[S \cup T \cup \{r\}]$ contains a directed path from r to every terminal $t \in T$ if and only if it contains a directed path from r to every source-terminal $t \in T_0$.*

The following is an important subroutine of our algorithm.

Lemma 2. *Let D be a digraph, $r \in V(D)$, $T \subseteq V(D) \setminus \{r\}$ and $T_0 \subseteq T$. There is an algorithm which can find a minimum size set $S \subseteq V(D)$ such that there is path from r to every $t \in T_0$ in $D[T \cup \{r\} \cup S]$ in time $O^*(2^{|T_0|})$.*

Proof. Nederlof [19] gave an algorithm to solve the STEINER TREE problem on undirected graphs in time $O^*(2^t)$ where t is the number of terminals. Misra et al. [17] observed that the same algorithm can be easily modified to solve the DST problem in time $O^*(2^t)$ with t being the number of terminals. In our case, we create an instance of the DST problem by taking the same graph, defining the set of terminals as T_0 and for every vertex $t \in T \setminus T_0$, *short-circuiting* across this vertex. Clearly, a k-sized solution to this instance gives a k-sized solution to the original problem. To actually find the set of minimum size, we can first find its size by a binary search and then delete one by one the non-terminals, if their deletion does not increase the size of the minimum solution. $\square$

We call the algorithm from Lemma 2, NEDERLOF(D, r, T, T_0). We also need the following structural claim regarding the existence of low degree vertices in graphs excluding K_h as a topological minor.

Lemma 3. *Let $G = (V, E)$ be an undirected graph excluding K_h as a topological minor[2] and let $X, Y \subseteq V$ be two disjoint vertex sets. If every vertex in X has at least $h - 1$ neighbors in Y, then there is a vertex in Y with at most ch^4 neighbors in $X \cup Y$ for some constant c.*

Proof. It was proved in [3,15] that there is a constant a such that any graph that does not contain K_h as a topological minor is $d = ah^2$-degenerate. Consider the graph $H_0 = G[X \cup Y] \setminus E(X)$. We construct a sequence of graphs $H_0, \ldots, H_l$, starting from H_0 and repeating an operation which ensures that any graph in the sequence excludes K_h as a topological minor. The operation is defined as follows. In graph H_i, pick a vertex $x \in X$. As it has degree at least $h - 1$ in Y and there is no K_h topological minor in H_i, it has two neighbors y_1 and y_2 in Y, which are non-adjacent. Remove x from H_i and add the edge (y_1, y_2) to obtain the graph H_{i+1}. By repeating this operation, we finally obtain a graph H_l where the set X is empty. As the graph H_l still excludes K_h as a topological minor, it is d-degenerate, and hence it has at most $d|Y|$ edges. In the sequence of operations, every time we remove a vertex from X, we added an edge between two vertices of Y. Hence, the number of vertices of X in H_0 is bounded by the number of edges within Y in H_l, which is at most $d|Y|$. As H_0 is also d-degenerate, it has at most $d(|X| + |Y|) = d(d+1)|Y|$ edges. Therefore, there is a vertex in Y incident on at most $2d(d+1) = 2ah^2(ah^2 + 1) \le ch^4$ edges where $c = 4a^2$. This concludes the proof of the lemma. $\square$

---

[2] As a graph G which exludes K_h as a minor also excludes K_h as a topological minor, the lemma also applies in the former case. While a stronger bound can be given for this case, stating the lemma this way allows us to use it in further sections and does not hurt the asymptotic running time.

Let (D, r, T, k) be a reduced instance of DST, $Y \subseteq V \setminus T$ be a set of non-terminals representing a partial solution and d_b be some fixed positive integer. We define the following sets of vertices

- $T_1 = T_1(Y)$ is the set of source terminals dominated by Y.
- $B_h = B_h(Y, d_b)$ is the set of non-terminals not in Y, which dominate at least $d_b + 1$ terminals in $T_0 \setminus T_1$.
- $B_l = B_l(Y, d_b)$ is the set of non-terminals not in Y, which dominate at most d_b terminals in $T_0 \setminus T_1$.
- $W_h = W_h(Y, d_b)$ is the set of terminals in $T_0 \setminus T_1$ which are dominated by B_h.
- $W_l = W_l(Y, d_b) = T_0 \setminus (T_1 \cup W_h)$ is the set of source terminals which are not dominated by Y or B_h.

Note that the sets $Y, T_1, B_h, B_l, W_h, W_l$ are pairwise disjoint. The constant d_b is introduced to describe the algorithm in a more general way so that we can use it in further sections of the paper. Throughout this section, we will have $d_b = h - 2$.

Lemma 4. *Let (D, r, T, k) be a reduced instance of* DST, *$Y \subseteq V \setminus T$, $d_b \in \mathbb{N}$, and T_1, B_h, B_l, W_h, and W_l as defined above. If $|W_l| > d_b(k - |Y|)$, then the given instance does not admit a solution containing Y.*

Proof. This follows from the fact that any non-terminal from $V \setminus (B_h \cup Y)$ in the solution, which dominates a vertex in W_l can dominate at most d_b of these vertices. Since the solution contains at most $k - |Y|$ such non-terminals, at most $d_b(k - |Y|)$ of these vertices can be dominated. This completes the proof.

Lemma 5. *Let (D, r, T, k) be a reduced instance of* DST, *$Y \subseteq V \setminus T$, $d_b \in \mathbb{N}$, and T_1, B_h, B_l, W_h, and W_l as defined above. If W_h is empty, then there is an algorithm which can test if this instance has a solution containing Y in time $O^*(2^{d_b(k-|Y|)+|Y|})$.*

Proof. We use Lemma 2 and test whether $|\text{NEDERLOF}(D, r, T \cup Y, Y \cup (T_0 \setminus T_1))| \leq k - |Y|$. We know that $|Y| \leq k$ and, by Lemma 4, we can assume that $|T_0 \setminus T_1| \leq d_b(k - |Y|)$. Therefore, the size of $Y \cup (T_0 \setminus T_1)$ is bounded by $|Y| + d_b(k - |Y|)$, implying that we can solve the DST problem on this instance in time $O^*(2^{d_b(k-|Y|)+|Y|})$. This completes the proof of the lemma. $\square$

We now proceed to the main algorithm of this subsection.

Theorem 2. DST *can be solved in time $O^*(3^{hk+o(hk)})$ on graphs excluding K_h as a minor.*

Proof. Let T_0 be the set of source terminals of this instance. The algorithm we describe takes as input a reduced instance (D, r, T, k), a vertex set Y and a positive integer d_b and returns a smallest solution for the instance which contains Y if such a solution exists. If there is no solution, then the algorithm returns a dummy symbol S_∞. To simplify the description, we assume that $|S_\infty| = \infty$. The algorithm is a recursive algorithm and at any stage of the recursion, the

```
Input   : An instance (D, r, T, k) of DST, degree bound d_b, set Y
Output: A smallest solution of size at most k and containing Y for the instance (D, r, T, k)
          if it exists and S_∞ otherwise
1  Compute the sets B_h, B_l, Y, W_h, W_l
2  if |W_l| > d_b(k − |Y|) then return S_∞
3  else if B_h = ∅ then
4  │    S ← NEDERLOF(D, r, T ∪ Y, W_l ∪ Y) ∪ Y .
5  │    if |S| > k then S ← S_∞
6  │    return S
7  end
8  else
9  │    S ← S_∞
10 │    Find vertex v ∈ W_h with the least in-neighbors in B_h.
11 │    for u ∈ B_h ∩ N^−(v) do
12 │    │    Y' ← Y ∪ {u},
13 │    │    S' ← DST-SOLVE((D, r, T, k), d_b, Y').
14 │    │    if |S'| < |S| then S ← S'
15 │    end
16 │    D' ← D \ (B_h ∩ N^−(v))
17 │    S' ← DST-SOLVE((D', r, T, k), d_b, Y).
18 │    if |S'| < |S| then S ← S'
19 │    return S
20 end
```

Algorithm 3.1. Algorithm DST-SOLVE for DST on graphs excluding K_h as a minor

corresponding recursive step returns the smallest set found in the recursions initiated in this step. We start with Y being the empty set.

By Lemma 4, if $|W_l| > d_b(k − |Y|)$, then there is no solution containing Y and hence we return $S_∞$ (see Algorithm 3.1). If B_h is empty, then we apply Lemma 5 to solve the problem in time $O^*(2^{d_b k})$. If B_h is non-empty, then we find a vertex $v \in W_h$ with the least in-neighbors in B_h. Suppose it has d_w of them.

We then branch into $d_w + 1$ branches described as follows. In the first d_w branches, we move a vertex u of B_h which is an in-neighbor of v, to the set Y. Each of these branches is equivalent to picking one of the in-neighbors of v from B_h in the solution. We then recurse on the resulting instance. In the last of the $d_w + 1$ branches, we delete from the instance non-terminals in B_h which dominate v and recurse on the resulting instance. Note that in the resulting instance of this branch, we have v in $W_l(Y)$.

Correctness. At each node of the recursion tree, we define a measure $\mu(I) = d_b(k − |Y|) − |W_l|$. We prove the correctness of the algorithm by induction on this measure. In the base case, when $d_b(k − |Y|) − |W_l| < 0$, then the algorithm is correct (by Lemma 4). Now, we assume as induction hypothesis that the algorithm is correct on instances with measure less than some $\mu \geq 0$. Consider an instance I such that $\mu(I) = \mu$. Since the branching is exhaustive, it is sufficient to show that the algorithm is correct on each of the child instances. To show this, it is sufficient to show that for each child instance I', $\mu(I') < \mu(I)$. In the first d_w branches, the size of the set Y increases by 1, and the size of the set W_l does not decrease. Hence, in each of these branches, $\mu(I') \leq \mu(I) − d_b$. In the final branch, though the size of the set Y remains the same, the size of the set W_l increases by at least 1. Hence, in this branch, $\mu(I') \leq \mu(I) - 1$. Thus, we have shown that in each branch, the measure drops, hence completing the proof of correctness of the algorithm.

Analysis. Since D exludes K_h as a minor, Lemma 3, combined with the fact that we set $d_b = h - 2$, implies that $d_w^{\max} = ch^4$, for some c, is an upper bound on the maximum d_w which can appear during the execution of the algorithm. We first bound the number of leaves of the recursion tree as follows. The number of leaves is bounded by $\sum_{i=0}^{d_b k} \binom{d_b k}{i} (d_w^{\max})^{k - \frac{i}{d_b}}$. To see this, observe that each branch of the recursion tree can be described by a length-$d_b k$ vector as shown in the correctness paragraph. We then select i positions of this vector on which the last branch was taken. Finally for $k - \frac{i}{d_b}$ of the remaining positions, we describe which of the first at most $d_w^{\max}$ branches was taken. Any of the first d_w^{max} branches can be taken at most $k - \frac{i}{d_b}$ times if the last branch is taken i times.

The time taken along each root to leaf path in the recursion tree is polynomial, while the time taken at a leaf for which the last branch was taken i times is $O^*(2^{d_b(k-(k-\frac{i}{d_b}))+k-\frac{i}{d_b}}) = O^*(2^{i+k})$ (see Lemmata 4 and 5). Hence, the running time of the algorithm is

$$O^* \left(\sum_{i=0}^{d_b k} \binom{d_b k}{i} (d_w^{\max})^{k - \frac{i}{d_b}} \cdot 2^{i+k} \right) = O^* \left((2d_w^{\max})^k \cdot \sum_{i=0}^{d_b k} \binom{d_b k}{i} \cdot 2^i \right).$$

For $d_b = h - 2$ and $d_w^{\max} = ch^4$ this is $O^*(3^{hk+o(hk)})$. This completes the proof of the theorem.   □

Theorem 2 has the following corollary.

Corollary 1. *If $\mathcal{C}$ is a class of digraphs excluding $o(\log n)$-sized minors, then DST parameterized by k is* FPT *on $\mathcal{C}$.*

DST on d-degenerate graphs. Since DST has a $O^*(f(k,h))$ algorithm on graphs excluding minors (the previous subsection) and topological minors, a natural question is- does DST have a $O^*(f(k,d))$ algorithm on d-degenerate graphs. However, we show that in general, we cannot expect an algorithm of this form even for an arbitrary 2-degenerate graph.

Theorem 3. DST *parameterized by k is* W[2]-*hard on 2-degenerate graphs.*

Proof. The proof is by a parameterized reduction from SET COVER. Given an instance $(\mathcal{U}, \mathcal{F} = \{F_1, \ldots, F_m\}, k)$ of SET COVER, we construct an instance of DST as follows. Corresponding to each set F_i, we have a vertex f_i and corresponding to each element $u \in \mathcal{U}$, we add a directed cycle C_u of length l_u where l_u is the number of sets in $\mathcal{F}$ which contain u. For each cycle C_u, we add an arc from each of the sets containing u, to a unique vertex of C_u. Since C_u has l_u vertices, this is possible. Finally, we add another directed cycle C of length $m + 1$ and for each vertex f_i, we add an arc from a unique vertex of C to f_i. Again, since C has length $m + 1$, this is possible. Finally, we set as the root r, the only remaining vertex of C which does not have an arc to some f_i and we set as terminals all the vertices involved in a directed cycle C_u for some u and all the vertices in the cycle C except the root r. It is easy to see that the resulting digraph has degeneracy 3. Finally, we subdivide every edge which lies in a cycle C_u for some u, or on the cycle C and add the new vertices to the terminal set. This results in a digraph D of degeneracy 2. Let T be the set of terminals as defined above. This completes the construction. We defer the straightforward proof that $(\mathcal{U}, \mathcal{F}, k)$ is a YES instance of SET COVER iff (D, r, T, k) is a YES instance of DST to the full version of the paper.

In the instance of DST obtained in the above reduction, it seems that the presence of directed cycles in the subgraph induced by the terminals plays a major role in the *hardness* of this instance. We formally show that this is indeed the case by presenting an FPT algorithm for DST for the case the digraph induced by the terminals is acyclic.

Theorem 4. DST *can be solved in time* $O^*(2^{O(dk)})$ *on d-degenerate graphs if the digraph induced by the terminals is acyclic.*

Theorem 4 has the following corollary.

Corollary 2. *If* $\mathcal{C}$ *is an* $o(\log n)$-*degenerate class of digraphs, then* DST *parameterized by* k *is* FPT *on* $\mathcal{C}$ *if the digraph induced by terminals is acyclic.*

Before concluding this section, we also observe that analogous to the algorithms in Theorems 2 and 4, we can show that in the case when the digraph induced by terminals is acyclic, the DST problem admits an algorithm running in time $O^*(2^{O(hk)})$ on graphs excluding K_h as a topological minor.

Theorem 5. DST *can be solved in time* $O^*(2^{O(hk)})$ *on graphs excluding* K_h *as a topological minor if the digraph induced by terminals is acyclic.*

Hardness of DST.   In this section, we show that the algorithm given by Theorem 4 is essentially the best possible with respect to the dependency on the degeneracy and the solution size. We begin by proving a lower bound on the time required by any algorithm for DST on $O(\log n)$-degenerate graphs.

Theorem 6. DST *cannot be solved in time* $f(k)n^{o(\frac{k}{\log k})}$ *on* $c \log n$-*degenerate graphs for any constant* $c > 0$ *even if the digraph induced by terminals is acyclic, where* k *is the solution size and* f *is an arbitrary function, unless* ETH *fails.*

In order to prove Theorem 6, we first prove the following lemma.

Lemma 6. *There is a constant* γ *such that* SET COVER *with size of each set bounded by* $\gamma \log m$ *cannot be solved in time* $f(k)m^{o(\frac{k}{\log k})}$, *unless* ETH *fails, where* k *is the size of the solution and* m *is the size of the family of sets.*

Proof of Theorem 6. The proof is by a reduction from the restricted version of SET COVER shown to be hard in Lemma 6. Fix a constant $c > 0$ and let

$(\mathcal{U} = \{u_1, \ldots, u_n\}, \mathcal{F} = \{F_1, \ldots, F_m\}, k)$ be an instance of SET COVER, where the size of any set is at most $\gamma \log m$, for some constant γ. For each set F_i, we have a vertex f_i. For each element u_i, we have a vertex x_i. If an element u_i is contained in set F_j, then we add an arc (f_j, x_i). Further, we add another vertex r and add arcs (r, f_i) for every i. Finally, we add $m^{2\gamma/c}$ isolated vertices. This completes the construction of the digraph D. We set $T = \{x_1, \ldots, x_n\} \cup \{r\}$ as the set of terminals and r as the root.

We claim that $(\mathcal{U}, \mathcal{F}, k)$ is a YES instance of SET COVER iff (D, r, T, k) is a YES instance of DST. Suppose that $\{F_1, \ldots, F_k\}$ is a set cover for the given instance. Clearly, the vertices $\{f_1, \ldots, f_k\}$ form a solution for the DST instance.

Conversely, suppose that $\{f_1, \ldots, f_k\}$ is a solution for the DST instance. Since the only way that r can reach a vertex x_i is through some f_j, and the construction implies that $u_i \in F_j$, the sets $\{F_1, \ldots, F_k\}$ form a set cover for $(\mathcal{U}, \mathcal{F}, k)$. This concludes the proof of equivalence of the two instances.

We claim that the degeneracy of the graph D is $c \log n_1 + 1$. First, we show that the degeneracy of the graph D is bounded by $\gamma \log m + 1$. This follows from that each vertex f_i has total degree at most $\gamma \log m + 1$ and if a subgraph contains none of these vertices, then it contains no edges. Now, n_1 is at least $m^{2\gamma/c}$. Hence, $\log n_1 \geq (2\gamma/c) \log m$ and the degeneracy of the graph is at most $\gamma \log m + 1 \leq c \cdot (2\gamma/c) \log m \leq c \log n_1$. Finally, since each vertex f_i is adjacent to at most $\gamma \log m + 1$ vertices, $n_1 = O(m \log m + m^{2\gamma/c})$ and, thus, it is polynomial in m. Hence, an algorithm for DST of the form $f(k) n_1^{o\left(\frac{k}{\log k}\right)}$ implies an algorithm of the form $f(k) m^{o\left(\frac{k}{\log k}\right)}$ for the SET COVER instance. This concludes the proof. $\square$

By Theorem 6 we get the following corollary.

Corollary 3. *There are no two functions f and g such that $g(d) = o(d)$ and DST has an algorithm on d-degenerate graphs running in time $O^*(2^{g(d)f(k)})$ unless ETH fails.*

To examine the dependency on the solution size we utilize the following theorem, leading to Corollary 4.

Theorem 7. ([12]) *There is a constant c such that DOMINATING SET does not have an algorithm running in time $O^*(2^{o(n)})$ on graphs of maximum degree $\leq c$ unless ETH fails.*

Corollary 4. *There are no two functions f and g such that $f(k) = o(k)$ and DST has an algorithm on d-degenerate graphs running in time $O^*(2^{g(d)f(k)})$, unless ETH fails.*

References

1. Alon, N., Gutner, S.: Linear time algorithms for finding a dominating set of fixed size in degenerated graphs. Algorithmica 54(4), 544–556 (2009)
2. Bern, M.W., Plassmann, P.E.: The Steiner problem with edge lengths 1 and 2. Inf. Process. Lett. 32(4), 171–176 (1989)

3. Bollobás, B., Thomason, A.: Proof of a conjecture of Mader, Erdös and Hajnal on topological complete subgraphs. Eur. J. Comb. 19(9), 883–887 (1998)
4. Charikar, M., Chekuri, C., Cheung, T.-Y., Dai, Z., Goel, A., Guha, S., Li, M.: Approximation algorithms for directed Steiner problems. J. Algorithms 33(1), 73–91 (1999)
5. Demaine, E.D., Hajiaghayi, M., Klein, P.N.: Node-weighted Steiner tree and group Steiner tree in planar graphs. In: Albers, S., Marchetti-Spaccamela, A., Matias, Y., Nikoletseas, S., Thomas, W. (eds.) ICALP 2009, Part I. LNCS, vol. 5555, pp. 328–340. Springer, Heidelberg (2009)
6. Downey, R.G., Fellows, M.R.: Parameterized Complexity. Springer, New York (1999)
7. Garey, M.R., Johnson, D.S.: The rectilinear Steiner tree problem is NP-complete. SIAM J. Appl. Math. 32(4), 826–834 (1977)
8. Garey, M.R., Johnson, D.S.: Computers and Intractability. Freeman, San Francisco (1979)
9. Grohe, M., Marx, D.: Structure theorem and isomorphism test for graphs with excluded topological subgraphs. In: STOC, pp. 173–192 (2012)
10. Halperin, E., Kortsarz, G., Krauthgamer, R., Srinivasan, A., Wang, N.: Integrality ratio for group Steiner trees and directed Steiner trees. SIAM J. Comput. 36(5), 1494–1511 (2007)
11. Hwang, F.K., Richards, D.S., Winter, P.: The Steiner Tree Problem. North-Holland (1992)
12. Impagliazzo, R., Paturi, R., Zane, F.: Which problems have strongly exponential complexity? J. Comput. Syst. Sci. 63(4), 512–530 (2001)
13. Jones, M., Lokshtanov, D., Ramanujan, M.S., Saurabh, S., Suchý, O.: Parameterized complexity of directed steiner tree on sparse graphs. CoRR, abs/1210.0260 (2012)
14. Klein, P., Ravi, R.: A nearly best-possible approximation algorithm for node-weighted steiner trees. Journal of Algorithms 19(1), 104–115 (1995)
15. Komlós, J., Szemerédi, E.: Topological cliques in graphs 2. Combinatorics, Probability & Computing 5, 79–90 (1996)
16. Korte, B., Prömel, H.J., Steger, A.: Steiner trees in VLSI-layout. In: Paths, Flows and VLSI-Layout, pp. 185–214 (1990)
17. Misra, N., Philip, G., Raman, V., Saurabh, S., Sikdar, S.: FPT algorithms for connected feedback vertex set. In: Rahman, M. S., Fujita, S. (eds.) WALCOM 2010. LNCS, vol. 5942, pp. 269–280. Springer, Heidelberg (2010)
18. Mölle, D., Richter, S., Rossmanith, P.: Enumerate and expand: Improved algorithms for connected vertex cover and tree cover. Theory Comput. Syst. 43(2), 234–253 (2008)
19. Nederlof, J.: Fast polynomial-space algorithms using möbius inversion: Improving on steiner tree and related problems. In: Albers, S., Marchetti-Spaccamela, A., Matias, Y., Nikoletseas, S., Thomas, W. (eds.) ICALP 2009, Part I. LNCS, vol. 5555, pp. 713–725. Springer, Heidelberg (2009)
20. Prömel, H.J., Steger, A.: The Steiner Tree Problem; a Tour through Graphs, Algorithms, and Complexity. Vieweg (2002)
21. Zelikovsky, A.: A series of approximation algorithms for the acyclic directed Steiner tree problem. Algorithmica 18(1), 99–110 (1997)
22. Zosin, L., Khuller, S.: On directed Steiner trees. In: Proc. SODA 2002, pp. 59–63 (2002)

Improved Approximation Algorithms for Projection Games[*]
(Extended Abstract)

Pasin Manurangsi and Dana Moshkovitz

Massachusetts Institute of Technology, Cambridge MA 02139, USA
{pasin,dmoshkov}@mit.edu

Abstract. The projection games (aka Label-Cover) problem is of great importance to the field of approximation algorithms, since most of the NP-hardness of approximation results we know today are reductions from Label-Cover. In this paper we design several approximation algorithms for projection games:

1. A polynomial-time approximation algorithm that improves on the previous best approximation by Charikar, Hajiaghayi and Karloff [7].
2. A sub-exponential time algorithm with much tighter approximation for the case of smooth projection games.
3. A PTAS for planar graphs.

Keywords: Label-Cover, projection games.

1 Introduction

The projection games problem (also known as LABEL COVER) is defined as follows.

INPUT: A bipartite graph $G = (A, B, E)$, two finite sets of labels Σ_A, Σ_B, and, for each edge $e = (a, b) \subset E$, a "projection" $\pi_e : \Sigma_A \to \Sigma_B$.

GOAL: Find an assignment to the vertices $\varphi_A : A \to \Sigma_A$ and $\varphi_B : B \to \Sigma_B$ that maximizes the number of edges $e = (a, b)$ that are "satisfied", i.e., $\pi_e(\varphi_A(a)) = \varphi_B(b)$.

An instance is said to be "satisfiable" or "feasible" or have "perfect completeness" if there exists an assignment that satisfies all edges. An instance is said to be "δ-nearly satisfiable" or "δ-nearly feasible" if there exists an assignment that satisfies $(1 - \delta)$ fraction of the edges. In this work we focus on satisfiable instances of projection games.

LABEL COVER has gained much significance for approximation algorithms because of the following PCP Theorem, establishing that it is NP-hard, given a satisfiable projection game instance, to satisfy even an ε fraction of the edges:

---

[*] This material is based upon work supported by the National Science Foundation under Grant Number 1218547.

H.L. Bodlaender and G.F. Italiano (Eds.): ESA 2013, LNCS 8125, pp. 683–694, 2013.

Theorem 1 (Strong PCP Theorem). *For every n, $\varepsilon = \varepsilon(n)$, there is $k = k(\varepsilon)$, such that deciding* SAT *on inputs of size n can be reduced to finding, given a satisfiable projection game on alphabets of size k, an assignment that satisfies more than an ε fraction of the edges.*

This theorem is the starting point of the extremely successful long-code based framework for achieving hardness of approximation results [6,11], as well as of other optimal hardness of approximation results, e.g., for SET-COVER [9,16,8].

We know several proofs of the strong PCP theorem that yield different parameters in Theorem 1. The parallel repetition theorem [19], applied on the basic PCP Theorem [5,4,3,2], yields $k(\varepsilon) = (1/\epsilon)^{O(1)}$. Alas, it reduces exact SAT on input size n to LABEL COVER on input size $n^{O(\log 1/\varepsilon)}$. Hence, a lower bound of $2^{\Omega(n)}$ for the time required for solving SAT on inputs of size n only implies a lower bound of $2^{n^{\Omega(1/\log 1/\varepsilon)}}$ for LABEL COVER via this theorem. A different proof is based on PCP composition [17,8]. It has smaller blow up but larger alphabet size. Specifically, it shows a reduction from exact SAT with input size n to LABEL COVER with input size $n^{1+o(1)}poly(1/\varepsilon)$ and alphabet size $exp(1/\varepsilon)$.

One is tempted to conjecture that a PCP theorem with both a blow-up of $n^{1+o(1)}poly(1/\varepsilon)$ and an alphabet size $(1/\varepsilon)^{O(1)}$ holds. See [16] for a discussion of potential applications of this "Projection Games Conjecture".

Finding algorithms for projection games is therefore both a natural pursuit in combinatorial optimization, and also a way to advance our understanding of the main paradigm for settling the approximability of optimization problems. Specifically, our algorithms help make progress towards the following questions:

1. Is the "Projection Games Conjecture" true? What is the tradeoff between the alphabet size, the blow-up and the approximation factor?
2. What about even stronger versions of the strong PCP theorem? E.g., Khot introduced "smooth" projection games [13] (see discussion below for the definition). What kind of lower bounds can we expect to get via such a theorem?
3. Does a strong PCP theorem hold for graphs of special forms, e.g., on planar graphs?

2   Our Results

2.1   Better Approximation in Polynomial Time

In 2009, Charikar, Hajiaghayi and Karloff presented a polynomial-time $O((nk)^{1/3})$-approximation algorithm for LABEL COVER on graphs of size n with alphabets of size k [7]. This improved on Peleg's $O((nk)^{1/2})$-approximation algorithm [18]. Both Peleg's and the CHK algorithms worked in the more general setting of arbitrary constraints on the edges and possibly unsatisfiable instances. We show a polynomial-time algorithm that achieves a better approximation for satisfiable projection games:

Theorem 2. *There is a polynomial-time algorithm that given a satisfiable instance of projection games on a graph of size n and alphabets of size k, finds an assignment that satisfies $O((nk)^{1/4})$ edges.*

2.2   Algorithms for Smooth Projection Games

Khot introduced "smooth" projection games in order to obtain new hardness of approximation results, e.g., for coloring problems [13]. In a smooth projection game, for every vertex $a \in A$, the assignments projected to a's neighborhood by the different possible assignments $\sigma_a \in \Sigma_A$ to a, differ a lot from one another (alternatively, form an error correcting code with high relative distance). More formally:

Definition 1. *A projection game instance is μ-smooth if for every $a \in A$ and any distinct assignments $\sigma_a, \sigma_a' \in \Sigma_A$, we have*

$$Pr_{b \in N(a)}[\pi_{(a,b)}(\sigma_a) = \pi_{(a,b)}(\sigma_a')] \leq \mu.$$

Intuitively, smoothness makes the projection games problem easier, since knowing only a small fraction of the assignment to a neighborhood of a vertex $a \in A$ determines the assignment to a.

Smoothness can be seen as an intermediate property between projection and uniqueness, with uniqueness being 0-smoothness. Khot's Unique Games Conjecture [14] is that the Strong PCP Theorem holds for unique games on nearly satisfiable instances for any constant $\varepsilon > 0$.

The Strong PCP Theorem (Theorem 1) is known to hold for μ-smooth projection games with $\mu > 0$. However, the known reductions transform SAT instances of size n to instances of smooth LABEL COVER of size at least $n^{O((1/\mu)\log(1/\varepsilon))}$ [13,12]. Hence, a lower bound of $2^{\Omega(n)}$ for SAT only translates into a lower bound of $2^{n^{\Omega(\mu/\log(1/\varepsilon))}}$ for μ-smooth projection games.

Interestingly, the efficient reduction of Moshkovitz and Raz [17] inherently generates instances that are not smooth. Moreover, for unique games it is known that if they admit a reduction from SAT of size n, then the reduction must incur a blow-up of at least $n^{1/\delta^{\Omega(1)}}$ for δ-almost satisfiable instances. This follows from the sub-exponential time algorithm of Arora, Barak and Steurer [1].

Given this state of affairs, one wonders whether a large blow-up is also necessary for smooth projection games. We make progress toward settling this question by showing:

Theorem 3. *There exists $c \geq 1$ for which the following holds: there is a randomized algorithm that given an μ-smooth satisfiable projection game in which $d_A \geq \frac{c\log n_A}{\mu}$, finds an optimal assignment in time $exp(O(\mu n_B \log |\Sigma_B|))poly(n_A, |\Sigma_A|)$ with probability $2/3$.*

There is also a deterministic $O(1)$-approximation algorithm for μ-smooth satisfiable projection games of any degree. The deterministic algorithm runs in time $exp(O(\mu n_B \log |\Sigma_B|))poly(n_A, |\Sigma_A|)$ as well.

The algorithms work by finding a subset of fraction μ in B that is connected to all, or most, of the vertices in A and going over all possible assignments to it.

Theorem 3 essentially implies that a blow-up of n/μ is necessary for any reduction from SAT to μ-smooth LABEL COVER, no matter what is the approximation factor ε.

2.3   PTAS For Planar Graphs

As the strong PCP Theorem shows, LABEL COVER is NP-hard to approximate even to within very small ε. Does LABEL COVER remain as hard when we consider special kinds of graphs?

In recent years there has been much interest in optimization problems over *planar graphs*. These are graphs that can be embedded in the plane without edges crossing each other. Many optimization problems have very efficient algorithms on planar graphs.

We show that while projection games remain NP-hard to solve *exactly* on planar graphs, when it comes to approximation, they admit a PTAS:

Theorem 4. *The following hold:*

1. *Given a satisfiable instance of projection games on a planar graph, it is NP-hard to find a satisfying assignment.*
2. *There is a polynomial time approximation scheme for satisfiable instances of projection games on planar graphs.*

The PTAS works via Klein's approach [15] of approximating the graph by a graph with constant tree-width.

3   Conventions

We define the following notation to be used in the paper.

- Let $n_A = |A|$ denote the number of vertices in A and n_B denote the number of vertices in B. Let n denote the number of vertices in the whole graph, i.e. $n = n_A + n_B$.
- Let d_v denote the degree of a vertex $v \in A \cup B$.
- For a vertex u, we use $N(u)$ to denote set of vertices that are neighbors of u. Similarly, for a set of vertex U, we use $N(U)$ to denote the set of vertices that are neighbors of at least one vertex in U.
- For each vertex u, define $N_2(u)$ to be $N(N(u))$. This is the set of neighbors of neighbors of u.
- Let σ_v^{OPT} be the assignment to v in an assignment to vertices that satisfies all the edges. In short, we will sometimes refer to this as "the optimal assignment". This is guaranteed to exist from our assumption that the instances considered are satisfiable.
- For any edge $e = (a, b)$, we define p_e to be $|\pi^{-1}(\sigma_b^{OPT})|$. In other words, p_e is the number of assignments to a that satisfy the edge e given that b is assigned σ_b^{OPT}, the optimal assignment. Define $\bar{p}$ to be the average of p_e over all e; that is $\bar{p} = \frac{\sum_{e \in E} p_e}{|E|}$.
- For each set of vertices S, define $G(S)$ to be the subgraph of G that contains all edges with at least one endpoint in S, i.e., the set of edges $E(S)$ of $G(S)$ is $\{(u, v) \in E \mid u \in S \text{ or } v \in S\}$.
- For each $a \in A$, let $h(a)$ denote $|E(N_2(a))|$. Let $h_{max} = max_{a \in A} h(a)$.

For simplicity, we make the following assumption:

- G is connected. This assumption can be made without loss of generality, as if G is not connected, we can always perform any algorithm presented below on each of its connected component and get an equally good or a better approximation ratio.
- For every $e \in E$ and every $\sigma_b \in \Sigma_B$, the number of pre-images in $\pi_e^{-1}(\sigma_b)$ is the same. In particular, $p_e = \bar{p}$ for all $e \in E$.

We only make use of the assumptions in the algorithms for proving Theorem 2. We defer the treatment of graphs with general number of pre-images to the journal version.

4 Polynomial-Time Approximation Algorithms for Projection Games

In this section, we present an improved polynomial time approximation algorithm for projection games and prove Theorem 2.

To prove the theorem, we proceed to describe four polynomial-time approximation algorithms. In the end, by using the best of these four, we are able to produce a polynomial-time $O\left((n_A|\Sigma_A|)^{1/4}\right)$-approximation as desired (See also figure below):

1. **Satisfy one neighbor $- |E|/n_B$-approximation.** Assign each vertex in A an arbitrary assignment. Each vertex in B is then assigned to satisfy one of its neighboring edges. This algorithm satisfies at least n_B edges.
2. **Greedy assignment $- |\Sigma_A|/\bar{p}$-approximation.** Each vertex in B is assigned an assignment $\sigma_b \in \Sigma_B$ that has the largest number of preimages across neighboring edges $\sum_{a \in N(b)} |\pi_{(a,b)}^{-1}(\sigma_b)|$. Each vertex in A is then assigned so that it satisfies as many edges as possible. This algorithm works well when Σ_B assignments have many pre-images.
3. **Know your neighbors' neighbors $- |E|\bar{p}/h_{max}$-approximation.** For a vertex $a_0 \in A$, we go over all possible assignments to it. For each assignment, we assign its neighbors $N(a_0)$ accordingly. Then, for each node in $N_2(a_0)$ we leave only the assignments that satisfy all the edges between it and vertices in $N(a_0)$.
   When a_0 is assigned the optimal assignment, the number of choices for each node in $N_2(a_0)$ is reduced to at most $\bar{p}$ possibilities. In this way, we can satisfy $1/\bar{p}$ fraction of the edges that touch $N_2(a_0)$. This satisfies many edges when there exists $a_0 \in A$ such that $N_2(a_0)$ spans many edges.
4. **Divide and Conquer $- O(n_A n_B h_{max}/|E|^2)$-approximation.** For every $a \in A$ we can fully satisfy $N(a) \cup N_2(a)$ efficiently, and give up on satisfying other edges that touch $N_2(a)$. Repeating this process, we can satisfy $\Omega(|E|^2/(n_A n_B h_{max}))$ fraction of the edges. This is large when $N_2(a)$ does not span many edges for all $a \in A$.

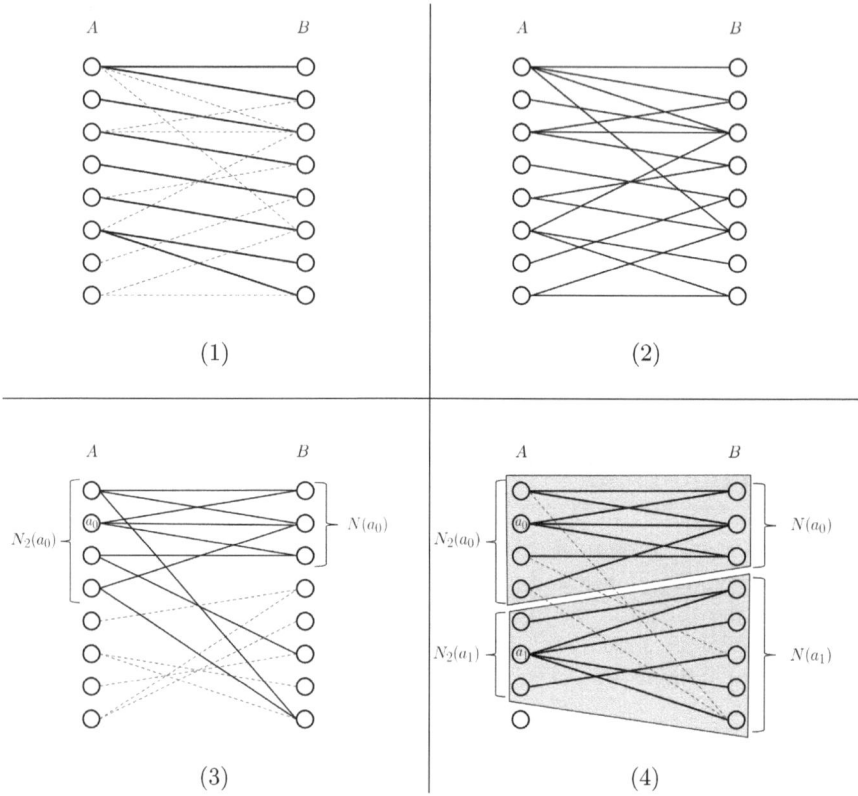

Fig. 1. The Four approximation algorithms

The largest of the four approximation factors is at least as large as their geometric mean, i.e.,

$$O\left(\sqrt[4]{\frac{|E|}{n_B} \cdot \frac{|\Sigma_A|}{\overline{p}} \cdot \frac{|E|\overline{p}}{h_{max}} \cdot \frac{n_A n_B h_{max}}{|E|^2}}\right) = O((n_A|\Sigma_A|)^{1/4}).$$

Due to space limitations, we only describe the fourth algorithm here, which is arguably the most complicated algorithm of the four.

We will present an algorithm that separates the graph into disjoint subgraphs for which we can find the optimal assignment in polynomial time. We shall show below that, if $h(a)$ is small for all $a \in A$, then we are able to find such subgraphs that contains most of the graph's edges.

Lemma 1. *There exists a polynomial-time* $O\left(\frac{n_A n_B h_{max}}{|E|^2}\right)$*-approximation algorithm for satisfiable instances of projection games.*

Proof. To prove lemma 1, it is enough to find an algorithm that gives an assignment that satisfies $\Omega\left(\frac{|E|^3}{n_A n_B h_{max}}\right)$ edges.

We use $\mathcal{P}$ to represent the collection of subgraphs we find. The family $\mathcal{P}$ consists of disjoint sets of vertices. Let $V_{\mathcal{P}}$ be $\bigcup_{P \in \mathcal{P}} P$.

For any set X of vertices, define G_X to be the graph induced by X with respect to G. Moreover, define E_X to be the set of edges of G_X. We also define $E_{\mathcal{P}} = \bigcup_{P \in \mathcal{P}} E_P$.

The algorithm works as follows.

1. Set $\mathcal{P} \leftarrow \emptyset$.
2. While there exists a vertex $a \in A$ such that $\left| E_{(N(a) \cup N_2(a)) - V_{\mathcal{P}}} \right| \geq \frac{1}{4} \frac{|E|^2}{n_A n_B}$:

    (a) Set $\mathcal{P} \leftarrow \mathcal{P} \cup \{(N_2(a) \cup N(a)) - V_{\mathcal{P}}\}$.

3. For each $P \in \mathcal{P}$, find in time $|\Sigma_A| \cdot |P|^{O(1)}$ an assignment to the vertices in P that satisfies all the edges spanned by P.

We will divide the proof into two parts. First, we will show that when we cannot find a vertex a in step 2, $\left| E_{(A \cup B) - V_{\mathcal{P}}} \right| \leq \frac{|E|}{2}$. Second, we will show that the resulting assignment from this algorithm satisfies $\Omega\left(\frac{|E|^3}{n_A n_B h_{max}}\right)$ edges.

We will start by showing that if no vertex a in step 2 exists, then $\left| E_{(A \cup B) - V_{\mathcal{P}}} \right| \leq \frac{|E|}{2}$.

Suppose that we cannot find a vertex a in step 2. In other words, $\left| E_{(N(a) \cup N_2(a)) - V_{\mathcal{P}}} \right| < \frac{1}{4} \frac{|E|^2}{n_A n_B}$ for all $a \in A$.

Consider $\sum_{a \in A} \left| E_{(N(a) \cup N_2(a)) - V_{\mathcal{P}}} \right|$. Since $\left| E_{(N(a) \cup N_2(a)) - V_{\mathcal{P}}} \right| < \frac{1}{4} \frac{|E|^2}{n_A n_B}$ for all $a \in A$, we have the following inequality.

$$\frac{|E|^2}{4 n_B} \geq \sum_{a \in A} \left| E_{(N(a) \cup N_2(a)) - V_{\mathcal{P}}} \right|.$$

Let $N^p(v) = N(v) - V_{\mathcal{P}}$ and $N_2^p(v) = N_2(v) - V_{\mathcal{P}}$. Similary, define $N^p(S)$ for a subset $S \subseteq A \cup B$. It is easy to see that $N_2^p(v) \supseteq N^p(N^p(v))$. This implies that, for all $a \in A$, we have $\left| E_{(N^p(a) \cup N_2^p(a))} \right| \geq \left| E_{(N^p(a) \cup N^p(N^p(a)))} \right| = \sum_{b \in N^p(a)} |N^p(b)|$.

Moreover:

$$\sum_{a \in A} \left| E_{(N(a) \cup N_2(a)) - V_{\mathcal{P}}} \right| = \sum_{a \in A} \left| E_{(N^p(a) \cup N_2^p(a))} \right|$$

$$\geq \sum_{a \in A} \sum_{b \in N^p(a)} |N^p(b)|$$

$$= \sum_{b \in B} \sum_{a \in N^p(b)} |N^p(b)|$$

$$= \sum_{b \in B} |N^p(b)|^2.$$

From Jensen's inequality, we have

$$\sum_{a \in A} \left| E_{(N(a) \cup N_2(a)) - V_{\mathcal{P}}} \right| \geq \frac{1}{n_B} \left(\sum_{b \in B} |N^{\mathcal{P}}(b)| \right)^2$$
$$= \frac{1}{n_B} \left| E_{(A \cup B) - V_{\mathcal{P}}} \right|^2 .$$

Since $\frac{|E|^2}{4 n_B} \geq \sum_{a \in A} \left| E_{(N(a) \cup N_2(a)) - V_{\mathcal{P}}} \right|$ and $\sum_{a \in A} \left| E_{(N(a) \cup N_2(a)) - V_{\mathcal{P}}} \right| \geq \frac{1}{n_B} \left| E_{(A \cup B) - V_{\mathcal{P}}} \right|^2$, we can conclude that

$$\frac{|E|}{2} \geq \left| E_{(A \cup B) - V_{\mathcal{P}}} \right|$$

which concludes the first part of the proof.

Next, we will show that the assignment the algorithm finds satisfies at least $\Omega\left(\frac{|E|^3}{n_A n_B h_{max}} \right)$ edges. Since we showed that $\frac{|E|}{2} \geq \left| E_{(A \cup B) - V_{\mathcal{P}}} \right|$ when the algorithm terminates, it is enough to prove that $|E_{\mathcal{P}}| \geq \frac{|E|^2}{4 n_A n_B h_{max}} \left(|E| - \left| E_{(A \cup B) - V_{\mathcal{P}}} \right| \right)$. Note that the algorithm guarantees to satisfy all the edges in $E_{\mathcal{P}}$.

We will prove this by using induction to show that at any point in the algorithm, $|E_{\mathcal{P}}| \geq \frac{|E|^2}{4 n_A n_B h_{max}} \left(|E| - \left| E_{(A \cup B) - V_{\mathcal{P}}} \right| \right)$.

Base Case. At the beginning, we have $|E_{\mathcal{P}}| = 0 = \frac{|E|^2}{4 n_A n_B h_{max}} \left(|E| - \left| E_{(A \cup B) - V_{\mathcal{P}}} \right| \right)$, which satisfies the inequality.

Inductive Step. The only step in the algorithm where any term in the inequality changes is step 2a. Let $\mathcal{P}_{old}$ and $\mathcal{P}_{new}$ be the set $\mathcal{P}$ before and after step 2a is executed, respectively. Let a be the vertex selected from step 2. Suppose that $\mathcal{P}_{old}$ satisfies the inequality.

From the condition in step 2, we have $\left| E_{(N(a) \cup N_2(a)) - V_{\mathcal{P}_{old}}} \right| \geq \frac{1}{4} \frac{|E|^2}{n_A n_B}$. Since $\left| E_{\mathcal{P}_{new}} \right| = \left| E_{\mathcal{P}_{old}} \right| + \left| E_{(N(a) \cup N_2(a)) - V_{\mathcal{P}_{old}}} \right|$, we have

$$\left| E_{\mathcal{P}_{new}} \right| \geq \left| E_{\mathcal{P}_{old}} \right| + \frac{1}{4} \frac{|E|^2}{n_A n_B} .$$

Now, consider $\left(|E| - \left| E_{(A \cup B) - V_{\mathcal{P}_{new}}} \right| \right) - \left(|E| - \left| E_{(A \cup B) - V_{\mathcal{P}_{old}}} \right| \right)$. We have

$$\left(|E| - \left| E_{(A \cup B) - V_{\mathcal{P}_{new}}} \right| \right) - \left(|E| - \left| E_{(A \cup B) - V_{\mathcal{P}_{old}}} \right| \right) = \left| E_{(A \cup B) - V_{\mathcal{P}_{old}}} \right| - \left| E_{(A \cup B) - V_{\mathcal{P}_{new}}} \right|$$

Since $V_{\mathcal{P}_{new}} = V_{\mathcal{P}_{old}} \cup (N_2(a) \cup N(a))$, we can conclude that

$$((A \cup B) - V_{\mathcal{P}_{old}}) \subseteq ((A \cup B) - V_{\mathcal{P}_{new}}) \cup (N_2(a) \cup N(a)) .$$

Thus, we can also derive

$$E_{(A \cup B) - V_{\mathcal{P}_{old}}} \subseteq E_{((A \cup B) - V_{\mathcal{P}_{new}}) \cup (N_2(a) \cup N(a))}$$
$$= E_{(A \cup B) - V_{\mathcal{P}_{new}}} \cup \{(a', b') \in E \mid a' \in N_2(a) \text{ or } b' \in N(a)\} .$$

From the definition of N and N_2, for any $(a', b') \in E$, if $b' \in N(a)$ then $a' \in N_2(a)$. Thus, we have $\{(a', b') \in E \mid a' \in N_2(a) \text{ or } b' \in N(a)\} = \{(a', b') \in E \mid a' \in N_2(a)\} = E(N_2(a))$. The cardinality of the last term was defined to be $h(a)$. Hence, we can conclude that

$$
\begin{aligned}
\left| E_{(A \cup B) - V_{\mathcal{P}_{old}}} \right| &\leq \left| E_{(A \cup B) - V_{\mathcal{P}_{new}}} \cup \{(a', b') \in E \mid a' \in N_2(a) \text{ or } b' \in N(a)\} \right| \\
&\leq \left| E_{(A \cup B) - V_{\mathcal{P}_{new}}} \right| + |\{(a', b') \in E \mid a' \in N_2(a) \text{ or } b' \in N(a)\}| \\
&= \left| E_{(A \cup B) - V_{\mathcal{P}_{new}}} \right| + |\{(a', b') \in E \mid a' \in N_2(a)\}| \\
&= \left| E_{(A \cup B) - V_{\mathcal{P}_{new}}} \right| + |E(N_2(a))| \\
&= \left| E_{(A \cup B) - V_{\mathcal{P}_{new}}} \right| + h(a) \\
&\leq \left| E_{(A \cup B) - V_{\mathcal{P}_{new}}} \right| + h_{max}.
\end{aligned}
$$

This implies that $\left(|E| - \left| E_{(A \cup B) - V_{\mathcal{P}}} \right| \right)$ increases by at most h_{max}.

Hence, since $\left(|E| - \left| E_{(A \cup B) - V_{\mathcal{P}}} \right| \right)$ increases by at most h_{max} and $|E_{\mathcal{P}}|$ increases by at least $\frac{1}{4} \frac{|E|^2}{n_A n_B}$ and from the inductive hypothesis, we can conclude that

$$
|E_{\mathcal{P}_{new}}| \geq \frac{|E|^2}{4 n_A n_B h_{max}} \left(|E| - \left| E_{(A \cup B) - V_{\mathcal{P}_{new}}} \right| \right).
$$

Thus, the inductive step is true and the inequality holds at any point during the execution of the algorithm.

When the algorithm terminates, since $|E_{\mathcal{P}}| \geq \frac{|E|^2}{4 n_A n_B h_{max}} \left(|E| - \left| E_{(A \cup B) - V_{\mathcal{P}}} \right| \right)$ and $\frac{|E|}{2} \geq \left| E_{(A \cup B) - V_{\mathcal{P}}} \right|$, we can conclude that $|E_{\mathcal{P}}| \geq \frac{|E|^3}{8 n_A n_B h_{max}}$. Since the algorithm guarantees to satisfy every edge in $E_{\mathcal{P}}$, we can conclude that the algorithm gives $O(\frac{n_A n_B h_{max}}{|E|^2})$ approximation ratio, which concludes our proof of Lemma 1.

5   Sub-exponential Time Algorithms for Smooth Projection Games

Due to space limitations we omit the proof of Theorem 3 from the current version.

6   PTAS for Projection Games on Planar Graphs

The NP-hardness of projection games on planar graphs is proved by reduction from 3-coloring on planar graphs. The latter was proven to be NP-hard by Garey, Johnson and Stockmeyer [10]. The reduction will appear in the full version of this paper.

Next, we will describe PTAS for projection game instances on planar graph and prove Theorem 4. We use the framework presented by Klein for finding PTAS for problems on planar graphs [15] to one for satisfiable instances of the projection games problem. The algorithm consists of the following two steps:

1. **Thinning Step:** Delete edges from the original graph to obtain a graph with bounded treewidth.
2. **Dynamic Programming Step:** Use dynamic programming to solve the problem in the bounded treewidth graph.

6.1   Tree Decomposition

Before we proceed to the algorithm, we first define *tree decomposition*. A *tree decomposition* of a graph $G = (V, E)$ is a collection of subsets $\{B_1, B_2, \ldots, B_n\}$ and a tree T whose nodes are B_i such that

1. $V = B_1 \cup B_2 \cup \cdots \cup B_n$.
2. For each edge $(u, v) \in E$, there exists B_i such that $u, v \in B_i$.
3. For each B_i, B_j, if they have an element v in common. Then v is in every subset along the path in T from B_i to B_j.

The *width* of a tree decomposition $(\{B_1, B_2, \ldots, B_n\}, T)$ is the largest size of $B_1, \ldots, B_n$ minus one. The *treewidth* of a graph G is the minimum width across all possible tree decompositions.

6.2   Thinning

Even though a planar graph itself does not necessarily have a bounded treewidth, it is possible to delete a small set of edges from the graph to obtain a graph with bounded treewidth. Since the set of edges that get deleted is small, if we are able to solve the modified graph, then we get a good approximate answer for the original graph.

Klein has proved the following lemma in his paper [15].

Lemma 2. *For any planar graph $G = (V, E)$ and integer k, there is a linear-time algorithm returns an edge-set S such that $|S| \leq \frac{1}{k}|E|$, a planar graph H, such that $H - S = G - S$, and a tree decomposition of H having width at most $3k$.*

By selecting $k = 1 + \frac{1}{\epsilon}$, we can conclude that the number of edges in $H - S = G - S$ is at least $\left(1 - \frac{1}{k}\right)|E| = \frac{1}{1+\epsilon}|E|$.

Moreover, since a tree decomposition of a graph is also a tree decomposition of its subgraph, we can conclude that the linear-time algorithm in the lemma gives tree decomposition for $G - S = H - S$ which is a subgraph of H with width at most $3k = 3(1 + \frac{1}{\epsilon})$.

6.3   Dynamic Programming

In this section, we will present a dynamic programming algorithm that solves the projection game problem in a bounded treewidth bipartite graph $G' = (A', B', E')$ and projections $\pi_e : \Sigma_A \to \Sigma_B$ for each $e \in E'$, given its tree decomposition $(\{B_1, \ldots, B_n\}, T)$ with a bounded width w.

The algorithm works as follows. We use depth first search to traverse the tree T. Then, at each node B_i, we solve the problem concerning only the subtree of T starting at B_i.

At B_i, we consider all possible assignments $\phi : B_i \rightarrow (\Sigma_A \cup \Sigma_B)$ of B_i. For each assignment ϕ and for each edge $(u, v) \in E'$ such that both u, v are in B_i, we check whether the condition $\pi_{(u,v)}(\phi(u)) = \phi(v)$ is satisfied or not. If not, we conclude that this assignment does not satisfy all the edges. Otherwise, we check that the assignment ϕ works for each subtree of T starting at each child B_j of B_i in T; this result was memoized when the algorithm solved the problem at B_j. The full algorithm will appear in the full version of the paper.

References

1. Arora, S., Barak, B., Steurer, D.: Subexponential algorithms for unique games and related problems. In: Proc. 51st IEEE Symp. on Foundations of Computer Science (2010)
2. Arora, S., Lund, C., Motwani, R., Sudan, M., Szegedy, M.: Proof verification and the hardness of approximation problems. Journal of the ACM 45(3), 501–555 (1998)
3. Arora, S., Safra, S.: Probabilistic checking of proofs: a new characterization of NP. Journal of the ACM 45(1), 70–122 (1998)
4. Babai, L., Fortnow, L., Levin, L.A., Szegedy, M.: Checking computations in poly-logarithmic time. In: Proc. 23rd ACM Symp. on Theory of Computing, pp. 21–32 (1991)
5. Babai, L., Fortnow, L., Lund, C.: Nondeterministic exponential time has two-prover interactive protocols. Computational Complexity 1, 3–40 (1991)
6. Bellare, M., Goldreich, O., Sudan, M.: Free bits, PCPs, and nonapproximability-towards tight results. SIAM J. Comput. 27(3), 804–915 (1998)
7. Charikar, M., Hajiaghayi, M., Karloff, H.: Improved approximation algorithms for label cover problems. In: Fiat, A., Sanders, P. (eds.) ESA 2009. LNCS, vol. 5757, pp. 23–34. Springer, Heidelberg (2009)
8. Dinur, I., Steurer, D.: Analytical approach to parallel repetition. Tech. Rep. 1305.1979, arXiv (2013)
9. Feige, U.: A threshold of ln n for approximating set cover. Journal of the ACM 45(4), 634–652 (1998)
10. Garey, M.R., Johnson, D.S., Stockmeyer, L.: Some simplified np-complete problems. In: Proceedings of the Sixth Annual ACM Symposium on Theory of Computing, STOC 1974, pp. 47–63. ACM, New York (1974)
11. Håstad, J.: Some optimal inapproximability results. Journal of the ACM 48(4), 798–859 (2001)
12. Holmerin, J., Khot, S.: A new PCP outer verifier with applications to homogeneous linear equations and max-bisection. In: Proceedings of the Thirty-Sixth Annual ACM Symposium on Theory of Computing, STOC 2004, pp. 11–20. ACM, New York (2004)
13. Khot, S.: Hardness results for coloring 3-colorable 3-uniform hypergraphs. In: Proc. 43rd IEEE Symp. on Foundations of Computer Science, pp. 23–32 (2002)
14. Khot, S.: On the power of unique 2-prover 1-round games. In: Proc. 34th ACM Symp. on Theory of Computing, pp. 767–775 (2002)

694   P. Manurangsi and D. Moshkovitz

15. Klein, P.N.: A linear-time approximation scheme for TSP for planar weighted graphs. In: Proceedings of the 46th IEEE Symposium on Foundations of Computer Science, pp. 146–155 (2005)
16. Moshkovitz, D.: The projection games conjecture and the NP-hardness of ln n-approximating set-cover. In: Gupta, A., Jansen, K., Rolim, J., Servedio, R. (eds.) APPROX/RANDOM 2012. LNCS, vol. 7408, pp. 276–287. Springer, Heidelberg (2012)
17. Moshkovitz, D., Raz, R.: Two query PCP with sub-constant error. Journal of the ACM 57(5) (2010)
18. Peleg, D.: Approximation algorithms for the label-cover max and red-blue set cover problems. J. of Discrete Algorithms 5(1), 55–64 (2007)
19. Raz, R.: A parallel repetition theorem. SIAM J. Comput. 27, 763–803 (1998)

The Compressed Annotation Matrix: An Efficient Data Structure for Computing Persistent Cohomology

Jean-Daniel Boissonnat[1], Tamal K. Dey[2], and Clément Maria[1]

[1] INRIA Sophia Antipolis-Méditerranée
{jean-daniel.boissonnat,clement.maria}@inria.fr
[2] The Ohio State University
tamaldey@cse.ohio-state.edu

Abstract. Persistent homology with coefficients in a field $\mathbb{F}$ coincides with the same for cohomology because of duality. We propose an implementation of a recently introduced algorithm for persistent cohomology that attaches annotation vectors with the simplices. We separate the representation of the simplicial complex from the representation of the cohomology groups, and introduce a new data structure for maintaining the annotation matrix, which is more compact and reduces substancially the amount of matrix operations. In addition, we propose a heuristic to simplify further the representation of the cohomology groups and improve both time and space complexities. The paper provides a theoretical analysis, as well as a detailed experimental study of our implementation and comparison with state-of-the-art software for persistent homology and cohomology.

1 Introduction

Persistent homology [10] is an algebraic method for measuring the topological features of a space induced by the sublevel sets of a function. Its generality and stability with regard to noise have made it a widely used tool for the study of data, where it does not need any knowledge a priori. A common approach is the study of the topological invariants of a nested family of simplicial complexes built on top of the data, seen as a set of points in a geometric space. This approach has been successfully used in various areas of science and engineering, as for example in sensor networks, image analysis, and data analysis where one typically needs to deal with big data sets in high dimensions. Consequently, the demand for designing efficient algorithms and software to compute persistent homology of filtered simplicial complexes has grown.

The first persistence algorithm [11,14] can be implemented by reducing a matrix defined by face incidence relations, through column operations. The running time is $O(m^3)$ where m is the number of simplices of the simplicial complex and, despite good performance in practice, Morozov proved that this bound is tight [13]. Recent optimizations taking advantage of the special structure of the

H.L. Bodlaender and G.F. Italiano (Eds.): ESA 2013, LNCS 8125, pp. 695–706, 2013.

matrix to be reduced have led to significant progress in the theoretical analysis [5,12] as well as in practice [1,5].

A different approach [7,8] interprets the persistent homology groups in terms of their dual, the persistent cohomology groups. The cohomology algorithm has been reported to work better in practice than the standard homology algorithm [7] but this advantage seems to fade away when optimizations are employed to the homology algorithms [1]. An elegant description of the cohomology algorithm, using the notion of annotations [3], has been introduced in [9] and used to design more general algorithms for maintaining cohomology groups under simplicial maps.

In this work, we propose an implementation of the annotation-based algorithm for computing persistent cohomology. A key feature of our implementation is a distinct separation between the representation of the simplicial complex and the representation of the cohomology groups. Currently the simplicial complex can be represented either by its Hasse diagram or by using the more compact simplex tree [2]. The cohomology groups are stored in a separate data structure that represents a compressed version of the annotation matrix. As a consequence, the time and space complexities of our algorithm depend mostly on properties of the cohomology groups we maintain along the computation and only linearly on the size of the simplicial complex.

Moreover, maintaining the simplicial complex and the cohomology groups separately allows us to reorder the simplices while keeping the same persistent cohomology. This significantly reduces the size of the cohomology groups to be maintained, and improves considerably both the time and memory performance as shown by our detailed experimental analysis on a variety of examples. Our method compares favourably with state-of-the-art software for computing persistent homology and cohomology.

Background: A *simplicial complex* is a pair $\mathcal{K} = (V, S)$ where V is a finite set whose elements are called the *vertices* of $\mathcal{K}$ and S is a set of non-empty subsets of V that is required to satisfy the following two conditions : 1. $p \in V \Rightarrow \{p\} \in S$ and 2. $\sigma \in S, \tau \subseteq \sigma \Rightarrow \tau \in S$. Each element $\sigma \in S$ is called a *simplex* or a *face* of $\mathcal{K}$ and, if $\sigma \in S$ has precisely $s + 1$ elements ($s \geq -1$), σ is called an *s-simplex* and its dimension is s. The dimension of the simplicial complex $\mathcal{K}$ is the largest k such that S contains a k-simplex. We define $\mathcal{K}^p$ to be the set of p-dimensional simplices of $\mathcal{K}$, and note its size $|\mathcal{K}^p|$. Given two simplices τ and σ in $\mathcal{K}$, τ is a subface (resp. coface) of σ if $\tau \subseteq \sigma$ (resp. $\tau \supseteq \sigma$). The *boundary* of a simplex σ, denoted $\partial\sigma$, is the set of its subfaces with codimension 1.

A *filtration* [10] of a simplicial complex is an order relation on its simplices which respects inclusion. Consider a simplicial complex $\mathcal{K} = (V, S)$ and a function $\rho : S \rightarrow \mathbb{R}$. We require ρ to be monotonic in the sense that, for any two simplices $\tau \subseteq \sigma$ in $\mathcal{K}$, ρ satisfies $\rho(\tau) \leq \rho(\sigma)$. We will call $\rho(\sigma)$ the *filtration value* of the simplex σ. Monotonicity implies that the sublevel sets $\mathcal{K}(r) = \rho^{-1}(-\infty, r]$ are subcomplexes of $\mathcal{K}$, for every $r \in \mathbb{R}$. Let m be the number of simplices of $\mathcal{K}$, and let $(\rho_i)_{i=1\cdots n}$ be the n different values ρ takes on the simplices of $\mathcal{K}$. Plainly $n \leq m$, and we have the following sequence of $n + 1$ subcomplexes:

$$\emptyset = \mathcal{K}_0 \subseteq \cdots \subseteq \mathcal{K}_n = \mathcal{K}, \quad -\infty = \rho_0 < \cdots < \rho_n, \quad \mathcal{K}_i = \rho^{-1}(-\infty, \rho_i]$$

Applying a (co)homology functor to this sequence of simplicial complexes turns (combinatorial) complexes into (algebraic) abelian groups and inclusion into group homomorphisms. Roughly speaking, a simplicial complex defines a domain as an arrangement of local bricks and (co)homology catches the global features of this domain, like the connected components, the tunnels, the cavities, etc. The homomorphisms catch the evolution of these global features when inserting the simplices in the order of the filtration. Let $H_p(\mathcal{K})$ and $H^p(\mathcal{K})$ denote respectively the homology and cohomology groups of $\mathcal{K}$ of dimension p with coefficients in a field $\mathbb{F}$. The filtration induces a sequence of homomorphisms in the homology and cohomology groups in opposite directions:

$$0 = H_p(\mathcal{K}_0) \to H_p(\mathcal{K}_1) \to \cdots \to H_p(\mathcal{K}_{n-1}) \to H_p(\mathcal{K}_n) = H_p(\mathcal{K}) \qquad (1)$$

$$0 = H^p(\mathcal{K}_0) \leftarrow H^p(\mathcal{K}_1) \leftarrow \cdots \leftarrow H^p(\mathcal{K}_{n-1}) \leftarrow H^p(\mathcal{K}_n) = H^p(\mathcal{K}) \qquad (2)$$

We refer to [10] for an introduction to the theory of homology and persistent homology. Computing the persistent homology of such a sequence consists in pairing each simplex that creates a homology feature with the one that destroys it. The usual output is a *persistence diagram*, which is a plot of the points $(\rho(\tau), \rho(\sigma))$ for each persistent pair (τ, σ). It is known that because of duality the homology and cohomology sequences above provide the same persistence diagram [8].

The original persistence algorithm [11] considers the homology sequence in Equation 1 that aligns with the filtration direction. It detects when a new homology class is born and when an existing class dies as we proceed forward through the filtration. Recently, a few algorithms have considered the cohomology sequence in Equation 2 which runs in the opposite direction of the filtration [7,8,9]. The birth of a cohomology class coincides with the death of a homology class and the death of a cohomology class coincides with the birth of a homology class. Therefore, by tracking a cohomology basis along the filtration direction and switching the notions of births and deaths, one can obtain all information about the persistent homology of the complex. The algorithm of de Silva et al. [8] computes the persistent cohomology following this principle which is reported to work better in practice than the original persistence algorithm [7]. Recently, Dey et al. [9] recognized that tracking cohomology bases provides a simple and natural extension of the persistence algorithm for filtrations connected with general simplicial maps (and not simply inclusion). Their algorithm is based on the notion of annotation [3] and, when restricted to only inclusions, is a re-formulation of the algorithm of de Silva et al. [8]. Here we follow this annotation based algorithm.

2   Persistent Cohomology Algorithm and Annotations

In this section, we recall the annotation-based persistent cohomology algorithm of [9]. It maintains a cohomology basis under simplex insertions, where representative cocycles are maintained by the value they take on the simplices. We rephrase the description of this algorithm with coefficients in an arbitrary field $\mathbb{F}$, and use standard field notations $\langle \mathbb{F}, +, \cdot, -, /, 0, 1 \rangle$.

Definition 1. *Given a simplicial complex $\mathcal{K}$, let $\mathcal{K}^p$ denote the set of p-simplices in $\mathcal{K}$. An annotation for $\mathcal{K}^p$ is an assignement $\mathsf{a}^p : \mathcal{K}^p \to \mathbb{F}^g$ of an $\mathbb{F}$-vector $\mathsf{a}_\sigma = \mathsf{a}^p(\sigma)$ of same length g for each p-simplex $\sigma \in \mathcal{K}^p$. We use a when there is no ambiguity on the dimension. We also have an induced annotation for any p-chain $c = \sum_i f_i \sigma_i$ given by linear extension: $\mathsf{a}_c = \sum_i f_i \cdot \mathsf{a}_{\sigma_i}$.*

Definition 2. *An annotation $\mathsf{a} : \mathcal{K}^p \to \mathbb{F}^g$ is valid if:*
1. $g = rank\, H_p(\mathcal{K})$ and 2. two p-cycles z_1 and z_2 have $\mathsf{a}_{z_1} = \mathsf{a}_{z_2}$ iff their homology classes $[z_1]$ and $[z_2]$ are identical.

Proposition 1 ([9]). *The following two statements are equivalent:*
1. An annotation $\mathsf{a} : \mathcal{K}^p \to \mathbb{F}^g$ is valid
2. The cochains $\{\phi_j\}_{j=1\cdots g}$ given by $\phi_j(\sigma) = \mathsf{a}_\sigma[j]$ for all $\sigma \in \mathcal{K}^p$ are cocycles whose cohomology classes $\{[\phi_j]\}_{j=1\cdots g}$ constitute a basis of $H^p(\mathcal{K})$.

A valid annotation is thus a way to represent a cohomology basis. The algorithm for computing persistent cohomology consists in maintaining a valid annotation for each dimension when inserting all simplices in the order of the filtration. Since we process the filtration in a direction opposite to the cohomology sequence (as in Equation 2), we discover the death points of cohomology classes earlier than their birth points. To avoid confusion, we still say that a new cocycle (or its class) is born when we discover it for the first time and an existing cocycle (or its class) dies when we see it no more.

We present the algorithm and refer to [9] for its validity. We insert simplices in the order of the filtration. Consider an elementary inclusion $\mathcal{K}_i \hookrightarrow \mathcal{K}_i \cup \{\sigma\}$, with σ a p-simplex. Assume that to every simplex τ of any dimension in $\mathcal{K}_i$ is attached an annotation vector a_τ from a valid annotation a of $\mathcal{K}_i$. We describe how to obtain a valid annotation for $\mathcal{K}_i \cup \{\sigma\}$ from that of $\mathcal{K}_i$. We compute the annotation $\mathsf{a}_{\partial\sigma}$ for the boundary $\partial\sigma$ in $\mathcal{K}_i$ and take actions as follows:

Case 1: If $\mathsf{a}_{\partial\sigma} = 0$, $g \leftarrow g+1$ and the annotation vector of any p-simplex $\tau \in \mathcal{K}_i$ is augmented with a 0 entry so that $\mathsf{a}_\tau = [f_1, \cdots, f_g]^T$ becomes $[f_1, \cdots, f_g, 0]^T$. We assign to the new simplex σ the annotation vector $\mathsf{a}_\sigma = [0, \cdots, 0, 1]^T$. According to Proposition 1, this is equivalent to creating a new cohomology class represented by $\phi(\tau) = 0$ for $\tau \neq \sigma$ and $\phi(\sigma) = 1$.

Case 2: If $\mathsf{a}_{\partial\sigma} \neq 0$, we consider the non-zero element c_j of $\mathsf{a}_{\partial\sigma}$ with maximal index j. We now look for annotations of those $(p-1)$-simplices τ that have a non-zero element at index j and process them as follows. If the element of index j of a_τ is $f \neq 0$, we add $-f/c_j \cdot \mathsf{a}_{\partial\sigma}$ to a_τ. Note that, in the annotation matrix whose columns are the annotation vectors, this implements simultaneously a series of elementary row operations, where each row ϕ_i receives $\phi_i \leftarrow \phi_i - (\mathsf{a}_{\partial\sigma}[i]/c_j) \times \phi_j$. As a result, all the elements of index j in all columns are now 0 and hence the entire row j becomes 0. We then remove the row j and set $g \leftarrow g - 1$. σ is assigned $\mathsf{a}_\sigma = 0$. According to Proposition 1, this is equivalent to removing the jth cocycle $\phi_j(\tau) = \mathsf{a}_\tau[j]$.

As with the original persistence algorithm, the pairing of simplices is derived from the creation and destruction of the cohomology basis elements.

3 Data Structures and Implementation

In this section, we present our implementation of the annotation-based persistent cohomology algorithm. We separate the representation of the simplicial complex from the representation of the cohomology groups.

Representation of the Simplicial Complex. We represent the simplicial complex $\mathcal{K}$ in a data structure $\mathcal{K}_{\text{DS}}$ equipped with the operation COMPUTE-BOUNDARY(σ) that computes the boundary of a simplex σ. We denote by $\mathcal{C}_\partial^p$ the complexity of this operation where p is the dimension of σ. Additionally, the simplices are ordered according to the filtration.

Two data structures to represent simplicial complexes are of particular interest here. The first one is the *Hasse diagram*, which is the graph whose nodes are in bijection with the simplices (of all dimensions) of the simplicial complex and an edge links two nodes representing two simplices τ and σ iff $\tau \subseteq \sigma$ and the dimensions of τ and σ differ by 1. The second data structure is the *simplex tree* introduced in [2], which is a specific spanning tree of the Hasse diagram. For a simplicial complex $\mathcal{K}$ of dimension k and a simplex $\sigma \in \mathcal{K}$ of dimension p, the Hasse diagram has size $O(k|\mathcal{K}|)$ and allows to compute COMPUTE-BOUNDARY(σ) in time $\mathcal{C}_\partial^p = O(p)$, whereas the simplex tree has size $O(|\mathcal{K}|)$ and allows to compute COMPUTE-BOUNDARY(σ) in time $\mathcal{C}_\partial^p = O(p^2 D_{\text{m}})$, where D_{m} is typically a small value related to the time needed to traverse the simplex tree. Both structures can be used in our setting. For readability, we will use a Hasse diagram in the following.

The Compressed Annotation Matrix. For each dimension p, the pth cohomology group can be seen as a valid annotation for the p-simplices of the simplicial complex. Hence, an annotation $\mathsf{a} : \mathcal{K}^p \to \mathbb{F}^g$ can be represented as a $g \times |\mathcal{K}^p|$ matrix with elements in $\mathbb{F}$, where each column is an annotation vector associated to a p-simplex. We describe how to represent this annotation matrix in an efficient way.

Compressing the annotation matrix: In most applications, the annotation matrix is sparse and we store it as follows. A column is represented as the singly-linked list of its non-zero elements, where the list contains a pair (i, f) if the ith element of the column is $f \neq 0$. The pairs in the list are ordered according to row index i. All pairs (i, f) with same row index i are linked in a doubly-linked list.

Removing duplicate columns: To avoid storing duplicate columns, we use two data structures. The first one, $\mathcal{AV}^p$, stores the annotation vectors and allows fast search, insertion and deletion. $\mathcal{AV}^p$ can be implemented as a red-black tree or a hash table. We denote by $\mathcal{C}^p_{\mathcal{AV}}$ the complexity of an operation in $\mathcal{AV}^p$. For example, if $\mathcal{AV}^p$ contains n elements and $c_{\max}$ is the length of the longest column, we have $\mathcal{C}^p_{\mathcal{AV}} = O(c_{\max} \log(n))$ for a red-black tree implementation and $\mathcal{C}^p_{\mathcal{AV}} = O(c_{\max})$ amortized for a hash-table. The simplices of the same dimension that have the same annotation vector are now stored in a same set and the various (and disjoint) sets are stored in a *union-find* data structure denoted $\mathcal{UF}^p$. $\mathcal{UF}^p$ is encoded as a forest where each tree contains the elements of a set, the root being the "representative" of the set. The trees of $\mathcal{UF}^p$ are in bijection with the different annotation vectors stored in $\mathcal{AV}^p$ and the root of each tree maintains a pointer to the corresponding annotation vector in $\mathcal{AV}^p$. Each node representing a p-simplex σ in the simplicial complex $\mathcal{K}_{\mathrm{DS}}$ stores a pointer to an element of the tree of $\mathcal{UF}^p$ associated to the annotation vector $\mathbf{a}_\sigma$. Finding the annotation vector of σ consists in getting the element it points to in a tree of $\mathcal{UF}^p$ and then finding the root of the tree which points to $\mathbf{a}_\sigma$ in $\mathcal{AV}^p$. We avail the following operations on $\mathcal{UF}^p$:

- CREATE-SET: creates a new tree containing one element.
- FIND-ROOT: finds the root of a tree, given an element in the tree.
- UNION-SETS: merges two trees.

The number of elements maintained in $\mathcal{UF}^p$ is at most the number of simplices of dimension p, i.e. $|\mathcal{K}^p|$. The operations FIND-ROOT and UNION-SETS on $\mathcal{UF}^p$ can be computed in amortized time $O(\alpha(|\mathcal{K}^p|))$, where $\alpha(\cdot)$ is the very slowly growing inverse Ackermann function (constant less than 4 in practice), and CREATE-SET is performed in constant time. We will refer to this data structure as the *Compressed Annotation Matrix*.

Operations: The compressed annotation matrix described above supports the following operations. We define $c_{\max}$ to be the maximal number of non-zero elements in a column of the compressed annotation matrix (or equivalently in an annotation vector) and $r_{\max}$ to be the maximal number of non-zero elements in a row of the compressed annotation matrix, during the computation. We will express our complexities using $c_{\max}$ and $r_{\max}$:

- SUM-ANN($\mathbf{a}_1, \mathbf{a}_2$): computes the sum of two annotation vectors $\mathbf{a}_1$ and $\mathbf{a}_2$, and returns the lowest non-zero coefficient if it exists. The column elements are sorted by increasing row index, so the sum is performed in $O(c_{\max})$ time.
- SEARCH-ANN/ADD-ANN/REMOVE-ANN ($\mathbf{a}$): searches, adds or removes an annotation vector $\mathbf{a}$ from $\mathcal{AV}^p$ in $O(\mathcal{C}^p_{\mathcal{AV}})$ time.
- CREATE-COCYCLE(): implements **Case 1** of the algorithm described in section 2. It inserts a new column in $\mathcal{AV}^p$ containing one element $(i_{\mathrm{new}}, 1)$, where i_{new} is the index of the created cocycle. This is performed in time $O(\mathcal{C}^p_{\mathcal{AV}})$. We also create a new disjoint set in $\mathcal{UF}^p$ for the new column. This is done in $O(1)$ time using CREATE-SET. CREATE-COCYCLE() takes $O(\mathcal{C}^p_{\mathcal{AV}})$ in total.
- KILL-COCYCLE($\mathbf{a}_{\partial\sigma}, c_j, j$): implements **Case 2** of the algorithm. It finds all columns with a non-zero element at index j and, for each such column A, it adds

to A the column $-f/c_j \cdot \mathsf{a}_{\partial\sigma}$ if f is the non-zero element at index j in A. To find the columns with a non-zero element at index j, we use the doubly-linked list of row j. We call SUM-ANN to compute the sums. The overall time needed for all columns is $O(c_{\max} r_{\max})$ in the worst-case. Finally, we remove duplicate columns using operations on $\mathcal{AV}^p$ (in $O(r_{\max}\, \mathcal{C}_{\mathcal{AV}}^{p-1})$ time in the worst-case) and call UNION-SETS on $\mathcal{UF}^{p-1}$ if two sets of simplices, which had different annotation vectors before calling KILL-COCYCLE, are assigned the same annotation vector. This is performed in at most $O(r_{\max}\, \alpha(|\mathcal{K}^{p-1}|))$ time. The total cost of KILL-COCYCLE is $O(r_{\max}(c_{\max} + \mathcal{C}_{\mathcal{AV}}^{p-1} + \alpha(|\mathcal{K}^{p-1}|)))$.

Computing Persistent Cohomology. Given as input a filtered simplicial complex represented in a data structure $\mathcal{K}_{\mathrm{DS}}$, we compute its persistence diagram.
Implementation of the persistent cohomology algorithm: We insert the simplices in the filtration order and update the data structures during the successive insertions. The simplicial complex $\mathcal{K}$ is stored in a simplicial complex data structure $\mathcal{K}_{\mathrm{DS}}$ and we maintain, for each dimension p, a compressed annotation matrix, which is empty at the beginning of the computation. For readability, we add the following operation on the set of data structures:

• COMPUTE-$\mathsf{a}_{\partial\sigma}(\sigma)$: given a p-simplex σ in $\mathcal{K}$, computes its boundary in $\mathcal{K}_{\mathrm{DS}}$ using COMPUTE-BOUNDARY (in $O(\mathcal{C}_{\partial}^p)$ time). For each of the $p+1$ simplices in $\partial\sigma$, it then finds their annotation vector using FIND-ROOT in $\mathcal{UF}^{p-1}$ (in $O(p\,\alpha(|\mathcal{K}^{p-1}|))$ time). Finally, it sums all these annotation vectors together (with the appropriate $+/-$ sign) using at most $p+1$ calls to SUM-ANN (in $O(p\, g_{\mathrm{m}})$ time). Note that, with the compression method, two simplices in $\partial\sigma$ may point to the same annotation vector; the computation is fasten by adding such annotation vector only once, with the appropriate multiplicative coefficient. The total worst case complexity of this operation is $O(\mathcal{C}_{\partial}^p + p\,\alpha(|\mathcal{K}^{p-1}|) + p\, g_{\mathrm{m}})$.

Let σ be a p-simplex to be inserted. We compute the annotation vector of $\partial\sigma$ using COMPUTE-$\mathsf{a}_{\partial\sigma}$. Depending on the value of $\mathsf{a}_{\partial\sigma}$, we call either CREATE-COCYCLE or KILL-COCYCLE. The algorithm computes the pairing of simplices from which one can deduce the persistence diagram. By reversing the pointers from the $\mathcal{UF}^p$s to the simplices in $\mathcal{K}_{\mathrm{DS}}$, one can compute explicitly the representative cocycles of the basis classes and have an explicit representation of the cohomology groups along the computation.
Complexity analysis: Let k be the dimension and m the number of simplices of $\mathcal{K}$. Recall that $c_{\max}$ and $r_{\max}$ represent respectively the maximal number of non-zero elements in an annotation vector and in a row of the compressed annotation matrix, along the computation. Recall that, in dimension p, $\mathcal{C}_{\partial}^p$ is the complexity of COMPUTE-BOUNDARY in $\mathcal{K}_{\mathrm{DS}}$ and $\mathcal{C}_{\mathcal{AV}}^p$ the complexity of an operation in $\mathcal{AV}^p$. $\alpha(\cdot)$ is the inverse Ackermann function.
The complexity for inserting σ of dimension p is:

$$O\left(\mathcal{C}_{\partial}^p + p(\alpha(|\mathcal{K}^{p-1}|) + c_{\max}) + \mathcal{C}_{\mathcal{AV}}^p + r_{\max}(c_{\max} + \mathcal{C}_{\mathcal{AV}}^{p-1} + \alpha(|\mathcal{K}^{p-1}|))\right)$$

Consequently, the total cost for computing the persistent cohomology is:

$$O\left(m \times \left[\mathcal{C}_{\partial}^k + k(\alpha(m) + c_{\max}) + r_{\max}(c_{\max} + \mathcal{C}_{\mathcal{AV}} + \alpha(m))\right]\right)$$

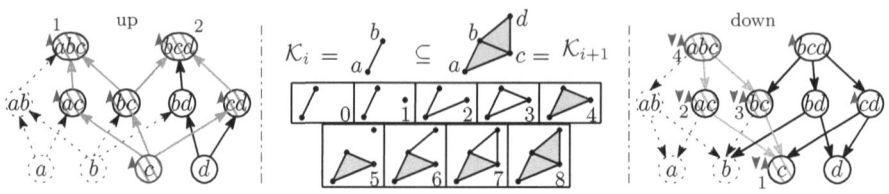

Fig. 1. Inclusion $\mathcal{K}_i \subseteq \mathcal{K}_{i+1}$. Left: upward traversal (in green) from simplex $\{c\}$. The ordering of the maximal cofaces appears in blue. Right: downward traversal (in orange) from simplex $\{abc\}$. The ordering of the subfaces appears in blue.

Specifically, if we implement $\mathcal{K}_{DS}$ as a Hasse diagram and the $\mathcal{AV}$s as hash-tables, we get $\mathcal{C}_\partial^k = O(k)$ and $\mathcal{C}_{\mathcal{AV}} = O(c_{max})$. If we consider $\alpha(m)$ as a small constant and remove it for readability, we get that the total cost for computing persistent cohomology is $O(m\,c_{max}(k + r_{max}))$. We show in section 5 that c_{max} and r_{max} remain small in practice. Hence, the practical complexity of the algorithm is linear in m for a fixed dimension.

4   Reordering Iso-simplices

Many simplices, called iso-simplices, may have the same filtration value. This situation is common when the filtration is induced by a geometric scaling parameter. Assume that we want to compute the cohomology groups $H^p(\mathcal{K}_{i+1})$ from $H^p(\mathcal{K}_i)$ where $\mathcal{K}_i \subseteq \mathcal{K}_{i+1}$ and all simplices in $\mathcal{K}_{i+1} \setminus \mathcal{K}_i$ have the same filtration value. Depending on the insertion order of the simplices of $\mathcal{K}_{i+1} \setminus \mathcal{K}_i$, the dimension of the cohomology groups to be maintained along the computation may vary a lot as well as the computing time. This may lead to a computational bottleneck. We propose a heuristic to reorder iso-simplices and show its practical efficiency in Section 5.

Intuitively, we want to avoid the creation of many "holes" of dimension p and want to fill them up as soon as possible with simplices of dimension $p + 1$. For example, in Figure 1, we want to avoid inserting all edges first, which will create two holes that will be filled when inserting the triangles. To do so, we look for the maximal faces to be inserted and recursively insert their subfaces. We conduct the recursion so as to minimize the maximum number of holes. In addition, to avoid the creation of holes due to maximal simplices that are incident, maximal simplices sharing subfaces are inserted next to each other. We can describe the reordering algorithm in terms of a graph traversal. The graph considered is the graph of the Hasse diagram of $\mathcal{K}_{i+1} \setminus \mathcal{K}_i$, defined in section 3 (see Figure 1).

Let $\sigma_1 \cdots \sigma_\ell$ be the iso-simplices of $\mathcal{K}_{i+1} \setminus \mathcal{K}_i$, sorted so as to respect the inclusion order. We attach to each simplex two flags, a flag F_{up} and a flag F_{down}, set to 0 originally. When inserting a simplex σ_j, we proceed as follows. We traverse the Hasse diagram upward in a depth-first fashion and list the inclusion-maximal cofaces of σ_j in $\mathcal{K}_{i+1} \setminus \mathcal{K}_i$. The flags F_{up} of all traversed nodes are set to 1 and the maximal cofaces are ordered according to the traversal. From each maximal coface in this order, we then traverse the graph downward and order the subfaces in a depth-first fashion: this last order will be the order of insertion of the simplices. The flags F_{down} of all traversed nodes are set to 1. We stop the upward (resp. downward) traversal when we encounter a node whose flag F_{up} (resp. F_{down}) is set to 1. We do not insert either simplices that have been inserted previously.

By proceeding as above on all simplices of the sequence $\sigma_1 \cdots \sigma_\ell$, we define a new ordering which respects the inclusion order between the simplices. Indeed, as the downward traversal starts from a maximal face and is depth first, a face is always inserted after its subfaces. Every edge in the graph is traversed twice, once when going upward and the other when going downward. Indeed, during the upward traversal, at each node N associated to a simplex σ_N, we visit only the edges between N and the nodes associated to the cofaces of σ_N and, during the downward traversal, we visit only the edges between N and the nodes associated to the subfaces of σ_N. If $\mathcal{K}_{i+1} \setminus \mathcal{K}_i$ contains ℓ simplices, the reordering takes in total $O(\ell \times (\mathcal{C}_\partial + \mathcal{C}_{co\partial}))$ time, where $\mathcal{C}_\partial$ (resp. $\mathcal{C}_{co\partial}$) refers to the complexity of computing the codimension 1 subfaces (resp. cofaces) of a simplex in the simplicial complex data structure $\mathcal{K}_{\text{DS}}$. The reordering of the filtration can either be done as a preprocessing step if the whole filtration is known, or on-the-fly as only the neighboring simplices of a simplex need to be known at a time. The reordering of a set of iso-simplices respects the inclusion order of the simplices and the filtration, and therefore does not change the persistence diagram of the filtered simplicial complex. This is a direct consequence of the stability theorem of persistence diagrams [6]. However, it may change the pairing of simplices.

5   Experiments

In this section, we report on the experimental performance of our implementation. Given a filtered simplicial complex as input, we measure the time taken by our implementation to compute its persistent cohomology, and provide various statistics. We compare the timings with state-of-the-art software computing persistent homology and cohomology. Specifically, we compare our implementation with the Dionysus library (www.mrzv.org/software/dionysus/) which provides implementation for persistent homology [11,14] and persistent cohomology [8] (denoted DioCoH) with field coefficients in $\mathbb{Z}_p$, for any prime p. We also compare our implementation with the PHAT library (version 1.0) (www.phat.googlecode.com) which provides an implementation of the optimized algorithm for persistent homology [1,4] (using the -twist option) as well as an implementation of persistent cohomology [1,7] (using the -dualize option), with coefficients in $\mathbb{Z}_2$ only. DioCoH and PHAT have been reported to be the most efficient implementation in practice [1,7]. All timings are measured on a Linux machine with 3.00 GHz processor and 32 GB RAM. Dionysus, PHAT and our implementation are written in C++ and compiled with gcc 4.6.2 with optimization level -O3. Timings are all averaged over 10 independent runs. The symbols T_∞ means that the computation lasted more than 12 hours.

Data	Cpx	$|\mathcal{P}|$	D	d	$\rho_{\max}$	k	$|\mathcal{K}|$	DioCoH		PHAT$^\perp$		PHAT		CAM	
								$\mathbb{Z}_2$	$\mathbb{Z}_{11}$	$\mathbb{Z}_2$	$\mathbb{Z}_{11}$	$\mathbb{Z}_2$	$\mathbb{Z}_{11}$	$\mathbb{Z}_2$	$\mathbb{Z}_{11}$
Cy8	Rips	6040	24	2	0.41	16	21×10^6	420	4822	44	—	5.3	—	6.4	6.5
S4	Rips	507	5	4	0.715	5	72×10^6	943	1026	95	—	3591	—	22.5	23.2
L57	Rips	4769	—	3	0.02	3	34×10^6	239	240	35.2	—	972	—	9.3	9.5
Bro	Wit	500	25	?	0.06	18	3.2×10^6	807	T_∞	6.3	—	0.88	—	2.7	2.9
Kl	Wit	10000	5	2	0.105	5	74×10^6	569	662	101	—	1785	—	19.7	19.9
L35	Wit	700	—	3	0.06	3	18×10^6	109	110	17.5	—	869	—	5.1	5.1
Bud	αSh	49990	3	2	∞	3	1.4×10^6	30.0	30.9	2.6	—	0.32	—	0.7	0.7
Nep	αSh	2×10^6	3	2	∞	3	57×10^6	T_∞	T_∞	163	—	33	—	39.5	40.2

Fig. 2. Data, timings (in seconds) and statistics

We construct three families of simplicial complexes [10] which are of particular interest in topological data analysis: the Rips complexes (denoted Rips), the relaxed witness complexes (denoted Wit) and the α-shapes (denoted αSh). These complexes depend on a relaxation parameter ρ. When the data points are embedded, the complexes are constructed up to embedding dimension, with euclidean metric. They are constructed up to the intrinsic dimension of the space with intrinsic metric otherwise. We use a variety of both real and synthetic datasets: **Cy8** is a set of points in $\mathbb{R}^{24}$, sampled from the space of conformations of the cyclo-octane molecule, which is the union of two intersecting surfaces; **S4** is a set of points sampled from the unit 4-sphere in $\mathbb{R}^5$; **L57** and **L35** are sets of points in the *lens spaces* $L(5,7)$ and $L(3,5)$ respectively, which are non-embedded spaces; **Bro** is a set of 5×5 *high-contrast patches* derived from natural images, interpreted as vectors in $\mathbb{R}^{25}$, from the Brown database; **Kl** is a set of points sampled from the surface of the figure eight Klein Bottle embedded in $\mathbb{R}^5$; **Bud** is a set of points sampled from the surface of the *Happy Buddha* (http://graphics.stanford.edu/data/3Dscanrep/) in $\mathbb{R}^3$; and **Nep** is a set of points sampled from the surface of the *Neptune statue* (http://shapes.aimatshape.net/). Datasets are listed in Figure 2 with details on the sets of points $\mathcal{P}$, their size $|\mathcal{P}|$, the ambient dimension D, the intrinsic dimension d of the object the sample points belong to (if known), the threshold $\rho_{\max}$, the dimension k of the simplicial complexes and the size $|\mathcal{K}|$ of the simplicial complexes.

Time Performance: As Dionysus and PHAT encode explicitly the boundaries of the simplices, we use a Hasse diagram for implementing $\mathcal{K}_{DS}$. We thus have the same time complexity for accessing the boundaries of simplices. We use the persistent homology algorithm of PHAT with options `-twist -sparse-pivot` and the persistent cohomology algorithm (noted PHAT$^\perp$) with option `-twist -sparse-pivot -dualize` as the `-sparse-pivot` representation of columns has been observed to be the most efficient in practice. As illustrated in Figure 2, the persistent cohomology algorithm of Dionysus is always several times slower than our implementation. Moreover, DioCoH is sensitive to the field used, as illustrated in the case of **Cy8** and **Bro**. On the contrary, CAM shows almost

Nep	$|M|$	#$\mathbb{F}op.$		Nep	average	maximum
Compression	126057	84×10^6		c_{av}, c_{max}	0.79	18
¬Compression	574426	3860×10^6		r_{av}, r_{max}	1.02	18

Bro	time		Bro	$\mathbb{Z}_{11}$			$\mathbb{Q}$		
Reordering	2.9 s.			M_{DS}	$a_{\partial\sigma}$	M_{op}	M_{DS}	$a_{\partial\sigma}$	M_{op}
¬Reordering	14.2 s.			71%	19%	10%	67%	21%	12%

Fig. 3. Statistics on the effect of the optimizations

identical performance for $\mathbb{Z}_2$ and $\mathbb{Z}_{11}$ coefficients on all examples. The persistent cohomology algorithm PHAT$^\perp$ performs better than DioCoH. However, CAM is still between 2.3 and 6.9 times faster.

The persistent homology algorithm of PHAT shows good performance in the case of the alpha shapes and on **Cy8** and **Bro**: CAM and PHAT have close timings. However, PHAT provides computation with $\mathbb{Z}_2$ coefficients only, whereas CAM computes persistence for general field coefficients and integrates no specific optimization for $\mathbb{Z}_2$. Moreover, CAM scales better to more complex examples (such as **S4**, **L57**, **Kl** and **L35**, which have higher intrinsic dimension and more complex topology). Indeed, the running time per simplex of CAM remains stable on all examples and for all field coefficients (between 2.7×10^{-7} and 9.1×10^{-7} seconds per simplex).

Statistics and Optimization: Figure 3 presents statistics about the computation. The top table presents, on the left, the effect of the compression (removal of duplicate columns) of the annotation matrix on the number of elements $|M|$ stored in the sparse representation and the number of changes #$\mathbb{F}op.$ in the matrix during the computation of the persistence diagram of **Nep**. We note a reduction factor of 4.5 for the size of the matrix, and we proceed to 46 times less field operations with the compression. Considering **Nep** is 57 million simplices, we proceed to less than 1.5 field operations per simplex on average. The right part of the table shows the average and maximum number of non-zero elements in a column when proceeding to a sum of annotation vectors (SUM-ANN) and the average and maximum number of non-zero elements in a row when proceeding to its reduction (KILL-COCYCLE). These values are key variables (c_{max} and r_{max} respectively) in the complexity analysis of the algorithm. We note that these values remain really small. The bottom table presents the effect of the reordering strategy on the example **Bro**. We note that reordering iso-simplices makes the computation 4.9 faster. Finally, the right side of the table presents how the computing time is divided into maintaining the compressed annotation matrix (noted M_{DS}), computing the annotation vector $a_{\partial\sigma}$ and modifying the values of the elements in the compressed annotation matrix (noted M_{op}). The percentage are given when computing persistent cohomology with $\mathbb{Z}_{11}$ and $\mathbb{Q}$ coefficients. The computational complexity of field operations $\langle \mathbb{F}, +, \cdot, -, /, 0, 1 \rangle$ depends on the field we use. For $\mathbb{Z}_{11}$, or any field of small cardinal, the operations can be precomputed and accessed in constant time. The field operations in $\mathbb{Q}$ are more costly. Specifically, an element q in $\mathbb{Q}$ is represented as a pair of coprime integers (r, s) such that $q = r/s$, and field operations may require gcd computation to ensure that nominator and denominator remain coprime. However, the computational time of CAM is quite insensitive to the field we use. Specifically, as it minimizes the number of matrix changes using the compression method, the computational time is only increased by 8% when computing persistence with $\mathbb{Q}$ coefficients instead of $\mathbb{Z}_{11}$, whereas the computation involving field operations takes 34% more time.

In all our experiments, the size of the compressed annotation matrix is negligible compared to the size of the simplicial complex. Consequently, combined with the simplex tree data structure [2] for representing the simplicial complex, we have been able to compute the persistent cohomology of simplicial complexes of several hundred million simplices in high dimension.

A public and fully documented version of our code will be released soon.

Acknowledgement. This research is partially supported by the 7th Framework Programme for Research of the European Commission, under FET-Open grant number 255827 (CGL Computational Geometry Learning). This research is also partially supported by NSF (National Science Foundation, USA) grants CCF-1048983 and CCF-1116258.

References

1. Bauer, U., Kerber, M., Reininghaus, J.: Clear and compress: Computing persistent homology in chunks. arXiv/1303.0477 (2013)
2. Boissonnat, J.-D., Maria, C.: The simplex tree: An efficient data structure for general simplicial complexes. In: Epstein, L., Ferragina, P. (eds.) ESA 2012. LNCS, vol. 7501, pp. 731–742. Springer, Heidelberg (2012)
3. Busaryev, O., Cabello, S., Chen, C., Dey, T.K., Wang, Y.: Annotating simplices with a homology basis and its applications. In: Fomin, F.V., Kaski, P. (eds.) SWAT 2012. LNCS, vol. 7357, pp. 189–200. Springer, Heidelberg (2012)
4. Chen, C., Kerber, M.: Persistent homology computation with a twist. In: Proceedings 27th European Workshop on Computational Geometry (2011)
5. Chen, C., Kerber, M.: An output-sensitive algorithm for persistent homology. Comput. Geom. 46(4), 435–447 (2013)
6. Cohen-Steiner, D., Edelsbrunner, H., Harer, J.: Stability of persistence diagrams. Discrete & Computational Geometry 37(1), 103–120 (2007)
7. de Silva, V., Morozov, D., Vejdemo-Johansson, M.: Dualities in persistent (co)homology. Inverse Problems 27, 124003 (2011)
8. de Silva, V., Morozov, D., Vejdemo-Johansson, M.: Persistent cohomology and circular coordinates. Discrete Comput. Geom. 45, 737–759 (2011)
9. Dey, T.K., Fan, F., Wang, Y.: Computing topological persistence for simplicial maps. CoRR, abs/1208.5018 (2012)
10. Edelsbrunner, H., Harer, J.: Computational Topology - an Introduction. American Mathematical Society (2010)
11. Edelsbrunner, H., Letscher, D., Zomorodian, A.: Topological persistence and simplification. Discrete Comput. Geom. 28(4), 511–533 (2002)
12. Milosavljevic, N., Morozov, D., Skraba, P.: Zigzag persistent homology in matrix multiplication time. In: Symposium on Comp. Geom. (2011)
13. Morozov, D.: Persistence algorithm takes cubic time in worst case. In: BioGeometry News, Dept. Comput. Sci., Duke Univ. (2005)
14. Zomorodian, A., Carlsson, G.: Computing persistent homology. Discrete & Computational Geometry 33(2), 249–274 (2005)

Approximation Algorithms for Facility Location with Capacitated and Length-Bounded Tree Connections

Jannik Matuschke, Andreas Bley, and Benjamin Müller

TU Berlin, Institut für Mathematik
{matuschke,bley,bmueller}@math.tu-berlin.de

Abstract. We consider a generalization of the uncapacitated facility location problem that occurs in planning of optical access networks in telecommunications. Clients are connected to open facilities via depth-bounded trees. The total demand of clients served by a tree must not exceed a given tree capacity. We investigate a framework for combining facility location algorithms with a tree-based clustering approach and derive approximation algorithms for several variants of the problem, using techniques for approximating shallow-light Steiner trees via layer graphs, simultaneous approximation of shortest paths and minimum spanning trees, and greedy coverings.

1 Introduction

We study a generalization of the uncapacitated facility location (UFL) problem where instead of directly connecting a client to an open facility, shared access trees connecting multiple clients to an open facility are allowed. Problems of this type arise in the planning of optical access networks in telecommunications. In these so-called fiber-to-the-home (FTTH) or fiber-to-the-curb (FTTC) networks, optical splitters are used to split a single fiber emanating from the central office into a fiber tree serving multiple clients. The main advantage of this technology, compared to using an individual fiber for each client, is a considerable reduction in the number of fibers and, more importantly, the active fiber termination equipment at the central office location. On the other hand, all clients in the same fiber tree share the limited transmission capacity of a single fiber. Thus, not too many clients may be aggregated into such a tree. Furthermore, the optical signal emitted at the central office must fulfill several power and quality requirements when reaching a client, in order to guarantee a reliable connection. These technical requirements imply an upper bound on the path length between the central office and any client within the fiber tree or, in other words, on the depth of the tree.

When planning the deployment of such a network, one generally has to decide where to set up central offices and how to connect the clients to these offices via fiber trees. One of the most important objectives in this planning is to minimize the total network cost, which is comprised by cost for setting up the offices and the cost for laying out the fibers.

H.L. Bodlaender and G.F. Italiano (Eds.): ESA 2013, LNCS 8125, pp. 707–718, 2013.

We model this planning problem as an uncapacitated facility location problem with capacitated and length-bounded tree connections (UFL-CLT), which is defined as follows. We are given a graph $G = (V, E)$ whose vertex set is partitioned into a set of n clients $\mathcal{C}$ and a set of possible facilities $\mathcal{F}$. The facilities correspond to the potential central office locations and the edges correspond to routes where fibers might be installed. Each edge $e \in E$ has length $\ell(e) \in \mathbb{Z}_+$ and cost $c(e) \in \mathbb{Z}_+$. Each facility $w \in \mathcal{F}$ has an opening cost $f(w) \in \mathbb{Z}_+$ and each client $v \in \mathcal{C}$ has a demand $d(v) \in \mathbb{Z}_+$. Finally, there is a cable capacity $U \in \mathbb{Z}_+$ and a length bound $L \in \mathbb{Z}_+$.

A solution to UFL-CLT consists of a set of open facilities $F \subseteq \mathcal{F}$, a set of (not necessarily disjoint) trees $\mathcal{T} \subseteq 2^E$, with each tree T being rooted at a facility $w_T \in F$, and an assignment ϕ that specifies for each client v a tree $\phi(v) \in \mathcal{T}$ that contains v. Let P_v denote the unique path from v to $w_{\phi(v)}$ in $\phi(v)$. The solution is feasible, if it respects the length bound

$$\ell(P_v) := \sum_{e \in P_v} \ell(e) \leq L \quad \forall v \in \mathcal{C} \tag{1}$$

and the capacity constraint

$$\sum_{v \in \mathcal{C}: \phi(v) = T, e \in P_v} d(v) \leq U \quad \forall e \in E, T \in \mathcal{T}. \tag{2}$$

Using a framework that combines approximation algorithms for various subproblems, we obtain bicriteria approximation algorithms for several variants of UFL-CLT, approximating length bound and optimal cost at the same time.

Definition 1. *An (α, β)-approximation algorithm for UFL-CLT is an algorithm that computes in polynomial time a solution fulfilling (2) and $\ell(P_v) \leq \alpha L$ for all $v \in \mathcal{C}$, with cost at most βOPT, where OPT denotes the minimum cost of a solution fulfilling both (1) and (2).*

1.1   Related Work

UFL-CLT is closely related to the *capacitated cable facility location* problem (CCFL), which corresponds to UFL-CLT without the length bound and where additionally the demand of a client can be split among several trees. Ravi and Sinha [13] devise a framework to combine a ρ_{ST}-approximation for Steiner tree and a ρ_{UFL}-approximation for UFL to obtain a $(\rho_{\mathrm{ST}} + \rho_{\mathrm{UFL}})$-approximation algorithm for CCFL. Our algorithms are largely based on a variant of their framework, incorporating a partitioning technique by Harks et al. [5] to ensure that demands do not have to be split. Similar clustering techniques have also been applied by Maßberg and Vygen [12] for a related problem in VLSI design, where capacities bound the total cost of a tree rather than the demand of its clients.

UFL-CLT combines two combinatorial optimization problems, uncapacitated facility location (UFL) and shallow-light Steiner tree (SLST), both of which have been studied extensively in literature. We give a short overview on approximability results for each problem, with a particular focus on those algorithms and special cases occurring in this paper.

Facility Location. The general (non-metric) version of UFL does not allow for an $o(\log(n))$-approximation unless $P = NP$. The greedy algorithm of Hochbaum [7] achieves a ratio of $O(\log(n))$. If the connection cost is metric, the best known constant factor approximation is a 1.5-approximation by Byrka and Aardal [2] that also allows for adjustable individual approximation ratios for connection cost and opening cost.

Shallow-light Trees. The *shallow-light Steiner tree* problem (SLST) asks for a Steiner tree with the property that the length of a path from each terminal to the given root obeys the length bound while minimizing the cost. It corresponds to the special case of UFL-CLT with $U = \infty$ and $|\mathcal{F}| = 1$. Using a matching augmentation technique, Marathe et al. [11] give a polynomial time $(O(\log(n)), O(\log(n)))$-approximation for SLST. While known lower bounds do not rule out the possibility of an $(O(1), O(1))$-approximation, the only improvement on the approximation factor in [11] so far is a parameterized $(O(p\log(n)/\log(p)), O(\log(n)/\log(p)))$-approximation for an input parameter p, $1 \le p < n$ by Kapoor and Sarwat [8].

The *directed Steiner tree* problem asks for a minimum cost arborescence in a directed graph rooted at a given node and spaninng all terminals. Charikar et al. [4] provide an algorithm for this problem that returns an $O(p^2 s^{1/p})$-approximation in time $O(s^{3p})$ for a parameter p, where s is the number of terminals. Setting $p = \log(s)$ yields an $O(\log^2(s))$-approximation in time $O(s^{3\log(s)})$. Using a condensed layer graph, this result immediately translates into a quasi-polynomial $(1 + \varepsilon, O(\log^2(n)))$-approximation for SLST.

The special case of SLST with $\ell \equiv 1$ is known as *hop-constrained Steiner tree* problem. For this problem, Kortsarz and Peleg [10] give a $(1, O(\log(n)))$-approximation, whose running time is polynomial when L is bounded by a constant. For the hop-constrained spanning tree problem with metric costs, Althaus et al. [1] provide a $(1, O(\log(n)))$-approximation using approximation of metrics by trees. Finally, for SLST instances with $c = \ell$, Khuller et al. [9] devise an algorithm that transforms a given tree T into a tree T' with cost $c(T') \le (1 + 2/(1 - \alpha))c(T)$ such that the path length of any vertex v to the specified root r in T' is a most α times the length of a shortest v-r-path in the graph. They call such trees *light approximate shortest-path trees* (LAST).

Remark 1. Some of the references cited above in fact study *bounded diameter* versions of the problem instead of shallow-light trees. However, the two problems are polynomially equivalent in the sense that any instance of SLST can be transformed into an equivalent instance of the diameter version, while any instance of bounded diameter Steiner tree can be solved by solving at most $|E|$ instances of SLST.

1.2   Contribution and Structure of the Paper

In this paper, we study a general framework for obtaining approximation algorithms to UFL-CLT with varying ratios for path lengths and cost. All our algorithms deviate from best-known approximation factors for the corresponding versions of SLST at most by a constant factor, showing the power of the chosen approach for resolving capacity restrictions in depth-bounded tree networks.

In Section 2, we establish hardness of approximation better than $(3, O(\log(n)))$ even for a very resticed special case of UFL-CLT and introduce two lower bounds on the cost of an optimal solution. We then describe the general algorithmic framework in Section 3 and use it to devise two algorithms for the general version of UFL-CLT, a polynomial $(O(\log(n)), O(\log(n)))$-approximation and a quasi-polynomial $(3 + \varepsilon, O(\log^2(n)))$-approximation. The remainder of the paper investigates two important special cases, for which we can obtain considerably better approximation factors. In Section 4, we show how to obtain a $(3\alpha, (1+2/(\alpha-1))O(1))$-approximation for a parameter $\alpha > 1$ in case c and ℓ are proportional to a common metric. In Section 5, we consider the hop-constrained version of UFL-CLT, i.e., $\ell \equiv 1$, and obtain a $(1 + \varepsilon, O(\log(n)))$-approximation algorithm.

2   Preliminaries

2.1   Basic Definitions and Notations

Given a rooted tree T and two vertices $v, w \in V(T)$, we denote by $T[v, w]$ the unique path from v to w in T and by $T[v]$ the subtree rooted at v. Let $x : E \to \mathbb{Q}_+$ be a function. For a vertex $v \in V$, and a set of vertices $W \subseteq V$, we define $x(v, W) := \min_{w \in W}\{x(vw)\}$. For $S \subseteq E$ we define $x(S) := \sum_{e \in S} x(e)$.

2.2   Hardness of Approximation

As already indicated in Section 1.1, UFL-CLT generalizes several problems that (unless $P = NP$) do not allow for an approximation ratio better than $O(\log(n))$, like (non-metric) UFL or SLST. This is even true for the special case $c = \ell$.

Theorem 1. *Unless $P = NP$, there is no $(3 - \varepsilon, o(\log(n)))$-approximation to UFL-CLT for any $\varepsilon > 0$, even when restricting to instances where cost and length are proportional to a common metric.*

2.3   Two Lower Bounds

Assume we are given an instance of UFL-CLT and let OPT be the cost of an optimal solution to this instance. The following two lower bounds are generalizations of the lower bounds used in [13].

Lemma 1. *For $v \in \mathcal{C}$ and $w \in \mathcal{F}$ define*

$$\tilde{c}(v, w) := \frac{d(v)}{U} \min\{c(P) : P \text{ is a } v\text{-}w\text{-path in } G \text{ with } \ell(P) \leq L\}.$$

Let $\tilde{F} \subseteq \mathcal{F}$ be an optimal solution to the UFL instance with facilities $\mathcal{F}$, clients $\mathcal{C}$, opening costs f and connection costs $\tilde{c}$. Then $\sum_{w \in \tilde{F}} f(w) + \sum_{v \in \mathcal{C}} \tilde{c}(v, \tilde{F}) \leq OPT$.

Lemma 2. *Let $G' := (V', E')$ with $V' := V \cup \{r\}$ and $E' := E \cup \{rw : w \in \mathcal{F}\}$. Extend c and ℓ by defining $c(rw) := f(w)$ and $\ell(rw) := 0$ for all $w \in \mathcal{F}$. Let $T \subseteq E'$ be a tree of minimal cost among all trees spanning $\mathcal{C} \cup \{r\}$ with $\ell(T[r, v]) \leq L$ for all $v \in \mathcal{C}$ (i.e., T is an optimal shallow-light Steiner tree on G'). Then $c(T) \leq OPT$.*

3   Approximation Algorithms for General UFL-CLT

In this section, we introduce the algorithmic framework. We apply it to the general version of UFL-CLT and derive two approximation algorithms. The first runs in polynomial time and approximates both length bound and optimal cost by a logarithmic factor. The second runs in quasi-polynomial time, violating the length bound only by a constant while approximating the cost by a polylogarithmic factor.

3.1   Algorithmic Framework

The algorithm consists of three main steps. In the first two steps, the facility location and shallow-light tree instances introduced in Lemmas 1 and 2 are constructed and approximately solved. The two solutions are merged in the third step, using opened facilities to relieve overloaded subtrees. In a fourth step, the final solution, consisting of the subtrees of the main tree rooted at facilities and the additional subtrees constructed in the third step, is returned.

Algorithm 1

Step 1: Construct the UFL instance described in Lemma 1 and compute a β_{UFL}-approximate solution $\tilde{F} \subseteq \mathcal{F}$ to this instance.

Step 2: Construct the graph G' as described in Lemma 2 and compute an $(\alpha_{\mathrm{SLST}}, \beta_{\mathrm{SLST}})$-approximate shallow-light Steiner tree T on G' spanning $\mathcal{C} \cup \{r\}$. Set $\phi(v) = T[w]$ for all $v \in \mathcal{C}$, where w is the last facility on $T[v, r]$.

Step 3: Let $\mathcal{T}'$ be an initially empty set of trees. While there is an edge $e \in T \cap E$ with $\sum_{v:\phi(v)=T(w), e \in T[r,v]} d(v) > U$, let v' be a client incident to exactly one such edge and call relieve(v').

Step 4: Let $F = \tilde{F} \cup \{w \in \mathcal{F} : rw \in T\}$ and $\mathcal{T} = \{T[w] : rw \in T\} \cup \mathcal{T}'$, and return $(F, \mathcal{T}, \phi)$.

relieve(v'): Let $S := \{v \in \mathcal{C} : v$ is a child of v' in $T\}$ and for each $v \in S$, let $S(v) := \{u \in V(T[v]) : \phi(u) = \phi(v)\}$, and let $S(v') := \{v'\}$. Partition $S \cup \{v'\}$ into groups $S_0, \ldots, S_k$ such that $\sum_{v \in S_0} d(S(v)) \leq U$ and $U/2 \leq \sum_{v \in S_i} d(S(v)) \leq U$ for all $i \in \{1, \ldots, k\}$. Then, for each $i \in \{1, \ldots, k\}$
  – find a client-facility pair v_i, w_i with $v_i \in \bigcup_{v \in S_i} S(v)$ and $w_i \in F'$ such that $c(v_i, w_i)$ is minimum among all such pairs,
  – add $T_i := \bigcup_{v \in S_i} (T[v] \cup \{v'v\}) \cup \{v_i w_i\}$ to $\mathcal{T}'$ and remove the edges of T_i from T,
  – set $\phi(u) = T_i$ for all $u \in \bigcup_{v \in S_i} S(v)$.

Remark 2. An approximate version of the UFL instance in Step 1 can be constructed in polynomial time by using an FPTAS for the length constrained shortest path problem yielding a path from v to w with length at most L and cost at most $1 + \varepsilon$ times the cost of an optimal length constrained path [6].

Lemma 3. *Let* $(F, \mathcal{T}, \phi)$ *be the solution computed by the algorithm. Let* $\hat{T}$ *be the tree computed in Step 2. Then the following three statements hold true.*

1. $\sum_{v \in \mathcal{C}:\, \phi(v) = T',\, e \in P_v} d(v) \leq U$ *for all* $T' \in \mathcal{T}$ *and all* $e \in T$
2. $\ell(P_v) \leq (2\alpha_{SLST} + 1)L$ *for all* $v \in \mathcal{C}$
3. $\sum_{w \in F} f(w) + \sum_{T' \in \mathcal{T}} c(T') \leq \sum_{w \in \tilde{F}} f(w) + c(\hat{T}) + 2\sum_{v \in \mathcal{C}} \tilde{c}(v, \tilde{F})$

Proof.   1. After termination of the while loop in Step 3, the capacity constraint is fulfilled for all edges $e \in T$. Every tree introduced in the relieve procedure only serves a demand of at most U and thus the capacity constraint is fulfilled in these trees as well.
  2. Let $v \in \mathcal{C}$. If $\phi(v) = T$, then $\ell(P_v) \leq \alpha_{\mathrm{SLST}} L$. If $\phi(v) \neq T$, then v has been removed from T in Step 3 when relieving the subtree $T[v']$ for some client v'. In this case, its demand was rerouted to some facility w_i via a client v_i and thus $\ell(P_v) \leq \ell(\hat{T}[v, v']) + \ell(\hat{T}[v', v_i]) + \ell(v_i w_i)$. While the first two summands each are bounded by $\alpha_{\mathrm{SLST}} L$ by construction of $\hat{T}$, the third summand is bounded by L by construction of the UFL instance in Step 1.
  3. Clearly, $\sum_{w \in F} f(w) \leq \sum_{w \in \tilde{F}} f(w) + \sum_{w:rw \in \hat{T}} c(rw)$. If $e \in T'$ for some $T' \in \mathcal{T}$, then either $e \in \hat{T}$, or $e = v_i w_i$ was introduced in the relieve procedure when connecting the clients in group S_i for some $i \in \{1, \ldots, k\}$. In the latter case, $c(e) \leq c(u, \tilde{F})$ for all $u \in U_i := \bigcup_{v \in S_i} S(v)$. Then $\sum_{v \in S_i} d(S(v)) \geq U/2$ implies $c(e) \leq 2\sum_{u \in U_i}(d(u)/U)c(u, F) = 2\sum_{u \in U_i} \tilde{c}(u, \tilde{F})$. As the sets U_i are disjoint, the third statement follows.   □

The algorithmic framework introduced above can be implemented using different variants of approximation algorithms for the UFL and SLST instances in Steps 1 and 2, resulting in different overall approximation factors and running times.

3.2   A Polynomial-Time $(O(\log(n)), O(\log(n)))$-Approximation

A natural choice for the UFL approximation of Step 1 is the greedy $O(\log(n))$-approximation by Hochbaum [7]. In Step 2, the $(O(\log(n)), O(\log(n)))$-approximation for diameter-constrained Steiner trees by Marathe et al. [11] can be used to approximate the SLST instance. Thus, Lemma 3 yields the following theorem.

Theorem 2. *Using the greedy UFL approximation [7] in Step 1 and the (diameter, cost)-algorithm of [11] in Step 2, Algorithm 1 computes in polynomial time an* $(O(\log(n)), O(\log(n)))$-*approximate solution to UFL-CLT.*

3.3   A Quasi-Polynomial $(3 + \varepsilon, O(\log^2(n)))$-Approximation

In order to improve the approximation guarantee for the length bound to constant factor, we employ a quasi-polynomial directed Steiner tree algorithm on the layer graph corresponding to the SLST instance in Step 2.

Definition 2 (Layer graph). *Given an instance of SLST on a graph $G = (V, E)$ with root $r \in V$, terminal set $S \subseteq V$, costs c, edge lengths ℓ, and length bound L, the corresponding* layer graph *is a directed graph G_L with node set V_L, arc set E_L, and costs c_L defined as follows. For every vertex $v \in V$ and every $t \in \{0, \ldots, L\}$, there is a node v_t in V^*. For every edge $e \in E$ with end points v and w, and every $t \in \{0, \ldots, L - \ell(e)\}$, there is an arc from v_t to $w_{t+\ell(e)}$ and an arc from w_t to $v_{t+\ell(e)}$ both with cost $c(e)$. Finally, there also is an arc from v_t to v_L for every terminal $v \in S$ and every $t \in \{0, \ldots, L - 1\}$ with cost 0.*

A directed Steiner tree on G_L is an aborescence rooted at r_0 that spans all vertices v_L for $v \in S$.

Lemma 4. *Given an instance of SLST on a graph G, the minimum cost of a directed Steiner tree in G_L is equal to the minimum cost of a shallow-light Steiner tree in G. Given a directed Steiner tree T_L in G_L, one can compute in time polynomial in $|T_L|$ a shallow-light Steiner tree T in G with $c(T) \le c_L(T_L)$.*

Lemma 5. *For every $\varepsilon > 0$, there is an $O(n^{3\log(n)})$-time algorithm that computes a $(1+\varepsilon, O(\log^2(n)))$-approximate solution to SLST (where n is the number of terminals).*

Proof. Let $\varepsilon > 0$ and let $(G = (V, E), S, r, c, \ell, L)$ be an instance of SLST. Define $\kappa := \varepsilon L / |V|$, $\tilde{L} := \lfloor (1/\kappa)L \rfloor$, and $\tilde{\ell}(e) := \lfloor (1/\kappa)\ell(e) \rfloor$ for all $e \in E$. Construct the layer graph $G_{\tilde{L}}$ w.r.t. $\tilde{\ell}$ and $\tilde{L}$ and apply the $O(p^2 s^{1/p})$-approximation for directed Steiner tree of [4] on $G_{\tilde{L}}$ with $p = \log(s)$. Observe that the number s of terminals in $G_{\tilde{L}}$ is n and that $G_{\tilde{L}}$ has at most $\varepsilon^{-1}|V|^2$ nodes and $\varepsilon^{-1}|E||V|$ arcs. Thus the algorithm computes an $O(\log^2(n))$-approximate directed Steiner tree $\tilde{T}$ in quasi-polynomial time (depending on $|V|$, $|E|$, and ε^{-1}, but not on L). Using Lemma 4, we can compute a Steiner tree T in G with cost at most $c_{\tilde{L}}(\tilde{T})$ and $\tilde{\ell}(T[v, r]) \le \tilde{L}$ for all $v \in S$. Thus

$$\ell(T[v, r]) \le \kappa \sum_{e \in T[v,r]} (\tilde{\ell}(e) + 1) \le \kappa(\tilde{L} + |V|) \le (1 + \varepsilon)L.$$

Furthermore, observe that if $\ell(P) \le L$ for any path P in G, then also $\tilde{\ell}(P) \le \tilde{L}$, implying that the cost of an optimal shallow-light Steiner tree in G w.r.t. $\tilde{\ell}$ and $\tilde{L}$ is at most the cost of a shallow-light Steiner tree w.r.t. ℓ and L. Thus, by Lemma 4, T is a $(1 + \varepsilon, O(\log^2(n)))$-approximate shallow-light Steiner tree to the original SLST instance.   □

Again using the greedy UFL approximation [7] in Step 1 and the quasi-polynomial algorithm for SLST in Step 2 of Algorithm 1, we can achieve a constant factor approximation of the length bound by Lemma 3.

Theorem 3. *For every $\varepsilon > 0$, there is an algorithm that computes a $(3 + \varepsilon, O(\log^2(n)))$-approximate solution to UFL-CLT in time $O(n^{3\log(n)})$.*

4   Length-Dependent Cost

Throughout this section, we assume G to be complete and both c and ℓ to be proportional to a common metric on G (w.l.o.g., this implies $c = \ell$). For this special case, a modified version of Algorithm 1, using the LAST algorithm of Khuller et al. [9] and a greedy covering, yields a solution that approximates both the length bound and the optimal cost by constants. We can adjust the two constants by modifying the parametric approximation factor of the LAST.

4.1 μ-Stretched Covers

Before applying the LAST algorithm, we need to ensure that for each client there is an open facility whose distance to the client does not exceed the length bound by more than a constant factor.

Definition 3. *Given an instance of UFL-CLT, a μ-stretched cover is a set of facilities $F \subseteq \mathcal{F}$ such that for every client $v \in \mathcal{C}$ there is a facility $w \in F$ with $\ell(vw) \leq \mu L$.*

Greedy 3-Streched Covering Algorithm.

> **Input**: an instance of UFL-CLT
> **Output**: a set of facilities $F' \subseteq \mathcal{F}$
> $\mathcal{C}' \leftarrow \mathcal{C}$; $F' \leftarrow \emptyset$
> **while** $\mathcal{C}' \neq \emptyset$ **do**
> > choose any $v' \in \mathcal{C}'$ and a let $\mathcal{F}_{v'} = \{w \in \mathcal{F} : \ell(v'w) \leq L\}$
> > choose $w' \in \mathcal{F}_{v'}$ minimizing $f(w')$
> > $F' \leftarrow F' \cup \{w'\}$
> > $\mathcal{C}' \leftarrow \mathcal{C}' \cap \{v \in \mathcal{C} : \ell(vw') > 3L\}$
>
> **end**
> **return** F'

Lemma 6. *The set of facilities F' returned by the greedy 3-streched covering algorithm is a 3-stretched cover and $\sum_{w \in F'} f(w) \leq OPT$.*

Proof. A client v is only removed from $\mathcal{C}'$ if $\ell(v, F') \leq 3L$, thus, when the algorithm terminates, F' is a 3-stretched cover. For any facility $w \in F'$, let $v_w \in \mathcal{C}$ be the client that was chosen in the first line of the while loop in the iteration where w was added to F', and let $o(w) \in \mathcal{F}$ be the facility that serves v_w in a fixed optimal solution. Clearly, $f(w) \leq f(o(w))$ as $o(w) \in \mathcal{F}_{v_w}$. We now show $o(w) \neq o(w')$ for $w \neq w'$, which implies the claim of the lemma. By contradiction assume $o(w) = o(w')$ for $w \neq w'$. W.l.o.g., w was added to F' before w'. Then $o(w) = o(w')$ implies $\ell(v_{w'}, w) \leq \ell(v_{w'}, o(w')) + \ell(v_w, o(w)) + \ell(v_w, w) \leq 3L$. Thus $v_{w'}$ was erased from $\mathcal{C}'$ when w was added to F'.   □

4.2   The Algorithm

The first three steps of Algorithm 2 resemble those of Algorithm 1 with the only difference that we ignore the length bound. We therefore can apply constant factor approximations for the resulting metric UFL and Steiner tree instances in

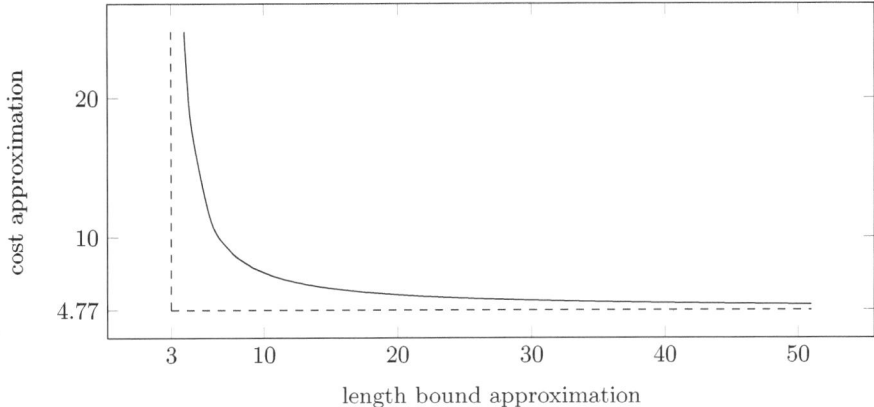

Fig. 1. Approximation factors achieved by Algorithm 2 for UFL-CLT with length-proportional cost, using the Steiner tree approximation by Byrka et al. [3] with $\beta_{\mathrm{ST}} = 1.39$

Step 1 and 2, respectively. After applying the relieve procedure, the algorithm computes a 3-stretched cover of facilities to be opened in addition to those stemming from the previous steps. Finally, each tree in the solution computed thus far is processed using the LAST algorithm, ensuring that the distance of each client to the root of its tree is at most α times the distance to the closest open facility (which is at most $3L$), while increasing the cost of the tree by a factor of at most $1 + 2/(\alpha - 1)$.

In order to improve the approximation guarantee of the algorithm, the connection and opening costs of the UFL approximation are balanced carefully: The UFL algorithm of Byrka and Aardal [2] takes as additional input a parameter γ, returning a solution whose opening cost is at most γ times the opening cost of an initial LP solution and whose connection cost is at most $1 + 2e^{-\gamma}$ times the connection cost of that LP solution. We denote the optimal choice of γ for a given α by γ_α. It is the unique solution to $(1 + 2/(\alpha - 1))(2 + 4e^{-\gamma}) = \gamma$, as can be derived from the proof of Theorem 4.

Algorithm 2

Step 1: Construct the UFL instance with facilities $\mathcal{F}$, clients $\mathcal{C}$, opening costs f and connection costs $\bar{c}(v, w) := (d/u)c(v, w)$. Compute an approximate solution $\tilde{F} \subseteq \mathcal{F}$ to this instance using the algorithm in [2] with $\gamma = \gamma_\alpha$.

Step 2: Construct the graph G' as described in Lemma 2 and compute w.r.t. c' a β_{ST}-approximate Steiner tree T with terminals $\mathcal{C}$ on G'.

Step 3: Let $\mathcal{T}'$ be an initially empty set of trees. While there is an edge $e \in T \cap E$ with $\sum_{v:e \in T[r,v]} d(v) > U$, let v' be a client incident to exactly one such edge and call relieve(v').

Step 4: Compute a 3-stretched cover $\bar{F}$.

Step 5: Let $F = \tilde{F} \cup \{w \in \mathcal{F} : rw \in T\} \cup \bar{F}$ and $\mathcal{T} = \{T[w] : rw \in T\} \cup \mathcal{T}'$. Temporally contract F to a single vertex r'. For each $T' \in \mathcal{T}$, apply the algorithm of Khuller et al. [9] using T' as initial tree and the star $\{r'v : v \in V(T')\}$ as shortest path tree, to compute an $(\alpha, 1 + 2/(\alpha - 1))$-LAST T^* w.r.t. c and root r'. Replace T' by T^*. Expand r' to F and return $(F, \mathcal{T}, \phi)$.

Theorem 4. *For every $\alpha > 1$ there is a polynomial-time algorithm that computes a $(3\alpha, (1 + 2/(\alpha - 1))\beta_{ST} + \gamma_\alpha + 1)$-approximate solution for UFL-CLT restricted to instances with length-proportional cost, where γ_α is the unique solution to $(1 + 2/(\alpha - 1))(2 + 4e^{-\gamma}) = \gamma$ and β_{ST} is the approximation factor of a Steiner tree algorithm.*

5   Hop Constraints

We now consider the case where $\ell \equiv 1$, the graph is complete and c is a metric. In this case, the length bound is also known as *hop constraint*. We will show how to adapt Algorithm 1 so as to approximate the hop constraint with arbitrary precision while still achieving a logarithmic cost approximation and polynomial run time. This is done by applying two different tree algorithms, depending on the number of hops allowed in the instance.

If L is large, we will use a $(1, O(\log(n)))$-approximation for the minimum hop-constraint spanning tree problem by Althaus et al. [1]. However, the transformation of the corresponding lower bound from Steiner tree to spanning tree incurs an increase in the number of hops by an additive constant. We will compensate for this by using a different algorithm for instances where L is small: The $(1, O(\log(n)))$-approximation for shallow light Steiner tree by Kortsarz and Peleg [10] runs in polynomial time for constant number of hops. The final ingredient is a slight modification of the relieve subprocedure, ensuring that the depth of newly created trees does not exceed that of the original tree.

Lemma 7 (Spanning tree variant of Lemma 2). *Let G' be the graph constructed in Lemma 2 and let $\hat{c}(rw) := 0$ for all $w \in \mathcal{F}$, $\hat{c}(vw) = c(vw) + f(w)$ for all $v \in \mathcal{C}$ and $w \in \mathcal{F}$, and $\hat{c}(e) = c(e)$ for all other edges. Then the cost of a minimum $(L + 2)$-hop spanning tree in G' w.r.t. $\hat{c}$ and root r is at most $2OPT$.*

Proof. Let $(F, \mathcal{T}, \phi)$ be an optimal solution to the UFL-CLT instance. For a tree T and a facility $w \in V(T)$, let $v_{w,T}^{\min}$ be a child of w in T that minimizes $c(vw)$ among all children v of w. We modify each tree $T \in \mathcal{T}$ by applying the following change for each facility $w \neq w_T$ in T. Let u be the parent of w in T. Remove $v_{w,T}^{\min}w$ and wu from T and insert $v_{w,T}^{\min}u$. Then, for each child v of w with $v \neq v_{w,T}^{\min}$, replace vw by $vv_{w,T}^{\min}$ in T. Observe that this modification does not increase the length of any client-root-path in the tree.

Thus, for each client $v \in \mathcal{C}$ there is a tree $T \in \mathcal{T}$ with a v-w_T-path of length at most L. The set $T^* := \bigcup_{T \in \mathcal{T}} T \cup \{rw : w \in \mathcal{F}\}$ spans G' and contains a path of length $L + 1$ from each client to r. If T^* contains a cycle, there must be at least one edge uv in the cycle such that $\mathrm{dist}_\ell(u, r) + 1 > \mathrm{dist}_\ell(v, r)$. Iteratively remove such edges until T^* contains no more cycles and is an $(L+1)$-hop spanning tree. If there is a facility $w \in \mathcal{F}$ that has more than one child in T^*, again for each child v of w with $v \neq v^{\min}_{w,T^*}$, replace vw by $vv^{\min}_{w,T^*}$ in T. This increases the length of any client-root-path by at most 1. Observe that by triangle inequality, every replacement of an edge vw by $vv^{\min}_{w,T}$ increases the cost of the corresponding tree w.r.t. c by at most $c(v^{\min}_{w,T}w) \leq c(vw)$. As this replacement happens at most once for every client, the total increase in cost is bounded by $\sum_{T \in \mathcal{T}} c(T)$. Also, every facility $w \in \mathcal{F} \setminus F$ is a leaf in T^* and every facility $w \in F$ has at most one child in T^*. Thus, T^* is an $(L+2)$-hop spanning tree and $\hat{c}(T^*) \leq c(T^*) + \sum_{w \in F} f(w) \leq \sum_{w \in F} f(w) + 2\sum_{T \in \mathcal{T}} c(T) \leq 2OPT$. □

Algorithm 3. Let $\varepsilon > 0$. Run Algorithm 1 with the following modifications.

- In Step 1, use the greedy $O(\log(n))$-approximation for (non-metric) UFL [7].
- In Step 2: If $L \leq \lceil 1/\varepsilon \rceil$, then use the algorithm of Kortsarz and Peleg [10] to compute an $O(\log(n))$-approximate $(L+1)$-hop Steiner tree in G' w.r.t. cost c, root r and terminals $\mathcal{C}$. Otherwise, use the algorithm of Althaus et al. [1] to compute an $O(\log(n))$-approximate $(L+2)$-hop spanning tree in G' w.r.t. cost $\hat{c}$ and root r.
- In relieve(v'): When creating tree T_i, insert edge $v'w_i$ instead of edge v_iw_i.

Theorem 5. *Algorithm 3 is a polynomial time $(1+\varepsilon, O(\log(n))$-approximation for instances of UFL-CLT where $\ell \equiv 1$, G is complete, and c is a metric.*

Proof. Statement 1 of Lemma 3 is still valid and guarantees that the solution respects link capacities. Using the triangle inequality, the modified relieve procedure exceeds the cost bound given in Lemma 3 at most by the cost of the initial tree. In conjunction with Lemma 7, the total costs are thus bounded by $O(\log(n)OPT)$.

Let D be the depth of the tree T computed in Step 2 (recall that $D \leq L+1$ if $L \leq \lceil 1/\varepsilon \rceil$ and $D \leq L+2$ otherwise). We now argue that for each client v, the client-facility-path P_v has length at most $D-1$. If $\phi(v) = T[w]$ for a tree that arose as a subtree of T rooted at facility w, then $P_v = T[v,r] \setminus \{wr\}$, so the assertion is true. Otherwise, v was reassigned to a newly created tree T_i with facility root w_i when relieving the subtree at client v'. In this case $|P_v| = |T_i[v, w_i]| \leq |T[v,v']| + 1 = |T[v,r]| - |T[v',r]| + 1$. As v' is not a facility, $|T[v',r]| \geq 2$ and the assertion is correct again. Thus, if $L \leq \lceil 1/\varepsilon \rceil$, then $|P_v| \leq L$. If $L > \lceil 1/\varepsilon \rceil$, then $|P_v| \leq L+1 \leq (1+\varepsilon)L$. □

6   Conclusion

In this article, we studied a framework for approximating the UFL-CLT problem achieving individual approximation guarantees for maximum connection length and cost. By choosing suitable shallow-light tree and facility location algorithms as subroutines and carefully adapting the algorithm, improved approximation guarantees can be achieved for two special cases of particular importance, demonstrating the power and flexibility of the combined approach.

In the practical application motivating this work, precise client demands are unknown during the early planning phase but can only be estimated roughly. However, fixing location decisions with sufficient lead time can reduce installment costs considerably. Designing approximation algorithms for a generalization of UFL-CLT that incorporates this uncertainty in a 2-stage optimization problem is an interesting subject of future research. Further, the suitability of the algorithmic approach in practical applications (providing initial solutions for improvement heuristics) should be investigated in a computational study.

Acknowledgements. This work was supported by Deutsche Forschungsgemeinschaft (DFG) as part of the Priority Program "Algorithm Engineering" (1307) and the DFG Research Center MATHEON "Mathematics for key technologies".

References

1. Althaus, E., Funke, S., Har-Peled, S., Könemann, J., Ramos, E.A., Skutella, M.: Approximating k-hop minimum-spanning trees. Operations Research Letters 33(2), 115–120 (2005)
2. Byrka, J., Aardal, K.: An optimal bifactor approximation algorithm for the metric uncapacitated facility location problem. SIAM Journal on Computing 39(6), 2212–2231 (2010)
3. Byrka, J., Grandoni, F., Rothvoß, T., Sanità, L.: An improved LP-based approximation for Steiner tree. In: Proceedings of the 42nd ACM Symposium on Theory of Computing, pp. 583–592 (2010)
4. Charikar, M., Chekuri, C., Cheung, T.-Y., Dai, Z., Goel, A., Guha, S., Li, M.: Approximation algorithms for directed Steiner problems. Journal of Algorithms 33(1), 73–91 (1999)
5. Harks, T., König, F.G., Matuschke, J.: Approximation algorithms for capacitated location routing. Transportation Science 47(1), 3–22 (2013)
6. Hassin, R.: Approximation schemes for the restricted shortest path problem. Mathematics of Operations Research 17(1), 36–42 (1992)
7. Hochbaum, D.S.: Heuristics for the fixed cost median problem. Mathematical Programming 22(1), 148–162 (1982)
8. Kapoor, S., Sarwat, M.: Bounded-diameter minimum-cost graph problems. Theory of Computing Systems 41(4), 779–794 (2007)
9. Khuller, S., Raghavachari, B., Young, N.: Balancing minimum spanning trees and shortest-path trees. Algorithmica 14(4), 305–321 (1995)
10. Kortsarz, G., Peleg, D.: Approximating the weight of shallow Steiner trees. Discrete Applied Mathematics 93(2), 265–285 (1999)
11. Marathe, M.V., Ravi, R., Sundaram, R., Ravi, S.S., Rosenkrantz, D.J., Hunt, H.B.: Bicriteria network design problems. Journal of Algorithms 28(1), 142–171 (1998)
12. Maßberg, J., Vygen, J.: Approximation algorithms for a facility location problem with service capacities. ACM Transactions on Algorithms (TALG) 4(4), 50 (2008)
13. Ravi, R., Sinha, A.: Approximation algorithms for problems combining facility location and network design. Operations Research 54(1), 73–81 (2006)

The Recognition of Simple-Triangle Graphs and of Linear-Interval Orders Is Polynomial*

George B. Mertzios

School of Engineering and Computing Sciences, Durham University, UK
george.mertzios@durham.ac.uk

Abstract. Intersection graphs of geometric objects have been extensively studied, both due to their interesting structure and their numerous applications; prominent examples include interval graphs and permutation graphs. In this paper we study a natural graph class that generalizes both interval and permutation graphs, namely *simple-triangle* graphs. Simple-triangle graphs – also known as *PI* graphs (for Point-Interval) – are the intersection graphs of triangles that are defined by a point on a line L_1 and an interval on a parallel line L_2. They lie naturally between permutation and trapezoid graphs, which are the intersection graphs of line segments between L_1 and L_2 and of trapezoids between L_1 and L_2, respectively. Although various efficient recognition algorithms for permutation and trapezoid graphs are well known to exist, the recognition of simple-triangle graphs has remained an open problem since their introduction by Corneil and Kamula three decades ago. In this paper we resolve this problem by proving that simple-triangle graphs can be recognized in polynomial time. As a consequence, our algorithm also solves a longstanding open problem in the area of partial orders, namely the recognition of *linear-interval orders*, i.e. of partial orders $P = P_1 \cap P_2$, where P_1 is a linear order and P_2 is an interval order. This is one of the first results on recognizing partial orders P that are the intersection of orders from two different classes $\mathcal{P}_1$ and $\mathcal{P}_2$. In contrast, partial orders P which are the intersection of orders from the same class $\mathcal{P}$ have been extensively investigated, and in most cases the complexity status of these recognition problems has been established.

Keywords: Intersection graphs, PI graphs, recognition problem, partial orders, polynomial algorithm.

1 Introduction

A graph G is the *intersection graph* of a family $\mathcal{F}$ of sets if we can bijectively assign sets of $\mathcal{F}$ to vertices of G such that two vertices of G are adjacent if and only if the corresponding sets have a non-empty intersection. It turns out that many graph classes with important applications can be described as intersection

* This work was partially supported by (i) the EPSRC Grant EP/K022660/1 and (ii) the EPSRC Grant EP/G043434/1.

H.L. Bodlaender and G.F. Italiano (Eds.): ESA 2013, LNCS 8125, pp. 719–730, 2013.
© Springer-Verlag Berlin Heidelberg 2013

graphs of set families that are derived from some kind of geometric configuration. One of the most prominent examples is that of *interval* graphs, i.e. the intersection graphs of intervals on the real line, which have natural applications in several fields, including bioinformatics and involving the physical mapping of DNA and the genome reconstruction[1] [3,6,7].

Generalizing the intersections on the real line, consider two parallel horizontal lines on the plane, L_1 (the upper line) and L_2 (the lower line). A graph G is a *simple-triangle* graph if it is the intersection graph of triangles that have one endpoint on L_1 and the other two on L_2. Furthermore, G is a *triangle* graph if it is the intersection graph of triangles with endpoints on L_1 and L_2, but now there is no restriction on which line contains one endpoint of every triangle and which contains the other two. Simple-triangle and triangle graphs are also known as *PI* and *PI** graphs, respectively [2,4,14], where PI stands for "Point-Interval". Such representations of simple-triangle and of triangle graphs are called *simple-triangle* (or *PI*) and *triangle* (or *PI**) *representations*, respectively. Simple-triangle and triangle graphs lie naturally between *permutation* graphs (i.e. the intersection graphs of line segments with one endpoint on L_1 and one on L_2) and *trapezoid* graphs (i.e. the intersection graphs of trapezoids with one interval on L_1 and the opposite interval on L_2) [2,14]. Note that, using the notation *PI* for simple-triangle graphs, permutation graphs are *PP* (for "Point-Point") graphs, while trapezoid graphs are *II* (for "Interval-Interval") graphs [4].

A *partial order* is a pair $P = (U, R)$, where U is a finite set and R is an irreflexive transitive binary relation on U. Whenever $(x, y) \in R$ for two elements $x, y \in U$, we write $x <_P y$. If $x <_P y$ or $y <_P x$, then x and y are *comparable*, otherwise they are *incomparable*. P is a *linear order* if every pair of elements in U are comparable. Furthermore, P is an *interval order* if each element $x \in U$ is assigned to an interval I_x on the real line such that $x <_P y$ if and only if I_x lies completely to the left of I_y. One of the most fundamental notions on partial orders is *dimension*. For any partial order P and any class $\mathcal{P}$ of partial orders (e.g. linear order, interval order, semiorder, etc.), the $\mathcal{P}$-*dimension* of P is the smallest k such that P is the intersection of k orders from $\mathcal{P}$. In particular, when $\mathcal{P}$ is the class of linear orders, the $\mathcal{P}$-dimension of P is known as the *dimension* of P. Although in most cases we can efficiently recognize whether a partial order belongs to a class $\mathcal{P}$, this is not the case for higher dimensions. Due to a classical result of Yannakakis [15], it is NP-complete to decide whether the dimension, or the interval dimension, of a partial order is at most k, where $k \geq 3$.

There is a natural correspondence between graphs and partial orders. For a partial order $P = (U, R)$, the *comparability* (resp. *incomparability*) *graph* $G(P)$ of P has elements of U as vertices and an edge between every pair of comparable (resp. incomparable) elements. A graph G is a *(co)comparability graph* if G is the (in)comparability graph of a partial order P. There has been a long

---

[1] Benzer [1] earned the prestigious Lasker Award (1971) and Crafoord Prize (1993) partly for showing that the set of intersections of a large number of fragments of genetic material in a virus form an interval graph.

line of research in order to establish the complexity of recognizing partial orders of $\mathcal{P}$-dimension at most 2 (e.g. where $\mathcal{P}$ is linear orders [14] or interval orders [9]). In particular, since permutation (resp. trapezoid) graphs are the incomparability graphs of partial orders with dimension (resp. interval dimension) at most 2 [5,14], permutation and trapezoid graphs can be recognized efficiently by the corresponding partial order algorithms [9,14].

In contrast, not much is known so far for the recognition of partial orders P that are the intersection of orders from different classes $\mathcal{P}_1$ and $\mathcal{P}_2$. One of the longstanding open problems in this area is the recognition of *linear-interval orders* P, i.e. of partial orders $P = P_1 \cap P_2$, where P_1 is a linear order and P_2 is an interval order. In terms of graphs, this problem is equivalent to the recognition of simple-triangle (i.e. PI) graphs, since PI graphs are the incomparability graphs of linear-interval orders; this problem is well known and remains open since the introduction of PI graphs in 1987 [4] (cf. for instance the books [2,14]).

Our Contribution. In this article we establish the complexity of recognizing simple-triangle (PI) graphs, and therefore also the complexity of recognizing linear-interval orders. Given a graph G with n vertices, such that its complement $\overline{G}$ has m edges, we provide an algorithm with running time $O(n^2m)$ that either computes a PI representation of G, or it announces that G is not a PI graph. Equivalently, given a partial order $P = (U, R)$ with $|U| = n$ and $|R| = m$, our algorithm either computes in $O(n^2m)$ time a linear order P_1 and an interval order P_2 such that $P = P_1 \cap P_2$, or it announces that such orders P_1, P_2 do not exist. Surprisingly, it turns out that the seemingly small difference in the definition of simple-triangle (PI) graphs and triangle (PI*) graphs results in a very different behavior of their recognition problems; only recently it has been proved that the recognition of triangle graphs is NP-complete [11]. In addition, our polynomial time algorithm is in contrast to the recognition problems for the related classes of *bounded tolerance* (i.e. *parallelogram*) graphs [12] and of *max-tolerance* graphs [8], which have already been proved to be NP-complete.

As the main tool for our algorithm we introduce the notion of a *linear-interval cover* of bipartite graphs. As a second tool we identify a new tractable subclass of 3SAT, called *gradually mixed* formulas, for which we provide a linear time algorithm. The class of gradually mixed formulas is *hybrid*, i.e. it is characterized by both *relational* and *structural* restrictions on the clauses. Then, using the notion of a linear-interval cover, we are able to reduce our problem to the satisfiability problem of gradually mixed formulas.

Our algorithm proceeds as follows. First, it computes from the given graph G a bipartite graph $\widetilde{G}$, such that G is a PI graph if and only if $\widetilde{G}$ has a linear-interval cover. Second, it computes a gradually mixed Boolean formula ϕ such that ϕ is satisfiable if and only if $\widetilde{G}$ has a linear-interval cover. This formula ϕ can be written as $\phi = \phi_1 \wedge \phi_2$, where every clause of ϕ_1 has 3 literals and every clause of ϕ_2 has 2 literals. The construction of ϕ_1 and ϕ_2 is based on the fact that a necessary condition for $\widetilde{G}$ to admit a linear-interval cover is that its edges can be colored with two different colors (according to some restrictions). Then the edges of $\widetilde{G}$ correspond to literals of ϕ, while the two edge colors encode the truth value of the corresponding variables. Furthermore every clause of ϕ_1 corresponds to the edges of an *alternating cycle* in $\widetilde{G}$ (i.e. a closed walk that alternately visits edges and non-edges) of length 6, while the clauses of ϕ_2 correspond to specific pairs of edges of $\widetilde{G}$ that are not allowed to receive the same color. Finally, the equivalence between the existence of a linear-interval cover of $\widetilde{G}$ and a satisfying truth assignment for ϕ allows us to use our linear algorithm to solve satisfiability on gradually mixed formulas in order to complete our recognition algorithm.

Organization of the Paper. We present in Section 2 the class of gradually mixed formulas and a linear time algorithm to solve satisfiability on this class. In Section 3 we provide the necessary notation and preliminaries on alternating cycles. Then in Section 4 we introduce the notion of a linear-interval cover of bipartite graphs to characterize PI graphs, and in Section 5 we translate the linear-interval cover problem to the satisfiability problem on a gradually mixed formula. Finally, in Section 6 we present our PI graph recognition algorithm.

2   A Tractable Subclass of 3SAT

In this section we introduce the class of *gradually mixed* formulas and we provide a linear time algorithm for solving satisfiability on this class. Any gradually mixed formula ϕ is a mix of binary and ternary clauses. That is, there exist a 3-CNF formula ϕ_1 (i.e. a formula in conjunctive normal form with at most 3 literals per clause) and a 2-CNF formula ϕ_2 (i.e. with at most 2 literals per clause) such that $\phi = \phi_1 \wedge \phi_2$, while ϕ satisfies some constraints among its clauses. Before we define gradually mixed formulas (cf. Definition 2), we first define *dual* clauses.

Definition 1. *Let ϕ_1 be a 3-CNF formula. If $\alpha = (\ell_1 \vee \ell_2 \vee \ell_3)$ is a clause of ϕ_1, then the $\overline{\alpha} = (\overline{\ell_1} \vee \overline{\ell_2} \vee \overline{\ell_3})$ is the dual clause of α.*

Note by Definition 1 that, whenever α is a clause of a formula ϕ_1, the dual clause $\overline{\alpha}$ of α may belong, or may not belong, to ϕ_1.

Definition 2. *Let ϕ_1 and ϕ_2 be CNF formulas with 3 literals and 2 literals in each clause, respectively. The mixed formula $\phi = \phi_1 \wedge \phi_2$ is gradually mixed if the next two conditions are satisfied:*

1. *Let α and β be two clauses of ϕ_1. Then α does not share exactly one literal with either the clause β or the clause $\overline{\beta}$.*

2. *Let $\alpha = (\ell_1 \vee \ell_2 \vee \ell_3)$ be a clause of ϕ_1. Then:*
   - *if $(\ell_0 \vee \ell_1)$ is a clause of ϕ_2, then ϕ_2 contains also (at least) one of the clauses $\{(\ell_0 \vee \overline{\ell_2}), (\ell_0 \vee \overline{\ell_3})\}$,*
   - *if $(\ell_0 \vee \overline{\ell_1})$ is a clause of ϕ_2, then ϕ_2 contains also (at least) one of the clauses $\{(\ell_0 \vee \ell_2), (\ell_0 \vee \ell_3)\}$.*

As an example of a gradually mixed formula, consider the formula $\phi = \phi_1 \wedge \phi_2$, where $\phi_1 = (x_1 \vee \overline{x_2} \vee x_3) \wedge (\overline{x_1} \vee x_2 \vee x_4) \wedge (x_5 \vee x_6 \vee \overline{x_7})$ and $\phi_2 = (x_8 \vee \overline{x_3}) \wedge (x_8 \vee x_1) \wedge (x_8 \vee x_4) \wedge (\overline{x_8} \vee x_9) \wedge (x_5 \vee x_{10}) \wedge (\overline{x_6} \vee x_{10})$.

Note by Definition 2 that the class of gradually mixed formulas contains 2SAT as a proper subclass, since every 2-CNF formula ϕ_2 can be written as a gradually mixed formula $\phi = \phi_1 \wedge \phi_2$ where $\phi_1 = \emptyset$. Furthermore the class of gradually mixed formulas ϕ is a *hybrid* class, since the conditions of Definition 2 concern simultaneously *relational* restrictions (i.e. where the clauses are restricted to be of certain types) and *structural* restrictions (i.e. where there are restrictions on how different clauses interact with each other). The intuition for the term *gradually mixed* in Definition 2 is that, whenever the sub-formulas ϕ_1 and ϕ_2 share more variables, the number of clauses of ϕ_2 that are imposed by condition 2 of Definition 2 increases. In the next theorem we use resolution to prove that satisfiability can be solved in linear time on gradually mixed formulas.

Theorem 1. *There exists a linear time algorithm which decides whether a given gradually mixed formula ϕ is satisfiable and computes a satisfying truth assignment of ϕ, if one exists.*

The conditions of Definition 2 which guarantee the tractability of gradually mixed formulas are *minimal*, in the sense that, if we remove any of these two conditions, the resulting subclass of 3SAT is NP-complete.

3 Preliminaries

Notation. In this article we consider finite, simple, and undirected graphs. An edge between two vertices u and v of a graph $G = (V, E)$ is denoted by uv, and in this case u and v are said to be *adjacent*. The *neighborhood* of a vertex $u \in V$ is the set $N(u) = \{v \in V \mid uv \in E\}$ of its adjacent vertices. The complement of G is denoted by $\overline{G}$, i.e. $\overline{G} = (V, \overline{E})$, where $uv \in \overline{E}$ if and only if $uv \notin E$. For any subset $E_0 \subseteq E$ of the edges of G, we denote for simplicity $G - E_0 = (V, E \setminus E_0)$. A subset $S \subseteq V$ of its vertices induces an *independent set* in G if $uv \notin E$ for every pair of vertices $u, v \in S$. Furthermore, S induces a *clique* in G if $uv \in E$ for every pair $u, v \in S$. A graph G is a *split graph* if its vertices can be partitioned into a clique K and an independent set I.

The smallest k for which there exists a proper k-coloring of G is the *chromatic number* of G, denoted by $\chi(G)$. If $\chi(G) = 2$ then G is a *bipartite* graph, i.e. its vertices are partitioned into two independent sets, the *color classes*. A bipartite graph G is denoted by $G = (U, V, E)$, where U and V are its color classes and E is the set of edges between them. For a bipartite graph $G = (U, V, E)$, its *bipartite complement* is the graph $\widehat{G} = (U, V, \widehat{E})$, where for two vertices $u \in U$ and $v \in V$, $uv \in \widehat{E}$ if and only if $uv \notin E$. A bipartite graph $G = (U, V, E)$ is a *chain graph* if the vertices of each color class can be ordered by inclusion of their neighborhoods, i.e. $N(u) \subseteq N(v)$ or $N(v) \subseteq N(u)$ for any two vertices u, v in the same color class. Note that chain graphs are closed under bipartite complementation, i.e. G is a chain graph if and only if $\widehat{G}$ is a chain graph.

For two graphs $G_1 = (V, E_1)$ and $G_2 = (V, E_2)$, we denote $G_1 \subseteq G_2$ whenever $E_1 \subseteq E_2$. Moreover, we denote for simplicity by $G_1 \cup G_2$ and $G_1 \cap G_2$ the graphs $(V, E_1 \cup E_2)$ and $(V, E_1 \cap E_2)$, respectively. Similarly, for any two partial orders $P_1 = (U, R_1)$ and $P_2 = (U, R_2)$, we denote $P_1 \subseteq P_2$ whenever $R_1 \subseteq R_2$. Moreover, we denote for simplicity $P_1 \cup P_2$ and $P_1 \cap P_2$ for the partial orders $(U, R_1 \cup R_2)$ and $(U, R_1 \cap R_2)$, respectively.

Alternating Cycles in a Graph. The next definition of an *alternating cycle* is crucial for our recognition algorithm for PI graphs.

Definition 3. *Let* $G = (V, E)$ *be a graph,* $\widetilde{E} \subseteq E$ *be an edge subset, and* $k \geq 2$. *A set of* $2k$ *(not necessarily distinct) vertices* $v_1, v_2, \ldots, v_{2k} \in V$ *builds an alternating cycle* AC_{2k} *in* $\widetilde{E}$, *if* $v_i v_{i+1} \in \widetilde{E}$ *whenever* i *is even and* $v_i v_{i+1} \notin E$ *whenever* i *is odd (where indices are* mod $2k$*). Furthermore, we say that* G *has an alternating cycle* AC_{2k}, *whenever* G *has an* AC_{2k} *in the edge set* $\widetilde{E} = E$.

For instance, for $k = 3$, there exist two different possibilities for an AC_6, which are illustrated in Figures 1(a) and 1(b). These two types of an AC_6 are called an *alternating path of length 5* or *of length 6*, respectively (AP_5 and AP_6 for short, respectively). Furthermore, note that for $k = 2$, a set of four vertices $v_1, v_2, v_3, v_4 \in V$ builds an alternating cycle AC_4 if $v_1 v_2, v_3 v_4 \notin E$ and $v_2 v_3, v_1 v_4 \in E$. There are three possible graphs on four vertices that build an alternating cycle, AC_4 which are illustrated in Figures 1(c)-1(e).

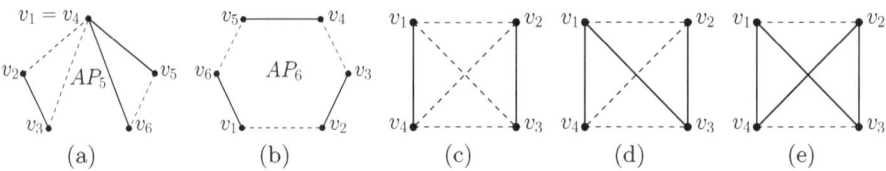

(a)   (b)   (c)   (d)   (e)

Fig. 1. The two possibilities for an AC_6: (a) an alternating path AP_5 of length 5 and (b) an alternating path AP_6 of length 6. Furthermore, the three possibilities for an AC_4: (c) a $2K_2$, (d) a P_4, and (e) a C_4. The solid lines denote edges of the graph and the dashed lines denote non-edges of the graph.

Alternating cycles can be used to characterize chain graphs as the bipartite graphs with no induced $2K_2$ [10]. We define now for any bipartite graph G the associated split graph of G, which we use extensively throughout of the paper.

Definition 4. *Let* $G = (U, V, E)$ *be a bipartite graph. The associated split graph of* G *is the split graph* $H_G = (U \cup V, E')$, *where* $E' = E \cup (V \times V)$, *i.e.* H_G *is the split graph made by* G *by replacing the independent set* V *of* G *by a clique.*

The next two definitions of a *conflict* between two edges and the *conflict graph* are essential for our results.

Definition 5. *Let* $G = (V, E)$ *be a graph and* $e_1, e_2 \in E$. *If the vertices of* e_1 *and* e_2 *build an* AC_4 *in* G, *then* e_1 *and* e_2 *are in* conflict, *and in this case we denote* $e_1 \| e_2$ *in* G. *Furthermore, an edge* $e \in E$ *is* committed *if there exists an edge* $e' \in E$ *such that* $e \| e'$; *otherwise* e *is* uncommitted.

Definition 6 ([13]). *Let $G = (V, E)$ be a graph. The conflict graph $G^* = (V^*, E^*)$ of G is defined by*

- *$V^* = E$ and*
- *for every $e_1, e_2 \in E$, $e_1 e_2 \in E^*$ if and only if $e_1 \| e_2$ in G.*

4   Linear-Interval Covers of Bipartite Graphs

In this section we introduce the crucial notion of a *linear-interval cover* of bipartite graphs (cf. Definition 9). Then we use linear-interval covers to provide a new characterization of PI graphs (cf. Theorem 3), which is one of the main tools for our PI graph recognition algorithm. First we provide in the next theorem the characterization of PI graphs using linear orders and interval orders.

Theorem 2. *Let $G = (V, E)$ be a cocomparability graph and P be a partial order of $\overline{G}$. Then G is a PI graph if and only if $P = P_1 \cap P_2$, where P_1 is a linear order and P_2 is an interval order.*

For every partial order P we define now the domination bipartite graph $C(P)$, which has been used to characterize interval orders [9]. Here "C" stands for "Comparable", since the definition of $C(P)$ uses the comparable elements of P.

Definition 7 ([9]). *Let $P = (U, R)$ be a partial order, where $U = \{u_1, u_2, \ldots, u_n\}$. Furthermore let $V = \{v_1, v_2, \ldots, v_n\}$. The domination bipartite graph $C(P) = (U, V, E)$ is defined such that $u_i v_j \in E$ if and only if $u_i <_P u_j$.*

Lemma 1 ([9]). *Let $P = (U, R)$ be a partial order. Then, P is an interval order if and only if $C(P)$ is a chain graph.*

Extending the notion of $C(P)$, we now introduce the bipartite graph $NC(P)$ to characterize linear orders (cf. Lemma 2). Here "NC" stands for "Non-strictly Comparable". Namely, this graph can be obtained by adding to the graph $C(P)$ the perfect matching $\{u_i v_i \mid i = 1, 2, \ldots, n\}$ on the vertices of U and V.

Definition 8. *Let $P = (U, R)$ be a partial order, where $U = \{u_1, u_2, \ldots, u_n\}$. Furthermore let $V = \{v_1, v_2, \ldots, v_n\}$. Then, $NC(P) = (U, V, E)$ is the bipartite graph, such that $u_i v_j \in E$ if and only if $u_i \leq_P u_j$.*

Lemma 2. *Let $P = (U, R)$ be a partial order. Then, P is a linear order if and only if $NC(P)$ is a chain graph.*

Now we introduce the notion of a *linear-interval cover* of a bipartite graph. This notion is crucial for our main result of this section, cf. Theorem 3.

Definition 9. *Let $G = (U, V, E)$ be a bipartite graph, where $U = \{u_1, u_2, \ldots, u_n\}$ and $V = \{v_1, v_2, \ldots, v_n\}$. Let $E_0 = \{u_i v_i \mid 1 \leq i \leq n\}$ and suppose that $E_0 \subseteq E$. Then, G is linear-interval coverable if there exist two chain graphs $G_1 = (U, V, E_1)$ and $G_2 = (U, V, E_2)$, such that $G = G_1 \cup G_2$ and $E_0 \subseteq E_2 \setminus E_1$. In this case, the sets $\{E_1, E_2\}$ are a linear-interval cover of G.*

Theorem 3. *Let* $P = (U, R)$ *be a partial order. In the bipartite complement* $\widehat{C}(P)$ *of the graph* $C(P)$*, denote* $E_0 = \{u_i v_i \mid 1 \leq i \leq n\}$*. The following three statements are equivalent:*

(a) $P = P_1 \cap P_2$*, where* P_1 *is a linear order and* P_2 *is an interval order.*
(b) $\widehat{C}(P) = \widehat{NC}(P_1) \cup \widehat{C}(P_2)$ *for two partial orders* P_1 *and* P_2 *on* V*, where* $\widehat{NC}(P_1)$ *and* $\widehat{C}(P_2)$ *are chain graphs.*
(c) $\widehat{C}(P)$ *is linear-interval coverable, i.e.* $\widehat{C}(P) = G_1 \cup G_2$ *for two chain graphs* $G_1 = (U, V, E_1)$ *and* $G_2 = (U, V, E_2)$*, where* $E_0 \subseteq E_2 \setminus E_1$*.*

Furthermore, a linear-interval cover of the bipartite graph $\widehat{C}(P)$ does not only guarantee that the input graph G is a PI graph, but it can be also used to efficiently compute a PI representation of G, as the next theorem states.

Theorem 4. *Let* G *be a cocomparability graph with* n *vertices and* P *be the partial order of* $\overline{G}$*. Let* $\{E_1, E_2\}$ *be a linear-interval cover of* $\widehat{C}(P)$*. Then we can construct in* $O(n^2)$ *time a PI representation* R *of* G*.*

5   Detecting Linear-Interval Covers Using Boolean Satisfiability

The natural algorithmic question that arizes from the characterization of PI graphs using linear-interval covers in Theorems 2 and 3, is the following: "Given a cocomparability graph G and a partial order P of $\overline{G}$, can we efficiently decide whether the bipartite graph $\widehat{C}(P)$ has a linear-interval cover?" We will answer this algorithmic question in the affirmative in Section 6. In this section we translate *every* instance of this decision problem (i.e. whether the bipartite graph $\widehat{C}(P)$ has a linear-interval cover) to a restricted instance of 3SAT (cf. Theorem 5). That is, for every such a bipartite graph $\widehat{C}(P)$, we construct a Boolean formula ϕ in conjunctive normal form (CNF), with size polynomial on the size of $\widehat{C}(P)$ (and thus also on G), such that $\widehat{C}(P)$ has a linear-interval cover if and only if ϕ is satisfiable. In particular, this formula ϕ can be written as $\phi = \phi_1 \wedge \phi_2$, where ϕ_1 has three literals in every clause and ϕ_2 has two literals in every clause. Moreover, as we will prove in Section 6, the satisfiability problem can be efficiently decided on the formula ϕ, by exploiting an appropriate sub-formula of ϕ which is gradually mixed (cf. Definition 2).

In the remainder of the paper, given a cocomparability graph G and a partial ordering P of its complement $\overline{G}$, we denote by $\widetilde{G} = \widehat{C}(P)$ the bipartite complement of the domination bipartite graph $C(P)$ of P. Furthermore we denote by H the associated split graph of $\widetilde{G}$ and by H^* the conflict graph of H. Moreover, we assume in the remainder of the paper without loss of generality that $\chi(H^*) \leq 2$, i.e. that H^* is bipartite. Indeed, as we can prove, if $\chi(H^*) > 2$ then $\widetilde{G}$ does not have a linear-interval cover, i.e. G is not a PI graph. Note that every proper 2-coloring of the vertices of the conflict graph H^* corresponds to exactly one 2-coloring of the edges of H that includes no monochromatic AC_4. We assume in the following that a proper 2-coloring (with colors blue and red) of the vertices of H^* is given as input; note that χ_0 can be computed in polynomial time.

Let $C_1, C_2, \ldots, C_k$ be the connected components of H^*. Some of these components of H^* may be isolated vertices, which correspond to uncommitted edges in H. We assign to every component C_i, where $1 \leq i \leq k$, the Boolean variable x_i. Since H^* is bipartite by assumption, the vertices of each connected component C_i of H^* can be partitioned into two color classes $S_{i,1}$ and $S_{i,2}$. Without loss of generality, we assume that $S_{i,1}$ (resp. $S_{i,2}$) contains the vertices of C_i that are colored red (resp. blue) in χ_0. Note that, since vertices of H^* correspond to edges of H (cf. Definition 6), for every two edges e and e' of H that are in conflict (i.e. $e||e'$) there exists an index $i \in \{1, 2, \ldots, k\}$ such that one of these edges belongs to $S_{i,1}$ and the other belongs to $S_{i,2}$. We now assign a *literal* ℓ_e to every edge e of H as follows: if $e \in S_{i,1}$ for some $i \in \{1, 2, \ldots, k\}$, then $\ell_e = x_i$; otherwise, if $e \in S_{i,2}$, then $\ell_e = \overline{x_i}$. Note that, by construction, whenever two edges are in conflict in H, their assigned literals are one the negation of the other.

Observation 1. *Every truth assignment τ of the variables $x_1, x_2, \ldots, x_k$ corresponds bijectively to a proper 2-coloring χ_τ (with colors blue and red) of the vertices of H^*, as follows: $x_i = 0$ in τ (resp. $x_i = 1$ in τ), if and only if all vertices of the component C_i have in χ_τ the same color as in χ_0 (resp. opposite color than in χ_0). In particular, $\tau = (0, 0, \ldots, 0)$ corresponds to the coloring χ_0.*

Description of the 3-CNF Formula ϕ_1: Consider an AC_6 in the split graph H, and let e, e', e'' be its three edges in H, such that no two literals among $\{\ell_e, \ell_{e'}, \ell_{e''}\}$ are one the negation of the other. Then the Boolean formula ϕ_1 has for this triple $\{e, e', e''\}$ of edges exactly the two clauses $\alpha = (\ell_e \vee \ell_{e'} \vee \ell_{e''})$ and $\alpha' = (\overline{\ell_e} \vee \overline{\ell_{e'}} \vee \overline{\ell_{e''}})$. It is easy to check by the assignment of literals to edges that the clause α (resp. the clause α') of ϕ_1 is false in a truth assignment τ of the variables if and only if all edges $\{e, e', e''\}$ are colored red (resp. blue) in the 2-edge-coloring χ_τ of H (cf. Observation 1).

Consider now another AC_6 of H on the edges $\{e_1, e_2, e_3\}$, in which at least one literal among $\{\ell_{e_1}, \ell_{e_2}, \ell_{e_3}\}$ is the negation of another literal, for example $\ell_{e_1} = \overline{\ell_{e_2}}$. Then, for *any* proper 2-coloring of the vertices of H^*, the edges e and e' of H receive different colors, and thus this AC_6 is not monochromatic.

Observation 2. *The formula ϕ_1 is satisfied by a truth assignment τ if and only if the corresponding 2-coloring χ_τ of the edges of H does not contain any monochromatic AC_6.*

Description of the 2-CNF Formula ϕ_2: Denote for simplicity $H = (U, V, E_H)$, where $U = \{u_1, u_2, \ldots, u_n\}$ and $V = \{v_1, v_2, \ldots, v_n\}$. Furthermore denote $E_0 = \{u_i v_i \mid 1 \leq i \leq n\}$. Let $E' = E_H \setminus E_0$ and $H' = H - E_0$, i.e. H' is the split graph that we obtain if we remove from H all edges of E_0. Consider now a pair of edges $e = u_i v_t$ and $e' = u_t v_j$ of E', such that $u_i v_j \notin E'$. Note that i and j may be equal. However, since $E' \cap E_0 = \emptyset$, it follows that $i \neq t$ and $t \neq j$.

Moreover, since the edge $u_t v_t$ belongs to E_H but not to E', it follows that the edges e and e' are in conflict in H' but not in H (for both cases where $i = j$ and $i \neq j$). That is, although e and e' are two non-adjacent vertices in the conflict graph H^* of H, they are adjacent vertices in the conflict graph of H'. For both cases where $i = j$ and $i \neq j$, an example of such a pair of edges $\{e, e'\}$ is illustrated in Figure 2. For every such pair $\{e, e'\}$ of edges in H, the Boolean formula ϕ_2 has the clause $(\ell_e \vee \ell_{e'})$. It is easy to check by the assignment of literals to edges of H that this clause $(\ell_e \vee \ell_{e'})$ of ϕ_2 is false in the truth assignment τ if and only if both e and e' are colored *red* in the 2-edge coloring χ_τ of H.

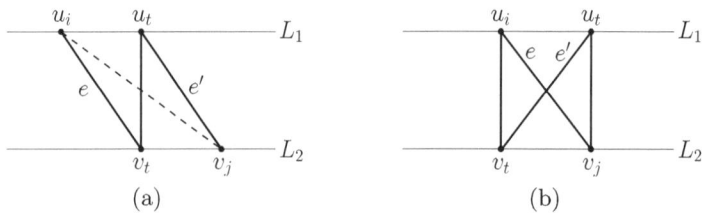

Fig. 2. Two edges $e = u_i v_t$ and $e' = u_t v_j$ of H, for which the formula ϕ_2 has the clause $(\ell_e \vee \ell_{e'})$, in the case where (a) $i \neq j$ and (b) $i = j$

Now we provide the main result of this section, which relates the existence of a linear-interval cover in $\widetilde{G} = \widehat{C}(P)$ with the satisfiability of the formula $\phi_1 \wedge \phi_2$.

Theorem 5. $\widetilde{G} = \widehat{C}(P)$ *is linear-interval colorable if and only if $\phi_1 \wedge \phi_2$ is satisfiable. Given a satisfying assignment τ of $\phi_1 \wedge \phi_2$, a linear-interval cover of $\widetilde{G}$ can be computed in $O(n^2)$ time.*

6   The Recognition of Linear-Interval Orders and PI Graphs

In this section we investigate the structure of the formula $\phi_1 \wedge \phi_2$ that we computed in Section 5. In particular, we first prove some fundamental structural properties of $\phi_1 \wedge \phi_2$, which allow us to find an appropriate sub-formula of $\phi_1 \wedge \phi_2$ which is gradually mixed (cf. Definition 2). Then we exploit this sub-formula of $\phi_1 \wedge \phi_2$ to provide an algorithm that solves the satisfiability problem on $\phi_1 \wedge \phi_2$ in time linear to its size, cf. Theorem 6. Finally, using this satisfiability algorithm, we combine our results of Sections 4 and 5 to recognize efficiently PI graphs and linear-interval orders.

The main structural properties of $\phi_1 \wedge \phi_2$ are proved in Lemmas 3 and 4. The proof of the next lemma is a based on the results of [13].

Lemma 3. *Let α and β be two clauses of ϕ_1. If α and β share at least one variable, then $\{\alpha, \overline{\alpha}\} = \{\beta, \overline{\beta}\}$.*

Algorithm 1. Recognition of PI graphs

Input: A graph $G = (V, E)$
Output: A PI representation R of G, or the announcement that G is not a PI graph

1: **if** G is a trapezoid graph **then**
2:     Compute a partial order P of the complement $\overline{G}$
3: **else return** "G is not a PI graph"

4: Compute the domination bipartite graph $C(P)$ from P
5: $\widetilde{G} \leftarrow \widehat{C}(P)$
6: Compute the associated split graph H of $\widetilde{G}$
7: Compute the conflict graph H^* of H

8: **if** H^* is bipartite **then**
9:     Compute a 2-coloring χ_0 of the vertices of H^*
10:     Compute the formulas ϕ_1 and ϕ_2

11:     **if** $\phi_1 \wedge \phi_2$ is satisfiable **then**
12:         Compute a satisfying truth assignment τ of $\phi_1 \wedge \phi_2$ by Theorem 6
13:         Compute from τ a linear-order cover of $\widetilde{G}$ by Theorem 5
14:         Compute a PI representation R of G by Theorem 4
15:     **else**
16:         **return** "G is not a PI graph"
17: **else**
18:     **return** "G is not a PI graph"

19: **return** R

Definition 10. *The clauses of ϕ_2 are partitioned into the sub-formulas ϕ_2', ϕ_2'', such that ϕ_2' contains all tautologies of ϕ_2 and all clauses of ϕ_2 in which at least one literal corresponds to an uncommitted edge, while ϕ_2'' contains all the remaining clauses of ϕ_2.*

Lemma 4. *Let $\{e_1, e_2, e_3\}$ be the three edges of an AC_6 in H, which has clauses in ϕ_1. Let e be an edge of H such that $(\ell_e \vee \ell_{e_1})$ is a clause in ϕ_2''. Then ϕ_2'' has also one of the clauses $\{(\ell_e \vee \overline{\ell_{e_2}}), (\ell_e \vee \overline{\ell_{e_3}})\}$.*

The next corollary, which follows easily by Definition 2 and by Lemmas 3 and 4, allows us to use the linear time algorithm for gradually mixed formulas (cf. Theorem 1) in order to solve the satisfiability problem on $\phi_1 \wedge \phi_2''$.

Corollary 1. $\phi_1 \wedge \phi_2''$ *is a gradually mixed formula.*

In the next theorem we use Corollary 1 to design an algorithm that decides satisfiability on $\phi_1 \wedge \phi_2$ in time linear to its size. This will enable us to combine the results of Sections 4 and 5 to recognize efficiently whether a given graph is a PI graph, or equivalently, due to Theorem 2, whether a given partial order P is the intersection of a linear order P_1 and an interval order P_2.

Theorem 6. $\phi_1 \wedge \phi_2$ *is satisfiable if and only if $\phi_1 \wedge \phi_2''$ is satisfiable. Given a satisfying truth assignment of $\phi_1 \wedge \phi_2''$ we can compute a satisfying truth assignment of $\phi_1 \wedge \phi_2$ in linear time.*

Now we are ready to present our recognition algorithm for PI graphs (Algorithm 1). Its correctness and timing analysis is established in Theorem 7. Due to characterization of PI graphs in Theorem 2 using partial orders, Theorem 8 follows also by Theorem 7.

Theorem 7. *Let $G = (V, E)$ be a graph and $\overline{G} = (V, \overline{E})$ be its complement, where $|V| = n$ and $|\overline{E}| = m$. Then Algorithm 1 constructs in $O(n^2 m)$ time a PI representation of G, or it announces that G is not a PI graph.*

Theorem 8. *Let $P = (U, R)$ be a partial order, where $|U| = n$ and $|R| = m$. Then we can decide in $O(n^2 m)$ time whether P is a linear-interval order, and in this case we can compute a linear order P_1 and an interval order P_2 such that $P = P_1 \cap P_2$.*

References

1. Benzer, S.: On the topology of the genetic fine structure. Proc. of the National Academy of Sciences (PNAS) 45, 1607–1620 (1959)
2. Brandstädt, A., Le, V.B., Spinrad, J.P.: Graph classes: a survey. In: Society for Industrial and Applied Mathematics (SIAM) (1999)
3. Carrano, A.V.: Establishing the order to human chromosome-specific DNA fragments. In: Biotechnology and the Human Genome, pp. 37–50. Plenum Press (1988)
4. Corneil, D.G., Kamula, P.A.: Extensions of permutation and interval graphs. In: Proceedings of the 18th Southeastern Conference on Combinatorics, Graph Theory and Computing, pp. 267–275 (1987)
5. Dagan, I., Golumbic, M.C., Pinter, R.Y.: Trapezoid graphs and their coloring. Discrete Applied Mathematics 21(1), 35–46 (1988)
6. Goldberg, P.W., Golumbic, M.C., Kaplan, H., Shamir, R.: Four strikes against physical mapping of DNA. Journal of Computational Biology 2(1), 139–152 (1995)
7. Golumbic, M.C.: Algorithmic graph theory and perfect graphs, 2nd edn. Annals of Discrete Mathematics, vol. 57. North-Holland Publishing Co. (2004)
8. Kaufmann, M., Kratochvil, J., Lehmann, K.A., Subramanian, A.R.: Max-tolerance graphs as intersection graphs: cliques, cycles, and recognition. In: Proceedings of the 17th Annual ACM-SIAM Symposium on Discrete Algorithms (SODA), pp. 832–841 (2006)
9. Ma, T.-H., Spinrad, J.P.: On the 2-chain subgraph cover and related problems. Journal of Algorithms 17, 251–268 (1994)
10. Mahadev, N., Peled, U.N.: Threshold Graphs and Related Topics. Annals of Discrete Mathematics, vol. 56. North-Holland Publishing Co. (1995)
11. Mertzios, G.B.: The recognition of triangle graphs. Theoretical Computer Science 438, 34–47 (2012)
12. Mertzios, G.B., Sau, I., Zaks, S.: The recognition of tolerance and bounded tolerance graphs. SIAM Journal on Computing 40(5), 1234–1257 (2011)
13. Raschle, T., Simon, K.: Recognition of graphs with threshold dimension two. In: Proceedings of the 27th ACM Symposium on Theory of Computing (STOC), pp. 650–661 (1995)
14. Spinrad, J.P.: Efficient graph representations. Fields Institute Monographs, vol. 19. American Mathematical Society (2003)
15. Yannakakis, M.: The complexity of the partial order dimension problem. SIAM Journal on Algebraic and Discrete Methods 3, 351–358 (1982)

Z-Skip-Links for Fast Traversal of ZDDs Representing Large-Scale Sparse Datasets

Shin-Ichi Minato

Graduate School of Information Science and Technology, Hokkaido University/
JST ERATO MINATO Discrete Structure Manipulation System Project
minato@ist.hokudai.ac.jp

Abstract. ZDD (Zero-suppressed Binary Decision Diagram) is known as an efficient data structure for representing and manipulating large-scale sets of combinations. In this article, we propose a method of using *Z-Skip-Links* to accelerate ZDD traversals for manipulating large-scale sparse datasets. We discuss average case complexity analysis of our method, and present the optimal parameter setting. Our method can be easily implemented into the existing ZDD packages just by adding one link per ZDD node. Experimental results show that we obtained dozens of acceleration ratio for the instances of the large-scale sparse datasets including thousands of items.

1 Introduction

A *set of combinations* is one of the most fundamental model of discrete structure for solving combinatorial problems in various applications. Binary Decision Diagram (BDD) [2], a state-of-the-art data structure of Boolean function representation, is sometimes used for solving combinatorial problems because n-input Boolean functions have one-to-one correspondence to the sets of combinations considering n items. Zero-suppressed BDD (ZDD) [5] is a variant of BDD, customized for manipulating sets of combinations. ZDDs have been successfully applied not only for VLSI design but also for various real-life applications, such as data mining, system diagnosis, and network analysis.

Recently, processing of "Big Data" have attracted a great deal of attention, and we often deal with a large-scale sparse dataset, which has more than thousand or ten thousands of items as the columns of a dataset. If we represent such data using a ZDD, the height of ZDD grows as large as the number of items, and the depth of recursive operations also becomes very large. Thus, ZDD-based manipulation is usually not very efficient for such large-scale sparse datasets.

In this paper, we propose an idea of attaching a "Z-Skip-Link" to each ZDD node for accelerating the traversal of ZDDs of large-scale sparse datasets. It consumes only a constant size of additional memory, and can easily be implemented into a conventional BDD/ZDD package. We also show the average-case computation time of traversing ZDDs, in order to evaluate the effect of Z-Skip-Links and their optimal setting of the skip length. In the practical case of sparse datasets with thousands of items, our experiments show that the use of Z-Skip-Links

H.L. Bodlaender and G.F. Italiano (Eds.): ESA 2013, LNCS 8125, pp. 731–742, 2013.

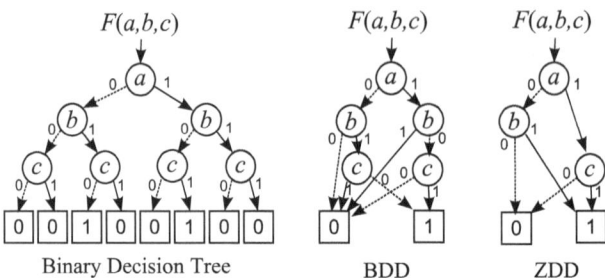

F(a,b,c)   F(a,b,c)   F(a,b,c)

Binary Decision Tree   BDD   ZDD

Fig. 1. Binary Decision Tree, BDD and ZDD

makes the membership operations 10 to 30 times faster than using conventional ZDD operations.

In the rest of this paper, we first explain the basic properties of ZDDs and what is the problem in manipulating large-scale sparse datasets. We then present an idea of Z-Skip-Links and average-case complexity analysis. Finally we describe the algorithm implementation and the experimental results.

2   Preliminary – Basic Properties of ZDDs

A Binary Decision Diagram (BDD) [2] is a graph representation for a Boolean function. As illustrated in Fig. 1, it is derived by reducing a binary decision tree, which represents a decision making process that depends on some input variables. In this graph, we may find the following two types of decision nodes:

(a) *Redundant node*: A decision node whose two child nodes are identical.
(b) *Equivalent nodes*: Two or more decision nodes having the same variable and the same pair of child nodes.

If we find such types of nodes, we can reduce the graph without changing the semantics (in other words, we can compress the graph). If we fix the order of input variables and apply the two reduction rules as much as possible, then we obtain a canonical form for a given Boolean function [1]. Such a data structure is called an Ordered BDD (OBDD), but in this article we will just call it a BDD.

The compression ratio of a BDD depends on the properties of Boolean function to be represented, but it can be 10 to 100 times more compact in some practical cases. In addition, we can systematically construct a BDD that is the result of a binary logic operation (i.e., AND or OR) for a given pair of BDDs. This algorithm is based on a recursive procedure with hash table techniques, and it is very efficient when the BDDs have a good compression ratio. The computation time is bounded by the product of the BDD sizes of the two operands, and in many practical cases, it is linearly bounded by the sum of input and output BDD sizes [13].

Zero-suppressed BDDs (ZDDs, or ZBDDs) [5] are a variant of BDDs, customized to manipulate sets of combinations. An example is shown in Fig.1. ZDDs are based on special reduction rules that differ from the ordinary ones. As shown

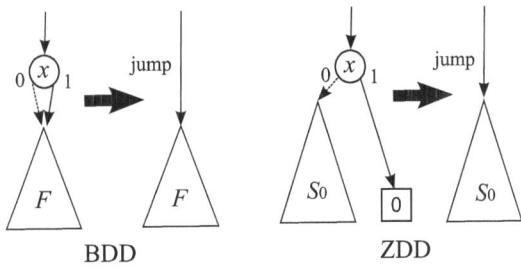

Fig. 2. ZDD reduction rule

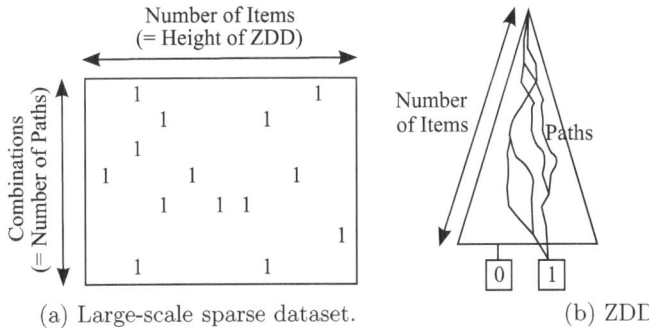

(a) Large-scale sparse dataset.   (b) ZDD

Fig. 3. Large-scale sparse dataset and ZDD

in Fig. 2, we delete all nodes whose 1-edge directly points to the 0-terminal node, but do not delete the nodes which would be deleted in an ordinary BDD. Similarly to ordinary BDDs, ZDDs give compact canonical representations for sets of combinations. We can construct ZDDs by applying algebraic set operations such as union, intersection and difference, which correspond to logic operations in BDDs.

The zero-suppressing reduction rule is extremely effective if we are handling a set of sparse combinations. If the average appearance ratio of each item is 1%, ZDDs are possibly up to 100 times more compact than ordinary BDDs. Such situations often appear in real-life problems, for example, in a supermarket, the number of items in a customer's basket is usually much less than the number of all the items displayed.

Recently, ZDD has become more widely known, since D. E. Knuth intensively discussed ZDD-based algorithms in the latest volume of his famous series of books [7]. The original BDD was invented and developed for VLSI logic design, but ZDD is now recognized as the most important variant of BDD, and is widely used in various kinds of problems in computer science [3,8,9,6].

3   Problem for Handling Large-Scale Sparse Datasets

Recently, some kinds of Big Data are regarded as a large-scale sparse dataset, which is a set of many combinations each of which selects a few items out of

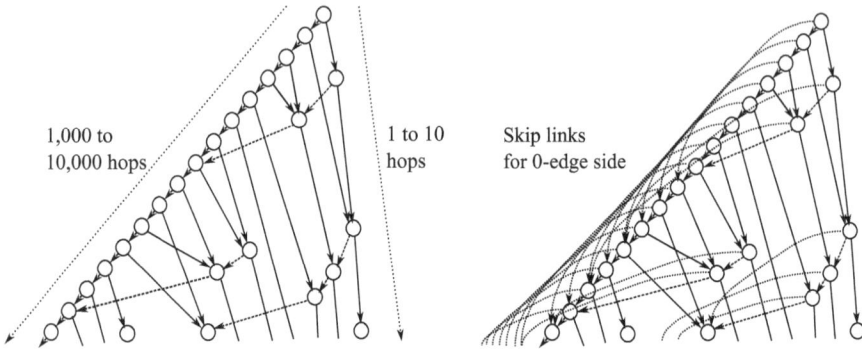

Fig. 4. Example of very unbalanced ZDD   **Fig. 5.** ZDD with Z-Skip-Links

thousands of ones, as illustrated in Fig. 3(a). There are so many practical examples, such as the basket data in a supermarket, the word-correlation dataset in a natural language, the key word lists in document database, grouping of Internet web pages, and SNPs (mutation points) of human gene data.

   If we represent such a large-scale sparse dataset using a ZDD, each path from the root node to the 1-terminal node corresponds to one combination in the dataset, as shown in Fig. 3(b). Namely, the number of such paths in the ZDD equals the number of combinations in the dataset, and the 1-edges on each path represent the occurrence of items in a combination. Since the zero-suppression rule automatically deletes the ZDD nodes corresponding to the items not occurring in the combination, the ZDD size is bounded by the total number of item occurrences in the datasets. If there are many partially similar patterns of combinations, the paths in the ZDD are shared with each other, and in such cases very large compression ratio is obtained. For example, *LCM-ZDD* method [11] efficiently generates nearly a billion patterns of frequent itemsets by using only thousands of ZDD nodes.

   Such data compression is one of the big advantage of using ZDDs, however, there is a hidden problem that the height of the ZDD must be n if the dataset contains n relevant items. If we represent a large-scale and sparse dataset using thousands of items, the ZDD becomes a very unbalanced form as shown in Fig. 4, such that we need more than thousand hops for 0-edge traversal while only a few hops needed for 1-edge side to reach a terminal node. Unfortunately, such type of datasets often appear in real-life applications. In general, ZDD-based operations require a computation time linear in the height of the ZDD. The membership testing operation is especially inefficient in this case because thousands of steps of ZDD traversal is needed to check the existence of the k-th item's decision node, even if the membership query has only a combination with a few items. It means that the ZDD-based operation could be hundreds or more times slower than a naive data structure based on arrays and linked lists.

   In this paper, we propose a practical method for addressing this problem for handling large-scale sparse datasets using ZDDs.

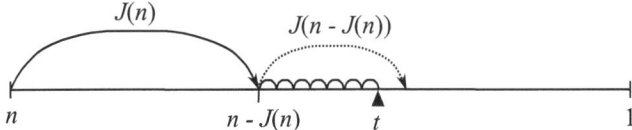

Fig. 6. Basic Model with a Number Line

4   Z-Skip-Links

Basic Idea

Conventional ZDD operation linearly traverses the cascade of 0-edges, and it requires $O(n)$ steps in average to check the existence of a ZDD node with a given item where n is the height of the very unbalanced ZDD. If we prepare a table of pointers to the descendant ZDD nodes indexed by the item IDs, we can directly jump to the destination node in $O(1)$ steps. However, we have to prepare such a pointer table for each ZDD nodes as the start point, so additional memory requirement becomes $O(n)$ words per ZDD nodes, and the total memory requirement becomes $O(nG)$, where G is ZDD size. This is an unacceptable memory increase for large n as hundreds or thousands.

Our basic idea is to attach only one pointer to each ZDD node, as illustrated in Fig. 5. The pointer, named *Z-Skip-Link*, indicates a descendant node reachable by a cascade of 0-edges of a given length. When we want to search a ZDD node with the t-th level, we first refer the Z-Skip-Link at the root node and check the level of the pointed node s. If s is still higher than t, we execute the jump by the Z-Skip-Link and continue the search from there. If s is exceedingly lower than t, we cancel the jump and descend the 0-edges for one step, and then continue the search at the next node.

Our method is quite simple, but the reachability is clearly guaranteed as ordinary linear search. The search time can be reduced as much as the total jump length of Z-Skip-Links. The additional memory requirement is just a constant factor for the ZDD size. Since the Z-Skip-Links do not have any side-effect to the ZDD operations, they can be easily implemented in a conventional BDD/ZDD package without any significant modification to the basic data structure and operation codes.

The key issue of this method is the design of the skip length of Z-Skip-Links. The longer skip length gets the more saving time, but a too long skip length may increase the probability of exceeding the target reducing the chance of speed up. In this article, we discuss the average case analysis of time complexity and the optimal skip length for a given n.

Basic Model and Average Search Cost

For analyzing average search cost, we consider the model of a number line as shown in Fig. 6. There are n positions on the line from 1 to n, where the start position is n, and the target position is t. We define a *skip length function* $J(x)$ to represent the skip length at the current position x $(1 \leq x \leq n)$. The first skip

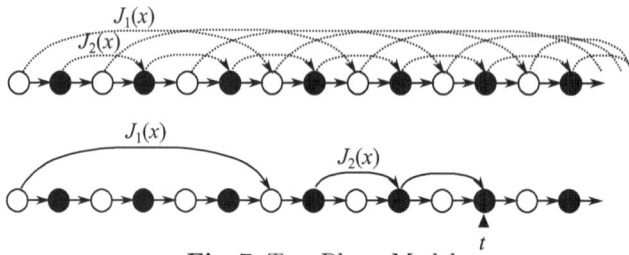

Fig. 7. Two-Phase Model

length should be $J(n)$ and the second length becomes $J(n - J(n))$. We repeat the jumps until passing through the target t, and after that we execute a linear search from the last position. In this model, we do not know t beforehand, thus we assume that t has a discrete uniform distribution between 1 and n. Here, we define $C_1(n)$ as the average number of jumps, and $C_2(n)$ as the average number of moves in the linear search. Our objective is to design a good skip length function $J(x)$ to minimize the total average cost $C(n) = C_1(n) + C_2(n)$. The function $J(x)$ cannot depend on t but may depend on n.

At first, we assume the simple skip length function with a constant $\alpha \geq 1$:

$$J(x) = \left\lceil \frac{1}{\alpha} x \right\rceil. \tag{1}$$

For example, $\alpha = 4$ means that we skip $1/4$ distance of the remaining search range. Then, we get the average search cost $C(n)$ as follows. (Detailed calculation is shown in Appendix of [10].)

$$C(n) = \alpha + \frac{n}{2(2\alpha - 1)} \tag{2}$$

If we set $\alpha = 16$, we get $C(10000) \approx 177.3$, which is 28.2 times faster than ordinary linear search.

Let us consider the optimal α for a given n. The derivative of $C(n)$ with respect to α can be written as:

$$\frac{d}{d\alpha} C(n) = 1 - \frac{n}{(2\alpha - 1)^2}$$

and it becomes zero, thus:

$$\alpha_{opt} = \frac{1}{2} \left(\sqrt{n} + 1 \right), \quad C_{opt}(n) = \sqrt{n} + \frac{1}{2} \tag{3}$$

are obtained. In other words, we can accelerate up to $\frac{\sqrt{n}}{2}$ times in average from linear search. When $n = 10000$, we get $\alpha_{opt} = 50.5$ and $C_{opt}(10000) = 100.5$, about 50 times faster than linear search.

Two-Phase Model

The above discussion assumed a uniform skip length function, but we may improve the performance if we use different forms of skip length functions for different positions. Figure 7 illustrates our idea of the two-phase model, using the two types of skip length functions $J_1(x)$ and $J_2(x)$ for an even number x and an odd number x, respectively. $J_1(x)$ provides long jumps and $J_2(x)$ provides middle length jumps. We first repeat long jumps with $J_1(x)$ until passing through the target t, next we repeat middle jumps with $J_2(x)$, and after that we execute a linear search. We set that both $J_1(x)$ and $J_2(x)$ return an even number of skip length, and thus destination of J_1 is always a position of J_1, and J_2 also going to a J_2 position.

We also analyzed the average case search cost in this two-phase model. We assume the same skip length as $1/\alpha$ of remaining search range for both $J_1(x)$ and $J_2(x)$. Then $C(n)$ is described as follows. (Detailed calculation is shown in Appendix of [10].)

$$C(n) = 2\alpha + \frac{n}{2(2\alpha - 1)^2} \tag{4}$$

If we set $\alpha = 16$, we get $C(10000) \approx 37.2$, which is about 134 times faster than ordinary linear search.

Similarly to the uniform model, we can calculate optimal α for a given n.

$$\alpha_{opt} = \frac{1}{2}\left(\sqrt[3]{n} + 1\right), \quad C_{opt}(n) = \frac{3}{2}\sqrt[3]{n} + 1 \tag{5}$$

When $n = 10000$, we get $\alpha_{opt} = 11.27$ and $C_{opt}(10000) \approx 33.3$, about 150 times faster than linear search.

In the above discussion, we ignored at most 2 steps which are required for changing J_1 and J_2 positions, so we must add a small constant factor for exact analysis. Anyway, we can significantly improve the performance only using two different skip length functions without any additional memory requirement.

k-Phase Model

Extending the two-phase model, we can consider k types of skip length functions, by classifying the position x into k groups by x modulo k. If we write $C^{(k)}(n)$ for the average search cost for k-phase model, it can be described as:

$$C^{(k)}(n) = k\alpha + \frac{n}{2(2\alpha - 1)^k} \tag{6}$$

and optimal α is obtained as:

$$\alpha_{opt} = \frac{1}{2}\left(n^{\frac{1}{k+1}} + 1\right), \quad C_{opt}^{(k)}(n) = \frac{k+1}{2}n^{\frac{1}{k+1}} + \frac{k}{2} \tag{7}$$

For example, when $k = 4$, we get

$$\alpha_{opt} = \frac{1}{2}\left(\sqrt[5]{n} + 1\right), \quad C_{opt}^{(4)}(n) = \frac{5}{2}\sqrt[5]{n} + 2. \tag{8}$$

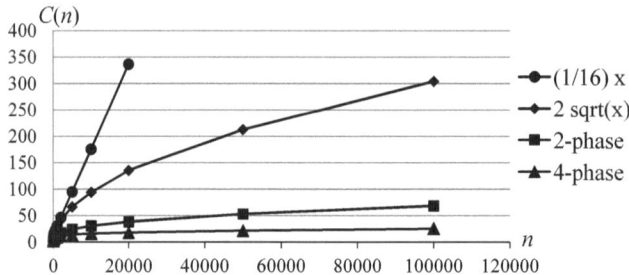

Fig. 8. Computational Experiments

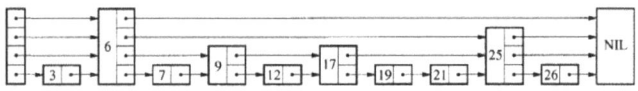

Fig. 9. An Example of Skip List [12]

Next, let us consider the optimal k for given n. The detailed calculation is shown in Appendix of [10] and we can obtain the following result.

$$k_{opt} = \frac{\ln n}{\ln c} - 1, \quad C_{opt}^{(k_{opt})}(n) = \left(\frac{c+1}{2 \cdot \ln c} \right) \ln n - \frac{1}{2}, \qquad (9)$$

where c is a constant nearby 3.6. Consequently, the average search cost of the Z-Skip-Links is theoretically shown as $O(\log n)$, if we use optimal k.

In the case of $n = 65536$, we get $k_{opt} \approx 8$ and $C_{opt}^{(k_{opt})}(65536) \approx 19.4$. However, even for $k = 4$ we can achieve $C_{opt}^{(4)}(65536) \approx 24.97$, which is more than 1300 times faster than ordinary linear search. We can conclude that, for up to $n = 100000$, 4-phase model is sufficiently effective for practical applications.

Computational Experiments
To confirm the above theoretical analysis, we conducted computational experiments. For a given n, we counted the number of moves to t, and computed average moves $C(n)$ for all t from 1 to n. In this experiments we tested four kinds of skip length functions as $\frac{1}{16}x$, $2\sqrt{x}$, the optimal 2-phase model, and the optimal 4-phase model. Figure 8 shows the results. We can observe that the results are very close to theoretical formulas up to $n = 100000$.

Related Work
Our method is related to *Skip List* proposed by Pugh [12] in 1990. Skip List is a quick access technique used for sorted linear linked list. It equips probabilistic distributed skip pointers as shown in Fig. 9. It is known that the average access time is $O(\log n)$. From this viewpoint, Z-Skip-Links can be regarded as a kind of Skip List technique deterministically attached to all ZDD nodes. However, our

Table 1. Look-up Table for Skip Length Function

range of x	$J_1(x)$	$J_2(x)$	$J_3(x)$	$J_4(x)$
$5 - 15$	4	—	—	—
$16 - 63$	8	4	—	—
$64 - 255$	64	16	8	4
$256 - 511$	128	32	8	4
$512 - 1023$	256	64	16	4
$1,024 - 2047$	512	128	32	8
$2,048 - 4095$	1,024	256	64	8
$4,096 - 8,191$	2,048	512	64	8
$8,192 - 32,767$	4,096	512	64	8
$32,768 - 65,535$	8,192	1,024	128	16

```
ZDescend(F, t)
{  while(F.top > t)
   {  G ← cache("ZSkip at F") ;
      if (G exists and G.top ≥ t)
          F ← G ;
      else F ← (0-Child of F) ;
   }
    return F ;
}
```

```
SetZSkip(F)
{  if (F.top ≤ 4) return ;
   if (cache("ZSkip at F") exists) return ;
   SetZSkip(0-Child of F) ;
   t ← (F.top − J(F.top)) ;
   G ← ZDescend(F, t) ;
   cache("ZSkip at F") ← G ;
   SetZSkip(1-Child of F) ;
}
```

Fig. 10. ZDD Traversal with Z-Skip-Links   **Fig. 11.** Construction of Z-Skip-Links

method is not limited to a simple linear list but applicable to the complex ZDD structure including a large number of paths shared with each other.

5   Algorithm Implementation

Implementation to Conventional BDD/ZDD Package

Most of conventional BDD/ZDD packages use short int (16 bit integer) for item IDs, so we assumed n up to 65535. For this range, we know that 4-phase is effective enough, so we adopted the 4-phase model. It is time consuming to calculate $\sqrt[5]{n}$ many times, so we approximate it by table look-up. The table is shown in Table. 1

Also, most of conventional BDD/ZDD Packages have an internal table called "operation cache," which stores the recent operations and their results by a hash-key with the pointers to the operand ZDDs and operation IDs. Our Z-Skip-Links can be easily implemented using the framework of the operation cache, and we needed only 100 lines of additional C++ codes for processing Z-Skip-Links.

Here we explain the algorithm of fast ZDD traversal using Z-Skip-Links, and the preprocessing algorithm for constructing Z-Skip-Links for all ZDD nodes.

ZDD Traversal Using Z-Skip-Links

Figure 10 shows pseudo code of ZDescend(F, t). For given root node of ZDD F and the target item-ID t, this algorithm returns the pointer to a ZDD node which is the first meet such that the item ID is equal or lower than t when descending 0-edges starting from F. In this codes, cache("ZSkip at F") means to refer the operation cache. This algorithm enjoys the benefit of Z-Skip-Links, however, even if Z-Skip-Lists are not prepared yet, it returns correct result by linear search.

Construction of Z-Skip-Links
Figure 11 shows pseudo codes of SetZSkip(F). This algorithm constructs Z-Skip-Links for all internal nodes of ZDD F by using a recursive procedure. At first, we check the operation cache and if a Z-Skip-Link is already constructed, the procedure terminates. Otherwise, it recursively calls itself on the 0-child node to construct Z-Skip-Links for all descendant nodes. After that, we compute the skip length and find the destination node by using ZDescend procedure, and then register it to the operation cache.

The number of recursive calls of SetZSkip is bounded by the number of F's ZDD nodes if operation cache works well. SetZSkip also calls ZDescend procedure, so we need several steps to reach the destination node. In the range of $n \leq 65536$, ZDescend procedure requires about 20 to 30 steps in average, therefore, the total computation time will be several ten times of the number of ZDD nodes. This would be feasible overhead if we repeat the search process many times for a large-scale sparse dataset.

6   Experimental Results

We implemented the algorithms and conducted experiments for performance evaluation. The specification of our PC is as follows. Intel Core i7 2700K 3.5GHz, 32GB memory, OpenSuSE Linux 12.1 (64bit), GNU C++ 4.6.2. We used our own BDD/ZDD package written in C and C++.

Membership Testing for Random-Generated Sparse Datasets
We applied our method to the randomly generated large-scale sparse datasets. We generated a ZDD for the dataset $D_{(n,m)}$ including m combinations each of which constructed randomly selecting two items out of n items. Next we also randomly generated a two item pattern p and evaluated the computation time for the membership testing if p is in $D_{(n,m)}$. We executed SetZSkip procedure for the ZDD as preprocessing and then repeated the membership testing 10000 times with different p's. We compared our method with conventional ZDD-based membership operation as $D_{(n,m)} \cap p_{(n)}$.

The experimental results are shown in Table 2. Here "pre-prc." means the time for preprocessing, "old" means conventional method. "accel." means the ratio of acceleration, "net" means ignoring preprocessing time, and "gross" means including preprocessing time. From the result, if we ignore the preprocessing time, the acceleration ratio reaches more than 500 times when $n = 50000$. Even including preprocessing time, we are more than 70 times faster than conventional method. The experimental results show that our method is effective when implementing on the practical BDD/ZDD package.

Table 2. Results for Random-Generated Sparse Datasets

		ZDD	CPU time(sec)			accel.	
n	m	nodes	pre- prc.	search x10000	old x10000	net	gross
1,000	10,000	10,011	0.004	0.012	0.158	13.2	9.9
2,000	20,000	20,181	0.005	0.013	0.390	30.0	21.7
5,000	50,000	50,683	0.007	0.018	1.025	56.9	41.0
10,000	100,000	101,521	0.026	0.021	2.055	97.9	43.7
20,000	200,000	203,272	0.058	0.029	4.499	155.1	51.7
50,000	500,000	508,335	0.170	0.028	14.393	514.0	72.7

Table 3. Results for Frequent Itemset Datasets

			ZDD	CPU time(sec)			accel.	
θ	n	m	nodes	pre- prc.	search x10000	old x10000	net	gross
100	932	11,928	2,298	0.001	0.009	0.094	10.4	9.4
50	1,597	70,713	6,059	0.001	0.011	0.169	15.4	14.1
20	2,434	63,4065	30,410	0.007	0.012	0.268	22.3	14.1
10	2,885	4,440,335	93,899	0.021	0.012	0.328	27.3	9.9
5	3,129	26,946,004	353,091	0.050	0.015	0.393	26.2	6.0

Membership Testing for Frequent Itemset Datasets

Next we applied our method to the frequent itemset data, which are dealt with in the basic problem of data mining. We used a dataset "BMS-WebView2" chosen from the KDD benchmark [4]. This data is known as the access log of web pages at an Internet shopping site, This dataset has 3340 items and 77512 transactions (combinations), but only 4.6 items appears in average per transaction, so it has very sparse combinations. We applied LCM-ZDD method[11] to this dataset and generated a ZDD D_θ including all frequent itemset patterns with a given minimum support θ. For example, D_5 includes 26946004 patterns and there are 3129 relevant items. The number of ZDD nodes are only 353091, so it has very good compression rate.

Then, we also randomly generated a two item pattern p, and similarly evaluated the computation time for the membership testing if p is in D_θ. Table 2 shows that our method is 10 to 27 times faster than conventional method without considering preprocessing time, and 6 to 14 times faster considering preprocessing time.

7 Summary

We proposed Z-Skip-Links to accelerate the traversal of ZDDs of large-scale sparse datasets. It consumes only a constant size of additional memory, and can easily be implemented into a conventional BDD/ZDD package. We have analyzed the average-case computation time and clarified the optimal settings. Our experiments show that the use of Z-Skip-Links makes the membership operations much faster than using conventional ZDD operations when handling large-scale sparse datasets. We expect that our method will widen the effective applications of ZDDs in the era of Big Data.

Acknowledgment. The author would like to thank a reviewer of ESA 2013 who corrected an error of calculation in the average-case analysis.

References

1. Akers, S.B.: Binary decision diagrams. IEEE Transactions on Computers C-27(6), 509–516 (1978)
2. Bryant, R.E.: Graph-based algorithms for Boolean function manipulation. IEEE Transactions on Computers C-35(8), 677–691 (1986)
3. Coudert, O.: Solving graph optimization problems with ZBDDs. In: Proc. of ACM/IEEE European Design and Test Conference (ED&TC 1997), pp. 224–228 (1997)
4. Goethals, B., Zaki, M.J.: Frequent itemset mining dataset repository. In: Frequent Itemset Mining Implementations (FIMI 2003) (2003), http://fimi.cs.helsinki.fi/
5. Minato, S.I.: Zero-suppressed BDDs for set manipulation in combinatorial problems. In: Proc. of 30th ACM/IEEE Design Automation Conference (DAC 1993), pp. 272–277 (1993)
6. Ishihata, M., Kameya, Y., Sato, T., Minato, S.: Propositionalizing the EM algorithm by BDDs. In: Proc. of 18th International Conference on Inductive Logic Programming (ILP 2008), p. 9 (2008)
7. Knuth, D.E.: The Art of Computer Programming: Bitwise Tricks & Techniques; Binary Decision Diagrams, vol. 4, fascicle 1. Addison-Wesley (2009)
8. Loekit, E., Bailey, J.: Fast mining of high dimensional expressive contrast patterns using zero-suppressed binary decision diagrams. In: Proc. The Twelfth ACM SIGKDD International Conference on Knowledge Discovery and Data Mining (KDD 2006), pp. 307–316 (2006)
9. Minato, S., Satoh, K., Sato, T.: Compiling bayesian networks by symbolic probability calculation based on zero-suppressed BDDs. In: Proc. of 20th International Joint Conference of Artificial Intelligence (IJCAI 2007), pp. 2550–2555 (2007)
10. Minato, S.-I.: Z-skip-links for fast zdd traversal in handling large-scale sparse datasets (revised ed.). Hokkaido University, Division of Computer Science. TCS Technical Reports, TCS-TR-A-13-66 (2013), http://www-alg.ist.hokudai.ac.jp/tra.html/
11. Minato, S.-I., Uno, T., Arimura, H.: LCM over ZBDDs: Fast generation of very large-scale frequent itemsets using a compact graph-based representation. In: Washio, T., Suzuki, E., Ting, K.M., Inokuchi, A. (eds.) PAKDD 2008. LNCS (LNAI), vol. 5012, pp. 234–246. Springer, Heidelberg (2008)
12. Pugh, W.: Skip lists: A probabilistic alternative to balanced trees. Algorithms and Data Structures 33(6), 668–676 (1990)
13. Yoshinaka, R., Kawahara, J., Denzumi, S., Arimura, H., Minato, S.-I.: Counterexamples to the long-standing conjecture on the complexity of bdd binary operations. Information Processing Letters 112(16), 636–640 (2012)

Optimal Color Range Reporting
in One Dimension

Yakov Nekrich and Jeffrey Scott Vitter

The University of Kansas
yakov.nekrich@googlemail.com, jsv@ku.edu

Abstract. Color (or categorical) range reporting is a variant of the orthogonal range reporting problem in which every point in the input is assigned a *color*. While the answer to an orthogonal point reporting query contains all points in the query range Q, the answer to a color reporting query contains only distinct colors of points in Q. In this paper we describe an $O(N)$-space data structure that answers one-dimensional color reporting queries in optimal $O(k + 1)$ time, where k is the number of colors in the answer and N is the number of points in the data structure. Our result can be also dynamized and extended to the external memory model.

1 Introduction

In the orthogonal range reporting problem, we store a set of points S in a data structure so that for an arbitrary range $Q = [a_1, b_1] \times \ldots \times [a_d, b_d]$ all points from $S \cap Q$ can be reported. Due to its importance, one- and multi-dimensional range reporting was extensively studied in computational geometry and database communities. The following situation frequently arises in different areas of computer science: a set of d-dimensional objects $\{ (t_1, t_2, \ldots, t_d) \}$ must be preprocessed so that we can enumerate all objects satisfying $a_i \le t_i \le b_i$ for arbitrary a_i, b_i, $i = 1, \ldots, d$. This scenario can be modeled by the orthogonal range reporting problem.

The objects in the input set can be distributed into *categories*. Instead of enumerating all objects, we may want to report distinct categories of objects in the given range. This situation can be modeled by the color (or categorical) range reporting problem: every point in a set S is assigned a color (category); we pre-process S, so that for any $Q = [a_1, b_1] \times \ldots \times [a_d, b_d]$ the distinct colors of points in $S \cap Q$ can be reported.

Color range reporting is usually considered to be a more complex problem than point reporting. For one thing, we do not want to report the same color multiple times. In this paper we show that complexity gap can be closed for one-dimensional color range reporting. We describe color reporting data structures with the same space usage and query time as the best known corresponding structures for point reporting. Moreover we extend our result to the external memory model.

H.L. Bodlaender and G.F. Italiano (Eds.): ESA 2013, LNCS 8125, pp. 743–754, 2013.
© Springer-Verlag Berlin Heidelberg 2013

Previous Work. We can easily report points in a one-dimensional range $Q = [a, b]$ by searching for the successor of a in S, $succ(a, S) = \min\{ e \in S \,|\, e \geq a \}$. If $a' = succ(a, S)$ is known, we can traverse the sorted list of points in S starting at a' and report all elements in $S \cap [a, b]$. We can find the successor of a in S in $O(\sqrt{\log N / \log \log N})$ time [4]; if the universe size is U, i.e., if all points are positive integers that do not exceed U, then the successor can be found in $O(\log \log U)$ time [22]. Thus we can report all points in $S \cap [a, b]$ in $O(\mathrm{tpred}(N) + k)$ time for $\mathrm{tpred}(N) = \min(\sqrt{\log N / \log \log N}, \log \log U)$. Henceforth k denotes the number of elements (points or colors) in the query answer. It is not possible to find the successor in $o(\mathrm{tpred}(N))$ time unless the universe size U is very small or the space usage of the data structure is very high; see e.g., [4]. However, reporting points in a one-dimensional range takes less time than searching for a successor. In their fundamental paper [14], Miltersen et al. showed that one-dimensional point reporting queries can be answered in $O(k)$ time using an $O(N \log U)$ space data structure. Alstrup et al. [1] obtained another surprising result: they presented an $O(N)$-space data structure that answers point reporting queries in $O(k)$ time and thus achieved both optimal query time and optimal space usage for this problem. The data structure for one-dimensional point reporting can be dynamized so that queries are supported in $O(k)$ time and updates are supported in $O(\log^\varepsilon U)$ time [16]; henceforth ε denotes an arbitrarily small positive constant. We refer to [16] for further update-query time trade-offs. Solutions of the one-dimensional point reporting problem are based on finding an arbitrary element e in a query range $[a, b]$; once such e is found, we can traverse the sorted list of points until all points in $[a, b]$ are reported. Therefore it is straightforward to extend point reporting results to the external memory model.

Janardan and Lopez [10] and Gupta et al. [9] showed that one-dimensional color reporting queries can be answered in $O(\log N + k)$ time, both in the static and the dynamic scenarios. Muthukrishnan [17] described a static $O(N)$ space data structure that answers queries in $O(k)$ time if all point coordinates are bounded by N. We can obtain data structures that use $O(N)$ space and answer queries in $O(\log \log U + k)$ or $O(\sqrt{\log N / \log \log N} + k)$ time using the reduction-to-rank-space technique. No data structure that answers one-dimensional color reporting queries in $o(\mathit{tpred}(N)) + O(k)$ time was previously known. A dynamic data structure of Mortensen [15] supports queries and updates in $O(\log \log N + k)$ and $O(\log \log N)$ time respectively if the values of all elements are bounded by N.

Recently, the one- and two-dimensional color range reporting problems in the external memory model were studied in several papers [11,18,12]. Larsen and Pagh [11] described a data structure that uses linear space and answers one-dimensional color reporting queries in $O(k/B + 1)$ I/Os if values of all elements are bounded by $O(N)$. In the case when values of elements are unbounded the best previously known data structure needs $O(\log_B N + k/B)$ I/Os to answer a query; this result can be obtained by combining the data structure from [2] and reduction of one-dimensional color reporting to three-sided[1] point reporting [9].

In another recent paper [5], Chan et al. described a data structure that supports the following queries on a set of points whose values are bounded by $O(N)$: for any query point q and any integer k, we can report the first k colors that occur after q. This data structure can be combined with the result from [1] to answer queries in $O(k+1)$ time. Unfortunately, the solution in [5] is based on the hive graph data structure [6]. Therefore it cannot be used to solve the problem in external memory or to obtain a dynamic solution.

Our Results. As can be seen from the above discussion and Table 1, there are significant complexity gaps between color reporting and point reporting data structures in one dimension. We show in this paper that it is possible to close these gaps.

In this paper we show that one-dimensional color reporting queries can be answered in constant time per reported color for an arbitrarily large size of the universe. Our data structure uses $O(N)$ space and supports color reporting queries in $O(k + 1)$ time. This data structure can be dynamized so that query time and space usage remain unchanged; the updates are supported in $O(\log^\varepsilon U)$ time where U is the size of the universe. The new results are listed at the bottom of Table 1.

Our internal memory results are valid in the word RAM model of computation, the same model that was used in e.g. [1,16,17]. In this model, we assume that any standard arithmetic operation and the basic bit operations can be performed in constant time. We also assume that each word of memory consists of $w \geq \log U \geq \log N$ bits, where U is the size of the universe. That is, we make a reasonable and realistic assumption that the value of any element fits into one word of memory.

Furthermore, we also extend our data structures to the external memory model. Our static data structure uses linear space and answers color reporting queries in $O(1 + k/B)$ I/Os. Our dynamic external data structure also has optimal space usage and query cost; updates are supported in $O(\log^\varepsilon U)$ I/Os.

In Section 2 we describe a static data structure for color reporting in one dimension. The key component of our solution is a data structure that supports *highest range ancestor* queries. In Section 3 we show how our static data structure can be adopted to the external memory model. We show how to dynamize our data structure in Sections 4, 5, and 6. Details of our dynamic solution and its modification for the external memory model are provided in the full version of this paper [19].

---

[1] A three-sided range query is a two-dimensional orthogonal range query that is open on one side. For instance, queries $[a, b] \times [0, c]$ and $[a, b] \times [c, +\infty]$ are three-sdied queries.

Table 1. Selected previous results and new results for one-dimensional color reporting. The fifth and the sixth row can be obtained by applying the reduction to rank space to the result from [17].

Ref.	Query Type	Space Usage	Query Cost	Universe	Update Cost
[1]	Point Reporting	$O(N)$	$O(k+1)$		static
[16]	Point Reporting	$O(N)$	$O(k+1)$		$O(\log^\varepsilon U)$
[9,10]	Color Reporting	$O(N)$	$O(\log N + k)$		$O(\log N)$
[17]	Color Reporting	$O(N)$	$O(k+1)$	N	static
[17]	Color Reporting	$O(N)$	$O(\log \log U + k)$	U	static
[17]	Color Reporting	$O(N)$	$O(\sqrt{\log N / \log \log N} + k)$		static
[15]	Color Reporting	$O(N)$	$O(\log \log N + k)$	N	$O(\log \log N)$
[5]+[1]	Color Reporting	$O(N)$	$O(k+1)$		
Our	Color Reporting	$O(N)$	$O(k+1)$		static
Our	Color Reporting	$O(N)$	$O(k+1)$		$O(\log^\varepsilon U)$

2   Static Color Reporting in One Dimension

We start by describing a static data structure that uses $O(N)$ space and answers color reporting queries in $O(k+1)$ time.

All elements of a set S are stored in a balanced binary tree $\mathcal{T}$. Every leaf of $\mathcal{T}$, except for the last one, contains $\log N$ elements, the last leaf contains at most $\log N$ elements, and every internal node has two children. For any node $u \in \mathcal{T}$, $S(u)$ denotes the set of all elements stored in the leaf descendants of u. For every color z that occurs in $S(u)$, the set $Min(u)$ ($Max(u)$) contains the minimal (maximal) element $e \in S(u)$ of color z. The list $L(u)$ contains the $\log N$ smallest elements of $Min(u)$ in increasing order. The list $R(u)$ contains the $\log N$ largest elements of $Max(u)$ in decreasing order. For every internal non-root node u we store the list $L(u)$ if u is the right child of its parent; if u is the left child of its parent, we store the list $R(u)$ for u. All lists $L(u)$ and $R(u)$, $u \in \mathcal{T}$, contain $O(N)$ elements in total since the tree has $O(N/\log N)$ internal nodes.

We define the middle value $m(u)$ for an internal node u as the minimal value stored in the right child of u, $m(u) = \min\{ e \mid e \in S(u_r) \}$ where u_r is the right child of u. The following *highest range ancestor* query plays a crucial role in the data structures of this and the following sections. The answer to the highest range ancestor query (v_l, a, b) for a leaf v_l and values $a < b$ is the highest ancestor u of v_l, such that $a < m(u) \leq b$; if $S \cap [a, b] = \emptyset$, the answer is undefined. The following fact elucidates the meaning of the highest range ancestor.

Fact 1. *Let v_a be the leaf that holds the smallest $e \in S$, such that $e \geq a$; let v_b be the leaf that holds the largest $e \in S$, such that $e \leq b$. Suppose that $S(v_l) \cap [a, b] \neq \emptyset$ for some leaf v_l and u is the answer to the highest range ancestor query (v_l, a, b). Then u is the lowest common ancestor of v_a and v_b.*

Proof: Let w denote the lowest common ancestor of v_a and v_b. Then v_a and v_b are in w's left and right subtrees respectively. Hence, $a < m(w) \leq b$ and w is not an ancestor of u. If w is a descendant of u and w is in the right subtree of u, then $m(u) \leq a$. If w is in the left subtree of u, then $m(w) > b$. $\qquad\square$

We will show that we can find u without searching for v_a and v_b and answer highest range ancestor queries on a balanced tree in constant time.

For every leaf v_l, we store two auxiliary data structures. All elements of $S(v_l)$ are stored in a data structure $D(v_l)$ that uses $O(|S(v_l)|)$ space and answers color reporting queries on $S(v_l)$ in $O(k+1)$ time. We also store a data structure $F(v_l)$ that uses $O(\log N)$ space; for any $a < b$, such that $S(v_l) \cap [a, b] \neq \emptyset$, $F(v_l)$ answers the highest range ancestor query (v_l, a, b) in $O(1)$ time. Data structures $D(v_l)$ and $F(v_l)$ will be described later in this section. Moreover, we store all elements of S in the data structure described in [1] that supports one-reporting queries: for any $a < b$, some element $e \in S \cap [a, b]$ can be found in $O(1)$ time; if $S \cap [a, b] = \emptyset$, the data structure returns a dummy element $\perp$. Finally, all elements of S are stored in a slow data structure that uses $O(N)$ space and answers color reporting queries in $O(\log n + k)$ time. We can use e.g. the data structure from [10] for this purpose.

Answering Queries. All colors in a query range $[a, b]$ can be reported with the following procedure. Using the one-reporting data structure from [1], we search for some $e \in S \cap [a, b]$ if at least one such e exists. If no element e satisfying $a \leq e \leq b$ is found, then $S \cap [a, b] = \emptyset$ and the query is answered. Otherwise, let v_e denote the leaf that contains e. Using $F(v_e)$, we search for the highest ancestor u of v_e such that $a \leq m(u) \leq b$. If no such u is found, then all e, $a \leq e \leq b$, are in $S(v_e)$. We can report all colors in $S(v_e) \cap [a, b]$ using $D(v_e)$. If $F(v_e)$ returned some node u, we proceed as follows. Let u_l and u_r denote the left and the right children of u. We traverse the list $L(u_r)$ until an element $e' > b$ is found or the end of $L(u_r)$ is reached. We also traverse $R(u_l)$ until an element $e' < a$ is found or the end of $R(u_l)$ is reached. If we reach neither the end of $L(u_r)$ nor the end of $R(u_l)$, then the color of every encountered element $e \in L(u_r)$, $e \leq b$, and $e \in R(u_l)$, $e \geq a$, is reported. Otherwise the range $[a, b]$ contains at least $\log N$ different colors. In the latter case we can use any data structure for one-dimensional color range reporting [10,9] to identify all colors from $S \cap [a, b]$ in $O(\log n + k) = O(k + 1)$ time.

Leaf Data Structures. A data structure $D(v_l)$ answers color reporting queries on $S(v_l)$ as follows. In [9], the authors show how a one-dimensional color reporting query on a set of m one-dimensional elements can be answered by answering a query $[a, b] \times [0, a]$ on a set of m uncolored two-dimensional points. A standard priority search tree [13] enables us to answer queries of the form $[a, b] \times [0, a]$ on m points in $O(\log m)$ time. Using a combination of fusion trees and priority search trees, described by Willard [23], we can answer queries in $O(\log m / \log \log N)$ time. The data structure of Willard [23] uses $O(m)$ space and a universal

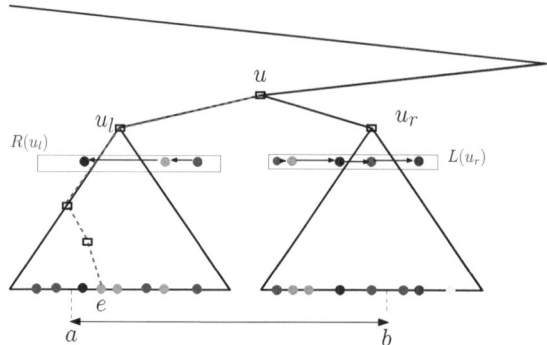

Fig. 1. Answering a color reporting query $Q = [a, b]$: e is an arbitrary element in $S \cap [a, b]$, u is the highest range ancestor of the leaf that contains e, the path from e to u is indicated by a dashed line. We assume $\log n = 5$, therefore $L(u_r)$ contains 5 elements and the yellow point is not included in $L(u_r)$. To simplify the picture, we assumed that each leaf contains only one point; only relevant parts of $\mathcal{T}$ are on the picture.

look-up table of size $O(\log^\varepsilon N)$ for an arbitrarily small ε. Updates are also supported in $O(\log m / \log \log N)$ time[2].

Since $S(v_l)$ contains $m = O(\log N)$ elements, we can answer colored queries on $S(v_l)$ in $O(\log m / \log \log N) = O(1)$ time. Updates are also supported in $O(1)$ time; this fact will be used in Section 4.

Now we describe how $F(v_l)$ is implemented. Suppose that $S(v_l) \cap [a, b] \neq \emptyset$ for some leaf v_l. Let π be the path from v_l to the root of $\mathcal{T}$. We say that a node $u \in \pi$ is a *left parent* if $u_l \in \pi$ for the left child u_l of u; a node $u \in \pi$ is a *right parent* if $u_r \in \pi$ for the right child u_r of u. If $S(v_l)$ contains at least one $e \in [a, b]$, then the following is true.

Fact 2. *If $u \in \pi$ is a left parent, then $m(u) > a$. If $u \in \pi$ is a right parent, then $m(u) \leq b$.*

Proof: If $u \in \pi$ is a left parent, then $S(v_l)$ is in its left subtree. Hence, $m(u)$ is greater than any $e \in S(v_l)$ and $m(u) > a$. If u is the right parent, than $S(v_l)$ is in its right subtree. Hence, $m(u)$ is smaller than or equal to any $e \in S(v_l)$ and $m(u) \leq b$.   □

Fact 3. *If $u_1 \in \pi$ is a left parent and u_1 is an ancestor of $u_2 \in \pi$, then $m(u_1) > m(u_2)$. If $u_1 \in \pi$ is a right parent and u_1 is an ancestor of $u_2 \in \pi$, then $m(u_2) > m(u_1)$.*

Proof: If u_1 is a left parent, then u_2 is in its left subtree. Hence, $m(u_1) > m(u_2)$ by definition of $m(u)$. If u_2 is a right parent, then u_1 is in its right subtree. Hence, $m(u_1) < m(u_2)$ by definition of $m(u)$.   □

---

[2] In [23], Willard only considered queries on N points, but extension to the case of any $m \leq N$ is straightforward.

Suppose that we want to find the highest range ancestor of v_l for a range $[a, b]$ such that $S(v_l) \cap [a, b] \neq \emptyset$. Let $\mathcal{K}_1(\pi)$ be the set of middle values $m(u)$ for left parents $u \in \pi$ sorted by height; let $\mathcal{K}_2(\pi)$ be the set of $m(u)$ for right parents $u \in \pi$ sorted by height. By Fact 3, elements of $\mathcal{K}_1$ ($\mathcal{K}_2$) increase (decrease) monotonously. By Fact 2, $m(u) > a$ for any $m(u) \in \mathcal{K}_1$ and $m(u) < b$ for any $m(u) \in \mathcal{K}_2$. Using fusion trees [7], we can search in $\mathcal{K}_1$ and find the highest node $u_1 \in \pi$ such that u_1 is a left parent and $m(u_1) \leq b$. We can also search in $\mathcal{K}_2$ and find the highest node $u_2 \in \pi$ such that u_2 is a right parent and $m(u_2) > a$. Let u denote the higher node among u_1, u_2. Then u is the highest ancestor of v_l such that $m(u) \in [a + 1, b]$.

Removing Duplicates. When a query is answered, our procedure returns a color z two times if z occurs in both $S \cap [a, m(u) - 1]$ and $S \cap [m(u), b]$. We can easily produce a list without sorting in which each color occurs exactly once. Let Col denote an array with one entry for every color that occurs in a data structure. Initially $\mathrm{Col}[i] = 0$ for all i. We traverse the list of colors $\mathcal{L}$ produced by the above described procedure. Every time when we encounter a color z in $\mathcal{L}$ such that $\mathrm{Col}[z] = 0$, we set $\mathrm{Col}[z] = 1$; when we encounter a color z such that $\mathrm{Col}[z] = 1$, we remove the corresponding entry from $\mathcal{L}$. When the query is answered, we traverse $\mathcal{L}$ once again and set $Col[z] = 0$ for all $z \in \mathcal{L}$.

Theorem 1. *There exists an $O(N)$-space data structure that supports one-dimensional color range reporting queries in $O(k + 1)$ time.*

3   Color Reporting in External Memory

The static data structure of Section 2 can be used for answering queries in external memory. We only need to increase the sizes of $S(v_l)$, $R(u)$, and $L(u)$ to $B \log_B N$, and use an external memory variant of the slow data structure for color reporting [2]. This approach enables us to achieve $O(1 + k/B)$ query cost, but one important issue should be addressed. As explained in Section 2, the same color can be reported twice when a query is answered. However, we cannot get rid of duplicates in $O(1 + k/B)$ I/Os using the method of Section 2 because of its random access to the list of reported colors. Therefore we need to make further changes in our internal memory solution. For an element $e \in S$, let $prev(e)$ denote the largest element $e' \leq e$ of the same color. For every element e in $L(u)$ and any $u \in T$, we also store the value of $prev(e)$.

We define each set $S(v_l)$ for a leaf v_l to contain $B \log_B N$ points. Lists $L(v)$ and $R(v)$ for an internal node v contain $B \log_B N$ leftmost points from $Min(v)$ (respectively, $B \log_B N$ rightmost points from $Max(v)$). Data structures $F(v_l)$ are implemented as in Section 2. A data structure $D(v_l)$ supports color reporting queries on $S(v_l)$ and is implemented as follows. We can answer a one-dimensional color reporting query by answering a three-sided point reporting query on a set Δ of $|S(v_l)|$ two-dimensional points; see e.g., [9]. If $B \geq \log_2 N$, $S(v_l)$ and Δ contain $O(B^2)$ points. In this case we can use the data structure from [2] that uses linear space and answers three-sided queries in $O(\log_B |S(v_l)| + k/B) = O(1 + k/B)$

I/Os. If $B < \log N$, $S(v_l)$ and Δ contain $O(\log^2 N)$ points. Using the data structure from [7], we can find the predecessor of any value v in a set of $O(\log^2 N)$ points in $O(1)$ I/Os. Therefore we can apply the rank-space technique [8] and reduce three-sided point reporting queries on Δ to three-sided point reporting queries on a grid of size $|\Delta|$ (i.e., to the case when coordinates of all points are integers bounded by $|\Delta|$) using a constant number of additional I/Os. Larsen and Pagh [11] described a linear-space data structure that answers three-sided point queries for m points on an $m \times m$ grid in $O(1 + k/B)$ I/Os. Summing up, we can answer a three-sided query on a set of $B \log_B N$ points in $O(1 + k/B)$ I/Os. Hence, we can also answer a color reporting query on $S(v_l)$ in $O(1 + k/B)$ I/Os using linear space.

A query $Q = [a, b]$ is answered as follows. We find the highest range ancestor u for any $e \in S \cap [a, b]$ exactly as in Section 2. If u is a leaf, we answer the query using $D(u)$. Otherwise the reporting procedure proceeds as follows. We traverse the list $R(u_l)$ for the left child u_l of u until some point $p < a$ is found. If $e \geq a$ for all $e \in R(u_l)$, then there are at least $B \log_B N$ different colors in $[a, b]$ and we can use a slow data structure to answer a query in $O(\log_B N + \frac{k}{B}) = O(1 + \frac{k}{B})$ I/Os. Otherwise we traverse $L(u_r)$ and report all elements e such that $prev(e) < a$. If $prev(e) \geq a$ for $e \in L(u)$, then an element of the same color was reported when $R(u_l)$ was traversed. Traversal of $L(u_r)$ stops when an element $e > b$ is encountered or the end of $L(u_r)$ is reached. In the former case, we reported all colors in $[a, b]$. In the latter case the number of colors in $[a, b]$ is at least $B \log_B N$. This is because every element in $L(u_r)$ corresponds to a distinct color that occurs at least once in $[a, b]$. Hence, we can use the slow data structure and answer the query in $O(\log_B N + \frac{k}{B}) = O(\frac{k}{B} + 1)$ I/Os.

Theorem 2. *There exists a linear-space data structure that supports one-dimensional color range reporting queries in $O(k/B + 1)$ I/Os.*

4   Base Tree for Dynamic Data Structure

In this section we show how the base tree and auxiliary data structures of the static solution can be modified for usage in the dynamic scenario. To dynamize the data structure of Section 2, we slightly change the balanced tree $\mathcal{T}$ and secondary data structures: every leaf of $\mathcal{T}$ now contains $\Theta(\log^2 N)$ elements of S and each internal node has $\Theta(1)$ children. We store the lists $L(u)$ and $R(u)$ in each internal non-root node of u. We associate several values $m_i(u)$ to each node u: for every child u_i of u, except the leftmost child u_1, $m_i(u) = \min\{e \mid e \in S(u_i)\}$. The highest range ancestor of a leaf v_l is the highest ancestor u of v_l such that $a < m_i(u) \leq b$ for at least one $i \neq 1$. Data structures $D(v_l)$ and $F(v_l)$ are defined as in Section 2. We also maintain a data structure of [16] that reports an arbitrary element $e \in S \cap [a, b]$ if the range $[a, b]$ is not empty.

We implement the base tree $\mathcal{T}$ as the weight-balanced B-tree [3] with the leaf parameter $\log^2 N$ and the branching parameter 8. This means that every internal node has between 2 and 32 children and each leaf contains between $2 \log^2 N$ and $\log^2 N$ elements. Each internal non-root node on level ℓ of $\mathcal{T}$ has between $2 \cdot 8^\ell \log^2 N$ and $(1/2) \cdot 8^\ell \log^2 N$ elements in its subtree. If the number of elements in some node u exceeds $2 \cdot 8^\ell \log^2 N$, we split u into two new nodes, u' and u''. In this case we insert a new value $m_i(w)$ for the parent w of u. Hence, we may have to update the data structures $F(v_l)$ for all leaf descendants of w. A weight-balanced B-tree is engineered in such a way that a split occurs at most once in a sequence of $\Omega(8^\ell \log^2 N)$ insertions (for our choice of parameters). Since $F(v_l)$ can be updated in $O(1)$ time, the total amortized cost incurred by splitting nodes is $O(1)$. When an element e is deleted, we delete it from the set $S(v_l)$. If $e = m_i(u)$ for a deleted element e and some node u, we do not change the value of $m_i(u)$. We also do not start re-balancing if some node contains too few elements in its subtree. But we re-build the entire tree $\mathcal{T}$ if the total number of deleted elements equals $n_0/2$, where n_0 is the number of elements that were stored in $\mathcal{T}$ when it was built the last time. Updates can be de-amortized without increasing the cost of update operations by scheduling the procedure of re-building nodes (respectively, re-building the tree) [3].

Auxiliary Data Structures. We implement $D(v_l)$ in the same way as in Section 2. Hence color queries on $S(v_l)$ are answered in $O(\log |S(v_l)| / \log \log N) = O(1)$ time and updates are also supported in $O(1)$ time [23].

We need to modify data structures $F(v_l)$, however, because $\mathcal{T}$ is not a binary tree in the dynamic case. Let π denote a path from v_l to the root for some leaf v_l. We say that a node u is an i-node if $u_i \in \pi$ for the i-th child u_i of u.

Fact 4. *Suppose that $S(v_l) \cap [a, b] \neq \emptyset$ and π is the path from v_l to the root. If $u \in \pi$ is an i-node, then $m_j(u) < b$ for $1 \leq j \leq i$ and $m_j(u) > a$ for $j > i$.*

We say that a value $m_j(u)$ for $u \in \pi$ is a *left value* if $j \leq i$ and u is an i-node. A value $m_j(u)$ for $u \in \pi$ is a *right value* if $j > i$ and u is an i-node.

Fact 5. *If $m_j(u_1)$ is a left value and $u_1 \in \pi$ is an ancestor of $u_2 \in \pi$, then $m_j(u_1) \leq m_f(u_2)$ for any f. If $m_j(u_1)$ is a right value and $u_1 \in \pi$ is an ancestor of $u_2 \in \pi$, then $m_j(u_1) > m_f(u_2)$ for any f.*

It is easy to check Facts 4 and 5 using the same arguments as in Section 2.

We store all left values $m_j(u)$, $u \in \pi$, in a set $\mathcal{K}_1$; $m_j(u)$ in $\mathcal{K}_1$ are sorted by the height of u. We store all right values $m_j(u)$, $u \in \pi$, in a set $\mathcal{K}_2$; $m_j(u)$ in $\mathcal{K}_2$ are also sorted by the height of u. Using fusion trees on $\mathcal{K}_1$, we can find the highest node u_1, such that at least one left value $m_g(u_1) > a$. We can also find the highest u_2 such that at least one right value $m_f(u_2) \leq b$. Since $\mathcal{K}_1$ and $\mathcal{K}_2$ contain $O(\log N)$ elements, we can support searching and updates in $O(1)$ time; see [7,21]. By Fact 4, $a < m_g(u_1) \leq b$ and $a < m_f(u_2) \leq b$. If u is the higher node among u_1, u_2, then u is an answer to the highest range ancestor query $[a, b]$ for a node v_l.

5   Fast Queries, Slow Updates

In this section we describe a dynamic data structure with optimal query time. Our improvement combines an idea from [15] with the highest range ancestor approach. We also use a new solution for a special case of two-dimensional point reporting problem presented in the full version of this paper [19] in Section A.1.

Let $height(u)$ denote the height of a node u. For an element $e \in S$ let $h_{\min}(e) = height(u')$, where u' is the highest ancestor of the leaf containing e, such that $e \in Min(u')$. We define $h_{\max}(e)$ in the same way with respect to $Max(u)$. All colors in a range $[a, b]$ can be reported as follows. We identify an arbitrary $e \in S \cap [a, b]$. Using the highest range ancestor data structure, we can find the lowest common ancestor u of the leaves that contain the successor of a and the predecessor of b. Let u_f and u_g be the children of u that contain the successor of a and the predecessor of b. Let $a_f = a$, $b_g = b$; let $a_i = m_i(u)$ for $f < i \le g$ and $b_i = m_{i+1}(u) - 1$ for $f \le i < g$. We can identify unique colors of relevant points stored in each node u_j, $f < j \le g$, by finding all $e \in [a_j, b_j]$ such that $e \in Min(u_j)$. This condition is equivalent to reporting all $e \in [a_j, b_j]$ such that $h_{\min}(e) \ge height(u_j)$. We can identify all colors of relevant points in u_f by reporting all $e \in [a_f, b_f]$ such that $h_{\max}(e) \ge height(u_f)$. Queries of the form $e \in [a, b]$, $h_{\min}(e) \ge c$, (respectively $e \in [a, b]$, $h_{\max}(e) \ge c$) can be supported using Lemma 1 (see Section A.1 in [19]). While the same color can be reported several times, we can get rid of duplicates as explained in Section 2.

When a new point is inserted into S or when a point is deleted from S, we can update the values of $h_{\min}(e)$ and $h_{\max}(e)$ in $O(\log \log U)$ time. We refer to [15,20] for details.

While updates of data structures of Lemma 1 are fast, re-balancing the base tree can be a problem. As described in Section 4, when the number of points in a node u on level ℓ exceeds $2 \cdot 8^\ell \log N$, we split it into two nodes, u' and u''. As a result, the values $h_{\min}(e)$ for e stored in the leaves of u'' can be incremented. Hence, we would have to examine the leaf descendants of u'' and recompute their values for some of them. Since the height of $\mathcal{T}$ is logarithmic, the total cost incurred by re-computing the values $h_{\min}(e)$ and $h_{\max}(e)$ is $O(\log N)$. The problem of reducing the cost of re-building the tree nodes is solved as follows. In Appendix A.2 in [19] we describe another data structure that supports fast updates but answering queries takes polynomial time in the worst case. In Section 6 we show how the cost of splitting can be reduced by modifying the definition of $h_{\min}(e)$, $h_{\max}(e)$ and using the slow data structure from [19] when the number of reported colors is sufficiently large.

6   Fast Queries, Fast Updates

Let $n(u)$ denote the number of leaves in the subtree of a node u. Let $Left(u)$ denote the set of $(n(u))^{1/2}$ smallest elements in $Min(u)$; let $Right(u)$ denote the set of $(n(u))^{1/2}$ largest elements in $Max(u)$. We maintain the values $\overline{h_{\min}}(e)$ and $\overline{h_{\max}}(e)$ for $e \in S$, such that for any $u \in \mathcal{T}$ we have: $\overline{h_{\min}}(e) = h_{\min}(e)$ if $e \in Left(u)$ and $\overline{h_{\min}}(e) \le h_{\min}(e)$ if $e \in S(u) \setminus Left(u)$; $\overline{h_{\max}}(e) = h_{\max}(e)$ if $e \in Right(u)$ and $\overline{h_{\max}}(e) \le h_{\max}(e)$ if $e \in S(u) \setminus Right(u)$. We keep $\overline{h_{\min}}(e)$ and $\overline{h_{\max}}(e)$ in data structures of Lemma 1. We also maintain the data structure described in Section A.2 in [19]. This data structure is used to answer queries when the number of colors in the query range is large. It is also used to update the values of $\overline{h}_{\min}(e)$ and $\overline{h}_{\max}(e)$ when a node is split.

To answer a query $[a, b]$, we proceed in the same way as in Section 5. Let u, u_f, u_g, and a_i, b_i, $f \leq i \leq g$ be defined as in Section 5. Distinct colors in each $[a_i, b_i]$, $f \leq i \leq g$, can be reported using the data structure of Lemma 1. If the answer to at least one of the queries contains at least $(n(u_i))^{1/2}$ elements, then there are at least $(n(u_i))^{1/2}$ different colors in $[a, b]$. The total number of elements in $[a, b] \cap S$ does not exceed $n(u) = 16n(u_j)$. Hence, we can employ the data structure from Section A.2 in [19] to report all colors from $[a, b]$ in $O(([a, b] \cap S)^{1/2} + k) = O(k)$ time. If answers to all queries contain less than $(n(u_i))^{1/2}$ elements, then for every distinct color that occurs in $[a, b]$ there is an element e such that $e \in Left(u_i) \cap [a_i, b_i]$, $f \leq i < g$, or $e \in Right(u_g) \cap [a_g, b_g]$. By definition of $\overline{h}_{\min}$ and $\overline{h}_{\max}$ we can correctly report up to $(n(u_i))^{1/2}$ leftmost colors in $Left(u_i)$ or up to $(n(u_i))^{1/2}$ rightmost colors in $Right(u_i)$.

When a new element e is inserted, we compute the values of $h_{\min}(e)$, $h_{\max}(e)$ and update the values of $h_{\min}(e_n)$, $h_{\max}(e_n)$, where e_n is the element of the same color as e that follows e. This can be done in the same way as in Section 5. When a node u on level ℓ is split into u' and u'', we update the values of $h_{\min}(e)$ and $h_{\max}(e)$ for $e \in S(u') \cup S(u'')$. If $\ell \leq \log \log N$, we examine all $e \in S(u') \cup S(u'')$ and re-compute the values of $prev(e)$, $h_{\min}(e)$, and $h_{\max}(e)$. Amortized cost of re-building nodes u on $\log \log N$ lowest tree levels is $O(\log \log N)$. If $\ell > \log \log N$, $S(u)$ contains $\Omega(\log^5 N)$ elements. We can find $(n(u'))^{1/2}$ elements in $Left(u')$, $Left(u'')$, $Right(u')$, and $Right(u'')$ using the data structure from Lemma 3 in [19]. This takes $O((n(u)^{1/2}) \log N + \log N \log \log N) = O(((n(u))^{7/10})$ time. Since we split a node u one time after $\Theta(n(u))$ insertions, the amortized cost of splitting nodes on level $\ell > \log \log N$ is $O(1)$. Thus the total cost incurred by splitting nodes after insertions is $O(\log \log N)$. Deletions are processed in a symmetric way. We obtain the following result

Theorem 3. *There exists a linear-space data structure that supports one-dimensional color range reporting queries in $O(k + 1)$ time and updates in $O(\log^\varepsilon U)$ amortized time.*

In [19] we also show how the result of Theorem 3 can be extended to the external memory model.

References

1. Alstrup, S., Brodal, G.S., Rauhe, T.: Optimal static range reporting in one dimension. In: Proc. 33rd Annual ACM Symposium on Theory of Computing (STOC), pp. 476–482 (2001)
2. Arge, L., Samoladas, V., Vitter, J.S.: On two-dimensional indexability and optimal range search indexing. In: Proc. 18th ACM SIGACT-SIGMOD-SIGART Symposium on Principles of Database Systems (PODS), pp. 346–357 (1999)
3. Arge, L., Vitter, J.S.: Optimal external memory interval management. SIAM J. Comput. 32(6), 1488–1508 (2003)

4. Beame, P., Fich, F.E.: Optimal bounds for the predecessor problem and related problems. J. Comput. Syst. Sci. 65(1), 38–72 (2002)
5. Chan, T.M., Durocher, S., Skala, M., Wilkinson, B.T.: Linear-space data structures for range minority query in arrays. In: Fomin, F.V., Kaski, P. (eds.) SWAT 2012. LNCS, vol. 7357, pp. 295–306. Springer, Heidelberg (2012)
6. Chazelle, B.: Filtering search: a new approach to query-answering. SIAM J. Comput. 15(3), 703–724 (1986)
7. Fredman, M.L., Willard, D.E.: Trans-dichotomous algorithms for minimum spanning trees and shortest paths. J. Comput. Syst. Sci. 48(3), 533–551 (1994)
8. Gabow, H.N., Bentley, J.L., Tarjan, R.E.: Scaling and related techniques for geometry problems. In: Proc. 16th Annual ACM Symposium on Theory of Computing (STOC 1984), pp. 135–143 (1984)
9. Gupta, P., Janardan, R., Smid, M.H.M.: Further results on generalized intersection searching problems: counting, reporting, and dynamization. Journal of Algorithms 19(2), 282–317 (1995)
10. Janardan, R., Lopez, M.A.: Generalized intersection searching problems. International Journal of Computational Geometry and Applications 3(1), 39–69 (1993)
11. Larsen, K.G., Pagh, R.: I/O-efficient data structures for colored range and prefix reporting. In: Proc. 23rd Annual ACM-SIAM Symposium on Discrete Algorithms (SODA), pp. 583–592 (2012)
12. Larsen, K.G., van Walderveen, F.: Near-optimal range reporting structures for categorical data. In: Proc. 24th Annual ACM-SIAM Symposium on Discrete Algorithms (SODA) (to appear, 2013)
13. McCreight, E.M.: Priority search trees. SIAM J. Comput. 14(2), 257–276 (1985)
14. Miltersen, P.B., Nisan, N., Safra, S., Wigderson, A.: On data structures and asymmetric communication complexity. J. Comput. Syst. Sci. 57(1), 37–49 (1998)
15. Mortensen, C.W.: Generalized static orthogonal range searching in less space. Technical report, IT University Technical Report Series 2003-33 (2003)
16. Mortensen, C.W., Pagh, R., Patrascu, M.: On dynamic range reporting in one dimension. In: Proc. 37th Annual ACM Symposium on Theory of Computing (STOC), pp. 104–111 (2005)
17. Muthukrishnan, S.: Efficient algorithms for document retrieval problems. In: Proc. 13th Annual ACM-SIAM Symposium on Discrete Algorithms (SODA), pp. 657–666 (2002)
18. Nekrich, Y.: Space-efficient range reporting for categorical data. In: Proc. 31st ACM SIGMOD-SIGACT-SIGART Symposium on Principles of Database Systems (PODS), pp. 113–120 (2012)
19. Nekrich, Y., Vitter, J.S.: Optimal color range reporting in one dimension. CoRR, abs/1306.5029 (2013)
20. Shi, Q., JáJá, J.: Optimal and near-optimal algorithms for generalized intersection reporting on pointer machines. Inf. Process. Lett. 95(3), 382–388 (2005)
21. Thorup, M.: Undirected single-source shortest paths with positive integer weights in linear time. J. ACM 46(3), 362–394 (1999)
22. van Emde Boas, P., Kaas, R., Zijlstra, E.: Design and implementation of an efficient priority queue. Math. Sys. Theory 10, 99–127 (1977)
23. Willard, D.E.: Examining computational geometry, van Emde Boas trees, and hashing from the perspective of the fusion tree. SIAM J. Comput. 29(3), 1030–1049 (2000)

Lagrangian Duality in Online Scheduling with Resource Augmentation and Speed Scaling

Kim Thang Nguyen*

IBISC, University of Evry Val d'Essonne, France

Abstract. We present an unified approach to study online scheduling problems in the resource augmentation/speed scaling models. Potential function method is extensively used for analyzing algorithms in these models; however, they yields little insight on how to construct potential functions and how to design algorithms for related problems. In the paper, we generalize and strengthen the dual-fitting technique proposed by Anand et al. [1]. The approach consists of considering a possibly non-convex relaxation and its Lagrangian dual; then constructing dual variables such that the Lagrangian dual has objective value within a desired factor of the primal optimum. The competitive ratio follows by the standard Lagrangian weak duality. This approach is simple yet powerful and it is seemingly a right tool to study problems with resource augmentation or speed scaling. We illustrate the approach through the following results.

1. We revisit algorithms EQUI and LAPS in Non-clairvoyant Scheduling to minimize total flow-time. We give simple analyses to prove known facts on the competitiveness of such algorithms. Not only are the analyses much simpler than the previous ones, they also explain why LAPS is a natural extension of EQUI to design a scalable algorithm for the problem.
2. We consider the online scheduling problem to minimize total weighted flow-time plus energy where the energy power $f(s)$ is a function of speed s and is given by s^α for $\alpha \geq 1$. For a single machine, we showed an improved competitive ratio for a non-clairvoyant memoryless algorithm. For unrelated machines, we give an $O(\alpha/\log \alpha)$-competitive algorithm. The currently best algorithm for unrelated machines is $O(\alpha^2)$-competitive.
3. We consider the online scheduling problem on unrelated machines with the objective of minimizing $\sum_{i,j} w_{ij} f(F_j)$ where F_j is the flow-time of job j and f is an arbitrary non-decreasing cost function with some nice properties. We present an algorithm which is $\frac{1}{1-3\epsilon}$-speed, $\frac{2K(\epsilon)}{\epsilon}$-competitive where $K(\epsilon)$ is a function depending on f and ϵ. The algorithm does not need to know the speed $(1 + \epsilon)$ a priori. A corollary is a $(1 + \epsilon)$-speed, $\frac{k}{\epsilon^{1+1/k}}$-competitive algorithm (which does not know ϵ a priori) for the objective of minimizing the weighted ℓ_k-norm of flow-time.

---

* Supported by the French National Agency (ANR) project COCA ANR-09-JCJC-0066-01, and the GdR RO.

H.L. Bodlaender and G.F. Italiano (Eds.): ESA 2013, LNCS 8125, pp. 755–766, 2013.

1 Introduction

We consider online scheduling problems where jobs arrive at unrelated machines over time. Each job j has release date r_j and its processing time p_{ij} and weight w_{ij} on machine i. At the arrival time r_j, job j becomes known to the scheduling algorithm. We distinguish two different models. At time r_j, in the *non-clairvoyant* model only the weights w_{ij}'s becomes known to the scheduler while in the *clairvoyant* model, all parameter of jobs j are available. A scheduler must determine how to process jobs in order to optimize a quality of service without the knowledge about future. In the paper, we study natural qualities of service related to the flow-times of jobs. The *flow-time* of a job is the total amount of time it spends in the system, i.e., the difference of its completion time and its release time.

A popular measure for studying the performance of online algorithms is *competitive ratio*. An algorithm is said to be *c-competitive* if for any instance its objective is within factor c of the optimal offline algorithm's objective. Unfortunately, for many problems, any online algorithm has large competitive ratio even that some heuristics have performance very close to the optimum in practice. To remedy the limitation of pathological instances in worst-case analysis, a popular relaxation *resource augmentation* model was introduced in [22]. In this relaxation, the online algorithm is given extra speed to process jobs and compared to the optimal offline algorithm. This model has successfully provided theoretical evidence for heuristics with good performance in practice. Besides, algorithms could be classified according to their competitive ratios in the model of resource augmentation for practical choices. We say an algorithm is *s-speed c-competitive* if for any input instance the objective value of the algorithm while running at speed s is at most c times the objective value of the optimal offline scheduler while running at unit speed. Ideally, we would like algorithms to be constant competitive when given $(1+\epsilon)$ times a resource over the optimal offline algorithm for any constant $\epsilon > 0$. Such algorithms are called *scalable*.

The most successful tool until now to analyze online scheduling algorithms with resource augmentation is the potential function method. Potential functions has been designed and show that the corresponding algorithms behave well in an amortized sense. Designing such potential functions is far from trivial and often yields little insight about how to design such potential functions and algorithms for related problems (a generalized variant with additional constraints for example).

Recently, Anand et al. [1] gave a more direct and interesting approach for analyzing online scheduling algorithms with resource augmentation based on the technique of dual fitting for convex programming relaxation. Informally, the technique could be described as follows. Consider a linear (convex) programming relaxation of a given problem and the dual linear program (or Lagrangian dual). Then construct a feasible solution for the dual (given an online algorithm) and prove that its objective value is close to that of the online algorithm. The main advantage of this technique is that the dual variables (which constitute the desired dual solution) often have intuitive interpretations and their construction could be naturally deduced from the algorithm. Consequently, the procedures of analyzing and designing algorithms are more interactive and could be done in a principled manner.

Independently, Gupta et al. [18] gave a principled method to design online algorithms for non-linear programs. Their approach could be seen as an extension of the online primal-dual method for linear programming [9]. Roughly speaking, in the method the dual variables are set in such a way that the increase rate in the dual objective is proportional to the one in the primal objective. This approach is particularly powerful while the primal objective function is convex.

1.1 Approach and Contributions

The main contribution of the paper is to show a principled approach to design/analyze online scheduling algorithms with resource augmentation (or speed scaling) by strengthening the dual fitting technique in [1]. The approach is sharply inspired by the one in [1]. First, consider a mathematical programming relaxation (associated with a given problem) which is *not* necessarily convex and its Lagrangian dual. Then construct dual variables such that the Lagrangian dual has objective value within a desired factor of the primal one (due to some algorithm). Then by the standard Lagrangian weak duality for mathematical programming, the competitive ratio follows.

Lemma 1 (Weak duality). *Consider a possibly non-convex optimization problem*

$$p^* := \min_x f_0(x) \ : \quad f_i(x) \leq 0, \quad i = 1, \ldots, m.$$

where $f_i : \mathbb{R}^n \to \mathbb{R}$ *for* $0 \leq i \leq m$. *Let* $\mathcal{X}$ *be the feasible set of* x. *Let* $L : \mathbb{R}^n \times \mathbb{R}^m \to \mathbb{R}$ *be the Lagragian function*

$$L(x, \lambda) = f_0(x) + \sum_{i=1}^m \lambda_i f_i(x).$$

Define $d^* = \max_{\lambda \geq 0} \min_{x \in \mathcal{X}} L(x, \lambda)$ *where* $\lambda \geq 0$ *means* $\lambda \in \mathbb{R}_+^m$. *Then* $p^* \geq d^*$.

Weak duality is indeed a direct consequent of the *minimax* inequality

$$\max_{\lambda \in \mathcal{Y}} \min_{x \in \mathcal{X}} L(x, \lambda) \leq \min_{x \in \mathcal{X}} \max_{\lambda \in \mathcal{Y}} L(x, \lambda)$$

where $\mathcal{X}$ and $\mathcal{Y}$ are feasible sets of x and λ. Intuitively, our approach could be considered as a one-shot game between an algorithm and an adversary. The algorithm chooses dual variables λ^* in such a way that whatever the choice of the adversary, the value $\min_{x \in \mathcal{X}} L(x, \lambda^*)$ is always within a desirable factor c of the objective due to the algorithm. In the model, the adversary has less resource than the algorithm. For example, if the algorithm processes jobs with unit rate then the adversary can run only with rate $(1 - \epsilon)$. We extensively use that advantage in proving bounds for the dual objective.

In high level, our approach is the same as the one in [1] except that the relaxation is possibly non-convex. However, the flexibility of our approach provides many advantages. First, a problem could be more directly and naturally formulated as a non-convex program. For example, the online scheduling problem to minimize total weighted flow time plus energy could be naturally formulated by a non-convex relaxation (Section 4) while it is unclear how to formalize the problem by a convex program. Consequently, the analysis is usually simpler, cleaner and the performance guarantee is improved. Inversely, the simplicity of the analysis gives insights on the problems and so (simple) algorithms could be designed. Second, as it is not constrained to be a convex optimization program, additional constraints for generalized variants of a problem could be easily incorporated (for example, from a single machine to unrelated machines). Thereby an algorithm for generalized variants could be derived based on the previous ones for the basic problem and the ideas of the analyses remain essentially the same.

We illustrate the advantages of the approach through the following results.

1. In Section 3, we revisit algorithms EQUI and LAPS$_\epsilon$ in Non-clairvoyant Scheduling to minimize total flow-time. We give simple analyses to prove known facts that EQUI is $\frac{1}{1/2-\epsilon}$-speed, $\frac{1}{\epsilon}$-competitive [14] and LAPSϵ is $\frac{1}{1-2\epsilon}$-speed, $\frac{2}{\epsilon^2}$-competitive [15]. Not only are the analyses much simpler than the previous ones, they also explain why LAPS$_\epsilon$ is a natural extension of EQUI to design a scalable algorithm for the problem.

2. In Section 4, we consider the online scheduling problem to minimize total weighted flow-time plus energy where the energy power $f(s)$ is a function of speed s and is given by s^α for $\alpha \geq 1$. For a single machine, we showed an improved competitive ratio $O(2^\alpha)$ for a non-clairvoyant memoryless algorithm (its performance was previously known to be $O(3^\alpha)$). For unrelated machines, we give an $O(\alpha/\log\alpha)$-competitive algorithm. This bound matches to the currently best algorithm for a single machine [5]. The currently best algorithm for unrelated machines is $O(\alpha^2)$-competitive [1].

3. In Section 5, we consider the online scheduling problem on unrelated machines with the objective of minimizing $\sum_{i,j} w_{ij} f(F_j)$ where F_j is the flow-time of job j and f is an arbitrary non-decreasing cost function with some nice properties (for example, f is in class $\mathcal{C}^1$ and f' is non-decreasing). We derive an algorithm which is $\frac{1}{1-3\epsilon}$-speed, $\frac{2K(\epsilon)}{\epsilon}$-competitive where $K(\epsilon)$ is a function depending on f and ϵ. The algorithm does not need to know the speed $(1 + \epsilon)$ a priori. A corollary is a $(1 + \epsilon)$-speed, $\frac{k}{\epsilon^{1+1/k}}$-competitive algorithm (which does not know ϵ a priori) for the objective of minimizing the weighted ℓ_k-norm of flow-time. That answers an open question in [19] and marginally improves the currently best known algorithm which is $(1 + \epsilon)$-speed, $\frac{k}{\epsilon^{2+1/k}}$-competitive [1].

Besides, using the approach, related problems and direct generalizations of the above problems could be proved.

1.2   Related Work

The online problems of minimizing objectives related to (weighted) flow-times of jobs have been extensively studying. For the basic problem of minimizing total flow-time on single machine, it is well-known that Shortest Remaining Processing Time (SRPT) is the optimal algorithm. However, that is the only constant competitive algorithm. Bansal and Chan [3] showed that no algorithm is constant competitive for the problem of minimizing total weighted flow-time on single machine. In fact, no bounded competitive ratio holds for parallel machines setting [13,17].

The strong lower bounds motivate the use of resource augmentation, originally introduced by Kalyanasundaram and Pruhs [22], which circumvents the persimist worst-case paradigm. In the same paper, the authors gave an $O(1/\epsilon)$-competitive algorithm, called SETF, for the objective of minimizing flow-time on a single machine in the non-clairvoyant setting. In this setting, without resource augmentation the competitive ratios of every deterministic and randomized algorithms are $\Omega(n^{1/3})$ and $\Omega(\log n)$, respectively [26]. Edmonds [14] considered algorithm EQUI and showed that it was $(2 + \epsilon)$-speed, $2/\epsilon$-competitive. Later on, Edmonds and Pruhs [15] proposed a generalized algorithm called LAPS$_\epsilon$. They proved that LAPS$_\epsilon$ is $(1 + 2\epsilon)$-speed, $4/\epsilon^2$-competitive for minimizing the objective of total flow-time (even with sublinear non-decreasing speedup curves).

In the clairvoyant setting, Bansal and Pruhs [6] proved that the Highest Density First (HDF) algorithm is $(1 + \epsilon)$-speed, $O(1/\epsilon)$-competitive for the objective of weighted ℓ_k-norm of flow-time on a single machine. Chadha et al. [10] gave the first $(1 + \epsilon)$-speed, $O(1/\epsilon^2)$-competitive algorithm for minimizing weighted flow time on unrelated machines. Recently, using the approach based on linear programming and dual-fitting, Anand et al. [1] derived another simple algorithm which is $(1+\epsilon)$-speed, $O(1/\epsilon)$-competitive. Moreover, the authors extended this to an $(1 + \epsilon)$-speed, $O(k/\epsilon^{2+1/k})$-competitive algorithm for the objective of weighted ℓ_k-norm of flow-time. Note that the latter needs to know the speed $(1 + \epsilon)$ a priori.

For the objective of total flow-time plus energy on a single machine, Bansal et al. [5] gave a $(3 + \epsilon)$-competitive algorithm. Besides, they also proved a $(2 + \epsilon)$-competitive algorithm for minimizing total *fractional* weighted flow-time plus energy. Their results hold for a general class of convex power functions. Those results also imply an $O(\alpha/\log\alpha)$-competitive algorithm for weighted flow-time plus energy when the energy function is s^α. Again, always based on linear programming and dual-fitting, Anand et al. [1] proved an $O(\alpha^2)$-competitive algorithm for unrelated machines. The total (weighted) flow-time plus energy in non-clairvoyant setting has been also considered [11,25]. Chan et al. [12] proved that a memoryless non-clairvoyant algorithm, which a variant of algorithm EQUI with a policy on speed, was $O(3^\alpha)$ competitive.

The objective of minimizing $\sum_{i,j} w_{ij} f(F_j)$ for general cost function f aims to capture multiple standard objectives in literature (weighted ℓ_k-norm of flow-time, weighted tardiness). A competitive algorithm for a general cost function could be useful particularly in scheduling with multiple objectives or in setting where objectives may compete with each other [2,24]. For the offline version on a single machine, Bansal and Pruhs [7] presented a polynomial time $O(\log \log P)$-approximation algorithm [7,8] where P is the ratio of the maximum to minimum job size. Im et al. [21] showed that the HDF algorithm is $(2 + \epsilon)$-speed, $O(1)$-competitive for arbitrary non-decreasing cost function f on a single machine. They also gave a scalable algorithm when f is a concave and twice differentiable.

Almost all of competitive algorithms with resource augmentation are proved by potential functions. Those clever functions are used to show that a particular algorithm is locally competitive in an amortized sense. Recently, a principle approach to construct potential functions for online scheduling has been systematically formalized and given in [20] for many problems. However, it does not apply to all, for example the problem we consider in Section 5. More importantly, that still yields little insight about how to design algorithms and construct potential functions for related problems or for non-trivial generalized variants.

Anand et al. [1] was the first who proposed studying online scheduling with resource augmentation by linear (convex) programming and dual fitting. By this elegant approach, they gave simple algorithms and simple analysis with improved performance for problems where the analyses based on potential functions are complex or it is unclear how to design such functions. Our approach is greatly inspired by the one in [1].

Independently, Gupta et al. [18] gave a principled method to design online algorithms for non-linear programs. They showed the application of the method to online speed-scaling problems. Subsequently, [23] have applied the method to design an α^{α}-competitive for the problem of minimizing the consumed energy plus lost values.

2   Preliminaries

In unrelated machine environment, we are given a set of m machines and jobs arrive over time. A job j is released at time r_j and requires p_{ij} units of processing time if it is scheduled on machine i. The machines are allowed to process jobs preemptively. The *flow-time* of a job j is $F_j = C_j - r_j$ where C_j is its the completion time. If a job j is assigned to machine i then its weighted flow-time is $w_{ij}F_{ij}$. Consider a scheduling algorithm. A job j is *pending* at time t if it is not completed by the algorithm, i.e., $r_j \leq t < C_j$. At time t, we denote $q_{ij}(t)$ the *remaining* processing time of job j on machine i. The *total weight* of pending jobs assigned to machine i at time t is denoted as $W_i(t)$. In case where all jobs have unit weight, we use $N_i(t)$ (number of pending jobs) instead of $W_i(t)$. The *residual density* of a pending job j assigned to machine i at time t is $\delta_{ij}(t) = w_{ij}/q_{ij}(t)$. The *density* of a job j on machine i is $\delta_{ij}(r_j)$. We distinguish two different models: the *non-clairvoyant* model in which at the arrival of job j, only the weights w_{ij}'s becomes known to the scheduler; and the *clairvoyant* model in which all parameter of jobs j are available at its release time. Note that when only a single machine is considered, for simplicity the notations remain the same except that the machine index (usually i) will be dropped.

3   Non-clairvoyant Scheduling

The Problem. In this section, we study the non-clairvoyant online scheduling problem with the objective of minimizing the total flow-time on a single machine. Let $x_j(t)$ be the variable that represents the processing rate of the machine on job j at time t for every job j. Let C_j be a variable representing the completion time of j. The relaxation could be formulated as the following mathematical program. We notice again that in our approach the programs do *not* need to be convex.

$$\min \quad \sum_j \frac{C_j - r_j}{p_j} \int_{r_j}^{C_j} x_j(t)dt$$

$$\text{subject to} \quad \int_{r_j}^{C_j} x_j(t)dt = p_j \qquad \forall j$$

$$\sum_{j=1}^{n} x_j(t) \le 1 \qquad \forall t$$

$$x_j(t) \ge 0 \qquad \forall j, t$$

$$x_j(t) = 0 \qquad \forall j, \forall t \notin [r_j, C_j]$$

Observe that the last constraints are redundant but they are kept in order to make the relaxation clear. The dual of that program is $\max_{\lambda,\gamma,\mu} \min_{x,C} L(x, C, \lambda, \gamma, \mu)$ where L is the Lagrangian

$$\sum_j \int_{r_j}^{C_j} \frac{C_j - r_j}{p_j} x_j(t)dt + \sum_j \lambda_j \left(p_j - \int_{r_j}^{C_j} x_j(t)dt \right)$$

$$+ \int_0^\infty \left(1 - \sum_j x_j(t) \right) \gamma(t)dt - \sum_j \int_0^\infty x_j(t)\mu_j(t)dt$$

$$= \sum_j \lambda_j p_j - \sum_j \int_0^\infty x_j(t) \cdot \left(\lambda_j + \gamma(t) - \frac{C_j - r_j}{p_j} \right) dt$$

$$+ \int_0^\infty \gamma(t)dt - \sum_j \int_0^\infty x_j(t)\mu_j(t)dt$$

Remark that the weak duality holds also for functions instead of variables. In the setting, one could see the dual $\max_{\lambda,\gamma,\mu} \min_{x,C} L(x, C, \lambda, \gamma, \mu)$ as an optimization problem over functions $x_j(t)$ and others (calculus of variations); or as an optimization over variables (x, t) and others (by a transformation $x_j(t) \mapsto (x_j, t)$).

3.1   EQUI

Algorithm EQUI. The processor shares its resource equally to the pending jobs.

Let $q_1 \leq \ldots \leq q_n$ be remaining processing times of pending jobs at some time t. Assume that no new job is released after t, then the remaining time before completion for the first job is nq_1, that for the second job is $nq_1 + (n-1)(q_2 - q_1)$. By recurrence, the remaining time before completion for job j is $q_1 + \ldots + q_{j-1} + (n-j)q_j$ for $1 \leq j \leq n$.

Suppose that at time t, a new job arrives with processing time q such that $q_k \leq q < q_{k+1}$ for some index k. Then the flow time of the new job, assuming that no new job is released after t, is $q_1 + \ldots + q_{k-1} + (n+1-k)q$. Moreover, due to the arrival of the new job, the completion time of job k' is increased by $q_{k'}$ for $k' \leq k$; and by q for $k' > k$. Hence, the marginal increase of the total flow time due to the arrival of the new job is bounded by twice the flow time of that job.

Dual Variables. Choose $\gamma(t) = 0$, $\mu_j(t) = 0$ for every j, t and $\lambda_j = \lambda_j^E$ such that $\lambda_j^E p_j$ equals the flow time of j (due to the algorithm) assuming that no new job arrives after r_j. By the observation on the flow time of jobs in EQUI, we have that $\sum_j \lambda_j^E p_j \leq \mathcal{F}^E \leq 2 \sum_j \lambda_j^E p_j$ where $\mathcal{F}^E$ is the total flow-time due to EQUI.

Lemma 2. *It holds that $\frac{1}{p_j} \left(\lambda_j^E p_j - (t - r_j) \right) \leq N^E(t)$ for $t \geq r_j$ where $N^E(t)$ is the number of pending jobs at time t by algorithm EQUI.*

Proof. Observe that if some request arrives between time r_j and t, the left-hand side remains unchanged while the right hand-side is non-decreasing. Therefore, it is sufficient to prove the inequality assuming that no job is released after r_j. Consider $t \leq C_j^E$ (since otherwise the inequality trivially holds since the left-hand side is negative). Rename jobs in non-decreasing order of the remaining processing times at r_j, i.e., $q_1(r_j) \leq \ldots \leq q_n(r_j)$. Note that $p_j = q_j(r_j)$. Suppose that k is the pending job with smallest index at time t, i.e., jobs $1, \ldots, k-1$ have been completed. We have that

$$\frac{1}{p_j} \left(\lambda_j^E p_j - (t - r_j) \right) = \frac{1}{p_j} \left(q_k(t) + \ldots + q_{j-1}(t) + (n-j)q_j(t) \right) \leq N^E(t)$$

where the last inequality follows since $q_k(t) \leq \ldots \leq q_{j-1}(t) \leq q_j(t) \leq p_j$. □

Theorem 1 ([14]). *Algorithm EQUI is $\frac{1}{1/2 - \epsilon}$-speed, $\frac{1}{\epsilon}$-competitive for the problem of minimizing total flow time.*

Proof. As the adversary has only the speed $(1/2 - \epsilon)$, the processing rate of adversary $\sum_j x_j(t) \leq 1/2 - \epsilon$ for all t. By the choice of dual variables corresponding to EQUI, we have

$$\min_{x,C} L \geq \frac{\mathcal{F}^E}{2} - \int_0^\infty \sum_j x_j(t) N^E(t) \geq \frac{\mathcal{F}^E}{2} - \left(\frac{1}{2} - \epsilon \right) \int_0^\infty N^E(t) = \epsilon \mathcal{F}^E$$

where the first inequality is due to Lemma 2; the second inequality follows by $\sum_j x_j(t) \leq 1/2 - \epsilon$. Hence, the competitive ratio of EQUI is at most $1/\epsilon$. □

3.2 LAPS$_\beta$

Inspecting the analysis of EQUI, one realizes that in order to get a scalable algorithm, the machine should share its power only to a small fraction of pending jobs instead of all such jobs. This observation naturally leads to algorithm LAPS introduced in [15].

Algorithm LAPS$_\beta$ Let $0 < \beta \leq 1$. The processor shares its resource equally to the $\beta N^L(t)$ jobs with the latest arrival times where $N^L(t)$ is the number of pending jobs at time t.

Note that in the definition of the algorithm, $\beta N^L(t)$ is not necessarily an integer. However, that algorithm is equivalent to the following procedure. First, choose the $\lceil \beta N^L(t) \rceil$ most recent jobs. Then among such jobs, the machine shares its power to the $\lfloor \beta N^L(t) \rfloor$ most recent ones proportional to 1 and to the last job proportional to $(\beta N^L(t) - \lfloor \beta N^L(t) \rfloor)$. For the ease and simplicity of the exposition, we consider the version described in the definition.

Theorem 2 ([15]). *Algorithm* LAPS$_\epsilon$ *is* $\frac{1}{1-2\epsilon}$*-speed,* $\frac{2}{\epsilon^2}$*-competitive for the problem of minimizing total flow time.*

4 Weighted Flowtime Plus Energy

The Problem. In this section, we study the online scheduling with the objective of minimizing the total weighted flow-time plus energy. The energy power function is given by s^α where s is the speed of the machine and $\alpha \geq 1$ is a constant. In Section 4.1, we consider non-clairvoyant algorithms on a single machine and Section 4.2, we consider algorithms on unrelated machines.

4.1 Non-clairvoyant Scheduling on Single Machine

Algorithm. At time t, the machine maintains a speed $s(t) = \beta W(t)^{1/\alpha}$ where $W(t)$ is the total weight of pending jobs and β is a constant to be defined later. At any time, the machine shares its resource to pending jobs proportional to their weights.

Theorem 3. *The algorithm is* 2^α*-competitive for* $\beta = 2$.

4.2 Clairvoyant Scheduling on Unrelated Machines

Scheduling Policy. At any time t, every machine i sets its speed $s_i(t) = \beta W_i(t)^{1/\alpha}$ where $W_i(t)$ is the total (integral) weight of pending jobs assigned to machine i; and $\beta > 0$ is a constant to be chosen later. At any time, every machine i processes the highest residual density job among the pending ones assigned to i.

Assignment Policy. At the arrival of a job j, assign j to machine i that minimizes the marginal increase (due to the scheduling policy) of the total weighted flow-time.

Theorem 4. *The algorithm is* $8(1+\frac{\alpha}{\ln \alpha})$-*competitive for* $\beta = \frac{1}{\alpha-1}(\alpha-1+\ln(\alpha-1))^{\frac{\alpha-1}{\alpha}}$.

5   Arbitrary Cost Functions of Flow-Time

The Problem. In this section, we study the online scheduling on unrelated machines to minimize a general objective $\sum_{i,j} w_{ij} f(F_j)$ where f is a function with certain properties (described below). At the arrival time of a job, the scheduler has to immediately assign it to a machine. Jobs will be entirely processed on their machines and the migration of jobs across machines is not allowed. (In practice, it is not desirable to migrate jobs from a machine to others.)

Properties $f(0) = f'(0) = 0$ and for any $\epsilon > 0$ arbitrarily small,

(P1) there exists a function $K_1(\epsilon)$ such that $f(z_1 + z_2) \leq \frac{1}{1-\epsilon} f(z_1) + K_1(\epsilon) f(z_2)$ $\forall z_1, z_2 \geq 0$;

(P2) $f'(z)$ is non-decreasing. By this property, we can deduce that

$$\sum_{\ell=1}^{k} a_\ell f'(A_{\ell-1}) \leq f(A_k) \leq \sum_{i=\ell}^{k} a_\ell f'(A_\ell)$$

where $A_\ell = a_1 + \ldots + a_\ell$ and $a_\ell \geq 0$ for every $1 \leq \ell \leq k$.

(P3) there exists a function $K_2(\epsilon)$ such that $f'(z_1+z_2) \leq \frac{1}{1-\epsilon} f'(z_1) + K_2(\epsilon) f'(z_2)$ $\forall z_1, z_2 \geq 0$;

(P4) there exists a function $K_3(\epsilon)$ such that $f'(z + \frac{z}{K_3(\epsilon)}) \leq \frac{1}{1-\epsilon} f'(z)$ $\forall z \geq 0$;

(P5) there exists a function $K_4 \geq 1$ such that $z f'(z) \leq K_4 f(z)$ $\forall z \geq 0$.

Scheduling Policy. At time t, every machine i schedules the highest residual density job among the ones assigned to i.

Assignment Policy. For a job j, recall that $q_{ij}(t)$ is the remaining processing time of j on machine i. Let $Q_j(t)$ be the remaining time of job j from t to its completion time by the algorithm. Let $U_i(t)$ be the set of jobs assigned to machine i and are still pending at t. At the arrival time r_j, job j is assigned to the machine i that minimize $\widetilde{\lambda}_{ij}$, which is defined as

$$\delta_{ij} f\left(\sum_{\substack{u \in U_i(r_j) \\ \delta_u(r_j) \geq \delta_{ij}}} q_u(r_j) + p_{ij} \right) + \sum_{\substack{u \in U_i(r_j) \\ \delta_u(r_j) < \delta_{ij}}} \frac{w_{iu}}{p_{ij}} \Big(f(Q_u(r_j) + p_{ij}) - f(Q_u(r_j)) \Big)$$

where δ_{ij} is the density of job j on machine i, i.e., $\delta_{ij} = \delta_{ij}(r_j)$. Note that $\widetilde{\lambda}_{ij} p_{ij}$ is the marginal increase of the objective function if job j is assigned to machine i.

Theorem 5. *The algorithm is $\frac{1}{1-3\epsilon}$-speed and $\frac{2K(\epsilon)}{\epsilon}$-competitive where $K(\epsilon) = \max\{K_1(\epsilon), 3K_2(\epsilon)K_3(\epsilon)K_4\}$.*

Corollary 1. *The algorithm is $\frac{1}{1-3\epsilon}$-speed $O(\frac{k}{\epsilon^{1+1/k}})$-competitive for the objective of weighted ℓ_k-norm of flow-time.*

6 Conclusion and Further Directions

In the paper, we have proved competitive algorithms in the resource augmentation/speed scaling models for different online scheduling problems using an unified approach. The approach is simple yet powerful in designing and analyzing algorithms. It seems to be a right tool to study problems in the resource augmentation/speed scaling models. Besides the extensions mentioned in previous sections, a future direction is to study online scheduling problems with the objectives of different nature, for example throughput-related objective. Moreover, different constraints might be incorporated, for example the bounded-speed model [4,25] or the capacitated machine model [16].

An interesting future direction is to investigate different online problems with resource augmentation using the approach. Moreover, the min max game between algorithms and adversaries may give insights not only for designing algorithms but also for constructing counter-examples.

Acknowledgment. We would like to thank Kirk Pruhs and anonymous reviewers for pointing out related references and useful comments.

References

1. Anand, S., Garg, N., Kumar, A.: Resource augmentation for weighted flow-time explained by dual fitting. In: Proc. 23rd ACM-SIAM Symposium on Discrete Algorithms, pp. 1228–1241 (2012)
2. Azar, Y., Epstein, L., Richter, Y., Woeginger, G.J.: All-norm approximation algorithms. J. Algorithms 52(2), 120–133 (2004)
3. Bansal, N., Chan, H.-L.: Weighted flow time does not admit $o(1)$-competitive algorithms. In: Proc. 20th ACM-SIAM Symposium on Discrete Algorithms, pp. 1238–1244 (2009)
4. Bansal, N., Chan, H.-L., Lam, T.-W., Lee, L.-K.: Scheduling for speed bounded processors. In: Aceto, L., Damgård, I., Goldberg, L.A., Halldórsson, M.M., Ingólfsdóttir, A., Walukiewicz, I. (eds.) ICALP 2008, Part I. LNCS, vol. 5125, pp. 409–420. Springer, Heidelberg (2008)
5. Bansal, N., Chan, H.-L., Pruhs, K.: Speed scaling with an arbitrary power function. In: Proc. 20th ACM-SIAM Symposium on Discrete Algorithms, pp. 693–701 (2009)
6. Bansal, N., Pruhs, K.R.: Server scheduling in the weighted ℓ_p norm. In: Farach-Colton, M. (ed.) LATIN 2004. LNCS, vol. 2976, pp. 434–443. Springer, Heidelberg (2004)
7. Bansal, N., Pruhs, K.: The geometry of scheduling. In: Proc. 51th Symposium on Foundations of Computer Science, pp. 407–414 (2010)

8. Bansal, N., Pruhs, K.: Weighted geometric set multi-cover via quasi-uniform sampling. In: Epstein, L., Ferragina, P. (eds.) ESA 2012. LNCS, vol. 7501, pp. 145–156. Springer, Heidelberg (2012)

9. Buchbinder, N., Naor, J.: The design of competitive online algorithms via a primal-dual approach. Foundations and Trends in Theoretical Computer Science 3(2-3), 93–263 (2009)

10. Chadha, J.S., Garg, N., Kumar, A., Muralidhara, V.N.: A competitive algorithm for minimizing weighted flow time on unrelatedmachines with speed augmentation. In: Proc. 41st ACM Symposium on Theory of Computing, pp. 679–684 (2009)

11. Chan, H.-L., Edmonds, J., Lam, T.W., Lee, L.-K., Marchetti-Spaccamela, A., Pruhs, K.: Nonclairvoyant speed scaling for flow and energy. Algorithmica 61(3), 507–517 (2011)

12. Chan, S.-H., Lam, T.W., Lee, L.-K., Ting, H.-F., Zhang, P.: Non-clairvoyant scheduling for weighted flow time and energy on speed bounded processors. Chicago J. Theor. Comput. Sci. (2011)

13. Chekuri, C., Khanna, S., Zhu, A.: Algorithms for minimizing weighted flow time. In: Proc. 33rd ACM Symposium on Theory of Computing, pp. 84–93 (2001)

14. Edmonds, J.: Scheduling in the dark. Theor. Comput. Sci. 235(1), 109–141 (2000)

15. Edmonds, J., Pruhs, K.: Scalably scheduling processes with arbitrary speedup curves. ACM Transactions on Algorithms 8(3), 28 (2012)

16. Fox, K., Korupolu, M.: Weighted flowtime on capacitated machines. In: Proc. 24th ACM-SIAM Symposium on Discrete Algorithms, pp. 129–143 (2013)

17. Garg, N., Kumar, A.: Minimizing average flow-time: Upper and lower bounds. In: Proc. 48th Symposium on Foundations of Computer Science, pp. 603–613 (2007)

18. Gupta, A., Krishnaswamy, R., Pruhs, K.: Online primal-dual for non-linear optimization with applications to speed scaling. In: Proc. 10th Workshop on Approximation and Online Algorithms, pp. 173–186 (2012)

19. Im, S.: Online Scheduling Algorithms for Average Flow Time and its Variants. PhD thesis, University of Illinois at Urbana-Champaign (2012)

20. Im, S., Moseley, B., Pruhs, K.: A tutorial on amortized local competitiveness in online scheduling. SIGACT News 42(2), 83–97 (2011)

21. Im, S., Moseley, B., Pruhs, K.: Online scheduling with general cost functions. In: Proc. 23rd ACM-SIAM Symposium on Discrete Algorithms, pp. 1254–1265 (2012)

22. Kalyanasundaram, B., Pruhs, K.: Speed is as powerful as clairvoyance. J. ACM 47(4), 617–643 (2000)

23. Kling, P., Pietrzyk, P.: Profitable scheduling on multiple speed-scalable processors. In: Proc. 25th Symposium on Parallelism in Algorithms and Architectures (2013)

24. Kumar, V.S.A., Marathe, M.V., Parthasarathy, S., Srinivasan, A.: A unified approach to scheduling on unrelated parallel machines. J. ACM 56(5) (2009)

25. Lam, T.W., Lee, L.-K., To, I.K., Wong, P.W.H.: Online speed scaling based on active job count to minimize flow plus energy. Algorithmica 65(3), 605–633 (2013)

26. Motwani, R., Phillips, S., Torng, E.: Non-clairvoyant scheduling. Theor. Comput. Sci. 130(1), 17–47 (1994)

Euclidean Greedy Drawings of Trees

Martin Nöllenburg and Roman Prutkin

Institute of Theoretical Informatics, Karlsruhe Institute of Technology, Germany

Abstract. Greedy embedding (or drawing) is a simple and efficient strategy to route messages in wireless sensor networks. For each source-destination pair of nodes s, t in a greedy embedding there is always a neighbor u of s that is closer to t according to some distance metric. The existence of Euclidean greedy embeddings in $\mathbb{R}^2$ is known for certain graph classes such as 3-connected planar graphs. We completely characterize the trees that admit a greedy embedding in $\mathbb{R}^2$. This answers a question by Angelini et al. (Graph Drawing 2009) and is a further step in characterizing the graphs that admit Euclidean greedy embeddings.

1 Introduction

Message routing in wireless ad-hoc and sensor networks cannot apply the same established global hierarchical routing schemes that are used, e.g., in the Internet Protocol. A family of alternative routing strategies in wireless networks known as *geographic routing* uses node locations as addresses instead. The *greedy routing* protocol simply passes a message at each node to a neighbor that is closer to the destination. Two problems with this approach are (i) that sensor nodes typically are not equipped with GPS receivers due to their cost and energy consumption and (ii) that even if nodes know their positions messages can get stuck at voids, where no node closer to the destination exists.

An elegant strategy to tackle these issues was proposed by Rao et al [12]. It maps nodes to virtual rather than geographic coordinates, on which the standard greedy routing is then performed. A mapping that always guarantees successful delivery is called a *greedy embedding* or *greedy drawing*.

The question about the existence of greedy embeddings for various metric spaces and classes of graphs has attracted a lot of interest. Papadimitriou and Ratajczak [11] conjectured that every 3-connected planar graph admits a greedy embedding into the Euclidean plane. Dhandapani [4] proved that every 3-connected planar triangulation has a greedy drawing. The conjecture by Papadimitriou and Ratajczak itself has been proved independently by Leighton and Moitra [9] and Angelini et al. [3]. Kleinberg [8] showed that every connected graph has a greedy embedding in the hyperbolic plane.

Since efficient use of storage and bandwidth are crucial in wireless sensor networks, virtual coordinates should require only few, i.e., $O(\log n)$, bits in order to keep message headers small. Greedy drawings with this property are called *succinct*. Eppstein and Goodrich proved the existence of succinct greedy drawings for 3-connected planar graphs in $\mathbb{R}^2$ [6], and Goodrich and Strash [7] showed

H.L. Bodlaender and G.F. Italiano (Eds.): ESA 2013, LNCS 8125, pp. 767–778, 2013.

it for any connected graph in the hyperbolic plane. Wang and He [14] used a custom distance metric and constructed convex, planar and succinct drawings for 3-connected planar graphs using Schnyder realizers [13].

It has been known that not all graphs admit a Euclidean greedy drawing in the plane, e.g., $K_{k,5k+1}$ ($k \geq 1$) has no such drawing [11], including the tree $K_{1,6}$. Leighton and Moitra [9] showed that a graph containing at least six *pairwise independent irreducible triples* (e.g., the complete binary tree containing 31 nodes) cannot have a greedy embedding. They used this fact to present a planar graph that admits a greedy embedding, although none of its spanning trees does. We show that there are trees with no greedy drawing that contain at most *five* such triples [10]. Further, some greedy-drawable trees have no succinct Euclidean greedy drawing [2].

Self-approaching drawings [1] are a subclass of greedy drawings with the additional constraint that for any pair of nodes there is a path ρ that is distance decreasing not just for the node sequence of ρ but for any triple of intermediate points on the edges of ρ. Alamdari et al. [1] gave a complete characterization of trees admitting self-approaching drawings. Since self-approaching drawings are greedy, all trees with a self-approaching drawing are greedy-drawable. However, there exist numerous trees that admit a greedy drawing, but no self-approaching one, and the characterization of those trees turns out to be quite complex.

Contributions. We give the first complete characterization of all trees that admit a greedy embedding in $\mathbb{R}^2$ with the Euclidean distance metric. This solves the corresponding open problem stated by Angelini et al. [2] and is a further step in characterizing the graphs that have greedy embeddings. We show that deciding whether T has a greedy drawing is equivalent to deciding whether there exists a valid angle assignment in a certain wheel polygon. This includes a non-linear constraint known as the *wheel condition* [5]. For most cases (all trees with maximum degree 4 and most trees with maximum degree 5) we are able to give an explicit solution to this problem, which provides a linear-time recognition algorithm. For trees with maximum degree 3 we give an alternative characterization by forbidden subtrees in the full version of this paper [10]. For some trees with one degree 5 node we resort to using non-linear solvers. For trees with nodes of degree ≥ 6 no greedy drawings exist.

Our proofs are constructive, however, we ignore the possibly exponential area requirements for our constructions. This is justified by the aforementioned result that some trees require exponential-size greedy drawings [2]. Due to space constraints several proofs are omitted; for details we refer to the full paper [10].

2   Preliminaries

In this section, we introduce the concept of the opening angle of a rooted subtree and present relations between opening angles that will be crucial for the characterization of greedy-drawable trees.

Let $T = (V, E)$ be a tree. A *straight-line drawing* Γ of T maps every node $v \in V$ to a point in the plane $\mathbb{R}^2$ and every edge $uv \in E$ to the line segment between its endpoints. We say that Γ is *greedy* if for every pair of nodes s, t there is a neighbor u of s with $|ut| < |st|$, where $|st|$ is the Euclidean distance between points s and t. To ease notation we identify nodes with points and edges with line segments. Furthermore we assume that all drawings are straight-line drawings.

It is known that for a greedy drawing Γ of T any subtree of T is represented in Γ by a greedy subdrawing [2]. We define the *axis* of an edge uv as its perpendicular bisector. Let h^u_{uv} denote the open half-plane bounded by the axis of uv and containing u. Let T^u_{uv} be the subtree of T containing u obtained from T by removing uv. Angelini et al. [2] showed that in a greedy drawing of T every subtree T^u_{uv} is contained in h^u_{uv}. The converse is also true.

Lemma 1. *Let Γ be a drawing of T with $T^u_{uv} \subseteq h^u_{uv} \; \forall uv \in E$. Then, Γ is greedy.*

Angelini et al. [2] further showed that greedy tree drawings are always planar and that in any greedy drawing of T the angle between two adjacent edges must be strictly greater than $60°$. Thus T cannot have a node of degree ≥ 6.

Let $\text{ray}(u, \vec{uv})$ denote the ray with origin u and direction $\vec{uv}$. For $u, v \in V$, let $d_T(u, v)$ be the length of the u-v path in T. For vectors $\vec{ab}, \vec{cd}$, let $\angle_{\text{ccw}}(\vec{ab}, \vec{cd})$ denote the counterclockwise turn (or turning angle) from $\vec{ab}$ to $\vec{cd}$.

Lemma 2 (Lemma 7 in [2]). *Consider two edges uv and wz in a greedy drawing of T, such that the path from u to w does not contain v and z. Then, the rays $\text{ray}(u, \vec{uv})$ and $\text{ray}(w, \vec{wz})$ diverge; see Fig. 1a.*

Lemma 3. *Let Γ be a greedy drawing of T, $v \in V$, $\deg(v) = 2$, $N(v) = \{u, w\}$ the only two neighbors of v, and $T' = T - \{uv, vw\} + \{uw\}$. The drawing Γ' induced by replacing segments uv, vw by uw in Γ is also greedy.*

Next we generalize some concepts from Leighton and Moitra [9]. For $k = 3, 4, 5$, we define an *irreducible k-tuple* as a k-tuple of nodes $(b_1, \ldots, b_k)$ in a graph $G = (V, E)$, such that $\deg(b_1) = k$, $b_1 b_2, b_1 b_3, \ldots, b_1 b_k \in E$ (we call these $k - 1$ edges *branches* of the k-tuple) and removing any branch $b_1 b_j$ disconnects the graph. A k-tuple $(b_1, \ldots, b_k)$ and an l-tuple $(x_1, \ldots, x_l)$ are *independent*, if $\{b_1, \ldots, b_k\} \cap \{x_1, \ldots, x_l\} = \emptyset$, and deleting all the branches keeps b_1 and x_1 connected.

Let Γ be a greedy drawing of T. We shall consider subtrees $T_i = (V_i, E_i)$ of T, such that T_i has root r_i, $\deg(r_i) = 1$ in T_i and v_i is the neighbor of r_i in T_i. We define the *polytope of a rooted subtree* T_i as $\text{polytope}(T_i) = \bigcap \{h^w_{uw} \mid uw \in E_i, uw \neq r_i v_i, d_T(w, r_i) < d_T(u, r_i)\}$.

Definition 1 (Extremal edges). *For $j = 1, 2$, let $a_j b_j \neq v_i r_i$ be an edge of T_i, $d_T(a_j, r_i) < d_T(b_j, r_i)$, such that $\angle_{\text{ccw}}(\vec{v_i r_i}, \vec{a_j b_j})$ is maximum for $j = 1$ and minimum for $j = 2$. We call edges $a_j b_j$ extremal.*

Note that by Lemma 2, $\text{ray}(a_j, \vec{a_j b_j})$ and $\text{ray}(v_i, \vec{v_i r_i})$ diverge. In the following two definitions, let $e_j = a_j b_j$, $j = 1, 2$ be the extremal edges of T_i.

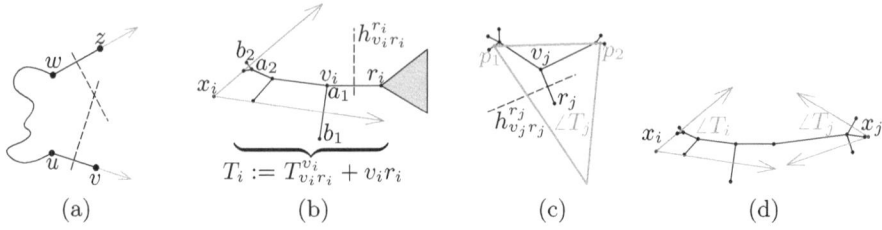

(a)   (b)   (c)   (d)

Fig. 1. (a) Sketch of Lemma 2. (b) Subtree T_i with opening angle $\angle T_i$ (orange), extremal edges a_1b_1, a_2b_2 (blue) and apex x_i (red). The subtree $T_{v_ir_i}^{r_i}$ (gray triangle) must be contained in the half-plane $h_{v_ir_i}^{r_i}$ and the cone $\angle T_i$. (c) Subtree T_j with closed angle $\angle T_j$ and boundary segment p_1p_2. (d) Open angles of independent subtrees must contain apices of each other.

Definition 2 (Open angle). *Let $\angle_{\text{ccw}}(a_1\vec{b}_1, a_2\vec{b}_2) > 180°$. Then, polytope($T_i$) is unbounded, and we say that T_i is drawn with an open angle.*

(a) *If $a_1b_1 \not\subseteq h_{a_2b_2}^{b_2}$ and $a_2b_2 \not\subseteq h_{a_1b_1}^{b_1}$, define $\angle T_i = h_{a_1b_1}^{a_1} \cap h_{a_2b_2}^{a_2}$. Let x_i be the intersection of* axis(a_1b_1) *and* axis(a_2b_2). *We set* apex($\angle T_i$) $= x_i$; *see Fig. 1b.*

(b) *If $a_jb_j \subseteq h_{a_kb_k}^{b_k}$ for $j = 1, k = 2$ or $j = 2, k = 1$, let $\angle T_i$ be the cone defined by moving the boundaries of $h_{a_1b_1}^{a_1}$, $h_{a_2b_2}^{a_2}$ to b_j ($b_k \in \angle T_i$), and apex($\angle T_i$) $= b_j$.*

We call $\angle T_i$ the opening angle of T_i in Γ (orange in Fig. 1b). We write $|\angle T_i| = \alpha$, where α is the angle between the two rays of $\angle T_i$.

Obviously, polytope(T_i) $\subseteq \angle T_i$ in (a). This is also true in (b) by Observation 1.

Observation 1 *Let h be an open half-plane and $p \notin h$. Let h' be the half-plane created by a parallel translation of the boundary of h' to p. Then, $h \subseteq h'$.*

Definition 3 (Closed and zero angle). *Let $\angle_{\text{ccw}}(a_1\vec{b}_1, a_2\vec{b}_2) < 180°$ (or $= 180°$). Let $C_i = h_{a_1b_1}^{a_1} \cap h_{a_2b_2}^{a_2}$, and let p_j be the midpoint of e_j. We denote the part of C_i bounded by segment p_1p_2 containing r by $\angle T_i$ and say that T_i is drawn with a closed (or zero) angle; see Fig. 1c. We write $|\angle T_i| < 0$ (or $= 0$).*

We say that two subtrees T_1, T_2 are *independent*, if $T_2 \setminus \{r_2\} \subseteq T_{v_1r_1}^{r_1}$ and $T_1 \setminus \{r_1\} \subseteq T_{v_2r_2}^{r_2}$. If T_1 and T_2 are independent, then $T_2 \setminus \{r_2\} \subseteq h_{v_1r_1}^{r_1}$ and $T_1 \setminus \{r_1\} \subseteq h_{v_2r_2}^{r_2}$ in Γ. Also, if $r_2 \notin T_{v_1r_1}^{r_1}$, then $r_2 = v_1$.

Lemma 4. *Let T_i and T_j be independent, $|\angle T_i|, |\angle T_j| > 0$ in Γ. Then, apex($\angle T_i$) $\in \angle T_j$ and apex($\angle T_j$) $\in \angle T_i$.*

Lemma 5 (generalization of Claim 4 in [9]). *Let T_i, T_j be two independent subtrees. Then, either $|\angle T_i| > 0$ or $|\angle T_j| > 0$.*

We shall use the following lemma to provide a certificate of non-existence of a greedy drawing.

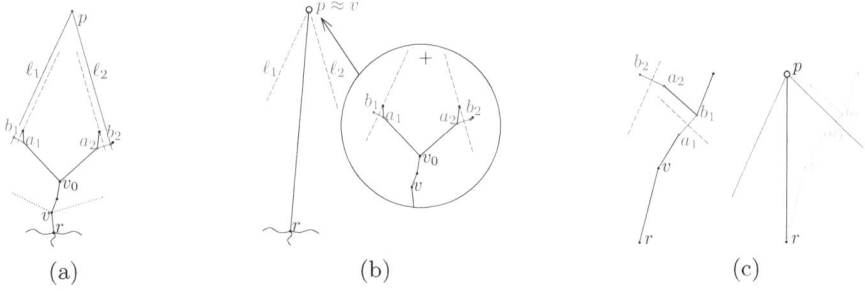

Fig. 2. Illustration of Lemma 8. (a) Greedy drawing Γ. Edges a_1b_1, a_2b_2 are extremal. Dotted blue: bounding cone of T'. (b) Greedy drawing Γ'. Subtree T_{rv}^v has been moved to a new point $p \notin V$ and drawn infinitesimally small. (c) Drawings Γ and Γ' for the case when a_1,b_1 lie on the r-b_2-path. Here, $p = b_2 \in V$.

Lemma 6. *Let T_i, $i = 1, \ldots, d$ be pairwise independent subtrees, and $\alpha_i = |\angle T_i|$. Then, $\sum_{i=1,\ldots,d,\alpha_i>0} \alpha_i > (d - 2)180°$.*

Let T contain a set of n_k irreducible k-tuples, $k = 3, 4, 5$, that are all pairwise independent. Leighton and Moitra [9] showed that for $n_3 \geq 6$ no greedy drawing of T exists. We generalize this result slightly:

Lemma 7. *No greedy drawing of T exists if $n_3 + 2n_4 + 3n_5 \geq 6$.*

2.1   Shrinking Lemma

We now present a lemma which is crucial for later proofs. Let the *bounding cone* of a subtree $T_{rv}^v + rv$ defined for an edge rv in a greedy drawing Γ of T be the cone with apex v and boundary rays $\mathrm{ray}(v, a_1\vec{b}_1)$ and $\mathrm{ray}(v, a_2\vec{b}_2)$ for extremal edges a_1b_1, a_2b_2 of $T_{rv}^v + rv$ that contains the drawing of T_{rv}^v.

Lemma 8. *Let $T = (V, E)$ be a tree and $T' = T_{rv}^v + rv$, $rv \in E$, a subtree of T. Let Γ be a greedy drawing of T, such that $|\angle T'| > 0$. Then, there exists a point p in the bounding cone of T_{rv}^v, such that shrinking T_{rv}^v infinitesimally and moving it to p keeps the drawing greedy, and $|\angle T'|$ remains the same.*

Proof. Let $e_i = a_ib_i$, $i = 1, 2$, be the two extremal edges of T' in Γ, ρ_i the r-b_i-path, and $a_i \in \rho_i$; see Fig. 2 for a sketch. We distinguish two cases:

(1) Edge e_1 is not on ρ_2 and edge e_2 is not on ρ_1. Then, $\{a_1, b_1\} \subseteq h_{a_2b_2}^{a_2}$, and $\{a_2, b_2\} \subseteq h_{a_1b_1}^{a_1}$. Let ℓ_i be the line parallel to $\mathrm{axis}(e_i)$ through b_i and p the intersection of ℓ_1 and ℓ_2; see Fig. 2a. Let $v_0 \in V$ be the last common node of ρ_1 and ρ_2, and let η_i be the v_0-b_i-path in T, $i = 1, 2$.

We now define three intermediate drawings. Let Γ_1 be the drawing gained by replacing T' in Γ by the edge rv_0 and the two paths η_1 and η_2, and let $\Gamma_2 = \Gamma_1 - \eta_1 - \eta_2 + \{v_0b_1, v_0b_2\}$; see Fig. 3a. By Lemma 3, both Γ_1 and Γ_2 are greedy. Let $\Gamma_3 = \Gamma_2 - \{v_0b_1, v_0b_2\} + \{v_0p\}$. Let V_1 be the node set of T_{vr}^r with addition of v_0. Note that the nodes in V_1 have the same coordinates in Γ, Γ_1, Γ_2 and Γ_3. Further, since $v_0 \in h_{a_ib_i}^{a_i}$ for $i = 1, 2$, p lies inside the angle $\angle b_1v_0b_2 < 180°$.

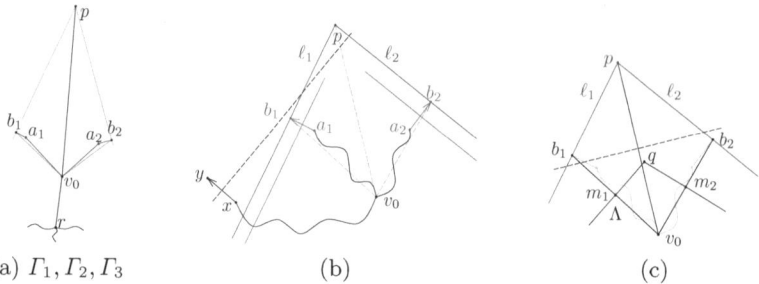

(a) $\Gamma_1, \Gamma_2, \Gamma_3$ (b)   (c)

Fig. 3. Proof of Lemma 8. (a): Intermediate drawings Γ_1 (black and red), Γ_2 (black and green) and Γ_3 (black and blue). (b): For an edge $xy \notin T_{rv}^v$, its axis doesn't cross $v_0 b_1$, $v_0 b_2$. It also doesn't cross $v_0 p$ due to Lemma 2. (c): It is $\Lambda \subseteq h_{v_0 p}^{v_0}$.

We have to prove the greediness of Γ_3. Since $p \notin V$, it doesn't follow directly from Lemma 3. We first show that for an edge xy in Γ_3, $xy \neq v_0 p$, where x is closer to v_0 in T than y, it holds $p \in h_{xy}^x$. Edge xy is also contained in Γ_1. Nodes x, v_0 and a_i lie on the y-b_i-path in T, $i = 1, 2$. Hence, $\{v_0, a_1, a_2, b_1, b_2\} \subseteq \eta_1 \cup \eta_2 \subseteq h_{xy}^x$, therefore, axis$(xy)$ doesn't cross edges $v_0 b_1$, $v_0 b_2$. Now assume $p \notin h_{xy}^x$. Then, axis(xy) must cross $v_0 p$, $b_1 p$ and $b_2 p$ (but not $v_0 b_i$); see Fig. 3b. This is only possible if for some $i \in \{1, 2\}$, rays ray$(x, \vec{xy})$ and ray$(a_i, \vec{a_i b_i})$ are parallel or converge, which is a contradiction to Lemma 2.

Next, we show that $V_1 \subseteq h_{pv_0}^{v_0}$. Without loss of generality, let $v_0 b_1$ be directed upwards to the left and $v_0 b_2$ upwards to the right. Note that a_1 lies to the right of $v_0 b_1$ and a_2 to the left of $v_0 b_2$ (otherwise, the edge $a_i b_i$ would not be extremal in T'). Hence, $\angle v_0 b_i p \geq 90°$. Let Λ be the opening angle of the subtree induced by edges $\{rv_0, v_0 b_1, v_0 b_2\}$ with root r in Γ_2 (blue in Fig. 3c). It is $\Lambda \subseteq h_{pv_0}^{v_0}$ (see [10]). Hence, $V_1 \subseteq \Lambda \subseteq h_{pv_0}^{v_0}$. This proves the greediness of Γ_3. Due to the extremality of $a_1 b_1$, $a_2 b_2$, p lies in the bounding cone of T'.

Removing v_0 and connecting r to p keeps the drawing greedy. Finally, we acquire Γ' by drawing the subtree T_{rv}^v of T infinitesimally small at p. Let C_1 be the cone $\angle T'$ in the original drawing Γ, and C_2 the cone bounded by ℓ_1 and ℓ_2, $a_i \in C_2$. By Observation 1, $C_1 \subseteq C_2$. Consider an edge e in T_{rv}^v, $e \notin \{e_1, e_2\}$ in Γ. Let ℓ be the line parallel to axis(e) through p.

Due to the extremality of e_1, e_2, cone C_2 lies on one side of ℓ. Therefore, since $V_1 \subseteq C_2$, the drawing Γ' is greedy, and it is $\angle T' = C_2$. Since ℓ_i is parallel to axis$(a_i b_i)$, $|\angle T'|$ in Γ' is as big as in Γ.

(2) Now assume $a_1 b_1$ lies on ρ_2. Let Γ_4 be the drawing obtained by replacing T' in Γ by edge rb_2. By Lemma 3, Γ_4 is greedy. It is $b_2 \in h_{a_1 b_1}^{b_1}$. Similar to (1), we acquire Γ' by drawing the subtree T_{rv}^v of T infinitesimally small at $p = b_2$. Then, $|\angle T'|$ remains the same as in Γ, see Fig. 2c. $\qquad\square$

3 Opening Angles of Rooted Trees

The main idea of our decision algorithm is to process the nodes of T bottom-up while calculating tight upper bounds on the maximum possible opening angles

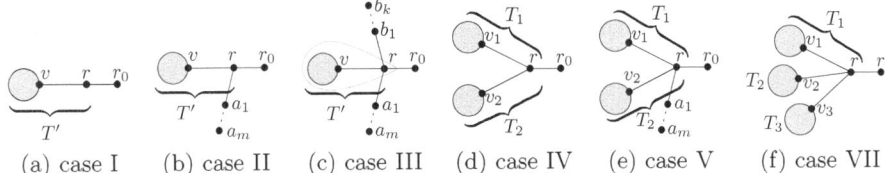

(a) case I   (b) case II   (c) case III   (d) case IV   (e) case V   (f) case VII

Fig. 4. (a)–(e): Possible cases when combining subtrees to maintain an open angle. Subtrees T_1, T_2 have opening angles $\in (90°, 120°)$. In case VII ((f)) or in case VI $(|\angle T_i| \leq 90°$ in IV or V for one $i \in \{1, 2\})$ no open angle is possible.

of the considered subtrees. If T contains a node of degree 5, it cannot be drawn with an open angle, since each pair of consecutive edges forms an angle strictly greater than 60°. In this section, we consider trees with maximum degree 4.

If a subtree T' can be drawn with an open angle $\varphi - \varepsilon$ for any $\varepsilon > 0$, but not φ, we say that it has opening angle φ^- and write $|\angle T'| = \varphi^-$. For example, a triple has opening angle 120^- and a quadruple 60^-. We call a subtree *non-trivial* if it is not a single node or a simple path. Figure 4 shows possibilities to combine or extend non-trivial subtrees T', T_1, T_2. We shall now prove tight bounds on the possible opening angles for each construction. As we shall show later, only cases I–V are feasible for the resulting subtree to have an open angle. To compute the maximum opening angle of the combined subtree $\overline{T}$ in cases I–V, we use the following strategy. We show that applying Lemma 8 to T' does not decrease the opening angle of $\overline{T}$ in a drawing. Hence, it suffices to consider only drawings in which T'^v_{rv} is shrunk to a point. We then obtain an upper bound by solving a linear maximization problem. Finally, we construct a drawing with an almost-optimal opening angle for $\overline{T}$ inductively using an almost-optimal construction for T'. We give a proof for case II, see [10] for the remaining cases.

Lemma 9. *Let T' be a subtree with $\angle T' = \varphi^-$, and consider the subtree $\overline{T} = T' + rr_0 + ra_1 + a_1a_2 + \ldots + a_{m-1}a_m$ in Fig. 4b. Then $|\angle\overline{T}| = (45° + \frac{\varphi}{2})^-$ if $\varphi > 90°$ (case (i)), and $|\angle\overline{T}| = \varphi^-$ if $\varphi \leq 90°$ (case (ii)).*

Proof. First, let $m = 1$. (i) Consider a greedy drawing Γ of $\overline{T}$. Let a_1r be drawn horizontally and v above it and to the left of $\text{axis}(ra_1)$; see Fig. 5a,b,d. Due to Lemma 2, the right boundary of $\angle\overline{T}$ is formed by $\text{axis}(ra_1)$. The left boundary is either formed by (1) the left boundary of $\angle T'$ (see Fig. 5a), or (2) by $\text{axis}(rv)$

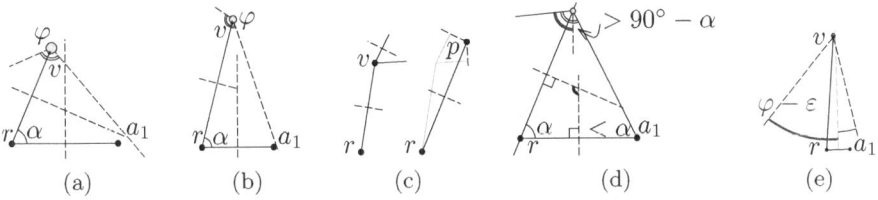

Fig. 5. Optimal construction and tight upper bound for case II

Table 1. Computing maximum opening angle of the combined subtree $\overline{T}$. Let $|\angle T_i| = \varphi_i^-$, $\varphi_i \geq \varphi_{i+1}$, and $|\angle T_i| = \varphi_i = 180°$ if T_i is a path.

case	φ_1	φ_2	φ_3	maximum $|\angle\overline{T}|$
I	$(0°, 180°]$	-	-	φ_1^-
II.i	$180°$	$(90°, 120°]$	-	$(\frac{\varphi_2}{2} + 45°)^- \in (90°, 120°)$
II.ii	$180°$	$(0°, 60°]$	-	$\varphi_2^- \in (0°, 60°)$
III	$180°$	$180°$	$(0°, 120°]$	$\frac{\varphi_3}{2}^- \in (0°, 60°)$
IV	$(90°, 120°]$	$(90°, 120°]$	-	$(\varphi_1 + \varphi_2 - 180°)^- \in (0°, 60°)$
V	$180°$	$(90°, 120°]$	$(90°, 120°]$	$(\frac{3}{4}\varphi_2 + \frac{1}{2}\varphi_3 - 112.5°)^- \in (0°, 60°)$
VI	$(0°, 120°]$	$(0°, 90°]$	-	$< 0°$
VII	$(0°, 120°]$	$(0°, 120°]$	$(0°, 120°]$	$< 0°$

(Fig. 5b). We apply Lemma 8 to $T_{rv}^{\prime v}$ in Γ and acquire Γ', in which $T_{rv}^{\prime v}$ is drawn infinitesimally small. In Γ', axis(ra_1) remains the right boundary of $\angle\overline{T}$. In case (1), the left boundary of $\angle\overline{T}$ is again formed by the left boundary of $\angle T'$, and $|\angle\overline{T}|$ remains the same. In case (2), the subtree $T_{rv}^{\prime v}$ must lie to the right of $r\vec{v}$ in Γ (since each edge in it is oriented clockwise relative to $r\vec{v}$), and so does the point p from Lemma 8. Thus, the edge rv is turned clockwise in Γ', and $|\angle\overline{T}|$ increases; see Fig. 5c. Thus, to acquire an upper bound for $|\angle\overline{T}|$ it suffices to only consider drawings in which $T_{rv}^{\prime v}$ is drawn infinitesimally small. Let $\alpha = \angle a_1 rv$. Then, for $\overline{\varphi} = |\angle\overline{T}|$ it holds: $\overline{\varphi} \leq 180° - \alpha$, $\overline{\varphi} < \varphi - 90° + \alpha$; see the blue and green angles in Fig. 5d. Thus, $\overline{\varphi}$ lies on the graph $f(\alpha) = 180° - \alpha$ or below it and strictly below the graph $g(\alpha) = \varphi - 90° + \alpha$. Maximizing over α gives $\overline{\varphi} < 45° + \frac{\varphi}{2}$. We can achieve $\overline{\varphi} = (45° + \frac{\varphi}{2})^-$ by choosing $\alpha = 135° - \frac{\varphi}{2} + \varepsilon'$ and drawing $T_{rv}^{\prime v}$ sufficiently small with $|\angle T'| = \varphi - \varepsilon$ for sufficiently small $\varepsilon, \varepsilon' > 0$.

(ii) Obviously, $|\angle T'| \geq |\angle\overline{T}|$. For the second part, see Fig. 5e. We choose $\angle a_1 rv = 90° - \frac{\varepsilon}{2}$ and draw ra_1 long enough, such that its axis doesn't cross $T_{rv}^{\prime v}$. We rotate $T_{rv}^{\prime v}$ such that the right side of the opening angle $\angle T'$ and rv form an angle $\frac{3\varepsilon}{2}$. Then, the opening angle φ' of the drawing is defined by the left side of $\angle T'$ and the axis of ra_1 and is $\varphi - \varepsilon$.

For $m \geq 2$, draw $a_2, \ldots, a_m$ collinear with ra_1 and infinitesimally close to a_1.

□

Tight upper bounds on opening angles of the combined subtree $\overline{T}$ for all possible cases are listed in Table 1. Note that no bounds in $(120°, 180°)$ and $(60°, 90°]$ appear. See [10] for the proofs of cases III–VII.

4   Arranging Rooted Subtrees with Open Angles

In this section, we consider the task of constructing a greedy drawing Γ of T by combining independent rooted subtrees with a common root. The following problem (restricted to $n \in \{3, 4, 5\}$) turns out to be fundamental in this context.

Problem 1. Given n angles $\varphi_0, \ldots, \varphi_{n-1} > 0°$, is there a convex n-gon P with corners $v_0, \ldots, v_{n-1}$ (in arbitrary order) with interior angles $\psi_i < \varphi_i$ for

maximize ε under:

$$\varepsilon, \alpha_i, \beta_i, \gamma_i \in [0, 180],$$
$$i = 0, \ldots, n-1;$$
$$\beta_i + \varepsilon \le \alpha_i, \ \gamma_i + \varepsilon \le \alpha_i,$$
$$\beta_i + \gamma_{i+1} + \varepsilon \le \varphi_i$$
$$\alpha_i + \beta_i + \gamma_i = 180,$$
$$\alpha_0 + \ldots + \alpha_{n-1} = 360,$$
$$\textstyle\prod_{i=0}^{n-1} \sin \beta_i = \prod_{i=0}^{n-1} \sin \gamma_i$$

(a) optimization problem $(*)$

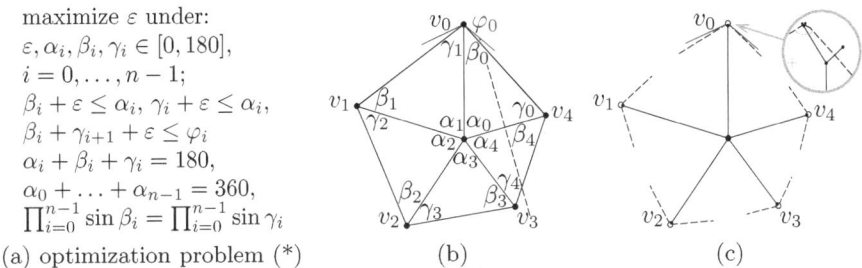

(b)   (c)

Fig. 6. (a) Optimization problem $(*)$; (b) sketch for $(*)$. (c) Solving $(*)$ lets us construct greedy drawings by placing sufficiently small drawings of subtrees at n-gon corners.

$i = 0, \ldots, n-1$, such that the star $K_{1,n}$ has a greedy drawing with root r inside P and leaves $v_0, \ldots, v_{n-1}$?

If Problem 1 has a solution we write $\{\varphi_0, \ldots, \varphi_{n-1}\} \in \mathcal{P}^n$. It can be solved using a series of optimization problems as in Fig. 6a (one for each fixed cyclic ordering of $(\varphi_1, \ldots, \varphi_n)$). The last constraint in $(*)$ follows from applying the law of sines and is known as the wheel condition in the work of di Battista and Vismara [5].

Lemma 10. *It is* $\{\varphi_0, \ldots, \varphi_{n-1}\} \in \mathcal{P}^n$ *if and only if there exists a solution of* $(*)$ *with* $\varepsilon > 0$ *for an ordering* $(\varphi_0, \ldots, \varphi_{n-1})$.

Deciding whether a solution of $(*)$ with $\varepsilon > 0$ exists is in fact equivalent to deciding whether the wheel condition can be satisfied in the interior of a $2n-1$-dimensional simplex; see [10] for more details.

Theorem 1. *For* $n = 3, 4, 5$, *consider trees* T_i, $i = 0, \ldots, n-1$ *with root* r, *edge* rv_i *in* T_i, $\deg(r) = 1$ *in* T_i, $T_i \cap T_j = \{r\}$ *for* $i \ne j$, *such that each* T_i *has a drawing with opening angle at least* $0 < \varphi_i - \varepsilon < 180°$ *for any* $\varepsilon > 0$. *Then, tree* $T = \bigcup_{i=0}^{n-1} T_i$ *has a greedy drawing with* $|\angle T_i| < \varphi_i$ *for all* $i = 0, \ldots, n-1$ *if and only if* $\{\varphi_0, \ldots, \varphi_{n-1}\} \in \mathcal{P}^n$.

Proof. First, consider a drawing of $K_{1,n}$ with edges rv_i that solves $\mathcal{P}^n$, and, without loss of generality, let the angles be ordered such that $\psi_i := \angle v_{i-1}v_iv_{i+1} < \varphi_i$. We create a greedy drawing Γ of T by drawing $(T_i)_{rv_i}^{v_i}$ infinitesimally small at v_i with opening angle $\varphi_i - \varepsilon > \psi_i$ for a sufficiently small $\varepsilon > 0$ and orienting it such that $v_j \in \angle T_i$ for all $j \ne i$; see Fig. 6c.

Now assume a greedy drawing Γ_0 of T with $|\angle T_i| < \varphi_i$, $i = 0, \ldots, n-1$ exists. For one i, it might be $|\angle T_i| < 0$ in Γ_0. Then, there also exists a greedy drawing Γ, in which $0 < |\angle T_j| < \varphi_j$, $j = 0, \ldots, n-1$. By Lemma 5, the subtree $\overline{T} = \{rv_i\} + \bigcup_{j \ne i} T_j$ must have an open angle in Γ_0. We then obtain Γ by making the edge rv_i sufficiently long inside $\angle \overline{T}$ and drawing T_i with $|\angle T_i| > 0$, such that $\overline{T} \subseteq \angle T_i$ and $T_i \subseteq \angle \overline{T}$.

We apply Lemma 8 to T_0, then to $T_1, \ldots, T_{n-1}$ and obtain a greedy drawing Γ' of T with opening angles $\angle T_i$ of same size, such that each subtree $(T_i)_{rv_i}^{v_i}$ is

Table 2. Solving non-linear problem $\mathcal{P}^n$ explicitly. Let $\varphi_i \geq \varphi_{i+1}$, $\varphi_i \in (0°, 60°] \cup (90°, 120°] \cup \{180°\}$, $\sum_{i=0}^{n-1} \varphi_i > (n-2)180°$. See the full version for the proofs.

n	case	$\{\varphi_0, \ldots, \varphi_{n-1}\} \in \mathcal{P}^n$ iff
3, 4		always
5	$\varphi_0 = \ldots = \varphi_3 = 180°$	always
5	$\varphi_0 \leq 120°$	always
5	$\varphi_0 = \ldots = \varphi_2 = 180°$	$\varphi_3 + \varphi_4 > 120°$
5	$\varphi_0 = \varphi_1 = 180°$	$\varphi_2 + \varphi_3 + \varphi_4 > 240°$
5	$\varphi_0 = 180°, \varphi_1, \varphi_2, \varphi_3 \in (90°, 120°], \varphi_4 \leq 60°$	?
5	$\varphi_0 = 180°, \varphi_1, \ldots, \varphi_4 \in (90°, 120°]$	?

drawn infinitesimally small at v_i. For $n = 4, 5$, for each pair of consecutive edges rv_i, rv_j in Γ' the turn from rv_i to rv_j is less than $180°$, so r lies inside the convex polygon with corners $v_0, \ldots, v_{n-1}$. Therefore, Γ' directly provides a solution of $\mathcal{P}^n$. For $n = 3$, v_1 might lie inside angle $\angle v_0 r v_2 \leq 180°$. However, since $\varphi_0 + \varphi_1 + \varphi_2 > 180°$, it is $\{\varphi_0, \varphi_1, \varphi_2\} \in \mathcal{P}^3$; see Table 2. □

Although Problem (*) is non-linear, we are almost always able to give tight conditions for the existence of the solution; see Table 2, which summarizes all possible cases. The last two cases for $n = 5$ are the only remaining ones to consider (for $\varphi_3 + \varphi_4 > 120°$, $\varphi_2 + \ldots + \varphi_4 > 240°$, $\varphi_1 + \ldots + \varphi_4 > 360°$). In practice, it is possible to either strictly prove $\{\varphi_0, \ldots, \varphi_4\} \notin \mathcal{P}^5$ or numerically construct a solution for many such sets of angles. If we drop the last constraint in (*), we acquire a linear program which has a constant number of variables and constraints and can be solved in $O(1)$ time. If it has no solution for any cyclic order of φ_i, neither has $\mathcal{P}^5$. For example, this is the case for $\{180°, 105°, 105°, 105°, 60°\}$. If this linear program has a solution, we can try to solve (*) using nonlinear programming solvers. However, if the non-linear solver finds no solution, we obviously have no guarantee that none exists. In [10], we present examples of trees for which we could prove the existence of a greedy drawing by solving $\mathcal{P}^5$ using MATLAB. Further, we formulate a sufficient condition for the first of the two above cases.

5   Recognition Algorithm

Maximum Degree 4. In this section we formulate Algorithm 1, which decides for a tree T with maximum degree 4 whether T has a greedy drawing. First, we describe a procedure to determine the tight upper bound for the opening angle of a given rooted subtree. Let $N(v)$ denote the neighbors of $v \in V$ in T. After processing a node v, we set a flag $p(v) = \text{true}$. Let $N_p(v) = \{u \mid uv \in E, p(u) = \text{true}\}$, and $\angle_{\text{optimal}}$ the new tight upper bound calculated according to Table 1.

Lemma 11. *Procedure getOpenAngle is correct and requires time $O(|V|)$.*

Proof. The algorithm processes tree nodes bottom-up. For $v \in V$, let π_v be the parent of v, $\deg(v) = d_v$, $T_v = T^v_{\pi_v v} + \pi_v v$ with root π_v. For a subtree with one

Procedure getOpenAngle(T,r)	**Algorithm 1.** hasGreedyDrawing(T)

Input : tree $T = (V, E)$,
      root $r \in V$, $d_r = 1$
 Result: tight upper bound on $|\angle T|$,
      0 if no open angle possible.
 $\mathrm{p}(r) \leftarrow$ false
 for $v \in V \setminus \{r\}$ **do**
      if $d_v = 5$ **then return** 0
      else if $d_v = 1$ **then**
          $\mathrm{p}(v) \leftarrow$ true
          $\angle(v) \leftarrow 180$
      else $\mathrm{p}(v) \leftarrow$ false
 while $\exists v \in V : \neg \mathrm{p}(v)$
    & $|\mathrm{N_p}(v)| = d_v - 1$
      if $\forall u \in \mathrm{N_p}(v) : \angle(u) = 180$ **then**
          $\angle(v) \leftarrow 180 - (d_v - 2) \cdot 60$
      else if *case I,..., V applicable*
      then $\angle(v) \leftarrow \angle_{\mathrm{optimal}}(\mathrm{N_p}(v))$
      else return 0
      $\mathrm{p}(v) \leftarrow$ true
 return $\angle(v)$ for $\{v\} = \mathrm{N}(r)$
	Input : tree $T = (V, E)$, max deg 4
	 Result: whether T has a greedy drawing
	 for $v \in V$ **do**
	      if $d_v = 1$ **then**
	          $\mathrm{p}(v) \leftarrow$ true; $\angle(v) \leftarrow 180$
	      else $\mathrm{p}(v) \leftarrow$ false
	 while $\exists v \in V : \neg \mathrm{p}(v)$ & $|\mathrm{N_p}(v)| \geq d_v - 1$
	      if $|\mathrm{N_p}(v)| = d_v$ **then**
	          return $\sum_{u,uv \in E} \angle(u) > (d_v - 2)180$
	      else if $\forall u \in \mathrm{N_p}(v) : \angle(u) = 180$ **then**
	          $\angle(v) \leftarrow 180 - (d_v - 2) \cdot 60$
	      else if *case I,..., V applicable* **then**
	          $\angle(v) \leftarrow \angle_{\mathrm{optimal}}(\mathrm{N_p}(v))$
	      else
	          $w \leftarrow \mathrm{N}(v) - \mathrm{N_p}(v)$
	          $\angle(w) \leftarrow$ getOpenAngle($T_{vw}^w + vw, v$)
	          return $\angle(w) > 0$
	          & $\sum_{u,uv \in E} \angle(u) > (d_v - 2)180$
	      $\mathrm{p}(v) \leftarrow$ true

or two nodes, define its opening angle as 180°. We prove the following invariant for the *while* loop: For each $v \in V$ with $\mathrm{p}(v) = $ true, $\angle(v) > 0$ stores a tight upper bound for the opening angle in a greedy drawing of T_v.

The invariant holds for all leaves of T after the initialization. The first *if*-statement inside the *while* body ensures that if all nodes in T_v except v have degree 1 or 2, then $\angle(v) = 180$ if $d_v = 1, 2$ in T, $\angle(v) = 120$ if $d_v = 3$ and $\angle(v) = 60$ if $d_v = 4$. Now consider the first *else* clause inside the *while* loop. Assume $\mathrm{p}(v) = $ false, $|\mathrm{N_p}(v)| = d_v - 1$ and the invariant holds for all subtrees T_u, $u \in \mathrm{N_p}(v)$. If one of the cases I–V can be applied to v and subtrees T_u, then, after the current loop, $\angle(v) > 0$ stores the tight upper bound for the opening angle in a greedy drawing of T_v; see Table 1. Otherwise, we have case VI or VII, and T_v cannot be drawn with an open angle. Each node v is processed in $O(d_v)$, and if for $u \in \mathrm{N}(v) - \mathrm{N_p}(v)$, it holds $|\mathrm{N_p}(u)| \geq d_u - 1$ after processing v, we put u in a queue. Hence the running time is $O(|V|)$. □

Algorithm 1 also requires $O(|V|)$ time and is similar to Procedure getOpenAngle, except that T now does not have a distinguished root. We proceed from the leaves of T inwards, until we reach some "central" node v with neighbors $\{u_1, \ldots, u_{d_v}\}$, such that a greedy drawing of T exists only if all tight upper bounds φ_i on $|\angle(T_{vu_i}^{u_i} + vu_i)|$ are positive. Then, we report true if and only if $\sum_{i=1}^{d_v} \varphi_i > (d_v - 2)180°$. See [10] for the formal correctness proof.

Maximum Degree 5 and Above. If T contains a node v with $\deg(v) \geq 6$, no greedy drawing exists. Also, a greedy-drawable tree can have at most one node of degree 5 by Lemma 7, otherwise, there are two independent 5-tuples. For unique $r \in V$, $\deg(r) = 5$, consider the five rooted subtrees $T_0, \ldots, T_4$ attached to it and the tight upper bounds φ_i on $|\angle T_i|$. If $\sigma = \sum_{i=0}^{4} \varphi_i \leq 540°$, T cannot be drawn greedily. The converse, however, does not hold. By Theorem 1, a greedy drawing exists if and only if $\{\varphi_0, \ldots, \varphi_4\} \in \mathcal{P}^5$. To decide whether $\{\varphi_0, \ldots, \varphi_4\} \in \mathcal{P}^5$, we apply the conditions from Table 2. If in the remaining case $\varphi_0 = 180°$, $\varphi_1, \ldots, \varphi_4 \leq 120°$ (i) the sufficient condition does not apply, (ii) the linear relaxation of Problem (*) has a solution, but (iii) the non-linear solver finds none, we report *uncertain*. The full formulation of this algorithm as well as uncertain examples can be found in [10].

References

1. Alamdari, S., Chan, T.M., Grant, E., Lubiw, A., Pathak, V.: Self-approaching graphs. In: Didimo, W., Patrignani, M. (eds.) GD 2012. LNCS, vol. 7704, pp. 260–271. Springer, Heidelberg (2013)
2. Angelini, P., Di Battista, G., Frati, F.: Succinct greedy drawings do not always exist. In: Eppstein, D., Gansner, E.R. (eds.) GD 2009. LNCS, vol. 5849, pp. 171–182. Springer, Heidelberg (2010)
3. Angelini, P., Frati, F., Grilli, L.: An algorithm to construct greedy drawings of triangulations. J. Graph Algorithms Appl. 14(1), 19–51 (2010)
4. Dhandapani, R.: Greedy drawings of triangulations. Discrete Comput. Geom. 43, 375–392 (2010)
5. Di Battista, G., Vismara, L.: Angles of planar triangular graphs. In: Proc. Symp. Theory of Computing (STOC 1993), pp. 431–437. ACM (1993)
6. Eppstein, D., Goodrich, M.T.: Succinct greedy geometric routing using hyperbolic geometry. IEEE Trans. Computers 60(11), 1571–1580 (2011)
7. Goodrich, M.T., Strash, D.: Succinct greedy geometric routing in the euclidean plane. In: Dong, Y., Du, D.-Z., Ibarra, O. (eds.) ISAAC 2009. LNCS, vol. 5878, pp. 781–791. Springer, Heidelberg (2009)
8. Kleinberg, R.: Geographic routing using hyperbolic space. In: Proc. Computer Communications (INFOCOM 2007), pp. 1902–1909. IEEE (2007)
9. Leighton, T., Moitra, A.: Some results on greedy embeddings in metric spaces. Discrete Comput. Geom. 44, 686–705 (2010)
10. Nöllenburg, M., Prutkin, R.: Euclidean greedy drawings of trees. CoRR arXiv:1306.5224 (2013)
11. Papadimitriou, C.H., Ratajczak, D.: On a conjecture related to geometric routing. Theor. Comput. Sci. 344(1), 3–14 (2005)
12. Rao, A., Ratnasamy, S., Papadimitriou, C., Shenker, S., Stoica, I.: Geographic routing without location information. In: Proc. Mobile Computing and Networking (MobiCom 2003), pp. 96–108. ACM (2003)
13. Schnyder, W.: Embedding planar graphs on the grid. In: Proc. Discrete Algorithms (SODA 1990), pp. 138–148. SIAM (1990)
14. Wang, J.J., He, X.: Succinct strictly convex greedy drawing of 3-connected plane graphs. In: Snoeyink, J., Lu, P., Su, K., Wang, L. (eds.) FAW-AAIM 2012. LNCS, vol. 7285, pp. 13–25. Springer, Heidelberg (2012)

Sparse Fault-Tolerant BFS Trees*

Merav Parter** and David Peleg

Department of Computer Science and Applied Mathematics,
The Weizmann Institute, Rehovot, Israel
{merav.parter,david.peleg}@weizmann.ac.il

Abstract. A *fault-tolerant* structure for a network is required to continue functioning following the failure of some of the network's edges or vertices. This paper considers *breadth-first search (BFS)* spanning trees, and addresses the problem of designing a sparse *fault-tolerant* BFS tree, or *FT-BFS tree* for short, namely, a sparse subgraph T of the given network G such that subsequent to the failure of a single edge or vertex, the surviving part T' of T still contains a BFS spanning tree for (the surviving part of) G. For a source node s, a target node t and an edge $e \in G$, the shortest $s - t$ path $P_{s,t,e}$ that does not go through e is known as a *replacement path*. Thus, our FT-BFS tree contains the collection of all replacement paths $P_{s,t,e}$ for every $t \in V(G)$ and every failed edge $e \in E(G)$. Our main results are as follows. We present an algorithm that for every n-vertex graph G and source node s constructs a (single edge failure) FT-BFS tree rooted at s with $O(n \cdot \min\{\texttt{Depth}(s), \sqrt{n}\})$ edges, where $\texttt{Depth}(s)$ is the depth of the BFS tree rooted at s. This result is complemented by a matching lower bound, showing that there exist n-vertex graphs with a source node s for which any edge (or vertex) FT-BFS tree rooted at s has $\Omega(n^{3/2})$ edges. We then consider *fault-tolerant multi-source BFS trees*, or FT-MBFS *trees* for short, aiming to provide (following a failure) a BFS tree rooted at each source $s \in S$ for some subset of sources $S \subseteq V$. Again, tight bounds are provided, showing that there exists a poly-time algorithm that for every n-vertex graph and source set $S \subseteq V$ of size σ constructs a (single failure) FT-MBFS tree $T^*(S)$ from each source $s_i \in S$, with $O(\sqrt{\sigma} \cdot n^{3/2})$ edges, and on the other hand there exist n-vertex graphs with source sets $S \subseteq V$ of cardinality σ, on which any FT-MBFS tree from S has $\Omega(\sqrt{\sigma} \cdot n^{3/2})$ edges. Finally, we propose an $O(\log n)$ approximation algorithm for constructing FT-BFS and FT-MBFS structures. The latter is complemented by a hardness result stating that there exists no $\Omega(\log n)$ approximation algorithm for these problems under standard complexity assumptions. In comparison with previous constructions our algorithm is deterministic and may improve the number of edges by a factor of up to $\sqrt{n}$ for some instances. All our algorithms can be extended to deal with one *vertex* failure as well, with the same performance.

* Supported in part by the Israel Science Foundation (grant 894/09), the I-CORE program of the Israel PBC and ISF (grant 4/11), the United States-Israel Binational Science Foundation (grant 2008348), the Israel Ministry of Science and Technology (infrastructures grant), and the Citi Foundation.
** Recipient of the Google European Fellowship in distributed computing; research is supported in part by this Fellowship.

H.L. Bodlaender and G.F. Italiano (Eds.): ESA 2013, LNCS 8125, pp. 779–790, 2013.

1   Introduction

Background and Motivation. Modern day communication networks support a variety of logical structures and services, and depend on their undisrupted operation. As the vertices and edges of the network may occasionally fail or malfunction, it is desirable to make those structures robust against failures. Indeed, the problem of designing fault-tolerant constructions for various network structures and services has received considerable attention over the years.

Fault-resilience can be introduced into the network in several different ways. This paper focuses on a notion of fault-tolerance whereby the structure at hand is augmented or "reinforced" (by adding to it various components) so that subsequent to the failure of some of the network's vertices or edges, the surviving part of the structure is still operational. As this reinforcement carries certain costs, it is desirable to minimize the number of added components. To illustrate this type of fault tolerance, let us consider the structure of graph k-spanners (cf. [19,20,21]). A graph spanner H can be thought of as a skeleton structure that generalizes the concept of spanning trees and allows us to faithfully represent the underlying network using few edges, in the sense that for any two vertices of the network, the distance in the spanner is stretched by only a small factor. More formally, consider a weighted graph G and let $k \geq 1$ be an integer. Let $\text{dist}(u, v, G)$ denote the (weighted) distance between u and v in G. Then a k-spanner H satisfies that $\text{dist}(u, v, H) \leq k \cdot \text{dist}(u, v, G)$ for every $u, v \in V$. Introducing fault tolerance, we say that a subgraph H is an f-*edge fault-tolerant* k-*spanner* of G if $\text{dist}(u, v, H \setminus F) \leq k \cdot \text{dist}(u, v, G \setminus F)$ for any set $F \subseteq E$ of size at most f, and any pair of vertices $u, v \in V$. A similar definition applies to f-vertex fault-tolerant k-spanners. Sparse fault-tolerant spanner constructions were presented in [6,10]. This paper considers *breadth-first search (BFS)* spanning trees, and addresses the problem of designing *fault-tolerant* BFS trees, or FT–BFS trees for short. By this we mean a subgraph T of the given network G, such that subsequent to the failure of some of the vertices or edges, the surviving part T' of T still contains a BFS spanning tree for the surviving part of G. We also consider a generalized structure referred to as a *fault-tolerant multi-source BFS tree*, or FT–MBFS *tree* for short, aiming to provide a BFS tree rooted at each source $s \in S$ for some subset of sources $S \subseteq V$.

The notion of FT–BFS trees is closely related to the problem of constructing *replacement paths* and in particular to its *single source* variant, studied in [13]. That problem requires to compute the collection $\mathcal{P}_s$ of all $s - t$ replacement paths $P_{s,t,e}$ for every $t \in V$ and every failed edge e that appears on the $s - t$ shortest-path in G. The vast literature on *replacement paths* (cf. [4,13,24,26,28]) focuses on *time-efficient* computation of the these paths as well as their efficient maintenance in data structures (a.k.a *distance oracles*). In contrast, the main concern in the current paper is with optimizing the *size* of the resulting fault tolerant structure that contains the collection $\mathcal{P}_s$ of all replacement paths given a source node s. A typical motivation for such a setting is where the graph edges represent the channels of a communication network, and the system designer would like to purchase or lease a minimal collection of channels (i.e., a subgraph $G' \subseteq G$) that maintains its functionality as a "BFS tree" with respect to the source s upon any single edge or vertex failure in G. In such a context, the cost of computation at the preprocessing stage may often be negligible compared to the purchasing/leasing cost of the resulting structure. Hence, our key cost measure in this paper is the *size* of the fault tolerant structure, and our main goal is to achieve *sparse* (or *compact*) structures. Most previous work on sparse / compact fault-tolerant structures and services concerned structures that are *distance-preserving* (i.e., dealing with distances, shortest paths or shortest routes), *global* (i.e., centered on "all-pairs" variants), and *approximate* (i.e., settling for near optimal distances), such as *spanners*, *distance oracles* and *compact routing schemes*. The problem considered here, namely, the construction of FT-BFS trees, still concerns a distance preserving structure. However, it deviates from tradition with respect to the two other features, namely, it concerns a "single source" variant, and it insists on exact shortest paths. Hence our problem is on the one hand easier, yet on the other hand harder, than previously studied ones. Noting that in previous studies, the "cost" of adding fault-tolerance (in the relevant complexity measure) was often low (e.g., merely polylogarithmic in the graph size n), one might be tempted to conjecture that a similar phenomenon may reveal itself in our problem as well. Perhaps surprisingly, it turns out that our insistence on exact distances plays a dominant role and makes the problem significantly harder, outweighing our willingness to settle for a "single source" solution.

Contributions. We obtain the following results. In Sec. 2, we define the *Minimum* FT-BFS and *Minimum* FT-MBFS problems, aiming at finding the minimum such structures tolerant against a single edge or vertex fault. We show that these problems are NP-hard and moreover, cannot be approximated (under standard complexity assumptions) to within a factor of $\Omega(\log n)$, where n is the number of vertices of the input graph G. Section 3 presents lower bound constructions for these problems. For the single source case, we present a lower bound stating that for every n there exists an n-vertex graph and a source node $s \subseteq V$ for which any FT-MBFS tree from s requires $\Omega(n^{3/2})$ edges. We then show that there exist n-vertex graphs with source sets $S \subseteq V$ of size σ, on which any FT-MBFS tree from the source set S has $\Omega(\sqrt{\sigma} \cdot n^{3/2})$ edges. These results are complemented by matching upper bounds. In Sec. 4, we present a simple algorithm that for every n-vertex graph G and source node s, constructs a (single edge failure) FT-BFS tree rooted at s with $O(n \cdot \min\{\text{Depth}(s), \sqrt{n}\})$ edges. A similar algorithm yields an FT-BFS tree tolerant to one vertex failure, with the same size bound. In addition, for the multi source case, we show that there exists a polynomial time algorithm that for every n-vertex graph and source set $S \subseteq V$ of size $|S| = \sigma$ constructs a (single failure) FT-MBFS tree $T^*(S)$ from each source $s_i \in S$, with $O(\sqrt{\sigma} \cdot n^{3/2})$ edges.

Note that while those algorithms match the worst-case lower bounds, they might still be far from optimal for certain instances, see [18]. Consequently, in Sec. 5, we complete the upper bound analysis by presenting an $O(\log n)$ approximation algorithm for the Minimum FT-MBFS problem. This approximation algorithm is superior in instances where the graph enjoys a sparse FT-MBFS tree, hence paying $O(n^{3/2})$ edges is wasteful. In light of the hardness result for these problems (in Sec. 2), the approximability result is tight (up to constants). All our results hold for directed graphs as well.

Related Work. To the best of our knowledge, this paper is the first to study the sparsity of fault-tolerant BFS structures for graphs. The question of whether it is possible to construct a sparse fault tolerant *spanner* for an arbitrary undirected weighted graph, raised in [8], was answered in the affirmative in [6], presenting algorithms for constructing an f-vertex fault tolerant $(2k - 1)$-spanner of size $O(f^2 k^{f+1} \cdot n^{1+1/k} \log^{1-1/k} n)$ and an f-edge fault tolerant $2k - 1$ spanner of size $O(f \cdot n^{1+1/k})$ for a graph of size n. A randomized construction attaining an improved tradeoff for vertex fault-tolerant spanners was shortly afterwards presented in [10], yielding (with high probability) for every graph $G = (V, E)$, odd integer s and integer f, an f-vertex fault-tolerant s-spanner with $O\left(f^{2-\frac{2}{s+1}} n^{1+\frac{2}{s+1}} \log n\right)$ edges. This should be contrasted with the best stretch-size tradeoff currently known for non-fault-tolerant spanners [25], namely, $2k-1$ stretch with $\tilde{O}(n^{1+1/k})$ edges. Fault tolerant spanners for the d-dimensional Euclidean case were studied in [8,16,17].

A related network service is the *distance oracle* [3,23,26], which is a succinct data structure capable of supporting efficient responses to distance queries on a weighted graph G. A distance query (s, t) requires finding, for a given pair of vertices s and t in V, the distance (namely, the length of the shortest path) between u and v in G. The query protocol of an oracle $\mathcal{S}$ correctly answers distance queries on G. In a *fault tolerant distance oracle*, the query may include also a set F of failed edges or vertices (or both), and the oracle $\mathcal{S}$ must return, in response to a query (s, t, F), the distance between s and t in $G' = G \setminus F$. Such a structure is sometimes called an *F-sensitivity distance oracle*. The focus is on both fast preprocessing time, fast query time and low space. It has been shown in [9] that given a directed weighted graph G of size n, it is possible to construct in time $\tilde{O}(mn^2)$ a 1-sensitivity fault tolerant distance oracle of size $O(n^2 \log n)$ capable of answering distance queries in $O(1)$ time in the presence of a single failed edge or vertex. The preprocessing time was recently improved to $\tilde{O}(mn)$, with unchanged size and query time [4]. A 2-sensitivity fault tolerant distance oracle of size $O(n^2 \log^3 n)$, capable of answering 2-sensitivity queries in $O(\log n)$ time, was presented in [11].

Recently, distance sensitivity oracles have been considered for weighted and directed graphs in the *single source* setting [13]. Specifically, Grandoni and Williams considered the problem of *single-source replacement paths* where one aims to compute the collection of all replacement paths for a given source node s, and proposed an efficient randomized algorithm that does so in $\tilde{O}(APSP(n, M))$

where $APSP(n, M)$ is the time required to compute all-pairs-shortest-paths in a weighted graph with integer weights $[-M, M]$.

A relaxed variant of distance oracles, in which distance queries are answered by *approximate* distance estimates instead of *exact* ones, was introduced in [26], where it was shown how to construct, for a given weighted undirected n-vertex graph G, an approximate distance oracle of size $O(n^{1+1/k})$ capable of answering distance queries in $O(k)$ time, where the *stretch* (multiplicative approximation factor) of the returned distances is at most $2k - 1$. An f-sensitivity approximate distance oracle $\mathcal{S}$ was presented in [5]. For an integer parameter $k \geq 1$, the size of $\mathcal{S}$ is $O(kn^{1+\frac{8(f+1)}{k+2(f+1)}} \log (nW))$, where W is the weight of the heaviest edge in G, the stretch of the returned distance is $2k - 1$, and the query time is $O(|F| \cdot \log^2 n \cdot \log \log n \cdot \log \log d)$, where d is the distance between s and t in $G \setminus F$. A fault-tolerant label-based $(1 + \epsilon)$-approximate distance oracle for the family of graphs with doubling dimension bounded by α is presented in [2]. Our final example concerns fault tolerant routing schemes. A fault-tolerant routing protocol is a distributed algorithm that, for any set of failed edges F, enables any source vertex $\hat{s}$ to route a message to any destination vertex $\hat{d}$ along a shortest or near-shortest path in the surviving network $G \setminus F$ in an efficient manner (and without knowing F in advance). Compact routing schemes are considered in [1,7,19,22,25]. Fault-tolerant routing schemes are considered in [5].

2 Preliminaries

Notation. Given a graph $G = (V, E)$ and a source node s, let $T_0(s) \subseteq G$ be a shortest paths (or BFS) tree rooted at s. For a source node set $S \subseteq V$, let $T_0(S) = \bigcup_{s \in S} T_0(s)$ be a union of the single source BFS trees. Let $\pi(s, v, T)$ be the $s - v$ shortest-path in tree T, when the tree $T = T_0(s)$, we may omit it and simply write $\pi(s, v)$. Let $\Gamma(v, G)$ be the set of v neighbors in G. Let $E(v, G) = \{(u, v) \in E(G)\}$ be the set of edges incident to v in the graph G and let $\deg(v, G) = |E(v, G)|$ denote the degree of node v in G. When the graph G is clear from the context, we may omit it and simply write $\deg(v)$. Let $\mathtt{depth}(s, v) = \mathrm{dist}(s, v, G)$ denote the *depth* of v in the BFS tree $T_0(s)$. When the source node s is clear from the context, we may omit it and simply write $\mathtt{depth}(v)$. Let $\mathtt{Depth}(s) = \max_{u \in V} \{\mathtt{depth}(s, u)\}$ be the *depth* of $T_0(s)$. For a subgraph $G' = (V', E') \subseteq G$ (where $V' \subseteq V$ and $E' \subseteq E$) and a pair of nodes $u, v \in V$, let $\mathrm{dist}(u, v, G')$ denote the shortest-path distance in edges between u and v in G'. For a path $P = [v_1, \ldots, v_k]$, let $\mathtt{LastE}(P)$ be the last edge of path P. Let $|P|$ denote the length of the path and $P[v_i, v_j]$ be the subpath of P from v_i to v_j. For paths P_1 and P_2, $P_1 \circ P_2$ denote the path obtained by concatenating P_2 to P_1. Assuming an edge weight function $W : E(G) \to \mathbb{R}^+$, let $SP(s, v_i, G, W)$ be the set of $s - v_i$ shortest-paths in G according to the edge weights of W. Throughout, the edges of these paths are considered to be directed away from the source node s. Given an $s - v$ path P and an edge $e = (x, y) \in P$, let $\mathrm{dist}(s, e, P)$ be the distance (in edges) between s and e on P. In addition, for an edge $e = (x, y) \in T_0(s)$, define $\mathrm{dist}(s, e) = i$ if $\mathtt{depth}(x) = i - 1$ and $\mathtt{depth}(y) = i$.

Definition 1. *A graph T^* is an edge (resp., vertex) FT-BFS tree for G with respect to a source node $s \in V$, iff for every edge $f \in E(G)$ (resp., vertex $f \in V$) and for every $v \in V$, $\mathrm{dist}(s, v, T^* \setminus \{f\}) = \mathrm{dist}(s, v, G \setminus \{f\})$.*

A graph T^* is an edge (resp., vertex) FT-MBFS tree for G with respect to source set $S \subseteq V$, iff for every edge $f \in E(G)$ (resp., vertex $f \in V$) and for every $s \in S$ and $v \in V$, $\mathrm{dist}(s, v, T^* \setminus \{f\}) = \mathrm{dist}(s, v, G \setminus \{f\})$.

For simplicity, we refer to edge FT-BFS (resp., edge FT-MBFS) trees simply by FT-BFS (resp., FT-MBFS) trees. Throughout, we focus on edge fault, yet the entire analysis extends trivially to the case of vertex fault as well.

Like other papers in this field [14,4], throughout, we assume without loss of generality that the shortest paths are unique since we can always add small perturbations to break any ties. Let W be a weight assignment that captures these symbolic perturbations.

The Minimum FT-BFS Problem. Denote the set of solutions for the instance (G, s) by $\mathcal{T}(s, G) = \{\widehat{T} \subseteq G \mid \widehat{T} \text{ is an FT-BFS tree w.r.t. } s\}$. Let $\mathtt{Cost}^*(s, G) = \min\{|E(\widehat{T})| \mid \widehat{T} \in \mathcal{T}(s, G)\}$ be the minimum number of edges in any FT-BFS subgraph of G. These definitions naturally extend to the multi-source case where we are given a source set $S \subseteq V$ of size σ. Then $\mathcal{T}(S, G) = \{\widehat{T} \subseteq G \mid \widehat{T} \text{ is a FT-MBFS with respect to } S\}$ and $\mathtt{Cost}^*(S, G) = \min\{|E(\widehat{T})| \mid \widehat{T} \in \mathcal{T}(S, G)\}$.

In the *Minimum* FT-BFS problem we are given a graph G and a source node s and the goal is to compute an FT-BFS $\widehat{T} \in \mathcal{T}(s, G)$ of minimum size, i.e., such that $|E(\widehat{T})| = \mathtt{Cost}^*(s, G)$. Similarly, in the *Minimum* FT-MBFS problem we are given a graph G and a source node set S and the goal is to compute an FT-MBFS $\widehat{T} \in \mathcal{T}(S, G)$ of minimum size i.e., such that $|E(\widehat{T})| = \mathtt{Cost}^*(S, G)$. We begin by establishing hardness (for missing proofs see full version [18]).

Theorem 1. *The Minimum FT-BFS problem is NP-complete and cannot be approximated to within a factor $c \log n$ for some constant $c > 0$ unless $\mathcal{NP} \subseteq \mathcal{TIME}(n^{poly \log(n)})$.*

3   Lower Bounds

We now present a lower bound for the case of a single source.

Theorem 2. *There exists an n-vertex graph $G(V, E)$ and a source node $s \in V$ such that any FT-BFS tree rooted at s has $\Omega(n^{3/2})$ edges, i.e., $\mathtt{Cost}^*(s, G) = \Omega(n^{3/2})$.*

Proof: Let us first describe the structure of $G = (V, E)$. Set $d = \lfloor \sqrt{n}/2 \rfloor$.

The graph consists of four main components. The first is a path $\pi = [s = v_1, \ldots, v_{d+1} = v^*]$ of length d. The second component consists of a node set $Z = \{z_1, \ldots, z_d\}$ and a collection of d disjoint paths of deceasing length, $P_1, \ldots, P_d$, where $P_j = [v_j = p_1^j, \ldots, z_j = p_{t_j}^j]$ connects v_j with z_j and its length is $t_j = |P_j| = 6 + 2(d - j)$, for every $j \in 1, \cdots, d$. Altogether, the set of nodes in these paths, $Q = \bigcup_{j=1}^d V(P_j)$, is of size $|Q| = d^2 + 7d$.

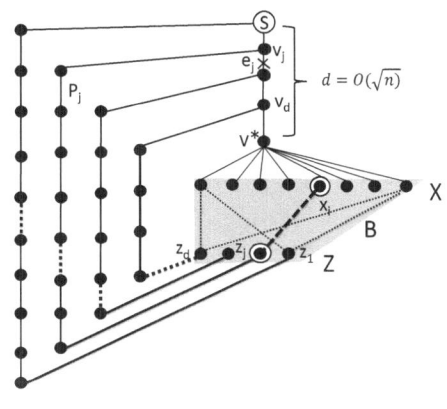

The third component is a set of nodes X of size $n - (d^2 + 7d)$, all connected to the terminal node v^*. The last component is a complete bipartite graph $B = (X, Z, \hat{E})$ connecting X to Z. Overall, $V = X \cup Q$ and $E = \hat{E} \cup E(\pi) \cup \bigcup_{j=1}^d E(P_j)$. Note that $n/4 \le |Q| \le n/2$ for sufficiently large n. Consequently, $|X| = n - |Q| \ge n/2$, and $|\hat{E}| = |Q| \cdot |X| \ge n^{3/2}/4$. A BFS tree T_0 rooted at s for this G (illustrated by the solid edges in the figure) is given by

$$E(T_0) = \{(x_i, z_i) \mid i \in \{1, \ldots, d\}\} \cup \bigcup_{j=1}^d E(P_j) \setminus \{(p_{\ell_j}^j, p_{\ell_j - 1}^j)\},$$

where $\ell_j = t_j - (d - j)$ for every $j \in \{1, \ldots, d\}$. We now show that every FT-BFS tree $T' \in \mathcal{T}(s, G)$ must contain all the edges of B, namely, the edges $e_{i,j} = (x_i, z_j)$ for every $i \in \{1, \ldots, |X|\}$ and $j \in \{1, \ldots, d\}$ (the dashed edges in the figure). Assume, towards contradiction, that there exists a $T' \in \mathcal{T}(s, G)$ that does not contain $e_{i,j}$ (the bold dashed edge (x_i, z_j) in the figure). Note that upon the failure of the edge $e_j = (v_j, v_{j+1}) \in \pi$, the unique $s - x_i$ shortest path connecting s and x_i in $G \setminus \{e_j\}$ is $P_j' = \pi[v_1, v_j] \circ P_j \circ [z_j, x_i]$, and all other alternatives are strictly longer. Since $e_{i,j} \notin T'$, also $P_j' \nsubseteq T'$, and therefore $\text{dist}(s, x_i, G \setminus \{e_j\}) < \text{dist}(s, x_i, T' \setminus \{e_j\})$, in contradiction to the fact that T' is an FT-BFS tree. It follows that every FT-BFS tree T' must contain at least $|\hat{E}| = \Omega(n^{3/2})$ edges. The theorem follows.   ∎

We next consider an intermediate setting where it is necessary to construct a fault-tolerant subgraph FT-MBFS containing several FT-BFS trees in parallel, one for each source $s \in S$, for some $S \subseteq V$. In the full version [18], we establish the following.

Theorem 3. *There exists an n-vertex graph $G(V, E)$ and a source set $S \subseteq V$ of cardinality σ, such that any FT-MBFS tree from the source set S has $\Omega(\sqrt{\sigma} \cdot n^{3/2})$ edges, i.e., $\text{Cost}^*(S, G) = \Omega(\sqrt{\sigma} \cdot n^{3/2})$.*

4   Upper Bounds

Single Source. In this section we consider the case of FT-BFS trees and establish the following.

Theorem 4. *There exists a polynomial time algorithm that for every n-vertex graph G and source node s constructs an FT-BFS tree rooted at s with $O(n \cdot \min\{\mathtt{Depth}(s), \sqrt{n}\})$ edges.*

To prove the theorem, we first describe a simple algorithm for the problem and then prove its correctness and analyze the size of the resulting FT-BFS tree. Using the sparsity lemma of [24] and the tools of [13], one can provide a randomized construction for an FT-BFS tree with $O(n^{3/2} \log n)$ edges with high probability. In contrast, the simple algorithm presented here is *deterministic* and achieves an FT-BFS tree with $O(n^{3/2})$ edges, matching exactly the lower bound established in Sec. 3. We note that known time-efficient (and rather involved) algorithms for constructing replacement paths and distance sensitivity oracles (cf., [14,24,4,28,13]) can be modified to construct sparse FT-BFS and FT-MBFS trees by breaking shortest path ties properly and maintaining the successors of the computed replacement paths. Since our focus here is on the size of the resulting FT-BFS trees, and not on optimizing the running time, we introduce the construction using a simple but slow ($O(nm + n^2 \log n)$ round) algorithm. In the analysis section we then show that as long as the collection of the single-source replacement paths are computed in a way that breaks shortest path ties properly, the total number of edges in this collection is bounded by $O(n^{3/2})$.

The Algorithm. Recall that W is a weight assignment that guarantees the uniqueness of the shortest paths, by introducing some symbolic perturbation to the edge lengths. Let $T_0 = BFS(s, G)$ be the BFS tree rooted at s in G, computed according to the weight assignment W. For every $e_j \in T_0$, let $T_0(e_j)$ be the BFS tree rooted at s in $G \setminus \{e_j\}$. Then the final FT-BFS tree is given by $T^*(s) = T_0 \cup \bigcup_{e_j \in T_0} T_0(e_j)$. The correctness is immediate by construction.

Observation 5. $T^*(s)$ *is an FT-BFS tree.*

It remains to bound the size of $T^*(s)$.

Size Analysis. We first provide some notation. For a path P, let $\mathtt{Cost}(P) = \sum_{e \in P} W(e)$ be the weighted cost of P, i.e., the sum of its edge weights. An edge $e \in G$ is defined as *new* if $e \notin E(T_0)$. For every $v_i \in V$ and $e_j \in T_0$, let $P_{i,j}^* = \pi(s, v_i, T_0(e_j)) \in SP(s, v_i, G \setminus \{e_j\}, W)$ be the optimal *replacement path* of s and v_i upon the failure of $e_j \in T_0$. Let $\mathtt{New}(P) = E(P) \setminus E(T_0)$ and

$$\mathtt{New}(v_i) = \{\mathtt{LastE}(P_{i,j}^*) \mid e_j \in T_0\} \setminus E(T_0)$$

be the set of v_i new edges appearing as the last edge in the replacement paths $P_{i,j}^*$ of v_i and $e_j \in T_0$. It is convenient to view the edges of $T_0(e_j)$ as directed away from s. We then have that

$$T^*(s) = T_0 \cup \bigcup_{v_i \in V \setminus \{s\}} \mathtt{New}(v_i).$$

I.e., the set of new edges that participate in the final FT–BFS tree $T^*(s)$ are those that appear as a last edge in some replacement path.

We now upper bound the size of the FT–BFS tree $T^*(s)$. Our goal is to prove that $\text{New}(v_i)$ contains at most $O(\sqrt{n})$ edges for every $v_i \in V$. The following observation is crucial in this context.

Observation 6. *If* $\text{LastE}(P^*_{i,j}) \notin E(T_0)$, *then* $e_j \in \pi(s, v_i)$.

Obs. 6 also yields the following.

Corollary 1. *(1)* $\text{New}(v_i) = \{\text{LastE}(P^*_{i,j}) \mid e_j \in \pi(s, v_i)\} \setminus E(T_0)$ *and* *(2)* $|\text{New}(v_i)| \le \min\{\text{depth}(v_i), \deg(v_i)\}$.

This holds since the edges of $\text{New}(v_i)$ are coming from at most $\text{depth}(v_i)$ replacement paths $P^*_{i,j}$ (one for every $e_j \in \pi(s, v_i)$), and each such path contributes at most one edge incident to v_i.

For the reminder of the analysis, let us focus on one specific node $u = v_i$ and let $\pi = \pi(s, u)$, $N = |\text{New}(u)|$. For every edge $e_k \in \text{New}(u)$, we define the following parameters. Let $f(e_k) \in \pi$ be the failed edge such that $e_k \in T_0(f(e_k))$ appears in the replacement path $P_k = \pi(s, u, T')$ for $T' = T_0(f(e_k))$. (Note that e_k might appear as the last edge on the path $\pi(s, u, T_0(e'))$ for several edges $e' \in \pi$; in this case, one such e' is chosen arbitrarily).

Let b_k be the *last* divergence point of P_k and π, i.e., the last vertex on the replacement path P_k that belongs to $V(\pi) \setminus \{u\}$. Since $\text{LastE}(P_k) \notin E(T_0)$, it holds that b_k is not the neighbor of u in P_k.

Let $\text{New}(u) = \{e_1, \ldots, e_N\}$ be sorted in non-decreasing order of the distance between b_k and u, $\text{dist}(b_k, u, \pi) = |\pi(b_k, u)|$. I.e.,

$$\text{dist}(b_1, u, \pi) \le \text{dist}(b_2, u, \pi) \ldots \le \text{dist}(b_N, u, \pi). \tag{1}$$

We consider the set of truncated paths $P'_k = P_k[b_k, u]$ and show that these paths are vertex-disjoint except for the last common endpoint u. We then use this fact to bound the number of these paths, hence bound the number N of new edges. The following observation follows immediately by the definition of b_k.

Observation 7. $(V(P'_k) \cap V(\pi)) \setminus \{b_k, u\} = \emptyset$.

Lemma 1. $(V(P'_i) \cap V(P'_j)) \setminus \{u\} = \emptyset$ *for every* $i, j \in \{1, \ldots, N\}$, $i \ne j$.

Proof: Assume towards contradiction that there exist $i \ne j$, and a node

$$u' \in (V(P'_i) \cap V(P'_j)) \setminus \{u\}$$

in the intersection. Since $\text{LastE}(P'_i) \ne \text{LastE}(P'_j)$, by Obs. 7 we have that $P'_i, P'_j \subseteq G \setminus E(\pi)$. The faulty edges $f(e_i), f(e_j)$ belong to $E(\pi)$. Hence there are two distinct $u' - u$ shortest paths in $G \setminus \{f(e_i), f(e_j)\}$. By the optimality of P'_i in $T_0(f(e_i))$, (i.e., $P_i \in SP(s, u, G \setminus \{f(e_i)\}, W)$), we have that $\text{Cost}(P'_i[u', u]) < \text{Cost}(P'_j[u', u])$. In addition, by the optimality of P'_j in $T_0(f(e_j))$, (i.e., $P_j \in SP(s, u, G \setminus \{f(e_j)\}, W)$), we have that $\text{Cost}(P'_j[u', u]) < \text{Cost}(P'_i[u', u])$. Contradiction. ∎

We are now ready to prove our key lemma.

Lemma 2. $|\texttt{New}(u)| = O(n^{1/2})$ *for every* $u \in V$.

Proof: Assume towards contradiction that $N = |\texttt{New}(u)| > \sqrt{2n}$. By Lemma 1, we have that $b_1, \ldots, b_N$ are distinct and by definition they all appear on the path π. Therefore, by the ordering of the P'_k, we have that the inequalities of Eq. (1) are strict, i.e., $\text{dist}(b_1, u, \pi) < \text{dist}(b_2, u, \pi) < \ldots < \text{dist}(b_N, u, \pi)$. Since $b_1 \neq u$ (by definition), we also have that $\text{dist}(b_1, u, \pi) \geq 1$. We Conclude that

$$\text{dist}(b_k, u, \pi) = |\pi(b_k, u)| \geq k . \tag{2}$$

Next, note that each P'_k is a replacement $b_k - u$ path and hence it cannot be shorter than $\pi(b_k, u)$, implying that $|P'_k| \geq |\pi(b_k, u)|$. Combining with Eq. (2), we have that

$$|P'_k| \geq k \quad \text{for every} \ \ k \in \{1, \ldots, N\} . \tag{3}$$

Since by Lemma 1, the paths P'_k are vertex disjoint (except for the common vertex u), we have that

$$\left| \bigcup_{k=1}^{N} (V(P'_k) \setminus \{u\}) \right| = \sum_{k=1}^{N} |V(P'_k) \setminus \{u\}| \geq \sum_{k=1}^{N} (k-1) > n,$$

where the first inequality follows by Eq. (3) and the last by the assumption that $N > \sqrt{2n}$. Since there are n nodes in G, we end with contradiction. ∎

Multiple Sources. For the case of multiple sources, in the full version [18], we establish the following upper bound.

Theorem 8. *There exists a polynomial time algorithm that for every n-vertex graph $G = (V, E)$ and source set $S \subseteq V$ of size $|S| = \sigma$ constructs an* FT-MBFS *tree $T^*(S)$ from each source $s_i \in S$, with a total number of* $n \cdot \min\{\sum_{s_i \in S} \texttt{depth}(s_i), O(\sqrt{\sigma n})\}$ *edges.*

We note that both our lower and upper bound analysis naturally extend to the case of directed and edge weighted graphs with integer weights in the range $[-M, M]$ by paying an extra factor of $O(\sqrt{M})$ in the size of the FT-MBFS trees.

5 $O(\log n)$-Approximation for FT-MBFS Trees

In Sec. 4, we presented an algorithm that for every graph G and source s constructs an FT-BFS tree $\widehat{T} \in \mathcal{T}(s, G)$ with $O(n^{3/2})$ edges. In Sec. 3, we showed that there exist graphs G and $s \in V(G)$ for which $\texttt{Cost}^*(s, G) = \Omega(n^{3/2})$, establishing tightness of our algorithm in the worst-case. Yet, there are also inputs (G', s') for which the algorithm of Sec. 4, as well as algorithms based on the analysis of [13] and [24], might still produce an FT-BFS $\widehat{T} \in \mathcal{T}(s', G')$ which is denser by a factor of $\Omega(\sqrt{n})$ than the size of the optimal FT-BFS tree, i.e., such that $|E(\widehat{T})| \geq \Omega(\sqrt{n}) \cdot \texttt{Cost}^*(s', G')$. For an illustration of such a case see [18]. Clearly, a universally optimal algorithm is unlikely given the hardness of approximation result of Thm. 1. Yet the gap can be narrowed down. The goal of this section is to present an $O(\log n)$ approximation algorithm for the Minimum FT-BFS Problem (hence also to its special case, the Minimum FT-BFS Problem, where $|S| = 1$).

Theorem 9. *There exists a polynomial time algorithm that for every n-vertex graph G and source node set $S \subseteq V$ constructs an* FT-MBFS *tree $\widehat{T} \in \mathcal{T}(S,G)$ such that $|E(\widehat{T})| \leq O(\log n) \cdot \mathtt{Cost}^*(S,G)$.*

To prove the theorem, we first describe the algorithm and then bound the number of edges. Let $\mathtt{ApproxSetCover}(\mathfrak{F}, U)$ be an $O(\log n)$ approximation algorithm for the Set-Cover problem, which given a collection of sets $\mathfrak{F} = \{S_1, \ldots, S_M\}$ that covers a universe $U = \{u_1, \ldots, u_N\}$ of size N, returns a cover $\mathfrak{F}' \subseteq \mathfrak{F}$ that is larger by at most $O(\log N)$ than any other $\mathfrak{F}'' \subseteq \mathfrak{F}$ that covers U (cf. [27]).

The Algorithm. Starting with $\widehat{T} = \emptyset$, the algorithm adds edges to $\widehat{T}$ until it becomes an FT-MBFS tree.

Set an arbitrary order on the vertices $V(G) = \{v_1, \ldots, v_n\}$ and on the edges $E^+ = E(G) \cup \{e_0\} = \{e_0, \ldots, e_m\}$ where e_0 is a new fictitious edge whose role will be explained later on. For every node $v_i \in V$, define

$$U_i = \{\langle s_k, e_j \rangle \mid s_k \in S \setminus \{v_i\}, e_j \in E^+\}.$$

The algorithm consists of n rounds, where in round i it considers v_i. Let $\Gamma(v_i, G) = \{u_1, \ldots, u_{d_i}\}$ be the set of neighbors of v_i in some arbitrary order, where $d_i = \deg(v_i, G)$. For every neighbor u_j, define a set $S_{i,j} \subseteq U_i$ containing certain source-edge pairs $\langle s_k, e_\ell \rangle \in U_i$. Informally, a set $S_{i,j}$ contains the pair $\langle s_k, e_\ell \rangle$ iff there exists an $s_k - v_i$ shortest path in $G \setminus \{e_\ell\}$ that goes through the neighbor u_j of v_i. Note that $S_{i,j}$ contains the pair $\langle s_k, e_0 \rangle$ iff there exists an $s_k - v_i$ shortest-path in $G \setminus \{e_0\} = G$ that goes through u_j. I.e., the fictitious edge e_0 is meant to capture the case where no fault occurs, and thus we take care of true shortest-paths in G. Formally, every pair $\langle s_k, e_\ell \rangle \in U_i$ is included in every set $S_{i,j}$ satisfying that

$$\mathrm{dist}(s_k, u_j, G \setminus \{e_\ell\}) = \mathrm{dist}(s_k, v_i, G \setminus \{e_\ell\}) - 1. \tag{4}$$

Let $\mathfrak{F}_i = \{S_{i,1}, \ldots, S_{i,d_i}\}$. The edges of v_i that are added to $\widehat{T}$ in round i are now selected by using algorithm $\mathtt{ApproxSetCover}$ to generate an approximate solution for the set cover problem on the collection $\mathfrak{F} = \{S_{i,j} \mid u_j \in \Gamma(v_i, G)\}$. Let $\mathfrak{F}'_i = \mathtt{ApproxSetCover}(\mathfrak{F}_i, U_i)$. For every $S_{i,j} \in \mathfrak{F}'_i$, add the edge (u_j, v_i) to $\widehat{T}$. In [18], we prove the correctness of this algorithm and establish Thm. 9.

Acknowledgment. We are grateful to Gilad Braunschvig, Alon Brutzkus, Adam Sealfon, Oren Weimann and the anonymous reviewers for helpful comments.

References

1. Awerbuch, B., Bar-Noy, A., Linial, N., Peleg, D.: Compact distributed data structures for adaptive network routing. In: STOC, pp. 230–240 (1989)

2. Abraham, I., Chechik, S., Gavoille, C., Peleg, D.: Forbidden-Set Distance Labels for Graphs of Bounded Doubling Dimension. In: PODC, pp. 192–200 (2010)
3. Baswana, S., Sen, S.: Approximate distance oracles for unweighted graphs in expected $O(n^2)$ time. ACM Trans. Algorithms 2(4), 557–577 (2006)
4. Bernstein, A., Karger, D.: A nearly optimal oracle for avoiding failed vertices and edges. In: STOC, pp. 101–110 (2009)
5. Chechik, S., Langberg, M., Peleg, D., Roditty, L.: f-sensitivity distance oracles and routing schemes. Algorithmica, 861–882 (2012)
6. Chechik, S., Langberg, M., Peleg, D., Roditty, L.: Fault-tolerant spanners for general graphs. In: STOC, pp. 435–444 (2009)
7. Chechik, S.: Fault-Tolerant Compact Routing Schemes for General Graphs. In: Aceto, L., Henzinger, M., Sgall, J. (eds.) ICALP 2011, Part II. LNCS, vol. 6756, pp. 101–112. Springer, Heidelberg (2011)
8. Czumaj, A., Zhao, H.: Fault-tolerant geometric spanners. Discrete & Computational Geometry 32 (2003)
9. Demetrescu, C., Thorup, M., Chowdhury, R., Ramachandran, V.: Oracles for distances avoiding a failed node or link. SIAM J. Computing 37, 1299–1318 (2008)
10. Dinitz, M., Krauthgamer, R.: Fault-tolerant spanners: better and simpler. In: PODC, pp. 169–178 (2011)
11. Duan, R., Pettie, S.: Dual-failure distance and connectivity oracles. In: SODA (2009)
12. Feige, U.: A Threshold of ln n for Approximating Set Cover. J. ACM, 634–652 (1998)
13. Grandoni, F., Williams, V.V.: Improved Distance Sensitivity Oracles via Fast Single-Source Replacement Paths. In: FOCS (2012)
14. Hershberger, J., Subhash, S.: Vickrey prices and shortest paths: What is an edge worth? In: FOCS (2001)
15. Hershberger, J., Subhash, S., Bhosle, A.: On the difficulty of some shortest path problems. In: Alt, H., Habib, M. (eds.) STACS 2003. LNCS, vol. 2607, pp. 343–354. Springer, Heidelberg (2003)
16. Levcopoulos, C., Narasimhan, G., Smid, M.: Efficient algorithms for constructing fault-tolerant geometric spanners. In: STOC, pp. 186–195 (1998)
17. Lukovszki, T.: New results on fault tolerant geometric spanners. In: Dehne, F., Gupta, A., Sack, J.-R., Tamassia, R. (eds.) WADS 1999. LNCS, vol. 1663, pp. 193–204. Springer, Heidelberg (1999)
18. Parter, P., Peleg, D.: Sparse Fault-Tolerant BFS Trees (2013),
http://arxiv.org/abs/1302.5401
19. Peleg, D.: Distributed Computing: A Locality-Sensitive Approach. SIAM (2000)
20. Peleg, D., Schäffer, A.A.: Graph spanners. J. Graph Theory 13, 99–116 (1989)
21. Peleg, D., Ullman, J.D.: An optimal synchronizer for the hypercube. SIAM J. Computing 18(2), 740–747 (1989)
22. Peleg, D., Upfal, E.: A trade-off between space and efficiency for routing tables. J. ACM 36, 510–530 (1989)
23. Roditty, L., Thorup, M., Zwick, U.: Deterministic constructions of approximate distance oracles and spanners. In: Caires, L., Italiano, G.F., Monteiro, L., Palamidessi, C., Yung, M. (eds.) ICALP 2005. LNCS, vol. 3580, pp. 261–272. Springer, Heidelberg (2005)
24. Roditty, L., Zwick, U.: Replacement paths and k simple shortest paths in unweighted directed graphs. ACM Trans. Algorithms (2012)
25. Thorup, M., Zwick, U.: Compact routing schemes. In: SPAA, pp. 1–10 (2001)
26. Thorup, M., Zwick, U.: Approximate distance oracles. J. ACM 52, 1–24 (2005)
27. Vazirani, V.: Approximation Algorithms. Georgia Inst. Tech. (1997)
28. Weimann, O., Yuster, R.: Replacement paths via fast matrix multiplication. In: FOCS (2010)

On the Most Likely Convex Hull
of Uncertain Points*

Subhash Suri, Kevin Verbeek, and Hakan Yıldız

University of California, Santa Barbara, USA

Abstract. Consider a set of points in d dimensions where the existence
or the location of each point is determined by a probability distribution.
The convex hull of this set is a random variable distributed over exponentially many choices. We are interested in finding the *most likely convex
hull*, namely, the one with the maximum probability of occurrence. We
investigate this problem under two natural models of uncertainty: the
point (also called the *tuple*) model where each point (site) has a fixed
position s_i but only exists with some probability π_i, for $0 < \pi_i \leq 1$, and
the *multipoint* model where each point has multiple possible locations
or it may not appear at all. We show that the most likely hull under
the point model can be computed in $O(n^3)$ time for n points in $d = 2$
dimensions, but it is NP–hard for $d \geq 3$ dimensions. On the other hand,
we show that the problem is NP–hard under the multipoint model even
for $d = 2$ dimensions. We also present hardness results for approximating
the probability of the most likely hull. While we focus on the most likely
hull for concreteness, our results hold for other natural definitions of a
probabilistic hull.

1 Introduction

We study the problem of computing the *most likely convex hull* of n uncertain points. The problem is fundamental in its own right, extending the notion
of minimal convex enclosure to probabilistic input, but is also motivated by a
number of applications dealing with noisy data. Before formalizing the problem, let us mention some motivating scenarios for our problem. In *movement
ecology* [12, 13], scientists track the movements of a group of animals using sensors with the goal of inferring their natural "home range". The ecologists have
long known that the smallest convex polygon containing all possible locations
visited by the animals is a gross overestimation of the home range, due to the
outlier problem, and instead have begun to consider probability-based isopleths.
The most likely hull is one possible tool in this analysis: use a discrete set of
landmarks (points), assign probability to each based on the frequency of the
animals' visits to the landmarks, and compute the most likely convex hull of
this probabilistic set of points as the most probable home range. As another
example, consider monitoring of a large geographic area for physical activity
(e.g., earthquake tremors). After collecting data over a period of time, we want

---

* This research was partially supported by the National Science Foundation grants
CCF-1161495 and CNS-1035917.

H.L. Bodlaender and G.F. Italiano (Eds.): ESA 2013, LNCS 8125, pp. 791–802, 2013.

to estimate the most likely region of activity. Since the value of a prediction decreases sharply with the rate of false positives, we want to find the tightest region for expected activity, and the most likely hull is a natural candidate. Finally, as a growing number of applications rely on machine learning and data mining for classification, we are inevitably forced to work with data whose attributes are inherently probabilistic. Computing meaningful geometric structures over these data is an interesting, and challenging, algorithmic problem. The most likely hull is a convenient vehicle to investigate these types of problems, although our methods and results are applicable more broadly, as discussed later.

In the *point* model of uncertain data,[1] the input is a pair (S, Π), where $S = \{s_1, s_2, \ldots, s_n\}$ is a set of n points (sites) in d-dimensional real Euclidean space, and $\Pi = \{\pi_1, \pi_2, \ldots, \pi_n\}$ is a probability vector with the interpretation that site s_i is active (namely, present) with probability π_i. The probabilities π_i are mutually independent. Thus, a random instance of (S, Π) includes each point s_i with an independent probability π_i. The convex hull of (S, Π) is a random variable, which assumes values over the convex hulls of the (at most) 2^n possible subsets. We are interested in computing the *most likely convex hull* for (S, Π).

The *multipoint* model generalizes the point model to incorporate *locational uncertainty*. The ith point of the input is described as $\left(\{s_i^1, \pi_i^1\}, \ldots, \{s_i^{k_i}, \pi_i^{k_i}\} \right)$, with the interpretation that the point appears at the position s_i^j with probability π_i^j, for $j = 1, 2, \ldots, k_i$. Different points can have a different number of possible locations k_i, but for simplicity we assume that the total number of locations is linear. Finally, we allow $\sum_{j=1}^{k_i} \pi_i^j < 1$ to include the possibility that the ith point does not exist at all, thus achieving a strict generalization of the point model.

Our first result shows that the most likely hull of points in 2 dimensions in the point model can be found in $O(n^3)$ time. We then show that the problem becomes NP-hard for dimensions $d \geq 3$. We also show that approximating the probability of the most likely hull is provably hard. In particular, computing a hull whose likelihood is within factor $2^{-O(n^{1-\epsilon})}$ of the optimal is NP–hard. This is nearly tight because a factor-(2^{-n}) approximate hull is easily computed by a simple greedy algorithm. Under the multipoint model, we show that the most likely hull problem is NP-hard even in two dimensions, and also inapproximable to a factor better than $2^{-O(n^{1-\epsilon})}$ unless P=NP. Note that in both models the problem is clearly in P for $d = 1$, since the number of distinct convex hulls in one dimension is only polynomial. While we focus on the most likely hull as a natural and concrete example, our algorithms and techniques apply more broadly to other possible ways of defining a probabilistic convex hull. Omitted proofs can be found in the full version of the paper.

Related Work. Uncertainty in geometric computing has been studied in a few different ways. In [16, 17], Löffler and van Kreveld have considered problems on "imprecise" objects: each object, such as a point, can be anywhere inside a simple

---

[1] The point model is also called the *tuple* model in database research, and has been used for studying clustering, ranking etc. of uncertain multi-attribute objects, modeled as points in d-space.

geometric region. For instance, given a set of imprecise points, one can ask for the maximum possible area of the convex hull of these points. However, this line of research looks at the worst-case behavior, and not the stochastic behavior, which is the main focus of our work. In a more closely related and interesting work [14], Jørgensen, Löffler and Phillips, develop a general framework for geometric shape-fitting problems and describe how the solutions to these problems vary with respect to the uncertainty in the points. Another line of research has focused on uncertainty caused by the finite machine precision [15, 18, 19]. The goal there is to achieve robustness under bounded precision, and not to compute structures that are most representative under a probability distribution. There also has been extensive research in the database community on clustering and ranking of uncertain data [4, 5, 10] and on range searching and indexing [1–3].

2   Two-Dimensional Most Likely Hull in the Point Model

In this section, we describe a dynamic programming algorithm for computing the most likely hull of n points in the plane under the point model of uncertainty. For simplicity, we assume that no three points are collinear, but the algorithm is easily modified to handle such degeneracies. We begin with some general technical facts related to convex hulls of uncertain points in the point model.

Let (S, Π) denote the input to the uncertain convex hull problem in d-space. A subset $A ß S$ occurs as an outcome of a probabilistic experiment with probability $\pi(A)$ given by

$$\pi(A) \;=\; \prod_{s_i \in A} \pi_i \;\times\; \prod_{s_i \notin A} \overline{\pi_i}$$

where we use the notation $\overline{\pi_i} = (1 - \pi_i)$. Given an outcome A, its convex hull is denoted as $\mathcal{CH}(A)$. For a convex polytope C, we define its *likelihood*, denoted $\mathcal{L}(C)$, as the probability that C is the convex hull of the random outcome of a probabilistic experiment on (S, Π). In other words,

$$\mathcal{L}(C) \;=\; \Pr\big[\mathcal{CH}(A) \equiv C\big] = \sum_{\substack{A ß S \\ \mathcal{CH}(A) \equiv C}} \pi(A)$$

The *most likely hull* of (S, Π) is the polytope C with the maximum value of $\mathcal{L}(C)$. Our first lemma shows that $\mathcal{L}(C)$ can be written as a product of two factors where the first factor involves only the *vertices* of C, and not all the sites that fall inside C.

Lemma 1. *Let C be a convex polytope, $V ß S$ be its vertex set, and $S_{out} ß S$ the set of sites lying outside C. Then, we have the following:*

$$\mathcal{L}(C) \;=\; \prod_{s_i \in V} \pi_i \times \prod_{s_i \in S_{out}} \overline{\pi_i},$$

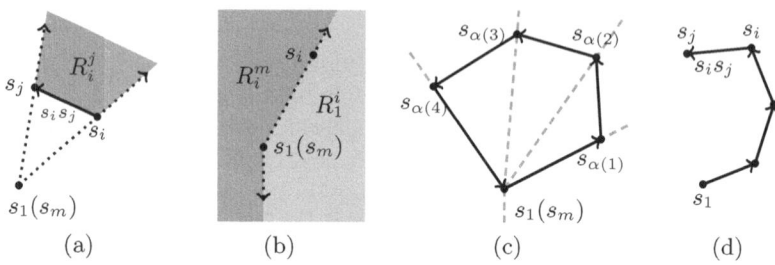

Fig. 1. Illustrations for the two-dimensional most likely hull in the point model

Likelihood Contributions of Edges. We now describe how to find the most likely hull for a 2-dimensional input under the point model. Our algorithm computes, for each site s_i, the most likely hull with s_i as its lowest (minimum y-coordinate) vertex, and then outputs the best hull over all choices of s_i. For ease of reference, let us call a convex polygon with s_i as its lowest vertex, a *hull rooted at s_i*. We decompose the likelihood of a convex hull into several components, each associated with an edge of the hull. The key to the computational efficiency is to ensure that the component associated with an edge *does not depend on the hull* in which the edge participates. Geometrically, we associate a *wedge shaped region* with each edge, depending only on the choice of the lowest vertex, and define the contribution based only on the sites contained in this wedge. We now discuss this in more details.

Suppose we want to compute the most likely hull rooted at s_1. Without loss of generality, let $s_2, \ldots, s_{m-1}$ be the sequence of sites (all lying above s_1) in the counter-clockwise order around s_1, for $(m-1) \le n$. Any hull rooted at s_1 has a subsequence of $s_1, \ldots, s_{m-1}$ as its vertex set. Finally, for notational convenience, we add an artificial site $s_m = s_1$ (a copy of the root point) with probability zero.

Given two sites s_i and s_j, with $1 \le i < j \le m$, we use $s_i s_j$ to denote the *directed edge* drawn from s_i to s_j. To each directed edge $s_i s_j$, we associate a region of space R_i^j. For an edge not involving s_1 or its copy s_m, namely $s_i s_j$, for $1 < i < j < m$, R_i^j is the region bounded by the segment $s_i s_j$ and the rays $\overrightarrow{s_1 s_i}$ and $\overrightarrow{s_1 s_j}$. See Figure 1a for illustration. For edges with the first endpoint at s_1, namely $s_1 s_i$, for $1 < i < m$, R_1^i is the region bounded (on its left) by the downward ray extending from s_1 and the ray $\overrightarrow{s_1 s_i}$. The complementary region of R_1^i is also important, and we call it R_i^m, associated with the edge $s_i s_m$, which is the reverse edge of $s_1 s_i$. See Figure 1b.

We now define the *contribution* of the directed edge $s_i s_j$, denoted $\mathcal{C}(s_i s_j)$, as π_i times the probability that none of the sites in the region R_i^j (except s_i and s_j) are present, including the sites that may lie below s_1. That is,

$$\mathcal{C}(s_i s_j) = \pi_i \times \prod_{s_k \in R_i^j} \overline{\pi_k}$$

The following lemma shows how these edge contributions help us compute the likelihood of a convex hull C.

Lemma 2. *Let C be a hull rooted at s_1, with vertices $s_1, s_{\alpha(1)}, \ldots, s_{\alpha(\ell)}$ in the counter-clockwise order. Then,*

$$\mathcal{L}(C) = \mathcal{C}(s_1 s_{\alpha(1)}) \times \mathcal{C}(s_{\alpha(1)} s_{\alpha(2)}) \times \cdots \times \mathcal{C}(s_{\alpha(\ell-1)} s_{\alpha(\ell)}) \times \mathcal{C}(s_{\alpha(\ell)} s_m)$$

Proof. Partition the space outside C into the regions $R_1^{\alpha(1)}, R_{\alpha(1)}^{\alpha(2)}, \ldots, R_{\alpha(\ell-1)}^{\alpha(\ell)}, R_{\alpha(\ell)}^m$ by drawing a downward ray from s_1 and drawing rays $\overrightarrow{s_1 s_{\alpha(j)}}$ for each $1 \leq j \leq \ell$. (See Figure 1c for an example.) Then, by Lemma 1, it is easy to see that the $\mathcal{L}(C)$ is the product of the contributions of the edges of C. □

The contribution of each edge can be computed in constant time after an $O(n^2)$-time preprocessing, using a modified version of a triangle query data structure of [11]. We give the details of this structure in the full version of the paper.

The Dynamic Programming Algorithm. Our dynamic programming algorithm computes, for each edge $s_i s_j$, the convex chain whose edges yield the *maximum product of contributions* under the following constraints:

1. The sequence of vertices in the chain is a subsequence of $s_1, \ldots, s_m$.
2. The first vertex of the chain is s_1.
3. The last edge of the chain is $s_i s_j$. (See Figure 1d for an example.)

We denote this maximum chain by $\mathcal{T}(s_i s_j)$. With a slight abuse of notation, we also use $\mathcal{T}(s_i s_j)$ to denote the product of the edge contributions of this chain. Clearly, all chains of the form $\mathcal{T}(s_i s_m)$ correspond to polygons rooted at s_1, and the one with the maximum contribution is the most likely hull we want. Our dynamic programming formulation is fairly standard, and similar style of algorithms have been used in the past for computing largest convex subsets [6, 9] and monochromatic islands [7].

We now describe an optimal substructure property crucial for our dynamic programming algorithm. Consider a chain $\mathcal{T}(s_i s_j)$. This, by definition, has the maximum likelihood of all chains terminating with the edge $s_i s_j$. If we remove the last vertex s_j of $\mathcal{T}(s_i s_j)$, and the corresponding edge $s_i s_j$, then the remaining chain should be the optimal chain terminating at s_i *that can be extended to s_j without violating convexity.* In other words, the remaining chain is the maximum among all chains $\mathcal{T}(s_k s_i)$ (where $1 \leq k < i$) such that the path $s_k \to s_i \to s_j$ is a left turn. This implies the following recurrence:

$$\mathcal{T}(s_i s_j) = \begin{cases} \mathcal{C}(s_1 s_j) & \text{if } i = 1 \\ \mathcal{C}(s_i s_j) \times \displaystyle\max_{\substack{1 \leq k < i \\ s_k \to s_i \to s_j \text{ is a left turn}}} \left(\mathcal{T}(s_k s_i) \right) & \text{otherwise} \end{cases}$$

We use this recurrence to compute all the chains $\mathcal{T}(s_i s_j)$ as follows. We begin by setting $\mathcal{T}(s_1 s_i)$ to $\mathcal{C}(s_1 s_i)$ for all $1 < i \leq m$. Then, we process all sites s_i in increasing order of i. When we process a site s_i, we compute all chains $\mathcal{T}(s_i s_j)$ by using the previously computed chains. This can be done in $O(n)$ time as follows. Let S_{prec} be the set of sites $\{s_1, \ldots, s_{i-1}\}$ and S_{succ} be the set $\{s_{i+1}, \ldots, s_m\}$. Let $s_{\beta(1)}, \ldots, s_{\beta(\ell)}$ be the sites in S_{prec} in counter-clockwise order around s_i, starting with s_1.[2] For each site $s_{\beta(u)}$ in S_{prec}, we define s_u^* to be the site s_k among the sequence $s_{\beta(1)}, \ldots, s_{\beta(u)}$ that maximizes $\mathcal{T}(s_k s_i)$. The site s_u^* can be computed for all sites $s_{\beta(u)}$ with a linear sweep of the sites in S_{prec} in order.

For each site $s_{\beta(u)}$ in S_{prec}, we set the value $\mathcal{T}(s_i s_j)$ to $\mathcal{C}(s_i s_j) \times \mathcal{T}(s_u^* s_i)$ for all sites s_j in S_{succ} inside the wedge bounded by the lines $\overleftrightarrow{s_{\beta(u)} s_i}$ and $\overleftrightarrow{s_{\beta(u+1)} s_i}$.[3] (See Figure 2.) Note that the sites in this wedge are the sites that form a left turn when connected to $s_{\beta(1)}, \ldots, s_{\beta(u)}$ through s_i (the condition in the recurrence relation). By considering the sites $s_{\beta(u)}$ in radial order around s_i, we can locate each site in the wedge of interest in constant time.

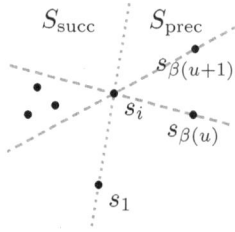

Fig. 2. Wedge for $\mathcal{T}(s_i s_j)$

The processing of a single point s_i takes $O(n)$ time, and thus we can find the most likely hull rooted at s_1 in $O(n^2)$ time, and the global most likely hull of P in $O(n^3)$ time. The algorithm needs $O(n^2)$ space, dominated by the storage of the $\mathcal{T}(\cdot)$ values.

Theorem 1. *The most likely convex hull of an uncertain point set defined by n sites in the point model can be computed in $O(n^3)$ time and in $O(n^2)$ space.*

3   Hardness of the 3-Dimensional Most Likely Hull

We now show that computing the most likely hull in 3 or more dimensions is NP-hard in the point model. In particular, we give a reduction from the vertex cover problem in penny graphs to the 3-dimensional most likely hull problem.

A *penny graph* is a graph $G = (V, E)$ along with an embedding $\rho : V \to \mathbb{R}^2$ such that $\|\rho(u) - \rho(v)\|_2 = 2$ if $(u, v) \in E$, and $\|\rho(u) - \rho(v)\|_2 > 2$ if $(u, v) \notin E$, where $\|.\|_2$ denotes the L_2 norm. In other words, a penny graph admits a planar drawing where vertices are represented as unit disks with pairwise disjoint interiors, and two disks make contact if and only if there is an edge between the two corresponding vertices. We denote the centers of the unit disks by the points $p_1, \ldots, p_n$, and the point of contact between two adjacent disks with centers p_i and p_j by p_{ij}. The following simple observation about the penny graph embedding will be critical in our reduction. See Figure 3 for an illustration.

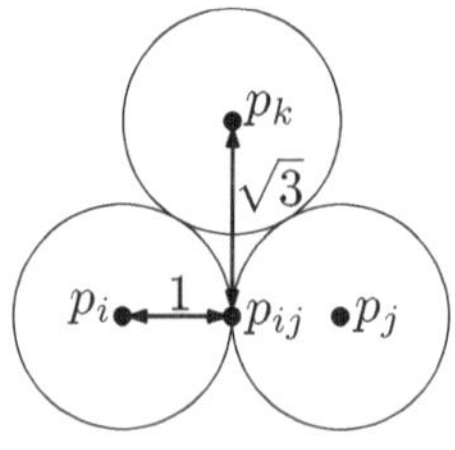

Lemma 3. $\|p_k - p_{ij}\|_2 \geq \sqrt{3}$, *for all* $k \neq i, j$.

Fig. 3. Lemma 3

---

[2] This counter-clockwise order for all sites s_i can be precomputed in $O(n^2 \log n)$ time.
[3] We also remember how $\mathcal{T}(s_i s_j)$ is computed, so the corresponding chain can be constructed later.

The vertex cover problem for penny graphs is to find the smallest subset $U \subseteq V$ of vertices such that every edge of the graph has an endpoint in U. This problem was shown to be NP-hard in [8]. Our reduction relies on the following simple but important property of the most likely hull in the point model.

Lemma 4. *Any point (s_i, π_i) with $\pi_i \geq 1/2$ is in the most likely hull.*

The Reduction. Consider an instance of the vertex cover problem for a penny graph G, with $p_1, \ldots, p_n$ being the disk centers of the embedding of G. We create an instance of the most likely hull problem in three dimensions, as follows. All the sites lie on one of the two paraboloids, $\mathcal{P}_1 : z = x^2 + y^2$ or $\mathcal{P}_2 : z = x^2 + y^2 - 2$. In particular, for each disk center p_i, we create a site u_i by vertically lifting p_i onto the paraboloid $\mathcal{P}_2$. All these points are assigned a fixed probability $\pi_i = \alpha < \frac{1}{2}$.

The sites on $\mathcal{P}_1$ are associated with the contact points p_{ij} but are not a direct lifting of the contact points themselves. Instead, for each contact point $p_{ij} = (x_{ij}, y_{ij})$, we define four new points $p_{ij}^N = (x_{ij}, y_{ij} + \delta)$, $p_{ij}^E = (x_{ij} + \delta, y_{ij})$, $p_{ij}^S = (x_{ij}, y_{ij} - \delta)$, and $p_{ij}^W = (x_{ij} - \delta, y_{ij})$, for some $\delta > 0$. (We set the value of δ later.) Next, we add a set X_{ij} of m arbitrary points inside the quadrilateral formed by p_{ij}^e ($e \in \{N, E, S, W\}$). We lift each of the p_{ij}^e onto $\mathcal{P}_1$ to obtain a site u_{ij}^e, for $e \in \{N, E, S, W\}$, and each of these points is assigned a probability of 1. Finally, the subsets X_{ij} are lifted onto $\mathcal{P}_1$ to get subsets Y_{ij}, and each of these points are assigned a fixed probability $\beta > \frac{1}{2}$. All these points, lying on the paraboloids $\mathcal{P}_1$ and $\mathcal{P}_2$, along with their associated probabilities form the input for our most likely hull problem.

The main idea of the reduction is that we want to "cover" each set Y_{ij} by putting either u_i or u_j on the most likely hull. In the penny graph, this corresponds to covering the edge associated with the contact point p_{ij} by the vertex associated with p_i or p_j. We now describe this relation in more depth, starting with a well-known lemma about the lifting transform.

Lemma 5. *Consider a point $p \in \mathbb{R}^2$, and let $u(p)$ be its vertical projection (lifting) onto the paraboloid $\mathcal{P}_1$, and $H(p)$ the hyperplane tangent to $\mathcal{P}_1$ at $u(p)$. Then, the vertical projections $u(p')$ of all points $p' \in \mathbb{R}^2$ at distance r from p lie on a hyperplane parallel to $H(p)$ whose vertical distance from $H(P)$ is r^2.*

The points u_i's (liftings of p_i's) lie on $\mathcal{P}_2$, which is a vertical downward shift of $\mathcal{P}_1$. Now, if u_{ij} is the point obtained by lifting p_{ij} to $\mathcal{P}_1$, then by Lemma 5 the points u_i and u_j are vertically 1 unit below the tangent plane of $\mathcal{P}_1$ at u_{ij}, while the points u_k ($k \neq i, j$) are at least vertically 1 unit above this plane by Lemma 3 (see Figure 4). If we treat $\mathcal{P}_1$ as an "obstacle", then u_i and u_j can "see" u_{ij} from below, while the points u_k ($k \neq i, j$) cannot. Thus there exists a small enough $\delta > 0$ such that Y_{ij} is contained in the convex hull of u_{ij}^e ($e \in \{N, E, S, W\}$) with either of u_i and u_j,

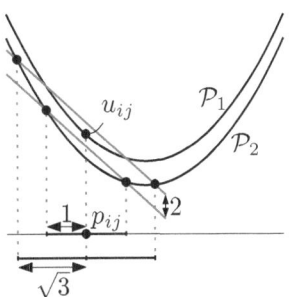

Fig. 4. Lift to $\mathcal{P}_1$ and $\mathcal{P}_2$ (vertically scaled)

but not with u_k $(k \neq i, j)$. The following lemma describes a sufficient upper-bound on δ.

Lemma 6. *If $\delta < \sqrt{3} - \sqrt{2}$, then the points u_i and u_j can see the entire quadrilateral on $\mathcal{P}_1$ formed by u_{ij}^e ($e \in \{N, E, S, W\}$) from below, but no u_k $(k \neq i, j)$ can see any part of the quadrilateral from below.*

Theorem 2. *Computing the most likely hull in three dimensions is NP-hard.*

Proof. We show that computing the likelihood of the most likely hull is NP-hard. Given an instance of the vertex cover problem for penny graphs, we construct an instance of the most likely hull problem in three dimensions as described above (e.g., with $\delta = 0.25$). We choose m, α, and β such that $\beta^m < \alpha$, and $\alpha < 0.5 < \beta$; e.g., $m = 3$, $\alpha = 0.25$, and $\beta = 0.6$. By Lemma 4 all points on $\mathcal{P}_1$ must be on or inside the most likely hull, and so we only need to choose which points u_i $(1 \leq i \leq n)$ are on the most likely hull. No point from a set Y_{ij} can be on the most likely hull because then we could add either u_i or u_j to the hull and increase the likelihood of the hull, since $\beta^m(1 - \alpha) < \alpha$. Thus, the likelihood of the most likely hull is determined by the number κ of points u_i $(1 \leq i \leq n)$ that are on the most likely hull, and its likelihood is $\alpha^\kappa(1 - \alpha)^{n-\kappa}$. Every point u_i on the most likely hull corresponds to a vertex of the penny graph, and by construction and Lemma 6, these vertices form a vertex cover of the penny graph. Thus the penny graph has a vertex cover of size κ if and only if the likelihood of the most likely hull is at least $\alpha^\kappa(1 - \alpha)^{n-\kappa}$. Finally, it is easy to see that the construction can be performed in polynomial time. $\square$

The proof above directly implies that there exists no polynomial-time $(\frac{\alpha}{1-\alpha})$-approximation algorithm to compute the likelihood of the most likely hull unless $P = NP$. Although we can change the value of α to obtain a stronger bound, we give a more general argument below.

Inapproximability. The likelihood of a hull is a product of terms. We show that, under mild conditions, NP-hard optimization problems of this form cannot be approximated well by a multiplicative factor, unless $P = NP$.

Let $\mathcal{O} = (\mathcal{I}, \mathcal{F}, f)$ be an optimization problem where $\mathcal{I}$ is the set of instances, $\mathcal{F}$ is a function over $\mathcal{I}$ such that $\mathcal{F}(I)$ describes the set of feasible solutions for instance I, and f is an optimization function over all feasible solutions. For an instance $I \in \mathcal{I}$, let $|I|$ denote the size of I. We say that $\mathcal{O}$ is *product composable* if, given any collection of problem instances $I_1, \ldots, I_k \in \mathcal{I}$, we can construct a new instance $I^* \in \mathcal{I}$ in polynomial time (w.r.t. $|I^*|$) satisfying the following:

1. $|I^*| = \sum_{i=1}^{k} |I_i|$.
2. There is a bijection between $\mathcal{F}(I^*)$ and $\mathcal{F}(I_1) \times \ldots \times \mathcal{F}(I_k)$ such that for each solution $S \in \mathcal{F}(I^*)$ with the matching tuple $(S_1, \ldots, S_k)$, $f(S) = \prod_{1 \leq i \leq k} f(S_i)$.
3. Given a solution $S \in \mathcal{F}(I^*)$, one can construct the solutions in its matching tuple in polynomial time.

In other words, we can form a new instance I^* by combining the instances $I_1, \ldots, I_k$ in an independent way.

Lemma 7. *If a maximization problem $\mathcal{O}$ is product composable and cannot be approximated within a constant $c < 1$ in polynomial time, then there exists no polynomial-time $2^{-O(n^{1-\epsilon})}$-approximation algorithm for $\mathcal{O}$, where n is the size of the instance and $\epsilon > 0$.*

Although the most likely hull problem is not product composable itself, this property only needs to hold for a subproblem. The subproblem formed by the instances used in our NP-hardness reduction is product composable, which easily follows from the construction. We defer a detailed explanation of this property to the full version of the paper.

Corollary 1. *For any $\epsilon > 0$, there exists no polynomial-time $2^{-O(n^{1-\epsilon})}$-approximation algorithm for the most likely hull problem in three dimensions, unless $P=NP$.*

Finally we observe that one can trivially achieve a 2^{-n}-approximation of the most likely hull problem as follows: simply take the convex hull of all sites with probability at least $\frac{1}{2}$. If $\pi_i < \frac{1}{2}$ for all i, then the convex hull is empty.

4   Most Likely Hull in the Multipoint Model

In this section, we show that computing the most likely hull in the multipoint model is NP–hard even for two dimensions. (The technical definition of the most likely hull under the multipoint model differs slightly from that of the point model, but the following abridged description should be accessible without a need for those details. A more complete formal description can be found in the full version of the paper.) Our proof uses a reduction from 3-SAT.

Consider a 3-SAT instance (V, U) where V is the set of the variables and U is the set of clauses. We first construct $6|U|$ points on the unit circle. We call these points the *anchors* and use them as permanent points (i.e., points with probability 1) in our hull problem instance. Between each pair of consecutive anchors, we place a single point on the unit circle that we call a *spike*. (See Figure 5a.) We assign an independent existence probability of $\frac{1}{2}$ to each spike. As we will explain shortly, the main idea of our construction is that the most likely hull includes all spikes in its interior if and only if the 3-SAT instance is satisfiable.

For each variable v, we construct two additional sets of points, one corresponding to the case that v is true and one corresponding to the case that v is false. In particular, for each clause u that v appears in positive form, we construct a point p_v^u covering a single spike, at the intersection of the lines tangent to the unit circle at the two anchors next to the spike. We assign each p_v^u a probability of $\frac{1}{2}$ but this probability is dependent, as we will put p_u^v in the same tuple with another point in the rest of the construction. We construct all points p_u^v for a single variable v over a consecutive sequence of spikes, and then put a single point t_v covering the constructed points. (See Figure 5b.)

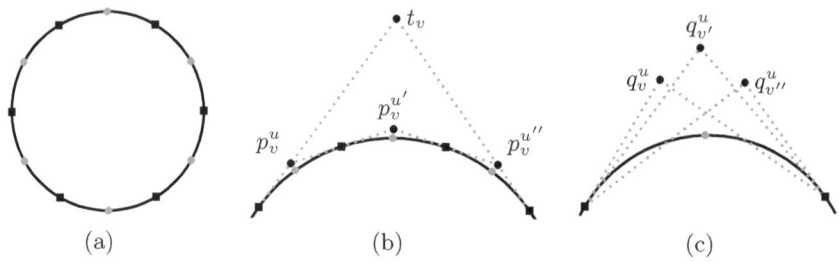

Fig. 5. (a) Anchors (black squares) and spikes (gray circles) on the unit circle. (b) Construction of t_v. (c) The three points constructed for clause u.

We apply the same construction for all clauses that v appears in negated form. This creates an additional set of points p_u^v, all of which we cover with a single point f_v as we did for t_v. We put t_v and f_v to the same probabilistic tuple and assign each a probability of $\frac{1}{2}$. That is, in a probabilistic experiment, either t_v or f_v is present (with equal probability), but not both. Existence of t_v is meant to imply that v is assigned true, whereas the existence of f_v is meant to imply that v is assigned false.

Finally, for each clause u, we construct three additional points covering a single spike. These points are constructed in such a way that: (1) they do not cover any other spike, and (2) they are in convex position with respect to each other and the two anchors next to the covered spike. Each of these points corresponds to a distinct variable v that appears in the clause. We denote the point associated with variable v by q_v^u. (See Figure 5c.) We put each q_v^u to the same probability tuple as the previously constructed point p_v^u and assign it probability $\frac{1}{2}$. That is, in an experiment, either q_v^u or p_v^u exists (with equal probability), but not both.

Lemma 8. *The most likely hull has likelihood $(1/2)^{3|U|+|V|}$ if and only if it contains all spikes in its interior. Otherwise, its likelihood is at most $(1/2)^{3|U|+|V|+1}$.*

We now describe how the satisfiability of the 3-SAT instance relates to our construction. Consider a variable v. Notice that, if the most likely hull covers all spikes below t_v, then either t_v or all points p_v^u below t_v appears in the hull as a vertex. If t_v appears in the hull, then the hull can pass through the points q_v^u (which are in the same probabilistic tuples with points p_v^u), and cover spikes representing the clauses that v appears in positive form. This corresponds to the case that v is assigned true and all corresponding clauses are satisfied. Similar notion also applies to f_v and the clauses that v appears in negated form. If all spikes are covered, then all clauses are satisfied and so is the 3-SAT instance. Combining this idea with Lemma 8, we deduce the following lemma.

Lemma 9. *The 3-SAT instance is satisfiable if and only if the most likely hull has likelihood $(1/2)^{3|U|+|V|}$.*

Theorem 3. *Computing the most likely hull in the multipoint model is NP-hard.*

Lemma 8 in fact implies a stronger result: It is NP-hard to compute the likelihood of the most likely hull within any factor $c > \frac{1}{2}$. By construction, the problem instances that we create are product composable. Then, by Lemma 7, we can state the following theorem.

Theorem 4. *For any $\epsilon > 0$, there exists no polynomial-time $2^{-O(n^{1-\epsilon})}$-approximation algorithm for the most likely hull problem in the multipoint model unless P=NP.*

5   Extensions and Concluding Remarks

Making sense of probabilistic (uncertain) data is a complex and challenging task. Even for simple numerical data, elementary statistics such as mean, median, or mode serve a useful first order approximation. For multi-dimensional spatial data, however, there are no universally agreed upon summaries of similar generality. Our work is an attempt to explore some natural geometric structures, and their complexity, over probabilistic data. For "convexity" of uncertain data, one possibility is to compute the distribution over the entire space: for each point of the space, compute the probability that it is inside the convex hull. In a different work, we are also exploring that direction but (i) a full distribution is inevitably quite expensive to compute (requiring a worst-case space complexity $\Omega(n^{d^2})$), and (ii) the distribution still does not lend itself to a simple and "intuitive" description of a convex hull.

Therefore, algorithms for computing or estimating succinct summary hulls are a useful tool in the analysis of uncertain geometric data. While we focused exclusively on the Most Likely Hull, our techniques are applicable to several other ways of defining the "best" hull. Any useful definition of the likely hull must include a penalty function for misclassifying points, *both false positives and false negatives*. If only false negatives (points outside the hull) are penalized, then the convex hull of *all* the points has the best score. Our dynamic programming algorithm for the point model in 2 dimensions can be extended for several natural scoring functions. Although entries may need to be computed differently, the subproblem structure utilized by the dynamic programming algorithm also applies to these other settings.

For instance, one simple scoring function measures the *agreement* on the "in" and "out" classification. A convex hull C splits the point set into two parts: inside and outside. We can measure the "quality" $Q(C)$ of a hull C by its expected agreement with a random hull's classification: the number of points of S whose classification (in or out) is the same for both C and the hull of a random outcome. Both our dynamic programming algorithm for computing the hull in 2 dimensions, and the hardness in 3 dimensions, under the point model carry over to this "Symmetric Difference Hull" definition. Similarly, another scoring function for measuring the fraction of points correctly classified counts the number of points in the random outcome that lie in C plus the number of non-sample points that lie outside C. All our results hold for this model as well.

In summary, we believe that the study of geometric structures over probabilistic data is a fundamental problem, and our results are only a first, but promising, step.

References

1. Afshani, P., Agarwal, P.K., Arge, L., Larsen, K.G., Phillips, J.M.: (Approximate) uncertain skylines. Theory Comput. Syst. 52(3), 342–366 (2013)
2. Agarwal, P.K., Cheng, S.-W., Tao, Y., Yi, K.: Indexing uncertain data. In: PODS, pp. 137–146 (2009)
3. Agarwal, P.K., Cheng, S.-W., Yi, K.: Range searching on uncertain data. ACM Transactions on Algorithms 8(4), 43 (2012)
4. Aggarwal, C.C.: Managing and Mining Uncertain Data. Advances in Database Systems, vol. 35. Kluwer (2009)
5. Aggarwal, C.C., Yu, P.S.: A survey of uncertain data algorithms and applications. IEEE Trans. Knowl. Data Eng. 21(5), 609–623 (2009)
6. Avis, D., Rappaport, D.: Computing the largest empty convex subset of a set of points. In: Proc. of the 1st Symp. on Comput. Geometry, pp. 161–167 (1985)
7. Bautista-Santiago, C., Díaz-Báñez, J.M., Lara, D., Pérez-Lantero, P., Urrutia, J., Ventura, I.: Computing optimal islands. Op. Res. Letters 39(4), 246–251 (2011)
8. Cerioli, M.R., Faria, L., Ferreira, T.O., Protti, F.: On minimum clique partition and maximum independent set on unit disk graphs and penny graphs: Complexity and approximation. Electronic Notes in Discrete Mathematics 18, 73–79 (2004)
9. Chvátal, V., Klincsek, G.: Finding largest convex subsets. Congresus Numeratium 29, 453–460 (1980)
10. Cormode, G., McGregor, A.: Approximation algorithms for clustering uncertain data. In: Proc. 27th Symp. on Principles of Database Systems, pp. 191–200 (2008)
11. Eppstein, D., Overmars, M., Rote, G., Woeginger, G.: Finding minimum area k-gons. Discrete & Computational Geometry 7(1), 45–58 (1992)
12. Getz, W.M., Fortmann-Roe, S., Cross, P.C., Lyons, A.J., Ryan, S.J., Wilmers, C.C.: Locoh: Nonparameteric kernel methods for constructing home ranges and utilization distr ibutions. PLoS ONE 2(2), 02 (2007)
13. Getz, W.M., Wilmers, C.C.: A local nearest-neighbor convex-hull construction of home ranges and utilization distributions. Ecography 27(4), 489–505 (2004)
14. Jørgensen, A., Löffler, M., Phillips, J.M.: Geometric computations on indecisive and uncertain points. CoRR, abs/1205.0273 (2012)
15. Kettner, L., Mehlhorn, K., Pion, S., Schirra, S., Yap, C.-K.: Classroom examples of robustness problems in geometric computations. Comput. Geom. 40(1) (2008)
16. Löffler, M.: Data Imprecision in Computational Geometry. PhD thesis, Utrecht University (2009)
17. Löffler, M., van Kreveld, M.: Largest and smallest convex hulls for imprecise points. Algorithmica 56, 235–269 (2010)
18. Salesin, D., Stolfi, J., Guibas, L.J.: Epsilon geometry: Building robust algorithms from imprecise computations. In: Symp. on Comput. Geom., pp. 208–217 (1989)
19. Yap, C.-K., Pion, S.: Special issue on robust geometric algorithms and their implementations. Comput. Geom. 33(1-2) (2006)

Top-k Document Retrieval in External Memory[*]

Rahul Shah[1], Cheng Sheng[2], Sharma V. Thankachan[1],
and Jeffrey Scott Vitter[3]

[1] Louisiana State University, USA
{rahul,thanks}@csc.lsu.edu
[2] The Chinese University of Hong Kong, China
csheng@cse.cuhk.edu.hk
[3] The University of Kansas, USA
jsv@ku.edu

Abstract. Let $\mathcal{D}$ be a given set of (string) documents of total length
n. The top-k document retrieval problem is to index $\mathcal{D}$ such that when
a pattern P of length p, and a parameter k come as a query, the index
returns those k documents which are most relevant to P. We present
the first non-trivial external memory index supporting top-k document
retrieval queries in optimal $O(p/B + \log_B n + k/B)$ I/Os, where B is the
block size. The index space is almost linear $O(n \log^* n)$ words.

1 Introduction and Related Work

The inverted index is the most fundamental data structure in the field of information retrieval [43]. It is the backbone of every known search engine today. For
each word in any document collection, the inverted index maintains a list of all
documents in that collection which contain the word. Despite its power to answer various types of queries, the inverted index becomes inefficient, for example,
when queries are phrases instead of words. This inefficiency results from inadequate use of word orderings in query phrases [37]. Similar problems also occur in
applications when word boundaries do not exist or cannot be identified deterministically in the documents, like genome sequences in bioinformatics and text in
many East-Asian languages. These applications call for data structures to answer
queries in a more general form, that is, (string) pattern matching. Specifically,
they demand the ability to identify efficiently all the documents that contain a
specific pattern as a substring. The usual inverted-index approach might require
the maintenance of document lists for all possible substrings of the documents.
This can take quadratic space and hence is neither theoretically interesting nor
sensible from a practical viewpoint.

The first frameworks for answering document retrieval queries were proposed
by Matias et al. [30] and Muthukrishnan [31]. Their data structures solve the *document listing problem*, where the task is to index a collection $\mathcal{D}$ of D documents,
such that whenever a pattern P of length p comes as a query, report all those
documents containing P exactly once. Muthukrishnan also initiated the study

[*] Work supported by National Science Foundation (NSF) Grants CCF–1017623 (R.
Shah and J. S. Vitter) and CCF–1218904 (R. Shah).

H.L. Bodlaender and G.F. Italiano (Eds.): ESA 2013, LNCS 8125, pp. 803–814, 2013.
© Springer-Verlag Berlin Heidelberg 2013

of relevance metric-based document retrieval [31], which was then formalized by Hon et al. [22] as follows:

Problem 1 (Top-k document retrieval problem). *Let $w(P, d)$ be the score function capturing the relevance of a pattern P with respect to a document d. Given a document collection $\mathcal{D} = \{d_1, d_2, .., d_D\}$ of D documents, build an index answering the following query: given P and k, find k documents with the highest $w(P, .)$ values in its sorted (or unsorted) order.*

Here, instead of reporting all the documents that match a query pattern, the problem is to output the k documents most relevant to the query in sorted order of relevance score. Relevance metrics considered in the problem can be either pattern-independent (e.g., PageRank) or -dependent. In the latter case one can take into account information like the frequency of the pattern occurrences (or *term-frequency* of popular tf-idf measure, which takes the number of occurrences of P in a document d as $w(P, d)$) and even the locations of the occurrences (e.g., *min-dist* [22] which takes proximity of two closest occurrences of pattern as the score). In general, we assume that other than a static weight which is fixed for each document d, $w(P, d)$ is dependent only on the set of occurrences of P in d. The framework of Hon et al. [22] takes linear space and answers the query in $O(p + k \log k)$ time. This was then improved by Navarro and Nekrich [33] to achieve $O(p + k)$ query cost. Both [22] and [33] reduced this problem to a 4-sided orthogonal range query in 3d, which is defined as follows: the data consists of a set S of 3-dimensional points and the query consists of four parameters x', x'', y' and z', and output is the set of all those points $(x_i, y_i, z_i) \in S$ such that $x_i \in [x', x'']$, $y_i \leq y'$ and $z_i \geq z'$. While general 4-sided orthogonal range searching is proved hard [9], the desired bounds can nevertheless be achieved by identifying a special property that one dimension of the reduced subproblem can only have p distinct values. Even though there has been series of work on top-k string, including in theory as well as practical IR [5, 10, 11, 13, 15, 17–25, 33–35, 37, 38, 42] communities, most implementations (as well as theoretical results) have focused on RAM based compressed and/or efficient indexes (See [32] for an excellent survey). We introduce an alternative framework for solving this problem and obtain the first non-trivial external memory [3] solution as follows:

Theorem 1. *In the external memory model, there exists an $O(nh)$-word structure that solves the top-k (unsorted) document retrieval problem in $O(p/B + \log_B n + \log^{(h)} n + k/B)$ I/Os for any $h \leq \log^* n$, where $\log^{(h)} n = \log \log^{(h-1)} n$, $\log^{(1)} n = \log n$ and B is the block size.*

For $h = \log^* n$, $\log^{(h)} n$ is a constant, and hence we have the following result.

Corollary 1. *There exists an $O(n \log^* n)$-word structure for answering top-k (unsorted) document retrieval problem in optimal $O(p/B + \log_B n + k/B)$ I/Os.*

Our framework can also be used for improving the existing internal memory results [22, 33] (see Theorem 2). In situations where the locus node can be computed in $o(p)$ time, our new index support faster queries. For example in cross-document pattern matching [26], the locus can be computed in $O(\log \log p)$ time. Another application is autocompletion search (like in Google Instant^{TM}), where multiple loci are searched with amortized constant time for each locus (see [27, 41] for other examples).

Theorem 2. *There exists an $O(n)$ word space data structure in word RAM model for solving (sorted) top-k document retrieval problem in $O(k)$ time, once the locus of the pattern match is given.*

A related but somewhat orthogonal line of research has been to get top-k queries on general array based ranges. In this, we are given array A of colors with each color is assigned a score, and for a range query (i, j), we have to output k highest scored colors in this range (with each color reported at most once). If the scoring criteria is based on frequency, for example say score of a color is its number of occurrences in $A[i..j]$, then lower-bounds on range-mode problem [8, 16] would imply no efficient (linear space and polylog time) data structures can exist. There are variants considered where each entry in the array has a fixed score or each color (document) has a fixed score, independent of number of occurrences. Recent [29] surprising result of achieving optimal I/Os with $O(n \log^* n)$ space has been for 3-sided categorical range reporting where each entry has another attribute called score, and the query specifies range as well as score threshold. We are supposed to output all colors whose at least one entry within the range satisfies the score criteria. There are easier variants where each entry of the same color gets the same score attribute like PageRank which have been shown to have efficient external memory results [36]. There are even simpler variants, where only top-k scores are to be reported [2, 28, 39] without considering colors or unique colors are to be reported without considering scores (as in *document listing*). Both these variants lead to 3-sided queries which are easier to solve in external memory. In internal memory, there exists optimal space and time data structures for outputting these scores in the sorted order [7].

2   Preliminary: Top-k Framework

This section briefly explains the linear space framework for top-k document retrieval based on the work of Hon et al. [22], and Navarro and Nekrich [33]. The generalized suffix tree (GST) of a document collection $\mathcal{D} = \{d_1, d_2, d_3, \ldots, d_D\}$ is the combined compact trie (a.k.a. Patricia trie) of all the non-empty suffixes of all the documents. Use n to denote the total length of all the documents, which is also the number of the leaves in GST. For each node u in GST, consider the path from the root node to u. Let $depth(u)$ be the number of nodes on the path, and $prefix(u)$ be the string obtained by concatenating all the edge labels of the path. For a pattern P that appears in at least one document, the *locus* of P, denoted as u_P, is the node closest to the root satisfying that P is a prefix of $prefix(u_P)$. By numbering all the nodes in GST in the pre-order traversal manner, the part of GST relevant to P (i.e., the subtree rooted at u_P) can be represented as a range.

Nodes are marked with documents. A leaf node ℓ is marked with a docu-
ment $d \in \mathcal{D}$ if the suffix represented by ℓ belongs to d. An internal node u
is marked with d if it is the lowest common ancestor of two leaves marked
with d. Notice that a node can be marked with multiple documents. For each
node u and each of its marked documents d, define a *link* to be a quadruple
$(origin, target, doc, score)$, where $origin = u$, $target$ is the lowest proper ancestor[1] of u marked with d, $doc = d$ and $score = w(prefix(u), d)$. Two crucial
properties of the links identified in [22] are listed below.

Lemma 1. *For each document d that contains a pattern P, there is a unique
link whose origin is in the subtree of u_P and whose target is a proper ancestor
of u_P. The score of the link is exactly the score of d with respect to P.*

Lemma 2. *The total number of links is $O(n)$.*

Based on Lemma 1, the top-k document retrieval problem can be reduced to
the problem of finding the top-k links (according to its score) stabbed by u_P,
where *link stabbing* is defined as follows:

Definition 1 (Link Stabbing). *We say that a link is* stabbed *by node u if it
is originated in the subtree of u and targets at a proper ancestor of u.*

If we order the nodes in GST as per the *pre-order traversal* order, these
constraints translate into finding all the links (i) the numbers of whose origins
fall in the number range of the subtree of u_P, and (ii) the numbers of whose
targets are less than the number of u_P. Regarding constraint (i) as a two-sided
range constraint on x-dimension, and regarding constraint (ii) as a one-sided
range constraint on y-dimension, the problem asks for the top-k weighted points
that fall in a three-sided window in 2d space, where weight of a point is the score
of the corresponding link [33].

3   External Memory Structures

This section is dedicated for proving Theorem 1. The initial phase of pattern
search can be performed in $O(p/B + \log_B n)$ I/O's using a string B-tree [12].
Once the suffix range of P is identified, we take the lowest common ancestor of
the left-most and right-most leaves in the suffix range of GST to identify the
locus node u_P. Hence, the first phase (i.e., finding the locus node u_P of P) takes
optimal I/O's and now we focus only on the second phase (i.e., reporting the
top-k links stabbed by u_P). Instead of solving the top-k version, we first solve a
threshold version in Sec 3.1 where the objective is to retrieve those links stabbed
by u_P with *score* at least a given threshold τ. Then in Sec 3.2, we propose a
separate structure that converts the original top-k-form query into a threshold-
form query so that the structure in Sec 3.1 can now be used to answer the
original problem. Finally, we obtain Theorem 1 via bootstrapping on a special
structure for handling top-k queries in lesser number of I/Os for small values of
k. We shall assume all scores are distinct and are within $[1, O(n)]$. Otherwise,
the ties can be broken arbitrarily and reduce the values into rank-space.

---

[1] Define a dummy node as the parent of the root node, marked with all the documents.

3.1   Breaking Down into Sub-Problems

Instead of solving the top-k version, we first solve a threshold version, where the objective is to retrieve those links stabbed by u_P with *score* at least a given threshold τ. We show that the problem can be decomposed into simpler subproblems, which consists of a 3d dominance reporting and $O(\log(n/B))$ 3-sided range reporting in 2d, both can be solved efficiently using known structures. The main result is captured in Lemma 3 defined below. From now onwards, the origin, target and score of a link L_i are represented by o_i, t_i and w_i respectively.

Lemma 3. *There exists an $O(n)$ space data structure for answering the following query: given a query node u_P and a threshold τ, all links stabbed by u_P with score $\geq \tau$ can be reported in $O(\log^2(n/B) + z/B)$ I/Os, where z is the number of outputs.*

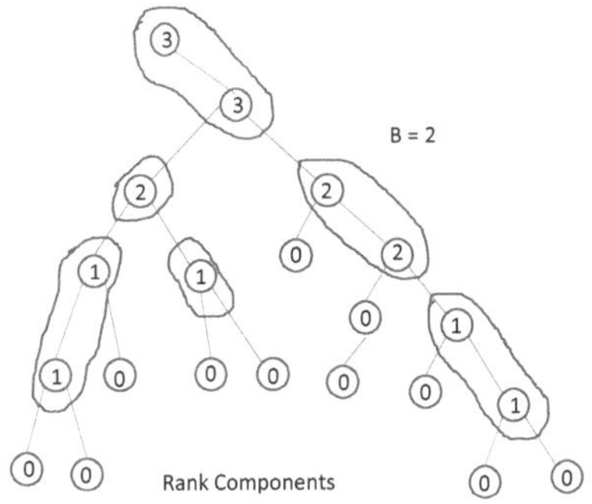

Fig. 1. Rank Components

Rank and Components. For any node u in GST, we use u to denote its pre-order rank as well. Let $size(u)$ denotes the number of leaves in the subtree of u, then we define its *rank* as:

$$rank(u) = \lfloor \log\lceil \frac{size(u)}{B} \rceil \rfloor$$

Note that $rank(.) \in [0, \lfloor\log\lceil\frac{n}{B}\rceil\rfloor]$. A contiguous subtree consisting of nodes with the same rank is defined as a *component*, and the *rank* of a component is same as the rank of nodes within it (see figure 1). Therefore, a *component* with $rank = 0$ is a bottom level subtree of size (number of leaves) at most B. From the definition, it can be seen that a node and at most one of its children can have the same *rank*. Therefore, a component with $rank \geq 1$ consists of nodes in a path which goes top-down in the tree.

   The number of links originating within the subtree of any node u is at most $2size(u) - 1$. Therefore, the number of links originating within a component with $rank = 0$ is $O(B)$. These $O(B)$ links corresponding to each component with $rank = 0$ can be maintained separately as a list, taking total $O(n)$ words space. Now, given a locus node u_P, if $rank(u_P) = 0$, the number of links originating within the subtree of u_P is also $O(B)$ and all of them can be processed in $O(1)$ I/O's by simply scanning the list of links corresponding to the component to which u_P belongs to. The query processing is more sophisticated when $rank(u_P) \geq 1$. For handling this case, we classify the links into the following 2 types based on the *rank* of its target with respect to the *rank* of query node u_P:

1. *equi-ranked links*: links with $rank(target) = rank(u_P)$
2. *high-ranked links*: links with $rank(target) > rank(u_P)$

Next we show that the problem of retrieving outputs among equi-ranked links can be
reduced to a 3d dominance query, and the
problem of retrieving outputs among high-
ranked links can be reduced to at most
$\lfloor \log\lceil \frac{n}{B}\rceil \rfloor$ 3-sided range queries in 2d.

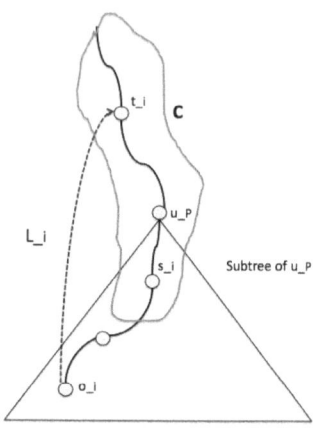

Processing Equi-Ranked Links. Let C be
a component and S_C be set of all links L_i,
such that its target t_i is a node in C. Also,
for any link $L_i \in S_C$, let pseudo_origin s_i be
the (pre-order rank of) lowest ancestor of its
origin o_i within C (see Figure 2). Then a link
$L_i \in S_C$ originates in the subtree of any node
u within C if and only if $s_i \geq u$. Now if the
locus u_P is a node in C, then among all equi-
ranked links, we need to consider only those

Fig. 2. Pseudo Origin

links $L_i \in S_C$, because the origin o_j of any
other equi-ranked link $L_j \notin S_C$, will not be in the subtree of u_P. Based on
the above observations, all equi-ranked output links are those $L_i \in S_C$ with
$t_i < u_P \leq s_i$ and $w_i \geq \tau$. To solve this in external memory, we treat each link
$L_i \in S_C$ as a 3d point (t_i, s_i, w_i) and maintain a 3d dominance query structure
over it. Now the outputs with respect to u_P and τ are those links corresponding
to the points within $(-\infty, u_P) \times [u_P, \infty) \times [\tau, \infty)$. Such a structure for S_C can
be maintained in linear $O(|S_C|)$ words of space and can answer the query in
$O(\log_B |S_C| + z_{eq}/B)$ I/O's using the result by Afshani [1], where $|S_C|$ is the
number of points (corresponding to links in S_C) and z_{eq} be the output size.
Thus overall these structures occupies $O(n)$-word space.

Lemma 4. *Given a query node u_P and a threshold τ, all the equi-ranked links
stabbed by u_P with score $\geq \tau$ can be retrieved in $O(\log_B n + z_{eq}/B)$ I/Os using
an $O(n)$ word space data structure, where z_{eq} is the output size.*   □

Processing High-Ranked Links. The following is an important observation.

Observation 1. *Any link L_i with its origin o_i within the subtree of a node u is
stabbed by u if $rank(t_i) > rank(u)$, where t_i is the target of L_i.*

This implies, while looking for the outputs among the high-ranked links, the
condition of t_i being a proper ancestor of u_P can be ignored as it is taken care
of automatically if $o_i \in [u_P, u'_P]$, where u'_P be the (pre-order rank of) rightmost leaf in the subtree rooted at u_P. Let G_r be the set of all links with rank
equals r for $1 \leq r \leq \lfloor \log\lceil \frac{n}{B}\rceil \rfloor$. Since there are only $O(\log(n/B))$ sets, we shall
maintain separate structures for links in each G_r by considering only *origin* and
score values. We treat each link $L_i \in G_r$ as a 2d point (o_i, w_i), and maintain

a 3-sided range query structure over them for $r = 1, 2, .., \lfloor \log\lceil \frac{n}{B}\rceil \rfloor$. All high-ranked output links can be obtained by retrieving those links in $L_i \in G_r$ with the corresponding point $(o_i, w_i) \in [u_P, u'_P] \times [\tau, \infty]$ for $r = rank(u_P)+1, .., \lfloor \log\lceil \frac{n}{B}\rceil \rfloor$. By using the linear space data structure in [4], the space and I/O bounds for a particular r is given by $O(|G_r|)$ words and $O(\log_B |G_r| + z_r/B)$, where z_r is the number of output links in G_r. Since a link can be a part of at most one G_r, the total space consumption is $O(n)$ words and the total query I/Os is $O(\log_B n \log(n/B) + z_{hi}/B) = O(\log^2(n/B) + z_{hi}/B)$, where z_{hi} represents the number of high-ranked output links.

Lemma 5. *Given a query node u_P and a threshold τ, all the high-ranked links stabbed by u_P with score $\geq \tau$ can be retrieved in $O(\log^2(n/B) + z_{hi}/B)$ I/Os using an $O(n)$ word space data structure, where z_{hi} is the output size.* □

By combining Lemma 4 and Lemma 5, we obtain Lemma 3.

3.2   Converting Top-k to Threshold via Logarithmic Sketch

Here we derive a linear space data structure, such that given a query node u and a parameter k, a threshold τ can be computed in constant I/Os, such that the number of links z stabbed by u with score $\geq \tau$ is bounded by, $k \leq z \leq 2k + O(\log n)$. Hence query I/Os in Lemma 3 can be modified as $O(\log^2(n/B) + z/B) = O(\log^2(n/B) + k/B)$. From the retrieved z outputs, the actual top-k answers can be computed by selection [6, 40] and filtering in another $O(z/B) = O(k/B + \log_B n)$ I/O's. We summarize our result in the following lemma.

Lemma 6. *There exist an $O(n)$ word data structure for answering the following query in $O(\log^2(n/B) + k/B)$ I/O's: given a query point u and an integer k, report the top-k links stabbed by u.* □

The details of top-k to threshold conversion are given below.

Marked Nodes and Prime Nodes in GST. We identify certain nodes in the *GST* as marked nodes and prime nodes with respect to a parameter g called the *grouping factor*. The procedure starts by combining every g consecutive leaves (from left to right) together as a group, and marking the lowest common ancestor (LCA) of first and last leaf in each group. Further, we mark the LCA of all pairs of marked nodes recursively. Additionally, we ensure that the root is always marked. At the end of this procedure, the number of marked nodes in *GST* will be $O(n/g)$ [22]. Prime nodes are those which are the children of marked nodes [2]. Corresponding to any marked node u^* (except root), there is a *unique prime* node u', which is its closest prime ancestor. In case u^*'s parent is marked then $u' = u^*$. For every prime node u' with atleast one marked node in its subtree, the corresponding closest marked descendant u^* is unique. If u' is marked then the closest marked descendant u^* is same as u'.

---

[2] Note that the number of prime nodes can be $\Theta(n)$ in the worst case.

Hon et al. [22] showed that, given any node u with u^* being its highest marked descendent (if it exists), the number of leaves in the subtree of u, but not in the subtree of u^* (which we call as fringe leaves) is at most $2g$. This means for a given threshold τ, if z is the number of outputs corresponding to u^* as the locus node, then the number of outputs corresponding to u as the locus is within $z \pm 2g$. This is because of the fact that the number of documents d with $w(prefix(u), d) \neq w(prefix(u^*), d)$ cannot be more than the number of fringe leaves. Therefore, we maintain the following information at every marked node u^*: the score of q–th highest scored link stabbed by u^* for $q = 1, 2, 4, 8, \ldots$ By choosing $g = \log n$, the total space can be bounded by $O((n/g) \log n) = O(n)$ words, and can retrieve any particular entry in $O(1)$ time.

Using the above values, the threshold τ corresponding to any given u and k can be computed as follows: first find the highest marked node u^* in the subtree of u ($u^* = u$ if u is marked). Now identify i such that $2^{i-1} < k + 2g \leq 2^i$ and choose τ as the score of 2^i-th highest scored link stabbed by u^*. This ensures that $k \leq z < 2k + O(g) = 2k + O(\log n)$.

3.3   Special Structures for Bounded k

In this section, we derive a faster data structures for the case when k is upper bounded by a parameter g. The main idea is to identify smaller sets of $O(g)$ links, such that top-g links stabbed by any node u are contained in one of such sets. Thus by constructing the structure described in Lemma 6 over the links in each such sets, the top-k queries for any $k \leq g$ can be answered faster as follows:

Lemma 7. *There exists a $O(n)$ word data structure for answering top-k queries for $k \leq g$ in $O(\log^2(g/B) + k/B)$ I/O's.*

Recall the definitions of marked nodes and prime nodes from Sec 3.2. Let u' be a prime node and u^* (if it exists) be the unique highest marked descendent of u' by choosing a grouping factor g (which will be fixed later). All the links originated from the subtree of u' are categorized into the following (Figure 3).

- *near-links:* The links which are stabbed by u^*, but not by u'.
- *far-link:* The links which are stabbed by both u^* and u'.
- *small-link:* The links which are stabbed neither by u^*, nor by u'.
- *fringe-links:* The links originated not from the subtree of u^*.

Lemma 8. *The number of* fringe-links *and the number of* near-links *of any prime node u' is $O(g)$.*

Proof. The number of leaves in $subtree(u') \backslash subtree(u^*)$ is at most $2g$ [22]. Therefore, the number of *fringe-links* can be bounded by $O(g)$. For every document d whose link originates from $subtree(u^*)$ going out of it ends up as a *near-link* if and only if d exists at one of the leaves of $subtree(u') \backslash subtree(u^*)$. Thus, this can also be bounded by $O(g)$. In the case where u^* does not exist for u', only fringe-links exist. More over the subtree size of u' is $O(g)$ there can be no more than $O(g)$ of these links.   □

Consider the following set, consisting of $O(g)$ links with respect to u': all *fringe-links*, *near-links* and g highest scored *far-links*. We maintain these *links* at u' (as a data structure to be explained later). For any node u, whose closest prime ancestor (including itself) is u', the above mentioned set is called *candidate links* of u. From each u, we maintain the pointer to its closest prime ancestor where the set of *candidate links* is stored.

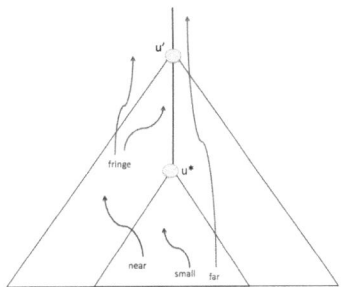

Lemma 9. *The* candidate links *of any node* u *contains top-g highest scored links stabbed by* u.

Fig. 3. Categorization of Links

Proof. Let u' be the closest prime ancestor of u. If no marked descendant of u' exist, then all the links are stored as candidate links. Otherwise, *small-links* can never be candidates as they never cross u. Now, if u lies on the path from u' to u^* then all *far-links* will satisfy both origin and target conditions. Else, *far-links* do not qualify. Hence, any link which is not among top-g (highest scored) of these far-links, can never be the candidate.   □

Taking a clue from Lemma 8 and 9, for every prime node u', we shall maintain a data structure as in Lemma 6 by considering only the links stored at u', and top-k queries can be answered faster when $k \le g$. For this we shall define a candidate tree $CT(u')$ of node u' (except the root) to be a modified version of subtree of u' in GST augmented with candidate links stored at u'. Firstly, for every candidate link which is targeted above u', we change the target to v, which will be a dummy parent of u' in $CT(u')$. Now $CT(u')$ consists of those nodes which are either origin or target (after modification) of some candidate link of u'. Moreover, all the nodes in $subtree(u') \backslash subtree(u^*)$ are included as well. Since only the subset of nodes is selected from $subtree(u')$, our tree is basically a Steiner tree connecting these nodes. Moreover, the tree is edge-compacted so that no degree-1 node remains. Thus, the size of the tree as well as the number of associated links is $O(g)$. Next we do a rank-space reduction of pre-order rank (w.r.t to GST) of the nodes in $CT(u')$ as well as the scores of candidate links.

The candidate tree (no degree-1 nodes) as well as the associated candidate links satisfies all the properties which we have exploited while deriving the structure in Lemma 6. Hence such a structure for $CT(u')$ can be maintained in $O(min(g, size(u')))$ words space and the top-k links in $CT(u')$ stabbed by any node u, with u' being its lowest prime ancestor can be retrieved in $O(\log^2(g/B) + k/B)$ I/O's. The total space consumption of structures corresponding all prime nodes can be bounded by $O(n)$ words as follows: the number of prime nodes with at least a marked node in its subtree is $O(n/g)$, as each such prime node can be associated with a unique marked node. Thus the associated structures takes $O(n/g \times g) = O(n)$ words space. The candidate set of a prime node u' with no marked nodes in its subtree consists of $O(size(u'))$ links, moreover a link cannot be in the candidate set of two such prime nodes. Thus the total space is $O(n)$ words in this case as well. Note that for $g = O(B)$, we need not store any structure on $CT(u')$, because such a candidate tree fits entirely in constant number of blocks which can be processed in $O(1)$ I/Os. This completes the proof of Lemma 7.

3.4 I/O-Optimal Data Structure via Bootstrapping

The bounds in Theorem 1 can be achieved by maintaining multiple structures as in Lemma 7. Clearly the structure in Lemma 6 is optimal for $k \geq B \log^2(n/B)$. However, for handling the case when $k < B \log^2(n/B)$, we shall choose the grouping factor $g_i = B(\log^{(i)}(n/B))^2$, for $i = 1, 2, 3, .., h \leq \log^* n$ and maintain h separate structures as in Lemma 7, occupying $O(nh)$ space. Thus top-k query for any $k \geq g_h$ can be answered by querying on the structure corresponding to the grouping factor g_j, where $g_j \geq k > g_{j+1}$ in $O(\log^2(g_j/B) + k/B) = O(g_{j+1}/B + k/B) = O(k/B)$ I/Os. For $k < g_h$, we shall query on the structure corresponding to the grouping factor g_h, and the I/Os are bounded by $O(\log^2(g_h/B) + k/B) = O(\log^{(h)} n + k/B)$. This completes the proof of Theorem 1.

4   Adapting to Internal Memory

Our external memory framework can be adapted to internal memory by choosing $B = \Theta(1)$, and by replacing the external memory substructures by the corresponding internal memory counterparts. Retrieving the outputs among high-ranked links is reduced to $O(\log n)$ 3-sided range reporting queries. By using an interval tree like approach, the problem of retrieving outputs among equi-ranked links also can be reduced to $O(\log n)$ 3-sided range reporting queries. By using the linear-space sorted range reporting structure by Brodal et al. [7] for 3-sided range reporting, the outputs can be obtained in the sorted order of score. Further, these sorted outputs from $O(\log n)$ different places can be merged using an atomic heap [14], which is capable of performing all heap operations in $O(1)$ time, provided the number of elements in the heap is $O(\log^{O(1)} n)$ as in our case. At the beginning of each of these $O(\log n)$ queries, we may need to perform a binary search for finding the boundaries, thus resulting in a total query time of $O(\log^2 n + k)$, which is $O(k)$ for $k \geq \log^2 n$. The space can be bounded by $O(n)$ words. For the case when $k < \log^2 n$, we obtain a linear space and $O(\log^2 \log n + k)$ query time structure by using the ideas from Sec 3.3 (here we choose grouping factor $g = \log^2 n$). Again, this structure can answer queries in $O(k)$ time for $k \geq \log^2 \log n$. We do not continue this bootstrapping further. Instead, we make use of the following observation: the candidate set of a node consists of only $O(g)$ links, hence a pointer to any particular link within the candidate set of any node can be maintained in $O(\log g) = O(\log \log n)$ bits. Thus, at every node u in GST, we shall maintain the top-$(\log^2 \log n)$ links stabbed by u in the decreasing order of score as a pointer to its location within the candidate set of u. This occupies $O(n \log^3 \log n)$ bits or $o(n)$ words and top-k queries for any $k \leq \log^2 \log n$ can be answered in $O(k)$ time by chasing the first k pointers and retrieving the documents associated with the corresponding links. This completes the proof of Theorem 2.

References

1. Afshani, P.: On dominance reporting in 3D. In: Halperin, D., Mehlhorn, K. (eds.) ESA 2008. LNCS, vol. 5193, pp. 41–51. Springer, Heidelberg (2008)
2. Afshani, P., Brodal, G.S., Zeh, N.: Ordered and unordered top-k range reporting in large data sets. In: SODA, pp. 390–400 (2011)
3. Aggarwal, A., Vitter, J.S.: The input/output complexity of sorting and related problems. Commun. ACM 31(9), 1116–1127 (1988)
4. Arge, L., Samoladas, V., Vitter, J.S.: On two-dimensional indexability and optimal range search indexing. In: PODS, pp. 346–357 (1999)
5. Belazzougui, D., Navarro, G., Valenzuela, D.: Improved compressed indexes for full-text document retrieval, vol. 18, pp. 3–13 (2013)
6. Blum, M., Floyd, R.W., Pratt, V.R., Rivest, R.L., Tarjan, R.E.: Time bounds for selection. J. Comput. Syst. Sci. 7(4), 448–461 (1973)
7. Brodal, G.S., Fagerberg, R., Greve, M., López-Ortiz, A.: Online sorted range reporting. In: Dong, Y., Du, D.-Z., Ibarra, O. (eds.) ISAAC 2009. LNCS, vol. 5878, pp. 173–182. Springer, Heidelberg (2009)
8. Chan, T.M., Durocher, S., Larsen, K.G., Morrison, J., Wilkinson, B.T.: Linear-space data structures for range mode query in arrays. In: STACS, pp. 290–301 (2012)
9. Chazelle, B.: Lower bounds for orthogonal range searching: I. the reporting case. J. ACM 37(2), 200–212 (1990)
10. Culpepper, J.S., Navarro, G., Puglisi, S.J., Turpin, A.: Top-k ranked document search in general text databases. In: de Berg, M., Meyer, U. (eds.) ESA 2010, Part II. LNCS, vol. 6347, pp. 194–205. Springer, Heidelberg (2010)
11. Culpepper, J.S., Petri, M., Scholer, F.: Efficient in-memory top-k document retrieval. In: SIGIR (2012)
12. Ferragina, P., Grossi, R.: The string b-tree: A new data structure for string search in external memory and its applications. J. ACM 46(2), 236–280 (1999)
13. Fischer, J., Gagie, T., Kopelowitz, T., Lewenstein, M., Mäkinen, V., Salmela, L., Välimäki, N.: Forbidden patterns. In: Fernández-Baca, D. (ed.) LATIN 2012. LNCS, vol. 7256, pp. 327–337. Springer, Heidelberg (2012)
14. Fredman, M.L., Willard, D.E.: Trans-dichotomous algorithms for minimum spanning trees and shortest paths. J. Comput. Syst. Sci. 48(3), 533–551 (1994)
15. Gagie, T., Navarro, G., Puglisi, S.J.: New algorithms on wavelet trees and applications to information retrieval. Theor. Comput. Sci. 426, 25–41 (2012)
16. Greve, M., Jørgensen, A.G., Larsen, K.D., Truelsen, J.: Cell probe lower bounds and approximations for range mode. In: Abramsky, S., Gavoille, C., Kirchner, C., Meyer auf der Heide, F., Spirakis, P.G. (eds.) ICALP 2010. LNCS, vol. 6198, pp. 605–616. Springer, Heidelberg (2010)
17. Hon, W.-K., Patil, M., Shah, R., Thankachan, S.V., Vitter, J.S.: Indexes for document retrieval with relevance. In: Munro Festschrift, pp. 351–362 (2013)
18. Hon, W.-K., Shah, R., Thankachan, S.V.: Towards an optimal space-and-query-time index for top-k document retrieval. In: Kärkkäinen, J., Stoye, J. (eds.) CPM 2012. LNCS, vol. 7354, pp. 173–184. Springer, Heidelberg (2012)
19. Hon, W.-K., Shah, R., Thankachan, S.V., Vitter, J.S.: String retrieval for multi-pattern queries. In: Chavez, E., Lonardi, S. (eds.) SPIRE 2010. LNCS, vol. 6393, pp. 55–66. Springer, Heidelberg (2010)

20. Hon, W.-K., Shah, R., Thankachan, S.V., Vitter, J.S.: Document listing for queries with excluded pattern. In: Kärkkäinen, J., Stoye, J. (eds.) CPM 2012. LNCS, vol. 7354, pp. 185–195. Springer, Heidelberg (2012)
21. Hon, W.-K., Shah, R., Thankachan, S.V., Vitter, J.S.: Faster compressed top-k document retrieval. In: DCC (2013)
22. Hon, W.-K., Shah, R., Vitter, J.S.: Space-efficient framework for top-k string retrieval problems. In: FOCS 2009, pp. 713–722 (2009)
23. Hon, W.-K., Shah, R., Vitter, J.S.: Compression, indexing, and retrieval for massive string data. In: Amir, A., Parida, L. (eds.) CPM 2010. LNCS, vol. 6129, pp. 260–274. Springer, Heidelberg (2010)
24. Karpinski, M., Nekrich, Y.: Top-k color queries for document retrieval. In: SODA, pp. 401–411 (2011)
25. Konow, R., Navarro, G.: Faster compact top-k document retrieval. In: DCC (2013)
26. Kucherov, G., Nekrich, Y., Starikovskaya, T.: Cross-document pattern matching. In: Kärkkäinen, J., Stoye, J. (eds.) CPM 2012. LNCS, vol. 7354, pp. 196–207. Springer, Heidelberg (2012)
27. Külekci, M.O., Vitter, J.S., Xu, B.: Efficient maximal repeat finding using the burrows-wheeler transform and wavelet tree. IEEE/ACM Trans. Comput. Biology Bioinform. 9(2), 421–429 (2012)
28. Larsen, K.G., Pagh, R.: I/o-efficient data structures for colored range and prefix reporting. In: SODA, pp. 583–592 (2012)
29. Larsen, K.G., van Walderveen, F.: Near-optimal range reporting structures for categorical data. In: SODA, pp. 265–276 (2013)
30. Matias, Y., Muthukrishnan, S.M., Şahinalp, S.C., Ziv, J.: Augmenting suffix trees, with applications. In: Bilardi, G., Pietracaprina, A., Italiano, G.F., Pucci, G. (eds.) ESA 1998. LNCS, vol. 1461, pp. 67–78. Springer, Heidelberg (1998)
31. Muthukrishnan, S.: Efficient algorithms for document retrieval problems. In: SODA, pp. 657–666 (2002)
32. Navarro, G.: Spaces, trees and colors: The algorithmic landscape of document retrieval on sequences. CoRR, abs/1304.6023 (2013)
33. Navarro, G., Nekrich, Y.: Top- k document retrieval in optimal time and linear space. In: SODA, pp. 1066–1077 (2012)
34. Navarro, G., Puglisi, S.J.: Dual-sorted inverted lists. In: Chavez, E., Lonardi, S. (eds.) SPIRE 2010. LNCS, vol. 6393, pp. 309–321. Springer, Heidelberg (2010)
35. Navarro, G., Valenzuela, D.: Space-efficient top-k document retrieval. In: Klasing, R. (ed.) SEA 2012. LNCS, vol. 7276, pp. 307–319. Springer, Heidelberg (2012)
36. Nekrich, Y.: Space-efficient range reporting for categorical data. In: PODS, pp. 113–120 (2012)
37. Patil, M., Thankachan, S.V., Shah, R., Hon, W.-K., Vitter, J.S., Chandrasekaran, S.: Inverted indexes for phrases and strings. In: SIGIR, pp. 555–564 (2011)
38. Sadakane, K.: Succinct data structures for flexible text retrieval systems. J. Discrete Algorithms 5(1), 12–22 (2007)
39. Sheng, C., Tao, Y.: Dynamic top-k range reporting in external memory. In: PODS, pp. 121–130 (2012)
40. Tao, Y.: Lecture 1: External memory model and sorting
41. Välimäki, N., Ladra, S., Mäkinen, V.: Approximate all-pairs suffix/Prefix overlaps. In: Amir, A., Parida, L. (eds.) CPM 2010. LNCS, vol. 6129, pp. 76–87. Springer, Heidelberg (2010)
42. Välimäki, N., Mäkinen, V.: Space-efficient algorithms for document retrieval. In: Ma, B., Zhang, K. (eds.) CPM 2007. LNCS, vol. 4580, pp. 205–215. Springer, Heidelberg (2007)
43. Zobel, J., Moffat, A.: Inverted files for text search engines. ACM Comput. Surv. 38(2) (July 2006)

Shell: A Spatial Decomposition Data Structure for 3D Curve Traversal on Many-Core Architectures

Kai Xiao[1], Danny Ziyi Chen[1], Xiaobo Sharon Hu[1], and Bo Zhou[2]

[1] Department of Computer Science and Engineering, University of Notre Dame
{kxiao,dchen,shu}@nd.edu
[2] Member of Technical Staff, Altera Corp.
allen.bo.zhou@gmail.com

Abstract. Shared memory many-core processors such as GPUs have been extensively used in accelerating computation-intensive algorithms and applications. 3D curve traversal is a fundamental process in many applications, and is commonly accelerated by spatial decomposition schemes captured in hierarchical data structures (e.g., kd-trees). However, using hierarchical structures requires repeated hierarchical searches, which are time-consuming on shared memory many-core architectures. In this paper, we propose a novel spatial decomposition based data structure, called Shell, which completely avoids hierarchical search for 3D curve traversal. In Shell, a structure is built on the boundary of each region in the decomposed space, which allows any curve traversing in a region to find the next neighboring region to traverse using table lookup schemes. While our approach works for other spatial decomposition paradigms and many-core processors, we illustrate it using kd-tree on GPU and compare with the fastest known kd-tree ray traversal algorithms. Experimental results show that our approach accelerates ray traversal considerably over the kd-tree approaches.

Keywords: Many-core architecture, GPU, data structure, spatial decomposition, 3D curve traversal.

1 Introduction

3D Curve traversal is a fundamental process in many applications, including graphics ray tracing [4], volume rendering [2], and radiation dose calculation [12]. Since applications using curve traversal often involve large numbers of curves and repeatedly conduct the traversal process, the execution speed of curve traversal is critical. Spatial decomposition based data structures have been developed on shared memory many-core architecture (e.g., general purpose graphics processing units (GPGPUs)) for accelerating curve traversal solutions. In this paper, a new efficient spatial decomposition based data structure for 3D curve traversal is proposed, which better exploits the characteristics of shared memory many-core architectures and avoids hierarchical searches that are commonly performed in known spatial decomposition data structures. We use ray traversal to illustrate the problem and our solution, but it can be easily extended to other curves, such as line segments or general algebraic curves.

H.L. Bodlaender and G.F. Italiano (Eds.): ESA 2013, LNCS 8125, pp. 815–826, 2013.

Shared memory many-core processors present great opportunities to speed up computation intensive applications. Recent advances on GPGPUs leverage massively parallel architectures based on single-instruction multiple-data (SIMD) processor cores to achieve high performance. However, the memory efficiency and execution divergence is often the major performance bottleneck for applications running on this architecture [1]. Typically, memory operations are much more expensive than computations in GPU. For example, in the NVIDIA GPU, the latency of an off-chip memory transaction is 400-600 times longer than the fastest computational instructions [9].

A number of data structures have been developed to partition and organize geometric objects in 3D scenes to improve the efficiency of ray traversal. One important class of such data structures is based on spatial decomposition (e.g., grids [8], octrees and kd-trees [4]), which partitions a geometric scene into a set of regions, each containing a small number of objects. The subdivided regions are normally organized by a hierarchical structure which represents the geometric relationship among those regions. Kd-tree is a commonly used hierarchical structure due to its ability of matching the subdivided regions with the distribution of geometric objects in the scene. It is also considered to be the data structure that provides the fastest known ray traversal speed in static scenes because of its efficient hierarchical search mechanism [13].

Quite a few algorithms have been developed for 3D ray traversal using kd-tree structures, in which repeated hierarchical searches in the tree are needed to find the neighboring regions for traversal [11]. For kd-tree based ray traversal on a traditional CPU architecture, a stack is normally used to store the tree nodes along a search path. On GPU, due to the lack of stack support and limited capacity of on-chip memory, a stack based approach typically allocates stacks to off-chip memory (e.g., global memory). The long access latency of the off-chip memory can easily become a performance bottleneck for ray traversal on GPUs [7]. To address this challenge, Foley et al. [3] and Horn et al. [6] proposed a restart traversal scheme on kd-trees which starts the tree search from the root, and uses a push-down method to move the "root" down to the minimum subtree to be searched. Horn et al. [6] also developed a short-stack approach which builds a small circular stack in the on-chip memory of GPU, and combines it with the push-down method (PD-SS). Popov et al. [10] proposed a kd-rope algorithm [5] using the concept of neighbor links, where each leaf node in the kd-tree contains not only the region R it represents but also a set of pointers, each pointing to a minimum subtree that contains all neighboring regions touching each of the boundaries (or faces) of R. Santos et al. [11] improved the kd-rope implementation, making it the fastest known kd-tree searching ray traversal approach on GPU (called kd-rope++).

Although these algorithms considerably improved the kd-tree searching based 3D ray traversal performance on GPU, all of them still rely on hierarchical search to find the next traversal region when a ray crosses a region boundary (i.e., a ray exits from a region). Each search operation incurs reading a tree node, and search for the next neighboring traversal region can visit $O(H)$ nodes in a kd-tree, where H is the height of the tree. Note that each reading of a visited tree node may take multiple memory transactions to obtain the entire information package for the node, because each GPU memory transaction has a limited size. Also, the threads for different rays may follow different sequences of visited tree nodes which can result in execution divergence.

We propose a new spatial decomposition based data structure, called Shell, which completely eliminates hierarchical search, hence leading to an efficient solution for 3D curve traversal on GPU. Shell provides a neighboring region locating mechanism based on table lookup techniques to replace hierarchical search and find the next region for any traversing ray, allowing the ray to directly access the next region.

Generally speaking, given the set of decomposed 3D regions in a hierarchical structure (say, a kd-tree), Shell focuses on the neighboring relationship among the regions for all leaf tree nodes (i.e., leaf regions). For each leaf region R, the information of its neighboring leaf regions is captured in Shell by a geometric structure called arrangements, with one arrangement per boundary face of R. An arrangement partitions a face F of R into a set of 2D regions called cells, such that each cell C covers an area of F touching a neighboring leaf region R' of R and contains information of R'. When a ray exits R by crossing F in C, it acquires the neighboring region information of R' by accessing C. The table lookup scheme in Shell is crucial for quickly finding C without tree searches and other performance bottlenecks.

There are two key factors in designing the table lookup schemes: the efficiency of cell locating mechanism (for performance) and the size of lookup table (for memory usage). Both factors are affected by the partition schemes for generating the arrangements. We seek to balance the ray traversal performance with good memory usage in Shell, and present a set of partition schemes, including (from simple to sophisticated) uniform grids, multi-level uniform grids, and compressed non-uniform grids, to deal with different neighboring region settings. Each such scheme can be viewed as an extension of the simpler ones for obtaining a good trade-off between ray traversal performance and memory usage for a more complicated neighboring region setting. In many applications, uniform grids are provided in the scenes, such as the CT images in radiation dose calculation and the data volumes of smog animation in volume rendering.

Given a geometric scene with N objects, suppose a kd-tree decomposition partitions it into M leaf regions with a tree height of $O(\log M)$ using $O(M + N)$ memory. A ray traversal in such a kd-tree takes $O(\log M)$ search steps (i.e., the number of visited tree nodes or memory transactions involved) each time a ray crosses a region boundary. In comparison, using Shell, only one step (or $O(1)$ memory transactions) is needed. The Shell memory usage is $O(M + N + U)$, where U is the total number of cells in the arrangements for all leaf regions. Using judicious partition and compression schemes, the $O(M + N + U)$ memory bound of Shell can be made in practice comparable to the $O(M + N)$ memory bound of the kd-tree. Since many applications use massive numbers of traversing rays each of which can cross many regions in the scenes, reducing the memory transactions from $O(\log M)$ to $O(1)$ per region crossing for each ray on GPU can be significant.

We implemented our Shell data structure based on kd-tree decomposition with all our arrangement schemes. Our experiments were conducted on ray traversals in several benchmark geometric scenes used in graphics rendering. The results show that our Shell approach outperforms the fastest known kd-tree searching ray traversal approaches on GPU by over 2X to 5X. Although ray traversals, kd-tree, and GPU are used to illustrate our method in this paper, our solutions can be easily extended for traversing other curves, using different spatial decompositions, and on other many-core architectures.

2 Ray Traversal Using Kd-tree Decomposition

Given a spatially decomposed scene, we can use a dual graph G to represent all leaf regions, as follows: Each vertex in G corresponds to exactly one leaf region, and two vertices are connected by an edge in G if and only if the two corresponding regions share any common boundary portion. With this graph model, the ray traversal problem for a ray r is to find a path (i.e., a sequence of vertices) in G determined by the trajectory of r (i.e., the sequence of regions intersected by r in the order of the corresponding vertices on the path). To obtain such a path in G, a key issue, which we call the *next-region issue*, is, at each vertex v (for a region R_v) of the path, to find the next vertex v' (for a region $R_{v'}$) on the path, i.e., after region R_v, the ray r enters the next region $R_{v'}$. This issue can be resolved by using the geometric location information of r traversing in R_v and the neighboring leaf regions of R_v. In implementation, resolving this issue means to map a point on a boundary face of R_v (from which r exits R_v) to a memory address (which stores the information of the neighboring region $R_{v'}$ traversed next by r). The efficiency of resolving the next-region issue is an essential task for ray traversal.

Since a kd-tree organizes the leaf regions in a tree structure, hierarchical search is the basic process for solving the next-region issue. However, the internal nodes accessed in the hierarchical search impacts ray traversal performance by incurring considerable memory transactions and divergent branches. Although considered as the most efficient kd-tree searching ray traversal algorithms on GPU, PD-SS and kd-rope inherit the hierarchical search mechanism from the kd-tree structure to find neighboring regions. In the worst case, both algorithms still visit $O(H)$ nodes to find a neighboring region, where H is the height of the kd-tree. Further, accessing each node takes multiple memory transactions, which is a considerable overhead. Moreover, the numbers of accessed nodes between different rays can be quite different, which provides the rays with different workloads. In parallel execution on GPU, the workload variance of the rays can cause execution divergence and hence deteriorate the ray traversal performance.

3 The Shell Data Structure

We propose the Shell data structure for directly finding the next neighboring regions without any tree search. By avoiding hierarchical search, Shell reduces both memory transactions and execution divergence. Given a kd-tree decomposed scene, based on our dual graph model, Shell builds on the leaf regions by putting a geometric structure (i.e., the arrangement) on the boundary of each leaf region R (intuitively, the "Shell" of region R), in order to capture the information of all neighboring leaf regions of R. When a ray r hits a boundary face F of R to exit from R, the arrangement for F allows r to find quickly the neighboring region of R to traverse next.

The arrangements are the key components of Shell, which essentially provide table lookup schemes for mapping the geometric information of a ray r traversing in a region R and the neighboring regions R's of R to a GPU memory address storing the information of the R' to be traversed next by r. For each face F of every leaf region R, an arrangement $A(F)$ partitions F into a set of 2D areas (i.e., the cells), such that each cell touches only one neighboring region of R on F. All cells of $A(F)$ are organized by a specific data structure such that given any point p on F, the cell of $A(F)$

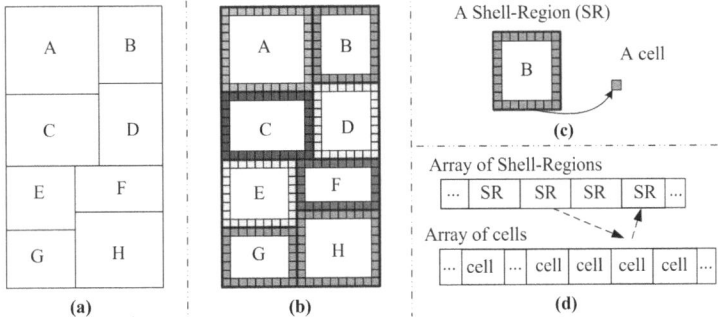

Fig. 1. (a) A 2D scene with 8 regions. (b) A uniform-grid based Shell structure for the scene in (a). (c) The Shell-Region for region B and one of its cells. (d) The memory layout of Shell.

containing p can be found quickly. As a table lookup scheme, the cells of $A(F)$ form a data table, and each cell gives a pointer to its corresponding neighboring region of R. Ideally, the following features are desired from good arrangements: (1) An arbitrary ray r can quickly find the cell C containing its hit point on a face of R (say, with only few memory transactions), and (2) the memory requirement for storing all cells is not too high. There is a trade-off between these two features. Depending on different settings of neighboring regions, we propose a set of partition and compression schemes for building arrangements to achieve good performance with respect to ray traversal speed and memory usage of Shell. Since the GPU memory architecture prefers simple memory layout for efficient addressing, we choose arrays as the main data structure in GPU for storing arrangements based on all our partition schemes.

Figure 1(b) illustrates the Shell structure based on the spatial decomposition in Figure 1(a). Here, to illustrate the idea of a partition scheme for arrangements of neighboring regions, we use a simple uniform grid as example to partition all region boundaries in the scene. Each leaf region is represented by a *Shell-Region* (SR) (e.g., see a Shell-Region in Figure 1(c)). Every Shell-Region uses an arrangement to partition each of its boundary faces into a set of cells (e.g., the shaded small boxes around the boundary of the Shell-Region in Figure 1(c)). Each cell covers an area on its boundary face and contains a pointer to a neighboring Shell-Region touching that cell on the face. The memory layout of Shell, shown in Figure 1(d), consists of an array of all Shell-Regions and an array of all cells stored in the GPU main memory. The cells of the same boundary face of a Shell-Region are all allocated as a group in consecutive memory space, and the memory address of the first cell in the group is stored in the corresponding Shell-Region structure. Thus, any cell can be addressed by computing its memory offset from the first cell in the group. When a ray r leaves a region through a face F, the neighboring region entered next by r is found by accessing the cell in which r crosses F.

3.1   The Structure of an Individual Shell-Region

A Shell-Region contains information of a 3D leaf region and the arrangements of its six 2D boundary faces. To ensure that a Shell-Region can be quickly loaded, its memory requirement should ideally be no bigger than the size of a single memory transaction in the hardware architecture (e.g., 128 consecutive bytes on NVIDIA GTX570 GPU). Essentially, the leaf region information of a Shell-Region R is reserved from the spatial decomposition, which consists of its geometric location and size, as well as a pointer to its associated scene objects. An arrangement consists of a geometric partition scheme subdividing a region face into a set of cells, and a data structure mapping the locations to the memory addresses of the cells. The partition scheme determines the speed and memory usage of a ray traversal algorithm using Shell. To achieve a good balance between these two factors, we propose a series of partition schemes, including (from simple to sophisticated) uniform grid, non-uniform grid, and grid compression. Each scheme is built on top of the simpler ones and reduces memory usage by using slightly more computational operations for locating cells (but not more memory transactions), to handle a more complicated neighboring region setting.

Uniform Grid Scheme. A uniform grid partitions each region face into a matrix of cells of the same size. This arrangement easily maps the geometric locations to the memory addresses of the cells. Each cell contains a pointer to the neighboring region touching this cell. Using a uniform grid for all region boundaries is the simplest approach to build a Shell structure (see Figure 1(b)). In many applications (e.g., radiation dose calculation and volume rendering), the uniform grid partition is provided with scenes, such as CT images and smog volumes. The uniform grid scheme provides a fast cell-accessing mechanism, but it can use a large amount of memory. Therefore, it is good when each region's face has a simple distribution of neighboring regions (say, for scenes containing well-shaped objects with a relatively regular distribution). However, for a complicated scene, the region faces often have a large number of neighbors with non-regular distributions. The uniform grid has to choose the finest resolution of the neighbors on all region faces as the cell size, and significantly increases the grid size.

To address this issue, we can use multiple uniform grids with different levels of resolutions to partition each region faces regarding with its neighboring region settings. This is called the *multi-level uniform grid* scheme, which provides some flexibility to the region faces with simple neighbor distributions so that they can use coarse grid resolutions and thus store fewer cells. For the region faces with complicated neighbor distributions, we still must use uniform cells of sufficiently fine resolutions. The resolution of each grid used in the multi-level uniform grid scheme needs to be stored for each face of every Shell-Region, which is used to calculate the cell hit by a ray on the face and the memory address of the corresponding cell. Comparing with the single uniform grid, the multi-level grid scheme can reduce the number of cells on the region faces and hence save memory space, with a very small time overhead.

Non-uniform Grid Scheme. The memory space requirement of the uniform grid often prevents its usage for complicated neighboring region settings. Observe that multiple cells in a uniform grid of a boundary face often touch the same neighboring region and hence store multiple copies of the same corresponding Shell-Region. We propose a non-uniform grid scheme which aims to merge cells of a uniform size on each face sharing the same neighboring region into a larger cell to save memory space.

For a boundary face F, a non-uniform grid is built on the partition of a uniform grid. Figure 2(a) shows a boundary face (for a 3D scene) with 16 neighboring regions

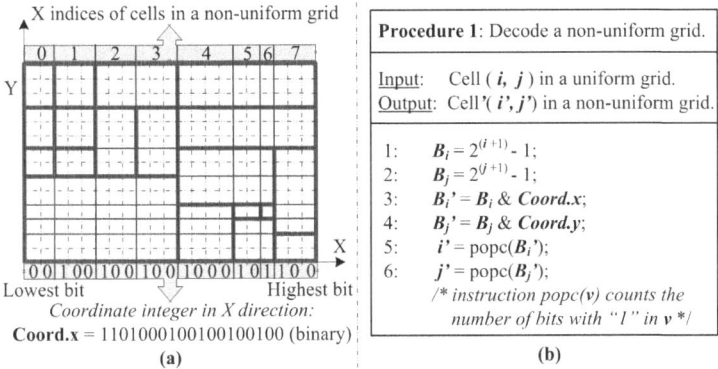

Fig. 2. (a) A non-uniform grid for a boundary face F with 16 neighboring regions. (b) The decoding procedure for converting the cell indices from a uniform grid to a non-uniform grid.

(marked by heavy lines) and originally partitioned into a 21×14 uniform grid. We use the set of vertical or horizontal lines along the projected boundaries of the neighboring regions on F (e.g., the solid bold red boxes in Figure 2(a)) to partition F. Note that only a subset of the lines for the uniform grid (e.g., the solid green lines in Figure 2(a)) is actually aligned with such projected region boundaries on F. A non-uniform grid uses those aligned lines to form a new 8×7 grid partition, whose cells (the boxes bounded by solid green lines in Figure 2(a)) may cover multiple cells from the uniform grid.

A decoding mechanism is needed to transit the indices from the uniform grid to the non-uniform grid, so that any cell of the non-uniform grid (hit by a ray) can be located. To support this mechanism, the partition lines in each axis of the uniform grid are indexed by a bit sequence, where each bit is set as 1 if the corresponding line is used in the non-uniform grid partition and 0 otherwise. Every bit sequence is then stored in the integer format, called a *coordinate integer*, in the Shell-Region (e.g., $Coord.x$ in Figure 2(a)). During ray traversal, suppose a ray r exits the Shell-Region through a cell $C(i, j)$ in the uniform grid for a face F. Then the following is done: (1) The two coordinate integers $Coord.x$ and $Coord.y$ for F are obtained; (2) using $Coord.x$ (resp., $Coord.y$), find the number of 1's in the bit sequence of $Coord.x$ (resp., $Coord.y$) from its left end up to position i (resp., j), denoted by i' (resp., j'). Then cell $C'(i', j')$ is the one in the non-uniform grid of F hit by the ray r. To avoid using any branch operations (which may cause execution divergence), we design a short procedure to accomplish Step (2) (see Figure 2(b)). Note that given the two coordinate integers, the decoding process uses no memory transaction.

Grid Compression Scheme. Even with a non-uniform grid scheme, the Shell structure still tends to use more memory than a kd-tree structure. A grid is commonly represented as a matrix, each its element storing a value. On a face F with K neighboring regions partitioned into a grid of size $m \times n$, its matrix representation stores $m \times n$ copies of K pointers (each for a neighboring region of F). To reduce such duplication, we employ a grid compression scheme similar to the sparse matrix compression which improves the memory usage for F from $O(m \times n + K)$ to $O(m + n + K) = O(m + n)$.

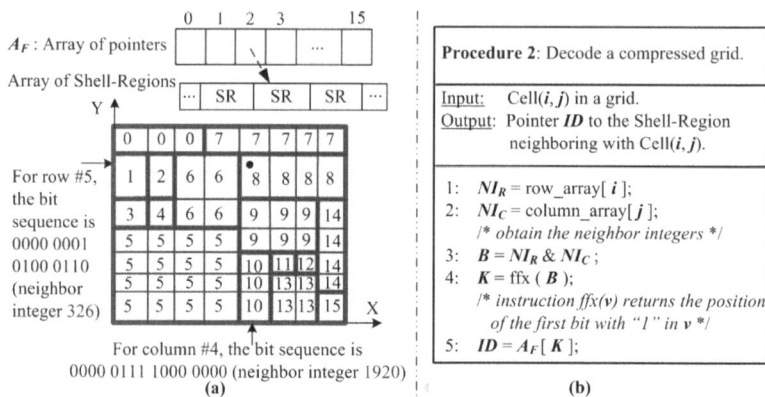

Fig. 3. (a) A compressed non-uniform grid for the boundary face F in Figure 2(a). (b) The decoding procedure for obtaining the pointer to the neighboring Shell-Region touching a given cell.

Note that for any boundary face F, the projected shape of each neighboring region on F is an axis-parallel rectangle. Consider a non-uniform grid G_n on F. The cells in each row or column of G_n contain pointers to a set of neighboring regions touching F. Observe that for any row R_i and any column C_j of G_n, the neighboring region R' pointed to by the cell $C(i, j)$ in G_n appears in both the set of neighboring regions pointed to by the cells of R_i and the set of neighboring regions pointed to by the cells of C_j; further, R' is the only neighboring region appearing in both these two sets (this follows from the rectangular shapes of the projected neighboring regions on F). Based on this observation, we use the following grid compression scheme. (1) Store pointers to all neighboring Shell-Regions of F in an indexed array A_F. (2) For each row (or column) of G_n, build a bit sequence as follows: the sequence has K bits, one for each neighboring Shell-Region of F, in their order as stored in the array A_F; each bit is set as 1 if and only if the corresponding neighboring Shell-Region appears in that row (or column) of G_n. Every such bit sequence is stored as an integer, called a *neighbor integer*. Thus, we have an array of pointers to the set of K neighboring Shell-Regions of F, and two sequences of neighbor integers (one for the rows and one for the columns of G_n). Since there are K neighboring Shell-Regions touching F and G_n has m rows and n columns on F, the memory usage of the above grid compression scheme for F is clearly $O(m + n + K)$. (Here, we assume that the number K of neighboring regions touching F is not too big, which is usually the case in practice.)

For a ray r hitting a point on F to exit, we use the following decoding process to find the next Shell-Region traversed by r: (1) Compute the cell indices in the uniform grid for the hit point of r on F; (2) find the indices (for a row R_i and a column C_j) in the non-uniform grid G_n; (3) take the neighbor integers for R_i and C_j, and perform a logic AND operation on them to identify the unique neighboring Shell-Region pointed to by the cells in both R_i and C_j. Note that both the uniform grid and non-uniform grid of F involved in our decoding process above are only used *conceptually*, i.e., they are only concepts for helping our computation but are not actual structures explicitly maintained in Shell. Figure 3 illustrates the grid compression scheme and decoding process.

3.2 GPU Memory Layout and Memory Bound of Shell

Since the GPU memory architecture prefers simple memory layout schemes in order to implement easy and efficient addressing, we use arrays to store the Shell data structure. The Shell memory layout consists of an array of Shell-Regions and an array of cells, addressed by their indices (e.g., see Figure 1(d)). Essentially, each Shell-Region is a 128 bytes aligned memory block and each cell stores a pointer point to a Shell-Region.

Suppose for a 3D scene with N objects (for specific applications) represented by Shell, we need to store M Shell-Regions and U cells besides the N objects, where M is the number of leaf regions in the spatially decomposed scene and U is the total number of cells. Then clearly Shell uses $O(M + N + U)$ memory. Note that in the non-uniform grid compression scheme, the number of cells on each boundary face is equal to the number of its neighboring regions, and the total size of the three arrays (for neighbor integers of the rows and columns of a non-uniform grid and for pointers to the neighboring Shell-Regions) is proportional to the total size of these neighboring regions. Thus, U is proportional to the number of neighboring leaf region pairs in the spatial decomposition (i.e., the number of edges in the dual graph G).

4 Ray Traversal Based on Shell

In this section, we show how to use the Shell data structure effectively in ray traversal, which is to efficiently decode the information of Shell so as to obtain the memory address of the neighboring region R' of R, when a ray r is exiting region R.

In the Shell data structure, the cell containing the hit point of a ray r on a boundary face F of a region R stores a pointer to (i.e., the memory address of) the neighboring region of R touching that cell. We call this cell a *target cell* and denote it by $cell_t$. Thus, locating the next region entered by r can be accomplished in three steps:

(I) Find the hit point p of r on F;

(II) determine the memory location of the target cell $cell_t$ containing p;

(III) access $cell_t$ to obtain the address of the Shell-Region for the next traversing region of r. Step (I) is taken by all ray traversal algorithms regardless of the data structure used and Step (III) is trivial. Below, we elaborate how Step (II) is done.

The cells of a boundary face F, which we call a *cell group*, are stored consecutively in the memory as part of an array. Hence, the memory location of $cell_t$ can be computed based on the memory address of the cell group for face F (denoted as M_{base}) and the offset of $cell_t$ from the first element in the cell group. Since M_{base} has already been read into the on-chip memory upon r entering the current region, the main hurdle is to determine the offset. The process of computing the offset is equivalent to decoding the Shell structure associated with the arrangement on F. The actual decoding method depends on the specific partition and compression schemes used for F. Our decoding solutions for two cases (the uncompressed uniform grid scheme and compressed non-uniform grid scheme) are given in Algorithm 1 and 2. Decoding methods for other cases can be easily derived from these two algorithms.

In an ideal case, locating the next traversing region using the Shell data structure takes only two memory transactions: one for accessing a Shell-Region and one for a

---

Algorithm 1. Locating the next traversing region using a uniform grid

1: **Input:** ray R, boundary face F, Shell-Region SR.
2: **Output:** target cell $cell_t$.
3: compute the point P where R intersects F and exits from SR;
4: obtain the uniform grid G for F;
5: M_{base} = the address of the cell group for F;
6: compute the location index (i, j) of the cell in G containing P;
7: $M_{off} = i \times (row\_size\_of\_G) + j$;
8: $M_{addr} = M_{base} + M_{off}$;   /* Compute the memory address M_{addr} for the target cell. */
9: $cell_t$ = cell_array[M_{addr}];   /* Load and return the target cell*/

Algorithm 2. Locating the next traversing region using a compressed non-uniform grid

1: same as Line 1-4 in Algorithm 1.
2: compute the location index (i, j) of the cell in G containing P;
3: obtain the non-uniform grid G_n for F, with $Coord.x$ and $Coord.y$;
4: compute index (i', j') by calling *Procedure 1* in Figure 2(b); /* Decode non-uniform grid. */

5: compute $cell_t$ by calling *Procedure 2* in Figure 3(b);   /* Decode compressed grid. */

---

cell (Line 9 of Algorithm 1, or Line 5 of Procedure 2 (see Figure 2(b))). But, the decoding for the more sophisticate schemes may need more than two memory requests. For example, the compressed non-uniform grid scheme incurs three additional memory requests: one access to the grid coordinates (Line 7 of Algorithm 2) and two accesses for decoding the target cell (Lines 1–2 of Procedure 2 (see Figure 2(b))). By properly aligning the memory layout of the Shell-Region structure and the coordinate integers for the non-uniform grid or the neighbor integers for a compressed grid, we can fulfill the five sequentially issued memory accesses by two memory transactions with cache.

The combined effort above guarantees that only two memory transactions are used for locating the next traversing region. For comparison, consider a geometric scene decomposed into M regions and a ray traversing through K leaf regions. A kd-tree ray traversal algorithm accesses $O(K \times \log M)$ tree nodes; each access takes 2–4 memory transactions depending on the implementation. With the Shell ray traversal algorithm, the ray accesses K Shell-Regions; each access needs only two memory transactions.

5   Evaluation

To evaluate the proposed Shell approach for ray traversal on GPU, we implemented our Shell data structure and Shell based ray traversal algorithm. We also implemented two state-of-the-art kd-tree searching ray traversal algorithms: PD-SS and kd-rope++. For PD-SS, although its performance is slightly lower comparing with kd-rope, it takes much less memory space. To ensure the quality of our PD-SS and kd-rope implementations, we tested them on graphics ray tracing applications and achieved statistics matching with those published in related works (e.g., [4,11]). The hardware platform used in our evaluation is an NVIDIA GTX570 graphics card (480 cores, 1.6GHz core frequency). Two different Shell data structures are built on the kd-tree decomposition of

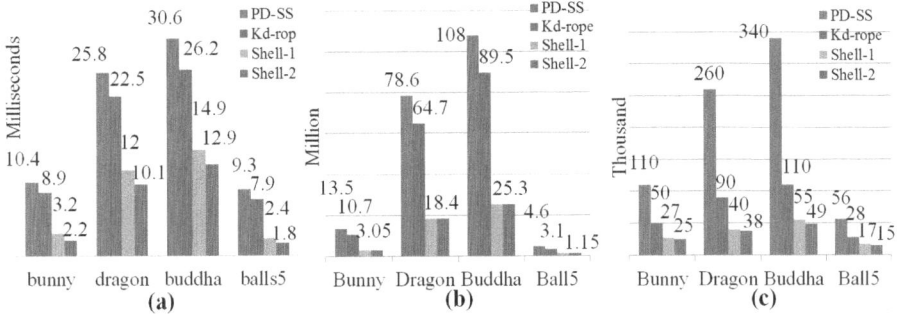

Fig. 4. The ray traversal performance evaluation for the PD-SS, kd-rope, and Shell based algorithms on the scenes. (a) Execution time. (b) The total numbers of nodes accessed by all rays. (c) The number of divergent branches.

each scene. Shell-1 aims to minimize the memory usage by adopting sophisticated arrangements if they can save memory space. Shell-2 tends to achieve a good balance between memory usage and ray traversal performance, which uses the sophisticated schemes only if they reduce a certain portion of memory usage (e.g., more than 20%).

Figure 4 shows the statistics of the ray traversal execution using PD-SS, kd-rope, and our Shell based algorithms. All data are obtained by traversing a set of rays in the process of rendering a 1024×1024 image for the scenes. Figure 4(a) shows that a speedup between 2.6X–5.1X can be achieved by using Shell comparing to PD-SS and 2.2X–4.3X to kd-rope. Furthermore, comparing the performance of Shell-1 and Shell-2, we see that a larger Shell data structure tends to lead to a faster ray traversal speed.

The Shell based ray traversal algorithm gains its performance advantage through removing many expensive memory accesses to internal nodes in the kd-tree searching methods. As demonstrated in Figure 4(b), Shell based ray traversal accesses on average 4.2X and 3.5X fewer nodes than PD-SS and kd-rope, respectively. Although the kd-rope approach uses neighbor links to reduce accessing internal nodes, it still needs to visit a number of internal nodes when a boundary face has multiple neighboring regions.

Another factor contributing to the performance improvement of the Shell based ray traversal is that Shell reduces a significant amount of execution divergence. Figure 4(c) summarizes the number of divergent branches for each algorithm, which shows Shell based ray traversal incurs on average 70% less divergent branches. But, Shell base ray traversal cannot completely eliminate execution divergence due to the different partition schemes involved. The more sophisticated schemes a Shell structure uses (e.g., compressed non-uniform grid), the more divergent branches it may have. This is shown by comparing the Shell-1 and Shell-2 results in Figure 4(c).

6   Extending Shell for Other Applications

Besides graphics ray tracing implemented in our experiments, the Shell data structure can also be applied to many other applications such as radiation dose calculation in radiation cancer treatment, where a uniform grid is "naturally" available and efficient traversal algorithms for computing the trajectories of huge numbers of radiation rays are needed. With some simple modifications, our Shell data structure and Shell based ray traversal algorithm as presented in this paper can be extended to tracing other types of 3D curves, such as line segments or general algebraic curves. Besides the GPU architecture discussed in this paper, the Shell based approach can also be applied to other types of shared memory many-core processors such as those using MIMD architecture, for reducing off-chip memory transactions in 3D curve traversal.[1]

Acknowledgements. This research was supported in part by NSF under Grants CCF-0916606 and CCF-1217906, and also in part by a research contract from the Sandia National Laboratories.

References

1. Aila, T., Laine, S.: Understanding the efficiency of ray traversal on GPUs. In: Proceedings of the 1st ACM Conference on High Performance Graphics, pp. 145–149 (2009)
2. Bethel, E., Howison, M.: Multi-core and many-core shared-memory parallel raycasting volume rendering optimization and tuning. International Journal of High Performance Computing Applications 26(4), 399–412 (2012)
3. Foley, T., Sugerman, J.: Kd-tree acceleration structures for a GPU raytracer. In: Proceedings of Graphics Hardware, pp. 15–22 (2005)
4. Hapala, M., Havran, V.: Review: Kd-tree traversal algorithms for ray tracing. Computer Graphics Forum 30(1), 199–213 (2011)
5. Havran, V., Bittner, J., Zara, J.: Ray tracing with rope trees. In: Proceedings of Spring Conference on Computer Graphics, pp. 130–139 (1998)
6. Horn, D.R., Sugerman, J., Houston, M., Hanrahan, P.: Interactive k-d tree GPU ray tracing. In: Proceedings of Symposium on Interactive 3D Games and Graphics, pp. 167–174 (2007)
7. Huges, D.M., Lim, I.S.: Kd-jump: A path-preserving stackless traversal for faster isosurface raytraing on GPUs. IEEE Transactions on Visualization and Computer Graphics 15(6), 1555–1562 (2009)
8. Kalojanov, J., Billeter, M., Slusallek, P.: Two-level grids for ray tracing on GPUs. Computer Graphics Forum 30(2), 307–314 (2011)
9. NVIDIA Corporation. NVIDIA CUDA C programming guide version 5.0 (2013),
    http://docs.nvidia.com/cuda/cuda-c-programming-guide/
    index.html
10. Popov, S., Gunther, J., Seidel, H.P., Slusallek, P.: Stackless kd-tree traversal for high performance GPU ray tracing. Computer Graphics Forum 26(3), 415–424 (2007)
11. Santos, A., Teixeira, J.M., Farias, T., Teichrieb, V., Kelner, J.: Understanding the efficiency of kd-tree ray traversal techniques over a GPGPU architecture. International Journal of Parallel Programming 40(3), 331–352 (2012)
12. Xiao, K., Zhou, B., Chen, D.Z., Hu, X.S.: Efficient implementation of the 3D-DDA ray traversal algorithm on GPU and its application in radiation dose calculation. Medical Physics 39, 7619–7626 (2012)
13. Zlatuska, M., Havran, V.: Ray tracing on a GPU with CUDA – Comparative study of three algorithms. In: Proceedings of Computer Graphics, Visualization and Computer Vision, pp. 69–75 (2010)

---

[1] Due to the length limitation of the conference, we omitted several non-essential implementation details in this paper. For the interested readers, those details are provided in the full version of this paper, which can be found in the technical report repository in University of Notre Dame (http://www.cse.nd.edu/Reports/2013/TR-2013-01.pdf).

Author Index